T5-ANW-729

PRODUCTION HANDBOOK

ADVISORY BOARD

PRODUCTION HANDBOOK

Fourth Edition

JOHN A. WHITE
Georgia Institute of Technology
Atlanta, Georgia

JOHN WILEY & SONS

New York · Chichester · Brisbane · Toronto · Singapore

Library of Congress Cataloging in Publication Data:

Production handbook.

Bibliography: p.
1. Production management—Handbooks, manuals, etc.
I. White, John A., 1939–

TS155.P747 1987 658.5 86-26796
ISBN 0-471-86347-5

Printed in the United States of America

10 9 8 7 6 5 4 3 2 1

CONTRIBUTORS

William J. Ainsworth
Peat, Marwick, Mitchell & Company
Chicago, Illinois

Jane C. Ammons
Georgia Institute of Technology
Atlanta, Georgia

David C. Anderson
Purdue University
West Lafayette, Indiana

Roger Anderson
Westinghouse/Unimation
Danbury, Connecticut

James M. Apple
SysteCon
Duluth, Georgia

C. Ray Asfahl
University of Arkansas
Fayetteville, Arkansas

Moshe M. Barash
Purdue University
West Lafayette, Indiana

Bruce Barger*
Modern Materials Handling
Newton, Massachusetts

William L. Berry
The University of Iowa
Iowa City, Iowa

J T. Black
Auburn University
Auburn, Alabama

Warren Boe
The University of Iowa
Iowa City, Iowa

Ronald W. Brockett
Kurt Salmon Associates, Inc.
Atlanta, Georgia

John A. Buzacott
University of Waterloo
Waterloo, Ontario, Canada

John R. Canada
North Carolina State University
Raleigh, North Carolina

G. Harlan Carothers, Jr.
Consultant
Knoxville, Tennessee

Kenneth E. Case
Oklahoma State University
Stillwater, Oklahoma

W. Colebrook Cooling
Consultant
South Orange, New Jersey

Robert E. Crowley
Electonic Data Systems
Indianapolis, Indiana

Louis E. Davis
University of California
Los Angeles, California

* Deceased.

Steven D. Duket
Pritsker and Associates, Inc.
West Lafayette, Indiana

Glenn C. Dunlap
ITT Advanced Technology Center
Shelton, Connecticut

Harry K. Edwards
University of Michigan
Flint, Michigan

Samuel Eilon
Imperial College of Science and Technology
London, England

Hamilton Emmons
Case Western Reserve University
Cleveland, Ohio

Richard L. Engwall
Westinghouse Electric Corporation
Columbia, Maryland

Orlando J. Feorene
Georgia Institute of Technology
Atlanta, Georgia

John K. Freund
Deere and Company
Moline, Illinois

Inyong Ham
The Pennsylvania State University
University Park, Pennsylvania

Catherine M. Harmonosky
Purdue University
West Lafayette, Indiana

Arnoldo C. Hax
Massachusetts Institute of Technology
Cambridge, Massachusetts

Robert H. Hayes
Harvard University
Boston, Massachusetts

James E. Heaton
Oracle, Inc.
Chelsea, Michigan

Russell G. Heikes
Georgia Institute of Technology
Atlanta, Georgia

John M. Hill
Logisticon, Incorporated
Santa Clara, California

Amit K. Jain
Professional Computer Resources, Inc.
Oakbrook Terrace, Illinois

David W. Jung
ITT Advanced Technology Center
Shelton, Connecticut

Leland R. Kneppelt
Management Science America, Inc.
Winston-Salem, North Carolina

Kenneth Knott
The Pennsylvania State University
University Park, Pennsylvania

Ronald A. Kohser
University of Missouri
Rolla, Missouri

Samuel B. Korin
IBM Corporation
Thornwood, New York

Ingram Lee II
Texas Instruments, Incorporated
Dallas, Texas

Robert N. Lehrer
PTY, Incorporated
Atlanta, Georgia

Ferdinand F. Leimkuhler
Purdue University
West Lafayette, Indiana

John D. Lorenz
GMI Engineering and Management Institute
Flint, Michigan

John C. Luber
IBM Corporation
Thornwood, New York

Hanan Luss
AT&T Bell Laboratories
Holmdel, New Jersey

Leon F. McGinnis
Georgia Institute of Technology
Atlanta, Georgia

Alfred H. McKinlay
Consultant
Pattersonville, New York

John A. Maddox
Consultant
Arlington, Texas

Steve Malkin
Israel Institute of Technology
Technion City, Haifa, Israel

John J. Mariotti
GMI Engineering and Management Institute
Flint, Michigan

Devendra Mishra
Creative Video Services
Newbury Park, California

Yasuhiro Monden
University of Tsukuba
Sakura, Japan

John W. Mueller
Litton, ASD
Hebron, Kentucky

J. Greg Murphy
Kurt Salmon Associates, Inc.
Atlanta, Georgia

Michael B. Murray
Xerox Corporation
Webster, New York

Ray M. Nagan
Arwood Corporation
Rockleigh, New Jersey

Rajeev V. Naik
Arwood Corporation
Rockleigh, New Jersey

Benjamin W. Niebel
The Pennsylvania State University
University Park, Pennsylvania

Shimon Y. Nof
Purdue University
West Lafayette, Indiana

Richard P. Paul
University of Pennsylvania
Philadelphia, Pennsylvania

David W. Poock
GMI Engineering and Management Institute
Flint, Michigan

A. Alan B. Pritsker
Pritsker and Associates, Inc.
West Lafayette, Indiana

Robert L. Propst
The Propst Company
Redmond, Washington

Jack B. ReVelle
Hughes Aircraft Company
Fullerton, California

Harvey Rickles
SysteCon
Duluth, Georgia

William B. Rouse
Georgia Institute of Technology
Atlanta, Georgia

Randall P. Sadowski
Purdue University
West Lafayette, Indiana

Gavriel Salvendy
Purdue University
West Lafayette, Indiana

Robert K. Shattuck
Xerox Corporation
Webster, New York

William T. Shirk
AT&T Technologies
Norcross, Georgia

James J. Solberg
Purdue University
West Lafayette, Indiana

Frederick T. Sparrow
Purdue University
West Lafayette, Indiana

Gerald J. Thuesen
Georgia Institute of Technology
Atlanta, Georgia

James A. Tompkins
Tompkins Associates, Inc.
Raleigh, North Carolina

Joseph Tulkoff
Lockheed-Georgia Company
Marietta, Georgia

Chester J. Van Tyne
Lafayette College
Easton, Pennsylvania

Robert B. Vollum
R. B. Vollum & Associates
Huntingdon Valley, Pennsylvania

Gerald J. Wacker
Xerox Corporation
Los Angeles, California

Steven C. Wheelwright
Stanford University
Stanford, California

David W. Whitmyre
Xerox Corporation
Webster, New York

Theodore J. Williams
Purdue University
West Lafayette, Indiana

Bernard Zimolong
University of Bochum
Bochum, West Germany

FOREWORD

The massive impact of global competition has combined with fast moving manufacturing technology to suddenly transform the production function from the role of expensive corporate necessity to a wholly new position as now the most critical management sector in American industry.

Even five years ago this was not so. The rapid advances in marketing management in the 1960s and financial management in the 1970s had so advanced the contributions of those functions as to largely replace production as a corporate competitive lever in the perceptions of top management. By contrast, production had regressed into a position of apparent maturity, dull stability, and a backwater of discouraged managers. It certainly was not the best place in business from which to move up to top management. More important, the level of intellectual vitality was low: few new, influential concepts and techniques were emerging from either practitioners or academics.

But, as always, economics and technology have their way. Competition from Japan, Korea, and other Asian nations, plus that of European manufacturers, has forced the production function in management to become the arena in which industrial firms must learn to survive by restoring a competitive edge. Companies of all sizes have learned that to compete in today's world markets, and indeed in most domestic markets, requires an expertise in producing at low cost with quality levels never demanded before, and to ever more stringent delivery requirements.

Against this stressful background, the production function has undergone a five-year period of extraordinary revitalization. Nearly every company has renewed its emphasis on productivity, using productivity committees, productivity staff functions, and a return to the basics of industrial engineering. Once again attention is on excellent processes, the best tools and equipment, straight line production flows, clear visual controls, and carefully engineered jobs. Further, there is a newly regained emphasis on efficiency and productivity via worker training, better supervision, and innovative experiments in the management of people in production operations.

On an amazingly broad scale there is a new and disciplined effort to vastly upgrade quality standards through total quality programs, statistical quality control, and awareness of the great cost of quality if the work is not done right the first time. This emphasis on quality is resulting in new attention on excellence in production from the handling of materials and equipment, the attention of workers and supervisors, and marketing recognition that quality can now be a competitive weapon.

The revitalization of production is also seen in a vast outpouring of new manufacturing technologies. CAD and CAM and flexible machining centers are leading the way with robotics to computer integrated manufacturing and the paperless factory. These technologies are being accompanied by innovations in management such as family vendor relationships, compressed and interactive new product and process development cycles, closer relationships with engineering, inventory controls to increase returns on assets employed, and new ideas in materials management such as "just in time."

All this energetic experimentation in production management and manufacturing technology will undoubtedly help to restore some of the lost competitive edge of American industry. The question, however, is when, for it is clear that to date relatively little progress has been made in reducing trade deficits and in regaining market shares in most markets. Our many sick industries are for the most part still sick, and overall productivity is moving up too slowly. Manufacturing is seldom a formidable competitive weapon.

This paradox is yet to be explained. My own sense is that the splendid array of new tools and concepts which are being deployed are not being coordinated and managed at the general management level. In these times of duress we have become fascinated with promising new programs and techniques, reaching out for most any program or technique which seems to offer relief. Few companies have a careful and coherent manufacturing strategy by which their managers can select the right techniques with the right emphasis. The popular focus on cost and efficiency has resulted in a kind of productivity paradox which has not yet been resolved.

For these reasons this new *Production Handbook* could not come at a more timely period in our industrial history. We have more change going on than ever before, changes in concepts, techniques, tools, new insights, new technologies, and new approaches to the management of quality, inventories, people, and production resources. It is difficult yet essential that production managers and engineers get on top of and capture the latest in these areas of production management competencies. Managers are under more pressure than ever before to learn and grow and to acquire both specific knowledge and a broad view of the new tools, techniques, and technologies available.

This handbook does an excellent job of covering the vast and changing field of production management in a new and exciting way. Its authors are outstanding in their fields. They have been asked to focus on both the basics and that which is new and changing; they have been asked to capture in effective prose exactly what is going on in the field and to make available those ideas and techniques for practicing managers.

This will be a broad source book, a reference for production managers from first level supervisor to the vice presidential level. The *Production Handbook* is a microcosm of the great energy and vitality going on in the production management field today.

WICKHAM SKINNER
James E. Robison Professor
of Business Administration, Emeritus
Harvard Business School

PREFACE

The design and management of production systems have undergone more change in the past five years than during any other period since the early 1900s. The breadth and scope of the field of production have changed dramatically, as has management's understanding of the role of production in today's competitive age. The emergence of global competition in a global economy has resulted in greater emphasis being placed on manufacturing excellence. The situation that exists can be summarized in the following way: the world is at war, but too few recognize it! The battle lines have been drawn on the production floors of organizations throughout the world.

Despite the fact that the survival of many firms depends on achieving manufacturing excellence, few seem to understand what is needed and how to make the necessary changes. Also, too many have a narrow view of manufacturing excellence. A broad systems perspective is presented in the handbook. For example, the reader will discover that manufacturing excellence is more than state-of-the-art production processes; it is more than record increases in production efficiency; it is more than having "zero defects"; it is more than being the high-quality, low-cost producer; it is more than having highly motivated and well managed workforces; it is more than producing products that were "designed for production"; it is more than achieving "economies of scale and scope" by using focused factories and flexible factories; it is more than short-term, bottom-line oriented performance; it is more than an emphasis on design simplicity, as opposed to design complexity; it is more than integrated systems versus "islands of automation"; it is more than non-stop, continuous flow manufacturing; it is more than computer integrated manufacturing; and it is more than just-in-time production. Manufacturing excellence includes most, if not all, of the above—and more!

Manufacturing excellence begins at the top of the organization, with a commitment to manufacturing evidenced by the development of a corporate manufacturing strategy. In organizing production for manufacturing excellence, consideration must be given to capacity planning, competitive planning, and performance and productivity measurement. Policies, procedures, and the production organization must be developed consistent with production's goals and objectives. An environment conducive to manufacturing excellence must be created and sustained and management's attention must be given to the design of the products to be manufactured. These subjects are addressed in Section I of the *Handbook*.

Following the introductory section, "Management: Organizing for Production," the *Production Handbook* is organized into seven sections, based on the 5Ms and 2Ss: manpower, methods, machines, material, money, space, and system. The authors include an international cross section of persons well qualified to address an important dimension of manufacturing excellence. They are recognized experts in their fields. Production managers, engineers, consultants, and educators have drawn on their experiences and abilities to communicate the essential elements of their subject concisely and effectively.

Several have focused on what is; others have focused on what can or should be; and some have focused on what will be.

The intended audience for this handbook includes managers at all levels, engineers, consultants, and educators involved in designing and/or managing production systems, as well as teaching courses on the subject. Because of the emphasis given to designing products, processes, and systems to facilitate production, this handbook will be of particular interest to electrical engineers, industrial engineers, manufacturing engineers, and mechanical engineers, as well as computer scientists, management scientists, and operations researchers.

The issues addressed are sufficiently generic as to include those from small and large organizations. Both high-volume and low-volume production needs are addressed. Although the handbook targets discrete parts production, several chapters also apply to continuous production.

Because of the diversity of needs, the breadth of issues, and the explosion of technologies supportive of the design and management of production systems, the *Handbook* resembles a buffet, rather than a fixed meal. A variety of topics are presented, and we recognize that no one will be interested in all of the aspects of production treated in the *Handbook*. The level of coverage of topics is quite diverse, ranging from philosophical to mathematical treatments of particular subjects.

The organization of the *Handbook* was the result of considerable input from the Advisory Board. The membership of the Advisory Board, collectively, spanned the production function. It included those whose interests were primarily with new and emerging technology, those who were concerned with management issues, those who were concerned with engineering issues, and those who were interested in the design of integrated production systems. The Advisory Board assisted in identifying topics and authors, as well as reviewing manuscripts.

The handbook required far longer to complete than any of us ever intended. The logistics associated with a handbook involving more than 60 chapters and 80 authors are immense. Yet, as Wickham Skinner noted in the Foreword, the Fourth Edition of the *Production Handbook* could not have been published at a better time. The need is great for a comprehensive and contemporary treatment of the subject. In spite of this need, there simply does not exist another book treating the subject of production management and design as comprehensively as does this edition. Issues are addressed that are important today and will continue to be important through the next decade. The design and management of future generation production systems will be based on manpower, methods, machines, material, money, space, and system aspects treated in the Fourth Edition.

In addition to the assistance of the authors, the members of the Advisory Board, and Michael Hamilton and Kathleen Kelly of John Wiley & Sons, Inc., the preparation of the handbook benefited greatly from the dedication of Freida A. Breazeal, Katharyn A. McLendon, Joene Owen, Nan C. Thomas, and Edward H. Frazelle of Georgia Institute of Technology. William T. Shirk of AT&T made numerous contributions, including extensive reviews of manuscripts submitted for publication. Finally, the efforts of Jaime Trevino, Assistant to the Editor, were essential to the successful completion of the handbook. I appreciate their commitment to the project. I know they were glad to see it completed!

JOHN A. WHITE, EDITOR
Regents' Professor
School of Industrial and
Systems Engineering
Georgia Institute of Technology
Atlanta, GA

May 1987

CONTENTS

PART 1. MANAGEMENT: ORGANIZING FOR PRODUCTION

1.1. **The Manufacturing Enterprise** 1 • 3
David W. Whitmyre
1.2. **Strategic Planning for Manufacturing** 1 • 11
Robert H. Hayes and Steven C. Wheelwright
1.3. **Capacity Planning** 1 • 23
Hanan Luss
1.4. **Manufacturing and Competitive Planning** 1 • 33
William J. Ainsworth
1.5. **Competitive Benchmarking: The First Step in Meeting the Competitive Challenge** 1 • 39
Michael B. Murray and Robert K. Shattuck
1.6. **Measures of Performance and Productivity** 1 • 45
Robert N. Lehrer
1.7. **Manufacturing Management** 1 • 58
Amit K. Jain
1.8. **Production Organization** 1 • 66
Orlando J. Feorene
1.9. **Manufacturing Environment** 1 • 72
Robert L. Propst
1.10. **Management of Design for Manufacture** 1 • 78
G. Harlan Carothers, Jr.

PART 2. MANPOWER: PEOPLE IN PRODUCTION

2.1. **Job Design** 2 • 3
Louis E. Davis and Gerald J. Wacker
2.2. **Industrial Safety** 2 • 33
Jack B. ReVelle
2.3. **Man–Machine Systems** 2 • 52
William B. Rouse
2.4. **Human Factors** 2 • 67
Bernhard Zimolong and Gavriel Salvendy
2.5. **Work Measurement** 2 • 97
Devendra Mishra
2.6. **Industrial Employee Training** 2 • 125
J. Greg Murphy and Ronald W. Brockett

PART 3. METHODS: DESIGNING, IMPROVING, AND CONTROLLING PRODUCTION SYSTEMS

3.1. **Design-to-Manufacture** 3 • 3
Benjamin W. Niebel
3.2. **Design-to-Assemble** 3 • 21
Roger Anderson
3.3. **Design-to Cost** 3 • 40
Ingram Lee II
3.4. **Production System Models** 3 • 49
John A. Buzacott
3.5. **Control Models for Production Systems** 3 • 68
Shimon Y. Nof and Theodore J. Williams
3.6. **Quality Control and Assurance** 3 • 96
Kenneth E. Case
3.7. **Aggregate Production Planning** 3 • 116
Arnoldo C. Hax
3.8. **Production Forecasts** 3 • 128
Russell G. Heikes
3.9. **Production and Project Scheduling** 3 • 137
Randall P. Sadowski and Catherine M. Harmonosky
3.10. **Scheduling and Sequencing Algorithms** 3 • 159
Hamilton Emmons
3.11. **Assembly Line Balancing** 3 • 176
John D. Lorenz and David W. Poock
3.12. **Production Activity Control** 3 • 191
Robert B. Vollum
3.13. **Factory Floor Information System (FFIS)** 3 • 203
David W. Jung and Glenn C. Dunlap
3.14. **Methods Improvement** 3 • 220
Kenneth Knott
3.15. **Group Technology** 3 • 237
Inyong Ham
3.16. **The Toyota Production System** 3 • 249
Yasuhiro Monden

PART 4. MACHINES: PRODUCTION PROCESSES AND EQUIPMENT

4.1. **Foundry Processes** 4 • 3
Ray M. Nagan and Rajeev V. Naik
4.2. **Machining Processes** 4 • 18
J T. Black
4.3. **Forming Processes** 4 • 38
Chester J. Van Tyne and Ronald A. Kohser
4.4. **Abrasive Processes** 4 • 61
Steve Malkin
4.5. **Industrial Robots** 4 • 74
C. Ray Asfahl

4.6. Materials Handling Equipment 4 • 100
Bruce Barger
4.7. Automatic Identification Systems 4 • 128
John M. Hill
4.8. Maintenance 4 • 143
W. Colebrook Cooling

PART 5. MATERIAL: ASSET MANAGEMENT

5.1. Material Requirements Planning 5 • 3
William L. Berry and Warren Boe
5.2. Material Handling and Storage 5 • 15
James M. Apple, Jr., and Harvey V. Rickles
5.3. Material Flow Control 5 • 38
William T. Shirk
5.4. Material Protection 5 • 72
Alfred H. McKinlay

PART 6. MONEY: FINANCIAL MANAGEMENT

6.1. Economic Analysis for Production Planning 6 • 3
Ferdinand F. Leimkuhler and Frederick T. Sparrow
6.2. Economic Analysis 6 • 23
Gerald J. Thuesen
6.3. Cost Control Systems 6 • 38
John R. Canada
6.4. Unit Cost Analysis 6 • 49
Samuel Eilon

PART 7. SPACE: THE PRODUCTION FACILITY

7.1. Facility Location 7 • 3
Leon F. McGinnis
7.2. Facilities Planning 7 • 18
James A. Tompkins
7.3. Facilities Layout 7 • 30
John J. Mariotti, David W. Poock, and Harry K. Edwards
7.4. Energy Management 7 • 58
John K. Freund

PART 8. SYSTEM: FACTORY INTEGRATION

8.1. Factories of the Future 8 • 3
James J. Solberg, David C. Anderson, Moshe M. Barash, and Richard P. Paul

8.2. Computer Aided Design and Computer Aided Manufacturing 8 • 15
Robert E. Crowley
8.3. Computer Aided Process Planning 8 • 21
Joseph Tulkoff
8.4. Computer Control: Software Design 8 • 41
Leland R. Kneppelt
8.5. Computer Control: Hardware Design 8 • 51
James E. Heaton
8.6. Simulating Production Systems 8 • 57
Steven D. Duket and A. Alan B. Pritsker
8.7. Flexible Manufacturing Systems 8 • 74
Jane C. Ammons and Leon F. McGinnis
8.8. Advanced Flexible Manufacturing Systems 8 • 90
Samuel B. Korin and John C. Luber
8.9. Controlling the Flexible Factory 8 • 96
John W. Mueller
8.10. Implementing the Flexible Factory 8 • 106
Richard L. Engwall
8.11. Managing the Flexible Factory 8 • 122
John A. Maddox

INDEX

PRODUCTION HANDBOOK

PART 1

MANAGEMENT: ORGANIZING FOR PRODUCTION

CHAPTER 1.1

THE MANUFACTURING ENTERPRISE

DAVID W. WHITMYRE

Xerox Corporation
Webster, New York

In the past, the *Production Handbook* was dedicated to the practitioners of factory engineering—the production engineer, production supervisors/managers, and industrial engineers who were charged with maintaining the throughput and productivity of our factories. However, during the past few years our preeminent manufacturing capability has been challenged by a manufacturing enterprise whose high technology was spurred by heavy investment, highly flexible manufacturing systems, new inventory management policies, and a highly dedicated work force focused on a common goal of quality.

This challenge has not gone unanswered. We stand on the brink of a new era that will link our ability to compete in world markets with the ability of our factories to be highly flexible. This flexibility will extend beyond the factory walls starting with DESIGN and linking CAD and CAM to the new manufacturing enterprise. Traditional boundaries of organization and mission will give way to automation and computerized systems. Those companies that do not seek flexibility as a goal are likely to find further deterioration of market share and slippage of profits.

Many of us are committed to the revitalization of our manufacturing capability. However, the path to recovery and to the goal of the automated factory appears awesome, not only from the perspective of knowing where to start, but also how to plan, organize, sell, design, build, and implement a complex system. In an era where large investments are required to maintain a competitive edge, manufacturing has been plagued by increasing complexity, making it even more important that we understand manufacturing.

1.1.1 FIVE M AND TWO S APPROACH

About three years ago, the Xerox Corporation undertook an in-depth comparison of our company versus our Japanese subsidiary. In this competitive benchmark test, it was clear that the Japanese firm achieved higher results in most categories. Puzzled by the firm's unparalleled success, a team of U.S. managers began to dissect the Japanese firm's approach to problem solving and the management process that was evidently so successful for them. They rediscovered, in fact, good management practices. By defining problems in simple terms, the Japanese management team was able to make more decisions in less time. That technique is called the "5Ms and 2Ss" of manaufacturing. The five Ms of manufacturing are:

Manpower
Methods
Machines
Material
Money

The two Ss of manufacturing are:

Space
Systems

With the 5Ms and 2Ss system (see Figure 1.1.1), one can describe all the facets of manufacturing in simple terms and with new understanding of the interfaces and overlaps within manufacturing.

Similarly, a look at each subset of the 5Ms and 2Ss reveals an interesting view of the infrastructure within manufacturing. Below is a listing of those subsets.

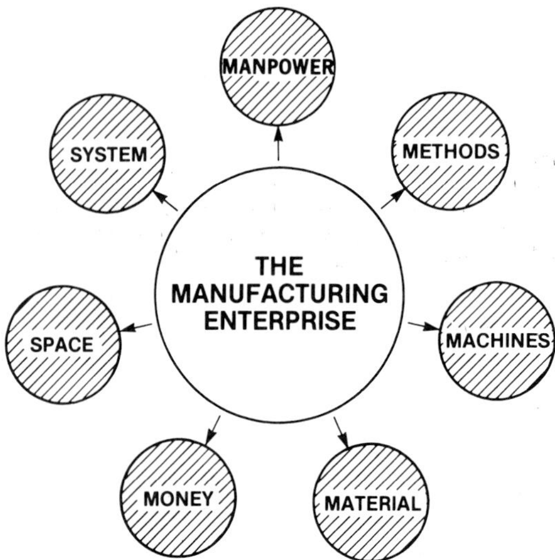

Fig. 1.1.1. The 5Ms and 2Ss.

Manpower. In Figure 1.1.2, job classification, line balance, training, standards, business requirements, safety, performance, and ergonomics are some of the elements of the subset Manpower.

Methods. In manufacturing, Methods is the focal point where many of the 5Ms and 2Ss intersect. Figure 1.1.3 defines Methods to include the critical sequence, safety, tooling, material, station layout, CAD, time standards, manpower, quality/parts inspection, and packaging.

Machines. Machines includes but is not limited to tool management, speeds and feeds, flexible robotics, hard automation, facilities, gaging, tool equipment, calibration, test equipment, ergonomics, and safety. Refer to Figure 1.1.4.

Material. While the value of material is usually disproportionately higher than the value added of labor, there are other considerations associated with material, as seen in Figure 1.1.5. They are material handling,

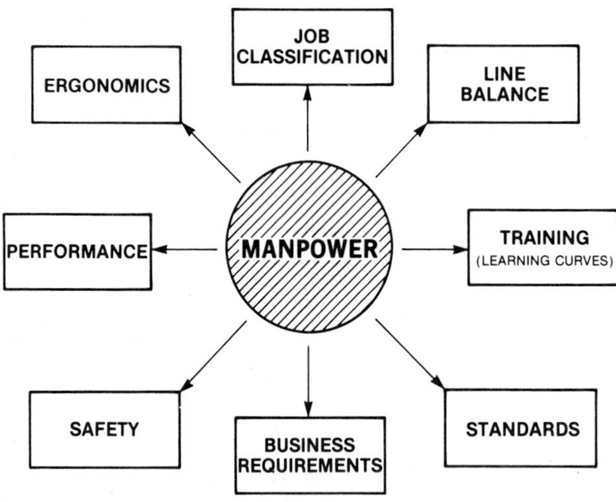

Fig. 1.1.2.

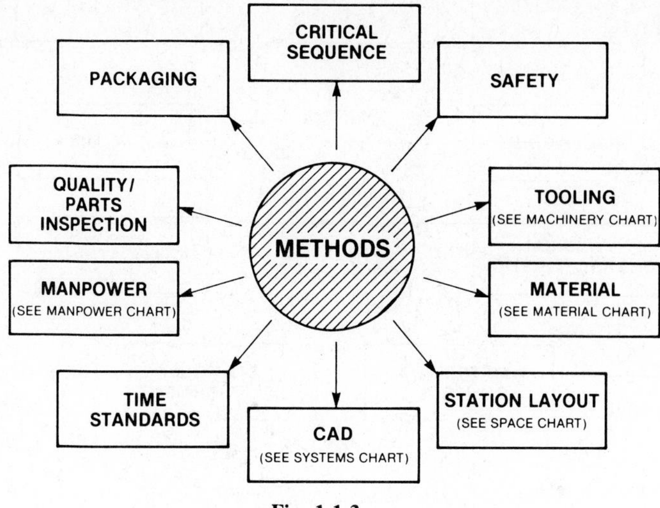

Fig. 1.1.3.

volume, bill of material/M.R.P., station location, usage, station replenishment, material control, physical characteristics, packaging, and shelf life.

Money. The fifth M is Money. Some of the familiar terms are budgets, ROI, capital, NPV, material cost, cost reduction, labor cost, payback, overhead cost, and expense. Refer to Figure 1.1.6.

Space. Space can be described in terms of factory layout, station layout, assembly line layout, safety, physical constraints, capacity, planning function, storage, space utilization, and energy management. Refer to Figure 1.1.7.

Systems. In any complex manufacturing enterprise there are many supporting systems, varying from informal to formal and from simple to sophisticated. Some systems perform tasks while others are used

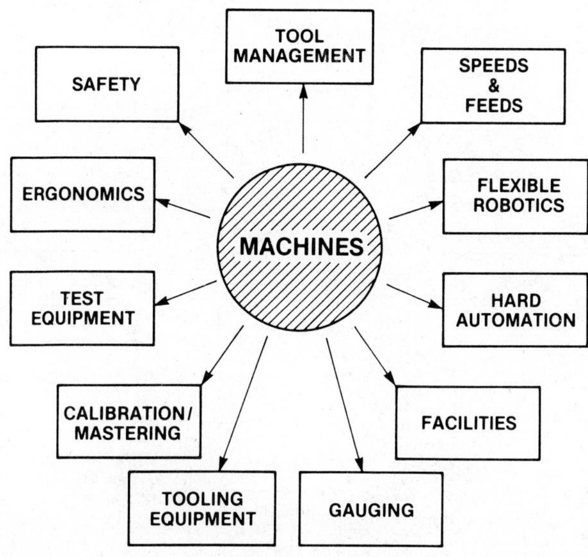

Fig. 1.1.4.

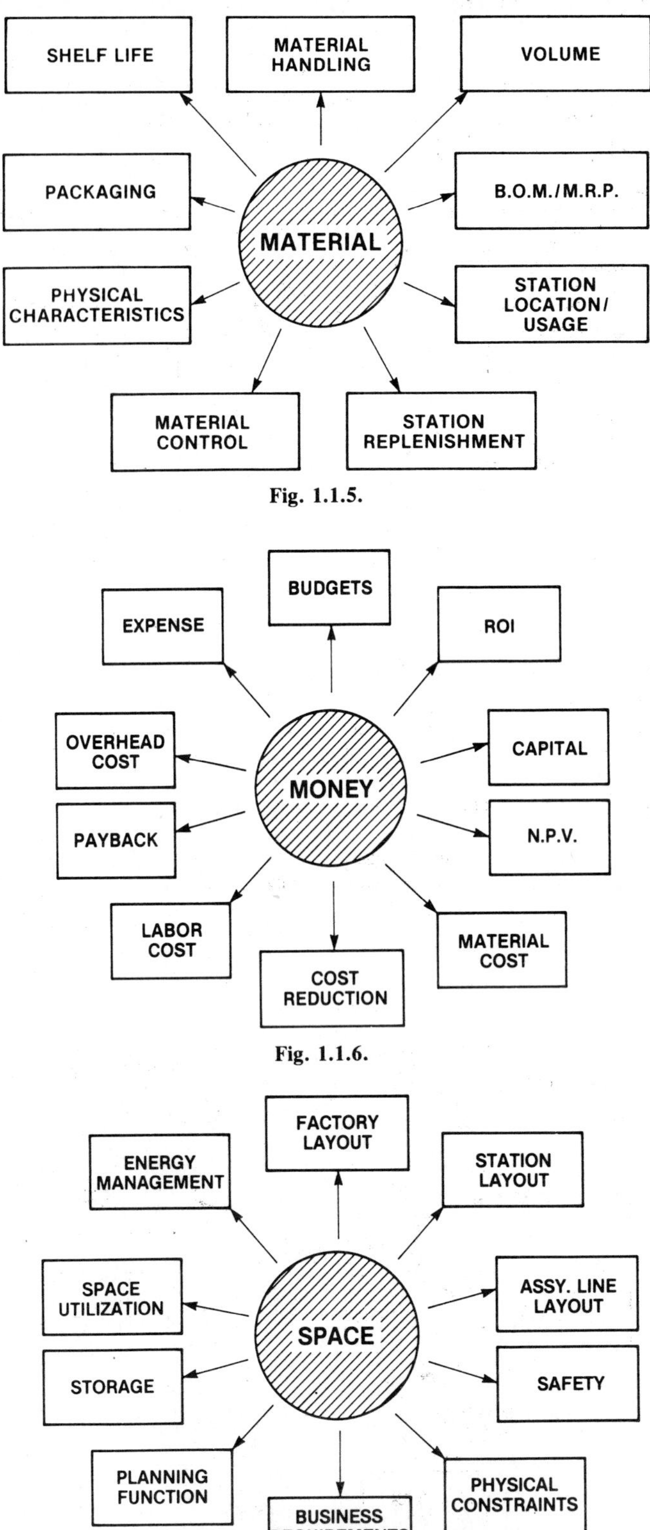

Fig. 1.1.5.

Fig. 1.1.6.

Fig. 1.1.7.

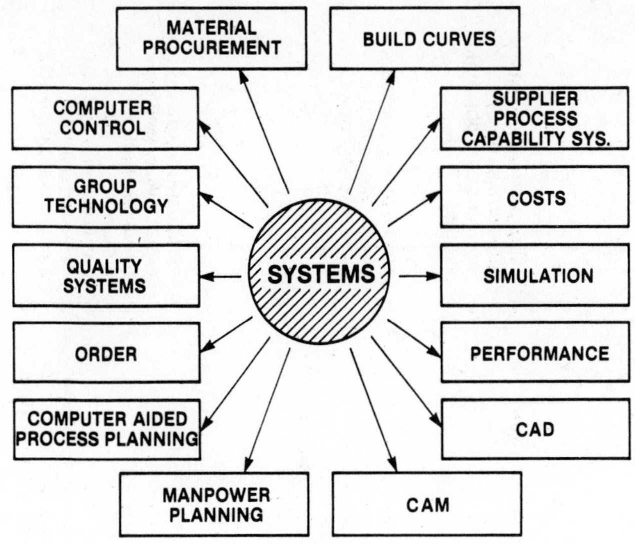

Fig. 1.1.8.

for measurement. Figure 1.1.8 describes such common manufacturing systems as material procurement, learning curves, build curves, supplier process capability system, cost, simulation, performance, CAD, CAM, manpower planning, computer-aided process planning, change order, quality systems, group technology, and computer control.

The 5Ms and 2Ss provide a taxonomy for classifying the complexities of manufacturing into several logical categories. When one applies this to a particular business, although some of the elements are likely to be different, the 5Ms and 2Ss structure will be the same. The 5Ms and 2Ss can be applied equally to discrete parts manufacturing for batch production, flexible manufacturing systems, and continuous production. Because the infrastructure of an organization is likely to vary between businesses, the 5Ms and 2Ss can be applied as a competitive benchmarking technique to compare the common elements of the business.

1.1.2 APPLYING THE 5Ms AND 2Ss

There has been much discussion about the automated factory and the recent realization that a corporation's strength is directly related to superior manufacturing presence and capability. In the race to the automated factory, the path is generally evolutionary as a firm develops the knowledge, the talent, and the strategic planning capability. In Figure 1.1.9, the path from Islands of Automation to Continents of Automation generally requires the integration of two major elements out of three—material handling, the process, and the control system. Whereas Islands of Automation were products of short-term paybacks with a high certainty of recovering the savings, Continents of Automation and Hemispheres of Automation require longer payback periods and have higher risk in realizing and measuring the savings. The evolutionary path to Hemispheres of Automation requires even greater planning, for this requires the automation of all three key elements—material handling, control system, and the process. Systems integration requires higher level control systems that are very complex in order to generate greater efficiency and throughput.

In Figure 1.1.10, the 5Ms and 2Ss system is used to compare Islands of Automation, Continents of Automation, and Hemispheres of Automation.

Figure 1.1.10 demonstrates how the 5Ms and 2Ss system can be used to describe a problem as complex as the automated factory. Use of the 5Ms and 2Ss system has also proven to be successful in other applications, including

defining the strengths and weaknesses of a corporation;

defining the strengths and weaknesses of the competition in a comparative analysis; and

defining the natural overlaps that exist between the elements of manufacturing as a result of its infrastructure.

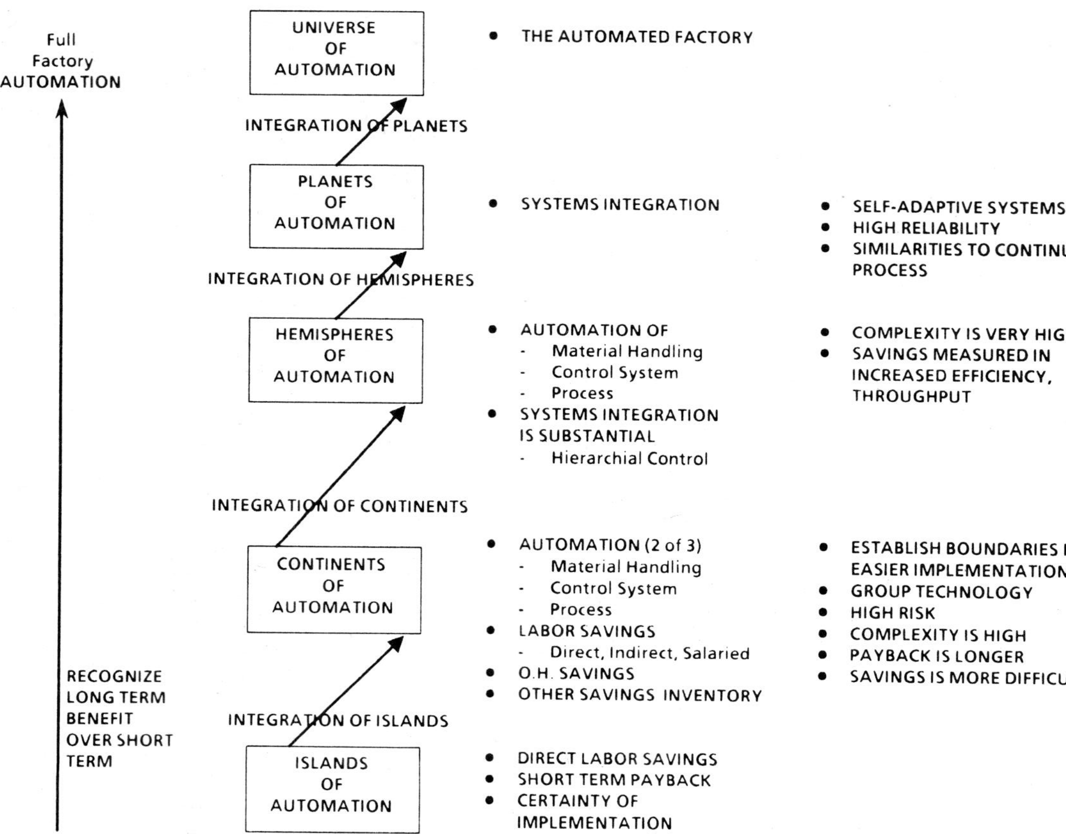
EVOLUTIONARY PATH TO THE AUTOMATED FACTORY
Full
Factory
AUTOMATION
RECOGNIZE
LONG TERM
BENEFIT
OVER SHORT
TERM
UNIVERSE
OF
AUTOMATION
• THE AUTOMATED FACTORY
INTEGRATION OF PLANETS
PLANETS
OF
AUTOMATION
• SYSTEMS INTEGRATION
• SELF-ADAPTIVE SYSTEMS
• HIGH RELIABILITY
• SIMILARITIES TO CONTINUOUS PROCESS
INTEGRATION OF HEMISPHERES
HEMISPHERES
OF
AUTOMATION
• AUTOMATION OF
- Material Handling
- Control System
- Process
• SYSTEMS INTEGRATION IS SUBSTANTIAL
- Hierarchial Control
• COMPLEXITY IS VERY HIGH
• SAVINGS MEASURED IN INCREASED EFFICIENCY, THROUGHPUT
INTEGRATION OF CONTINENTS
CONTINENTS
OF
AUTOMATION
• AUTOMATION (2 of 3)
- Material Handling
- Control System
- Process
• LABOR SAVINGS
- Direct, Indirect, Salaried
• O.H. SAVINGS
• OTHER SAVINGS INVENTORY
• ESTABLISH BOUNDARIES FOR EASIER IMPLEMENTATION
• GROUP TECHNOLOGY
• HIGH RISK
• COMPLEXITY IS HIGH
• PAYBACK IS LONGER
• SAVINGS IS MORE DIFFICULT
INTEGRATION OF ISLANDS
ISLANDS
OF
AUTOMATION
• DIRECT LABOR SAVINGS
• SHORT TERM PAYBACK
• CERTAINTY OF IMPLEMENTATION
• LOW RISK

Fig. 1.1.9. Evolutionary path to the automated factory.

	ISLAND OF AUTOMATION	CONTINENTS OF AUTOMATION	HEMISPHERES OF AUTOMATION
MANPOWER	• Direct labor base may require learning new skills or improving existing skills	• Support personnel requires higher skills base to support increased level of technology • Training/retraining of personnel for new skills • Training at start-up • Special skills required to support system in on-going state • Safety	• Support personnel requires higher skills base to support increased level of technology • Training/retraining of personnel for new skills • Training at start-up • Special skills required oto support system in on-going state • Safety
METHOD	• May be manual/semi-automatic or automatic	• Manual/semi-automatic/or automatic - May require product enablers such as Design for Assembly • Quality Control	• Manual/semi-automatic/or automatic - May require product enablers such as Design for Assembly • Quality Control • Begins to simulate a continuous flow process
MACHINES	• Special purpose, stand-alone machines	• Machine cells, perhaps F.M.S. system, or flexible assembly system • Quick change-over of tooling highly desirable	• Machine cells, perhaps F M.S. system, or flexible assembly system • Need for highly reliable equipment and machinery • Quick change-over of tooling required
MATERIALS	• May be more efficient with family of materials	• MRP and material control • Requires standardization of materials • Semi-auto/automatic material handling	• Most efficient with standardization of materials • Automated material handling
MONEY	• Direct labor saving • Short term payback • Certainty of implementation	• Direct/indirect/salaried savings • Reduction in factory overhead • Inventory savings • Higher risk • Longer payback	• Savings realized through further systems integration
SPACE	• Layout	• Layout • Energy management	• Layout • Energy management
SYSTEM	• System can exist stand-alone	• Partial systems integration is achieved • Systems simulation	• System integration is key • Systems throughput becomes key system indicator • Systems simulation

Fig. 1.1.10. Applying 5Ms and 2Ss to the automated factory.

1.1.3 CONCLUSION

In summary, the 5Ms and 2Ss system describes all the elements of the *manufacturing enterprise*. During the next decade our ability to solve complex problems may directly relate to our ability to clearly define the problems and systematically seek a solution. The 5Ms and 2Ss system provides an organized approach and a tactical analysis whose simplicity can offer a new approach to the *manufacturing enterprise*.

CHAPTER 1.2

STRATEGIC PLANNING FOR MANUFACTURING

ROBERT H. HAYES

Harvard University
Cambridge, Massachusetts

STEVEN C. WHEELWRIGHT

Stanford University
Stanford, California

Intensified competition in a number of global manufacturing industries recently has triggered renewed interest in the manufacturing function and the contribution it can make to a company's overall competitive success. There has been a growing recognition that manufacturing can be a formidable competitive weapon if equipped and managed properly, and that a key to doing that is the development of a coherent manufacturing strategy.

Most managers believe they have a fairly good idea of what a *business strategy* consists of, and they are generally familiar with the basic issues and concerns associated with the terms *marketing strategy* and *financial strategy*. But the fact that there is such a thing as a *manufacturing strategy*, beyond simply "doing whatever is required in order to carry out our other strategies" or pursuing "improved efficiency," comes as a surprise to many people, even those within the manufacturing function. The continual pressure for quick decisions tends to stifle strategic thinking and compels manufacturing managers to adopt stopgap measures that are drawn from a variety of concepts, techniques, and approaches. As a result, these measures are likely to lack clear purpose and lead to inconsistent results. Moreover, their selection and implementation are often disjointed and only indirectly linked to broader issues of general management concern.

The purpose of this chapter is twofold. In the first portion, we present a conceptual framework that we have found useful in thinking about how manufacturing capabilities can be used to achieve a competitive advantage. We begin by reviewing the link between a firm's overall *business philosophy* and its *competitive strategy*, and then describe the relationship between its competitive strategy and its *manufacturing* (functional) *strategy*. Next we discuss the kinds of decisions that are involved in implementing a manufacturing strategy. Finally, we consider alternative approaches to defining and developing a "corporate manufacturing strategy" and recommend one that we have found particularly useful.

In the second portion of the chapter we describe some of the different ways a firm might pursue a competitive advantage and indicate where and how its manufacturing function can make an important contribution to creating that advantage. We also discuss the links between manufacturing strategy and what we refer to as a firm's enduring characteristics—the basic attitudes and preferences that shape the way it manages itself and competes in the marketplace.

1.2.1 COMPANY PHILOSOPHY AND MANUFACTURING STRATEGY

For reasons we explain shortly, the formulation and implementation of an effective manufacturing strategy normally takes several years, requires the support and coordinated efforts of many people throughout an organization, and once in place is difficult to change. Therefore, it is essential that such a strategy be based

Taken, with permission of the authors and publisher, from Robert H. Hayes and Steven C. Wheelwright, *Restoring our Competitive Edge: Competing through Manufacturing*, New York: Wiley, 1984.

on a set of organizational values and preferences that are expected to endure for a long time and that are widely shared. Perhaps this explains why the companies that have built the most formidable manufacturing organizations and are most adept at translating their manufacturing capabilities into competitive success—in short, the companies that have developed and implemented the most effective manufacturing strategies—are those that are characterized by a strongly held set of values and beliefs: a "philosophy of doing business."

We define a company philosophy as *the set of guiding principles, driving forces, and ingrained attitudes that help communicate goals, plans, and policies to all employees and that are reinforced through conscious and subconscious behavior at all levels of the organization.** Such a philosophy ties people together and gives meaning and purpose to their everyday working lives. When an organization comes under attack or finds its environment changing rapidly, a consistent set of values enables people at all levels to contribute usefully, though independently, to its overall objectives.

A "set of common values" is *not* a strategy but it can have similar impact, in that a set of shared objectives also serves to guide decisions and efforts throughout an organization. It encourages certain modes of behavior within the organization and suggests how that organization ought to behave toward its own people, its customers, its suppliers, and the community it serves.

A number of highly successful U.S. firms[1] have developed very explicit, well-thought-out philosophies that help communicate what is and is not important to the firm and, more significantly, what is and what is not "right." For example, there have traditionally been three tenets of IBM's company philosophy: (1) respect for the dignity of the individual, (2) first-rate customer service, and (3) excellence. Hewlett-Packard's philosophy encompasses both business practice and people management:

1. Company-related philosophies

   Pay as you go (no long-term borrowing)

   Market expansion and leadership based on new product contributions

   Customer satisfaction second to none

   Honesty and integrity in all matters

2. People-related philosophies

   Belief in our people

   Emphasis on working together and sharing rewards (teamwork/partnership)

   A superior working environment

These and other companies, such as Timken and S. C. Johnson, believe that their philosophies are so powerful and important that they take precedence over, and serve to screen, all strategic plans and decisions. In particular, they feel their philosophy should infuse all their product/market strategies.

Whether stated explicitly or only implied, such a philosophy can have a profound impact on an organization. It serves as an umbrella over the various corporate, business, and functional strategies adopted by different groups within the company. It not only establishes the context within which day-to-day operating decisions are made but also sets bounds on the strategic options available to the firm. Further, an organization's philosophy guides it in making tradeoffs not only among competing performance measures (such as flexibility, delivery, cost, and quality) but also between short-term and long-term goals. Finally, the degree to which a philosophy is able to achieve consistency among diverse activities tends to be proportional to its coherence and the extent to which that philosophy is shared throughout the organization.

It is not uncommon in firms that have developed strong philosophies to find that the allegiance of managers and workers to their company supersedes their allegiance to their professions. Employees tend to use a common vocabulary, and a set of common examples (a sort of folklore) guides their behavior. They pride themselves on the "company way" of doing things and are imbued with a sense of tradition and continuity. Former employees retain a strong sense of pride in and loyalty to the company and feel a common bond with other "alumni."

As will be described later in this chapter, a manufacturing strategy consists of a pattern of decisions affecting the key elements of a manufacturing system. Because the choices that constitute this pattern are affected by, and should be reflective of, the company's philosophy, making fundamental changes in a manufacturing strategy requires careful attention to its potential interaction with that philosophy (both implied and explicit) and the associated driving forces within the organization. Moreover, companies do themselves a disservice if they fail to exploit the contribution that the manufacturing organization can make as an integrating, coordinating, and communicating force. If managed properly, the manufacturing function can play a unique role in helping to define, support, and enhance the philosophy *and* competitive success of the business, operating in concert with all its functions. This is the theme of this chapter.

* During the early 1980s, the concept of a "corporate culture" gained widespread attention. Generally, culture has been used to describe the rules, norms, and expectations of an organization with regard to the behavior of its members. Such behavior, in turn, reflects the basic values, principles, and philosophies shared by the organization. In our discussions we do not separate such aspects of behavior from the philosophy and values behind them.

1.2.2 CHARACTERISTICS OF A STRATEGY

The word *strategy* (derived from the Greek military term *strategos*, meaning, literally, "the general's art") has been used so extensively in the past decade that it has lost much of its unique meaning when applied to the practice of management. Most definitions of strategy, however, include such elements as establishing purpose, setting direction, developing plans, taking major actions, and securing a distinctive advantage. (See, for example, Ref. 2, p. 93.) At least five important characteristics are common to the use of the term in business:

1. *Time horizon.* Generally, *strategy* describes activities that involve an extended time horizon, both with regard to the time it takes to carry out such activities and the time it takes to observe their impact.
2. *Impact.* Although the consequences of pursuing a given strategy may not become apparent for a long time, their eventual impact will be significant.
3. *Concentration of effort.* An effective strategy usually requires concentrating one's activity, effort, or attention on a fairly narrow range of pursuits. Focusing on these chosen activities implicitly reduces the resources available for other activities.
4. *Pattern of decisions..* Although some companies need to make only a few major decisions in order to implement their chosen strategy, most strategies require that a series of decisions be made over time. These decisions must be supportive of one another, in that they follow a consistent pattern.
5. *Pervasiveness.* A strategy embraces a wide spectrum of activities ranging from resource allocation processes to day-to-day operations. In addition, the need for consistency over time in these activities requires that all levels of an organization act, almost instinctively, in ways that reinforce the strategy.

Because the word *strategy* is used in a variety of settings and has such a range of definitions, it is useful to identify and contrast different types of management-related strategies. As outlined in Figure 1.2.1, business organizations, especially those structured around functionally organized business units, develop and pursue strategies at three levels. At the highest level, *corporate strategy* specifies two areas of overall interest to the corporation: the definition of the businesses in which the corporation will participate (and by omission, those in which it will *not* participate), and the acquisition and allocation of key corporate resources to each of those businesses.

The definition of the businesses in which a corporation will engage can be based on any of several dimensions. For example, some corporations refer to themselves as steel companies, others as glass companies, and still others as aluminum companies. All such companies tend to focus their activities around a particular material and its associated production processes. Others, such as consumer products firms, use a market segment or group of consumers to guide their selection of activities to engage in. We expand on this topic in Section 1.2.5.1.

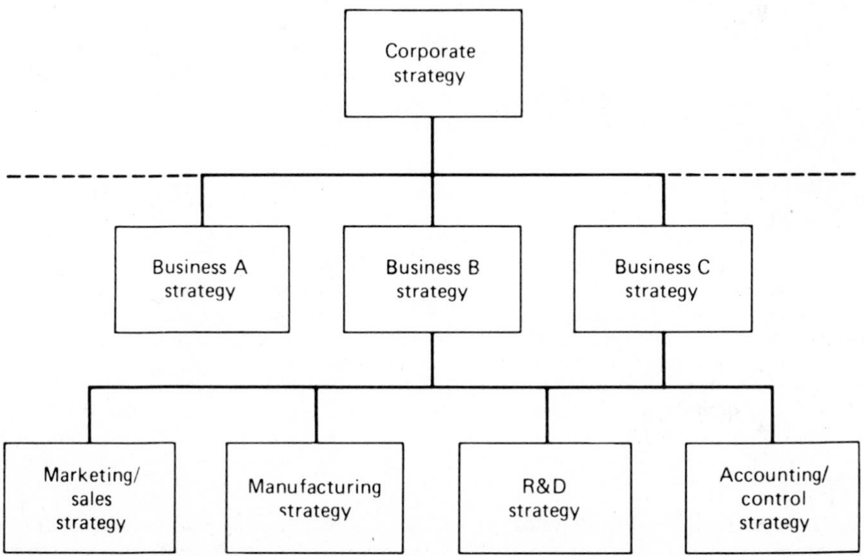

Fig. 1.2.1. Levels of strategy. First level might also refer to the group or sector level in a large, diversified organization. Second level usually refers to a division or strategic business unit (SBU).

The second element of corporate strategy—the acquistion and deployment of resources—is usually dominated by a strong finance staff at the corporate level. This group typically is given primary responsibility for acquiring financial resources and, through a capital budgeting system, allocating them to various aspects of the firm's businesses. Human resources are just as important, of course, and many firms recently have begun strengthening their corporate-level personnel activities in the hope that their acquisition and deployment of valuable human resources can become as sophisticated and well integrated as their management of capital.

The second major level of strategy identified in Figure 1.2.1 is that associated with a strategic business unit (SBU) or strategic planning unit (SPU), which is usually a subsidiary, division, or product line within the firm. Many U.S. corporations have moved to an organizational structure that is composed of a number of relatively autonomous SBUs, each with its own business strategy. A *business strategy* specifies (1) the scope of that business, in a way that links the strategy of the business to that of the corporation as a whole, and (2) the basis on which that business unit will achieve and maintain a competitive advantage.

Specifying the scope of a business requires a statement of the product/market/service subsegments to be addressed. Such a statement is necessary not only to prevent direct competition among the firm's business units (several of which may operate within the same industry, such as "consumer products"), but also to focus the efforts of each unit on activities that are likely to enhance its competitive position in that business. A given SBU might achieve a defensible competitive advantage using one of a variety of approaches, including such generic ones as "low cost/high volume," "product innovation and unique features," or "customized service in selected niches." To be effective, such an advantage must be sustainable using the unit's own resources, take into account competitors' strategies, and fit the customer segments being pursued.

The third level is comprised of *functional strategies*. Once a business unit has developed its business strategy, each functional area must develop strategies that support this strategy. As illustrated in Figure 1.2.1, Business B might have four such functional strategies: a marketing/sales strategy, a manufacturing strategy, a research and development strategy, and a financial/control strategy. In another business unit different functions such as distribution, field maintenance, or quality assurance might be involved.

To be effective, each functional strategy must support, through a specific and consistent pattern of decisions, the competitive advantage being sought by the business strategy. For example, decisions in such areas as pricing, packaging, distribution, and field service—all subparts of the marketing functional strategy—would be very different if the desired competitive advantage were high volume/low cost rather than, say, unique features/ customized service. Similarly, such research and development decisions as the selection of which technologies to pursue, whether to be a technological leader or follower, and whether to emphasize basic research or developmental engineering/manufacturability of designs constitute subparts of the R&D functional strategy. It cannot be overemphasized that it is the *pattern of decisions* actually made, and the degree to which that pattern supports the business strategy, that constitutes a function's strategy, not what is said or written in annual reports or planning documents.*

An important aspect of functional strategy that is often overlooked is the difference between horizontal and vertical activities. Briefly, vertical activities are those that relate a single function to the business-level strategy or relate a subfunction to the overall functional-level strategy. These are more easily tackled, by academics as well as practitioners, because they follow classic hierarchical organizational relationships, as seen from the perspectives of both the manager and the technical specialist. Horizontal activities are those that cut across multiple functions at fairly low levels. Examples are quality improvement, product development/manufacturing startup, and large-scale engineering projects. These require much more coordination and consistency among functions than do vertical activities and should also be addressed when different functional groups formulate their strategies.

1.2.3 DEFINING MANUFACTURING'S STRUCTURE, INFRASTRUCTURE, AND STRATEGY

As implied by the preceding discussion of different kinds of strategies, an effective manufacturing operation is not necessarily one that promises the maximum efficiency or engineering perfection, but rather one that fits the needs of the business—that strives for consistency between its capabilities and policies and the competitive advantage being sought. Translating the business strategy into an appropriate collection of bricks and mortar, equipment, people, and procedures requires resources, time, and management perseverance to ensure that literally hundreds of manufacturing decisions, large and small, are mutually supportive.

Because of the diversity of manufacturing decisions that must be made over time, an organizing framework that groups them into categories is useful both in identifying and in planning a firm's manufacturing strategy. A framework that we have found particularly helpful in working with a variety of firms uses eight major categories, as summarized in Table 1.2.1.

* Some writers and managers distinguish between an "enunciated" (or planned) strategy and an "implemented" strategy. We do not make that distinction because it suggests that developing a strategy and then implementing it are somehow separable. We think the development of a manufacturing strategy is an interactive process involving planning and execution at various levels and in a variety of areas. In the end, it is the pattern of decisions actually pursued that determines the firm's manufacturing capabilities.

A second, and we think more useful, definition of corporate manufacturing strategy can be stated using Table 1.2.3. In one sense, this second definition simply clarifies and corrects the first, but it does so in a manner consistent with the levels of strategy identified in Fig. 1.2.1. In Table 1.2.3, each business unit is permitted to have its own manufacturing strategy, as indicated by the different symbols (000000, //////, and ++++++) that are found within the eight major decision categories under each of the three business units (A, B, and C).

Each SBU within the corporation may pursue similar policies within certain subcategories even though its complete manufacturing strategy is unique. If so, it may be possible to identify a few corporate-wide policies, regarding certain types of manufacturing decisions, that are common across all (or most) of the businesses in which the company engages. In Table 1.2.3 these are indicated by the symbol ××××××. The column on the far right contains examples of such common policies for a hypothetical corporation. For example, a firm might specify certain size and location characteristics for its individual manufacturing facilities, or certain personnel policies, independent of the particular business or division involved. In fact, many of these common policies are likely to arise out of the philosophy and values that the corporation shares with all its business units.

The corporate manufacturing strategy under this definition consists of those subparts of each of the eight decision categories that are governed by common policies across all the firm's businesses. The choice of the decisions to be held constant in this manner can vary significantly for different companies. In some corporations, for example, decisions regarding work-force policies may be left entirely to the individual business unit. In others, a strong corporate culture may dictate that a common set of work-force policies be applied to all business units, whereas quality decisions may be left entirely to individual divisions.

Determining which decisions should be held constant for a given firm can be a difficult problem, because it involves tradeoffs between *ease of control* and *focus*. Some people believe that all subelements should be adapted to local circumstances; this, of course, thwarts attempts to identify the commonalities (and to develop associated controls) that could be the basis of a corporate manufacturing strategy. Most companies, however, lean in the opposite direction: they require too much commonality.

A third definition is sometimes implied in discussions of "corporate manufacturing strategy." It simply identifies those areas where a corporate-wide perspective and focused efforts would be preferable to letting each business unit develop its own manufacturing policies. In essence, this simply raises division level issues to the corporate level for analysis and solution. For instance, when the U.S. Occupational Safety and Health Act (OSHA) was first passed, a number of companies established a corporate-level position to monitor adherence to the Act because few of their operating units had enough experience (or motivation) to do so properly. Later many of those efforts were decentralized and eventually became embedded in the manufacturing strategy adopted by each individual business unit.

Another area where it is often useful to adopt a corporate-wide perspective is in basic R&D, particularly that relating to manufacturing processes. A company may decide that it will need a certain manufacturing capability in the future, even though none of its business units has an immediate need for it. Therefore it may choose to develop an in-house capability in that technology so it will be available when needed. For some period of time, that capability would be found only at the corporate level, although subsequently it would show up in one or more of the business units.

This third definition of corporate manufacturing strategy is very consistent with that outlined in Table 1.2.3. For this reason it can be considered simply an extension of the second definition, which is the one we prefer.

1.2.5 SEEKING COMPETITIVE ADVANTAGE THROUGH A MANUFACTURING STRATEGY

If the manufacturing function assembles and aligns its resources through a cohesive strategy, it can play a major role in helping a business attain a desired competitive advantage. The notion that manufacturing can be a competitive weapon rather than just a collection of rather ponderous resources and constraints is not new, although its practice is not very widespread. Most manufacturing-based businesses simply seek to minimize the negative impact that manufacturing can have. Even in many well-managed firms, manufacturing plays an essentially *neutral* role, reflecting the view that marketing, sales, and R&D provide better bases for achieving a competitive advantage.

To understand the potential contribution that the manufacturing function can make in strengthening a firm's competitive position, it is useful to look more closely at the relationship between manufacturing strategy and business strategy, and how the attitudes and preferences that underlie a business strategy also shape its manufacturing strategy. Identifying such preferences, which relate to specific aspects of the "company philosophy" we discussed earlier, can assist a business unit in setting priorities, making necessary tradeoffs, and developing more effective functional strategies. In this section, four general types of preferences—dominant orientation, pattern of diversification, attitude toward growth, and choice of competitive priorities—are described and their implications for manufacturing strategy discussed.

1.2.5.1 The Dominant Orientation of the Business Unit

Some companies have a strong market orientation. They consider their primary expertise to be the ability to understand and respond effectively to the needs of a particular market or consumer group. In exploiting this

Table 1.2.3. The Concept of a Corporate Manufacturing Strategy

Dimensions of a Manufacturing Strategy	Individual Business Strategies			Examples of Generic (Corporate-Wide) Policies and Guidelines
	Business A[a]	Business B[a]	Business C[a]	
Capacity[b]	××××××	××××××	××××××	A common set of criteria to be used in developing/presenting an investment proposal
	000000	//////	++++++	
	000000	//////	++++++	Policies for the economic or competitive conditions required to plan/start/postpone capacity changes
	000000	//////	++++++	
Facilities[b]	××××××	××××××	××××××	Parameters governing the size and location of individual facilities
	××××××	××××××	××××××	
	××××××	××××××	××××××	Guidelines for permanent reductions in capacity at mature facilities
	000000	//////	++++++	
	000000	//////	++++++	
Technology[b]	××××××	××××××	××××××	Policies for the organization and layout of production processes
	000000	//////	++++++	
	000000	//////	++++++	Criteria for equipment selection and the levels of automation to be pursued
	000000	//////	++++++	
	000000	//////	++++++	
Vertical Integration[b]	××××××	××××××	××××××	Policies for make/buy analysis and changes in backward integration
	000000	//////	++++++	
	000000	//////	++++++	Rules for establishing internal transfer prices

Work force[b]	××××××	××××××	××××××	Establishment of benefit packages and pay scales
	××××××	××××××	××××××	Policies on unionization, hiring, promotion, and employment stability
	××××××	××××××	××××××	
	000000	//////	++++++	
Quality[b]	××××××	××××××	××××××	Standardized reports, reporting relationships and job definitions
	××××××	××××××	××××××	
	000000	//////	++++++	Guidelines on performance measures such as the cost of quality, field failures, and expected quality levels
	000000	//////	++++++	
Production planning/ Materials control[b]	××××××	××××××	××××××	Parameters for manufacturing system specifications and hardware approval
	××××××	××××××	××××××	
	××××××	××××××	××××××	Rules for measuring and evaluating inventory performance
	000000	//////	++++++	
	000000	//////	++++++	
Organization[b]	××××××	××××××	××××××	Definitions for job classifications and direct/indirect staffing levels
	000000	//////	++++++	
	000000	//////	++++++	Policies regarding manufacturing engineering support levels and use of outside services

[a] Each column represents the manufacturing strategy (pattern of manufacturing decisions) that complements a specific business strategy.
[b] Each row represents behavior, practices, and policies in that decision category that are consistent across business (indicated by ××××××), and those not consistent across all businesses.

market knowledge, they use a variety of products, materials, and technologies. Gillette and Procter & Gamble come to mind. Other companies have a material, or product, orientation. They develop multiple uses for their product or material and follow those uses into a variety of markets. Corning Glass, Goodyear, DuPont, and Shell Oil are examples of companies with such a dominant orientation. Still other companies and businesses are technology oriented (most electronics and pharmaceutical companies fall into this class), and tend to follow the lead of their technological expertise into various materials and markets.

In essence, such companies believe that by confining themselves to activities that allow their various business units to exploit similar capabilities, they can do much better than if they tried to pursue a full range of business strategies in a very diverse set of business activities. A company that defines itself in this way often finds it very difficult to venture outside its dominant orientation. An example is provided by Texas Instruments' decision, in the early 1970s, to produce consumer products such as electronic calculators, digital watches, and home computers. Although TI may continue in some of these consumer markets, its announcement in mid-1981 that it was abandoning the digital watch market, followed by a similar announcement two years later concerning home computers (its explanation in each case was that this would allow it to devote more resources to its primary business, integrated circuits) illustrates how difficult even a "natural" diversification move can be for such a company.

Since most dominant orientations incorporate implicit, if not explicit, judgments as to the relative importance of various functions and their roles in achieving a competitive advantage, they establish strong mind-sets within an organization as to the role that manufacturing should play in its competitive strategy. Those mind-sets are very real constraints. If they are negative or neutral, they must be identified and addressed if manufacturing is to make a significant positive contribution to the company's competitive success.

1.2.5.2 The Pattern of Diversification

A second, and related, preference is the pattern of diversification a company follows. Diversification can be accomplished in several ways: (1) product diversification within a given market; (2) market diversification (geographic or consumer group) with a given product line; (3) process, or vertical, diversification (increasing the span of the process so as to gain more control over vendors or customers) with a given mix of products and markets; and (4) unrelated (horizontal) diversification, as exemplified by conglomerates. These patterns of diversification are closely interrelated with a company's dominant orientation. They also reflect the company's preference to concentrate on a relatively narrow set of activities, products, or markets rather than spread itself broadly over many.

Generally speaking, the greater the variety in a company's businesses, the more likely it is that there will be variety in the business strategies pursued and thus in the manufacturing strategies adopted. Unfortunately, our observations suggest that the greater the variety of manufacturing strategies, the less likely it is that senior-level managers will develop a detailed understanding of manufacturing's potential contribution. This is due both to the fact that there are not enough common threads for them to exploit from their position and because they are unlikely to be familiar with all the technologies involved.

1.2.5.3 The Attitude Toward Growth

The role and importance of growth is a third factor influencing the competitive role of manufacturing. For some companies, growth represents an *input* to the company's or business unit's planning process; for others it is an *output*. Every company continually confronts a variety of growth opportunities. Its decision to accept some and to reject others signals, in a fundamental way, the kind of company it prefers to be. Some companies, for example, in their concentration on a particular market, geographic area, or material, accept the growth permitted by that market or area or material. Other companies, however, are managed so that a certain rate of growth is required if they are to function properly.

Within companies that require a certain rate of growth, two different approaches for transmitting that orientation to individual business units can be observed. Under one approach the corporation requires each business unit to meet that growth rate. Under the other, each business unit is assigned a mission within the "corporate portfolio" that encourages it to pursue a specified rate of growth (as well as other dimensions of expected performance).

The firm's attitude toward growth has a powerful influence on its attitude toward manufacturing as a competitive weapon. In businesses in which high growth is considered essential, the primary task assigned manufacturing is often simply to keep up with that growth—"get the product out the door!" This need tends to take precedence over establishing a competitive advantage on other dimensions of manufacturing effectiveness. Divisions or businesses in which growth is not the primary motivating factor, on the other hand, are more likely to assign a larger and richer strategic role to manufacturing.

1.2.5.4 The Choice of Competitive Priorities

Another set of enduring preferences—that act, in a sense, to integrate and summarize the others—is embodied in the way a company chooses to compete in the marketplace and the types of markets it pursues. In its simplest form, a company's competitive posture indicates whether it prefers to seek high profit margins and low volumes or high volumes and low margins. Some companies, for example, consistently seek out high-volume products, even when this decision subjects them to severe cost-reduction pressure or low margins. This sort of competitive preference often is pervasive throughout a company; that is, most of its business units adopt a similar competitive posture.

Price (and therefore cost) is the most familiar competitive dimension, but it is not the only basis on which a business can compete. A business can configure its competitive strategy around several different modes of competition, and the clearer the priorities placed on these modes, the greater the role that manufacturing can play in supporting that strategy.

In some businesses the basis of competitive advantage is superior *quality*, achieved either by providing higher product reliability or performance in a standard product (for example, Mercedes-Benz) or by manufacturing a product with features or capabilities that are unavailable in competing products. The cost of providing higher quality, as defined this way, must be balanced against the market's willingness to pay for it, of course. The nature of this balance has powerful implications for the role of manufacturing in the business.

Another competitive dimension that some businesses use to differentiate themselves in the marketplace is *dependability*. Although the products of such firms may be priced higher than the products of others, and they may not offer the highest performance or the latest technology, they do work as specified, they are delivered on time, and the company stands ready to mobilize its resources instantly to ensure that any failures are corrected immediately. IBM, Caterpillar, and Sysco are often cited as examples of companies whose business strategies emphasize such "peace of mind."

Still another important basis for competitive advantage is *flexibility*. There are at least two important types of flexibility: product flexibility and volume flexibility. A business that competes on the basis of product flexibility emphasizes its ability to handle difficult, nonstandard orders and to take the lead in new produce introduction. Smaller companies often make this their primary basis of competitive advantage. Other businesses compete through volume flexibility, emphasizing their ability to accelerate or decelerate production very quickly and juggle orders so as to meet demands for unusually rapid delivery. Successful companies in highly cyclical industries, like housing or furniture, often make volume flexibility a primary priority.

Within a given industry, different companies or business units give different emphases to each of these four competitive dimensions: price, quality, dependability, and flexibility. It is difficult (if not impossible), and potentially dangerous, for a company to try to compete by offering superior performance along *all* of these dimensions simultaneously, since it will probably end up second best on each dimension to some other company that devotes more of its resources to developing that competitive advantage. Instead, a business must attach clear priorities to each dimension, and these priorities will determine how that business positions itself relative to its competitors. Specifying and clarifying these priorities is the first step in formulating manufacturing's role in the business, since the acid test of whether a business has a strategy is whether it displays a consistent set of preferences through the pattern of decisions it makes over time.

1.2.6 MANUFACTURING'S ROLE AS A COMPETITIVE WEAPON

In the context of this web of preferences, attitudes, and competitive priorities, the task for manufacturing is to structure and manage itself so as to mesh with, reinforce, and enhance its SBU's competitive strategy. Manufacturing should be capable of helping the business do the things it considers essential without wasting resources in lower-priority pursuits, because if resources are diverted to low-priority tasks, some of the activities that are really important simply will not get done. Understanding the company's biases and preferences, and their implications for manufacturing/decision patterns, is a prerequisite to realizing the full potential of a manufacturing organization.

Up to this point we have concentrated on what a manufacturing strategy is and how it can support a given business strategy. We now want to challenge the idea that the role of the manufacturing function is simply to assist in implementing the strategy that others have developed. We argue that manufacturing should take a more proactive role in defining the competitive advantage that is to be pursued. It must communicate clearly to top management the constraints it operates under, the capabilities it can exploit, and the options available to it. And it must seek collaborative relationships with other functions. In fact, manufacturing cannot become a significant competitive weapon unless it takes this kind of role.

Common sense suggests that manufacturing should play an equal role with other functions in defining the competitive strategy for the business. That is, top management should consult manufacturing to get its perspective on the major issues facing the business, the strategies being proposed by other functional heads, and the options open to manufacturing. If such an approach is to work, the status and credibility of the firm's manufacturing managers must be made equal to those of other functional managers. This may require changes in how potential manufacturing managers are selected and trained, the career paths they follow, and the way their performance is measured and evaluated. Helpful though such actions might be, unfortunately, they will not be truly successful unless the equality of manufacturing becomes embedded in the "culture" of the organization.

If manufacturing's role in the firm is made credible by its culture, moreover, the firm is less likely to experience the type of disaster that can occur when manufacturing falls out of step with the other functions. It helps managers make the appropriate tradeoffs between short-run and long-run interests and provides a mechanism for viewing decisions that are often regarded as merely tactical (work force, quality, production planning/materials, etc.) in their strategic context.

The firm's philosophy (and derived culture) specifies the kind of organization it wishes to be, how it is viewed by competitors, stockholders, employees, and the public, and the common values these groups share. The implications of this are particularly important for manufacturing, because a philosophy is effective only to the extent that it is widely shared by the people in the organization, and the majority of a company's people usually are members of its manufacturing organization. Recoginizing that the manufacturing organization is the major "keeper" of the company's philosophy is an important step in expanding the contribution that manufacturing can make.

Finally, strategy development should be an ongoing, interactive process, with inputs and perspectives contributed by all functions, rather than a single, all-encompassing (and inevitably dated and incomplete) statement at a given moment. It is easier for manufacturing to play an active, rather than reactive, role in companies that encourage creativity and consistency across functions and over time.

REFERENCES

1. Thomas J. Peters and Robert H. Waterman, Jr. *In Search of Excellence*, Harper & Row, New York, 1982.
2. C. Roland Christensen, Kenneth A. Andrews, Joseph L. Bowers, Richard G. Hamermesh, and Michael E. Porter. *Business Policy*, 5th ed. Irwin, Homewood, IL, 1982.

BIBLIOGRAPHY

Andrews, K. R. *The Concept of Corporate Strategy*, rev. ed. Homewood, IL: Dow Jones-Irwin, 1980.

Grant, J. H., and W. R. King, *The Logic of Strategic Planning*. Boston: Little, Brown, 1982.

Hofer, Charles W., and Dan Schendel. *Strategy Formulation: Analytical Concepts*. St. Paul: West Publishing Co., 1978.

Pascale, Richard T., and Anthony G. Athos, *The Art of Japanese Management*. New York: Simon & Schuster, 1981.

Peters, Thomas J. "Putting Excellence into Management." *Business Week* (July 21, 1980), pp. 196–205.

Peters, Thomas J., and Robert H. Waterman, Jr. *In Search of Excellence*. New York: Harper & Row, 1982.

Porter, Michael E. *Competitive Strategy*. New York: Free Press, 1980.

Skinner, Wickham. *Manufacturing in the Corporate Strategy*. New York: Wiley, 1978.

Wheelwright, Steven C. "Japan—Where Operations Really Are Strategic." *Harvard Business Review* (July–August 1981), pp. 67–74.

Wheelwright, Steven C., "Reflecting Corporate Strategy in Manufacturing Decisions." *Business Horizons* (February 1978), pp. 57–66.

CHAPTER 1.3

CAPACITY PLANNING

HANAN LUSS

AT&T Bell Laboratories
Holmdel, New Jersey

1.3.1 INTRODUCTION

Planning for the expansion of production capacity is of vital importance in both the private and public sectors. Examples can be found in heavy process industries, telecommunications networks, electric power services, and water resource systems. Since the late 1950s many quantitative studies of capacity expansion problems have been conducted and a variety of models and computational algorithms have been developed.

In the first comprehensive treatment of the subject, Manne[1] edited a book on capacity expansion problems that includes case studies for heavy process industries in India, as well as theoretical papers. This book has provided an excellent stimulus to researchers and practitioners, who have developed and used capacity expansion models. Recently, Freidenfelds[2] gave an excellent exposition of various capacity expansion problems, and Luss[3] published a state-of-the-art survey on operations research models and algorithms for capacity expansion problems. Many of these models require some familiarity with basic operations research techniques covered in an introductory book such as Hillier and Lieberman.[4]

Rather than rigorously define what is meant by capacity expansion problems, several representative examples will be described.

Heavy Process Industries

Suppose the demand for aluminum is forecast to grow over the next 30 years. To satisfy the growing demand, it is necessary to expand the production capacity over the years. The capacity expansion costs have substantial economies of scale (the average cost per capacity unit decreases with the expansion size). A sound economic policy must therefore consider the trade-off between the economies of scale of large expansion sizes and the cost of installing capacity before it is needed. The basic capacity expansion problem consists of determining the size of facilities to be added and the associated times at which the facilities should be added so that the present worth cost of these expansions is minimized.

In other applications, like those in the cement and fertilizer industries, products are shipped to various demand locations at substantial transportation costs. Hence, in addition to the sizing and timing decisions described above, the appropriate location for an expansion becomes an important factor in the capacity expansion planning process.

Telecommunications Networks

Overall capacity planning of telecommunications networks has additional complex facets. The demand is specified in terms of point-to-point requirements for all relevant nodes of the network. Given the demand for each period over the planning horizon, the objective is to find the capacity expansion policy that minimizes the total discounted cost, subject to satisfying the demand at all periods. In order to find such a policy, one should plan not only the expansion of each link over time, but also design the routing of demand throughout the network in each period. Obviously, the routing decisions affect the capacity expansion decisions and vice versa.

Consider expansion policies for a single network link that is used to satisfy two types of demand, e.g., demands that require cables of different quality. Suppose that a standard cable can be used to satisfy only one type of demand, whereas a more expensive cable can be used to satisfy both types of demand. Due to the economies of scale of large expansion sizes, it may be beneficial to use the expensive cable to satisfy both demands. Thus, given the demands over time, the expansion planning consists of determining the sizes, timing, and type of cables to be installed.

Electric Power, Water Resources, and Other Public Services

Electric power utilities must expand power generation facilities in order to meet growing demands. Typically, there are various types of power generation equipment with different capital and operating costs. The optimal mix of equipment types for any given period depends on the Load Duration Curve, a function that describes the demand for power within the period. The capacity expansion planning consists of timing, sizing, and location decisions, as well as decisions regarding the optimal mix of equipment types.

The expansion of water resource services often requires the development of a river basin. Typically, it consists of implementing over time a subset of projects selected from a finite number of possible projects. For example, several dams can be built in different locations and various canals may be constructed. Optimal project selection and sequencing decisions may have to consider various dependencies; e.g., a dam utilization may depend on the set of previously constructed canals.

In growing communities, various public services (schools, libraries, roads, jails, etc.) require the expansion of facilities over time. Unfortunately these expansion decisions are often influenced by political considerations and short-term budgetary constraints. Taxpayers' money is therefore not always spent in an optimal way.

Due to the large sums of money required for capacity expansion, capacity planning is done very carefully by professionals familiar with mathematical modeling, sophisticated algorithms, and existing software packages. The decision making is typically done by middle and upper management. This chapter addresses problems that arise in the applications above and explores appropriate models and algorithms. The most important issue in capacity planning is the trade-off between economies-of-scale savings of large expansion sizes and the cost of installing capacity before it is needed. The next section describes this and a variety of other key issues.

Certain capacity expansion problems are not discussed here, e.g., capacity planning of production machines within a factory. The difficult part consists of identifying the true bottlenecks under ever-changing production requirements. Manufacturing Resource Planning (MRP) systems, discussed in Chapter 5.1, may be used to help identify such bottlenecks. Other more sophisticated production scheduling packages that employ bottleneck optimization methods or just-in-time systems can also be used. Once bottlenecks are identified, the decision whether to add one or more machines can be addressed using simple economic analysis or, if needed, via the models described in this chapter.

In Section 1.3.2 key modeling issues are described. In Sections 1.3.3–1.3.6, capacity expansion models and algorithms are presented: single-facility problems, capacity expansion via a finite set of projects, two-facility problems, and multifacility problems. Final remarks are given in Section 1.3.7.

1.3.2 KEY ISSUES IN CAPACITY EXPANSION PROBLEMS

In all the applications described above, the expansion of production capacity requires the commitment of substantial capital resources over long time periods. The major decisions are the selection of the following:

Expansion sizes
Expansion times
Facility locations
Facility types
Project selections

Given the pattern of demand over time, the most common objective is to minimize the total discounted cost of the expansion process. Management often imposes budgetary constraints and restrictions on acceptable policies.

1.3.2.1 Fundamental Modeling Issues

The simplest capacity expansion model is for a deterministic demand that grows linearly over time with a rate of δ per year. Suppose that the capacity, once installed, has an infinite economic life and that whenever demand reaches the existing capacity, x units of capacity are added (all expansion sizes are the same). Figure 1.3.1 shows the pattern of demand and capacity over time. Suppose $f(x)$ denotes the cost of an expansion of size x, x/δ is the time between successive expansions, and r is the discount rate of money ($f(x)$ and r should be on a comparable basis; e.g., net of inflation). Then the discounted cost $C(x)$ of all expansions over an infinite horizon is

$$C(x) = \sum_{k=0}^{\infty} \exp(-rkx/\delta)f(x) = f(x)/[1 - \exp(-rx/\delta)] \ .$$

In many problems the expansion size is taken to be a continuous variable, i.e., the expansion size may take on any value. This is a common convention that is quite realistic when the number of possible expansion sizes is large. Of course, when this number is small, the feasible expansion sizes should be considered explicitly.

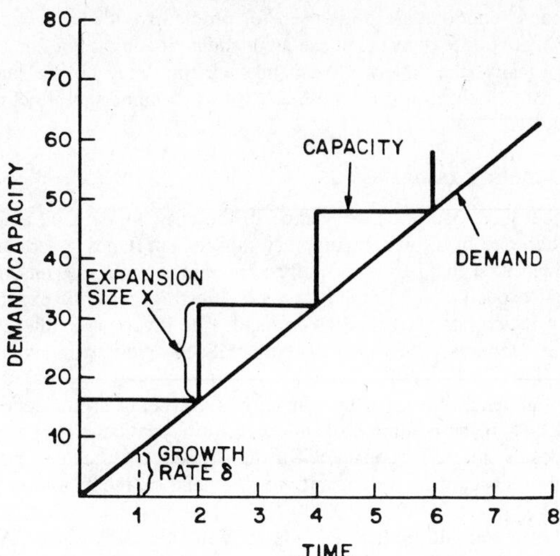

Fig. 1.3.1. The capacity-expansion process. *Source*: Ref. 3, p. 909. Reprinted with permission from *Operations Research*, **30**(5), 1982, Operations Research Society of America. No further reproduction permitted without the consent of the copyright owner.

The discounted cost of all expansions depends on the expansion cost function $f(x)$, the discount rate r, and the growth of demand over time. The expansion cost function is often concave, exhibiting economies of scale. Common cost functions are the power cost function,

$$f(x) = Kx^{\alpha}\ (0 < \alpha < 1) \qquad x \geqslant 0$$

the fixed charge cost function,

$$f(x) = \begin{cases} 0 & \text{if } x = 0 \\ A + Bx & \text{if } x > 0 \end{cases}$$

and some combination of the two. Suppose that different technologies are used to add facilities of different sizes. The cost function is then usually piecewise concave, i.e., it is concave in the range covered by any single technology, but not over all values of x.

As in most investment problems, the discount rate r has a significant effect on the optimal policy. Unfortunately, the estimation of r is generally subjective; the discount rates used by government agencies are usually smaller than those used by the private sector. Further, since the planning horizon for capacity expansion problems is usually long, it is virtually impossible to forecast the "appropriate" discount rate over that period. An attempt is often made to adjust the discount rate to reflect reductions in expansion costs due to future technological innovations. For example, the cost per capacity unit in communication networks is significantly smaller today than it was 20 years ago and it may decrease further during the next 20 years.

Finally, the discounted cost is affected by demand growth. The most commonly used demand functions $D(t)$ over time t are

$$D(t) = \mu + \delta t \tag{1.3.1}$$

$$D(t) = \mu e^{\delta t} \tag{1.3.2}$$

and

$$D(t) = \beta[1 - e^{-\delta t}] \tag{1.3.3}$$

where $\mu, \delta, \beta \geqslant 0$ are given parameters. The linear function (Eq. 1.3.1) represents demand with a constant growth rate over time. It is used quite often since it simplifies the mathematical analysis. The exponential function (Eq. 1.3.2) represents demand with a growth rate that is proportional to the demand volume at any point in time. The last function represents demand with a decreasing growth rate over time with a saturation level of β.

General nonlinear demand functions are often used for problems with finite and relatively short planning horizons. In these problems, it is important to choose an adequate planning horizon and assess its effect on the optimal expansion policy. Finally, the pattern of demand over time may not be completely known. Treating the demand function $D(t)$ as a stochastic process allows one to examine the effect of demand uncertainty on the optimal expansion policy.

1.3.2.2 Additional Modeling Issues

So far, we have assumed that the existing capacity exceeds the growing demand at any given time. However, capacity expansion can be deferred by allowing temporary capacity shortages at certain shortage costs. Allowing for capacity shortages implies either that part of the demand remains temporarily unsatisfied, or that it is satisfied temporarily by "importing" capacity. A different way of deferring capacity expansion is by accumulating inventory during periods in which capacity exceeds demand. The inventory is then used to satisfy demand in periods of capacity shortage. Obviously, part of the saving realized by deferring expansion is spent on capacity shortages and inventory build-up.

Facilities may require maintenance even if the capacity is not being used. Hence, a holding cost is often associated with excess capacity. If the holding costs are high, it may be beneficial to dispose of some capacity. In particular, capacity disposals should be considered if the demand is expected to decrease for a long period of time. Further, the operating costs of active capacity may depend on the technology available at the selected expansion times.

Management is often concerned with indices associated with expansion policy. A frequently cited index is capacity utilization over time, defined as the demand divided by the existing capacity. Since large expansions may in fact be optimal, higher utilization does not necessarily imply better planning. Furthermore, the utilization index fluctuates within a cycle (time interval between two successive expansions) and from cycle to cycle. For example, in Figure 1.3.1, the utilization varies from 0 to 100% within the first cycle and from 50% to 100% within the second cycle.

In some applications, capacity is increased by implementing several projects chosen from a set of available projects. Further, the capacity associated with any project may depend on previously established projects. For example, this would be the case if the capacity of a dam can be fully utilized only if certain canals have already been constructed. Similarly, the operating cost at any time may depend on existing projects. In particular, if projects are implemented in different locations, the interlocation transportation costs depend on the existing facilities.

Multifacility problems raise additional issues. We distinguish between *multilocation* and *multitype* problems. *Multilocation* problems are those in which production facilities may be installed in various locations and products are shipped from any production facility to various demand locations at specified transportation costs. *Multitype* problems are those in which several capacity types are being used to satisfy various types of demand. Each type of capacity is designed to satisfy a specific type of demand, but it can be converted (perhaps at some cost) to satisfy another type of demand.

As a final point, many capacity expansion models require sophisticated algorithms to generate optimal policies. However, the uncertainties regarding the data, such as future demands and appropriate discount rates, often make it necessary to conduct extensive sensitivity studies. Hence, practitioners often resort to heuristic methods that are computationally more efficient and still provide "good" solutions.

1.3.3 SINGLE-FACILITY PROBLEMS

Single-facility problems address many of the issues described above. We categorize these problems into those with an infinite planning horizon and those with a finite planning horizon.

1.3.3.1 Infinite-Horizon Problems

Figure 1.3.1 exhibits a simple capacity expansion process in which identical expansion sizes (and identical time intervals between successive expansions) are used to satisfy a linearly growing demand. From the discounted cost expression $C(x)$ derived previously, the optimal expansion size can be readily found.

Let t_k $(k = 1,2,3, \ldots)$ denote the expansion times. Suppose the discounted cost incurred from an expansion at t_k until the next expansion epoch at t_{k+1} can be expressed as $\exp(-\gamma t_k)g(t_{k+1} - t_k)$ where γ is a constant and $g(\cdot)$ depends only on $t_{k+1} - t_k$. Then it can be shown that there is an optimal policy with equal time intervals between successive expansions, i.e., $t_{k+1} - t_k = \tau$ for all k. For example, this would be the case when $D(t) = \mu e^{\delta t}$ and $f(x) = Kx^{\alpha}$. The discounted cost from t_k to t_{k+1} is then

$$e^{-rt_k}K(\mu e^{\delta t_{k+1}} - \mu e^{\delta t_k})^{\alpha} = K\mu^{\alpha}e^{-(r-\delta\alpha)t_k}(e^{\delta(t_{k+1}-t_k)} - 1)^{\alpha}$$

Thus, $\gamma = r - \delta\alpha$ and $g(\tau) = K\mu^{\alpha}(e^{\delta\tau} - 1)^{\alpha}$.

Other cases for which the optimal policy consists of equal time intervals between expansions include $f(x) = Kx^{\alpha}$ when $D(t)$ is given by any of the functions displayed before, and $f(x) = A + Bx$ when $D(t)$ is linear. Although this policy is not always optimal, it may often provide very good solutions. Details can be found in Sinden[6] and Smith[7].

An optimal expansion policy does not always have a simple structure. Suppose the discounted cost from t_k to t_{k+1} is given as $g_k(t_k, t_{k+1})$. By differentiating the sum of the g_k's with respect to t_k and equating the result to zero, t_{k+1} can sometimes be expressed in terms of t_k and t_{k-1}. Thus, by varying t_1, the total discounted cost can be plotted as a function of t; see Sinden[6]. Although this approach is interesting, it is often impractical from a computational point of view. In general, it appears easier to find the optimal policy by one of the known dynamic programming algorithms for finite-horizon problems, to be discussed later. Of course, then the sensitivity of the optimal policy to the chosen planning horizon needs to be examined.

More complex expansion policies consist of cycles in which the capacity does not always exceed the demand. During periods of capacity shortages, the demand is satisfied either by using accumulated inventory or by temporarily importing capacity. An optimal capacity expansion policy then consists of determining expansion sizes and time intervals for inventory accumulation and for capacity shortages. All these decisions are interrelated, and finding an optimal policy is quite involved; see Erlenkotter.[8]

All the models above are for deterministic demands. However, sometimes the stochastic nature of the demand has significant impact on optimal expansion policies and should be explicitly considered. Typical approaches include the use of birth and death processes and diffusion approximations; see Freidenfelds,[2] Chapter 7.

1.3.3.2 Finite-Horizon Problems

Many production models with a finite planning horizon can also be used for capacity expansion problems by an appropriate interpretation of the parameters and variables. We start with a fundamental model with discrete time periods $t = 1, 2, \ldots, T$ where T is the finite horizon. Let $r_t \geqslant 0$ be the demand increment for additional capacity in period t, let x_t be the expansion size in period t, and let I_t be the excess capacity at the end of period t. The costs incurred in period t include the expansion cost plus the holding cost of excess capacity. The objective is to find an expansion policy that minimizes the total discounted cost incurred over the T periods provided that capacity shortages are not allowed. Specifically,

$$\underset{x_t}{\text{Minimize}} \sum_{t=1}^{T} [f_t(x_t) + h_t(I_t)]$$

subject to

$$\begin{aligned} I_t &= I_{t-1} + x_t - r_t, & t &= 1, 2, \ldots, T \\ I_t &\geqslant 0, & t &= 1, 2, \ldots, T-1 \\ I_0 &= I_T = 0 \end{aligned}$$

where $f_t(\cdot)$ and $h_t(\cdot)$ are the expansion and holding cost functions in period t, respectively. (The appropriate discount factors are assumed to be included in these functions.)

This problem can be reformulated as a standard dynamic programming problem. Let w_t be the optimal discounted cost over periods $0, 1, \ldots, t$ given $I_t = 0$, and let d_{uv} be the discounted cost associated with a subproblem that finds an optimal plan over periods $u + 1, \ldots, v$ given $I_u = I_v = 0$ and $I_t > 0$ for $t = u + 1, \ldots, v - 1$. These definitions lead to the forward dynamic programming equations.

$$\begin{aligned} w_0 &= 0 \\ w_v &= \underset{0 \leqslant u < v}{\text{minimum}} \, [w_u + d_{uv}], \qquad v = 1, 2, \ldots, T. \end{aligned}$$

These equations can be solved easily if the costs d_{uv} are known. Thus the primary issue is, under what conditions can the costs d_{uv} be obtained with moderate effort?

In many applications the cost functions $f_t(\cdot)$ and $h_t(\cdot)$ are concave (exhibiting economies of scale). In this case there exists an optimal solution that is an extreme point solution; i.e., it cannot be expressed as a convex combination of two other distinct feasible solutions. Using a network flow approach, this property can be exploited to find the cost values d_{uv} of the subproblems. Figure 1.3.2 gives a network representation of the problem in which all possible expansions are represented by arcs emanating from a single source. An extreme flow, which corresponds to an extreme point solution, implies that there are no cycles in which all arcs have nonzero flows. This in turn implies that there is only a single positive flow into each node. Therefore, only a single capacity expansion within the subproblem can be positive (If $r_{u+1} > 0$, then $x_{u+1} = \Sigma_{t=u+1}^{v} r_t$, and $x_t = 0$ for $t = u + 2, \ldots v$.) Hence the costs d_{uv} can be obtained in a straightforward manner.

Various extensions can readily be incorporated into the model and the dynamic programming algorithm. These include negative demand increments, capacity disposals, capacity shortages, and upper bounds on excess

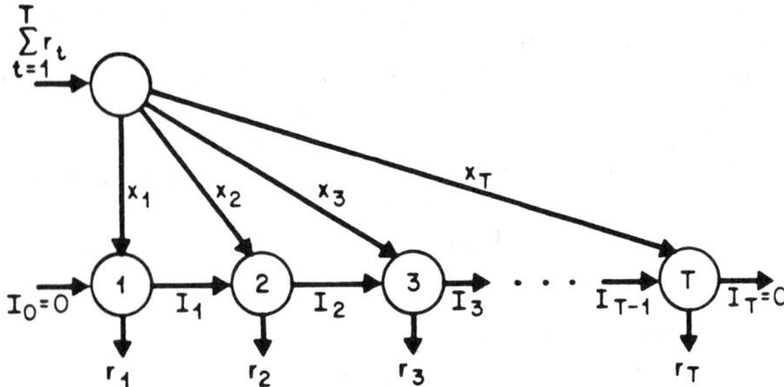

Fig. 1.3.2. A network flow representation of a single-facility model. *Source*: Ref. 3, p. 920. Reprinted with permission from *Operations Research*, **30**(5), 1982, Operations Research Society of America. No further reproduction permitted without the consent of the copyright owner.

capacity and capacity shortages. More details can be found in Manne and Veinott[9] and Love[10]. Upper bounds can also be imposed on the expansion sizes, but the problem becomes more difficult; set Lambert and Luss[11].

1.3.4 CAPACITY EXPANSION VIA A FINITE SET OF PROJECTS

In some applications capacity is expanded by implementing several projects selected from a finite set of possible projects.

Consider a situation in which each project s is defined by its construction cost c_s and capacity x_s, $s = 1, 2, \ldots, S$. Given the demand over time, $D(t)$, the simple expansion problem is to select the time (if any) at which each project should be implemented in order to satisfy the demand at any point and minimize the overall discounted cost. Let X be the set of implemented projects. The set X is described by a vector with S binary digits, where the s-th digit from the right represents the s-th project and all digits that are equal to one represent implemented projects. For example, for $S = 3, X = (110)$ indicates that projects 2 and 3 have been implemented. *Let* $\tau(X)$ be the earliest time at which demand exceeds the available capacity. (For $X = (110)$ the available capacity is $z_2 + z_3$.) Since only the projects' construction costs c_s are considered, it is optimal to implement new projects only when the existing capacity is exhausted; i.e., if project 1 is constructed first, then the next project would be implemented at time $\tau(001)$. Of course, if the capacity z_1 exceeds all future demands, then $\tau(001) = \infty$, and projects 2 and 3 will never be implemented.

Figure 1.3.3 shows a network representation of the capacity expansion problem (known also as the project selection and sequencing problem) for $S = 3$. Each node represents a state X with a corresponding fill time $\tau(X)$. The arcs represesnt the possible transitions from a given value of X; i.e., from $X = (001)$ only nodes (011) and (101) can be reached. The cost associated with each link is the discounted cost of implementing the project; i.e., the cost associated with the link that connects nodes (001) and (101) is $c_3 \exp[-r\tau(001)]$. The problem is to find the shortest path from node (000) to node (111). This can be solved readily by a simple

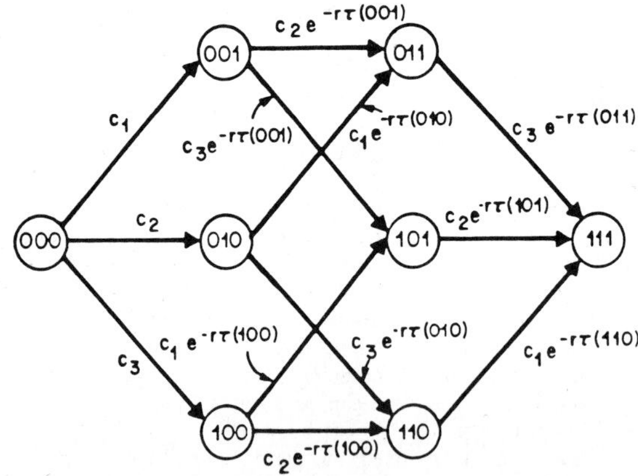

Fig. 1.3.3. A capacity-expansion problem with three possible projects.

dynamic programming algorithm or by any other shortest path method. Since the number of possible nodes grows exponentially with S, this approach is practical only for problems with a relatively small number of projects (up to about 20). Still, this approach is very useful since a complete enumeration of all possible $S!$ sequences is a formidable task ($20! = 2.4 \times 10^{18}$).

For linear demand, $D(t) = \delta t$, the optimal policy can be obtained without using dynamic programming. Let the equivalent cost rate ρ_s of project s be defined implicitly by $c_s = \int_0^{\tau'} \rho_s \exp(-rt)\,dt$, where τ' is the time it takes to use up the capacity associated with project s. Thus $\tau' = z_s/\delta$, yielding $\rho_s = rc_s/[1 - \exp(-rz_s/\delta)]$. The optimal policy satisfies $\rho_s \leqslant \rho_{s+1}$ for all s. Since ρ_s does not depend on the implementation time of project s, the optimal sequence is readily obtained. However, when the demand is nonlinear, the equivalent cost rate of any project depends on its implementation time (τ' will depend on X). So a simple ranking of these rates is not possible. The interested reader is referred to Erlenkotter[12].

Extensions of the basic sequencing model include models with: (1) several capacity levels for each project; (2) dependencies among capacities; (3) dependencies among costs; and (4) operating costs that depend on X. The last requires the explicit consideration of timing decisions, since it may then be attractive to introduce a new project before the existing capacity of all previously implemented projects has been exhausted.

1.3.5 TWO-FACILITY PROBLEMS

Two-facility problems introduce another factor into the capacity expansion process, namely, the location or type of each expansion. Recall that in the two-location problem, the capacity at location i ($i = 1, 2$) can be used to satisfy demand at the other location, but a transportation cost will be incurred. In the two-facility type problem, two capacity types are used to satisfy two types of demand. The *deluxe* capacity is designed primarily to satisfy demand type 1, but it can be converted to satisfy demand type 2 as well. In contrast, the *standard* capacity can accommodate only demand type 2. In some models it is assumed that converted capacity can be rearranged at any time to the deluxe capacity at no cost. In other models it is assumed that converted capacity becomes an integral part of the standard capacity and cannot be rearranged.

Here we shall concentrate on two-facility-type problems. Suppose the demand for type i ($i = 1, 2$) grows linearly with a rate of δ_i over an infinite planning horizon. Let $I_1(t)$ and $I_2(t)$ be the capacity surpluses of the deluxe and standard types respectively at time t, and suppose that rearrangements of capacity can be done at any time at no cost. Since both demands must be satisfied at all times, I_1 and I_2 (the argument t is deleted) must always satisfy $I_1 \geqslant 0$ and $I_1 + I_2 \geqslant 0$. Further, if only expansion costs are considered, an optimal policy consists of expanding only when $I_1 = 0$ or when $I_1 + I_2 = 0$. Let $C(I_1,I_2)$ be the discounted cost of optimally satisfying all future demands starting with spare capacities I_1 and I_2. Further, let x_i be the expansion size ($i = 1, 2$), let $f_i(x_i)$ be the expansion cost function, and let τ be the time interval until the next expansion:

$$\tau = \min\,[(I_1 + x_1)/\delta_1,\ (I_1 + x_1 + I_2 + x_2)/(\delta_1 + \delta_2)]\ .$$

Then, the functional dynamic programming equation is

$$C(I_1,I_2) = \underset{x_1 x_2}{\text{minimum}}\ [f_1(x_1) + f_2(x_2) + C(I_1',I_2')e^{-r\tau}]\ ,$$

where I_1' and I_2' are the resulting capacity surpluses at the next expansion instant. Since the relevant values for the state variables are only those with $I_1 = 0$ or $I_1 + I_2 = 0$, the formulation above reduces to a single state variable problem. The dynamic programming equations can be solved, for example, by policy iteration methods.

Other approaches for solving this problem include heuristics that limit the policy space to policies in which $I_1 + I_2 = 0$ periodically. One simple policy consists of installing the deluxe capacity at time 0 to satisfy both demands until time τ'. At time τ' the standard capacity is installed so that $I_1 + I_2 = 0$ at time τ ($\tau \geqslant \tau'$, where τ' and τ are the decision variables). More complicated "coordinated policies" include several expansions of the standard capacity from time 0 to time τ. Details can be found in Freidenfelds,[2] Ch.6.

As in the case of single-facility problems, the two-facility problems with nonlinear demands have been examined for finite planning horizons with discrete time periods. We now describe a model in which a joint set-up cost is incurred at any period in which at least one expansion has taken place.

Let x_{it} be the expansion size of capacity type i at period t ($t = 1, 2, \ldots, T$); let z_t be the sum of the expansion sizes for both types at period t; let y_t be the amount of capacity type 1 converted to type 2 at period t (once converted, it is assumed to be an integral part of type 2 and cannot be rearranged); let I_{it} be the surplus of capacity type i at the end of period t; and let $r_{it} \geqslant 0$ be the additional demand for capacity type i at period t. The costs incurred include those for expansions, joint set-ups, conversions, and holding excess capacities. The corresponding cost functions $f_{it}(x_{it})$, $d_t(z_t)$, $g_t(y_t)$, and $h_{it}(I_{it})$ are assumed to be equal to zero at zero, nondecreasing, and concave. The model is

$$\text{Minimize} \sum_{t=1}^{T} \left[\sum_{i=1}^{2} [f_{it}(x_{it}) + h_{it}(I_{it})] + d_t(z_t) + g_t(y_t) \right]$$

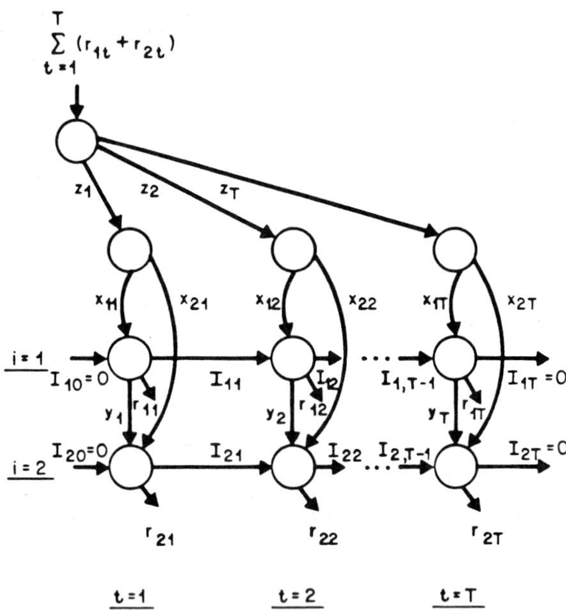

Fig. 1.3.4. A network representation of the two-facility model with joint expansion set-up costs.

subject to

$$
\begin{aligned}
I_{1t} &= I_{1,t-1} + x_{1t} - y_t - r_{1t} & & \\
I_{2t} &= I_{2,t-1} + x_{2t} + y_t - r_{2t} & \quad i &= 1, 2 \\
z_t &= x_{1t} + x_{2t} & \quad t &= 1, 2, \ldots, T \\
x_{it} &\geqslant 0, y_{it} \geqslant 0, I_{it} \geqslant 0 & & \\
I_{i0} &= I_{iT} = 0 & &
\end{aligned}
$$

Since the objective function is concave, there exists an optimal solution that is an extreme point. A network flow representation of the model is given in Figure 1.3.4. Using the network flow representation, properties of extreme flows are derived; for example, only a single incoming flow is positive at any node.

The problem can be reformulated as a dynamic programming problem that searches for the best extreme point solution. The state variables are I_{1t} and I_{2t}, where at most one of them is positive. Hence, this formulation is in fact a single-state problem. Most of the computational effort is spent on deriving a large number of cost values, which are required for the dynamic programming algorithm. These values are computed using the properties associated with extreme flows. Luss[5] proposed this algorithm for problems with two or more facility types. He describes both a heuristic method and an algorithm that provides optimal solutions.

There are similar models and algorithms for two-location problems, but they will not be described here. In general, two-facility problems are primarily solved by dynamic programming methods. For problems with m facility types or locations, the dynamic programming algorithms use $m - 1$ state variables. Though these algorithms can readily be applied to two-facility problems, they are not very practical for multifacility problems with $m > 2$, and heuristic methods are often needed.

1.3.6 MULTIFACILITY PROBLEMS

For two-facility problems we have concentrated on models with two facility types. To balance the presentation, we now discuss multilocation problems.

Suppose capacity expansions may take place in production locations $i = 1, 2, \ldots, m$. Furthermore, products often have to be shipped to different geographical regions, $j = 1, 2, \ldots, n$, at substantial transportation costs. These costs should be considered explicitly while making decisions regarding the timing, size, and location of capacity expansions.

Let $Z = (Z_1, Z_2, \ldots Z_m)$ be the vector of existing capacity levels; let y_{ijt} be the amount shipped at time t from i to j at a price of p_{ij} per unit; and let $D_j(t)$ be the demand at location j at time t. The total transportation cost $P(Z,t)$ as a function of Z and t (but not discounted) is:

$$P(Z,t) = \underset{y_{ijt}}{\text{minimum}} \sum_{i=1}^{m} \sum_{j=1}^{n} p_{ij} y_{ijt} ,$$

subject to

$$\sum_{j=1}^{n} y_{ijt} \leq Z_i \qquad i = 1, 2, \ldots, m$$

$$\sum_{i=1}^{n} y_{ijt} = D_j(t) \qquad j = 1, 2, \ldots, n$$

$$y_{ijt} \geq 0 \qquad \text{for all } i \text{ and } j$$

This is the well-known transportation problem, which can be solved parametrically as a function of the time index t. The capacity expansion problem is that of finding the sequence of expansion times, sizes, and locations so that the total discounted expansion and transportation costs are minimized. We assume that demand in any location must be satisfied at all times. The planning horizon may be finite or infinite.

Various heuristics were developed in which the key assumptions are that additional capacity is added to any location only if all the existing capacity has been exhausted, and that only a single expansion may take place at any given time.

Suppose x_i is installed at location i when the existing capacity $\Sigma_{i=1}^{m} Z_i$ is exhausted. The total discounted cost $g_i(Z,x_i)$ incurred during the fill time $\tau(Z,x_i)$ associated with the added capacity is

$$g_i(Z,x_i) = f_i(x_i) + \int_0^{\tau(Z,x_i)} P(Z', t)e^{-rt}dt ,$$

where $f_i(x_i)$ is the expansion cost and Z' is the resulting capacity vector.

A simple and efficient heuristic is the Minimum Annual Cost (MAC) algorithm, in which expansion sizes and locations are decided myopically. These decisions are based on the equivalent cost rate, $\rho_i(Z,x_i)$, defined implicitly by $\int_0^{\tau(Z,x_i)} \rho_i(Z,x_i)\exp(-rt)\,dt = g_i(Z,x_i)$. For a given capacity vector Z, one can compute for each location i the values of $\rho_i(Z, x_i)$ and $\rho_i(Z)$, where $\rho_i(Z)$ is the minimum over x_i of $\rho_i(Z, x_i)$. The next expansion takes place in the location that provides the smallest $\rho_i(Z)$. Thereafter, the capacity vector Z is updated and the computations are repeated until a desired planning horizon is reached. This heuristic and a more complex "incomplete" dynamic programming heuristic are described in Erlenkotter[13].

If transportation costs are relatively high, restricting the policies to those in which capacity expansions are delayed as long as possible may result in unsatisfactory solutions. Several algorithms have been developed that allow for expansions before all the existing capacity has been exhausted, e.g., Fong and Srinivasan[14].

Erlenkotter[15] compared many of the existing methods for multilocation problems. The comparisons are based on two examples taken from studies done for the heavy process industries in India.

1.3.7 FINAL REMARKS

In this chapter, we described possible applications, basic capacity expansion problems, and various solution techniques. Several useful references were also given. As noted, capacity planning is a complex task and involves careful analysis of expansion decisions over long planning horizons. Finding optimal solutions is often impractical, and instead, heuristic methods are employed. Many of the basic models described here can be used as building blocks for such heuristics. Finally, due to data inaccuracies, sensitivity studies should always be done.

REFERENCES

1. A. S. Manne, Ed., 1967. *Investments for Capacity Expansion: Size Location and Time-Phasing*, Cambridge, MA: MIT Press, 1967.
2. J. Freidenfelds, *Capacity Expansion, Analysis of Simple Models with Applications*, New York: Elsevier, 1981.
3. H. Luss, "Operations Research and Capacity Expansion Problems: A Survey," *Operations Research* **30** (1982), 907–947.

4. F. S. Hillier, and G. J. Lieberman, *Operations Research*, 3rd ed., Holden-Day, San Francisco: 1980.
5. H. Luss, "A Multifacility Capacity Expansion Model with Joint Expansion Set-up Costs," *Naval Research Logistics Quarterly* **30** (1983), 97–111.
6. F. W. Sinden, "The Replacement and Expansion of Durable Equipment," *Journal of SIAM* **8** (1960), 466–480.
7. R. L. Smith, "Optimal Expansion Policies for the Deterministic Capacity Problem," *Engineering Economist* **25** (1980), 149–160.
8. D. Erlenkotter, "Capacity Expansion with Imports and Inventories," *Management Science* **23** (1977), 694–702.
9. A. S. Manne, and A. F. Veinott, Jr., "Optimal Plant Size with Arbitrary Increasing Time Paths of Demand." In *Investments for Capacity Expansions: Size, Location and Time-Phasing*, pp. 178–190, A. S. Manne (Ed.), Cambridge, MA: MIT Press, 1967.
10. S. F. Love, "Bounded Production and Inventory Models with Piecewise Concave Costs," *Management Science* **20** (1973), 313–318.
11. A-M. Lambert, and H. Luss, "Production Planning with Time-Dependent Capacity Bounds," *European Journal of Operational Research* **9** (1982), 275–280.
12. D. Erlenkotter, "Sequencing Expansion Projects," *Operations Research* **21** (1973), 542–553.
13. D. Erlenkotter, "Capacity Planning for Large Multi-Location Systems: Approximate and Incomplete Dynamic Programming Approach," *Management Science* **22** (1975), 274–285.
14. C. O. Fong and V. Srinivasan, "The Multiregion Dynamic Capacity Expansion Problem, Parts I and II," *Operations Research* **29** (1981), 787–816.
15. D. Erlenkotter, "A Comparative Study of Approaches to Dynamic Location Problems," *European Journal of Operational Research*, **6** (1981), 133–143.

CHAPTER 1.4

MANUFACTURING AND COMPETITIVE PLANNING

WILLIAM J. AINSWORTH

Peat, Marwick, Mitchell & Co.
Chicago, Illinois

1.4.1 MANUFACTURING AND COMPETITIVE PLANNING

In recent years many industrial companies have adopted strategic planning to position themselves better for optimum long-term growth and profitability. In highly competitive industrial markets, this can be achieved through competitive planning, which is based on securing and maintaining an increased product value or advantage over market competitors, as perceived by end users. Competitive planning calls for an increased involvement of the manufacturing function in the planning process.

Industrial companies have inherent limitations in strategic decision-making since they are highly labor- and capital-intensive. Manufacturing is in a unique position to determine the product mix that will optimize production resources (labor and capital). Increased production efficiency may give the company a marketplace advantage over its competitors through lower price, higher quality, or improved service. The improvement or expansion of existing production resources can also result in increased efficiency and economy of scale, but this simply underscores the importance of integrating capital planning into the strategic planning process. It is not suggested that manufacturing be moved ahead of the functions that have traditionally governed the strategic planning process, but that it be given increased involvement in jointly developing corporate objectives and strategies.

This chapter will review how this joint responsibility may be performed as part of the traditional strategic planning process based on this concept of competitive planning. The primary objective of the process is a strategy in which the company achieves and maintains a competitive advantage for the planning period. This is the product of a careful assessment of company strengths and weaknesses and the opportunities and threats existing in the marketplace. The long-term viability of a company is often dependent on its ability to effectively perform this strategic assessment of its marketplace. The increased involvement of manufacturing not only impacts strategy formulation positively, but also facilitates the successful implementation and execution of the plan, because manufacturing often has more leverage to control cost, and consequently, profitability. This chapter focuses on applying competitive planning, specifically, the strategic assessment of a company's production resources, competition and market, and the formulation of the strategic plan.

1.4.2 THE STRATEGIC PLANNING PROCESS

The traditional strategic planning process incorporates a number of steps in achieving its primary objective of positioning the company for maximum long-term growth and profitability.

State corporate mission
State corporate objectives
Establish strategic resources framework
Identify strategies
Formulate strategic plan
Monitor and review results

The strategic audit of the external environment, generally the responsibility of the marketing function, involves review of the economic environment, government and societal issues, and, perhaps most important, the market/competitive environment. A joint effort by marketing and manufacturing in evaluating each of these areas provides an added perspective. The contribution is likely to be greatest in the area of technology, suppliers, and market/competitive environments.

1.4.3 COMPETITIVE PLANNING CONCEPT

Competitive planning is a concept that establishes a focus for all the planning systems that the company may use in positioning itself in the future. The objectives of competitive planning are both to position the company at a comparative advantage relative to its competitors and to facilitate quick response to competitors' actions in the marketplace. Manufacturing can play a significant role in defining corporate strategy based on its understanding of what is required to successfully implement the strategy.

To accomplish these objectives, it is necessary to apply the competitive planning concept during both the formulation of the strategic plan and its implementation and execution. Implementation and execution represent the programs for putting into action the elements of the strategic plan that impact the current operating periods.

In applying competitive planning to the audit, certain considerations should be kept in mind. Due to the inherent limitations mentioned earlier, a manufacturing company cannot readily expand or improve its production capabilities. Recognizing this, the first step is determining the optimal configuration of existing production and distribution systems that achieves maximum cost efficiencies based on various historical product volume/mixes. The end product of this exercise would be a ranking of product volume/mix alternatives based on maximum profit contribution. These alternatives are then reviewed with marketing, which has concurrently developed alternative product strategies based on pricing, competitive, and market factors that they feel will offer maximum profit contribution. The major product volume/mix strategy is then agreed upon jointly by marketing and manufacturing. What often becomes evident at this point is the number of products providing negative contribution. It may also become apparent in the analysis that the company needs to reposition itself in more attractive markets by changing its product mix to products that it does not currently produce. This may lead to planned changes in production and distribution capacity contingent upon the availability of capital.

Also derived from this analysis are the key corporate strategies that the company may pursue and that may vary on a product basis. They may include any one or a combination of the following strategies:

Low-cost producer
Market niche
Differentiation/value added

After formulating the strategies, it is manufacturing that accepts the major responsibility for the successful execution of the plan. Such responsibility often requires painstaking process and quality control as well as monitoring the achievement of economies of scale. The remainder of this chapter discusses the role of manufacturing management in utilizing the competitive planning concept in achieving the corporate objectives.

Table 1.4.1 presents the competitive planning approach that can be taken by manufacturing and marketing in formulating corporate strategy as part of the traditional strategic planning.

As noted earlier, competitive planning is also intended to facilitate planned reaction to competitive and market actions. This means anticipating what competitors will do and developing strategies to counter their actions. An example is anticipating sudden changes in demand or supply balances in the marketplace and being prepared to respond quickly. This concept is similar to contingency planning in operations management.

Competitive planning at the execution stage must remain flexible in order to respond to short-term changes in the competitive marketplace. However, if such competitive or market actions evolve into a longer-term trend, a review of the strategic plan and a change in strategy may be in order.

1.4.4 ELEMENTS OF A COMPETITIVE PLAN

This section shows how the competitive planning process may be applied to a typical manufacturing company—ABC Manufacturing Company. The company and situation are, of course, fictitious and are intended only to provide a framework for reviewing the application of the planning concept.

ABC is a medium-sized, privately held manufacturer and fabricator located on the East Coast of the United States. Last year, its sales were $200 million and it made a small after-tax profit on operations. The company manufactures a variety of products for the electronic and electric aftermarket that it sells to commercial and domestic users throughout the United States and Canada. The products range from a line of simple solderless terminals to a sophisticated line of "high-tech" electronic circuit breakers and voltage controllers. The company is a recognized leader in a narrow range of voltage controllers for use with data processing and other sensitive equipment. Its base-business products do not command a significant market share, nor are they recognized industry leaders. The base-business products account for 50 percent of its sales and the high-tech products the other 50 percent. The company has shown a slow to moderate rate of growth for the past several years, barely ahead of inflation, with sales increasing an average of 5 to 8 percent, and profits keeping pace.

Table 1.4.1. Competitive Planning Approach to the Formulation of Corporate Strategy

Manufacturing	Marketing
Determine product mix offering maximum contribution based on historical volume levels.	Obtain competitive pricing for all products produced by the company.
Determine production costs for any new market products identified by market not currently produced by the company and factor into revised product mix.	Obtain competitive pricing on any new market products not currently produced by the company.
Determine optimum production and distribution configuration and develop summary of capital necessary for improvements and/or expansion. Also take into account increased capabilities to meet existing and projected new product volumes.	Rank each existing and new product by the following criteria: Sales equity/supplier relationship Service and development costs Geographic advantages Working capital costs Inventory capital costs Account receivables position Competitive resistance
Assess competitive environment: Determine strengths and weaknesses of competitive production and distribution, sales, and marketing capabilities; Develop pro forma product costs by competitor; and Factor resistance to market-share loss by competitor. Rank alternative strategies by product lines according to maximum long-term growth, profitability, investment leverage, and probability of success.	

ABC management has begun to express some concern at the slow rate of growth and equally disappointing return on equity and has decided that the time is right to promote more rapid growth. Among management's concerns are safeguarding the market share of its more profitable high-tech business. That market is extremely competitive and the lead times to develop and introduce new products are long. ABC executives worry that at any time a competitor might introduce a new product that is either technically superior or offers a low-cost alternative to their product. If this occurs, they will no longer be on the cutting edge of technology in their business and will lose the competitive advantage it offers. They are also concerned about the base-business low-tech products and the increasing costs of maintaining the older plants and equipment that manufacture these products.

Management has decided to apply competitive planning concepts to this year's strategic planning process. The approach would require more involvement from manufacturing and its related functions.

The ABC Company established two primary corporate objectives as part of its strategic planning process:

Attain a 20 percent return on equity (ROE), a 15 percent sales growth per year, and operating margins of 20 percent.

Finance growth internally in line with debt/equity targets.

1.4.4.1 Strategy Formulation

Using the corporate objectives, the manufacturing and marketing functions proceed with the competitive planning process of developing strategies to achieve the objectives. As the process begins, it is important to reflect on the participation necessary. Marketing collects the following information based on both current and prospective markets:

Market size and growth projections

Characteristics of product users

Sales projections by product and product line

Product evolution plans (extensions to current product lines and new products)

Although marketing has a major role in this process, emphasis is placed on the roles of manufacturing and its related functions. Manufacturing and distribution must review aspects of their functions relative to competition and marketplace. A sampling of these reviews is summarized below.

The Product

A complete understanding of the existing product and the planned future product is essential. There should be ample information on how the existing product is produced at its present volume. However, in competitive planning it is necessary to detail exactly how the product will be made and distributed at the projected plan volume. Information about how a competitor makes and distributes products is invaluable, particularly if the competitor is the low-cost and market-share leader in ABC's base business. The analysis is important to all ABC products, but especially to the high-tech portion of the product line because of the volatility of the market and the need to stay ahead of competition.

ABC management also recognizes that whether the company is developing a new product, or merely improving an existing one, a measure of success of product planning and development is the ability to efficiently manufacture the product. Maintaining a competitive advantage demands that manufacturing requirements developed during the original product design be part of any subsequent product evolution or design change.

The Process

The most overlooked element in the development and marketing of a product is how it will be made and distributed. Significant competitive advantage can be gained if the process is thoroughly analyzed at the time the product is conceived. The aspects of the process that can be used to achieve and maintain a competitive advantage are:

Cost-effectiveness of the use of machinery, facilities, and material;

Extent of customer satisfaction to be gained by effective distribution and service; and

Ability of the machinery and facilities to support extensive growth in product sales and evolution in product type.

An analysis of the competitive advantage of the appropriate manufacturing and distribution processes will always show results on the profit line of operations.

Research and Development

This area is concerned with the *manufacturing* technology that will be used to develop and manufacture the product, as distinguished from product research and development. It requires analysis of ABC's own technology and the possible technologies used by competing firms. At this stage, the competitive plan will also include an assessment of the risks of new technology and a comparison of ABC's technology to existing and future technologies that others may employ. ABC wants to assure that if others are working on similar products and product-manufacturing technology, it will maintain a sufficient lead to beat them to product introduction.

Products Evolution

ABC is aware that competitive advantage can be maintained only if it has a plan to keep the company's products and product technology responsive to market conditions. As ABC's competition develops and introduces new products, ABC must be prepared to introduce either a new product or an improved version of an existing product. ABC recognizes that significant time is required to get the product through pilot manufacturing, into production, and to the customers. The steps required to do this must be planned in advance as part of a competitive planning strategy designed to minimize the time from announcement to delivery.

At this point, it should be made clear that competitive planning is *not* a substitute for strategic planning. It is intended to be used *in conjunction with* strategic planning, making the overall process more effective by shaping both strategy and execution to achieve a competitive advantage in the marketplace. The financial, human resource, and other implications are given more emphasis in the overall strategic planning process.

The competitive planning process is intended to identify the major business strategies necessary for ABC to achieve and maintain an advantage in the marketplace in order to meet its corporate objectives. The sequence of planning steps as they may apply to ABC manufacturing, complete with possible findings, are shown in the following summary. The findings are intended to provide a framework for describing the process and are not intended to be all-inclusive.

Planning Step	Finding
Manufacturing determines product mix providing maximum contribution.	Large percent of current base product mix is run at negative contribution. Idle and obsolete base product capacity. Large number of similar high-tech products with low order quantities diminish operating efficiencies.
Marketing obtains competitive pricing for all products produced by the company.	Both product groups are in highly price competitive markets, although both also have product segments with diminished competition. Growing foreign competition, particularly in the base products.
Marketing identifies new products meeting long-term profitability and growth goals and obtains competitive pricing.	Opportunities in a number of attractive new markets and products, particularly where the number of suppliers are decreasing.

Planning Step	Finding
Manufacturing determines production costs of new products and factors into product-mix alternatives.	Many new products offer greater contribution than existing products.
Marketing ranks existing and new products according to selected company, market, and competitive criteria.	Many existing products in both business segments do not offer long-term growth and profitability compared to new products.
Manufacturing determines optimum production configuration and estimates capital required to make changes.	Combination of new and existing products offers product mix with the maximum profit contribution.
Marketing and Manufacturing assess competitive environment.	Many base products exist where ABC will have an increasing advantage over competitors due to diminishing competition.
Research and development provides technical and competitor data and refines product concepts and process techniques.	Clear advantage over competitors in high-tech business. Inventory and receivables terms given some customers make many products marginally profitable.
Develop and rank alternative strategies.	ABC selects following strategies to achieve corporate objective. Market niche/specialization High-tech producer Low-cost producer Market dominance

The reader may recognize that each step in the process required considerable preparation and analysis in order to arrive at the findings indicated. For example, to identify an optimum product mix based on contribution entails a complete breakdown of ABC's and competitors' pro forma costs by product, along with the quantification of other important sales and market criteria to facilitate ranking. In addition, in order to maintain a competitive advantage, it is necessary to consider the same elements in positioning the product for the future. The successful completion of this analysis provides the most effective basis for the formulation of ABC's or any company's business strategies. This effectiveness is based on the fact that if the process is properly performed, the strategies that evolve will be *market-driven* and provide the end user an increased value in utilizing ABC's products versus competitors'. Understandably, the absence of these two characteristics is often cited as the primary reason businesses fail or fall prey to takeover attempts resulting from declining shareholder value as reflected in stock prices.

1.4.4.2 Implementation and Execution of the Plan

We have discussed the first phase of competitive planning as it applies to strategy formulation. The second phase deals with implementing and executing the plan. Aside from its application in formulating strategy, the market and competitive intelligence collected during the first phase provides the basis for the development of contingency plans to respond to market and competitive actions. Competitive planning at this phase requires the review of major actions or strategies which key competitors may take, affecting our product's position in the marketplace. These prospective actions are ranked based on both probability and impact on ABC Company. The market conditions most likely to evoke these competitive actions are also defined and monitored on an ongoing basis. For each action, a set of response plans involving the necessary functions of ABC are developed in preparation.

The possible competitive actions that may affect ABC's market include:

Introduction of a new or improved product by a competitor.

Shift in market emphasis to the use of a new technology or the obsolescence of ABC's technology.

Entry of a new and major competitor into the market (possible foreign competition).

Changes in the cost and pricing structure of the products in the market (new production equipment, etc.).

Changes in ABC's ability to produce and deliver its product (strikes, equipment breakdown, raw material availability, etc.).

Introduction of a new product line by ABC.

Acquisition by ABC of another company.

To provide a ranking of the likelihood of a given action or combination of actions taking place, probability studies can be done for each of the foregoing actions. These are examples of major competitive actions for which reactive strategies must be developed. Such actions are important to companies in a high-tech environment, but in no way represent all of the possible competitive actions that can affect a company. The character of the industry in which the company competes will define the set of competitive actions that must be analyzed for any particular organization.

Competitive Reactions

When an analysis of possible competitive actions has been completed, the next step is to plan possible reactions to each of the actions. The extent of the planned reaction is dependent upon determining the effect of each action upon the well-being of the company. For ABC, the most significant actions are likely to be those that revolve around new products or new technology, since they affect basic survivability in a high-tech market and require extremely rapid reaction for protection. Some competitive planning actions (reactions) might include:

Advance studies of possible new product introductions and the probable next step in product evolution. Ongoing R&D (in both product and process) is necessary so that ABC will be in a position to introduce its own new or improved product and capitalize on the marketing efforts of its competitors.

Development of new requirements and uses for ABC's product lines to extend product life and provide time to develop and introduce new product lines. These new uses can be communicated to customers in a positive manner after new product introduction by a competitor.

Anticipation and upstaging of a competitor's introduction with the introduction of an even more extensive product line. This has been most effectively used in the computer industry when announcement follows announcement about new computers, software, and peripheral devices. It has, however, fallen into some disrepute because new products introduced to counter competitive introductions have not been available or have fallen far short of expectations. Nonetheless, this remains the most effective countermeasure in ABC's high-tech market, and management will include this strategy in its competitive planning process.

The entry into the market of a major new competitor does not usually call for extensive reaction on the part of the manufacturing department. The primary reaction strategy will be developed by marketing and sales; manufacturing strategy should ensure that operations has the ability to execute the strategy.

Cost and pricing changes, as well as unexpected problems in product-delivery capability, are actions that may be included in competitive planning. Pricing changes can be particularly troublesome if initiated by a competitor, because ABC, except for its superior electronic voltage regulators, operates in a very price-sensitive marketplace. Effective monitoring of price and cost activity is an inherent part of competitive planning. Some management tools by which manufacturing costs may be monitored are:

Cost control studies

Process R&D

Labor relations studies

Alternative supplier identification

Other appropriate and ongoing measures

Thus, competitive planning for the events mentioned above is nothing more than prudent business management.

The remaining competitive actions—new ABC product introduction or acquisition of the company—are initiated and largely controlled by ABC. They are themselves competitive strategies and can be used as mechanisms to deal with other events requiring a reaction. They are also actions for which ABC's manufacturing management must plan, as well as participate in their implementation and execution.

1.4.5 CONCLUSION

The competitive planning approach presented, with its emphasis on the participation of manufacturing in the process, is aimed primarily at the industrial markets. The inherent capital and labor constraints associated with companies in these markets require that they be aware of what optimum leverage their production resources may provide in securing a competitive advantage in the marketplace. In competitive planning, manufacturing and market-based strategies combine to keep a firm viable and provide for long-term value to its shareholders.

CHAPTER 1.5

COMPETITIVE BENCHMARKING: THE FIRST STEP IN MEETING THE COMPETITIVE CHALLENGE

MICHAEL B. MURRAY

Xerox Corporation
Webster, New York

ROBERT K. SHATTUCK

Xerox Corporation
Webster, New York

Competitive benchmarking is a natural process that we all use in our everyday lives. When comparison shopping for a house, a car, a television set, clothing, in fact, anything we buy, we are using a benchmarking philosophy. When we look for the most effective way to modify a room in our house, pick the best way to travel to a distant destination, or follow the path of an opponent's putt, we are benchmarking. Unfortunately this process has not followed us into our work environment. This has not been so with our Japanese counterparts and is a key contributor to the rate at which they have caught up to and passed many United States and European industries.

The basic premises underlying competitive benchmarking are:

In a competitive market, the more efficient manufacturer will succeed.

To stay competitive, real year-over-year productivity is required.

Your competition is doing something to achieve this productivity.

By understanding your competitor's actions, you can make better decisions on how to achieve your productivity goal.

The only companies who do not need to utilize competitive benchmarking are those who know everything about everything, have unlimited resources, or have no competition. Very few companies fit this description.

We believe that although competitive benchmarking is something we all do at home, we do not do it at work. Yet the need to use it is almost universal. Now let us discuss what competitive benchmarking is:

An ongoing process of measuring methods/practices, products, and/or philosophies of companies involved in equivalent or comparable activities.

There are three key points in this definition. The first is that this is an ongoing process. Second, almost everything can be benchmarked. And third, the benchmark need not necessarily be made against a direct competitor. We will discuss these points in more detail throughout this chapter.

Before moving on, let us give an example of where competitive benchmarking has been applied. United States automotive companies, having discovered their cost disadvantage vis-à-vis their Japanese counterparts, have taken significant action to close this gap through new product designs, shortened product development process, capital investments in their factories, new working relationships with the workforce, etc. They established a goal to be competitive and have put in place the actions to achieve it, based upon what their competitors were doing.

1.5.1 PROCESS

The process of competitive benchmarking can be applied to all areas of our job, whether we provide a product or a service. Some examples follow:

Specific face-off of products or services
Business practices of the organization
Technical advancements/applications
Organizational structure
Productivity levels
Quality, cost, and delivery levels

While it is useful to understand how a direct competitor operates in a specific area of interest, it is by no means the only basis for comparison. The word *competitive* in competitive benchmarking does not necessarily imply a direct competitor. There are many well-run organizations performing functions similar to those of your company, against whom meaningful comparisons can be made. For instance, comparing your distribution function to that of another company may provide valuable insights into how you manage your process. The actual product flowing through the system may be irrelevant. The objective of competitive benchmarking is to make *you* more competitive. This means that you are limited only by your imagination as to what you want to evaluate and who will be used for comparison.

The information you require for your benchmark can often be obtained more easily from similar companies than from direct competitors. However, initially you will increase your chances for success by concentrating on direct competitors. This will generally be more meaningful to management and get their attention more quickly. You avoid the problem of management responding, "I see what you're saying, but they are in a different business than we are."

The process of competitive benchmarking is intended to answer four questions:

Who is the best?
Why are they the best?
How much better are they?
What do I need to do to become the best?

To accomplish this, there are a number of steps that must be followed (shown in Appendix 1.5.1). By proceeding through this process you will be able to effectively answer these four questions.

1.5.1.1 Planning

The process begins by identifying the organization's customers and identifying the output that is provided to those customers. Every organization provides an output to a customer, whether that customer is internal or external. You must establish some measurement criteria that can be applied to these outputs. Quantifying the outputs enables comparison and should be done from the customer's perspective to determine how he or she would measure the effectiveness of your organization. Many companies continually survey their customer base as well as those of their competitors to monitor their success in delivering their product or services.

The next step of the process is to identify the best competitors. Information can be gathered by scanning trade or technical journals, talking to consultants, specialists, or academics familiar with the subject, attending trade shows and seminars, and participating in tours. You will begin to develop your own competitive benchmarking network. From these activities, some organizations/companies will emerge as being more successful and effective than others in your area of interest. To make this process more manageable, you may want to limit yourself initially to 4–6 candidates for evaluation.

Once you have identified the best competitor(s), it then becomes necessary to quantify how well they are delivering their particular output to their customer. A number of tactics may be used at this point. The previously used sources, which helped you define the best competitor, can also be used to obtain more detailed information. As data needs become more specific, so do the approaches for obtaining them. Face-to-face discussions can be held, direct mailing of questionnaires may be considered, consultants may be contracted, etc. An office-products company in Europe employed graduate students for a benchmarking study. The results were valuable to the company's management and an excellent business experience for the students. Similar information must also be gathered for your own organization in order to do the comparison.

1.5.1.2 Analysis

After you acquire the data you must analyze it. The first requirement is to assure that you are comparing data on an "apples-to-apples" basis. In some cases this means that data must be normalized. Often it is easier to obtain data on a relative basis than on an absolute basis. For instance, a competitor or other leading company may be willing to discuss a range or relative size of marketing expenses but not to reveal the actual amount spent. The analysis should focus on two items—determination of the current competitive gap and a projection of future trends. Many people have difficulty at this point in trying to determine whether or not they are better than the benchmark. Is more or less of something (i.e., manpower, capital investments, etc.) better? What is important is how your customer rates your output, how well you are satisfying his or her needs. Companies will apply resources in different ways, but that is not what is being compared. The *effectiveness* of the output or customer satisfaction is being compared. For instance, a United States computer manufacturer analyzed its product and found that a comparable Japanese product cost only half as much as its planned product. Several airline companies that began after the industry was deregulated provided their services at a substantial cost advantage to the customer. The existing airlines responded with cost reductions and improved services to counter these new competitors.

Second, performance will change over time. You need to understand not only your position relative to your competitor today, but also in the future. The analysis must comprehend not only the present gap but the competitor's potential for changing the dimensions of that gap. The initial sources for your information should therefore be ones that can be used in the future.

1.5.1.3 Implementation

Ongoing management involvement in the process will improve the acceptance of the final results. Be prepared to be challenged as you proceed through the competitive benchmarking process. No one likes to be told that somebody else is performing a function better than they themselves, especially their direct competitor. Demonstrating the reliability of the data, using a sound analytical methodology, and understanding the methods being employed to achieve the results, improve acceptance. Present the information as evidence of the need to take positive actions to remain or to achieve a leadership position rather than criticizing existing practices. As the completed analysis will indicate a gap between yourself and a competitor, actions should be taken to close an unfavorable gap or to continue a favorable one. Achievement of the goals will assure a leadership position. Remember that you are using the competitive benchmarking process to establish achievable goals and actions to reach the goals.

In the long run, this information must be included in the organization's overall planning process. It is an additional source of information that can be used to set long-range strategy for the organization. These plans must clearly communicate to all members of the organization the steps that will be taken to overcome the identified shortfalls. The actions you take to become effective must complement your own organizational culture. This may result in a totally different approach from your benchmark to solve the same problem, but when effectiveness is measured, you have achieved your goals. Again, the intent is not to copy someone else's practices but to compare their business practices to yours, determine where they are more effective, and establish goals for your organization to become as effective. Often these actions demand innovative organizational responses or a technology breakthrough to achieve success. This can motivate employees to develop the appropriate response.

After action plans have been implemented, progress must be monitored. This requires checkpoints to measure both internal progress toward the original goals as well as your ongoing relationship to the benchmark. You should recalibrate on a periodic basis—say, once a year. The recalibration should involve both how the benchmarks are changing over time as well as understanding if the original outputs being benchmarked are the ones still important when viewed from the customer's perspective. Information on progress against the goals should be reported on a regular basis to all employees.

1.5.1.4 Maturity

After the competitive benchmarking process has been started, what then? To be considered truly effective, it must be viewed as an "evergreen" process. It requires an ongoing commitment to benchmark your business practices and update them based upon changing conditions. Once success has been demonstrated in one area of the business, there will be interest in expanding it into other areas.

Maturity is reached only when you have truly achieved a leadership position in all areas of your business. This obviously takes time. To maintain this leadership position requires that competitive benchmarking be integrated into the everyday business practices and the management process. Competitive benchmarking forces you to expand your perspective and consider other ways of doing business or to search out people who may be doing business in a better manner. It encourages continual searching of comparative information to be able to respond to the dynamics of a changing world.

1.5.2 CRITICAL SUCCESS FACTORS AND STARTUP

We have now discussed the benchmarking process and many of the things that can/must be done to have an efficient program in your company. We must emphasize that there are several factors that are critical to the success of any benchmarking activity.

1. *Problem Recognition.* Unless you are willing to accept the fact that you have a problem or that other companies may be more effectively executing activities similar to those of your company, then you will never benchmark successfully.

2. *Patience.* The process is a learned activity in which you are likely to meet dead ends before you find a successful approach. Also, what works in one area may not work in another. Both the people involved in the process and management must have patience.
3. *Commitment from Top Management.* This is more than the patience factor. Top management must be open-minded and must listen to the findings. If bad news about your company is discovered, those who found it are not at fault. Do not shoot the messengers. Rather, support and reward them for a job well done. The willingness of top management to change and to support needed changes leads to the fourth factor.
4. *Willingness to Change.* Change is never easy to implement. But the fact that change is necessary and will happen must be accepted by all members of the organization. Competitive benchmarking in many ways represents an attitude to drive constantly toward excellence. Change is a permanent member of the team that plays in a dynamic, competitive marketplace.
5. *Followup/Feedback.* Benchmarking is a closed-loop process. Your current strategies and benchmarks must be challenged on a regular basis, and your people must be kept informed to ensure their continued commitment.

With this understanding, let us now briefly discuss some of the factors involved in implementing a Competitive Benchmarking process. The first idea revolves around critical success factors 1 and 2—problem recognition and management committment.

To determine whether you have a problem, you must find out whether your performance is acceptable vis-à-vis your competition, using such quantifiable measures or metrics as:

Market share
Product selling price
Margins
Quality levels
Manpower ratios

A more complete list of metrics that you might use is contained in Appendix 1.5.2. Whatever metrics you do use, you must ensure that you are measuring performance that directly or indirectly impacts customers.

The second key is that top management must support the benchmarking activity. In all probability, the solutions to the performance problems identified will require solutions that cross organizational boundaries or require policy revisions.

Approaches to staffing a benchmarking activity are as large as your imagination, but can generally be categorized as follows:

Task force
Staff assignment
Special line management assignment
Combinations of the above

These can be single-function or multifunctional, single-level or multileveled. Use the approach that is most easily initiated within your company. You must keep in mind, however, that the results must be accepted across a wide range of the organization. The "multi" approaches tend to facilitate this acceptance and will probably give you a better set of benchmarking results.

1.5.3 SUMMARY

Competitive benchmarking is a tool to identify, establish, and achieve standards of excellence. It provides insight to help set performance goals and to develop action plans to achieve those goals. From this arises an increased awareness of costs, products, services, competitors, and markets. Competitive benchmarking drives a continual introspection and self-improvement process. It ultimately tests your performance against your competitors' for satisfying your customers' requirements, rather than improvement against your own past performance. Given today's fast-changing and highly competitive world, competitive benchmarking is necessary for long-term survival.

APPENDIX 1.5.1 PROCESS STEPS FOR COMPETITIVE BENCHMARKING

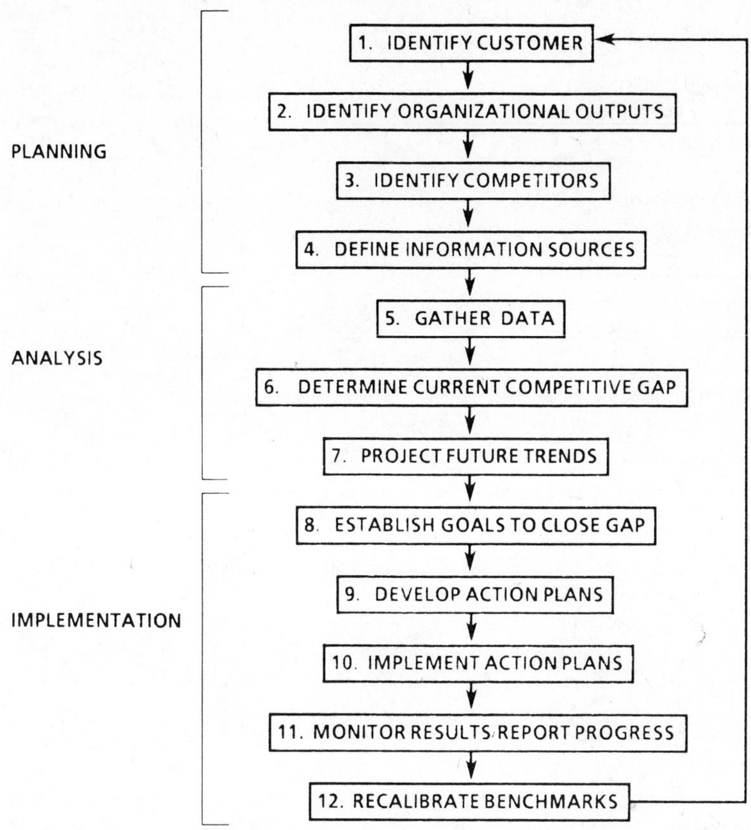

AT MATURITY YOU WILL HAVE:

- ATTAINED A LEADERSHIP POSITION
- INTEGRATED THE PROCESS INTO YOUR BUSINESS PRACTICES

APPENDIX 1.5.2 COMMONLY USED ORGANIZATIONAL OUTPUTS

Quality

Percentage of defective parts
Percentage of products delivered without defects
Reliability mean time between product failures
Billing error rate
Software code error rate
Percent of process steps with demonstrated process capability
Percentage of tools in certification

Cost

Cost as percentage of revenue for:
—Service
—Manufacturing

—Customer Administration
—Distribution
—Engineering
Return on assets
Labor overhead rate (percent)
Material overhead rate (percent)
Inventory turns per year
Cost per order
Cost per engineering drawing
Output per man year

Delivery
Service response time
Percent of product downtime
Percentage of parts/product delivered on time
Number of telephone rings before answer
Achievement of planned internal schedules

APPENDIX 1.5.3 KEY QUESTIONS

Are the critical success factors in place?

How rapidly can the competitive benchmarking philosophy and process be integrated into the normal business activities of your organization?

Are your reward and recognition systems in conflict with benchmarking and change or supportive of it?

Are organizational goals and management measurement tools in line with the benchmarking metrics?

What are the major barriers or inhibitors to a competitive benchmarking process and to change within the company?

Is there clear accountability with the benchmark measurements or metrics?

Are you measuring a cause or an effect?

What are the tradeoffs between metrics? Do they work together or against each other?

Will responses that improve your overall effectiveness drive the metrics and indicators in the "right" direction?

Are all functions equally able to work together to identify and solve problems?

How rapidly is competition improving their performance?

How much time do you have and/or need to improve your performance?

How can you involve your total organization in the benchmarking process? in the solution identification process? in the solution implementation process?

Have you properly identified solution strategies and their tactical steps?

Are recalibration steps in place?

CHAPTER 1.6

MEASURES OF PERFORMANCE AND PRODUCTIVITY

ROBERT N. LEHRER

PTY, Incorporated
Atlanta, Georgia

The objective of this chapter is to provide a philosophy of measurement related to production systems. Most organizations have various measures of performance and productivity, and associated "control" systems, that can be made more effective by addressing the objectives of measurement and alternative ways to better achieve the desired results. Basic concepts and alternative approaches are presented. Detailed how-to presentations are reserved for other chapters.

> *I often say that when you can measure what you are speaking about, and express it in numbers, you know something about it; but when you cannot express it in numbers, your knowledge is of a meager and unsatisfactory kind; it may be the beginning of knowledge, but you have scarcely, in your thoughts, advanced to the stage of science, whatever the matter may be. (Lord Kelvin, 1883)*

The statement above is frequently used to set the stage for discussions of measurement, and it expresses a fitting philosophy for this presentation. However, one should recognize that the present subject is not "science" in the sense Lord Kelvin used the word. The intricacies of production (human) organizations are not yet well enough understood to provide the basis for a truly scientific measurement of performance and productivity. How scientific one wants measurements to be depends upon their use. Useful measures can be developed, and Lord Kelvin's philosophy can be a useful guide.

The introduction of the concept of "science" to the management of production systems is associated with "scientific management," developed largely on the basis of Fredrick Taylor's pioneering work of the late 1800s. Use of piece-rate incentive compensation plans and of time study (work measurement), a Taylor innovation, are usually thought to be the sum and substance of scientific management. The use of these concepts has been dubbed "Taylorism."

Use of work measurement and financial incentive compensation have achieved many worthwhile economic benefits. However, there have also been adverse consequences, so much so that "Taylorism" is currently used as a term of derision to describe unenlightened management that exploits and demeans its labor resource. This is unfortunate, for it tends to depreciate the utility of Taylor's concepts. But there is a brighter side. Recognition that both beneficial and dysfunctional consequences have resulted from their use has sparked a search for other, more beneficial (or less objectionable), approaches. We are now much better informed about why dysfunctional results occur, and how they can be avoided. Basic work measurement, developing from Taylor, still has validity and utility. However, there are many other "measures" equally valid and useful. Also, the objectives of measuring—for "control and manipulation" or to help people "grow" and to contribute better to the mission of the organization—are now recognized as being critically important.

1.6.1 BASIC MEASURES OF PERFORMANCE AND PRODUCTIVITY FOR DIRECT PRODUCTION

The key concept upon which scientific management was based was "plan the work—then work the plan." Planning work—identifying the kind, amount, and timing of resources required to achieve a defined and desired end result—and then using that plan to guide and "control" the actual use of resources is eminently logical and is still an appropriate cornerstone for developing a family, or hierarchy, of measures of performance and productivity.

The usual approach is to start with the business plan, that is, what is to be produced, when, and in what quantities. Bills of materials are "exploded" to identify individual piece parts and to define what resources are required, when, and in what quantities. Detailed production plans and schedules are developed that define what is to happen and when, and what flow rates are associated with each production step or operation. Work measurement in some form is usually used to establish a quantitative basis for labor requirements and flow rates. Other measurements are required to quantify the requirements for such other resources as materials, equipment, inventory, working capital, and energy. Measurement techniques are varied and well documented, both in the open literature as well as in later sections of this volume.

The developed plan is used as the basis for execution and management of the use of resources—for acquiring facilities and machinery, ordering materials and supplies, acquiring necessary labor skills and services, for initiating work activities—and for monitoring accomplishment. Labor productivity, machine utilization, adherence to quality and design specifications, materials utilization and yields, meeting of schedules, inventory control, etc., are evaluated in comparison with the plan. If they conform to the plan, progress is satisfactory. If the "norms," or standards of performance, are not achieved as the plan is being executed, causes for deviation are sought out and corrected.

"Plan the work—then work the plan" is what is involved. The plan must have adequate specifications and measures, and the system of feedback must be sensitive and appropriate in order to achieve good results. Some incentive or motivational influences, not necessarily direct financial incentives, are required to encourage those involved to help achieve planned performance.

The structure of this basic plan and the feedback control loops should provide performance indicators that are sensitive to facilitating control of current operations and supporting key decisions that influence use of resources and achievement of desired outputs. The correct "controllables" are measured with minimum time lag between action and corrective action.

In Taylor's day the emphasis was upon facilitating effective use of the labor resource. This is still important, but no longer sufficient. All resources required to accomplish the production called for by the basic business plan, both those which directly contribute to production and those which are supportive or facilitative in nature, should be included, but even this is not sufficient. So far we have only been concerned with "production." We should be concerned with all the resources necessary to achieve the mission of the organization as well as with total organizational performance.

Other measures must account for the many support services that are needed by the organization in order to achieve its mission but which do not make direct contributions to "producing."

1.6.2 WHY MEASURE?

To plan work, some quantification of the resources required to achieve a desired output is essential. In many situations, estimates based upon past experience may be satisfactory for developing a work plan, which is then acceptable as a guide for use of these resources. One could argue that in these cases, if overall performance is acceptable as measured, say, by profitability, then there is no need for additional measurement. This may be valid. What would be the benefits of measurement, and what would be the cost of the development, maintenance, and use of the measurements? Would the benefits outweigh the costs?

Evaluation of the benefit/cost relationship for additional measurement of performance and productivity is complex, but usually more measurement is advantageous, except in those cases where the organization is highly successful in self-directing innovation and improvement, or where extensive formalized measurement has become bureaucratized.

Evaluation of the benefit/cost relationship for additional measurement of performance and productivity is complex, but usually more measurement is advantageous, except in those cases where the organization is highly successful in self-directing innovation and improvement, or where extensive formalized measurement has become bureaucratized.

In the former case, people throughout the organization are well in tune with the mission and objectives of the organization and clearly recognize what they can do to make a maximum contribution to organizational performance. This is highly desirable, but very unusual.

In the second case, measurement is extensive and oppressive, and tends to prevent people in the organization from recognizing and wanting to make their maximum (or even "normal") contribution to organizational performance. A rethinking of the organization's philosophy of performance and productivity measurement and a fresh approach are indicated. This can be done best by focusing on what can be done to motivate and make possible maximum job performance by everyone within the organization, thus contributing to better overall organizational performance. Measurement is a means to an end—the end being good and continually improving organizational performance.

1.6.3 OBJECTIVES OF MEASUREMENT

The basic objectives of performance and productivity measurement is to establish a relationship between individual and group performance and its contribution to achieving the mission and objectives of the organization, thus assisting individuals and groups to improve their contribution to organizational performance. This basic objective can be restated: to define a norm or standard for the resources required to achieve a standard output, so that the production system can be effectively operated, managed, and improved.

Organizational performance is only partially evaluated in profitability terms, even though this element of performance is extremely important. Other elements of mission and organizational objectives tend to place profitability in a secondary role. However, profitability will naturally follow when other elements of mission are adequately addressed.

Measurement allows quantification of the resources required to produce a defined output, which provides a plan for acquiring and using those resources. If the plan is accurate and is followed, the results should be favorable, providing the defined output contributes to organizational mission in a desirable way. If the elements of planning and use of the plan are motivational to all those affected, measurement is the basis for developing and achieving improvement. Measurement defines an achievable norm or base level of performance that is a challenge to be improved upon.

Therefore, measurement should be viewed as a tool for helping people throughout the organization do their jobs effectively—to make maximum individual and group contributions to organizational performance both in regular performance and in developing improvements to further enhance performance.

Measures should provide a "score card" evaluation of individual, group, and organizational performance—the use of resources and the achievement of units of desired output, referenced to some base of expectancy or standard. This helps individuals and groups do their job well and also informs the next higher level of individuals and groups how well things are going up to their level of responsibility, and indicates areas of superior and/or deficient performance.

1.6.4 MEASUREMENT NEEDS AT DIFFERENT LEVELS

The ideal system of performance and productivity measures should provide every level and every individual within the organization with indicators that help them to plan and guide their own efforts in order to maximize their contribution to achieving the organizational mission. The needs at various levels and within various areas of the organization for these purposes are quite different. They are a reflection of the formal organizational structure and associated allocation of responsibility, and of organizational culture. Each individual, ideally, should be provided with measures of what he or she is to achieve, and of the resourses that are required. These measures reflect a standard, or norm, for achievement and use of resources that is straightforward in most production operations, but becomes ambiguous with support activities.

Measures needed by top management are concerned with the generalized, overall operation of their organization, with considerable attention to the interplay between short- and long-term needs and performance. Appropriate measures of performance and productivity, which must integrate with long-range planning and financial profitability measures, are aggregate indicators of how the organization is performing.

Measures needed by operational levels of management are less generalized and more detailed, reflecting their more direct responsibility for the use of resources.

Measures useful to direct production workers and their staff counterparts are restrictive and reflect the assigned responsibilities at these levels for specific task assignments.

The ideal system should be integrative in an upward direction, providing each successive level with appropriate aggregations of performance and productivity, with supportive detailed information available if desired. The time lag between "occurrence" and availability of measurement by those who "control or perform" should be minimized.

1.6.5 PERFORMANCE AND PRODUCTIVITY

Performance and productivity measures are distinctively different in characteristics and in the ways in which they can be measured. They are also interrelated.

Performance has to do with achievement of a goal or result. *Productivity* has to do with the relationship between the resources used and the output achieved.

Measurement of organizational performance is usually (almost always) done in financial terms, most frequently profitability. Various other financial measurements are used to reflect performance of contributional factors. While these financial measurements are both useful and necessary, they are not sufficient.

Financial accounting is geared to historical perspective and not to planning. Even though profitability includes the effect of changes in productivity, the effect and cause are masked by changes in price/cost relationships, staffing patterns, indirect/overhead expenses, product mix, quality, and many other significant aspects that should be directly managed. Management of these other resources can best be approached by measurement, in order to evaluate their contribution to overall performance.

Productivity measurement is usually (almost always) done in terms of output per labor input. While this partial measure of productivity is useful, the other resources required to produce output—capital in various forms and materials—are also important and should be included in productivity measurement.

An approach to more comprehensive management of all resources and their contribution to overall profitability performance is to supplement financial accounting with productivity measurement. This helps to achieve a complete measurement of performance in terms of profitability as a consequence of price/cost changes and productivity in the use of all resources.

1.6.6 PROFITABILITY = PRODUCTIVITY × PRICE RECOVERY

One of the commonly used systems for relating productivity and profitability is the American Productivity Center (APC) Firm-Level Total Productivity Measurement System. The approach is one of integrating upward various measures of resource usage and adjusting for the effect of cost/price, product mix, and volume changes, so that productivity and profitability can be planned and managed at all levels of the organization.

A base period is used for comparison, so that subsequent performance can be evaluated to determine which changes in productivity and cost/price relationships account for the achieved profit. For example, profitability performance over several periods for the two organizations shown in Figure 1.6.1 is the same, but for vastly different reasons. Company A has achieved its profitability by improved productivity. Inflation-induced increases in the cost of resources have not been offset by increases in product/service prices, as shown by the declining price recovery line. Productivy increases, as shown by the productivity line, have been sufficient to more than offset this additional cost and to contribute to improved profitability. Company B, on the other hand, with an identical profitability performance over the same period, has achieved its profitability by aggressively increasing the prices for its products/services. Price increases have offset cost increases and the effect of declining productivity.

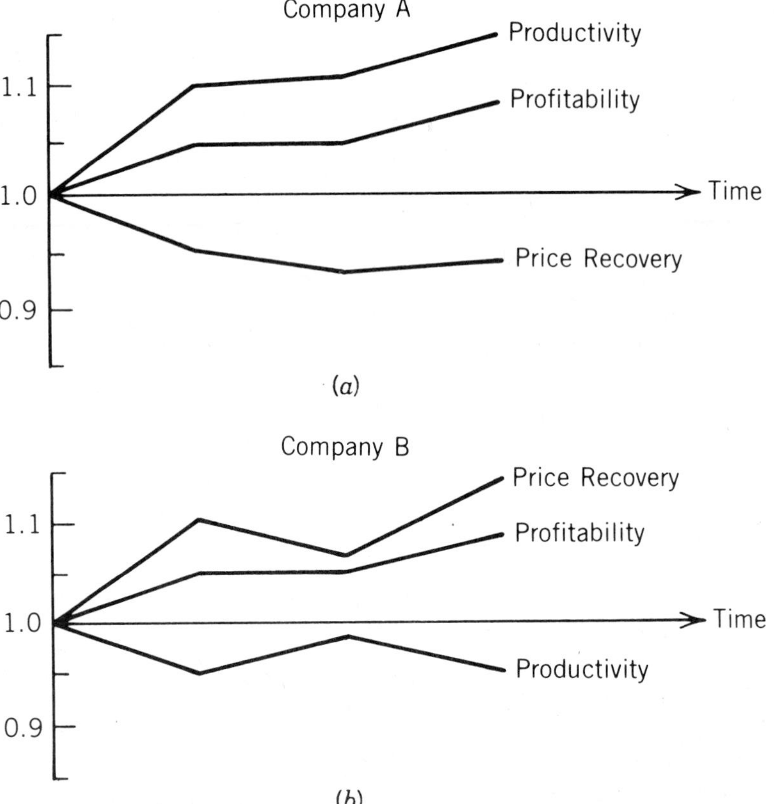

Fig. 1.6.1. Total productivity measurement provides for tracking, over time, changes in the component effects of productivity and price recovery on probability. *Source*: Carl G. Thor, "Productivity Measurement in White Collar Groups," in *White Collar Productivity*, Robert N. Lehrer, Ed., McGraw-Hill, New York, 1983.

Capital, depreciation

Inputs	Change in Profitability	Change in Productivity	Change in Price Recovery
Labor 1			
Labor 2			
Labor 3			
Total labor			
Materials 1			
Materials 2			
Total materials			
Capital, leases			
Capital, depreciation			
Capital, return			
Total capital			
Energy 1			
Energy 2			
Total energy			
Total all inputs			

Fig. 1.6.2. The APC performance measurement system results in a matrix that connects changes in profitability, productivity, and price recovery with each of the input factors. Results are expressed in both rate of change and dollar effect. *Source*: Carl G. Thor, "Productivity Measurement in White Collar Groups," in *White Collar Productivity*, Robert N. Lehrer, Ed., McGraw-Hill, New York, 1983.

Overall performance, relating profit performance to productivity and price recovery performance is presented graphically in Figure 1.6.1. Detailed supportive data starting at the production level and integrating upward and encompassing the entire organization is required to develop the overall financial, productivity, and price recovery measurements.

Base-period measurements for the quantity and cost of all resources and the units of output and their values are first established starting at the production level by organizational unit, then extending to incorporate the entire organization. Subsequent period measurements are compared with base-period measurements to establish values for changes in profitability associated with each resource and the associated productivity and price recovery contributions.

Summary data for a production unit might be in a format similar to Figure 1.6.2. Changes in profitability, productivity, and price recovery are expressed as ratios and in dollars, providing a comparison with base-period values. Supportive detailed data is needed to develop the summary data, and are useful for analysis of causes of superior and/or deficient performance indicated by the summary data. Data for each lower unit are aggregated with other unit data to provide appropriate summary performance data at each level of aggregation up to the top of the organization.

Not only can performance be evaluated in comparison with a base period, but period performance can be evaluated in comparison with hypothetical norms or goals, such as a budget, business plan, or target level of achievement.

Total productivity measurement requires slightly different input data than that provided by most accounting systems, particularly those related to the evaluation of capital resources. All capital components must be evaluated to reflect current-period opportunity costs and values, without direct consideration of normal depreciation or capital recovery procedures. Product changes, quality changes, and the determination of equivalent values may also be beyond the scope of accounting data.

1.6.7 A HIERARCHY OF WORK-UNITS

Another approach to total productivity measurement within an organization or an organizational unit is represented by the hierarchy of work-units and work-unit analysis developed by Mundel.[2]

A work-unit is an amount of work (input or resource) or the result of an amount of work (output or accomplishment), which is related to the hierarchy of the organization and responsibilities/functions/activities that interlink each element of the organization.

The highest level is the 8th-order work-unit concerned with results—what is achieved because of the output of the activity of the organization or organizational unit being considered. Each lower level work unit is directed to less general output and input measures, increasing in specificity and narrowness of scope to the 1st-order work-unit, human motion.

The 7th-order work-unit, gross output, deals with a large group of end products or services of a working group having some common affinity.

The 6th level deals with programs, a group of outputs or completed services that is part of a 7th-order work-unit, grouped according to some aspect of "alikeness."

The 5th level is directed toward end products, units of output (used outside the organization or the organizational unit being considered) that contribute to the objectives (mission) of the organization and/or the organizational unit being considered without further work being done on that output.

Additional and lower levels of work-units deal with intermediate products (or services), tasks, task elements, and individual human motions.

Mundel's work-unit structure can be used for measuring the output/input relationships within an organization or within an organizational unit, as a total productivity measurement system of the organization or any organizational unit. It does not provide the interlinking of productivity and price recovery as they affect profitability, as does the APC system, but it does provide a more comprehensive evaluation of output related to organizational mission and objectives, and of the use of resources to achieve desired outputs.

Developing the structure for a specific situation starts with definition of objectives and/or mission, and exhaustive development of each lower level work-unit structure until a level is reached that provides measures useful for forecasting, planning, and control, and suitable for some type of measurement. During this process, clear relationships are retained between objectives (and/or mission) and all progressively lower level work-units, and the list of work-units at each level is developed to be individually mutually exclusive and collectively all-inclusive.

Figure 1.6.3 summarizes 8th-order work-unit analysis for a plant-engineering organizational unit of a fiber mill, defining the type of unit output (facilitative or support, as opposed to production), mission areas, intent, dimensions, goals, limitations, and freedoms. More detailed work-units, carried to the 7th, 6th, and 5th levels, are developed in Figure 1.6.4. Figure 1.6.4 defines measurable elements of accomplishment, which are derived from the 8th-order work-unit definitions. They are mutually exclusive units that completely define the discharge of responsibilities of the organizational unit.

The 8th-order work-unit structure develops measurable goals for guiding and evaluating performance of the plant-engineering unit of the organization, goals that can be related to the unit's contribution to organizational mission.

The 7th-order work-unit structure defines program areas and provides a basis for strategic planning and resource allocation to major responsibility areas.

The 6th-order work-unit structure provides a more detailed definition of activities required to discharge program responsibilities, useful for a finer level of resource allocation, and provides the basis for definition of measurable units of activity or tasks at the 5th-order level. The 5th-order work-unit structure, of which the figure presents only a portion, provides the basis for measurement of specific tasks and the specification of what is to be achieved, when, and with what resources. It is the base from which measures of performance and productivity are developed, and integrates upward to define a hierarchy of measures to facilitate planning and to guide execution of the plan.

1.6.8 SUPPORT ACTIVITIES

In most organizations, support activities, usually regarded as indirect work or overhead, consumes relatively large and increasing amounts of total resources. Effective measurement and management of these activities—defining contribution to organization mission, and quantifying needed accomplishments and required resources—is critical to overall organization performance.

Both the APC and Mundel's work-unit structure approaches to performance and productivity measurement can encompass support activities as well as direct production activities. Several other approaches to development of performance and productivity measures for support type activities, less comprehensive but often useful, should be considered.

Establishing appropriate measures of performance and productivity for support activities is inherently more complex than doing so for direct production activities. Most direct production activities are fairly easy to measure, for the tasks are well-structured, limited in complexity, fairly consistent, repetitive, of short duration, and with definable output. Further, their contribution to organization mission can be evaluated reasonably easily.

Most support activities are difficult to measure, for they are composed of ambiguous tasks, which may be done in a variety of ways (frequently at the discretion of the individual), with a range of results of varying but still acceptable quality, complexity and nonrepetitiveness, of relatively long duration, the outcome of which is frequently difficult to define precisely and to relate to fulfilling the organization mission, and which may have a considerable time lag and/or interaction with other activities before achieving any impact.

Support activities may be performed along with production activities by the same individuals or organizational units, or they may be separated from production activities and performed by different individuals or groups of individuals that are part of production or line groups, or they may be allocated to independent organizational units, departments, or divisions within the organization. The allocation of responsibility for support activities influences the approach to developing and using measures of performance and productivity. Definition of the

TYPE: Output facilitative
MISSION AREA:

1. All physical plant
2. All production equipment
3. All materials handling equipment

INTENT:

1.1 Maintain intact
1.2 Upgrade as scheduled, within planned costs
2.1 Maintain available for on-line use
2.2 Reduce accident rate caused by machine-correctable faults
3.1 Maintain available for on-line use
3.2 Optimize economy of operation

DIMENSION:

1.1 Cost of damages attributable to prior, uncorrected defects
1.2 Conformance to upgrading schedules and costs
2.1 Percent of time production equipment is available for use during prime shift
2.2.1 Frequency of lost-time accidents per 1000 hours of direct labor, machine correctable
2.2.2 Non-lost-time accidents per 1000 hours of direct labor, machine correctable
3.1 Percent of time available for on-line use
3.2 Reduction of M-H costs while maintaining current service level to production and warehouse operation

GOALS:

1.1 Not in excess of what prevention costs would have been
1.2 Not in excess of ±5 percent deviation
2.1 97 percent
2.2.1 Reduce to 1 in 10,000 or less
2.2.2 Reduce to 1 in 5000 or less
3.1 90 percent
3.2 5 percent cost reduction

LIMITATIONS:

1. Low educational level of labor force
2. Old plant
3. No current repair parts usage data
4. Rising wage costs

FREEDOMS:

1. Can use budget inventory money as desired
2. Adequate spare parts storage area
3. M-H equipment relatively new

Fig. 1.6.3. Work-unit structure, 8th-order, for a mill plant engineering function. *Source*: "Work Unit Analysis," Marvin E. Mundel, *ACSE Spring Annual Conference Proceedings*, 1977, © American Institute of Industrial Engineers, Inc., 25 Technology Park, Norcross, GA.

7th-order work-units
01 Grounds and roadway maintenance provided
02 Building maintenance provided
03 Production equipment maintenance provided
04 Materials handling service provided
05 Materials handling equipment maintenance provided

6th-order work-units
0101 Main roadway maintenance provided
0102 Feeder roadway maintenance provided
0103 Decortification areas maintenance provided

0201 Production building maintenance provided
0202 Warehouse building maintenance provided
0203 Office building maintenance provided

0301 Decortification production equipment maintenance provided
0302 Spinning shed production equipment maintenance provided
0303 Weaving shed production equipment maintenance provided
0304 Dyeing shed production equipment maintenance provided
0305 Finishing and inspection area production equipment maintenance provided

0401 In-plant decortification M-H service provided
0402 In-plant spinning M-H service provided
0403 In-plant weaving M-H service provided
0404 In-plant dyeing M-H service provided
0405 In-plant finishing and inspection M-H service provided
0406 Between-plant M-H service provided

0501 In-plant push truck maintenance provided
0502 In-plant lift truck maintenance provided
0503 In-plant hoist equipment maintenance provided
0504 Between-plant lift truck maintenance provided
0505 Truck maintenance provided
0506 Maintenance equipment maintained

5.5th-order work-units for 0301
0301 Decortification production equipment maintained

030101 Supply tanks inspection completed
030102 Supply lines inspection completed
030103 In-plant tank inspection completed
030104 In-plant piping inspection completed
030105 In-plant electrical system inspection completed
030106 Decortification beater system inspected
030107 Supply tank repair completed, prior to failure
030108 Supply tank repair completed, after failure
030109 Supply line repair completed, prior to failure
030110 Supply line repair completed, after failure
And so forth

Fig. 1.6.4. Expansion of work units for 7th-, 6th-, and 5th-orders. *Source*: "Work Unit Analysis," Marvin E. Mundel, *ACSE Spring Annual Conference Proceedings*, 1977, © American Institute of Industrial Engineers, Inc., 25 Technology Park, Norcross, GA.

activity and its significance, what must be accomplished and its contribution to organization mission, and the resources that are required to produce the desired result are needed, but made more difficult when support activities are decentralized. Centralization, however, may be counterproductive.

The ease or difficulty with which support activities and their significance can be defined and the required resources specified also have a bearing on measurement. A variety of measurement techniques are available, including estimation, scoring models, predetermined motion standards, work sampling, statistical modeling, and group participative approaches. Each has its place and associated limitations. Selecting the most appropriate techniques depends upon intended use, benefit/cost relationships, and necessary accuracy and precision of resulting measures.

What to measure and why are also important issues. Is measurement for control? If so, who is it intended to help in exercising control, and does this individual or group have control over that which is imputed by the measures? Do the measures include elements over which this individual or group has no control, but for which they are held accountable? Or is measurement intended to help individuals and groups do their jobs well, and to improve? Are individuals and groups provided with measures that reflect performance and productivity over which they have control and influence? Have measures and their use been approached from a bureaucratic base or do they reflect a self-control attitude? Have measures and the systems for using them been imposed in order to manipulate people and to require them to behave in ways which they do not desire? Do the measures provide a positive basis for motivating and helping the individual grow?

Is it necessary to define and measure all outputs and all inputs, or is it appropriate to concentrate on selected aspects of productivity and performance? Should control be directed toward use of resources, accomplishment of specific results, or to the value of contributions to achieving organization mission? Each alternative and various combinations may be appropriate, depending upon the relative significance of the specific element ot the overall, the potential for improvement, and the ease or difficulty of developing and using measures. What is appropriate is "situation specific," dependent upon what is practicable in developing and using measures and the associated benefits/cost relationship. Partial productivity measures are particularly appropriate where they are perceived by the individuals involved as relevant to their own personal views and where they are compatible with higher level aggregate measures.

Who should develop measures? How involved should the individuals and groups who are directly affected by measures be in their definition and development? What should be the role of the "expert" in development of definitions and measurements?

Should measurement of performance and productivity be done on a continuing basis, or only periodically? What are appropriate intervals for measurement? Should the same cycles and intensity of measurement apply to all measures, or should they be used selectively?

There are many issues to ponder when considering measures of performance and productivity for production systems. These multiply as attention is shifted away from or beyond direct production to include support activities as well. The following sections amplify on some of these issues.

1.6.9 PARTICIPATIVE DEVELOPMENT OF MEASURES

Participative development of measures of performance and productivity is potentially useful for any activity, for those who do the work are directly involved in the process of definition and measurement. Participation allows clarification of expectations and responsibilities of all who are involved—individuals, groups, those who do the work and those to whom they are responsible—and for a sense of ownership of the measures and their use. Participative approaches are potentially useful for any area of activity, but are particularly appropriate in the support-activity area. Those who are doing the work frequently have almost exclusive insight into the matter of what measures are most meaningful, both to those directly involved and to the organization, and of how appropriate measures can be accurately, reliably, and economically developed and used.

Use of participative approaches does not mean that management is soft and abandoning its responsibilities. Properly used, participative approaches insure the recognition of mutuality of concern by all parties for the well-being of the organization and its members, and result in meaningful and realistic measures that help individuals and the organization achieve acceptable and improving performance.

Participative approaches may include measurement specialists as a resource, to provide guidance in measurement techniques and to insure that workable results are achieved. They may be very informal, associated with an open style of management and used in conjunction with more formal techniques, or used in a formalized way by themselves. Various approaches may be appropriate and useful. They all share common characteristics when used successfully. Participation is open and guided. It is made meaningful by providing individuals with opportunity for growth in acquiring the necessary understanding and skill to use participation as an effective way to help achieve task accomplishment. All affected individuals and resources are utilized. The focus is upon mutuality of concern, and how to best "integrate" the interests of all parties. Consensus determinations are sought.

One formalized participative approach often used for developing performance and productivity measures, but useful for many other purposes, is the Nominal Group Technique (NGT). The name derives from the concept of a participant group being a group only in the sense of individual's participation in the group. This may be the case where individuals from different areas and responsibility levels come together only because they represent a cross-section of resources, or knowledge and insight that would be useful for dealing with a specific issue. Or a group might be formed to include a portion, or even all, of an organizational entity or unit. The important point is that the group is constituted on the basis of bringing together all the resources that are appropriate to effectively address the issues at hand.

When dealing with definition, development, and use of performance and productivity measures, various "stakeholders" are involved and various skills are required. The quality aspect of the outcome is important, but acceptance by all stakeholders is critically important to best results.

The Nominal Group Technique (NGT), briefly outlined, operates in this way: (a) issues or objectives are defined; (b) appropriate resources (people) are identified; (c) the group is assembled; (d) the issues have been communicated to the group with a request that each individual come prepared with written recommendations, or this is done at the meeting. This phase is usually referred to as *silent generation*. (e) Each individual, in turn, is invited to contribute one item, which is listed for all to see. The next individual contributes an item, and so on around the group for as many cycles as necessary to include all contributions. An individual may pass his or her turn at any time; (f) The list of contributions is then edited by the group to combine like items, remove redundancy, obtain clarification if necessary, and possibly to add additional items. The items are numbered; (g) Each member of the group is then asked to select a number of items that are most significant in his or her judgment, say five or maybe seven; (h) Cards are furnished to each person on which to record the number and statement of each item selected as particularly significant, and then arranged in order of importance, from most important to least. Each person then numbers the cards to indicate the rank-ordering, 5 for the most important (if five items are being ranked, 6 if six items are involved, etc.) and successively to 1 for the least important of the selected group; (i) The leader then collects the cards, tallies the points for each item, and lists the items in rank-order of total assigned points along with the point totals. This scoring identifies the items that are considered by the group to be important, and provides relative scores for the degree of importance.

The NGT approach provides an opportunity for all members of the group to participate and to contribute in a protected atmosphere. Ideas are individually generated, shared as the individual wishes, and each one has an opportunity to be stimulated to further creativity by listening to the item contributions of others. Everyone has an equal chance to participate. Power plays, dominance, and use of authority are prevented by the process. If an individual wants to contribute an idea that he or she thinks may offend or involve personal risk, it can be held back until someone opens up the issue, or it can be held back altogether. Usually group members are open and candid if encouraged to be, and a held-back item is likely to be duplicated by someone else's contribution. Evaluative comments as such are not permitted, but clarifying comments are.

An example of the results of NGT to establish performance measures for a computer service unit is shown in Figure 1.6.5. Five measures, selected from a listing of some 40 suggested by the group, are used as five independent measures of unit performance, and combined in a weighted fashion to develop a composite performance index.

Not everyone in the computer group was involved in the NGT development of performance measures. The group included representatives from sectors of responsibility of providers and users of computer services. The group was sufficiently representative of the whole that meaningful measures were developed, and accepted not only by the group but also by nonparticipants in the provider- and user-groups. Involving the broader group of stakeholders in the development of performance measures may be desirable in order to insure that all affected parties view the final measures as "theirs."

Measure	Range, Worst to Best	Weighting	Measure Value		Scaled Value		Weighted and Scaled Value	
			Last Quarter	Current Quarter	Last Quarter	Current Quarter	Last Quarter	Current Quarter
Downtime	40–15	0.3	20	17	80.0	92.0	24.0	27.6
Meet user deadline	58–98	0.3	87	95	71.8	92.3	21.5	27.7
Rerun time	500–200	0.2	221	250	93.0	83.3	18.6	16.7
Repeat complaints	30–10	0.1	13	11	85.0	95.0	8.5	9.5
On-line response	4.0–2.0	0.1	2.3	2.7	85.0	65.0	8.5	6.5
						Composite index	81.1	88.0

Fig. 1.6.5. A sample performance report for a computer center, showing the 5 performance measures and their relative weightings as developed by the nominal group process. The actual range over which each measure varies, from *worst* to *best*, is converted to a 100-point scale so that the worst value equals 0 and the best value equals 100. The raw measure values for last quarter and for the current quarter are listed under *measure value*. The scaled values, i.e., the raw values converted to the 100-point scale for each measure, are listed under *scaled value*. Those listings show that relative performance for each measure on its own 100 points equal 100% scale. The final columns present the weighted and scaled values on a composite basis, whereby 100% performance for all individual measures adds to a composite index of 100 points, or 100%. *Source*: Carl G. Thor, "Productivity Measurement in White Collar Groups," in *White Collar Productivity*, Robert N. Lehrer, Ed., McGraw-Hill, New York, 1983.

1.6.10 PRODUCER/USER RELATIONSHIPS IN SUPPORT SERVICES

Standard costing and transfer pricing are used by some organizations to better identify measures of performance and productivity for individual production departments or cost centers. Some organizations emphasize marketplace economics to the extent that outside suppliers may be selected by the user department in preference to internal providers if service and cost are more attractive. The same concept can be applied to support services in a variety of ways. Support services can be priced by the provider, budgeted for and purchased by the user, and controlled by the marketplace. The concepts outlined earlier concerning measures of performance and productivity can be applied.

Less elaborate evaluation, involving a different approach to measurement, may aso be appropriate, as illustrated by Overhead Value Analysis (OVA) developed by McKinsey and Company.[4] OVA involves analysis and measurement of all the individual services provided by an organizational unit relative to the cost incurred by the providing unit and relative to the value of each service to the user unit. The relationship between benefits and costs is then assessed, and alternative ways to improve this relationship by modification of needs and services or ways to provide services are identified. In addition to directly involving the providers and users of support services, upper levels of management are involved in reviewing recommendations, obtaining compromise agreement, and evaluating the impact of potential improvements on the organization.

The essence of the OVA philosophy may be used as a point of departure for measuring the costs and values of support services, establishing improved benefit/cost relationships, and to establish measures of performance and productivity for support services that are meaningful to both user and provider, can be used to track performance against target values, and to evaluate trends over time.

1.6.11 MEASURING FUNCTIONAL ACTIVITIES

Support services may be provided from within various production units or departments, grouped in separate organizational units, or both. Within any ogranization it is useful to have some measures for relating support activities and their resources to the operation of the "line" elements of the organization. These measures may be very broad, such as total indirect manufacturing costs/total direct manufacturing costs, or rather detailed, say maintenance costs per line by shift per unit of production, or full-time equivalent order-entry personnel head count per 100 orders received, or secretarial head count per 100 indirect plant personnel. These measures can be the basis for establishing norms, or standards, can be tracked over time to show trends, and in many cases can be used for intra- and interindustry comparisons.

Identification of groupings of support activities in various organizational units that are similar or identical in function may be useful as the basis for developing measures that relate support-service resource requirements to need or output. One system, the Common Staffing System (CSS) developed by IBM, has been used to establish relationships between 115 functional activities and 40 indicator measures of the cause or reason for the activities being required. The CSS has been used by its originators to "measure" all nondirect activities, providing periodic summaries of performance for various groupings of functional activities at various levels of detail and aggregation, and showing trends and inter-/intraplant comparisons in order to stimulate improvement in utilization of resources and performance.[5]

1.6.12 MEASURES OF ORGANIZATIONAL STRUCTURE

Organizational structure may impose severe limitation on the productivity and performance of both line and support services. A deep organization usually represents more support and staff personnel and more managers than a shallow organization. A centralized organization is more bureaucratic than a decentralized one. A narrow span of responsibility or control is likely to have greatly different impacts on costs and performances than a broad one. Concentrating on secondary activities, rather than those which are primary contributors to helping the organization achieve its mission, may be wasteful and ineffective use of resources.

Measurement of organizational structure, perhaps on an occasional or periodic basis, may be a useful way to insure that organizational barriers do not limit performance and productivity, particularly in the area of staff and support services.

There are many measurements of organizational structure, for example: span of control, span of responsibility, allocation of effort to primary and secondary tasks, mix of activities, relationship of activities to organizational mission, duplication of effort, fragmentation of effort, support costs relative to business volume, cost to manage a unit of business volume[6]. A few key measurements usually suffice.

Standards or norms for measures of organizational structure are highly "situation specific," but general values are recommended in management literature and can be used as the basis for determining the general magnitude of base values and the desirable direction of change.

```
3080 DATA CASES PRODUCED,LINE UTILIZATION
3090 DATA LINE MECHANICAL EFFICIENCY
3100 DATA LINE CAPACITY UTILIZATION,LINE TIME UTILIZATION
3110 DATA SYRUP YIELD
3120 DATA UTILIZATION OF TOTAL PERSONNEL (PLANT & WAREHOUSE)
3130 DATA PRODUCTION DIRECT PERSONNEL
3140 DATA PRODUCTION INDIRECT PERSONNEL
3150 DATA PRODUCTION SUPERVISION AND CLERICAL PERSONNEL
3160 DATA TOTAL PRODUCTION PERSONNEL,PRODUCTION OVERTIME HOURS
3170 DATA WAREHOUSE CASE HANDLING PERSONNEL
3180 DATA WAREHOUSE BOTTLE SORTING PERSONNEL
3190 DATA WAREHOUSE PALLET MAKEUP PERSONNEL
3200 DATA WAREHOUSE TOTAL DIRECT PERSONNEL
3210 DATA SUPERVISION AND CLERICAL PERSONNEL
3220 DATA TOTAL WAREHOUSE PERSONNEL,WAREHOUSE OVERTIME HOURS
3230 DATA ABSENTEEISM
3240 DATA CONCENTRATE/BEVERAGE BASE YIELD,SUGAR 1 YIELD
3250 DATA SUGAR 2 YIELD,SUGAR 3 YIELD,ROLL ON 1 YIELD
3260 DATA ROLL ON 2 YIELD,LABEL YIELD,CROWN YIELD,CAN END YIELD
3270 DATA OTHER A YIELD,OTHER B YIELD,OTHER C YIELD
3280 DATA OTHER D YIELD,CO2 USAGE,CAUSTIC USAGE,WATER USAGE
3290 DATA OTHER A USAGE,OTHER B USAGE,OTHER C USAGE,OTHER D USAGE
3300 DATA PRODUCTION BOTTLE BREAKAGE,WAREHOUSE BOTTLE BREAKAGE
3310 DATA PRODUCTION CAN WASTAGE,WAREHOUSE CAN WASTAGE
3320 DATA CARTON REPLACEMENT,MT CASE REPLACEMENT
3330 DATA PRODUCT LOSS (CASES LOST FOR LOW QUALITY)
3340 DATA STORAGE UTILIZATION,ENERGY CONSUMPTION ELECTRICITY
3350 DATA ENERGY CONSUMPTION OIL,ENERGY CONSUMPTION GAS
3360 DATA ENERGY CONSUMPTION OTHER,ENERGY CONSUMPTION TOTAL
3370 DATA FREQUENCY OF INJURIES
3380 DATA NUMBER OF HOURS LOST DUE TO ACCIDENTS
3390 DATA PRODUCT-MATERIAL-OR-PROPERTY LOSS DUE TO ACCIDENTS
```

Fig. 1.6.6. A listing of measures from which plant personnel can select to obtain daily and/or weekly performance reports.

1.6.13 MICROCOMPUTERS

Most organizations utilize computers to facilitate the summary and reporting of productivity and performance data. Many systems are centralized, with attendant delays in processing of data. Where appreciable delay separates actual performance and the availability of reports to those who are in the position to directly control and influence performance, the value of the data for guiding actual performance is lessened.

The advent of microcomputers, particularly those with capacity for business applications, has opened up many new opportunities to couple more closely performance and the availability of evaluative data to those directly involved in the execution of work and the use of resources. The current generation of microcomputers rivals the power of earlier large-scale computer systems at very modest costs, allowing key personnel to monitor performance and productivity on nearly a real-time basis.

Figure 1.6.6 is from a microcomputer application, showing a listing of measures from which plant personnel in one multiplant organization can select for daily and/or weekly report summaries to keep themselves informed of how well they are performing and managing the use of resources. The microcomputer program was designed for ease of use by plant personnel, and so that it would integrate upward to provide summary data for period productivity and performance reports processed by the organization's main computer.

1.6.14 SUMMARY

Measures of performance and productivity for production systems should be designed to help people within the organization do their jobs effectively and to cultivate improvement to enhance the performance of the production system and the organization. Not only should measures be available to indicate expected accomplishments and resource requirements for direct production work, but also for required support services that facilitate production accomplishment indirectly.

REFERENCES

1. Additional information about the APC Firm-Level Total Productivity Measurement System can be obtained from the American Productivity Center, 123 Post Oak Lane, Houston, TX 77024.
2. Mundel's work-unit structure is fully described in Marvin E. Mundel, *Improving Productivity and Effectiveness*, Prentice-Hall, Englewood Cliffs, NJ, 1983.
3. Use of the nominal group technique is covered in: Andre L. Delbecq, Andrew H. Van de Ven, and David H. Gustafson, *Group Techniques for Program Planning*, Scott, Foresman, Glenview, IL, 1975.

Stewart, William, *Performance Measurement and Improvements in Common Carriers*, Purdue University, Lafayette, IN, 1980.

Morris, William T., Implementation Strategies for Industrial Engineers, Grid Publishing, Columbus, OH, 1979.

4. For further information on Overhead Value Analysis see John L. Neuman, "Making Overhead Cuts That Last," *Harvard Business Review*, May–June 1975 (reproduced as Chapter 4 in Robert N. Lehrer, *White Collar Productivity*, McGraw-Hill, New York, 1983), and Neuman and Hardy's chapter "Gaining a Competitive Edge through Participative Management of Overhead" in *The Handbook of Planning and Budgeting*, Van Nostrand Reinhold, New York, 1985.
5. The Common Staffing System is described by David L. Conway in Ch. 9 of *White Collar Productivity* (see Ref. 4).
6. The Introspect process of analysis and measurement of organizational structure is described by Ralph A. Johnson, Ch. 5, in *White Collar Productivity* (see Ref. 4).
7. Various publications of the Institute of Industrial Engineers, 25 Technology Park, Norcross, GA, 30092, are excellent sources for information dealing with measures of performance and productivity.

CHAPTER 1.7

MANUFACTURING MANAGEMENT

AMIT K. JAIN

Professional Computer Resources, Inc.
Oakbrook Terrace, Illinois

In recent times, manufacturing management systems have gained a lot of management attention. For a growing number of companies, this is one system that is relied upon for day-to-day management of operations and is the primary source of information used to make strategic decisions. The initial benefits derived from the implementation of such a system should be reduced inventory and improved customer service. A well-designed and well-managed computer-aided manufacturing control system permits both these goals. The inventories will be of the proper mix, and through control of work-in-process, manufacturing lead times will be minimized.

The effective utilization of manpower and equipment will result in improved productivity and efficiency. Time and material will no longer have to be applied to parts that are not needed. Similar improvements in the utilization of machinery and equipment can be realized, offering potential elimination of, and/or delays in, capital investment. The manufacturing control system can forecast needs for capacity, labor, and cash. This allows for increased bookings through improved on-time delivery performance, effective utilization of capacity, and better planning for necessary expansion.

The manufacturing management system will become a company system that will improve not only profitability but also the management process. The decisions that are made will be based on sound information and will be communicated to all interested users. Rapidly expanding computer technology has paved the way for more extensive use of the computer in the control of manufacturing operations. Similarly, this has permitted the use of manufacturing systems techniques that were impractical on a manual basis. In this chapter, we will be concentrating on the more significant application areas in a manufacturing control system, the associated practical technical features and functions available, and the selection of these techniques for various manufacturing environments.

1.7.1 OPERATIONS ORGANIZATION

Let's first define the word *manufacturing* as we use it in the context of a manufacturing control system: "The manufacturing function is responsible for the acquisition of facilities, manpower, and materials, and for coordinating their use in the production of goods."

"Facilities" in a manufacturing company is a plant, along with the machines and equipment that are used to make the product. Manufacturing is also responsible for hiring and supervising the numerous employees who assist in the manufacture of products and for acquiring the raw materials that are converted into the product.

The manufacturing functions are the responsibilities of the Vice-president of Engineering, Manufacturing, Materials, and Quality Assurance. Part of the responsibilities of the Vice-president of Sales and of the Controller can also be viewed as manufacturing control functions.

Specifically, the Vice-president of Engineering is responsible for both creation and maintenance of design, testing, and engineering records. In addition to designing new products, engineering is also responsible for improving the design of current products and designing custom products to satisfy certain customers' requirements.

The manufacturing engineering function reports to the Vice-president of Manufacturing and is responsible for designing the process to build the end-product and for designing the tooling and gauges that will be required. It is also responsible for setting the time standards for the manufacturing operations.

Production control is responsible for translating the sales plan into production plans that best utilize the manufacturing capacity; it also schedules operations in accordance with the plan.

One of the primary responsibilities of the materials function is to control raw material components, in-process and finished-goods levels, and to determine the requirements for the individual components of the end product and to ensure the availability of material through maintenance of inventory records. Material management is also responsible for control of the materials in storage.

Inspection's responsibility is the all-important one of assuring material and product quality. There is a move in American industries for the "quality-at-source" concept, where each vendor will be responsible for the quality, and the customer will eliminate incoming inspection. A final function within the manufacturing area is accounting for manufacturing cost and reporting of performance.

The organizational units that have been described may vary from enterprise to enterprise in terminology and responsibilities, but the functions they perform must exist to collectively perform the manufacturing functions. These units all have a common goal: to satisfy the demand for quality products, delivered on time, produced at lowest cost, with minimum capital investment.

With so many organizational units involved, it is not surprising that there would be conflicts in short-term goals. For example, the objective of maintaining a level manufacturing schedule conflicts with the objective of minimum inventory investment. To properly balance these conflicting objectives, a series of integrated systems modules covering all aspects of manufacturing is necessary. These will be discussed in detail in the next section.

1.7.2 MANUFACTURING MANAGEMENT SYSTEMS

The following pages will provide an overview of the scope of a typical manufacturing management system in terms of the functions it serves, the systems utilized, and the benefits that can be realized. In the design and installation of manufacturing management systems, each application has unique characteristics that vary by industry, product type, enterprise size, nature of assembly and fabrication operations, and management capability. "Fitting" the system to the enterprise's environment is one of the most important aspects of manufacturing planning, procedures, and policies. A management approach to the planning, execution, and feedback processes can be improved through effective use of the Closed-Loop Manufacturing System shown in Figure 1.7.1. In the following pages, a detailed description of these processes will be provided.

1.7.2.1 Strategic Planning

In the 1960s, business schools emphasized strategic elegance over hands-on experience and well-managed line operations. The decision-making process focused on short-term results and was mainly influenced by the

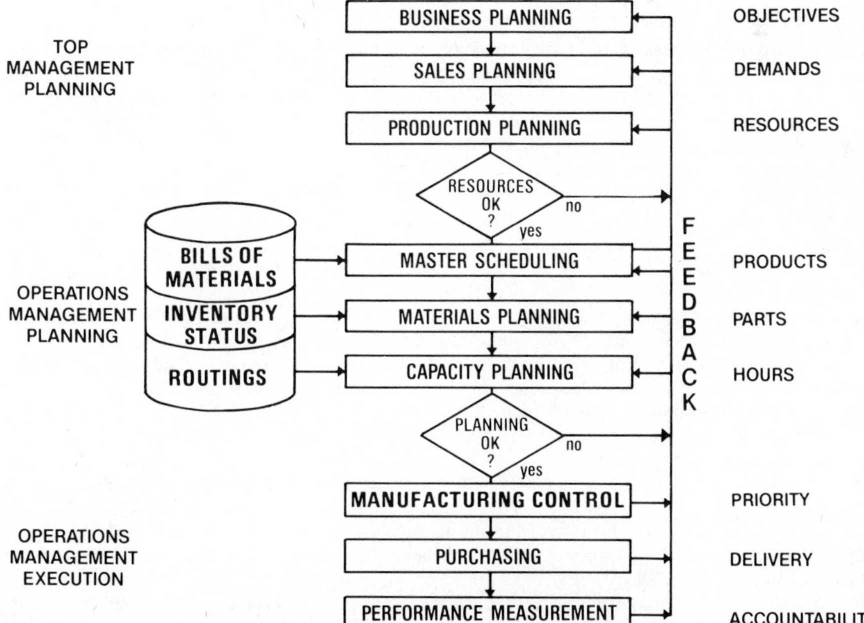

Fig. 1.7.1. Closed loop manufacturing system.

marketing and financial managements, at the expense of manufacturing and technological needs. However, manufacturing strategic planning should be an equal partner with finance and marketing in formulating corporate strategic plans.

Manufacturing Strategy

The driving force needed to develop a strategy should include clear understanding on at least the following three fundamental elements:

A. Dominant Business of Company
   1. Market oriented.
   2. Technology dominated.
   3. Conglomerate.

B. Diversification
   1. Develop new products for new markets.
   2. Develop new products for existing markets.
   3. Develop new markets for existing products.
   4. Process diversification.

C. Growth Strategy
   1. Growth in market share without concern for productivity and profitability.
   2. Growth permitted only if it will help earnings and profitability.

An effective manufacturing strategy is not necessarily one that provides the maximum efficiency, but one that "fits" the needs of the business, perspective, and background of key managers in the operational capability. The area of manufacturing strategy, although important, has received very limited attention in the past. There are at least the following four explanations:

1. Role of manufacturing managers in most companies is reactive rather than strategic.
2. Most companies are looking for short-term gains.
3. Difficulty in measuring results and improvement in performance.
4. Inability to communicate and execute manufacturing strategy.

Strategic planning initiates the closed-loop manufacturing process. During strategic planning, a company's long-range direction is established through the development of objectives, goals, and strategies based on the industry and competitive marketplace in which the company operates.

1.7.2.2 Market and Sales Planning

The next step in this framework is to take the long-range direction that has been established by the top management and convert it into a sales and marketing plan. The marketing plan establishes the target markets and market shares the company must have to move in the direction specified by the strategic plan.

The primary function of the marketing plan is to develop products in response to existing market demand or to create demand for new or existing products. This entails:

Identifying target markets, market segments, and market share desired

Determining the right type of promotion and advertising to maintain demand for existing products and accomplish established market-share targets. Pricing is important competetive strategy that affects both new and existing products.

The pricing and promotion of products must be oriented toward the target market, which requires knowledge of the customers' characteristics and knowledge of the competition's strengths and weaknesses.

One of the important considerations in market planning is successful new-product introduction. In the next decade, along with an increasing number of new product introductions, companies will also be under pressure to phase out unprofitable products, which is very difficult due to the reluctance of Sales to decrease the number of products it can offer to customers. But it will become essential, because continuing with declining products diverts resources from more profitable uses.

The sales plan is derived from the market plan. It consists of a specific unit- and dollar-forecast by product line, which relates to the strategic plan. The sales plan must relate the annual-sales dollar forecast to a monthly unit forecast to determine whether manufacturing has the capability to produce the planned products. To remain competitive, Sales must be knowledgeable about customers' needs to be able to evaluate the level of service that manufacturing must provide.

The sales plan, properly developed, becomes the basis for the production plan, which is what manufacturing must build to meet the strategic, marketing, and sales plans. It is the production plan that drives the manufacturing resource-planning process, which we will discuss later. One of the steps in the sales-planning process is the development of a sales forecast.

Sales Forecast

Forecasting and demand planning cover the process of estimating expected future demand. Accurate forecasts and demand plans are essential for the effective planning and control of manufacturing operations. Figure 1.7.2 provides a synopsis of the types of sales forecasts and their implications.

Long-term forecasting for long-range planning purposes typically uses many external variables and employs sophisticated analytical techniques. However, for production and materials planning purposes only a short-term forecast is necessary. "Short-term" in this context is defined as equal to or greater than the cumulative product lead time. Time-series forecasting is the preferred approach for short-term forecasting. The data from the forecasting system are modified based on delivery schedule status and plant conditions. This modification results in a *time-phased* sales plan.

The sales plan is reviewed by a team consisting of the top functional executives in the company. This process in itself may highlight problems or impracticalities. The planning team then passes the recommended sales plan to the president for approval, after which it is forwarded to the production department. After orders have been processed and forecasts updated, finished-goods inventory balances are evaluated and replenishment decisions are developed considering the on-hand inventory.

1.7.2.3 Production Planning

Unlike sales forecasting, which may or may not be required, some form of production planning is required in almost every manufacturing enterprise. The production plan takes into account the constraints on key facilities, material, major components, production capacity, manpower, and funds. Each alternative production plan must be tested to ascertain whether it can be accomplished within these resource constraints.

The production plan not only tests the resources but also emphasizes different business strategies. For example, a company may wish to:

Minimize inventory cost by lowering inventories which degrade delivery service.

Increase customer deliveries by increasing inventories.

Minimize production changes and costs by planning long production runs, which usually result in higher or unbalanced inventories.

Adopt a "Just-in-time" environment that reduces inventory but requires tight control of output and accompanying plans.

In performing this exercise, it is important to remember that there is need to check out only the critical or bottleneck facilities, "key facilities," which act as the tightest constraints on production, along with major components with long lead times.

The first step in developing a production plan is the sales plan. For the purposes of production planning, the sales plan can be summarized into product categories (also called product *lines*) consisting of groupings of similar products with similar production characteristics. This reduces the production planning workload and makes it easier to forecast the load for future time periods. Many alternative production plans that emphasize different business strategies may be evaluated and passed through the planning cycle and review process. Once the production plan has been approved, a master schedule is prepared. This schedule identifies product-category requirements in the short range, usually covering the period required for final product manufacturing lead time.

Master Scheduling

Production planning is a process of producing high-level schedules in gross terms. When management has reached a decision as to the general plans, this plan is translated into a specific master schedule used to schedule all manufacturing activity. This is where top management involvement is essential.

	Long Range	Medium Range	Short Range
Horizon	3-5 Years	1-2 Years	3-12 Months
Planning Interval	Year	Quarters/ Months	Months/ Weeks/ Days
Detail	Aggregate	Families	Code by Plant Location
Use	Strategic	Tactical	Operations
Technique	Macro	Extrapolation	Statistical

Fig. 1.7.2. Forecasting.

What is master scheduling? Typically, master scheduling involves developing daily or weekly schedules for final assembly. In some manufacturing companies, other key production-pacing facilities are master scheduled. Two examples of other key facilities are a feeder subassembly line and the paper machine in a paper-product manufacturer.

The master schedule is not a high-level generalized plan, but rather must be a specific schedule that can be used to drive all manufacturing and procurement activities. The master schedule should be the best possible approximation of what will actually be produced based on planned capacity.

Successful master scheduling is dependent on the accurate and timely status of components in inventory, which is handled by inventory management.

Inventory Management

Inventory control is basically concerned with answering the questions of what, when, and how much to order. These inventory control questions are for both independent and dependent demand items. Independent demand items are usually controlled based on history, such as order point or sales forecast, whereas dependent demand items are ordered using a material requirements planning (MRP) procedure, which will be discussed later. A basic difference between order-point and MRP systems is that order point utilizes an independent forecast for each part, whereas MRP recognizes that the demand for component parts is a result of the demand for higher-level items. Both systems, order point and MRP, utilize certain inventory control techniques in the process of determining when and how much to order.

When-to-Order Techniques. There are four techniques that are usually considered for making the when-to-order decisions:

1. Variable reorder point
2. Fixed reorder point
3. Replenishment cycle
4. As required

The reorder index is another way of presenting information developed using the variable reorder point, fixed reorder point, or replenishment-cycle techniques. The variable reorder point is a sophisticated, flexible approach that places orders based on the most current demand projections, lead times, and stock status.

The fixed reorder point does not require a complex demand projection and involves infrequent review of the reorder point, typically using historical usage data. It is not nearly as flexible as the variable reorder point and does not respond to rapidly changing requirements or lead times.

The replenishment-cycle approach is useful when frequent review of stock status is not necessary, for example, in warehouse resupply, where the vehicle delivery schedule may determine when to review stock status. The as-required approach places one order for each demand as the demand becomes known. It should be considered for custom products, items produced on a make-to-order basis, and items with a high risk of obsolescence.

How Much to Order. Practical limitations or constraints should be placed upon any order quantity technique. Minimum order quantities may be set by vendors or by the supply operation in order to improve their operations. Maximum order quantities may be set by management. Space constraints may limit the maximum order quantity that can be accommodated.

The concept of blanket orders permits large quantities to be purchased with attendant price savings. This approach allows the ordering of one or more items and the scheduling of multiple deliveries for these items. Blanket orders are negotiated in advance by the purchasing department. The inventory ordering techniques discussed will not produce the desired results without reasonable inventory record accuracy. The next section is concerned with physical inventory management, which has a large impact on record accuracy. A storeroom with restricted access is a potential improvement in the accuracy of perpetual inventory balances, along with well-designed transaction systems to record receipts and issues from storeroom.

Physical Counts. Over time, inventory records become inaccurate as counting and clerical errors occur. To remedy inaccuracy, physical counts and reconciliation to records becomes necessary. The alternatives are:

1. To count everything periodically (annually, for example). This can be quite expensive because it is usually necessary to stop production until the counting is completed.
2. To continuously cycle count. While the count effort, if everything is counted at least once a year, could be equal to the count-everything alternative, production can continue while the count is being taken.
3. Some combination of the above.

1.7.2.4 Design Engineering

Design engineering information is the foundation of any integrated manufacturing planning and control system. It provides the base information that details the parts included in each assembly. The major use of this information is for:

Production planning
Procurement
Operation scheduling
Product costing

Because of the importance of design engineering, a great deal of attention has been given to techniques of storing and accessing engineering data. Four basic files are normally utilized in an engineering information system. These files are:

Item master file
Product structure file
Production router file
Work center file

These files, and the basic information they contain, are essential whether we are discussing manual or computerized systems.

The *item master file* contains the data on raw material to finished-product items used in the manufacturing process. The file also contains stock status, order status, and requirements data by time period.

The *product structure file* defines the components used in each manufactured item through a parent-component relationship. It is usually designed on a single-level concept in which a particular structure is defined only once, regardless of how many higher-level assemblies it may be part of. This concept is the basis upon which most modern data base management systems are built. Among other uses, this file structure enables a new bill of material to be constructed by making slight changes to an existing one. This procedure is useful when designing custom products that are essentially minor variations of standard products. The use of such a maintenance aid is a necessary capability in any system.

1.7.2.5 Manufacturing Engineering

Manufacturing engineering decides how the product will be made and sets standard times for operations. The data are stored in a production-router file. For each manufactured item, each manufacturing process is described by sequence of work center it will have to go through and what type of tools and fixtures will be required.

The *production router file* is used in a number of ways. One important use is to multiply the quantity of items to be produced by the standard times shown and then to post, or "load," the aggregate times into each *work center file*. This load is compared with the predetermined capacity of each work center, based on man and machine in the center, adjusted by historical utilization and efficiency factor, and stored in the work center file. This permits the user to determine whether plant resources are adequate to meet production plans. This file is also used to document shop-floor-processing flows for instructing employees on standard manufacturing methods, for cost accounting, and for a variety of other purposes.

1.7.2.6 Material Requirements Planning

A sound master schedule is the foundation for requirements planning. Material requirements planning ranks as one of the most important operating tools available to manufacturing management. The objective of requirements planning is to determine which parts must be purchased, what assemblies must be placed into production, and when these events must occur so that the master schedule can be met. This requires identification of all components used in manufacturing of the end product or subassemblies. This component breakdown, level by level, is referred to as the *bill of material explosion*.

Numerous systems alternatives exist and should be thoroughly explored and adapted depending on user requirements. Following are some examples in connection with the explosion process.

Should the explosion be gross requirements at all levels or netting at each level?

Should requirements be regenerated each time the plan is modified or only adjusted for changes?

Should provision be made for both normal and reduced lead time offset between levels in order to allow schedules to be compressed?

Should engineering changes be controlled on a due-date basis or on a quantity basis or both?

How should unfilled issues of components allocated to released manufacturing orders be handled?

Similarly, there are several alternative methods of determining order quantities:

Discrete quantities may be ordered weekly to correspond to calculated net requirements.

Fixed quantities may be released at variable time intervals. This may be the case where standard lot sizes are a factor.

Another alternative is annualized economic order quantities.

Order certain period requirements, such as a six-weeks' supply.

Finally, dynamic order quantities might be applicable. These are economic order quantities, which will be recalculated for each week in the future based on planned future requirements. They are effective when demand is highly variable.

There are many more systems alternatives in other areas of a requirements planning system, and each should be explored thoroughly to evaluate its usefulness in a particular situation. Finally, a requirements planning system should generate a number of inventory performance reports, including those showing overall inventory status, overall inventory performance, shortages, and excess and obsolete stock. This area will be covered in more detail later.

In total, a requirements planning system specifies the quantity of materials and components needed to meet a particular production plan. It also specifies the time within which action must be taken to acquire or produce these materials and components. Any deficiencies in this area will be magnified in subsequent operation of the manufacturing control system.

Once specific requirements are known, we are ready to release orders for the purchase of materials and the manufacture of components. Even though our framework defines purchasing as part of logistics, we will discuss it here because of its impact on the manufacturing control system.

1.7.2.7 Purchasing

The purchasing function can contribute significantly to improve profitability, particularly where purchased materials contribute a large portion of the product cost. In making final decisions in the purchasing process, certain supplemental information is provided to the purchasing agent over and above the technical description of the materials required.

This data relates to the procurement decision itself and includes information on vendor reliability, quotes and price structures, freight considerations, and other similar matters. Based on the quantities needed and other vendor-related considerations, the purchase orders are finalized and released to vendors.

1.7.2.8 Manufacturing Control

A companion action has to do with the release of planned manufacturing orders:

Operation scheduling
Capacity planning and loading
Material and component availability

Before planned manufacturing orders can be released, they must be scheduled in terms of time, usually weeks, in each work center or operation. They must then be loaded against available capacity. The purpose of the capacity plan is to give management an indication of short-range capacity so overtime or necessary changes in employee assignments can be made to meet the schedules. They must be checked for material and component availability by reference to shortage reports or status information in the item master file. Finally, they must be scheduled in a sequence for release to production.

The operation scheduling process may be performed with widely varying degrees of sophistication, depending on the nature of the product and the production process. Reports generated will provide data needed to start and maintain factory production, including, for each order, descriptions of the production process, material required, work standards, and production lead time. This data, when combined with procedures to report the status of orders on the production floor, permits the preparation of key control reports to monitor the status of work stations and individual production orders and to identify delays and errors.

Using logic similar to that described in the production planning process, projections can be made on the effect on production facilities of an order to be released. Capacity limitations thus can be detected and corrective action taken before they become serious. This process can operate on either a finite or infinite concept. The infinite concept assumes unlimited production capacity and develops reports which compare those instances where planned production exceeds capacity.

If an overload condition is detected, manual intervention may be needed to revise plans so that they fit within the capacity constraints. The finite concept does not permit planned production to exceed capacity. When excesses do occur, production plans are corrected based on the logic of priority rules designed into the system. In practice, most companies use the infinite concept. Prior to the release of the order for the end product or subassembly, material and component availability must be checked to ensure that the shop floor is not flooded with paperwork or partially completed work-in-process. Shop packets, including production orders, material requisitions, move tickets, and completed production reports, are issued to the shop floor.

The feedback of data from the shop floor and the reporting of manufacturing order status represent a major area of manufacturing control. The number of scheduling and reporting points and the frequency of reporting are closely related to the capacity planning and operation scheduling system, so the two areas are typically treated as one from a design viewpoint. The objective in all cases, however, is to provide status reports, on an exception basis, to facilitate expediting action.

Finally, purchasing and manufacturing transaction data, including material movements, provide the basis for cost and performance reporting. In this connection, it is important to note that the quality of system design and the discipline underlying the execution of the shop-floor procedures system directly affect the quality of the cost and performance reporting effort.

In addition to the cost and performance reporting, two other modules are necessary to manufacturing operations. These modules are quality control and maintenance. Both areas are critical to a manufacturing operation, yet they are frequently overlooked.

1.7.3 PERFORMANCE MEASUREMENT

Performance measurement is management's responsibility. Management can effectively guide or direct people and other company resources toward company goals only if it first understands the company's current status. Therefore, management must be kept continually abreast of the company's status as it guides the company toward its objectives. Performance measurement provides management with status information that is accurate, timely, useful, and meaningful. Management must then select the performance criteria, measure performance, and take appropriate corrective action. Successful performance measurement requires serious consideration of what performance is to be measured and how. Managers should be involved in this process. The following are important:

- Functions to be measured clearly identified
- Responsibility and accountability defined
- Frequency of measurement
- Realistic definition of goals/targets
- Reward system for exceptional performance

We have now completely developed all the modules comprising an integrated manufacturing planning and reporting system that serves as a base point for tailoring systems. The number of modules required and degree of their use will vary by industry and enterprise. The planning and control features incorporated into these systems provide for large-scale cost savings, typically in the areas mentioned. Based on the number and magnitude of costs involved, substantial opportunities for profit improvement are also available.

Finally, a word on the application of computers to manufacturing systems: The features of an integrated system, coupled with the power of the computer, enable the manufacturers to deal rapidly and effectively with the large volume of data and the complex interrelationships which exist. The results are:

- Better delivery service
- More complete control over resources and costs
- The ability to control the expansion of facilities and markets

In the case of the large amounts of data and coordination required in the manufacture of a typical product, it would be difficult to effectively and efficiently utilize the necessary manufacturing control systems without computer assistance. It should also be recognized that the manufacturing organization will have to change in accordance with the procedures developed.

In summary, the closed-loop manufacturing system, which consists of an integration of top management planning, operations management planning, operations execution, and feedback, takes a data-driven approach. It recognizes the importance of information flowing between various levels of management and various departmental entities of manufacturing companies to achieve overall coordination. Even though the concepts are sound, the results of the closed-loop system have often been disappointing to companies. There has been a tendency to believe that if these concepts were put in place, the benefits would automatically be forthcoming. Perhaps there has not been sufficient recognition of the need to work for the benefits. When installing such systems, the attention has been on achieving clean bills of material, accurate inventory status, and clean routings, as well as the disciplines required for timely and accurate information from the shop floor. However, not enough emphasis is placed on questioning the real cost generators and obstacles to customer service and quality. Questions are not asked regarding why we have inventory in the first place, why our lead times are as long as they are, and why it takes so much time to process information from one department to another. Problems in these areas are not automatically addressed by the closed-loop manufacturing system. In addition, the closed-loop manufacturing system does not address key areas that are critical to achieving overall productivity improvement. In effect, the closed-loop manufacturing system does not integrate the shop floor. It has not encouraged people to manage and improve productivity. This approach recognizes a number of departmental barriers that tend to increase lead time, add redundancy, and result in inefficiencies. The closed-loop can be a powerful tool if the information developed in the various departments is used for pointing out where excessive costs are generated or where there are opportunities for improvement. That is when we will start seeing real benefit through an essential interplay between information, technology, and management itself.

CHAPTER 1.8

PRODUCTION ORGANIZATION

ORLANDO J. FEORENE

Georgia Institute of Technology
Atlanta, Georgia

1.8.1 INTRODUCTION

Technological developments are changing not only the way in which work is performed in offices and factories but also how it is organized and managed. At the same time, the design and implementation of modern production facilities are heavily influenced by changing societal needs, aspirations, and values that cannot easily be ignored by managers in the pursuit of productivity gains and the improvement of quality. Organizational structures suitable for an earlier, less complex manufacturing environment are inadequate in managing the highly integrated and interdependent processes that characterize the production of goods and services in modern factories and plants. It is important that planners and designers as well as managers understand the impact of technologically innovative methods and processes on the communication and decision-making channels, and on the roles traditionally assigned to workers, supervision, and to managers.

The computerization and automation of manufacturing processes has actually made possible more efficient organization structures for management. Parallel although independent information processing developments within the office environment have introduced the possibility of an essentially instantaneous communication capability across organizational lines, and between facilities literally anywhere on the face of the globe. However, the dramatic technical innovations in controlling machine tools, in monitoring processes, and in the automation of the physical plant has captured most of the attention of management. The human and organizational implications of these technical changes, although not completely ignored, have suffered by comparison and have yet to be fully understood at all levels of management. Organizational development must progress hand-in-hand with technological developments. The organizational structure cannot be left to the last, any more than integrated manufacturing systems can be forced into an organization based on departmentalization of production facilities. Independent design processes destroy any possibility for a synergistic synthesis of an optimal production capability.

Integrated manufacturing systems will require workers with generalized skills that traditionally have been categorized into separate crafts or disciplines. These workers will be a new breed of operators with monitoring responsibilities and decision-making assignments. The lines of demarcation between blue-collar and white-collar workers is certain to become increasingly less definable or discernible. Interactions between line and staff at both the technical level and the management level will gravitate toward a more closely knit shared responsibility as the complexity and integration of operations increase. The ability to widely and instantaneously propagate easily accessible, massive amounts of information contributes to a democratization of information flow that in itself is bringing about changes in organizational behavior and responsibilities. The future is here—technically speaking.

This chapter first briefly profiles the early departmentalization of functions that effectively served the needs of management for nearly a century of industrial progress. The factors contributing to the need for a change in methods of organizing resources are then described, along with the inadequacies of the earlier forms of pyramidal organizational structures. Finally, a rationale is offered for developing methods to more effectively manage the orchestration of the many skills and disciplines required to design, implement, and operate a modern production facility.

1.8.2 IN THE BEGINNING

Organizations came into being to provide a structure that would allow enterprises to efficiently marshal resources toward some common goal. Since people were the means to this end it was only natural to place them in groups in which they held something in common—a craft or skill, for example—in order to employ and direct their activities more effectively.

This natural grouping of resources or specialized capabilities was pervasive and widely practiced. Some managers adopted the principle of occupational specialty in creating separate work sections and departments; others relied on process differentation or on specific product groupings to determine the boundary lines of the organizational subdivisions. In shop areas, machine tools tended to be grouped by functional families: punch presses were separated from turret lathes, shapers, and planers, for example, while plating operations in a metal producing plant were conducted away from rolling or forming operations.

Clerical and stenographic activities were all grouped to take full advantage of specialized skills or office equipment. Functional activities such as production planning, labor scheduling, purchasing, accounting, and materials control tended to be located in facilities often quite removed from production operations. Information was usually compiled and batched, processed off-line and delivered at some later time to plant supervisors for action.

Technical support staffs experienced similar specialized grouping arrangements. It is not uncommon even today for product designers to be in one department while process-development engineers are in another; information and computer specialists are still separate from industrial and systems engineering groups in many companies. These few examples illustrate the wide acceptance of functional groupings as a method for organizing operations.

The resulting departmentalization required the delegation of a certain amount of authority and responsibility to leaders, usually selected from the ranks of the specialized groups. These leaders were chosen probably because they knew more than their coworkers about a particular craft or function. With some justification, expertise and demonstrated skills were deemed adequate prerequisites for a supervisory role. The now familiar hierarchical pyramid came into being. Procedures were developed to facilitate horizontally related activities and the vertical relationships necessary to integrate the efforts of the individual decision makers with those of the enterprise as a whole. In retrospect, the leaders of these functional groups formed a rather effective managerial network and the bureaucratic pyramid has held up over time as a viable structure for managing and controlling resources. The functional and physical departmentalization of work has, for the most part, satisfactorily met the needs of industrial, commerical, and civil service management.

Two recent developments challenge the effectiveness of organizations based on occupational specialties and the subdivision of work. The first is the growing realization that the maximization of output of individual activities in a series of related operations can often lead to suboptimal results for the enterprise as a whole. Computer assisted analytical and modeling techniques have extended management's ability to assess the impact of incremental changes on organizational goals and objectives. This improved capability has tended to raise the level at which decisions are made within the organization structure to minimize the impact of well-meaning but nearsighted changes contemplated at the lower decision-making levels. However, a second series of related developments have combined to form what is probably an even more compelling reason for reexamining how work is to be accomplished in the future. The successful exploitation of the technical advances made in microelectronics, software engineering, and sensor technology have significantly and irrevocably affected how products and processes will be designed, produced, and distributed. As a result these developments will also affect how people will solve problems in a multidisciplined technical environment and how they will interface with the computerized tools of production. The traditional role of the factory worker, the supervisor, and the knowledge worker will be different in this new work environment. Societal and behavioral developments combined with technological advances have generated interactions not easily defined, since human needs, expectations, and lifestyles are involved as well as technical and economic factors. There is a compelling need to examine critically and to understand the implications of these changes on existing and future organizational structures. A not-too-hypothetical description of a modern factory may help.

1.8.3 A CHANGING WORLD OF WORK

A computerized and automated manufacturing plant consists of production equipment with integral sensors that continuously monitor the performance of the machines. Electronic links to programmable logic controllers and microprocessors actuate servomechanisms to automatically correct performance deviations. More sophisticated machines have built-in preventive maintenance capability and provision for automatic adjustments for tool wear and bearing degradations. Each machine is programmed for minimum downtime and maximum yields. Upon command, video terminals display diagnostic graphics to assist in correcting aberrant behavior. The level of autonomy assigned to each production tool is a balance between human capabilities versus machine capabilities, and the size of the knowledge base implanted in the memory of the machine. These options in turn reflect a trade-off between design factors and the economics of operation.

An electronic network connects these fully automated machines to automatic transfer and transport devices and to other computerized production tools. A minicomputer in the network optimally balances the entire sequence of events and scans the entire cycle for the early identification and correction of quality degradation. Human interrogation and intervention is available at a number of predetermined stages in the process. Quality, yield, and machine performance data are continuously displayed in real time at strategic points for human review.

A computerized decision center dominates the hierarchical computer networks that optimize production and scheduling within the plant. The decision center is in constant electronic communication with the sensors and logic controllers of the automated machines. Line balancing is a real-time process for all but the most exceptional of operating conditions. Replacement components or tools are dispatched via robotic vehicles that arrive just in time to minimize downtime, production disruptions, or quality degradation. The decision center assumes overriding responsibility for operating conditions that exceed the programmed logic of the production machines. Humans, in turn, monitor the behavior of the autonomous work stations and introduce a level of refinement that transcends both the optimizing algorithms in the individual machine tools and the higher level logic vested in the knowledge base in the electronic decision center. At any moment three levels of intelligence are available: programmed machine automaticity, intelligent (expert) systems, and human judgment—a new approach to Frederick Taylor's principle of productivity improvement through subdivision of work.

In some manufacturing plants a variety of features may be desired in a product line having certain basic common components. Computer aided design terminals in the engineering or product planning organizations electronically linked with the production machines and conveyance systems allow real-time interaction in modifying assemblies or fabricating components. A high level of flexibility in utilizing production facilities can be realized. Touch and voice-data entry systems directly coupled to computerized numerical control machines (CNC) networked to tactile sensors and vision devices will enhance opportunities for innovative approaches to more flexible manufacturing methods.

Integrated planning, scheduling, and production-control systems will link marketing and distribution information functions directly to the manufacturing plants, minimizing inventories of raw materials, work in process, and finished goods. In an increasing number of instances the electronic network will include both suppliers and consumers. For the supplier, on-line product performance information, consumption rates, and inventory levels will allow more efficient operations through reduced lead times and an improved ability to react quickly to variations in materials or parts delivered to the production facilities. Both the suppliers and the final customer will be able to exploit to their advantage the potential of being linked to a subsystem of an integrated telecommunication capability regardless of where in the world—or in space—they may be.

Other managers will take advantage of the potential of the electronic aids spawned by the computer age to improve the quality and the productivity of their tasks. Computer aided decision support systems will facilitate financial analyses and strategic planning processes. Research scientists will multiply their ability for innovative work with the aid of computers in controlling bench experiments, and in the execution of arithmetic, analytical, and computational tasks. Computer aided literature and file searches for relevant scientific research alone will extend the capability for intellectual work. Engineers and scientists will be able to undertake a higher order of mental and judgmental tasks while sensors and microprocessors recognize and respond to the many functions that are beyond human capability.

The technical aspects of this scenario are not fanciful and have already been achieved. The human and organizational components of complex systems, however, need definition and restructuring to fully realize the potential of these innovative technological developments.

1.8.4 A RATIONALE FOR CHANGE

The decision to design a new manufacturing operation should signal the need to establish an organization that can cope with the problems of creating and implementing a technologically advanced facility.

The conceptualization of an automated process, the design of the software architecture, the human tradeoffs in optimizing the man/machine interactions are complexities far beyond the body of knowledge held by any single technical discipline.

The knowledge workers—the engineers, the scientists, and the computer specialists—need to be brought together into new and different relationships that support the synergism necessary to generate innovative solutions to complex problems. The contributions of many disciplines are necessary ingredients. The unilateral dominance of any single discipline will produce biased and suboptimal design configurations.

The organizational entity that needs to be developed cannot survive in an environment suited for an earlier and different generation of manufacturing technology. A force fit is not tenable; an entrepreneurially oriented organization spinoff will have more of a chance for success.

The practice of grouping professional and technical personnel into separate departments creates a barrier to the orchestration of scarce resources. In far too many instances the concentration of highly specialized technologies has been carried to the point of creating an isolated subculture within the company. In such an environment the interchange of technical developments with other departments becomes increasingly difficult. A doctor/patient relationship develops with managerial clients that excludes participative opportunities and the possibility of generating alternative options for management consideration.

There is a very real danger that left to his or her own initiative, a technical specialist working within the boundaries of a discipline could initiate concepts that would predetermine the form and nature of the manufacturing process and even how it will be managed. Computer systems specialists, for example, will be influenced by the unique determinants of programming logic in designing the information and communications channels for controlling the manufacturing process. In so doing they will be affecting the configuration of the machine tools, the sequence of operations, and even the job content for the operator. A total systems strategy will inevitably lead the designer to consider methods of linking the manufacturing control systems to the management information systems. At this stage managerial involvement is mandatory, since it is management's responsibility to select the nature and the form of the controls that must be in place to manage the company, not the responsibility of some technical specialist. In the absence of any other direction, systems designers could easily program into the information systems decision-support algorithms that might not reflect the wishes and needs of the responsible managers.

In like manner, systems definition that does not involve the active participation of the engineers responsible for the design of the process, the production facilities, and the maintenance procedures will result in design configurations constrained by the systems architecture. There will be no opportunity for synergistic "tradeoffs."

Therefore it is important to identify the architects of a complex manufacturing process and include them in an organizational structure composed of both technical and managerial leaders. This new group should be charged with the responsibility for acquiring and managing the resources required to design and implement the new facility. In doing so, the group should be separated from the ongoing activities of the parent organization, but should be able to rely on it for technical and administrative support.

The emergence of multidiscipline groups or project teams is one response to the need for a different process to manage the design and implementation of complex systems. Quite often a series of meetings or workshops is held at the beginning of the program to define the mission of the team and to clarify the roles of the participants. Issues such as relations with functional groups in the parent organization should be discussed and resolved during this formative phase of the program.

Particular attention should be paid at these sessions to the concerns and needs of the technical personnel assigned to the newly formed team; questions touching on technical responsibilities, authority relationships, and reporting channels are very important to them since these are issues that could affect performance appraisals, salary reviews, and career development plans. Teams of highly competent individuals are quite likely to generate a host of problems as they attempt to function both within and outside a more structured organization. These problems will challenge all levels of management and can be "show-stoppers" in the absence of a process and a plan for their resolution.

The value of early identification of areas of potential conflict within mixed-group settings should be self-evident. Conducting participative problem-solving meetings to discover these areas can be facilitated with the help of a third party skilled in organizing and directing such meetings. However, project leaders should not hesitate to conduct these meetings personally, with or without the help of a facilitator. The group must have confidence in its ability to function as a team; the leader is the central figure in creating the environment that will develop this confidence.

It is quite possible that the design team may need to enlist experts from outside the organization. Not all companies have internal technical resources capable of designing a computer integrated manufacturing system. A number of enterprises can and do provide this resource. In addition to private consultants and research institutions, a high level of technical expertise has been developed by vendors representing the manufacturers of high technology equipment. The engagement of outside experts does not alter the need for a team-building process. The players may change but the issues will remain and should therefore be anticipated.

A synthesis of the experience of managers who have successfully directed and led multidiscipline teams in the design and operation of complex systems is yet to be written. However, there is mounting evidence supporting the need to hold design discussions with a relatively broad group of designers and managers. Typical representation should include machine designers, process development engineers, manufacturing engineers, industrial engineers, and computer technologists. The managers who are to assume the responsibility for operating and maintaining the new facility should also be invited to participate in the initial sessions.

As the group develops, the technical aspects of the broad representation of disciplines present insures the interweaving of human factors into the systems design. The level of "intelligent" control to be assigned to each production unit, the availability of information to allow human intervention, and the kind of performance diagnostics to be made available are typical tradeoffs affecting costs, quality, and yields. Parenthetically, the stage is being set for the organizational structures necessary to manage the new facility. The level of decision-making responsibility to be assigned to operators and supervisors will begin to be established at these early technical sessions as byproducts of the systems configuration. This is the proper time and place for management participation. The process of determining which tasks are to be assigned to machines and which ones are better assigned to humans, and the degree of operator freedom in accessing production information normally shared only with supervision and management are issues requiring the attention and judgment of both managers and technologists. The design system finally accepted will have a profound affect on future organizational and staffing plans. The analysis and resolution of the human and organizational implications of alternative technical designs are as critical to the ultimate productivity of the facility as the investment in capital.

A byproduct of the shared responsibility in resolving the human/machine interactions will be the benefits that joint ownership brings in developing a strong committment on everyone's part to launch a successful program. Since a new manufacturing facility will affect such matters as inventory policies, quality assurance, procurement practice, and staffing needs, some thought must be given to a method of informing the managers of other staff and service functions of the plans and progress being made. A series of monthly or quarterly informational meetings, for example, could provide an opportunity for the exchange of ideas and the sharing of mutual concerns.

Too often the workers who are expected to operate and maintain a new plant are introduced to their new work assignments after the facilities are all in place. Workers today are better educated and more knowledgable about the technological developments available to management. They are also aware of the impact of these developments on their own future. Their participation in some manner needs to be considered to minimize their concerns and to enlist their assistance in the design of the new process. Discussion sessions with the operators, conducted by supervisory personnel and attended by design engineers, provide an open environment for considering questions and evaluating suggestions. The value of these sessions will be apparent in the improvements in the learning curves as the new plant starts operations. The level of committment of the workers should match or surpass that of the designers and managers.

As a matter of design necessity workers will have direct access to information that will allow them to make informed decisions at the operating level at the right time to favorably affect quality, yield, and costs. Video display facilities and interactive programs will greatly enhance their ability to manage a fairly well defined area of responsibility. The ability of workers to manage an increasingly greater portion of their job content is a fundamental change in industrial relations practice. Since workers are already receiving an increasingly larger portion of their indoctrination and training through computerized programs rather than from their foreman or group leader, the inevitable change in the worker/supervisor relationship is all too evident.

The feasibility of shedding or cascading decision-making responsibilities to a lower level of an organization as a result of technological developments is bound to create behavioral reactions up the organizational ladder. Middle managers will feel threatened and insecure if the responsibilities that have migrated to the worker are not replaced by equal or more important tasks—preferably tasks or responsibilities held by their supervisors. A chain-like reaction will impact a series of supervisory and managerial positions. The possibility of combining or perhaps even eliminating positions will be difficult to resist and perhaps may even be desirable in some instances through planned attrition programs. However, the opportunity for even greater productivity gains by restructuring overhead functions should not be overlooked. At the very least an *overhead* functional analysis—one that parallels the design of the physical plant—will suggest innovative and effective organizational relationships that capture and more fully exploit the potential of computer-integrated operations.

The automated facilities will require innovative layouts for the office and production-floor areas. Efficient layouts should package the computerized system and, like a tailored suit of clothes, should be designed to uniquely fit the design configuration. The degree of fit will directly affect the work content and the methods of performing the work. One result of more efficient layouts will be a blurring of the lines of demarcation between office functions and production tasks, which presents one more opportunity to reassess indirect functions and where they can best be performed.

Since many decision-supporting transactions will be directly and immediately available in the production area, the need for off-line compilation and analysis will either be eliminated or reduced. The production, office, and laboratory settings will evolve to a different configuration of functions and communication as well as layout. Facilities for knowledge workers are already being redesigned to exploit the productivity of information processing technologies; distributed information systems and local area networks facilitate the reassignment of functions once considered to be exclusively the domain of the office or the test laboratory. The early warning advantage of information centers in the operations units permits out-of-control corrective action in real time, thereby maximizing yields while minimizing resource expenditures.

1.8.5 A NEED TO MODEL

The assessment of manufacturing alternatives in advance of the implementation of a complex system requires a mechanism for testing the completeness of the logic, especially when it is impossible to experiment with real systems. The tradeoffs between human and machine tasks and the degree of automation in integrating the manufacturing process from raw materials to finished products are typical problems ideally suited for modeling and simulation techniques. However, there is a much higher level of value in using simulations and models in the design of new systems. The collaborative construction of the model by the project team members is an effective vehicle for developing group understanding of the process under design.

In addition, the design of a model that lends itself to *what if* questions brings to the surface decision opportunities to bridge inconsistencies between long-range or strategic plans, and immediate, short-range operational goals. The troublesome inability to pinpoint a proper planning horizon can be greatly facilitated through the joint construction of a model of the system under study.

Complex systems bring together many factors that have traditionally been categorized into separate bodies of knowledge. The barriers between disciplines impede the systems analyst trained or directed to follow an area of specialization. In real life, within the same organizational structure, there is an interaction of physical, technical, economic, behavioral, and cultural factors. These interactions are complex and very important. A separation of the total system into one or more of these component systems will destroy the ability to discover, define, and analyze these interactions.

Simulations are not expected to deliver solutions, but they can be extremely effective in bringing together the separate disciplines essential to the design process. This in itself is sufficient reason for their use by a project team.

Models can be quite simple—flow charts, process charts, two-dimensional layouts—or they may be complex mathematical expressions dependent on computerized algorithms. The essential ingredient for a project team is the completeness of the logic so that it can be manipulated. It does not have to contain all of the specific information regarding the system, but it should be structurally complete.

Modeling and simulations lend themselves to the definition and analysis of the many variables and factors in a complex system, and that is a useful end in itself. But the far greater value to a multidisciplinary group is the means they provide for a common focal point for the synthesis of ideas.

1.8.6 CONCLUSION

The age of microelectronics is young and robust. There will be increases in the number of industrial and commercial operations installing computerized facilities. The world of work is moving toward more complexities, not fewer. Organizational processes will need to change with this changing environment.

The organization selected to design a complex manufacturing process will of necessity undertake a total systems approach. The usual departmentalization of resources does not lend itself to the orchestration of the many disciplines required to design complex systems.

The transfer of human tasks to automated machines, readily accessible information for the operator, and the migration of off-line functions from the office to the production areas will change organizational patterns. Inevitably there will be an upgrading of the responsibilities assigned to all levels as the roles of operators and supervisors are redefined. A significant gain in productivity should result from the increased time made available, time that can be spent on the more demanding intellectual pursuits of strategizing and planning. However, to achieve this goal the design of a complex manufacturing process cannot be left to the technologists and scientists alone. The involvement in the technical-design process by those who must implement and operate the new facility is necessary in order to fully realize more efficient production organizations.

BIBLIOGRAPHY

Drucker, Peter F., *Management: Tasks, Responsibilities, Practices*, New York: Harper & Row, 1973.

Kahn, Herman, *The Coming Boom*, New York: Simon & Schuster, 1982.

Morris, William T., *Implementation Strategies for Industrial Engineering*, Columbus, OH: Grid Publishing, 1979.

Naisbett, John, *Megatrends*, New York: Warner Books, 1982.

Ross, Joel E. and Robert G. Murdock, *Management Update*, New York: AMACOM, 1973.

Salvendy, Gavriel, Ed., *Handbook of Industrial Engineering*, New York: Wiley, 1982.

CHAPTER 1.9
MANUFACTURING ENVIRONMENT

ROBERT L. PROPST

The Propst Company
Redmond, Washington

Every living environment is an expression of intentions and purpose. The messages of the environment are always there, shaped by purpose or accident. The influence is undeniable. We tend to behave as our surroundings direct.

The manufacturing environment is not outside this phenomenon. The physical workplace explains with inescapable candor the true nature of an organization. A place may show pride, purpose, and productive intention. It may be bright, clean, and energizing. It can show process logic. It may provide information and direction, it may have style and provide a sense of *belonging*.

Unfortunately, manufacturing environments frequently convey powerful negative images. Often they provide, dull, blank, enervating surroundings that debilitate the morale and interest of employees. Many facilities by concept and construction are doomed to early shabbiness, illogical to keep clean, and with no method to deal with obsolescence—a long, grey life. The place as a whole may even be in serious conflict with the manufacturing process, with a kind of hopeless immutability, resisting thoughts of change, growth, or improvement. Too often a kind of "war is hell" climate has dominated the character of factory environments. The message transmitted has been "bear down, get on with the job, and don't complain about a little discomfort; it may help keep you awake."

In the end such negative surroundings stymie the best-laid plans and intentions of an organization. A poor environment reflects the organization's failure to see itself in a holistic, integrated manner.

Meanwhile, in the last 20 years workers have been undergoing a change in the perception of their role. The 2000 hours a worker spends each year at work is more than the waking hours spent in the home, which is about 1200. Thus the largest single investment in living time is at the job.

In any organization the physical and mental health of employees must be regarded as more than a maintenance factor. Members of a better educated, more participatory workforce no longer view the places where they work as the sole possession of the bosses and owners. A growing consensus views the place that consumes a major portion of an individual's life as requiring more positive attributes. Management, where aware of the environmental potential, is making it an important new tool.

In 1957 Joe Blinka, a Procter & Gamble manager, took over an old, decrepit paper mill with a dispirited, demoralized workforce. Procter & Gamble was entering the tissue market and had acquired the Charmin plant in Wisconsin.

Mr. Blinka recalls his experience,

> *It was a Company which had a great need for improved products, lower costs, and obviously a boost in morale. It was very obvious that money had been "short" for a long time and that the Company was on the way to nonexistence. I was assigned the job of (a) finding out what the processes across the plant were all about, then (b) setting up a program to do something about the problems. The Charmin management was retained and I had to work through these people.*
>
> *It was my practice to arrive at Green Bay on Sunday in the late afternoon, then to tour the plant to talk to operators and also to supervisors. This process allowed me to get to know the people well enough so that they would talk to me candidly when I later talked to them about the processes. There were a number of conflicting crosscurrents in existence. For example: "We're glad that P&G bought Charmin because now we probably can have a secure job. Without P&G, Charmin would not exist" and, "These big companies, with their ideas about efficiency, will rip us off.*
>
> *I can recall many discussions, at about 9:00* PM *Sunday nights, during which we discussed the changes that should be expected and the apprehensions, on the part of the workers, about these changes.*

Faced with the rigorous new production goals set by the company, many managers would have first replaced the worn-out equipment. In a paper mill it might have seemed unimportant that the employees were coming to work with several days' growth of beard, slovenly dressed, and with manure on their shoes. Joe Blinka decided that he had to retrieve the people first. He bet the game on building back pride, purpose, and a sense of value.

Mr. Blinka did not have time to renovate the whole plant so he worked on another scale—he worked to alter every employee's perception of his or her direct personal surroundings. He replaced broken windows and rickety stairs. He added power tools to relieve manual labor. He built clean, bright work stations. He provided a soundproof phone station. He cleaned and repaired the washrooms. In some cases he even built little second roofs over work areas to shed the leaks from the roof.

Mr. Blinka recalls,

> *The little changes that were initiated had a great effect in that they reflected the standards that are expected, but these changes were assimilable and did not create fear. Instead they gradually created pride in the job and very good suggestions concerning improvements were volunteered at a greater and greater frequency.*
>
> *As these changes (and more) were made, the operators' appearances changed—clean clothes, clean shoes, daily shaves, etc. Believe it or not, but key operators bought white lab coats (like some chemists use) and referred to one another as Professor so and so. One of the Union stewards even asked me to visit with his son to talk him into going to college, and I did.*

Much of a subsequent exceptional turnaround in the plant productivity is credited by Mr. Blinka to changing the direct environment. He proved that traditional manufacturing can be influenced by surroundings that motivate, identify, and build a sense of possession and ownership for the employee.

But the manufacturing environment is being revolutionized by other factors driven by the demands of the new high-tech industries. This might be defined as an effort to design, install, and run totally integrated manufacturing facilities.

This new thinking requires for the first time a wedding of performance areas that in the past existed somewhat oblivious of each other. This synergistic effort is by no means an easy new relationship and does not succeed without a high level of endorsement and authority. But it recognizes that real productivity depends on a continuing concert in the concept, organization, and continuing management of the whole manufacturing environment.

The electronic industry has led the way in advancing a new attitude about manufacturing environments. Pat Haggerty expressed TI's corporate position on facilities:

> *in a very real sense what I hope we are accomplishing at TI in the buildings we erect, is an expression of a continuous self-renewal of ourselves as an institution. The structures we build are, or can be, one expression of this self-renewal. And, if we do manage structures which are both contemporary and distinguished, it is only because we have also managed a continuous self-renewal of ourselves and of Texas Instruments.*

Dale Cunningham, a senior Texas Instruments executive, heavily involved in their facility development, reflects on how their corporate facility philosophy has evolved.

> *In the old days a factory was traditionally made up of two or three separate and distinct areas basically. The front of a plant usually contained the office space, which was probably air-conditioned, fixed up with very nice offices, private dining room, executive washroom, this was where the office people associated with the factory worked. The next section of the facility was the manufacturing portion, which was frequently a different style or designed structure attached to the rear of the offices, sometimes with or without windows, and usually not air-conditioned. It was normally fairly rough space in the sense that it was a floor, four walls, a roof, and some lights.*
>
> *The third section of the factory was generally the warehouse, which may or may not have been a part of the manufacturing building. Quite frequently it was a separate structure some distance from the manufacturing facility.*
>
> *Usually the environments in these areas were quite different, being totally unlike the offices. There were obviously varying degrees of housekeeping, work rules, discipline of operation, which usually relegated the manufacturing people to be what might be called "second class citizens" within the company. These inconsistent environments could and did cause labor problems, quality problems, production and scheduling problems.*
>
> *We at Texas Instruments have never agreed with that type of operation and have always strived to maintain as nearly as possible an equal and similar environment for both our office administration/engineering space as well as within our manufacturing and production areas. For years the majority of our facilities have been designed as, what we refer to, universal type buildings. Exactly the same building and structure can be used for many, many different operations.*

For instance, in one of our main buildings on the Expressway Site in Dallas, Texas, we have our Corporate offices with Board Room, and a few feet down the hall within exactly the same structure we have engineering design offices, a machine shop, painting and plating operations, electronic assembly, testing, and even warehousing. In the highly competitive electronics business that we are in, we are constantly shifting and changing operations within a given facility. Some of our major buildings have started as predominantly manufacturing facilities, and within a few years have become predominantly office/engineering/administration facilities, and then a few years later reconverted to manufacturing space. In this type of facility environment, we do not have to do any modifications to the buildings, the only thing we do is change the internal furnishings and equipment as required.

Prior to the mid-1970s, we used what I would call conventional office equipment in the offices—desks, chairs, stand-alone tables, filing cabinets, etc., and we used traditional grey colored metal work benches, metal table, racks, storage bins, etc., in our assembly and test areas.

Material was always a problem and was handled from the warehouse to the manufacturing area usually on skids, either by fork lift or hand truck. In the mid-1970s we started to convert all of our office space over to the use of modular furniture and the open-type layouts with bright colors. We found that the office layout utilizing modular-type equipment was very efficient and flexible in that almost any configuration could be made from components easily and quickly, and even those who had a special requirement could be satisfied using standard modular components. At about the same time, we asked ourselves why couldn't the same type of equipment be used in the manufacturing area?

At that time we were building a new modular facility in Lubbock, Texas, so we decided to give it a try. All of the assembly work stations in the new facility were constructed using basic modular equipment. Several benefits came out of this. One was the environment of our manufacturing area not only looked similar to the office environment, but in fact, was made up of practically the same components. Our people liked it and we had great flexibility in configuring work stations to suit the operation that was going on at that particular place. To fill out the requirements for the manufacturing operation, a number of new components were developed and added to the normal modular line so that now we have just about everything necessary for electronic-type manufacturing requirements.

Another benefit from doing this was the fact that if we needed to shrink manufacturing and expand office and administration, we had many of the components already available to convert to office space, and obviously the opposite if we had to go from office to manufacturing space, without the need of buying 100% new equipment. A work surface in an office is the same component as a work surface in an assembly manufacturing area.

The next step following this was our effort to try to totally integrate our manufacturing operation. What we mean by integration is to not only integrate one manufacturing process to the next manufacturing process, but also to integrate manufacturing areas with office area. In a number of electronic manufacturing operations, a significant amount of office or engineering support is necessary to go along with the manufacturing function. We find that through the use of modular equipment layouts, these two areas can be effectively integrated within a very close proximity.

We were also interested in integrating the people in the manufacturing area with the material which they require for manufacturing. A number of different types of material handling equipment were developed that integrated with the modular furniture system which allowed, with great ease, the movement of materials from permanent storage area to manufacturing line, then from one process to next process, and ultimately back to the packing and warehousing function.

One of the overall effects of this concept is that when you look at an assembly and test area within Texas Instruments today, at first glance it does not look like a manufacturing operation, but almost like an office-type operation. We feel this overall concept has done a lot to not only improve the environment for our people, but also give people a pride in their work space and pride in producing quality products, and productivity has gone up. Generally the same work rules apply in both manufacturing areas and office areas and, in fact, in many cases work rules and disciplines are now much more strict in the manufacturing areas.

Texas Instruments and other electronic companies have made a sustained effort to realize such new aspirations, and they have made notable progress. They have been able to overcome the classical difficulty of most process environments, namely, arrangements of a great many specialized products with improvised interfaces. Thus, the manufacturing process often is a bumping-along action where one special process after another makes a cumbersome transition to the next.

In response to the need, a new generation of modular systems equipment is now being supplied to high-tech industries. These responsive systems can express a broad range of office and factory production environments. They also provide systems of integrated containers and transporters. Using these new system parts vocabularies, a factory manager can engage in a tactically fast, ongoing reformulation of production schemes and the scale and configuration of process areas. A broader dialogue with suppliers of equipment and systems will help to make more progress in integrated systems.

However, the real gain lies ahead, where user organizations become more involved in setting criteria for whole-facility performance. The seed of generating this more comprehensive grip on the productive environment will start with a form of all-events accountability.

Robert Knight, who has been extensively involved in developing new manufacturing facility concepts at IBM, describes their effort to provide a truly integrated process environment.

With each new high technology product workers, whose interests are remote from the product originators, are enlisted to make the devices, and to them must be conveyed an impression of the importance of their tasks by the order, design and sense of purpose implicit in the total facility. The need exists to reconcile the great difference in kind and performance between facility and equipment, and to integrate the worker in a way that compels his best and most interested efforts.

To do this is to increase total facility performance while reducing cost in the same way as the integrated circuit did. This has not yet happened to any appreciable extent but has been demonstrated with indisputable results in limited cases.

By the technology advance which led from circuits containing discrete devices—transistors, resistors, capacitors, to integrated circuits in which devices are created by the thousands, already in place within the circuit, the cost of electronics has been reduced by hundreds, even thousands of times, giving rise to throw-away watches and five dollar calculators able to extract irrational roots of ten digit numbers.

In a way this parallels the phenomenon of the factory in which these devices are made. The disparate origins of its two major constituents, the physical facility and the major process equipment attest to opportunities for creating a single, coherent system. *The assembly worker for whose aid these components exist is among the last consideration in factory design despite the fact that his leverage on the total product investment is greater than any other, and so he, too, must be included under the integrating principle.*

Ten years ago at an IBM plant, saturated with product demand and overburdened with facility changes, an inconspicuous backroom project began which produced several such demonstrations. There was a need for an integrated circuit facility to produce an experimental computer memory device in quantities sufficient to build several large machines, after which to expand it to commerical production capacity.

At this plant it had long been observed that the high technology facility, the immediate environmental surrounding of the manufacturing process, set up inside the factory building but distinct from it, resembled, by its structure and function, manufacturing equipment more than it did the factory building.

From this view, the way facilities came into existence began to appear a little strange. It seemed possible to build a facility almost entirely without assistance from the construction industry, following the conventional manner in which units of manufacturing equipment were collected on the shop floor once the facility was completed. What was needed was the selection of appropriate components which performed facility functions in the same specific manner as manufacturing equipment performed manufacturing functions.

This was not conventional thinking, for the construction industry at this time dominated the high technology facility commercially and technically, often preempting decisions belonging to the technologist by its sheer size and apparent authority. Builders were able to impose their schedule as well, even though their logistically driven thinking had long been out of touch with the realities of the client's market.

Had the project that followed this unconventional line not enjoyed tacit management support in virtually all quarters from almost the beginning, it would have died an early death at the hands of the stewards of conventional construction wisdom. Management did, however, see the necessity for the competition which evolved these new strategies and designs to take place at the lowest organizational levels and so they remained as remote as possible to avoid determining the outcome.

At the outset components were needed from which to assemble a freestanding interior environment which was structurally independent of the host building and had every utility and amenity required by the process.

The filling in of the missing pieces between directly useable purchased hardware began (a job hardly as difficult as designing and specifying a special process tool to be built in a commercial prototype shop) and soon a parts list of dozens, then hundreds, of items accumulated. While most were catalog hardware, many were variations on commerical items to be factory ordered as volume warranted but now modified by commercial shops, and a few were unique special purpose designs, again, built in local shops.

Partitions and material management facility system components now available were modified or used as delivered for four side enclosure, work-in-process storage and work surfaces. Doors and frames, ceiling and partitions with openings or services—electrical or bulk chemical and gas delivery, were designed to integrate with the off-the-shelf product.

Altogether, the parts could be called up from a CAD program for a layout on the screen, itself composed of standard plan modules and dimensions, to instantly print a bill of materials and a cost/schedule estimate.

Initial applications used class 100 particulate environments having close temperature and R/H control requiring air filtering and conditioning. This was built into the individual manufacturing clean air work stations and could be added or removed without disturbing operations within the clean room.

Chemical distribution, in addition to deionized water and gases, and chemical waste collection were installed overhead allowing the integrity of the floor to be completely undisturbed during construction and obviated the need for basements. Chemical waste management has the potential to be particularly overbearing in the manufacturing environment. Gravity collection demands graded subfloor piping in covered trenches or in basements, direct burial being forbidden by local ordinance for the protection of ground water. It is, at its best, disruptive at this most critical interface, the floor, when access or modification is needed. Vacuum waste collection, installed overhead using flexible, continuous polymer piping on the other hand, is compliant to any change with little disturbance.

A vacuum waste collection unit, the utility behind the vacuum waste lines, was designed with the ability to serve twenty to thirty points of discharge within distances up to fifty feet. It was able, in turn, to discharge through plant or flexible piping to a tank farm a thousand feet or more distant. The unit was a cube only twenty inches on a side, highly portable, weighing less than two hundred pounds.

Chemical flow into and out of the facility was controlled by a single unit, the Chemical Distribution Terminal, which acted as the interface between plant piping and facility piping, and was soon recognized as the jurisdictional boundary of the construction industry.

In the past it has been very difficult to know, in detail sufficient for design, the future course of a rapidly developing technology in a growing market, thus the great majority of added features to serve this future were either inappropriate or unnecessary, and when the inevitable expansion or modification came, were stripped out unused.

Now, components were assembled according to design directly determined by the immediate needs of the manufacturing process, the need for far-seeing plans were precluded by the quick, easy and non-disruptive ability of the composition to be changed.

In four areas roughly equivalent to four thousand square foot increments, the final semiconductor fabrication line was occupied during about a nine month period. Confidence grew with experience to the point that we were willing to undertake a time trial with the last increment, to attempt completion in thirty working days. *As evidence, a time lapse videotape was produced.*

We saw for the first time the possibility of what was absolutely impossible by conventional means. With the succession of trades, each competing for space at a limited worksite, and the need to fabricate much of the facility at the site, as rapid a response, by even half, was physically impossible. By adding in parallel, the factory resources used in creating the manufactured components, which required only minimal assembly effort at the site, time was effectively compressed.

The product was successfully manufactured in volume at product yield levels far above expectation and during this time the performance of the facility was scrutinized by many prospective users.

The microelectronic production worker feels severe stress from several sources: he is constantly reminded of the critical nature of each production operation, the extreme vulnerability and high value of the product, and the loss of physical freedom restrained by clean room disciple and garments. The need exists for some form of environmental compensation for these necessary losses.

With this as our general design consideration, we were still surprised by the attachment that the workers formed for this facility and the competition among them for work assignments to it. We have attempted to explain this as the result of two factors: the clear and evident subservience of the facility to the worker and his needs, and the visual effect of the interior environment.

As shown by these pioneering efforts in leading corporations, an accountability that incorporates all the contributing elements as part of a manufacturing environment—the buildings, the process systems, the material handling systems, the container systems, the man/machine interface systems, the communictions systems—can happen.

The charter for this activity must be a willingness to account for all events and transactions and see that they meet the performance criteria. Such an approach will deliver a new wave manufacturing environment.

REFERENCES

1. Pat Haggerty, *Management Philosophies and Practices of Texas Instruments Inc.*, Texas Instruments Incorporated, Dallas, TX, 1976.

CHAPTER 1.10

MANAGEMENT OF DESIGN FOR MANUFACTURE

G. HARLAN CAROTHERS, JR.

Consultant
Knoxville, Tennessee

There is presently a great emphasis on the cost and quality of goods manufactured in the United States. Discussions about manufacturing include cost of labor, automation options, retooling, cost of quality, cost ratios, and supplier performance. This chapter suggests that these discussions miss the critical issue. There is only one crucial issue, *product cost management*.

Instead of defining the problem as management of *cost of manufactured goods*, let us define the key issue as management of *product cost*. Although one might consider them to be the same, a closer examination will reveal the differences.

For purposes of this chapter, *product cost* will be defined as the product cost agreed upon by Marketing and Engineering. Cost of manufacture is the cumulative cost incurred by the manufacturing/installation/service functions in producing goods. Missing in most organizations is management of product cost. This management begins with the Marketing/Engineering agreement. It ends only when Manufacturing has analytically demonstrated that it can repeatedly produce the product performance at the decided product cost. Engineering is principally responsible for management of product cost.

Major deficiencies in designing for Manufacturing result if Engineering does not accept accountability for management of product cost. We will consider the design process with and without this accountability.

For brevity and simplicity, we will consider manufacturing to be the production of products comprised of electronic and/or mechanical parts. Those manufacturers concerned with production of processes, i.e., chemical, electro-chemical, photographic, are asked to develop the analogy.

1.10.1 WITHOUT ENGINEERING ACCOUNTABILITY FOR PRODUCT COST

With agreement between Marketing and Engineering on the performance specification, design engineers put pencil to paper and detail the system, subsystem, and component specifications. Component design follows. Component design results in dimensioned drawings, material specifications, and component selection, including part characteristics, quantities, tolerances, types, etc. Manufacturing and assembly methods may also be provided. Significantly, the designers first describe the ideal part parameters. Realizing that the ideal is not always available, acceptable tolerances are chosen. Ideally, the establishment of tolerances is done based on legitimate analysis. In reality it is often superficial. Many tolerance selections are made strictly on the basis of supplier data sheets. This may lead to the design description being established by hearsay or the commitment of supplier sales personnel.

The more conscientious designer does a "worst case analysis" following the initial design and tolerancing. Supposedly, this step insures that variation such as component-part variation is taken into account. An example of component-part variation is the change in performance due to environment, application, part derating, interface variation, product shelf life, product planned, service life, user skills, etc. Proper worst case analysis involves a very comprehensive simulation of what might take place.

Up to this point we can certainly agree that all steps in this design process are appropriate. Designers follow these responsible methods to *choose and describe a specific design*.

However, the actual organizational objective of design is not only to choose and describe a specific design but also to *guarantee that the design can be predictably reproduced at the agreed product cost*.

In an organization that lacks product-cost accountability, only the section concerning "*choose and describe a specific design*" is given attention. The responsibility for *predictable reproduction* and *product cost* is shirked, forgotten, passed over, abdicated. In the final analysis, product-cost accountability is lost. Unless this significant omission is stopped, design choices, design descriptions, and design confirmations are out of context. The unchecked design, unbridled of cost constraints, moves quickly to the next step. Later, when Design is asked about the product cost, the question is met with a blank stare, a puzzled look, or worse, a statement that Manufacturing has not yet been brought into the engineering phase. So much for product-cost accountability.

Free of product-cost responsibility, designers hasten on. Supposedly, prototyping confirms the paper analysis and component descriptions. Design, via drawing, directs the purchase or fabrication of parts, assembly and test of a *limited* number of units. The purpose of limited units needs to be examined closely. Often the choice of parts for the initial units is made, forgetting that the problem is not one of producing a limited number, but rather of reproducing the desired units in volume. In most cases prototype parts are selected with no concern for ideal characteristics, or may be hand-selected for the first prototype. This gives no assurance that identical parts can be reproduced for volume production. Forgetting the organizational objective of design, the designer might even choose parts never before produced in volume. Obviously, cost of these parts is unknown. The product cost and product performance becomes increasingly unreliable.

A closer inspection of design confirmation shows glaring deficiencies. The agenda of the individual designers is to prove performance once. Again, this is not the organizational objective of design. Often the designers test and "fix" the design with an unpredictable and possibly unreproducible array of parts. There is no regard for *predictable reproducibility at the product cost*. The designer's misunderstanding of the organization objective for design leads to unreproducible products with unpredictable cost. This is not the worst of what one finds. In most prototyping phases, parts are not characterized. This results in the assembly and testing of prototypes with no knowledge of what specifically has been put into them. It raises the question: what is the prototype? Does it reflect a collection of parts with ideal (nominal) characteristics? Supplied parts do not always meet published specifications. Is it safe then to assume that the parts in the prototype correspond to the characteristics analyzed and chosen? If the performance of a prototype is not up to expectation, is the offending part what we originally analyzed or what the supplier provided? Why is performance different from the desired? With this much doubt, what do we have when we finally succeed in "checking out and making the prototype play"? Suppose we test the prototype—what did we confirm will work?

Ironically, in many cases where a prototype is made to work, we declare the design *good* and "release it to manufacturing." It is frightening to think about our confidence in predictable reproducibility at product cost.

With a sigh of relief, the designers release the drawings, bills of material, and process specifications/methods to Manufacturing. Little do they know, much less care, that the drawings may not reflect the prototype. They are even less concerned with the predictability of reproducing the characteristics of the components in the prototypes. Typically, we have a working prototype but sadly do not confirm that the drawings represent it.

Now the drawings are released to fabrication and procurement. The predictability of parts is neither specified, questioned, nor confirmed. No one asks: "Are there part types that have never been produced in quantity before? Can our selected suppliers assure predictable part parameters? What is the confidence in cost of parts if parts are not predictable in the quantities desired?" Worse yet, the drawings make little or no mention of required predictability or sourcing for predictable reproduction.

The worry at this point should not be with credible product cost only, but also with product performance predictability. Without product performance predictability, product cost is thrown out the window.

The story gets worse. Purchasing and fabrication, without the requirement for predictable parts characteristics, are unleashed. Lowest price becomes the criterion for sourcing. Source inspection and vendor qualification do not confirm predictable reproduction. Incoming inspection falls short of insuring predictability. Product manufacturability is adrift. Fabrication, raw materials, and processing methodology are not geared to predictability, so repeatability at assembly is jeopardized.

So far, our discussion concerns steps prior to manufacture. If we have typified the content of the steps leading to manufacture, is it any wonder product costs are suspect? Along with its other problems of labor costs, inefficiencies, scrap/rework, machining practices, etc., Manufacturing has inherited a questionable design.

Shortcomings in the front-end design process manifest themselves in manufacture. Good parts are produced, which do not work in assemblies. Some units work while others do not in spite of "worst case analysis." Product performance varies without anyone understanding why. Engineering support to Manufacturing continues for long periods. Engineering change orders abound. The product-development cycle gets longer and longer. Manufacturing requests for corrective action pile up. Field retrofits become a way of life. Field repair kits do not work. Product reliability in customer hands varies greatly. Cost variation in manufacturing is not predictable. Planned margins wax and wane as do planned profits.

So much for the condition; what about an alternative?

1.10.2 WITH ENGINEERING ACCOUNTABILITY FOR PRODUCT COST

As before, Marketing and Engineering agree on performance specification. Now, however, they also agree on *product cost, unit sales volume*, time of market introduction, and duration. This properly defines the design task for Engineering. Engineering now chooses and describes the specified design, which can be predictably reproduced at the product cost to match the market needs.

All subsequent steps are subservient to and must be tested against the responsibility of assuring reproducible performance and adhering to product cost.

The paper design must be targeted at detailing system, subsystem, and component specifications, which can be predictably reproduced for allocated portions of product cost.

Component design after specification analyzes and determines part types, part characteristics, quantity, and tolerances. The ultimate measure in final selection of components, however, is predictable reproducibility. This critical characteristics dictates which components will be used in the design. Insured predictability will be required before any decisions on parts are made. Part predictability is the first block of credibility in product cost management.

Specification of predictability must be identified on drawings and parts specifications. A measure of the reliability of predictions will be assigned to each part. This measure reflects the amount of acceptable variation.

With these variations determined or known, the worst case analysis will be directed toward a simulation of these variations. In addition, statistical tolerance analysis will be added as a confirmation of reproducibility of design.

Here also, there must be a prototype validation phase. However, the units prototyped will be built or purchased to the ideal (nominal) part characteristics specified in the design analysis. Each part characteristic will be measured and recorded. The designers will know exactly what was put into the prototype. Therefore, when the design works or is made to work, the designer will know what works and how to replicate the unit and insure the same performance.

By having a confirmed nominal design, the designer can proceed to validate the predictable reproducibility of each component critical to the performance. This can be done only with actual data (i.e., appropriate sampling of parts) and confidence-interval calculations. If parts are predictable, their cost can be calculated with confidence. The cost variability of final assembly and final test are the only open issues. These can be estimated, with the final confirmation after the pilot manufacturing run.

Now a set of drawings, material lists, and methods exist that insure a predictable product cost and product performance. The nominal design works and the distribution of predictability on parts is specified. The design description clearly describes what needs attention to insure manufacturability.

Given the clout of design in today's society, engineering management, by designing for manufacture, could greatly reduce the cost of manufacture. Cost of manufacturing would begin to approach product cost.

Demanding responsibility for product cost and thus predictable product reproduction will force behavioral changes in Purchasing, Incoming Inspection, Fabrication, and Manufacturing.

Purchasing will stop buying from sources with questionable performance predictability. Source qualification will change. Vendors will be required to demonstrate statistical capability. Vendors will know cost and price, surprises will be reduced. Vendor quality and predictability will be assured. Product cost and price will be credible.

Fabricators, working toward reliable parts, will make what is needed when it is needed. Assemblers will assemble, not refabricate. Test will test, not redesign through selective acceptance during test.

Product-cost confidence will increase. Assembled and tested product performance will vary less. Product quality delivered to the customer will improve. Predictable product cost is the objective: cost of manufacture will be product cost!

The management of product cost is quite different from the management of cost of manufactured goods. The discipline of management of product cost is more far-reaching than it at first appears. With the crucial issue clarified, American manufacturers can take appropriate action toward management of product cost. America's product competitiveness is threatened and requires *proper* attention.

PART 2

MANPOWER: PEOPLE IN PRODUCTION

CHAPTER 2.1

JOB DESIGN

LOUIS E. DAVIS

University of California
Los Angeles, California

GERALD J. WACKER

Xerox Corporation
Los Angeles, California

2.1.1 INTRODUCTION

At the heart of any organization are its jobs. The division of the work to be done by the organization into jobs sets forth relationships among people and between people and technology. This chapter examines the factors and decision processes that lead to the design of jobs. Emphasized are approaches to the design of effective and satisfying jobs, which meet both the organization's needs for effectively achieving its goals through the use of its human resources and the individual's needs, expectations, and goals.

2.1.2 SYSTEMS APPROACH TO JOB DESIGN

Research into the factors underlying job design,[1,2] and the effects of job designs on productivity and employee satisfaction,[3–7] suggests that jobs can be understood only in relation to a whole enterprise or agency as a system. This section reviews the systems perspective for the design of jobs.

2.1.2.1 Multiple Dimensions of Jobs

An organization, whether a manufacturing plant, a service firm, or a government agency, exists in several dimensions. First, it is a *production entity*, consisting of buildings, equipment, and technology (technical systems) that have been brought together for the purpose of transforming certain inputs into desired outputs. Of course an organization is more than just a production entity. It is also a *social or institutional entity*, or "miniature society," consisting of roles, traditions, conflicts, long-term strategies, and relationships with other institutions. Finally, an organization is a *collection of particular individuals*. Each employee has goals, commitments, and life-styles that only partially intersect with the organization. To design jobs that are effective, account must be taken of the production dimension, the organizational dimension, and the individual dimension as they pertain to the various kinds of decisions that go into job design.

2.1.2.2 Definition of Organization and Its Jobs

An organization is a system of roles or jobs and a structure of role relationships deliberately brought together at a particular time to accomplish desired outcomes. To achieve its desired outcomes, an organization does work. To do work, an organization brings together people (social system) and the means (technical system) through which they can jointly accomplish the desired outcomes. To perform effectively, these subsystems are required to act as a joint system, frequently referred to as a sociotechnical system.

Reprinted, with permission, from Louis E. Davis and Gerald J. Wacker, "Job Design," in Gavriel Salvendy, Ed., *Handbook of Industrial Engineering*, Wiley, New York, 1982.

2.1.2.3 Definition of Job

A job is a cluster of tasks assigned to a role in an organization. It is a part of the organization with which its members are identified. Over the last 100 years, jobs were defined as clusters of work tasks performed by individuals, whereas roles were defined as clusters of work tasks *plus* other kinds of tasks, which needed to be done because work was done not simply by an individual but by an individual in an organization. In organizations designed on the basis of contemporary organization theory, the difference between job and role has disappeared. An organization is functioning when its roles are occupied by people carrying out the requirements of each role or job. An organization remains virtually unchanged even though its roles or jobs are occupied by different people at different times.

2.1.2.4 Definition of Job Design

Job design is the process of making the decisions that determine (1) what tasks are to be performed by the work force, (2) what tasks are to be clustered into what jobs, and (3) how the jobs are to be linked together.

2.1.2.5 Technical and Social Support Systems

Job design decisions are interdependent with decisions regarding technical systems and with personnel policy. A number of technical systems alternatives may be derived from a conversion technology. A technical system is a set of procedures, techniques or methods, instructions, equipment, tools, and layout chosen for carrying out a particular process of converting material, information, or human input into a desired output. Choices made regarding the content and configuration of a technical system are in part job design decisions. How technical system choices create tasks and influence the clustering of tasks into jobs is reviewed in the remainder of this section.

The design of jobs involves not only technological choices but also social choices. The technical system used to convert material, information, or human input into desired outputs generates tasks that must be performed. Similarly, the social system within which the technical system is embedded also generates tasks that must be performed. Both kinds of tasks, to different extents, come to be included in jobs. However, what the tasks are and how they are configured originate with the choices made regarding the design of the social system, also referred to as the social support system.

For machines to function properly, they must be accommodated to the organization. Technical support systems include data processing devices, dust collectors, dehumidifiers, spare parts inventory systems, technical manuals, and even access to technical consultants. Analogously, a social support system helps people to adjust to the organization so that they can function properly in their jobs. Social support systems include aids for recruitment, training, safety, and security; arrangements for counseling, skill maintenance, discipline, and justice; facilities for rest, recreation, eating, and hygiene; reward systems for compensation and advancement; employee benefits; provisions for connecting employees' work and nonwork roles (such as the availability of telephones for personal calls, car pools, etc.); and provisions for managing the organization as a miniature society.

Because of the interdependence of technical system design and social support system design with job design, these decision-making areas are included within the realm of job design (see Fig. 2.1.1). Decision-making areas that lead to the establishment of work roles include those that determine

- The materials, equipment, information, and resources to be used by jobholders.
- The latitudes of decision-making authority awarded to jobholders.
- The means for adapting to difficulties or unexpected events.
- The means for scheduling, supervising, evaluating, and controlling work.
- Paths of advancement and promotion.
- The provisions for maintaining and elaborating the organizational structure.

2.1.2.6 Job Design under Conditions of Human and Technological Uncertainty

Underlying job design are assumptions about human and technological certainty and predictability. People differ from one another in size, skill, experience, viewpoint, and taste. In addition, the availability and alertness of any one person can vary greatly from day to day, because of health, mood, outside commitments, and so on. Designers can generally predict the actions and capabilities of a machine with much greater certainty than those of a person. However, this fact can be quite misleading in job design, as several authors have noted.

Jordan[8] pointed to the futility of trying to categorize routine tasks into those best suited to machines versus those best suited to humans. Tasks that are so routine as to be reduced to a formula are almost always more efficiently done by machine. The primary advantage of humans is their ability to deal with *technological* uncertainty, such as mechanical breakdowns, unusual production requirements, disruptive events, and processes that involve value judgments. Davis and Taylor[2,9,10] discussed the job design implications of automated technology. Human interactions with automated machinery are less routine and less predictable than interactions with older technologies. When situations arise that require responses beyond the capability of the technical system itself, then the presence of humans, with their versatile response capability, is essential to providing responses that

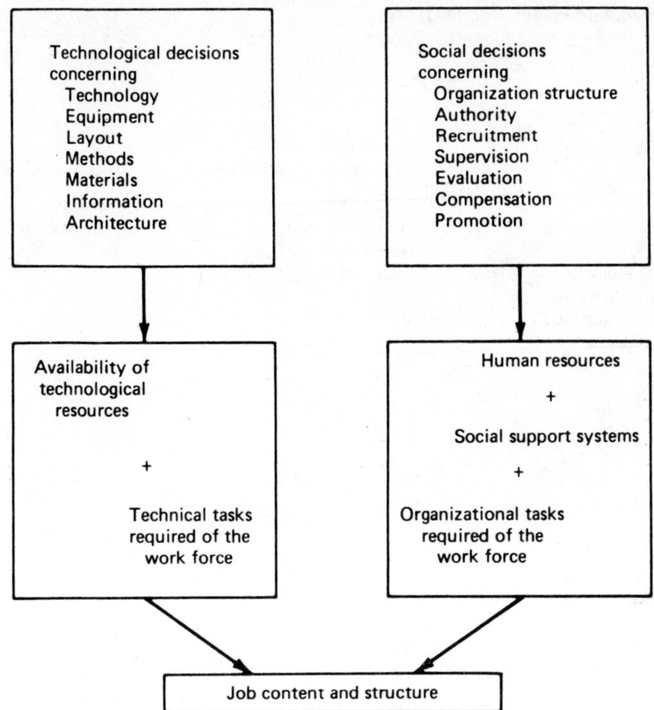

Fig. 2.1.1. Social and technological decisions underlying job design.

the organization needs if it is to succeed. The same applies when the social system or organization has been upset by absence of people or by other events. "The human system is far overdesigned, overengineered for simple jobs," wrote Waddel in the 1956 edition of the *Industrial Engineering Handbook*.[11]

Additionally, people have a perennial tendency to form social bonds and to help each other. If jobs are designed in recognition of this tendency, then some of the human uncertainty due to individual differences, health, and mood can be counteracted by mutually supportive relationships among employees. It is important that the job designer give maximal consideration to the social and creative characteristics of humans, especially when technical systems do not behave with complete certainty.

An anecdote may serve as an object lesson. A computerized information system for housekeeping and accounting was installed in a hospital. Soon afterward, the chief of surgery requested of the night receptionist that a particular patient be assigned a particular room for the next day. Unfortunately, the computer system had already reserved that room for someone else, and no provision had been made for the receptionist to reexamine the decision and exercise her own judgment.

Two scenarios can be envisioned. First, the receptionist could shrug her shoulders and tell the surgeon, "I only work here; if you want something, then *you* can try talking to the computer." The surgeon might have to wait until morning and then spend time searching for the proper forms and approvals. The second scenario is what actually occurred. The receptionist had previously discovered that, after five successive invalid entries into the computer, it would display the message "revert to manual." The old manual procedure involved calling a floor nursing supervisor and arranging room assignments with her. So the receptionist "subverted" the computer system in order to respond to an unpredicted event. Had the receptionist not discovered how to control her technical system accidentally, had she not been familiar with the manual procedure and acquainted with the floor supervisor, and had she not been motivated, then an ineffective outcome for the organization would have occurred. In this case the technical system for room assignment was unable by itself to respond to an unusual and unpredicted event and thus required an adaptation by the organization's social system through the coordinated actions of the surgeon, the receptionist, and the floor supervisor.

2.1.2.7 Summary

Jobs contain three dimensions: a production dimension, an organizational dimension, and an individual dimension. Each dimension should be considered in all job design decisions. These decisions determine what tasks are to be performed by the work force, how tasks are clustered into jobs, and how jobs are linked together. This framework is outlined in Table 2.1.1.

Table 2.1.1. Systemic Factors in Job Design

Sources of Tasks	Task Clusters	Job Linkages
	As a Production Entity[a]	
Production and Technological Needs	*Imposed by the Technological Design*	*Technological Resources*
Operation Maintenance Quality assurance Control of variance Information processing Reliability of equipment Variability of inputs and outputs Predictability of events Complexity of the technology Level of automation Scale of production Rate of technological innovation	Availability of data Timeliness of data Location of readouts Location of controls Layout Manual overrides Ability of equipment to be adjusted by its operators	Examples–conveyor belt, telephone
	As an Organizational Entity[a]	
Organizational Needs	*Imposed by the Organization Design*	*Organizational Resources*
Hiring Training Coordination Social maintenance Adaptation to change Organizational learning Exchange of information Conflict resolution Flexibility Labor turnover rate Absence rate Union-management requirements Stability of goals Long-term goals Financial requirements Legal/regulatory requirements	Reward opportunities Career paths Who reports to whom Latitudes of authority and responsibility	Reporting relationships Promotion paths Meetings Newsletters Records
	As a Collection of Individuals[a]	
Needs of Individuals	*Carried into the Enterprise by Individuals*	*Personal and Interpersonal Resources*
Physical needs Psychological needs Social needs Economic needs Individual differences Experiences Expectations Values Culture Other need providing institutions	Existing clusters of skill and experience Expectations about jobs and organizations Work habits Lifestyles	Acquaintanceships Acceptance of goals Social skills Understandings of others' tasks

[a] Dimension of the enterprise.

An organization is a joint system, composed of people and equipment, tools, and so on. A technical support system enables the equipment to function properly in the organizational setting, and a social support system enables employees to adjust to the the organization. Because the behavior of humans and of technical systems is not totally certain and predictable, jobs should be designed to enable the organization to deal with uncertainty. Whereas machines can perform routine tasks more predictably than humans, people have the potential to act more creatively and to coordinate their actions in adaptive response to nonroutine events.

2.1.3 SOURCES OF TASKS

Jobs are composed of tasks and the conditions under which they can be accomplished. These tasks derive from several sources, which reflect the three dimenions of an organization.

2.1.3.1 Production and Technological Needs

The most obvious source of tasks derives from the production and technological needs of the organization. From this source one can identify tasks relating to the performance, inspection, and evaluation of work; the preparation, operation, and maintenance of equipment; the recording, processing, and retrieval of technical information; the transportation and care of materials; and the control of technological variables that could adversely affect the production process.

Several mediating factors can affect production and technical system task requirements. Technical system uncertainty or instability places demands on the role of people in the production system. Technological uncertainty is affected by the reliability of equipment, the variability of inputs and outputs, and the unpredictability of events. In addition, the rate of technological innovation introduces long-term uncertainty about the particular tasks that will be required in the future of the work force. The complexity of the technology, the degree of automation, and the scale of production (from custom to mass production) are other mediating factors.

2.1.3.2 Organizational Needs

In addition to tasks that contribute directly to the production of goods and services, another source of tasks derives from the needs of the organization. These needs include the coordination of activities, the maintenance of the social order, the management of the organization as a miniature society, and the short- and long-term adaptations of the organization to changing conditions in its environment. Organizational needs call for tasks relevant to the hiring and training of a work force, communication, the discovery and exchange of knowledge, and the constructive resolution of interpersonal and organizational conflicts.

Organizational task requirements are affected by the following factors: labor turnover rate, absentee rate, union-management requirements, short- and long-term goals of the organization, the stability of these goals, financial requirements, and legal/regulatory requirements. The first two variables have both mediating and outcome effects. On the one hand, poorly designed jobs are likely to exacerbate labor turnover, absenteeism, and grievances. On the other hand, frequent turnover and absenteeism—regardless of cause—give rise to greater need for hiring, training, coordination, and evaluation.

2.1.3.3 Individual Needs

A third source of tasks derives from the needs of individuals. People have physical, psychological, social, and economic needs. The particular directions and strengths of each of these needs differ greatly from individual to individual. Needs are a function of personality and of the commitments, requirements, and life-style preferences that come from outside the organization. For example, need for elbow room or self-control expressed by younger workers influence demands for providing areas of decision making that a worker can call his or her own. These needs are influenced by past experiences,[3] expectations,[12] social reference groups,[13] fads, fashions, values, and the culture. The organization is not alone in providing for individual needs; other institutions in society help people to satisfy their various needs.

One of the most obvious needs of individuals that needs to be addressed in job design is that of financial compensation. In addition, organizations must concern themselves with employees' needs for safety, health, comfort, promotion and career advancement, socialization with co-workers, and job satisfaction. These needs have been discussed in the job design literature under the rubric of "quality of working life."[14] A comprehensive checklist of quality-of-working-life criteria, to serve as an aid to designers, is provided in Table 2.1.2. Job designs directed toward providing a high quality of working life incorporate various attempts to provide better job security, equity, and rewards and to satisfy the increasingly articulated psychological needs of all workers. These needs have been stated before by Englestad,[15] and the list provided here was enlarged by Davis[16] to include satisfaction of individual differences.

1. The need for the content of the work to be *reasonably demanding* in terms other than of sheer endurance, yet to provide a minimum of *variety* (not merely novelty).
2. The need to be able to learn on the job and to go on learning. At a minimum, this requires outcome specifications, standards of performance, and knowledge of results (feedback).

Table 2.1.2. Quality-of-Working-Life Criteria Checklist

Physical Environment

Safety
Health
Attractiveness
Comfort

Compensation

Pay
Benefits

Institutional Rights and Privileges

Employment security
Justice and due process
Fair and respectful treatment
Participation in decision making

Job Content

Variety of tasks
Feedback
Challenge
Task identity
Individual autonomy and self-regulation
Opportunity to use skills and capacities
Perceived contribution to product or service

Internal Social Relations

Opportunity for social contact
Recognition for achievements
Provision of interlocking and mutually supportive roles
Opportunity to lead or help others
Team morale and spirit
Small-group autonomy and self-regulation

External Social Relations

Job-related status in the community
Few work restrictions on outside lifestyle
Multiple options for engaging in work (e.g., flexible work hours, part-time options, shared jobs, and subcontracting)

Career Path

Learning and personal development
Opportunities for advancement
Multiple career path possibilities

3. The need for some area of decision making that an individual can call his or her own. Within this area one can exercise one's own discretion and be evaluated on the basis of objective outcomes.
4. The need for social support in the workplace: the need for an individual to know that he or she can rely on others for help when needed in performing the job and that sympathy and understanding will be available when needed.
5. The need for recognition, within the organization, of one's performance and contributions.
6. The need to be able to relate what one does and what one produces to one's social life outside the organization.
7. The need to feel that one's job leads to some sort of desirable future, that is, no dead-end jobs, but available career paths.
8. The need to know that choices are available in the organization by which one can satisfy one's needs and achieve one's objectives—that is, options as to kinds of jobs, localized structuring of work, many career paths, options for progressing at different rates, and so on.

Some tasks, such as oiling a machine or filling out a form for vacation, can be clearly identified with a single type of need. Other tasks serve several needs. For example, the task of planning a weekly work schedule has to take into account the satisfaction of production needs, organizational needs, and individuals' needs. It is important that the design of jobs enhance the ability of the jobholder to take all relevant needs into account when performing his or her assigned tasks.

2.1.4 SOCIAL AND TECHNICAL SYSTEMS DECISIONS THAT CREATE TASKS

Figure 2.1.1 shows various social and technological decisions that underlie job design. How these decisions create tasks are reviewed here; how they influence the clustering of tasks into jobs and how they link jobs together are discussed later.

2.1.4.1 Technological Decisions

A technology consists of the physical and informational resources by which people can systematically bring about some desired result. Throughout human history there have been incidents of both the accidental discovery of technology and the purposeful, painstaking development of technology. In the modern era more than ever, technological development is influenced by hidden assumptions about people and jobs.

For example, Davis and Taylor[2] reported a technical system design, comprised of machine and "human machine" elements, in which there were a number of parallel, partially automated machines for filling and capping aerosol spray cans. Each machine, with one operator in attendance, was placed so that there was no communication among the operators. Most of the operator's time was spent inserting a small plastic tube into the large hole in the top of each upright can, brought to the operator on a circular conveyor belt. The second, less frequent intervention—and the basic reason for the presence of the operator at all—was to press, when needed, a stop switch on a post directly in front of him or her. In the event of perceptible trouble anywhere in the machine, the operator was expected to shut off the machine and to seek help to resolve the problem.

The job design called for workers to be human machine elements in this production system. They performed in isolation the technologically unnecessary and tedious task of inserting tubes, which could have been easily done by the machine. The human task of inserting tubes into cans was developed simply because the primary task of sensing and diagnosing a problem required human eyes and ears, and by hiring those, the company acquired a set of hands, which were not to be left idle. It is difficult to know whether this "truncated" technological development was based on the state of the art or on assumptions held by the designers about people. Failure to develop servomechanical devices and/or remote sensing and control mechanisms for automatic shutoff was a major factor leading to an extremely poor job design. Perhaps if the designers had taken into account the needs of the social being to whom the eyes, ears, and hands were attached, a different technological design would likely have emerged.

In contrast to the preceding case is one involving the design of a food processing plant.[17] With the existing technology, there was a task that involved the manual unloading of 100-lb. sacks of raw material from boxcars and the stacking of the sacks on pallets, to be stored until the dumping of their contents into hoppers. From that point, automated equipment sorted, cooked, processed, and packaged the product. In existing plants the unloading and dumping tasks were performed by men, whereas the operation of the automatic equipment was done primarily by women. This was unacceptable for a number of reasons. One reason was the desire of all individuals to have a chance to learn all phases of the operation. With the backing of top management, the designers initiated an industrywide conversion to bulk packaging of the raw material so that mechanical equipment could handle it. Additional automated machinery was developed so that all the routine human tasks in the plant could be performed by either men or women.

2.1.4.2 Social System Decisions

Perhaps the most dramatic instance of a set of social decisions creating tasks was the framing of the U.S. Constitution. The then-radical criterion that government representatives were to be popularly elected created the multitude of tasks involved with conducting elections. On a smaller scale, a food processing plant[17] was designed so that its social support system included peer performance appraisal. This created organizational task pertaining to the administration and facilitation of the performance evaluation process.

Decisions about methods of recruitment not only make human resources available to fill jobs but also stimulate employee expectations regarding job security, career opportunities, and interpersonal relationships. Choices of the methods for supervision, compensation, promotion, and conflict resolution set forth their own tasks. At the food processing plant, a set of social decisions led to the identification of the following tasks to be performed by workers, which became part of their jobs: problem solving, counseling, task assignment scheduling, record keeping, communicating between work teams and between shifts, convening and leading meetings, safety coordination, security coordination, and training.

2.1.5 SOCIAL AND TECHNOLOGICAL DECISIONS THAT CONSTRAIN THE CLUSTERING OF TASKS

2.1.5.1 Technological Decisions

Technological decisions can add or eliminate options in designing jobs. Even such an apparently simple decision as the location of a switch or meter, or the mode of input and output for a computer, allocates technological resources to particular workstations and thus constrains the clustering of tasks into jobs through the physical locations of the means for controlling the production process. It is typically thought to be effective, for example, to construct a railroad spur and loading dock of a plant so that shipping and receiving tasks are performed in one physical location. In one recently designed paper mill, however, this arrangement was deemed inappropriate to the job design, since it would not have permitted the group of employees responsible for converting the raw material to control its inputs as to quality and time in order to achieve the outputs desired. From a purely technical viewpoint, it would have been desirable to build a single railroad spur and loading dock for both receiving and shipping. The broader systemic viewpoint prevailed, and the extra cost was more than repaid by building separate spurs and docks.

All too often, engineers have designed technical systems with hardly a glance at job design criteria. Underlying this neglect is an ideology of "technological determinism," meaning that technology does—and should—develop in isolation of and unconstrained by social considerations and that social structure is—and should be—fitted to the technical system whether or not the consequences may be dysfunctional. The folly of technological determinism is often illustrated by citing the case of the Chevrolet Vega plant at Lordstown, Ohio, in 1972. Although hailed in 1970 as "the world's most technologically advanced assembly line," the plant was soon gripped by labor discontent, sabotage, absenteeism, and carelessness. This culminated in a strike in 1972, which attracted national attention because it centered on the consequences of the job designs of a young, rural work force whose members said money alone was not enough and whose local union leaders publicly decried the dehumanized jobs and dizzying speeds. Since then, several GM divisions have sought ways to develop technical systems designs that meet job design requirements as well as more conventional criteria. In many industries, assembly technologies once characterized by long conveyors have been succeeded by U-shaped configurations, buffer stages, and self-propelled assembly carriers.[18] These technical innovations were stimulated in part by the need to make jobs more flexible and appealing.

Reviewing the empirical literature regarding the relationship between technology and job structure, Davis and Taylor[2] concluded that a self-fulfilling prophecy appears to be operating. Production systems that are designed in accordance with assumptions that people are reliable and intelligent and that they desire variety and challenge yield jobs that reflect those assumptions. Similarly, when technical systems are designed on the basis of assumptions that people are unreliable, that they should be isolated, and that they have few tasks to do, these assumptions lead to technological choices that ensure their own fulfillment, thus reinforcing the original unfounded assumptions.

2.1.5.2 Social Decisions

Social decisions can bias the allocation of tasks to jobs. The designers of a new food processing plant[17] incorporated both training and operational tasks into production jobs, so that workers could cross-train each other. In other plants this would have been constrained by a social support system that did not encourage workers to share their skills and knowledge other than informally. In the new design, pay levels were graded according to the number of skills and tasks an individual had mastered, regardless of the individual's particular work assignment at a particular time. Under such a reward structure, workers meet their own needs for pay and advancement while simultaneously meeting the organization's needs for flexibility, by cross-training each other.

2.1.6 CLUSTERING CONSTRAINTS CARRIED INTO THE ORGANIZATION BY INDIVIDUALS

Job design takes account of the existing clusters of skills and experience in the labor market. This in itself can be a social decision, since designers often choose a particular segment of the labor market for which to design jobs.

More subtle are requirements or constraints coming from individuals' expectations, values, and past experience. After World War II many of the operations in the coal mines of Durham, England, were mechanized, replacing manual methods of coal mining.[19] Accompanying the mechanization was job redesign. For generations miners in that region had worked in small, relatively autonomous crews. The dangers and production uncertainties inherent in coal mining required high levels of mutual dependence, cooperation, and trust among co-workers. The new job design called for a one-man/one-job mode of working that did not fit the mining culture. Each miner was accustomed to performing all the tasks required to work a section of the mine. The miners were unwilling to accept a narrowing of task assignment or to give up control over the selection of their work mates. Subsequently, a crew type of job design was reinstated, still retaining the new technology, and productivity vastly improved.

A lively controversy has developed around the question of employees' desires for variety, challenge, and autonomy in their jobs. Some research suggests that deep-seated personality traits may underlie employee preferences for one or another type of job.[20] Other experiments and field studies indicate that employee desires and expectations are simply reflections of culture, familiar experiences, and reference group values.[5,7,12,13,21] Many job design endeavors have proceeded on the negative principle of the self-fulfilling prophecy or on the positive principle of rising expectations, namely, that employees' expectations tend to grow as society makes more opportunites available inside and outside the workplace.

2.1.7 RESOURCES FOR LINKING JOBS TOGETHER

Technological resources help transfer materials and information from person to person. The conveyor belt, for example, is one device that links jobs by virtue of transporting materials between work locations. Information is transferred by such technical devices as telephone, teletype, and computer.

Organizational resources help transmit decisions, instructions, feedback, and rewards among jobholders. Communication, promotion and career paths, meetings, and newsletters all serve these functions.

Personal and interpersonal resources are frequently overlooked linkages between jobs. Acquaintanceships, personal commitments to organization goals, social skills, and familiarity with tasks outside one's own job all help to integrate the many jobs into a single system.

Technological and social decisions create job boundaries by creating tasks and constraining clustering of tasks. These decisions also provide linkages across job boundaries. Some boundary predispositions are brought into the enterprise from outside. It is the task of the job designer to ensure that the technological, organizational, and individual boundaries complement and reinforce each other. After reviewing the main types of job designs, we present some methods of analysis to aid in the alignment of job boundaries.

2.1.8 TYPES OF JOB DESIGNS

2.1.8.1 Undesigned Jobs

Certainly one way to design jobs is to let them "grow like Topsy." This laissez-faire approach to job design characterized production systems before the Industrial Revolution (1780). Jobs, or rather "trades," or sets of skills, evolved slowly, changing little from generation to generation. The particular allocation of tasks to jobs, the tools and techniques for task performance, and the quality standards were based on traditions and rules of thumb put forth by those who had mastered the skills. They were enforced and regulated by associations or guilds of master craftsmen. Even today many job designs derive from tradition. As an illustration, the roles of physician and nurse are *not* designed simply as an extension of the technology of medicine. These job designs also rest on social traditions that are now protected by legislation and by guilds of doctors and nurses.

Probably the best-known critic of undesigned jobs was Frederick W. Taylor.[22] He urged that traditional rules of thumb give way to more deliberate methods for making job design decisions. Traditions, argued Taylor, are rarely tested by controlled experimentation, and they are improved only by a very slow and haphazard process of trial and error. Taylor attributed this to the tradesman's lack of scientific education as well as to social support systems that did not encourage improved work methods. From 1890 to 1910 Taylor went on to develop another approach to job design as a part of his system of scientific management.

2.1.8.2 Machine Model of Jobs

The machines that emerged during the Industrial Revolution had two major impacts on job design. The first impact was technological. The proliferation of new machines meant wholly new tasks and thus new job designs. The second impact was social. Engineers and inventors were the heroes of the day, and optimism ran high that the perspectives of engineering and physical science could be fruitfully applied to the design of social systems. Organizations came to be visualized as precise, clockworklike mechanisms, which contained human cogs as well as mechanical cogs. The most descriptive metaphor is that of the organization as an elaborate clockworks.

Functional Specialization

The machine metaphor of organization led to the job design principle of functional specialization. This principle originated with Adam Smith[23] and Charles Babbage in 1790[24] and was reified by Taylor 100 years later.[22] It derived from the then-radical proposition, breaking with past tradition, that only the economic criterion should be used in assigning work to people. This meant, as with other single, variable answers, that work was to be done only on the basis of the cheapest way (in the short run), which was to break up or fragment skills.

Smith, Babbage, and later Taylor urged owners and managers to design each job as narrowly and precisely as possible. Precise prescription was necessary so that workers could be controlled by managers and supervisors. By logical extension of this principle, all the tasks deriving from the needs of individuals become vested in specialized "personnel" jobs, the tasks deriving from organizational needs are given to specialized clerical and staff jobs, and all tasks derived only from production needs are given to production jobs. Specialization is taken even further: all maintenance needs are assigned only to maintenance jobs, all inspection tasks are clustered into the job of inspector, and so on. A principle related to functional specialization, which has guided clustering of tasks, is that all tasks within a job should be at the same skill level.[1]

The alleged advantage of the functional specialization principle is that scarce knowledge and skill can be put to its widest use. Its disadvantages are a highly fragmented work force, which is costly to establish and coordinate; the possible loss of information between job boundaries when many specialties are needed in a single production process; and jobholders who are alienated by the narrowness and dead-endedness of their jobs, which reduces their willingness to contribute maximally to the organization.

These disadvantages can lead to what is sometimes called "job myopia," or the "It's not my job" syndrome. Some needs go unmet because those who are aware of a problem shrug it off as not part of their narrow jobs, which is likely the case. Those whose jobs call for them to deal with the needs do not learn of them until a problem of sufficient magnitude has developed. Additionally, specialists ignore and/or create problems according to their own workloads rather than according to the real needs of the organization. Another symptom of job myopia is that the human energy that would ideally be devoted to cross-training and spontaneous problem solving is instead diverted to protecting and/or aggrandizing individual specialties.

Deterministic Job Specification

Another principle to come out of the machine model of organizations and jobs is that of deterministic job specification. In the design of a machine nothing can be left to change; engineers specify every detail. By analogy, in the design of jobs a designer may see the need to specify every task, movement, and interaction, thus prescribing everything in the job, leaving no discretion to the employee. The analogy is fallacious. Unanticipated events often require employees purposely to deviate from their job specifications, that is, to use their discretion. Incidents of "working to rule" or "malicious obedience" show that discretion may be the most important ingredient that people bring to their jobs, even to routine jobs.

In Britain, railroad workers, instead of striking a few years ago, critically impeded train service by doing exactly what their job descriptions and standard operating procedures called for, no more and no less. They stopped the railroad from operating by leaving discretion at home when they came to work. Similarly, in New Jersey, police enforced all traffic laws without any exercise of discretion and thus managed to cripple traffic flow on key arteries.

Of course, no model of job design can prevent employees from expressing their grievances in one way or another; however, that is no reason for limiting discretion. A set of jobs will result in the achievement of an organization's goals not only because of the specified or prescribed aspects of jobs but also because of the ability and willingness of employees to use discretion.

Taylor's Approach

The machine model of organizations and jobs was developed to its most elaborate form by Frederick W. Taylor. He based his approach on the vision that a value-free science of human work existed and that it was his mission to develop that science. Taylor confidently foretold of a time when questions about the amount of work expected of a particular jobholder would be determined no differently than the calculation of the rising and setting of the sun, that is, scientifically.

The mainstay of Taylor's "science" was the fragmentation of tasks or skills into their simplest elements. These were then measured in units of time. Using Adam Smith's and Charles Babbage's economic criterion for assigning work to people, the sole criterion for creating and clustering tasks became minimum time per task or movement. This practice enabled the principles of functional specialization and deterministic job specification to be carried to their ultimate.

Taylor further insisted that the organizational tasks required to administer his scientific system be allocated to a new cadre of specialist-supervisors, thus adding the further organizational principle of separation of planning (thinking) from doing. Taylor's job designs thus separated the planning and conceptualization of work from its execution. To Taylor's critics that separation constituted a fundamental dehumanization of work. Nevertheless, after 70 years of widespread application of Taylor's principles of scientific management,[22] the discretionary element of jobs has not been replaced, and recognition of its need is increasing. Taylor's misconceived concept of a value-free science of human work became the basis for dehumanized work, which is so roundly rejected now.

Social and Technological Requirements

In order to design jobs according to the machine model, a number of social and technological decisions are required. Emery[25] listed four. The first is the development and installation of transfer devices, the classic example being Henry Ford's automobile assembly conveyors. Transfer devices move each object requiring work to each worker, making it feasible for each worker to perform but a single task, and enable managers to control directly the behavior of each worker, in this instance the pace of performing tasks.[26]

The second requirement is the standardization of both the product (or service) and the means to be used to make it. Since a production process is fragmented into many tasks, and since each task becomes a job assigned to a worker, workers thus become the interchangeable parts of the organization. Standardization requires an elaborate planning process so that the established specifications or norms will be applicable to all the cases that workers, customers, and clients will encounter. If customized products and services are needed, or if conditions of technological uncertainty prevail, then standardization will more than likely not be cost-effective.

Third are decisions inherent in "balancing the line." A set of single-task jobs is designed so that an exact number of human components will produce an exact number of product or service outputs per unit of time. These numbers cannot be expanded or contracted without a major system redesign, constraining the flexibility of the organization. Any breakdown, human or technical, that "unbalances" the system leads to downtime for the whole process. With downtime comes enforced idleness of production workers, and because it is caused by the design itself, such idleness is excused as "unavoidable delay" in the short run. In the long run come layoffs, with costly unemployment insurance claims and social disruptions. A tightly balanced production line is also vulnerable to unanticipated absences, thus requiring a surplus of workers to be available.

The fourth requirement is to develop elaborate supervisory jobs. Coercive supervision is needed to make people adhere to the machine model of organizations and jobs. Since only supervisors may undertake organizational tasks while workers perform single production task jobs, many supervisors and layers of supervisors are needed to coordinate the work and to deal with the human and technological uncertainty that inevitably arises.

2.1.8.3 Job Enlargement and Job Enrichment

During World War II the massive amounts of military material to be produced, including complex equipment, required rapid, large-scale expansion of the labor force, bringing an influx of people who were unskilled and inexperienced in industrial work. Widespread reliance on the machine model of job design in war-related industry led to discoveries of the limits of that model. Researchers developed new approaches to job design, called "job enrichment" and "job enlargement."

In job enlargement, the principles of functional specialization and deterministic job specification are somewhat moderated. Instead of striving to have each job consist of a single task or as small a fragment of work as possible, job enlargement clusters larger groups of tasks and allows jobholders certain degrees of discretion. Conant and Kilbridge[27] reported that job enlargement was beneficial, since it reduced nonproductive work and assembly line balance delays, increased the quality of the product and worker satisfaction, and provided savings in labor costs and greater production flexibility. Given the needs and expectations of workers today, it is doubtful that job enlargement can be adequate to the challenges of the 1980s as it was to those of the 1950s.

From psychological studies of worker motivation, Frederick Herzberg[28] developed job enrichment as an approach to worker satisfaction. According to this concept, a single job should encompass not only production tasks but also many of the setup, scheduling, maintenance, and control tasks related to the operation. In addition, each jobholder should be given the resources and responsibility for results relevant to his or her operation. Herzberg emphasized that job enrichment entails a "vertical" as well as a "horizontal" clustering of tasks, so that a job contains not just a variety of tasks, but planning and control tasks and a range of skill levels. The central criteria for job enrichment are that the job offer a challenge to the jobholder and that the jobholder be provided with feedback to gauge his or her own personal accomplishment.

A plastic bag plant[29] tripled its number of automatic bag making machines while enriching the jobs in its bag-making department. The unenriched jobs (Fig. 2.1.2) clustered the tasks associated with the front end of the machines into one set of jobs, and back-end tasks into another set of jobs. Since the front-end jobs required more skill, the positions, entitled "operators," were filled by men at higher pay. The back-end jobs, entitled "inspector-packers," were filled by women at lower pay. Much of the work on the machines involved dealing with breakdowns and errors. It was felt that these could be greatly reduced if each operator were assigned to perform all the tasks associated with one particular machine. In addition to this redesign of the jobs (Fig. 2.1.3), a new compensation plan was introduced in which all workers were paid a monthly salary rather than an hourly wage. Both men and women held the newly enriched operator jobs, and the company reported an increase in both productivity and job satisfaction.

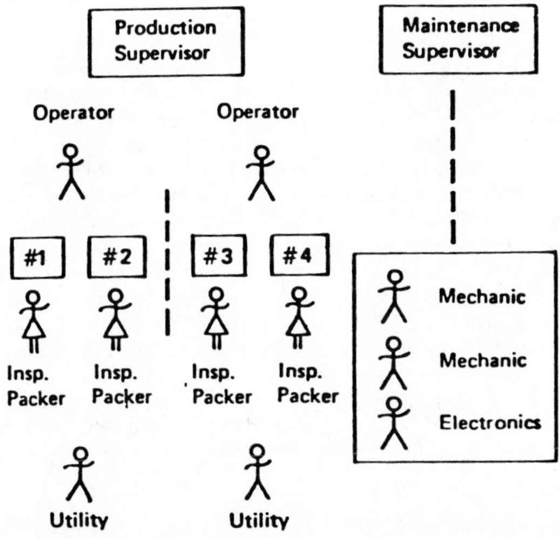

Fig. 2.1.2. Staffing of bag-making shift with initial fragmented jobs.

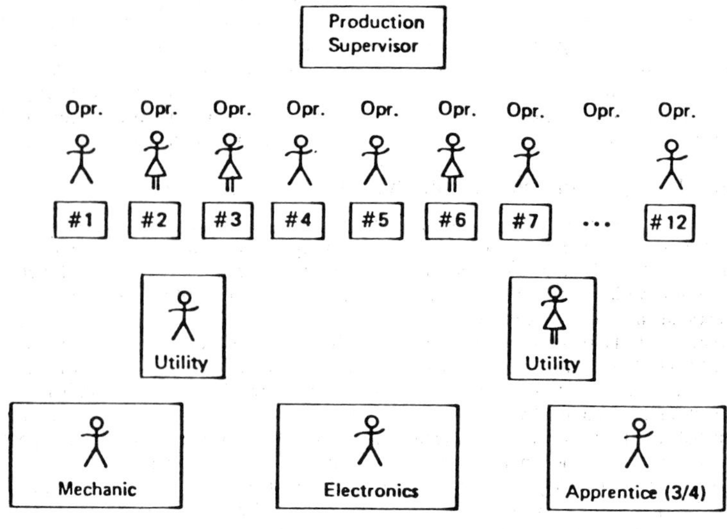

Fig. 2.1.3. Staffing of bag-making shift with current enriched jobs.

Although job enlargement and job enrichment represented major steps toward more efficient uses of human resources as well as increased humanization of work, each has significant shortcomings. These approaches do not take explicit consideration of the relationships between jobs. The one-person/one-job unit of analysis is retained from the machine model of organization. Neither do they deal with the needs and expectations that people have about their jobs (as distinct from work tasks), about coercive working environments, and about their futures.

2.1.8.4 Self-Maintaining Work Teams

Beginning with studies of English coal mines in the 1950s, researchers at the Tavistock Institute, London, developed a body of literature[30] aimed at incorporating groups as well as individuals into job design. That literature draws on parallel work in the United States[3,6] and has become known as sociotechnical systems theory. It has led not only to the modern job design option of self-supervising teams but more importantly to the underlying principle of *self-maintaining work units*.

The organization is divided into segments that can operate as fairly autonomous service or production units, or "minibusinesses." Each of these organizational segments or units is designed to contain all the tasks and resources necessary to meet its service or production and organization needs as well as many of its members' needs. In some cases all the tasks of a self-regulating work unit can be incorporated into one job, which may be the ultimate in modern job design. However, designing a work unit around a single person leaves it vulnerable to human variability in terms of the health, personality, availability, and so on, of the person. In many, if not most, cases it is preferable to design self-regulating work units around small groups of people, thus making them into self-maintaining organizational units. This enables the natural forces of group cooperation and cohesiveness to counteract both the human and the technological uncertainty that arises within the work unit.

A second principle of sociotechnical systems theory is *minimal critical specification*, the exact antithesis of the machine model's principle of deterministic complete job specification.[31] The self-maintaining work unit is designed so as to allow it as much flexibility and discretion as feasible. Job designers specify only the minimum that is absolutely critical for the work unit to function as a part of the whole organization. Other job design decisions are left as options for the work team to exercise, based on the experiences of its members. In this way, production, organizational, and individual needs can be integrated into local decision-making processes. In some of the latest organization job designs, self-maintaining work teams take responsibility for organizational learning, cross-training, peer counseling, internal coordination, and adaptation to uncertainty. The job design decisions that are typically specified and those that are typically left as options for work teams are listed in Table 2.1.3.

Illustration of Self-Regulating Team

The use of self-regulating work teams is illustrated by a field experiment conducted in a textile plant in India following installation of automatic looms. Rice[32] reports that the engineers conducted an intensive time study

Table 2.1.3. Job Design Decisions for Self-Maintaining Work Teams

Typically Specified by Job Designers, With Participation by Workers When Available	Typically Left for Team Decision Making, With Manager Consultation When Needed
Measurable input specifications	Task assignment scheduling
Measurable output specifications–quantity and quality	Work methods
	Work pace
Equipment and resources	Work hours
Work station layout options	Counseling and discipline
Compensation and aavancement plan	Internal information flows
External information flows	Internal leadership
	Team membership

of the automatic looms in order to lay out equipment and assign workloads to people on the basis of narrow tasks. These looms did not produce the quantity and quality levels already attained by the nonautomatic looms, let alone attain the improvements expected.

The automatic weaving sheds contained 240 looms, and work to be done was divided into 10 one-task jobs:

A weaver tended approximately 30 looms.

A battery filler served about 50 looms.

A smash hand tended about 70 looms.

A gater, cloth carrier, jobber, and assistant jobber were assigned to 112 looms.

A bobbin carrier, feeler-motion fitter, oiler, sweeper, and humidification fitter were each assigned to 224 looms.

These tasks were highly interdependent, and the utmost coordination was required to maintain continuity of production. However, the worker-machine assignments created organizational confusion. Each weaver had to relate to five-eighths of a battery filler, three-eighths of a smash hand, one-fourth of a gater, one-eighth of a bobbin carrier, and so on. The jobbers who carried out online maintenance reported to shed management through a separate supervisory channel from weavers, and there were no criteria to establish whose looms should have priority when breakdowns and other trouble occurred.

A redesign was undertaken to provide a self-regulating group responsible for the operation and maintenance of a specific bank of looms. Geographic, rather than functional, division of the weaving shed produced interaction patterns that enabled regularity of relationships among those with interrelated jobs. Individuals could now be held responsible for the production of their teams. The redesign was suggested by the workers themselves based on discussions with a consultant, supervisors, and managers. The consolidated teams reported to a single shift supervisor, who reported to an overall shed manager.

As a result of these changes, efficiency rose from an average of 80% to 95%, and damage dropped from a mean of 32% to 20% after 60 working days. In the adjacent part of the weaving shed, where job design changes were not made, efficiency dropped for a while to 70% and never rose above 80%, while damage continued at an average of 31%. The whole shed was then converted, and the improvements were permanently maintained. When it became clear that there was improvement, a way was found to introduce consolidated loom groups throughout the large number of nonautomatic sheds. A third shift, which had been previously resisted by the union, could then be introduced. Within loom groups, status differences were reduced; the less skilled were given opportunities to learn the roles of the more skilled, so that a promotion path was created. Wages were increased as substantially as costs were decreased.

Compatible Social and Technological Decisions

Self-maintaining work teams require compatible social and technological decisions. At Volvo's Kalmar plant in Sweden, a new technology for automobile assembly was developed to support the team form of job design.[18] Computer-monitored carriers enable each work group to lay out its own work and to control the pace of work. Volvo estimated that the new technology would cost 10% more than a conventional assembly line but expected to recover the costs in increased flexibility and higher worker motivation. At Philips in the Netherlands, televisions are assembled in "production islands," which are U-shaped assembly lines that enable assembly workers to interact and cooperate.[33] In a new polypropylene plant,[34] a process control computer displays decision-aiding information rather than making decisions. The responsibility for final decisions rests with work teams. This design was chosen because of the technological uncertainty contained in the large number of uncontrollable variables flowing in the system. If these could come to be controlled, large economic advantages would accrue. In this instance economic success was seen to be correlated with learning: the more variables that workers could learn to control, the higher the efficiency.

Compensation Plans

Three types of compensation plans have been used to support team job designs. The first, used primarily in Europe, is group piecework payment. The whole team is paid an amount to be disbursed among individual members according to internal decisions.

The second type of compensation plan is called the "group bonus," and if applied to an entire plant, it is the Scanlon plan.[35,36] Here individuals are paid on a conventional basis, but they receive bonuses when their respective teams exceed a certain production or cost target or invent improved methods that result in cost savings. Under the Scanlon plan the sharing of cost savings is plantwide.

The third type of compensation plan is called "pay by skill" or "pay by knowledge." Tasks and skills—not whole jobs—are graded, and then standards for training and testing are developed. Individuals' pay levels are determined according to their demonstrated skills and knowledge. This system serves to qualify an individual to perform certain tasks for his or her team and to train other team members in those tasks. The pay differentials do not, however, bestow authority over other team members, since the team operates as a group of peers and makes decisions by consensus.

Advantages and Disadvantages of Work Teams

One of the major advantages of self-maintaining work teams is the ability of a small group to accommodate variations in individual and organizational needs. Some members may prefer complex, multiskilled tasks; others, simple tasks; and still others, a variety of tasks. Team members, if they have adequate knowledge, skills, resources, flexibility, and social support, can usually develop mutually acceptable solutions, which often include job rotation and flexible working arrangement. This localized, participatory decision-making process increases the likelihood that decisions will be accepted by employees without the need for coercion.

For self-maintaining work teams to function at their potential, their members must possess considerable social skills and have attained a fairly high degree of emotional maturity. The present-day labor force, both white and blue collar, does largely possess the social and emotional sophistication needed for team job design. In addition, a team job design should provide in its social supplant to undergo 40 hours or more of training in social and organizational skills and team development prior to startup. Refresher sessions and problem-solving workshops are frequently provided.

One of the pitfalls of team job design is inadequate attention to team makeup. For team members to deal with intrateam and interteam problems, action should be taken to minimize the entry of conflicts present in the larger society into the organization or team. In one new plant in the southeastern United States, subteams were self-selected along clique lines, which also happened to be racial lines. Within days after startup, before the whole team could develop either cohesiveness or complete understanding of its tasks, problems within the team began to take on racial overtones. The team recognized that its makeup enabled old, deep-seated external societal conflicts to impede its internal problem-solving processes. It reassigned its members to subteams so that its internal organization cut across racial and clique lines, thus creating work-centered communication links between its parts.

It is important that a self-maintaining work team develop a practical understanding of consensus decision making. One of the most critical and difficult areas of team decision making is that of dealing internally with members who do not meet team membership norms or standards (excessive absence, unwillingness to help others, etc.). For teams to develop a style of consensus decision making, managers must deal with teams in a manner that is neither neglectful nor interfering. A facilitative management style is one of the social system requirements in support of team job design.

2.1.8.5 Matrix Structure

The concept of matrix organization originated in large aerospace enterprises.[37,38] The complexity of interrelationships among the different parts of an aerospace project, coupled with the great technological uncertainty faced by aerospace firms, may make it infeasible to segment the organization into autonomous minibusinesses. In the matrix structure, human resources are managed by two superimposed organizational structures. One of these is based on functional specialty or technology and the other on particular projects or interdisciplinary problems. This is illustrated in Figure 2.1.4.

Kingdon[39] reported a matrix job design in which the "workers" were engineering analysts with doctoral degrees and computer programmers with master of science degrees. Although a functional supervisor was nominally responsible for homogeneous groups of specialists, project assignments were determined by project managers who "hired" specialists from the functional departments for as long as they were needed. This practice was called the "job shop." Heterogeneous groups of specialists often worked together for several weeks in a place known as the "bat cave," from which they would emerge only when a critical problem had been solved.

Several recently designed manufacturing plants have instituted matrix structures within their production work forces. In one instance clusters of technical tasks and production responsibilities were allocated to five self-maintaining work teams, consisting of 4 to 28 members per team. The matrix concept was used to bring representatives of each team together to deal with the organizational tasks that required interteam coordination. On each team one member was elected to fill each of several "coordinator" roles, such as quality-control coordinator, communication coordinator, and safety coordinator. These were not full-time duties; it was each

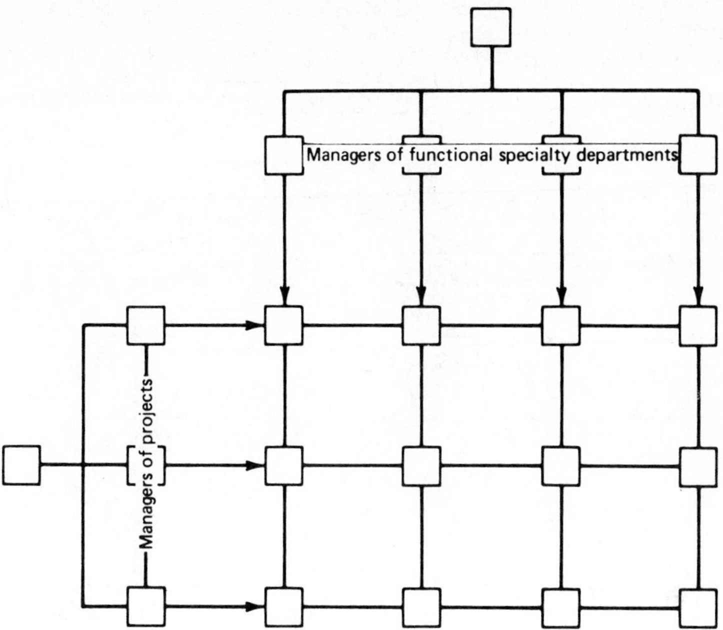

Fig. 2.1.4. Matrix job structure.

team's collective responsibility to allocate members' time so that both organizational and production needs would be met. From time to time all teams' quality-control coordinators would meet with the plant's quality-control manager, safety coordinators with the personnel manager, and so on. Communicators from each team jointly met with the operations manager every day. Special task forces composed of ad hoc representatives from all teams (including management) were created to solve a variety of plantwide problems.

Matrix structure is the most flexible form of job design, but it makes very great demands on both workers and managers. These demands stem from the ambiguity of intersecting channels of authority. Whereas many people thrive on the freedom and responsibility accorded to jobholders in a matrix structure, others prefer more rigidly structured and less responsible jobs. Managers sometimes lack the social skills necessary to manage a matrix job structure or are uncomfortable in a system in which they do not have continuing authority over a particular fixed group of subordinates. Despite the great demands of matrix structure, careful recruitment methods have turned up more than enough qualified applicants for those organizations in which jobs were so designed.

2.1.9 ANALYTIC METHODS IN SUPPORT OF JOB DESIGN

The aim of this section is to present some methods for identifying tasks and task clusters. The overriding objective of these methods of analysis is to help job designers to align the technological and organizational boundaries between jobs.

2.1.9.1 Transformation Flowchart

To conceptualize the organization as a production entity, job designers may wish to draft a flowchart showing how the product or service progresses from its input state to its output state. Industrial engineering flowcharts tend to break down actions too finely, often presuming a prior selection and layout of equipment and sometimes even a prior design of jobs. The purpose of a transformation flowchart is to enable the designers to conceptualize production requirements without being locked into specific technological or social decisions.

The transformation flowchart, an example of which is given in Figure 2.1.5, shows the progressive states of the product or service and the "unit operations" by which the product or service is transformed from one state into another. Product or service states are of three types: inputs, throughputs, and outputs. An input is a raw material or entry state that is to be transformed by the organization. A throughput is an intermediate state of work within the process. An output is a finished product, final state, or waste material that leaves the process.

Unit operations are self-contained segments of process describing the transformation of an object, in the form of material, information, or person, from an input to an output state. They describe state changes of the

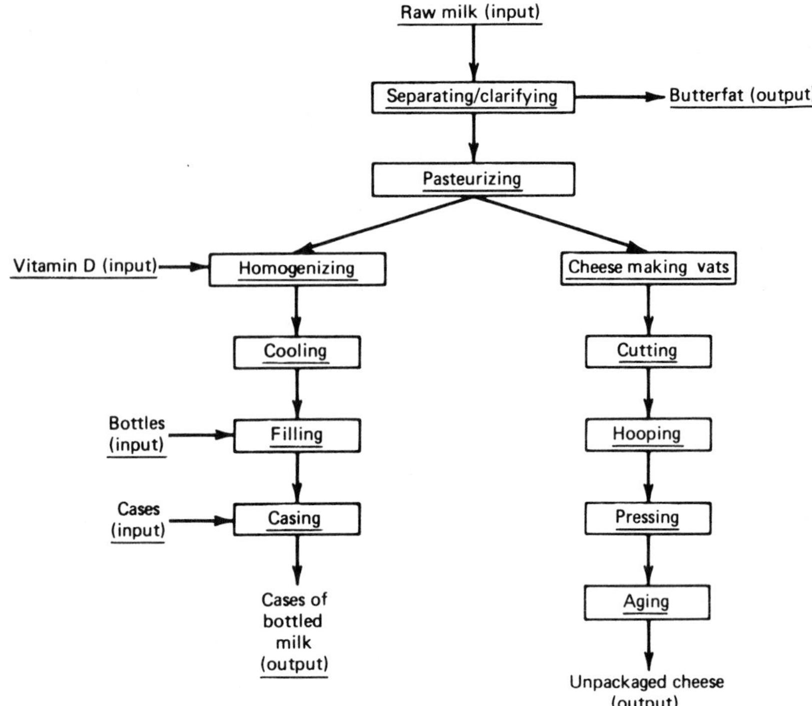

Fig. 2.1.5. Transformation flowchart for a dairy.

object being transformed. To carry out the state change, tasks need to be performed. Thus the unit operation indicates the tasks to be performed, which the job designer then clusters or assigns to a job.

Unit operations are expressed in terms of the transformation or service being rendered. If particular machines connote particular changes in product state, then it may be useful to express the unit operations in terms of these machines. Care must be taken not to exclude technical system alternatives. Inspection is usually not shown as a separate unit operation if it serves merely to verify that a transformation occurred. On the other hand, creating or updating permanent records, such as a medical history, that are not used merely to verify transformation or to control the production process may be considered separate unit operations. Storage is usually not considered a unit operation unless some change of product state—whether desired or undesired—could occur during the storage. Decision making, calculation, setup, positioning, and so on, are not expressed as unit operations by themselves but are considered tasks belonging to a unit operation.

The selection and layout of equipment and the design of jobs should enable employees to visualize the transformation flowchart as they work. Job or team boundaries should encompass one or more unit operations, so that each job or team is allocated a distinct piece of the transformation process, one having identifiable and measurable outcomes, and is responsible for an actual transformation of the product or process from one state to another.

2.1.9.2 Variance Analysis

With complete technological certainty, there would be no deviation or disturbance in the throughput states. Most organizations, however, are faced with pockets of technological uncertainty where human intervention is critical for controlling variability. The purpose of variance analysis is to identify the areas of technological uncertainty in the transformation or production process and ultimately to design jobs so that jobholders, individually or in a team or group, can effectively control variability where it arises.

A "variance" is an unwanted discrepancy between a specification or desired state and an actual state. Often specifications appear as ranges of tolerance within which deviation is permissible. A variance then occurs only when a specified range of tolerance is exceeded. There are two types of variances: product state (or service state) variances, which pertain to input, throughput, and output specifications, and system variances, which pertain to equipment and the ambient physical environment (heat, dust, humidity, etc.).

Although the concept of variance could easily be extended to discrepancies in human performance, skill, motivation, and so on, this would confound too many variables together. The concept of variance is therefore

Unit Operation	Product State Variances	System Variances

Fig. 2.1.6. Variance list format.

applied only to the technical system. The process of variance analysis first identifies the kinds and locations of human interventions needed to control technological variability at its source and, second, identifies the information, skills, and decisions needed for successful control or regulation of variances (discrepancies or disturbances).

The first step in variance analysis is to list all variances that could impede the production or service process. Trivial variances can be ignored, but when in doubt, variances should be included on the list. Often it is useful to organize the variance list in the format shown in Figure 2.1.6.

Here are some guidelines for identifying variances:

A variance should be listed with reference to the unit operation in which it would occur, regardless of where the variance is ultimately detected.

It is sometimes desirable to express variances as deviations rather than merely as variables, for example, "water too cold" rather than "water temperature." Some analysts, however, prefer simply to list variable names.

Variances are better expressed in state terms rather than in process terms, for example, "dirty dishes" rather than "unwashed dishes."

It is useful to express variances so as to reflect the degree of precision and objectivity of the specifications. For example, temperature can be a subjective variance, states as "water too cold," or an objective one, stated as "water less than 92°C."

The second step in variance analysis is to identify the dependency or causal relationship among the variances. An aid for this is a variance interrelatedness matrix, an example of which is shown in Figure 2.1.7, used in the redesign of jobs in a paper mill. The format of the matrix is analogous to that of an intercity mileage chart on a road map. Each cell of the matrix shows the degree of relatedness between two variances. A blank cell indicates no relatedness, while a "3" indicates relatedness of great importance, such as between the variances numbered 22 and 42 in Figure 2.1.7. The matrix helps the job designers to understand how the parts of the transformation or production system are dependent on each other. In designing jobs, tasks should be clustered and jobs linked so as to reflect the dependency relationships among variances.

The third step is the identification of "key variances," those whose control is most critical to the successful outcome of the production system. Job design should focus on control of key variances. Paraphrasing the Pareto principle, a large percentage of the problems in a production system are caused by a small percentage of the variances. A key variance has the following attributes: its actual, not theoretical, occurrence can seriously impair the quantity, quality, or cost of production, human resources, or technological resources; it interacts with, or causes disturbances in, many other variables; it occurs stochastically—the time, place, frequency, or intensity of occurrence cannot be predicted with certainty; and it can be detected, prevented, corrected, or otherwise controlled by timely appropriate human action.

The fourth step in variance analysis is to construct a table of key variance control, the format of which is shown in Table 2.1.4. This should contain brief descriptions of how, where, and by whom each key variance can occur and can be detected, corrected, and prevented. It should also describe how information about each key variance can be transmitted. "Feedforward" is the information that is transmitted from the point where a variance is detected to the point where it can be corrected or kept under control. "Feedback" is the information that is transmitted from the point of detection to the point where further occurrence can be prevented.

The fifth step is to construct a table of the skills, knowledge, information, and authority required for people to control key variances. The format for such a table is shown in Figure 2.1.8. Jobs should be designed so that jobholders have the skills, knowledge, information, and authority needed to control key variances.

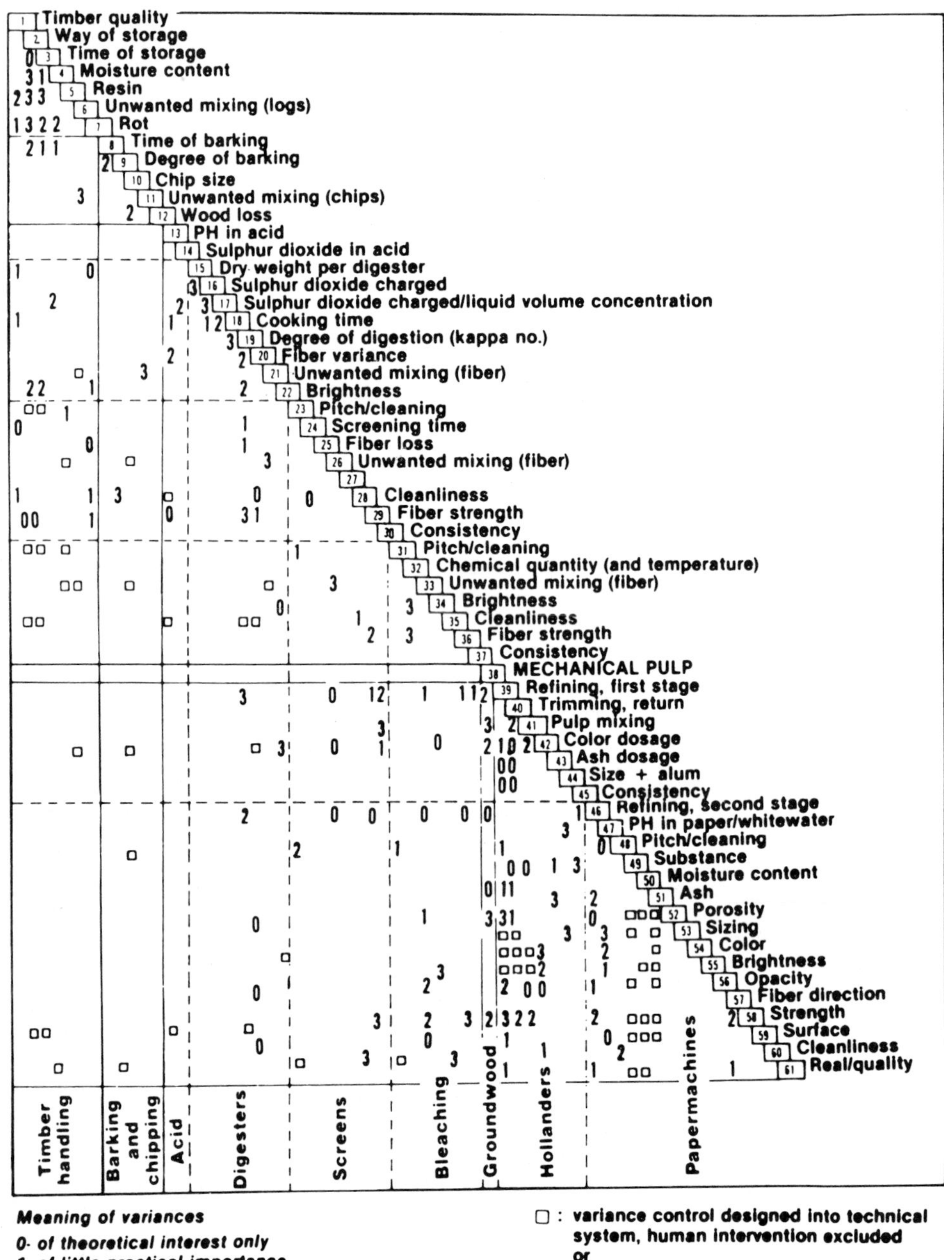

Fig. 2.1.7. Variance interrelatedness matrix. From Englestad.[41]

2.1.9.3 Technological Assessment

"Technology" has been defined as the physical and informational resources by which people can systemtically bring about some desired result. It has been noted that one of the primary functions of people in an organization is to deal with technological uncertainty. Whereas variance analysis can help to identify specific areas of technological uncertainty, technological assessment seeks a broader description of the general characteristics

Table 2.1.4. Format of Key Variance Control Table

Key Variance	Occurrence	Detection	Feedforward	Correction	Feedback	Prevention	Suggestions for Better Control	
							Technical	Social
Unit operation: Variance:	Where: How: By whom:	Where: How: By whom:	Channels: Lead time:	Where: How: By whom:	Channels: Lag time:	Where: How: By whom:		
Unit operation: Variance:	Where: How: By whom:	Where: How: By whom:	Channels: Lead time:	Where: How: By whom:	Channels: Lag time:	Where: How: By whom:		

Key Variance to be Controlled	Skills Needed	Knowledge Needed	Information Needed	Local Authority Needed

Fig. 2.1.8. Format of table of skills, knowledge, information, and authority needed to control key variances.

of the technology and their implications for job design. A number of dimensions of technology relevant to job design are described here. A suggested format for this analysis is shown in Figure 2.1.9.

Automation

Automation is technology that does not require human assistance. There are three phases of automation. The first is the operational phase, in which human physical movement is replaced or extended by machine. Second is the sensory phase, in which human senses are replaced by equipment sensors. The third phase of automation is the logic phase, in which informtion processing tasks that exceed the response capability internal to the technology itself, constitute the production-related tasks that are clustered into jobs. A typical effect of automation is a decrease in employee time spent on operational tasks and a relative increase in the proportion of employee time spent on maintenance, regulation, planning, and control.[2,40] Some unit operations may require primarily operational tasks of people and others primarily regulatory tasks. This may suggest ways in which team job designs can make flexible use of human resources.

Programmability

Some technologies are based on an exact science, so there is complete information about what needs to be done in order to achieve a specified result; there is a recipe to be followed. In other technologies the underlying science is inexact or incomplete, and the necessary actions entail some intuition or trial and error. The job designer should assess the areas of high and low programmability of action so that jobs contain elements of each. Human discretion is required when tasks are nonprogrammable.

Subjectivity

With some technologies, output quality can be determined objectively, but with others, specifications are vague and variance detection is subjective. A food processing engineer remarked casually, "We could almost train monkeys to operate the equipment; what really takes skill is knowing how to recognize the color and taste of a good product and what to do to get it consistently good." The jobs needed to be designed to provide feedback between equipment operation and product results so that operators could develop a continuous sensitivity to product quality.

Stability

Although most variances occur during the performance of some unit operation, some can occur "spontaneously" during storage, transport, waiting, and so on. Variance control is more critical when input material, product states, or equipment are unstable.

Equivalency

Engelstad[41] reported on a paper mill in which four pulp digesters were treated as equivalent or interchangeable pieces of equipment. In fact, however, one of the digesters was particularly efficient with certain kinds of wood fiber. A few operators discovered this, but their jobs were designed so that they had no incentive for sharing this information with others. On the other hand, jobs in an aluminum smelter[42] were redesigned to take account of the fact that each smelting furnace tended to develop its own "personality." Workers performed a wide range of tasks for a small group of furnaces in order to learn the peculiarities of each one.

Scale

In general, technologies designed for producing small lots or batches call for different job designs than those designed for continuous mass production.[43] Indeed, one of the ways to generate alternative technical system choices is to design two hypothetical production systems, the first for producing the product in a lot of one and the second for automated mass production.

Degradation

Some technologies function in either an "up" or "down" state. A computer, for example, rarely makes random errors; either it is not functioning at all or if it makes errors when functioning, they are made consistently, given the same inputs and programs. Other technologies function in intermediate states, such as an automobile in need of a tune-up. When many diverse technical components are united into a technical system, a breakdown in one of the components can degrade the whole system. Under some conditions the whole system will go down. Under other conditions the degradation is less severe—the system can continue to function, albeit in a slower or less powerful mode.

People, of course, usually degrade quite gracefully; rarely do they abruptly stop functioning.[8] The job designer should place people in such a way as to enable the system as a whole to degrade less abruptly in the case of a breakdown in one part of the system. The conditions of abrupt degradation should be identified and social system backup procedures designed. Jobs should be designed to take advantage of people's flexibility. One example is the integration of both operation and maintenance activities into jobs; thus when operations cannot be performed, jobholders can engage in maintenance activities.

Unit Operation	Automation	Programmability	Subjectivity	Stability	Equivalency	Scale	Degradation	Implications for Job Design

Fig. 2.1.9. Technological assessment format.

2.1.9.4 Task Ratings

The principal tasks to be performed by the work force can be subjectively rated according to the quality-of-working-life criteria shown in Table 2.1.2 (also see Hackman and Oldham[44]). These ratings can be used to guide the clustering of tasks into jobs so that each job or team contains a reasonable balance of both undesirable and desirable tasks, isolated work and teamwork, simple and complex tasks, and so on.

Task ratings can be misleading in job design. A job is not simply the sum of its tasks but the interrelationship among the tasks so as to form a meaningful whole job. Attempting to design jobs solely according to the summation of task ratings, ignoring task interrelationships, is contrary to the systems perspective of job design.

2.1.9.5 Mobility Analysis

An important source of job design information resides in the actual capabilities of employees. Mobility analysis addresses a simple question: Who can move to what other jobs should the need arise? The format for mobility analysis is shown in Figure 2.1.10. The example shown suggests that jobs C, D, E, and F might possibly be linked into a team job design.

2.1.9.6 Responsibility Analysis

Concomitant with the clustering of tasks into jobs is the allocation to each role of certain responsibilities for making decisions, for use of resources, and for achieving results. The allocation of these responsibilities should be explicitly considered in job design. Figure 2.1.11 illustrates a format for compiling such information.

In a study of the redesign of supervisory jobs,[45] two modifications in supervisors' responsibilities were introduced separately into a number of aircraft instrument repair shops. One of the modifications was to allocate to the supervisor's unit responsibility for all tasks required to complete the instruments processed in his or her shop. This represented a change from a function-based to a product-based division of the organization into smaller unit shops. The other modification involved not only the shift to a product-based organizational unit but also the addition of quality responsibility. Inspection tasks and authority for final acceptance of the instruments were added to the supervisor's unit. The supervisors under the second condition initially performed the inspection themselves and later delegated it to subordinates. Compared with control shops in which no changes took place, attitudes, productivity, and quality improved in the modified shops. As supervisors concerned themselves with planning and controlling their increased responsibility, their subordinates took on some of the characteristics of self-regulating work teams.

2.1.9.7 Cross-Boundary Interaction Analysis

In an existing job structure, the goodness of fit between actual job boundaries and organization needs can be investigated by analyzing the interactions across job and group boundaries. A format for such an analysis is

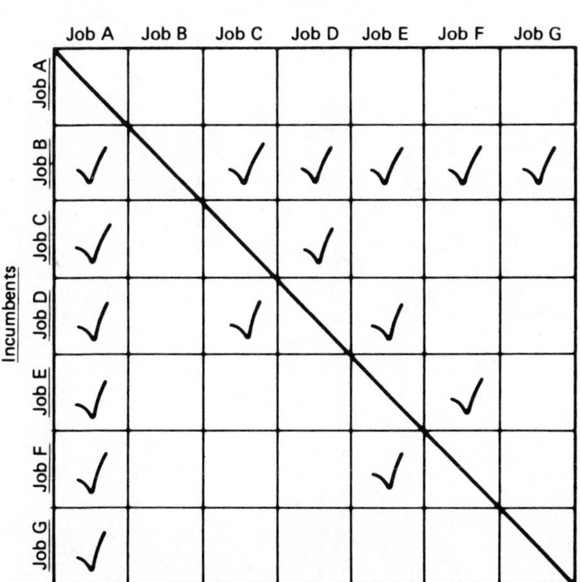

Fig. 2.1.10. Mobility analysis format.

Job or Team	Tasks	Discretionary Aspects of Task Performance	Latitude for Decisions	Available Resources	Accountability for Results	Sources of Feedback

Fig. 2.1.11. Responsibility analysis format.

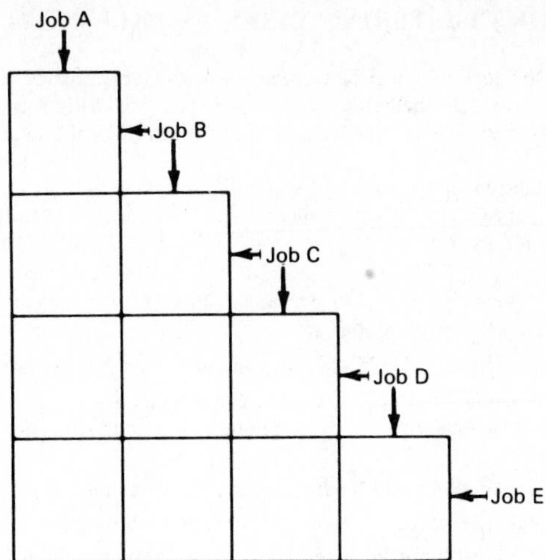

Fig. 2.1.12. Format of cross-boundary interaction analysis.

shown in Figure 2.1.12. Jobholders are asked to describe the frequencies and reasons for work-related interactions with others in the organization. These responses are summarized and recorded in the appropriate cells in the table. This analysis may provide information on task interdependence and variance control as well as on social relationships that have evolved among jobholders.

2.1.9.8 Loci of Organizational Stability and Instability

Social system stability indicators, such as labor turnover rate, absenteeism rate, grievance rate, incidents of antisocial behavior, and attitude measures, can point to special needs for job design. The increasing heterogeneity of the work force (age, sex, ethnicity, and education dimensions), together with the faster rates of social change and shorter lead times of market shifts and technological innovations, has focused attention on internal stability. There is the desire both to preserve employee skills and commitments within the organization and to adapt to changing conditions. Social system instability presents a dual challenge for job design: first, to design jobs that abate the causes of instability and, second, to design jobs that enable the system to cope with instability.

The analysis of turnover is both important and problematic. On the one hand, turnover has a negative connotation in its disruption of continuity of role incumbents and interpersonal relations. On the other hand, it has a positive connotation if individuals move to better jobs along a career path. A format for the analysis of labor turnover is presented in Figure 2.1.13. Exit interviews provide a source of data. Turnover analysis can suggest needs for more challenging jobs and for better paths of advancement.

Job	Jobholder	Dates Job Held	Prior Job	Successive Job	Reasons for Leaving

Fig. 2.1.13. Labor turnover analysis format.

2.1.10 CRITERIA FOR CLUSTERING TASKS AND ALIGNING BOUNDARIES

Following the analysis of the technological and organizational characteristics of an organization, tasks are clustered into jobs, and provisions are made for linkages between jobs. This is in essence a partitioning of the organization, which requires an alignment of technological and organizational boundaries, as discussed previously in this chapter.

A number of criteria should be considered in this partitioning process. It is unlikely that all criteria can be optimized; rather, job designers must deal with tradeoffs. The criteria for clustering tasks into jobs and teams and for aligning boundaries are as follows:

Each task cluster forms a meaningful unit of the organization—a minibusiness or an identifiable craft.

Jobs or teams are separated by stable buffer areas.

Each job or team has definite, identifiable, and measurable inputs and outputs.

Each job or team has appropriate units of performance evaluation—such as production, quality, maintenance, cost, waste, errors, absenteeism, turnover, cross-training, skill acquisition, customer complaints, machine utilization, and downtime.

Timely feedback about output states and feedforward about input states are available.

Each job or team is provided resources to measure and control throughput states.

Short feedback and feedforward loops localize variance control within the boundaries of jobs or teams.

There is more interdependence among tasks clustered together within a boundary than across the boundary. The same holds for jobs linked together.

Tasks are clustered around mutual cause–effect relationships.

Tasks are clustered around common skills, knowledge, or data bases.

Jobs are linked together to facilitate cross-training.

Jobs incorporate opportunities for skill acquisition relevant to career advancement.

Jobs or teams each contain a balance of group and individual tasks, desirable and undesirable tasks, and simple and complex tasks.

Jobs or teams each contain a balance of stressful and comfortable work environments.

Jobs or teams are capable of self-regulation and adaptation to variations in their environments.

The work stations of a team are geographically and/or temporally proximate to each other.

A team can maintain social relationships with face-to-face interactions.

2.1.11 PROCESSES AND STRATEGIES FOR DESIGNING NEW JOBS

When a new organization is designed, or when an existing organization is markedly expanded, the design of its jobs should be systematically integrated into other design decisions. As a general rule, these design projects pass through four phases.

2.1.11.1 Phase 1: Formation of the Steering Committee

The design of an organization cannot be undertaken solely by design specialists. All parties who may be affected by the design should be involved in design decisions.

Previously we reviewed the case of a food-processing plant whose designers initiated an industry-wide conversion to bulk packaging of their input material in order to accommodate job design criteria. This decision was beyond the latitude initially given to the project engineer and required the involvement of several corporate executives. Other decisions for that plant's design required the involvement of quality control executives, employee relations executives, and others who were members of a steering committee.

To set the stage for wide involvement in the design process, a steering committee is formed, consisting of key individuals with a direct stake in the new organization. The functions of the steering committee are to provide the necessary response, support, and protection to the design project team and the resources for implementation; to engender cooperation between the design task force and the parties who are affected by its design decisions; to bring design objectives into focus and to assist in future diffusion; and to oversee and guide the design process.

2.1.11.2 Phase 2: Design Task Force

Data gathering, analysis, and generation of design alternatives is done by a design task force. The task force often includes outside architects, engineers, and job design consultants. The design progresses by identifying the technological and environmental characteristics of the enterprise, identifying some "sketches" of various job design alternatives, and then considering more specific technical and social design decisions. Sometimes certain technical and social design decisions confront existing institutional precedents, constraints, or restrictions. It is then that the philosophy statement is given its test as a viable document.

2.1.11.3 Phase 3: Philosophy Statement

The initial task of the design task force is to draft a statement of organizational philosophy that acts as a charter for the design project once it is approved by the steering committee. The philosophy statement addresses the following issues:

Stakeholders in the project and their interests in it
Purposes of the design project
Production and technological needs and requirements
Organizational needs and requirements
Explicit recognition of the needs and requirements of individuals who, as members of the organization, will fill jobs
Values and assumptions about the organization's environment and future
Values and assumptions about people and work
Values and assumptions about management and managers
Relationship between the organization to be designed and existing institutions
Processes for decision making during the design
Processes for decision making during and following startup
Evolutionary aspects of the project—what is not to be designed until after startup
Resources for implementation
Risk bearing and responsibility for results
Expectations of participation, cooperation, and support

There are various methods for drafting the philosophy statement, depending on the degree to which values are already shared.

2.1.11.4 Phase 4: Evolution

In job redesign in an existing organization, and after managers and supervisors have been appointed in the design of a new organization, the process of job design shifts to a more participatory mode. The underlying strategy for this shift was discussed previously as the principle of minimal critical specification. One member of a steering committee expressed it more colloquially: "We want to avoid dictating other people's work lives. If we were designing this plant for ourselves to work in, we would want to leave a lot of options open." As the organization evolves, its jobholders use their experiences to help perfect its design. Continual redesign of jobs becomes one of the organizational tasks of the work force.

2.1.12 PROCESSES AND STRATEGIES FOR REDESIGNING EXISTING JOBS

Although existing organizations might do well to begin job redesign with a reexamination of managerial strategy, technology, and organizational structure, there are usually a number of constraints that cannot be removed. Job redesign is often just part of a continuous stream of changes the organization makes to meet requirements coming from its environment. One of the objectives of job redesign is to provide an experience that sets a precedent for future redesign needs.

The designer should identify all relevant change forces, both internal and external to the organization and both past and future. Usually there is a complex system of forces that underlies the job redesign endeavor. In an aluminum smelter,[42] rising labor turnover rates, increasing difficulty in recruiting workers, technological obsolescence, and more aggressive government regulation of occupational safety and health conditions and equal employment opportunities for minorities and women led to efforts to automate some of the most physically demanding and undesirable tasks. The new automated technology required less manpower and new task clusters. But the new technology alone did not solve the organization's problems of turnover, recruitment, and efficiency. Thus the job redesign was directed not only at suiting the new technology but also at creating more attractive and challenging jobs, attaining work force flexibility so that the organization could function with a broad range of turnover rates, and training and advancement geared to individual readiness, regardless of vacancies created by turnover.

Employee suggestions for, and their reactions to, redesigned jobs are influenced by their past experiences and their expectations for the future. In one plant employees saw redesign as being accomplished by petitioning for a second time clock so that they would not have to lose time in clocking out for lunch. That there were more fundamental aspects to an improved quality of working life was beyond their experiences. In the aluminum smelter cited previously, the job redesign resulted in fewer status differences between smelter jobs. Although most of the employees welcomed the new opportunities and new technology, some complained about the loss of their former perquisites. Attitudes improved among both the employees whose jobs were changed and those whose jobs had not been changed, because the latter group looked forward to an imminent changeover; their jobs now had a future.

The phases of redesign are analogous to those of new design, with the following additions:

Representatives of existing jobholders, or of their union, are usually included in the redesign project task force. Sometimes the redesign team's membership represents a "diagonal slice" of the organization, with various levels of the hierarchy represented such that no one is on the committee along with his or her direct superior.

If several separable redesign projects are considered, each has its own design task force; a single steering committee coordinates the projects.

Implementation planning is more critical in redesign, and implementation issues are considered concomitantly with design decisions. Probably the most important aspect of implementation is *ownership of change*. Those responsible for carrying out and living with proposed redesigns should feel that the changes have come out of their own interests and efforts. A widespread commitment to a proposal is a vital factor to its ultimate success, for two reasons: emotional identification with the redesign and the desire to put forth the effort to make it work; and understanding the details of the redesign sufficiently to estimate and minimize the effects, that is, to reduce the uncertainty over the consequences for the individual.

A number of methods can enable jobholders to contribute to the redesign. These include "nominal group technique"[46]; sociotechnial analysis approaches, including variance analysis; redesign principles; and quality-of-working-life characteristics.

2.1.12.1 Role of Consultants

In both new design and redesign, the job design consultant has a very different role from that of a conventional engineering consultant. Because the design process is multiphasic and includes many parties, the consultant performs five basic functions:

Training and Guidance

The principles of job design and the methods of analysis are taught to members of design task forces and steering committees, who, in turn, make design decisions. The consultant does not design the jobs but guides the principals through the job design endeavor.

Process Facilitation. The consultant helps plan and facilitate the processes in all phases of the design. for example, steering committee and task force members may be unacquainted with their roles, and the consultant helps clarify these issues as the need arises and aids them in working together.

Mediating. When several distinct institutions are party to a redesign project, for example, company and union, the consultant may act as a neutral broker and go-between in order to ensure consensual agreement for design and implementation.

Research. The assessment of job design results may require research methods for which the consultant has special training and experience. In such cases he or she may collect and analyze data and present findings useful for the teams.

Change Agent. As innovative job designs are developed, the consultant acts as a carrier of these innovations to other parts of the institution of which the organization is a part. The role here is not one of "expert" so much as "idea broker."

2.1.13 CONCLUSION

Job design is a process establishing the content of roles or jobs, which are then assigned to individuals to perform. In the design of jobs for new organizations, the process is inseparable from organization design. In the redesign of jobs, organizational boundary issues as well as technical system design issues must be examined, since they may place unnecessary constraints on the design. The designer must remember that the organization is a system and that the jobs are component parts of this system. Further, the system embraces both the organization and its environment, bringing into focus the fact that job design has to concern itself not only with the technical and economic needs of the organization but with the needs and expectations of the jobholders on whose task performances the organization depends for the achievement of its goals.

REFERENCES

1. L. E. Davis, R. Canter, and J. Hoffman. "Current Job Design Criteria." *Journal of Industrial Engineering*, **6**(2) (1955), 5–11.

2. L. E. Davis and J. C. Taylor. "Technology and Job Design." In L. E. Davis and J. C. Taylor, Eds., *Design of Jobs*, 2nd ed. Santa Monica, CA: Goodyear, 1979.
3. L. E. Davis and R. R. Canter. "Job Design Research." *Journal of Industrial Engineering*, **7** (1956), 275–282.
4. L. E. Davis and R. Werling. "Job Design Factors." *Occupational Psychology*, **34** (1960), 109–120.
5. G. I. Susman. "Job Enlargement: Effects of Culture on Worker Responses." *Industrial Relations*, **12** (1973), 1–15.
6. L. E. Davis and E. L. Trist. "Improving the Quality of Working Life: Sociotechnical Case Studies." In J. O'Toole, Ed., *Work and the Quality of Life*. Cambridge, MA: MIT Press, 1974.
7. R. Dunham. "Reactions to Job Characteristics: Moderating Effects of the Organization." *Academy of Management Journal*, **20** (1977), 42–65.
8. N. Jordan. "Allocation of Functions Between Man and Machine in Automated Systems." *Journal of Applied Psychology*, **47** (1963), 161–165.
9. L. E. Davis. "The Coming Crisis for Production Management: Technology and Organization." *International Journal of Production Research*, **9** (1971), 65–82.
10. J. C. Taylor. "Some Effects of Technology in Organizational Change." *Human Relations*, **24** (1971), 105–123.
11. H. L. Waddel. "The Fundamentals of Automation." In H. B. Maynard, Ed., *Industrial Engineering Handbook*. New York: McGraw-Hill, 1956, pp. 325–331.
12. C. Orpen. "The Effects of Job Enrichment on Employee Satisfaction, Motivation, Involvement, and Performance: A Field Experiment." *Human Relations*, **32** (1979), 189–217.
13. G. R. Oldham and H. E. Miller. "The Effect of Significant Other's Job Complexity on Employee Reactions to Work." *Human Relations*, **32** (1979), 247–260.
14. L. E. Davis and A. B. Cherns, Eds. *The Quality of Working Life*, Vols. 1 and 2. New York: Free Press, 1975.
15. P. H. Engelstad. "Sociotechnical Approach to Problems of Process Control." In L. E. Davis and J. C. Taylor, Eds., *Design of Jobs*, 2nd ed. Santa Monica, CA: Goodyear, 1979.
16. L. E. Davis. "Evolving Alternative Organization Designs." *Human Relations*, **30** (1977), 261–273.
17. L. E. Davis and G. J. Wacker. *Comprehensive Socio-Technical System Design: A Case Study*, in preparation.
18. P. G. Gyllenhammar. *People at Work*. Reading, MA: Addison-Wesley, 1977.
19. E. L. Trist, G. W. Higgin, H. Murray, and A. B. Pollock. *Organizational Choice*. London: Tavistock, 1963.
20. J. W. Lorsch and J. J. Morse. *Organizations and Their Members*. New York: Harper & Row, 1974.
21. R. M. Kanter. *Men and Women of the Corporation*. New York: Basic Books, 1977.
22. F. W. Taylor. *The Principles of Scientific Management*. New York: Harper & Row, 1911.
23. A. Smith. *The Wealth of Nations*. Penguin, London, 1970 (originally published 1776).
24. C. Babbage. *On the Economy of Machinery and Manufacturers*. New York: Augustus M. Kelly, 1965 (4th ed. enlarged; originally published 1835).
25. F. E. Emery. "The Assembly Line—Its Logic and Our Future." In L. E. Davis and J. C. Taylor, Eds., *Design of Jobs*, 2nd ed. Santa Monica, CA: Goodyear, 1979.
26. L. E. Davis. "Pacing Effects on Manned Assembly Lines." *International Journal of Industrial Engineering*, **4** (1966), 171–180.
27. E. H. Conant and M. D. Kilbridge. "An Interdisciplinary Analysis of Job Enlargement: Technology, Costs, and Behavioral Implications." *Industrial and Labor Relations Review*, **18** (1965), 377–390.
28. F. Herzberg. *Work and the Nature of Man*, Cleveland: World, 1966.
29. L. E. Davis and A. B. Cherns. "Transition to More Meaningful Work—A Job Design Case." In L. E. Davis and A. B. Cherns, Eds., *The Quality of Working Life*, Vol. 2. New York: Free Press, 1975.
30. F. E. Emery, Ed. *Systems Thinking*. London: Penguin, 1969.
31. P. G. Herbst. *Socio-Technical Design*. London: Tavistock, 1974.
32. A. K. Rice. *Productivity and Social Organization: The Ahmedabad Experiment*. London: Tavistock, 1958.
33. J. F. den Hertog. "The Search for New Leads in Job Design." *Journal of Contemporary Business*, **6**(2) (1977), 49–66.
34. L. E. Davis and C. S. Sullivan. "A Labour-Management Contract and Quality of Working Life." *Journal of Occupational Behavior*, Vol. 1 (1980), pp. 29–41.
35. J. N. Scanlon. "Profit Sharing Under Collective Bargaining: Three Case Studies." *Industrial and Labor Relations Review*, Vol. 2 (1948), p. 58 ff.
36. F. G. Lesieur and E. S. Puckett. "The Scanlon Plan Has Proved Itself." *Harvard Business Review*, **47**(5) (1969), 109–118.
37. D. I. Cleland and W. R. King. *Systems Analysis and Project Management*. New York: McGraw-Hill, 1968.
38. J. Galbraith. *Designing Complex Organizations*. Reading, MA: Addison-Wesley, 1973.
39. D. R. Kingdon. *Matrix Organization*. London: Tavistock, 1973.

40. R. J. Hazlehurst, R. J. Bradbury, and E. N. Corlett. "A Comparison of the Skills of Machinists on Numerically Controlled and Conventional Machines." *Occupational Psychology*, **43**(3) (1969), 169–182.
41. P. H. Engelstad. "Sociotechnical Approach to Problems of Process Control." 1979.
42. G. J. Wacker. "Evolutionary Job Design: A Case Study," working paper, Department of Industrial Engineering, University of Wisconsin, Madison, 1979.
43. J. Woodward. *Industrial Organization: Theory and Practice*. New York: Oxford University Press, 1965.
44. J. R. Hackman and G. R. Oldham. "Development of the Job Diagnostic Survey." *Journal of Applied Psychology*, **60** (1975), 159–170.
45. L. E. Davis and E. S. Valfer. "Supervisory Job Design." *Ergonomics*, **8** (1965), 1.
46. A. L. Delbeq, A. H. van de Ven, and D. H. Gustafson. *Group Techniques for Program Planning*. Glenview, IL: Scott, Foresman, 1976.

BIBLIOGRAPHY

Csikszentmihalyi, M. *Beyond Boredom and Anxiety*. San Francisco: Jossey-Bass, 1976.

Ford, R. N. *Why Jobs Die and What to Do About It*. New York: AMACOM, 1979.

Herzberg, F. *The Managerial Choice: To Be Efficient and Human*. New York: Dow Jones, 1976.

Hill, P. *Towards a New Philosophy of Management*. London: Gower, 1976.

Mumford, E., and H. Sackman, Eds. *Human Choice and Computers*. Amsterdam: North Holland, 1975.

Nadler, D. A., J. R. Hackman, and E. E. Lawler. *Managing Organizational Behavior*. Boston: Little, Brown, 1979.

Schon, D. A. *Beyond the Stable State*. New York: Random House, 1971.

Sheppard, H. L., and N. Q. Herrick. *Where Have All the Robots Gone?* New York: Free Press, 1972.

CHAPTER 2.2

INDUSTRIAL SAFETY

JACK B. REVELLE
Hughes Aircraft Company
Los Angeles, California

2.2.1 POLICY

An important policy and goal of any company should be to maintain a safe and healthful working environment for the well-being of its employees. This can be achieved through an industrial safety program aimed at the prevention of occupational illness and injury, the elimination or control of occupational health and safety hazards or nuisances, the prevention of property damage, and, thereby, the prevention of work interruptions.

2.2.2 RESPONSIBILITIES

2.2.2.1 General Responsibilities

Management is responsible for providing and maintaining safe and healthful working conditions for all employees. Supervisors are responsible for training and directing employees in the safe performance of their work assignments and enforcing all company rules and regulations pertaining to health and safety.

Supervisors in all activities must maintain a constant state of alertness to prevent accidents and eliminate environmental health hazards. Supervisors should listen carefully and establish the facts when an employee reports a safety or health hazard. In each case, the supervisor should communicate immediately with the industrial safety manager for assistance in recognizing, evaluating, and controlling hazardous conditions and, when necessary, in alleviating the concern of the employee.

2.2.2.2 Legal Responsibilities

Management and supervisory responsibility, according to federal and some state laws, can extend to personal liability on the part of the supervisor.

2.2.2.3 Accident Reporting

Employees should report all injuries, no matter how slight, to their supervisors and obtain treatment at the nearest first aid dispensary. Prompt and adequate medical treatment helps to prevent further complications and assures the timely administration of workers' compensation benefits. Each supervisor and employee should become familiar with the location of the nearest first aid station. Emergency phone numbers and the location of telephones should be listed in the front of the plant telephone book.

Supervisors are expected to make a timely report of any accident. The report is to be made by the most rapid means available, either in person or by telephone. All accidents should be reported to the industrial safety office.

2.2.3 FUNCTIONAL OPERATIONS

2.2.3.1 Industrial Safety Office

The staff of the industrial safety office should be available to establish, direct, and maintain environmental health and safety programs as a support function for supervisors and managers.

2.2.3.2 Industrial Safety Functions

The functions of an industrial safety office may include

Administering area inspections and surveys (preventive)

1. General safety audits and specific safety surveys
2. Noise survey
3. Ventilation surveys
4. Radiation survey (ionizing, nonionizing, infrared, microwave, etc.)
5. Monitoring dusts, fumes, and solvent vapors
6. General housekeeping

Investigating accident/incidents
Consulting with supervisors/managers
Interpreting standards and regulations
Interfacing with local/state/federal agencies
Training supervisors and employees in safety procedures
Reviewing plans and procedures
Regulating radioactive materials
Interfacing with customers
Administering safety glass program
Defining uses of protective equipment
Coordinating meetings of safety committees
Administering and coordinating workers' compensation
Publishing industrial safety bulletins and procedures
Issuing and enforcing industrial safety directives

2.2.3.3 Supervisory Functions

The functions of supervisors and managers may include

General Functions

1. Providing a proper and safe working environment
2. Correcting hazards
3. Conducting safety inspections
4. Maintaining a high level of good housekeeping
5. Reporting and investigating all accidents and taking corrective action to prevent recurrence
6. Maintaining adequate fire prevention equipment

Training. Insure that employees are knowledgeable in

1. Safe and proper performance of their jobs
2. Proper use of respiratory equipment
3. Safe use of hazardous materials
4. Environmental pollution control
5. Electrical lockout procedures
6. Use and maintenance of proper eye protection
7. Proper care, maintenance, and use of machine guards
8. Proper use of hand and portable power tools
9. Safe methods of work in confined spaces
10. Proper use of exhaust systems
11. Proper lifting techniques
12. Certification requirements

Rule Enforcement. All safety rules must be consistently enforced, including those that relate to

1. Use of respiratory devices
2. Environmental safety program
3. Pollution control
4. Regulatory control
5. Use of eye protection devices
6. Assignment of certification tasks to "only" qualified and certified employees

7. Modification of exhaust systems
8. De-energizing and locking out of electrical equipment prior to work or repair
9. Use of personal protective devices
10. Proper placement and maintenance of machine guards

Communicating Company/OSHA Requirements. Inform employees about

1. Potential safety hazards to which they are exposed
2. Safety procedures to be practiced on the job
3. Symptoms of overexposure to hazards
4. Emergency treatment for accidents
5. Dangers of hazardous materials
6. Purpose, use, value, accessibility, and limitations of personal protective devices
7. Wearing apparel requirements/restrictions
8. Hazards of confined spaces

Maintenance. Insure proper and regular maintenance of

1. Facilities and equipment in the work area
2. Facilities and equipment used for the storage, disposal, and use of hazardous materials
3. Exhaust systems
4. Facilities and equipment safeguards
5. Machine guards
6. Hand and portable power tools

Liaison. Consult with industrial safety officer for information/assistance to

1. Interpret company/OSHA rules, regulations, and safety standards
2. Obtain approval to purchase machines/materials
3. Evaluate conditions/operations that could produce harmful atmospheres or working conditions
4. Determine/investigate need for exhaust systems
5. Determine need/requirements for exhaust system repair
6. Identify areas with hazards to the eyes
7. Conduct employee work limitation review

2.2.4 MANAGEMENT PRINCIPLES*

Regardless of the industry or the process, the role of managers and supervisors in any safety program takes precedence over any of the other elements. This is not to say that the managerial role is necessarily more important than the development of safe environments, but without manager and supervisor participation, the other elements have a lukewarm existence. There is a dynamic relationship between management and the development of safe working conditions, and management and the development of safety awareness, and the relationship must not be denied.

Where responsibility for preventing accidents and providing a healthful work environment is sloughed off to the safety department or a safety committee, any reduction in the accident rate is minimal. To reduce the accident rate, and in particular, to make a good rate better, both senior and line managers must be held responsible and accountable for safety. Every member of the management team must have a role in the safety program. Admittedly, this idea is not new, but application of the concept still requires crystal-clear definition and vigorous promotion.

Notwithstanding the many excellent examples of outstanding safety records that have been achieved because every member of management had assumed full responsibility for safety, there are still large numbers of companies, particularly the small establishments, using safety contests, posters, or safety committees as the focal point of their safety programs—but with disappointing results because full management support is lacking. Under such circumstances safety is perceived as an isolated aspect of the business operation with rather low ceiling possibilities at best. But there are some who feel that gimmicks must be used because foremen and managers don't have time for safety.

As an example of the case in point, a handbook on personnel contains the statement that "A major disadvantage of some company-sponsored safety programs is that the supervisor can't spare sufficient time from his regular duties for running the safety program." Significantly, this was not a casual

* This section is taken, with permission of the author and publisher, from De Reamer.[1]

comment in a chapter on safety. It was indented and in bold print for emphasis. Yet it is a firmly accepted fact that to achieve good results in safety, managers and supervisors must *take the time to fulfill their safety responsibilities. Safety is one of their* regular *duties.*

2.2.5 ELEMENTS OF ACCIDENT PREVENTION*

Safety policy must be clearly defined and communicated to all employees.

The safety record of a company is a barometer of its efficiency. An American Engineering Council Study revealed "maximum productivity is ordinarily secured only when the accident rate tends toward the unreducible minimum."

Unless line supervisors are accountable for the safety of all employees, no safety program will be effective. Top management must put the cards on the table and let all managers and supervisors know what is expected of them in safety.

Periodic progress reports are required to let managers and employees know what they have accomplished in safety.

Meetings with managers and supervisors to review accident reports, compensation costs, accident-cause analysis, and accident-prevention procedures are an important element of the overall safety program.

The idea of putting on a big safety campaign with posters, slogans, and safety contests without full management support is wrong. The Madison Avenue approach doesn't work over the long run without management guidance and involvement.

Good housekeeping and the enforcement of safety rules show that management has a real concern for employee welfare. They are important elements in the development of good morale. (A U.S. Department of Labor study has revealed that workers are vitally concerned with safety and health conditions of the work place. A surprisingly high percentage of workers ranked protection against work-related injuries and illness and pleasant working conditions as having a priority among their basic on-the-job needs. In fact, they rated safety higher than fringe benefits and steady employment.)

The use of personal protective equipment (safety glasses, safety shoes, hard hats, etc.) must be a condition of employment in all sections of the plant where such protection is required.

Safety files must be complete and up-to-date to satisfy internal information requirements as well as external inspection by OSHA compliance officers and similar officials. [Table 2.2.1.]

2.2.6 ELIMINATING UNSAFE CONDITIONS†

2.2.6.1 Steps for Eliminating Unsafe Conditions

The following steps should be taken to effectively and efficiently eliminate an unsafe condition:

Remove. *If at all possible, have the hazard eliminated.*

Guard. *If danger point (i.e., high tension wires) can't be removed, see to it that hazard is shielded by screens, enclosures, or other guarding devices.*

Warn. *If guarding is impossible or impractical, warn of the temporarily unsafe condition. If a truck must back up across a sidewalk to a loading platform, the sidewalk cannot be removed or a fence built around the truck. All that can be done is to warn that a temporarily unsafe condition exists. This is done by posting a danger sign or making use of a bell, horn, whistle, signal light, painted striped lines, red flag, or other device. Where appropriate, a flag-person can be stationed to provide needed warnings.*

Recommend. *If you cannot remove or guard an unsafe condition on your own, notify the proper authorities about it. Make specific recommendations as to how the unsafe condition can be eliminated.*

Follow up. *After a reasonable length of time, check to see whether the recommendation has been acted on, or whether the unsafe condition still exists. If it remains, the person or persons to whom the recommendations were made should be notified.*

2.2.6.2 Factors Providing Maximum Productivity and Well-Being

The following factors should be considered in organizing a plant that provides for maximum productivity and employee well-being:

* This section is taken, with permission of the author and publisher, from De Reamer.[1]

† This section is taken, with permission of the author and publisher, from ReVelle.[5]

Table 2.2.1. Requirements for Safety Files[a]

Action Required	Action Completed Yes	No
1. Is there a separate section for safety-related files?		
2. Are the following subjects provided for in the safety section of the files:		
a. Blank OSHA forms?		
b. Completed OSHA forms?		
c. Blank company safety forms?		
d. Completed company safety forms?		
e. Blank safety checklists?		
f. Completed safety checklists?		
g. Agendas of company safety meetings?		
h. Minutes of company safety meetings?		
i. Records of safety equipment purchases?		
j. Records of safety equipment checkouts?		
k. Incoming correspondence related to safety?		
l. Outgoing correspondence related to safety?		
m. Record of safety projects assigned?		
n. Record of safety projects completed?		
o. Record of fire drills (if applicable)?		
p. Record of external assistance used to provide specialized safety expertise?		
q. Record of inspections by fire department, insurance companies, state and city inspectors, company safety personnel, and OSHA compliance officers?		
r. National Safety Council catalogs and brochures for films, posters, and other safety-related materials?		
3. Are the files listed in item 2 reviewed periodically to insure that they are current and to retire material over five years old?		
4. Are safety-related files reviewed periodically to determine the need to eliminate selected files and add new subjects?		
5. Is the index to the file current, so that an outsider could easily understand the system?		

[a] The following items are presented for your convenience as you review your administrative storage index to determine the adequacy of your safety-related files.
Source: Jack B. ReVelle, *Safety Training Methods,* John Wiley and Sons, 1980.

The general arrangement of the facililty should be efficient, orderly, and neat.

Work stations should be clearly identified so employees can be assigned according to the most effective working arrangement.

Material flow should be designed to prevent unnecessary employee movement for given work.

Materials storage, distribution, and handling should be routinized for efficiency and safety.

Decentralized tool storage should be used wherever possible. Where centralized storage is essential (e.g., general supply areas, locker areas, and project storage areas), care should be given to establish a management system that will avoid unnecessary crowding or congested traffic flow. (Certain procedures, such as time staggering, may reduce congestion.)

Time-use plans should be established for frequently used facilities to avoid having workers wait for a particular apparatus.

A warning system and communications network should be established for emergencies such as fire, explosion, storm, injuries, and other events that would affect the well-being of employees.

2.2.6.3 Checklist of Unsafe Conditions

The following unsafe conditions checklist presents a variety of undesirable characteristics to which both employers and employees should be alert.

Unsafe Conditions—Mechanical Failure. *These are types of unsafe conditions that can lead to occupational accidents and injuries.* Note. *Keep in mind that unsafe conditions often come about as a result of unsafe acts.*

Lack of Guards. *This applies to hazardous places like platforms, catwalks, or scaffolds where no guard rails are provided; power lines or explosive materials that are not fenced off or enclosed in some way; and machines or other equipment where moving parts or other danger points are not safeguarded.*

Inadequate Guards. *Often a hazard that is partially guarded is more dangerous than it would be if there were no guards. The employee, seeing some sort of guard, may feel secure and fail to take precautions that would ordinarily be taken if there were no guards at all.*

Defects. *Equipment or materials that are worn, torn, cracked, broken, rusty, bent, sharp, or splintered; buildings, machines, or tools that have been condemned or are in disrepair.*

Hazardous Arrangement (housekeeping). *Cluttered floors and work areas; improper layout of machines and other production facilities; blocked aisle space or fire exits; unsafely stored or piled tools and material; overloaded platforms and vehicles; inadequate drainage and disposal facilities for waste products.*

Improper Illumination. *Insufficient light; too much light; lights of the wrong color; glare; arrangement of lighting systems that result in shadows and too much contrast.*

Unsafe Ventilation. *Concentration (build-up) of vapors, dusts, gases, fumes; unsuitable design, capacity, location, or arrangement of ventilation system; insufficient air changes, impure air source used for air changes; abnormal temperatures and humidity.*

2.2.6.4 Conditions to Be Inspected

In describing conditions for each item to be inspected, terms such as the following should be used:

Broken	*Leaking*
Corroded	*Loose (or slipping)*
Decomposed	*Missing*
Frayed	*Rusted*
Fuming	*Spilled*
Gaseous	*Vibrating*
Jagged	

2.2.6.5 Problem Sources to Be Inspected

An alphabetized listing of possible problems to be inspected is presented in Table 2.2.2.

2.2.6.6 Classification of Hazards

It is important to differentiate the degrees of severity *of different hazards. The commonly used standards are given below.*

Class A Hazard. *Any condition or practice with* potential *for causing* loss *of life or body part and/ or extensive loss of structure, equipment, or material.*

Class B Hazard. *Any condition or practice with* potential *for causing serious injury, illness, or property damage, but less severe than Class A.*

Class C Hazard. *Any condition or practice with* probable potential *for causing* nondisabling *injury or illness or* nondisruptive *property damage.*

2.2.7 POTENTIAL UNSAFE CONDITIONS INVOLVING MECHANICAL OR PHYSICAL FACILITIES*

The total working environment must be under constant security because of changing conditions, new employees, equipment additions and modifications, and so on. The following checklist is presented as a guide to identify potential problems.

2.2.7.1 Building

Correct ceiling height

Correct floor type, in acceptable condition

* This section in taken, with permission of the author and publisher, from ReVelle.[5]

Table 2.2.2. Problem Sources to Be Inspected

Acids	Dusts	Railroad cars
Aisles	Electric motors	Ramps
Alarms	Elevators	Raw materials
Atmosphere	Explosives	Respirators
Automobiles	Extinguishers	Roads
Barrels	Flammables	Roofs
Bins	Floors	Safety devices
Blinker lights	Forklifts	Safety glasses
Boilers	Fumes	Safety shoes
Borers	Gas cylinders	Scaffolds
Buggies	Gas engines	Shafts
Buildings	Gases	Shapers
Cabinets	Hand tools	Shelves
Cables	Hard hats	Sirens
Carboys	Hoists	Slings
Catwalks	Hoses	Solvents
Caustics	Hydrants	Sprays
Chemicals	Ladders	Sprinkler systems
Claxons	Lathes	Stairs
Closets	Lights	Steam engines
Connectors	Mills	Sumps
Containers	Mists	Switches
Controls	Motorized carts	Tanks
Conveyers	Piping	Trucks
Cranes	Pits	Vats
Crossing lights	Platforms	Walkways
Cutters	Power tools	Walls
Docks	Presses	Warning devices
Doors	Racks	

Source: Principles and Practices of Occupational Safety and Health: A Programmed Instruction Course. OSHA 2213, Student Manual Booklet 1, U.S. Department of Labor, Washington, D.C., p. 40.

Adequate illumination

Adequate plumbing and heating pipes and equipment

Windows with acceptable opening, closing, and holding devices; protection from breakage

Acceptable size doors with correct swing and operational quality

Adequate railing and nonslip treads on stairways and balconies

Adequate ventilation

Adequate storage facilities

Adequate electrical distribution system in good condition

Effective space allocation

Adequate personal facilities (restrooms, drinking fountains, washup facilities, etc.)

Efficient traffic flow

Adequate functional emergency exits

Effective alarms and communications systems

Adequate fire prevention and extinguishing devices

Acceptable interior color scheme

Acceptable noise absorption factor

Adequate maintenance and cleanliness

2.2.7.2 Machinery and Equipment

Acceptable placement, securing, and clearance

Clearly marked safety zones

Adequate nonskid flooring around machines

Adequate guard devices on all pulleys

Sharp, secure knives and cutting edges

Properly maintained and lubricated machines in good working condition

Functional, guarded, magnetic-type switches/interlocks on all major machines

Properly wired and grounded machines

Functional hand and portable power tools in good condition and grounded

Quality machines adequate to handle the expected workload

Conspicuously posted safety precautions and rules near each machine

Guards for all pinch points within 7 feet of the floor

2.2.8 PROCEDURE AND BULLETIN TOPICS

2.2.8.1 Industrial Safety Procedures

Each company should prepare and distribute industrial safety procedures or directives that outline in the most explicit terms both acceptable and unacceptable behavior for its employees. The following list provides a point of departure regarding topics that your company should address in industrial safety procedures.

Administration

1. Industrial safety
2. Outside contractors
3. Corporate industrial safety committee
4. Accident/incident reporting
5. Personal injury accident—investigation and analysis
6. Property accident—investigation and analysis
7. Procedure supplements
8. Recordkeeping
9. Development of standard operating procedures for potentially hazardous operations
10. Guidance for the preparation of standard operating procedures for potentially hazardous operations
11. Inspections or investigations by outside agencies
12. OSHA injury/illness recordkeeping and reporting requirements

Environmental health

1. Environmental health program
2. Radio-frequency radiation
3. Ionizing/nonionizing radiation
4. Respiratory protective equipment program
5. Hazardous materials control
6. Hazardous waste disposal form
7. HAZMAT data request form
8. Environmental pollution control
9. Industrial ventilation
10. Hearing conservation program
11. Noise control
12. Temperature stress

Safety

1. Occupational safety program
2. Laser safety
3. Electrical safety
4. Electrical safety requirements for test equipment and laboratories
5. Explosive safety
6. Applicable codes and regulations

7. Receipt and use log
8. Personal protective equipment
9. Wearing apparel
10. Facility design review
11. Fleet safety
12. Eye protection—safety glasses program
13. Machine guarding
14. Hand and portable power tool safety
15. General facility and equipment standards
16. Laboratory and specialized task safety
17. Confined space entry
18. Pressure systems
19. Pressure vessels
20. Hazard identification systems
21. Housekeeping program

Workers' compensation

1. Workers' compensation administration
2. Workers' compensation—rehabilitation

Health services

1. Health services program
2. Nursing program
3. Physical examination/biological monitoring program
4. Pregnancy
5. Sanitation
6. Alcohol and drug abuse
7. Toxic substances
8. Pulmonary function testing (prior to respirator use)

Fire safety

Material handling

1. Material handling
2. Manual material handling
3. Cranes and hoists
4. Lifting devices and special fixtures
5. Powered industrial trucks
6. Storage compatibilities
7. Truckjacks and wheel chucks
8. Tagging and removal of damaged containers

Education and training

1. Industrial safety bulletins
2. Industrial safety training

2.2.8.2 Industrial Safety Bulletins

Industrial safety procedures are relatively stable, that is, once established, major changes are not likely to occur frequently. Industrial safety bulletins are intended to provide current information that supplements or temporarily supersedes the procedures. The followintg list, while not necessarily all inclusive, should provide a basis for the development of your company's industrial safety bulletins.

Program policy statement
Abstracts of supervisors' responsibilities
Lifting devices and special fixtures
General health and safety regulations
Foot protection
Pregnancy
Food, beverage, and tobacco consumption

Safety glass program
Lead standard
Mechanical ventilation systems
Noise standard
Earthquake preparedness
Earthquake rules
Crane, hoist, and sling inspections
Special safety
Lift truck operator certification
Statement of health and safety responsibilities
Standby industrial safety engineer
Eye hazard in the home environment
Smoking in conference rooms
Blocking of aisleways
Storage and handling of flammable liquids
Shipment of hazardous materials/dangerous goods
Disposal of hazardous materials/dangerous goods
Movement of hazardous material/dangerous goods
Suspended ceiling panels
Good housekeeping and personal fire prevention habits
Think and practice life safety
Caution: aerosol spray cans
Fire prevention
Standards for storage, handling, and dispensing of flammable liquids
Fire corridors
Fire prevention—miniature immersion heaters
Machine guarding

2.2.9 PROCEDURE ABSTRACTS

Section 2.2.8.1 presented a list of subjects that your company should consider as industrial safety procedure topics. This section provides abstracts of what some of these procedures should include.

2.2.9.1 Accident/Incident Reporting

Supervisors directly responsible for work, property, or personnel should immediately report, by the most expeditious means, any accident to the industrial safety office. A follow-up form, such as the "Supervisor's Accident/Incident Notification," must be sent to the industrial safety office within 24 hours.

Accidents of a minor or marginal nature, which do not fall within the spirit and intent of this procedure (e.g., normal property attrition, manufacturing errors, functional part failure, scrap of low-value items, or minor occupational injuries/illness requiring only routine first aid which are recorded on medical dispensary records), should still be reported. The industrial safety office determines the necessity for further reporting and investigation.

2.2.9.2 Personal Injury Accident—Investigation and Analysis

Insures that all occupational injuries are promptly reported to the industrial safety office

Investigates accidents and takes necessary corrective action

Documents the investigation of individual accidents ("Supervisor's Report of Accident Investigation" portion of the "Employee's Report of Industrial Injury or Illness") when requested by the industrial safety manager

Assesses actual or potential accidents or accident conditions within areas of responsibility and assures their resolution

2.2.9.3 Property Accident—Investigation and Analysis

Insures that all accidents/incidents are promptly reported to the industrial safety office by the most expeditious means.

Investigates accidents/incidents ("Property Accident Investigation/Analysis") when requested by the industrial safety manager

Documents the investigation of accidents ("Property Accident Investigation/Analysis")

Assesses actual or potential accidents or accident conditions within their areas of responsibility and assures their resolution

2.2.9.4 OSHA Injury/Illness Recordkeeping and Reporting Requirements

Insures that all occupational injuries/illnesses are promptly reported to the industrial safety office

2.2.9.5 Environmental Health Program

Insures compliance with applicable regulatory environmental health standards and company environmental health procedures. In addition, the industrial safety office is consulted concerning interpretation of environmental health standards applicable to specific operations. It is the requesting organization's responsibility to obtain required approval from the industrial safety office regarding the purchase of all equipment or devices used for personnel health and safety and for prevention of property damage

Insures that facilities and equipment within their jurisdiction are maintained in safe operating condition and that appropriate environmental health safeguards are provided and utilized

Insures that employees under their supervision are properly trained in the safe performance of their duties; provides written departmental environmental health practices if required; requires that employees perform their duties in accordance with safe practices and initiates appropriate disciplinary action for violations of environmental health rules and regulations

Informs employees, by the use of labels, signs, posted information, group meetings, or other appropriate forms of warning, of all hazards to which they are exposed, relevant symptoms, appropriate emergency treatment, and proper conditions and precuations for personnel safety

2.2.9.6 Radio-Frequency Radiation

Insures that the industrial safety office is advised and consulted prior to the procurement of any radio-frequency (RF) generating equipment that can produce potentially hazardous fields of radiation

Advises the industrial safety office of equipment and/or procedural changes to existing operations that might affect employee exposure to RF radiation

Prepares standard operating procedures as required by the industrial safety office

2.2.9.7 Ionizing Radiation

Coordinates with the industrial safety office the procurement, storage, use, transfer, and disposal of sources of ionizing radiation

Requires, through the industrial safety office, a preexposure physical examination and dosimetry services for each employee subject to work in controlled radiation areas

2.2.9.8 Respiratory Protective Equipment Program

Requests assistance of the industrial safety office for evaluation of all operations or conditions that may produce potentially harmful atmospheres to which employees may be exposed. It is essential that evaluations be performed during pre-operational planning stages.

Insures that approved administrative and engineering controls are fully employed and employees are physically capable and trained to use and maintain appropriate personal respiratory protective equipment

Develops written procedures and obtains necessary approvals covering the use of respiratory protective devices

2.2.9.9 Hazardous Materials Control

Assures that employees are trained in safe work procedures and that they are adequately informed of the hazardous nature of the materials with which they work

Assures that adequate facilities and equipment are provided and maintained for the safe storage, use, and disposal of hazardous materials

Assures that all containers of hazardous materials, regardless of size, are adequately labeled in accordance with this procedure

2.2.9.10 Environmental Pollution Control

Assures that all operations are conducted in accordance with applicable environmental pollution control regulations

Supervision/training department performs necessary employee training in environmental pollution control with the assistance of the industrial safety office

2.2.9.11 Industrial Ventilation

Advises the industrial safety office of plans for use of materials that may require new or modified exhaust systems so that system requirements may be evaluated in relation to chemicals and equipment used and work processed.

Insures that use of local exhaust systems are limited to the purpose for which they were designed, installed, and approved by the industrial safety office. This includes insuring that exhaust systems are not modified or adjusted in ways that may affect proper operation of the system

Develops necessary safety and health operation practices in the use of specific local exhaust systems; provides procedures that forewarn maintenance and other personnel of the dangers associated with the performance of tasks at or near roof-level exhausts that may discharge hazardous levels of gases; vapors, mists, or fumes; insures that employees are adequately instructed concerning such practices

Notifies the maintenance department and the industrial safety office if it is known or suspected that a local system is not operating correctly

2.2.9.12 Occupational Safety Program

Insures compliance with applicable regulatory occupational safety standards and company industrial safety procedures. Industrial safety is consulted whenever interpretation of occupational safety standards applicable to specific operations is required. It is the requesting organization's responsibility to obtain required approval from the industrial safety office regarding the purchase of all equipment or devices used for personnel occupational safety and for prevention of property damage

Insures that facilities and equipment within their jurisdiction are maintained in safe operating condition and that appropriate occupational safeguards are provided and utilized

Insures that employees under their supervision are properly trained in the safe performance of their duties and provides written copies of departmental safety practices if required. Supervisors are also responsible for requiring employees to perform their duties in accordance with safe practices and initiate appropriate disciplinary action for violations of published company safety rules and regulations

Informs employees, by the use of labels, signs, posted information, group meetings, or other appropriate forms of warning, of occupational safety hazards to which they may be exposed and instructs them in precautions for personnel safety and emergency safety procedures

2.2.9.13 Laser Safety

Prepares laser safety standard operating procedures and precautions, as required, prior to laser installations

Coordinates facility equipment designs and laser safety standard operating procedures for the fabrication of new systems and changes in present systems/environments with the industrial safety office

Assures that all personnel assigned to work in areas where harmful laser radiation exposure may occur are laser safety-certified and include a baseline eye examination

Notifies immediately the cognizant medical service unit and industrial safety office of all injuries or suspected injuries from laser exposure

Notifies the industrial safety office prior to the transfer or termination of laser-certified personnel

Registers and licenses laser devices, systems, or installations if such is required by government agencies and coordinates such registration and licensing with the corporate industrial safety office

Assure that the design, installation, and operation of all laser systems and work areas are in accordance with the *American National Standard for the Safe Use of Lasers*, Z136.1-1973, and all applicable governmental regulations

2.2.9.14 Electrical Safety

Assures that electrical equipment and circuits are de-energized and locked out prior to any normal electrical maintenance or repair work

Assures that all qualified employees are provided an individually keyed lock and are properly trained in lock-out procedures

Assures, when it is necessary to work on energized systems, conductors, or equipment, that only qualified persons are assigned to do this work and that applicable rules and regulations are followed

Assures that when personal protective equipment is used or required, it meets the applicable codes

Assures that work on high-voltage electrical systems is performed by a qualified high-voltage electrical worker with an observer present who is a qualified high-voltage electrical worker or who is a trainee for such a position

Assures that trainees be permitted to work on high-voltage electrical systems only when specifically assigned by the electrical supervisor and that such trainees are continuously supervised by a qualified high-voltage electrical worker

2.2.9.15 Electrical Safety Requirements for Test Equipment and Laboratories

Assures that employees assigned to work with energized systems or equipment are trained in the hazards associated with such equipment

Assures that set procedures are followed

Assures that test or experimental operations are not left unattended while energized in such a manner that inadvertent contact would result in injury

2.2.9.16 Explosives Safety

Assures that explosives received are safely transported and stored in accordance with requirements of this procedure and local operating regulations

Maintains appropriate records of receipt and issue of explosives

Assures that shipping and receiving departments are notified of any explosives being shipped so that applicable regulations are met

Assures that explosives are issued only to authorized personnel who are properly trained in the use and handling of explosives

2.2.9.17 Personal Protective Equipment

Insures that employees use personal protective equipment and are informed of the specific requirements, value, accessibility, and proper use of such equipment

Insures that employees fully understand the purpose and limitations of the equipment and that employees use the equipment as directed

2.2.9.18 Wearing Apparel

Assures that employees are informed of specific wearing apparel requirements. Assures the enforcement of these requirements

2.2.9.19 Eye Protection—Safety Glasses Program

Identifies, in cooperation with the industrial safety office, potential eye hazard areas and work operations that require eye protection

Insures that employees receive and use appropriate eye protection when required

Instructs employees in the proper use and maintenance of eye protection equipment

2.2.9.20 Machine Guarding

Assures that all equipment in use has proper guarding

Assures that adequate periodic maintenance programs and recordkeeping systems are established and maintained

Assures that deficiencies are promptly corrected

Assures that regulatory requirements and company practices are incorporated and enforced

Assures that all machine guards in use are in proper working order

Assures that employees are properly trained in the care, use, and maintenance of machine guarding systems

Assures that employees who circumvent machine guarding systems are properly disciplined

2.2.9.21 Hand and Portable Power Tool Safety

Assures that employees are properly trained and proficient in the use of hand and portable power tools, that only tools specified are used, that tools are maintained in good working order, and that where required tools receive periodic servicing or calibration

2.2.9.22 Confined Space Entry

Develops specific procedures, for confined space entry

Insures that only trained personnel enter confined spaces or perform the safety standby function

Informs employees of the hazards of the specific confined space

Does not permit entry into a confined space until all provisions of this procedure have been met

2.2.9.23 Pressure Systems

Assures that only qualified operators operate pressure systems

Assures that adequate facilities and equipment are provided in areas where pressure systems are employed

Assures that all employees working with or around pressure systems are provided with and use personal protective equipment

Assures that standard operating procedures are prepared and used by personnel engaged in pressure system operation

Coordinates with the industrial safety office the acquisition, design, installation, and modification of pressure systems

Assures no maintenance or modification is performed on high-pressure systems or systems containing toxic gases until such work is adequately defined by documentation and approved by the industrial safety office

2.2.9.24 Pregnancy

Informs cognizant medical and/or industrial safety office of pregnant employees under their supervision

Assures compliance with subsequent work limitations established

Notifies industrial relations of all changes including termination of pregnancy and tentative return to work date

2.2.9.25 Material Handling

Assures that all material delivered is maintained in an appropriate location for the protection of property and employees

Assures that material handling equipment and facilities are suitable and safe for the purposes for which they are operated or used, maintained in safe operating condition, and in compliance with this procedure

Assures that all employees assigned material handling tasks are properly trained and provided with any required personal protective equipment

2.2.9.26 Industrial Safety Training

Insures that all tasks under their jurisdiction are performed by qualified persons and that tasks requiring performance by a certified person are performed only by certified persons

Arranges for necessary safety and health training and certificates for employees

Conducts on-the-job training of employees under their direction in the specific health and safety requirements and potential hazards associated with their duties

2.2.10 NOTICE TO SUPERVISION

One of the industrial safety bulletin topics listed in Section 2.2.8.2 is General Health and Safety Regulations. This particular bulletin should be distributed to all levels of supervision to insure their awareness regarding these employee responsibilities. A typical bulletin about this subject appears in Table 2.2.3.

Any questions on or interpretation of the above items should be directed to the industrial safety office.

2.2.11 HOUSEKEEPING PROGRAM

One of the industrial safety procedure topics listed in Section 2.2.8.1 is the Housekeeping Program. This program affects every employee, without exception, and therefore should be widely distributed among supervisors. Tables 2.2.4 and 2.2.5 provide examples of procedures your company could use with modifications needed to meet your particular circumstances.

REFERENCES

1. Russell De Reamer. *Modern Safety and Health Technology*. New York: Wiley, 1980.
2. Ted S. Ferry. *Safety Management Planning*. Santa Monica, CA: The Merritt Company, 1982.
3. William G. Johnson. *MORT Safety Assurance Systems*. New York: Dekker, 1979.
4. Ross A. McFarland. "Application of Human Factors Engineering to Safety Engineering Problems," *National Safety Congress Transactions*, 12, Chicago: National Safety Council, 1967.
5. Jack B. ReVelle, *Safety Training Methods*. New York: Wiley, 1980.

Table 2.2.3. General Health and Safety Regulations

1. Employees must report all accidents to their supervisors and obtain first aid promptly.
2. Employees must report unsafe conditions immediately to their supervisors. The supervisor-in-charge must take necessary action to correct hazardous conditions in a timely fashion and take appropriate action to protect personnel and equipment.
3. Employees will not willfully or knowingly perform an unsafe act or create an unsafe condition that exposes or potentially exposes himself/herself or other employees to a hazard or that will result in a work interruption or property damage.
4. Employees must wear clothing appropriate for the work being performed. Complete arm and/or leg covering may be required in some areas. Where protective clothing is required, its use is mandatory.
   - 4.1. All employees must wear eye protection where there is danger of foreign substances entering the eye. Eye protection is mandatory in all designated eye hazard areas and where specific machines are posted. Contact lenses may not be worn in eye hazard areas. Temporary issue safety glasses must be made available at eye hazard areas for use by casual vistors. These glasses should be issued from the area supervision.

     Industrial safety glasses should be available to all employees at an optical center (see Industrial Safety Bulletin "Obtaining Safety Glasses" for details)
   - 4.2. Employees performing work that exposes their feet to injury from falling materials or objects must wear adequate foot protection. Open-toed shoes, high heels, tennis (running) shoes, and cloth shoes should not be allowed in shop, assembly, and lab areas.
   - 4.3. Employees working in locations where there is a hazard from falling objects must wear approved head protection. Where there is a risk of injury from hair entanglement, employees must confine their hair to eliminate the hazard.
   - 4.4. Employees whose work exposes them to potentially harmful dusts, mists, or vapors must wear approved respirators. Respirators for the specific hazard involved should be available from the industrial safety office.

     NOTE: Personnel protective equipment supplied by the company should be obtained by supervisors from the company store. The need for protective equipment is determined by supervisors in conjunction with the industrial safety office. Respirators require prior approval, fitting, and training, which should be provided by the industrial safety office.
5. Employees are not permitted to wear loose clothing, wrist watches, rings, bracelets, or other jewelry when working with moving machinery or exposed, energized electrical circuits.
6. Employees must not be permitted to operate machines or perform potentially hazardous work until they are properly trained by responsible supervision. Equipment posted for use by authorized operators must not be used without the approval of the supervisor in charge.
7. Only qualified and authorized employees shall modify or make repairs to machinery or equipment.
8. No employee shall remove, or make ineffective any lockout/tagout, safeguard, safety device or appliance, except as required for authorized maintenance or repair.
9. Use of compressed air for for removing dust or other material from clothing or body is forbidden.
10. Employees must not use chairs, boxes, or other improvised platforms to stand on in place of approved ladders, work stands or scaffolding.
11. Flammable liquids used in work operations shall be kept in approved flammable containers and labeled properly.
12. No material or equipment shall be stored in aisles or corridors within 36 inches of power panels, within 18 inches of fire sprinkler heads, or in such manner as to obstruct access to fire extinguishers, stretchers, or other emergency equipment.
13. Flexible extension cords may not be used as a substitute for permanent wiring. Extension cords over six feet in length may not be used under any circumstances without prior approval of the fire department and industrial safety office.
14. Employees are not to consume food, beverages, or tobacco where hazardous materials (e.g., flammable materials, hazardous chemicals, dusts, fumes) are used.
15. The in-plant speed limit for all vehicles is 10 mph.
16. The use of seat belts (restraint devices) is mandatory in all company vehicles operated on public streets and highways.

Table 2.2.4. Company Housekeeping Program

1. Purpose

   This practice establishes the company housekeeping program, standards, and responsibilities for providing and maintaining a clean and orderly working environment.

2. Housekeeping program/responsibilities
   - 2.1. The housekeeping program consists of a set of standards that are applied and enforced by all levels of management and supervision to insure compliance (see housekeeping standards attachment following).
   - 2.2. In order to assist management in implementing and auditing the program, housekeeping committees are established to implement the program and provide for routine inspection of work areas.
   - 2.3. In the interest of promoting efficiency, protecting sensitive information, and maintaining a proper working environment, supervisors are responsible for the following:
     - a. Indoctrinating employees in housekeeping standards and procedures,
     - b. Maintaining assigned areas in proper/good housekeeping conditions,
     - c. Verifying that documents and equipment are stored in appropriate areas or containers when not in use,
     - d. Ensuring that classified and company private documents are properly stored and adequately protected according to established security requirements,
     - e. Taking corrective action when housekeeping violations are observed or reported, and
     - f. Submitting violation notices or inspection reports directly to the company housekeeping coordinator.

3. Company housekeeping committee
   - 3.1. A company housekeeping committee is established to implement the housekeeping program by providing a forum for
     - a. exchanging ideas,
     - b. reviewing significant problems and their solutions, and
     - c. defining and developing policies and standards.
   - 3.2. The company committee chairman
     - a. Is appointed by the president's office,
     - b. Establishes an overall housekeeping program,
     - c. Determines the representation required from each functional area,
     - d. Schedules and presides over periodic committee meetings,
     - e. Conducts program audits for the company and for each functional area at least once a year, and
     - f. Reports program deficiencies to management.
   - 3.3. Committee representatives
     - a. Are appointed by each functional manager, and
     - b. Serve as coordinator for each area's committee.

4. Area housekeeping committees
   - 4.1. Area housekeeping committees are directly responsible for implementing the housekeeping program in their area.
   - 4.2. Area managers act as chairs and appoint a committee coordinator to manage the committee.
   - 4.3. An area committee coordinator
     - a. Establishes an area housekeeping program,
     - b. Schedules and presides over periodic meetings,
     - c. Determines required representation,
     - d. Conducts monthly inspections,
     - e. Serves as a representative to the company housekeeping committee,
     - f. Reports housekeeping violations to supervision and management,
     - g. Distributes copies of the inspection record or equivalent form as follows:

Original	Area supervisor
Duplicate	Area manager
Duplicate	Retained for recordkeeping

     - h. Maintains area housekeeping records and makes records available to the company committee chair during audits,
     - i. Provides a monthly summary of activities to the company committee,
     - j. Arranges to have company material handling/packaging engineering perform a parts protection audit once a year (this is applicable for all areas that process deliverable materials),
     - k. Arranges to have other support functions, such as an industrial safety engineer or fire protection officer, assist in inspections, and
     - l. Responds to housekeeping violations brought to his/her attention by industrial safety or fire protection.

Table 2.2.5. Housekeeping Standards Attachment

1. General areas
   1.1. Areas are maintained in a clean and orderly manner. Floors, storage areas, and stairways are maintained in a safe condition and free of debris.
   1.2. Aisles are kept clear of clutter, protruding objects, and safety hazards, such as chairs, stools, wastebaskets, cartons, pallets. Boxes and cartons for janitorial pickup are plainly marked "trash" and placed in aisles not more than 30 minutes before the end of work shift.
   1.3. Outside storage areas are maintained in an orderly, clean condition.
   1.4. Walls, partitions, doors, and windows are kept clean and free of reference data, notes, charts, and nonjob-related material. Reference data is utilized in accordance with approved standards, that is, in picture frames, on bulletin boards, or affixed to tack strips.
   1.5. Calculators, typewriters, and other office machines are dusted by the operator at periodic intervals. Machines are promptly covered when not in use.
   1.6. Furniture is arranged in such a manner that it does not obstruct aisles, doorways, or building service units, such as fire extinguishers, switch panels, and controls.
   1.7. Individual file cabinets and desk drawers are kept closed when not in use and attended. Materials are not stored on top of file cabinets.
   1.8. Chairs and stools are placed under or adjacent to a desk or table when not in use.
   1.9. Desk tops or conference tables are kept reasonably clear of materials. During nonwork hours, reference manuals, books, desk trays, calendar pads, and ash trays are neatly arranged on the desk. At the end of the working day, letters and documents are placed in desk drawers and file cabinets.
   1.10. Workstations are cleared of all material except that which is in immediate use. Storage of parts, supplies, jigs, and other materials is permitted on shelves under workbenches.
   1.11. Articles of clothing are placed on coat racks and are not hung on stools or chairs.
   1.12. Reference materials, charts, and company tools are returned to approved storage areas when no longer required for immediate use.
   1.13. Safety hazards, such as nails, jagged corners, improper extension cords, hanging wires, and unsanitary conditions noted in the restrooms and other areas, are immediately reported to plant maintenance.
   1.14. Chalkboards are used for work purposes only.
   1.15. Decorative plants are acceptable in office areas as long as they do not detract from or interfere with the work environment or create a safety hazard.
2. Drafting and technical illustration areas
   2.1. Storage of material on the tops of cabinets or under drafting tables is not permitted. Covers are stored neatly when the tables are in use.
   2.2. Rolled drawings are stored in a neat and orderly manner.
   2.3. During nonworking hours, drafting tables are cleaned of all items other than unclassified drawings, drafting machines, and table lamps.
3. File areas
   3.1. File cabinet tops are kept clean and clear of all material and are not to be used for storage of material. Files are arranged in a neat and orderly manner and as uniform in size and color as possible.
   3.2. Counter tops are kept clean and clear of material other than that which is in immediate use or is standard for counter installation.
   3.3. Cabinets, files, desks, and other equipment are kept reasonably clear and free of dust.
4. Manufacturing work areas
   4.1. Equipment, both stationary and mobile, is kept clean and free of dirt, dust, grease, and scrap material.
   4.2. Bin tops, racks, and other equipment are kept clean and clear at all times. Hazardous protrusions from racks, bins, trucks, et cetera, are immediately corrected.
   4.3. Parts, jig boards, detailed assemblies, and other items not in use are stored neatly and in such a manner that damage to the unit is prevented.
   4.4. Decorative plants are not allowed in manufacturing work areas.
5. Engineering work areas
   5.1. Laboratory areas are cleared of excess supplies, equipment, and empty containers.
   5.2. Laboratory equipment is kept reasonably clean and free of dust and residue.
   5.3. Reference data, books, log books, and other documents are properly placed in approved storage areas when not in use.
   5.4. Test equipment and product items are packaged, stored, transported, and used in a manner that prevents damage or degradation.
   5.5. Decorative plants are not allowed in lab areas.
6. Conference rooms
   6.1. Conference rooms are maintained in a clean and orderly manner.
   6.2. Plastic cups, stirring sticks, et cetera, are disposed of in trash containers.
   6.3. Empty ash trays are stacked on the top of the conference table.
   6.4. When no longer required, information appearing on chart boards is erased and chart presentations and other reference material are removed from easels, walls, and partitions.

BIBLIOGRAPHY

"Accident Prevention: Your Key to Controlling Surging Workers' Compensation Costs," *Occupational Hazards* (November 1979), p. 35.

"Accident: Related Losses Make Cost Soar," *Industrial Engineering* (May 1979), p. 26.

"Analyzing a Plant Energy-Management Program: Part I—Measuring Performance," *Plant Engineering* (October 30, 1980), p. 59.

"Analyzing a Plant Energy-Management Program: Part II—Forecasting Consumption," *Plant Engineering* (November 13, 1980), p. 149.

"Anatomy of a Vigorous In-Plant Program," *Occupational Hazards* (July 1979), p. 32.

"A Shift Toward Protective Gear," *Business Week* (April 13, 1981), p. 56H.

"A Win for OSHA," *Business Week* (June 29, 1981), p. 62.

"Computers Help Pinpoint Worker Exposure," *Chemecology* (May 1981), p. 11.

"Complying With Toxic and Hazardous Substances Regulations—Part I," *Plant Engineering* (March 6, 1980), p. 283.

"Complying With Toxic and Hazardous Substances Regulations—Part II," *Plant Engineering* (April 17, 1980), p. 157.

"Conserving Energy by Recirculating Air From Dust Collection Systems," *Plant Engineering* (April 17, 1980), p. 151.

"Control Charts Help Set Firm's Energy Management Goals," *Industrial Engineering* (December 1980), p. 56.

"Controlling Noise and Reverberation With Accoustical Baffles," *Plant Engineering* (April 17, 1980), p. 131.

"Controlling Plant Noise Levels," *Plant Engineering* (June 24, 1976), p. 127.

"Cost–Benefit Decision Jars OSHA Reform," *Industry Week* (June 29, 1981), p. 18.

"Cost Factors for Justifying Projects," *Plant Engineering* (October 16, 1980), p. 145.

"Costs, Benefits, Effectiveness, and Safety: Setting the Record Straight," *Professional Safety* (August 1975), p. 28.

"Costs Can Be Cut Through Safety," *Professional Safety* (October 1976), p. 34.

"Elements of Effective Hearing Protection," *Plant Engineering* (January 22, 1981), p. 203.

Engineering Control Technology Assessment for the Plastics and Resins Industry, NIOSH Research Report Publication No. 78-159.

"Engineering Project Planner: A Way to Engineer out Unsafe Conditions," *Professional Safety* (November 1976), p. 16.

"EPA Gears up to Control Toxic Substances," *Occupational Hazards* (May 1977), p. 68.

"Fume Incinerators for Air Pollution Control," *Plant Engineering* (November 13, 1980), p. 108.

"Groping for a Scientific Assessment of Risk," *Business Week* (October 20, 1980), p. 120J.

"Hand and Body Protection: Vital to Safety Success," *Occupational Hazards* (February 1979), p. 31.

"Hearing Conservation—Implementing an Effective Program," *Professional Safety* (October 1978), p. 21.

"How Do You Know Your Hazard Control Program Is Effective?" *Professional Safety* (June 1981), p. 18.

"How to Control Noise," *Plant Engineering* (October 5, 1972), p. 90.

"Human Factors Engineering—A Neglected Art," *Professional Safety* (March 1978), p. 40.

"New OSHA Focus Led to Noise-Rule Delay," *Industry Week* (June 15, 1981), p. 13.

"OSHA Communique," *Occupational Hazards* (June 1981), p. 27.

"OSHA Moves Health to Front Burner," *Purchasing* (September 16, 1979), p. 46.

"OSHA to Analyze Costs, Benefits of Lead Standard," *Occupational Health & Safety* (June 1981), p. 13.

Patty's Industrial Hygiene and Toxicology. Vol. I, 3rd rev. ed. New York: Wiley-Interscience, 1978.

"Practical Applications of Biomechanics in the Workplace," *Professional Safety* (July 1975), p. 34.

"Private Sector Steps up War on Welding Hazards," *Occupational Hazards* (June 1981), p. 50.

"Reducing Noise Protects Employee Hearing," *Chemecology* (May 1981), p. 9.

"Regulatory Relief Has Its Pitfalls, Too." *Industry Week* (June 29, 1981), p. 31.

"ROI Analysis for Cost-Reduction Projects," *Plant Engineering* (May 15, 1980), p. 109.

"Safety & Profitability—Hand in Hand," *Professional Safety* (March 1976), p. 36.

"Safety Managers Must Relate to Top Management on Their Terms," *Professional Safety* (November 1976), p. 22.

"The Cost/Benefit Factor in Safety Decisions," *Professional Safety* (November 1978), p. 17.

"The Economics of Safety . . . A Review of the Literature and Perspective," *Professional Safety* (December 1977), p. 31.

"The Design of Manual Handling Tasks," *Professional Safety* (March 1980), p. 18.

"The Hidden Cost of Accidents," *Professional Safety* (December 1975), p. 36.

"The Human Element in Safe Man–Machine Systems," *Professional Safety* (March 1981), p. 27.

"The Problem of Manual Materials Handling," *Professional Safety* (April 1976), p. 28.

"Tips for Gaining Acceptance of a Personal Protective Equipment Program," *Professional Safety* (March 1976), p. 20.

"Were Engineering Controls 'Economically Feasible'?" *Occupational Hazards* (January 1981), p. 27.

"Were Noise Controls 'Technologically Feasible'?" *Occupational Hazards* (January 1981), p. 37.

"What Are Accidents Really Costing You?" *Occupational Hazards* (March 1979), p. 41.

"Where OSHA Stands on Cost–Benefit Analysis," *Occupational Hazards* (November 1980), p. 49.

"Worker Attitudes and Perceptions of Safety" (Part 1). *Professional Safety* (December 1981), p. 28.

"Worker Attitudes and Perceptions of Safety" (Part 2). *Professional Safety* (January 1982), p. 20.

CHAPTER 2.3

MAN–MACHINE SYSTEMS

William B. Rouse

Georgia Institute of Technology
Atlanta, Georgia

2.3.1 INTRODUCTION

Humans become involved with machines in a variety of ways. They *operate* machines such as bicycles, drill presses, automobiles, and computers. Humans also *maintain* these machines. Beyond operations and maintenance, humans *manage* organizations of humans and machines. And, of course, humans also *design* and *fabricate* machines.

Considering the range of human activities that fall within the categories of design, fabrication, operations, maintenance, and management, it seems quite reasonable to claim that virtually any problem is, at least in part, a man–machine problem. As a result, in order to understand fully the nature of most problems, devise possible solutions, and successfully implement the chosen solution, one must consider the potential impact of a variety of man–machine issues. This chapter discusses these issues and how they can be addressed.

2.3.1.1 System Perspective*

The nature of man–machine interaction is dominated by the fact that humans and machines typically interact to form a system whose purpose is to satisfy a system-oriented requirement. The behavior of humans and the operation of machines involved in this process are only fully meaningful in the context of the man–machine system. Thus, we are not so much interested in understanding humans or machines individually but understanding a system within which humans represent one or more components.

While the system perspective is necessary to understanding man–machine interaction, it is not sufficient to assure success. One also must keep in mind that a human is not "just another component." Humans exhibit a very high degree of adaptability and tend to modify their behavior to meet the demands of the environment. As a result, one cannot describe human characteristics in the static manner that is possible for electrical and mechanical components. A human can be appropriately described as an information processor that dynamically adapts its processing to reflect the requirements dictated by its tasks and overall environment.

In order to understand man–machine systems, one must understand the abilities and limitations of the human information processor as well as the demands placed on the processor by the machines in the system and how environmental factors affect both humans and machines. This chapter discusses the state of the art of knowledge in these areas.

2.3.1.2 Overall Issues

One can approach man–machine systems issues on several levels, as shown in Figure 2.3.1. At the lowest level, one is concerned with *environmental compatibility*, insuring that humans can survive and function successfully within the man–machine systems environment. Short-term and long-term occupational safety and health issues are paramount at this level. The field of *ergonomics* is traditionally associated with this level.

At the next highest level, one is interested in insuring that the *input/output* requirements of man–machine systems are compatible with human abilities and limitations. Thus, displays and controls within man–machine systems must be consistent with sensing (input) and affecting (output) characteristics of humans. These types of issue are usually studied within the field of *human factors*.

Information processing, the next level, involves what happens between input and output. Here the concern is with humans' abilities to solve problems, make decisions, and execute appropriate responses. One is also

* For additional information, see references 33, 40, 55, and 64.

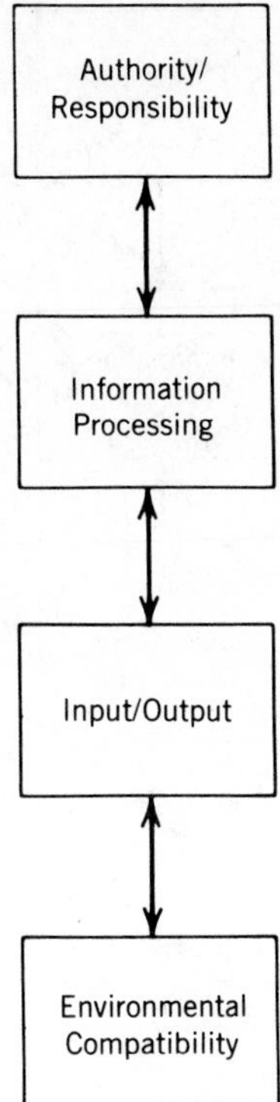

Fig. 2.3.1. Levels of man–machine systems issues.

interested in the way in which task requirements affect the availability of humans' information processing resources. These topics usually are pursued within the field of *experimental psychology*, in general, or *engineering psychology*, in particular.

The highest level in Figure 2.3.1, *authority/responsibility*, deals with who is in control, man or machine? This issue reflects trends in computer technology that have resulted in "intelligent machines" that can perform many tasks once thought to be solely the province of humans. It is now possible to delegate authority for certain tasks to computers; but is it possible, or will it be possible, to delegate responsibility to computers? Probably not, but, then, how can one be certain that humans who are divested of authority will continue to accept responsibility? These are difficult questions, which are being addressed in many fields of study. These questions are reconsidered later in this chapter.

The study of man–machine systems is concerned with all of the issues shown in Figure 2.3.1, not in and of themselves but to the extent that each has a practically significant impact on system functions. From this perspective, any particular issue may or may not be relevant to a specific system and problem. Nevertheless, it is important for those who pursue man–machine systems problems to have a broad appreciation and understanding of the range of issues that are potentially relevant, as presented in this chapter.

2.3.1.3 Overview

The topics discussed in this chapter are summarized in Figure 2.3.2. Section 2.3.2 discusses input/output in terms of sensing, affecting, and endogenous (internal) and exogenous (external) factors. This section primarily considers the environmental compatibility and input/output levels.

Section 2.3.3 discusses information processing from both a basic psychological viewpoint and a task-oriented perspective. Tasks discussed include control, decision making, problem solving, and those associated with human–computer interaction. The considerations in this section are mainly on the information processing level.

Section 2.3.4 reconsiders many of the topics discussed in Sections 2.3.2 and 2.3.3 but emphasizes system design rather than human abilities and limitations. In a sense, this section describes the "technology" of man–machine systems. This technology is essential if the "science" discussed in Sections 2.3.2 and 2.3.3 is to have an impact on actual man–machine systems.

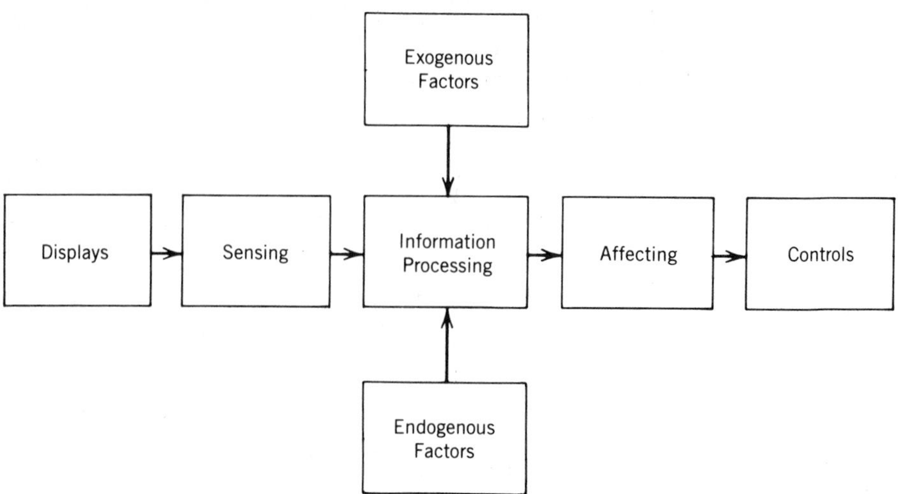

Fig. 2.3.2. Relationships of man–machine systems topics.

The fifth and final section of this chapter summarizes the state of the art and what appear to be inherent limitations to progress in man–machine systems. Also, the highest level issues in Figure 2.3.1 are reconsidered, and prospects for dealing with these issues are discussed.

2.3.2 INPUT/OUTPUT

This section is concerned with human abilities and limitations in sensing inputs via displays and affecting outputs via controls and how endogenous and exogenous factors affect human performance.

2.3.2.1 Sensing: Input*

There are a variety of sensing systems that enable humans to receive inputs from their environment. *Vision* and *hearing* are the primary means, and *touch*, *smell*, *taste*, and *kinesthetic sense* are also important.

Vision is probably the most extensively studied human sense. A considerable amount of practical information is available on visual acuity, image quality and legibility, eye movements and visual search, motion perception, size and distance perception, dark adaptation, and color vision. Kantowitz and Sorkin[13] have recently summarized the knowledge in this area.

Hearing has also been studied extensively. A significant amount is known about sensitivity and discriminability; effects of loudness, pitch, distortion, and nonlinearity; and the nature of binaural hearing. Kantowitz and Sorkin[13] also provide a recent summary of knowledge in this area.

The other human senses have received less attention, at least in a man–machine systems context, because they are not relied upon as heavily as vision and hearing. A slight exception is kinesthetic sense (i.e., sensation of position, movement, etc.), which is important in vehicle systems such as aircraft and spacecraft. One important concern in this area is insuring that humans do not become kinesthetically disoriented.

2.3.2.2 Affecting: Output†

The primary task-oriented outputs produced by humans are *speech* and *motion*. Other possible human "outputs" include electromyographic signals and cortical responses, but the practical use of such outputs is, at this point, very specialized and often speculative. Therefore, this section only considers speech and motion.

Speech has been reasonably well studied. Articulation and intelligibility of both human and synthetic speech are fairly well understood, as are the effects of intensity, distortion, and channel quality. The effects of context, predictability, and familiarity are more subtle and quite important to understanding speech as opposed to just hearing words. Kantowitz and Sorkin[13] provide a recent review of these topics.

There are several aspects of motion that are important. *Anthropometrics* is concerned with the space required to accommodate humans, both statically and functionally. Extensive data are available on the distribution of

* For additional information, see references 1, 13, 15, 16, 17, 26, 55, 61, and 69.

† For additional information, see references 1, 9, 13, 15, 16, 17, 26, 28, 37, 61, and 69.

static and functional dimensions of the human body. Beyond having the space to execute required motions, humans also have to have the strength, which has implications for energy requirements. Information relevant to these various aspects of motion is reviewed by Huchingson[9] and Konz[15].

2.3.2.3 Endogenous Factors*

Beyond task-oriented abilities and limitations of humans, other factors intrinsic to humans that affect their performance include *aptitude*, *ability*, *style*, *attitude*, and *motivation*. These human characteristics tend to modulate (either positively or negatively) humans' behaviors.

Aptitude refers to natural inclinations or the potential to acquire ability. Ability denotes both natural and acquired skills. Style, often referred to as "cognitive" style, is the way in which someone approaches a task (e.g., impulsively vs. reflectively). Reasonably standard tests exist for assessing aptitude, ability, and style; however, interpretation of results can be difficult.

Attitude refers to a human's disposition and is a broader concept than style in that it need not be task-oriented. Motivation can relate to internally generated incentives or motives or it can result from reactions to externally defined goals. Bailey[1] provides a recent review of motivation as it relates to man–machine systems.

The endogenous factors noted in this section are generally recognized to be of potential importance in many man–machine system contexts; but very little is usually done to consider these types of factor at a level of rigor commensurate with the other factors discussed in this chapter. This is perhaps a shortcoming of the field but also may reflect an inadequate scientific base upon which to build.

2.3.2.4 Exogenous Factors†

The environment within which a man–machine system functions often includes factors that are not intrinsic to the tasks being pursued by the humans and machines. Nevertheless, these factors can substantially affect humans' abilities to perform. The machines also can be affected, but this is not the concern of this chapter.

Inappropriate levels of *noise*, *illumination*, and *temperature* can substantially degrade human performance and reduce the amount of time that humans can continue to perform tasks. And, of course, extremes can make it impossible for humans to function in the environment. Kantowitz and Sorkin[13] provide a thorough review of these factors.

Vibration, *motion*, *acceleration*, and *weightlessness* are factors that are often important in vehicle systems. Effects range from disorientation to complete inhibition of performance to biomechanical damage. Huchingson[9] discusses many of the issues related to these factors.

Safety is a general concern in man–machine systems, with *biochemical* and *radiological hazards* being particularly problematic. While factors such as these tend to be long-term medical issues rather than human performance concerns, it is very important for designers of man–machine systems to be aware of their potential effect. Huchingson[9] and Konz[15] review the effects of these factors.

Another exogenous factor is *organization*. The nature of the organization within which a man–machine system functions can substantially affect the humans in the system and is closely related to the endogenous factors of attitude and motivation.

Exogenous factors are of particular concern to the extent that limits may be reached beyond which performance degrades or is inhibited. Within these limits, the effects of exogenous factors are often ignored. This can be problematic when the limits for negative long-term effects are much narrower than the limits for short-term effects.

2.3.2.5 Displays‡

Most displays in man–machine systems are of three types: visual, auditory, and tactile. Visual displays are, by far, the most common and most thoroughly researched.[1,3,9,13,16,17,26,28,61] Bailey[1] and Kantowitz and Sorkin[13] provide recent overviews of the wealth of material available on visual displays.

One can consider visual displays in terms of four general attributes. The first, *display technology*, can range from traditional, hardwired electromechanical displays to a variety of computer-generated displays. The second category of attributes is *physical characteristics* of a display, including size, resolution, luminance, contrast, and color. The last three items are particularly important for computer-generated displays.

A third attribute is *display type*, which may be variable-oriented or pattern-oriented displays. Variable-oriented displays are used to show specific temperatures, pressures, flow rates, and so forth. In contrast, pattern-oriented displays emphasize relationships and trends. Computer-generated displays are particularly useful for pattern-oriented displays.

Displays may also be static or dynamic. Man–machine systems in the domains of vehicle control and process control are such that their states change with time, often very rapidly. Such dynamic environments require displays appropriate for indicating changes and trends.[14,30,37,55]

* For additional information, see references 1, 5, and 37.

† For additional information, see references 1, 5, 9, 13, 15, 16, 17, 26, 28, and 61.

‡ For additional information, see references 1, 3, 9, 13, 14, 16, 17, 26, 28, 30, 37, 52, 56, and 61.

Finally, displays may be analog or digital. This distinction does not refer to the physical nature of the display device but to the way information is displayed. An excellent example of this distinction, taken from everyday life, is the wristwatch. Traditionally, watches provided analog displays, and one could tell, at a glance, the approximate time. More recently, watches with digital displays have allowed one to tell, with somewhat more effort, the exact time. In general, digital displays are good for exact readings of static or slowly changing variables. For somewhat qualitative, and often pattern-oriented readings, analog displays are better.

The fourth attribute of interest is *display format*. This includes coding display elements via color, size, shape, or mnemonic (i.e., memory aids) and scaling and labeling display elements. The arrangement of display elements also is important. Display integration[13] can be very useful for economizing on display space and making arrangement easier. This is particularly important for computer-generated displays where, in contrast to traditional display panels, the number of display elements that can be viewed simultaneously is limited by the size of the display surface.

While there has been considerable research on auditory displays,[1,9,13,16,17,26,61] they are used much less frequently than visual displays. This is due to the limited (or underdeveloped) capacity of the auditory channel (i.e., one picture can be worth very many spoken words). However, for simple "messages," auditory displays can be very appropriate. For example, auditory caution, warning, and alarm indicators are very good for attracting a human's attention. Synthesized speech can provide richer auditory inputs and, of course, auditory communication between humans is a central activity in many man–machine systems. Kantowitz and Sorkin[13] provide a recent overview of auditory displays.

Tactile displays are a rarity outside of developmental laboratories. This is due to the fact that the capacity of the cutaneous (i.e., touch) channel is even more limited or underdeveloped than the auditory channel. Nevertheless, there are some common applications, namely, braille communication of language and, to a lesser extent, aircraft applications such as the "stick shaker" for warning of possible stalls. Kantowitz and Sorkin[13] provide a brief review of tactile displays.

2.3.2.6 Controls*

While displays are the means by which humans sense inputs, controls are their means for affecting outputs. There are three general types of controls: tools, dedicated controls and control systems, and general-purpose input devices. There are important human factors issues associated with the geometry, grips, and so forth of hand tools,[1,13,15,17,26,61] but these issues are seldom, if ever, central from an overall man–machine systems perspective.

Until recently, dedicated controls and control systems were the norm in man–machine systems. Alternative controls include knobs, levers, switches, and buttons. Shape coding, size coding, labeling, and protection from inadvertent operation are considerations in choosing controls. Selection of controls also depends on whether inputs are to be discrete or continuous, linearly or rotationally oriented, and single or multidimensional and on any accessibility limitations where the controls are to be installed. Kantowitz and Sorkin[13] provide a fairly thorough overview of dedicated controls.

Control systems provide the means by which a human affects the state of a dynamic system, such as a vehicle or a production process.[14,30,37,55] As elaborated in Section 2.3.3.2, in a dynamic system human control actions are affected by the rates of change of displayed variables (e.g., position, velocity, and acceleration of temperature, pressure, and flow rates). In order for the human control actions to affect directly these rates of change, these actions as well as the display variables may have to be transformed to compensate for human limitations in sensing rates of change and implementing complicated control rules. In addition, special controls, such as joysticks or trackballs, are often necessary. Kelley[14] and Sheridan and Ferrell[55] provide thorough treatments of control systems.

General-purpose input devices are increasingly common, primarily due to the widespread use of general-purpose computer systems. Keyboards are the most common, but touch displays, tablets, mice, and light pens are also frequently used. Common problems with multipurpose devices such as these include their relative inefficiency for any particular purpose and also the possibility that the human will confuse one purpose with another and make inappropriate choices as a result. General-purpose input devices are discussed by Huchingson[9] and Kantowitz and Sorkin.[13]

2.3.3 INFORMATION PROCESSING

Section 2.3.3 discussed many "classical" issues associated with man–machine systems, namely, inputs, outputs, and endogenous and exogenous factors that may affect the relationships between inputs and outputs. Most of these issues are in the bailiwick of ergonomics, human factors, and to an extent, engineering psychology. These issues are important, but are not usually the central focus of most efforts in man–machine systems. Instead, the primary emphasis in man–machine systems tends to be the process whereby inputs from the machine are transformed into outputs by the human.

* For additional information, see references 1, 9, 13, 14, 15, 16, 17, 26, 28, 30, 37, 55, 59, and 61.

More specifically, man–machine systems are concerned with the processing of information by humans in order to produce behavior that achieves system performance objectives. This topic can be approached on several levels. At the physiological level, one can consider neural information processing; however, it is difficult to assess or predict man–machine system performance at this level. At the higher level of basic psychology, one can think in terms of elements of human information processing. Finally, one can take a task-oriented perspective and consider information processing in the context of manual and supervisory control, decision making and problem solving, and human–computer interaction. This section addresses the basic psychology and task-oriented levels of information processing.

2.3.3.1 Elements of Human Information Processing*

For the first half of this century, psychology frowned on investigations that attempted to go beyond simple stimulus–response mappings. Speculation about internal information processing mechanisms was considered unscientific. Fortunately, the last two or three decades have seen a rapid erosion of this strict behaviorist point of view. Perhaps the best exposition of the current state of thinking, particularly as it relates to man–machine systems, is the recent book by Wickens.[62]

The currently accepted view is that human information processing involves several stages. Inputs, or stimuli, are temporarily held in *short-term sensory storage*. The next stage is perception, which involves the use of *long-term memory* for interpretation of stimuli. The outputs of perception serve as inputs for *decision and response selection*, a stage of processing that relies heavily on *short-term memory*. The final stage is *response execution*, which, as the phrase implies, produces outputs such as motion and speech. The processes of perception, decision and response selection, and response execution, as well as the use of short-term memory, are all affected by the human's allocation of *attentional resources*.

Beyond being intellectually more satisfying than the behaviorists' strict stimulus–response perspective, the above view of human information processing has important practical implications. For example, when considering human performance in multitask situations, it is important to be able to predict whether tasks will be independent, conflicting, or perhaps supportive of each other.[41] Research in information processing psychology, using the above conceptual framework, has provided important insights into the subtleties of attentional resources and the impact on multitask performance.[13,62]

A very important consideration in the design of most man–machine systems is workload. While ergonomics and traditional industrial engineering have focused on *physical workload*, current trends in man–machine systems interpret the humans' roles as being more cognitive, which shifts the emphasis to *mental workload*.[5,13,24,25,47,62,65,67] The most comprehensive source on this topic is Moray,[24] with more recent but briefer treatments available in Wickens[62] and Kantowitz and Sorkin.[13] While there is a variety of views on mental workload and no clear consensus (except that mental workload is multidimensional and complicated), virtually all of the available conceptualizations have borrowed heavily from information processing psychology. Further, the most viable progress is being made by investigations working within human information processing frameworks.

While the information processing perspective outlined above appears to be sound theoretically, it does have practical limitations. The most important of these is the difficulty of applying its fine-grained perspective to realistically complex tasks. Further, there is no convenient way to predict the effects of the task characteristics and demands in a contextually meaningful manner. To overcome these limitations, a more task-oriented perspective must be integrated with the more elemental view of human information processing.

2.3.3.2 Manual and Supervisory Control†

Control is the process of causing the state of a system (or a subset of the state) to follow a desired trajectory as closely as possible considering the effort and costs involved. The word "state" refers to a set of variables (e.g., positions, velocities, temperatures, and pressures) whose *current* values are sufficient to allow prediction of *future* values if one has a knowledge of the system dynamics and inputs.[40,55] Control of a system involves using *feedback* of the current state to synthesize a control input to the system that will cause the system to reach a desired future state. If the process of observing the system state and generating a control input is performed by a human, the task is termed *manual control*. In contrast, if this process is performed by a machine, it is called *automatic control*.

Humans' abilities in manual control have been studied in great depth. The effects of system complexity, response time, input characteristics, and alternative displays are well documented and readily available for the practitioner.[13,30,55,62] In fact, manual control is so well codified that an impressive set of quantitative models of performance are available and have been extensively used in design.[39,40,55]

Somewhat ironically and perhaps inevitably, the increasingly refined knowledge base on manual control behavior generally is becoming less important as many manual activities in engineering systems are becoming automated. As a result, humans are now monitoring automatic control systems, intervening only when problems occur, such as failures or unforeseen events. This type of role for humans is termed *supervisory control*. It is a very active research area with emphasis on issues related to levels of automation, decision-making and problem

* For additional information, see references 1, 4, 13, 16, 55, and 62.

† For additional information, see references 5, 10, 13, 14, 16, 26, 30, 39, 40, 46, 55, 56, 61, and 62.

solving aids, and display design.[35,46,56] A variety of mostly qualitative models have thus far emerged for applications in vehicle and process systems as well as command and control.[40,46,55]

Most manual and supervisory control models and data are related to vehicle and process systems. Most vehicle systems efforts have been for aircraft, automobiles, and ships, with the majority of the effort having been focused on aircraft systems.[2,9,37,39,41,56] Process systems efforts have included both continuous and batch production processes and power plant operations.[2,6,7,27,31,36,56,62] In general, there is a rather impressive body of knowledge available for manual control of vehicle and process systems and a substantial but still emerging body of knowledge for supervisory control in these and other domains.

2.3.3.3 Decision Making and Problem Solving*

While manual control tasks rely primarily on humans' psychomotor abilities, supervisory control tasks rely more heavily on humans' decision-making and problem-solving abilities. Therefore, progress in supervisory control depends not only on a better understanding of the nature of supervisory tasks (as noted above) but also on increased knowledge of human decision making and problem solving.

Human decision making has been a topic of study for a very long time. For centuries, researchers have been concerned with how humans choose among alternatives and make judgments. Utility theory resulted from this work, and more recently, subjective expected utility theory was developed to account for humans' deviations from normative utility theory. Very recently, prospect theory has provided psychological explanations for these deviations. Considerable related work has been done in areas such as probability estimation, probability updating, and other aspects of statistical inference. Wickens[62] and Kantowitz and Sorkin[13] provide recent reviews of this work; an older but more thorough treatment of this material can be found in Sheridan and Ferrell.[55]

The research noted in the above paragraph considers human decision making in general. If one focuses on human decision making in the context of man–machine systems, the general research results can be structured in a more useful manner. Several taxonomies, or classification schemes, have recently emerged that take advantage of the goal-oriented nature of man–machine systems. Further, these taxonomies serve to broaden the range of issues covered to include human problem solving, which can be viewed as a much more general type of task than decision making in the sense of simply selecting among alternatives. This expansion of scope is very important because the trend in man–machine systems is to view humans as "ultimate backup systems," who intervene, take control, and find solutions whenever problems arise that exceed the capabilities of the automatic systems. From this perspective, humans not only have to select among alternatives; they also have to generate and evaluate alternatives.

Building on the work of Rasmussen,[32–34] Wohl,[68] and many others, Rouse[48,52] has shown how decision making and problem solving includes three general tasks: *situation assessment*, *planning and commitment*, and *execution and monitoring*. Situation assessment is composed of two phases, information seeking and explanation. These four general tasks can be decomposed into 13 more specific tasks, 11 of which involve generation of, evaluation of, and selection among alternative *information sources*, *explanations*, *courses of action*, and *acceptability of state deviations*. The remaining two specific tasks involve implementing and monitoring plans. This taxonomy, or classification, of decision-making and problem-solving tasks has been used as a basis for developing a methodology for the design of decision aids in general[48] and adaptive aids in particular.[52] Applications thus far have focused on process control[48] and command and control systems.[52]

If the predominant role of humans in emerging and future man–machine systems is that of problem solver, it is useful to focus on that role. This allows the task taxonomy to be much more specific and include *detection*, *diagnosis*, *compensation*, and *correction*. A considerable amount of research has been directed toward human abilities in detection and diagnosis, particularly when the "problems" of interest are system failures.[35,43] Compensation, in the sense of normal manual control, has been studied as a task itself; however, the much more difficult multitask problem of coordinating compensation and diagnosis (i.e., simultaneously maintaining operation and attempting to solve problems) has only recently received much attention.[43]

As trends in communications and computer technology enable large-scale, integrated systems that are highly dependent on automation, the tasks faced by the humans who monitor and supervise these systems have the potential for occasionally being very complex. It may not be reasonable to expect humans to cope with this complexity, particularly in light of the low frequency with which they will have to intervene in the highly reliable automatic operations. It is quite easy to imagine human performance being substantially degraded when a very complex task is rarely practiced. This situation is providing difficult challenges for those who design training programs and operational aids, and the man–machine systems community is currently very active in these areas.[35,52]

2.3.3.4 Human–Computer Interaction†

Computer technology is not only affecting humans' role in man–machine systems but also the way humans interact with their machines. Increasingly, humans are sitting in front of visual display units (VDUs) and inputting via keyboards and, occasionally, mice or touch panels, or voice. This type of human–machine

* For additional information, see references 1, 5, 13, 19, 35, 40, 54, 55, and 62.

† For additional information, see references 1, 4, 5, 6, 13, 18, 23, 26, 38, 57, and 61.

interface allows considerable flexibility and adaptability in that a single type of workstation can be used for many tasks.

There are also disadvantages. Succinctly, a general solution to many problems is not always the best solution for any particular problem. As a result, users may have to learn about many modes, conventions, and pitfalls before they can be at all productive using the system. This can be, at the very least, frustrating and may lead to inefficiencies and errors.

Computer technology may have a solution for these problems via artificial intelligence. Much current research is focusing on systems that can explain themselves as well as identify and provide context-specific explanations for human inefficiencies and errors. While this trend is needed, it does not come without potential problems. It is quite possible that natural intelligence (i.e., humans) will come to depend too heavily on artificial intelligence, making it impossible for humans to fill their role as ultimate backup systems. This question is currently of great concern.[63]

A variety of guidelines is available for design of the human-computer interface. Three types of tasks have received considerable attention: data entry,[1,9,13,26,61] text editing,[4,13,18,23] and programming.[5,13,18,23,57] Less frequently researched tasks include data base searching,[18,49,61] management information systems,[18,53] computer-based instruction,[12,18,42] and natural language interfaces.[18,57] Broadly applicable guidelines are provided by Smith[60] and Williges and Williges.[66]

2.3.4 SYSTEM DESIGN

In the context of man–machine systems, the primary reason for attempting to understand human input/output characteristics and information processing abilities and limitations is to be able to design such systems better. The material discussed in Sections 2.3.2 and 2.3.3 should be important ingredients in the design process. However, this information is not inherently structured in a process-oriented manner. As a result, the products of research in man–machine systems tend to be answers to questions that designers seldom realize they should ask. To minimize the possibility of this unfortunate result, methodologies are needed whereby consideration of man–machine systems issues is systematically integrated within the design process. This section discusses these types of methodology.

2.3.4.1 The Design Process

Figure 2.3.3 illustrates the design process that serves as the basic structure for much of the discussion of system design in this section. Before elaborating on the elements of this structure, two general points should be made. First, the orderly process depicted in Figure 2.3.3 is very much a simplification; in practice, there is much iteration among the steps of the process. Second, it is a bit of a misnomer to label this figure "the" design process, since many authors have used different terminology and, occasionally, different orderings of the steps. Despite these two caveats or precautions, the process shown in Figure 2.3.3 provides an excellent basis for integrating man–machine systems considerations into system design.

A system is designed in order to achieve one or more *objectives*. Defining objectives involves answering several questions: What is the purpose of the system? Who will be involved with the system? Where will the system operate? When will the system be operated? For many engineering systems, particularly in the military, these questions are all combined into a single question: What are the mission requirements?

While all systems are designed to achieve objectives, these objectives are not always well defined and operationally useful. In some cases the objectives are not even documented. These types of deficiency can lead to a rather inefficient design process at best and possibly to design solutions that are inadequate or inappropriate. Obviously, the design process is enhanced when everyone involved clearly knows what the design effort is trying to accomplish.

Once objectives have been defined, the next step is identification of *constraints*. The achievement of objectives can be constrained by the availability of technology, limitations on resources, schedule requirements, and organizational factors. These considerations may limit the range of alternative design solutions and may require the modification of objectives if there is no feasible solution within the boundaries of the constraints.

Objectives are achieved by synthesizing a set of *functions* that produce the desired system behaviors.[20] For example, minimal functional requirements for a personal transportation system (e.g., an automobile) might include propulsion, braking, and steering. In general, functions specify *what* must be provided to achieve objectives; the questions of *how* and *who* are not resolved until after functions are determined.

Tasks are the sequences of activities that provide required functions.[1,4,19,59] For example, to provide the steering function in an automobile, one must sense the state of the automobile relative to the highway, determine an appropriate sequence of control actions, and execute these actions. These tasks (which might be characterized as situation assessment, planning and commitment, and execution and monitoring) describe *how* objectives are achieved.

Once tasks are determined, the next step involves analysis of *information and control requirements*. This involves determining necessary inputs and outputs for each task as well as how inputs will be sensed and outputs affected. For example, in order to steer an automobile, lateral deviations, relative heading, velocity, and so on, must be sensed and the steering mechanism and accelerator pedal affected appropriately.

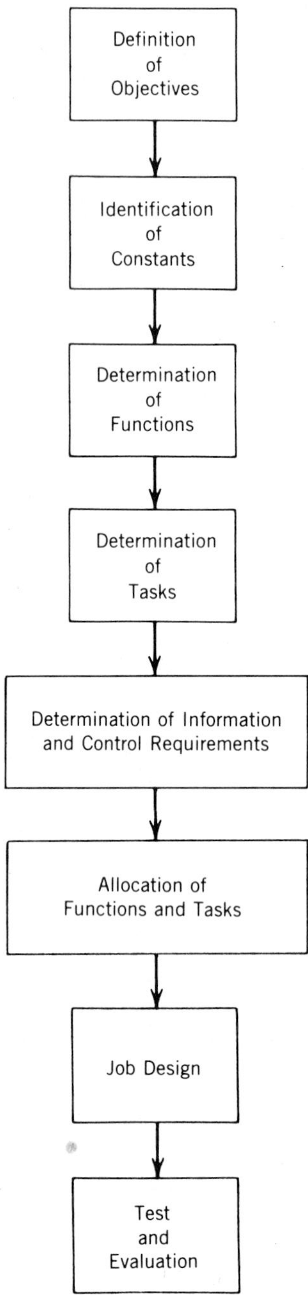

Fig. 2.3.3. The design process in man–machine systems.

The next step in the design process involves the question of *who* will perform each task. The *allocation of functions and tasks* among humans and machines is a central issue in the design of man–machine systems.[1,41,59] Two concerns govern the resolution of this issue. First, one must assess or predict the relative abilities of humans and machines to perform each task. The material discussed in Sections 2.3.2 and 2.3.3 can be quite helpful in this regard. Further, many tasks simply must be allocated to machines (e.g., in an automobile, those associated with the propulsion function); for others, there is no technological alternative to humans (e.g., steering in traffic). Nevertheless, considering the "intelligence" of many forthcoming machines, there is an increasingly larger set of tasks that could be performed by either humans or machines.

The question, therefore, becomes one of choosing the best mix of tasks for humans. This leads to the second concern: information processing workload. While humans may be able to perform acceptably a large variety of individual tasks, the number of tasks that humans can perform simultaneously is much smaller due to the fact that multiple tasks are likely to compete for the same limited human information processing resources.[41,62] Thus, one should assess or predict the workload associated with any candidate mix of tasks to insure that acceptable performance is achievable and that excessive workload does not potentially lead to problems such as fatigue and stress.

Given one or more candidate mixes of tasks to be allocated to humans, the next step of the process is *job design*. This involves the integration of functions and tasks allocated to humans into one or more jobs.[15] If there is more than one job, then more than one human is required and the organizational structure of this crew must be determined.

At this point in the design process, one should know what tasks each human in the system will perform and the information and control requirements for those tasks. The next consideration is *workspace layout*.[13,17,26,28,61,69] This aspect of job design involves determining approximate locations for displays and controls so as to produce an appropriate workflow within the physical constraints imposed by the environment. One particularly important consideration in this analysis is the need for crew members to share information, perhaps verbally.

With the workspace layout roughly defined, one then can proceed to *workstation design*.[1,8,9,13,26,28,61,69] This involves determining the dimensions of each workstation so as to accommodate both static and functional anthropometric characteristics of the population of humans involved and also considering physical limitations inherent in the environment. Workstation design also includes selection and arrangement of display and control elements, which involves determining the *how* and *where* of these elements rather than the *what* that should previously have emerged from the analysis of information and control requirements.

The final step of the design process is *test and evaluation*.[20,21,44,48] If one views design as a process of developing a plan to solve a problem, then test and evaluation can be considered to have two phases: verification and validation. *Verification* is the process of determining that the plan has been fabricated (or perhaps programmed) as intended. *Validation* involves determining whether the plan is an acceptable solution of the design problem. In other words, verification is concerned with insuring that the system was constructed as intended, and validation focuses on the extent to which the intentions were appropriate. While verification is important and can be time consuming, validation is a more difficult evaluation problem. Rouse has recently considered this problem from both an empirical[44] and analytical[48] perspective and suggested a multilevel approach to validating system designs.

2.3.4.2 Major Design Issues

There is a tendency to view the design of a man–machine system, as well as the design of the machine itself, in isolation from several considerations that are, in fact, crucial to the success of the system. For example, design quite naturally focuses on *operability*, since that is the way in which design objectives are achieved. Of equal importance, however, at least in the sense of potentially being an inhibiting factor, is *maintainability* in particular and *supportability* in general.[5,13,26,35,59,61] The aerospace industry and the military refer to the "life cycle" of a system and attempt to account for all of the activities and costs associated with the system. It is not unusual to find that life-cycle costs are higher than procurement costs, to a great extent due to supportability costs.

System *staffing*[5,19] is also a factor that is likely to receive insufficient attention. While function and task allocation results in a determination of what humans will do in the system, this does not directly determine the necessary characteristics of these humans. These characteristics include knowledge and skills required for certain tasks and physical attributes (e.g., strength or visual acuity) dictated by these tasks. Once these characteristics are determined, the question then becomes one of finding (via selection) or creating (via training) such individuals.

Selection is the process of recruiting, evaluating, and placing personnel.[1,15,16] The purpose of selection is to insure that those selected inherently have the required characteristics or, if that is not feasible, that they have the potential to develop the necessary characteristics. A variety of testing and evaluation methods are available; unfortunately, their diagnosticity is not as great as one would hope.

Training involves providing personnel with the knowledge and skills necessary to perform the tasks associated with their jobs.[1,5,12,15,16,26,35,37,42,59,61] Designing a training program involves determining both content and methods. *Content* defines what will be taught; *methods* define *how* it will be taught. Determining the knowledge and skills required by personnel is not as straightforward as one might imagine. For example, a recurring controversy concerns the extent to which operations and maintenance personnel need to understand the physical theories (e.g., chemistry and physics) that formed the basis for design of the system with which they are involved. Intuition and empirical results in this area strongly conflict, and training practices, therefore, are inconsistent across organizations and even among instructors within a single organization.

Considering training methods, traditional classroom lectures are more and more heavily augmented by simulator methods in many industries. Until recently, emphasis has been on full-scope, high-fidelity simulators. However, the costs of such devices have grown substantially. Further, there are serious doubts about the usefulness of such simulators for all aspects of training. As a result, interest has increased in using lower-fidelity devices in conjunction with high-fidelity simulators or real equipment to achieve a mixed-fidelity approach to training.[42]

Aiding is an important alternative (or complement) to training.[1,35,52,59] Rather than attempting to insure that humans have all necessary knowledge and skills prior to actual job performance, it is possible to assist them on the job with approaches that range from job performance aids, such as procedures, to "intelligent" assistants, such as provided by "expert systems." Another possibility is *adaptive aiding*, which modifies the level and nature of the assistance depending on the state of the task and/or human being trained.[52] Yet another possibility is to combine training and aiding to provide online instruction and assistance. This concept is referred to as *embedded training*.

Given that one has limited resources and can invest them in selection, training, or aiding, how should the allocation be determined? While this seems like an obvious question, it has, until recently, gone unaddressed. This is due to the fact that these issues cut across organizational boundaries, at least in terms of different departments and often in terms of different institutions. An approach to dealing with this large-scale man–machine systems problem has recently been proposed,[45] but much work needs to be done to refine and validate this formulation.

2.3.4.3 Design Methods*

As noted in the introduction of Section 2.3.4, the area of man–machine systems tends to be "long" on data and "short" on methodology. Nevertheless, there are a variety of methodological components available that can be used within the framework provided by Figure 2.3.3.

The empirical tradition is very strong within man–machine systems; hence, a variety of *empirical methods* have been developed and used extensively.[15,20–22] The range of methods includes those for field studies, simulator studies, and laboratory investigations as well as the classical methods of time and motion study. A recent analysis of the wealth of empirical studies of man–machine systems provided several insights regarding the frequency with which different methods are associated with successful studies.[51] In general, methods that allow more experimental control and measures that provide more fine-grained assessments tend to be more frequently associated with definitive results.

Modeling methods are also frequently employed in the design of man–machine systems. A model is a representation of how a system is assumed to function and enables one to answer questions about how that system will respond to various situations and events. As a design method, models are used to predict the performance of a man–machine system as a function to design parameters of interest.

The process of using a model to predict performance can be pursued either analytically or with simulation. Simulation approaches fall into two classes.[22,58] There are simulations, such as the Human Operator Simulator (HOS), that require only specification of requirements and model parameters, and there are simulation languages, such as SAINT or SLAM, that require one to develop the model from basic modeling primitives. Simulations tend to be very useful over rather limited ranges of applicability. In contrast, simulation languages have a much wider range of applicability but require much more effort on the user's part.

Analytical approaches to modeling are much "cleaner" and avoid the overhead of repetitive Monte Carlo runs to obtain statistical accuracy. Many problems, however, are analytically intractable. Further, analytical approaches usually require considerable experience on the part of the user. Nevertheless, when possible and practical, analytical models are usually preferred to simulation. There are a wide range of analytical models available for predicting biomechanical response, human behavior and performance, and human cognitive processing.[11,22,29,35,40,55]

Designers in many domains are concerned with finding the "best" design solution. As a result, many *optimization techniques* have emerged. While such techniques have only seen limited applications to man–machine systems, there have been notable exceptions with efforts to integrate link analysis and mathematical programming[22] and development of control theoretic approaches to display design.[39] The previously cited effort on optimal resource allocation among manpower-oriented activities[45] also fits into this category. While the complexity of man–machine systems may preclude the formulation of all aspects as an optimization problem, there is definitely a need for more systematic and analytical approaches for managing and resolving the many tradeoffs associated with the design of a man–machine system.

Test and evaluation methods are particularly important in the design of man–machine systems.[1,5,16,20–22,26,30,44,48,57,61] While the basic concepts and practice of experimental design are very important and essential to evaluation, additional methods are needed to deal with the nature of man–machine systems. Recent work[44,48] has viewed the evaluation problem as actually being three problems: compatibility, understandability, and effectiveness. *Compatibility* is concerned with insuring that the human–machine interface is compatible with human input/output abilities and limitations. *Understandability* reflects a requirement for meaningful "messages" between humans and machines. *Effectiveness* involves assessing whether the information transmitted between humans and machines is useful. Depending on which of these three problems is of interest, different evaluation methods are appropriate.[44,48]

* For additional information, see references 16, 18, 22, and 59.

2.3.4.4 Measurement Techniques*

Test and evaluation as well as the validation of models requires that measurements be made. There is a wide range of measurement techniques available. Due to the nature of human cognition and behavior, many of these techniques provide only surrogates for what one would actually like to measure.

An exception to this generality is *activity measurement*. If one is solely concerned with how the human moves his or her limbs, or which displays and controls they access, then direct measurement is possible. On the other hand, eye movements and voice recordings ae typically employed to supply information that will allow one to infer the variables of interests.

The indirect nature of measurement is even more evident with *physiological measurement*. Heart rate, skin resistance, EMG, and EEG are common choices for physiological measures. Using such measures to infer workload, stress, and attention provides, at best, very indirect insights into these constructs.

Subjective assessments are frequently used. Alternative approaches include questionnaires, interviews, observations, verbal protocols, and rating scales. For some of these alternatives, the word "subjective" refers to assessments provided by the person being studied; in others, it refers to assessments by an outside observer. Bailey[1] provides a recent review of many of the issues involved in using these methods.

One area in which measurement is a particularly important issue is *workload assessment*.[24,25,65,67] There is a continuing debate concerning the usefulness of various objective measures and the validity of subjective measures. Indeed, with all of the measures, it is not exactly clear how workload relates to what is measured. This is a good example of how an intuitively clear concept becomes rather complicated when one attempts to address it scientifically. This is a common problem in man–machine systems.

2.3.5 CONCLUSIONS

This chapter has reviewed the state of the art in man–machine systems, with emphasis on the current understanding of humans' abilities and limitations in information processing and the design of man–machine systems. In general, the field has produced a wealth of data and a variety of useful design concepts. However, the development of methodology has not been sufficiently emphasized, often with the result that excellent and relevant data are ignored by designers.

While there are many traditional man–machine systems issues that have not been totally resolved, there are several new issues that demand immediate and considerable attention. The impact of computer and communications technology has been an accelerating trend toward large-scale, complex systems whose normal operations are almost toally automated. Many off-normal operations are also automated, and humans are required only when something unusual occurs. However, it is not clear that humans can fill this role. It is likely that humans will have to be more involved if they are to feel responsible for system operations and be able to intervene and assume control authority when necessary. From this perspective, it may be "optimal" to use less automation than is technologically feasible. The question of what should and should not be automated, therefore, becomes a central issue.[63]

Many efforts in automation have been motivated by a desire to eliminate "human error." Indeed, for most engineering systems, human error accounts for 60% to 90% of all consequential incidents.[50] With a little reflection,[45] it is easy to show that the line of reasoning which leads to this conclusion is untenable. Certainly, human operators and maintainers are often "agents" of errors; however, only occasionally are they the causes of errors.

But the issue goes beyond who or what is to be blamed. The notion of eliminating error is fundamentally counterproductive. Humans are included in systems because they are flexible, adaptive information processors who can innovate when necessary. This enables humans to fulfill their role as ultimate backup systems.

Unfortunately, humans' innovations are not always appropriate, and occasionally a "human error" results. These errors can be eliminated by establishing fixed operational procedures and providing interlocks that prohibit unusual uses of displays and controls. However, this approach will also prevent humans from innovating when necessary.

To avoid having to forego this essential human ability, the emphasis should shift from error reduction to error tolerance. Systems should be designed to provide sufficient feedback to humans to enable them to detect and reverse inappropriate actions or, for irreversible situations, to enable them to "try out" their intentions before implementing them. With this approach, the occurrence of errors will not be as important because their consequences will not be allowed to propagate.

In summary, three issues are currently of much interest and deserve concerted attention. First, man–machine systems greatly need more and better design methodology. Second, a broader and longer-term perspective needs to be taken on the question of what is the appropriate level of automation. Finally, human error needs to be reconsidered and the emphasis shifted to design of error-tolerant systems.

* For additional information, see references 1, 20, 21, 22, and 44.

REFERENCES

1. R. W. Bailey. *Human Performance Engineering, A Guide for Systems Designers*. Englewood Cliffs, NJ: Prentice-Hall, 1982.
2. S. Baron, Ed. "Special Issue on Man–Machine Systems." *Journal of Cybernetics and Information Science* (Summer–Fall–Winter 1980).
3. R. Bernotat and K. P. Gaertner. *Displays and Controls*. Amsterdam: Swets and Zeitlinger, 1972.
4. S. K. Card, T. P. Moran, and A. Newell. *The Psychology of Human–Computer Interaction*. Hillsdale, NJ: Lawrence Erlbaum Associates, New York: 1983.
5. K. B. De Greene, Ed. *Systems Psychology*. New York: McGraw-Hill, 1970.
6. E. Edwards and F. P. Lees. *Man and Computer in Process Control*. London: Institute of Chemical Engineers, 1973.
7. E. Edwards and F. P. Lees. *The Human Operator in Process Control*. London: Taylor & Francis, 1974.
8. P. R. Frey, W. H. Sides, R. M. Hunt, and W. B. Rouse. *Computer-Generated Display System Guidelines*, Volume I: *Display Design*, Electric Power Research Institute, Palo Alto, CA, 1984, Report No. NP 3701, Volume 1.
9. R. D. Huchingson. *New Horizons for Human Factors in Design*. New York: McGraw-Hill, 1981.
10. G. Johannsen, H. E. Boller, E. Donges, and W. Stein. *Der Mensch in Regelkreis, Lineare Modelle*. Munich: R. Oldenbourg Verlag, 1977.
11. G. Johannsen and W. B. Rouse. "Mathematical Concepts for Modeling Human Behavior in Complex Man–Machine Systems." *Human Factors* (December 1979), pp. 733–747.
12. W. B. Johnson, M. E. Maddox, W. B. Rouse, and G. C. Kiel. *Diagnostic Training for Nuclear Plant Personnel*. Electric Power Research Institute, Palo Alto, CA, 1984, Report No. NP 3829, Volume 1.
13. B. H. Kantowitz and R. D. Sorkin. *Human Factors: Understanding People–System Relationships*. New York: Wiley, 1983.
14. C. R. Kelley. *Manual and Automatic Control*. New York: Wiley, 1968.
15. S. Konz. *Work Design*. Columbus, OH: Grid Publishing, 1979.
16. K. F. Kraiss and J. Moraal, Eds. *Introduction to Human Engineering*. Federal Republic of Germany: Verlag TUV, 1976.
17. E. J. McCormick and M. S. Sanders. *Human Factors in Engineering and Design*. New York: McGraw-Hill, 1982.
18. C. T. Meadow. *Man–Machine Communication*. New York: Wiley-Interscience, 1970.
19. D. Meister. *Human Factors: Theory and Practice*. New York: Wiley-Interscience, 1971.
20. D. Meister. *Behavioral Foundations of Systems Development*. New York: Wiley-Interscience, 1976.
21. D. Meister and G. F. Rabideau. *Human Factors Evaluation in System Development*. New York: Wiley, 1965.
22. J. Moraal and K. F. Kraiss, Eds. *Manned Systems Design: Methods, Equipment, and Applications*. New York: Plenum Press, 1981.
23. T. P. Moran, Ed. "Special Issue: The Psychology of Human–Computer Interaction." *ACM Computing Surveys* (March 1981).
24. N. Moray, Ed. *Mental Workload, Its Theory and Measurement*. New York: Plenum Press, 1979.
25. N. Moray. "Subjective Mental Workload." *Human Factors* (February 1982), pp. 25–40.
26. C. T., Morgan, J. S. Cook III, A. Chapanis, and M. W. Lund, Eds. *Human Engineering Guide to Equipment Design*. New York: McGraw-Hill, 1963.
27. N. M. Morris. *The Human Operator in Process Control: A Review and Evaluation of the Literature*, Center for Man–Machine Systems Research, Georgia Institute of Technology, Atlanta, GA, 1982, Report No. 82-1.
28. K. F. H. Murrell. *Human Performance in Industry*. New York: Reinhold Publishing, 1963.
29. R. W. Pew, S. Baron, C. E. Feehrer, and D. C. Miller. *Critical Review and Analysis of Performance Models Applicable to Man–Machine Systems Evaluation*, Bolt Beranek and Newman, Cambridge, MA, 1977, Report No. 3446.
30. E. C. Poulton. *Tracking Skill and Manual Control*. New York: Academic Press, 1974.
31. *Proceedings of Workshop on Cognitive Modeling of Nuclear Plant Control Room Operators*, Dedham, MA, August 15–18, 1982, Oak Ridge National Laboratory, Oak Ridge, TN, 1982, NUREG/CR-3114.
32. J. Rasmussen. *On the Structure of Knowledge—A Morphology of Mental Models in a Man–Machine Context*, Riso National Laboratory, Roskilde, Denmark, 1979, RISO-M-2192.
33. J. Rasmussen. "The Human as a System Component." In H. T. Smith and D. Green, Eds., *Man–Computer Research*, New York: Academic Press, 1979.

34. J. Rasmussen. "Skills, Rules, and Knowledge; Signals, Signs, and Symbols and Other Distinctions in Human Performance Models." *IEEE Transactions on Systems, Man, and Cybernetics* (May/June 1983), pp. 257–266.

35. J. Rasmussen and W. B. Rouse, Eds. *Human Detection and Diagnosis of System Failures*. New York: Plenum Press, 1981.

36. J. E. Rijnsdorp and W. B. Rouse. "Design of Man–Machine Interfaces in Process Control." In *Digital Computer Applications to Process Control*, H. R. Van Nauta Lemke and H. R. Verbruggen, Eds. New York: North Holland, 1977.

37. S. N. Roscoe, Ed. *Aviation Psychology*. Ames: Iowa State University Press, 1980.

38. W. B. Rouse. "Design of Man–Computer Interfaces for On-Line Interactive Systems." *Proceedings of the IEEE*, June 1975, pp. 847–857.

39. W. B. Rouse, Ed. "Special Issue on Applications of Control Theory in Human Factors." *Human Factors* (August and October 1977).

40. W. B. Rouse. *Systems Engineering Models of Human–Machine Interaction*. New York: North-Holland, 1980.

41. W. B. Rouse. "Human–Computer Interaction in the Control of Dynamic Systems." *ACM Computing Surveys* (March 1981), pp. 71–100.

42. W. B. Rouse. "A Mixed-Fidelity Approach to Technical Training." *Journal of Educational Technology Systems* (1982), pp. 103–115.

43. W. B. Rouse. "Models of Human Problem Solving: Detection, Diagnosis, and Compensation for System Failures." *Automatica* (November 1983), pp. 613–625.

44. W. B. Rouse. *Computer-Generated Display System Guidelines*, Volume II: *Developing an Evaluation Plan*, Electric Power Research Institute, Palo Alto, CA, 1984, Report No. 3701, Volume 2.

45. W. B. Rouse. "Optimal Allocation of System Development Resources to Reduce and/or Tolerate Human Error." *IEEE Transactions on Systems, Man, and Cybernetics*, 1985.

46. W. B. Rouse, Ed. *Advances in Man–Machine Systems Research*, Volume 1. Greenwich, CT: JAI Press, 1984.

47. W. B. Rouse, Ed. *Advances in Man–Machine Systems Research*, Volume 2. Greenwich, CT: JAI Press, 1985.

48. W. B. Rouse, P. R. Frey, and S. H. Rouse. *Classification and Evaluation of Decision Aids for Nuclear Power Plant Operators*, Search Technology, Norcross, GA, 1984, Report No. 8303-1.

49. W. B. Rouse, J. M. Hammer, and D. R. Morehead. *Modeling of Human Behavior in Seeking and Generating Information*, Center for Man–Machine Systems Research, Georgia Institute of Technology, Atlanta, GA, 1983, Report No. 83-1.

50. W. B. Rouse and S. H. Rouse. "Analysis and Classification of Human Error," *IEEE Transactions on Systems, Man, and Cybernetics* (July/August 1983), pp. 539–549.

51. W. B. Rouse and S. H. Rouse. "A Note on Evaluation of Complex Man–Machine Systems," *IEEE Transactions on Systems, Man, and Cybernetics* (July/August 1984), pp. 633–636.

52. W. B. Rouse and S. H. Rouse. *A Framework for Research on Adaptive Decision Aids*, Air Force Aerospace Medical Research Laboratory, Wright-Patterson Air Force Base, OH, 1983, Report No. AFAMRL-TR-83-082.

53. W. B. Rouse and S. H. Rouse. "Human Information Seeking and Design of Information Systems," *Information Processing and Management* (1984), pp. 129–138.

54. A. P. Sage. "Behavioral and Organizational Considerations in the Design of Information Systems and Processes for Planning and Decision Support," *IEEE Transactions on Systems, Man, and Cybernetics* (September 1981), pp. 640–678.

55. T. B. Sheridan and W. R. Ferrell. *Man–Machine Systems: Information, Control, and Decision Models of Human Performance*. Cambridge, MA, MIT Press, 1974.

56. T. B. Sheridan and G. Johannsen. *Monitoring Behavior and Supervisory Control*. New York: Plenum Press, 1976.

57. B. Shneiderman. *Software Psychology*. Cambridge, MA: Winthrop Publishers, 1980.

58. A. I. Seigel and J. J. Wolf. *Man–Machine Simulation Models*. New York: Wiley, 1969.

59. W. T. Singleton, R. S. Easterby, and D. C. Whitfield, Eds. *The Human Operator in Complex Systems*. London: Taylor & Francis, 1967.

60. S. L. Smith. *User–System Interface Design for Computer-Based Information Systems*, Electronic Systems Division, Air Force Systems Command, Hanscom Air Force Base, MA, 1982, Report No. ESD-TR-82-132.

61. H. P. Van Cott and R. G. Kinkade. *Human Engineering Guide to Equipment Design*. Washington, DC: U. S. Government Printing Office, 1972.

62. C. D. Wickens. *Engineering Psychology and Human Performance*. Columbus, OH: Charles E. Merrill, 1984.

63. E. L. Wiener and R. E. Curry. "Flight-Deck Automation: Promises and Problems." *Ergonomics* (1980), pp. 95–1101.
64. N. Wiener. *Cybernetics or Control and Communication in the Animal and the Machine*. New York: Wiley, 1948.
65. W. W. Wierwille. "Physiological Measures of Aircrew Mental Workload." *Human Factors* (October 1979), pp. 575–593.
66. B. H. Williges and R. C. Williges. *User Considerations in Computer-Based Information Systems*, Virginia Polytechnic Institute and State University, Blacksburg, VA, 1982, Report No. CSIE-81-2.
67. R. C. Williges and W. W. Wierwille. "Behavioral Measures of Aircrew Mental Workload." *Human Factors* (October 1979), pp. 549–574.
68. J. G. Wohl. "Force Management Decision Requirements for Air Force Tactical Command and Control." *IEEE Transactions on Systems, Man, and Cybernetics* (September 1981), pp. 618–639.
69. W. E. Woodson and D. W. Conover. *Human Engineering Guide for Equipment Design*. Berkeley: University of California Press, 1964.

CHAPTER 2.4

HUMAN FACTORS

BERNHARD ZIMOLONG*

University of Bochum
Bochum, West Germany

GAVRIEL SALVENDY

Purdue University
West Lafayette, Indiana

The purpose of this chapter is to acquaint the practicing engineer with the field of human factors, to present basic human factors data, and to show how to apply them to the design and management of production systems.

The first section provides an overview of the human factors profession, then reviews basic human capabilities. This is followed by design principles of visual displays and controls, task analysis, workplace design, and job design. In addition, the effects of environmental variables and the nature of human reliability on design and process variables are also discussed. Further, the application of these design principles to computer-aided design (CAD) and manufacturing (CAM) as well as to robotics is outlined. A bibliography is provided at the end of the chapter to enable the reader to obtain an in-depth understanding of the various methodologies and principles discussed here.

2.4.1 HUMAN FACTORS: DISCIPLINE, PROFESSION, AND GOALS

The human factors field emerged as a composite of specialists in enhancing human performance, training, safety, and the design of man–machine interfaces. Initially, the emphasis was on cockpits, consoles, controls, and displays in the military, and on work efficiency (time and motion) in industry. The emphasis changed to people in the system as system engineering. Recently a great deal of interest in the human within computer-based systems has developed.[1]

Two major disciplines contributed to the development of human factors: industrial engineering, concerned with the measurement of work output, and psychology, interested in the factors of information input, decision making, and motor performance output. Present-day human factors, or *ergonomics*, as it is called in most European countries, includes medicine, physiology, psychology, anthropology, physics, computer science, and engineering. The objectives of human factors are

1. to enhance the effectiveness and efficiency of human performance at work, and
2. to maintain or enhance certain desirable human values for people at work: health, safety, satisfaction, personal challenge, and growth.

The human factors specialist, working as a professional in the design process, has to take into account the following four major areas for improving tasks and jobs:

Equipment. Design and creation of the physical characteristics of the equipment to which humans have to respond (e.g., displays and controls in aircraft, surface transportation, industrial systems).

* This chapter was written while the author was a Heisenberg Fund (Deutsche Forschungsgemeinschaft) recipient and a visiting scholar in the School of Industrial Engineering at Purdue University.

Environment. Physical and organizational surroundings in which the equipment must be operated and maintained (e.g., illumination, temperature, noise, vibration).

Task. Special organizational and social characteristics of the jobs that people must perform to accomplish performance goals (e.g., job demand, workloads, work time and hours, salary system).

Person. The capabilities and limitations of equipment operators and *maintenance* personnel themselves (e.g., abilities, skills, training, motivation).

Several books and handbooks offer detailed and comprehensive treatments of the field of human factors. These are listed in the bibliography at the end of this chapter. Human factors and ergonomics are two terms for the same field, one of U.S. origin and the other of European origin. The major human factors journal in the United States is *Human Factors*, the official journal of the Human Factors Society. The corresponding European journal is *Ergonomics*. *Applied Ergonomics* features mostly articles on design problems. In *Ergonomics Abstracts* the reader may find current information about human factors and related issues.

2.4.2 HUMAN CAPABILITIES AND LIMITATIONS

2.4.2.1 Information Processing in Man–Machine/Computer Systems

Most work takes place in man–machine or man–computer systems, e.g., operators using visual display terminals (VDT) to control batch production. A machine is considered to be virtually any type of physical object, device, equipment, facility, or whatever to achieve some desired purpose. The human factors specialist tries to optimize the interactions between people and machine elements of the system, while taking the environment into account. A typical schematic of a person–machine system is shown in Figure 2.4.1.

Visual, auditory, and tactile displays represent the internal status of the machine subsystem. As an example, the operator gets information on the VDT about the present state of the production process by alphanumeric symbols, graphics, and other coded material. According to the encoded information, the operator starts or stops the process by using the attached keyboard or changes specific process parameters such as temperature, pressure, and velocity. Motor actions for control activities are the final result of human information processing and decision making. Information from displays and environment is perceived, processed, and decisions are made in order to maintain or alter control settings. Information from the control activities are fed back to the machine subsystem indicated in the diagram as a closed-loop system.

To optimize the information flow and operating characteristics of the whole system, the human factor specialist is especially interested in the design of the human–machine/computer interface. From the understanding of human capabilities and limitations, the specialist creates the information control and supplementary aid devices.

Visual Characteristics

The human receives information from the outside world through the use of the various sensory modalities, such as visual, auditory, and tactile. By far the most important modality is vision.

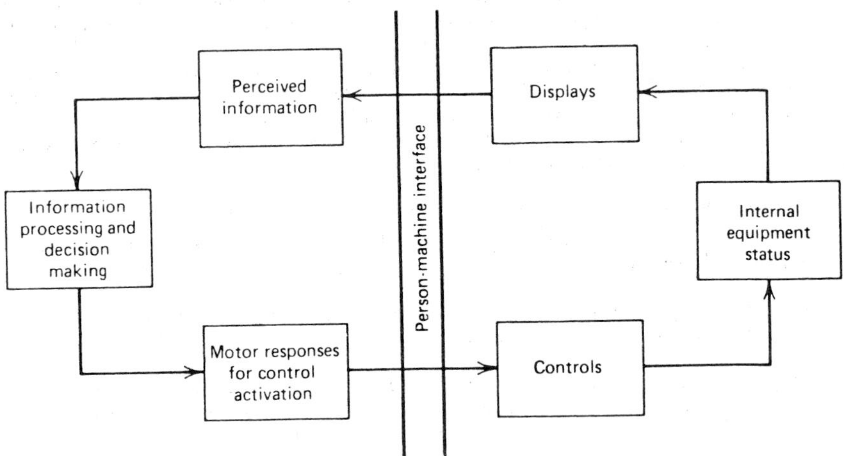

Fig. 2.4.1. Components of the man–machine system. *Source*: Ref. 1. Reprinted by permission of John Wiley & Sons, Inc.

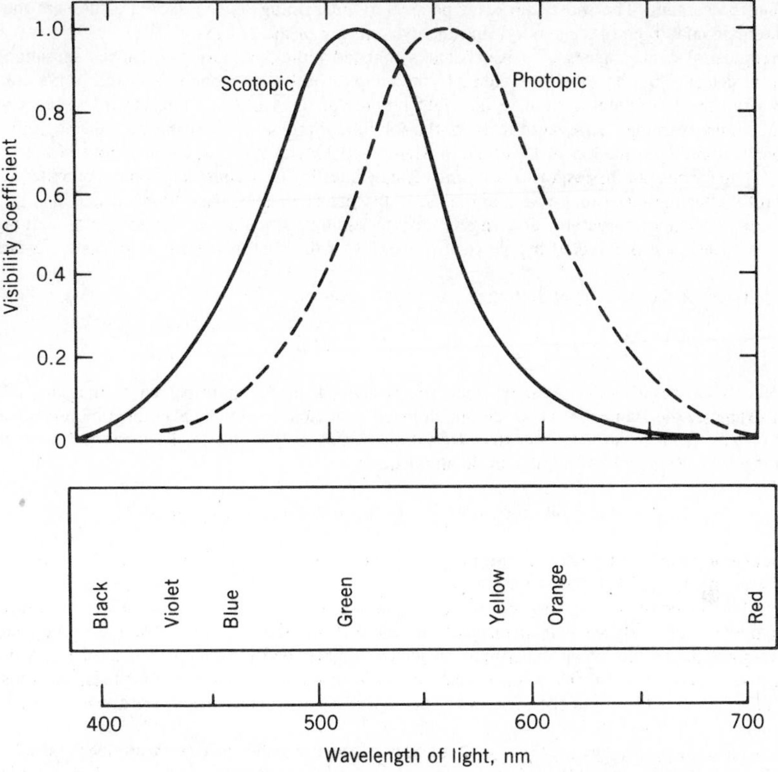

Fig. 2.4.2. The spectral sensitivity of the dark- and light-adapted eye. *Source*: Ref. 2.

The term *sensitivity* refers to the ability of the eyes to detect and correctly interpret the light signals that enter the eye. The eye is not equally sensitive to all wavelengths of light. Figure 2.4.2. shows how the spectral sensitivity of the eye varies according to whether the eyes are light- or dark-adapted, i.e., for photopic and scotopic vision.

Under the very best condition of adaptation, at least 40 minutes in the dark, the eye is most sensitive for green light with a wavelength of about 525 nanometers (nm). The light-adapted eye (adaptation from darkness to light takes place in a few seconds) is most sensitive to a more yellow-green shade with a wavelength of about 555 nm. In both cases, the sensitivity to red, about 650 nm, is only about 10% of the maximum sensitivity. This means that the quantity of radiation in the red region of the spectrum would have to be about 10 times greater than that in the green region in order to appear equally bright.

Visual acuity is the ability of the eyes to differentiate among the detailed features of a target. The most commonly used measure of acuity, minimum separable acuity, refers to the smallest space between the parts of a target that the eye can detect. Standard optometric techniques such as letter charts and checkerboard patterns are employed to calculate the resolution capability. Technically it is measured in terms of the reciprocal of the visual angle α subtended at the eye. Given target diameter or character height (H) and distance to the target (D), the tangent of the visual angle is the ratio of the target height and target distance: $\tan \alpha = H/D$.

The limit of acuity in detecting a single line (detection acuity) is about 2 seconds (2″) of arc or 0.01 milliradians (mrad) of visual angle. Minimal acuity in tests measuring how well closely spaced lines can be seen as separate (resolution acuity) is about 30″ of visual angle.

Visual acuity depends on a number of parameters including character size, shape, contrast, luminance, and color. In seeking to optimize character size, e.g. on a video screen, the lower limits of acuity are less important. Despite the ability of the eye to identify on a video screen stimuli that fall within a visual angle of between 1 and 2 seconds of arc, results from investigations of VDT operators indicate an optimal character height corresponding to 16–20 seconds of arc.[2]

Information Processing

The human information processing system can be divided into three interacting subsystems: (1) the perceptual system, (2) the cognitive or decision-making system, and (3) the motor system. Each system has its own memories and processors. The most important parameters concerning human factors issues are the storage capacity, the main code type of a memory, and the cycle time t of the processors.

The perceptual system consists of sensors and associated buffer memories. It carries sensations of the physical world detected by the sensory system. In the simplest tasks, the cognitive system serves to connect information from the perceptual system to the right outputs of the motor system. But most tasks are more complex and require learning, retrieval of facts, or the solution of problems. In such cases, the cognitive system uses previously stored information in Long Term Memory (LTM) to make decisions about how to respond. The motor system carries out the response, e.g. answering a question or pushing keys on a typewriter keyboard.

Very shortly after the onset of a visual stimulus, it appears in Visual Image Store. It is coded physically, that is, in terms of shape, curvature, and angles. For an auditory stimulus, a corresponding auditory buffer exists. The physical code is affected by physical properties of the stimulus, such as intensity, contrast, and size.

The cycle time[3] of the perceptual system, t_p, is on the order of:

$$t_p = 100 \qquad (50-200) \text{ ms}$$

Perceptual events occurring within a single cycle are combined into a single percept if they are sufficiently similar. For example, the frame rate at which an animated image on a video display must be refreshed to give the illusion of movement must be greater than 1/(100 ms/frame) = 10 frames/s. By using the lower bound of the cycle time, t_p = 50 ms, the solution can be augmented:

$$\text{Max frame rate for fusion} = 1/(50 \text{ ms/frame}) = 20 \text{ frames/s}$$

This calculation is in general accord with the frame rates commonly employed for motion picture cameras (18 frames/s for silent and 24 frames/s for sound).

Working Memory holds the information under current consideration. It is part of LTM; the information in Working Memory is commonly *symbolic acoustically* coded, that is, items are more liable to acoustic interference. Table 2.4.1 shows the results of an experiment in which subjects had to remember lists of five words, then recall them twenty seconds later. They made many errors with the acoustically similar lists, but substantially fewer with the semantically similar list. This was true regardless of whether they were given the lists orally or visually.

For application purposes, the user of a computer is especially liable to key in numbers mistakenly that sound like the numbers he has just looked up. In order to avoid confusion in the usage of error indicators (mnemonics) in computer systems, similar-sounding codes should therefore be avoided.[3]

The activated elements of LTM, which define Working Memory, consist of symbols, called *chunks*, which may themselves be organized into larger units. A chunk can be regarded as any familiar unit of information, regardless of size, that can be stored and recalled as one entity, given a single relevant cue. The effective capacity of Working Memory (e.g., the longest number that can be repeated) is the familiar 7 ± 2 chunks.[5] Apparently, there is no erasure from LTM. Successful retrieval of a chunk from LTM depends on whether associations to it can be found. The predominant code type is semantic. Items in LTM are more sensitive to

Table 2.4.1. Acoustic vs. Semantic Interference in Working Memory

	Experiment I (Spoken)				Experiment III (Visual)	
	Group A (N = 20)		Group S (N = 21)		Group AV (N = 10)	
	Acoustically Similar	Control	Semantically Similar	Control	Acoustically Similar	Control
Word set	*mad, man, mat, map, cad, can, cat, cap*	*cow, day, far, few, hot, pen, sup, pit*	*big, long broad, great, high, tall, large, wide*	*old, deep, foul, late, safe, hot, strong, thin*	Same as Expt. I plus *cab, max*	Same as Expt. I plus *rig, day*
Percentage correctly recalled	10	82	65	71	2	58

Subjects studied 25 five-word lists. The words in the lists were either acoustically similar, semantically similar, or unrelated (control condition). The numbers in the table are the proportion of lists recalled entirely correctly and in the proper order.

Source: Ref. 4.

Table 2.4.2. Cognitive Processing Rates

Rate at which an item can be matched against Working Memory	
Digits	33 [27–39] ms/item
Colors	38 ms/item
Letters	40 [24–65] ms/item
Words	47 [36–52] ms/item
Geometrical shapes	50 ms/item
Random forms	68 [42–93] ms/item
Nonsense syllables	73 ms/item
Range = 27–93 ms/item	
Rate at which four or fewer objects can be counted	
Dot patterns	46 ms/item
3-D shapes	94 [40–172] ms/item
Range = 40–172 ms/item	
Perceptual judgment	92 ms/inspection
Choice reaction time	92 ms/inspection
	153 ms/bit
Silent counting rate	167 ms/digit

Selected cycle times (ms/cycle) that might be identified with the cognitive processor cycle time.
Adapted from Ref. 3.

semantic interference than to acoustic interference, e.g., they are more confused with other items with similar meaning.

The basic cycle time[3] of the cognitive system is around one-tenth of a second:

$$t_c = 70 \qquad (25\text{–}170)\ \text{ms}$$

As with the perceptual system, the cycle time is not constant, but can be shortened by practice, task pacing, greater effort, or reduced accuracy. Table 2.4.2 shows selected cycle times (ms/cycle) that might be identified with the cognitive system cycle time.

The motor system can issue commands for continuous movements. The commands consist of a series of discrete movements, each requiring about

$$t_m = 70 \qquad (30\text{–}100)\ \text{ms}$$

The feedback loop from action to perception is sufficiently long (200–500 ms) that rapid behavioral acts such as typing and speaking must be executed in bursts of preprogrammed motor instructions.

2.4.2.2 Psychomotor Skill and Choice Reaction Time

Movement time is a function of movement distance and target width. Since movement consists of a series of microcorrections, each with a certain accuracy, the time to perform one correction is one cycle of the perceptual system to observe the hand, one cycle of the cognitive system to decide on the correction, and one cycle of the motor system to perform the correction, or $t_p + t_c + t_m$. The time to move the hand to the target is then the time to perform n of these corrections, or $n(t_p + t_c + t_m)$. Since the sum of the cycle times equals 240 ms, n is the number of roughly 240 ms intervals it takes to point to the target. Given the distance D to the target and its size S, the total movement time Mt is given by

$$Mt = K_m \log_2 (2D/S)$$

where

$$K_m = -(t_p + t_c + t_m)/\log_2 e$$

The constant e has been found to be about 0.07.[6] The equation above is called Fitt's law. It says that the time to move the hand to a target depends only on the relative precision required, that is, the ratio between the target's distance and its size. Measurements of K_m determined directly from experiments give somewhat higher values than those computed from the equation. They center around

$$K_m = 100 \qquad (50\text{–}120)\ \text{ms/bit}$$

where the bit is the unit of measurement based on information theory.

Attneave[7] presents a simple illustration of the information theory concept. By asking a series of questions each answerable only by yes or no, any square in an 8 × 8 checkerboard may be located with a minimum of six questions (by successively dividing regions of the checkerboard into halves). Hence, six bits of information are required to locate a square when the events are equally probable. For an 8 × 8 checkerboard, the amount of information, H, is $H = \log_2$ alternatives (64) = 6 bits. Similary, the amount of information in the throw of a die is $H = \log_2 6 = 2.58$ bits. Since most events are not equally probable, the more general form of the Shannon-Wiener information measure H is

$$H = p_i \log 1/p_i$$

which is equivalent to

$$H = p_i \log/p_i$$

Each alternative is weighted by the particular probability of its occurrence, $\log 1/p_i$, and alternative probabilities are summed.

The other global issue that controls reaction time is practice. The time t_n to do a task decreases with practice and is approximately proportional to a power of the amount of practice: $t_n = t_1 n^{-a}$, where t_1 is the time to do a task on the first trial, n is the trial number, and a is a constant, ranging between 200 and 600 ms. Figure 2.4.3 shows the results of an experimental study of the Power Law of Practice. A control panel had ten keys located under ten lights. Subjects had to press a subset of the keys in direct response to whatever subset of lights was illuminated. As can be seen, the data are well fit by a power law, except at the ends.

The Power Law of Practice applies to all skilled behavior, both cognitive and sensory-motor. However, it does not describe the acquisition of knowledge or apply to changes in the quality of performance.

Reaction-time calculations for simple decision tasks, as, for example, the time needed to press the space bar on a VDT whenever any symbol appears, requires the application of the appropriate cycle times: $t_p + t_c + t_m$. If there are a larger number of choices, the situation is more complicated. The task still can be analyzed as a sequential set of decisions made by the Cognitive System, each adding a nominal $t_c = 70$ ms to the response. But the relationship between time required and number of alternatives is not linear, because people apparently can arrange the processing hierarchically. The minimum number of steps necessary to process the alternatives can be derived from information theory. To a first-order approximation, decision time increases with uncertainty about the judgment to be made. The relation is often referred to as Hick's law.[9]

$$D_t = K_p + bH_s$$

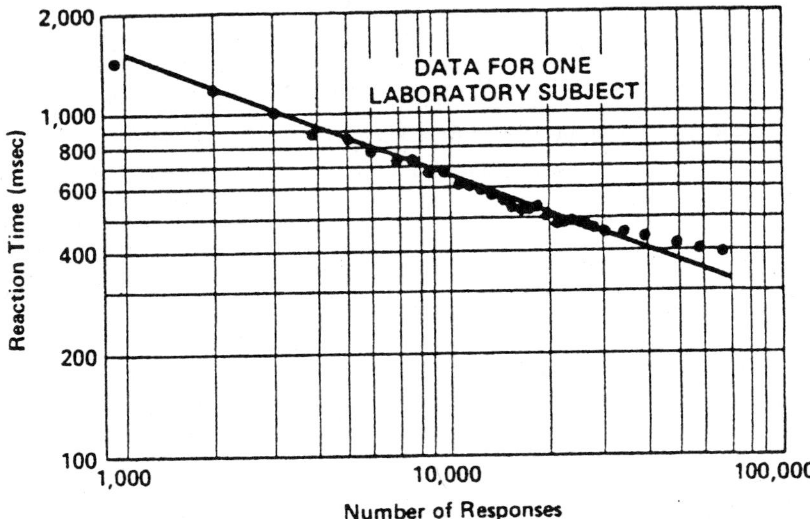

Fig. 2.4.3. The power law of practice. Improvement of reaction time with practice on a 1023-choice task. Subjects pressed keys on a ten-finger chordset according to pattern of lights directly above the keys. *Source*: Ref. 3, after Ref. 8.

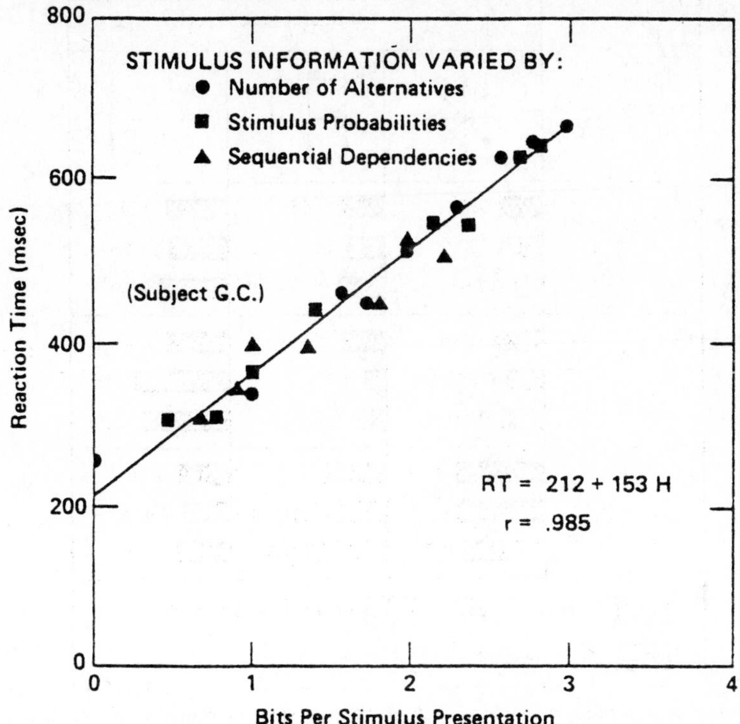

Fig. 2.4.4. Choice reaction time for three different ways of manipulating the stimulus information. *Source*: Ref. 10.

where H_s is the transmitted information of the decision in bits, b is the time needed to process one bit of information, and K_p is the time taken to encode the stimulus and to execute the response, $K_p = t_p + t_c$. Figure 2.4.4 shows the results of inducing three different ways of uncertainty. As can be seen, all the different ways of inducing uncertainty fit the same curve. It takes $b = 150$ ms/bit of uncertainty, above a base of about $K_p = 200$ ms, which could be identified as the processing time of the perceptual and motor systems: $t_p + t_m = 200$ ms.

There are many aspects of stimuli, responses, and the relationship between them that can alter reaction time. The physical features of stimuli such as size, shape, color, intensity, and discriminability as well as the control, e.g., switch, lever, knob, wheel, voice key, play a crucial role in establishing speed of reaction. One global issue that controls reaction time is stimulus–response compatibility. The classic demonstration of stimulus–response compatibility effects was published in 1953 by Fitts and Seeger.[11] Compatibility between the geometry of displays and response arrays was changed by altering spatial relationships. Reaction time was fastest and error rates lowest when the geometry of stimulus and response arrays corresponded directly (see Fig. 2.4.5). Nevertheless, after several days of practice, the speed and accuracy of incompatible stimulus–response mappings approaches that of compatible mappings. In emergency cases, however, as well as in other stress situations, response errors are more likely to occur in incompatible mappings.

2.4.2.3 Engineering Anthropometry and Biomechanics

Engineering anthropometry is defined as the application of scientific physical measurement methods to the human body in order to optimize the interface between humans and machines and other manufactured products. Worker performance problems are often caused by a mismatch of workplace design measures and anthropometric requirements. As an example, it was not long ago that a job designer would assume an easy overhead grasp type of reach of about 77 in. (195 cm). Today, with more than 43% of the labor force being female, approximately 73 in. (185 cm) is recommended in order to accommodate 95% of women.[12]

The anthropometric variations in the work force today dictate a great concern to ensure that the workplace layout can be physically fitted to the majority of people. An important aid is the percentile.

The percentile is the percentage of the population having a body dimension equal to or less than that indicated in the anthropometric task. For example, the 95th percentile for overhead reach is 87.6 in. for Air Force officer

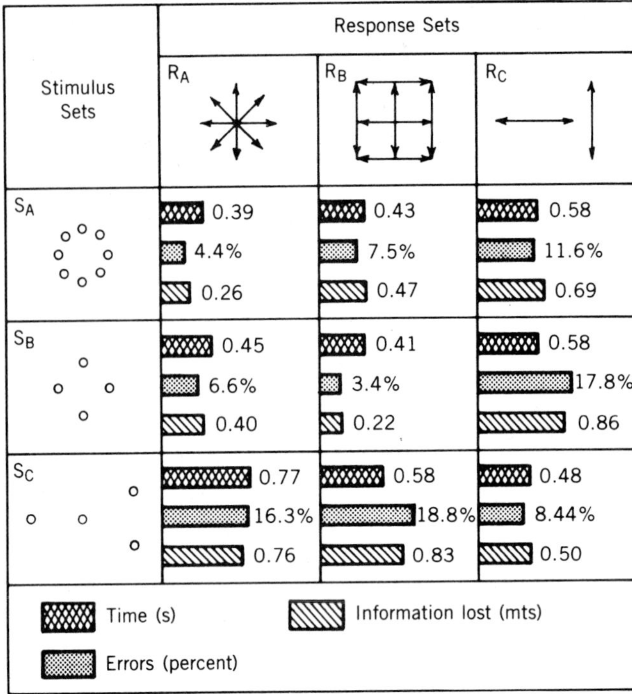

Fig. 2.4.5. Stimulus response compatibility. Each of the three stimulus panels on the left was assigned to one of the three response panels across the top. The natural compatibility assignments are seen down the negative diagonal. *Source*: Ref. 11.

flying personnel. The officer whose overhead reach is 87.6 in. can reach higher than 95% of his fellow officers. Comprehensive anthropometric data are still missing for population groups such as females, children, the elderly, and the handicapped. Most anthropometric data are based on young males in military service.

Static anthropometric dimensions are taken with the body of the subjects in fixed, standardized positions. For instance, to estimate overhead reach while seated, the lengths of the upper extremity and the torso can be combined to provide the necessary prediction (see Fig. 2.4.6). However, the prediction may have considerable error, since the amount of upward rotation of the shoulder and torso erection of the spine would not be depicted; and thus the estimate might be too short. The practical limit of arm reach is not the sole consequence of arm length. It is also affected in part by shoulder movement, partial trunk rotation, possible bending of the back and the function that is to be performed by the hand. If the task is disparic (e.g., moving or reaching) then dynamic anthropometric data need to be used; however, for static work (e.g., holding an item in position) static anthropometric data are needed. There are distinct differences between these two data bases. The use of static data tends to focus on clearance of body dimensions with the surroundings, whereas the use of dynamic measurements tends to focus on the functions of the operations involved. Because of this, the anthropometric considerations for each design must be based on the anthropometric data for the population that performs the task.*

Biomechanics is the study of mechanical reactions of the body to external or inertial loadings. This includes studies of range, strength, endurance, and speed of movement, and mechanical responses to such physical forces as acceleration and vibration. The human musculoskeletal system is a kinematic linkage. The ability to tolerate forces applied by loads to the linkage varies greatly between individuals and from joint to joint. The maximum occasional load, i.e., strength, that can be produced safely by a person performing a specific act has been assessed by several methods. Snook[13] has published limits to manual materials handling tasks based on psychophysical examinations. General rules have emerged for specific physical acts such as lifting, pushing and pulling carts, and control movement.[14] (see Fig. 2.4.7). The National Institute of Occupational Safety and Health (NIOSH) has suggested a preliminary guideline for lifting weights for both men and women. It is believed to be the most measurable design reference for occasional, two-handed symmetric lifts from near the floor to 30 in. (76 cm).

* For additional information on this subject, see reference 12a.

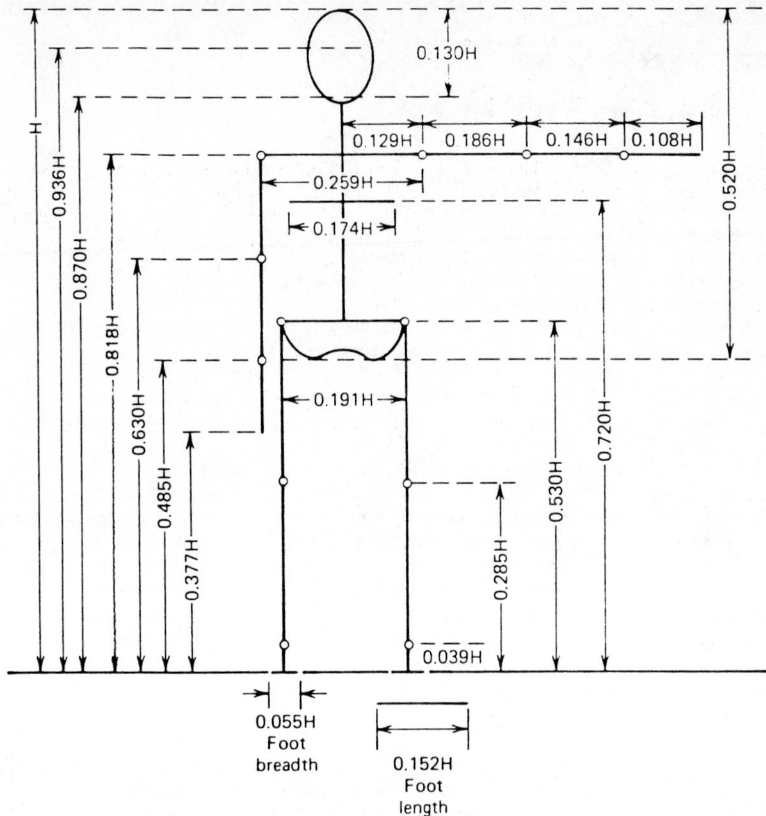

Fig. 2.4.6. Segment length as proportion of stature. *Source*: Ref. 12. Reprinted by permission of John Wiley & Sons, Inc.

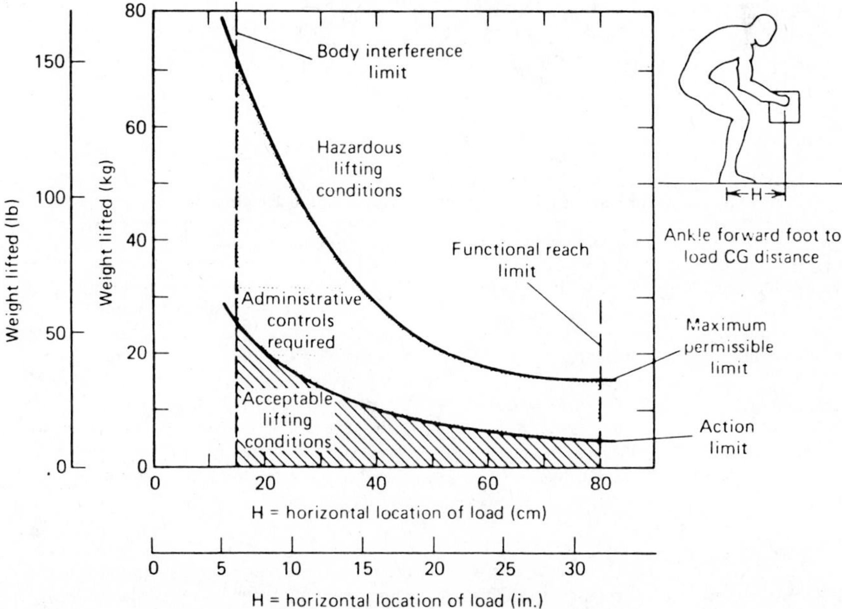

Fig. 2.4.7. Recommended weight and size limits for occasional lifts (less than once every 5 min.). *Source*: Ref. 14. Reprinted by permission of John Wiley & Sons, Inc.

2.4.3 DESIGN ERGONOMICS

2.4.3.1 Displays, Controls, and Layouts

Displays

The purpose of a display in a production system is to give information to the operator about the functional condition of the equipment or the process. Information is generally brought to the operator's attention by visual or auditory cues. Tactile cues may be used in situations where ambient noise levels interfere with auditory signals or where vision is reduced, such as in low ambient illumination. Visual presentation is preferred for complex messages in a noisy environment; auditory presentation is preferred for simple messages in areas where people move around frequently and where response time must be rapid.

There are several ways in which visual displays may be used in a production system. Table 2.4.3 includes some of these uses and suggests the display type most appropriate for each.

A digital readout should be used for quantitative readings. A moving pointer display is best for qualitative or check readings and some adjustments. A graph, e.g., pen recording, is best suited for detecting trends and qualitative readings. The annunciator light is optimal for giving operating instructions on a control panel where many functions are monitored.

The design and installation of visual displays will affect the performance of the operator of a production system. Factors such as the distance of an operator from a display when it is read, the number of displays on a single console, the readability of the dials, and the ambient illumination should be considered when selecting and installing displays. Guidelines are offered in McCormick and Sanders[16] and in Eastman Kodak.[15]

A general trend in display technology is the increasing use of electronic displays, including LEDs (Light-Emitting Diodes), liquid crystals, and characters generated in dot matrix. While these devices may save considerable space in a display panel, legibility may be a problem. Sequenced numbers are less legible and may be misread in situations where fast readings are needed, for instance, legibility problems often emerge between the numbers 6 and 8. The guidelines given in Table 2.4.4 specify some characteristics of electronic displays and suggestions for their installation.

Principles for Controls and Layouts

A control is any device that allows a human to transmit information to a man–machine system. Although many different types of controls are available, there are certain features that are common to all. Control knobs must have a shape and a size. Controls must offer some resistance to movement. There must be some label on or adjacent to a control. And at least, the designer must face the issue of protecting a control from accidental activation.

Control coding is an identification aid for operators. The primary methods of coding include color, shape, texture, size, locating, operational method, and labeling. Each of these methods has various desirable and undesirable features and principles of application.[16]

The following factors will determine which control is most suitable for a given application:

speed and accuracy of response needed
space available

Table 2.4.3. Types of Information Displayed and Recommended Displays for Each

Information Type	Preferred Display	Comments	Examples in Industry
Quantitative reading	Digital readout or counter	Minimum reading time Minimum error potential	Numbers of units produced on a production machine
Qualitative reading	Moving pointer or graph	Position easy to detect, trends apparent	Temperature changes in a work area
Check reading	Moving pointer	Deviation from normal easily detected	Pressure gauges on a utilities console
Adjustment	Moving pointer or digital readout	Direct relation between pointer movement and motion of control, accuracy	Calibration charts on test equipment
Status reading	Lights	Color-coded, indication of status (e.g., "on")	Consoles in production lines
Operating instructions	Annunciator lights	Engraved with action required, blinking for warnings	Manufacturing lines in major production systems

Source: Ref. 15, adapted from Grether and Baker.[17]

Table 2.4.4. Some Characteristics of Electronic Displays and Suggestions for Their Installation

Characteristics	Installation
1. Use a dot matrix character style, at least 5 × 7 and preferably 7 × 7 or 7 × 9, for the most accuracy	1. Provide for wide-range viewing angles to ensure full visibility of all characters without any background noise. Be sure obstructions do not prevent characters from being seen from all angles.
2. Provide the following geometry for the numerals and letters displayed: A width-to-height ratio of about 0.6 to 0.8 A distance between digits of 1.1 to 1.4 times the stroke width. Vertical numbers rather than slanted ones. A dot spacing of about 0.4 to 0.6 mm (0.02 to 0.025 in.).	2. Minimize internal reflections, unlit images, or distractions from the background of the display unit.
3. Select a display that does not persist so long that an operator is unable to read current values of numbers if they are changing rapidly.	3. Minimize glare on the display by adjusting the direction of ambient illumination, using shields or filters or both, and locating the displays away from glare sources.
4. Choose a display with lines for describing the characters that are sharp, not diffused, and have equal brightness throughout.	

Source: Adapted from Reference 15.

ease of use
readability in an array of similar controls
demands of other tasks performed simultaneously with control operation

Table 2.4.5 rates some of the more common controls for several of those factors. The ratings are based on typical examples of each control type, not on the extremes of performance in each range.

The location of individual controls and displays in relation to each other and to the operator is an important factor. However, in practice, an optimal layout of a man–machine interface is often difficult or impossible to design. Various competing principles and goals have to be taken into account:

Table 2.4.5. Characteristics of Common Controls

Control	Suitability Where Speed of Operation Is Required	Suitability Where Accuracy of Operation Is Required	Space Required to Mount Control	Ease of Operation in Array of Like Controls	Ease of Check Reading in Array of Like Controls
Toggle switch (On-Off)	Good	Good	Small	Good	Good
Rocker switch	Good	Good	Small	Good	Fair[1]
Push button	Good	Unsuitable	Small	Good	Poor[1]
Legend switch	Good	Good	Small	Good	Good
Rotary selector switch (discrete steps)	Good	Good	Medium	Poor	Good
Knob	Unsuitable	Fair	Small–Medium	Poor	Good
Crank	Fair	Poor	Medium–Large	Poor	Poor[2]
Handwheel	Poor	Good	Large	Poor	Poor
Lever	Good	Poor (H) Fair (V)	Medium–Large	Good	Good
Foot pedal	Good	Poor	Large	Poor	Poor

The types of controls that should be considered when the design or work situation has certain requirements, such as speed, accuracy, ease of operation, ease of reading, or limited space on a control panel, can be determined from this summary. *Source:* Ref. 15.

1. *Importance principle.* The most important instruments should be located in the optimal position in terms of convenient access and good visibility.
2. *Frequency-of-use principle.* The most frequently used instruments should be placed in the optimal location.
3. *Sequence-of-use principle.* Instruments should be laid out in the order of their usage.
4. *Functional principle.* Instruments with related functions should be grouped together.
5. *Compatibility principle.* Emphasizes the importance of population stereotypes.

People expect things to behave in certain ways when they are operating controls or when they are in certain environments. Such expectancies are called *population stereotypes*. Some examples are:

Knobs on electrical equipment are expected to turn clockwise for *on*, to increase current, and counterclockwise for *off*, to decrease current.

Wheels or cranks to control direction of a moving vehicle are expected to use clockwise rotation to make a right turn and counterclockwise rotation to make a left turn.

Pulling a control such as a throttle outward from a panel signifies that it has been activated (on).

Pushing it in disengages it (off).

The design and selection of controls to fit population stereotypes is particularly important for infrequently used emergency controls. Such controls should fit the population stereotype, should be simple to use, and require minimal decision making by the operator.

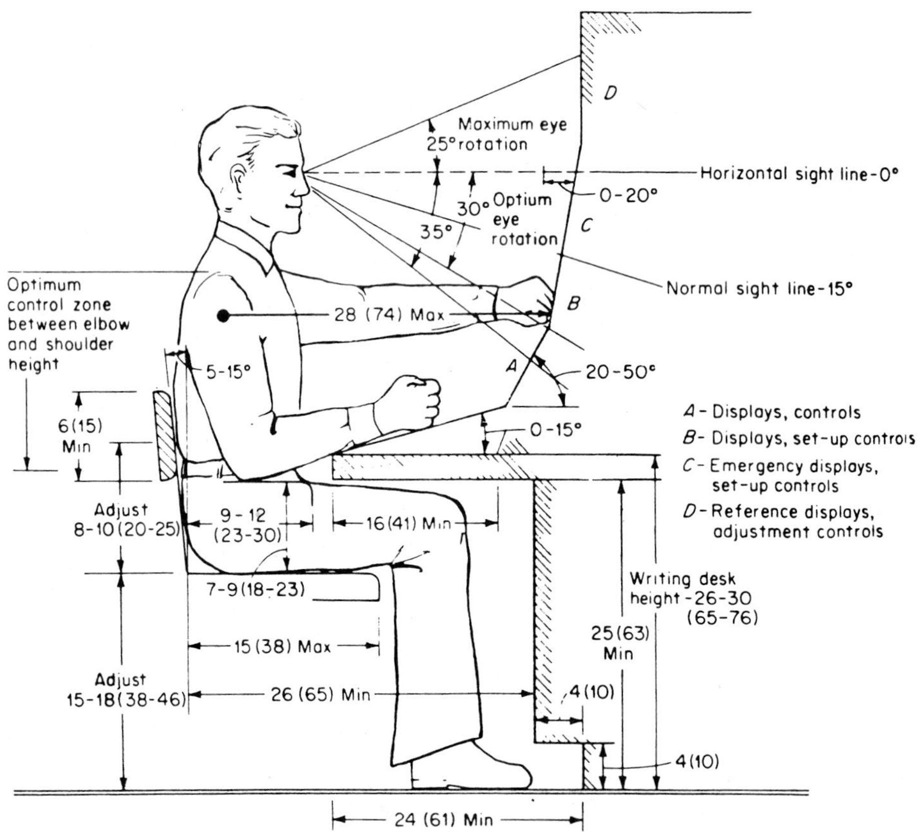

Fig. 2.4.8. Recommended design features of consoles for seated operators (5th–95th percentile of an Air Force population). *Source*: Ref. 16, adapted from Van Cott and Kinkade.[18]

2.4.3.2 Workplace Design

In designing the workplace, some compromises are almost inevitable due to competing priorities. Work should be arranged so that it can be seen without lowering the line of sight more than 30° below the horizontal and without raising it more than 5° above the horizontal. For fine detail, e.g., book or newspaper work, it should be between 100 in. (250 mm) and 140 in. (350 mm) from the eyes. Recommended design features of consoles for seated operators are illustrated in Figure 2.4.8. If the leg space permits a worker to cross the legs, this is an advantage. The space should be wider than just the width across the the worker's knees.

Industrial seats should be designed according to the physical structure and biomechanics of the human body. They must be adjustable over a range of about 6 in. (150 mm). This adjustment must be simple, or it will not be used. Supporting the chair on five legs adds to stability. The seat and back should have resilient coverings that can be readily wiped. Resilience rather than softness is the requirement; the seat should have a rounded front edge and backrest to keep the back in a state of balance. Arm rests are generally desirable unless the tasks require free mobility of the trunk, shoulder, and arms. Figure 2.4.9 shows recommended chair characteristics, using both front and side views.

The illumination level and direction from which the light travels in relation to the viewer affects contrast and glare and are crucial aspects of adequate vision.

The quantity of light that is emitted from a light source per unit time is called luminous flux. It is expressed in units of lumens (lm). The luminous flux is calculated in accordance with the spectral sensitivity of the "standard" human eye. *Luminance* is a measure of the intensity of light emitted from a light source per unit surface area normal to the direction of the light flux. The light source includes not only self-luminous sources such as lamps, but also any light transmitting or reflecting surface such as a wall, desktop, or a display screen. Luminance is expressed in units of candela/m^2 (cd/m^2). *Illuminance* is that part of the total luminous flux that is incident on a given surface. In practice, it is a measure of the quantity of light with which, for instance, a desktop is illuminated. The measure of illuminance is the footcandle, and 1 footcandle is equal to 10.8 lumens per square meter (lux) in the International System of Units.

Illuminance is usually measured in both the horizontal and vertical planes. Those values that are normally found in lighting tables are usually the values of horizontal illuminance. They do not adequately describe the illumination of objects in the vertical plane, e.g., objects on storage shelves, bookcases, and display screens. At a VDT workplace, a document lying on the desktop is illuminated by the horizontal illuminance, whereas the display screen is illuminated by the vertical illuminance. The ratio between these two quantities is usually between 0.3 and 0.5. As an example, if the illuminance in the room is quoted as 500 lux (lx), the illuminance of the desktop is 500 lx, whereas the illuminance of the display screen might be somewhere in the range of 150 to 250 lx. Table 2.4.6 lists recommended ranges of illuminance from various types of workplaces and tasks.[19]

In the selection of artificial light sources for illuminating work areas and workplaces, the two most important considerations are efficiency, in lumens per watt (lm/W), and color rendering. Color rendering is the degree to which the perceived colors of objects illuminated by various light sources match the perceived color of the

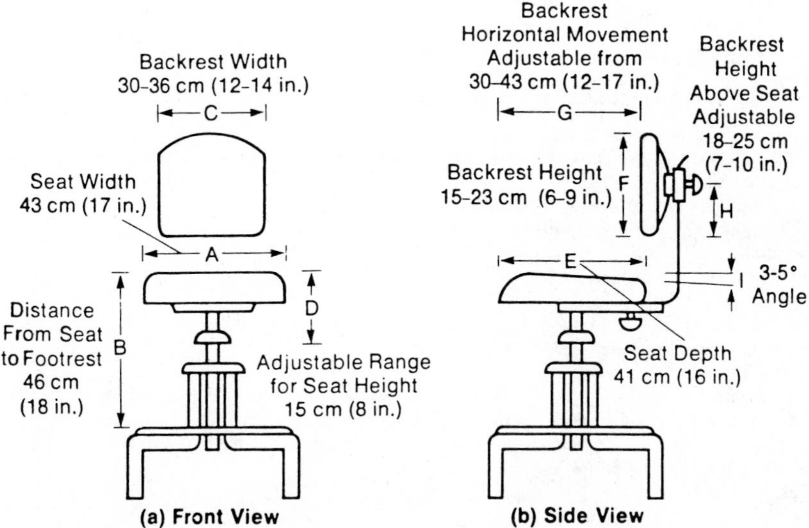

Fig. 2.4.9. Recommended dimensions for adjustable chair. *Source*: Ref. 15.

Table 2.4.6. Recommended Ranges of Illuminance

Category Letter	Range of Illuminances in Footcandles (lux)	Type of Activity or Area
A	2 to 3 to 5 (20 to 30 to 50)	Public areas with dark surroundings
B	5 to 7.5 to 10 (50 to 75 to 100)	Areas for brief visits
C	10 to 15 to 20 (100 to 150 to 200)	Working spaces where visual tasks are only occasionally performed
D	20 to 30 to 50 (200 to 300 to 500)	Performance of visual tasks of high contrast or large size–for example, reading printed material, typed originals, handwriting in ink, and good xerography; rough bench and machine work; ordinary inspection; rough assembly
E	50 to 75 to 100 (500 to 750 to 1000)	Performance of visual tasks of medium contrast or small size—for example, reading medium-pencil handwriting or poorly printed or reproduced material; medium bench and machine work; difficult inspection; medium assembly
F	100 to 150 to 200 (1000 to 1500 to 2000)	Performance of visual tasks of low contrast or very small size—for example, reading handwriting in hard pencil on poor-quality paper and very poorly reproduced material; highly difficult inspection
G	200 to 300 to 500 (2000 to 3000 to 5000)	Performance of visual tasks of low contrast and very small size over a prolonged period—for example, fine assembly; very difficult inspection; fine bench and machine work
H	500 to 750 to 1000 (5000 to 7500 to 10,000)	Performance of very prolonged and exacting visual tasks—for example, the most difficult inspection; extrafine bench and machine work; extrafine assembly
I	1000 to 1500 to 2000 (10,000 to 15,000 to 20,000)	Performance of very special visual tasks of extremely low contrast and small size—for example, surgical procedures

Source: Ref. 19.

same object when illuminated by standard light sources. Unfortunately, the more efficient light sources are often not suitable for tasks requiring color discrimination because of their poor color rendering. This is indicated in Table 2.4.7. The best way to determine the optimum tradeoff between efficiency and color rendering is through empirical testing.

2.4.3.3 Job Analysis and Design

There are three approaches to job analysis: the psychological, physiological, and industrial engineering methods. Psychological job analysis typically relates to such topics as job satisfaction and motivation, analyzing job demands and job skills, as well as psychologically and socially caused strain and stress. The physiological method of job analysis typically seeks input on fatigue, stress, overexertion, and anthropometric considerations in design. The industrial engineering method of job analysis seeks answers to such questions as, Does it have to be done at all? and If so, can it be done differently so that a higher production output could be realized? Well-known methods are time and motion studies. Job-analysis methods are discussed in detail in Gael[20]; some of them are reviewed briefly below.

There are various methods of collecting job information. The most common are interview and observation; sometimes group discussions are used. In recent years various types of structured job-analysis questionnaires have been developed.

The limitations of conventional job descriptions arise primarily from their dependence upon verbal material that is largely in essay form. Job-analysis methods that tend to be more systematic, standardized, and quantitative, are referred to as *structured job-analysis methods*. The job elements in the Position Analysis Questionnaire (PAQ) by McCormick[21] provide for analyzing jobs in terms of 187 job elements. They are organized in six divisions as follows:

1. *Information input.* Where and how does the worker get information? (e.g., from display, written material)

2. *Mental processes.* What reasoning and decision making processes are involved? (e.g., coding/decoding, problem solving)
3. *Work output.* What physical activities does the worker perform, what tools or devices are used? (e.g., lifting of weights, keyboard punching)
4. *Relationships with other person.* What kind of relationships are required by the job? (e.g., instructing workers, contacts with public customers)
5. *Job context.* What are the influencing factors of the physical and/or social context? (e.g., high temperature, interpersonal conflict situations)
6. *Other job characteristics.* What characteristics other than those already described are relevant to the job?

The analysis of jobs with the PAQ is typically carried out by job analysts, personnel officers, supervisors, or sometimes the job incumbents are asked to analyze their own jobs. A similar job-analysis technique has been developed by Rohmert and Landau.[22] It consists of a task- and job-analysis procedure and emphasizes the analysis of work demands, based on a physical and psychological stress/strain concept.

To diagnose existing jobs according to their motivational power and to get a job satisfaction index, the Job Diagnostic Survey (JDS) by Hackman and Oldham[23] has been used extensively in research and organizational development projects.

Table 2.4.7. Efficiency and Color Rendering of Artificial Light Sources

Type	Efficiency (lm/W)	Color Rendering	Comments
Incandescent	17–23	Good	Incandescent is a commonly used light source but is the least efficient. Lamp cost is low. Lamp life is typically less than 1 year.
Fluorescent	50–80	Fair to Good	Efficiency and color rendering vary considerably with type of lamp: cool white, warm white, deluxe cool light. Significant energy cost reductions are possible with new energy-saving lamps and ballasts. Lamp life is typically 5–8 years.
Mercury	50–55	Very Poor to Fair	Mercury has a very long lamp life (9–12 years), but its efficiency drops off substantially with age.
Metal halide	80–90	Fair to Moderate	Color rendering is adequate for many applications, but metal halide has more infrared than a tungsten lamp. Lamp life is typically 1–3 years.
High-pressure sodium	85–125	Fair	This lamp is a very efficient light source. Lamp life is 3–6 years at average burning rates, up to 12 hours per day.
Low-pressure sodium	100–180	Poor	This lamp is the most efficient light source. Lamp life is 4–5 years at average burning rate of 12 hours per day. Mainly used for roadways and warehouse lighting.

Higher values for efficiency indicate longer lamp life and better energy conservation.
Source: Ref. 15.

Table 2.4.8. The Properties of Motivating Jobs

Critical Psychological States	Outcomes
Experienced meaningfulness of the work	High internal work motivation
Experienced responsiblity for outcomes of the work	
Knowledge of the actual results of the work activities	

Source: Ref. 23.

Job Design

Jobs are composed of tasks together with conditions under which they can be accomplished. Job design is the process (1) of allocating the task to the worker, (2) of determining what tasks are to be clustered into what jobs, and (3) how the jobs are to be linked together.

The most commonly used principle in work design relates to functional specialization and deterministic job specification. This "scientific management" approach was introduced by Taylor.[24] Functional specialization refers to job design as narrowly and precisely as possible. Deterministic job specification relates to the issue of specifying every task, movement, and interaction by time and motion studies, predetermined motion–time systems, and work standards.

Highly functionalized and specialized work creates strong motivational problems. Job holders who are alienated by the simplicity and "dead-endedness" of their jobs are less willing to contribute maximally to the organization. Due to problems of labor turnover, absenteeism, grievances, and poor product qualities, alternative forms of job design have been introduced. They try to motivate employees by complex, meaningful, and challenging tasks instead of simplistic and repetitive ones.

Motivating properties of jobs, besides payment, are listed in Table 2.4.8. Experienced meaningfulness, responsibility for outcomes of the work, and knowledge of results improve internal work motivation. Three basic concepts of job design emerged as a result of these considerations: job enlargement, job enrichment, and autonomous work team. *Job enlargement* refers to improvements in task variety and related job skills, and in task identity and significance (meaningful, substantial piece of work). Improvements in task variety, autonomy, and responsibility refers to *job enrichment.* A self-maintaining work team is a fairly *autonomous* service or production unit. It is designed to maintain all the tasks and resources necessary to meet its service or production and organization needs as well as many of its members' personal and social needs.

These concepts are mainly employed in European countries. As an example, the Volvo Skovde plant for engine assembly in Sweden introduced job enlargement and self-maintaining work teams in 1974. The concepts involved were self-propelled assembly carts, ergonomically designed workstations and handtools, and a flow layout and plant architecture that supported the teamwork concept. The Saab-Scania plant for engine assembly has gone even further and has actually abandoned the moving belt and also altered the teamwork concept. Due to group pressures on workers who fell behind in production, the manual assembly is carried out in individual, ergonomically well-designed workstations. A computer-aided assembly carrier transports products to be assembled according to individual workplaces.[25]

2.4.4 ENVIRONMENT AND HUMAN RELIABILITY

2.4.4.1 Noise and Vibration

Noise is conveniently and frequently defined as "unwanted sound." In the frequency range of about 15 to 16,000 Hz, sound is audible and sensed mainly by the ear. Noise has several undesirable effects. In industry, continuous exposure to intense noise can result in deafness. Industrial noise interferes with communication and warning signals. Contrary to popular opinion, the mere physical intensity of noise is not a sufficient criterion for annoyance. Annoyance occurs when the noise interferes with a person's ability to carry out some desired activity, such as writing, planning, talking, or sleeping.

Because of the ear's large range of sensitivity, acousticians normally use a logarithmic sound pressure scale. Sound pressure level (SPL) is defined as:

$$\text{SPL} = 10 \log_{10} (p^2_{\text{rms}}/p^2_{\text{ref}}) \quad \text{dB}$$

where p_{rms} is the root-mean-square sound pressure and p_{ref} is 0.00002 Pascal, a reference sound approximately equal to the smallest sound an average young person can hear. The measurement units of sound pressure level most commonly used to assess the noise exposure of workers are decibels on the A-weighted scale (dBA). They deemphasize frequencies less than 1000 Hz.

Table 2.4.9. Noise Exposures Permitted by OSHA

Duration Per Day (hours)	Sound Level (dBA)
8	90
6	92
4	95
3	97
2	100
1.5	102
1	105
0.5	110
0.25 or less	115

Extremely intense continuous or impulsive noise (gun-fire or intense impacts over 120 dB) can cause permanent damage, although some partial healing may occur. Industrial noise at levels of about 90 to 110 dB also causes permanent hearing loss if experienced over a period of months or years.

Most modern industrialized countries have produced legally enforceable maximum noise levels for workers. The Occupational Safety and Health Act of 1970 (OSHA) developed maximum noise exposure standards for all employees, with the exception of farmers and construction workers (Table 2.4.9). Vibration is characterized by its frequency, acceleration, and direction. The resonance band for the body is 4–8 Hz. Long-term exposure to one gravity unit (1 g) of vibration at this frequency can affect people's health.

The following forms of vibration are of most concern to industry:

1. *Whole body vibration.* Most heavy construction equipment, trucks, and buses produce vibrations, predominantly vertical, with frequencies in the 0.1–20 Hz range and with accelerations generally less than 0.2 g but with peaks up to 0.4 g (26). These values would be acceptable for an 8-hour continuous exposure, but longer exposures might be associated with discomfort and physiological responses. Some of the known effects of vibration in the range of 2–20 Hz at 1 g acceleration (one g is equivalent to an acceleration of about 10 m/s^2) include abdominal pain, loss of equilibrium, nausea, muscle contractions, and shortness of breath. The vibration characteristics can be measured by mounting accelerometers on the vibrating surface.
2. *Sequential vibrations: hands.* For sequential vibrations to the hands, accelerations in the range of 1.5 g–80 g and frequencies from 8–5000 Hz are of concern. Raynaud's phenomenon is a vibration-associated circulatory disturbance of the fingers that results in stiffness, numbness, pain, and blanching of the fingers and a loss of strength. People who develop this problem are unable to do fine manipulative tasks in cool or cold environments and may lose endurance for sustained-holding tasks in warmer temperatures.

Any noise and vibration control problem may be expressed as a source, path, and receiver problem. The best approach is to reduce the dominant noise or vibration source; next best is to interfere with the paths of noise and vibration. Examples of this are the use of enclosures, barriers, absorbing material, vibration isolators, and vibration damping material. Figure 2.4.10 shows a comparison range of noise and vibration reduction methods applied to a typical machine. Approximate A-weighted sound level (SL) reductions are given.

Personal protective equipment, such as acoustic booths, earplugs, earmuffs, and helmets may be necessary to consider in cases where noise reduction at the source or along the path is difficult or too expensive. However, personal protective equipment is uncomfortable to wear, easy to take off, and should be used only as a last resort after engineering controls to reduce noise have failed.

2.4.4.2 Occupational Stress and Shiftwork

Occupational Stress

Stress is defined medically as the nonspecific response of the body to any demands. Useful for the human factors specialist is the more psychologically oriented definition: stress is a substantial imbalance between the demands of the tasks and/or the environment and the human's capability to handle those demands successfully. Some examples of demands or stressors are:

Physical stressors: temperature, climate, noise, vibration.

Task stressors: information overload, work underload, time pressure, physical dangers, shift work.

Social and organizational stressors: work-group conflicts, work-role ambiguity, job uncertainty.

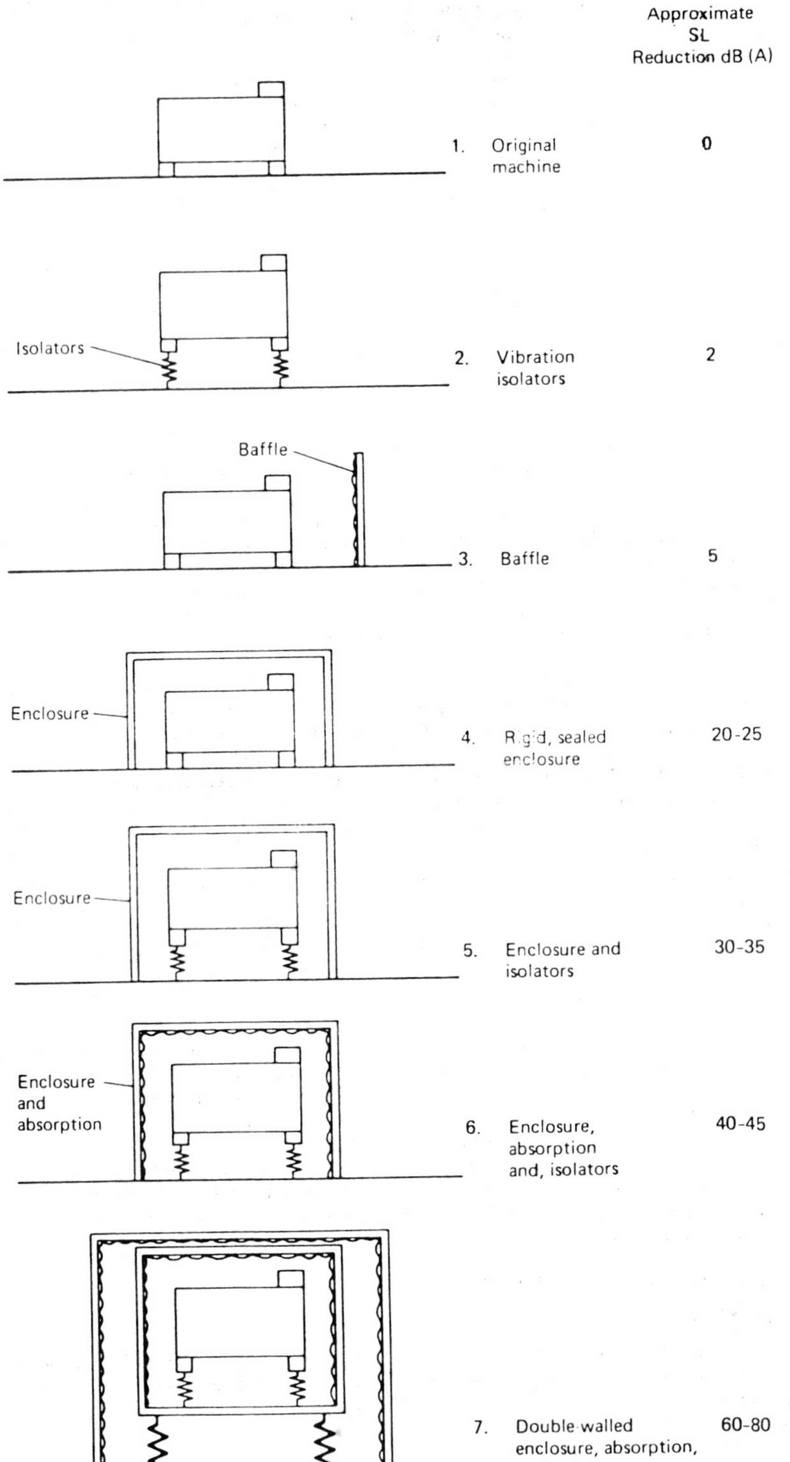

Fig. 2.4.10. Noise and vibration reduction methods. *Source*: Ref. 27. Reprinted by permission of John Wiley & Sons, Inc.

Strain is the effect of stressors on the worker, mediated by the person's ability to cope with the demands. While strain in physical work can be related to physiological measures such as heart rate and oxygen consumption, strain associated with mental workload is more difficult to measure.

The most common concept links stress to arousal, a kind of physiological alertness. Increasing arousal increases performance to an optimum and then decreases it. Although the empirical validity of the Yerkes-Dodson[28] law is widely accepted, it is not universally supported experimentally. It offers only general guidance to the design of a task. The main question is, where exactly is this optimal level? This mostly depends on task difficulty. A difficult task requires more arousal than an easy one but performance on an easy task would not be optimum with increased levels of arousal.

Information load is becoming a more important stressor as more and more jobs require cognitive effort. The main effects of informational and mental overload, such as in air-traffic control, are a systematic performance decrement, especially with tasks that require selection and execution of responses. Effects often include narrowing the focus of attention, decreasing flexibility of visual scanning, and rigidity of control motor activities. Effects of information underload, e.g., in repetitive tasks or in long-distance truck driving create physiological and performance symptoms consistent with stress and arousal.

Stress in Shiftwork. Shift work refers to work periods that fall largely after 4:00 pm and before 8:00 am. In 1978, U.S. evening shifts accounted for 4.9 million workers, and night shifts tallied another 2.1 million workers. Overall, 16% of all workers were shift workers.

Stress in shift work may result from the discrepancy between the time-structure of behavior (work, sleep) and the physiological functions geared to the normal daily routine. The circadian rhythm of physiological functions are typically recorded by heart and respiration rates, electroencephalographic (EEG) waves, hormone levels, and most commonly by body temperature. Shift work initiates adaptation processes of the circadian rhythm. The re-entrainment of most physiological functions takes place within a period of 3–14 days.[29] Of special interest is the duration of re-entrainment from night shift to day shift. In experiments with 21 night shifts, re-entrainment had not, in fact, been completed even after 4 days (Fig. 2.4.11).

The question of long-term adaptation of shift work is yet unsolved. There seems to be evidence of a long-term adjustment relative to part-timers. Personality differences in adaptation to shift work have been found.

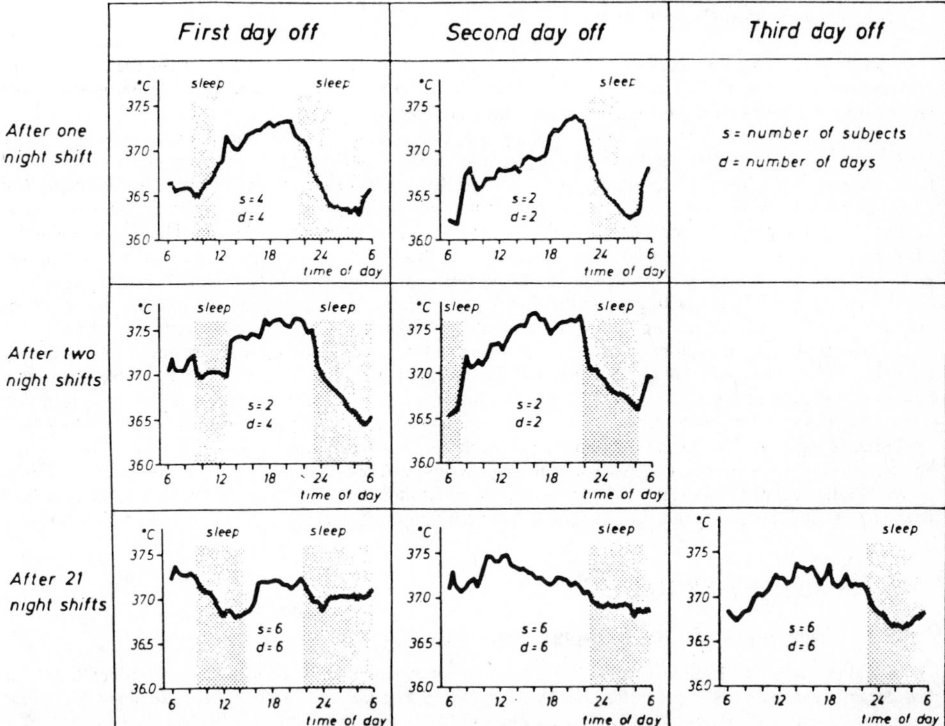

Fig. 2.4.11. Re-entrainment of circadian rhythm (rectal temperature) after 1, 2, and 21 experimental night shifts. *Source*: Ref. 29.

Ostberg[30] made the distinction between morning and evening types of subjects; the former has more pronounced difficulties in adapting to night work.

Another source of stress is social life. Most shift workers consider the disturbances of their family and social life to be more serious than the physiological or organizational disturbances.

The following recommendations made by Rutenfranz et al.[29] may be of assistance in constructing shift schedules:

1. Single night shifts are better than consecutive night shifts. If consecutive night shifts are required, short sequences are recommended. Due to re-entrainment of circadian rhythms, a sequence of night shifts longer than seven days would be acceptable. However, for psychosocial reasons, most workers need either to change their shift or to have some rest days.
2. For preventing the harmful effects of sleep deprivation, a substantial recovery period of at least 24 hours should be allowed after each night shift.
3. The cycle of a shift system should not be too long; four weeks are better than forty weeks. Short cycles and regular system of rotation helps workers and their families to plan their social life.

2.4.4.2 Human Error and Product Quality

Traditional approaches to reducing error in production rely mainly on personnel selection, placement, and training, supplemented by motivational campaigns to eliminate production defects. But production quality relates to two basic sources: human reliability and the characteristics of the work situation that predispose to error.

Workers in industrial settings will average about one error per 1000 or 10,000 acts. With highly intensive inspection, about 80–98% of their errors will be detected and corrected. The reason that errors seem to be so frequent and are actually a major problem in industry is due to the large amount of potential for error in the production process. External factors influencing human performance reliability can be readily modified by taking into account human capabilities and design ergonomics. These factors relate to

adequate job and/or task design,
adequate work space and optimal work layout,
adequate methods of handling, transportation, storing, or inspecting equipment, and
adequate environmental conditions

Two methods to identify existing or potential human performance problems are the man–machine systems analysis and the Technique for Human Error Rate Prediction (THERP). The former relies mostly on an ergonomic approach to checking various aspects of human–machine interfaces.

THERP is a technique that uses fault-tree analysis to calculate error probabilities for accomplishing certain kinds of tasks, with special emphasis on nuclear power plant applications.[31] For each task, technical and performance components involved are analyzed. The probability of the occurrence of any failure is determined as a joint probability of those basic system components. Probability estimates for technical failures of tools, equipment, and facilities resulted mostly from empirical findings or from expert judgment. The error probabilities of operator's performance such as errors of commission in check reading or use of wrong switches were mostly experimentally determined or also rated by human factors experts.

The most effective strategy to prevent the occurrence of errors emphasizes worker's training and selection, and human engineering of the work situation. The Error Cause Removal (ECR) program[31] is similar to the Quality Control (QC) circle program developed in Japan in 1963 and found highly successful in solving quality control problems. The essence of QC circles is participative problem solving. A group of 8–10 people performing similar or interrelated tasks forms a circle on a voluntary basis. They are given special training in quality control techniques. The group identifies specific causes of poor quality and recommends possible solutions.

Certain elements of the ECR team and the QC circles are much the same, e.g. the problem responsibility, and the crossover among levels of management, engineering, and production. Both teams use the motivational power of self-regulated work teams in order to achieve better product quality. Similar attempts to improve both production quality and quantity are used by various European companies, including Philips, Volvo, and Volkswagen.

2.4.5 COMPUTER ERGONOMICS

2.4.5.1 Ergonomics of Visual Display Terminals

Gantz and Peacock[32] estimate that the total computer power available to U.S. business increased tenfold in the last decade, and that it is expected to double every two to four years. Currently there are about 15 million computers, terminals, and electronic office machines in the United States[33].

Computers are still not as widely accepted as generally believed. In a study by Zoltan and Chapanis[34] on what professionals think about computers, the loudest complaints were related to such topics as dehumanizing, depersonalizing, and impersonal workplace. The respondents thought that computers were difficult and complicated to handle and that computing languages were not simple to learn and to understand.

The main sources of complaints of visual display terminal (VDT) operators are poorly working equipment, screen reflections, low screen luminance, flicker of the display, low contrast between characters and background, working postures, and job design problems. The following outlines provide some human factor considerations in order to overcome these problems. They are based on various experimental and study results[35]:

1. *Environmental design.* General room illumination should be betwen 500–700 lux, if screen and hard-copy work is required. Pure screen activities require about 300 lux of general room illumination. Individually adjustable workplace illumination is an advantage.

   Glare reduces contrast and increases the amount of visual effort. Proper placement of the screen is the most effective means of glare control. Screen filters can be added, although they diminish contrast. Flat keyboards with nonreflecting surfaces and the possibility of tilting the screen and moving it freely on its support may be a help in avoiding glare. Positive presentation (dark symbols on a light background) reduces sensitivity to discomfort glare, diminishes reflections on the screen, and improves adaptation conditions for the eye, when screen and hard-copy work is required.[36]
2. *Workstation design.* The screen should be approximately at eye level and tilted away from the user only slightly. A distance of about 450–500 mm (18–20 in.) is recommended. The possibility of tilting the screen has advantages in fitting the requirements of the operator as well as avoiding reflections.

   The keyboard should be detachable from the screen, so that each can be independently adjusted in height. Desks with independent adjustments for the keyboards and screen are helpful. The recommended dimensions for a seated VDT workplace without a footrest are those given in Fig. 2.4.8, section workplace design.

2.4.5.2 Computer Input and Output Devices

Virtually all keyboard-type data entry devices use the standard QWERTY typewriter layout. The 19th-century layout was developed in order to balance the mechanical loads. Dvorak et al.[37] criticized the standard QWERTY typewriter layout because it assigned most of the work to the left hand and had unbalanced finger loads. He developed his own keyboard (Fig. 2.4.12) in which the more common letters were on the home row. In tests with inexperienced or specially trained operators, productivity was up to 74% higher on the Dvorak keyboard as compared with the QWERTY keyboard.[38] The all-encompassing use of the QWERTY keyboard has made implementation of the Dvorak keyboard difficult, however. It takes approximately 28 days for QWERTY-trained typists to reach the same keying rates on the Dvorak keyboard.[38]

The digits 0–9 of a 10-key keyboard may be laid out in several ways. The layout of keys on a standard push-button telephone shows the lower numbers at the top, and the higher numbers in the bottom row nearest 0. Unlike the push-button telephone, on a calculator keyboard the higher numbers are in the top row. The sequence of numbers moves from 0 upward to 9; the more frequently keyed numbers (0, 1, 2, 3) are closer to the home row of a computer console keyboard.

The population stereotype favors the telephone arrangement. This arrangement has also been used for television channel selection and many appliances. But there is little empirical evidence that one is better than the other.[39] However, the calculator arrangement was established by tradition and has been employed on almost every electronic calculator keyboard in existence. It would be difficult to change for the reasons cited for the typewriter keyboard.

Entering and manipulating text, programs, and data requires position definition by cursor movement in two-dimensional space. Card et al.[3] studied the relative efficiency of four devices for text editing: the mouse, an isometric rate-controlled joystick, and conventional keys such as the step keys and text keys on teleprinters. The mouse is a small box with two little wheels beneath it. As the user rolls the mouse around on a flat surface, the cursor moves correspondingly. The findings, shown in Table 2.4.10, suggest that the mouse was slightly faster and less subject to errors as compared with all other entry devices.

Performance models of each of these devices showed that printing time for the analogue devices (mouse and joystick) is proportional to the log of the ratio of target distance and target size, as given by Fitt's law. The pointing time for the key devices is proportional to the number of keystrokes. Further analysis showed that the time to print using the mouse is not limited by the device itself, but by the information processing rate of the human eye–hand coordination system.

In an experiment[40] comparing positioning via light pen, joystick, and mouse, the critical measure was the time to move from the keyboard, indicate position with one of the devices, and then return to the keyboard. Results indicate that experienced subjects were fastest with the mouse, a little slower with light pen, and a lot slower with the joystick. Inexperienced subjects were fastest with the light pen, a little slower with the mouse, and a lot slower with the joystick. Thus it appears that the mouse and light pen are both good positioning devices, but that the joystick seems more time consuming. Most users prefer the mouse for long data entry session because it avoids arm fatigue as compared with the use of the light pen for long periods of time.

The current state of automated speech recognition limits its use to relatively simple input tasks. Even in these there are problems with different speakers, noise, and vocabularies of restricted size. Speech input seems

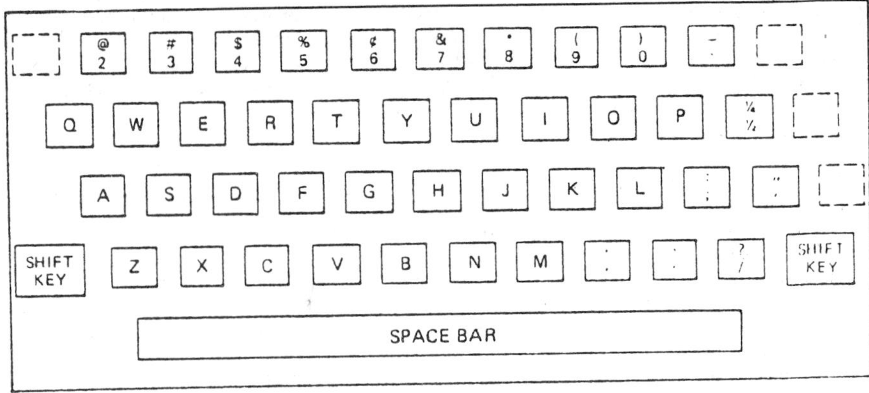

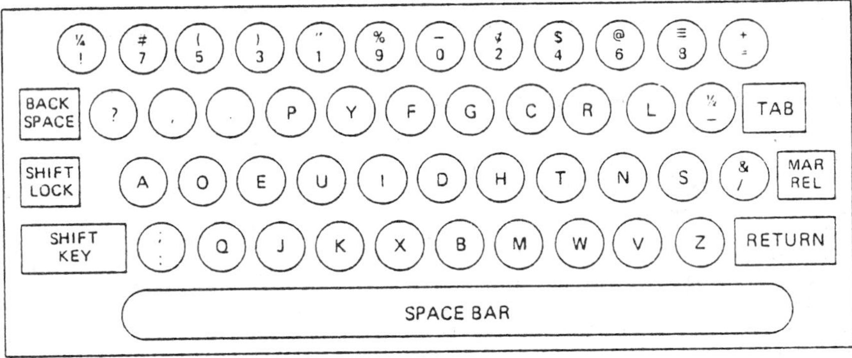

Fig. 2.4.12. The QWERTY keyboard, above, and the Dvorak arrangement, below. Lower figure from Ref. 37.

like a very desirable and natural input mode, but it is not clear whether it will prove to be widely applicable for human–computer interaction tasks.[41]

Although teletypewriters and alphanumeric CRT displays are the most common forms of output devices used in computer systems, there are numerous other possibilities: plasma displays, light-emitting devices (LEDs), liquid crystal displays (LCDs), tactile displays, and audio displays, including synthesized speech. Two handbooks of CRT displays cover a substantial number of design problems related to output devices.[42,43]

Table 2.4.10. Overall Printing Times for Four Data Entry Devices

		Movement Time for Nonerror Trials							
		Homing Time		Positioning Time		Total Time		Error Rate	
Device	Trials *N*	*M* (s)	*SD* (s)	*M* (s)	*SD* (s)	*M* (s)	*SD* (s)	*M*	*SD*
Mouse	1973	0.36	0.13	1.29	0.42	1.66	0.48	5%	22%
Joystick	1869	0.26	0.11	1.57	0.54	1.83	0.57	11%	31%
Step Keys	1813	0.21	0.30	2.31	1.52	2.51	1.64	13%	33%
Text Keys	1877	0.32	0.61	1.95	1.30	2.26	1.70	9%	28%

Source: Ref. 3.

2.4.5.3 Human–Computer Interaction

The task, the user, and the computer are the structural components of a human–computer system. They vary in many different aspects. Computer systems address different task domains, and they have different models of tasks in any given domain. Users vary widely in general intellectual ability, experience with the particular computer system, knowledge of the topic domain of the system, cognitive style, and perceptual-motor skills. User-interface aspects of computers vary in system architecture, dialogue style, command syntax, and input devices.

Task, user, and computer determine the performance of a system. The basic performance variables of a human–computer system are concerned with what tasks the system can do (functionality), how long it takes to acquire the functionality (learning), how long it takes to accomplish tasks (time), how frequently errors occur, and how consequential they are, how well tasks are done, and how robust (quality) the system is in the face of unexpected conditions. Card et al.[3] investigated computer text-editing as a prototypical human–computer interaction task. They developed a set of quantitative design rules for human–computer interaction, which will be discussed in more detail. Norman[44] suggested a quantitative scheme for evaluating the tradeoffs in design, e.g., to assess the relative psychological importance of user satisfaction with the use of an analogue pointing device against a digital pointing device.

The process of computer system design consists of a set of design functions. Some functions are:

Evaluation of a text-editor system against other systems.

Analysis of actual user performance with a specific keyboard in order to optimize time parameters.

Structural improvement of an existing pointing device.

Table 2.4.11 lists several design functions. Evaluation refers to the situation in which the structure of the system or part of the system has been specified and its performance needs to be understood. Parametric design refers to situations where the structure of the human–computer system is relatively fixed and there are a set of quantitative parameters of the structure to be determined, e.g., time to set up a text editor or maximum speed of data entry performed by a user. Structural design refers to the process in which a part of the system is configured or restructured to satisfy specific requirements.

A study to evaluate nondisplay and display editors was performed by Roberts.[45] She employed four evaluation scores. *Functionality* is measured by having expert users rate whether each task out of 212 tasks can be accomplished with the editor. *Learning* is measured as the average learning time per task for four novices. The *time* it takes experts to perform error free a benchmark set of about 60 tasks is also measured experimentally. The *error score* is the average percentage of time the four expert users spent correcting errors. In Table 2.4.12 the score is given as a percentage of the error-free time.

This methodology provides a multidimensional evaluation of editors. As Table 2.4.12 indicates, the major differences are between the nondisplay editors (TECO, WYLBUR) and the display editors (all the others). With the exception of NLS on error time and GYPSY on functionality, the display editors are better on all

Table 2.4.11. Classification of Design Function for Human Computer Systems

Design Function	Examples
Evaluation	
Compare systems	Compare on given performance variables, e.g., editor comparison, data entry, device evaluation.
Evaluate systems	Compare against some standard, e.g., layout system, editor evaluation.
Parametric Design	
Optimize parameter	Find best values on given performance variables, e.g., editor vs. typewriter, CRT panel layout.
Analyze sensitivity	Relate parameter value to performance variables, e.g., speed, accuracy, error, learning.
Structural Design	
Identify opportunity	Find place where system can be improved, e.g., installation of new display, alteration of character size, contrast.
Diagnose problem	Pinpoint structural component causing problem, e.g., low information rate, slow data entry device.
Generate improvement	Find structural change.

Adapted from Card et al.[3]

Table 2.4.12. Evaluation Summary of Eight Text Editors

	Evaluation Dimensions			
Editor	Time $M \pm CV$ (s/task)	Errors $M \pm CV$ (% time)	Learning $M \pm CV$ (min/task)	Functionality (% tasks)
TECO	49 ± .17	15 ± .70	19.5 ± .29	39
WYLBUR	42 ± .15	18 ± .85	8.2 ± .24	42
EMACS	37 ± .15	6 ± 1.2	6.6 ± .22	49
NLS	29 ± .15	22 ± .71	7.7 ± .26	77
BRAVOX	29 ± .29	8 ± 1.0	5.4 ± .08	70
WANG	26 ± .21	11 ± 1.1	6.2 ± .45	50
BRAVO	26 ± .32	8 ± .75	7.3 ± .14	59
GYPSY	19 ± .11	4 ± 2.1	4.3 ± .26	37
M(M)M(CV)	32 .19	11 1.0	8.1 .24	53
CV(M)	.30	.54	.58	.28

Two of these are nondisplay editors (TECO, WYLBUR), all the others are display-based editors. The Coefficient of Variation (CV) = Standard Deviation/Mean is a normalized measure of variability. The CVs on the individual scores indicate the amount of between-user variability. The M(CV)s give the mean between-user variability on each dimension, and the CV(M)s give the mean between-editor variance on each dimension. *Source:* Ref. 45.

performance dimensions. They are up to twice as fast to use and have about 50% more functionality. On learning, TECO stands out as taking nearly three times as long to learn as the others.

To predict the performance of a system, the designer must construct a specific performance model from the system's structural specification and then use the model to generate a prediction. Detailed performance models are typically based on an information processing analysis as outlined in Section 2.4.2. The following parametric design example uses a simple cognitive model. The structural design example for the maximum velocity of an analogue pointing device employs an information processing model.

A study to predict the sequence and time consumption of user actions in a computer text-editing system was performed by Card et al.[3] Cognitive models composed of a small set of Goals, Operators, Methods, and Selection rules gave a reasonable quantitative account for the behavior. The GOMS model for the manuscript-editing task predicted the sequences of user actions in the task reasonably well. It predicted a user's choices of methods about 80–90% of the time and the actual operators in sequence 80–100% of the time.

The time prediction error for subtasks of a text-editing task gave a root mean square (RMS) error of 33% of the mean observed time. If the unit of prediction were the whole manuscript rather than a subtask, the RMS error for predicting the time to edit the whole manuscript (70 subtasks) would be 4%. The error thus seems to be in the same range (about 5%) as that sometimes cited by industrial engineers for predetermined time system predictions of invariable-sequence physical activity.

The next example refers to structural design and stresses an information processing model.[3] The support hardware of an analogue pointing device, such as a mouse, should set the system parameters so that the maximum velocity will not be exceeded by a fast user. In Section 2.4.2, Fitt's law was derived from the assumption that macroscopic movements toward a target are made up of micromovements with constant error. The maximum velocity v_{max} will be reached on the first cycle. The average velocity on the first cycle is[3]:

$$v_{\max} = (1 - e)/(t_p + t_c + t_m)D$$

where D is the total distance to the target from the starting point and t_p, t_c, and t_m are the information processing time, respectively. Using Vince's[6] estimate of $e = 0.07$ and the fastest value for $t_p + t_c + t_m = 190$ ms, maximum velocity on a $D = 35$ cm screen diagonal will be

$$v_{\max} = 4.9\ D \text{ cm/s}$$

$$v_{\max} = 171 \text{ cm/s}$$

Figure 2.4.13 gives the median values recorded from an experiment. The median values are plotted as a straight line; the dotted lines indicate the ranges in which the points are predicted using the information processing time ranges of 190 and 260 ms. Step-tracking data are from Craik and Vince[6], maximum velocity data for mouse from Card et al.[3]

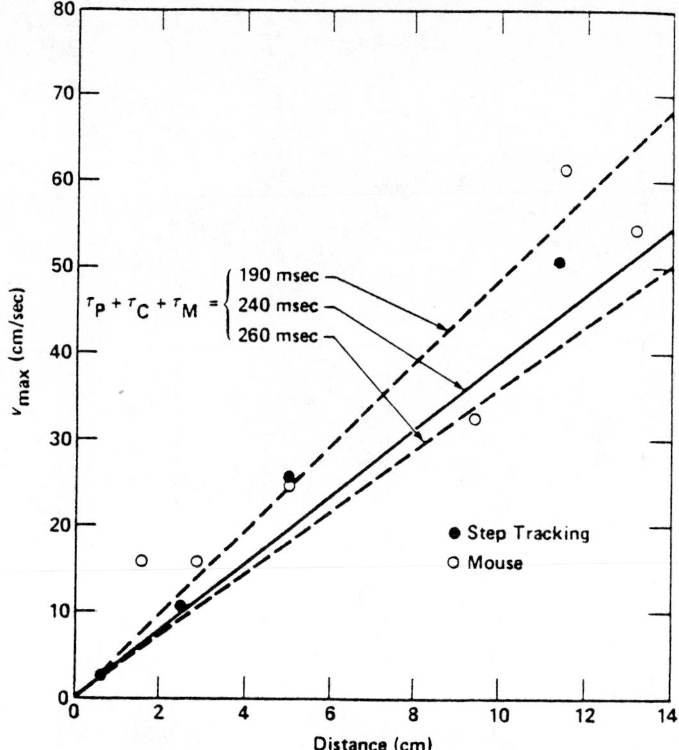

Fig. 2.4.13. Maximum velocity of cursor as a function of distance in a positioning task. *Source*: Ref. 3.

As the results indicate, positioning time for continuous movement devices is predicted by Fitt's law and derived from an information processor model. For key devices, positioning time is proportional to the number of keystrokes.

2.4.5.4 Robots and Expert Systems

The Electric Machinery Law of 1971 in Japan defined an industrial robot as an all-purpose machine, equipped with a memory device, and a terminal device capable of rotation and of replacing human labor by automatic performance of movements.

Industrial working robots are classified in Japan into six types. Production figures in 1982 are in parentheses[46]:

1. Manual manipulator—operated by a human (1942 units).
2. Fixed sequence control robot—manipulator whose control motions and preset information are fixed (13,483 units).
3. Variable sequence control robot—a manipulator with fixed control motions but adjustable preset information (1343 units). This is called a nonservo, pick-and-place robot, or a bang-bang robot in the United States.
4. Playback robot—a manipulator that can be taught by being moved through the motions. Welding and painting robots are of this type (2027 units). The U.S. term is programmable robot.
5. NC robot—operates under numerical control by paper tape or cards. Transfer robots incorporated into flexible manufacturing systems (FMS) belong to this type (992 units). The U.S. term is computerized robot.
6. Intelligent robot—has sensory (visual and/or tactile) and decision-making functions. Inspection robots with an "eye" at the end of the arm are of this type (131 units).

The Robotics Institute of America (RIA) defines a robot as a "manipulator designed to move material, parts, tools, or specialized devices through variable programmed motions for the performance of a variety of tasks." Thus, the U.S. definition of robots eliminates the manual manipulators and fixed sequence machines. In 1982, there were 6301 installed robots in U.S. industry as compared to 33,961 units in Japan. Robot production in

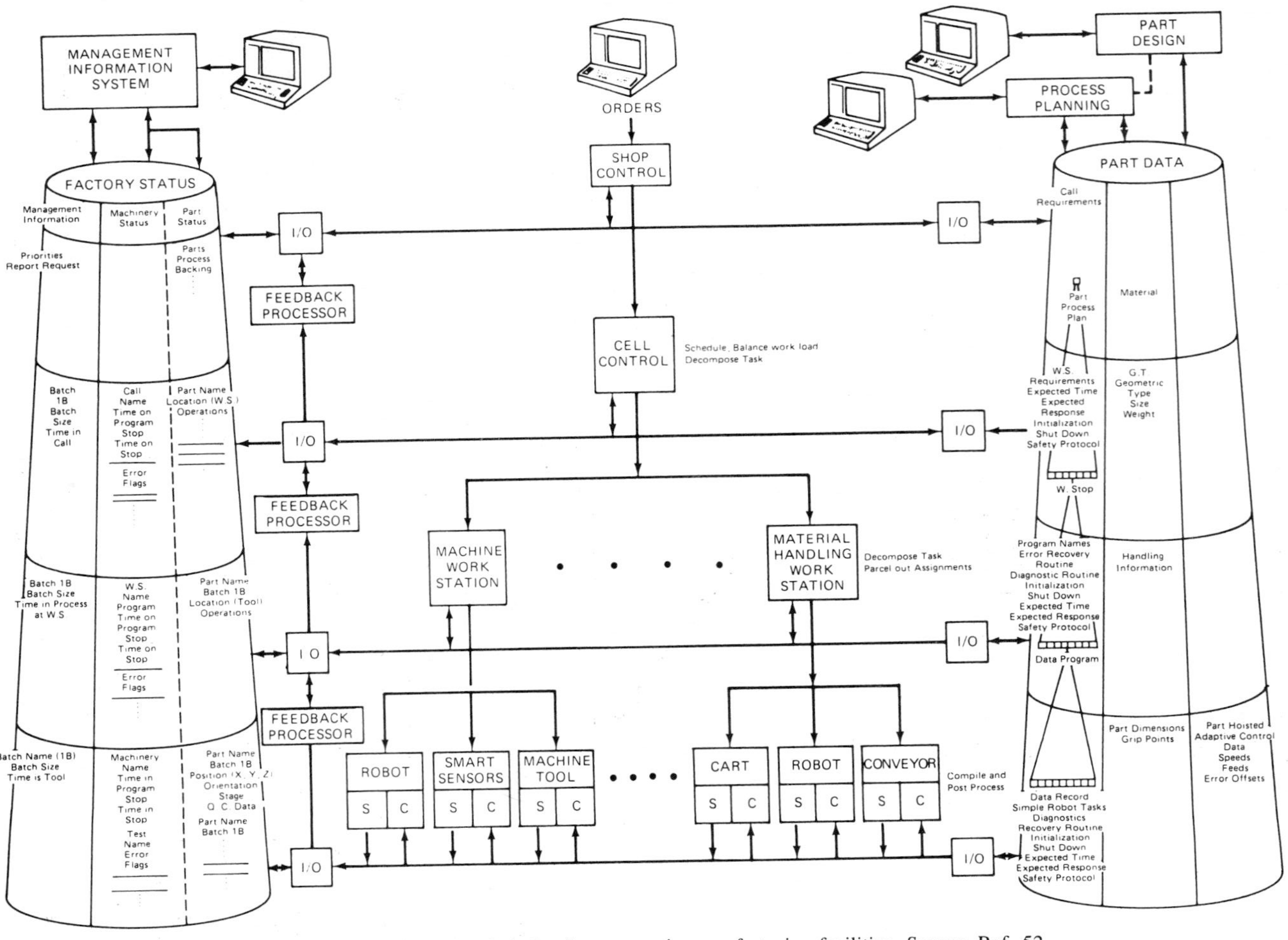

Fig. 2.4.14. Organizational design for automatic manufacturing facilities. *Source*: Ref. 52.

Table 2.4.13. Social Impacts of Robot Diffusion

Positive Impacts	Negative Impacts
1. Improvement of productivity and quality	Elimination of pride in old skills; unemployment problems
2. Increase in health and safety of workers by eliminating unhealthy and dangerous jobs	Safety and psychological problems of robot interaction with human
3. Elimination of short-cycled, repetitive, machine-paced tasks for workers	Nonproportional production capacity to the size of the labor force
4. Ease of keeping quality standards Ease of production scheduling	Shortage of engineers and newly trained skilled workers
5. Creation of new high-level jobs	Great movement of labor population from the second to the third sector of industry

the United States increased from 1981 to 1982 from 1269 to 1601 units, in Japan from 8182 to 14,937 units. These figures are based on the U.S. robot definition.[47]

Chief motives for the introduction of robots in industry are the reduction of labor cost, enhancement of efficiency, improvement of quality, and avoidance of dangerous or dirty jobs. There exists a long-range impact on employment. The loss of jobs due to the introduction of industrial robots in a German automobile plant was at a rate of 4 workers per industrial robot. The number of jobs created by maintenance and repair was 0.3 places, by investment in robots 0.5 workplaces per industrial robot. This results in a loss of 5 jobs to 1 gained.[48]

There are three main human issues affecting the productive use of industrial robots:

1. Allocation of functions between humans and robots and the use of industrial methods utilize robots effectively. These methodologies have been detailed elsewhere,[49] as have the methodologies for allocating functions between automation, robots, and humans.[50]
2. In planning for robotics systems there are many human aspects to be considered,[51] including the impact of robotics on information flow in an organization and the social effects of robot diffusion.
3. The human aspects of using robots in flexible manufacturing systems.[52] In this case the major issues are allocation of tasks between humans and computer; input as to when decision aids are helpful to the operator supervising an FMS; how and where it is desirable to substitute an expert system using artificial intelligence principles for the human supervisor.[53]

As illustrated in Figure 2.4.14, in an automatic manufacturing system a complex network of process-control computers, numerically controlled robots and machines, and automated material handling and storage systems is introduced. To aid decision making in this environment, computer aided manufacturing systems were developed. Although these systems address decision making for anticipated problems, where an appropriate response can be programmed, such systems provide little or no support for unstructured, unforeseen problems. An information system that may handle such unstructured decision problems is often called "a decision support system" or more generally an "expert system."[54]

Decision support systems are developed by formulating the application problem, designing and constructing the knowledge base of expertise, developing schemes of inference and problem solving, winning the confidence of experts, and evaluating the programs for production versions. A summary of recent research related to expert systems can be found in Hayes-Roth et al.[55] Some of the major features of these systems, including the schemes or internal models of problem solving approaches used in defining structures for the knowledge bases, are reviewed by Feigenbaum.[56] Examples of these systems include assisting users in such tasks as (1) deducing molecular structures from the output of mass spectrometers, (2) consulting for mineral exploration, and (3) diagnosing blood infections.

2.4.6 FUTURE TRENDS

With the increased utilization of CAD/CAM, robotics and other computer-based manufacturing systems in industry, the role of the human factors specialist is expanding and becomes more critical. There are a number of reasons for this. One is that intelligent manufacturing systems are emerging, which are based on the notion of how humans functioned in the production of these systems. This requires the extraction of expert knowledge, the understanding of human decision making, and understanding of computer programming languages to model human decision making. The human decision-making process is task specific, hence artificial intelligence and expert systems are also task specific. A critical aspect for the successful building of these systems is cognitive engineering—a domain of experimental psychologists.

A second key area that emerges for the human factors specialists is design for ease of use and reduced errors. Since there are frequently a number of computer-based manufacturing systems on the market, the one that can be learned in the shortest possible time will frequently be in the greatest demand by consumers. Likewise, since humans in the future will have control over larger systems than they do now, both the economic and social impact of human error will increase significantly. In order to reduce system errors, good human factors will have to be applied to the design, planning, operation, and control of all manufacturing systems.

A third factor is that the conventional ergonomics issues that emerged around World War II are still critical issues in a sound manufacturing work environment.

REFERENCES

1. D. Meister, *Human Factors: Theory and Practice*. New York: Wiley, 1971.
2. A. Cakir, D. J. Hart, and T. F. M. Stewart, *Visual Display Terminals*. Chichester: Wiley, 1980.
3. S. K. Card, T. P. Moran, and A. Newell, *The Psychology of Human–Computer Interaction*. Hillsdale, NJ: Erlbaum, 1983.
4. R. C. Calfee, *Human Experimental Psychology*. New York: Holt, Rinehart & Winston, 1975.
5. G. A. Miller, "The Magical Number Seven Plus or Minus Two: Some Limits on Our Capacity for Processing Information," *Psychological Review*, **63** (1956), 81–97.
6. K. J. Craik and M. Vince, "Psychological and Physiological Aspects of Control Mechanism," *Ergonomics*, **6** (1963) 419–440.
7. F. Attneave, *Applications of Information Theory to Psychology*. New York: Holt–Dryden, 1959.
8. E. T. Klemmer, "Communication and Performance," *Human Factors*, **4** (1962) 75–79.
9. W. E. Hick, "On the Rate of Gain of Information," *Quarterly Journal of Experimental Psychology*, **4** (1952) 11–26.
10. R. Hymann, "Stimulus Information as a Determinant of Reaction Time," *Journal of Experimental Psychology*, **45** (1953) 188–196.
11. P. M. Fitts and C. M. Seeger, "S-R Compatibility: Spatial Characteristics of Stimulus and Response Codes," *Journal of Experimental Psychology*, **46** (1953), 199–210.
12. J. A. Roebuck, K. H. E. Kroemer, and W. G. Thomson, *Engineering Anthropometry Methods*, New York: Wiley, 1975.

12a. NASA, *Anthropometric Source Book; Vol. 1: Anthropometrics for Designers; Vol 2: Handbook of Anthropometric Data; Vol. 3: Annotated Bibliography of Anthropometry*, NASA Reference Publication 1024, Scientific and Technical Information Office, July 1978.

13. S. H. Snook, "The Design of Manual Handling Tasks," *Ergonomics*, **12** (1978), 963–985.
14. D. B. Chaffin, "Engineering Anthropometry and Occupational Biomechanics," in G. Salvendy, Ed., *Handbook of Industrial Engineering*. New York: Wiley, 1982.
15. Eastman Kodak Company, *Ergonomic Design for People at Work*, Vol. 1. Belmont, CA: Lifetime Learning Publications, 1983.
16. E. J. Mc Cormick and M. S. Sanders, *Human Factors in Engineering and Design*, 5th ed. New York: McGraw-Hill, 1982.
17. W. F. Grether and C. A. Baker, "Visual Presentation of Information," in H. P. van Cott and R. G. Kinkade, Eds., *Human Engineering Guide to Equipment Design*. Washington, DC: U.S. Superintendent of Documents, 1972.
18. H. P. Van Cott and R. G. Kinkade, Eds., *Human Engineering Guide to Equipment Design*. Washington, DC: U.S. Superintendent of Documents, 1972.
19. C. A. Bennett, "Lighting," in G. Salvendy, Ed., *Handbook of Industrial Engineering*. New York: Wiley, 1982.
20. S. Gael, *Job Analysis Handbook*. New York: Wiley, 1987.
21. E. J. McCormick, "Job and Task Analysis," in G. Salvendy, Ed., *Handbook of Industrial Engineering*. New York: Wiley, 1982.
22. W. Rohmert and K. Landau, *A New Technique for Job Analysis*. New York: Taylor and Francis, 1983.
23. J. R. Hackman and G. R. Oldham, *Work Redesign*. Reading, MA: Addison-Wesley, 1980.
24. F. W. Taylor, *The Principles of Scientific Management*. New York: Harper & Row, 1911.
25. O. Ostberg and J. Enqvist, "Robotics in the Workplace: Robot Factors, Human Factors and Humane Factors," in H. W. Hendrick and O. Brown, *Proceedings of the First International Symposium on Human Factors in Organizational Design and Management*. Honolulu, Hawaii, New York: Elsevier, 1984.
26. D. E. Wassermann and D. W. Badger, "Vibration and the Worker's Health and Safety," *Technical Report No. 77*, U.S. Department of Health, Education and Welfare, National Institute for Occupational Safety and Health, Washington, DC, 1973.
27. M. J. Crocker, "Noise and Vibration," in G. Salvendy, Ed., *Handbook of Industrial Engineering*. New York: Wiley, 1982.

28. R. M. Yerkes and J. D. Dodson, "The Relation Strength of Stimulus to Rapidity of Habit Formation," *Journal of Comparative and Neurological Psychology*, **18** (1908), 459–482.

29. J. Rutenfranz, P. Knauth, and D. Angersbach, "Shift Work Research Issues," in L. C. Johnson, D. I. Tepas, W. P. Colquhoun, and M. J. Colligan, Eds., *The Twenty-Four Hours Workday: Proceedings of a Symposium on Variations in Work–Sleep Schedules*, U.S. Department of Health and Human Services, National Institute of Occupational Safety and Health, Cincinnati, 1981.

30. O. Oestberg, "Circadian Rhythms of Food Intake and Oral Temperature in Morning and Evening Groups of Individuals," *Ergonomics*, **16** (1973), 203–209.

31. A. D. Swain and H. E. Guttmann, *Handbook of Human Reliability Analysis with Emphasis on Nuclear Power Plant Applications*, U.S. Nuclear Regulatory Commission, NRC FIN A 1188, Washington, DC, June 1983.

32. J. Gantz and J. Peacock, "Computer Systems and Services for Business and Industry," *Fortune*, **103** (1981), 39–84.

33. U.S. Bureau of the Census, *Statistical Abstracts of the United States*, 100th ed.: Table Number 685, p. 415, Washington, DC: U.S. Government Printing Office, 1984.

34. E. Zoltan and A. Chapanis, "What Do Professional Persons Think about Computers?" *Behavior and Information Technology*, **1** (1982), 55–68.

35. M. J. Smith, "An Overview of the Health Effects Associated with the Automated Work Station and Results of Recent Research," National Institute for Occupational Safety and Health, Cincinnati, 1983.

36. G. W. Radl, "Visual Display Units: Present Status in Ergonomics Research and Applications in the Federal Republic of Germany," in E. Grandjean and W. Vigliani, Eds., *Ergonomic Aspects of Visual Display Terminals*. London: Taylor and Francis, 1980.

37. A. Dvorak, N. L. Merrick, W. L. Dealey, and G. C. Ford, *Typewriting Behavior*. New York: American Book, 1936.

38. E. P. Strong, *A Comparison Experiment in Simplified Keybord Retraining and Standard Keybord Supplementary Training*. Civil Services Administration, Washington, DC, 1956.

39. R. Conrad and A. Hull, "The Preferred Layout for Numeric Data Entry Keysets," *Ergonomics*, **11** (1968), 165–173.

40. W. K. English, D. C. Engelbart, and M. L. Bermann, "Display Selection Techniques for Text Manipulation," *IEEE Transactions on Human Factors in Electronics*, **8** (1967), 5–15.

41. B. Shneiderman, *Software Psychology*. Boston: Little, Brown, 1980.

42. D. A. Shurtleff, *How to Make Displays Legible*. La Mirada, CA: Human Interface Design, 1980.

43. W. O. Galitz, *Handbook of Screen Format Design*. Wellesley, MA: Q.E.D. Information Science, 1981.

44. D. A. Norman, "Design Principles for Human Computer Interfaces," in *Proceedings of the CHI '83 Conference on Human Factors in Computing Systems*, Boston, 1983.

45. T. Roberts and T. Moran, "Evaluation at Text Editors," in *Proceedings, Human Factors in Computer Systems*, Maryland, 1982.

46. K. Noro and Y. Okada, "Robotization and Human Factors," *Ergonomics*, **26** (1983) 985–1000.

47. P. H. Aron, *The Robot Scene in Japan: An Update*. New York: Daiwa Securities America, Ind., 1983.

48. P. Kalmbach, R. Kasiski, F. Manski, O. Mickler, W. Pelull, and W. Wobbe-Ohlenburg, *Robots' Effect on Production, Work and Employment*, Frankfurt (M): Campus Verlag, 1981 (in German).

49. S. Nof, J. Knight, and G. Salvendy, "Effective Utilization of Industrial Robots: A Job and Skills Analysis Approach," *AIIE Transactions*, **12** (1980), 216–225.

50. J. Kamali, C. L. Moodie, and G. Salvendy, "A Framework for Integrated Assembly Systems: Humans, Automation and Robots," *International Journal of Production Research*, **20** (1982), 431–448.

51. G. Salvendy, "Review and Reappraisal of Human Aspects in Planning Robotics Systems," *Behavior and Information Technology*, **2** (1983), 263–287.

52. J. Albus, "Industrial Robot Technology and Productivity Improvement," in *Exploration Workshop on the Social Impact of Robotics*, Office of Technology Assessment, U.S. Congress No. 30-240 0-82-2.

53. S. L. Hwang, J. Sharit, and G. Salvendy, "Optimal Management Strategies in the Design and Operation of Flexible Manufacturing Systems," in *Proceedings of the 27th Annual Meeting of the Human Factors Society*, Norfolk, Virginia, 10–14 October 1983.

54. S. L. Hwang, W. Barfield, Jr., T. C. Chang, and G. Salvendy, "The Role of the Human in the Operation and Control of Flexible Manufacturing Systems," in *Proceedings of the 7th International Conference on Production Research*, Windsor, Ontario, Canada, 22–24 August 1983. A modified version of proceedings paper appeared in *International Journal of Production Research*, **22**(5) (1984), 841–856.

55. S. Y. Nof, A. B. Whinston, and W. I. Bullers, "Control and Decision Support in Automatic Manufacturing Systems," *AIIE Transactions*, **12** (1980), 156–169.

56. F. Hayes-Roth, D. A. Waterman, and D. B. Lenat, *Building Expert Systems*. Reading, MA: Addison-Wesley, 1983.

57. E. A. Feigenbaum, "The Art of Artificial Intelligence—Themes and Case Studies of Knowledge Engineering," in S. P. Ghosh and L. Y. Liu, Eds., *Proceedings of the American Federation of Information Processing Societies*, **47**, Montvale, NJ: AFIPS Press, 1978.

BIBLIOGRAPHY

Eastman Kodak Company. *Ergonomic Design for People at Work*. Belmont, CA: Lifetime Learning Publications, 1983.

Kantowitz, B. H. and R. D. Sorkin, *Human Factors: Understanding People–System Relationships*. New York: Wiley, 1982.

McCormick, E. J. and M. S. Sanders. *Human Factors in Engineering and Design*, 5th ed. New York: McGraw-Hill, 1982.

Salvendy, G., Ed. *Handbook of Industrial Engineering*. New York: Wiley, 1982.

Salvendy, G., Ed. *Human–Computer Interaction*. Amsterdam: Elsevier, 1984.

Salvendy, G., Ed. *Handbook of Human Factors*. New York: Wiley, 1987.

CHAPTER 2.5

WORK MEASUREMENT

DEVENDRA MISHRA

Creative Video Services
Newbury Park, California

The measurement of work has been a primary pursuit of industrial engineers and their predecessors since Frederick W. Taylor utilized the stopwatch in 1883. Prior to the scientific measurement of work, tasks were standardized by the use of historical production data or estimates. *Work measurement* is the application of systematic techniques to determine the work content of a defined task and the time required for its completion by a qualified worker. The result of work measurement is a time standard or a work standard, which may be expressed as pieces per minute (or hour or day), hours to produce 1000 pieces, minutes per piece, or in some other useful form. In other instances, a work standard may apply to a group performing a function. Over the years, new techniques have evolved for work measurement: work sampling, predetermined motion, and time studies.

2.5.1 HISTORY OF WORK MEASUREMENT

In 1912, Frank B. and Lillian M. Gilbreth expounded the principles of motion economy, the systematic reduction of inefficiencies, and the concept of determining performance times by analyzing the motions required to perform work.[1] Gilbreth defined 18 fundamental motions called "Therbligs" ("Gilbreth" in reverse). These were the motions of the hands and the arms, some mental reactions, and other periods of rest and delays. Gilbreth pioneered the measurement of time required to perform an operation. In 1922, A. B. Segur began development of predetermined motion time systems (PMTS) by analyzing micromotion films taken of expert operators during World War I.[2] The objective of the research was to discover a means of training blind and handicapped workers. Segur called his system "Motion Time Analysis" (MTA). In the period 1934–38, Joseph H. Quick developed a very comprehensive work-measurement system called WORK-FACTOR®. It was the 1100 workers performing diverse operations of machine shops, punch-press shops, assembly plants, wood mills, plastic works, plating shops, and offices.[3] Joseph Quick also pioneered measurement of mental processes. Later, in 1946, Harold B. Maynard began the development of "Methods Time Measurement" (MTM) at the Westinghouse Electric Corporation in collaboration with Gustave J. Stegemerten and John L. Schwab.[4] His research consisted of taking motion pictures to describe and evaluate methods of a drill-press operation and analyzing them in terms of Therbligs. These pictures were subsequently rated by a number of engineers known for their experience in pace rating. Plotting of the leveled film times in relationship to distance for the cases of "transport empty" and "transport loaded" indicated a high coefficient of correlation. Thus, MTM became a Predetermined Motion Time System (PMTS) extremely useful for establishing a consistent time standard for any given method, in addition to being an approach for method evaluation.

In the late 1960s and 1970s, the application of computers to work measurement was begun. The headquarters Industrial Engineering Department of the Westinghouse Electric Corporation developed the Micro-Matic Methods and Measurement (4-M) system, which is now available through the nonprofit MTM Association.[5] Rath and Strong created a computer package called Computerized Standard Data, which aids in the analysis, computation, documentation, and maintenance of industrial engineering time standards.[6] It also serves as a system for data entry and file maintenance. The WOFAC Company, a division of Science Management Corporation, developed WOCOM®, computerized work measurement using Detailed and Ready Work-Factor and MTM.[7] In 1974, H. B. Maynard introduced Maynard Operation Sequence Technique (MOST) in the United States after it had been successfully developed in Sweden during 1969–1972.[8] It provided another alternative for utilization of predetermined data using a unique sequence model aimed at simplifying computation of time standards. Management Science, Inc., of Appleton, Wisconsin, marketed computer software for industrial engineering applications and called it UniVation Systems®. The systems utilize universal formulas to generate precise elemental standard times.[9] In the area of clerical operations, the Nolan Company has come up with the Automated Advanced Office Controls (AUTO-AOC) system.[10] Today, the work measurement system is emerging as a fundamental element of a management control system.

2.5.3 TECHNIQUES OF WORK MEASUREMENT

Several techniques are used for the measurement of work, depending on the accuracy required for the labor standard, availability of resources, nature of the activity, and management objectives. Some of the technqiues are historical data, estimates, time study, predetermined times, standard data, work sampling, and mathematical tools.

Figure 2.5.1 illustrates an idealized cost–benefit relationship of the various work-measurement techniques. Curve A represents the sum of the following basic costs:

1. Amortization of the cost of development and implementation of the work measurement system. Useful life of a labor work measurement is expected to be three to eight years, depending on the rate of capital investment made into a function.
2. Annual operating cost of maintenance of labor standards and management control system.

Curve B represents the cost of inaccuracy in labor standards or the opportunity cost for an operation standardized with a given work-measurement technique.

Curve C is the total cost curve, sum of curves A and B. The most economical strategy for standardizing and controlling an operation is indicated by the least-cost portion of Curve C. Although the cost-benefit ratio is the ideal selection criterion for the best work-measurement technique, other factors can prove to be overriding, such as the industrial engineering skills available, caliber of supervision, accuracy of labor standards required for disciplinary actions, and management urgency for program savings.

Work measurement encompasses all activities performed by an individual or a team in manufacturing, warehousing, and administrative operations. Therefore, as illustrated in Figure 2.5.2, produced by WOFAC, activities in all operations are measurable using the various techniques enumerated above. As the type of work changes from repetitive high volume to nonrepetitive low volume, the measurement technique changes from the more familiar predetermined time standards and stopwatch to the less familiar multiple regression analysis,

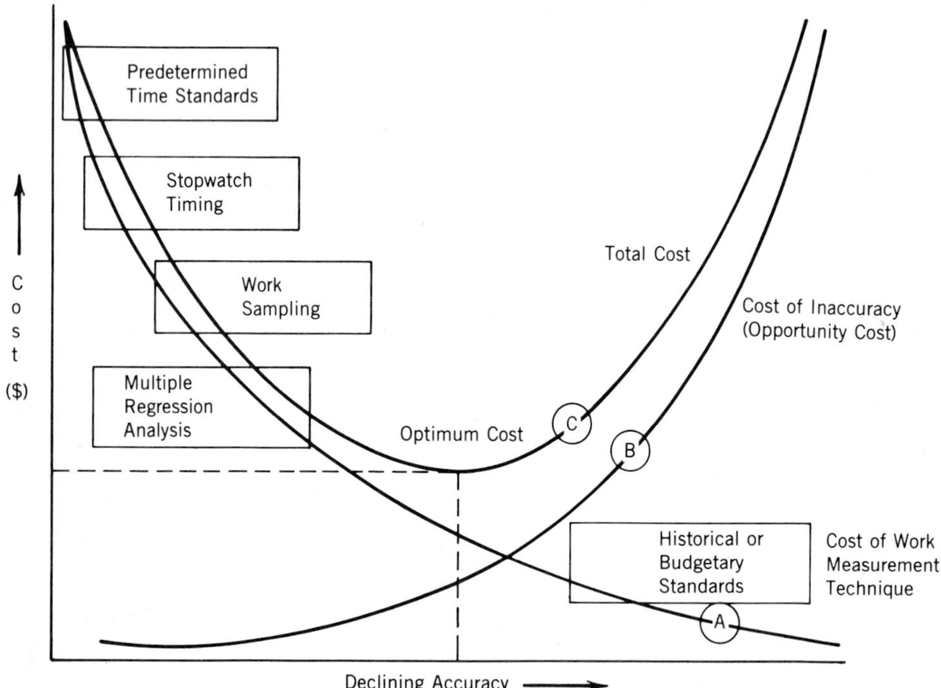

Fig. 2.5.1. Work measurement techniques: cost vs. inaccuracy.

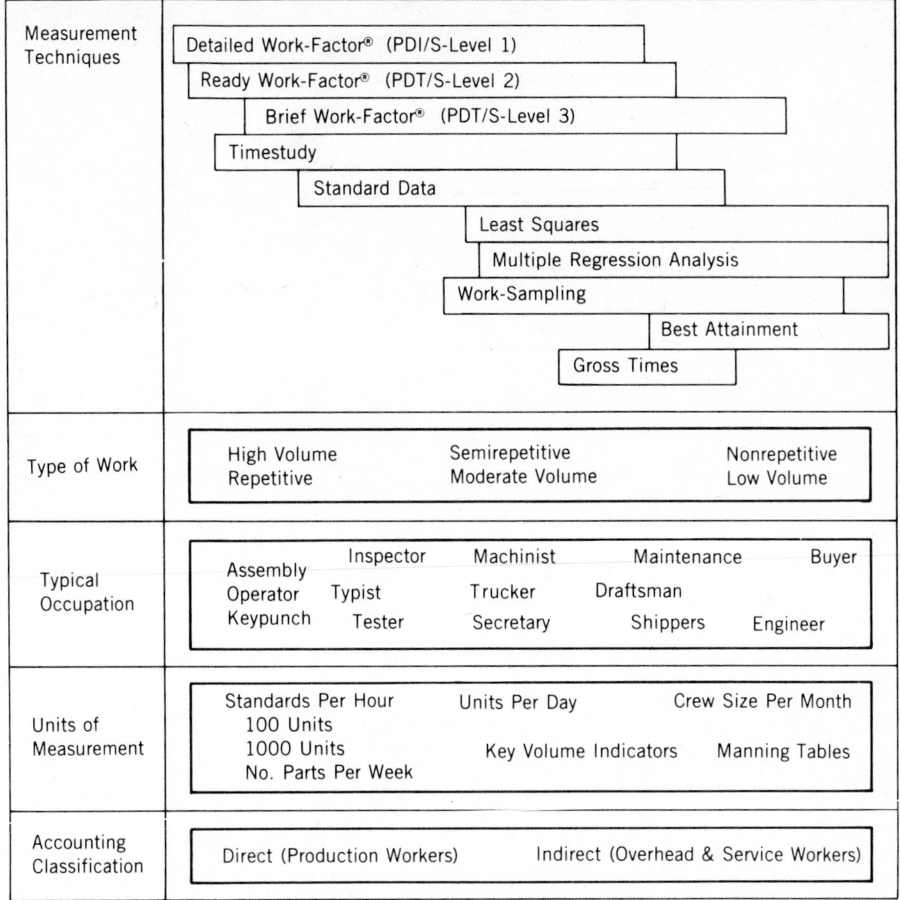

Fig. 2.5.2. Common applications of work measurement techniques.

work sampling, etc. In addition, you will notice that the unit of measurement changes from units or parts per hour to crew size or manning tables.

The common work measurement techniques are elaborated below.

Historical Data. Labor standards based on actual historical production data are determined by dividing the total pieces produced by the manhours expended in the given time period. Such standards are established quickly, but do not correct for poor methods, low performance, and poor reporting practices. These do not indicate the time that *should* be taken for a given task.

Estimates. In case engineered standards are not available, standards based on estimates are sometimes used. This is utilized particularly in the case of non-repetitive work that occurs infrequently. In other cases, estimates are also used as temporary standards for operations that may be studied in detail later or may be used to yield a cost estimate. The degree of accuracy of the estimate is dictated by the accuracy of the components of the data used.

Stopwatch Time Study. Time study is a procedure used to measure the time required by a qualified operator working at the normal performance level to perform a given task in accordance with a specified method. Figure 2.5.3 is a graphic analysis of the steps required in establishing a time standard, as presented by William Antis.[11] practice, an industrial engineer studies the method before standardizing it.

The tools of the time-study engineer are (1) time-study watch, (2) time-study form, (3) observation board, (4) pencil, (5) ten-foot tape, (6) speed indicator, and (7) calculator. The three main types of stopwatch used are the decimal minute watch, the decimal hour watch, and the split-second watch. Figures 2.5.4 and 2.5.5

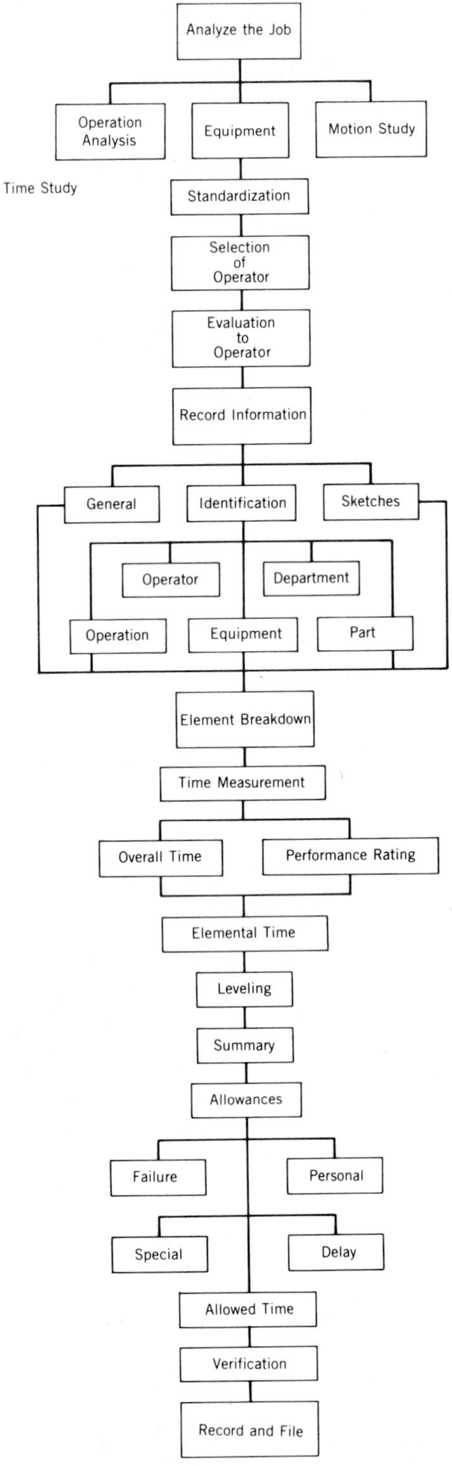

Fig. 2.5.3. Procedure for establishing time standards.

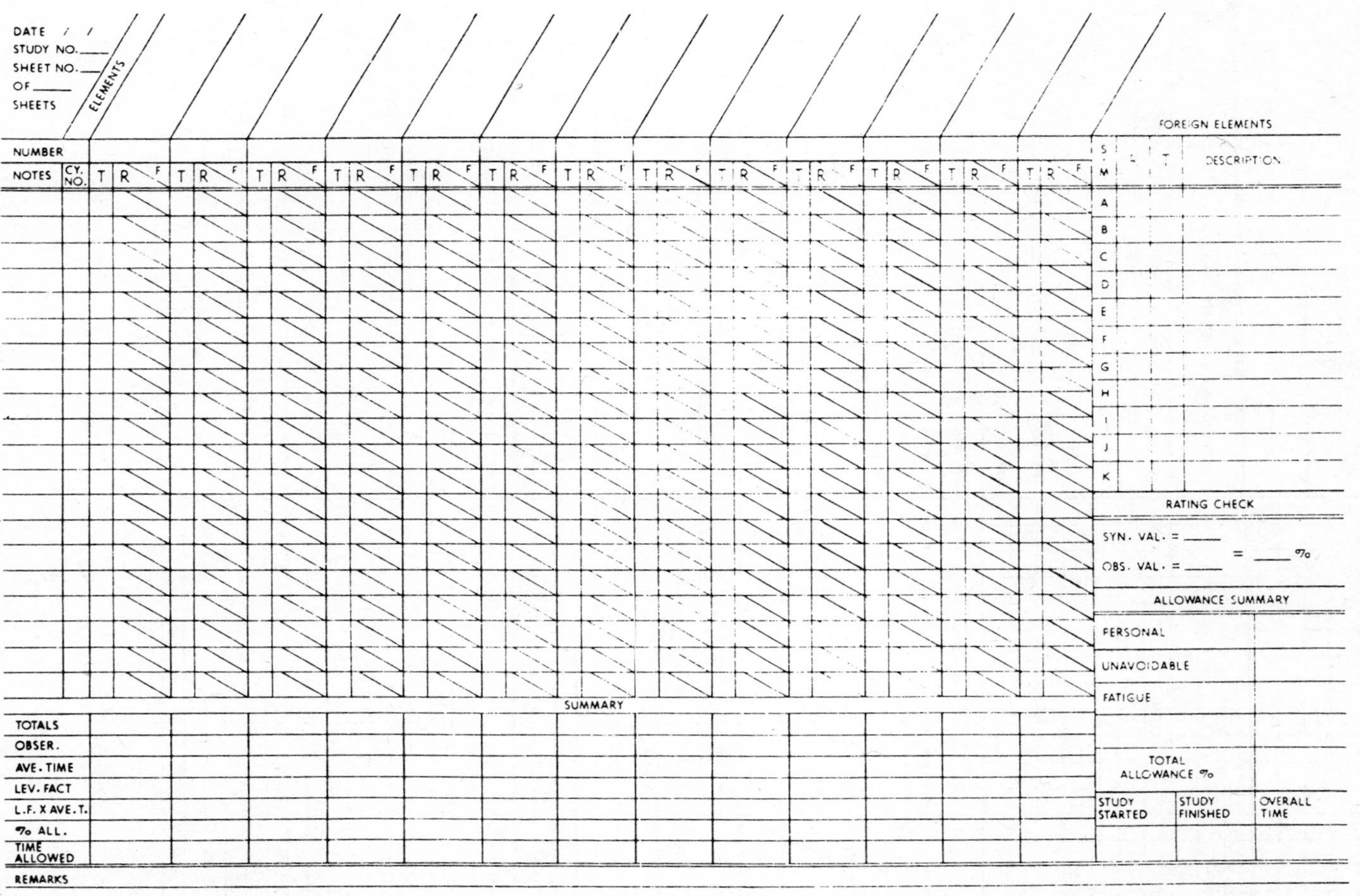

DATE / /
STUDY NO.____
SHEET NO.____
OF ____
SHEETS

ELEMENTS

NUMBER																																									
NOTES	CY. NO.	T	R	F	T	R	F	T	R	F	T	R	F	T	R	F	T	R	F	T	R	F	T	R	F	T	R	F	T	R	F	T	R	F	T	R	F	T	R	F

SUMMARY

TOTALS
OBSER.
AVE. TIME
LEV. FACT
L.F. X AVE. T.
% ALL.
TIME ALLOWED
REMARKS

FOREIGN ELEMENTS

S.M		T	DESCRIPTION
A			
B			
C			
D			
E			
F			
G			
H			
I			
J			
K			

RATING CHECK

SYN. VAL. = ____
= ____%
OBS. VAL. = ____

ALLOWANCE SUMMARY

PERSONAL
UNAVOIDABLE
FATIGUE
TOTAL ALLOWANCE %

STUDY STARTED	STUDY FINISHED	OVERALL TIME

Fig. 2.5.4. Front of time-study form.

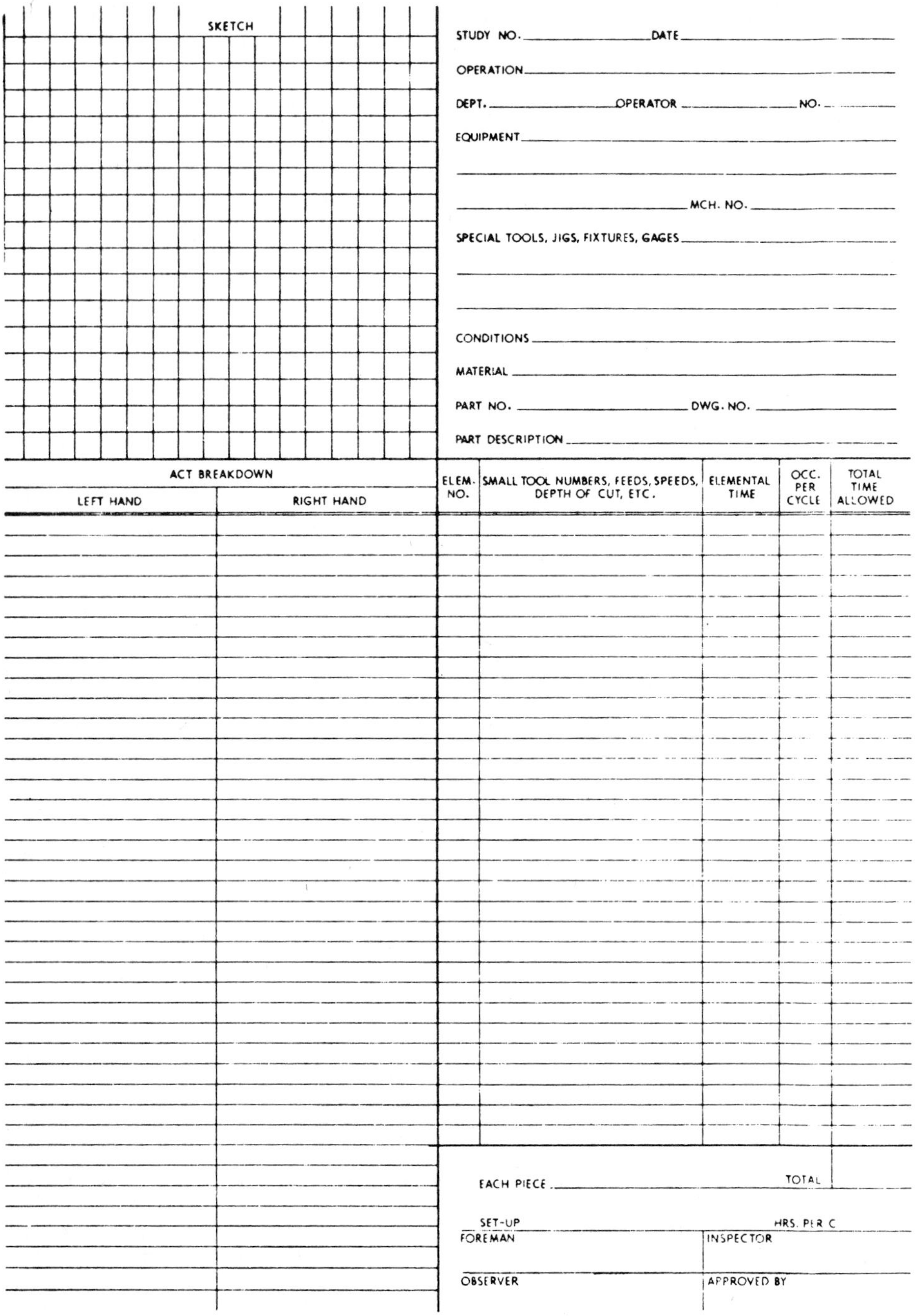

SKETCH

STUDY NO. ______ DATE ______

OPERATION ______

DEPT. ______ OPERATOR ______ NO. ______

EQUIPMENT ______

______ MCH. NO. ______

SPECIAL TOOLS, JIGS, FIXTURES, GAGES ______

CONDITIONS ______

MATERIAL ______

PART NO. ______ DWG. NO. ______

PART DESCRIPTION ______

ACT BREAKDOWN	
LEFT HAND	RIGHT HAND

ELEM. NO.	SMALL TOOL NUMBERS, FEEDS, SPEEDS, DEPTH OF CUT, ETC.	ELEMENTAL TIME	OCC. PER CYCLE	TOTAL TIME ALLOWED

EACH PIECE ______ TOTAL

SET-UP ______ HRS. PER C

FOREMAN | INSPECTOR

OBSERVER | APPROVED BY

Fig. 2.5.5. Back of time-study form.

are the time-study forms used. The observation board is used to hold the forms and the watch while the analyst moves around making observations.

The time-study procedure begins after the method has been established, working conditions standardized, and the operator trained in the proper method.

1. The information about the operator and the operation is recorded to include date and time of the study and the name of the time study engineer.
2. The operation is enumerated in terms of its elements and is fully described. It is necessary for the observer to know what elements probably will occur throughout the study and to maintain the same elements throughout the study.
3. The time taken by the operator to perform each element is observed and recorded. Three timing techniques are used:
   a. Continuous timing: the watch is started when the study begins and is allowed to run continuously during the study.
   b. Snapback timing: the watch is read and restarted (snapped back) at the end of each element.
   c. Accumulative: more than one watch is used to obtain the time. Continuous timing is the most popular approach, as it records everything an operator does. Element times are recorded without the decimal point, e.g., 335 represents 3.35 minutes.
4. It is vital for the observer to record the performance of the operator when the study is made. Based on the performance rating system universally used by work measurement specialists, the rating, or level, of operator performance is noted. If the study covers only a few cycles, one rating covering the whole operation is used. For longer studies, one rating may be used or the operator performance may be rated several times during the period of the study.
5. The data are summarized and analyzed. The abnormal values are removed and the total elapsed time for each element is determined. Based on the number of observations taken for each element, the average element time is computed. The leveling factor multiplied by the average time for the element gives the time for the element.
6. Next, to cover the major types of interruptions that the average operator experiences, allowances are applied to the standard time for the element. These are allowances for operator fatigue, personal needs, and unavoidable delays, to be discussed later.
7. Finally, the standard time is established by multiplying the element time by one plus the allowance percentage. At this stage, the procedure is written on a routing sheet or manufacturing information sheet. With this information, the computed standard can be verified with some degree of accuracy. The efficiency of the operator using the standard as the reference is compared with the performance rating, and a reasonable match confirms its validity. The number of studies to be taken is determined by the basic laws of statistics.

Predetermined Motion Time System

Predetermined Motion Time Systems (PMTS) may be classified as generic, functional, or specific. Basically, all predetermined motion time systems are based on the assumption that the time to make a manual motion having a specific purpose is constant. Tables and rules of application have been developed so that all manual motions encountered in productive work can be classified and time values assigned to each one.

A generic system is the one used by all users of work measurement and is not restricted in application. Examples of such a system are activities such as *reach*, *transport*, *grasp*, and *select*. Among the generic or universal systems in use today are MTM, Work-Factor Systems, Maynard Operation Sequence Technique, and Master Standard Data (MASD). A functional system has particular applications, such as in clerical work, microassembly, and tool usage. For example, *sweep* and *file* are action words in clerical work measurement, whereas *measure* is used in the machine shop. Finally, the third type is the specific system, which is proprietary and may have been developed for a specific industry. This also includes proprietary standard data systems.

PMTS can also be classified by the level of element complexity. The first category is the basic-level system, where elements consist mostly of single motions that cannot be further divided. As a result, the analysis of an operation becomes very time consuming. The second category is higher-level systems, where two or more of the single elements of a basic-level system are combined into a multimotion element. These are faster and easier to apply because the number of variables has been reduced. Some of the basic PMTS are described below.

Work-Factor System

The Work-Factor System has as its objective the analysis and cataloging of work-factors as they affect the time to make manual motions. A *work-factor unit* is equal to 0.0001 work-factor time minute. *Work-factor time* is the time required for an average, experienced operator working with good skill and good effort under standard conditions to perform one cycle of work. There are four basic types of work-factor systems. *Detailed work-factor* is the generic, basic-level system developed from original motion-time studies using stopwatches, phototimers, films, and fast film snapshots. The detailed time unit is 0.0001 minute. The *Mento-Factor System* is a basic-level system developed to determine the time required for measuring mental processes. The *Ready Work-Factor System* is a second-level system developed by simplifying detailed work-factor time values. Its time unit is 0.001 minute. This system is useful for measuring medium to long-run operations with cycles of 0.15 minute and greater. The fourth type is *Brief Work-Factor*, which was developed to measure nonrepetitive work. Figure 2.5.6 illustrates a labor standard developed with Detailed Work-Factor.

Methods Time Measurement System

The MTM-1 system is a generic, basic-level system with a time measurement unit (TMU) of 0.00001 hour. The second level of the family is MTM-2, which has a speed of analysis twice as great as the basic level, but has lower accuracy. The third level is MTM-3, whose speed of analysis is seven times greater than MTM-1. It is used where a less detailed methods description is required. Then there are special types for specific applications in clerical, machine shop, and microscope assembly operations. Figure 2.5.7 presents a labor standard developed using the MTM system.

Another variant of MTM is the Modular Arrangement of Predetermined Times (MODAPTS), developed as a generic and functional second-level system.

Master Standard Data

Serge A. Birn Company developed the generic, second-level MSD system in the late 1950s. It was used to set standard MTM-based data on manually controlled operations where production was less than 100,000 units per year or a few thousand units per week. Between production runs, the operator would lose most of the skill he or she has developed. Statistically, a very high percentage of work in industry falls within this limited-practice category.

Maynard Operation Sequence Technique (MOST)

The MOST technique, an MTM-based system, is applicable for any cycle length and repetitiveness, as long as there are variations in the motion pattern from one cycle to another. The system employs a small number of predetermined models of fixed activity sequences that cover practically all aspects of manual activity. It appears that there is a large category of operations for which the consistent behavior of MOST will produce standards with accuracy approximately equivalent to that of higher-level systems, but with far less effort and in far shorter time.

Work Sampling

A very popular technique of work measurement has been *work sampling*. Snap readings, ratio delay, group timing techniques, work-measurement sampling, and activity sampling are synonymous terms applied to what has come to be recognized as work sampling. Nearly fifty years ago, L. H. C. Tippett discovered the technique by randomly photographing the scene observed.

Work sampling is a method of randomly observing work to obtain information about an activity. This technique serves the needs of work measurement, problem analysis, method analysis, and so on. Basically, it provides a means to:

1. Gather information that cannot be obtained practically by continuous monitoring.
2. Obtain critical data, traditionally generated through continuous observation, in an economical manner.
3. Establish standards for long cycle work.
4. Identify opportunities for methods improvement.
5. Economically and speedily accumulate information that characterizes the nonrepetitive elements of a job for use in the establishment of standards.

Work sampling is based on the laws of probability. It works because a small number of chance occurrences tends to follow the same distribution pattern as a large one. Work is defined as an activity being studied. The work is subdivided into categories that are mutually exclusive and exhaustive.

Let

$$K = \text{total number of categories into which one activity is classified}$$

$$P^1 = \text{true proportion of time that worker or machine is engaged in a specific category of work}$$

$$P_1^1 + P_2^1 + \ldots + P_K^1 = 1$$

$$N = \text{total number of observations made in order to estimate } P^1$$

22D36

OPERATION NAME OBTAIN GOOD CASSETTES AND ASIDE TO TRAY 4-21-81 Sheet 1 of 1

LEFT HAND				Cumulative Time		RIGHT HAND			
No.	Elemental Description	Motion Analysis	Elem. Time			Elem. Time	Motion Analysis	Elemental Description	No.
1						61	VA10D	R TO CASSETTE	1
2					81	20	SOL VO 4	GR	2
3					142	61	A10D	M TO INSPECT	3
4					192	50	FO + I	INSPECT	4
5					224	32	PP-M-50%	PREPOSITION	5
6	WAIT OR HOLD	BD	250	250	250	26	VA4	M TO OTHER HAND	6
7	GR (TRANSFER)	½FI + ½FI	16	266	266	16	½FI + ½FI	RL (TRANSFER)	7
8				1330	1330	1064	266 X 4	REPEAT STEPS 1-7 4 TIMES	8
9	M STACK TOWARD TRAY	VA8 41.5	16	1346	1357	27	VA12D 41.5%	R TO TRAY	9
10					1360	3	½FI 41.5%	GR	10
11					1373	13	VA8 33%	PULL TRAY OUT	11
12	WAIT	BD	29	1375	1375	2	VA4 8.5%	PULL TRAY OUT	12
13	M STACK TO TRAY	VA4D 41.5%	16	1413					13
14	M STACK TO TRAY	VA12D 58.5%	38	1451					14
15	RL STACK	½FI	8	1459					15
16	R TO BACK OF STACK	A2D	29	1488					16
17	GR	GR-CT	—	1488					17
18	PUSH TO SEAT	A1	18	1506					18
19	RL	RL-CT	—	1506	1506	131	BD	HOLD TOTE OR WAIT	19
20	M HAND AWAY	VA8	38	1544	1519	13	VA8 33%	PUSH TRAY IN	20
21					1521	2	VA4 8.5%	PUSH TRAY IN	21
22					1524	3	½FI 41.5%	RL	22
23					1543	19	VA12 41.5%	M HAND AWAY	23
24									24
25									25
26									26
27									27
28									28
29									29
30									30

Fig. 2.5.6.

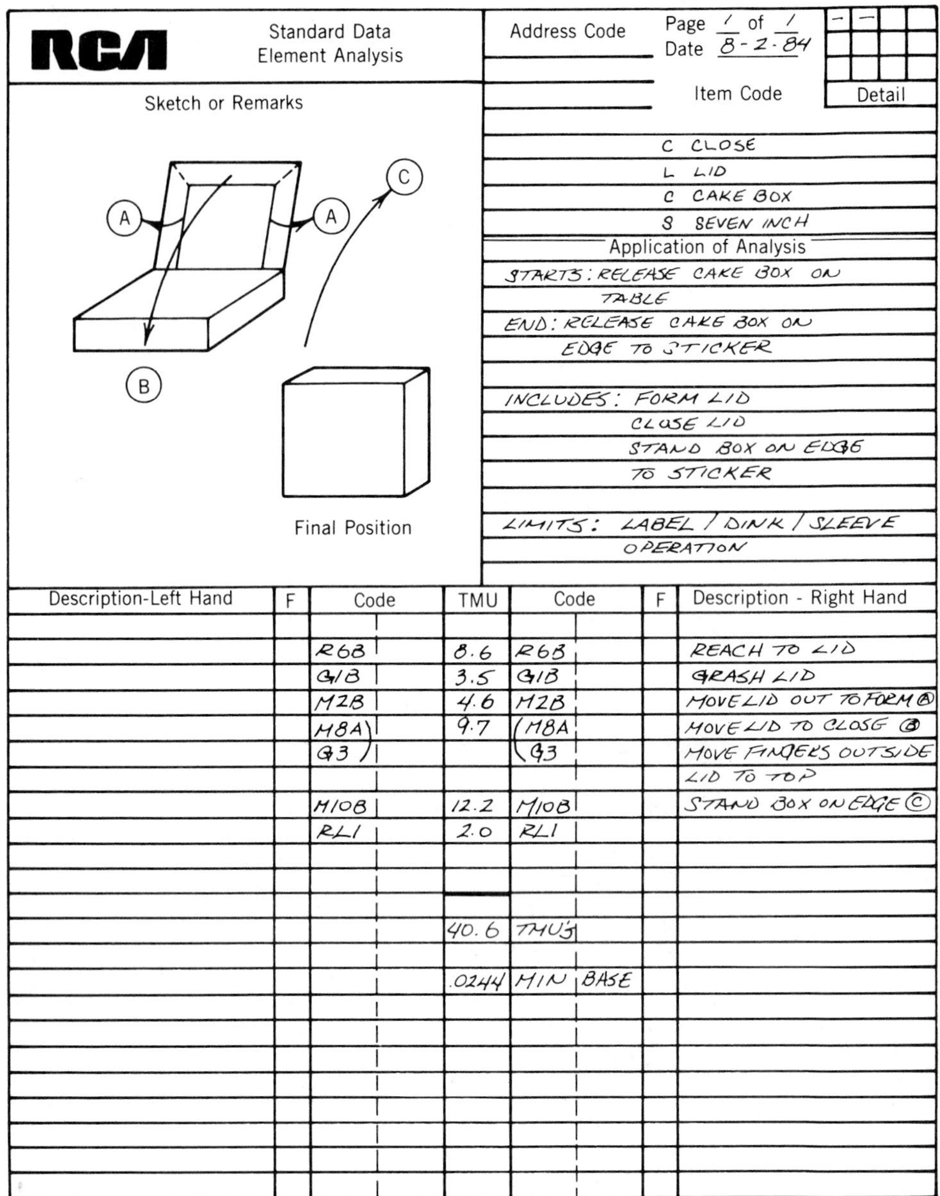

RCA Standard Data Element Analysis

Address Code

Page 1 of 1

Date 8-2-84

Item Code | Detail

C CLOSE
L LID
C CAKE BOX
S SEVEN INCH

Sketch or Remarks

Application of Analysis

STARTS: RELEASE CAKE BOX ON TABLE

END: RELEASE CAKE BOX ON EDGE TO STICKER

INCLUDES: FORM LID
CLOSE LID
STAND BOX ON EDGE
TO STICKER

LIMITS: LABEL / DINK / SLEEVE OPERATION

Description-Left Hand	F	Code	TMU	Code	F	Description - Right Hand
		R6B	8.6	R6B		REACH TO LID
		G1B	3.5	G1B		GRASP LID
		M2B	4.6	M2B		MOVE LID OUT TO FORM (A)
		M8A	9.7	M8A		MOVE LID TO CLOSE (B)
		G3		G3		MOVE FINGERS OUTSIDE
						LID TO TOP
		M10B	12.2	M10B		STAND BOX ON EDGE (C)
		RL1	2.0	RL1		
			40.6	THU'S		
			.0244	MIN BASE		

Fig. 2.5.7. MTM analysis.

X = total number of observations that find a worker or machine in a specific category

$X_1, X_2, X_3, \ldots, X_K$ = total number of observations of category 1, 2, 3, . . . , K

$$X_1 + X_2 + X_3 \ldots + X_K = N$$

P = an estimate of P^1 found by X/N

$P_1, P_2, \ldots, P_K$ = estimates of $P_1{}^1, P_2{}^1, \ldots, P_K{}^1$ found by $X_1/N, X_2/N, \ldots, X_K/N$.

The objective of a work sampling study is to generate a good estimate of a proportion p^1 such that there is no bias and variance is low. The sample size is statistically determined, based on the nature of the distribution of the work elements and the degree of accuracy required.

A step-by-step approach for work sampling has been outlined by Richardson and Pope,[12] as follows:

1. Define objective(s) of the work study.
2. Establish measure of output of the activity being sampled.
3. Determine who will collect data—an industrial engineer, a technician, or a supervisor.
4. Obtain approval of supervision and announce study.
5. Determine the general method to be used to record data—a pencil-and-paper form, punched or mark-sense cards, or electronic clipboards with built-in watches.
6. Train observers in observing activity and collecting/recording data.
7. Classify activity into categories.
8. Prepare for the physical aspect of data collection.
9. Decide on the number of observations needed to yield statistical validity.
10. Decide on the sampling method to be used. Several methods of systematic, stratified, and random sampling are in use, depending on the number of observations to be made, characteristics of the area and the workers, etc.
11. Select necessary random times for making observations.
12. Set up the necessary clerical or data entry manpower for processing data.
13. Make a short test study and remove bias that may creep in. Discard the results.
14. Make the intended study for the designed time span.
15. Set up control charts for the estimate of the true proportion of time that a worker or a machine is engaged in a specific category of work.
16. Re-evaluate precision of the estimate obtained in step 15.
17. Analyze and report results.

Advantages of work sampling are enumerated by Weaver as follows:[13]

Observers do not require special skill and training in work measurement. Thus, supervisors and others can be used as observers.

Properly used (making the proper number of observations, etc.), work sampling can provide accuracy and confidence in the information obtained.

Work sampling makes it practical to collect facts that wouldn't otherwise be obtained.

It produces few complaints from operators than does continuous study.

It produces less distraction in the operator's normal routine.

Several areas can be studied at the same time.

The observer can determine the distribution of work activities among members of a crew such as multiman changeovers and material handling operations.

The observer can determine the percent utilization of a piece of equipment, a group of similar machines, or a group of people.

Studies can be conducted over a long period of time, thus minimizing or eliminating day-to-day or seasonal effects.

Studies can be interrupted, then started again without affecting the results.

Accuracy of work sampling is determined by the number of observations. The greater the number of observations, the greater the accuracy and confidence. But nearly all problems have a point beyond which greater accuracy of data isn't worthwhile.

Mathematical Techniques

The techniques of operations research and statistics can be used to establish relationship between independent variables, which characterize output of work, and the dependent variable, which is manhours required. These methods are very appropriate for manpower staffing as well as labor standard estimation. This technique of work measurement is widely used in manufacturing, distribution, and office operations. In addition, it has proved to be a powerful tool for analyzing activities and problems.

One group of crewing problems is described by customers arriving for service and servers meeting their demands. Many service functions, such as maintenance personnel and tool-crib attendants, experience random calls for service. Due to the unanticipated nature of arrivals for service, the available manpower support can be insufficient, causing a delay in waiting time, or be surplus, resulting in wasted manpower. In these cases, the principles of Queuing or Waiting Line theory are utilized to determine the staffing level for the optimum cost of surplus manpower and idle machine time.

Another group of manpower estimation problems arises when many job arrivals determine the operator's output. For instance, the number of customer orders, number of line items, and number of units determine the output of an order picker in a warehouse. In these cases, a regression analysis can produce an equation that predicts the operator's output for different input variables. Using statistical techniques, it is possible to determine the correlation between the inputs and the output, the significance of any variable in predicting the output, and the amount of experimental error.

Master Standard Data

Standard data, also called basic data, is the compilation of all elements used for performing a specific task or operation with the elemental time value for each portion of the task. The essential differences between standard data and the techniques of time study, PMTS, and work sampling are the following:[14]

1. Standard data are a set of synthesized time values or a mathematical model, either of which may use time values established by one or more of the fundamental techniques.
2. Standard data in the form of a mathematical model use other parametric data and may also use time values that have not been established by one or more of the fundamental techniques.
3. Standard data are used to establish standards without the further use of any fundamental technique or timing device.

The primary purpose of using standard data is to minimize the expense of setting time standards. There are basically two kinds of standard data, namely synthetic or analytical. In the former, one utilizes the elemental data from time studies, basic motions or groups of motions from PMTS, and work-sampling studies. The smaller pieces of time data are added together to achieve larger blocks of time. The building block concept employed here combines first-order work units (motions) to achieve second-order work units (elements), and then additional second-order time study elements can be added. This set of second-order data can be used to set standards at the fourth-order work unit level. In case of analytical standard data, a mathematical technique is utilized. Multiple regression analysis is used to develop elemental data, such as "get part," where parameters such as the location of the part or its weight have been mathematically related to the time to get one part. The elemental data is then used with other data to develop the synthetic standard data.

Comparison of Techniques

A comparison of the common work measurement techniques is presented in Table 2.5.1. The markedly divergent methods of measurement indicate the inherent advantages in terms of costs and benefits. While predetermined time standards offer greatest accuracy and a basis for methods improvement, they are the most complex and expensive to establish. On the other hand, historical production standards provide a crude basis for work measurement.

2.5.3 MAJOR CONSIDERATIONS IN WORK MEASUREMENT

In 1914, Robert Franklin Hoxie, Professor of Economics at the University of Chicago, was appointed by the Federal Commission on Industrial Relations to investigate scientific management. The investigation included 35 workshops and people from all walks of industrial life, including labor unions. The result was a list of 17 factors considered variables in setting a good standard for a work-measurement program.

1. The general attitude, ideals, and purposes of the management and the consequent general instruction given to the time-study analyst.
2. The character, intelligence, training, and ideals of the time-study analyst.
3. The degree to which the job to be timed and all its appurtenances have been studied and standardized, considering uniform conditions for all workers.
4. The amount of change thus made from old methods and conditions of performance, e.g., the order of performance, the motions eliminated, and the degree of habituation of the workers to the old and the new situation when the task is set.
5. The mode of selection of the workers to be timed and their speed and skill relative to the other members of the group.
6. The relative number of workers timed and the number of readings considered sufficient to secure the result desired.
7. The atmospheric conditions, time of day, time of year, the mental and physical condition of the workers when timed, and the judgment exercised in reducing these matters to the "normal."

Table 2.5.1. Evaluation of Work Measurement Techniques for Manufacturing

Criteria	Historical Production Standards	Multiple Regression Analysis	Work Sampling	Stopwatch Timing	Predetermined Time Standards
1. Engineering effort required for					
System development and implementation	Minimal	Low–medium	Medium	Medium	High
System maintenance	Minimal	Low–medium	Medium	Medium	High
2. Complexity of labor standard	None	Low	Medium	Medium high	
3. Degree of work measurement accuracy	Low	Medium	High	High	Very high
4. Provide a basis for methods improvement	None	None	Low	Medium	Very good
5. Training and skill required for operating management	Low	Low	Medium	Medium high	High
6. Employee and management acceptability of standards (ease of understanding and administration)	Fair	Fair–good	Fair–good	Very good	Good
7. Ease of labor standards enforcement	Low	Fair	Good	Good	Good
8. Effectiveness in realizing maximum savings	Very low	Fair	Fair	Good	Very good
9. Program implementation	Very short	Short	Medium	Medium–long	Very long
10. Rate of savings realization (How quickly?)	Very short	Short	Medium–long	Long	Long
11. Effectiveness in					
Manpower forecasting and scheduling Performance appraisal Management reporting	Fair	Average	Average–high	High	High

8. The character and amount of special instruction and special training given the selected workers before timing them.
9. The instructions given to them by the time-study analyst as to care and speed, etc., to be maintained during the timing process.
10. The attitude of the time-study analyst toward the workers being timed and the secret motives and aims of the workers themselves.
11. The judgment of the time-study analyst as to the pace maintained under timing relative to the "proper," "normal," or maximum speed that should be demanded.
12. The checks on the actual results used by the time-study analyst in this connection.
13. The method and mechanism used for observing and recording times and the degree of accuracy with which actual results are caught and put down.
14. The judgment exercised by the time-study analyst in respect to the retention or elimination of possibly inaccurate or "abnormally" high or low readings.
15. The method used in summing up the elementary readings to get the "necessary" elementary time.
16. The method employed in determing how much should be added to the "necessary time" as a human allowance.
17. The method of determining the machine allowance.

2.5.3.1 Allowances in the Standard Time

In the establishment of production standards, it is necessary to apply certain allowances to the labor content that has been computed by any of the previously discussed techniques.

Allowances in labor standards are provided to cover time taken for personal needs, unavoidable delays, and a slowdown of output because of fatigue. To get the standard time, the allowance for personal, fatigue, and delay needs is applied to the predetermined or normal time. In general, the applied allowance will not be the same percentage for all elements of the job that has been studied. Certain allowance factors will apply to all the elements, while other factors will apply to the human efforts only, and others will apply to the machine-controlled elements only. A common technique for establishing the allowances is work sampling.

Personal time is the time the operator spends to maintain his/her well-being on the job. These times are used for trips to the restroom and drinking fountain, for example. It covers the time the machine is down because of the operator's personal need. The amount of time needed by the operator for personal reasons will depend upon the working conditions, the class of work performed, facility layout, etc. Generally, no personal time is allowed in the standard if (1) relief operators are provided; (2) the machine can be left unattended for periods of time of at least 3 to 5 minutes in length; (3) the operator can be relieved by someone else in his crew or by another operator doing similar work.

Fatigue, which may be defined as a lessening in the capacity for work, is most difficult to measure and, consequently, controversial. The amount of fatigue experienced varies significantly, not only from one person to the next, but also for the same person from one day to the next. It ranges from being strictly physical to purely psychological and, as a result, combines both the physical and the psychological. Research indicates that three major factors cause fatigue. The first is the general working environment, including the amount of light, the temperature level, the relative humidity, the noise level, the air freshness, etc. The second factor is the nature of the work being performed. Degree of monotony in the performance of the job can create fatigue along with the physical effort demanded. Finally, the overall health of the employee, mental as well as physical, has a bearing on fatigue. It is recommended that an allowance for fatigue should be given for the working conditions and job design. Fatigue allowance should not be made for general health factors or if relief operators are provided.

Miscellaneous unavoidable delays during the course of the day are too short and too unpredictable to warrant measuring and are not included in the standard as downtime or non-cyclic elements of work. This category includes brief instructions from the supervisor, minor machine malfunction, etc. On the other hand, this type of delay does not include clean-up, machine setup or changeover, handling bad parts, filling out time tickets, etc. In such situations, delays are measured and recognized in the standard on a frequency basis or as an element of the operation. In cases of major machine downtime or material shortage, the employee is assigned to another job or is given the actual downtime. Then there are instances where an operator is responsible for more than one machine and experiences machine interference. Such an allowance provides for time when one machine or more must wait until the operator completes the assigned work on other equipment. The allowance of machine interference increases with the degree of machine assignment and is determined using statistical techniques.

Finally, in companies that use incentive payment systems, provisions are made through allowances or other adjustments so that earnings above base pay can be obtained. This allowance results in incentive bonus or pay.

In conclusion, the basic allowances may be the result of studies or may be unilaterally established by management or may be determined by negotiation between management and union.

2.5.3.2 Recognition of Learning Curve

In the world of work measurement, the learning curve is an important phenomenon to be recognized. The learning curve has been known by various names, such as progress curve, improvement curve, cost curve, manufacturing progress function, experience curve, etc. It is the statement of the relationship between manhours and cumulative production or experience. Its fundamental principle is the simple fact that a worker learns as he or she works. Consequently, in the procedure for the establishment of labor standards, it is imperative that an experienced operator serve as a basis. In addition, learning cycles for different jobs can be scientifically determined and the rate of an employee's progress in terms of labor efficiency can be predicted.

The basic learning model can be expressed as follows:

$$\bar{y} = ax^b$$

where

$\bar{y}$ = cumulative average manhours for any production quantity x

a = the number of manhours required to build the first unit

x = any number of completed units

b = the slope of the learning curve (usually a negative number)

A learning curve is frequently designated by its percentage slope. Thus, a curve with $b = -.0322$ is an 80% curve because $2^{-0.322} = 0.80$. Figure 2.5.8 illustrates a learning curve with a 90% slope.

The application of work measurement requires that the learning curve phenomenon be recognized in order to provide adequate training time and to establish a fair expectation of employee performance. For example, studies at the Somerville plant of RCA classified jobs into four categories (A, B, C, and D) according to the time required for a new employee to obtain a high level of efficiency.

Training Period	*Classification*
13–16 weeks	A
9–12 weeks	B
5–8 weeks	C
1–4 weeks	D

Then a learning curve, produced in Table 2.5.2, was utilized to measure performance of employees. In the case of a current employee transferred from one job to another or a person who has been rehired after a time lapse of more than a month off the job, retraining factors were applied.

The application of learning curve is very prevalent in such industries as aircraft, automotive, housing, machine tools, and shipping. A work-measurement program in this environment of long cycle times, frequent

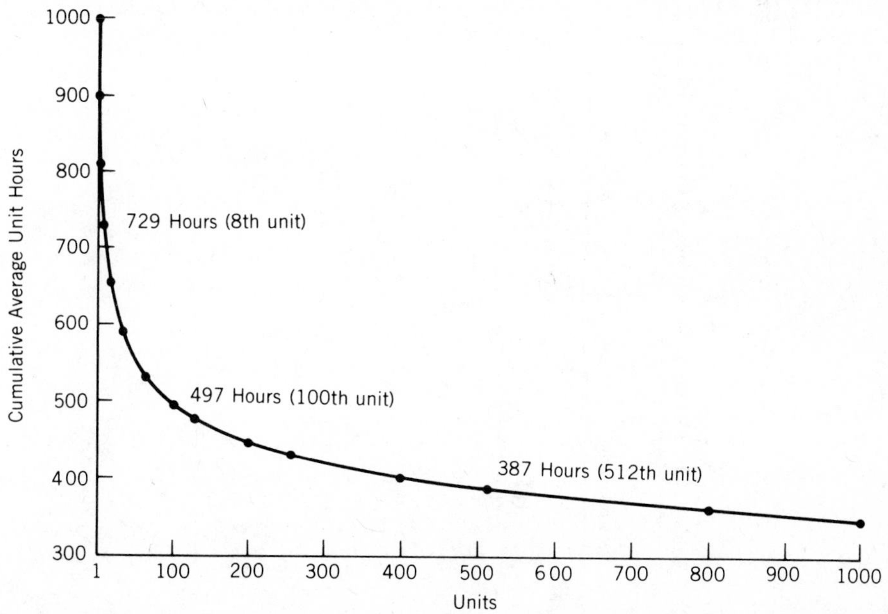

Fig. 2.5.8. Plot of 90% learning curve.

Table 2.5.2. **Weekly Efficiency during Learning (in %)**

	Total Training Period															
Weeks on Job	1	2	3	4	5	6	7	8	9	10	11	12	13	14	15	16
1	100	75	54	35	23	17	13	10	8	7	6	5	5	4	4	4
2		100	87	75	64	52	44	35	28	23	19	17	15	13	11	10
3			100	91	83	75	68	60	52	46	41	35	30	25	23	20
4				100	93	87	81	75	70	64	58	52	48	44	40	35
5					100	95	89	84	80	75	71	67	62	57	54	49
6						100	96	91	87	83	79	75	72	68	64	60
7							100	96	92	89	85	81	78	75	72	69
8								100	97	93	90	87	83	81	78	75
9									100	97	94	91	88	85	83	80
10										100	97	95	92	89	87	84
11											100	98	95	93	90	88
12												100	98	96	93	91
13													100	98	96	94
14														100	98	96
15															100	98
16																100

tooling and design changes, and relatively small quantity production, has to adequately incorporate the learning curve.

2.5.4 COMPUTERIZED WORK MEASUREMENT

Computerization allows the efficient application of work measurement techniques. Some of the enhancements through computerization can be broadly classified into the following areas:

1. Automation of data gathering in time study, work sampling, etc.
2. Mathematical and statistical analysis of work-measurement data.
3. Development and organization of data.
4. Development of standard data using mathematical and statistical techniques.
5. Documentation of methods of operations.
6. Audit control, work-measurement inputs, standard data, etc.
7. Storage and retrieval of standard data.
8. Establishment of labor standards for assembly lines.
9. Determination of indexes for optimization of methods.
10. Maintenance and updating.
11. Management control utilizing work measurement system, etc.

Today, several well known companies are offering computerized work-measurement systems with growing acceptance from users. Some of these systems are described below in terms of inputs, outputs, and operating system details. Bear in mind that these are only some of the better-known proprietary systems available today, and they do not represent an exhaustive list.

2.5.4 Micro-Matic Methods and Measurement Data Systems (4M Data)

The 4M Data System acquired its name as follows:

Micro. The system retains all the variables incorporated in the MTM procedures of reach, move, grasp, and position.

Matic. The methods analysts need no longer refer to a data card or table of values since the simple codes required to use the system are based on easily remembered categories. In addition, the system automatically determines a near-optimum method, subject only to restrictions imposed by the analyst in describing the general method.

Methods. The ability to develop and define sound manual methods is inherent in MTM-1.

Measurement. While "get" and "place" input codes are utilized to minimize the analysis time, the codes are designed so that the computer can interpret them, recognize their basic level components, and correctly apply them for the two hands.

The MTM motions that constitute 90% of the manual work performed in the common industrial situation are reach, grasp, move, position, and release. These motions are combined into the motion aggregates of GET and PLACE, as indicated in Figure 2.5.9. The computer program accepts simple GET or PLACE notations that describe the general method and then converts these motions to equivalent MTM-1 notations. An extensive electronic data processing logic program then utilizes the increments to develop near-optimum methods and standards.

Some of the standard reports from the system are reproduced in Figures 2.5.10–2.5.13. The Element Analysis is listed as an output. The Operation Analysis report contains the complete detail for a specific operation standard. The Operation Standard is enumerated in another report. Finally, the Operator Instruction Report provides the procedure for the employee to follow.

A major benefit of the 4M System is the computation of method improvement indexes. Five indexes are developed for a given method, showing opportunities for improvement. The Motion Assignment Index shows how completely the hands have been assigned work during the cycle. An index of 50% represents the effective use of only one hand at a time, while 100 percent would indicate perfect utilization of both hands. The GRA Index yields the percentage of cycle time that deals with grasping, releasing, and applying pressure. If high, it would suggest that improvements might be made by simplifying grasp requirements. The POS Index is the percentage of total cycle involved with positioning. A high value suggests that improvements might be possible in or relief of tolerances. The RMB Index expresses the degree to which cycle time can be reduced by shortening distances and reducing body motions. Finally, the Waiting-On-Process Intervals Index calculates the amount of process time remaining after internal motions have been performed. This leads to the reduction of high waiting intervals.

GET

G XXX xxx

Reach Distance

0	1		Reach	A	No Grasp
	2		″	B	″
	3		″	E	″
	4		″	D	″
1	1		″	A	Easy Pickup
	2		″	B	″ ″
	3		″	D	Flat ″
	C	1	″	B	Interference, Lg.
		2	″	B	″ , Med.
		3	″	B	″ , Sm.
2			Regrasp, No Distance		
3			Transfer Hand-to-Hand		
4	1		Select	Lg.	
	2		″	Med.	
	3		″	Sm.	

PLACE

P XXX xxx ww

Weight per hand
(If over 2.5#/1kg)

Move Distance

0	1		Move A, No Position
	2		″ B, ″
	2	T	Toss ″
	3		Move C, ″
1	☐	☐	Clearance $< .700 > .300$
2	☐	☐	″ $< .300 > .050$
3	☐	☐	″ $< .050 > .010$
☐	1	☐	Symmetrical
☐	2	☐	Semi-symmetrical
☐	3	☐	Non-symmetrical

Insertion

☐	☐	A Surface Alignment
☐	☐	0 0⅛″ Easy
☐	☐	1 ½″ Easy
☐	☐	2 1″ Easy
☐	☐	3 1½″ Easy
☐	☐	5 0⅛″ Difficult
☐	☐	6 ½″ Difficult
☐	☐	7 1″ Difficult
☐	☐	8 ½″ Difficult

Fig. 2.5.9. 4M data system notation.

```
4M ELEMENT ANALYSIS                    REQUESTED BY JPO   TIME 11.05.25  DATE 02/21/80        PAGE 1

TS, PLACE NEXT COIL AND IRON IN FIXTURE                 LEARNING LEVEL 100
                                                            TOTAL MU - MANUAL        943
UNITY                              UPDATED                           - PROCESS       190

                                                                   PROCESS                   NET
                                     RH OR BODY MOTIONS      FREQ.   TIME      LH     RH    MANUAL

N      G01 -9       G01 -9    REACH    CONTROL BUTTON                          83     83      83
       P01 -1       P01 -1    MOVE     BUTTON                                  25     25      25
                              PROCESS  HOLD BUTTONS                  190
                              PROCESS  COMPLETE PRESS ACTION         250*
                    G12 -14   GET      COIL, IRON IN TRAY                            173P    173
       G3  -12      P01 -26   MOVE     PART TO LH                             152    239P    295
                    G11 -8    GET      COIL, IRON IN FIXTURE                         109P    109
  RE   P121-6       P03 -24   MOVE     TO TRAY POCKET                         238    227     258

                                              TOTAL MU               190      496    856     943

   PERCENT   GRA   9 PERCENT   POS 10 PERCENT        PROC 17 PERCENT
```

Fig. 2.5.10. 4M element analysis.

The 4M Operating System, written in 1974 ANSI COBOL, is an interactive, on-line system capable of running on IBM, Honeywell, and other hardware.

2.5.4.2 The WOCOM System

The basic-level WOCOM system provides computerized application of the two most recognized systems of predetermined times: Work-Factor and MTM. It is designed for simple operation and flexibility in meeting the diverse needs of industry. The programs are available on the General Electric time-sharing network or can be run on in-house computers. WOCOM consists of eight modular computer programs that can be utilized for the analysis of human and machine work, alternative work methods, maintenance of labor standards, and design of assembly line. A brief description of the modules available in both Detailed and Ready Work-Factor are given below:

Work Measurement by Work-Factor (Modules 1 and 2)

The manual or machine operation is identified by the user in a simple input format, and the program determines the motions required to perform the operation, along with the times for each. The user receives the detailed motion analysis in Work-Factor and a manufacturing instruction sheet containing each major step in the operation. This module is available in both Detailed and Ready Work-Factor.

Work Measurement by MTM

This module is identical to module one, except that the application rules and time data of MTM are employed.

Application of Learning Allowances

The learning curve allowance is established for the operation variables of the cycle time.

Analysis of Mental and Perpetual Work

Repetitive mental operations, such as inspection of clerical work, are analyzed, and time values are determined.

Measurement of Highly Variable Operations

This module uses the tool of multivariate analysis for complex operations that are not amenable to analysis by the other work-measurement techniques. In this case the user has to provide times required to perform the operations under varying conditions, which are defined.

Interactive Line Balancing

By varying the allocation of work elements to the work stations of the assembly line, this module generates alternative configurations for it.

Batch Assembly Line Balancing

Based on user-supplied inputs, consisting of operation precedence relationships, operation grouping parameters, and desired workstation cycle time, the module provides an assembly line design.

```
4M DATA SYSTEM     R0820        4M OPERATION ANALYSIS              REQUESTED BY RE   TIME 09.05.50 DATE 04/11/80  PAGE 1

PART  2476SP8X-5                  WATT-HOUR METER                                              ORIGINATED  04/11/80 BY MT
OPERATION  09           ASSEMBLE BRACKETS TO RING                                              REVISION      / /    BY
DEPT. 605G       MACHINE 24         TOOLING             PRESS DIE           STANDARD HOURS - SETUP  .00000     RUN  .00625
                                                  ALLOWANCE - MANUAL  .080                          UNITS PER HOUR     160.08
                                                            - PROCESS .080

                                                                                               PROCESS                NET
             LH MOTIONS                                       RH OR BODY MOTIONS     FREQ.      TIME    LH    RH    MANUAL

                                        --- RUN ---

010    112-01    PLACE RING ON FIXTURE                                PO

 001                                        G12 -12  GET       RING TO BAR                                               B
 002 RECEIVE   SIDE OF RING       G3   -24  P01 -10  MOVE      RING TO LH                             205   271       327B
 003                                        G3   -3  RECEIVE   RING FOR BETTER GRASP                        109        109
 004 REACH     TO CONTACT RING  G02 -4      P02 -4   MOVE      RING TOWARD FIXTURE                     64    69         69
 005 ASSIST    RIGHT HAND         SP        P232-1   PLACE     RING ON FIXTURE                        294   294        294
 006 ASSIST    RIGHT HAND         SP        APA      PRESS     RING IN FIXTURE                        106   106        106

                                                                 ELEMENT SUBTOTAL                                      905


PART  2476SP8X-5                 WATT-HOUR METER
OPERATION  09           ASSEMBLE BRACKETS TO RING

                                                                                               PROCESS                NET
             LH MOTIONS                                       RH OR BODY MOTIONS     FREQ.      TIME    LH    RH    MANUAL

002 WITHDRAW RING FROM FIXTURE  026                                                                     75              75
003 MOVE     RING ON CONVEYOR   P02 -20  P231-2   PLACE      BKT, NUT IN FIXTURE                       183   301        301

                                                                 ELEMENT SUBTOTAL                                      467

                                             TOTAL RUN MU - MANUAL      5784  WITH ALLOWANCE    6247
                                                          - PROCESS        0  WITH ALLOWANCE       0
                                                                 TOTAL STANDARD RUN TIME        6247 MU
                                                                          CYCLE QUANTITY        1.00
                                                                  UNIT STANDARD RUN TIME        6247 MU
                                                                                              .00625 HOURS
                                                                                                .375 MINUTES
                                                                          UNITS PER HOUR      160.08

MAI  71 PERCENT          RMB  43 PERCENT    GRA  22 PERCENT    POS  34 PERCENT     PROC  0 PERCENT
```

Fig. 2.5.11. 4M operational analysis.

Revision and Maintenance of Labor Standards

This ninth module permits selective or across-the-board change to the system of standards as a result of a revision to an element. In addition, an analyst can examine in a test mode the effects of contemplated changes and can update commands identifying all standards that have changed by more than a specified percentage from their previous base times.

Figures 2.5.14 and 2.5.15 illustrate the typical WOCOM input and analysis.

2.5.4.3 Maynard Operation Sequence Technique (MOST)

In 1974 H. B. Maynard and Company introduced the MOST, a predetermined motion time system that was faster than previous techniques, including the MTM series on which it was based. It assumes that all manual work is performed with a standard sequence of activities. Motion times for each of the activities are predetermined and assigned to a "sequence model." For example, the first part of many manual jobs is transporting a workpiece to the workplace, which may consist of walking a few steps, bending, grasping the part, carrying it back, and placing it on the work surface. The MOST defines each of these actions by a letter, followed by a number denoting its complexity. Adding the numbers and multiplying by 10 gives the time standard in TMU for that operation.

The MOST work-measurement system is applicable for any cycle length and degree of repetitiveness, as long as there are variations in the motion pattern from one cycle to another. Based on the structure and theory of MTM-1 and MTM-2, this system can be applied to direct productive work such as machining, fabrication, assembly, material handling, distribution, maintenance, and clerical activities.

MOST is the integration of six self-contained modules. The main module is MOST analysis, which carries out the basic work measurement, including the workplace layout, method description, and determination of standard time. The second module is the suboperations data base for each job analyzed. The third module performs the calculation of operation time. With the fourth module, the standards are stored by part number, part noun, cost center, machine number, etc. The fifth module provides mass updating and maintenance of standards, and the final one interrelates the other five.

```
4M DATA SYSTEM    R0815     OPERATION STANDARD REPORT       REQUESTED BY KE     TIME 09.05.50  DATE 04/11/80   PAGE 1

PART  2476SP8X-5              WATT-HOUR METER                                          ORIGINATED  04/11/80 BY MT
OPERATION  09        ASSEMBLE BRACKETS TO RING                                         REVISION       /  /  BY
DEPT. 605G      MACHINE 24        TOOLING            PRESS DIE
                                                ALLOWANCE - MANUAL  .080    STANDARD HOURS - SETUP    .00000  RUN    .00625
                                                          - PROCESS .080                          UNITS PER HOUR    160.08

   ELEMENT                                                                  PRACTICE                                 EXTENDED
     CODE            DESCRIPTION                                          OPPORTUNITY    TIME   FREQUENCY           TIME

                                        -- RUN --

010  112-01     PLACE RING ON FIXTURE                                          Y            905     1.0000              905
020  112-02     ASSEMBLE RIVETS, NUT, BRACKET                                  Y           1675     2.0000             3350
030  112-03     REMOVE RING, PLACE BRACKET, NUT IN FIXTURE, REPLACE RING       Y           1062     1.0000             1062
040  112-04     ASIDE COMPLETED ASSEMBLY, PLACE BRACKET, NUT IN FIXTURE        Y            467     1.0000              467

                                                 TOTAL RUN MU - MANUAL          5784  WITH ALLOWANCE    6247
                                                              - PROCESS            0  WITH ALLOWANCE       0
                                                                            TOTAL STANDARD RUN TIME    6247
                                                                                     CYCLE QUANTITY    1.00
                                                                             UNIT STANDARD RUN TIME    6247 MU
                                                                                                     .00625 HOURS
                                                                                                       .375 MINUTES
                                                                                     UNITS PER HOUR 160.08

MAI  71 PERCENT             RMB  43 PERCENT    GRA  22 PERCENT    POS  34 PERCENT    PROC   0 PERCENT
```

Fig. 2.5.12. 4M operation standard report.

2.5.4.4 Automated Advanced Office Controls System

The widely used Advanced Office Controls (AOC) system, based on predetermined time standards for clerical operations, has been automated by the Nolan Company, Inc. The new version, AUTO-AOC, is a complete package of hardware and software. The simple microcomputer-based system utilizes two programs for maintaining the master file and establishing standards.

2.5.4.5 UniVation Systems

This is an integrated system consisting of a manufacturing planning data base, bills of materials, routings, operation method instructions, costs, assembly precedence relationships, facilities loading, and scheduling

```
4M DATA SYSTEM   R0840           OPERATOR INSTRUCTION REPORT                      REQUESTED

PART  2476SP8X-5                WATT-HOUR METER
OPERATION  09          ASSEMBLE BRACKETS TO RING
DEPT. 805G      MACHINE 24           TOOLING            PRESS DIE
                                        STANDARD HOURS - SETUP     .00000    RUN      .00625
                                        CYCLE QUANTITY  1.00    UNITS PER HOUR        160.08

                    LH MOTIONS                                    RH OR BODY MOTIONS

                                                                       --- RUN ---

010      PLACE RING ON FIXTURE

         001                                             GET      RING ON BAR
         002    RECEIVE    SIDE OF RING                  MOVE     RING TO LH
         003                                             RECEIVE  RING FOR BETTER GRASP
         004    REACH      TO CONTACT RING               MOVE     RING TOWARD FIXTURE
         005    ASSIST     RIGHT HAND                    PLACE    RING ON FIXTURE
         006    ASSIST     RIGHT HAND                    PRESS    RING IN FIXTURE

040      ASIDE COMPLETED ASSEMBLY.  PLACE BRACKET, NOT IN FIXTURE

         001    GET        RING IN FIXTURE
         002    WITHDRAW   RING FROM FIXTURE
         003    MOVE       RING ON CONVEYOR              PLACE    BRT, NOT IN FIXTURE

                                                     APPROVED BY:
  * TO BE DONE DURING PROCESS TIME                                ______________________________

                                                                  ______________________________
```

Fig. 2.5.13. Operator instruction report.

```
                              MOTION ANALYSIS

        LEFT HAND          TIME     LH      RH    TIME       RIGHT HAND

 1 + ASSEMBLE LITTLE PEOPLE TO FIXTURE ON CONVEYOR
 2 PICKUP PART
   GET HAIR                                      GET HAIR
    A8D                     54      54      54      54       A8D
    GR-R             S      48     102     102      48       GR-R             S
   PP HAIP                                       PP HAIR
    PP-0-75%         S      54     156     156      54       PP-0-75%         S
   M   HAIR TO FIXTURE                           M  HAIR TO FIXTURE
    A8SD                    70     226     226      70       A8SD
 3 ASY HAIR TO FIXTURE                           SIMO
    CT0.95080.985    S       5     231     231       5       CT0.95080.985    S
    ALN V0.2581S             7     238     238       7       ALN V0.25A1S
    DB 3.00                  2     240     240       2       DB 3.00
    UP A1S                  26     266     266      26       UP AIS
    IND F1S                 23     289     289      23       IND F1S
    INS A1P                 26     315     315      26       INS A1P
 4 PL  HAIR                                      SIMO
    0.50F1                   8     323     323       8       0.50F1
 5 PICKUP PART
   GET FACE                                      GET FACE
    A8D                     54     377     377      54       A8D
    GR-R             S      52     429     429      52       GR-R             S
   PP  FACE                                      PP FACE
    PP-0-75%         S      54     483     483      54       PP-0-75%         S
   M    FACE TO FIXTURE                          M    FACE TO FIXTURE
    A8SD                    70     553     553      70       A8SD
 6 ASY FACE TO HAIR AT FIXTURE                   SIMO
    CT0.34380.910    S      26     579     579      26       CT0.34380.910    S
    ALN V1.5081S            39     618     618      39       ALN V1.50A1S
    DB 3.00                 12     630     630      12       DB 3.00
    UP A1S                  26     656     656      26       UP A1S
    INS A1                  18     674     674      18       INS A1
 7 M    FACE TO SEAT TIMES  2.00                 SIMO
    A1-X2.00                36     710     710      36       A1-X2.00
 8 PL  FACE                                      SIMO
    0.50F1                   8     718     718       8       0.50F1
 9 PICKUP PART
   GET BODY                                      GET BODY
    A8D                     54     772     772      54       A8D
    GR-R             S      52     824     824      52       GR-R             S
   PP  BODY                                      PP BODY
    PP-0-50%         S      36     860     860      36       PP-0-50%         S
   M    BODY TO FIXTURE                          M    BODY TO FIXTURE
    A8SD                    70     930     930      70       A8SD
10 ASY BODY TO HAIR AT FIXTURE                   SIMO
    CT0.39080.800    S       9     939     939       9       CT0.39080.800    S
    ALN V0.50A1S            13     952     952      13       ALN V0.50A1S
    DB 3.00                  4     956     956       4       DB 3.00
     INS A1                 18     974      94      18       INS A1
11 M    BODY TO SEAT TIMES  2.00                 SIMO
    A1-X2.00                36    1010    1010      36       A1-X2.00
```

Fig. 2.5.14. WOCOM input.

information. Instead of using standard data, the system uses mathematical formulas to generate precise elemental standard times. As a result, only numerical input data are needed in the system. This eliminates the need for commonly used alphanumeric coding for storage of data. The UniVation systems are made up of nine modules, which are briefly outlined:

The UnivEl® System

This module generates precise elemental time standards using mathematical relationships and method instruction. In addition, it creates an integrated data base for a manufacturing planning and control network.

The Uni-CAM™ System

This module provides for the interactive development of process-planning information utilizing advanced group technology and parts classification concepts. A stand-alone minicomputer system is utilized. The system creates mathematically generated elemental methods and standards, routings, and an entire manufacturing data base. Mass updating of the entire manufacturing data base is also a feature of the Uni-CAM System. Automated process planning without the use of computer graphics is accomplished by Uni-CAM. Precosting of routings and methods are a by-product of the Uni-CAM System.

The UniComp® System

Using an algebraic language, this module feeds and maintains the mathematical relationships and allowances characterizing processes in the UnivEl System.

The VariComp™ System

This module is the software in the intelligent computer terminal for interactive development of input data to the UnivEl time, methods, and data base generator system. It edits the inputs and provides an efficient terminal operation.

The MultiComp™ System

This module is the workhorse responsible for the maintenance and mass updating of the overall manufacturing planning and control data base. A simplified data base language allows a product- or operation-oriented change to be reflected in the relevant data base with simple input instruction. This ensures a single source, or responsibility, for the maintenance of the manufacturing data base.

```
C PICKUP PART
GET -1 -2 -3 -4
PP -1 -5
M -1 to -6 -7 -8 -9
SAVE T 1
FR
C GET MORE PARTS
SAVE T 236 1530
FR
C FILL OUT PRODUCTION REPORT
SAVE T 237 648
FR
OBS
 1 30 HAIR
 2 29 FACE
 3 84 BODY
 4  1 FIXTURE
00
C ASSEMBLE LITTLE PEOPLE TO FIXTURE ON CONVEYOR
T 1 1 8 R 0 75 4 8 1 0 / SIMO
ASY 1 to 4 C .936 .950 DB 3 INDEX / SIMO
RL 1 / SIMO
T 1 2 8 R 0 75 4 8 1 0 / SIMO
ASY 2 TO 1 AT 4 C .312 .343 DB 3 / SIMO
M 2 TO SEAT 1 VE x 2 / SIMO
PL 2 / SIMO
T 1 3 8 R 0 50 4 8 1 0 / SIMO
M 3 TO SEAT 1 VF x 2 / SIMO
RL 3 / SIMO
C GET MORE PARTS AS REQUIRED
T 236 x .012
T 237 x .004
DETL
PCS 2
ALLOW 19
MIS
```

Fig. 2.5.15. WOCOM analysis.

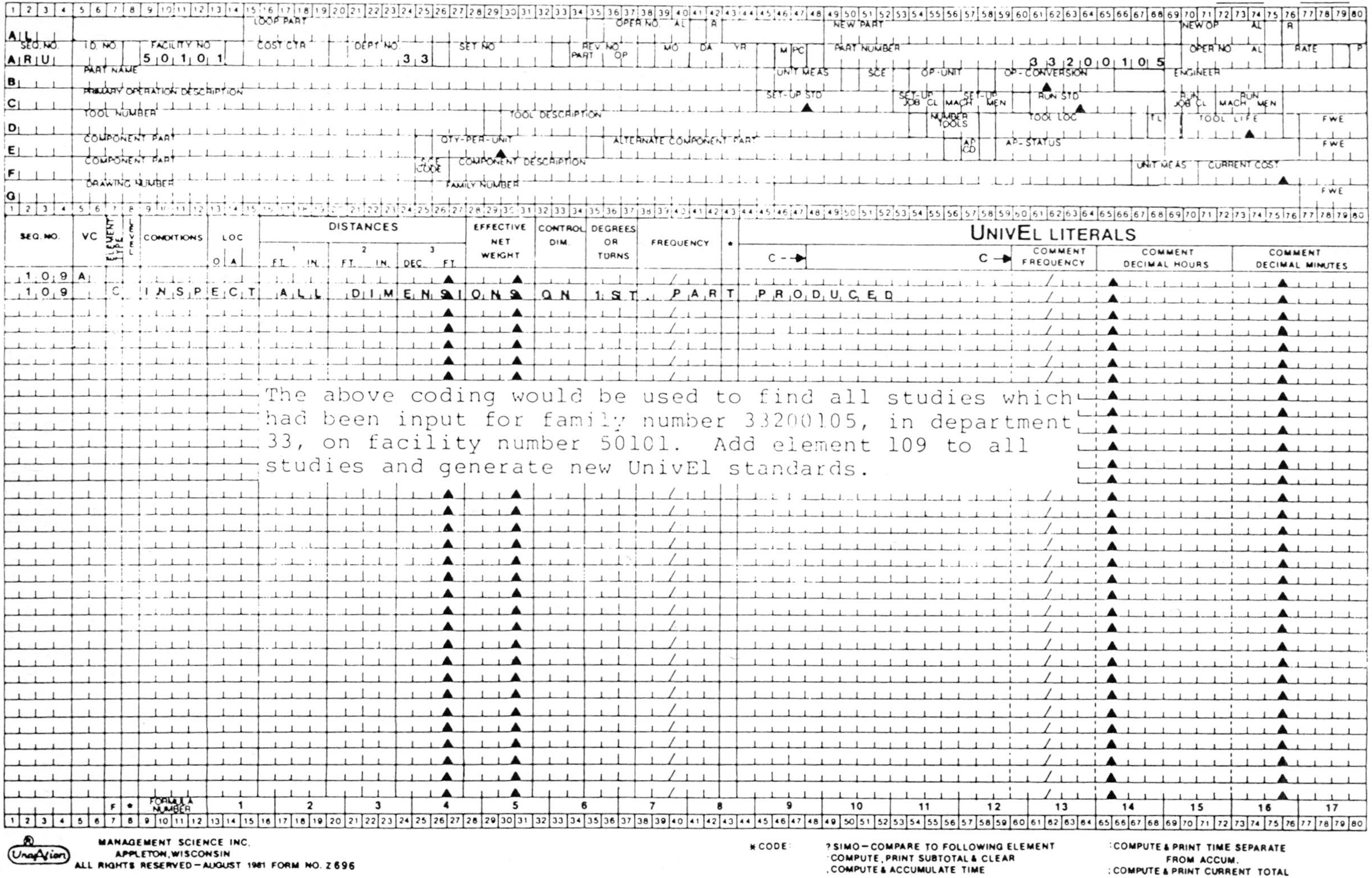

Header card fields (A–G):

Card	Field	Entry
A L	SEQ. NO. / LOOP PART / OPER NO. AL R / NEW PART / NEW OP AL R	
A R U	I.D. NO / FACILITY NO	5 0 1 0 1
	COST CTR / DEPT NO	3 3
	SET NO / REV NO PART OP / MO DA YR / M PC / PART NUMBER	3 3 2 0 0 1 0 5
	OPER NO AL / RATE / P	
B	PART NAME / UNIT MEAS / SCE / OP-UNIT / OP-CONVERSION / ENGINEER	
C	PRIMARY OPERATION DESCRIPTION / SET-UP STD / SET-UP JOB CL MACH MEN / RUN STD / RUN JOB CL MACH MEN	
D	TOOL NUMBER / TOOL DESCRIPTION / NUMBER TOOLS / TOOL LOC / TL / TOOL LIFE / FWE	
E	COMPONENT PART / QTY-PER-UNIT / ALTERNATE COMPONENT PART / AP CD / AP-STATUS / FWE	
F	COMPONENT PART / ACE CODE / COMPONENT DESCRIPTION / UNIT MEAS / CURRENT COST	
G	DRAWING NUMBER / FAMILY NUMBER / FWE	

Element lines:

SEQ. NO.	VC	ELEMENT TYPE	FREQ.	CONDITIONS	LOC O A	DISTANCES 1 FT. IN.	2 FT. IN.	3 DEC. FT.	EFFECTIVE NET WEIGHT	CONTROL DIM.	DEGREES OR TURNS	FREQUENCY	*	UNIVEL LITERALS C →	C → COMMENT FREQUENCY	COMMENT DECIMAL HOURS	COMMENT DECIMAL MINUTES
1 0 9	A																
1 0 9		C		INSPECT ALL DIMENSIONS ON 1ST. PART PRODUCED													

The above coding would be used to find all studies which had been input for family number 33200105, in department 33, on facility number 50101. Add element 109 to all studies and generate new UnivEl standards.

F • FORMULA NUMBER

* CODE: ? SIMO—COMPARE TO FOLLOWING ELEMENT
' COMPUTE, PRINT SUBTOTAL & CLEAR
. COMPUTE & ACCUMULATE TIME
: COMPUTE & PRINT TIME SEPARATE FROM ACCUM.
; COMPUTE & PRINT CURRENT TOTAL

Fig. 2.5.16. UnivEl coding sheet.

The UniPlan™ System

This module provides computerized assembly line balancing. Utilizing the time standards generated by UnivEl, this program develops the optimal design of the assembly line. Its outputs include detailed method instructions for the assembly line operators; tool, drawing, and parts lists for each work station; and revised assembly line configurations when labor crew or production requirements change.

The Routing Data Base® (RTG)

This module, generated automatically from data transferred by the UnivEl System, includes network and scheduling information, standard cost data, materials list and usage points, engineering change control information, tooling and process routing data, inventory information, and manufacturing operation measurement standards.

The UniCost™ System

This module computes standard and current costs at the operational level in terms of material, labor, and overhead. It offers the capability for costing work-in-process, scrap, and proposed changes.

The Performance Audit and Review (PAR)™ System

This module generates a labor performance report, based on actual operating data from the shop, by department, shift, and employee. Audit reports are also generated to identify discrepancies in data entry and calculation.

Inputs and Outputs. The outputs of the UniVation System can be seen vividly by examining the diverse management tools it provides. Figure 2.5.16 shows a UnivEl coding sheet consisting of constant data, such as weight and distance. Figure 2.5.17 depicts all the data sent to the computer for generating the standard, the method instruction sheet, and the routing. Figure 2.5.18 presents the standard and the method of instruction produced by the system. The file routing, consisting of tooling and material-used information, is displayed in Figure 2.5.19.

Work measurement is a fundamental science in the measurement and improvement of productivity. For the production engineer, it has widespread applications:

1. A standard for performance of all measurable jobs.
2. Utilized in product design to ensure labor efficiency.
3. Cost estimation, standard cost system, and budgetary control.

```
1...5...10...15...20...25...30...35...40...45...50...55...60...65...70...75...80  MESSAGE
ACOO    50101           33                A 082080                  52957 0020  RUM
BC000105 FEED LOWER HALF                            EACH MM      0000000    R P
CCOOTRIM                                            00000000    0000               0101
GCOO               E-08884                 33200105  470000  200000  187500
0100  S 0613 TRIM USING TRIM DIE & PRESS
0102  111   3    8             1  1         ?PART,SHELF PAN                    •
0104  111    4   22            1  1         ,PART WITH OTHER HAND,POSITION IN DIE
0106  111   2   22             1  1         ,HANDS,TRIP PRESS
0108  C.PRESS CYCLE                                                              00417
0109  C INSPECT ALL DIMENSIONS ON 1ST. PART PRODUCED
0110  C STOCK UP PARTS- CLEAN DIE AND OIL DIE
0114  112   2   42            10  1     1300?SCOOP,WOODEN BOX
0116  112   2   42            10  1     1300,RAKE,WOODEN BOX
0118  111    1   18           10 10     1300,RAKE,PULL PARTS INTO SCOOP
0120  111   2   12            10 10     5300,RAKE,REPOSITION IN PARTS
0122  111   1   12            10 10     5300,RAKE,PULL PARTS INTO SCOOP
0124  112   3   24            50 10     6300,SCOOP OF PARTS,SHELF PAN
0126  C,MOVE SCOOP TO POUR PARTS INTO SHELF PAN          6300                  01000
0128  112   2   24            20 10     5300,EMPTY SCOOP BACK,PARTS BOX
0130  112   2   24            20 10     1300,SCOOP,WOODEN BOX TO ASIDE
0132  112   1   24             1  1     1300,BACK,WORK AREA
0134  1222  33   24  28        3  6     1 25,OIN CAN,WOODEN BOX,IT,DIE
0136  C,MOVE OIL CAN DIE                                 1 25                  02500
0138  112   2   28             3  6     1 25,OIL CAN,WOODEN BOX
0140  112   1   24             1  1     1 25,BACK,WORK AREA
0142  1211  22   14  10        3  6     1100,AIR HOSE,HOLDER,IT,DIE
0144  C,MOVE AIR NOZZLE TO BLOW SCRAP FROM DIE           1100                  03000
0146  111   3   10             3  6     1100?AIR HOSE,HOLDER
0148  111   3   22             1  1     1100,PART,SHELF PAN
0150  F,9030
9999  E

1...5...10...15...20...25...30...35...40...45...50...55...60...65...70...75...80  MESSAGE
```

Fig. 2.5.17. UnivEl study listing.

PGM-V697 MANAGEMENT SCIENCE INC. ENG

PART NAME COVER SOURCE CODE MM PART-NO. 100-13022.000M

OPER DESC. NOTCH FAC. 80101 CTR. PCH17 DEPT. 42 OPER. 0030

METHOD SET A101 REV. NO. A1/ JOB CL CREW 1 MACH. 1 MAN PRT E ENGINEER B. K.

TOOL NO. PP-100-13022 DESC. NOTCHING DIE QUANTITY 001

DWG. NO. FAMILY NO. RATE TYPE R U 1 P L 09/26/79

NO.	ELEMENT DESCRIPTION	FREQ.	HRS/PC	MIN/PC
1	OBT. COVER AT SIDE FROM LOC 18 IN DIST, MOVE COVER TO SIDE TO POS IN DIE 25 IN DIST.	1/ 1	.0006365	.03819
2	MOVE HAND TO DEPRESS START BUTTON 12 IN DIST.	1/ 1	.0001390	.00834
3	PRESS CYCLE TIME	1/ 1	.0006583	.03950
4	OBT. ITEM AT SIDE FROM TABLE 15 IN DIST, MOVE IT TO SIDE TO POS IN DIE 34 IN DIST.	1/ 1	.0007220	.04332
5	OBT. SCRAPER AT SIDE FROM MACH 24 IN DIST, MOVE IT TO SCRAPE 45 IN DIST.	1/ 10	.0000842	.00505
6	MOVE SCRAPER TO REMOVE SCRAPES 6 IN DIST.	1/ 4	.0000536	.00322
7	MOVE SCRAPER TO ASIDE 45 IN DIST.	1/ 10	.0000589	.00353
		TOTAL –	.0023525	.14115
	F # 9030 15.0 PER CENT ALLOW.		.0003528	.02117
	.00271 HRS/ PC 369.0 PCS/HOUR	TOTAL	.0027053	.16232

RUN DATE 8/27/82 TIME 09:16:45 PG. 1

STANDARD

OUTPUT AT 100%

Fig. 2.5.18. UnivEl standard and method instruction.

```
PGM-U160                    MANAGEMENT SCIENCE, INC.                DATE 09/08/82 07:26:54
                         FILE ROUTING AND ASSEMBLY PARTS LIST
REVISION-NUMBER   -                                  PART-NUMBER              X0E005B MM

REVISION-DATE  07/10/80              PART-DESCRIPTION  BASIC SHORT PULL ON
                                     FAMILY--NUMBER  X
                                     DRAWING-NUMBER                         PAGE NUMBER--01
***************************************************************************************

OPER FACILITY COST  DEPT OP              OPERATION                  SETUP      RUN    ASSGN   REV
 NO.    NO.   CTR.  NO. CD              DESCRIPTION                 TIME       TIME   MH-MN CB NO

 10   SL620   SL    2   EDGE STITCH CREASE IN FRONT PANNELS 2      .00000   2.13549  1- 1 UC
                                     JOB CLASS  SETUP 000 RUN AOS   46.8  PCS/HR      L

                       - - - - -MATERIAL- - - - -
                     DK320 PP      .55000 YARD DOUBLE KNIT FABRIC
                     TH320 PP    11.00000 YARD THREAD

 20   SL160   SL    2   SEW DARTS ON BACK PANELS (2)               .00000   1.15558  1- 1 UC
                                     JOB CLASS  SETUP 000 RUN AOS   86.5  PCS/HR      L

                       - - - - -MATERIAL- - - - -
                     TH320 PP     3.00000 YARD THREAD

 30   02340   OL    2   JOIN FRONT & BACK INSEAM & RISE            .00000   1.66776  1- 1 UC
                                     JOB CLASS  SETUP 000 RUN HOS   60.0  PCS/HR      L

                       - - - - -MATERIAL- - - - -
                     TH320 PP    22.00000 YARD THREAD

 40   OL590   OL    4   SET ELASTIC ON WAIST                       .00000   2.88958  1- 1 UC
                                     JOB CLASS  SETUP 000 RUN AOS   34.6  PCS/HR      L

                       - - - - -MATERIAL- - - - -
                     EL100 PP      .66667 YARD ELASTIC 1 IN WHITE
                     TH320 PP    11.00000 YARD THREAD

 50   HB600   HB    3   BLIND HEM LEGS 1 1/2 IN                    .00000    .94240  1- 1 UC
                                     JOB CLASS  SETUP 000 RUN HOS  106.1  PCS/HR      L

                       - - - - -MATERIAL- - - - -
                     TH320 PP    10.00000 YARD THREAD

 60   HH500   HH    5   TOP STITCH WAIST CELASTIC BIBBINS TURN     .00000    .77301  1- 1 UC
```

Fig. 2.5.19. Routing sheet.

4. Methods improvement.
5. Capacity planning.
6. Manpower forecasting, scheduling, and control.
7. Work system design, such as an assembly line or man–machine assignment.
8. Layout and facility design.
9. Make-vs.-buy decisions, evaluation of alternatives, pricing decisions, etc.
10. Incentive wage payment system.
11. Serves to motivate employees.

In recent years we have witnessed the evolution of computerized work measurement techniques, which have minimized the need for human effort in the areas of data collection, analysis, and reporting. In addition, mathematical techniques have provided great accuracy in estimation of labor at a minimal expense for work measurement. We have also widened the application of work measurement from manufacturing to include service functions. Manpower utilization of office personnel and professionals through work measurement is becoming increasingly prevalent. Work measurement is finally beginning to encompass all activities of an enterprise.

REFERENCES

1. A. G. Shaw, "Motion Study," in *Industrial Engineering Handbook*, 3rd ed., H. B. Maynard, Ed. New York: McGraw-Hill, 1971.
2. A. B. Segur, "The Use of Predetermined Times," in *Industrial Management Society Clinic Proceedings*, Chicago, 1964.

3. J. H. Quick, J. H. Duncan, and J. A. Malcolm, *Work Factor Time Standards*. New York: McGraw-Hill, 1962, pp. 157–221.
4. H. B. Maynard, G. J. Stegemerten, and J. L. Schwab, *Methods Time Measurement*, **15**(3), 4–10, 1948.
5. J. C. Martin, "The 4M Data System," *Industrial Engineering*, **16**(3), 32–38, 1974.
6. P. Murphy, "A Computerized Standard Data System," *Industrial Engineering*, **12**(10), 10–17, 1970.
7. R. F. Weaver and E. A. Boepple, "WOCOM and Quick Work-Factor: The State-of-the-Art in Predetermined Systems," A.I.I.E. Fall Conference, 1979.
8. K. P. Zandin, "Relieving the Productivity Shortage," paper presented at Industrial Management Society Meeting, Arlington Heights, November 1973.
9. UniVation Systems, *General Specifications*, Management Science, Appletown, WI.
10. *Introducing Auto-A.O.C., General Specifications*, Robert E. Nolan Co., Simsbury, CT.
11. W. Antis, "Stopwatch Time Study," in *Industrial Engineering Handbook*, 3rd ed., H. B. Maynard, Ed. New York: McGraw-Hill, 1971.
12. W. J. Richardson and E. S. Pape, "Work Sampling," in *Handbook of Industrial Engineering*, G. Salvendy, Ed. New York: Wiley, 1982.
13. R. F. Weaver, M. D. Talerico, and E. A. Boepple, "Work Measurement Standards and Control."
14. A. W. Cywar, "Development and Use of Standard Data," in *Handbook of Industrial Engineering*, G. Salvendy, Ed. New York: Wiley, 1982.

CHAPTER 2.6

INDUSTRIAL EMPLOYEE TRAINING

J. GREG MURPHY
RONALD W. BROCKETT

Kurt Salmon Associates, Inc.
Atlanta, Georgia

The need for effective training in industry is a critical but often neglected area of management. Formal training programs can produce significant results when designed to meet a company's specific needs and can substantially reduce operating costs.

Poor training, with ensuing high turnover, is expensive. The cost to train a new employee can run as high as several thousand dollars, considering trainee wages, lost production, overhead, payroll costs, defective work, and other factors. But proper training procedures can reduce these costs to a fraction of what they would be otherwise.

The concepts of training that will be developed here have been successfully applied in a wide variety of manufacturing situations with resulting training improvements of the following magnitude:

Up to 60% reductions in training times.

Up to 75% reductions in training costs.

Up to threefold increases in the number of successful trainees.

Up to 30% increases in learner productivity during the training period.

2.6.1 INDUSTRIAL TRAINING—A HISTORICAL PERSPECTIVE

The origins of industrial training can be traced as far back as ancient Egypt with its craftsman training. We know a great deal about the system of apprenticeship training that grew out of the European craft guilds during medieval times.

A young man was "apprenticed" to work under the guidance of a master craftsman. The young apprentice went through a training period of two to seven years. At the end of this time, the apprentice became a journeyman of his craft. For a journeyman to become a master craftsman, he had to pass an examination or make a product that would be judged as a "masterpiece" by other masters. This was a very formal, laborious, and time-consuming approach to training.

Beginning in the early 1800s, the Industrial Revolution was in full bloom, and as machines were developed to do much of the work formerly done by skilled tradesmen, the apprenticeship system fell into a period of rapid decline.

Factories needed large numbers of workers to meet increasing production demands, but the mechanization of the Industrial Revolution allowed companies to employ previously unskilled workers. Because of new technology, operating methods had developed to a point where unskilled labor could replace skilled craftsmen. Mass migration of farm workers to the cities provided an abundant supply of unskilled labor.

Workers could easily be replaced, due to the surplus of available labor, so the primary technique of training became hard work and fear of replacement. The training that was done consisted largely of a foreman turning a new employee over to an experienced worker to observe how a job was done. If the new employee did not seem to catch on rapidly, there were other potential employees waiting at the door.

Despite advances in manufacturing concepts during the early 1900s by advocates of "scientific management," the general approach to training industrial employees remained largely unchanged for most of the first half of the 20th century. The most common technique was still "watch your neighbor and learn from him."

This traditional approach was quickly found to be inadequate as industry moved into the second half of the 20th century. The outbreak of World War II and the economic recovery following the war years focused new management interest on the area of industrial training.

Although progress was still being made, the levels of manufacturing improvement through mechanization had slowed, compared to the giant strides of the Industrial Revolution. Industries began to find that the key to increased productivity was falling back into the hands of individual workers. Their skill and knowledge in effectively using modern machinery and production techniques would become the critical factor in many companies' ability to grow and prosper in the competitive postwar industrial climate.

Clearly, modern industry was faced with a challenge. Traditional training methods were no longer adequate or effective. This realization on the part of management gave rise to the concepts of employee training that will be covered.

2.6.2 BASIC CONCEPTS OF TRAINING

The form of employee training we will describe has been given many names, including analytical training, scientific training, formal training, and structured training. Each name carries the same implication—a systematic approach to training employees.

It is doubtful that any one universal training program is totally applicable to all manufacturing situations. Effective training procedures must be tailored to meet specific needs. Some key variables that can affect the design of a training program include:

Type of Product. Is the nature of the product heavy manufacturing or light manufacturing (tractors vs small electronic components)? Are the jobs to be taught relatively stable (same product, same operation) or is there a significant style factor for an employee to contend with (multiproducts, multioperations)?

Type of Operations. Do the jobs primarily require manual skills or is there a significant amount of technical knowledge to be acquired? Are the jobs of a repetitive nature or do they require performance of nonroutine activities?

Complexity of Machinery. Do the jobs require simple machinery (basic hand tools) or complicated equipment (operating a continuous mining machine)?

Entry Level of New Employees. Is the incoming new employee likely to be unskilled, or a graduate of a technical school?

These variables need to be weighed before designing specific training programs. There are, however, a number of basic concepts that will apply, regardless of the specific training approach that is taken, and will form the basis for our system of employee training:

1. The mastery of an industrial task involves three separate areas: the development of skill, the development of knowledge, the development of job motivation. *Skill* refers to the physical ability to perform the movements necessary to do a job. *Knowledge* refers to the mental ability to retain information necessary to do a job. *Job motivation* refers to the desire on the part of the new employee to succeed on a given job.
2. Any new worker left alone to experiment with job methods, quality standards, and productive output will more than likely fail to become a satisfactory employee. The development of a training course covering the above job factors is essential to the planned and controlled training of new employees.
3. The training period should be viewed as the time required to develop a new employee to job standard performance. It has been found that trainees left on their own below this level have difficulty achieving full productivity.
4. People learn by doing. Training consisting solely of the presentation of information is incomplete. Practice must be provided to allow the new employee to put into effect what has been taught. The transition of theory into work reality under the direction of the teacher is a key element in effective training.
5. Learning separate parts of a job and then gradually combining them into the whole job often leads to more rapid overall mastery than attempting to learn the whole job from the outset. The more complicated the job, the more valid the principle. The frequent repetition of a skill on a short-term basis is more effective than periodic performance of the skill.
6. Some phases of training are best accomplished in a training center environment, away from the distractions of the production area.
7. Training is most effective when approached from a stair-step perspective (i.e., basic skills, parts of whole job, whole job, stamina development).

Any specific application of training should incorporate these seven basic concepts.

2.6.3 PRINCIPLES OF LEARNING

2.6.3.1 General Factors

In an industrial environment there are two basic ways in which an employee can learn to perform a particular job. The first way is through self-experimentation or trial and error. The second way is to learn within a controlled environment where the material to be taught is presented in a systematic manner and learner progress is carefully monitored. This second option is the one that has proven to be most successful and effective in industry.

A simple statement of sound industrial training procedure is:

Instruction is given to the learner.

The learner receives practice in the application of the instruction.

Instruction is reinforced through correction, monitoring and feedback.

Industrial training can be thought of as the process by which people retain and learn to apply job-related skills and information. In the development of instructional procedures, consideration must be given to how people learn. Studies on learner retention show that learners

retain 10% of what they read,

retain 20% of what they hear,

retain 30% of what they see,

retain 50% of what they see and hear, and

retain 90% of what they experience.

These statistics suggest a logical approach to the presentation of instruction.

The recommended steps in the presentation of instruction to a new employee are:

1. Demonstrate at full speed. Although seeing a full-speed demonstration of the task will only result in 30% retention, it is an attention getter for the learner and results in the learner paying close attention to more detailed instruction.
2. Explain method and demonstrate slowly. This step will increase retention to 50% and will give the learner the opportunity to closely observe the job method and ask questions.
3. Have learner practice. The learner is now experiencing the task with a move up to the 90% retention level. The person instructing should closely observe the learner and immediately correct mistakes in method, quality, or general application.
4. Follow-up. A continuing process of follow-up, monitoring, correction, and feedback will lead to maximum retention.

These instructional steps are applicable for an entire job or any component job part.

2.6.3.2 Learner Motivation

Although an applicant looking for work is motivated, this motivation is often in the form of securing employment for economic reasons or social prestige. There may be little or no initial motivation to learn a particular job or to acquire a specific skill. This motivation must be generated and fostered during the training process so that the learner will maintain a high degree of interest in the mastery of the specific job to be taught.

The motivation of industrial workers is the subject of vast research. We can break motivation down into two parts: during the learning process and continued motivation after job mastery is achieved.

Our main concern is the former—motivation during training.

There are a number of proven motivational principles that apply during training and can be built upon.

1. Motivation is closely related to understanding the job to be done.
2. Motivation is closely related to knowledge of results of job performance.
3. Motivation is closely related to rapport with the one giving instruction.
4. Motivation is closely related to experiencing success.
5. Motivation is closely related to competition with peers.

2.6.3.3 *Practical Applications of Motivational Concepts*

1. *Job Understanding.* The development of training courses provides the background to ensure thorough job understanding. The clear definition and analysis of the job provides the basis for learner understanding.

2. *Knowledge of Results.* The training targets developed as part of course design provide the basis for feedback of results in terms of achievement of time goals, production goals, and quality goals throughout the learning process.
3. *Rapport with Instructor.* Although personality differences are inevitable, there are several facets of instruction that should become "givens." The instructor will be successful only if the learner becomes successful. (Nothing has been taught until it has been learned.) The instructor must put the development of the learner ahead of any inherent personality differences that may exist. The instructor must be prepared to devote as much time as necessary (within practical limits) to the needs of assigned learners. This leads to the development of an instructor/learner ratio above which effective learner coverage is not possible. This will be discussed in more detail later in the chapter.
4. *Experience of Success.* The initial development of basic job skills provides the opportunity for the achievement of early success, even though that success may be on a limited part of the overall job. The idea that "success breeds success" has been demonstrated time and time again as learners move successfully through the simple aspects of a job into the more complicated stages.
5. *Competition.* It has been found that groups of learners starting out together often progress at a faster rate than do individual trainees. Competitiveness is a natural byproduct when two or more learners start out at the same time. It has also been found that public or semipublic posting of training results help develop pride in performance.

2.6.4 IMPLEMENTATION DECISIONS

The implementation of a formal approach to employee training presents a numer of options to be evaluated prior to starting a program. Formal training consists of a specified plan to train a new employee. The absolutes of a formal system are:

A training course to be followed.
Targets and goals to be measured against.
Initial emphasis on basic training and skill development.

The options to be considered do not center around whether or not to do something, but rather how best to do something.

2.6.4.1 Who Trains New Employees?

The options here are generally three:

1. A full-time instructor who specializes in employee training.
2. A supervisor who conducts training along with his/her other responsibilities.
3. A part-time instructor who otherwise performs operational functions.

The chosen option will normally be based on the frequency with which training situations arise and the technical complexity of the jobs to be taught.

In a situation of high turnover and job complexities requiring a training period of several weeks to achieve standard performance, full-time training instructors are usually found to be the superior approach.

In situations of low turnover with infrequent training required, the use of full-time instructors may not be economically justifiable. In this case, the department supervisor may be able to adequately handle the volume of training to be done.

In situations of relatively high turnover, but at the same time simple jobs with short training periods, the use of another production employee as a part-time instructor may be sufficient.

If the use of full-time instructors is justified, the question of training staff size must be addressed. We have found that a ratio exceeding eight to ten trainees per instructor significantly reduces training effectiveness. The recommended guideline is, therefore, one full-time instructor for each eight to ten trainees.

2.6.4.2 Where to Train?

The training of industrial workers generally takes place

in a vestibule or training center,
directly in the production department, or
a combination of vestibule and production department.

Each location offers some advantages. In a training center, the atmosphere can be controlled to make it as conducive to learning as possible. The instructor can always be in close proximity to the trainee and the unnerving distraction of the production floor can be avoided.

A criticism of vestibule-type training is that it often presents a too comfortable and unrealistic environment when compared to the normal activities of the production department and is therefore artificial. Training directly on the production line immediately exposes the learner to the real world of manufacturing, but this often leads to the confusion and frustration of the trainee.

The combined approach has proven to be a successful option. Basic training and skill development are best done in a training-center environment. When the learner achieves some degree of job proficiency, the continuation of training within the production department can normally be accomplished with avoidance of the confusion and frustration that occurs if the learner is placed directly in the production department.

2.6.4.3 Teaching

The training technique of breaking a complicated job down into component parts, learning one part at a time, and then combining the separate parts into the whole cycle of the operation has proven valid in many manufacturing situations.

The isolation of a particular part of the job cycle and the repetitive practice of the isolated part has proven to be a faster way to learn than attempting the job in its entirety from the outset. This is particularly true of an operation involving a series of pick-up, positioning, and repositioning elements.

On the other hand, relatively simple operations (with only one or two clearly discernible job parts) may be most effectively taught as a whole job cycle.

In many industrial training programs the range of jobs to be taught will lead to the development of both methodologies within the training framework.

2.6.5 DESIGNING A TRAINING COURSE

The end result of a training course is a road map to be followed by the trainee, culminating in job mastery. In industrial training, this journey often starts with the new employee possessing only raw, entry level skills and ends with the employee performing an operation at job standard.

We believe the best approach in designing a training course is to start with this idea in mind—the new employee begins with no appreciable entry level skill and must consequently learn it all. If a new employee turns out to possess some prior learning or if the course is to be used for retraining, then adjustments can be made to bypass some of the course materials.

2.6.5.1 General Contents of Training Course

A completed training course should include the following:

1. An outline of the sequential steps to be followed in taking a new employee up to the level of job standard.
2. A description of the methods and key points required for successful performance of the job to be taught.
3. Performance targets in terms of time, quality, and unit production for each of the steps in the training process.
4. A description of special training exercises designed to facilitate learning of the job in question.

While on the surface this may sound imposing, an effective training course for a relatively simple operation may consist of no more than a few pages, whereas a complicated operation to be taught may require a full manual of training course material.

2.6.5.2 Job Analysis

The first step in the training course design process is to perform a job, or task analysis. As a general rule, job instruction consists of teaching a set of distinct tasks. A job analysis begins with defining the particular items of skill and knowledge that must be maintained to lead to successful performance.

As an example of the process, we will develop the job analysis and training course of an industrial operation in a sewing plant. The operation is *Attach Yokes* on a man's dress shirt. This is considered a job of medium difficulty and is fairly easy to visualize.

The operator works at the sewing machine with three component parts to be assembled; the back of the shirt, the inside yoke containing the shirt label, and the outside yoke (Fig. 2.6.1). The two yokes are joined to the back to form a completed back assembly. At the next operation, the shirt fronts are joined to the completed back assembly to form the shell of the shirt.

The job analysis begins with an investigation of the general conditions of this operation.

Work comes to the station in cut stacks of 25 units each of the same shirt size (i.e., 25 backs, 25 inside yokes, 25 outside yokes for size 16–34 shirts).

The operation is performed on a single-needle lockstitch machine with bobbin.

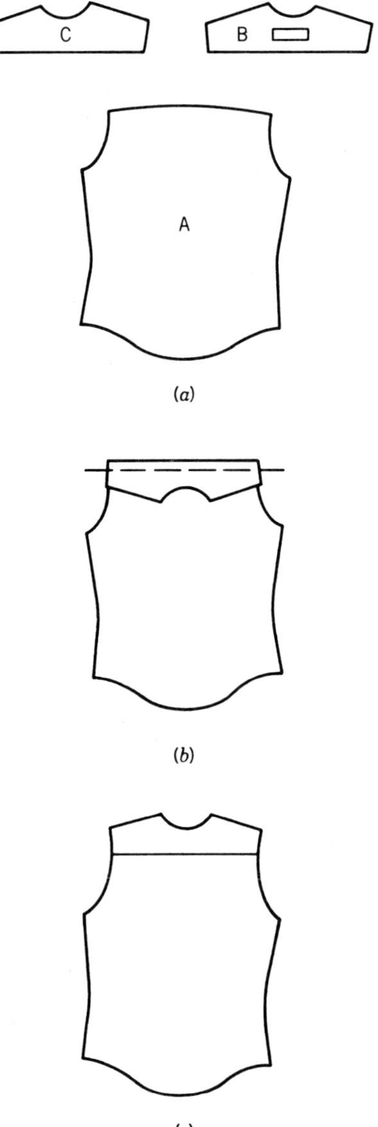

Fig. 2.6.1. (*a*) Component parts: A—back of shirt; B—inside yoke with label; C—outside yoke. (*b*) Position during operation. Back sandwiches between yokes with inside yoke face down on machine table. Dotted lines indicate stitching. (*c*) Completed back assembly.

The established job standard for the operation is 1450 units a day.

The operator must assemble the three component parts so that the inside and outside yokes are in correct position on the completed assembly.

After determining the general conditions affecting the operation, the method of job performance must be developed. There are generally two options available on job methods:

1. If the operations are highly engineered, the engineering department will probably have documented the method that corresponds to the established job standard.
2. If no engineering documentation is available, the course designer will study the methods of current workers, confer with the department supervisor, and then develop the method to be taught new employees.

Regardless of which approach is used, there are two key methods issues important to the training course:

A standard method for training is essential.

The key points of method should be clearly outlined. A *key point* is a technique of method that, while not affecting the end result of the product, is in fact a superior method of job performance. For example, a simultaneous pick-up of two parts will not affect the appearance or quality of the finished product, but it is a superior method in terms of efficiency.

The basic method for *Attach Yokes* requires the operator to:

1. Place the inside yoke face down (label down) on the machine table.
2. Align the back over the inside yoke.
3. Align the outside yoke over the back.
4. Sew across assembly to join parts together.
5. Dispose assembly to rear of machine table.
6. Repeat process above for each individual shirt.

Table 2.6.1 shows the operation analysis performed by the course designer. This analysis identifies the key factors of *Attach Yokes* and will be used as the basic working document for the design of the actual training course.

2.6.5.3 Training Course

At this point, the course designer must determine the most logical training elements of the operation. A training element may be different from the normal time study elements that are used to establish the job standard. A training element is a part of the complete job cycle that allows the trainee to practice an isolated component of the operation.

The separate mastery of isolated components of the overall job cycle has been proven to lead to a faster achievement of job standard than attempting to master the complete operation at once.

In our example, the training elements are defined as follows:

1. As left hand disposes the previous shirt to the rear of the machine table, right hand picks up the inside yoke, and places in position on machine table. .0.33 min
2. Left hand picks up the back as right hand picks up the outside yoke. Both hands align outside yoke to back while moving to table. Both hands position assembly to inside yoke on table and align for machine entry. 0.073 min
3. Sew without stopping to mid-point. Pause, align assembly at finishing end, sew to end without stops. 0.117 min

Total 0.223 min

The standard time for each of these training elements is shown in Table 2.6.2.

OPERATION: ATTACH YOKES

Job Conditions and Layout

The backs and yokes come to this operation in stacks of 25 units. A double cut-off yoke is attached to the back, and the back assembly then goes to join shoulders. The assembly is backstitched at both start and finish and the shirts are left in chain to be clipped apart at completion of stack.

The backs are positioned on table to left of operator with right side up and sewing edge toward machine. The inside yokes are placed in operator's lap with labels up and sewing edge toward machine. The outside yokes are placed to right of machine with right side up and sewing edge away from machine. The shirts are disposed in bin at rear of table (Fig. 2.6.2).

Method

1. RH picks up inside yoke as LH controls sew off of previous shirt.
2. LH pushes previous shirt to rear of machine as RH moves inside yoke to table.
3. LH picks up back as RH picks up outside yoke.
4. LH and RH align outside yoke to back in midair.
5. LH and RH place assembly on inside yoke and align three pieces along sewing edge.
6. Enter machine, backstitch, and sew to midpoint.

Table 2.6.1. Operation Analysis: Attach Yokes

Element	Left Hand	Right Hand	Foot or Knee	Skills	Mental Operation	Key Points
Sew off previous shirt pick up, position first yoke	Control backstitch and sew off of assembly in machine with L1, 2, 3, 4, LT on second yoke palm down	Grasp first yoke at back right edge with R1 and R2 on top right underneath	Raise presser foot slightly for backstitch	Foot-knee for backstitch(5)		RH grasps first yoke as LH controls sew off
	Push completed assembly to rear of machine	Move first yoke to table from label side up to label down. Place on table with label and sewing edge aligned to enter machine	Brake	Finger dexterity separate yokes(3) Manual dexterity move to machine(2)	Label must be turned face down	RH moves first yoke to table as LH disposes previous shirt
Pick up, position second yoke and back, align to first yoke	Move to back, grasp top sewing edge with LT on top, L1 and L2 underneath	Move to second yoke, grasp top sewing edge with R1 and R2 on top, RT underneath		Finger dexterity separate yokes and backs(3) Manual dexterity move to machine(2)		RH picks up second yoke as LH picks up back
	Move back to front of machine	Move second yoke to front of machine				
	Position back to second yoke in midair with back under yoke	Position second yoke to back in midair with yoke over back		Hand and eye-alignment of pieces(4)		Position yoke to back in midair
	Place assembly on top of first yoke	Place assembly on top of first yoke		Finger sen.-position to first yoke(4)		
	Align sewing edge of assembly to sewing edge of first yoke at starting end with LT on top at second yoke and L1 and L2 under first yoke	Align sewing edge of assembly to sewing edge of first yoke at midpoint with RT on top of second yoke and R1 and R2 under first yoke		Hand and eye-alignment of pieces(4)	Sewing edges of yokes and back must be equal length	
Enter machine, sew to midpoint	With RT, R1 and R2 pinching together to maintain alignment, enter machine	With RT, R1 and R2 pinching together to maintain alignment, enter machine				
	Move to palm down position on second yoke to left of needle during backstitch	Maintain alignment at mid-point during backstitch	Raise P. F. slightly for backstitch	Foot-knee for backstitch(5)		
	Keeping assembly against gauge, sew to midpoint	Maintaining alignment, sew to midpoint	Accelerate		All three pieces must be kept against gauge	Sew to midpoint without stops
Align at end, sew to end	Align second yoke and back to first yoke at end	Align second yoke and back to first yoke at end	Brake	Hand and eye alignment of pieces(4)		
	Keeping assembly against gauge, sew to end	Maintaining alignment at end, sew to end	Accelerate			RH maintains alignment with R1, R2 on top, RT underneath

Table 2.6.2. Training Course: Attach Yokes; Basic Machine Control Maintenance

	Target
1. Machine threading	0.250 min
2. Needle changes	0.175 min
3. Bobbin changes	0.333 min
4. Sew on fabric	0.417 min
Skill Development	
1. Pick up, position inside yoke	0.033 min
2. Pick up, position outside yokes and back	0.073 min
3. Attach yokes to back	0.117 min
4. Complete job cycle	0.223 min
Stamina Development	
1. Two cycles	0.446 min
2. Four cycles	0.892
3. Ten cycles	2.230 min
4. 25 cycles and setup	6.900 min
5. 1 hour	182 units
6. 2 hours	363 units
7. 4 hours	725 units
8. 8 hours	1,450 units

7. LH and RH align to end.

8. Sew to end, backstitch, and sew off.

Key Points

1. RH grasps inside yoke as LH controls sew off.

2. RH moves inside yoke to tableas Lh disposes previous shirt.

3. RH picks up outside yoke as LH picks up back.

4. Position outside yoke to back in midair.

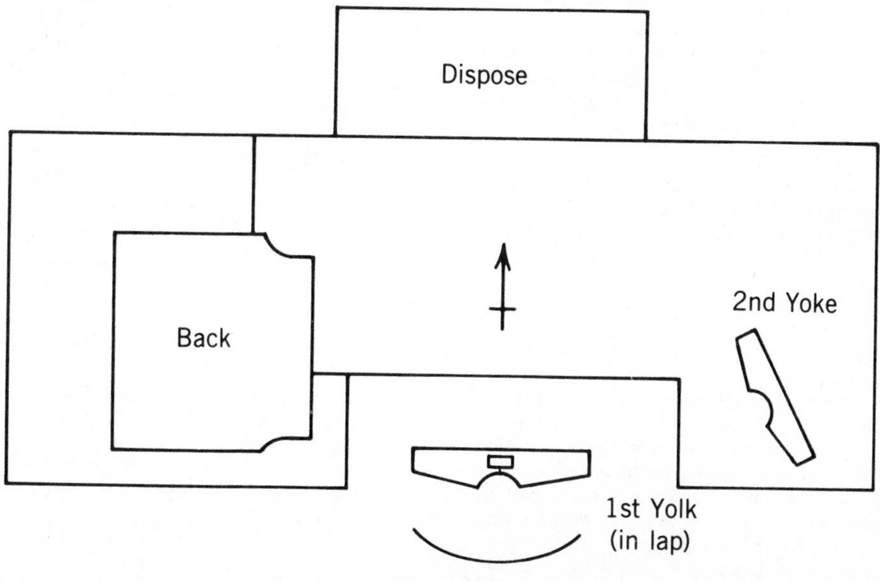

Fig. 2.6.2.

5. Sew to midpoint without stops.
6. RH holds end alignment with R1 and R2 on top and RT underneath.

Quality Specifications

12 SPI + 1.
Seam width: 3/8" ± 1/16".
Alignment at each end: even ± 1/8".
Alignment along sewing edge: even ± 1/8".

This course provides a step-by-step guideline for trainee development in the key areas of method, quality, and skill up to achievement of job standard.

2.6.6 TRAINING TOOLS

Although not complete training programs in themselves, there are a number of useful tools that can be incorporated into an overall employee training system.

2.6.6.1 Visual Aids

A *visual presentation* of job methods is a valuable enhancement of written material. Some of the more common forms include motion pictures, videotape recordings, and slide/sound presentations. All three offer learners the advantage of studying correct job methods without the distractions that are often present when viewing the job "live" in an actual production environment.

2.6.6.2 Programmed Instruction

Programmed instruction is a means of presenting written material in small bites of information and requiring the learner to make active responses that demonstrate understanding of the presented material before advancing to more information. In other words, programmed instruction allows the learner to actively ________ in order to verify that the presented information has been ________ .

. . .

The correct response to the above are *respond* and *understood*.

2.6.6.3 Computer-Assisted Instruction

A more sophisticated variation on the programmed instruction theme is interactive *computer-assisted instruction* (CAI). CAI allows the development of learning material that is presented on an interactive computer terminal. This offers the advantages of individual instruction, self-pacing, immediate feedback, and can be developed into a more interesting and entertaining format than traditional training material. A creative programmer/author can develop work simulations that allow the learner to test out newly acquired knowledge in a realistic work setting. Successful simulations lead to learner confidence in the actual job setting.

2.6.6.4 Pacing Devices

For production work that requires the achievement of an engineered job standard performance level, a variety of *pacing devices* are available that allow the trainees to pace themselves with 100% work levels for progressively longer periods of time. These devices have proven to be valuable aids when the learner's ability to meet or exceed job quotas influence wage levels within a production incentive system.

2.6.7 TRAINING CONTROLS

Essential to the success of any training program are controls that will allow management to evaluate and monitor program results.

In the training of production workers, it is important to carefully program trainee development so the learner successfully completes each prerequisite step before advancing to the next. In the sample training course presented earlier, the basic training steps should be completed before advancing to the job-parts step, and job parts should be completed before advancing to the stamina-development phase.

Targets and goals should be used for each step in the training process and their progressive achievement kept in summary form as part of a trainee's permanent record.

Individual performance graphs that measure a trainee's actual progress against an expected learning curve should be kept. In a program involving a number of trainees, a weekly management report should be compiled, summarizing the performance of each trainee in the program against expectation. This report highlights the success being achieved during the training process and also points out those individuals in need of special assistance.

At the completion of training, the tangible costs of employee training (for both successful and unsuccessful trainees) should be developed to provide an overall cost expenditure per success. This allows management to determine the cost effectiveness of its training effort and also provides a benchmark against which to measure in the future.

2.6.8 SUPERVISORY TRAINING

Effective first-line supervision is a vital cog in the efficiency of manufacturing operations. In labor intensive industries, well-trained supervisors play an especially pivotal role in the organization. By proper attention to supervisory training and development, a company is protecting its substantial investment in people, equipment, and materials.

The development of effective supervisors is best accomplished in an atmosphere wherein the supervisor is placed in a role of department or section manager rather than as a go-between from management to employees. As a department manager, a supervisor should be responsible for the work flow, cost, quality, and worker performance in his/her section.

This viewpoint of supervision leads to a rather clear analysis of the job to be done and allows the development of a specific training program for supervisors along the same lines covered for production employees. The main difference lies in the fact that training for supervisors will cover areas of job knowledge rather than production skills.

In the design of a supervisory training program, a logical first step is the development of specific supervisory duties and then a clear separation of primary duties and secondary duties. A frequently encountered problem in manufacturing supervision is the discovery that a supervisor may be spending the majority of his/her time performing secondary duties.

An inadequately trained supervisor will often use a "brush-fire" approach to solving problems, and will become heavily involved in activities that could be better delegated to someone else. One reason this happens is that people have a strong tendency to do the things they know well and are comfortable with and not necessarily the things that need to be done to genuinely solve an existing problem. The remedy is thorough training of supervisors in the essential areas of their jobs to the point that they will feel comfortable in fulfilling their roles as department managers.

A representative supervisory training curriculum outline might look like Table 2.6.3.

Two factors are felt to be of vital importance in a successful supervisory training effort. The first is that the subject covered be specific and relevant to the actual manufacturing situations faced by a particular company. There are many "canned" supervisory training programs available, but the most successful ones are those designed to meet the specific needs of a particular organization.

The second key factor is that classroom presentations and reference reading materials should be supplemented with actual on-floor project work to reinforce presented ideas with practical applications. This ensures that what is learned in a classroom is not left in the classroom.

Table 2.6.3. A Representative Supervisory Training Curriculum

		Presentation Method		
Subject Area	Specific Topics	Classroom Presentation	Reference Reading	On-Floor Project
Cost control	Manufacturing budgets	×	×	
	Control reports	×		×
	Department cost goals	×		×
	Cost control techniques	×	×	
Work flow	Production goals	×		×
	Staffing analysis	×		×
	Cross training	×	×	×
	Scheduling	×	×	×
Quality	Quality goals	×	×	
	Quality specifications	×		×
	Inspection procedures	×	×	×
	Defective work	×		×
Worker performance	Production standards	×		×
	Training techniques	×	×	×
	Performance evaluation	×	×	×
	Employee motivation	×	×	×

2.6.9 EMPLOYEE TRAINING: AN INTANGIBLE EXPENSE OR A MEASURABLE PROFIT GENERATOR?

Cost justification is an area of some controversy related to training. Is training an expense item producing intangible results, or is it truly a cost saver and profit generator? The training of industrial production workers and supervisors provides the opportunity for measurable results that can clearly identify the savings potential of effective training procedures.

Most companies have a continuing need to train or retrain employees, either production workers or production supervisors. Training is a continuous manufacturing expense. How large an expense it is depends on the effectiveness of the training procedures.

2.6.9.1 Production Workers

In the training of production workers, a number of quantitative measures can be established and equated to dollar values. Some of the most common measures are:

1. *"Make-Up" Wages.* A primary objective of training is for a new employee to produce at 100% of the job standard, but obviously this will not be the case in the early stages of training.

   If a new employee is guaranteed a wage of $4.00 per hour, but only produces at 50% of standard performance, then in reality only $2.00 per hour is earned and the other $2.00 becomes a training loss. In effect, $2.00 per hour is "made up" to the employee to meet the wage guarantee.

   To illustrate a typical training cost, consider a new employee at a $4.00 per hour guarantee who follows this progression to standard performance:

Week	1	2	3	4	5	6	7	8	9	10
Performance	20%	35%	45%	60%	65%	75%	85%	90%	95%	100%
Make-Up @ 40 h/wk	$128	$104	$88	$64	$56	$40	$24	$16	$8	—
Total Make-up Cost = $528										

   During training, this new employee was paid $528 in make-up (or unearned) wages.

2. *Training Time.* The training time for a production worker is the amount of time required to reach job standard. In our above example, the new employee was trained in 10 weeks.
3. *Trainee Productivity.* The productivity of a new employee is the average performance achieved during the training period. In our example above, the average performance over the 10-week period required to achieve job standard was

$$20\% + 35\% + 45\% + 60\% + 65\% + 75\% + 85\% + 90\% + 95\% + 100\% = 670\% \div 10 = 67\%$$

4. *Rate of Success.* Ideally, every new employee should eventually achieve job standard performance. Realistically, this is seldom the case. Some new employees will be released due to their inability to meet production or quality requirements. Others will quit out of job frustration or other reasons. The rate of success is the percentage of new employees who are able to successfully achieve job standard performance. If, for example, 5 out of every 10 new employees are successful, the rate of success is 50%.

Table 2.6.4. Weekly Performance (Percent)

Week	1	2	3	4	5	6	7	8	9	10
Employee A	20	35	45	60	65	75	85	90	95	100
Employee B	12	16	21	18	Released					
Employee C	31	28	36	42	48	51	65	67	70	73
Employee D	18	36	45	58	64	Quit				
Employee E	21	36	53	52	50	56	59	53	58	51
Employee F	50	Quit								
Employee G	28	33	40	43	46	52	53	61	69	72
Employee H	18	30	26	25	Quit					
Employee I	36	49	62	59	68	73	86	82	88	88
Employee J	10	20	30	29	26	28	29	Released		

5. *Cost per Success.* The cost per success is the most powerful measure of training effectiveness because it takes into account the combined effect of make-up wages, training time, and rate of success of new employees.

   Table 2.6.4 is an example of the cost per success calculation for ten new employees at a guaranteed wage level of $4.00/hour:

   The total make-up cost for this group of 10 new employees is $6298. 3 of the 10 achieved 100% of job standard (a 30% rate of success). The cost per success is $6298 ÷ 3 = $2099.

   Counting the cost of failures, $2,099 was spent for each new employee who achieved job standard.

6. *Other Quantitative Measures.* The statistical measures mentioned above are the most common ones for the training of production employees.

   However, two other cost factors are available:

   a. *Defective work produced.* It is generally accepted that a trainee will produce a higher rate of defective work than an experienced worker. This increase in defective work can be measured in terms of rework cost, scrapped material, and second-quality production.

   b. *Overall turnover cost.* This builds upon the cost per success measure by adding all other cost factors that can be attributed to a training situation created by an employee turnover. This computation typically includes:

      make-up wages,
      lost production cost (lost profits),
      instruction cost,
      cost of defective work,
      personnel department costs: paperwork, recruiting, signup, etc.

2.6.9.2 Production Supervisors

As with production workers, a number of quantitative measures are available to determine the effectiveness of the training of production supervisors. Among the most common are:

1. *Departmental Efficiency.* It is generally possible to measure the standard work content of a given amount of production coming from a supervisor's department in a given time period. For example, 5,000 units at a standard work content of 1.5 hours per unit should require 7,500 manhours to produce. If the actual manhours required to produce these 5,000 units amounts to 8,500, then the department efficiency is 88.2% (output ÷ input or 7,500 ÷ 8,500).
2. *Departmental Cost.* Similar to departmental efficiency, the standard labor cost (assuming no labor losses) can be established. In reality, some losses will occur due to low employee performance, rework, machine downtime, product changeovers, etc. It is common practice to establish an overcost budget (for example, overcost not to exceed 10% of standard cost). This provides the basis for measuring supervisory cost-control effectiveness.
3. *Departmental Quality.* The actual versus expected quality level of a department provides a measure of supervisory performance in this critical area of management.
4. *Other Measures.* Some other measurable performance indices for supervisors include department turnover, the number of labor grievances, and equipment downtime.

These factors for production workers and supervisors are both measurable and cost quantifiable. They provide the basis for balancing the cost of formal training against the savings generated.

11	12	13	14	15	16	17	18	19	20	Total Make-up Cost ($)
										528
										533
76	70	78	83	86	80	89	92	96	100	1,022
										446
59	57	Released								960
										80
86	93	90	98	Quit						858
										482
90	91	93	96	99	103					544
										845
										Total 6,298

2.6.10 THE BOTTOM LINE—PAYOFF TO MANAGEMENT

The best way to illustrate the impact of an effective approach to employee training is through a case study of an actual implementation.

2.6.10.1 Background

ABC Company is a manufacturing facility whose employees are predominantly female production workers. The plant had been in existence for five years and was a clean, modern facility. ABC had received some state training assistance during its first year start-up, mainly in the form of financial subsidies to absorb training cost.

The company had no formal system of training for its production workers, and most of its supervisors started as production workers during the plant start-up and worked themselves up through the ranks.

ABC employed 350 production workers. The annual labor turnover was 127% and this heavy loss rate had led the company to assign four of its employees to the role of "trainer," but these four had no formal program with which to work. The nature of the company's product required versatility and frequent changeovers, although no one product or operation by itself was extremely difficult.

Only 37% of the work force was performing above standard. The overall plant efficiency was only 75% and labor losses ran 29% of standard cost.

An analysis of production-worker training showed that the cost per success ran $1603 and the rate of success was only 20%. It took in excess of 50 weeks for the average new employee to achieve standard performance consistently.

2.6.10.2 Implementation

Based on an analysis of the aforementioned conditions, a training program was designed and implemented as follows:

1. *For Production Workers*

a. It was determined that due to the style factors involved, the trainee/instructor ratio would be set at 8 to 1.

The initial staffing was set to cover the current turnover rate (127%), although it was expected that improved training would reduce turnover.

Staffing was calculated as follows:

350 employees @ 127% turnover = 445 hires annually

It was estimated the rate of success could be improved to at least 40%, with training time per successful trainee at 12 weeks and time per unsuccessful trainee at 6 weeks.

445 hires @ 40 percent success rate:

178 successful @ 12 weeks	=	2136 weeks
267 unsuccessful @ 6 weeks	=	1602 weeks
Training weeks		3736

One instructor with 8 trainees for 49 work weeks covers 392 (8 × 49) of the needed training weeks.

Therefore, 3738 ÷ 392 = 9.5, or approximately 9 instructors.

Five additional instructors were selected to join the original four, and the group of nine was given a structured program within which to operate.

b. The training procedures designed were primarily aimed at new employees but contained enough flexibility to be useful in working with current employees who were classified as low performers in need of retraining.

c. The procedural aspects of the program gave each instructor a structured plan to follow with each operator (trainee or retrainee). Learning curves were established for each operation with trainees plotted from zero performance up to full job standard, and retrainees plotted from their current performance level up to job standard.

d. Training courses were developed for each operation. The courses included job methods, target times for various elements of each job, step-by-step progression from basic training through skill development to stamina buildup, and training times.

e. A separate training area was established and new hires spent approximately one to two weeks in this vestibule to complete their basic training and move into skill development on actual production work. When the trainee moved from the vestibule to the production line, the instructor continued the same work although now in a different location.

f. Training was continued until the employee (new hire or retrainee) achieved job standard on the particular operation.

2. *For Supervisors*

a. An evaluation of supervisory performance revealed that most of the supervisors received little, if any, formal training prior to assuming control of their departments. As a result, most supervisory

Table 2.6.5. Results

	Before Training	After Training	Improvement
New Employees			
Cost per success	$1,603	$ 672	58%
Rate of success	20%	4%	110%
Average weeks	50+	13	74%
Retrainees			
Average performance	82%	106%	29%
Turnover	127%	70%	45%
Employees above standard	37%	65%	76%
Plant efficiency	75%	94%	25%
Overcost (labor cost)	29%	13%	55%

time was spent "putting out fires" with little time given to any real attempt to manage the department. A typical solution to any product problem was for the supervisor to step in and operate the machinery as a production worker or to repair the defective work produced by operators.

b. A training program was designed for supervisors consisting of classroom sessions followed by actual on-floor work to implement the material covered in class. Specific areas covered included production flow techniques, cost control, quality control, employee performance, and human relations skills.

c. The supervisors were taught techniques to establish their own departmental performance goals (under the guidance of management) and also skills to achieve these goals.

d. The company's on-site data processing center ran both daily and weekly performance analyses for all production workers, but it was done on a plant-wide basis. The computer run was broken down by supervisory section to give each supervisor a daily and weekly analysis of both individual operator and overall department performance.

e. The nine training instructors attended the classroom sessions with the supervisors to blend the effort of both groups into a cohesive approach to improving plant performance.

2.6.10.3 Results

Within one year of the implementation of the training program for operators and supervisors, the following changes had occurred (Table 2.6.5). The combination of reduced labor loss and fixed overhead recovery on increased productivity amounted to an annual savings of $531,564. The implementation costs were $46,800 annually for five additional instructors, $15,000 for a training director on a half-time basis, and a one-time startup cost of $45,000. The program paid for itself within the first year and provided an 8 to 1 return on investment thereafter.

2.6.11 SUMMARY

The primary objective of industrial employee training is quite simple: to produce the greatest number of trained employees at the least possible cost. The experience of most manufacturing concerns has shown clearly that leaving new employees basically on their own to learn a job is not the way to achieve this objective.

While a formal program of employee training is costly, the savings achieved will far outweigh the expenditures.

The concepts of industrial training presented here have proven themselves effective over a wide range of manufacturing conditions and have stood the test of time. In today's competitive manufacturing environment, a commitment to sound principles of employee training is not a luxury item, but rather a basic necessity of industrial life.

PART 3

METHODS: DESIGNING, IMPROVING, AND CONTROLLING PRODUCTION SYSTEMS

CHAPTER 3.1

DESIGN-TO-MANUFACTURE

BENJAMIN W. NIEBEL

The Pennsylvania State University
University Park, Pennsylvania

3.1.1 OBJECTIVES OF DESIGN-TO-MANUFACTURE

The principal objective of design-to-manufacture is to produce a design that will satisfy both functional and physical requirements at a cost that is compatible to the user. Thus the design must be producible at a cost that will permit the product to be introduced competitively in the marketplace.

In order to achieve the principal objective, consideration should include but not necessarily be limited to the following areas:

1. To maximize
   a. simplicity of design,
   b. use of economical materials that will satisfy the functional design requirements,
   c. use of economical manufacturing tooling, methods, and procedures,
   d. standardization of materials and components,
   e. ease of assembly,
   f. ease of inspecting and testing the product, and
   g. maintainability of the design.
2. To minimize
   a. use of critical processes,
   b. generation of scrap, chips, or waste,
   c. procurement lead time,
   d. energy consumption,
   e. special manufacturing tests and inspection procedures,
   f. generation of pollution,
   g. specification of proprietary items, and
   h. limited availability materials, items, and processes.

Design-to-manufacture implies the incorporation of producibility not only during the conception and development of the design but during the production phases and throughout the entire life cycle of the product. However, it is during the functional design of the product that producibility should be incorporated as the design progressively takes shape.

3.1.2 ENGINEERING DRAWINGS

The principal communication tool between the functional product designer and those responsible for producing the product is the engineering drawing. Whether the drawing is prepared by the draftsman using traditional methods or is the result of computer-aided drafting where the computer terminal is the exclusive tool used to develop drawings of mechanical parts and structures and for the development of the cutter-line program to be used to machine the parts, it is the engineering drawings alone that control and completely delineate size, shape, form, fit, finish, function, and interchangeability requirements that lead to the most competitive procurement.

An accurate engineering drawing, together with reference specifications and standards, will permit a qualified manufacturer to produce the design within the dimensional and surface tolerance specifications provided. The drawing will convey to the creative manufacturing planner how the design can best be produced. Since producibility should be incorporated in the design, the drawings will reflect this characteristic. Frequently, in view of the space constraints, product specifications such as quality-assurance check points and inspection procedures will be separately summarized, but they should always be cross-referenced on the engineering drawing.

The following points are characteristic of sound design-to-manufacture engineering drawings:

1. The design as depicted on the engineering drawings is conducive to the application of economic processing.
2. Group technology has been incorporated in all components that comprise the design. Thus, "sameness" of the various components with similar or like components produced in the past are specified so as to take advantage of common tooling, facilities, and the economics of quantity production.
3. Every component in the design can be readily produced with the current state-of-the-art manufacturing.
4. All design specifications are compatible with the performance specifications.
5. Critical location surfaces and/or points are identified on the drawing.
6. Standard components are specified that are available from two or more sources.
7. All specifications shown on the design are definitive. For example, if chamfers are indicated they will be dimensioned.
8. All dimensions and associated tolerances, parallelism, perpendicularity, etc. should be shown so they can be easily measured and/or inspected.
9. Tolerances and specifications are no more restrictive than necessary.
10. The design permits easy and rapid assembly of mating parts.
11. All materials selected are readily available and are appropriate to product function and ease of processing.
12. Size, geometry, and finish specifications may be achieved with standard tools.
13. Requirements for wiring clearance, tool clearance, component space, and clearance for joining connectors have been met.
14. All welds points are accessible and have been identified with the correct symbols.
15. The design identifies limited-life materials, such as fluids, and provides for their replacement without difficulty.
16. The design provides for easy maintenance, repair, and overhaul.
17. The design is simple—overdesign is avoided.

3.1.3 THE PRODUCT DESIGN PROCESS

To help assure successful product designs, the design process must systematically incorporate the design-to-manufacture concept. Figure 3.1.1 illustrates the six steps associated with the new product design process: initial conception and evaluation, analysis, development of general design, development of detailed design, hardware development, and development of test and production models. Although this systematic procedure will be simpler in the case of certain elementary new product designs and extended in the case of more complex designs, the steps enumerated are characteristic of the procedure that should be followed in the majority of cases. From the first step (initial conception and evaluation) where a preliminary design is developed, until the final step where production models are produced, design-to-manufacture is considered as part of the design criteria to be evaluated for cost-effectiveness and ease of manufacture vs. the degree of compliance with the functional requirements defined in the first step.

Note that feedback paths must exist, not only within the six design steps, but between the design steps throughout the design continuance. In concert with design compliance, consideration is given to the classic manufacturing resources: people, facilities, technology, raw material, capital, and design. Thus, preliminary analyses will be made to tentatively select components, configurations, materials, and processes, without locking onto the design of any tentative selections. This initial selection will provide a basis for the functional designer to evaluate. If an approach appears to be confined to a single material, component, process, geometry, etc., it should suggest additional analysis, resulting in a competitive approach that offers more flexibility in materials or another process that will provide a more cost-effective means of achieving the required performance objectives. Thus, the design process may be thought of as a series of sequential steps. The various approaches are analyzed and either rejected or accepted tentatively. The design process is not a one-pass operation but is a chain of iterative loops and interactions that may be traveled a number of times.

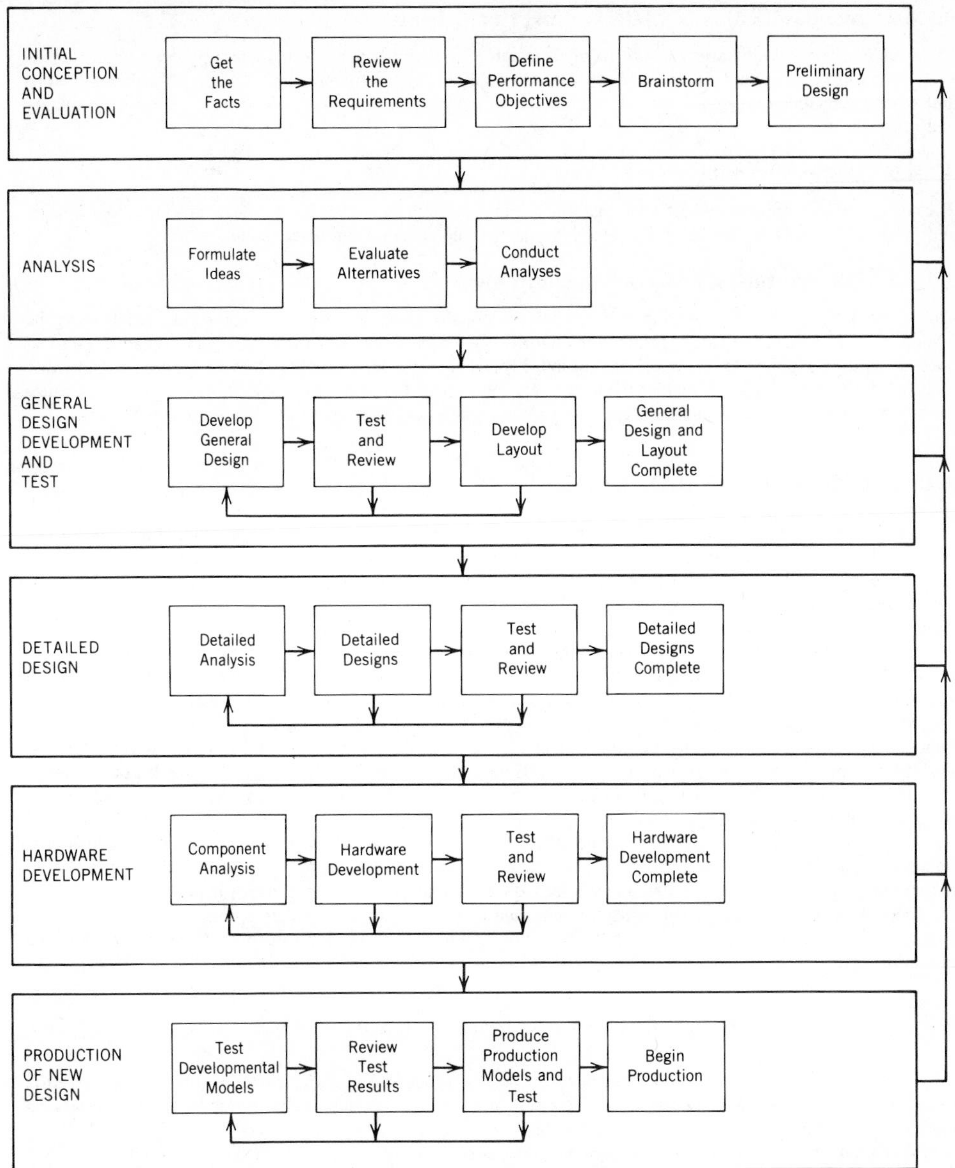

Fig. 3.1.1. The design process.

3.1.4 FUNDAMENTAL DESIGN PRACTICES

The following design-to-manufacture practices can make a substantial contribution so that the final effort will reflect a design that can be economically produced.

3.1.4.1 Simplicity of Design

Perhaps the most common deficiency of new design is excessive complexity. Simpler designs invariably lead to more economical production, are more reliable, are more easily maintained, and typically will exceed the performance of the more complex design. Simplicity is frequently reached by eliminating components of an assembly by building their function into other components.

3.1.4.2 Standardization of Materials and Components

Today an extremely wide variety of off-the-shelf materials and components is available. By designing to utilize such items, cost is generally reduced and spare parts availability is greatly increased. Utilizing standard materials and components frequently will provide higher reliability, since standard components have a known quality and reliability record.

3.1.4.3 Design Flexibility

The design should allow several different materials and manufacturing processes in order to produce an acceptable end product. In this way the flexibility of the design will enhance design-to-manufacture analysis.

3.1.4.4 Manufacturing Process Capability Analysis

This analysis should be performed in concert with the functional design of each component. The talents of the manufacturing engineering and/or production engineer should be utilized by the functional designer throughout the design process to avoid production restrictions. Typical errors in design that fail to consider process capability include specifying a threaded inside diameter to the bottom of the blind hole, specifying a ground tolerance on an outside diameter to a shoulder, and specifying a small hole (less than 0.015″) to too great a depth.

3.1.5 SELECTION OF MATERIALS

Design-to-manufacture embraces the selection of the most appropriate materials and processes for the economic production while maintaining product quality and reliability. Consideration of both materials and processes need to be made together, since it is only after materials are processed that the total cost (cost of materials and cost of processing) is known. An inexpensive material that is difficult to process can be an expensive finished material.

Usually there are several alternatives for the functional designer to consider in the selection of materials for a specific application. To simplify the selection process, the designer should organize information related to possible materials into three categories: properties, specifications, and procurement data.

The *property* category will provide the information that suggests the most desirable material from the standpoint of product quality and reliability over a predicted life. Alternative materials under consideration should be compared by their property profiles. Such important properties as yield point, modulus of elasticity, resistance to corrosion, conductivity, toughness, durability, and machinability should be compared. Those materials that qualify because of their properties will stand out.

The *specifications* category, where material information is organized, will include unique specifications, standard grades, and cost. This information will assist the designer to make a final decision between two or more materials that have survived an analysis based on the data from the properties category.

The last category, *procurement data*, provides data that are needed to place an order for the material. Here is recorded availability, alternative sources of supply, minimum-order size, quantity break points, and the economic order quantity.

3.1.6 SELECTION OF MANUFACTURING PROCESSES

Design-to-manufacture includes the concept of the lowest processed material costs, which summarizes costs for all the processing steps including setup and production-time costs along with the preprocessed material cost and the cost of scrap, chips, and waste generation.

Processes may be broken down into four categories: basic processes, secondary processes, finishing processes, and inspection and quality control processes. Basic processes include those in the liquid, plastic, and solid state.

3.1.6.1 Liquid Material Basic Processes

In the development of the preliminary design, consideration must be given to whether to start with a basic process that uses materials in the liquid state, such as a casting, or in a plastic state, such as a forging, or in a solid state, such as a fabrication. If the designer makes the decision to use a casting, he then must decide which alloy and what casting process can most nearly meet the required geometrical configuration, mechanical properties, dimensional tolerances, reliability, and production rate, at the least cost.

The family of processes designated as castings has several distinct positive characteristics: the ability to conform to a complex shape, economy, and a wide choice of alloys. The principal inherent problems that may exist include internal porosity, dimensional variations resulting from shrinkage upon cooling, and solid or gaseous inclusions that result from the molding operation. These problems can be minimized by sound design-for-manufacture.

All casting processes are basically similar, in that the metal being formed is initially in a liquid or highly viscous state and is poured or injected into a cavity of a desired configuration.

The following casting design guides will lead to economic productions:

1. When changes in section are required, use smoothly tapered sections to reduce stress concentration. Where sections join, use generous radii that blend in with joined sections.
2. The casting drawing should show machining allowances so that patterns are produced to ensure adequate stock and yet avoid excessive machining.
3. Avoid concave notches in casting design, as the metal mold or die is difficult to produce. Convex product designs permit easy milling in the making of the mold or die.
4. Specify raised lettering rather than depressed lettering in the casting.
5. Avoid thin sections in the design as they are difficult to fill completely.
6. Cored holes that will subsequently be drilled and/or reamed and/or tapped should have countersinking on both ends to facilitate the secondary operations.
7. To avoid warpage and distortion, break up large plain surfaces with ribs or serrations.
8. For maximum strength, keep material away from the neutral axis. Endeavor to keep plates in tension and ribs in compression.

Table 3.1.1 provides the important design parameters associated with the principal casting processes and provides those limitations and constraints that should be observed by the functional designer to assure economic production.

3.1.6.2 Plastic Material Basic Processes

The functional designer may consider using a forging rather than a casting because of improved mechanical properties associated with forgings. The hot forging process will break up the large dendritic grain structure characteristic of castings, thus giving the grain structure of the metal a refinement, and inclusions will be stretched out in the direction in which plastic flow occurs. In order to acquire the best load-carrying ability, forgings should be designed so that the flow lines run in the direction of the greatest load during service.

Guidelines that the designer should observe in the design of forgings in order to help assure economic production and reliability include the following:

1. The unsupported length of bar should not be longer than three times the diameter of the bar or distance across the flats. This is the maximum length of bar that can be upset in a single stroke without some buckling of the unsupported portion.
2. Recesses perpendicular to the forging plane with depth up to the diameter of the recess can be incorporated in either or both sides of a section. Secondary piercing operations to remove the residual web should be utilized on through-hole designs.
3. Draft should be added to all surfaces perpendicular to the forging plane to permit easy removal of the forged part. Outside draft can be less than inside draft, since the outside surfaces will shrink toward bosses in the die.
4. Draft for difficult forging materials such as titanium and nickel base alloys should be larger than that for easy forging materials. Similarly, deep die cavities need more draft than shallow ones.
5. Uniform draft results in more economic dies. One draft should be specified on all outside surfaces and one greater draft on all inside surfaces.
6. Corner and fillet radii should be large in order to facilitate metal flow and provide long die life. Usually 0.25 in. (6 mm) is the minimum radius for parts forged from high-temperature alloys, stainless steels, and titanium alloys.
7. In order to simplify the die design and keep its cost of construction low, endeavor to keep the parting line in one plane.
8. To avoid deep impressions with accompanying high die wear and difficulty in removing the forged part from the die, locate the parting line on a central element of the part.

Table 3.1.2 provides helpful design information for the economic production of quality forgings.

3.1.6.3 Solid State Basic Processes

There are several other basic metal shaping processes in addition to casting and forging where the material is formed in the solid state. The material may be in powdered, sheet, rod, or billet form. The more important of these processes that impart the approximate finished geometry include: powder metallurgy, cold heading, extrusion, roll forming, press forming, spinning, electroforming, and automatic screw machine work.

Powdered metallurgy may be considered a basic process, although parts from this process often do not require any secondary operations. In this process, powdered metal is placed in a die and compressed under high pressure. The resulting cold formed part is then sintered in a furnace to a point below the melting point

Table 3.1.1. Design Parameters Associated with the Principal Casting Processes

Design Parameter	Sand Casting		Shell	Plaster (preheated mold)	Investment (preheated mold)	Permanent Mold (preheated mold)	Die (preheated mold)	Centrifugal
	Green	Dry/Cold Set						
Weight	3.5 oz to 440 tons (100 g to 400 metric tons)	3.5 oz to 440 tons (100 g to 400 metric tons)	3.5 oz to 225 lb (100 g to 100 kg)	3.5 oz to 225 lb (100 g to 100 kg)	Fraction of oz to 110 lb (1g to 50 kg)	3.5 oz to 55 lb (100 g to 25 kg)	Fraction of oz to 65 lb (g to 30 kg)	0.25 oz to 440 lb (7 g to 200 kg)
Minimum section thickness	0.120″ (3 mm)	0.120″ (3 mm)	0.06″ (1.5 mm)	0.04″ (1 mm)	0.02″ (0.5 mm)	0.12″ (3 mm)	0.03″ (0.75 mm)	0.25″ (6 mm)
Allowance for machining	Ferrous: 0.10 to 0.375″ (2.5 to 9.5 mm) Nonferrous: 0.06 to 0.25″ (1.5 to 6.5 mm)	Ferrous: 0.10 to 0.375″ (2.5 to 9.5 mm) Nonferrous: 0.06 to 0.25″ (1.5 to 6.5 mm)	Often not required When required 0.10 to .375″ (2.5 to 6.5 mm)	0.03″ (0.75 mm)	0.01″ to 0.03″ (0.25 to 0.75 mm)	0.030″ to 0.12″ (0.80 to 3 mm)	0.030″ to 0.060″ (0.80 to 1.60 mm)	0.030″ to 0.060″ (0.75 mm to 1.5 mm)
General tolerance	±0.015″ to ±0.25″ (±0.4 to 6.4 mm)	±0.015″ to ±0.25″ (±0.4 to 6.4 mm)	±0.003″ to ±0.063″ (±0.08 to ±1.60 mm)	±.005″ to ±0.010″ (±0.13 to ±0.26 mm)	±0.002″ to ±.006″ (±0.75 to ±1.5 mm)	±0.010″ to ±0.06″ (±0.25 to ±1.5 mm)	±0.001″ to ±0.005″ (±0.025 to ±0.125 mm)	±0.031″ to 0.138″ (±0.80 to ±3.5 mm)
Surface finish (microns, RMS)	6.0 to 24.0	6.0 to 24.0	1.25 to 6.35	0.8 to 1.3	0.5 to 2.2	2.5 to 6.35	0.8 to 2.25	2.5 to 13.0
Process reliability (%)	90	90	90	90	90	90	95	90
Cored holes	Holes as small as 0.25″ (6 mm)	Holes as small as 0.25″ (6 mm)	Holes as small as 0.25″ (6 mm)	Holes as small as 0.50″ (12 mm)	Holes as small as 0.02″ (0.5 mm) diam.	Holes as small as 0.20″ (5 mm) diam.	Holes as small as 0.031″ (0.80 mm) diam.	Holes as small as 1″ (25 mm) in diameter. No undercuts
Minimum lot size	1	1	100	1	20	1000	3000	100
Draft allowances	1° to 3°	1° to 3°	0.25° to 1°	0.5° to 1°	0° to 0.5°	2° to 3°	2° to 5°	0° to 3°

Table 3.1.2. Design Parameters Associated with the Principal Forging Processes

Design Parameter	Open Die	Conventional Utilizing Preblocked	Closed Die	Upset	Precision Die
Size or weight	1.1 to 1100 lb (500 g to 5000 kg)	0.5 oz to 45 lb (14 g to 20 kg)	0.5 oz to 45 lb (14 g to 20 kg)	.75″ to 10″ bar (20 mm to 250 mm bar)	0.5 oz to 45 lb (14 g to 20 kg)
Allowance for finishing machining	0.08″ to 0.40″ (2 to 10 mm)	0.08″ to 0.40″ (2 to 10 mm)	0.04″ to 0.20″ (1 to 5 mm)	0.20″ to 0.40″ (5 to 10 mm)	0″ to 0.12″ (0 to 3 mm)
Thickness tolerance	+.024″–.008″ to +0.12″–0.04″ (+0.6 mm–0.2 mm to +3.00 mm–1.00 mm)	+0.016″–0.008″ to +0.08″–0.03″ (+0.4 mm–0.2 mm to +2.00 mm–0.75 mm)	+0.012″–0.006″ to +0.06″–.02″ (+0.3 mm–0.15 mm to +1.5 mm–0.5 mm)	—	+0.008″–0.004″ to +0.04″–0.008″ (+0.2 mm–0.1 mm to +1 mm–0.2 mm)
Fillet and corners	0.2″ to 0.5″ (5 to 13 mm)	0.12″ to 0.28″ (3 to 7 mm)	0.078″ to 0.16″ (2 to 4 mm)	—	0.04″ to 0.079″ (1 to 2 mm)
Surface finish (microns, RMS)	3.8 to 4.5	3.8 to 4.5	3.2 to 3.8	4.5 to 5.0	1.25 to 2.25
Process reliability	95	95	95	95	95
Minimum lot size	25	1000	1500	25	2000
Draft allowance	5° to 10°	3° to 5°	2° to 5°	—	0° to 3°
Die wear tolerance	±0.0014″ per lb (±0.075 mm per kg weight of forging)	±0.0014″ per lb (±0.075 mm per kg weight of forging)	±0.0014″ per lb (+0.075 mm per kg weight of forging)	—	±0.0014″ per lb (+0.075 mm per kg weight of forging)
Mismatching tolerance	±.0.0015″ to +0.00006″ per lb (±.25 mm to +0.01 mm per 3 kg weight of forging)	±0.0015″ to +0.00006″ per lb (±0.25 mm to +0.01 mm per 3 kg weight of forging)	±0.0015″ to +0.00006″ per lb (±0.25 mm to +0.01 mm per 3 kg weight of forging)	—	±0.0015″ to +0.00006″ per lb (±0.25 mm to +0.01 mm per 3 kg weight of forging)
Shrinkage tolerance	±0.003″ (±0.08 mm)	±0.003″ (±0.08 mm)	±0.003″ (±0.08 mm)	—	±0.003″ (±0.08 mm)

of its major constituent. Complex parts such as gears and self-lubricating bearing sleeves are made by powder metallurgy.

Cold heading involves striking a segment of cold material up to one inch (25 mm) in diameter in a die so that it is plastically deformed to the die configuration. Cold heading, as in powdered metallurgy, frequently does not require secondary operations. The heads of bolts and shafts with enlarged ends may be formed by this type of metal working.

Extrusion is performed by forcing heated metal through a die having an aperture of a specified geometry. The extruded shapes are then cut into the desired length. The following design features applicable to extrusion should be incorporated to assure sound design for economic production and reliability:

1. Thin sections with large circumscribed area should be avoided.
2. Any heavy wedge section that tapers to a thin edge should be avoided.
3. Thin sections that have close space tolerance should be avoided.
4. Do not include sharp corners in the design.
5. Do not design semiclosed shapes that require dies with long thin projections. Such dies are expensive to make and have a short life.
6. When a thin member is attached to a heavy section, the length of the thin member should not exceed ten times its thickness.

Examples of extrusions include the side rails and rungs of aluminum ladders, steps for escalators, forming for windows, and seamless tubing.

Roll forming is a process where strip metal passes through a series of rolls that progressively change the shape of the metal by stretching it beyond its yield point. In designing for the amount of bend caused by each roll, allowance must be made for springback. The final rolls will bring the material to its desired geometrical shape. As in extrusion, roll-formed lengths are cut into the desired length.

Forming in presses (hydraulic, pneumatic, and mechanical) is a process applied to sheet metal where the material is stressed beyond its yield point in a metal die. The original material may be reduced in thickness by drawing and/or ironing. The process is based upon two principles:

1. Compressing and stretching a material beyond its elastic limit on the inside and outside of a bend.
2. Compressing a material beyond the elastic limit without stretching or stretching the material beyond the elastic limit without compression.

Press forming, although classified here as a basic forming operation will produce parts that require no additional work.

Spinning is a process applied to sheet metal where the work is formed over a symmetrical pattern, often made of hard wood, epoxy or urethane resin plastics, or metal. The pattern and material are spun while a lubricated rounded tool is forced against the material, bringing it in contact with the pattern and finally taking shape conforming to the shape of the pattern. The process will result in thinning the material as it is formed over the pattern. This process is limited to symmetrical shapes and often requires no secondary operations.

The *electroforming* process involves the development of a mandrel having the desired inside geometry of the part. The mandrel is then plated with the appropriate material to the desired thickness. The completed part is then stripped from the mandrel and any necessary secondary operations are performed.

Automatic screw-machine work will often produce components without the necessity of performing any secondary operations. Here bar stock is fed and cut with multiple tools to obtain the desired configuration.

Table 3.1.3 outlines important design information for the utilization of these processes in an economic fashion.

3.1.7 DESIGNING FOR SECONDARY METAL FORMING AND SIZING OPERATIONS

Once the basic process for producing a given functional design is selected, consideration must be given to the selection of the most favorable secondary forming and sizing operations. Invariably there is more than one competing way to produce a part. A number of factors relating to a given design that need be considered in secondary process selection include: shape desired, the characteristics of the alloy, the tolerances required, the surface finishes required, the quantity to be produced, and the cost.

Some design considerations that should be observed so that secondary operations may be economically performed include:

1. Provide flat surfaces for the entering and exit of drilled holes.
2. On long members, design so that male threads can be machined between centers as opposed to female threads, where it would be difficult to support the work.

Table 3.1.3. Design Parameters Associated with Certain Basic Manufacturing Processes

Design Parameter	Powder Metallurgy	Cold Heading	Extrusion	Roll Forming	Press Forming	Spinning	Electroforming	Automatic Screw Machine
Size	Diam: 0.06″ to 12″ (1.5 to 300 mm) Length: 0.12″ to 9″ (3 to 225 mm) *L*/*D* <2 (usually)	Diam: 0.03 to 0.75″ (0.75 to 20 mm) Length: 0.06 to 10″ (1.50 to 250 mm)	0.06″ to 10″ diameter (1.5 mm to 250 mm diameter)	Up to 80″ (to 2000 mm)	Up to 20 ft (to 6 m) in diameter	0.25 in. to 13 ft in diameter (6 mm to 4000 mm)	Limited to size of plating tanks	0.031″ (0.8 mm) diam. by 0.06″ (1.5 mm) length to 8″ (200 mm) diam. by 35″ (900 mm) length
Minimum thickness	0.040″ (1 mm)	—	0.040″ (1 mm)	0.003″ (0.075 mm)	0.003″ (0.075 mm)	0.004″ (0.1 mm)	0.0001″ (0.0025 mm)	—
Allowance for finish machining	To size	To size	To size	To size	To size	To size	To size	To size
Tolerance	Diam: ±0.001 to ±0.005″ (±0.025 to 0.125 mm) Length: ±0.01″ to ±0.02″ (±0.25 to 0.50 mm)	Diam: ±0.002 to 0.005″ (±0.05 to 0.125 mm) Length: ±0.03″ to ±0.09″ (±0.75 to 2.25 mm)	Flatness: ±0.0004″ (0.01 mm) per in. of width; wall thickness: ±0.006″ to ±0.01″ (0.15 to ±0.25 mm) Cross section: ±0.006″ to ±0.008″ (±0.15 to ±0.20 mm)	Cross section: ±0.002″ to 0.01″ (±0.050 to 0.35 mm) Length: ±0.06″ (1.5 mm)	±0.01″ (±0.25 mm)	Length: ±0.005″ Thickness: ±0.002″ (length ±0.12 mm) (thickness ±0.05 mm)	Wall thicknesses: ±0.001″ Dimension ±0.0002″ (wall thicknesses: ±0.025 mm) (dimension ±0.005 mm)	Diam: ±0.0004″ to 0.002″ (±0.01 to 0.06 mm) Length: ±0.0016″ to 0.004″ (±0.04 to 0.10 mm) Concentricity: ±0.002″ (0.06 mm)
Surface finish								
Microinches (μin)	5 to 50	85 to 100	100 to 120	85 to 100	85 to 160	15 to 85	5 to 10	12 to 100
Microns (μm)	0.125 to.25	2.2 to 2.6	2.6 to 3	2.2 to 2.6	2.2 to 4.0	0.4 to 2.2	0.125 to 0.250	0.30 to 2.5
Process reliability	95	99	99	99	99	90–95	99	98
Minimum lot size	10,000	5000	1500 ft	10,000 ft	1500	5	25	1000
Draft allowancce	0	—	—	—	0°–0.25°	—	—	—
Bosses permitted	Yes	Yes	Yes	Yes	Yes	No	Yes	Yes
Undercuts permitted	No	Yes	Yes	Yes	Yes	Yes	Yes	Yes
Inserts permitted	Yes	No	No	No	No	No	No	No
Holes permitted	Yes	No	Yes	Yes	Yes	No	Yes	Yes

3. Parts to be machined should be designed with gripping surfaces so that the work can be held securely while machining takes place.
4. Parts that are held during machining must be sufficiently rigid to withstand the forces of machining.
5. In the design of mating parts, avoid double fits. It is easier to maintain close tolerances when a single fit is specified.
6. To minimize tooling costs when quantity requirements are small, avoid the design of special contours that would require a form tool.
7. When shearing sheet-metal work, avoid feather edges. Internal edges should be rounded, while corners along the edge of the strip stock should be sharp.
8. The flat blanks from which press formed parts are made should have straight edges.
9. Allow a distance of at least 1.5 times the thread pitch between the last thread and the bottom of tapped blind holes.
10. All blind holes should end with a conical geometry in order to allow the use of standard drills.
11. Work should be designed so that diameters of external features increase from the exposed face, while diameters of internal features decrease.
12. Internal corners of the work piece should be dimensioned with a radius machinable by a simple cutting tool.
13. Consider designing the work so that all secondary operations can be performed while holding the work in a single jig or fixture.

Table 3.1.4 provides pertinent information in connection with the principal secondary operations performed by machine tools.

3.1.8 DESIGNING FOR THE SHAPING OF PLASTIC COMPONENTS

When considering plastics as a product material, the functional designer has more than thirty distinct families to consider. From these families evolve thousands of specific formulations that might be considered for each component of a design. Although the number of plastic material alternatives is large, the number of basic processes utilized in the shaping of plastics is limited. The principal production processes used to shape plastic materials include: compression molding, transfer molding, injection molding, extrusion, casting, cold molding, thermoforming, and blow molding. In order to design effectively for economic production, the functional designer must consider the capabilities of competing processes in concert with the unique physical and mechanical properties of each plastic material under consideration.

3.1.8.1 Compression Molding

This process is compatible for both thermosetting and thermoplastic materials. Here an appropriate amount of the plastic compound (usually in power form) is placed into a heated metallic mold. The mold is closed under pressure and the molding material becomes soft by heat and is formed into a continuous mass having the shape of the mold cavity. If the material being processed is thermoplastic, the mold is cooled and the hardened product is removed. If the material is thermosetting, further heating will result in the hardening of the material.

Some considerations for the selection of compression molding in order to design for economic production include:

1. Parts with thin walls (less than 0.060 in.) can be molded with complete filling of the mold cavity, little part warpage, and good dimensional stability.
2. Since the material is placed directly into the mold cavity, no gate markings will be made.
3. Less material shrinkage and a more uniform shrinkage is characteristic of compression molding.
4. The process is economical for large parts (parts weighing more than 2 lb).
5. Mold costs are usually less than either transfer or injection molds designed to produce the same part.
6. When reinforcing fibers are used, the end product is usually stronger and tougher than when either transfer or injection molding is used, since in these closed mold processes the fibers are broken up.

3.1.8.2 Transfer Molding

In this process, plastics material temporarily stored in an auxiliary chamber is transferred under pressure through an orifice into the closed mold. The formed part along with the residue, referred to as the *cull*, is removed when the mold is opened. Unlike compression molding, there is no flash to trim. Only the cull and the runners between the mold cavities in multiple-cavity molds need to be removed.

Table 3.1.4. Design Parameters Associated with Secondary Forming and Sizing Operations

Process	Shape Produced	Machine	Cutting Tool	Tolerance	Surface Finish (Microinches, μin) (Microns, μm)	Relative Motion: Tool	Relative Motion: Work
Turning (external)	Surface of revolution (cylindrical)	Lathe, boring machine	Single point	±0.001″ to ±0.003″ (±0.025 mm to ±0.075 mm)	30 to 250 0.8 to 6.4		
Boring (internal)	Cylindrical (enlarges holes)	Boring machine, Lathe	Single point	±0.0005″ to ±0.003″ (±0.025 mm to ±0.075 mm)	15 to 200 0.4 to 5.0		
Shaping and planing	Flat surfaces or slots	Shaper, planer	Single point	±0.002″ to ±0.004″ (±0.050 mm to ±0.100 mm)	30 to 250 0.8 to 6.4		
Milling (end, form, slab)	Flat and contoured surfaces and slots	Milling machine, horizontal, vertical, bed type	Multiple points	±0.001″ to ±0.003″ (±0.025 mm to ±0.075 mm)	30 to 250 0.8 to 6.4		
Drilling	Cylindrical (originating holes 0.1 to 100 mm in diameter)	Drill press, lathe	Twin edged drill	±0.002″ to ±0.004″ (±0.050 mm to ±0.100 mm)	100 to 250 2.5 to 6.4		or Fixed
Grinding (cylindrical, surface, plunge)	Cylindrical, flat, and formed	Grinding machine, cylindrical, surface, thread	Multiple points	±0.0001″ to ±0.001″ (±0.005 mm to ±0.025 mm)	8 to 100 0.2 to 2.5		or Fixed
Reaming	Cylindrical (enlarging and improving finish of holes)	Drill press, turret lathe	Multiple points Expandable	±0.0005″ to ±0.002″ (±0.0125 mm to ±0.0500 mm) ±0.002″ ±0.050 mm	30 to 100 0.8 to 2.5		Fixed
Broaching	Cylindrical, flat, slots (constant cross section)	Broaching machine, press	Multiple points	±0.0002″ to ±0.0006″ (±0.005 mm to ±0.0150 mm)	30 to 100 0.8 to 2.5		Fixed

(*Table continues on next page.*)

Table 3.1.4 (*Continued*)

Process	Shape Produced	Machine	Cutting Tool	Tolerance	Surface Finish (Microinches, μin) (Microns, μm)	Relative Motion	
						Tool	Work
Electric discharge machining	Variety of shapes depending on shape of electrode	Electric discharge machine	Single point electrode	±0.002″ (±0.050 mm)	30 to 200 0.8 to 5.0 (pulse-energy dependent)	↕ Spark erosion	Fixed
Electrochemical machining	Variety of shapes; usually odd shaped cavities of hard material	Electrochemical machine	Dissolution process	±0.002″ (±0.050 mm)	12 to 60 0.2 to 1.5	Andoic dissolution; tool is cathode	Workpiece is anode
Chemical machining	Variety of shapes; usually blanking of intricate shapes, printed circuit etching or shallow cavities	Chemical machining machine	Chemical attack of exposed surfaces	±0.002″ (±0.050 mm)	24 to 70 0.6 to 1.8	Chemical attack of exposed surfaces	Fixed
Laser machining	Cylindrical holes as small as 5 μm	Laser beam machine	Single wave length beam of collimated light	Holes are reproducible within ±3% of diam (for thin materials)	24 to 100 0.6 to 2.5	Fixed vaporization or melting	Fixed

Ultrasonic machining	Same shape as tool	Machine equipped with magnetostrictive transducer, generator power supply	Shaped tool and abbrasive grit	±0.001″ (±0.025 mm)	12 to 35 0.3 to 0.9 (grit-size dependent)	Sonic erosion	Fixed
Electron beam machining	Cylindrical slots	Electron beam machine equipped with vacuum of 10^{-4} mm of mercury	High velocity electrons focus on workpiece	±0.001″ (±0.025 mm)	24 to 70 0.6 to 1.8	Vaporization or melting	Fixed
Gear generating	Eccentric cams, ratchets, gears	Gear shaper	Single point reciprocating	±0.0005″ to ±0.001″ (±0.013 mm to ±0.025 mm)	70 to 150 1.8 to 3.8		
Hobbing	Any form that regularly repeats itself on periphery of circular part	Hobbing machine	Multiple points	±0.0005″ to ±0.001″ (±0.013 mm to ±0.025 mm)	70 to 150 1.8 to 3.8		
Trepanning	Large through holes, circular grooves	Lathe-like	One or more single point cutters revolving around a center	±0.005″ (±0.13 mm)	100 to 250 2.5 to 6.4		

3.1.8.3 Injection Molding

More thermoplastic compounds are molded by this process than any other. Here the plastic material in the form of pellets or grains is placed in a hopper above a heated cylinder called the barrel. From the hopper, an appropriate amount of material is metered into the barrel in every cycle. The plastic material, under pressures up to 600 psi, is forced into the closed mold. When molding thermoplastics, the cooled mold is opened and the molded parts are ejected. When molding thermosets, such as phenolic resins, the barrel temperatures are considerably lower than when molding thermoplastics. Thermoset barrel temperatures range from 150 to 250°F, while thermoplastic barrel temperatures typically range from 350 to 600°F.

3.1.8.4 Extrusion

This process provides a continuous cross-sectional shape by forcing softened plastic material through an orifice having approximately the geometrical profile of the cross section of the work. The extruded plastic form now passes through a cooling cycle where it is hardened. Such products as filaments, tubes, rod, and uniform cross-sectional shapes are economically produced using the continuous extrusion process.

3.1.8.5 Casting

In this process, plastic materials in the liquid form are introduced into a mold that has the geometrical configuration of the desired part. The mold material is often flexible such as rubber, or at times made of nonflexible materials such as plaster or metal. Typical plastics families that can be produced as castings include the epoxies, phenolics, and polyesters.

3.1.8.6 Cold Molding

Cold molding is a process that is used for the production of thermosetting parts. The molding material is placed into a mold at room temperature (unlike compression molding). The mold is then closed under pressure. Subsequently the mold is opened and the formed part is removed to a heating oven where it is baked until the thermosetting material becomes hard.

3.1.8.7 Thermoforming

Under thermoforming, sheets of thermoplastic material are heated to a plastic condition and then drawn over a mold contour, where upon cooling the sheet takes the shape of the mold. This process can also use a sequence of rolls to produce the desired form from plastic sheet. Typically the plastic sheet is brought to the forming temperature (275 to 425°F) by infrared radiant heat, electrical resistance heating, or ovens using gas, fuel oil, or coal.

3.1.8.8 Calendering

This process involves the passing of thermoplastic compounds between a series of heated rolls in order to produce a sheet of uniform thickness. The thickness of the sheet is governed by adjusting the distance between mating rolls. After the sheet passes through the final sizing rolls, it is cooled in order to solidify the material preparatory to winding into large rolls.

3.1.8.9 Blow Molding

This process is used in the economic production of hollow, bottle-shaped thermoplastic products. Here a tube of plastic material called the parison is extruded over an apparatus called the blow pipe and then is encased in a split metallic mold. Air is now injected into this hot plastic section of extruded stock through the blow pipe and the parison is blown outward against the contour of the mold. The part is then cooled (in heavy sections liquid nitrogen or carbon dioxide may be used to hasten the cooling), the mold is opened, and the solidified part is removed.

3.1.9 FACTORS AFFECTING THE SELECTION OF THE MOST FAVORABLE PROCESS IN THE MOLDING OF PLASTICS

The selection of the most favorable molding process for plastic components can have a significant impact on the quality, reliability, life, and the cost of the design. Those parameters that need to be evaluated for the process selection include the specific plastic material used in the design, the geometry of the part, the quantity to be produced, and the cost of the process including the die or mold cost.

Usually the functional designer, in view of the physical and/or mechanical property requirements, is able to identify whether a thermoplastic or thermosetting resin should be used in the design. This information will be helpful in reducing the number of molding processes to be considered. If a thermosetting resin is selected, both thermoforming and blow molding can be eliminated, since these processes are usually restricted to thermoplastics. Because of economics, if the quantity to be produced is large, compression and transfer molding should be restricted to thermosetting resins. In both simple and complex shapes, where the quantity is large and the material is thermoplastic, injection molding will generally prove to be the most competitive.

After considering the material to be used, the functional designer will study the geometry of the part in order to help identify the most favorable process. If the part has a continuous cross section, extrusion is the logical selection. If the design is bottle-shaped and thin-walled, it probably should be blow molded. Flat sheet designs are limited to calendering.

Quantity also has an important impact on the process choice. Compression molding is economical for many thermoplastic and thermosetting designs where the quantity required is small or modest, yet would not be economical if the quantity required were large.

In order to design plastic components for producibility, the following guidelines should be observed:

1. Small holes (less than 1/16 in. diameter) should be drilled after molding.
2. Molded blind holes should be limited in depth to twice their diameter.
3. All holes should be located perpendicular to the parting line to allow easy removal of the part from the mold.
4. Undercuts should be avoided, since they require a more costly mold. The mold must either be split or have a removable core section.
5. The distance between adjacent holes should be greater than 1/8 in.
6. The height of bosses should be less than twice their diameter.
7. In the design of bosses, taper of at least 5° on each side should be incorporated to assure easy withdrawal from the mold.
8. Radii at both the top and the base should be included in the design of bosses and ribs. These radii should not be less than 1/32 in.
9. Ribs should be designed with 2° to 5° taper on each side and the rib width at the base should be at least 1/2 the wall thickness; the height should be limited to 1 1/2 times the wall thickness.
10. At the parting line, outside edges should be designed without a radius.
11. The design should permit both ends of inserts to be supported in the mold.
12. Inserts should be at right angles to the parting line.
13. A taper of 1° to 2° should be specified on those vertical surfaces parallel with the direction of mold pressure.
14. Concave numbers and lettering should be engraved in the mold. Letters should be approximately 3/32 in. high and 0.007 in. deep.
15. Threads that are less than 5/16 in. in diameter should be cut after molding.

Table 3.1.5 provides helpful design information related to the principal processes used in the fabrication of thermoplastic and thermosetting resins.

3.1.10 GROUP TECHNOLOGY IN FUNCTIONAL DESIGN

Group technology implies the classification and coding of the various components utilized in a company's products, so that parts similar in shape and processing sequence are numerically identified. When all of a company's products are so classified and coded, it is an easy matter to retrieve similar designs for review and study when beginning a new design. In this way not only will design time be reduced by capitalizing on existing designs, but design quality including designing for economic production will be enhanced. New designs can be developed to take advantage of existing facilities equipped with universal-type quick-acting jigs and fixtures.

Where group technology is practiced, frequently the functional designer will find it unnecessary to design a completely new part. There will be three choices for design rationalization: use a former design as is, modify an existing design, or design a new part.

3.1.11 COMPUTER-AIDED MANUFACTURING PLANNING

Manufacturing planning is an important step in the transformation of a new functional design into a marketable product. Manufacturing planning involves the identification of the best ways to produce the product so that quality and reliability are assured at a competitive profitable price. Good manufacturing planning will help assure the success of the enterprise, while poor manufacturing planning may result in costs that are so prohibitive that the product may not sell.

Manufacturing planning must consider several alternatives in order to select the best procedure to produce a part. Furthermore, there is seldom much time to perform this function. The product usually is required by the customer shortly after design approval. The computer permits process planning rapidly while considering all possible alternatives.

Table 3.1.5. Basic Processes for Fabricating Plastics and Their Principal Parameters

Process	Shape Produced	Machine	Mold or Tool	Material	Typical Tolerance	Minimum Wall Thickness	Ribs	Draft	Inserts	Minimum Quantity Requirements
Calendering	Continuous sheet or film	Multiple roll calendar	None	Thermoplastic	0.002″ to 0.008″ (0.05 mm to 0.200 mm) depending upon material		None	None	None	Low
Extrusion	Continuous form as rods, tubes, filaments, and simple shapes	Extrusion press	Hardened steel die	Thermoplastic	0.004″ to 0.012″ (0.10 mm to 0.30 mm) depending upon material		None	None	Possible to extrude over or around wire insert	Low (tooling is expensive)
Compression molding	Simple outlines and plain cross sections	Compression press	Hardened tool steel mold	Thermoplastic or Thermosetting	0.0016″ to 0.01″ (0.04 mm to 0.25 mm) depending upon material	0.05″ (1.25 mm)			Yes	Low
Transfer molding	Complex geometries possible	Transfer press	Hardened tool steel mold	Thermosetting	0.0016″ to 0.01″ (0.04 mm to 0.25 mm) depending upon material	0.06″ (1.5 mm)	3° to 5° taper Height < 3 wall thickness	½° to 5°	Yes	High
Injection molding	Complex geometries possible	Injection press	Hardened tool steel mold	Thermoplastic or Thermosetting	0.0016″ to 0.01″ (0.04 mm to 0.25 mm) depending upon material	0.03″ (0.75 mm) (exceptions down to 0.010 in)	½° to 5° taper Height = 1½ to 5 wall thickness Width = ½ wall	¼° to 4°	Yes	High

Casting	Simple outlines and plain cross sections	None	Metal mold or epoxy mold	Thermosetting	0.004″ to 0.02″ (0.10 mm to 0.50 mm) depending upon material	0.08″ (2.0 mm)	Yes	Low to medium depending on mold
Cold molding	Simple outlines and plain cross sections	None	Mold of wood, plaster, or steel	Thermosetting	0.004″ to 0.020″ (0.10 mm. to 0.50 mm) depending upon material	0.08″ (2.0 mm)	Yes	Low
Thermoforming	Thin walled and cup shaped	Thermo-forming machine	Suitable form	Thermoplastic			No	Low
Blow molding	Thin walled and bottle shaped	Pneumatic blow molding machine	Tool steel mold	Thermoplastic	0.012″ to 0.024″ (0.30 mm to 0.60 mm)	0.12″ (3.0 mm)	No	High
Rotational molding	Full or semi-enclosures (hollow objects)	Roto molding system	Cast aluminum or fabricated metal	Thermoplastic, limited Thermosetting	0.012″ to 0.024″ (0.30 mm to 0.60)	0.12″ (3.0 mm)	No	Medium
Filament winding	Tubes, piping, tanks	Filament winding machine	Must have axis about which the filament can be wound	Single-end continuous strand glass fiber and thermoplastic	0.008″ to 0.020″ (0.20 mm to 0.50 mm)	0.12 (3.0 mm)	No	Medium

To plan the complete processing of a product, four main classes of processes should be considered: basic processes, secondary processes, finishing processes, and quality control processes. Different parameters need to be considered in connection with each of these four families.

1. *Basic processes.* Those processes that are initially used to approximate the geometrical configuration of the component. Typical basic processes include casting, forging, extrusion, and roll forming.
2. *Secondary processes.* Those processes that follow the basic processes to bring the product to its final dimensional accuracy and form exclusive of the protective and/or aesthetic coatings. Representative secondary processes include drilling, reaming, tapping, turning, facing, grooving, grinding, broaching, honing, and heat treating.
3. *Finishing processes.* Those processes applied, usually after the completion of all secondary processes, to bring the product to its final specifications by the application of a protective or decorative coating. Finishing processes include plating, anodizing, enameling, and painting.
4. *Quality control processes.* Those processes applied to assure the quality and reliability of the product. Typical quality control processes include inspection and packing for shipment.

With reference to the selection of the optimum basic process, the computer program would need to consider the following parameters: size of part, geometry or configuration of part, material being utilized, microstructure resulting from the process, relative cost of secondary process that will need to be performed, quantity to be produced, and cost.

All the parameters above can serve as constraints. Size has a limiting effect on several basic processes. For metals, die castings larger than 75 lb are seldom produced, sand castings less than 2 oz usually are not made, and extrusions with cross-sectional areas of more than 140 in^2 are generally not produced.

Geometry, too, is an important constraint. For example, complex geometries can be cast but cannot be forged, nonsymmetrical bowl-shaped parts cannot be spun, and designs with undercut or re-entrant angles are not acceptable to the powdered metal process.

Material is a very important parameter related to process selection. Ferrous metals generally are not pressure-die cast because of economics. Plaster-mold casting is limited to nonferrous metals. Compression molding is the usual technique for producing thermoplastic plates less than 0.25 in. thick.

The microstructure resulting from the process may be a consideration. For example, controlled grain flow may be required for end-use products subject to dynamic or impact loading.

Quantity is a most important consideration. A permanent mold such as a die-casting die would not be made for the production of a small number of parts. Similarly, a dozen parts from a given bar stock design would be produced on a lathe not on an automatic screw machine.

The cost of the secondary operations must also be considered in selecting the most favorable basic operations. Secondary operation cost will vary considerably depending upon what basic operation was used. For example, the secondary operations required on a ferrous metal gear made of powdered metal would be much less than a similar gear made as a ferrous sand casting.

Of course, cost of the basic operation must be considered. If the quality and product reliability resulting from competing processes are equal and assuming the delivery time is the same, we will want to select the process that will result in the lowest product cost.

Software can be developed that considers the above parameters so that all alternatives are considered and the most favorable process may be selected for new designs during initial design phases. In this way the best process can be selected well in advance of completion of the design and the functional designer will be able to incorporate design requirements characteristic of the processes selected.

BIBLIOGRAPHY

Alting, Leo, *Manufacturing Engineering Processes*. New York: Dekker, 1982.

Bralla, James G., *Product Design for Manufacturing*. New York: McGraw-Hill, 1986.

DeGarmo, E. Paul, *Materials and Processes in Manufacturing*, 5th ed. New York: Macmillan, 1979.

Doyle, Lawrence E., *Manufacturing Processes and Materials for Engineers*, Englewood Cliffs, NJ: Prentice-Hall, 1969.

Greenwood, Donald D., *Mechanical Details for Product Design*, New York: McGraw-Hill, 1964.

LeGrand, Rupert, *Manufacturing Engineers' Manual*, New York: McGraw-Hill, 1971.

Niebel, Benjamin W. and Alan B. Draper, *Product Design and Process Engineering*, New York: McGraw-Hill, 1974.

Niebel, Benjamin and Edward N. Baldwin, *Designing for Production*, Homewood, IL: Irwin, 1963.

Trucks, H. E., *Designing for Economical Production*, Dearborn, MI: Society of Manufacturing Engineers, 1974.

Yankee, Herbert W., *Manufacturing Processes*, Englewood Cliffs, NJ: Prentice-Hall, 1979.

Defense Systems Management College, *Manufacturing Management Handbook for Program Managers*, Fort Belvoir, VA, 1982.

CHAPTER 3.2

DESIGN-TO-ASSEMBLE

ROGER ANDERSON

Unimation
Danbury, Connecticut

3.2.1 INTRODUCTION

When a decision is being made whether to automate the assembly of a particular product or subassembly or to make the assembly a manual operation, many factors have to be considered. Do the projected production requirements demand automation? Is it economically justified in terms of the reduced labor content, or is the local rate sufficiently low and stable to make a proautomation decision questionable? Is the expected life of the product long enough to justify automation? These questions and others demand a business analysis that the manufacturer must make.

In this chapter, it will be assumed that the economics of semi- or full automation have been resolved in favor of the mechanization of an assembly process. This text will deal with how the assembly process might be automated, what are the preferred methods and machine formats for an assembly system, and how they might be expected to perform.

Since the proper selection of machinery for automation is as important as the design of the components to be assembled, a general review of methods or options is included here. It is important for the design engineer to realize what choices are available in machinery to help insure compatibility between the assembling machines and the individual parts. After a review of a typical assembly system, each machine component will be examined in some detail along with possible alternatives.

To allow more space for a discussion of what "tools" should be used, this chapter will assume the design engineer has a fundamental knowledge of good design practice.

This chapter not only discusses machines associated with "hard" or "fixed" automation but reviews "flexible" automated assembly, including robots.

3.2.2 GENERAL AUTOMATIC ASSEMBLY SYSTEM

An automatic assembly machine in general is composed of parts feeders, magazines, transfer devices, and indexers. Various other miscellaneous equipment is required, but these machines are the major ones. A standard configuration is represented by Figure 3.2.1. Automatic assembly of any product made up of several parts requires that all parts be separated into individual containers (parts feeders). In a common assembly machine (Fig. 3.2.1), a series of feeders with orienters for each part is located around a rotary indexing turntable (dial) in order of assembly. The parts are transferred from the feeders to stations on the dial by "pick and place" machines, or "placers," in the preferred sequence for ease of assembly. During the dwell period, all stations "work" in either an active or passive manner. There are many variations on this format, but only the more conventional ones are discussed here.

3.2.2.1 Parts Feeders

In the typical assembly of a product (Fig. 3.2.1), the first component moves into position to be placed on the indexing turntable, or dial. Commonly the prime mover is a vibratory bowl that literally vibrates the part up a spiral ramp on the inside (or outside) wall of the bowl from a collection of similar parts at the bottom of the bowl. Approximately 90% of small assembly components can be oriented and fed with a vibratory feeder system.

Appropriate tooling placed along the path of bowl-fed parts orients them for transfer to the appropriate location on the dial. For example, if an assembly is held together with a series of nuts and screws, it would

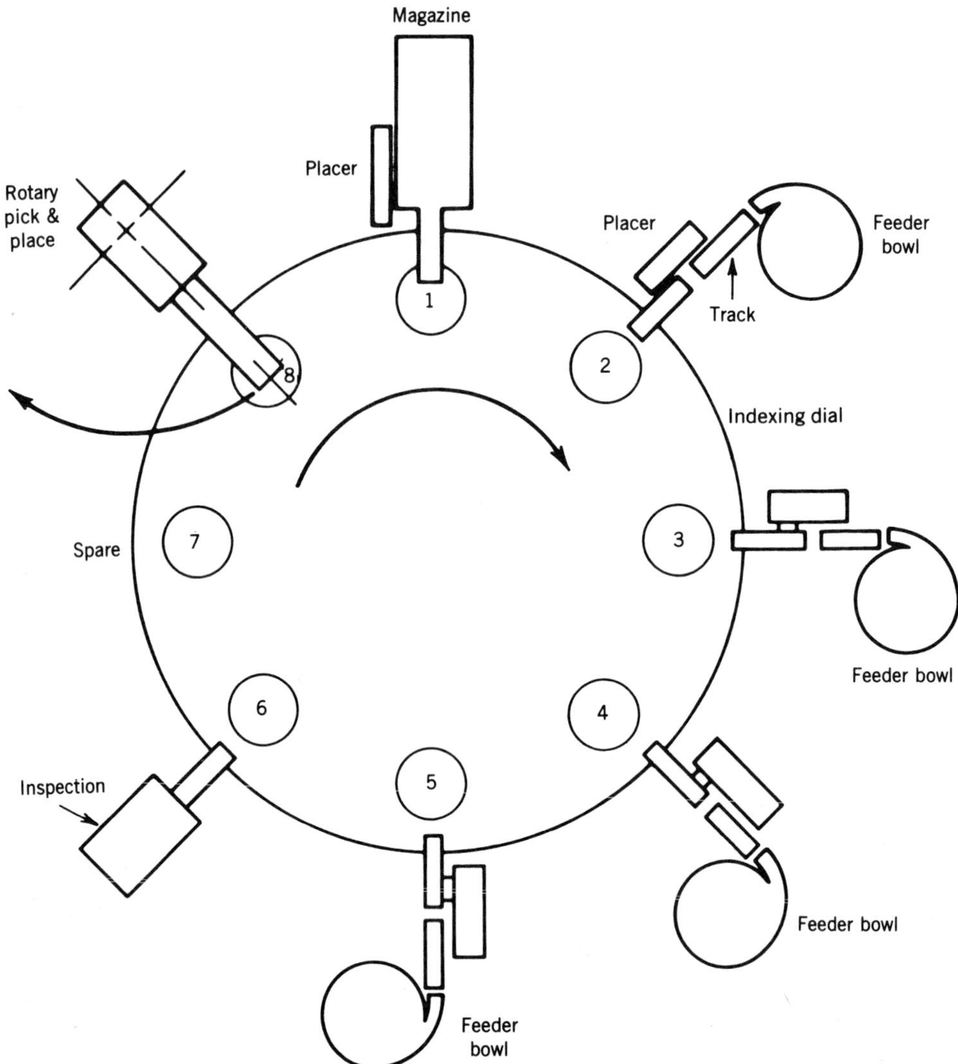

Fig. 3.2.1. Typical automatic assembly machine.

be common practice to orient and maintain alignment of the nuts in a feeder track, by capturing two sides of each nut within the track walls. When the nuts reach a dead plate at the end of the track, they would be picked up and placed by a suitably equipped transfer device to a dial station with sockets to hold the nuts from turning when driving the screws. Orientation is maintained during the transfer by designing the fingers of the gripper to receive the flats or corners of the nuts.

A vibratory feeder bowl and track can be arranged to deliver multiple nuts simultaneously on parallel tracks. Several of the same parts, such as the nuts, can just as readily be transferred simultaneously to a dial station (Fig. 3.2.2).

3.2.2.2 Magazines

Parts that cannot be readily bowl-fed may be stored in an oriented fashion in a magazine (see station 1 in Fig. 3.2.1). Magazines are normally hand-loaded by an operator at rates up to 120 parts a minute. A magazine is a simple and economical storage system but requires considerably more operator attention than vibratory feeder bowls.

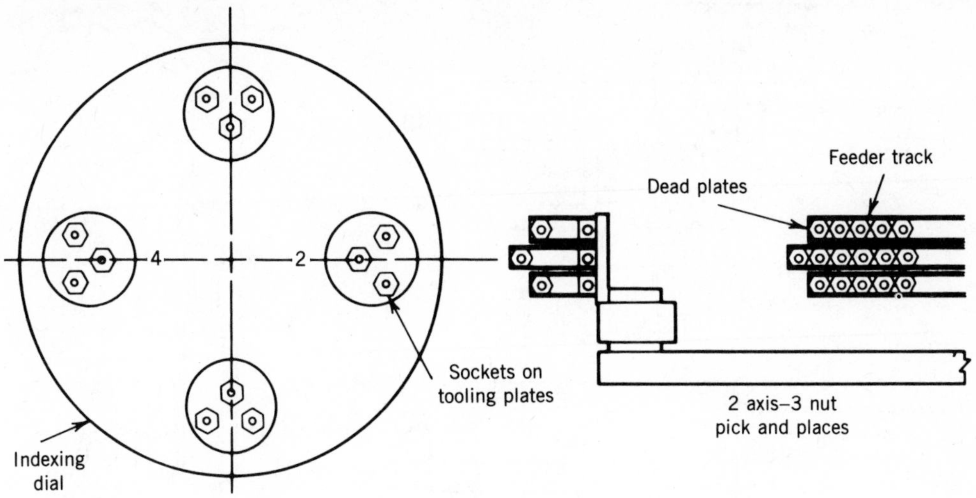

Fig. 3.2.2. Transferring multiple products to rotary dial.

3.2.2.3 Transfer Devices

Once a component of an assembly has reached the end of the track of a feeder bowl, or is at the bottom of a magazine, it is usually oriented and in a position to be transferred to the dial. It then must be separated from the rest of the components on the track when a feeder bowl is used or from the ones above it when a magazine is employed. The component can be lifted from the track and over to the dial with a transfer device, or it can be freed from the stack or column with an escapement.

A transfer device, commonly called "pick and place," or "placer," usually has a a roof over the track that prevents the second component in line from being affected by the removal of the first, regardless of the line pressure; that is, only the first part is displaced. A design of this sort has its own built-in static escapement. (Fig. 3.2.3).

Most simple "placers" are two-axis slide devices that are mechanically or pneumatically operated in X and Z modes. They typically travel up vertically ($+ Z$) for 2 or more inches to clear the track, after closing a gripper on the product, out ($+ X$) for 8–10 inches to reach the work station, and down vertically ($- Z$) to place the object in the proper position at the dial work station. A complete cycle returning to the starting point takes between 1½ and 2 seconds if air-actuated but considerably less if cam-operated (Fig. 3.2.4).

Placers can be used in more sophisticated ways, if required, such as rotating the grippers at the work station to tighten or loosen an object or rotating the tooling itself. Placers may have as many as four or more axes. If the placer is programmable in one axis or more, it fulfills the accepted defintion of a robot.

True or active escapements are often used in feeding balls or headed products, such as screws, to an automatic screwdriver. In an automatic screwdriver, the part is removed from the end of a feeder bowl with a

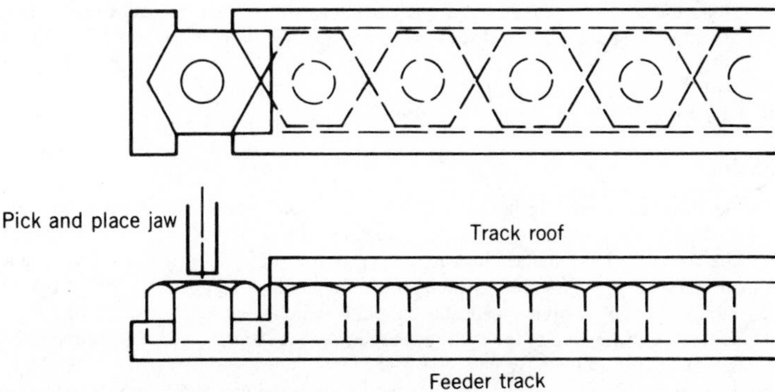

Fig. 3.2.3. Feeder track as static escapement.

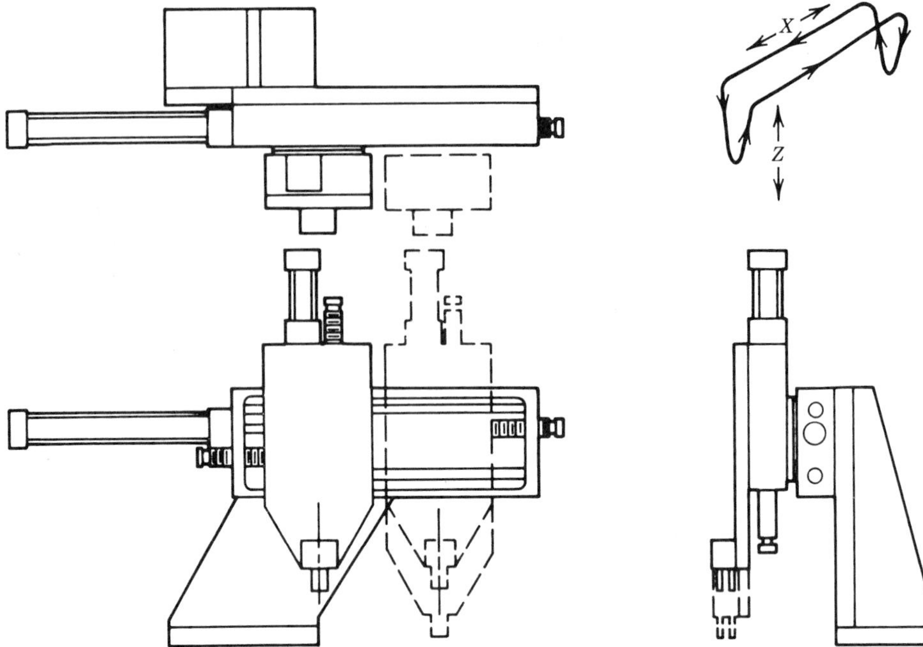

Fig. 3.2.4. Z-axis-($X + Z$) pick and place.

shuttle type escapement to a tube, usually plastic, and pneumatically forced down the tube into position at the chuck.

In many instances, components such as balls can be delivered directly to the dial workstation in a tube, using gravity or air pressure, thereby avoiding the need for a pick and place unit.

3.2.2.4 Rotary and In-Line Indexers

When a component has been successfully transferred from the parts feeder to a rotary indexing turntable, or dial, as in Figure 3.2.1, it is placed in parallel with all other operations that are occurring simultaneously at the other stations. In other words, during the dwell of the indexer, the dial will receive product parts from stations 1 through 5 while products are inspected at station 6 and removed at 8. Thus, the principal advantage of a multistation indexing assembly machine is that all operations are in parallel. An eight-station dial is illustrated in Figure 3.2.1, a standard format, but 32- or 48-station dials that assemble three or four identical products simultaneously are also possible (Fig. 3.2.5).

An in-line indexing system with an over-and-under construction is common when operators have to be interspersed between feeders, placers, and other equipment along the length of the machinery (Fig. 3.2.6). When an in-line indexer, which is conventionally driven by a chain or timing belt in an over and under configuration, is closed in a horizontal loop, it is referred to as a carousel. The advantage of a rotary indexing turntable over an in-line indexing conveyor is that one operator can load and unload at the same position.

Other types of assembly systems will be discussed in more detail in later sections.

3.2.3 PARTS FEEDERS IN DETAIL

3.2.3.1 Magazines

Magazines, while the simplest of feeding devices, require labor for loading. Besides low cost, the overriding advantage of a magazine is the ability to store parts that cannot be fed any other way except by manual orientation. When the parts are hand-loaded in a magazine to the preferred orientation, the loaded magazine can be stored for later inclusion in an assembly system. To eliminate loading operators for long periods, multiple magazines can be mounted on a revolving pedestal. A photoswitch, or other sensor, will indicate an empty magazine, and the pedestal will revolve to the next magazine. An inexpensive air-actuated rotary indexing table is an acceptable base for an array of magazines.

Because magazines are simple in construction, they can easily be adjusted. An array of adjustable magazines becomes a low-cost flexible system for assembling batch orders. No other machine feeder is as versatile as a well-designed magazine. It can feed most parts in an oriented fashion and has a built-in escapement (Fig. 3.2.7).

3.2.3.2 Vibratory Feeder Bowls

Vibratory feeder bowls are among the most commonly used method for feeding and orienting parts. The bowls themselves are usually constructed of stainless steel with a ramp spiraling up the inside. Although they are manufactured in many sizes, an outside diameter of ten times the part length is standard. Many bowls are constructed with vibratory in-line feeder tracks adjacent to the bowl that convey the part to a transfer device after it has left the bowl. Both track and bowl vibrations are the result of electromagnetic-induced motion through the leaf springs that support them. The vibration is regulated by the electrical input to the driving magnet. It is also affected by the stiffness, length and number of springs, so either the bowl or track is readily tuned to the product being fed. Bowl manufacturers also frequently add or remove weight on the bowl to increase the feed rate of a particular part (Fig. 3.2.8).

Feeding a part requires proper settings for ramp angle of the bowl, the amplitude of the vibration, the acceleration of the track (ramp), angle of the supporting springs, coefficient of friction between the part and the ramp, and other factors. While standard bowls may be purchased that are stripped of tooling for orienting a part, they have enough tunable features to drive most parts up the ramp. Design engineers must regard feeder bowl design and tooling with caution, however. Although a number of books analyze the interactions of a feeder bowl and part, it is advisable to rely on manufacturers specializing in feeder bowls. This is an area that remains an art. Some feeder bowl manufacturers train apprentices for years before they are entrusted with the design and construction of a feeder bowl for a customer.

Some parts, because of their surface or mass, will not progress up a feeder bowl ramp. Sponges, felt washers, and other light or resilient products, can be forced up a feeder bowl ramp with "pusher blocks." These blocks of plastic, or whatever the bowl design requires, are mixed in with the particular part and serve to force

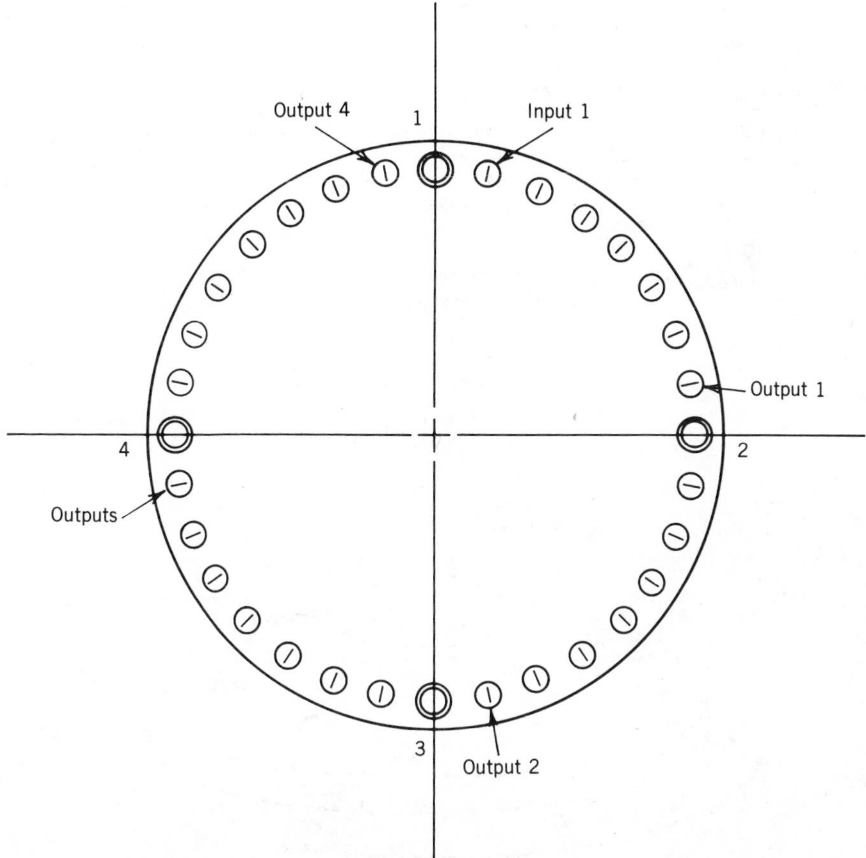

Fig. 3.2.5. Thirty-two station dial that can assemble four identical products simultaneously.

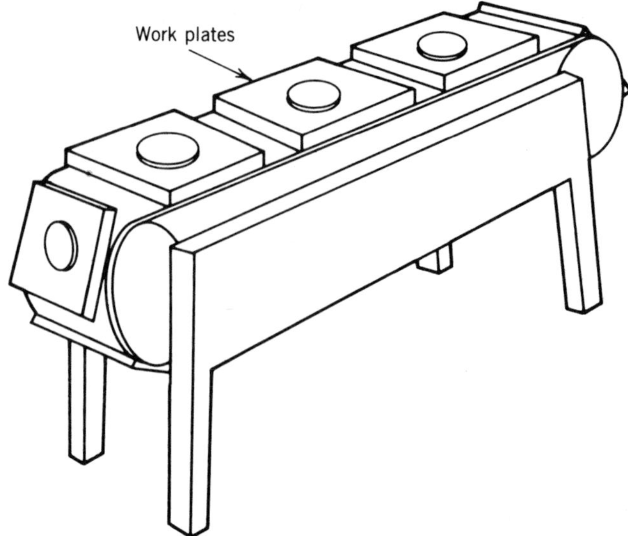

Fig. 3.2.6. Indexing in-line over and above assembly machine.

the washer or sponge up the ramp while they (the blocks) vibrate and feed independently. At a convenient place at the top of the travel, or along the feeder track, the blocks can be tooled or escaped out of the product path and returned to the bowl.

Part Orientation

While feeding a part with a vibratory bowl is relatively standard, the tooling required for orienting is not. It is here that the art of the bowl designer is in the most demand.

An orienting system is required to align and position a part by the time it is at the end of the track or has reached an escapement mechanism. Orienting in most cases is done with tooling along the side of the bowl or

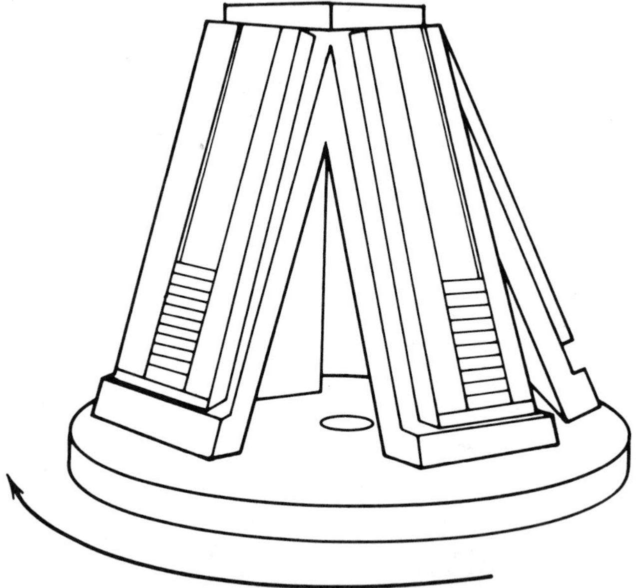

Fig. 3.2.7. Adjustable magazine on indexing base.

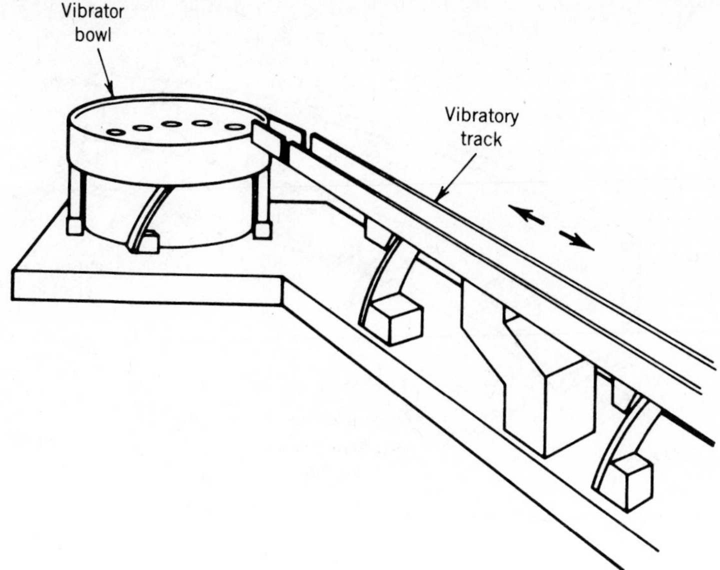

Fig. 3.2.8. Vibratory feeder bowl and track.

on the ramp. As the part progresses up the bowl ramp, it is intercepted by and interacts with whatever tooling has been designed to control and position the part.

Most tooling is static and is of a simple form that works on parts in "go" or "no go" manner. If the part is oriented correctly, it is a "go" and passes through. If it is a "no go," the part is returned to the bowl supply. For this type of tooling, an adequate feeding rate, enough to supply the demand, is all that is required. If a higher rate is needed, either a second bowl may be added or a station that can actively orient the part is inserted in the path. In this manner, 100% of the fed product is oriented. In the case of a part that can have many possible positions on the ramp but only one correct one, a cam, air jet, or projection can be introduced to insure more "go" product (Fig. 3.2.9).

Auxiliary Orienter

Auxiliary orienting tools are used with parts that are relatively easy to feed but virtually impossible to orient with conventional means. A typical one would be a flat disc with a notch or hole that requires orienting in a particular direction.

A common approach is to vibrate the part to a dead nest that will rotate the part under a sensor, then to remove it with a placer or other device that can maintain the orientation. The mechanism for halting the rotation of the part when the hole or lot has been sensed may be an air- or spring-operated pin or latch. A rotary air cylinder with flow controls will serve as a rotation device (Fig. 3.2.10).

There are components that can be fed and oriented in a feeder bowl to a particular position but because of a peculiarity in shape or distribution of mass they do not respond to tooling for orientation to an acceptable position for assembly. It is here that a robot can be employed not only to pick the part but also to orient it to the required position on the way to the assembly station. A robot can become a pick-and-place machine as well as an "in flight" orienting tool.

Semiflexible Bowl Feeders

Batch assembly has always been a large part of manufacturing and is becoming increasingly important as the emphasis on flexible automation continues to grow. (Flexible assembly will be covered in some detail in Section 3.2.6.) Vibratory feeder bowls, while considered an indispensible tool of "fixed" automation, have sometimes been equipped with semiprogrammable tools to make them more compatible with batch manufacturing.

Since the majority of tools that orient parts in a feeder bowl have relatively uncomplicated shapes themselves, it is feasible in many cases to fasten the various shapes on individually adjustable mountings to allow the "programmer" to introduce whatever shape is required for a particular part.

The adjustable feature permits a limited range of sizes of parts to be handled by one bowl. This flexibility is somewhat restricted to parts that are similar. Also, because of the changeover time, it is not widely used despite the fact that it has been available for many years. A variation on this concept is replaceable tooling, which permits one section of bowl track containing all tooling to be exchanged for another.

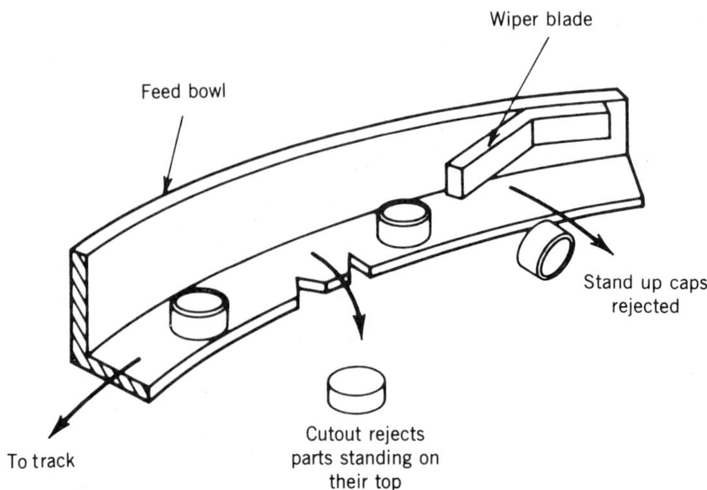

Fig. 3.2.9. Feeder bowl tooling example.

Coatings for Vibratory Feeder Bowls

There are several disadvantages to vibratory feeder bowls beyond the difficulty in tooling them. In most cases, these disadvantages may be overcome or at least modified.

Noise is a major factor, especially in a plant with multiple bowl installations where operators have to be present. One method of noise reduction is to coat the inside of the bowl with a plastic surface bonded to the original metal one. Urethane is most commonly used, and it not only reduces noise but part deterioration caused by vibration. Certain parts, because of their composition, are too brittle to be fed in even a urethane-coated bowl but may require a bowl coated with man-made fibers. In this case, the feeder bowl or feeder track is lined with a fiber that vibrates with the bowl and helps protect the part during transport.

3.2.3.3 Nonvibratory Feeders

Many nonvibratory feeders exist, and drawings of them may be found in various texts. However, the centrifugal, centerboard, and elevator feeders are of most interest to the design engineer.

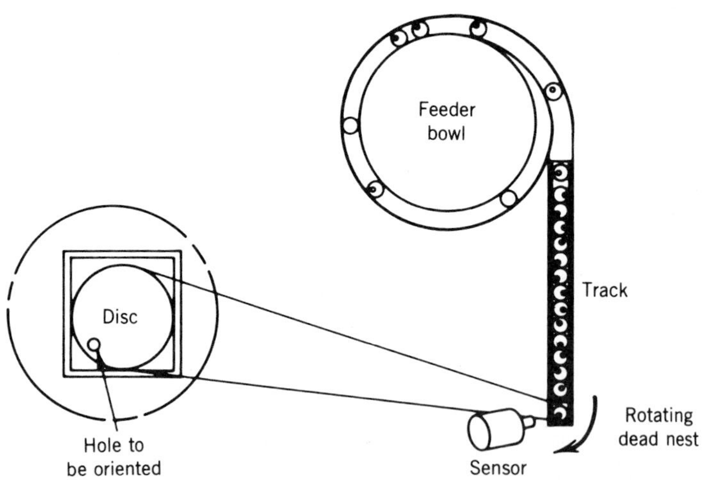

Fig. 3.2.10. Auxiliary orientating tool.

Centrifugal Feeders

If a particular component will not respond to vibratory feeding for reasons other than fragility, it may have to be hand-fed, and oriented or delivered with a centrifugal feeder, the most modern of the nonvibratory feeders.

Several types of centrifugal feeders are available. A typical one has a spinning disc in the middle of a surrounding bowl. The parts are moved by centrifugal force to the bowl wall, where appropriate tooling may be placed to orient the part. Most orienting tooling for this type of a feeder is relatively simple. Parts best suited to this machine range from large and light, such as a rectangular sponge, and from cylindrical to bullet-shaped.

Centrifugal feeders have both a relatively fast feed rate (one manufacturer claims up to 2,000 bullets per minute) and low noise levels (because it is nonvibratory). One disadvantage is cost; nonvibratory centrifugal feeders cost approximately 50% more than vibratory feeders. Nonvibratory feeders also are not as universal as vibratory feeder.

Centerboard Feeders

Another nonvibratory feeder-orienter is the centerboard type. These are among the oldest of all feeding systems, predating the vibratory systems, which emerged after World War II. The name *centerboard* is derived from the design of a moveable boat-type centerboard, which is contained within the stationary trunk of the hull. When the feeder hopper is full and the centerboard is dropped, parts automatically fall in over the top of the board. When the board is raised, and assuming there is a minimum amount of clearance between the board and trunk, some of the parts that do not fall back into the hopper will fall on top of the board. A properly designed groove along the top edge of the board usually will be filled with parts and automatically oriented. If the board pivots around a pin while raising, the part will slide down the groove to whatever stop has been provided (Fig. 3.2.11).

Centerboard feeders are ideal for headed pieces, such as screws, rivets, and nails. The body hangs in the centerboard groove suspended by the head. Pins without heads can be captured and fed horizontally as can almost any other cylindrical shape. Spheres or balls are the simplest shapes with which to work. A full track is a virtual certainty with every cycle of the proper part.

While feeding is reasonably simple with centerboards, orienting parts after the "pick" is not. Once the board has been raised from a hopper by an actuator, the parts are captured in the board groove and cannot be reoriented without an external tool. The parts must be oriented in the desired manner by the centerboard the first time.

Advantages of centerboard feeders are many if they are used with suitable parts. They are noiseless, inexpensive, easily altered for new parts (by substituting centerboards) and less damaging to parts than vibratory feeders. Disadvantages include use limited to certain part shapes, low "pick" content when hopper is low, the need for air jets or brushes when the *L/D* ratio is low, and tooling restrictions, that is, centerboards are oriented to one design.

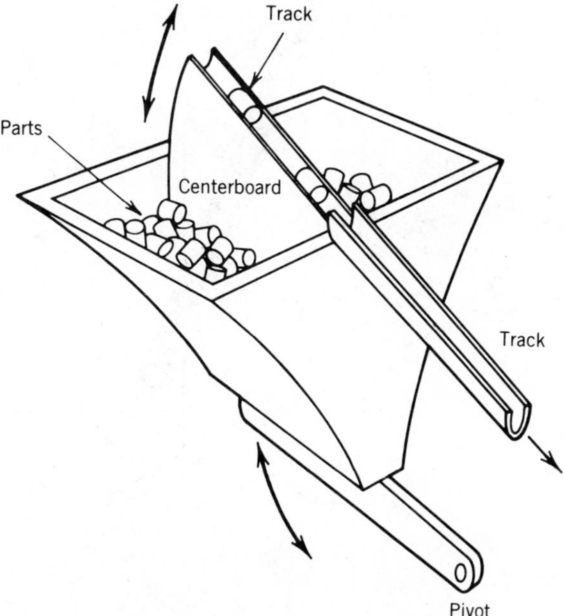

Fig. 3.2.11. Centerboard hopper feeder.

Half-Centerboards

Some parts can be fed and oriented in a "go," "no go" fashion with a half-centerboard, which uses a shelf or ledge on one side of the board to lift parts. A design of this type will pick parts with a tapered base (toward the board) and discard others that are not so oriented (Fig. 3.2.12).

Elevator Feeders

Another important form of nonvibratory feeder is the elevator. This classic form picks parts from a hopper with ledges attached to an endless belt. The ledges orient simple parts while they are rising through the hopper. Angling the ledges appropriately assures that the parts will slide down a delivery chute (Fig. 3.2.13).

If gravity alone acts on the parts, they must be compatible in material, coefficient of friction, and *L/D* ratio for efficient delivery. Any problem with feeding can usually be alleviated with a wiper or brush. In extreme cases, the belt may be indexed upward and the products cleared out of the ledge during a dwell period.

A variation on the ledge, which often is used in fairly high-speed automated stamping factories, is to place magnets behind the primary belt that will pick the parts from the hopper as determined by the magnet spacing. Because properly sized magnets will only pick one part at a time, simple ferrous parts are automatically oriented and spaced. At the top of the travel, the parts may be separated from the belt with a "doctor" blade combined with a chute.

Because of its inherent design, the elevator is useful for only the simplest products, usually cylinders or spheres.

3.2.4 TRANSFER DEVICES IN DETAIL

Once a part has reached the end of the feeder track, it presumably has been oriented and is ready for transfer to a work station. Parts can be moved pneumatically through a flexible tube, a gravity chute, a pick-and-place device (placer), or a robot.

Placers are equipped with grippers in two to four or more axes and cover ranges from a fraction of an inch to several feet and more. The longer distances are generally handled with a conventional pneumatic cylinder and sometimes with a more compact rodless variety.

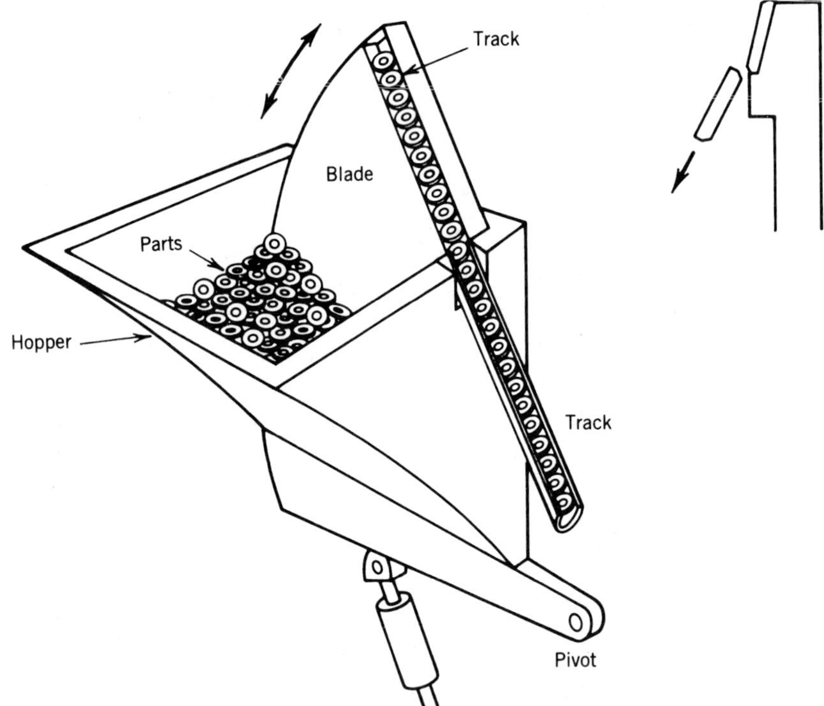

Fig. 3.2.12. Half-centerboard feeder.

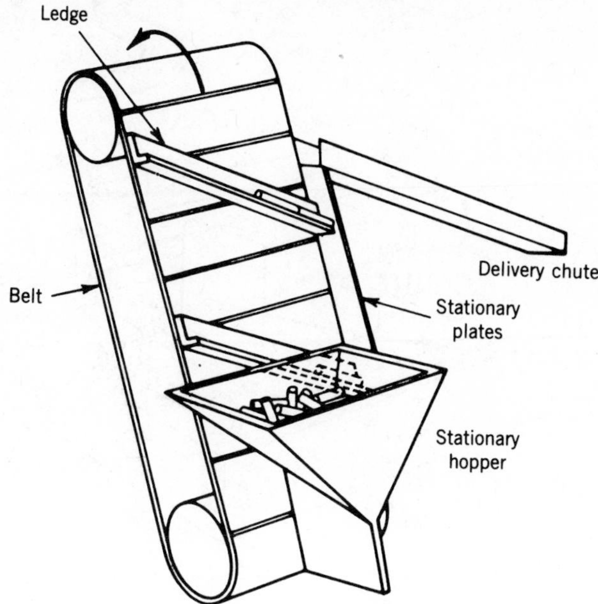

Fig. 3.2.13. Elevator feeder.

For high-speed operation, a cam-operated placer is preferred because the accelerations can be controlled and the cycle times are therefore lower. Some manufacturers have packaged their mechanical placers with the *X* and *Z* motions combined in one input, which simplifies the mechanism and eliminates the timing and contact problem associated with two inputs that have to be synchronized.

When a decision has been made to operate the placer pneumatically because the balance of the system is pneumatic, the placer may either be purchased or designed in-house. An in-house design has a few advantages over commerical units that might outweigh the high cost. Virtually all commerical units, while accurate at the ends of their strokes because the end is a hard stop, suffer from compliance in a lateral (rotary) direction if two support rods are used and in all directions if only one support rod is used.

If accuracy on the order of several thousandths in all directions is required in placing, the design has to be inherently rigid and requires a slide that is carried on linear ball bearings supported by rods anchored to the frame on both ends. This construction lends itself to adding adjustable shock absorbers at both ends of the travel, thus correcting other faults of most commercial placers, which have hard end stops and inadequate cylinder cushions, if they exist at all. In addition, no provisions are made for adjusting the stroke of most commercial placers. See Figure 3.2.14 for several examples of placers.

When designing a pneumatic placer, air cylinder sensors should be used to indicate each end of each stroke. This will keep the motions in sequence and insure that a stroke is completed before the second one starts. Many cylinders are designed for add-on sensors.

If only timers are used, there is inherently a greater possibility of a collision if one cylinder stroke should be incomplete, for whatever reason, while the second one is in operation.

3.2.5 ASSEMBLY MACHINES IN DETAIL

3.2.5.1 Rotary Assembly Machines

When a pick-and-place unit, or a robot, has placed a part on a work station of any assembly machine, often a rotatary indexing dial machine, the input motions normally take place during the dwell period of the indexer. Exceptions to this are continuous motion machines that operate at a constant speed and input products without a dwell period. Rotary assembly machines are commonly composed of a flat circular table, from one to eight feet in diameter mounted on top of and bolted to an indexer housing. A mechanical overload clutch is sometimes placed between the indexer and dial. The indexer is powered by an AC or DC motor with variable speed capability operating through a reducer. If the ratio of the dwell segment of the total cycle is a certain value with respect to the index time, the motor is allowed to run continuously (the time segments are found in manufacturers' catalogs). If the dwell time required is outside the specifications of the supplier, the unit becomes a demand system that has a clutch-brake combination added to the power train. With a demand arrangement,

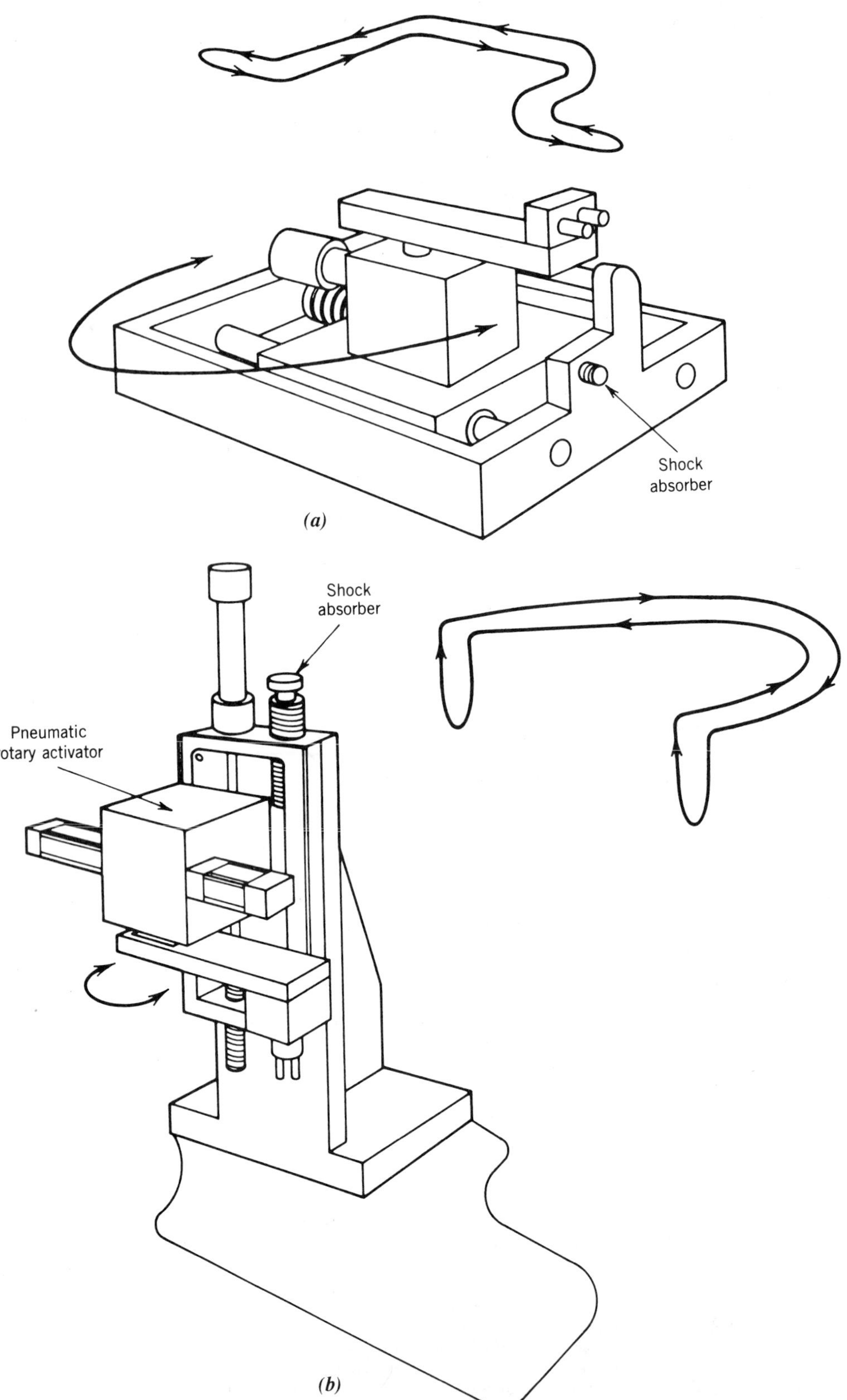

Fig. 3.2.14. Transfer devices. (a) Pick and place. (b) Rotary pick and place.

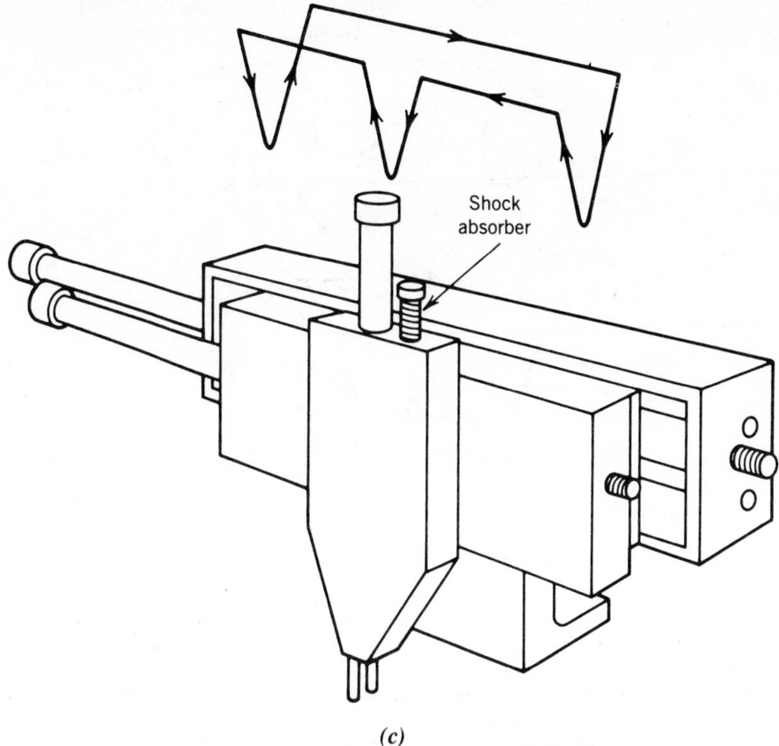

(c)

Fig. 3.2.14. *(Continued)* (c) Pick and place.

the indexer will wait for all functions to occur before a "complete" signal is processed to release the brake and engage the clutch.

Indexers began with a rachet and pawl, or a walking beam system, that frequently operated in-line machines. Later, they progressed to Geneva drives, then to the various roller and cam drives presently associated with both in-line and rotary machines.

Rachet and pawl systems, because of inflexibility, are generally found on older machines. They require low-speed operation because of overshoot. Rachet and pawl devices are also found on small mechanisms and sometimes on very large assembly in-line machines with inherently high friction.

There is a tendency away from the Geneva mechanism for indexing drives because of the high accelerations, wear characteristics, and attendant inaccuracies that develop. When the number of slots is low, all negative characteristics are attenuated, and unless special attention is given to maintenance, the accuracy will deteriorate until floating tool plates and shot pins are required.

A cam and pin drive is the most common form of indexing drive presently in use. The larger the pin circle diameter, the more readily the dial (turntable) accuracy can be maintained. Pin-to-pin distances can be measured very accurately for special indexers and with selective assembly; positional locations of ± .002 inches are possible with a 2-foot dial radius. This type of precision assumes a jig-bored dial with the appropriate quality levels for the rest of the parts (Fig. 3.2.15).

Indexers are manufactured in a variety of shapes to help the design-engineer package his or her machine. It is sometimes expedient to place all tooling on a secondary stationary dial mounted above the indexing one. This construction requires slip rings and rotary manifolds with a conventional indexer housing. However, some indexer manufacturers can provide a housing with a stationary main tube on the axis of the output that not only holds the stationary dial but permits all lines, electrical and pneumatic, to be led up the tube from below the indexer and support table. A design of this type provides a direct connection to the tooling above the rotary dial without resorting to rotary joints.

Another form of rotary indexing assembly machine positions a vertically reciprocating upper dial mounted over a rotary indexing lower dial. Common practice is to mount mechanically actuated placers that are driven by the cam action of the vertical dial on the upper dial. To enable the machine to survive a possible collision during operation, an air spring is made part of the linkage system.

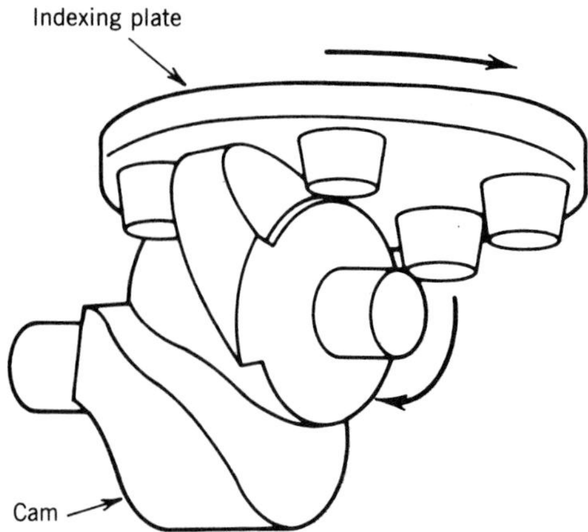

Fig. 3.2.15. Cam indexing unit.

3.2.5.2 In-Line Indexing Assembly Machines

In-line assembly machines are often of the over-and-under type, with an indexer at one end and a pair of chains that can carry tool plates. A device for tensioning the chains to compensate for wear and stretch is located at the end opposite the indexer. Matched chains can also be purchased in a prestretched condition.

Timing belts are used as a chain substitute for light to moderate loads because they are quieter and do not require lubrication. They are of reinforced rubber composites at the tooth portion with a section of flexible steel cables inside the belt for torque transmission.

When accuracy is required above that which the chains or belt provides, tooling plates are attached to chains with a "floating" connection. When a pair of shot pins anchored on the frame is forced in the tooling plates at the stations that require it, the plates can be positioned to within 0.002 to 0.003 inches.

Over-and-under in-line indexers are desirable for assemblies that require operator assisted stations. They consume little space and lend themselves to progressive assembly lines that are directed across the factory.

3.2.5.3 Carousel

When more space is required for assembly than either rotary or in-line machine will yield, a horizontal form of the over and under indexing machine is often used. Sometimes called a carousel, this type of construction usually has a rectangular loop configuration with assembly machines and/or operators placed on the inside or outside of the loop (Fig. 3.2.16).

Indexing the belt or chain is conventionally done with a motor-reducer-indexer package, as in the other indexing machines. Because of the corners, however, the workplates are sometimes attached on one side to a chain or belt and, if accuracy is required, are shot pinned into position. If the belt is composed of plates attached in the middle to a chain, as in "table-top" construction, the plates are reduced on the ends to negotiate the corners.

3.2.5.4 In-Line Free-Transfer Machines

A more flexible indexing system, found in many factories, is the "in-line free transfer," "power and free," or "power on, power off" machine. In-line free-transfer machines, found in all automotive plants, give a measure of flexibility to the operator-worker. When a task has been completed at a work station, the operator releases the pallet or workplate, which then descends to engage a main drive chain and be carried to the next station (Fig. 3.2.17).

A build-up of workplates or buffers allows the operator to release old plates or accept new ones at a rate partially established by the compentence of the individual. It is recognized, however, that the flexibility of the system is necessarily very limited because the remainder of the line would be under stress if one station was consistently underperforming.

In some automotive plants, the length of free-transfer system loops may be 100 feet or more, with 30 to 60 assembly workers and assembly machines interspersed around the outside.

Another version of the free-transfer machine, which is used in large plants, is an overhead version using the same principles as the floor-mounted type but carrying the product in hanging devices. The more extensive arrangements travel from floor to floor as well as plantwide. A hanging system is often found in assembly line painting.

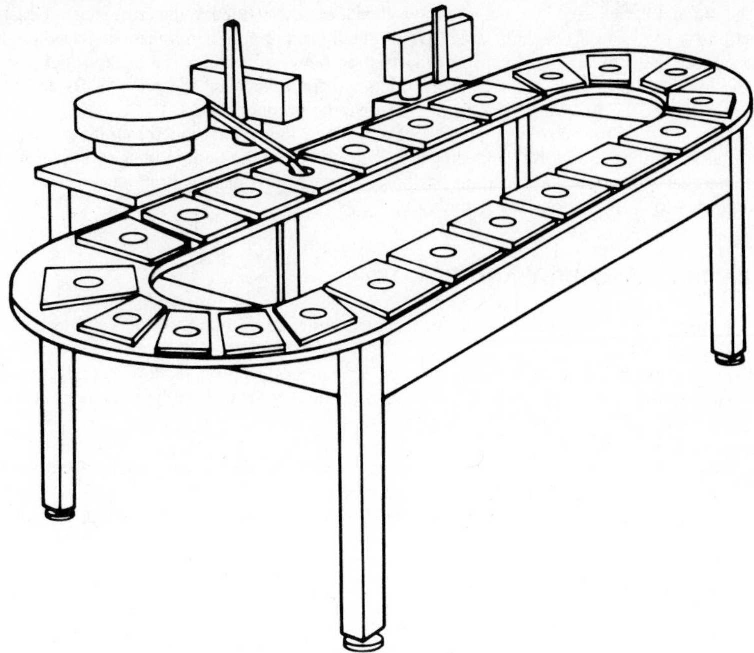

Fig. 3.2.16. Carousel assembly machine.

For small products requiring more flexibility during assembly than would be feasible in an automotive free-transfer line, there is a system that offers advantages of greater flexibility than the above designs and at a lower cost.

Several manufacturers produce a generic accumulating conveyor, which with adaptations may be made a part of a flexible automatic assembly system a level above a conventional free-transfer line. The principle involves a continually running main shaft of a conveyor system linked to a series of cross shafts carrying rollers

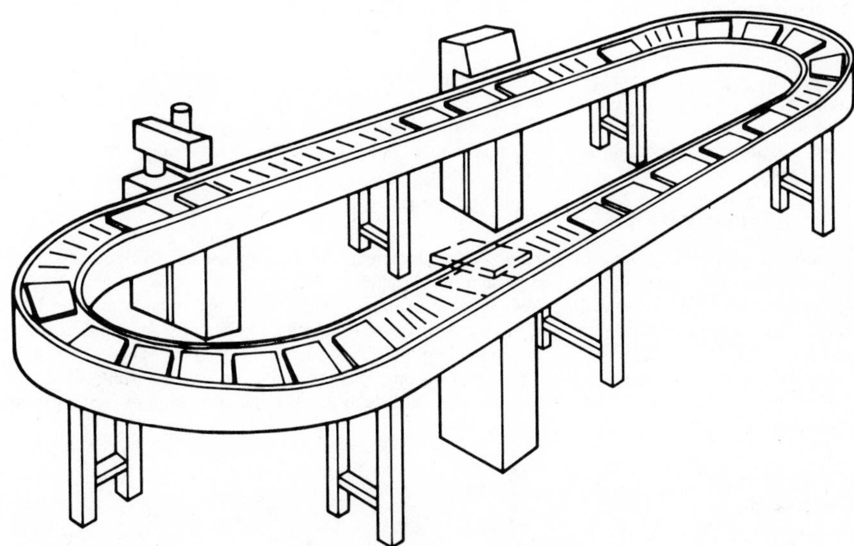

Fig. 3.2.17. In-line free transfer machine.

that support the product to be moved. All supporting rollers are loosely fitted to the cross shafts, which, because of the connections to the main drive shaft, are also continually running. When a mass is placed on the rollers, they rotate in concert with the shafts because of the friction between the material of the roller and the shaft. The greater the mass on the roller, the greater the frictional force between the cross shafts and rollers. The rollers, therefore, follow the rotation of the shaft and move the product.

When a gate is lowered to impede the progess of a product, the rollers underneath the product also stop, but the cross shafts continue to revolve, and slippage occurs between the two. With a design of this type, any small product may be stopped at any position on the conveyor, thus making it an ideal vehicle to transport pallets or workplates to work stations for assembly.

3.2.6 FLEXIBLE ASSEMBLY AND ROBOTS

Because of a growing need for more flexibility in manufacturing, assembling, testing, and inspecting, robots are becoming a major factor.

Flexibility is demanded for economic reasons, that is, for product life, model changes, and short runs. These are all putting an emphasis on the ability of a manufacturer to change tooling, manufacturing methods, assembly sequence and product as rapidly as possible.

Robots that can be be programmed and reprogrammed can affect the economies of a machine or machine system. Dedicated or fixed automation systems frequently have to be discarded if a manufacturing process or a product goes through a change, revolutionary or evolutionary.

This portion of the chapter discusses the role of the robot in assembly and how it might contribute in other aspects of manufacturing that leads to assembly.

3.2.6.1 Inspection

One of the primary input requirements for successful assembly in a modern plant are parts that reflect the design engineers' drawings of a proven design. An assembly system will not function if the components of the product to be assembled do not correspond to the drawings of the part and are randomly out of tolerance. Thus, inspection is necessary, and robots can help to handle this process.

A robot may be used to depalletize parts that are supplied in a suitable format. Without palletizing, it is possible to identify certain parts under some restrictive conditions with a vision system connected to the robot, but pallets with oriented and compartmented product are preferred. Pallets may be brought to the robot on a conveyor and up to a registration point.

If a component to be assembled is a complex casting, the robot is capable of positioning the part on a preset fixture where it can be automatically clamped. All measurements can be taken by a universal measuring machine, a servocontrolled device for automatically moving in all planes and taking measurements of an object with a load-sensitive probe. When required, the robot could turn all surfaces to the tooling of the measuring machine. The robot has the ability to alter the fixturing to accommodate the positional moves demanded for complete measurements.

If a component should fail to pass inspection, it could be unclamped and discarded or directed to a repair station under robotic control.

If the part is simple in shape, it can be placed before a laser gauge and measured where required by a laser without tooling or placing in a fixture. More complex shapes are more easily gauged when placed in a series of fixtures by the robot and presented to a laser gauge.

Several new-old approaches that could contribute considerably to automatic assembly with a robotic input are being integrated into modern manufacturing. These "new" approaches are in reality not new but are old methods that have been revived to increase productivity.

3.2.6.2 Selective Assembly

Before the principle of interchangeable parts was an acceptable manufacturing assembly method, all components that were assembled to make a product or portion of a product were individually measured and catagorized to enable the asembler to find or manufacture a suitable mating component. At that time, before the 1800s, the machinery was inadequate to manufacture mating parts every time, on demand.

Selective assembly has been used in the automobile industry for many years in the manufacture of engines. Optimum fits are possible between wrist pins and pistons, pistons and cylinders, connecting rod and main bearings to the crankshaft. The selection process also extends to the weights of rotating of reciprocating parts.

The advantages of selective assembly are

Fewer rejected parts
Better operating subassemblies
Longer wear of components
Less rework of out-of-tolerance parts
Lower warranty costs on products

Products are replaceable as a unit, not repairable. (High labor costs make this a preferred condition for most manufacturers.)

Selective assembly is dependent on very rapid and accurate measurements to properly match components. It also requires the ability of a machine to automatically categorize parts at high speed. Both operations are within the capabilities of most robots, and with the growth of selective assembly as a manufacturing method, robots may play an increasingly important role.

3.2.6.3 Zero Defects

Zero defects, or 100% acceptable parts, is a principle based on measuring all incoming components before they are assembled and modifying the design when necessary as well as using selective assembly to insure that after inspection all parts will come under a range of acceptable tolerances for the particular function. To implement this process, tolerances on drawings of components for mature products are reviewed and revised to the broadest range at which the product would function satisfactorily. Rejection of components at inspection stations are reduced to nearly zero because any part outside the tolerance range can be matched with a mating part through selective assembly. New products should undergo similar evaluation during the decision phase. This revised approach will result in higher productivity at a lower manufacturing cost.

3.2.6.4 Robotic Feeding, Orienting, Transferring, Assembling

After a component of a subassembly or product has been inspected and accepted, it must be fed, oriented, and transferred or placed at the work station of an assembly machine.

Earlier, this chapter was concerned with feeding, orienting, transferring, and assembling with fixed automation. The robot, however, is programmable, which adds flexibility. If after robotic-aided inspection, the part is placed on an assembly station from a compartmented pallet, there is no reason for a feeder-orienter system. A robot becomes the orienter and feeder. Since the robot inspected the part and placed it in an oriented position on the pallet, it only has to be transferred by the robot on demand. If the parts to be assembled are the product of selective assembly as a result of inspection, a computer would store the size category required from the previous part inspection stations and the robot then would transfer the correct size part from the appropriate compartment to the assembly machine.

At this time, robotic vision systems are not reliable enough for typical factory installation. Most current vision systems rely on back lighting of contrasting parts in a clean environment on a lighted belt. Additionally, the parts should not be touching one another. Conditions of this sort are found in laboratories but not factories other than possibly food processing plants. Sophisticated vision systems will be available in a few years and possibly will be of the laser scanning type for accuracy, fast rate of information processing, and the ability to work under adverse conditions.

The universal gripper is another robotic device that has not appeared in a reliable, economical, or practical enough form to be commercialized. One substitute for the universal gripper is the rack-mounted gripper, which can be exchanged automatically by the robot in approximately a three-second cycle. Any number of grippers may be racked to be exchanged when required. Another variation is to mount a series of grippers of varying designs on a rotating member attached directly to the robot. Weight is a factor for some robots, but for machines with the capacity, this method of gripper presentation would lower exchange times to under one second.

A number of manufacturers offer a line of compliant grippers, which are useful in many assembly situations but especially when a shaft is to be inserted in a hole. The design of the mechanism permits the shaft to be out of alignment with the hole in both angular and parallel modes. If the shaft is chamfered, and preferably the hole also, forces on the shaft as the robot exerts an inserting force deflect the compliant springs enough to allow the chamfers to guide the shaft into the hole.

Experience has shown that programming the robot to move in X, Y, and Z modes during shaft-to-hole insertion is often more effective than using a compliant mechanism. The motions would usually have to conform to required cycle times and, therefore, be extremely small, on the order of 0.010 to 0.015 inches in each direction, but the results are successful in most cases. This is known as a "dither" hand.

A combination of the programmed dither and compliant hands would appear to be a logical optimum assembly tool, but, in practice, this does not work well because the mass of the component added to that of the gripper deflects the gripper springs when the excursions are programmed. There is a point where the entire mass will resonate and make impossible the completion of the task. This point of resonance varies with the mass of each component and the accelerations of the dither.

A transfer move by a robot implies either preorientation from pallets or orientation during transfer. If the part is then assembled on a work station pallet that is an integral part of a flexible assembly conveyor, most of the elements of a flexible assembly system are present.

In hard or fixed automation, when parts are assembled around an indexing table, it is generally thought that the more work stations around the table, the longer the downtime and that this results in an impractical, uneconomical situation when there are more than ten stations. Actually, when the work stations are properly designed and monitored, many subassemblies, or components, can be ejected automatically during a jam-up. For assemblies to which it is difficult to apply an ejection system, however, a robotic alternative is possible. The core of this "bypass" system is an accumulating conveyor, which is traversed with work pallets. As the

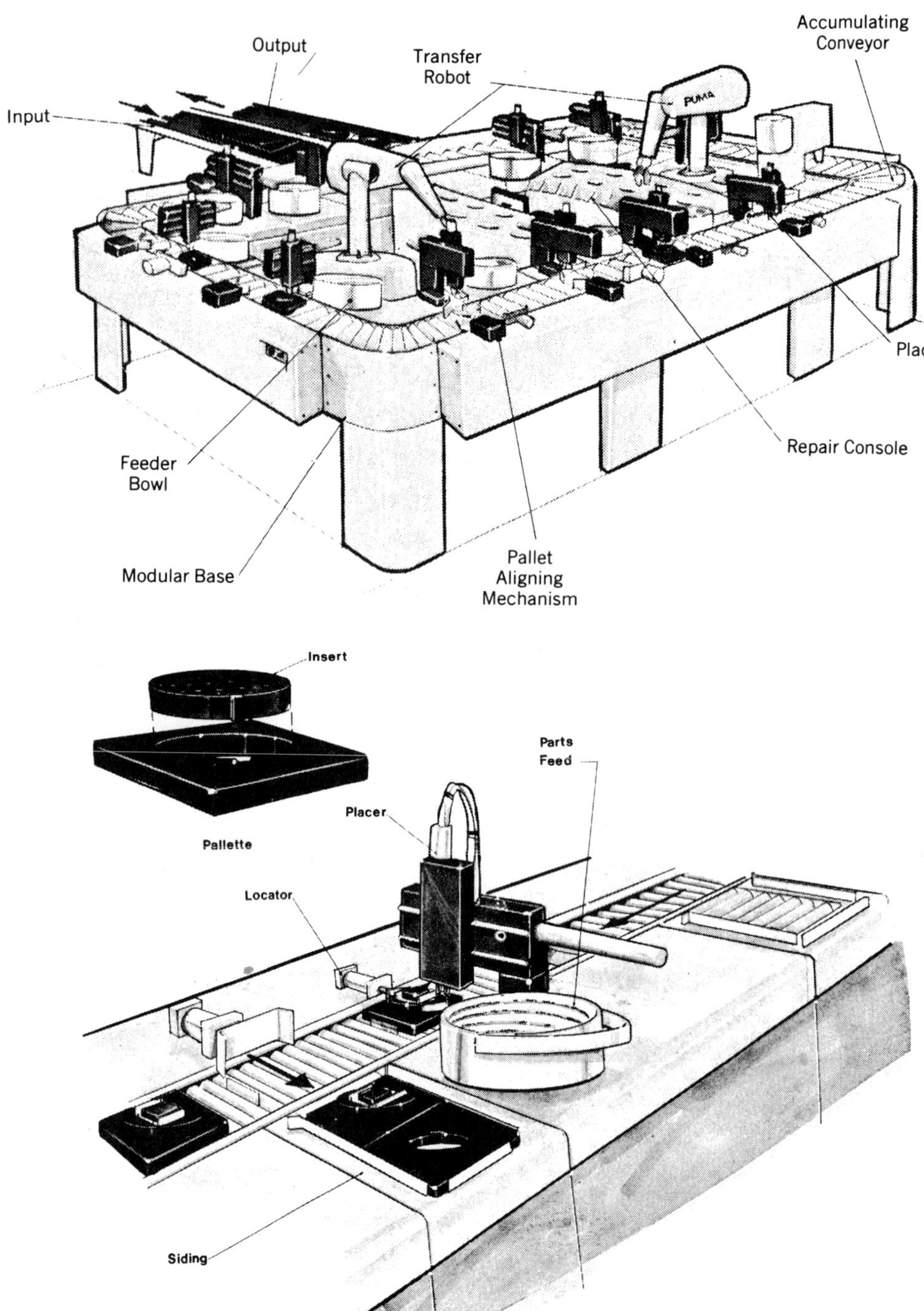

Fig. 3.2.18. (*a*) Typical bypass system. (*b*) Typical station bypass.

components are progressively assembled by either placers or robots, a system of sensors determines at each position if the assembly has been successful. If the particular stage is acceptable, the work pallet will move to the next gate. If a station fails to operate, the work pallet will be moved from the main conveyor to an accumulation spur, and the subassembly will be picked up by a robot and presented to a repair person located in the center of the assembly loop. After repair, the robot returns the subassembly to the proper spur, and the work pallet enters the main stream automatically when the opportunity is presented. In this manner, the assemblies are constructed at the required rate, the repairs are virtually instantaneous, and 100% throughput becomes feasible (Fig. 3.2.18).

Many variations of this system are possible, but a main component remains the flexibility of a robot that can be programmed to perform most tasks.

3.2.7 MODULAR FLEXIBLE AUTOMATED ASSEMBLY MACHINES

Another system is available for automated assembly that is between hard and flexible automation. The principle involves groups of modules that are identical in framing and similar in approach to tooling and transfer devices. Most of the installations contain four or more modules bolted together, carrying a cam shaft that actuates the transfer devices, inserters, presses, and other equipment. Parts reach multiple stations on work plates attached to an indexing chain that circles the machine on the outside. Conventional feeders are used, but robots can be incorporated if the cost is justified. Standardization through modular construction is the principal advantage of this system.

3.2.8 SUMMARY

Two important areas of assembly that are not discussed here but deserve mention are design of the components of a product to be automatically assembled and the means with which assemblies are held together, that is, which fastening methods are best suited to automatic assembly. The opportunity for the design engineer to develop a new product to be automatically assembled are virtually zero. Products have to be prototyped, tested many times, and redesigned before the economic justification for automatic assembly may be addressed. Once the decision to automate has been made, many texts are available that cover design-for-automation in detail.

Fasteners and joining methods are outlined in various texts, but the most complete and up-to-date compilations are found in trade journals. A number of the trade periodicals have devoted entire issues to an annual survey of fastener and joining methods.

CHAPTER 3.3

DESIGN-TO-COST

INGRAM LEE II

Texas Instruments, Incorporated
Dallas, Texas

3.3.1 INTRODUCTION

Design-to-cost is both a business and an engineering discipline that couples cost and design performance to customer acceptance throughout the life cycle of a product. It begins with the acceptance of the fact that *markets* set prices. Prices established by successful manufacturers merely reflect value as perceived by the customer.

The recognition that markets set prices must extend throughout the major areas of any industrial organization. Design, manufacturing, and marketing functions all share a responsibility in customer acceptance of a product. The more sophisticated the product, the more important it is that engineering and manufacturing accept this responsibility.

Too often, particularly in high-technology companies, the design of a product centers on advancing performance capabilities, totaling the cost, marking up the costs by the normal margins, and then handing the product over to marketing with instructions to sell it. When designers ignore the market price demand, potential buyers simply seek out a competitive product that meets their *combined* price and performance requirements.

3.3.2 THE DESIGN-TO-COST PROCESS

3.3.2.1 Identifying Customer Needs and Values

Design-to-cost is a logical process (Fig. 3.3.1) that starts with the identification of customer needs and values. Although a challenging undertaking, it is marketing's responsibility to determine the customer's needs and his perception of value.

3.3.2.2 Establishing Price/Volume Profit and Cost Goals

The customer's perception of value is expressed in terms of how much he is willing to pay, and this leads to the establishment of price and cost targets, which is the second step in the design-to-cost process flow. The concept of the relationship between price and volume is essential, since in most marketplaces as price is lowered, more customers are willing to purchase and volume therefore increases. This results in the classic market elasticity curve illustrated by Figure 3.3.2. Typical curves such as this demonstrate a knee at which point the potential market volume goes up rapidly for relatively small changes in price. This characteristic is more pronounced in the consumer marketplace than in industrial and institutional fields but exists in all markets, including governmental areas.

Closely related to the market elasticity curve is the determination of the volume and price at which a significant market share can be obtained. This is particularly difficult to define in new marketplaces, since consideration must be given to the rate of increase of market growth. While these are difficult subjects to measure, they must be evaluated in order to establish a viable strategy in the marketplace.

The foregoing discussion may seem strange when related to design, and it is not the intent of this section to make a marketeer out of the engineer. It is, however, essential that the engineer understand and appreciate these principles, since proper marketing requires a constant interchange of information between marketing and engineering. It is generally incumbent on the engineer to communicate with marketing in nontechnical terms.

After careful market study has established the probable relationship between price and unit volume, a target selling price to be achieved at a particular production level must be determined. A target profit margin also must be established. This profit margin is essential to finance growth and reward stockholders and must be assigned in the beginning.

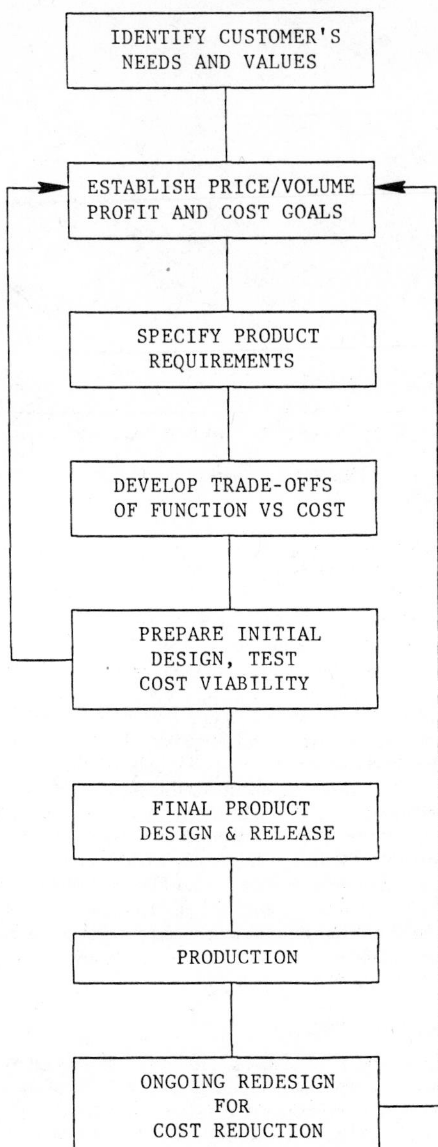

Fig. 3.3.1. The design-to-cost process.

3.3.2.3 Product Requirements

After the identification of customer needs and values and the establishment of price and cost targets, the design of the product may be undertaken. No detail design should be initiated until the product is defined and customer requirements are translated into a set of specifications that clearly state the *minimum acceptable* requirements for the product. The necessity for such requirements may seem obvious, but all too frequently this key item is overlooked or is not adequately defined by marketing and considered by engineering.

The concept of *minimum acceptable requirements* is key to the specifications and must be comprehended in design. Failure to clearly establish this level almost invariably results in overdesign, which in turn creates excess costs. While this is a difficult parameter to assess, it must be thoroughly understood by both marketing and design engineering. Customers may be pleased with products that exceed their expectations, but in general they will not pay for any significant excess of capability.

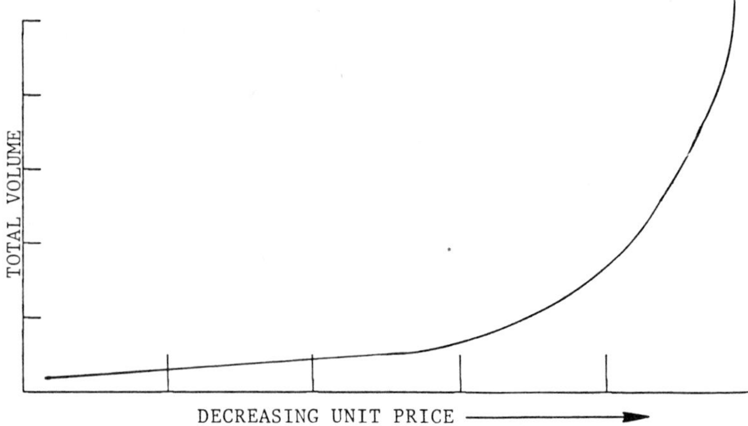

Fig. 3.3.2. Market elasticity.

Some of the factors that must be agreed upon between marketing and design are appearance, size, weight, reliability, operating environment, ease of maintenance, difficulty in learning to use, and possible compatibility with other equipment.

The importance of quality must not be underestimated. To most customers, the relationship between product reliability and life cycle performance is becoming increasingly critical. The purchaser has a right to expect some useful period of life, the length of which varies with the product and the price. A proper marketing plan must define this anticipated useful life, estimated mean time between failures, frequency and type of repair, and warranties involved. All of these are considerations that enter into the ultimate "cost of ownership."

The simplest definition for cost of ownership, and sometimes the one most commonly considered, is the price at the time of purchase. For many products, this is the only consideration; but for more sophisticated items, the cost of ownership is more far reaching, and marketing must define and specify the emphasis on the different aspects of cost of ownership. For every product there exists a life cycle; that is, the interval between acquisition and disposal. Within this life cycle, the initial purchase price is only one consideration. The frequency of repair, cost of maintenance over the lifetime, and possible technical obsolescence may play a key role in its marketability. These are factors that can be assessed and a life cycle cost determined.

Closely related to the reliability definition is environment definition. The key question is, what are the conditions under which the product must function for a defined period of time? The marketeer and the design engineer must agree on such essential items as temperature ranges, vibration, shock, humidity, electromagnetic interference or radiation, exposure to chemicals, sand, dust, and altitude and pressure variations.

3.3.2.4 Trade-Offs of Function versus Cost

Following proper product definition and specifications, preliminary design efforts may be undertaken, with engineering and marketing working together in determining trade-offs of function and weighing the value of functions versus costs. Each design element should be considered in terms of the following: What is it? What does it do? What does it cost? What is it worth to the customer? What alternatives exist and at what cost?

3.3.2.5 Cost Viability

In testing the cost/profit viability of the preliminary design, the impact of volume must be taken into consideration. Relating volume and cost is key to the success of a program and must employ the principle of the "improvement curve." The importance of this management principle is so key that its concept and application will be explored in detail here.

The improvement curve is sometimes referred to as the *volume learning curve*, but in the broader sense it encompasses far more than the process of learning. Since learning is an element, however, this will be examined first.

The learning curve is predicated on the fact that people learn with repetition and do so in a generally predictable pattern. The learning curve takes into consideration that improvement occurs within a work group with beneficial results.

Since learning follows regular patterns, it is possible to plot and then project progressive and cumulative improvements on a graph. With reasonable certainty, one may forecast the decline in labor content per manufactured unit over time.

The basics of the learning curve were developed in the late 1930s by the aircraft industry. At that time, aircraft manufacturers learned that building a second plane typically required only 80% of the direct labor needed to build the first plane. The fourth plane needed only 80% of the man-hours required for the second, and the eighth could be build 20% faster than the fourth.

During World War II the need for constantly increasing production by work forces with limited training resulted in the refinement of this managerial and scheduling tool. It was essential to be able to predict the number and cost of aircraft available at a point in the future in order to develop strategic planning.

In the prior discussion, it is evident that given unit time was reduced to 80% each time the quantity doubled. This is termed an 80% learning curve, which is typical of production processes that are labor intensive and illustrated by the plot shown in Figure 3.3.3. The "percent" of learning for different products will differ, with some curves as shallow as 95% mature products or as steep as 60% for products new to the production facility in question.

The graph of Figure 3.3.3 is plotted on linear scales such that all divisions are equal and represent a constant value. From a user standpoint, a logarithmic scale is of more value. In this case, equal distances on the scale represent a constant proportion, and the learning curve plots as a straight line reflecting a constant "percentage" rate of reduction. This makes cost reduction forecasts more understandable and easier to project. An example is shown in Figure 3.3.4.

A composite curve may be used to describe finished product costs, which include material, labor, and overhead. Characteristically, the curve is used for labor, but it applies with different slopes to any of the cost elements in the completed product, such as raw material, supplies, or other purchased items that might "learn" with volume discounts, increased purchasing sophistication, and so on. Similar patterns exist for reject rates and scrap losses, since these two variables are predictable.

Note that "learning curve" cost reduction slopes are plotted against total cumulative units produced and that time plots keyed to volume must be added to produce the projected time-phased product cost impacts.

Establishment of the proper slope for learning curves requires imagination, engineering judgment, data on performance curves achieved on prior and related equipments, and a willingness to accept risks. Prediction of too steep a slope results in unrealistic cost goals and the strong likelihood that projected profits will not be realized. On the other hand, the curve with too shallow a slope may result in overpriced products that fail to achieve sales potential or an unrealized profit potential.

There are many factors that must be considered in predicting the slope of an improvement curve. Some of these include the freedom of management to make changes without intervention by other parties, the quality and turnover of the work force, the condition of manufacturing facilities, and inflationary pressures. Despite all these variables, it must be recognized that improvement curves exist and must be aggressively used in design-to-cost.

The concept of the improvement or learning curve merely indicates opportunity. Achievement of this opportunity is attained only by dedicated effort and discipline. In the manufacturing areas where a repetitive product is produced, plans must be established and executed to increase production rates if the manufacturing work force is held constant or to remove people from the manufacturing area if the production level remains the same. Failure to do so will result in failure to achieve the proper curve. This discipline must be imposed on all levels of supervision to constantly reduce the labor input at a predicted rate.

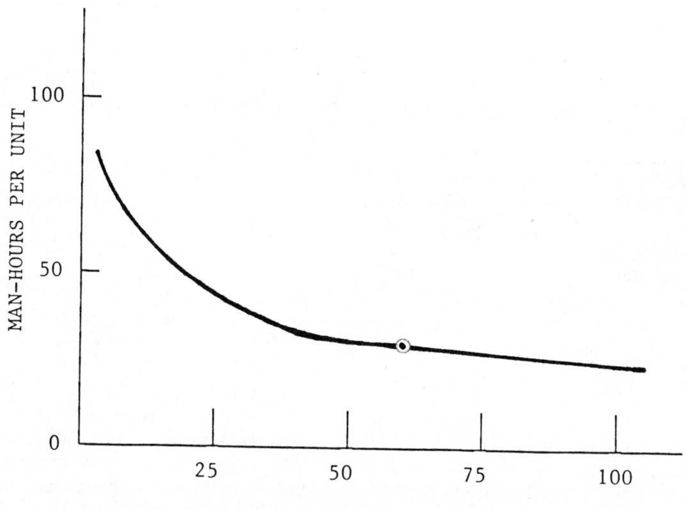

Fig. 3.3.3. The 80% learning curve on arithmetic scales.

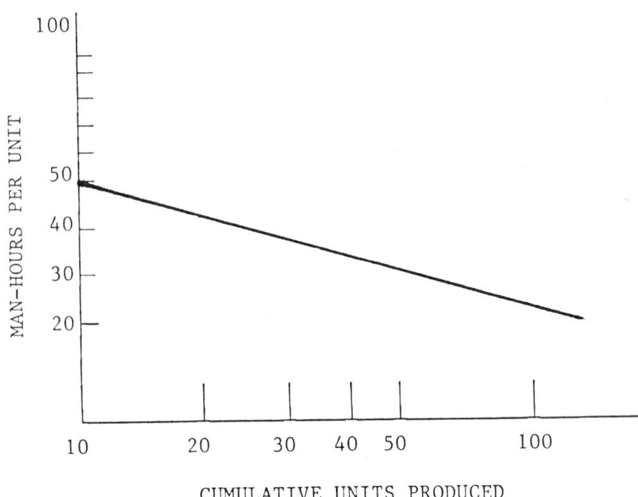

Fig. 3.3.4. The 80% learning curve on double logarithmic scales.

Cost forecasting and improvement predictions may indicate that the design does not result in a cost that will yield the desired margin when the product is taken to the marketplace. In short, it is not cost viable. As shown in the flow chart in Figure 3.3.1, a series of iterative steps must be taken to develop a design that satisfies all requirements, both financial and technical. At that point, a final product design may be completed and released to production.

Plans must be established in the beginning to measure the progress along such improvement curves. All too frequently, inadequate attention is given to the means of measuring various elements of cost, and it is realized too late that a large cost package defies analysis. The engineer who is truly working with design-to-cost concepts in mind must include the appropriate financial people in the planning of the program. Working with the financial organization, discrete points can be established for measuring the cost of material and labor and associated overhead. These points may relate to production quantities or intervals of time. At any rate, it is essential that points be identified early enough in the program so that corrective action can be taken in the event the desired slope is not achieved.

In the broadest sense, the achievement of the improvement curve in repetitive manufacturing areas is a joint effort between management and workers. The successful engineer or manager will recognize the importance of indoctrinating his entire organization with the philosophy of improvement. Further, much of the improvement will come about through the ideas of the people actually performing the work. Recognizing and incorporating these ideas result in more ideas with attendant improvement in labor performance. It is characteristic that people like to know how they are performing both individually and in groups, and graphic portrayal of the improvement curve can be a powerful motivator.

A typical improvement curve, plotted as the best fit along several points, is given in Figure 3.3.5. The first point to be plotted is theoretical and represents the required cost derived from the selling price at a specific quantity. An initial run has been made, which falls above the curve. A line between this point and the required cost point may indicate an unrealistic improvement curve and that changes in the design or method of manufacturing need to be examined. Subsequent releases to production show the first and second significant runs lying on a curve that approximates the forecast improvement curve.

The requirement for constantly reducing costs as cumulative quantities increase is not limited to the production organization but must be extended to the various suppliers and vendors who furnish material, components, subassemblies, and so forth. Each of these vendors experiences a learning curve. The more unique the vendor's product, the greater the opportunity for material cost reduction.

Reduction of vendor prices through establishing vendor cost goals requires a dedicated effort by design engineers, buyers, procurement specialists, and contract negotiators. When a large company is purchasing from a small vendor, it may become apparent that the supplier does not know about the learning curve or lacks the means to implement it. Procurement specialists can assist such a supplier in defining and projecting his learning curve and providing him the techniques for measurement. This concept must be imposed on suppliers to ensure that the total product cost is driven downward with fixed objectives in mind.

In the previous paragraphs, emphasis has been given to the obvious areas of direct material and direct labor input. The learning or improvement process must not be restricted to the production organization, since the opportunities in design engineering, production engineering, and management are far greater. The effect of changes in these areas appears as step functions in the improvement curve. The changes will come about only if management recognizes the necessity, sets the goals, establishes the points of measurement, and makes proper adjustments. All too frequently, products are launched using teams of top-level design and production engineers to guide the program through the initial phases. As soon as some minimum production rate is achieved, victory is celebrated and the forces are dispersed and assigned to different efforts. Their replacements generally do not feel the same responsibility or commitment to attainment of cost goals. Further, the new team frequently lacks the depth of technical knowledge necessary to make certain decisions that involve a technical risk. When this circumstance arises, the opportunities for dramatic changes in the slope of the improvement curve may be overlooked.

The design engineer who thoroughly comprehends the design requirements is best suited to effect the redesign that eliminates components, eases tolerances, and changes materials without detriment to function and ultimate customer satisfaction. The ability to take the risks associated with such changes and determine the proper trade-offs requires a depth of technical knowledge of the highest order. The designer must maintain a close relationship with the marketing organization to ensure that the changes contemplated do not result in a perceived decrease in quality by the ultimate customer.

One of the barriers to incorporating necessary changes for continually driving costs is "pride of invention." Narrow-minded individuals sometimes view changes to their original design as an indication that they did not do it right the first time and resist such changes. Objectivity, therefore, plays a part in the ability to achieve specific cost goals. The truly capable engineer recognizes that it is virtually impossible to create the perfect design in the beginning and that the search for improvement and subsequent cost reduction can be as challenging as the creation of the original design. In an ideal design environment, the engineer seeks out specialists who can provide information leading to lower cost design.

As a beginning, cost analysis must be made of each of the individual components of the product. Frequently, it will be found that a significant portion of the cost is associated with a relatively small number of the parts. According to Pareto's principle, 80% of the cost will be found in 20% of the parts. This provides a good starting point for challenging each of the high-cost parts.

The first objective is to eliminate components without quality deterioration. To do so, the function of each item must be examined. If customer satisfaction does not depend on a particular item, then steps to eliminate it should be taken. It may be possible to eliminate a variety of individual items by creative combination resulting from a different manufacturing technique. For example, quantity production can provide the opportunity for investing in tooling required for plastic molding or in castings.

Reduction of the number of parts can result in both direct and indirect savings. The indirect savings result from a reduction in paperwork associated with procurement, control, scheduling, inspection, and so on, of a variety of parts, reduction in stocking and material handling requirements, and reduced likelihood for disruptive shortages. The direct savings associated with such combinations result in the step functions in the improvement curve.

In achieving design cost goals, the design engineer should make full use of the procurement function. The decision to make or buy must be examined, and the procurement function should have established goals for its activity. Frequently, in obtaining information relative to making or buying parts, suppliers can suggest improvements and subsequent cost reduction to elements on which they are bidding.

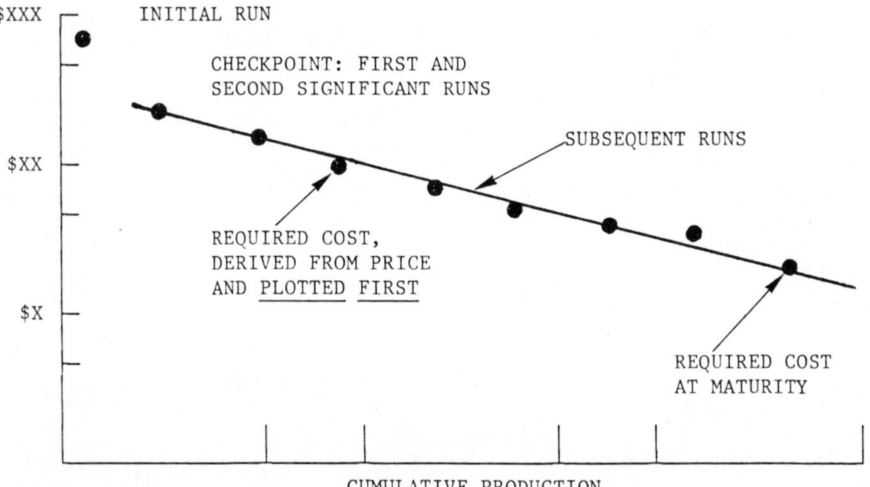

Fig. 3.3.5. Cost history and forecasts.

Further step functions in the improvement curve can be effected by the continued application of production engineering principles. Even operations that are largely automated can be improved throughout the life of the project. Application of new types of tooling, coolants, lubricants, and cutting tool materials all result in further progress down the improvement curve.

One area frequently overlooked is in the support or overhead activities. These, too, should exhibit reduction in costs as products mature. Frequently, such reductions are in the form of improved material handling techniques, simplified packaging, and product protection to reduce damage. Even the flow of paperwork can be streamlined as production stabilizes.

3.3.3 DESIGN-TO-COST FOR SMALL LOT MANUFACTURING

From the foregoing discussion, the conclusion might be drawn that only high-volume manufacturing operations are suitable for design-to-cost effort and improvement curve predictions. Similar opportunities, however, exist in small lot manufacturing. The principal difference is the necessity for applying cost reduction techniques from the very inception of the program. Failure to do so may well result in lost profit opportunities, since the life cycle of the product ends before normal cost reduction techniques can be employed.

As depicted in Figure 3.3.6, technology governed by the design-to-cost discipline reduced the price of the hand-held calculator to attract immense market areas. At the left on the log scale is the retail price of complete calculators. The key element has been cost reduction on semiconductor devices, and the dollar figures inside the graph on the right indicate cost changes for these elements.

Twenty years ago calculators were bulky electromechanical devices that could perform only the four basic math functions and calculate square roots. In 1962, calculators occupied half of a desktop and cost approximately $2,000.

This large $2,000 calculator was greatly simplified with the introduction of integrated circuits in 1968, which not only reduced the cost of parts but had a substantial impact on parts handling and assembly. In 1968 electronic calculators retailed at a little over $1,000.

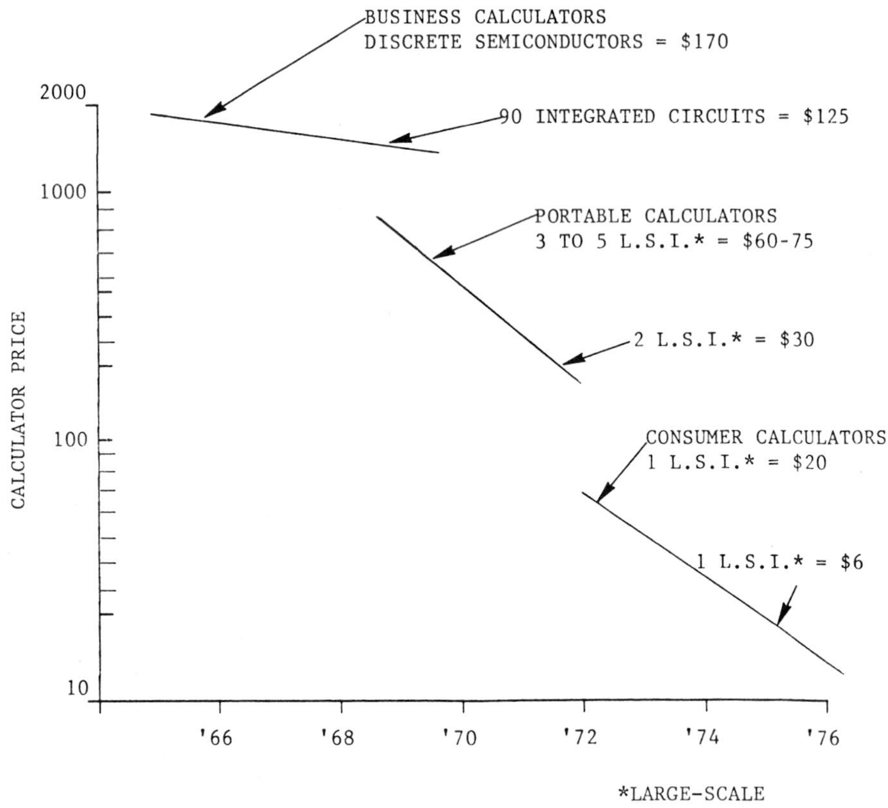

Fig. 3.3.6. Evolution of electronic hand-held calculators.

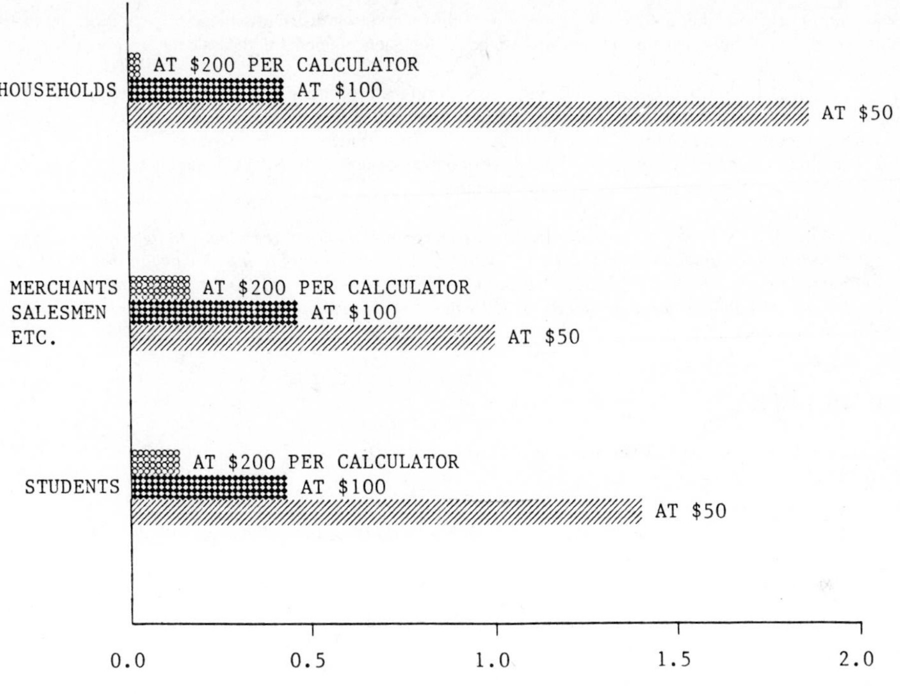

Fig. 3.3.7. Potential consumer market for hand-held calculators.

As the 1970s dawned, early "portable calculators" were made possible by further application of large-scale integrated circuits, and calculators became available to almost any businessman. This design-to-cost effort affected both the number of parts (20 to 1) and the value of the parts.

Still driven by design-to-cost and the goal of penetrating another marketplace, the development of a large-scale integrated circuit performing all logic and memory functions resulted in another step function. This drove the price below $200, and a consumer-oriented market was underway.

Over the short period of seven years, the decline in the number of integrated circuit elements and their price was very dramatic. From 90 units per assembly costing $125, the number of parts was reduced to one costing approximately $20. Several calculator product cycles were generated in this time period.

In 1971, an objective was established to enter the consumer marketplace. Careful study was given to the concept of market elasticity for various sectors of the consumer market, as shown in Figure 3.3.7. Note that each segment shows a marked increase in potential at around $50, which corresponds to the knee of the market elasticity curve shown in Figure 3.3.2.

The strategy for entering the consumer marketplace is shown in Figure 3.3.8. Key to this was establishing sales price goals and a timetable to achieve each. In order to relate market price to factory selling price, the

I. DESIGN REQUIREMENTS
    ENTER HIGH VOLUME CONSUMER MARKET RETAIL FOR LESS THAN $100
    IN FIRST YEAR OF PRODUCTION

    MARKET THROUGH CHANNELS REQUIRING 40% MARGIN
    POTENTIAL TO REACH $50 PRICE MARKET WITHIN TWO YEARS
    USE ADVANCED TI TECHNOLOGY PRODUCTS IN:
        DISPLAYS
        SEMICONDUCTORS
        KEYBOARDS

Fig. 3.3.8. Strategy for entering the consumer marketplace with hand-held calculators.

initial requirement was that a 40% margin be allocated to marketing distribution channels. In turn, the factory selling price was subdivided into design-to-cost goals for each element of the calculator over its projected product life.

A number of trade-off decisions were necessary during the program's early history. Each trade-off was worked backward toward redesign and forward toward consumer acceptance. Emphasis on the design-to-cost discipline forced new findings, and these findings forced these trade-offs.

This real life example highlights the philosophy of design-to-cost, which may be summarized by a quotation from J. Fred Bucy, former president of Texas Instruments:

> *The application of the design-to-cost concept must become second nature not only to those who design, but to those who manufacture and market. It must be applied not only to high volume products, but—in adapted form—to limited volume products. It must be applied not only to the products and services we sell, but to our internal functions as well. Design-to-cost principles, creatively adapted, will make a major contribution to the achievement of . . . objectives.*

BIBLIOGRAPHY

Corporate Engineering Council, Texas Instruments, Incorporated. *Design-To-Cost—An Introduction*, pp. 1–6 (June 1977).

Texas Instruments, Incorporated. *The Learning Curve and Design-To-Cost*, pp. 1–55 (1976).

John Whitmarsh. "Writing the Learning Curve," *Purchasing World* (February 1979).

CHAPTER 3.4

PRODUCTION SYSTEM MODELS

JOHN A. BUZACOTT

University of Waterloo
Waterloo, Ontario, Canada

3.4.1 INTRODUCTION

To design and improve production systems, it is necesssary to have some means of predicting their performance and identifying the effect of the key design parameters. Traditionally production system designers have relied on extrapolation of past experience or have used simple rules of thumb. However, the increasing complexity of modern automated systems with their high capital cost has meant that designers have come to recognize that better means of predicting performance are needed in order to avoid the financial and other consequences of the installed system failing to perform as originally expected. The use of formal models to assess performance has come to be seen as an invaluable aid to the system designer in choosing the main parameters of the design and in forcing a disciplined approach to the development of cost minimizing and performance maximizing production systems. The models can also be of great value in assessing the performance of systems once they are installed because they enable the sources of loss of productivity to be identified. Models provide understanding and insight and enable a wide variety of "What would happen if?" questions to be posed and answered.

The purpose of this section is to describe a wide range of production models that have been developed in order to predict the performance of production systems. The focus of the models is on predicting what the production rate of a system should be after allowing for disturbances such as machine breakdown, human operator performance variability, and quality problems.

3.4.2 HISTORICAL REVIEW

Because the production models to be described in this section focus on the effect of disturbances on the production rate, an essential mathematical tool is probability and statistics. Apart from the study of the "machine interference" problem by engineers in the 1930s, with the first adequate model due to Palm (1943),[1] it was not until the late 1950s that any reasonably correct models were suggested.[2,3,4,5] Then in the early 1960s came the development of various simulation languages such as GPSS and SIMSCRIPT. These languages provided a means by which modelers could describe complex manufacturing systems and simulate their behavior. Since that time production models have been developing in their comprehensiveness and usefulness, and the respective roles of simulation models and analytical (probability) models have come to be appreciated. Starting in the late 1970s production models have come to be seen by practicing engineers as having more than just an academic interest, and they are being used as a valued design tool.

3.4.3 PRODUCTION MODELS

To the user, a production model of a particular class of production system is either a mathematical formula or a computer program into which can be fed numerical values of the parameters of a system. The model can estimate the values of various performance indicators of the system. By changing the values of the parameters, it is possible for the user to gain insight into their influence on the system performance. In conjunction with an optimizing procedure, the model can be used to find the value of design parameters that maximize the system performance or minimize its cost subject to performance targets being met.

3.4.4 SYSTEM PERFORMANCE

The performance indicators of interest to the user are usually some or all of the following:

Gross production rate. The long-run average production rate of a system subject to no disturbance in the form of breakdowns, variability in task time performance, or quality problems.

Net production rate. The long-run average production, taking into account breakdowns and task time performance variability but ignoring quality problems.

Actual production rate. The long-run average production of items meeting quality standards.

Efficiency. The ratio of the net production rate to the gross production rate. That is,

$$\text{net production rate} = \text{efficiency} \times \text{gross production rate}$$

Net yield. The ratio of the number of items produced meeting quality standards to the total number produced. That is,

$$\text{actual production rate} = \text{net yield} \times \text{net production rate}$$

Most production models can predict a wide variety of other parameters, such as the time taken for a job to go through the system, the queue lengths in front of machines, and the average inventory levels.

3.4.5 THE DEVELOPMENT OF PRODUCTION MODELS

Development of a production model of a particular class of production system involves making a compromise between the detail of system description, the computational requirements of model, and the time and difficulty of model development. Simple models are easy to develop and easy to use. However, simple models do not include much of the reality of system behavior. Yet to capture all the features of a system would generally impose an impossible modeling task. Model development thus always involves making compromises. The skill of the modeler lies in choosing compromises that do not distort the reality of the representation of the system by the model yet enable results to be obtained in a reasonable time. Often, rather than having one model, it is desirable to have a set of models, each focusing on different aspects of system performance. Simpler models are more likely to lead to insight being gained.

3.4.6 TYPES OF MODELS

As mentioned in Section 3.4.2 there are two basic types of production models: Analytic models and Simulation models.

Simulation modeling techniques are described in Chapter 8.6. A simulation model is usually able to describe a system in considerable detail, but this means that it does not give general results. Simulation models are particularly useful for fine tuning a design, but they are not so useful at the initial stages when the designer is trying to determine the order of magnitude of key parameters.

Analytical models can be further subidivided into exact and approximate models, exact meaning that the equations describing the system can be solved and approximate meaning that, owing to the number of equations or their numerical characteristics, no solution can be found within a reasonable amount of computer time or space. Thus some assumptions are made that change the model in some way so that solutions can be obtained. Testing the validity of an approximate model requires the use of a simulation model, so developing approximate models is only worthwhile if the approximate model is easy to use and by its structure provides insight into the system behavior. Some exact models provide algebraic formulae, while others require a computational algorithm to be used in order to obtain numerical results.

3.4.7 BASIC APPROACH TO MODEL DEVELOPMENT

Model development requires the following steps:

1. Ascertain the needs of the user. What is the problem? How will the model be used? When is it needed?
2. Identify the components of the system. What machines, what material handling system, what storage locations will be used?
3. Determine the pattern of workflow through the system. Do all jobs follow the same path? What diversity in routing is there?
4. Find typical values of key parameters. What are the production targets? What sort of availability and yield is typical of systems?

5. Choose the modeling approach. Use simulation or analytic models? If analytic, do models already exist? If not, where do existing models fail?
6. Identify the performance measures of interest to the user. Which aspects of performance are most important, which are less so? What are the economic and social consequences of the system failing to meet its owners' goals?
7. Develop a preliminary model in order to obtain some results quickly and to develop the communication with the user that a model demands in order for the user to be convinced that the modeler understands the situation.
8. Produce some results using the model and assess their accuracy. This step will also serve to make the user aware of the data which use of the model will require.
9. Develop a more detailed model and collect the necessary data. Several iterations may be required in order to devise an adequate model.
10. Document, implement, and educate the user in the use of the model. What can it do? What can it not do? How accurate is it likely to be?

3.4.8 DESCRIPTION OF AVAILABLE MODELS

The remainder of this chapter will be devoted to a description of available analytical models of production systems. They are grouped according to the type of disturbance they seek to describe, which may be machine breakdowns, task time variability, or defective parts and quality problems. There are some models that attempt to capture more than one of these effects; however, these will not be described here. The discussion will emphasize analytical models that yield a formula-type result, as such results are most readily applicable. When no formula exists, we will outline the general principles of the computational algorithm.

3.4.9 ASSUMPTIONS

Every model is based on certain assumptions about the composition of the system, the workflow through it, and the availability of resources, such as the number of operators, the number of repair crews, and the way in which these resources are allocated when there are competing demands for them. These assumptions describe the essential features of the system. Then there are further assumptions that relate to the nature of the disturbances to the system operation, such as the distribution of time between successive failures of a machine, the distribution of time to repair a machine, the distribution of time to perform a task, and the pattern of occurrence of defective parts. In some models the results depend critically on these distributions, while in others the mean of the distribution is all that is required. Usually it is easier to derive results for certain distributions than from others; in particular it is often possible to derive formulae for exponential distributions but not for other distributions. Fortunately, in many cases the results are not particularly sensitive to the form of the distribution or, at least, they give the right general shape of the relationship. Thus, the results can be used to give a general insight, which is often the main use of analytical models.

Almost all analytical models make the following assumptions:

Distributions are stationary (e.g., process parameters do not change with time or cumulative production).

Successive events occur at intervals which are independent of each other (i.e., there is no serial correlation).

Events at one machine are independent of events at another machine.

These assumptions are often not strictly correct in reality. However, it would be difficult to derive results that would relax them. In most real systems, when the assumptions fail, the system would be considered out of control and managerial action would be taken to restore control (e.g., a significant worsening in quality with time would demand managerial intervention).

3.4.10 MODELS OF MACHINE BREAKDOWNS

3.4.10.1 Machine Interference

Consider a number of machines under the care of a single operator. The operator is required to repair or adjust each machine as the need arises. However, two or more machines could require service at the same time. Since the operator can only work on one at a time, the other machine requiring attention will be idle, with consequent loss of production. This is the classical machine interference problem.

This model also represents a variety of other situations. Suppose each member of a group of workers requires a particular facility from time to time, such as a grinding wheel to sharpen tools. Then this can be modeled in the same way as the earlier situation, with the grinding wheel being equivalent to the operator and the workers to the machines. Another situation to which this model has been applied arises in the steel industry.

There are a number of soaking pits in which steel ingots are heated until they have a uniform temperature. Then when the contents of the soaking pit are ready, the ingots are removed one at a time for rolling. The contents of one pit are rolled before the next pit is opened. In this case the interest is in "nothing hot" delays, that is, times when no ingots are ready for rolling, resulting in the rolling mill being idle. The soaking pits are equivalent to the machines and the rolling mill to the operator.[6]

It can be seen that the focus of machine interference models is on determining how the performance of a system of "machines" is affected if they share a common facility and only one machine at a time can use that facility. Otherwise the machines operate independently.

The Model

The assumptions being made in the above machine interference problem are the following:

1. There are N identical machines and a single repairman.
2. Time between failures of each machine is exponentially distributed with parameter λ; that is, λ is the failure rate of a machine.
3. Repair time of a machine has an exponential distribution with parameter μ.
4. Machines are served in order of failure (FCFS, first come first served, queue discipline).
5. The repairman is always available to repair failures, and there is no delay between the occurrence of a failure and the beginning of repair unless the repairman is working on another failed machine.

Let

$$x = \frac{\lambda}{\mu}$$

$$p_n = \text{probability that } n \text{ machines are down}$$

It can be shown that[7]

$$p_n = \frac{N!}{(N-n)!} x^n p_0, \qquad p_0 = \frac{1}{F(x, N)} \tag{3.4.1}$$

where

$$F(x, N) = \sum_{n=0}^{N} \frac{N!}{(N-n)!} x^n$$

$F(x,N)$ can be calculated using the recursive formula

$$\begin{aligned} F(x,N) &= 1 + NxF(x,N-1) \qquad (N \geqslant 1) \\ F(x,0) &= 1 \end{aligned} \tag{3.4.2}$$

Given the p_n, the performance measures can be obtained, for example,

$$\begin{aligned} \text{operator utilization } O &= \text{proportion of time repairman is busy} \\ &= 1 - p_0 \\ &= Nx \frac{F(x,N-1)}{F(x,N)} \end{aligned} \tag{3.4.3}$$

$$\begin{aligned} \text{machine efficiency } E &= \text{fraction of time a machine is working} \\ &= \frac{\sum_{n=0}^{N}(N-n)p_n}{N} \\ &= \frac{F(x,N-1)}{F(x,N)} \end{aligned} \tag{3.4.4}$$

A general relationship exists between E and O:

$$\text{number of repair completions per unit time} = O\mu$$

$$\text{number of machine failures per unit time } \lambda^* = NE\lambda$$

Since these must be the same,

$$E = \frac{O}{Nx} \tag{3.4.5}$$

a result that is true irrespective of the distributions of time to failure and repair time.

If R is the production rate of a machine per unit working time, then

$$\text{gross production rate of system } G = NR \tag{3.4.6}$$

$$\text{net production rate } P = NER \tag{3.4.7}$$

Other performance measures can also be obtained using simple arguments, for example,

Mean duration of down time of a machine W = mean waiting time + mean repair time:

$$W = \frac{1 - E}{E\lambda} \tag{3.4.8}$$

Mean number of machines down L ($L = \lambda^* W$):

$$L = N(1 - E) \tag{3.4.9}$$

Mean number of machines waiting for repair L_q:

$$\begin{aligned} L_q &= N(1 - E) - O \\ &= N[1 - E(1 + x)] \end{aligned} \tag{3.4.10}$$

Naturally the appropriate performance measure must be selected for the system represented by the model. For example, in the steel mill system

$$\begin{aligned} \text{net production rate} &= \text{number of repair completions per unit time in model} \\ &= O\mu \end{aligned}$$

where the units would be in pits of steel rolled per hour, with $1/\mu$ the mean time to roll a pit of steel and $1/\lambda$ the mean time to heat a soaking pit filled with steel ingots.

Modifications to the Model

1a. Machines not identical but repair time the same for all machines. The results depend on the order in which failed machines are repaired. For FCFS (first come first served) or RO (random order) of repair and denoting by $P(i_1,i_2,\ldots,i_k)$ the probability machines $i_1,\ldots,i_k$ are failed

$$P(i_1,i_2,\ldots,i_k) = k!\frac{\prod_{u=1}^{k}\lambda_{i_u}}{\mu^k}P(0) \tag{3.4.11}$$

where λ_{i_u} is the failure rate of machine i_u and $P(0)$ is the probability no machine is failed. $P(0)$ must be determined from the requirement that all probabilities sum to 1.

1b. Number of repairmen $= s$.
It can be shown that

$$p_0 = \frac{1}{G(x,s,N)}$$

$$G(x,s,N) = \sum_{n=0}^{s-1} \binom{N}{n} x^n + \sum_{n=s}^{N} \frac{N!}{(N-n)!} \frac{s^s}{s!} \left(\frac{x}{s}\right)^n \tag{3.4.12}$$

$$E = \frac{G(x,s,N-1)}{G(x,s,N)}$$

$$O = \frac{Nx}{s} E$$

2. General distribution of time to failure. The results for O and hence E remain unchanged if λ is now interpreted as the inverse of the mean time to failure.[8]
3. General distribution of repair time. The above results are no longer valid; see Barlow[8] for the appropriate results. Ashcroft[9] has results for deterministic repair time. In general, the change from exponential to deterministic repair time does not make a large difference to the results.
4. Machines serviced according to some priority rule. Various results are available for different priority rules; for example, the repairman patrols the machines and serves the closest machine first.[10] Another set of models has considered what is the optimal order in which the repairman should serve the machines waiting for repair. If the machines are not identical, then the machine with the lowest failure rate should be served first.

Extensions of the Model

Suppose that there are s repairmen and that the maximum number of machines permitted to work at any instant is w. That is, there are $N - w$ spare machines that can be used in order to maintain production in the event of machine failure.

This system can be represented as a two-stage cyclic queue. Stage 1 corresponds to the in-service position and consists of w servers in parallel. Stage 2 corresponds to the repair position and consists of s servers in parallel. The customers are the machines that go from stage 1 (working) to stage 2 (under repair) and then back to stage 1. A customer queueing in front of stage 1 is a spare machine; a customer in front of stage 2 is a machine waiting repair.

This is a special case of a general cyclic queue model in which operation and service may go through a number of stages. It has been used as a model of a variety of systems:

Aero engines. These require service in a number of phases. w corresponds to the number of aircraft required to operate; s corresponds to the capacity of the service system.

Mining. Stage 1 corresponds to the loading point; stage 2 corresponds to the unload or dump point; and the customers are the trucks or other material handling devices.

3.4.10.2 Parallel and Standby Systems

The next group of breakdown models looks at the effect of using parallel or standby machines. Sometimes such an arrangement is used to increase production because the gross production rate of a single machine is insufficient. In other cases, while the gross production rate is adequate, the net production rate is not, so standby or other arrangements are used.

Parallel Machines

Let G_i be the gross production rate of machine i. The net production rate of machine i will be P_i. Then the efficiency of the set of parallel machines will be

$$E = \frac{\Sigma_{i=1}^N P_i}{\Sigma_{i=1}^N G_i} = \Sigma_{i=1}^N f_i E_i \tag{3.4.13}$$

where

$$f_i = \frac{G_i}{\Sigma_{i=1}^N G_i}$$

is the fraction of production capacity contributed by machine i and E_i is the efficiency of machine i.

Standby Machines

In a standby system, one machine is producing at a time. If it fails, then another machine, the standby machine, takes over. In normal standby, this machine would continue to operate until it fails and production is switched back to the first machine. In priority standby, switchover would occur as soon as the first machine is repaired.

Normal Standby. If the time to failure and the repair time are exponentially distributed with parameters λ and μ, then it can be shown that with identical machines and $x = \lambda/\mu$.

$$E = \frac{1 - x^2}{1 - x^3} \tag{3.4.14}$$

If repair time is constant, then

$$E = \frac{1}{x + e^{-x}} \tag{3.4.15}$$

Priority Standby. In this case, it is likely that the priority unit and the standby unit would have different parameters. Such an arrangement is used when the priority unit is in some sense better than the standby.[11] Thus if machine 1 is the priority machine and machine 2 the standby machine,

$$E = E_1\left(1 + \frac{\lambda_1/\lambda_2}{1 + \mu_1/\lambda_2 + \lambda_1/\mu_2}\right) \tag{3.4.16}$$

where E_1 is the efficiency of machine 1 on its own, and λ_i and μ_i are the failure rates and repair rates of machine i. This result is not sensitive to the form of the distributions of time to failure and repair time.

Splitting. Another possible arrangement is used when a given production rate from the parallel machines is required. When all parallel machines are available, then each machine will slow down. As machines fail, then the remaining machines will speed up production in order to maintain a constant overall production rate. If the failure rate of a machine depends on its production rate, then this arrangement will result in improved efficiency.[11]

3.4.10.3 Transfer Line Models

If the machines of a production system are connected in series so that work flows from one machine to the next in the line, then the failure of any machine will result in flow being interrupted.

If in a transfer line, where all jobs transfer simultaneously, there is no inventory bank between machines, then failure of one machine stops all other machines in the line at the next transfer instant. This linkage between machines has a critical effect on the overall performance of the system.

No Inventory Banks

The assumptions being made in the transfer line problem are the following:

1. Jobs transfer simultaneously if all machines are working. If any machines fail, no transfer can occur.
2. No jobs are scrapped. All jobs entering the line eventually complete processing.
3. Time to failure has a general distribution.
4. Time to repair has a general distribution.
5. A forced down machine cannot fail (operation-dependent failures).

Then

$$E = \frac{1}{1 + \sum_{i=1}^{N} x_i} \tag{3.4.17}$$

where

$$x_i = \frac{\text{mean repair time}}{\text{mean time to failure}}$$

If assumption (5) does not apply and a forced down machine *can* fail (time-dependent failures), then

$$E = \prod_{i=1}^{N} E_i \tag{3.4.18}$$

where

$$E_i = \frac{1}{1 + x_i}$$

Effect of Inventory Banks

Infinite Capacity Banks. If the line can be subdivided into sections with infinite capacity banks between them, then

$$E = \min_i E_i \tag{3.4.19}$$

where E_i is the efficiency of the section viewed as an isolated line.

Finite Banks. The zero and infinite banks results provide useful upper and lower bounds on the effect of finite banks on the line efficiency. However, to get more precise results it is necessary to make assumptions about the form of the distributions involved. Even then it is only in the simplest case of a two-stage (section) line that formula-type results exist.

Two-Stage Line. Let the following assumptions apply:

1. The line is divided into two stages or sections separated by an inventory bank of capacity z.
2. The line processes discrete parts with a constant cycle time between the beginning of successive transfers.
3. There is no scrapping. All parts entering the line eventually leave with processing complete.
4. The probability is constant of failure per cycle of a stage when it is processing parts (a_i for stage i).
5. The probability is constant of failure per cycle of a stage that is forced down because the other stage is failed and, if it is the downstream stage, the bank is empty, or if it is the upstream stage the bank is full (c_i for stage i).
6. The probability is constant of completion of repair per cycle of a stage (b_i for stage i).

The solution involves the following derived parameters:

$$A_1 = a_1 + a_2 - a_1a_2 - b_1a_2$$

$$A_2 = a_1 + a_2 - a_1a_2 - a_1b_2$$

$$B_1 = b_1 + b_2 - b_1b_2 - a_1b_2$$

$$B_2 = b_1 + b_2 - b_1b_2 - b_1a_2$$

$$C = (B_1A_2)/(A_1B_2)$$

$$x_1 = a_1/b_1, \qquad x_2 = a_2/b_2$$

$$r = \frac{(c_1 + b_1)(c_2A_1 + a_2B_1)}{(c_2 + b_2)(c_1A_2 + a_1B_2)}$$

The line efficiency for bank size z, $E(z)$, is given by

$$E(z) = \frac{1 - rC^z}{1 + x_1 - (1 + x_2)rC^z} \qquad x_1 \neq x_2$$

$$= \frac{1 + m - (c_2 + b_2) + (c_2 + b_2)z(1 + x)}{(1 + m)(1 + x) - (c_2 + b_2)(1 + x)^2 + (c_2 + b_2)(b_1 + b_2)/b_1b_2 + (c_2 + b_2)z(1 + x)^2}$$

$$x_1 = x_2 = x \qquad (3.4.20)$$

where

$$m = \frac{c_2 + b_2}{c_1 + b_1}$$

Operation-Dependent versus Time-Dependent Failures. If failures are operation-dependent, then $c_i = 0$; if they are time dependent, $c_i = a_i$. Thus if $x_1 = x_2 = x$ and the stages are identical ($a_1 = a_2 = a$, $b_1 = b_2 = b$), then the above formula for efficiency becomes

$$E(z) = \frac{2 - b + bz(1 + x)}{2(1 + 2x) + b(z - 1)(1 + x)^2} \qquad (3.4.21)$$

for operation-dependent failures and

$$E(z) = \frac{2 - b(1 + x) + bz(1 + x)^2}{2(1 + x)^2 + b(z - 1)(1 + x)^3} \qquad (3.4.22)$$

for time-dependent failures.

Table 3.4.1 shows the difference between the two formulae for the case where $x = 0.5$ and $b = 0.1$. It can be seen that at small z, there can be a significant difference in the predictions made by the two formulae. Thus, it is necessary to decide which is the more reasonable assumption in the given context. Usually operation-dependent failures would be the appropriate assumption.

Other Two-Stage Line Formulae

DETERMINISTIC REPAIR TIME. Formulae only exist if the repair times for the two stages are identical and

1. bank capacity z is an integer multiple of the repair time T in cycles (i.e., $z = LT$ where L is an integer);
2. bank capacity z is less then the repair time in cycles ($z < T$).

In case (1), the formula involves the following derived parameters:

$$r^* = \frac{a_2(1 - a_1)}{a_1(1 - a_2)}$$

$$D = \frac{a_2(1 - a_1)^{T+1}}{a_1(1 - a_2)^{T+1}}$$

Table 3.4.1. Comparison of Operation-Dependent and Time-Dependent Failure Assumption ($x = 0.5$, $b = 0.1$)

	Efficiency $E(z)$	
z	Operation-Dependent Failure	Time-Dependent Failure
0	0.50	0.44
5	0.54	0.51
10	0.56	0.54
20	0.59	0.58
50	0.63	0.62
100	0.64	0.64
∞	0.67	0.67

For operation-dependent failures, then,

$$E(z = LT) = \frac{1 - r^*D^L}{1 + a_1T - (1 + a_2T)\, r^*D^L}, \qquad a_1 \neq a_2$$
$$= \frac{1 + L(1 + aT)}{1 + 2aT - a^2T + L(1 + aT)^2}, \qquad a_1 = a_2 = a \tag{3.4.23}$$

Note that when the operation-dependent formulae for geometric and deterministic repair times are compared, it can be seen that a given bank size is twice as effective in improving efficiency with deterministic repair time; that is,

$$E_D(z) \approx E_G(2z)$$

where the suffixes D and G denote deterministic and geometric repair times, respectively.

It follows that it is desirable to reduce the variability of repair times in order to improve the effectiveness of a given bank.

The formulae for case (2) ($0 < z < T$) are given by Dudick.[12]

DETERMINISTIC TIME TO FAILURE AND GEOMETRIC REPAIR TIME. If the stations are identical, it is possible to obtain formula-type solutions. It can be shown that two alternative formulae exist. The difference between them arises because with deterministic time to failure, it is possible that the two stages cannot ever be simultaneously under repair. If when the bank inventory is zero, the downstream station has an age (working time since last failure) that is M cycles greater than the age of the upstream station and $z < M < S$ (with S equal to the time to failure of a station), then the stations will never be simultaneously down.

If the repair probability per cycle is b, then

$$E(z) = \frac{1 + (z - 1)b}{(1 - b)(1 + 2/Sb) + zb(1 + 1/Sb)} \tag{3.4.24a}$$

if simultaneous failure is not possible and

$$E(z) = \frac{2 + (z - 1)b}{(2 - b)(1 + 2/Sb) + zb(1 + 1/Sb) - 1/Sb} \tag{3.4.24b}$$

if simutaneous failure is possible.

SINGLE REPAIRMAN. If there is only one repairman for the whole line, then it is possible that, as in the machine interference problem, both stages can be down and only one can be under repair. Thus it is necessary to specify which stage the repairman would work on if both are down.

A variety of assumptions are possible. One reasonable possibility is that the repairman would start repairing any failure as soon as it occurred and would continue on a repair until it was complete. However, it is possible for both stages to fail in the same cycle. It then would be necessary to specify on which stage the repairman would work first. He could choose a stage at random, or he could choose a stage depending on the inventory level in the bank. For example, if the bank were almost empty, it would seem better to work on the upstream stage. If he worked on the downstream stage, then once repair was complete, the stage would soon be forced down by the empty bank, and little benefit would be obtained by the repair. Working on the upstream stage would enable extra production to occur once its repair was complete.

Formulae have been obtained for such a situation with inventory level y such that if both stages fail in the same cycle and the inventory is at or below y, then the repairman works on the upstream stage. If the inventory is above y, then he works on the downstream stage.[13]

An alternative single repairman model assumes that the repairman begins repair of a stage as soon as it fails. However, if the other stage fails, then he may switch to it, depending on the inventory level in the bank. If the level is above some level y', he repairs the downstream stage; if it is at or below y', he repairs the upstream stage. The relevant formulae are given by Dudick.[12]

Other Two-Stage Line Models. Apart from the models outlined earlier, there are virtually no other models of two-stage lines processing discrete parts that give formula-type solutions. For other models the usual approach is to develop a Markov chain model giving an equation for each possible state and then to solve the equations directly using a standard routine for solving simultaneous equations. See Okamura and Yamashina[14] for an example of the use of this approach. Because the equations of the Markov model usually have a special structure with many zero elements in the coefficient matrix, recent research has attempted to develop procedures for solving the equations that take account of their structure (the so-called matrix geometric approach[15]).

Shanthikumar and Tien[16] describe the application of the matrix geometric approach to a transfer line where the work piece in process at a stage when failure occurs may be scrapped.

Three-Stage Line Models. While it is possible to develop Markov models of a three-stage line, these models do not have a closed-form solution. This means that a variety of other approaches have to be attempted. The number of states in the Markov chain model of the three stage line is approximately $8(z_1 + 2)(z_2 + 2)$, where z_1 and z_2 are the bank capacities. Thus one approach is to solve these equations directly, and this is appropriate if the bank capacities are small or a large computer is available (see Haussman and Buzacott[17] for an example of this approach).

Another approach is to use a method that exploits the structure of the equations in some way. Because inventory levels in a bank can only change by ± 1, the coefficient matrix is sparse and has a regular structure. One such method was proposed by Gershwin and Schick,[18] but it converts the set of linear equations into min $[4(z_1 + z), 4(z_2 + 2)]$ nonlinear equations so the method does not seem to yield significant computational benefits. While other methods of exploiting the structure exist, they have not been applied to three-stage transfer lines.

Rather than seeking numerical methods, the main approach has been to seek approximate methods that can be extended to more than three stages. One approach is the "loss transfer" method of Sevastyanov.[19] This method assumes that the stage repair times are identical and geometrically distributed with repair rate b per cycle. Stage failure rates are a_1, a_2, a_3 per cycle.

Then the three-stage line could be viewed as a two-stage line in two ways: from bank 1 and from bank 2. Viewed from bank 1, the upstream stage has failure rate $a_1' = a_1$ and the downstream stage has failure rate a_2'' given by

$$a_2'' = a_2 + a_3\delta_{32}$$

Viewed from bank 2, the upstream stage has failure rate a_2' given by

$$a_2' = a_2 + a_1\delta_{12}$$

and the downstream stage has failure rate $a_3'' = a_3$.

The δ_{ij} are the loss transfer coefficients and measure the fraction of stoppages of stage i which result in stoppage of stage j, where $j = i \pm 1$. If $j = i + 1$, then stage j is stopped because the bank between stages i and $i + 1$ is empty; if $j = i - 1$, stage j is stopped because the bank between stages $i - 1$ and i is full. The δ_{ij} can be obtained from the solution to the two-stage model. Using the same symbols as Eq. 3.4.20,

$$\delta_{(i-1)i} = \begin{cases} \dfrac{(1 - r)}{1 - rC^{z_{i-1}}}, & x_1 \neq x_2 \\[2ex] \dfrac{2 - b}{2 - b + z_{i-1}b(1 + x)}, & x_1 = x_2 = x \end{cases} \tag{3.4.25}$$

$$\delta_{(i+1)i} = \begin{cases} \dfrac{C^{z_i}(1 - r)}{1 - rC^{z_i}}, & x_1 \neq x_2 \\[2ex] \dfrac{2 - b}{2 - b + z_ib(1 + x)}, & x_1 = x_2 = x \end{cases} \tag{3.4.26}$$

where in calculating r and C the values of a_1 and a_2 to be used are the apparent failure rates of the upstream and downstream stages, respectively; that is, in calculating $\delta_{(i-1)i}$, set $a_1 = a_{i-1}'$ and $a_2 = a_i''$, while in calculating $\delta_{(i+1)i}$, set $a_1 = a_i'$ and $a_2 = a_{i+1}''$.

An iterative method can be used to solve the equations and determine the line performance. Let $\delta_{12}(j)$, $\delta_{32}(j)$, $a_2'(j)$, $a_2''(j)$ be the values of δ_{12}, δ_{32}, a_2', a_2'' calculated at step j of the iterative procedure.

PROCEDURE

STEP 0. Set $a_2''(0) = a_2$.

STEP $2j-1$. Consider the two-stage line seen by bank 1. Calculate $\delta_{12}(2j-1)$ using the formula for the two-stage line with $a_1 = a_1$ and $a_2 = a_2''(2j - 2)$. Then set

$$a_2'(2j - 1) = a_2 + a_1\delta_{12}(2j - 1).$$

STEP $2j$. Consider the two-stage line seen by bank 2. Calculate $\delta_{32}(2j)$ using the above formula for the two-stage line with $a_1 = a_2'(2j - 1)$ and $a_2 = a_3$. Then set

$$a_2''(2j) = a_2 + a_3\delta_{32}(2j).$$

Continue until the differences $a_2'(2j-1) - a_2'(2j-3)$ and $a_2''(2j) - a_2''(2j-2)$ are less than some allowed ϵ.

The line efficiency can then be found from the two-stage line efficiency formula either by considering the two-stage line seen by bank 1 with $a_1 = a_1$, $a_2 = a_2''$ or by considering the two-stage line seen by bank 2 with $a_1 = a_2'$, $a_2 = a_3$. In practice, convergence is very rapid and the method seems to give good approximations.

Extension to More Stages. The basic idea of the above approximation can be extended to more stages if the repair times at each stage are identical and geometrically distributed. Again the line is viewed as a two-stage line from each inventory bank. Suppose bank j is between stages j and $j+1$. Viewed from this bank the upstream stage has failure rate a_j' and the downstream stage has failure rate a_{j+1}''. Then

$$a_j' = a_j + a_{j-1}'\delta_{j-1j}$$

$$a_{j+1}'' = a_{j+1} + a_{j+2}''\delta_{j+2j+1} \qquad (3.4.27)$$

Also, there are the formulae for δ_{j-1j} and δ_{j+2j+1} given above. With M banks, there are thus $2M$ equations defining the a' and a'' and $2(M-1)$ equations defining the δs. Solving the equations by an iterative scheme based on the above, will give the a' and a''.

The line efficiency can then be found from the two-stage formula with the line divided at any bank j into two stages with $a_1 = a_j'$, $a_2 = a_{j+1}''$. For the extension of this approach to lines where the stages have unequal repair rates, see Gershwin.[20]

Optimal Line Division

Sevastyanov[19] showed that when the loss transfer method is used to calculate line efficiency, the optimal line division with M banks must be such that

$$a_j' = a_{j+1}'' \qquad \text{for } 1 \leq j \leq M$$

With all bank capacities equal, at the optimum the division will be such that

$$a_2 = a_3 = \cdots = a_M = a_1(1-\delta) = a_{M+1}(1-\delta) \qquad (3.4.28)$$

That is, the first and last stages should have higher failure rate.

3.4.11 MODELS OF TASK TIME VARIABILITY

Another source of disturbances is variability in the time required to perform a task. This can arise for a variety of reasons. If tasks are performed by human operators, then there will be an inherent variability in the time required to do them. Alternatively, due to the variety of jobs, each with its own processing time, it may be appropriate to treat the set of jobs as having some probability distribution of task time.

Most solvable models of the effect of variability in task time assume that the time at a machine or work station has an exponential distribution, that is, probability density $f(t) = \mu_i e^{-\mu_i t}$ where $1/\mu_i$ is the mean time at station i.

3.4.11.1 No Limit on In-Process Inventories

This system is represented by a queueing network with jobs queueing at servers that represent the machines and work stations. The classical job shop model is due to Jackson.[3] Assumptions follow:

1. There are M machines.
2. All jobs processed by machine i have an exponential service time with parameters μ_i.
3. Jobs are served at machine i using FCFS queue discipline.
4. There is no limit on the number of jobs allowed in the system.
5. There are no limits on the size of queue at any machine.
6. Jobs arrive at the system with exponential interarrival time with mean time between arrivals $1/\lambda$.
7. The first operation of an arriving job is at machine i with probability q_i.
8. The routing of jobs between machines is described by a routing matrix P with element p_{ij} = probability that on completion of a job at machine i its next operation will be at machine j.

The idea behind assumption (8) is to aggregate the routes of all jobs through the system and determine the p_{ij} by looking at the frequency of machine sequence ij.

The model determines the probability of a given set of queue sizes at the machines, that is, $\Pr(n_1,n_2, \ldots, n_M)$ where n_i is the queue size at machine i.

$$\Pr(n_1,n_2, \ldots, n_M) = \prod_{i=1}^{M} (\lambda a_i)^{n_i} \Pr(0,0, \ldots, 0) \quad (3.4.29)$$

where $a_i = e_i/\mu_i$ = expected total time a job will spend in process at machine i, and e_i = number of visits of a job to machine i.

The e_i are found by solving the equations

$$e_i = q_i + \sum_{j=1}^{M} p_{ji} e_j \quad (3.4.30)$$

Also,

$$\Pr(0,0, \ldots, 0) = \prod_{i=1}^{M} (1 - \lambda a_i) \quad (3.4.31)$$

Solutions of the form above are known as product-form solutions, since

$$\Pr(n_1,n_2, \ldots, n_M) = Q_1(n_1) \cdot Q_2(n_2) \cdot \ldots \cdot Q_M(n_M) \cdot K \quad (3.4.32)$$

Other parameters of interest can be found, for example,

$$U_i = \text{utilization of machine } i = \lambda a_i$$

$$L = \text{expected number of jobs in system} \quad (3.4.33)$$

$$= \sum_{i=1}^{M} \frac{\lambda a_i}{1 - \lambda a_i} = \sum_{i=1}^{M} L_i$$

where L_i is the expected number of jobs at machine i. Also

$$W = \text{expected time a job spends in the system} = \sum_{i=1}^{M} \frac{a_i}{1 - \lambda a_i} = \sum_{i=1}^{M} e_i W_i \quad (3.4.34)$$

where W_i is the expected time a job spends at machine i. And

$$\text{PC} = \text{production capacity, that is, the maximum arrival rate such that the system will be able to process all arriving jobs} \quad (3.4.35)$$

$$= \min_i 1/a_i$$

3.4.11.2 Extensions of the Model

State-Dependent Arrival and Service Rates

If the arrival rate is a function of the total number of jobs in the system, that is, $\lambda(N)$, and the service rate at machine i is a function of queue length n_i, that is, $\mu_i(n_i)$, then a product-form solution still applies.

$$\Pr(n_1,n_2, \ldots, n_M) = \prod_{j=1}^{\Sigma n_i} \lambda(j - 1) \prod_{i=1}^{M} e_i^{n_i} \prod_{r=1}^{n_i} (1/\mu_i(r)) \Pr(0,0, \ldots, 0) \quad (3.4.36)$$

$\Pr(0,0, \ldots, 0)$ must be found from the normalizing condition that the sum of all state probabilities equals 1.

This model includes a special case in which there are multiple servers at machine or station i; for example, if there are s servers each with service rate μ_i

$$\mu_i(r) = r\mu_i, \qquad 0 < r < s$$
$$\mu_i(r) = s\mu_i, \qquad r \geqslant s \quad (3.4.37)$$

Multiple-Job Classes

Suppose that there are t job classes and that the routing of class k job is described by a routing matrix P^k with p_{ij}^k being the probability that on completion of a class k job at machine i, it goes next to machine j. Let q_i^k be the probability that an arriving job is class k with its first operation at machine i.

Then determine e_i^k, the expected number of visits of a job to machine i with class k from

$$e_i^k = q_i^k + \sum_{j=1}^{M} p_{ji}^k e_j^k \tag{3.4.38}$$

and

$$e_i = \sum_{k=1}^{t} e_i^k \tag{3.4.39}$$

If all job classes have the same exponentially distributed time at machine i, with parameter μ_i, then the basic product-form solution still applies.

3.4.11.3 Relaxing the Assumptions

2′. There is general service time distribution.

An approximate approach is described in Shanthikumar and Buzacott.[21] This approximation works best if the squared coefficient of variation of the service time is less than one and the routing of jobs is fairly random.

3′. There are other scheduling rules besides FCFS.

The approximate approach in Shanthikumar and Buzacott[21] also enables other queueing disciplines to be examined, for example, shortest processing time (SPT), provided a single server queue model for the queue discipline exists.

4′. There is a limit on the number of jobs allowed in the system.

The state-dependent arrival rate model allows a special case to be readily treated. If C is the maximum number of jobs allowed in the system, then the product-form solution applies by setting $\lambda(N) = 0$ for $N \geqslant C$. However, this means that if jobs are arriving from some source with rate λ, irrespective of the number in this system, all jobs that arrive when $N \geqslant C$ are in fact assumed to be lost (or go to some other system to be processed).

An alternative approach to modeling a limit on the number of jobs allowed in the system is to assume that the system operates with the number of jobs constrained to be always equal to C. This gives rise to a closed queueing model.

3.4.11.4 Closed Queues

In this case the assumptions above apply except that assumption (6) becomes

6′. There are no external job arrivals.

The solution is then

$$\Pr(n_1, n_2, \ldots, n_M) = \prod_{i=1}^{M} (e_i/\mu_i)^{n_i} \Pr(0,0,\ldots,0) \tag{3.4.40}$$

with e_i given by the solution to

$$e_i = \sum_{j=1}^{M} e_j p_{ji} \tag{3.4.41}$$

where it will be necessary to set one e_i equal to 1; that is, set $e_1 = 1$.

$\Pr(0,0,\ldots,0)$ is determined by the normalizing conditions that the sum of all $P(n_1, \ldots, n_M)$ such that $\Sigma n_i = C$ must be 1.

Again the extensions to state-dependent service rates and multiple-job classes apply.[22]

3.4.11.5 Limits on Inventory at Machines

If the space or inventory at a machine is limited, then a variety of results exist.

Two-Stage Flow Line

If μ_1, μ_2 are the processing rates of the two machines (task times assumed to have an exponential distribution) and z the capacity of the inventory bank, then[2]

$$\Pr(n) = \left(\frac{\mu_1}{\mu_2}\right)^{n+1} \Pr(I), \qquad 0 \leq n \leq z \tag{3.4.42}$$

where $\Pr(I)$ is the probability that the bank is empty and the second machine is idle. The probability the bank is full and the first machine is blocked, $\Pr(B)$, is given by

$$\Pr(B) = \left(\frac{\mu_1}{\mu_2}\right)^{z+2} \Pr(I) \tag{3.4.43}$$

It follows that the production capacity $PC(z)$ is given by

$$\begin{aligned} PC(z) &= \mu_1(1 - \Pr(B)) = \mu_2(1 - \Pr(I)) \\ &= \begin{cases} \mu_1 \dfrac{(1 - (\mu_1/\mu_2)^{z+2})}{(1 - (\mu_1/\mu_2)^{z+3}}, & \mu_1 \neq \mu_2 \\ \mu \dfrac{z+2}{z+3}, & \mu_1 = \mu_2 = \mu \end{cases} \end{aligned} \tag{3.4.44}$$

Multistage Flow Line

Two approaches can be used. One is to set up a Markov process model of the line and solve the resultant set of equations; however, the number of equations is of order $\Pi_{i=1}^{M-1} (z_i + 2)$, where z_i is the capacity of the bank between stages i and $i + 1$ ($i = 1,2, \ldots, M - 1$). An alternative approximate approach due to Hillier and Boling[21] is to view the multistage line as a two-stage line for each inventory bank. For bank i, the upstream stage will have apparent processing rate μ_i' given by $\mu_i' = \mu_i[1 - P_{i-1}(I)]$, where $P_{i-1}(I)$ is the probability that the downstream stage is blocked in the two-stage line divided at bank $i - 1$. Again, for bank i the downstream stage will have apparent processing rate μ_{i+1}'' given by $\mu_{i+1}'' = \mu_{i+1}[1 - P_{i+1}(B)]$, where $P_{i+1}(B)$ is the probability that the upstream stage is blocked in the two-stage line divided at bank $i + 1$.

$P_i(I)$ and $P_i(B)$ can be calculated from the two-stage line formula as

$$P_i(I) = \begin{cases} \dfrac{1 - \mu_i'/\mu_{i+1}''}{1 - (\mu_i'/\mu_{i+1}'')^{z+3}}, & \mu_i' \neq \mu_{i+1}'' \\ \dfrac{1}{z+3}, & \mu_i' = \mu_{i+1}'' \end{cases} \tag{3.4.45}$$

$$P_i(B) = \begin{cases} \dfrac{(\mu_i'/\mu_{i+1}'')^{z+2}\,(1 - \mu_i'/\mu_{i+1}'')}{1 - (\mu_i'/\mu_{i+1}'')^{z+3}}, & \mu_i' \neq \mu_{i+1}'' \\ \dfrac{1}{z+3}, & \mu_i' = \mu_{i+1}'' \end{cases} \tag{3.4.46}$$

Hillier and Boling[23] observe that the flow through each inventory bank of the line will be the same and equal to, say, R. Thus alternatively, one can set

$$\mu_{i+1}'' = \frac{R}{1 - P_i(I)}$$

$$\mu_M'' = \mu_M$$

and

$$R \doteq \mu_M\,[1 - P_{M-1}(I)]$$

This method gives good prediction of the performance of a multistage flow line.

Two Machine Job Shops

If the jobs can follow random routings between the machine, then with finite inventory banks at the machines, it is necessary to take account of the external store where jobs will be kept waiting for space in the queue at the machine for their first operation. If $z_1 = z_2 = Z$, then the maximum number of jobs in the system is $2Z$. It can be shown that it is then desirable to restrict the maximum number of jobs to be less than $2Z$ in order to reduce the likelihood of blocking, that is, forcing a machine down because there is no space in the queue at the other machine for the job just completed service. If the maximum number of jobs allowed in the system is h, then it can be shown that maximum capacity is achieved when $h = z + 1$. For a randomly routed job shop, that is, $p_{12} = p_{21} = \frac{1}{2}$, $q_1 = q_2 = \frac{1}{2}$, with identical machines and FCFS queue discipline,

$$\begin{aligned} \mathrm{PC}(h) &= \mu \frac{h}{h+1}, & 0 < h \leqslant z \\ \mathrm{PC}(h) &= \mu \frac{4z - 2h + 3}{4z - 2h + 4}, & z + 1 \leqslant h \leqslant 2z \end{aligned} \qquad (3.4.47)$$

PC(h) is a maximum at $h = z + 1$, and it is then equal to $\mu(2z + 1) / (2z + 2)$.

It can be shown that the production capacity is increased through the use of alternative release rules from the queue of jobs waiting entry into the system. The optimum rule based on information on queue lengths at machines is the idle machine rule: release jobs only when the machine for the first operation on the job would otherwise be idle. It can be shown that for the above parameters

$$\mathrm{PC}(z) = \mu[1 - (1/2)^{z+1}] \qquad (3.4.48)$$

Multimachine Job Shops

No results are available for multimachine job shops except some where a different blocking mechanism is assumed. If, instead of having blocked job force the machine down (block and hold), the blocked job is returned to the beginning of the queue at the machine (block and recirculate), then for certain special routing matrixes, closed-form product-form solutions exist.[24]

3.4.11.6 Flexible Manufacturing System Models

Flexible manufacturing systems (FMS) are essentially automated job shops with the following special features:

1. Limited storage at machines (often no storage space for waiting jobs).
2. Central storage for work in process.
3. Combined with (2) may be storage for arriving jobs not yet processed at any machine.
4. Limited number of pallets to hold work in process.

Various approaches have been suggested for modeling FMS.[24,25] There are two main types of models, the closed queueing and open queueing networks.

Closed Queueing Network

The closed queueing network has an equal number of customers and pallets. The servers are the machine or work centers of the FMS; in addition, material handling between machines may be represented by one or more servers. The best-known approach is CAN-Q due to Solberg.[26]

Open Queueing Network

The open queueing network has a limited number of customers in the system with an external queue of jobs waiting release to the system. Were it not for the external queue, the system could be modeled as an open queueing network with a limit on the number of customers allowed in it. An approximate method is to analyze the system, including the queue of jobs waiting release, as a closed queue in which departures are immediately fed back to the beginning of the release queue.

Once PC(n) is determined for n customers in the closed queue, then it can be shown that

$$\mathrm{PC} = \lim_{n \to \infty} \mathrm{PC}(n)$$

and the other parameters, such as expected total number of jobs in the FMS plus release queue, can be found from analyzing a single server queue with state-dependent service rate $\mu(n) = \mathrm{PC}(n)$.[27]

The block and recirculate models by Buzacott and Yao[24] can also be used for some routing matrices. Approximate methods for the case of general service time distributions are also described by Buzacott and Yao.[24]

3.4.12 QUALITY MODELS

3.4.12.1 Assembly Process

If an assembly is made up of s parts and q_i is the probability part i is not defective, then the probability that the assembly is not defective q is given by

$$q = \prod_{i=1}^{s} q_i \tag{3.4.49}$$

Alternatively, if there are s stations on an assembly line and at each station a part is added, with probability of no errors in the assembly process at station i being q_i, then the yield of the line will be

$$Y = \prod_{i=1}^{s} q_i \tag{3.4.50}$$

3.4.12.2 Assembly Line with Rework

Consider a line where the probability of no error at station i is q_i. Suppose that when an error occurs, rework is necessary and the duration of rework is τ_i, where τ is geometrically distributed. Then this system can be modeled as a multistage transfer line in which the failure probability at station i is $a_i = 1 - q_i$ and the repair probability at station i is $b_i = 1/\tau_i$.

Assembly Line with Scrapping

Instead of rework, suppose that when an error occurs, the assembly is scrapped. However, scrapping of the assembly takes extra time τ_i, which is geometrically distributed. In the two-station case, the transfer line with scrapping on failure model due to Shanthikumar and Tien[16] can be used. Such a model would be appropriate for an automatic assembly machine.

3.4.12.3 Test and Rework Loops

Suppose assemblies arrive at a test station with rate R. Let f be the fraction that are defective. Suppose the defective assemblies go to a rework station and then back to the test station. If the rework is perfect, then the total load on the test station will be $L = R + Rf$ and the yield of the test station Y, defined as the number of good assemblies leaving it, divided by total load L will be

$$Y = \frac{R}{R + Rf} = \frac{1}{1 + f} \tag{3.4.51}$$

Note that $L = R/Y$.

Suppose instead that there is probability p that the rework station will be able to fix the defect. This means that an assembly will go around the loop several times. Then

$$\begin{aligned} L &= R + Rf + Rf(1 - p) + Rf(1 - p)^2 + \cdots \\ &= R + Rf/p \end{aligned}$$

and

$$Y = \frac{R}{R + Rf/p} = \frac{1}{1 + f/p} \tag{3.4.52}$$

Note that if $p = 1 - f$, that is, the rework process is no better than the original process (possible if rework means replacing defective components by the same quality of components originally installed), then $Y = 1 - f = p$.

3.4.13 OTHER USES OF PRODUCTION MODELS

The models described in this chapter have all focused on calculating efficiency and yield. Production models also can be used for optimization, that is, finding what combination of parameters yield maximum production rate or minimum cost. It is also possible to develop models that can determine the optimal method of controlling an operation of a production system, but as yet few such models exist. Some very simple models are described by Buzacott.[13] Kimemia[28] has a much more complex model for determining the optimal control of release of work to an FMS conditional or the inventory levels of the parts made by it.

REFERENCES

1. C. Palm. "Assignment of Workers in Servicing Automatic Machines." *Journal of Industrial Engineering*, **9**, 28–42 (1958). Originally appeared in Swedish in *Industritidningen Norden*, **75**, 75–80, 90–94, 119–123 (1947).
2. G. C. Hunt. "Sequential Arrays of Waiting Lines." *Operations Research*, **4**, 674–683 (1956).
3. R. R. P. Jackson. "Random Queueing Processes with Phase Type Service." *Journal of the Royal Statistical Society, Series B*, **18**, 129–132 (1956).
4. J. R. Jackson. "Networks of Waiting Lines." *Operations Research*, **5**, 518–521 (1957).
5. A. P. Vladzievskii. "The Probability Law of Operation of Automatic Transfer Lines and Internal Storage in Them." *Avtomatika i Telemekhanika*, **13**, 227–281 (1952).
6. J. A. Buzacott and J. R. Callahan. "The Capacity of the Soaking Pit-Rolling Mill Complex in Steel Production." *INFOR*, **9**, 87–95 (1971).
7. D. R. Cox and W. L. Smith. *Queues*. London: Methuen, 1961.
8. R. E. Barlow. "Repairman Problems." In K. J. Arrow, S. Karlin, and H. E. Scarf, Eds., *Studies in Applied Probability and Management Science*. Palo Alto: Stanford University Press, 1962.
9. H. Ashcroft. "The Productivity of Several Machines under the Care of One Operator." *Journal of the Royal Statistical Society, Series B*, **12**, 145–151 (1950).
10. G. H. Reynolds. "An M/M/m/n Queue for the Shortest Distance Priority Machine Interference Problem." *Operations Research*, **23**, 325–341 (1975).
11. J. A. Buzacott. "Prediction of the Efficiency of Production Systems without Internal Storage." *International Journal of Production Research*, **6**, 173–188 (1968).
12. A. Dudick. "Fixed-Cycle Production Systems with In-Line Inventory and Limited Repair Capability." Doctor of Engineering Science thesis, Columbia University, New York (1979).
13. J. A. Buzacott. "Optimal Operating Rules for Automated Manufacturing Systems." *IEEE Transactions on Automatic Control*, **AC-27**, 80–86 (1982).
14. K. Okamura and H. Yamashina. "Analysis of the Effect of Buffer Storage Capacity in Transfer Line Systems." *AIIE Transactions*, **9**, 127–135 (1977).
15. M. F. Neuts. *Matrix-Geometric Solutions in Stochastic Models—An Algorithmic Approach*. Baltimore: Johns Hopkins University Press, 1981.
16. J. G. Shanthikumar and C. C. Tien. "An Algorithmic Solution to Two Stage Transfer Lines with Possible Scrapping of Units." *Management Science*, **29**, 1069–1086 (1983).
17. W. Haussman and J. A. Buzacott. "Analysis of a Three-Stage Production Line with Two Finite Buffer Storages." Paper presented at ORSA/TIMS Meeting, Chicago, April 1983.
18. S. B. Gershwin and I. C. Schick. "Modelling and Analysis of Three-Stage Transfer Lines with Unreliable Machines and Finite Buffers." *Operations Research*, **31**, 354–380 (1983).
19. B. A. Sevastyanov. "Influence of Storage Bin Capacity on the Average Standstill Time of a Production Line." *Theory of Probability Applications*, **7**, 429–438 (1962).
20. S. B. Gershwin. *An Efficient Decomposition Method for the Approximate Evaluation of the Production Lines with Finite Storage Space*. Report LIDS-P-1308, Laboratory for Information and Decision Systems, Massachusetts Institute of Technology, Cambridge, Massachusetts, July 1983.
21. J. G. Shanthikumar and J. A. Buzacott. "Open Queueing Network Models of Dynamic Job Shops." *International Journal of Production Research*, **19**, 255–266 (1981).
22. F. Baskett, K. M. Chandy, R. R. Muntz, and R. G. Palacios. "Open, Closed and Mixed Networks of Queues with Different Classes of Customers." *Journal of the ACM*, **22**, 248–260 (1975).
23. F. S. Hillier and R. W. Boling. "Finite Queues in Series with Exponential or Erlang Service Time—A Numerical Approach." *Operations Research*, **15**, 286–303 (1967).
24. J. A. Buzacott and D. D. W. Yao. "On Queueing Network Models of Flexible Manufacturing Systems." *Queueing Systems: Theory and Applications*, **1**, 5–27 (1986).

25. J. A. Buzacott and D. D. W. Yao. "Flexible Manufacturing Systems—A Review of Analytical Models." *Management Science*, **32**, 890–905 (1986).

26. J. J. Solberg. "A Mathematical Model of Computerized Manufacturing Systems." *Proceedings of the 4th International Conference on Production Research*, Tokyo, 1977, 22–30.

27. J. A. Buzacott and J. G. Shanthikumar. "Models for Understanding Flexible Manufacturing Systems." *AIIE Transactions*, **12**, 339–350 (1980).

28. J. G. Kimemia. "Hierarchical Control of Production in Flexible Manufacturing Systems." Ph.D. thesis, Massachusetts Institute of Technology, Cambridge, Massachusetts, 1982.

CHAPTER 3.5

CONTROL MODELS FOR PRODUCTION SYSTEMS

SHIMON Y. NOF
THEODORE J. WILLIAMS

Purdue University
West Lafayette, Indiana

3.5.1 INTRODUCTION

Control is the fundamental engineering and managerial function whose major purpose is to measure, evaluate, and adjust the operation of a process, machine, or system under dynamic conditions so that it achieves desired objectives within its planned specifications under cost and safety considerations. A well-planned system can perform well without any control only as long as no variations are encountered in its own operation and in its environment. In reality, however, many changes occur over time. Machine breakdown, human error, variable material properties, and faulty information are a few examples of why systems must be controlled.

When a system is more complex and there are more potential sources of dynamic variations, a more complicated control is required. Particularly in automatic systems, in which human operators are replaced by machines and computers, a thorough design of control responsibilities and procedures is necessary. Control activities include automatic control of individual numerical control (NC) machines, materials handling equipment, and manufacturing processes. They also include control of production, inventory, quality, labor performance, and cost. Careful design of correct and adequate controls that continually identify and trace variations and disturbances, evaluate alternative responses, and result in timely and appropriate actions is therefore vital to the successful operation of a system.

The purpose of this chapter is to describe control models and explain their engineering use in the study and design of control systems. These models represent temporal relationships between the input and the resulting output or state of a system. Whether they specify the control functions in a small, self-contained mechanism or organization or in a multiechelon, distributed organization, they are based on common principles of control. Therefore, explanations and examples of control principles throughout the chapter, which, for clarity, may refer specifically to a mechanical system, should be considered true for organizational systems as well. As their name implies, control models describe the control activity, and their purpose is to predict the behavior of the system being modeled over a wide range of conditions.

3.5.1.1 Why Use Control Models?

There are three main motivations for using control models. First, in the design of a production system and its controls, prediction of performance is necessary in order to evaluate alternative proposals and select the best among them. A control model provides designers with a tool *to analyze control methods and requirements and to estimate the impact and economics of the control system.* Use of control models will guide designers in structuring the designed system and in specifying its control system. For instance, analysis of a model of the control system for a machine will yield the specification of characteristics of the controller for that machine. It may also have an impact on the design of the machine itself, for example, by pointing to circumstances under which the machine can or cannot operate safely. Similarly, a manual control model that represents the man–machine performance of an aircraft can be applied to design the aircraft such that its performance criteria are met while under pilot control.

The authors gratefully acknowledge permission by the publisher and editor of the *Handbook of Industrial Engineering* (G. Salvendy, Ed.), Wiley-Interscience, 1982, to use their Chapter 13.10, "Control Models," from which this chapter has been developed.

Second, during the operation of a working production system, control models can *serve as tools in the actual control activity*. The ability of the models to predict the output state of a system can now be applied to particular actual conditions as they occur. For instance, in controlling an automated production facility, different control policies may have to be compared periodically, often just one step before operations are being performed. Each time one or several control models may have to be analyzed with the actual current state of the facility in order to decide which policy is optimal.

The third motivation for using control models is the need *to investigate the control ability and attributes of systems and their components*. Although such an investigation is closely related to the previous two items, control design and control implementation, it is viewed as a more research-oriented task. Of interest here are fundamental attributes such as decision logic, sensing, response time, and control quality, which characterize various types of systems. Good examples are manual control models, whose purpose is to investigate and establish the capabilities of humans as controllers.

3.5.2 DEFINITIONS, TERMINOLOGY, AND TOOLS

3.5.2.1 Fundamentals of Control

Automatic control, as the term is commonly used, is "self-correcting," or feedback, control; that is, some control instrument is continuously monitoring certain output variables of a controlled process and is comparing this output with some preestablished desired value. The instrument then compares the actual and desired values of the output variable. Any resulting error obtained from this comparison is used to compute the required correction to the control setting of the equipment being controlled. As a result, the value of the output variable will be adjusted to its desired level and maintained there. This type of control is known as a *servomechanism*.

At this point the reader should distinguish servomechanism control from common on-off control. An on-off control acknowledges the presence or absence of an error, with no concern for the magnitude of that error, to govern equipment. For example, a thermostatic control of a home refrigerator or furnace is an on-off operation. The design and use of a servomechanism control system, which is an essential component of any truly automatic operation of complicated, modern production facilities, requires a knowledge of every element of the control loop. For example, in Figure 3.5.1, the engineer must know the dynamic response, or complete operating characteristics, of each pictured device: the sampler, or indicator, which senses and measures the actual output; the controller, including both the error detector and the correction computer, which contain the decision-making logic; the control actuator and the transmission characteristics of the connecting lines, which communicate and activate the necessary adjustment; and the operating characteristics of the plant, which is the process or system being controlled. *Dynamic response* or *operating characteristics* refer to a mathematical expression, for example, differential equations for the transient behavior of the process or its actions during periods of change in operating conditions. From it, one can develop the *transfer function* of the process (another, often simpler representation of the differential equations) or prepare an experimental or empirical representation of the same effects. For convenience, a control engineer usually lumps all of the devices into a so-called block diagram for convenience, such as the one shown in Figure 3.5.1.

Because of time lags due to the long communication line (typically pneumatic or hydraulic) from sensor to controller and other delays in the process, some time will elapse before knowledge of changes in an output process variable reaches the controller. When the controller notes a change, it must compare it with the variable value it desires, compute how much and in what direction the control actuator must be repositioned, and then activate this correction. Some time is required, of course, to make these decisions and to correct the actuator position.

Some time will also elapse before the effect of the actuator correction on the output variable value can reach the output itself and thus be sensed. It is only then that the controller will be able to know whether its first correction was too small or too large. At that time it makes a further correction, which will, after a time, cause another output change. The results of this second correction will be observed; a third correction will be made; and so on.

This series of measuring, comparing, computing, and correcting actions will go around and around through the controller and through the process in a closed chain of actions until the actual process value is finally balanced again at the value desired by the operator. Since from time to time there are disturbances and modifications in the desired level (or type) of the output, the series of control actions never ceases. This type of control is aptly termed *feedback control*. Figure 3.5.1 shows the direction and path of this closed series of control actions. The closed-loop concept is fundamental to a full understanding of automatic control.

Although the preceding example illustrates the basic principles involved, the actual attainment of automatic control of almost any industrial process or other complicated device will usually be much more difficult because of the speed of response, multivariable interaction, nonlinearities, response limitations, or other difficulties that may be present, as well as the much higher accuracy or degree of control that is usually desired beyond that required for the simple process just mentioned.

As defined here, automatic process control always implies the use of a feedback. This means that the control instrument is continuously monitoring certain output variables of the controlled process, such as a pressure or a composition, and is also comparing this output with some preestablished desired value, which is considered

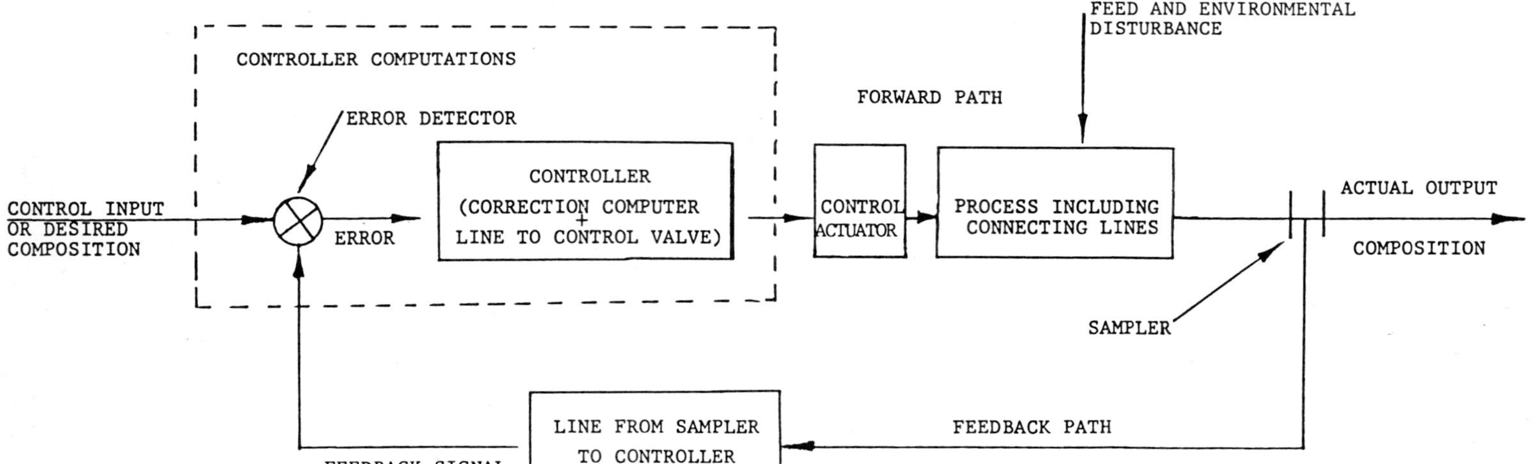

Fig. 3.5.1. Block diagram of a typical simple, single control loop of a process control system.

a *reference* or *set point* of the controlled variable. An error that is indicated by the comparison is used by the instrument to compute a correction to the setting of the process control valve or other final control element in order to adjust the value of the output variable to its desired level and to maintain it there.

If the set point is altered, the response of the control system to bring the process to the new operating level is brought about by what is termed a *servomechanism*, or *self-correcting device*. The action of holding the process at a previously established level of operation in the face of external disturbances operating on the process is the function of a *regulator*.

3.5.2.2 Implementation of an Automatic Control System

The large number of variables in a typical industrial plant constitutes a wide variety of flows, levels, temperatures, compositions, and other parameters to be measured by the sensor elements of the control system. Such devices sense some physical, electrical, or other informational property of the variable under consideration and use it to develop an electrical, mechanical, or pneumatic signal representative of the magnitude of the variable in question. The signal is then acted upon by a *transducer* to convert it to one of the standard signal levels used in industrial plants (3 to 15 psig for pneumatic systems and 1 to 4, 4 to 20, or 10 to 50 mA or 0 to 5 V for electrical systems). Signals may also be digitized at this point if the control system is digital. The important sensing and conversion techniques are well covered in the literature.[1]

The signals that are developed by many types of sensors are continuous representations of the sensed variables and as such are called *analog signals*. When analog signals have been operated upon by an analog-to-digital converter, they become a series of bits, or on-off signals, and are then called *digital signals*. Several bits must always be considered together in order to represent properly the converted analog signal (typically 10 to 12 bits).

As stated previously, the resulting sensed variable signal is compared at the controller to a desired value or set point for that variable. The set point is established by the plant operator or by an upper-level control system. Any error (difference) between these values is used by the controller to compute the correction to the controller output, which is transmitted to the valve or other actuator of the system's parameters.

A typical algorithm by which the controller (either analog or digital) computes its correction is as follows:[2] Suppose a system comprises components that convert inputs to outputs according to relationships, called *gains*, of three types, that is, proportional, derivative, and integral gains. Then the controller output is

$$\text{output} = K_p e_n + \sum_{i=1}^{n} K_R e_i + K_D(e_n - e_{n-1}) + K_1$$

where K_p, K_D, and K_R are the proportional, derivative, and integral gains, respectively, of the controller; K_1 is a midrange constant to allow proportional control only (i.e., $K_D = 0$, $K_R = 0$); n is the last sample obtained by the controller; and $(n-1)$ is the next to last sample. The summation for the integral gain term is over all the sampling iterations until the last one, n. The error e_n is calculated as

$$e_n = \pm(\text{set point} - \text{controlled variable})$$

The option of using either a plus or minus sign in the error term depends on the particular control loop.

3.5.2.3 Conventional Control Methods

Analog Control

Both theoretical and practical studies of automatic control were confined principally to single-loop studies similar to that in Figure 3.5.1, until the mid-1950s. Work involved continued development of the methods first proposed by researchers in the 1930s and used with important results in the succeeding two decades. Particularly important here were studies of stability (Fig. 3.5.2) and design methods of feedback control systems, which are discussed extensively in the classical control literature: root-locus method, Bode-plot representations, Nyquist diagrams, and Nicols charts.[3]

All of the preceding work involved the use of analog control systems, that is, mechanical or electrical devices that simulated the action desired. They were *analogs* of the action that a human operator would take in the same circumstance. Because of the limitations in complexity of the mechanical and electrical devices that were built in large quantities during this period, and because of the limitations in the control systems theory existing at the time, nearly all of these systems were built to control single loops.

Computer Control

Almost as soon as digital computers were developed, researchers began to consider them as candidates for automatic control systems because of their potential for breaking the limitations on complexity imposed by the earlier analog devices.

Thus, following the development of capable, inexpensive, and reliable digital computers in the late 1960s, these devices became very popular elements of control systems. A digital computer can easily store the values

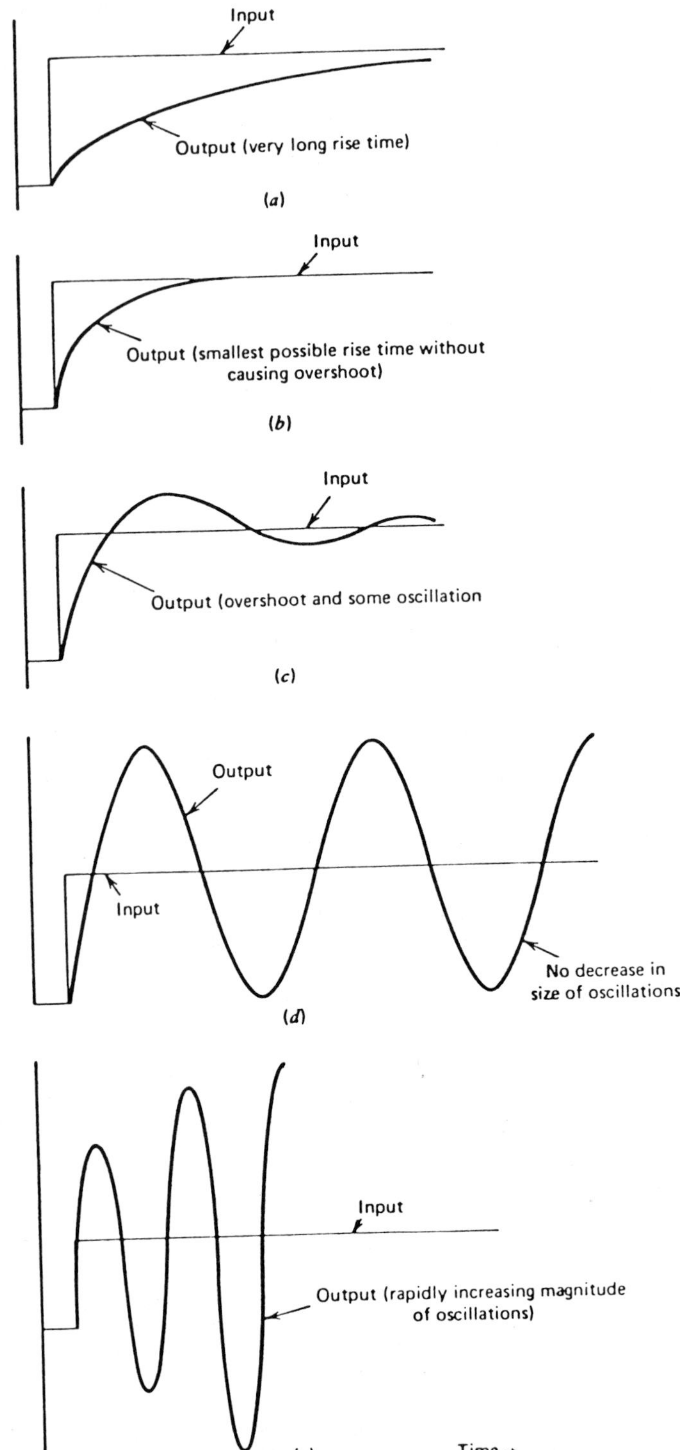

Fig. 3.5.2. Various degrees of stability in transient response as achieved by automatic control systems. (*a*) Stable and overdamped. (*b*) Stable and critically damped. (*c*) Stable and underdamped. (*d*) Oscillatory or threshold of stability. (*e*) Unstable.

of set points for a large number of variables; computer programs can define and quickly prepare highly complex computations that are necessary for decision making; a large number of sensors can transmit signals to the computer at the same time. Advances in computer technology have also enabled the design of hierarchical and distributed control systems. Here reliable communication between computers provides for the control of a network of machines, stations, departments, or even remote plants. In parallel, control models have been developed and advanced to represent, analyze, and design the more complex control systems.

3.5.3 BASIC CONTROL MODELS

3.5.3.1 Control Modeling

Four types of modeling methodologies have been employed to represent physical components and relationships in the study of control systems:

1. Mathematical equations, in particular, differential and difference equations that are the basis of classical control theory (transfer functions are a common form of these equations).
2. Mathematical equations that are based on state variables of multivariable systems and associated with modern control theory.
3. Block diagrams.
4. Signal flow graphs.

Mathematical models are employed when detailed relationships are necessary. To simplify the analysis of mathematical equations, we usually approximate them by linear, ordinary differential equations. For instance, a characteristic differential equation of a control loop model may have the form

$$\frac{d^2x}{dt^2} + 2\alpha \frac{dx}{dt} + \beta^2 x = f(t)$$

with initial conditions of the system given as

$$x(0) = X_0$$

$$x'(0) = V_0$$

where $x(t)$ is a time function of the controlled output variable; its first and second derivatives over time specify the temporal nature of the system; α and β are parameters of the system properties; $f(t)$ specifies the input functions; and X_0 and V_0 are specified constants.

Mathematical equations such as this example are developed to describe the performance of a given system. Usually an equation or a transfer function is determined for each system component. Then a model is formulated by appropriately combining the individual components. This process is often simplified by applying Laplace and Fourier transforms. A graphic representation by block and signal flow diagrams (see Fig. 3.5.3) is usually applied to define the connections between components.

Once a mathematical model is formulated, the control system characteristics can be analytically or empirically determined. The basic characteristics that are the object of the control system design are response time, relative stability, and control accuracy. They can be expressed either as functions of frequency, called *frequency domain specifications*, or as functions of time, called *time domain specifications*. To develop the specifications, the mathematical equations have to be solved, a procedure that is often complicated. A relatively simpler approach can be employed by using graphic methods such as those mentioned in Section 3.5.2.3 on conventional control methods.

Numerical analysis methods and computer simulation methods may also be used for the solution. When systems are too complex to model mathematically, the graphic methodologies are applied to describe the numerous control details and relationships. In this case computer simulation is usually employed to evaluate the performance characteristics of the system.

3.5.3.2 Open-Loop, Feedback, and Feedforward Models

Control models can be either the open-loop or the feedback type. Since automatic control always require feedback, systems in practice as well as their models involve combinations of the two types. It is quite common to find that the introduction of more automation and computer control into a system is accompanied by a process of "closing the loops," that is, providing more feedback whenever the performance of an open-loop operation is deemed unsatisfactory.

When certain knowledge about expected future input or operating conditions is available, it can be utilized by a control system to improve performance. This type of advanced preparation is called *feedforward compensation*. For instance, an automatic processing station can be alerted to expected changes in material composition in

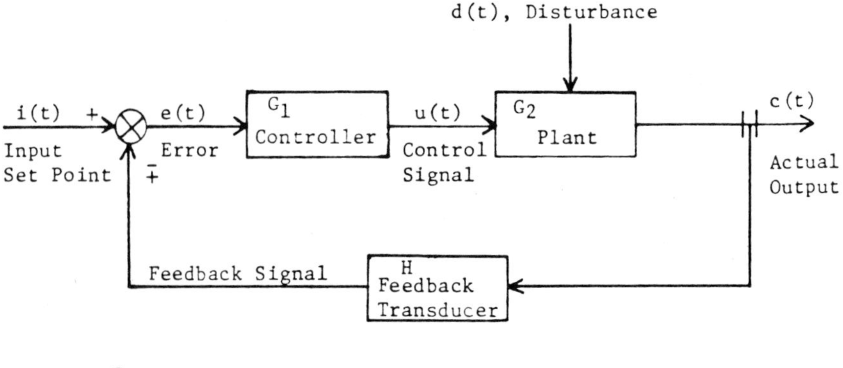

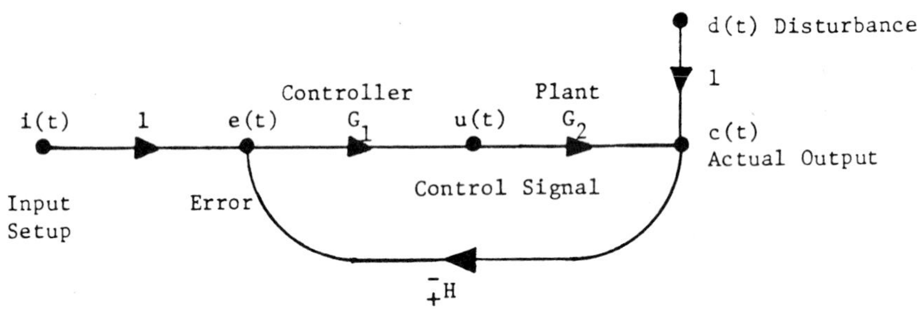

Fig. 3.5.3. Graphic techniques for control models. (*a*) Block diagram model of a feedback loop (similar to Fig. 3.5.1). (*b*) Signal flow graph of the same feedback loop.

order to actuate proper process variations. The feedback portion of the control system in this example will continue to be used for fine adjustment, but the overall control activity will be significantly improved.

A number of basic control models follow the three classes already described (see Fig. 3.5.4). With regard to computer process control, we can define three classes of application: *supervisory, or optimizing, control,* as exemplified in Figure 3.5.5; *direct digital control* (Fig. 3.5.6); and *hierarchy control* (Fig. 3.5.7), which is a combination of the others to effect all levels of decision making simultaneously in a plant.

As shown in Figure 3.5.5, supervisory, or optimizing, control puts the computer in an external or secondary control loop to the primary plant control system, which remains as the conventional plant instruments and individual electronic or pneumatic analog controllers discussed previously. The computer merely changes the set point, or level of control governed by the controllers, either directly or through manual intervention. Its task then is to "trim" the plant operation to improve the economic return from its operation and not to affect its dynamic control. At the same time, a malfunction of the computer cannot adversely affect the plant control. When the computer performs the set point computation just mentioned but does not itself adjust the plant controller and instead depends on a plant operator for this final action, we have what is called *open-loop control.* When the computer merely samples the process variables and determines their correct operating levels but does not perform the optimization calculations, we have what is called *process monitoring.*

Direct digital control (see Fig. 3.5.6) uses a computer to replace a group of single-loop analog controllers with a digital computer. This is done especially when the single computer will be less expensive than the many controllers it can supplant.[4]. The digital computer's computational ability also makes possible the application of more complex advanced control logic.

Hierarchy control, the most ambitious of models, as shown in Figure 3.5.7, attempts to apply computers to all plant control situations at the same time. Level 1 and 2 machines are very similar in function, differing only in the extent of the control functions assigned to each. The dedicated digital controller of Level 1 handles complex devices, such as chemical analyzers (e.g., chromatographs), and specialized feedforward and noninteracting multivariable control setups. The direct digital controller of Level 2 handles a much larger number of three-mode and related types of control loops, as in Figure 3.5.1. It also communicates with plant operators through a console, probably composed of CRTs and a keyboard. Communication between all digital computers and with all consoles is by digital signals. Connections to analog signals are required only of the Level 1 and Level

2 machines. Level 3 machines serve as supervisory computer control systems, as previously outlined. Level 4 machines carry out production scheduling and management information functions.

When the control system in the hierarchy is implemented through a number of computer controllers in each level, a *distributed control system* results. The rapid development of microprocessor technology has made the concept of distributed control quite attractive. The use of reliable communication between a larger number of individual controllers, each responsible for its own tasks rather than for the complete operation, improves the response of the total system.

3.5.3.3 Tasks at Each Level of The Hierarchy

In the context of large industrial plants, or of a complete industrial company based in one location, the tasks that would be carried out at each level of the hierachy are described in Tables 3.5.1 to 3.5.5.[6] These tasks subdivided within each table into the following areas: (I) production scheduling, (II) control enforcement, (III)

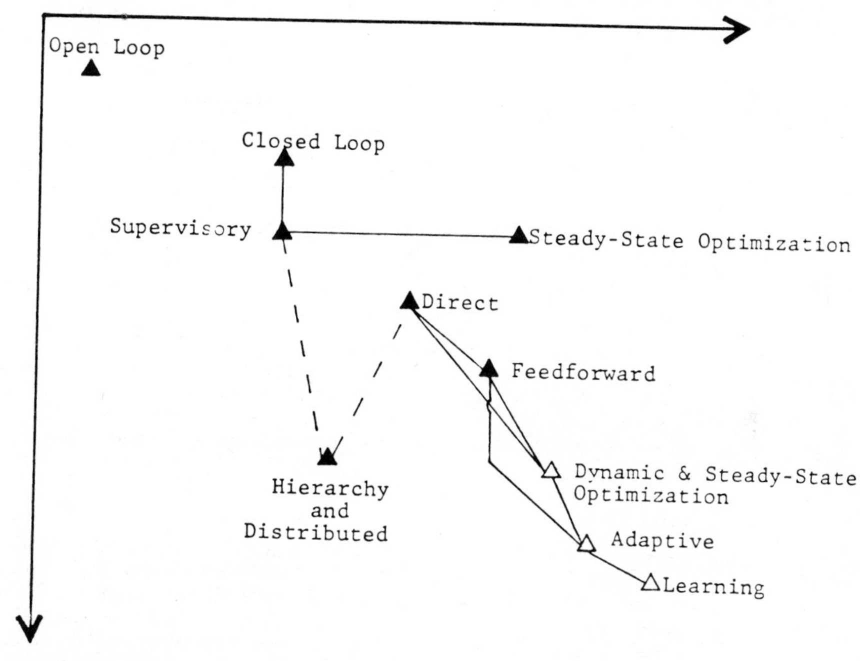

Fig. 3.5.4. Summary of control models, performance, and complexity.

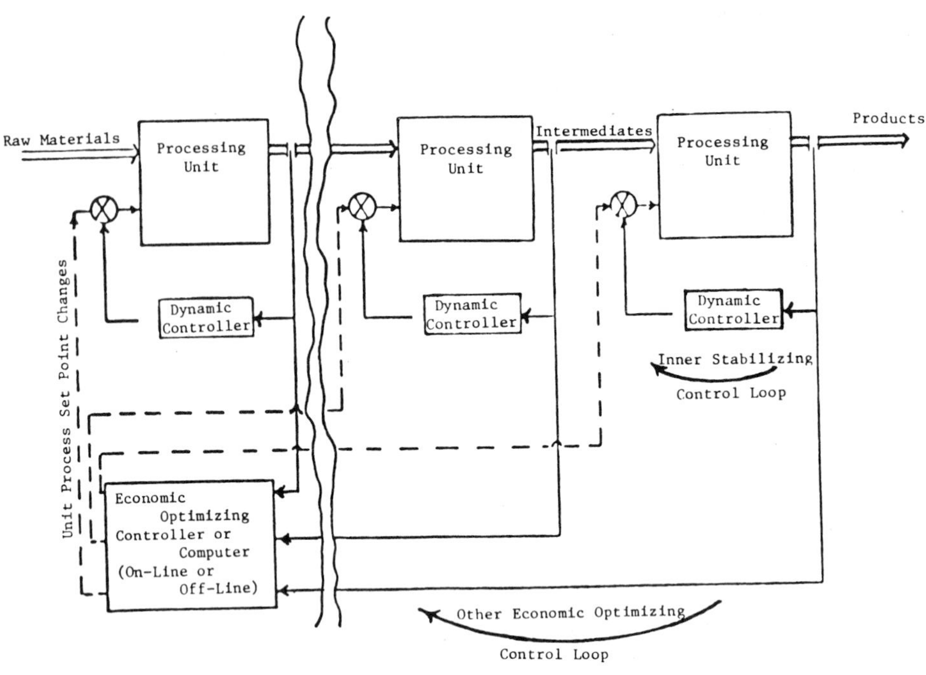

Fig. 3.5.5. A model of computer control as a plant optimizer and supervisory controller.

system coordination and operational data reporting, and (IV) reliability assurance. It is the authors' contention that these tables outline the tasks that *must* be carried out in any industrial plant, particularly at the upper levels of the hierarchy.[9, 10] Details of how these operations are actually carried out may vary drastically, particularly at the lowest levels, because of the nature of the actual production process being controlled, but the operations themselves remain the same in concept. The tasks of the hierarchy computer system are diagrammed in Figure 3.5.8.[5,6,7,8]

3.5.3.4 Manual Control Models

Before concluding this discussion on basic models, we must include certain model types that are particularly useful in manual control modeling.[11] These are the *preview model* and the *cognitive model* (Fig. 3.5.9). Both include feedforward compensation that depends on advanced knowledge of the human controller about the system. In the preview model a true display of future input set points provides the human controller with information well in advance so that he or she can effectively control the operation. In the cognitive model the human controller can plan operations well ahead of time based on his or her precognition of the system characteristics and idiosyncrasies. Since such cognitive knowledge is usually gained by experience and is subject to random variations, precognition may improve the response on the average but may fall in the extremes.

3.5.4 ADVANCED CONTROL MODELS

In the late 1950s it became readily evident that there were severe limitations on the ability of the classical control methods, as discussed previously and exemplified by the single-loop controllers of Figure 3.5.1, whether implemented by analog or digital methods. The appearance of the digital computer with its vastly increased computational ability provided incentive for the beginning of a whole new era of development in automatic control theory. This class of developments has subsequently been labeled *modern control*.

Important here was the application of the variational calculus to dynamic optimization problems along with the concept of dynamic programming as formulated by Bellman[12] and associates in 1956. The "maximum principle" of Pontryagin,[13] developed in 1959, and the identification techniques of the filter theory of Kalman[14] were other important discoveries during this period. This work has continued to the present day along with studies of the other important advanced control topics that are mentioned here. These developments have vastly extended the capabilities of industrial control systems in recent years. As indicated, a computer is almost always necessary to make desired implementations.

3.5.4.1 Adaptive Control

Adaptive control is the capability of the control system to modify its own operation to achieve the best possible mode of operation.[15] A general definition of adaptive control implies that an adaptive system must be capable of performing the following functions: providing continuous information about the present state of the system or identify the process; comparing present system performance to the desired or optimum performance and making a decision to change the system so as to achieve some previously defined optimum performance; and initiating a proper modification to drive the control system to the optimum. These three principles—identification, decision, and modification—are inherent in any adaptive system.[16,17] Figure 3.5.10 illustrates the adaptive control model with the adaptive controller in a secondary loop, where it modifies the operation of the regular systems controller, which is located in the primary process control loop.

3.5.4.2 Optimal Control

Compared to adaptive control, further progress toward optimal response can be achieved by methods and models of optimal control.

The *steady state optimization method* allows a computer to determine for the process a new best operating level if external conditions require such changes in order to maintain the process operation at some optimum, usually relative to economic, criteria. Optimizations are computed under the assumption that the process is at a steady state and can be instantaneously transferred from one steady state to another, hence, the name of this control method. Such an assumption is necessary to transform all process operating equations to algebraic form for ready solution by a computer. In most supervisory control systems (Fig. 3.5.5), this method can be implemented.

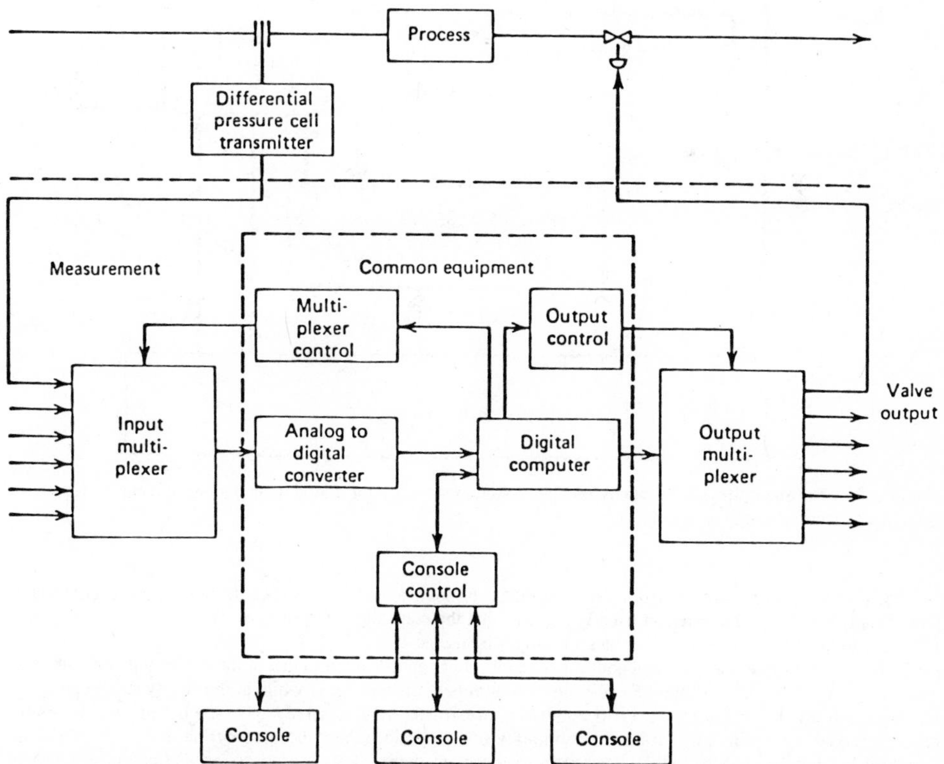

Fig. 3.5.6. Block diagram of a direct digital control system.

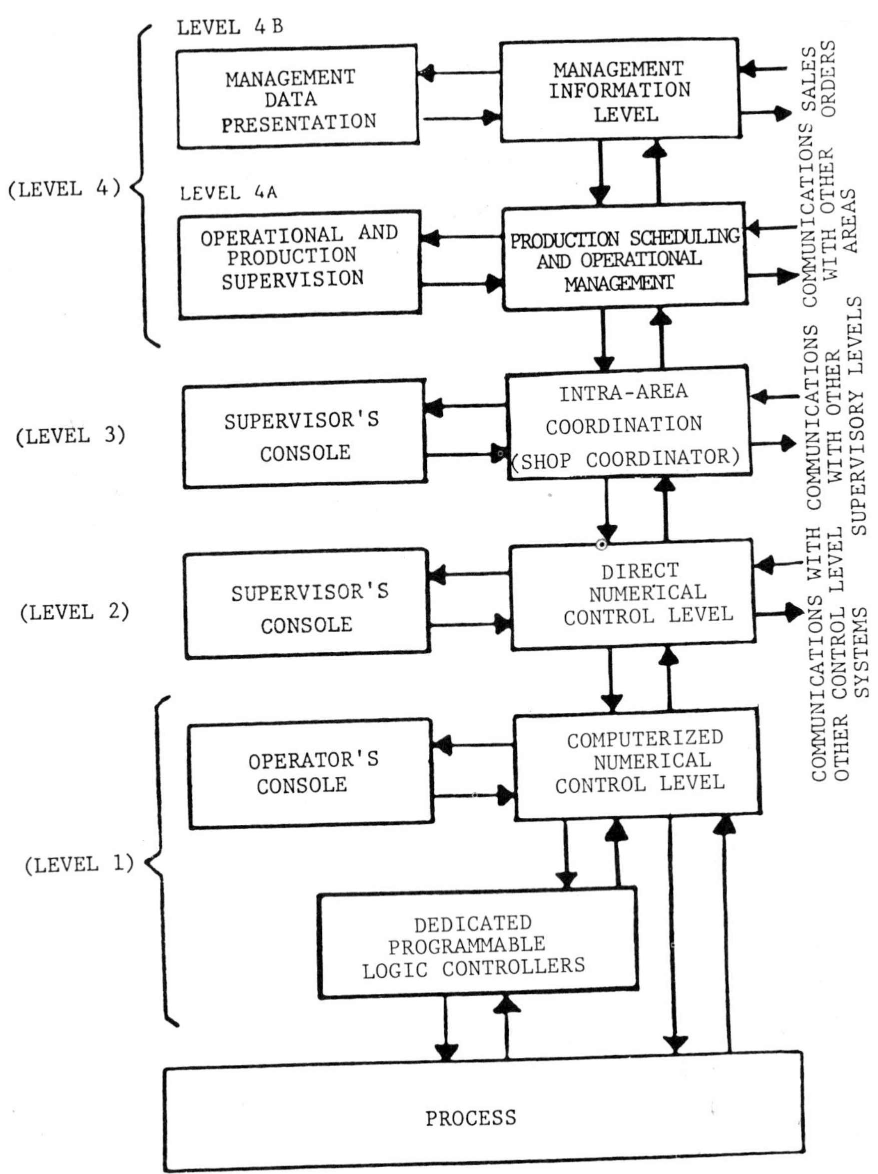

Fig. 3.5.7. Assumed physical hierarchy computer system structure for a large manufacturing complex [computer integrated manufacturing system (CIMS)].

The *dynamic optimizing control* system operates in such a way that a specific performance criterion is dynamically satisfied. The criterion usually requires that the controlled system move from the original to a new state in a minimum possible time, or at a minimum total cost.[18]

Dynamic optimization and control adds one more level of sophistication to that of the steady state optimization. Thus, not only is the process maintained at its optimum performance level while in the steady state regime but the change from one operating level to another is also made to best satisfy the established overall control criteria. Models of optimal control are now mainly of academic interest because of the extremely large and powerful computing capacity that is required to implement them. However, practical application is certainly feasible in the foreseeable future. When it does occur, it will be best suited in systems that are highly sensitive to the process path being followed, for example, cyclical catalytic processes and batch processing of plastics. A very large computing capacity, both in speed and memory, is necessary, since the equations to be solved in

Table 3.5.1. Tasks of the Intracompany Communications Control System (Level 4B of Fig. 3.5.7)

III. System Coordination and Operational Data Reporting
   1. Maintain interfaces with
      a. Plant and company management
      b. Sales personnel
      c. Accounting and purchasing departments
      d. Production scheduling level (Level 4A)
   2. Supply production and status information as needed (in the form of regular production and status reports and on-line inquiries) to
      a. Plant and company management
      b. Sales personnel
      c. Accounting and purchasing departments
      d. Personnel department
   3. Supply order status information as needed to sales personnel

IV. Reliability Assurance
   4. Perform self-check and diagnostic checks on itself

Table 3.5.2. Tasks of the Production Scheduling and Operational Management Level (Level 4A of Fig. 3.5.7)

I. Production Scheduling
   1. Establish basic production schedule
   2. Modify the production schedule for all units per order stream received, energy constraints and power demand levels
   3. Determine the optimum inventory level of goods in process at each storage point. The criteria to be used will be the trade-off between customer service (i.e., short delivery time) versus the capital cost of the inventory itself and the trade-offs in operating costs versus costs of carrying the inventory level (this is an off-line function).
   4. Modify production schedule as necessary whenever major production interruptions occur in downstream units and whenever such interruptions will affect prior or succeeding units

III. System Coordination and Operational Data Reporting
   5. Collect and maintain raw material use and availability inventory and provide data for purchasing for raw material order entry
   6. Collect and maintain overall energy use for transfer to accounting
   7. Collect and maintain overall goods in process and production inventory files
   8. Collect and maintain the quality control file
   9. Maintain interfaces with management interface level function and with area level systems

IV. Reliability Assurance
   10. Run self-check and diagnostic routines on self and lower-level machines

Table 3.5.3. Tasks of the Area Level (Level 3 of Fig. 3.5.7)

I. Production Scheduling
   1. Establish the immediate production schedule for its own area, including transportation needs
   2. Locally optimize the costs for its individual production area while carrying out the production schedule established by the production control computer system (Level 4A) (i.e., minimize energy usage or maximize production)

III. System Coordination and Operational Data Reporting
   3. Make area production reports
   4. Use and maintain area practice files
   5. Collect and maintain area data queues for production, inventory, raw materials usage and energy usage
   6. Maintain communication with higher and lower levels of the hierarchy
   7. Operations data collections and off-line analysis as required by engineering functions
   8. Service the man–machine interface for the area
   9. Carry out needed personnel functions
      a. Vacation schedule
      b. Work force schedules
      c. Union line of progression

IV. Reliability Assurance
   10. Diagnostics of self and lower level functions

Table 3.5.4. Tasks of the Supervisory Level (Level 3 of Fig. 3.5.7)

II. Control Enforcement
1. Respond to any emergency condition that may exist in its region of plant cognizance
2. Optimize the operation of units under its control within limits of established production schedule
3. Carry out all established process operational schemes or operating practices in connection with these processes

III. System Coordination and Operational Data Reporting
4. Collect and maintain data queues of production, inventory, and raw material and energy usage for the units under its control
5. Maintain communication with higher and lower levels
6. Service the man–machine interfaces for the units involved

IV. Reliability Assurance
7. Perform diagnostics on itself and lower level machines
8. Update all standby systems

the mathematical model are now sets of simultaneous differential equations as opposed to the algebraic equations in steady state optimization.

3.5.4.3 Learning Control

Learning control implies that the control system contains sufficient computational ability so that over time it can develop specific representations of the control model of the system being controlled and can modify its own operation to compensate for this newly developed knowledge. The learning control system is a further development of the adaptive controller and can apply artificial intelligence techniques.

3.5.4.4 Multivariable Noninteracting Control

Multivariable noninteracting control concerns large systems whose internal variables depend on the values of other, related variables of the process. The single-loop techniques of classical control theory will therefore not suffice. More sophisticated techniques must be used to develop appropriate control systems for such processes.[19,20] Figure 3.5.11 shows a model of such a system. The feedforward controller compensates for the dynamics and process delays of the process itself and helps the unit process controllers compensate for input upsets. The multivariable noninteracting controller compensates for the effect of variations in flow of product A due to action of unit controller 1 on product B and thus prevents interaction of the output variables of the process.

3.5.4.5 Control by Expert Systems and Machine Intelligence

The use of artifical intelligence techniques for manufacturing planning and control is described by Bullers et al.[21]

Traditional methods of control are often ill-suited for the increased complexity and timeliness requirements for decision making in an integrated, computerized production environment. Often, production objectives are conflicting, concurrent tasks have to be mitigated, and the control has to be based on numerous yet uncertain and incomplete data. Under such circumstances judgment, adaptation, and even intuition may be vital for necessary control functions. A promising artificial intelligence approach to control in such environments is by an *expert system*. An expert system is defined as a problem solving computer program that achieves a high level of intelligent performance in an area that is considered to be difficult and requiring specialized knowledge and skill. In terms of control, an expert control system, including its human users, will adaptively govern the behavior of a production system. The control operation must repeatedly perform several knowledge-based tasks,

Table 3.5.5. Tasks of the Control Level (Level 1 of Fig. 3.5.7)

II. Control Enforcement
1. Maintain direct control of the plant units under its control
2. Detect and respond to any emergency condition that may exist in these plant units

III. System Coordination and Operational Data Reporting
3. Collect information on unit production, raw material, and energy use and transmit to higher levels
4. Service the operator's man–machine interface

IV. Reliability Assurance
5. Perform diagnostics on itself
6. Update any standby systems

including[22]: interpretation of the current situation, prediction of future events, diagnosis of the causes of anticipated problems, formulation of remedial plans, and monitoring execution to ensure success. A model of problem solving in dynamic environments that may be useful for a control expert system is shown in Figure 3.5.12. Another example of an expert system developed to control the dynamic reconfiguration of a programmable production facility is described by Nof.[23]

3.5.5 APPLICATION AND SELECTION OF CONTROL MODELS IN PRODUCTION SYSTEMS

The purpose of this section is to describe several typical applications of control models in production systems engineering functions. The reader can by now realize that control always implies that there is a system—large or small, simple or complex—to be controlled. Three important system areas in which production engineers have traditionally been working with control models are manufacturing systems, man–machine systems, and information systems.

3.5.5.1 Control Models of Manufacturing Systems

Manufacturing systems probably provide the richest area for applications of control models. They include the self-contained machine or process for which a precise mathematical model can be formulated. They also require the many control activities that are less amenable to precise modeling, such as production and inventory control

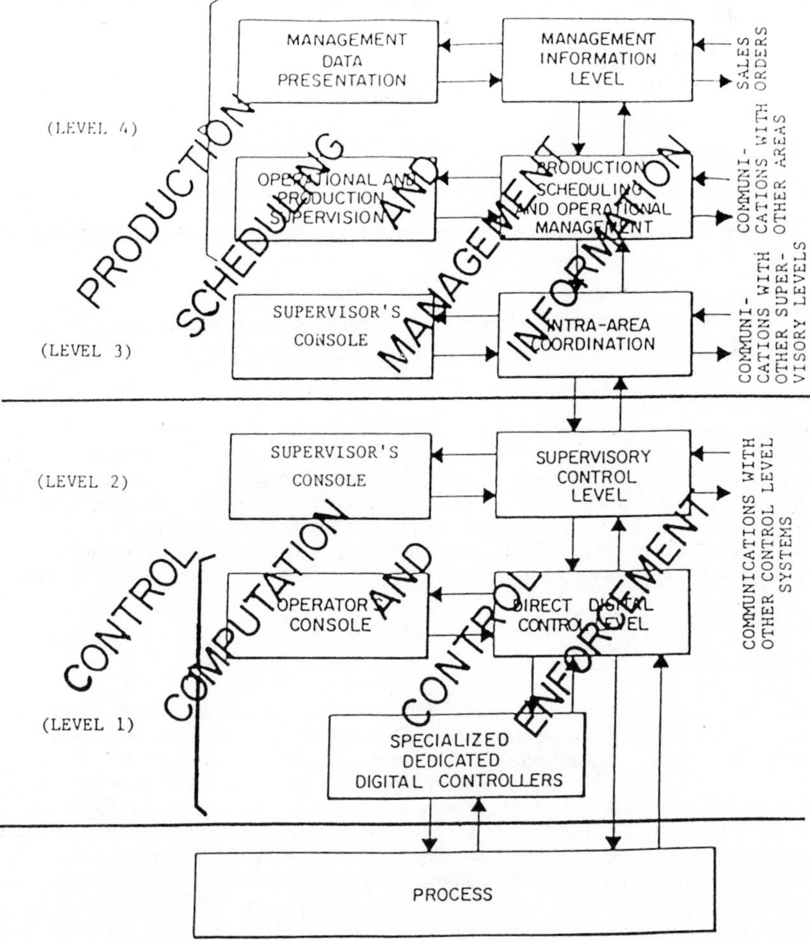

Fig. 3.5.8. Definition of the real tasks of the hierarchy computer control system.

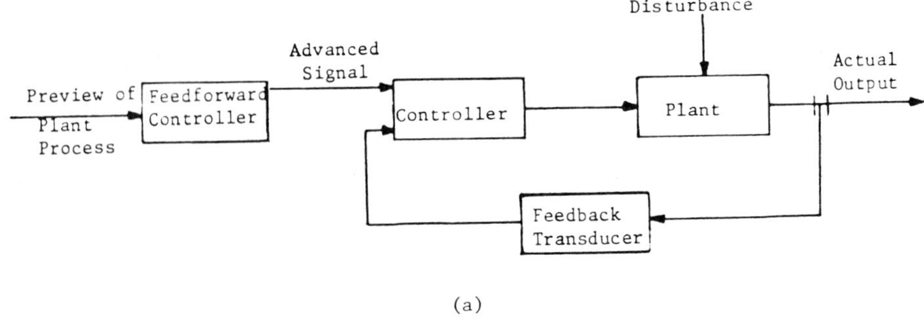

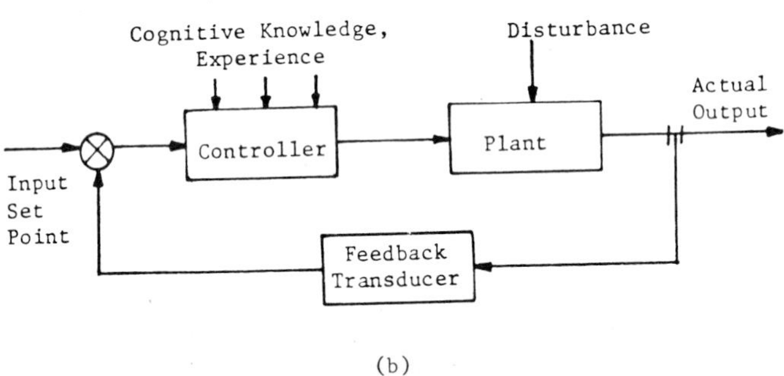

Fig. 3.5.9. Control models with feedforward compensation: (*a*) Preview control model and (*b*) Precognitive control model.

or quality control, for which operations research and simulation techniques are applied to evaluate and plan the control system.

Models of NC Machines

A typical example of a model of an NC machine is illustrated in Figure 3.5.13.[25] In this model, G is the transfer function (gain) of the position transducer. The velocity transducer is designed to have a transfer function of $G\alpha s$, where α specifies control parameters (damping and natural frequency) of the transducer and s is the Laplace transform variable. An input digital signal θ_i is passed as e_i to an amplifier with gain m, which controls a servomotor with gain G_2. This motor produces a torque, which turns a drive mechanism to a specific position displacement θ_0 at a specified velocity. With certain simplifying assumptions (e.g., that the input in this example is a step function), a control model for the torque output can be formulated, using Laplace transform, as

$$\theta_0 = m \left[\frac{G_2(e_i - e_0)}{s^2 J + G_1 \alpha s} \right]$$

In the equation, J is the system inertia. Analysis of the model can identify the specific values of the control parameters that yield the best response. For instance, the value of α determines the damping of the system, namely, how fast after every displacement command the machine can stabilize at the specified position.

Whereas the model in Figure 3.5.13 is based strictly on the closed-loop, feedback method, Figure 3.5.14 shows an adaptive control model for NC. As explained in Section 3.5.4, the adaptive control model applies a secondary loop that obtains feedback information beyond that obtained by the primary loop. In this example the primary loop involves position and velocity feedback, as in Figure 3.5.13, plus feedback on cutting forces, which provides for further corrections in the machine operation.

Models of Robot Control

Since robots are increasingly becoming an integral part of production systems, it is worthwhile to examine their control. Actually, robot control can involve four main levels, including the internal machine level; adaptive control level, reacting to sensory input; intelligent robot control; and hierarchical-distributed control of several robots integrated to other production equipment. These different levels are depicted in Figures 3.5.15, 3.5.16, and 3.5.17. In Figure 3.5.15 five control levels of a robot are shown. Axis control of Level 1 in Figure 3.5.15 is shown in Figure 3.5.16. The higher control levels, involving interaction with the outside dynamic world, can be handled by an activity controller system,[28] which is depicted in Figure 3.5.17. The design of robot control systems is discussed in detail in Luh[29] and Paul.[30]

Models of Overall Plant and Company Control Systems

Automatic control of a large, modern industrial plant, whether achieved by a computer-based system or by conventional means, involves an extensive system for the automatic monitoring of a large number of different variables operating under a very wide range of process dynamics. It requires the development of a large number of quite complex, usually nonlinear relationships for the translation of the plant variable values into the required control correction commands. Finally, these control corrections must be transmitted to another very large set of widely scattered actuation mechanisms of various types, which, because of the nature of manufacturing processes, may, and often do, involve the expenditure of very large amounts of energy. Also, plant personnel, both operating and management, must be kept aware of the current status of the plant and each of its processes.

In addition, industrial plants are faced with the continual problem of adjusting the production schedule to match new customer orders while maintaining a high plant productivity at the lowest practical production costs. This is handled in most cases at present through a manual, although computer-aided, production control information system along with an in process and finished goods inventory judged adequate by plant personnel.[31,32]

As a result of the circumstances cited, a precise, mathematical control model of an overall plant control is impractical. However, as the overall requirements for both energy savings and productivity gains become more complex, more and more sophisticated and capable control systems are necessary. To achieve this objective, control systems are gravitating toward large, digital computer-based systems. Proper design and implementation of such systems necessitate careful modeling and analysis. To obtain the necessary control responses, an overall control system must have the following capabilities:

1. A tight control of each operating unit of the plant to ensure that it is operating at its maximum efficiency of energy utilization and/or production capability. The operation is based on the production level set

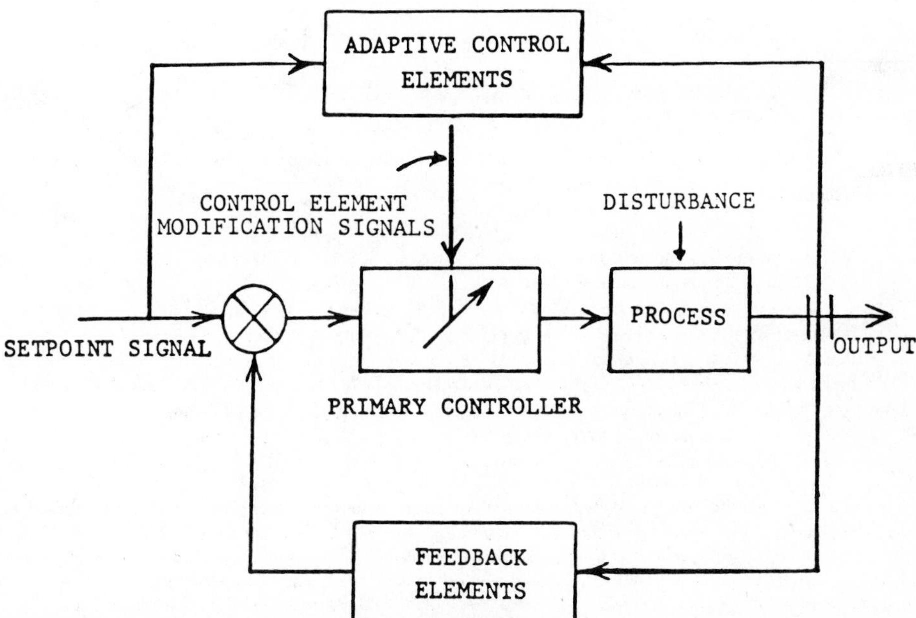

Fig. 3.5.10. Adaptive control model. Adaptive control elements form a secondary control loop around the conventional process control loop and change the characteristics of the primary controller to provide desired system response.

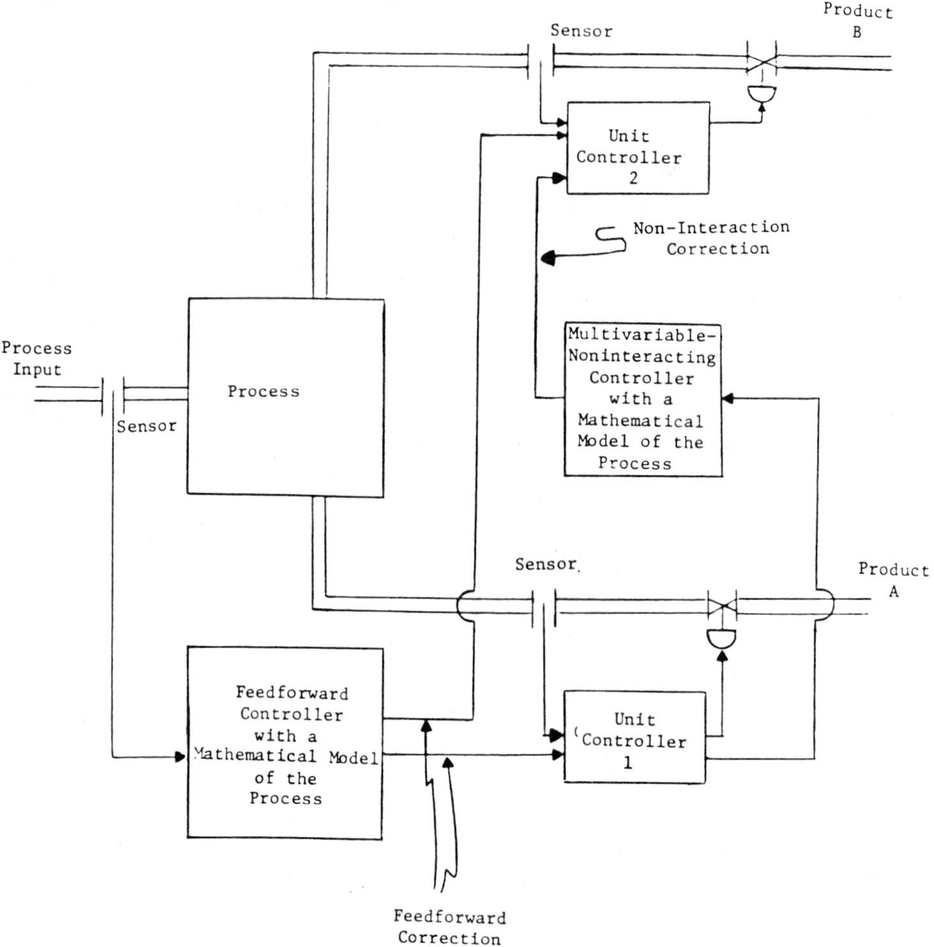

Fig. 3.5.11. Model of a feedforward and multivariable noninteracting controller system.

by the coordinating and scheduling functions listed in (2) and (3). This control reacts directly to any emergencies that may occur in its own unit.

2. A coordination system that determines and sets the production level of all units working together between inventory locations. This system ensures that no unit is exceeding the general area level and thus using excess energy or raw materials. It responds to emergenices or upsets in any of the units under its control by shutting down or systematically reducing the output in these and related units.
3. A system capable of carrying out the scheduling function for the plant from customer orders or management decision in order to produce the required products at the optimum combination of time, energy, and raw materials suitably expressed as cost functions.

Because of the ever-widening scope of authority of each of these requirements, in turn, they effectively become the distinct and separate levels of a superimposed control structure, one on top of the other. Also, in view of the amount of information that must be passed back and forth among these three "levels" of control, a distributed computational capability organized in a hierarchical fashion is necessary (see Fig. 3.5.7). A specific implementation of this model is shown in Figures 3.5.18 and 3.5.19 for the overall control of a steel mill.[33]

Other Models of Production Control

An example of the study of control models in computer-controlled production by operations research techniques is given in Nof et al.[34] A queuing network analysis and a simulation package are applied to evaluate alternatives

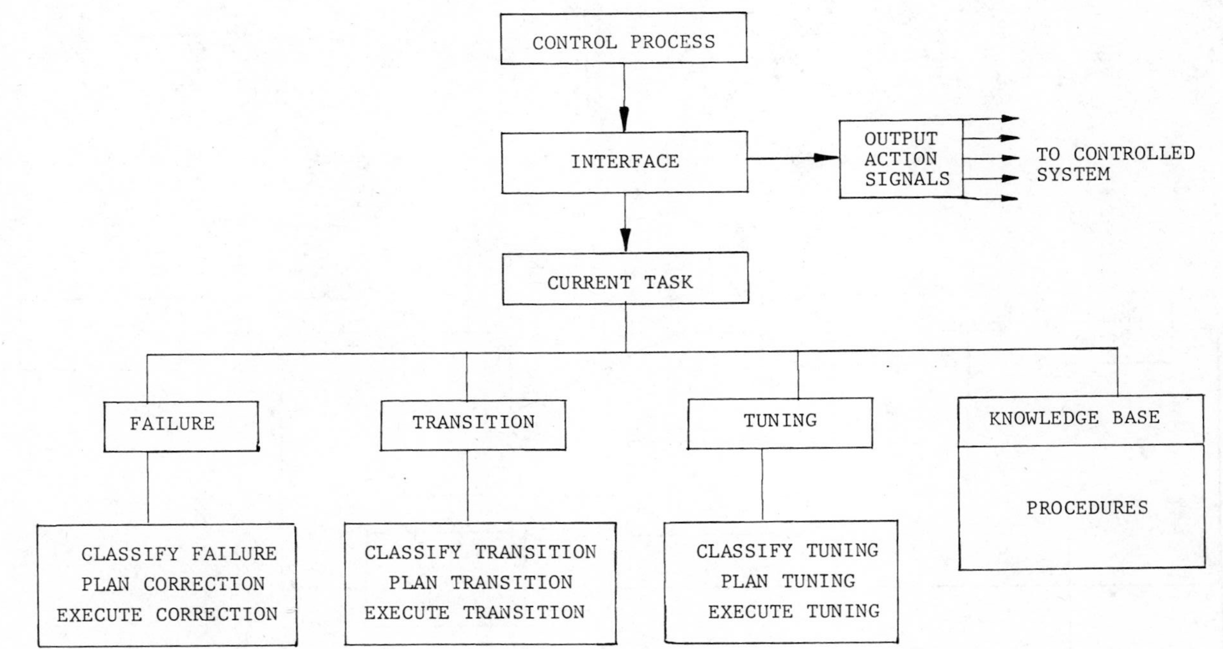

Fig. 3.5.12. A model of knowledge-based problem solving in dynamic environments (following Knaepuer et al.[24]).

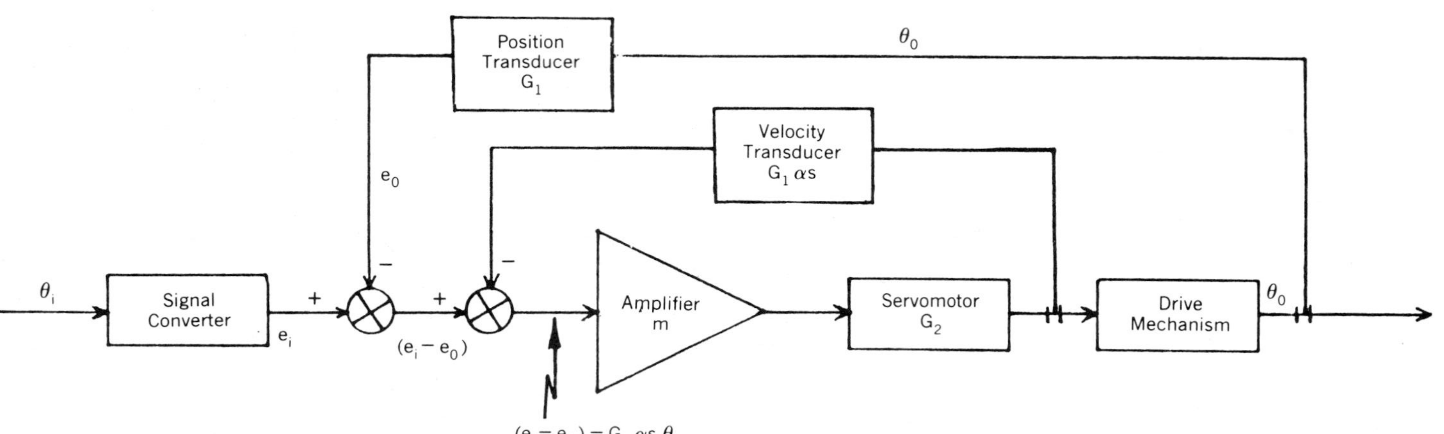

Fig. 3.5.13. A block diagram for position and velocity control in NC machines.

Input
Set Up
Machine
Control Unit
Control
Signals
Controlled
Machine
Position and
Velocity
Feedback
Adaptive
Controller
Cutting Forces

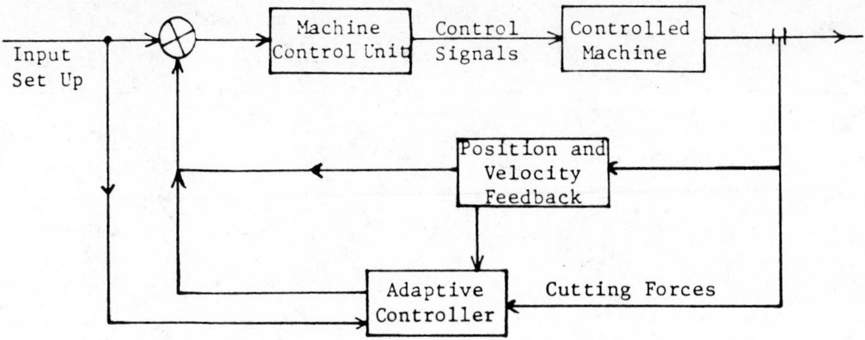

Fig. 3.5.14. An adaptive control model for NC machines.

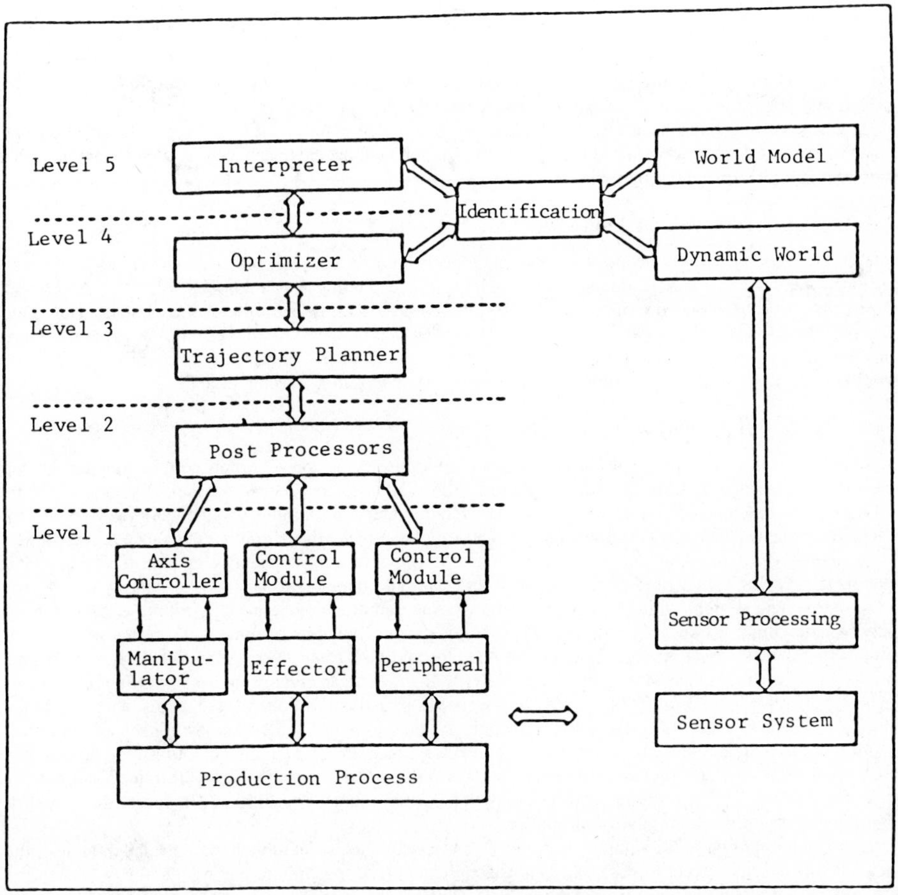

Level 5
Interpreter
Identification
World Model
Level 4
Optimizer
Dynamic World
Level 3
Trajectory Planner
Level 2
Post Processors
Level 1
Axis
Controller
Control
Module
Control
Module
Manipu-
lator
Effector
Peripheral
Sensor Processing
Sensor System
Production Process

Fig. 3.5.15. Hierarchical control models of a real-time robot control system.[26]

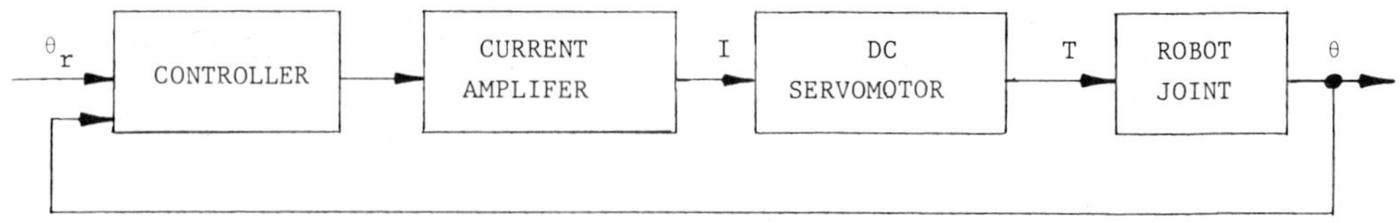

(a) TORQUE CONTROL OF ROBOT ARM

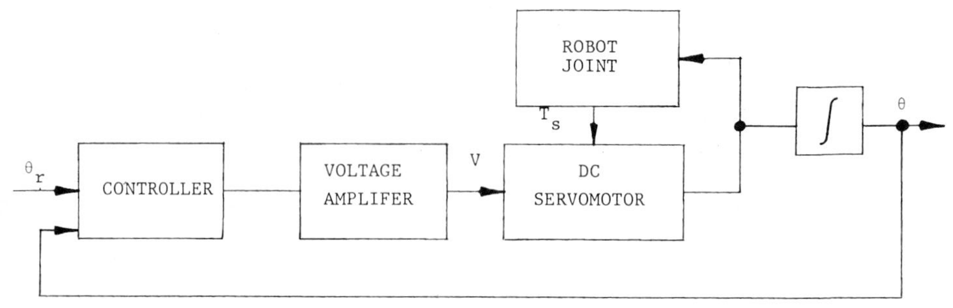

(b) SPEED CONTROL OF ROBOT ARM

Fig. 3.5.16. Control models of a robot arm.[27]

of operational control, including issues of how parts are loaded into the facility, which manufacturing process to select, and how to assign parts to particular machines in the facility.

Learning control models have also been suggested for overall plant control. In Nof et al.[35] the concept of the manufacturing operating system is described for the control of automatic manufacturing with certain intelligence to improve control decisions progressively.

3.5.5.2 Control Models in Man–Machine Systems

In the area of manual control, control models have been used to evaluate total man–machine performance, particularly for vehicular applications, for example, airplane piloting and automobile driving. Additionally, these models have been tested empirically in order to evaluate and measure the characteristics of the human as a controller. A large number of such models are reviewed by Sheridan and Ferrell.[11] A general review of control modeling of human behavior is provided in Rouse and Gopher.[36] Integration of human operators in the operation and control of flexible manufacturing systems is analyzed by Hwang et al.[37]

Control Models of Display Design

Any system that involves people such as operators, supervisors, or controllers in control functions requires some type of a display facility in order to present information to the decision maker. Examples of typical displays are instrument panels in airplanes and other vehicles and control lights and indicators in a control room of a chemical plant. A correctly designed display will significantly aid the overall performance of the controlled system and reduce the workload of the human operators. The design must specify a number of display factors, including the display format, information content and rate, and interaction patterns.

Display system design, which has been receiving increased attention, especially with the development of highly complex flight vehicles, has traditionally employed control models.[11] Some work in this area was based on frequency domain descriptions of the human operator and his or her workload, particularly with regard to visual scanning of the display. More recent works have been based on an *optimal control of the human operator*,[38] which is shown in Figure 3.5.20. In the previous section, optimal control models for steady state and dynamic optimization were defined. They represent automatic systems in which the controller utilizes an optimization algorithm to calculate and actuate an optimal response for dynamically changing conditions. In the case of the optimal control model for a human operator, it is hypothesized that the operator is well trained and motivated and will make an effort to respond as well as possible within human capabilities. However, the model does not imply that an optimization algorithm is used.

As shown in Figure 3.5.20, a human operator is assumed by the model to monitor a display or an instrument panel. The operator is subject to observation noise, for example, errors in interpreting the diplay, and to motor noise, which affects neuromuscular behavior. Additionally, the model assumes a certain time delay in the human response. A typical mathematical form of the model is

$$x'(t) = Ax(t) + Bu(t) + v(t)$$

where A = a matrix of parameters of the controlled process B = a matrix of parameters of the controlled output $x(t)$ = a time vector of state variables $u(t)$ = a time vector of control variables $v(t)$ = a vector of the noise variables

Different variants of this model have been applied for the design and evaluation of a variety of display issues.[11,39,40] One such model has been used in the design of vertical takeoff and landing aircraft. The tradeoff between the level of automation in the flight control system and the display system sophistication was analyzed. The model incorporated simultaneous monitoring and control by the pilot for various design combinations, ranging from complete manual control with no monitoring to fully automatic control with monitoring only.

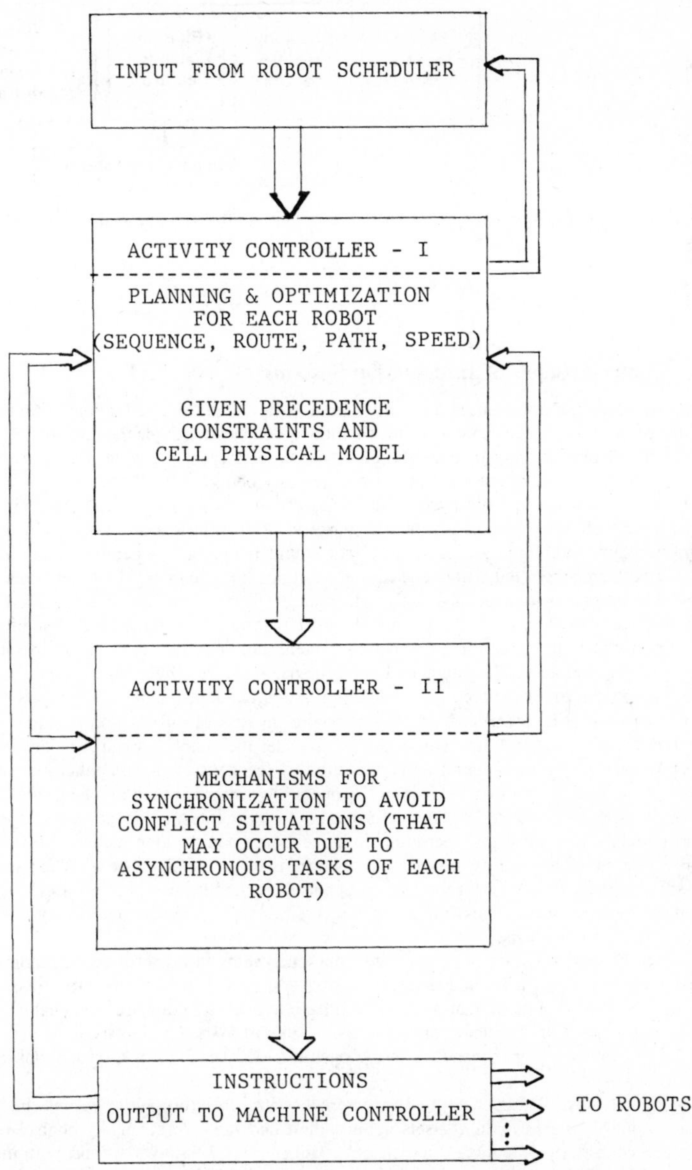

Fig. 3.5.17. Model of an on-line activity control in a multirobot operation.[28]

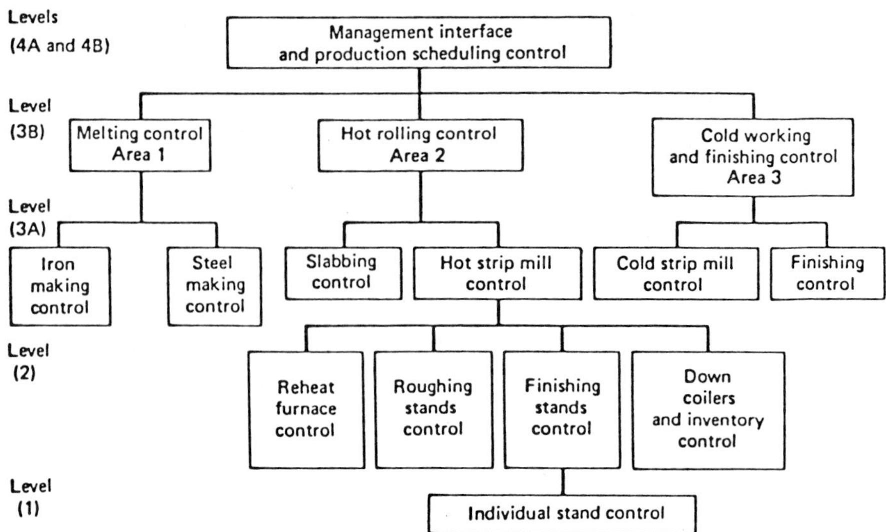

Fig. 3.5.18. Hierarchical arrangement of the steel plant control (an application of the model in Fig. 3.5.7).

3.5.5.3 Control Models of Information Systems

The numerous industrial applications of control models in computer information systems can be classified as being of one of two types: development of information systems that provide the information to control operations or maintenance of internal control over the quality and security of the information itself. Since information systems are usually complex, graphic models are typically being used.

Any of the control models surveyed in this chapter can essentially incorporate an information system as indicated in some of the examples given. The purpose of an information system is to provide useful, high-quality information; therefore, it can be used for sound planning of operations and for preparing realistic standards of performance. Gathering, classifying, sorting, and analyzing large amounts of data can provide timely and accurate measurement of actual performance. This can be compared to reference information and standards that are also stored in the information system in order to immediately establish discrepancies and initiate corrective actions. Thus an information system can improve the control operation in all its major functions by measuring and collecting actual performance measures, by analyzing and comparing the actual to the desired set points, and by directing or actuating corrective adjustments.

As in other application areas, the purpose of control models of information systems is to aid in the design and analysis of control capabilities. For instance, consider the adaptive control model in Figure 3.5.21. The actual performance of a certain operation is measured by feedback elements, either manually or mechanically. This feedback, together with other data about the operation or needed by it and goals and standards established for it, is processed and analyzed by the information system. Information produced can be used directly as signals or indirectly to control the operation. In a secondary control loop, adaptive control elements measure the actual performance as well as the external conditions of the whole system. When external conditions are significantly modified, as defined for the adaptive controller, modifications in the information processing system are indicated. Some of the changes can be implemented directly, whereas others may require the involvement of information system analysts.

The degree of sophistication of a given information system will determine the amount of human operator involvement. For instance, a limited data processing system will produce reports, possibly concentrating on highlighting exceptions, but a human operator will have to analyze, compare, and evaluate the information and decide how to respond. On the other hand, a decision support system will analyze the data, evaluate alternative actions, and recommend proper response to the operator. Chapter 3.13 of the handbook includes a discussion of information systems.

Control models have also been applied for internal control of information systems. Internal control includes the physical control of information assets against theft and tampering, quality control of data accuracy and integrity, and control of operation efficiency.[41,42] Figure 3.5.22 shows a model of internal control in an information system. Such a model is used mainly to explain the different types of internal control. More detailed models are typically employed to specify completely the necessary control functions in each particular area.

3.5.6 SOME FUTURE EXPECTATIONS

There are many possible paths that the development of control systems, and hence the development and use of control models, can take in the future. However, in many cases the paths now appear clear.

A major concern of all installations of computer control systems has been the reliability of the overall system. Not that computers are themselves any less reliable than the alternative analog systems, but their involvement in overall plant control systems in contrast to the multiple single loops of the analog system makes them more vulnerable. This will be countered by the use of multiple-redundant computer systems in place of single computers and of fault-tolerant and fault-remedying techniques whenever possible.

Computer programming complexity is the major bottleneck in computer control system development. Company proprietary considerations of both users and vendors will hold up the development of remedies to this problem. However, programming aids and software sharing must materialize because of the extreme economic incentives involved.

Electronic techniques have the potential, so far unexploited, of drastically reducing the cost of industrial control systems. The implementation of these techniques, however, will be slow because of developmental capital needed by vendor companies. There is also a lack of push on the part of user companies because of the relatively small percentage that instrumentation and control constitutes of the total plant cost. Therefore,

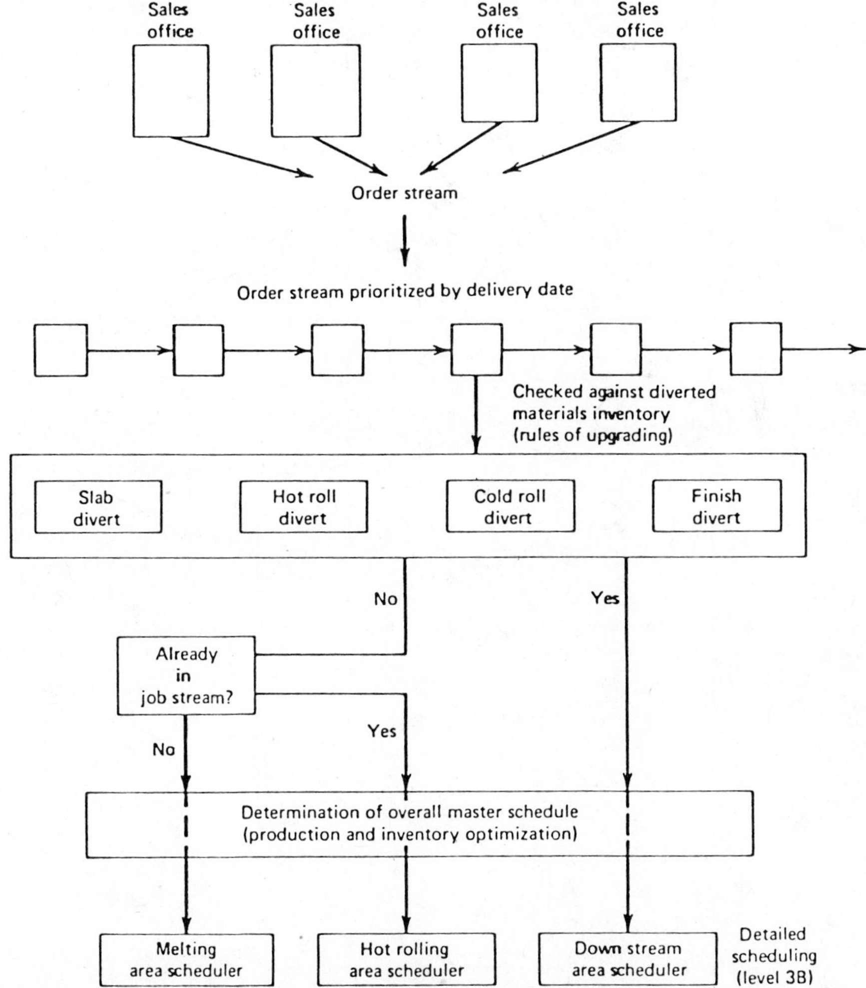

Fig. 3.5.19. A model of the steel mill scheduling procedure at the overall production scheduling level (Level 4A in Fig. 3.5.18).

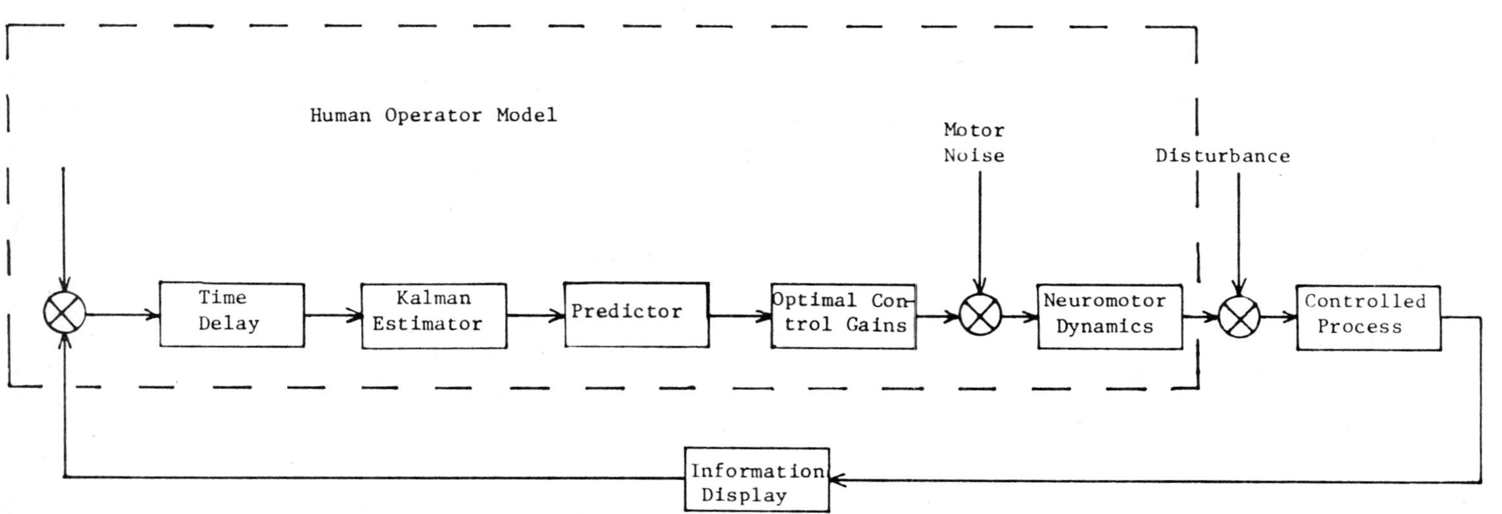

Fig. 3.5.20. Optimal control model of the human operator.[38]

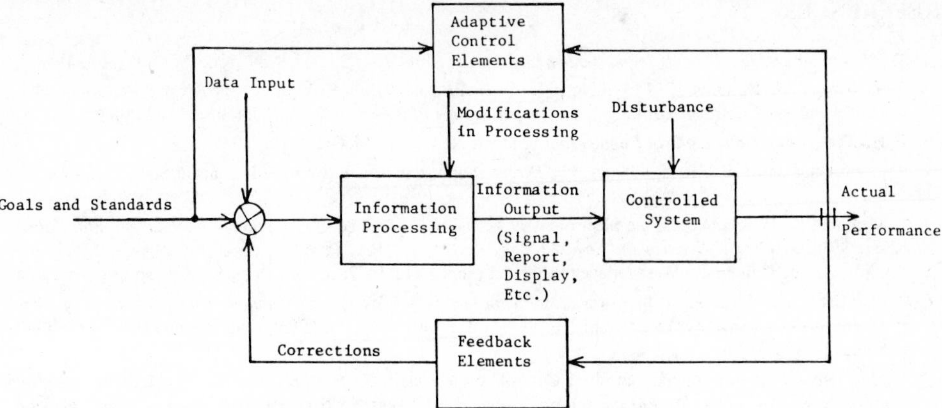

Fig. 3.5.21. An adaptive control model of an information system (application of the model in Fig. 3.5.9).

it would be highly desirable for vendors to compensate for lowered electronic component cost with a greater system complexity to increase reliability or to decrease programming costs.

Hierarchical and distributed computer systems are the wave of the future. Their acceptance is driven by the low cost and relatively low capability of the microprocessor. At the same time they are promoting a modularity and simplification of systems programming that should be retained. Likewise, they tend to promote reliability by subdividing system tasks among several computers, thus lessening the impact of the failure of any one computer.

Use of artificial intelligence techniques is expected to provide new approaches to control logic specification, particularly in adaptive and learning control, once these methods become practical.

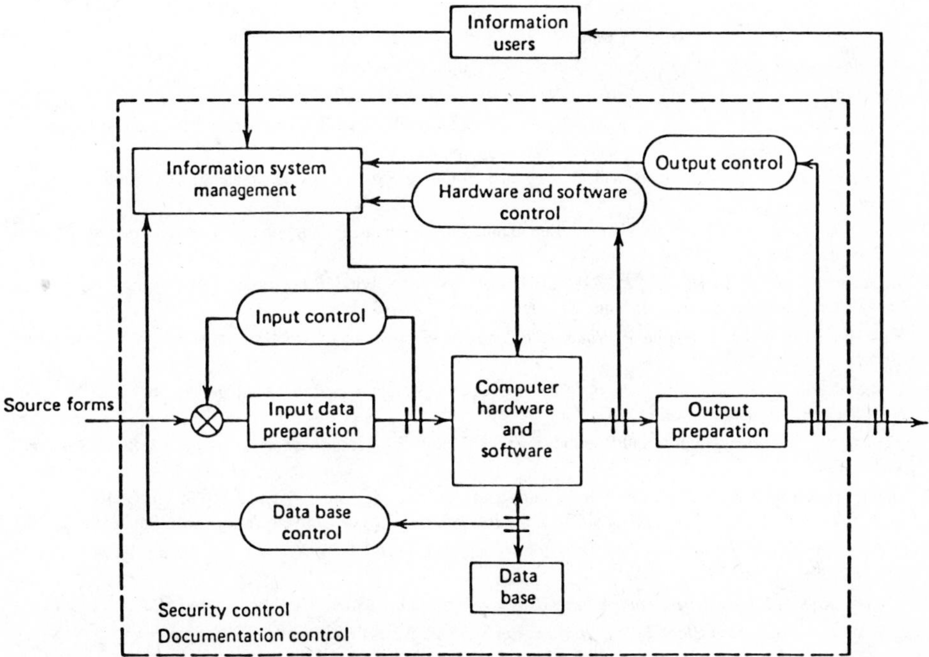

Fig. 3.5.22. Internal control of an information system.

REFERENCES

1. D. M. Considine, Ed. *Process Instruments and Controls Handbook*, 2nd ed. New York: McGraw-Hill, 1974.
2. J. B. Cox, L. J. Hellums, T. J. Williams, R. S. Banks, and G. J. Kirk, Jr. "A Practical Spectrum of DDC Chemical-Process Control Algorithms." *Instrument Society of America Journal*, **13**(10), 65–72 (1966).
3. F. H. Raven. *Automatic Control Engineering*, 2nd ed. New York: McGraw-Hill, 1975.
4. T. J. Williams and F. M. Ryan. *Progress in Direct Digital Control*. Pittsburgh: Instrument Society of America, 1969.
5. L. C. Long, J. H. Schunk, and the Steel Industry Advisory Committee. *Description of the Example Flat Rolled, Sheet, Strip, Steel Mill Used in the Steel Project*. Report No. 105, Purdue Laboratory for Applied Industrial Control, Purdue University, West Lafayette, Indiana (September 1977; revised October 1979 and October 1983).
6. Purdue Laboratory for Applied Industrial Control, project staff. *Tasks and Functional Specifications of the Steel Plant Hierarchy Control System*. Report No. 98, Purdue University, West Lafayette, Indiana (September 1977; revised June 1982).
7. Purdue Laboratory for Applied Industrial Control, project staff. *Systems Engineering of Hierarchy Computer Control Systems for Large Steel Manufacturing Complexes*. Report No. 100, Purdue University, West Lafayette, Indiana (September 1977).
8. P. Uronen and T. J. Williams. *Hierarchical Computer Control in the Pulp and Paper Industry*. Report No. 111, Purdue Laboratory for Applied Industrial Control, Purdue University, West Lafayette, Indiana (September 1978).
9. T. J. Williams, "Computer Control and Its Affect on Industrial Productivity." A. I. Johnson Memorial Lecture, National Research Council of Canada, Montreal (May 24, 1979).
10. T. J. Williams. "The New Process Control Hardware and Its Effect on Industrial Control of the Future." *Proceedings PROMECON I*, Institute of Measurement and Control, London, England, 173–186 (June 16–18, 1981).
11. T. B. Sheridan and W. R. Ferrell. *Man-Machine Systems: Information, Control, and Decision Models of Human Performance*. Cambridge, MA: MIT Press, 1974.
12. R. E. Bellman. *Dynamic Programming*. Princeton, NJ: Princeton University Press, 1957.
13. L. S. Pontryagin, V. G. Boltyanskii, R. V. Gamrelidze, and E. F. Mischenko. *The Mathematical Theory of Optimal Processes*. New York: Wiley-Interscience, 1962.
14. R. E. Kalman. *Transactions of the American Society of Mechanical Engineers*, **82D**, 35–45 (1960).
15. I. D. Landau. *Adaptive Control*. New York: Academic Press, 1979.
16. G. N. Saridis. *Self-Organizing Control of Stochastic Systems*. New York: Dekker, 1977.
17. R. M. Glorioso and F. C. Osorio. *Engineering Intelligent Systems*. Billerica, MA: Digital Press, 1980.
18. M. Athans and P. L. Falb. *Optimal Control*. New York: McGraw-Hill, 1966.
19. P. S. Buckley. *Techniques of Process Control*. New York: Wiley, 1964.
20. F. A. Shinskey. *Process Control Systems*. New York: McGraw-Hill, 1967.
21. W. I. Bullers, S. Y. Nof, and A. B. Whinston. "Artificial Intelligence in Manufacturing Planning and Control." *AIIE Transactions*, 351–363 (December 1980).
22. F. Hayes-Roth, D. A. Waterman, and D. B. Lenat. *Building Expert Systems*. Reading, MA: Addison Wesley, 1983.
23. S. Y. Nof. "An Expert System for Planning/Replanning Programmable Production Facilities." *International Journal of Production Research* (1984).
24. A. Knaepuer and W. B. Rouse. "A Model of Human Problem Solving in Dynamic Environments." *Proceedings of the Human Factors Society*, 27th annual meeting, 695–699 (1983).
25. R. S. Pressman and J. E. Williams. *Numerical Control and Computer-Aided Manufacturing*. New York: Wiley, 1977.
26. U. Rembold, R. Dillman, and P. Levi. "The Role of the Computer in Robot Intelligence." In S. Y. Nof, Ed., *Handbook of Industrial Robotics*. New York: Wiley, 1985.
27. Y. Koren. "Numerical Control and Robotics." In S. Y. Nof, Ed., *Handbook of Industrial Robotics*. New York: Wiley, 1985.
28. O. Z. Maimon and S. Y. Nof. "Activity Controller for a Multiple Robot Assembly Cell." In D. E. Hardt and W. J. Book, Eds., *Control of Manufacturing Processes and Robot Systems*. New York: ASME Publications, 1983.
29. J. Y. S. Luh. "Design of Control Systems for Industrial Robots." In S. Y. Nof, Ed., *Handbook of Industrial Robotics*. New York: Wiley, 1985.
30. R. Paul. *Robot Manipulators, Mathematics, Programming, and Control*. Cambridge, MA: MIT Press, 1981.
31. J. J. Verzijl. *Production Planning and Information Systems*. New York: Halsted, 1976.
32. A. K. Kochhar. *Development of Computer-Based Production Systems*. New York: Halsted, 1979.
33. T. J. Williams. "Hierarchy Computer Control Systems for Overall Industrial Plant Control." *Proceedings of the 13th International Conference on Systems Sciences*, University of Hawaii, Honolulu (1980).

34. S. Y. Nof, M. M. Barash, and J. J. Solberg. "Operational Control of Item Flow in Versatile Manufacturing Systems." *International Journal of Production Research*, **17**(5), 479–489 (1979).

35. S. Y. Nof, W. I. Bullers, and A. B. Whinston. "Decision and Control in Automatic Manufacturing." *AIIE Transactions*, **12**(2), 156–169 (June 1980).

36. W. B. Rouse and D. Gopher. "Estimation and Control Theory: Application to Modeling Human Behaviour." *Human Factors*, **19**(4), 315–330 (1977).

37. S. L. Hwang, et al. "Integration of Humans and Computers in the Operation and Control of Flexible Manufacturing Systems. *International Journal of Production Research* (1984).

38. D. L. Kleinman, S. Baron, and W. H. Levison. "A Control Theoretic Approach to Manned-Vehicle Systems Analysis." *IEEE Transactions on Automatic Control*, **AC-16**, 824–832 (1971).

39. S. Baron and W. H. Levison. "Display Analysis with the Optimal Control Model of the Human Operator." *Human Factors*, **19**(5), 437–458 (1977).

40. R. E. Curry, D. L. Kleinman, and W. C. Hoffman. "A Design Procedure for Control/Display Systems." *Human Factors*, **19**(5), 421–436 (1977).

41. D. W. Kroeber and H. J. Watson. *Computer-Based Information Systems*. New York: Macmillan, 1984.

42. G. B. Davis. *Computers and Information Processing*. New York: McGraw-Hill, 1978.

BIBLIOGRAPHY

Bailey, S. J., and K. Pluhar. "Flexible Manufacturing Systems, Digital Control and the Automatic Factory." *Control Engineering*, 59–64 (September 1979).

Bibbero, R. J. *Microprocessors in Instruments and Control*. New York: Wiley-Interscience, 1977.

Distefano, J. J., A. R. Stubberud, and I. J. Williams. *Feedback and Control Systems* (Schaum's Outline Series). New York: McGraw-Hill, 1967.

R. C. Dorf. *Modern Control Systems*, 3rd ed. Reading, MA: Addison-Wesley, 1980.

Y. Koren. *Computer Control of Manufacturing Systems*. New York: McGraw-Hill, 1983.

Michie, D., Ed. *Expert Systems in the Micro-Electronic Age*. Edinburgh University Press, 1979.

Nilsson, N. J. *Principles of Artificial Intelligence*. Tioga, 1980.

Nof, S. Y., Ed. *Handbook of Industrial Robotics*. New York: Wiley, 1985.

Schoeffler, J. D., and R. H. Temple. *Minicomputers: Hardware, Software, and Applications*. New York: IEEE Press, 1972.

CHAPTER 3.6

QUALITY CONTROL AND ASSURANCE

KENNETH E. CASE

Oklahoma State University
Stillwater, Oklahoma

3.6.1 QUALITY, QUALITY CONTROL, AND QUALITY ASSURANCE

Quality is what the customer says it is. Simple as this sounds, it is true that *satisfaction of needs* and *fitness for use* are ultimately judged by the customer whether that "customer" is a downstream production operation, a merchant, a maintenance shop, or the final consumer. Both terms are used in well-known definitions of quality. ANSI/ASQC Standard A3 defines quality as "the totality of features and characteristics of a product or service that bear on its ability to *satisfy given needs*."[1] Juran and Gryna say quality is simply "fitness for use."[2] These definitions of quality are excellent for describing overall intent but are inoperable for day-to-day production use.

Quality parameters and *quality characteristics* are useful in establishing a link between the generic definitions of quality and a set of operational surrogate measures useful for real-time quality control and assurance. Quality parameters can be thought of as customer expectations *as perceived by the producer*. Typical quality parameters include input and output criteria (often many), efficiency, reliability, maintainability, producibility, ease of use, and safety, most of which can be quantified during design and measured sometime during the life cycle. Quality characteristics are specific properties that ultimately dictate the extent to which quality parameters are achieved. They include, for example, dimensions, materials, properties, colors, textures, and contents, most of which are governed by explicit specifications and may be measured prior to release to the customer.

It is appropriate to conceptually divide quality into four elements:

Quality of design.
Quality of production.
Quality of sales, use, and service.
Quality of residual effects.

It is also useful to recognize that the "product" includes not just the physical entity itself but the tangible and intangible support items. These include sales materials, instructions, warranties, options, programs, spares, transport, installation, and service as illustrated in Figure 3.6.1

Quality of design involves "front end" analyses, development, decision making, and planning, which establish the "grade," or potential quality, of the product. It is here that the links between fitness for use, selection of quality parameters, and specification of quality characteristics is established. Decisions and plans made here affect quality of production, sales, use and service, and residual effects to follow. Quality of design is eventually assessed by the degree to which total production design is fit for use and satisfies needs.

Quality of production relates to the production of tangible portions of the product and how they conform to design. It is usually measured in terms of the degree to which specifications are met by the physical entity and its tangible support items (e.g., instruction manuals, programs, spares), which are produced for use by the customer.

Quality of sales, use, and service relates to the delivery of intangible portions of the product and their integration with the tangible portions of the product and the customer. It is measured in terms of how effectively and efficiently the tangible support items are accomplished compared to company policy and expectations.

Quality of residual effects relates to the extent of direct or indirect effects of the product during production, use, or disposal. Resource consumption, safety, pollution, and effects on producers, users and the environment in general, both short-term and long-term, are included. This may take months or years to measure, thus requiring special care and consideration during the design phase to minimize deleterious effects.

There are interrelationships between each of the elements of this conceptual division of quality. These links, illustrated in Figure 3.6.2, enable dynamic adjustment, permitting quality parameters or characteristics to be modified, or better permitting attainment of quality parameters or characteristics. The effectiveness and efficiency of these links are dependent on the integrity and timeliness of the information system and the ability of the organizational structure, policies, and procedures to respond to needed changes. This is where *quality control* and *quality assurance* are important.

Quality control and quality assurance are often used interchangeably, and their meanings are often confused. *Quality control* (QC) is the set of operational techniques and the activities that sustain a quality of product or service that will satisfy given needs; it is also the use of such techniques and activities. *Quality assurance* (QA) is all the planned or systematic actions necessary to provide adequate confidence that a product or service will satisfy given needs.[2] Quality control aims to provide quality through excellent integration of the controllable aspects of policies, planning, and administration; design; procurement; production; customer experience and feedback; corrective action; and employee selection, training, and motivation. Quality assurance aims to provide confidence that quality is what it should be and involves verifications, audits, and evaluations of products, processes and programs. Both QC and QA are necessary; together they might be thought of as "total quality control" or a modern comprehensive quality program.

3.6.2 THE IMPORTANCE OF QUALITY CONTROL AND ASSURANCE

Financial success and even survival have been conclusively linked to understanding, controlling, and assuring quality. The effects of quality improvement or degradation are pervasive, but nowhere are they seen more tangibly than in statistics on *market share, quality costs,* and *productivity.* The quality of design, production, sales, use and service, and residual effects are all important factors affecting market share, quality costs, and productivity, which all relate to the profit line.

Mounting evidence shows that quality has replaced price as the key to improving market share and profit margins. A recent study of 2,000 businesses showed that improving product quality is an effective way to gain market share.[3] Further, companies with high quality and high market share typically have profit margins five times greater than those at the other extreme. When a reputation for high quality is earned and maintained, the company can charge more than its competitors and continue to increase market share.

Quality costs are often described as the "expense of doing things wrong." A *quality cost system,* one tool used in quality assurance, monitors changes in quality costs and helps to identify high cost areas requiring improvement. Quality costs are typically broken down into four major categories as follows, and then into finer accounts as needed:[4]

Prevention. Costs associated with designing, implementing, and maintaining the quality program, including program audit.

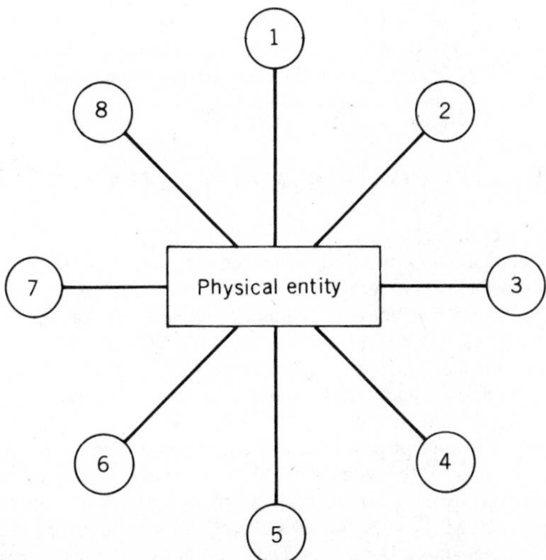

Fig. 3.6.1. The "product." (1) Sales materials, (2) instructions, (3) warranties, (4) options, (5) programs, 96) spares, (7) transport and installation, (8) service.

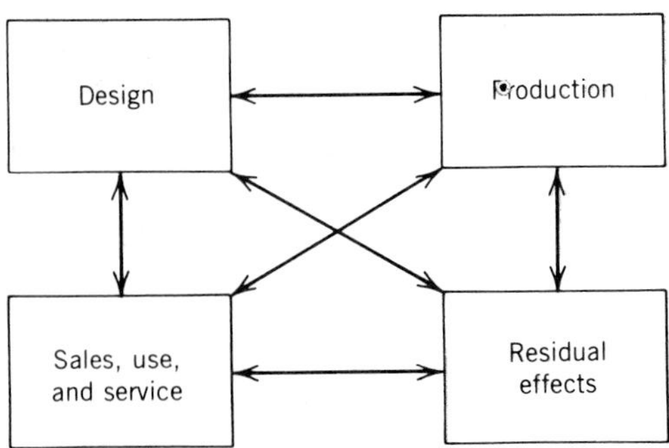

Fig. 3.6.2. Links between elements of quality.

Appraisal. Costs associated with measuring, evaluating, or auditing materials, components, or products to assure conformance with specifications.

Internal failure. Costs associated with nonconforming materials, components, or products that cause losses due to rework, repair, retest, scrap, sorting, and so on, prior to release to the customer.

External failure. Costs associated with nonconforming products that cause losses due to warranties, returns, allowances, and so on, after release to the customer.

The important concept underlying quality cost systems is that investments in prevention and, to a lesser extent, appraisal will be leveraged and result in vastly reduced internal and external failure costs. This philosophy has worked repeatedly in practice and is illustrated in Figure 3.6.3.

It is common to inquire about industry-standard quality costs as a percent of, say, sales. Or, the appropriate breakdown of the quality cost dollar into prevention, appraisal, internal failure, and external failure costs is often desired. Unfortunately, reliable estimates such as these do not exist, despite industry surveys.[5,6,7] What *is* known is that

Quality costs for a company just beginning a quality program are often in the range of 15–18% of sales.

Quality costs for a company with a well-developed quality program can be as low as 3–5% of sales.

Companies regularly operate on the "poorer quality" side of optimum total cost.

Each of these points leads to the conclusion that increased investment in prevention costs, or quality control and quality assurance, will result in substantial net profit increases through reduced failure costs.

Productivity improvement is a natural by-product of sound quality control. Reducing rework, repair, scrap, and the replacement of nonconforming items results in a direct productivity increase. The old quality saying "If you make it right the first time, you don't have to make it over" relates well to productivity. Likewise, many efforts to improve productivity, such as the innovative techniques utilized by Sink, result in quality improvement by-products.[8] Certainly, quality and productivity are not synonymous; however, they are inexorably linked to each other and to financial success and survival.

3.6.3 ACHIEVING QUALITY CONTROL AND QUALITY ASSURANCE

Communication is by far the key to achieving a quality product. Strong quality programs exhibit open and well-structured lines of communication between top management, design, purchasing, production, marketing, and training, with the quality organization serving as a facilitator. In addition, strong quality programs have equally open and well-structured external lines of communication with customers, vendors, regulators, and so on. Designing and producing a product that is not well accepted is indicative of failure of the potential customer, marketing, and design to communicate. Receipt of poor materials or components is indicative of communication failure between the vendor, purchasing, and design. Numerous examples can be given involving poor quality, nearly all of which involve poor communication.

Two menus for structuring a quality program are presented below. One is paraphrased from the 14-point program of Dr. W. Edwards Deming;[9,10] the other is based on ANSI/ASQC Z-1.15 entitled *Generic Guidelines for Quality Systems*.[11] Excellent sources of practical quality philosophy, tools, and techniques are the selected writings of J. M. Juran and Frank M. Gryna, Jr., whose text and handbook are indispensable for designing, implementing, and maintaining a quality program.[2,12] The importance of communications pervades each of the sources just named.

Deming's 14-point special program, adopted as a philosophy and implemented by many companies, may be paraphrased as follows[9] (an earlier version exists[10]):

1. Create constancy of purpose toward improvement of product and service, with the aim to become competitive, to stay in business, and to provide jobs.
2. Adopt the new philosophy. We are in a new economic age. We can no longer live with our commonly accepted style of American management, nor with commonly accepted levels of delays, mistakes, and defective products.
3. Cease dependence on inspection to achieve quality. Eliminate the need for inspection on a mass basis by building quality into the product in the first place.
4. End the practice of awarding business on the basis of price tag. Instead, minimize *total* cost.
5. Improve constantly and forever the system of production and service to improve quality and productivity and thus constantly decrease costs.
6. Institute training on the job.
7. Institute supervision [see point (12)]. The aim of supervision should be to help people and machines and gadgets to do a better job. Supervision of management, as well as supervision of production workers, is in need of overhaul.
8. Drive out fear, so that everyone may work effectively for the company. Fundamental changes are required, such as elimination of the annual rating and management by objectives.
9. Break down barriers between departments. People in research, design, sales, and production must work as a team to foresee problems of production and in-use that may be encountered with the product or service.
10. Eliminate slogans, exhortations, and targets for the work force asking for zero defects and new levels of productivity without providing road maps.
11. Eliminate work standards that prescribe numerical quotas for the day. Substitute aids and helpful supervision.
12. Remove barriers that stand between the hourly worker and his right to pride of workmanship. The responsibility of supervisors must be changed from sheer numbers to quality. The same statement holds for people in management.
13. Institute a vigorous program of education and retraining.
14. Put everyone in the company to work, to accomplish the transformation. The transformation is everyone's job.

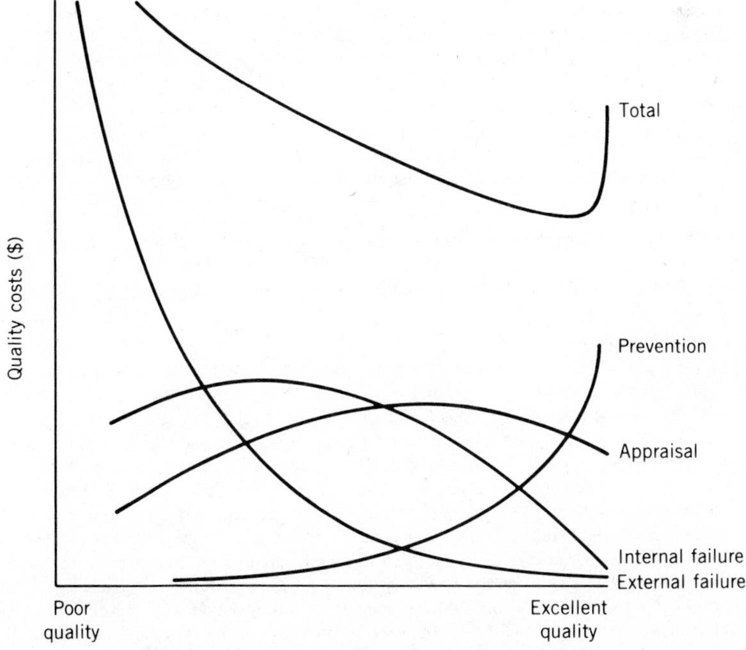

Fig. 3.6.3. Quality cost tradeoffs.

Generic Guidelines for Quality Systems[11] is an excellent aid for structuring or evaluating the quality program for a manufactured product. Its content will not be repeated here; however, major aspects of a sound quality program are outlined:

Policies, Planning, and Administration

Quality policies. A top management responsibility to establish the quality philosophy or "conscience" of the organization with respect to safety, product liability, legal requirements, fitness for use, and other major quality parameters.

Quality objectives. Written assignments providing specific operational plans of action and target levels consistent with the quality policies.

Quality program. A system that implements the quality policies and objectives.

Quality planning. Defines important quality parameters and characteristics, including clear specifications and acceptance standards. Ensure inspections, tests, and audits at appropriate stages using equipment meeting accuracy and precision requirements.

Quality manual. A document containing quality policies, quality program elements, and planning requirements.

Administration. Accomplishes the quality program through clear lines of authority, quality responsibility, quality performance reporting, management of quality costs, and audits of the quality program.

Design Assurance and Change Control

Drawings and specifications. Clear documents containing all applicable requirements.

Measurement control. Establishes measurement capability and uses calibrated equipment for design testing.

Design validation. Evaluates the design and development process of new or redesigned products. It includes one or more formal design reviews of quality characteristics, drawings, specifications, documents, and so on, embracing quality of production, quality of sales, use and service, and quality of residual effects. It also includes design qualification testing to assure the quality of design, market readiness review to ensure adequate production and field support, and design documentation thoroughly defining the product readied for release to production.

Design requalification. Used periodically to review and modify the current design.

Design change control. Ensures that documents and specifications are formally changed and promptly disseminated to all work areas. Obsolete documents should likewise be removed.

Purchased Material Control

Supplier quality management. Ensures that correct purchased materials, fit for use, arrive on time. Supplier selection methods utilize vendor surveys, quality history, and delivery results. Material quality requirements are clear and understood by both supplier and purchaser. Requirements of the supplier's quality system must be established. Material quality assessment may include detailed first production inspection, supplier quality evidence, inspection at source, and/or inspection at receipt. Supplier inspection facilities and procedures must be adequate. Evaluation of supplier includes audit visits and formal ratings. Nonconforming material control includes segregation, information feedback to purchasing, and corrective action by suppliers. Assistance to suppliers includes training, motivation, and open communication.

Supplier quality information. May be used, as practical, to avoid redundant inspection and testing.

Source quality control or surveillance. Specific quality checks of the supplier's quality program, processes, product submitted, acceptance procedures, and quality information generated.

Receiving inspection. Examination of submitted items upon receipt to determine acceptability.

Production Quality Control

Process planning and control. Evaluates the production system and quality program to ensure that they are capable of controlling quality during production. Process capability studies confirm that the inherent process variability is capable of meeting specifications. Operations control ensures that operating procedures and sequences are followed. In process inspection confirms that the product and process conform to requirements. Special process control is applied to difficult or touchy processes using periodic verification of equipment accuracy and variability, operators, environments, times, temperatures, and so on. Inspection and test station control confirms the quality program's ability to identify the inspection status of material and assemblies throughout production. Measurement equipment control ensures that gauges and measurement tools are the right type, range, accuracy, and precision and are calibrated, repaired, and adjusted or replaced as needed. Nonconforming material control clearly identifies nonconforming material, segregates it, ensures review and disposition by proper authority, and initiates corrective action. Document control sees that production and inspection instructions, specifications, and drawings are the proper types and are issued to the proper work station, with prompt recall of obsolete documents. Stock control assures identification, protection, and replacement of materials, components, and product, with attention given to shelf life and age control.

Completed item inspection. Acceptance inspection using 100% or sampling inspection and/or product quality audit may be used to measure overall quality. Important characteristics are inspected or tested. Records are maintained. Reinspection of reworked, repaired, or modified product is performed. Responsibility for product acceptance and product audit rests with the quality function, while responsibility for producing conforming product rests with production.

Handling, storage, and shipping. Proper containers, environments, and vehicles are used to prevent handling damage. Item audit ensures against deterioration. Shipping method control confirms that transit and shipping document requirements are met. Environment control reviews procedures for special protective environments. Carton identification verifies marking and labeling instructions and regulatory and safety compliance.

Product and carton identification. Proper identification provides traceability, safety of handling and unpacking, and ability to issue warnings or recalls when required. Forms identify stock or model numbers, production location, time of production, serial numbers, and warning labels and markings. Location of markings may be on the product itself, packaging, outer wrap, and on transit vehicles. Legibility in terms of size, permanence, and contrast should be considered.

Quality information. The quality program identifies, corrects, and prevents quality deficiencies and provides objective evidence as to whether quality policies and objectives are being achieved. A quality information policy specifies the type and quantity of information to be gathered, reported, and retained. Types of production quality records include product/production process identification, inspection and quality control procedures, inspection and test records, rejected product records, scrap and spoilage reports, and quality cost reports. Quality records should be current, accurate, legible, pertinent, dated; traceable to product, process, or production run, identifiable as to their originator; show quantity, type, and severity of defects; be retained and protected according to policy; and be analyzed and fed back in usable form to appropriate parts of the organization.

User Contact and Field Performance

Market information. Closes the loop on the quality program and verifies its effectiveness.

Product objectives and specifications. Clarifies customer needs, establishes quality parameters, specifies quality characteristics in advance, and measures the degree of success achieved after the product has been marketed.

Advertising and promotion. Reviewed for information on product performance, quality, safety, reliability, and maintainability.

Sales and service. Information on user product acceptance or field failure should be fed back as appropriate. Field failure data includes failure mode, failure frequency, and repair requirements, whether in or out of warranty, in a format that permits efficient analysis of failure data.

Installation, service, and use. Quality, reliability, safety, and performance should not be degraded during installation or use. User literature should include essential information in language suitable for the user. Field installation and service tooling, measuring and test equipment, purchased materials and components, and procedures should receive the same scrutiny as similar items documented "Design," "Purchased Materials," and "Production Quality Control."

User/consumer feedback. Provides data that may be analyzed to ensure that a viable product is in the market or that corrective action must begin. Product return and failure analysis provides data on failure modes and performance acceptance. Product failure early warning system expedites reporting, handling, and diagnosis aimed at rapid corrective action on new products. Complaints generally understate the magnitude of a problem but should be included in a feedback system. Product acceptance surveys solicit the user's reaction to the product. User review panels provide a structured group format for in-depth analysis of product performance criteria. Product safety performance may merit separate tracking. Other external feedback may be available from industry associations, government data, or insurance firms.

Corrective Action

Creation of change. Some element(s) of design, production, or distribution of the product must be changed to reduce or eliminate the probability that a failure will recur. Merely correcting the failed item is insufficient. The quality information system should collect, classify, and rank failure data and then ensure that corrective action is taken on the major causes of failure.

Detection and documentation. All data related to a problem and its effects, from diverse sources, should be used to analyze the product for corrective action.

Quality reports leading to corrective action. Different levels of reporting focus attention on different aspects of a problem. Each level of the organization needs to take action in its own area of responsibility.

Initiating corrective action. Anyone seeing such a need is responsible for initiating corrective actions which may include a written corrective action request.

Evaluating corrective action needs. Quality-related data must be categorized and analyzed to identify recurring problems for which corrective action is feasible and economically justified.

Responsibility for corrective action. Specifically assigned, and includes coordinating, evaluating, recording, and monitoring such efforts. Specific functions, such as design, purchasing, and production, found to have deficiencies that cause problems should be responsible for planning and implementing corrective action.

Determining the cause. Determining cause and effect relationships is essential to corrective action. Impact of a problem should be assessed in terms of quality costs, performance, safety, and customer satisfaction. Each significant variable should be classified as operator controllable or management controllable.[2] Important variables affecting the process capability to meet product specifications should be identified and quantified. New controls should be instituted and regularly audited to assure that production processes can meet product specifications.

Nonconforming product review and disposition. Nonconforming items should be reviewed for disposition and, perhaps, additional findings.

Recall of product items. Evaluation of similar product may be required during a nonconformance analysis. Such product may be in stock, transit, distribution, or the consumer's location, and voluntary audit ranging to mandatory recall may be required.

Recurrence file. Upon completion of corrective action, a final report should be circulated to relevant parties. The entire file should be kept so that suspected problem recurrences can be referred quickly to the file.

Employee Selection, Training and Motivation

Work and product quality. Areas in which personal work affects product quality should be identified and controlled.

Management responsibilities. Include identifying quality-related responsibilities and duties of personnel and state them in job descriptions to help ensure proper employee selection. Provides work environment, materials, equipment, instructions, and supervision capable of meeting requirements; linking piece rate systems to number of conforming items.

Employee selection and training. Selection based on capability, experience, and/or potential. Training helps develop skill competence. Qualification and certification of employees' skill may be required.

Employee motivation. Links job performance and its importance to other employees, customer satisfaction, market share, and job security. Individual or group quality measures give visibility to employees, their supervision, and to management. Visible recognition for achievement of individual or group quality goals can be a stronger motivator.

Clearly, these guidelines for a quality program, as written, are quite terse. They do, however, reflect the broad scope of a modern quality program.

3.6.4 NOTATION

In order to facilitate the presentation of subsequent sections related to statistical quality control, the following notation is presented. In the text, much of this notation is subscripted.

X	Single observation or measurement
$\overline{X}$	Sample average
$\overline{\overline{X}}$	Average of the sample averages
R	Range, or largest measurement minus smallest
$\tilde{X}$	Sample median
$\overline{\tilde{X}}$	Average of the sample medians
s	Sample standard deviation
$\overline{s}$	Average sample standard deviation
μ	Process mean
σ	Process standard deviation (or standard deviation of the sample in σ chart)
MA	Moving average
MR	Moving range
m	Number of subgroups selected
n	Subgroup or sample size
x	Number of nonconformances
p	Fraction nonconforming
$\overline{p}$	Average fraction nonconforming
c	Number of nonconformities (c chart). Allowable number of nonconformances for lot acceptance (in acceptance sampling)
$\overline{c}$	Average number of nonconformities
u	Number of nonconformities per unit
$\overline{u}$	Average number of nonconformities per unit

$\overline{\sigma}$	Average of the standard deviation of the samples in σ chart
$\overline{R}$	Average range
D_4	Factor for R chart upper control limit
D_3	Factor for R chart lower control limit
B_4	Factor for s and σ chart upper control limits
B_3	Factor for s and σ chart lower control limits
$\overline{\text{MR}}$	Average moving range
U	Upper specification limit
L	Lower specification limit
A_1, A_2, A_3	Factors for $\overline{X}$ chart control limits
E_2	Factor for individual (I) control chart limits.
$\overline{\overline{A}}_2$	Factor for median chart control limits
d_2, c_4, c_2	Factors used to estimate the process standard deviation in process capability analyses
$\hat{\sigma}$	Estimate of the process standard deviation.
Z	Standard normal deviate.
z	Stabilized p chart statistic.
α	Type 1 error; producer's risk; probability of rejecting a good lot
β	Type 2 error; consumer's risk; probability of accepting a bad lot
r	Minimum number of nonconformances requiring lot rejection (in acceptance sampling)
P_a	Probability of acceptance

3.6.5 STATISTICAL PROCESS QUALITY CONTROL

Process control is essential if a production operation is to consistently meet demanding specifications. Control charts have been used as a diagnostic and maintenance tool in the control of production processes for several decades since first introduced by Shewhart.[13] During this time there have been numerous refinements and additions to the art and science of control charting. By far the most popular control charts are the $\overline{X}$, R, p, and c charts. The $\overline{X}$ and R charts are used together on variables data to analyze the central tendency and dispersion of a process on a single measurable characteristic. The p chart is used to analyze attributes data such as the fraction nonconforming of a process. The c chart is used when attributes data such as a count of the number of nonconformities per sample unit is the measure of interest. There are many variations of (or alternatives to) these control charts, including the median, s, σ, moving average, moving range, individual, cusum, np, standardized p, u, and other control charts, plus other techniques such as narrow limit gauging (NLG).

Control charts can be used as a management tool to help bring a process into control, keep a process in control, and determine process capability to meet specifications. Output from a process is sampled and a statistic is calculated that bears the name of the control chart on which it is to be plotted (e.g., the sample average $\overline{X}$ is plotted on the $\overline{X}$ chart). When process variability is due to only *chance* causes, the process is said to be in a *state of statistical control* (SOSC) and the plotted pattern of statistics over time is reasonably well behaved, falling within predictable limits and according to predictable patterns. When process variability is due to *assignable* causes such as people, equipment, and/or materials, the process is said to be *out of control* (OOC) and nonrandom patterns appear on the control chart.

Control charts have a center line plus upper and lower control limits (UCL and LCL). These control limits are usually spaced three standard deviations (of the sample statistic being plotted) above and below the center line. As such, it is highly unlikely that a plotted point will fall outside control limits so long as an SOSC exists. In addition to points outside control limits, the *pattern* of plotted points within control limits is equally important in identifying an OOC condition and its cause. Therefore, sample statistics are *always* to be plotted in time order.

3.6.5.1 $\overline{X}$ and R Charts

The *$\overline{X}$ and R control charts* are used together in analyzing a single measurable characteristic. Approximately $m = 20$ or 30 subgroups of size n each are selected. Typical subgroup sizes are $n = 4$ or 5, selected consecutively from the process or produced under as nearly identical conditions as possible. Time between subgroups is dependent on judgment and may be once per hour, twice per day, once per shift, etc. The idea is to have *within* subgroup variation be as small as possible, with production process differences, if any, showing up over time *between* subgroups. For each subgroup i, the average $\overline{X}_i$ and range R_i statistics are calculated.

Once the desired number of subgroups m has been inspected, the $\overline{X}$ and R chart center line and control limits (UCL, LCL) can be calculated, as given in Table 3.6.1. Tables of values for factors A_2, D_4, and D_3 are readily available in statistical quality control texts and standards.[14,15,16] An abbreviated set of these and other factors is presented in Table 3.6.2. These control chart factors are all based on a three standard deviation spread of control limits above and below the center line.

Table 3.6.1. Variables Control Chart Formulas

Control Chart	Center Line	Upper Control Limit (UCL)	Lower Control Limit (LCL)	Comments
$\overline{X}$	$\overline{\overline{X}} = \sum_{i=1}^{m} \overline{X}_i / m$	$\overline{\overline{X}} + A_2\overline{R}$	$\overline{\overline{X}} - A_2\overline{R}$	When used with R chart
R	$\overline{R} = \sum_{i=1}^{m} R_i / m$	$D_4\overline{R}$	$D_3\overline{R}$	
$\overline{X}$	$\overline{\overline{X}} = \sum_{i=1}^{m} \overline{X}_i / m$	$\overline{\overline{X}} + A_3\overline{s}$	$\overline{\overline{X}} - A_3\overline{s}$	When used with s chart
s	$\overline{s} = \sum_{i=1}^{m} s_i / m$	$B_4\overline{s}$	$B_3\overline{s}$	
$\overline{X}$	$\overline{\overline{X}} = \sum_{i=1}^{m} \overline{X}_i / m$	$\overline{\overline{X}} + A_1\overline{\sigma}$	$\overline{\overline{X}} - A_1\overline{\sigma}$	When used with σ chart
σ	$\overline{\sigma} = \sum_{i=1}^{m} \sigma_i / m$	$B_4\overline{\sigma}$	$B_3\overline{\sigma}$	
Median	$\overline{\tilde{X}} = \sum_{i=1}^{m} \tilde{X}_i / m$	$\overline{\tilde{X}} + \overline{A}_2\overline{R}$	$\overline{\tilde{X}} - \overline{A}_2\overline{R}$	
MA	$\overline{X} = \sum_{i=1}^{m} X_i / m$	$\overline{X} + A_2\overline{\text{MR}}$	$\overline{X} - A_2\overline{\text{MR}}$	
MR	$\overline{\text{MR}} = \sum_{i=n}^{m+n-1} \text{MR}_i / (m + n - 1)$	$D_4\overline{\text{MR}}$	$D_3\overline{\text{MR}}$	
I	$\overline{X} = \sum_{i=1}^{m} X_i / m$	$\overline{X} + E_2\overline{R}$	$\overline{X} - E_2\overline{R}$	$\overline{R}$ based upon MR of size n (n=2 often used)
Modified Control Limits	No center *line*	$U - 3\sigma + 3\sigma / \sqrt{n}$	$L + 3\sigma - 3\sigma / \sqrt{n}$	

Table 3.6.2. Factors for Variables Control Charts

Observations in Subgroups	$\overline{X}$ and MA Charts			Median Chart	I Chart	R and MR Charts		s and σ Charts	
	A_1	A_2	A_3	$\tilde{A}_2$	E_2	D_3	D_4	B_3	B_4
2	3.760	1.880	2.659	1.88	2.660	0	3.267	0	3.267
3	2.393	1.023	1.954	1.19	1.772	0	2.574	0	2.568
4	1.880	0.729	1.628	0.80	1.457	0	2.282	0	2.266
5	1.595	0.577	1.427	0.69	1.290	0	2.114	0	2.089
6	1.410	0.483	1.287	0.55	1.184	0	2.004	0.030	1.970
7	1.277	0.419	1.182	0.51	1.109	0.076	1.924	0.118	1.882
8	1.175	0.373	1.099	0.43	1.054	0.136	1.864	0.185	1.815
9	1.095	0.337	1.032	0.41	1.010	0.184	1.816	0.239	1.761
10	1.028	0.308	0.975	0.36	0.975	0.223	1.777	0.284	1.716
11	0.972	0.285	0.927		0.945	0.256	1.744	0.321	1.679
12	0.925	0.266	0.886		0.921	0.283	1.717	0.354	1.646
13	0.885	0.249	0.850		0.899	0.307	1.693	0.382	1.618
14	0.848	0.235	0.817		0.881	0.328	1.672	0.406	1.594
15	0.817	0.223	0.789		0.864	0.347	1.653	0.428	1.572
16	0.788	0.212	0.763		0.849	0.363	1.637	0.448	1.552
17	0.762	0.203	0.739		0.836	0.378	1.622	0.466	1.534
18	0.739	0.194	0.718		0.824	0.391	1.608	0.482	1.518
19	0.717	0.187	0.698		0.813	0.403	1.597	0.497	1.503
20	0.698	0.180	0.680		0.803	0.415	1.585	0.510	1.490
21	0.679	0.173	0.663		0.794	0.425	1.575	0.523	1.477

EXAMPLE 3.6.1

The coded data in Table 3.6.3 represent measurements of a shaft diameter taken in 20 subgroups of 4 shafts each. The coded data replace *xx* in the dimension 2.54*xx* cm. For example, a data value of 13 represents 2.5413 cm.

Calculating center lines and control limits results in

$$\text{Center line}_{\overline{X}} = \overline{\overline{X}} = 679.00/20 = 33.95$$
$$\text{Center line}_R = \overline{R} = 424/20 = 21.20$$
$$\text{UCL}_{\overline{X}} = 33.95 + .729(21.20) = 49.40$$
$$\text{LCL}_{\overline{X}} = 33.95 - .729(21.20) = 18.50$$
$$\text{UCL}_R = 2.282(21.20) = 48.38$$
$$\text{LCL}_R = 0$$

The $\overline{X}$ and R control charts are plotted in Figure 3.6.4.

When used alone, Shewhart control charts are not sensitive to relatively small process shifts. A scheme consisting of four checks is recommended for making a control chart ($\overline{X}$, R, p, etc.) much more sensitive to OOC conditions. The steps of the scheme are illustrated in Figure 3.6.5

1. Divide the area on either side of the control chart center line into three equal zones, A, B, and C.
2. *Check 1.* If one point falls outside 3σ control limits (beyond zone A), mark each such point with an X slightly above or below the point, away from the center line.
3. *Check 2.* If two out of any three successive points fall in or beyond zone A of the same side, mark the second of the two points with an X. The "other" point may be anywhere.
4. *Check 3.* If four out of any five successive points fall in or beyond zone B of the same side, mark the fourth of the four points with an X. The "other" point may be anywhere.
5. *Check 4.* If eight successive points fall in or beyond zone C of the same side, mark the eighth such point with an X.

These checks were applied to the $\overline{X}$ and R charts of Figure 3.6.4, which clearly shows that a shift in process centering occurred toward the end of the data collection process. On an $\overline{X}$ chart, with its underlying normal

Table 3.6.3. Coded Data-Shaft Diameter

Subgroup Number i	Data X_{i1}	X_{i2}	X_{i3}	X_{i4}	Sample Average $\overline{X}_i$	Sample Range R_i
1	13	32	38	44	31.75	31
2	36	11	11	27	21.25	25
3	24	35	20	41	30.00	21
4	31	30	18	38	29.25	20
5	43	34	41	49	41.75	15
6	34	32	29	43	34.50	14
7	34	20	23	44	30.25	24
8	23	26	29	26	26.00	6
9	38	39	38	27	35.50	12
10	43	31	30	25	32.25	18
11	26	9	21	23	19.75	17
12	13	33	39	34	29.75	26
13	37	23	27	28	28.75	14
14	25	49	34	39	36.75	24
15	47	18	24	29	29.50	29
16	33	39	40	49	40.25	16
17	63	47	40	39	47.25	24
18	46	55	38	27	41.50	28
19	46	46	38	57	46.75	19
20	49	27	68	41	46.25	41
Totals					679.00	424

distribution, there is about a 68.3% chance that a point will fall in zone C, 27.2% in zone B, 4.2% in zone A, and 0.3% outside control limits when a process is actually in an SOSC. As a result, a process in an SOSC should incur few if any Xs indicating lack of process control.

Once a control chart has been set up, including application of the four checks, it must be interpreted in order to determine whether an OOC condition exists and, if so, to help diagnose the problem for corrective action. Pattern recognition within control limits is critical in establishing the reason(s) for an OOC condition. Excellent guidelines for interpreting patterns are available.[17] One of the most common patterns encountered in practice, the *stable mixture* pattern, is presented in Figure 3.6.6, which shows an $\overline{X}$ chart that has no points beyond control limits. Yet, the Xs marked and the void of points in zone C clearly indicate a two-cause system (different operators, processes, materials, and so on), as conceptually illustrated by the two overlapping process distributions.

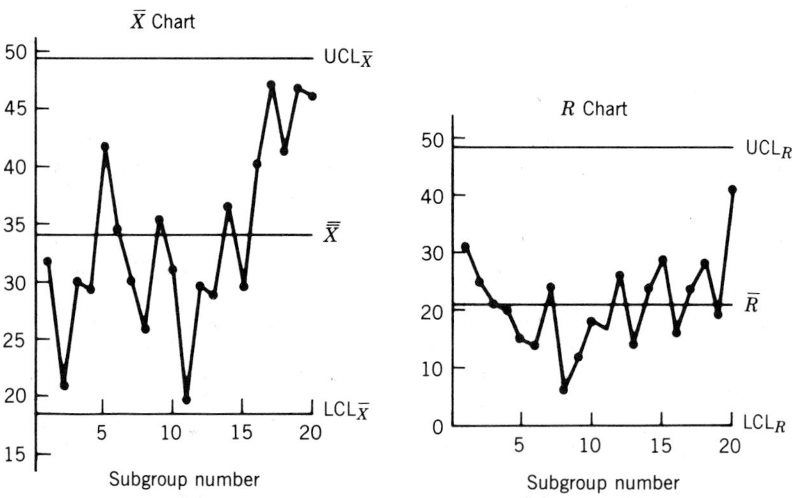

Fig. 3.6.4. $\overline{X}$ and R charts of coded data for shaft diameter.

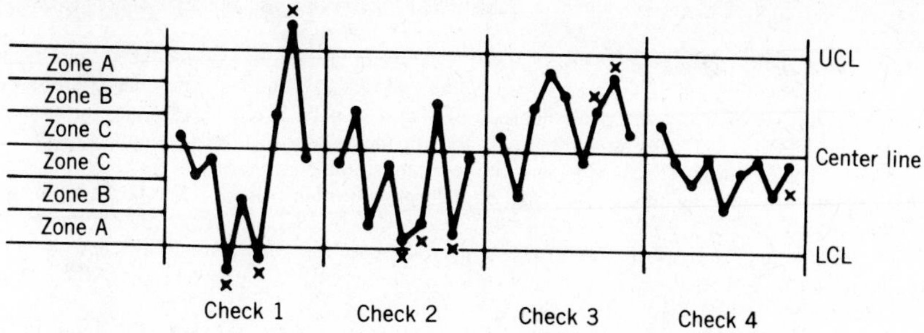

Fig. 3.6.5. Scheme for sensitizing Shewhart control charts.

3.6.5.2 Other Charts for Variables

Many other charts exist for statistical process control using variables data. None of these enjoys the extensive use that the $\bar{X}$ and R charts have received. The most prominent of these other charts are briefly described here.

The *median chart* may be used to replace the $\bar{X}$ chart. Individual values from the n items in a sample are plotted, with the middle (or median) value circled and evaluated in light of control limits. Calculations are not necessary; within-sample dispersion is readily apparent; and specification limits can be superimposed on the chart against which the individual item performances can be compared. The usual R chart is continued. The center line and control limit formulas appear in Table 3.6.1. Setup and use of the median chart are explained in reference 18.

The *s chart* may be used to replace the R chart as a measure of within-sample variation. Use of the s chart is recommended when the subgroup size n is greater than 10. For each subgroup i, the statistic

$$s_i = \left[\left(\sum_{j=1}^{n} X_{ij}^2 - \left(\sum_{j=1}^{n} X_{ij}\right)^2 \Big/ n\right) \Big/ \left(n - 1\right)\right]^{1/2}$$

is calculated. The center line and control limits are given in Table 3.6.1. Setup and use of the s chart are explained in references 14 and 18.

The *σ chart* may also be used to replace the R chart. The appropriate subgroup statistic is

$$\sigma_i = \left[\left(\sum_{j=1}^{n} X_{ij}^2 - \left(\sum_{j=1}^{n} X_{ij}\right)^2 \Big/ n\right) \Big/ n\right]^{1/2}$$

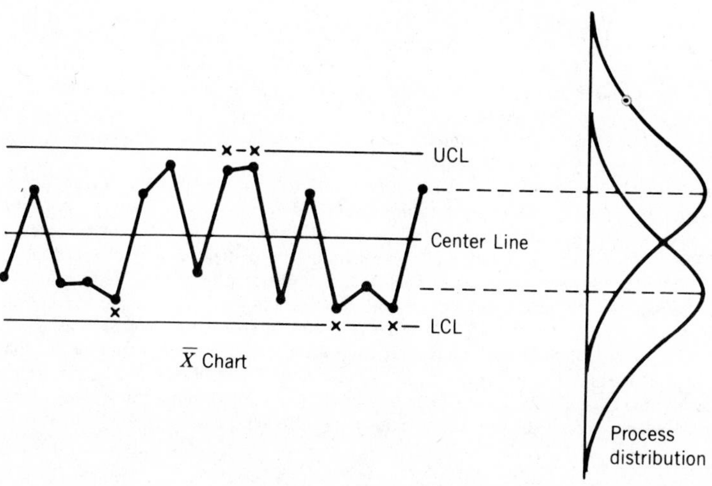

Fig. 3.6.6. Stable mixture pattern indicated on $\bar{X}$ chart.

The center line and control limits are given in Table 3.6.1. Setup and use of the σ chart are explained in reference 15.

The *moving average* (MA) and *moving range* (MR) *charts* can be used in place of the $\bar{X}$ chart when mixes or blends, gas compositions, temperatures, viscosities, and so on, are sampled. These items are not discrete and do not lend themselves to a traditional subgroup of size $n = 4$ or 5; rather, only one (or an average of several) measurement X_i is available for sample i. In the chemical industries, it is common to plot a chart of moving averages. The MA and MR charts permit one to establish the concept of process control charting. The size of the MA and MR moving subgroup n must be determined; often it is $n = 3$ or 4. The MR statistic is calculated as $MR_i = \max\{X_{i-n+1}, X_{i-n+2}, \ldots, X_i\} - \min\{X_{i-n+1}, X_{i-n+2}, \ldots, X_i\}$. The center line and control limits are given in Table 3.6.1. Rules for reading and interpreting the MA and MR charts do not follow those presented for the $\bar{X}$ chart since the points plotted are interdependent. Setup and use of the MA and MR charts are explained in reference 15.

The *individual* (I) *chart* is used under similar circumstances as for the MA and MR charts. In addition, the I chart has found use in tracking "administrative" measures such as production rates, costs per unit, and so on. The actual individual measurements X_i are plotted on this chart. Its center line and control limits are given in Table 3.6.1. Note that a moving range chart of size $n = 2$ is often used in conjunction with the I chart. Setup and use of the I chart are explained in reference 15.

The *acceptance control chart* (ACC) may be used to replace the $\bar{X}$ chart; however, its underlying philosophy is quite different. An acceptable process level (APL) and rejectable process level (RPL) are defined, usually based on location of the specification limits and the proportions of nonconforming product deemed acceptable and rejectable, respectively. Specific probabilities of not rejecting a process at the APL and declaring a process out of control at the RPL are used to design the control limits and sample size used. The ACC is most often used only when the specification limits are considerably wider than the natural tolerance limits ($\mu \pm 3\sigma$). The net effect is to have control chart with a center *zone* of acceptability, permitting some deviation in centering, rather than a single center line. A standard R chart is used in conjunction with the ACC. Setup and use of the ACC are explained in reference 14.

Modified control limits on the $\bar{X}$ chart may be used in lieu of the regular $\bar{X}$ chart. They are similar to the acceptance control chart in that they are applicable when the specification limits are considerably wider than the natural tolerance limits, resulting in a center zone of acceptability for the process. The upper control limit is placed $3\sigma_{\bar{X}}$ up from the highest allowable process center, which itself is 3σ below the upper specification limit; a similar statement describes placement of the lower control limit. Unlike the ACC, modified control limits are designed with emphasis on not rejecting an acceptable process level; there is no provision for ensuring, with a specified probability, rejection of a certain rejectable process level. Control limits are given in Table 3.6.1. Setup and use of modified control limits are explained in reference 15.

The *cusum chart* is alternatively known as the V-mask chart and may be used to replace the $\bar{X}$ chart. The cusum chart is particularly useful for detecting small sustained shifts in process centering that may be more difficult to detect on an $\bar{X}$ chart, especially if the sensitivity checks are not employed. Design of the cusum chart is slightly more involved than the charts mentioned earlier; also, its parameters are situation-specific rather than generic as are most other charts. Setup and use of the cusum chart are explained in reference 14.

3.6.5.3 *p* Charts

The *p chart* is an attributes data chart based on the fraction of product rejected. A sample of n items is selected from a process and each is inspected for nonconformities on multiple characteristics. Any item found to have one or more nonconformities is considered a reject. The number of rejects x divided by the sample size n is the fraction rejected.

The p chart, like $\bar{X}$ and R charts, is very popular in industry. It requires only attributes data, so go/no go gauging, visual inspections, and the like yield appropriate data. The p chart, unlike $\bar{X}$ and R charts, is particularly attractive where combining several characteristics into one chart is desired. Unfortunately, common sample sizes are often in the range of 50 to 300 items.

Approximately $m = 20$ or 30 subgroups of size n are selected from the process, keeping in mind the desire for within-subgroup homogeneity, with production process differences, if any, showing up over time between subgroups. For each subgroup i, the statistic $p_i = x_i/n_i$ is calculated. The p chart center line and control limits are given in Table 3.6.4. The center line and control limits are plotted on graph paper, and the individual statistics p_i are plotted in chronological order. The same four-check scheme presented for the $\bar{X}$ chart may be applied. Also, the pattern recognition methods discussed for reading control charts are applicable.[17]

Two additional items are important to note. First, the sample size n_i may not be constant from subgroup to subgroup on the p chart. In this case, individual control limits for each point may be calculated using the formulas presented in Table 3.6.4, with the appropriate subgroup size inserted for n_i. Second, it may be desired for later troubleshooting to keep a record of which characteristic nonconformities are observed during inspection, even though a knowledge of specific nonconformities is not needed for the p chart.

3.6.5.4 *c* Charts

The *c chart* is based on the number of nonconformities observed per sample group. The size of the sample group must first be established and can, for example, consist of an airplane skin section, five stereo receivers,

Table 3.6.4. Attributes Control Chart Formulas

Control Chart	Center Line	Upper Control Limit (UCL)	Lower Control Limit (LCL)
p	$\bar{p} = \sum_{i=1}^{m} x_i \Big/ \sum_{i=1}^{m} n_i$	$\bar{p} + 3\sqrt{\bar{p}(1 - \bar{p})/n_i}$	$\bar{p} - 3\sqrt{\bar{p}(1 - \bar{p})/n_i}$
np	$n_i\bar{p}$	$n_i\bar{p} + 3\sqrt{n_i\bar{p}(1 - \bar{p})}$	$n_i\bar{p} - 3\sqrt{n_i\bar{p}(1 - \bar{p})}$
Stabilized p	0	3	−3
c	$\bar{c} = \sum_{i=1}^{m} c_i \Big/ m$	$\bar{c} + 3\sqrt{\bar{c}}$	$\bar{c} - 3\sqrt{\bar{c}}$
u	$\bar{u} = \sum_{i=1}^{m} c_i \Big/ \sum_{i=1}^{m} n_i$	$\bar{u} + 3\sqrt{\bar{u}/n_i}$	$\bar{u} - 3\sqrt{\bar{u}/n_i}$

three fork-lift trucks, ten pairs of overalls, and so on. The size of the sample group must remain fixed from sample to sample in order to maintain constant the area of opportunity for the occurrence of nonconformities. The c chart, like the p chart, requires only attributes data. It is attractive when a simple *count* of nonconformities (deviant visuals, speeds, frequencies, dimensions, etc.) is the performance measure of interest.

Approximately $m = 20$ or 30 sample groups are selected over time. The items selected for a sample group should be homogeneous in the sense that they are produced together or under similar conditions. Once again, production process differences, if any, should show up over time between sample groups. For each sample group, the statistic c_i is kept, representing the total count of nonconformities observed over all units in the sample group. The c chart center line and control limits are given in Table 3.6.4. The center line and control limits are plotted on graph paper, and the individual statistics c_i are plotted in chronological order. The same four-check scheme as well as the pattern recognition methods discussed previously are applicable.[17]

3.6.5.5 Other Charts for Attributes

The *np chart* may be used in place of the p chart when it is desired to plot *numbers* of nonconformances x_i rather than fraction nonconforming. The primary disadvantage to this chart is that the center line shifts as a function of sample size n_i. The center line and control limits for this chart are given in Table 3.6.4. Setup and use of the *np* chart are explained in references 15 and 18.

The *stabilized p chart* may be used in lieu of the p chart when it is desired to have constant center line and control limits, even when the subgroup size n_i is changing from sample to sample. The statistic plotted is

$$z_i = (p_i - \bar{p})/\sqrt{\bar{p}(1 - \bar{p})/n}$$

The center line is set at zero and control limits are set at ± 3. Setup and use of the stabilized p chart are explained in reference 14.

The *u chart* is applicable in place of the c chart when the sample group consists of multiple units and it is desired to express results on a per unit basis. It *must* be used in place of the c chart when the number of units n_i in the sample group cannot be maintained constant. The statistic plotted is $u_i = c_i/n_i$. The center line and control limits are given in Table 3.6.4. Setup and use of the u chart are explained in references 15 and 18.

3.6.6 PROCESS CAPABILITY ANALYSIS

In using $\bar{X}$ and R (or s or σ) charts, an SOSC or OOC condition is determined without respect to the specification limits on the measurable dimension of interest. Once the process is in an SOSC, however, it is desirable to determine the capability of the process to meet specifications. This procedure is easily performed and often permits the estimation of process yields.

Steps of a process capability analysis include the following:

1. Establish an SOSC on both the $\bar{X}$ and R charts. Failure to do this will invalidate the procedures to follow.
2. Estimate the process standard deviation $\hat{\sigma} = \bar{R}/d_2$ (or $\hat{\sigma} = \bar{s}/c_4$ or $\hat{\sigma} = \bar{\sigma}/c_2$), where d_2 (or c_4 or c_2) is a function of the subgroup size. Table 3.6.5 presents values of d_2 (and c_4 and c_2). This estimate of $\hat{\sigma}$

Table 3.6.5. Factors for Estimating Process Standard Deviation

Observations in Subgroups	d_2 for $\hat{\sigma} = \bar{R}/d_2$	c_4 for $\hat{\sigma} = \bar{s}/c_4$	c_2 for $\hat{\sigma} = \bar{\sigma}/c_2$
2	1.128	.7979	.5642
3	1.693	.8862	.7236
4	2.059	.9213	.7979
5	2.326	.9400	.8407
6	2.534	.9515	.8686
7	2.704	.9594	.8882
8	2.847	.9650	.9027
9	2.970	.9693	.9139
10	3.078	.9727	.9227
11	3.173	.9754	.9300
12	3.258	.9776	.9359
13	3.336	.9794	.9410
14	3.407	.9810	.9453
15	3.472	.9823	.9490
16	3.532	.9835	.9523
17	3.588	.9845	.9551
18	3.640	.9854	.9576
19	3.689	.9862	.9599
20	3.735	.9869	.9619
21	3.778	.9876	.9638

is based on the within-sample variation, assuming this is typical of the process as a whole and that there is negligible between-sample variation, since the $\bar{X}$ chart indicates control.

3. If $6\hat{\sigma} \leq$ UCL, the process is said to be *capable* of meeting specifications. Centering of the process on the target value may still be a problem.
4. Fraction nonconforming and process yield may be estimated if the process distribution is normally distributed by determining $z_L = (L\text{-}\bar{\bar{X}})/\hat{\sigma}$ and $z_U = (U\text{-}\bar{\bar{X}})/\hat{\sigma}$ and using the standard normal table to estimate the fraction of product less than L and greater than U.

EXAMPLE 3.6.2

Using the data of Table 3.6.3, suppose it is discovered that subgroups 16–20 were collected during an assignable cause consisting of improper equipment setup. Having discovered and eliminated that cause, subgroups 16–20 are discarded as not being representative of process capability. Considering only subgroups 1–15, $\bar{\bar{X}}$ = 30.47 and R = 19.73, and the process is believed to be in an SOSC reflected by these values. Further, plotting a histogram, plotting the data on normal probability paper, or performing a goodness of fit test, yield confidence that the process is normally distributed. Estimating, $\hat{\sigma} = \bar{R}/d_2$ = 19.73/2.059 = 9.58. If specifications are L = 20 and U = 80, the process is capable, since 6(9.58) $\leq$ 80–20 or 57.48 $\leq$ 60. At present, there is a centering problem and Z_L = (20.00 − 30.47)/9.58 = −1.09 and Z_u = (80 − 30.47)/9.58 = 5.17. Using the standard normal table indicates that fraction nonconforming is 13.79%, all below the lower specification limit.

Control charting and determining whether an SOSC exists can be done regardless of whether the process has an underlying normal distribution; that is, those procedures are generally robust. Process capability on significantly nonnormal processes requires fitting of a distribution to the data followed by probability paper plotting or numerical integration to estimate process yields. Use of the outlined procedures for estimating and calculating process yields is not proper for nonnormal processes.

A "quick and dirty" approach to process capability analysis may be used (with care) if a preliminary opinion is needed. It does not involve the use of control charts; therefore, it cannot be certain whether the process is in an SOSC.

The steps to this approach include the following:

1. Sample 50 or more items from the process and measure the appropriate dimension of each.
2. Plot a histogram of the results. Often, the scaling for a histogram is done in advance and the process checker need only put a check mark in the appropriate histogram cell.
3. Overlay specification limits on the histogram. If the sample histogram extends beyond these limits in both directions, conclude that the process is OOC and/or not capable. If the histogram extends beyond only one specification limit, conclude that the process is in need of centering, OOC, and/or not capable.

If the histogram is well within both specification limits, conclude that the process at the time of sampling is capable; however, it may not be in an SOSC.

The "quick and dirty" procedure is often used to determine priorities for those processes and operations to receive a thorough control chart analysis.

3.6.7 ACCEPTANCE SAMPLING

Acceptance sampling is a systematic procedure for deciding on product disposition based on conclusions drawn from inspecting a fraction of the product. It is applicable to incoming, in process, and finished product inspection. Acceptance sampling is soundly based in mathematics and does work well when properly implemented; it was accepted in principle by much of the industrial community in the 1940s. It diminishes the need for 100% inspection, which can be far from perfect due to fatigue, monotony, and inherent inspector limitations. Also, it permits the inspector to be more thorough due to often-reduced time pressures.

Acceptance sampling plans exist for both inspection by attributes and by variables. Attributes acceptance sampling is by far the more commonly used of the two. In attributes acceptance sampling, multiple characteristics can be considered by one sampling plan; variables acceptance sampling considers only one characteristic per plan. Acceptance sampling can be used for product in finite lots, continuous production processes, or in bulk.

Although sampling plans can be custom-designed using appropriate mathematics, several established and accepted sampling schemes exist that provide coordinated families of plans. These sampling schemes are sufficiently rich that a plan can be selected to accomplish most any sampling objective. Two principles must be adhered to without exception, however, for proper acceptance sampling performance. First, pertinent performance measures should be investigated to ensure that the plan is capable of performing as desired. Second, all rules for the designated sampling plan must be followed explicitly. For example, resampling a barely rejected lot while using a single sampling plan is prohibited.

3.6.7.1 Attributes Lot Acceptance Sampling

A *single sampling* plan is defined by the sample size n and the acceptance number c. The n items are drawn at random (each item is equally likely to be selected) from the lot and inspected; the number of rejected items x is determined. If $x \leqslant c$, the lot is accepted; otherwise, it is rejected.

A *double sampling* plan is defined by two sample sizes, n_1 and n_2, two acceptance numbers, c_1 and c_2, and two rejection numbers, r_1 and $r_2 = c_2+1$. A random sample of size n_1 is drawn; the number of rejected items is x_1. If $x_1 \leqslant c_1$, the lot is accepted; if $x_1 \geqslant r_1$, the lot is rejected; if $c_1 < x_1 < r_1$, a second sample of size n_2 is drawn at random from the rest of the lot. The number of rejected items is x_2. If $x_1 + x_2 \leqslant c_2$, the lot is accepted; if $x_1 + x_2 \geqslant r_2$, the lot is rejected.

Multiple sampling is an extrapolation of double sampling in which more than two levels of sampling may be used. Again, there are acceptance and rejection numbers that dictate whether acceptance, rejection, or another sample will take place. *Sequential sampling* is a further extrapolation in which items are taken from a lot one at a time. At each sampling step, the results are compared to an acceptance number and a rejection number that dictate whether the lot should be accepted or rejected or whether another item should be sampled. The primary advantage of double, multiple, and sequential sampling over single sampling is that a smaller average sample size may be realized. Their primary disadvantage is the administrative and "bookkeeping" complexity involved.

Operating Characteristic (OC) Curve

The OC curve is the most important performance measure used in lot acceptance sampling. A typical OC curve is illustrated in Figure 3.6.7. The OC curve shows, for a given sampling plan, probability of acceptance versus incoming quality expressed in percent nonconforming. As such, the OC curve describes the risks of rejecting good quality or accepting poor quality in using a particular sampling plan; if the user finds the OC curve unacceptable, it follows that the sampling plan is unacceptable.

The *Acceptable quality level* (AQL) is the maximum percent nonconforming (or the maximum number of nonconformities per hundred units) that, for purposes of sampling inspection, can be considered satisfactory as a process average.[19] A "producer's risk" α of having lots of such quality rejected is often associated with the AQL.

The *Limiting quality* (LQ) is the percentage of nonconforming units (or nonconformities per hundred units) in a lot for which for purposes of acceptance sampling, the consumer wishes the probability of acceptance to be restricted to a low value. A "consumer's risk" β of having lots of such quality accepted is often associated with the LQ.

Actually, there are two different OC curves in use. The Type A OC curve shows the probability of acceptance for isolated lots of some specified quality. The more commonly used Type B OC curve shows the probability of acceptance for a series of lots formed randomly from a process operating at some specified quality.[14,15] Further reference will be to the Type B OC curve only. The probability of acceptance for an attribute's single sampling plan n, c when using a Type B OC curve is given by

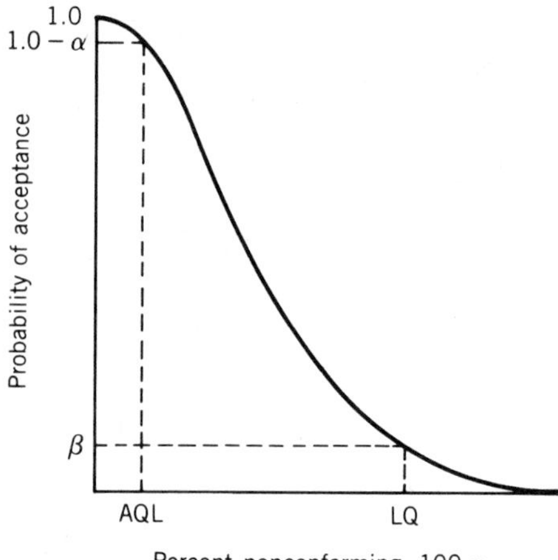

Fig. 3.6.7. Operating characteristic curve.

$$P_a = \sum_{x=0}^{c} \binom{n}{x} p^x(1 - p)^{n-x}, \qquad 0 \leq p \leq 1$$

Other Performance Measures

Average outgoing quality (AOQ) is an important performance measure applicable when rejected lots are rectified or screened. When lots are accepted on the basis of sampling inspection, they are passed on with, perhaps, some nonconforming items contained in the uninspected portion of the lot. If rejected lots are 100% inspected (rectified or screened), they should theoretically be free of nonconforming items. Any rejectables found during sampling and screening are dispositioned and may or may not be replaced. The AOQ is a measure of the average fraction of nonconforming items outgoing from the inspection process. It is a function of incoming percent nonconforming and is given by

$$\text{AOQ} = pP_a(N - n)/N \qquad \text{(single sampling with replacement)}$$

$$= pP_a(N - n)/[N - pn - p(1 - P_a)(N - n)] \qquad \text{(single sampling without replacement)}$$

A typical AOQ curve is shown in Figure 3.6.8. Note that the maximum AOQ is called the *Average outgoing quality limit* (AOQL), reflecting the worst average fraction nonconforming outgoing from the inspection process *regardless* of incoming quality levels. Formulas for the AOQ under many different conditions, for single or double sampling, are given in Beainy and Case.[20]

Average total inspection (ATI) reflects the average amount of sampling *and* screening inspection per lot when rejected lots are rectified. Of course, every lot is sampled; only rejected lots are 100% inspected. The ATI is a function of incoming percent nonconforming and is given by

$$\text{ATI} = [n + (1 - P_a)(N - n)]/(1 - p) \qquad \text{(single sampling with replacement)}$$

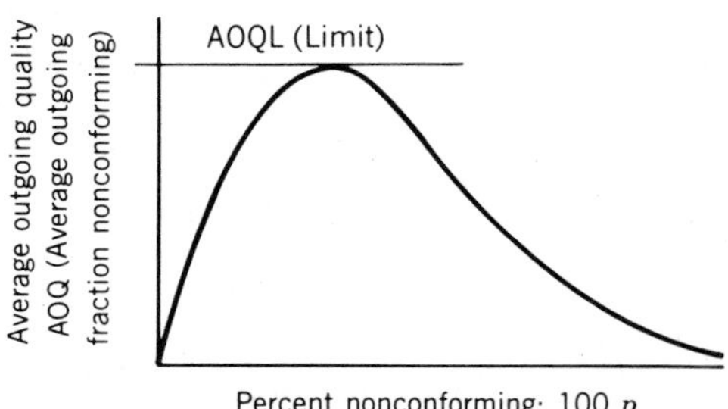

Fig. 3.6.8. Average outgoing quality curve.

$$= n + (1 - P_a)(N - n) \qquad \text{(single sampling without replacement).}$$

A typical ATI curve is shown in Figure 3.6.9. Formulas for the ATI under many different conditions, for single or double sampling, are given in Beainy and Case.[20]

The *average sample number* (ASN) shows the sampling economies achieved using double, multiple, or sequential sampling, relative to single sampling. This measure is applicable regardless of whether rejected lots are screened. Typical ASN curves for single, double, and multiple sampling are shown in Figure 3.6.10. Formulas for the ASN are given in reference 14.

ANSI/ASQC Z1.4

This standard, *Sampling Procedures and Tables for Inspection by Attributes,*[19] is essentially a slight revision of MIL-STD-105D. This sampling system presents a wide range of coordinated sampling plans designed to induce a supplier to maintain a process average at least as good as a specified AQL. It has the following major features:

Rules for switching between normal, tightened, and reduced inspection as dictated by inspection results.
Seven different inspection levels for discrimination between good and bad lots.
Acceptable quality levels (AQLs) ranging from 0.01% to 10% to 1,000 nonconformities per 100 units.
Provisions for lots of size 2 and over.
Single, double, and multiple sampling.
Average outgoing quality limit (AOQL) factors.
Provision for plan selection based on limiting quality (LQ).
Average sample number (ASN) curves.
Operating characteristic (OC) curves and tables.
Overall scheme performance curves and tables.
Instructions for proper use of the standard.

3.6.7.2 Variables Lot Acceptance Sampling

When actual measurements are taken on a particular characteristic, variables acceptance sampling may be used. Since variables data contain more information than attributes data, smaller sample sizes than required for an attributes plan can achieve the same OC curve. Unfortunately, a given variables sampling plan is applied to only one characteristic at a time. Also, most formulas, instructions, and sampling schemes for variables sampling assume that the distribution of characteristic dimensions is normal. Variables sampling may be carried out on nonnormal distributions; however, only those having a thorough understanding of the underlying mathematics should attempt to do so.

If the characteristic of interest has a normal distribution with known standard deviation σ and it is desired to have a sampling plan that fits an OC curve with risks α and β at the AQL and LQ, respectively, sampling plans are easily calculated as follows:

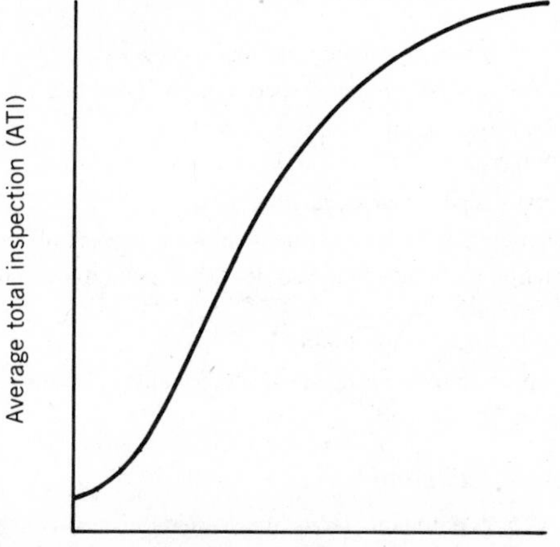

Fig. 3.6.9. Average total inspection curve.

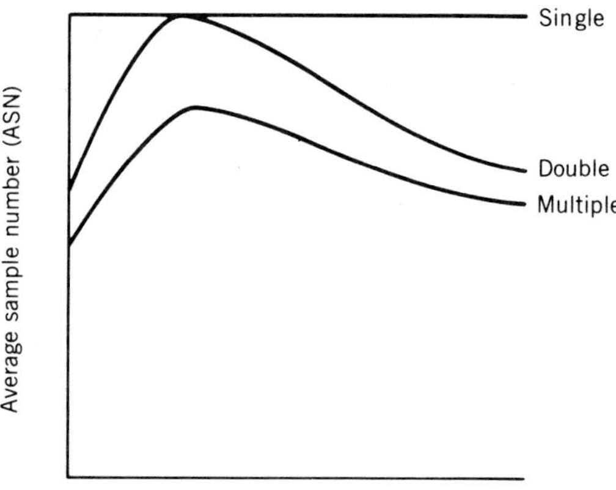

Fig. 3.6.10. Average sample number curve.

$$n = [(Z_\alpha + Z_\beta)/(Z_{AQL} - Z_{LQ})]^2$$

$$k = (Z_{AQL} - Z_\alpha + Z_{LQ} + Z_\beta)/(2\sqrt{n})$$

A random sample of n items is selected from the lot and the measurements on the appropriate characteristic are averaged ($\overline{X}$). For a single lower specification limit L, if $(\overline{X} - L)/\sigma \geqslant k$, the lot is accepted. Likewise, for the upper specification limit U, if $(U - \overline{X})/\sigma \geqslant k$, the lot is accepted.

If the standard deviation is unknown and σ is estimated by s, the equations for n and k become

$$k = (Z_\alpha Z_{LQ} + Z_\beta Z_{AQL})/(Z_\alpha + Z_\beta)$$

$$n = [(1+k^2)/2)((Z_\alpha + Z_\beta)/(Z_{AQL} - Z_{LQ})]^2$$

For a single lower specification limit L, if $(\overline{X} - L)/s \geqslant k$, the lot is accepted. Likewise, for the upper specification limit U, if $(U - \overline{X})/s \geqslant k$, the lot is accepted. For a discussion of variables plans for double specification limits, see Duncan.[14]

ANSI/ASQC Z1.9

This standard, *Sampling Procedures and Tables for Inspection by Variables for Percent Nonconforming,* is essentially a moderate revision of MIL-STD-414. ANSI/ASQC Z1.9 is roughly matched to ANSI/ASQC Z1.4, allowing inspection under either standard for stated AQLs and inspection levels with reasonably equivalent protection. It has the following major features:

Rules for switching between normal, tightened, and reduced inspection as dictated by inspection results.

Five different inspection levels for discrimination between good and bad lots.

Acceptable quality levels (AQLs) ranging from 0.10% to 10%.

Provision for lots of size 2 and over.

Operating characteristic curves for normal inspection.

Assumption that the underlying distribution of individual measurements is normal in shape.

Provision for standard deviation unknown (standard deviation method), standard deviation unknown (range method), and (3) standard deviation known.

Provision for single or double specification limits.

Two different but consistent procedures for use, one of which yields an estimate of lot percent nonconforming.

Instructions for proper use of the standard.

Other Methods of Acceptance Sampling

Continuous sampling is used with a continuous production system when it is unnatural or undesirable to "lot" items for acceptance sampling. The simplest formal continuous sampling plan is CSP-1, which requires the process to "qualify" for sampling inspection by achieving i consecutive items free of nonconformities.[21] Once achieved, 100% inspection is discontinued in favor of inspecting only a fraction f of the items. When a sampled unit is nonconforming, the process reverts to 100% inspection, and so on. There are numerous versions of continuous sampling plans available, including CSP-1, 2, 3, A, M, T, F, V, and R. The best source of information on implementing CSP is Stephens.[21]

Skip-lot sampling is an attributes sampling procedure for reducing the inspection effort on product submitted by suppliers who consistently produce superior quality material. It is intended only for a continuing series of lots or batches, with some lots in a series being accepted without inspection. Lots to be inspected are chosen randomly in accordance with a specified "skip-lot frequency," provided that a stated number of immediately preceding lots have met certain criteria. A related ANSI/ASQC standard is in preparation under the guidance of the American Society for Quality Control.

Chain sampling is effective when tests are costly or destructive, necessitating a small attributes sampling plan having an acceptance number of zero. Plans having $c = 0$ are well known to be relatively harsh on good lots, since just one nonconformance can cause lot rejection. Therefore, chain sampling plans have been developed for use when there is continuing production under similar conditions and lots are offered for acceptance in essentially the order of their production. Lots having samples with no nonconforming items are accepted. Also, lots having samples with just one nonconforming item are accepted if the previous i consecutive samples had no nonconformances. Two or more nonconformances is grounds for lot rejection. A primary reference is Dodge.[22]

Bulk sampling is necessary when product is not in the form of discrete items but exists in bulk. Typically, two forms of bulk material exist. One form is segmented bulk materials, such as powder in barrels and fertilizers produced in batches. Another form is bulk material moving in a continuous stream, such as a chemical ingredient being introduced to a continuous mixing or blending operation. Concern usually centers around quality differences over time, from batch to batch, and even within a batch. An excellent reference for sampling design techniques is Juran et al., Chapter 25A.[12]

REFERENCES

1. American Society for Quality Control. *ANSI/ASQC Standard A3—Quality Systems Terminology.* Milwaukee, WI, 1978.
2. J. M. Juran and Frank M. Gryna, Jr. *Quality Planning and Analysis,* 2nd ed. New York: McGraw-Hill, 1980.
3. "Quality: The U.S. Drives to Catch Up." *Business Week* (November 1, 1982), 66.
4. American Society for Quality Control. *Quality Costs—What & How,* 2nd ed. Milwaukee, WI, 1971.
5. Harold L. Gilmore. "Product Conformance Cost." *Quality Progress* (June 1974), 16.
6. "Quality Cost Survey." *Quality Progress,* **16**(6), 20–22 (June 1977).
7. "Quality Cost Survey." *Quality Progress,* **19**(7), 16–17 (July 1980).
8. D. S. Sink. *Productivity Management: Planning, Measurement, Evaluation, Control and Improvement.* New York: Wiley, 1984.
9. W. Edwards Deming. "Condensation of the 14 Points for Management." Office copy, Washington, DC (November 1983).
10. W. Edwards Deming. *Quality, Productivity and Competitive Position.* Cambridge, MA: Massachusetts Institute of Technology, Center for Advanced Engineering Study, 1982.
11. American Society for Quality Control. *ANSI/ASQC Z1.15—Generic Guidelines for Quality Systems.* Milwaukee, WI, 1979.
12. J. M. Juran, Frank M. Gryna, Jr., and R. S. Bingham, Jr. *Quality Control Handbook,* 3rd ed. New York: McGraw-Hill, 1974.
13. W. A. Shewhart. *Economic Control of Quality of Manufactured Product.* Princeton, NJ: Van Nostrand Reinhold, 1931.
14. Acheson J. Duncan. *Quality Control and Industrial Statistics,* 4th ed. Homewood, IL; Richard D. Irwin, 1974.
15. Eugene L. Grant and Richard S. Leavenworth. *Statistical Quality Control,* 5th ed. New York: McGraw-Hill, 1980.
16. American Society for Quality Control. *ANSI/ASQC A1—Definitions, Symbols, Formulas and Tables for Control Charts.* Milwaukee, WI, 1978.
17. AT&T (previously Western Electric Company, Inc.). *Statistical Quality Control Handbook,* 2nd ed. Easton, PA: Mack Printing Company, 1958 (reissued under AT&T name in 1984).
18. Ford Motor Company. *Continuing Process Control and Process Capability Improvement,* July 1983.
19. American Society for Quality Control. *ANSI/ASQC Z1.4—Sampling Procedures and Tables for Inspection by Attributes.* Mikwaukee, WI, 1981.
20. Ilham Beainy, and Kenneth E. Case. "A Wide Variety of AOQ and ATI Performance Measures With and Without Inspection Error." *Journal of Quality Technology* **13**(1), 1–9 (January 1981).
21. Kenneth S. Stephens. *How to Perform Continuous Sampling.* Milwaukee, WI: American Society for Quality Control, 1979.
22. H. F. Dodge. "Chain Sampling Inspection Plan." *Journal of Quality Technology* **9**(3) (July 1977).

CHAPTER 3.7

AGGREGATE PRODUCTION PLANNING

ARNOLDO C. HAX

Massachusetts Institute of Technology
Cambridge, Massachusetts

3.7.1 WHEN TO USE AGGREGATE PLANNING

Production planning is concerned with the determination of production, inventory, and work-force levels to meet fluctuating demand requirements. Normally, the physical resources of the firm are assumed to be fixed during the planning horizon of interest, and the planning effort is oriented toward the best utilization of those resources given the external demand requirements. A problem usually arises because the times and quantities imposed by the demand requirements seldom coincide with the time and quantities that use the firm's resources efficiently. Whenever the conditions affecting the production process are not stable in time (due to changes in demand, cost components, or capacity availability), production should be planned in an aggregate way to obtain effective resource utilization. The time horizon of this planning activity is dictated by the nature of the dynamic variations; for example, if demand seasonalities are present, a full seasonal cycle should be included in the planning horizon. Commonly the time horizon varies from 6 to 18 months, 12 months being a suitable figure for most planning systems.

Since it is usually impossible to consider every fine detail associated with the production process while maintaining such a long planning horizon, it is necessary to *aggregate* the information being processed. This aggregation can take place by consolidating similar items into product groups, different machines into machine centers, different labor skills into labor centers, and individual customers into market regions. The type of aggregation to be performed is suggested by the nature of the planning systems to be used, and the technical as well as managerial characteristics of the production activities. Aggregation forces the use of a consistent set of measurement units. It is common to express aggregate demand in production hours.

Once the aggregate plan is generated, constraints are imposed on the detailed production scheduling process that decide the specific quantities of each individual item to be produced. These constraints normally specify production rates or total amounts to be produced per month for a given product family. In addition, crew sizes, levels of machine utilization, and amounts of overtime to be used are determined.

When demand requirements do not change with time, and costs and prices are also stable, it may be feasible to bypass the aggregate planning process entirely, provided the resources of the firm are well balanced to absorb the constant requirements. However, when these conditions are not met, serious inefficiencies or even infeasibility might result from attempting to plan production, responding only to immediate requirements and ignoring the future consequences of present decisions. To illustrate this point, consider what happens when an order point-order quantity inventory control system, which treats every item in isolation, is applied in the presence of strong demand seasonalities. First, at the beginning of the peak season, demand starts increasing rapidly and a large number of items simultaneously trigger the order point, demanding production runs of the amount specified by the order quantities. Being unable to satisfy all these orders while maintaining an adequate service level, management may react by reducing the length of production runs, thereby creating multiple changeovers of small quantities. This in turn reduces the overall productivity (because of the high percentage of idle machine time due to the large number of changeovers), increases costs, and deteriorates customer service levels. Second, items at the end of the season are produced in normal order quantities (typically large). Since demand is low,

Taken, with permission of the authors and publisher, from Arnoldo C. Hax and D. Candea, *Production and Inventory Management*, Englewood Cliffs, NJ: Prentice-Hall, 1984.

capacities tend to be idle, replenishment times decrease, order points get lower, fewer items trigger, leading to increasing idle time, and so on. Inventory is created that is inactive until the beginning of the next season or that must be liquidated at salvage values. An effective aggregate capacity planning system would prevent such inefficiencies.

3.7.1.1 Ways to Absorb Demand Fluctations

There are several methods that managers can use to absorb changing demand patterns. These can be combined to create a large number of alternative production planning options.

1. Management can change the size of the work force by hiring and laying off, thus allowing changes in the production rate to take place. Excessive use of these practices, however, can create severe labor problems.
2. While maintaining a uniform regular work force, management can vary the production rate by introducing overtime and/or idle time, or relying on outside subcontracting.
3. While maintaining a uniform production rate, management can anticipate future demand by accumulating seasonal inventories. The trade-off between the cost incurred in changing production rates and holding seasonal inventories is the basic question to be resolved in most practical situations.
4. Management can also resort to planning backlogs whenever customers may accept delays in filling their orders.
5. An alternative that has to be resolved at a higher planning level is the development of complementary product lines with demand patterns that are counterseasonal to the existing products. This alternative is very effective in producing a more even utilization of the firm's resources, but it does not eliminate the need for aggregate planning.

3.7.1.2 Costs Relevant to Aggregate Production Planning

Relevant costs of aggregate production planning can be categorized as follows:

1. *Basic production costs.* Included here are material costs, direct labor costs, and overhead costs. It is customary to divide these into variable and fixed costs, depending on whether the amount of the incurred cost over a certain time span is or is not a function of the corresponding production volume.

2. *Costs associated with changes in the production rate.* Typically, this category includes the costs of increasing the production rate above the level that can be achieved during the regular work schedule with the available work force. Since this objective can be accomplished either by varying the size of the work force and/or working overtime, this category includes costs involved in hiring, training, and laying off personnel, as well as overtime compensations. For some companies outside subcontracting represents a third alternative, in which case the cost of subcontracting will also fall in this category.

3. *Inventory related costs.* In the most general sense, inventories can take on both positive and negative values. In the first case, inventory holding costs will be incurred, in which the major component is the cost of capital tied up in inventory; other components are storing, insurance, taxes, spoilage, and obsolescence. In the second case, shortage costs occur; they are usually very difficult to measure and include costs of expediting, loss of customer goodwill, and loss of sales revenues resulting from the shortage situation.

For more extensive discussions on the above cost elements, the reader can consult McGarrah,[24] Holt et al.,[17] and Buffa.[8]

3.7.1.3 The Role of Models in Aggregate Production Planning

Models have played an important role in supporting management decisions in aggregate production planning. Anshen et al.[1] indicate that models are of great value in helping managers to do the following:

1. Quantify and use the intangibles that are always present in the background of their thinking, but which are incorporated only vaguely and sporadically in scheduling decisions.
2. Make routine the comprehensive consideration of all factors relevant to scheduling decisions, thereby inhibiting judgments based on incomplete, obvious, or easily handled criteria.
3. Fit each scheduling decision into its appropriate place in the historical series of decisions and, through the feedback mechanism incorporated in the decision rules, automatically correct for prior forecasting errors.
4. Free themselves from routine decision-making activities, thereby giving themselves greater freedom and opportunity to dealing with extraordinary situations.

Models that support aggregate production planning decisions can be classified according to the structure of their cost components into four major categories: linear cost models, quadratic cost models, fixed cost models, and nonlinear cost models. Because of space limitations as well as their practical importance, we will restrict our presentation to linear cost models only. For a discussion of the other types of aggregate production planning models, the reader is referred to Hax[13] and Hax and Candea.[14]

3.7.2 LINEAR COST MODELS

Some of the very first models proposed to guide aggregate planning decisions assume linearity in the cost behavior of the decision variables. Such models are very popular even today because of the computational conveniences associated with linear programming. Moreover, these models are less restrictive than they first appear, because nonlinear cost functions can be approximated to any degree of accuracy by piecewise linear segments (Reference 16, Ch. 13.5); these approximations result in linear programs whenever a convex cost function is minimized or a concave cost function is maximized.

The linear cost models can be divided into two categories, depending on how management goals are incorporated into the formulation:

Classical linear programming formulations, where the objective function represents a unique goal (such as minimizing the total production cost).

Goal programming formulations, in which several goals appear explicitly, and trade-offs among these goals are worked out based on a set of priorities, rather than specific numerical cost factors.

3.7.2.1 Classical Linear Programming Models

Depending on whether hiring and firing are considered to be decision variables, two types of models will result: fixed-work-force and variable-work-force models.

Fixed-Work-Force Models

The case where the work force is fixed will be considered first. Hirings and layoffs to absorb demand fluctuations during the planning horizon are disallowed. Production rates can fluctuate only by using overtime from the regular work force.

The following notation is used to describe the model in mathematical terms.

Parameters

v_{it} = unit production cost for product i in period t (exclusive of labor costs)
c_{it} = inventory carrying cost per unit of product i held in stock from period t to $t + 1$
r_t = cost per manhour of regular labor in period t
o_t = cost per manhour of overtime labor in period t
d_{it} = forecast demand for product i in period t
k_i = manhours required to produce one unit of product i
$(rm)_t$ = total manhours of regular labor available in period t
$(om)_t$ = total manhours of overtime labor available in period t
I_{io} = initial inventory level for product i
W_o = initial regular work force level, in manhours
T = time horizon, in periods
N = total number of products.

Decision variables

X_{it} = units of product i to be produced in period t
I_{it} = units of product i to be left over as inventory in period t
W_t = manhours of regular labor used during period t
O_t = manhours of overtime labor used during period t.

A simple version of the fixed-work-force–linear-cost model is

$$\text{minimize } Z = \sum_{i=1}^{N} \sum_{t=1}^{T} (v_{it}X_{it} + c_{it}I_{it}) + \sum_{t=1}^{T} (r_tW_t + o_tO_t) \tag{3.7.1}$$

subject to

$$X_{it} + I_{i,t-1} - I_{it} = d_{it}, \quad t = 1, \ldots, T;\ i = 1, \ldots, N \tag{3.7.2}$$

$$\sum_{i=1}^{n} k_iX_{it} - W_t - O_t = 0, \quad t = 1, \ldots, T \tag{3.7.3}$$

$$0 \leqslant W_t \leqslant (rm)_t, \quad t = 1, \ldots, T \tag{3.7.4}$$

$$0 \leqslant O_t \leqslant (om)_t, \quad t = 1, \ldots, T \tag{3.7.5}$$

$$X_{it}, I_{it} \geqslant 0, \quad i = 1, \ldots, N;\ t = 1, \ldots, T \tag{3.7.6}$$

The objective function (3.7.1) expresses the minimization of variable production, inventory, and regular and overtime labor costs. If the marginal production costs are invariant over time, $v_{it} = v_i$, the terms $v_i X_{it}$ do not need to be included in the objective function. Due to the cost minimization objective, there will be no inventory left at the end of period T; therefore, total production over the planning horizon is a constant equal to the total demand less the initial inventory:

$$\sum_{i=1}^{N} \sum_{t=1}^{T} v_i X_{it} = \sum_{i=1}^{N} v_i \left(\sum_{t=1}^{T} d_{it} - I_{io} \right) = \text{constant}$$

Similarly, if the payroll of the available regular work force $(rm)_t$ constitutes a fixed commitment (that is, employees get full pay whether they are fully employed or idle), the term $\Sigma_{t=1}^{T} r_t W_t$ becomes $\Sigma_{t=1}^{T} r_t (rm)_t =$ constant, hence it should be deleted from (3.7.1).

Constraints (3.7.2) represent the typical production–inventory balance equation, in which both the amount of inventory I_{it} to be left in stock at the end of period t and the demand d_{it} in period t are supplied by the amount $I_{i,t-1}$ of product i in stock at the end of period $(t - 1)$ and the production X_{it} in period t. Notice that (3.7.6) implies that no backordering is allowed. The next model will show how backorders can be incorporated. Moreover, (3.7.2) assumes a deterministic demand d_{it} for every item in every time period. One way to allow for uncertainties in the demand forecast is to specify a lower bound for the ending inventory in each period, that is, $I_{it} \geqslant (ss)_{it}$, where $(ss)_{it}$ is the safety stock associated with item i in period t (safety stocks depend on the demand forecast errors and the level of customer service to be provided).

Constraints (3.7.3) define the total manpower to be used in every period. This model formulation assumes that manpower availability is the only constraining resource of the production process. It is straightforward to expand the number of resources being considered, provided that the linearity assumptions are maintained.

Constraints (3.7.4) and (3.7.5) impose lower and upper bounds on the use of regular and overtime manhours in every time period.

It has been indicated already how constraints (3.7.6) could be changed to incorporate safety stocks. One should bear in mind that if no terminal conditions are imposed upon the inventories at the end of the planning horizon, the model will drive them to zero. If total depletion of inventories is undesirable, a target inventory constraint should be added in the model. An additional constraint should also be attached if there are storage requirements that cannot be exceeded; for example, the constraint

$$\sum_{i=1}^{N} I_{it} \leqslant (sc)_t , \qquad t = 1, \ldots, T$$

implies that the total inventory in each period cannot be greater than the total storage capacity $(sc)_t$.

When it is necessary to assign products to different work centers with limited capacities, the decision variables are redefined to identify those decisions explicitly. For example, X_{ict} may denote the amount of product i produced at working center c during period t. It is straightforward to carry out the resulting transformations in the overall model.

Even the very simple model described by expressions (3.7.1)–(3.7.6) could present enormous computational difficulties if the individual items to be scheduled are not grouped in broad product categories. If one ignores constraints (3.7.4), (3.7.5), and (3.7.6), which merely represent upper and lower bounds for the decision variables, the model consists of $T(N + 1)$ effective constraints. When dealing with complex production situations, the total number of individual items, N, may be several thousands. For example, if the planning model has 12 time periods and the production planning process involves 5000 items, the model would have about 60,000 constraints, which exceed the capabilities of a regular linear programming code.

In most practical applications, however, it would not be functional to plan the allocations of the production resources at this level of detail. First, a detailed scheduling program should take into account a large number of technological and marketing considerations that cannot be included in the overall model because of their highly qualitative nature. Second, as has been mentioned before, many of the planning issues to be resolved within the model deal with broad allocations of resources, and excessively detailed information would obscure rather than enlighten these decisions. Third, aggregate forecasts are more accurate than detailed forecasts.

It is common practice, therefore, to aggregate items in product types. The criteria for aggregation are evident from the model structure: members of a single product type should share similar demand patterns (d_{it}), have similar cost characteristics (v_{it}, c_{it}), and require similar unit production times (k_i). Once the aggregate planning decisions are made, these decisions impose constraints that must be observed when performing detailed item scheduling (see Reference 14, Ch. 6).

Notice that this model, as well as any other dynamic planning model, requires the definition of a planning horizon T and the partitioning of this time horizon into multiple time periods. One might assume that this partitioning results in T equally spaced time periods; this does not need to be the case. Many operational planning systems are better designed if this partitioning generates uneven time periods, so that the more recent time periods carry more detailed information.

Because of the uncertain environment in which this planning effort is being conducted, a widely adopted strategy is a rolling horizon. Under this strategy the T period model (3.7.1)–(3.7.6) is repeatedly solved, usually at the end of every time period, as new information becomes available and is used to update the model parameters. Although in principle the solution to model (3.7.1)–(3.7.6) provides decisions to be carried out over all periods in the model, only the plan for the upcoming period is actually implemented before the model is rerun. Of course, questions could be raised with respect to how good a finite-horizon strategy is in an infinite horizon environment. There are two reasons why the infinite-horizon strategy is appropriate[4]:

In general, the optimal solution to an infinite-horizon model requires parameter estimations for an infinite number of future periods. The only exception occurs when planning-horizon theorems are applicable (see Reference 14, Sections 3.4.1 and 3.5.1.1). Because of the limited information available about the future and the computational problems associated with increasingly large models, in the general case the finite-horizon strategy imposes itself as the only viable alternative.

The quality of forecasts tends to deteriorate with the distance into the future of the period for which they are made. Therefore, the value to the planning process of forecasts made for periods beyond a certain planning horizon T is questionable.

Broad technological, institutional, marketing, financial, and organizational constraints can also be included in the model formulation. This flexibility, characteristic of the linear programming approach to problem solving, has made this type of model very useful and popular.

A simple version of the fixed-work-force linear programming model, having a transportation problem structure, was first proposed by Bowman.[5]

Variable-Work-Force Models

Whenever it is feasible to change the work force during the planning horizon as a way to counteract demand fluctations, the composition of the work force becomes a decision variable whose values can change by hiring and laying off personnel. Therefore, the corresponding hiring and layoff costs should be part of the objective function. Moreover, the corresponding model presented below allows for shortages to be included; thus a backordering cost is also part of the formulation. Besides the decision variables and parameters introduced in the previous model, the following additional notations are needed:

H_t = manhours of regular work force hired in period t
F_t = manhours of regular work force laid off in period t
I_{it}^+ = units of ending inventory of product i in period t
I_{it}^- = units of product i backordered at the end of period t
b_{it} = cost per unit of backorder of product i carried from period t to $t + 1$
h_t = cost of hiring one manhour in period t
f_t = cost of laying off one manhour in period t
p = overtime allowed as a fraction of the regular hours

A simple version of the variable-work-force model can be formulated as follows:

$$\text{minimize } Z = \sum_{i=1}^{N}\sum_{t=1}^{T}(v_{it}X_{it} + c_{it}I_{it}^+ + b_{it}I_{it}^-) + \sum_{t=1}^{T}(r_tW_t + o_tO_t + h_tH_t + f_tF_t) \tag{3.7.7}$$

subject to

$$X_{it} + I_{i,t-1}^+ - I_{i,t-1}^- - I_{it}^+ + I_{it}^- = d_{it}\,, \quad i = 1, \ldots, N;\ t = 1, \ldots, T \tag{3.7.8}$$

$$\sum_{i=1}^{N} k_iX_{it} - W_t - O_t \leq 0\,, \quad t = 1, \ldots, T \tag{3.7.9}$$

$$W_t - W_{t-1} - H_t + F_t = 0\,, \quad t = 1, \ldots, T \tag{3.7.10}$$

$$-pW_t + O_t \leq 0\,, \quad t = 1, \ldots, T \tag{3.7.11}$$

$$X_{it}, I_{it}^+, I_{it}^- \geq 0\,, \quad i = 1, \ldots, N;\ t = 1, \ldots, T \tag{3.7.12}$$

$$W_t, O_t, P_t, H_t, F_t \leq 0\,, \quad t = 1, \ldots, T \tag{3.7.13}$$

The objective function is self-explanatory.

Constraints (3.7.8) represent the production–inventory balance equation. Notice that this is equivalent to the old balance equation

$$X_{it} + I_{i,t-1} - I_{it} = d_{it}$$

except that now

$$I_{it} = I_{it}^{+} - I_{it}^{-} , \qquad t = 1, \ldots ,T$$

In the present model the ending inventory I_{it} can be either positive ($I_{it}^{+} > 0$ indicates that stock remains at the end of the period) or negative ($I_{it}^{-} > 0$ indicates an accumulation of backorders at the end of the period). Since there is a cost attached to both I_{it}^{+} and I_{it}^{-}, those variables will never be positive simultaneously. Note the real meaning of I_{it}^{-} in models (3.7.7)–(3.7.13): I_{it}^{-} represents backorders that accumulate from period to period, which means that if demand cannot be satisfied in some period it can be satisfied at a later time; of course, a penalty will be incurred for any backordering of this sort. A different situation, which can also be easily modeled, arises when stockouts result in lost sales rather than backorders; in this case constraints (3.7.8) would become $X_{it} + I_{i,t-1}^{+} - I_{it}^{+} + I_{it}^{-} = d_{it}$, where I_{it}^{-} is the amount of demand for item i in period t that cannot be served, resulting in lost sales.

Constraints (3.7.9) limit production to available manpower. Since a cost is associated with hirings and layoffs it is possible that at some point the regular work force W_t will be partially idle, therefore the $\leqslant$ in (3.7.9) is perfectly justified.

Constraints (3.7.10) define the change in the work force size during period t, that is, $W_t - W_{t-1} = H_t - F_t$. Labor is added whenever $H_t > 0$, or is subtracted whenever $F_t > 0$. Once again, since costs are attached to both hirings and layoffs, H_t and F_t will never have positive values simultaneously in a given time period.

Constraints (3.7.11) impose an upper bound on the total overtime available in period t as a function of the regular-work-force size; that is, $O_t \leqslant pW_t$, where p is the overtime allowed as a fraction of the regular hours.

Many of the comments made for the fixed-work-force model regarding ways to expand or simplify the models and ways to aggregate items in product types are applicable here and are not repeated. One remark, however: if safety stocks are imposed ($I_{it}^{+} \geqslant (ss)_{it}$), situations could develop in which backorders are planned along with carrying positive inventories in the form of safety stocks. This turns out to be no contradiction when one thinks that safety stocks are not intended to be used inside the planning model (3.7.7)–(3.7.13) to serve forecasted demand or to fill in accumulated backorders; rather, safety stocks are held to meet unexpected contingencies arising outside the model from uncertain demand.

The first of this type of models was proposed by Hanssman and Hess.[12] Several alternative approaches have been suggested, particularly those by Haehling von Lanzenaur,[11] and O'Malley, Elmaghraby, and Jeske.[25]

Lippman et al.[22] have analyzed the form of the optimal policies for a single-product problem assuming convex production costs, V-shaped manpower fluctuation costs, and increasing holding costs. In Reference 23, the authors provide an efficient algorithm to solve the special case where all the cost functions are linear and demand requirements are either monotone decreasing or increasing. The algorithm is an iterative procedure that starts by guessing the value of W_T, the regular manpower at the end of the planning horizon. It provides next an optimum policy for this value of W_T, and checks this policy against an optimality test. If an improvement is possible, the algorithm yields a better value of W_T and the process is repeated. Convergence is guaranteed in a finite number of iterations.

Whenever costs are linear and demand requirements are nondecreasing, there exists an optimum policy such that

$$W_{t+1} \geqslant W_t , \qquad t = 1, \ldots ,T - 1$$

$$O_{t+1} \geqslant O_t , \qquad t = 1, \ldots ,T - 1$$

$$O_{t+1}/W_{t+1} \geqslant O_t/W_t , \qquad t = 1, \ldots ,T - 1$$

$$(W_T - W_t)O_t = 0 , \qquad t = 1, \ldots ,T$$

This result is used throughout the computational process. Yuan[26] extended this approach to a multiproduct problem.

In an early work, Hoffman and Jacobs[16] and Antosiewicz and Hoffman[3] considered a linear cost model for a single product, allowing for changes in the production rate to be represented in the objective function. They analyzed the qualitative properties of the optimum solution, and proposed simple procedures to compute that solution when demand requirements are monotone increasing. This work was extended by Johnson and Dantzig.[19]

Linear programming models can be explained easily to cover production processes with several stages. A comprehensive discussion of multistage linear programming models including multiple routings, multiple sources, product mix decisions, and multiple production and distribution decisions is presented by Johnson and Montgomery[18]; a survey and a reformulation of the multistage production planning problem are given by Candea.[9]

3.7.2.2 Goal Programming Models

One important shortcoming of the classical linear programming formulations is that they can treat explicitly only one objective. This objective is expressed as the optimization of a function that must be homogeneous; this means that all relevant decision variables have to be converted so as to become measurable by a common unit, most often the dollar.

In many cases, however, managerial decisions require the considerations of several goals, which can often conflict with each other and might be incommensurable. There are various ways of approaching the problem:

Select the most important of these objectives, make it into the objective function of the optimization model, and incorporate the others into the constraint set to generate a minimum level of achievement of those objectives.

Establish a global objective function by heuristic means: weight the various objectives and add them up to obtain the global objective. A method for performing this unification of multiple objectives is suggested by Briskin.[7]

These two approaches, although relatively straightforward to implement, are arbitrary and pragmatic, and do not necessarily resolve the basic issues behind the multiple objective question.

A third approach is to use utility theory, which requires the direct assessment of the multiattribute preference function of the decision maker (Keeney and Raiffa[20]). Conceptually, this is extremely attractive, but requires a great deal of work to implement (Hax and Wiig[15]); also, there are still methodological limitations in developing utility functions for either individuals or groups.

Another approach is goal programming, which will be discussed in this section.

Goal programming, as defined by Lee (Reference 21, p. 21), is "a modification and extension of linear programming. The goal programming approach allows a simultaneous solution of a system of complex objectives rather than a single objective."

To formulate a problem as a goal programming model one has to specify a set of goals along with the priority level associated with each of them. The set of constraints is composed, in general, of two kinds of relationships: regular constraints (as in any LP) representing dependencies among decision variables and parameters, and goal constraints relating the decision variables to goals. The deviational variables, which are always part of the goal constraints, will show the extent to which the goals are met in the optimal solution. To complete the model the objective function has to be defined. The objective is to achieve to the fullest possible extent every specified goal, in order of its priority. Therefore, the objective function has a multilevel structure; each level contains the sum (possibly weighted) of the deviational variables associated with the goals of that priority level.

During the solution procedure, first the deviations from the goal with the highest level of priority are minimized as much as possible, then the deviations for the goal with the second priority level, and so on. Once a higher priority goal is optimized, the optimization of a lower priority goal cannot possibly improve the higher goal, nor is it allowed to worsen the accomplishment level of the higher priority goal.

It is clear from the considerations above that a goal programming model is, by its nature, always a minimization problem.

The general form of a goal constraint is

$$\sum_{j=1}^{n} a_{ij}X_j + D_i^- - D_i^+ = b_i , \qquad i = 1, \ldots ,m$$

where b_i can be regarded as the target level of the ith goal, and

X_j is the jth decision variable, $j = 1, \ldots, n$;
a_{ij} is the coefficient relating the jth decision variable to the ith goal;
D_i^- deviational variable denoting the amount by which the ith goal is underachieved; and
D_i^+ is the deviational variable denoting the amount by which the ith goal is overachieved.

With respect to the ith goal, three important types of objective functions can be distinguished:

1. Minimize $(D_i^+ + D_i^-)$. This minimizes the absolute value of $(\Sigma_{j=1}^n a_{ij}X_j - b_i)$, by searching for the vector of X_j's that would meet the goal $\Sigma_{j-1}^n a_{ij}X_j = b_i$ exactly, in which case $D_i^+ = D_i^- = 0$. It is not always possible to find such X_j's; therefore, either D_i^+ or D_i^- (but not both) can be positive. If, in the optimal solution $D_i^+ > 0$, then $D_i^- = 0$ and $\Sigma_{j=1}^n a_{ij}X_j > b_i$; similarly, if $D_i^- > 0$, then $D_i^+ = 0$ and $\Sigma_{j=1}^n a_{ij}X_j < b_i$.
2. Minimize D_i^+. This minimizes the positive deviation from the target value, which is equivalent to stating that deviations of $\Sigma_{j=1}^n a_{ij}X_j$ above b_i are undesirable. However, no penalty is attached to underachieving

the ith goal, which implies that underachievements are regarded as acceptable. For example, overtime operation hours should be limited, if possible, to no more than a specified level.

3. Minimize D_i^-. This minimizes the negative deviation from the target level of the goal. Deviations of $\Sigma_{j=1}^n a_{ij}X_j$ below b_i are not desirable, while overachievements are not penalized. For example, it is wished that profits be at least as high as a prespecified figure.

Other objective functions are also possible, although not very often used:

4. Minimize $(D_i^- - D_i^+)$. It is equivalent to the minimization of D_i^- with b_i set sufficiently large.
5. Minimize $(D_i^+ - D_i^-)$. This is equivalent to the minimization of D_i^+ and if b_i is made very small.

A variant of "Minimize $(D_i^+ + D_i^-)$" is obtained when different weights are attached to the two variables:

$$\text{minimize } (w_1 D_i^+ + w_2 D_i^-)$$

For example: Suppose the target inventory is zero, with D_i^+, D_i^- representing inventory and shortage, respectively; then, the two weights are either the costs or are proportional to the costs of carrying inventory and incurring shortages, respectively.

To illustrate the goal programming approach, consider an aggregate production planning problem with a fixed work force. N products are manufactured, the planning horizon is T periods long, there is only one binding resource, the work force is fixed, and regular hours have to be paid for whether employees are idle or not. Suppose the management has specified the following priorities associated with its aggregate planning goal structure:

P_1 = the highest priority assigned to the satisfaction of demand; the fewer the backorders, the higher the service level.

P_2 = the second priority level associated with the minimization of both overtime hours and idle time. The minimization of overtime is desired, since one hour of overtime costs 50% more than one hour of regular time. The minimization of idle time is sought because, although the regular hours constitute a fixed commitment, idle time is considered detrimental to workers' discipline and morale.

P_3 = the third priority assigned to the minimization of end-of-the month inventories.

The model is:

$$\text{minimize } Z = P_1 \sum_{i=1}^{N} \sum_{t=1}^{T} p_i D_{it}^- + P_2 \sum_{t=1}^{T} (1.5R_t^+ + R_t^-) + P_3 \sum_{i=1}^{N} \sum_{t=1}^{T-1} c_i D_{it}^+ \tag{3.7.14}$$

subject to:

$$I_{i0} + X_{i1} - D_{i1}^+ + D_{i1}^- = d_{i1}, \quad i = 1, \ldots, N \tag{3.7.15}$$

$$D_{i,t-1}^+ + X_{it} - D_{it}^+ + D_{it}^- - D_{i,t-1}^- = d_{it}, \quad t = 2, \ldots, T-1;\ i = 1, \ldots, N \tag{3.7.16}$$

$$D_{i,T-1}^+ + X_{iT} + D_{iT}^- - D_{i,T-1}^- = d_{iT}, \quad i = 1, \ldots, N \tag{3.7.17}$$

$$\sum_{i=1}^{N} k_i X_{it} + R_t^- - R_t^+ = (rm)_t, \quad t = 1, \ldots, T \tag{3.7.18}$$

$$R_t^+ \leqslant (om)_t, \quad t = 1, \ldots, T \tag{3.7.19}$$

$$X_{it}, D_{it}^-, R_t^+, R_t^- \geqslant 0, \quad i = 1, \ldots, N;\ t = 1, \ldots, T \tag{3.7.20}$$

$$D_{it}^+ \geqslant 0, \quad i = 1, \ldots, N;\ t = 1, \ldots, T-1 \tag{3.7.21}$$

The notations we have used are:

Parameters

p_i = profit per unit of product i
c_i = unit cost of product i
d_{it} = forecast demand for product i in period t
k_i = manhours required to produce one unit of product i
$(rm)_t$ = total manhours of regular labor available in period t

$(om)_t$ = total manhours of overtime labor available in period t
I_{i0} = initial inventory of product i.

Decision variables

X_{it} = amount of product i to be produced in period t
D^+_{it} = ending inventory of product i in period t
D^-_{it} = backorders of product i at the end of period t
R^+_t = hours of overtime in period t
R^-_t = hours of idle time in period t

The objective function (14) shows that the highest priority goal is the minimization of the backorders. Since the N products differ in terms of profitability, their backorders have been weighted by the corresponding profit margins.

The second level of priority involves the minimization of overtime and idle time. The overtime variable is weighted by its relative cost compared to regular time.

The lowest priority goal aims at minimizing ending inventories. It has been considered appropriate to weight the inventory variables by the unit costs of the items because the unit cost determines the investment in inventories; also, in most cases, the unit inventory carrying cost is proportional to the unit cost.

Constraints (3.7.15)–(3.7.17) are the inventory balance equations in which the deviational variables indicate by how much the availability of product i in period t exceeds or misses the forecast level of demand. In period 1 there are no initial backorders. In constraint (3.7.17) the deviational variable D^+_{iT} is omitted in order to avoid piling up excess inventory solely for the sake of minimizing idle time.

Constraints (3.7.18) stipulate that, if sufficient, the available regular manhours can be supplemented by overtime R^+_t. It is also possible, however, to underutilize the regular manhours by R^-_t.

Constraint (3.7.19) places an upper bound on the amount of overtime that can be scheduled.

P_1, P_2, P_3 are nonquantitative parameters. They are ranking coefficients that show goal priorities. The meaning of this hierarchy is that the decision variables have to be chosen to have the backorders minimized, disregarding the second and third priority goals. Only after the highest priority goal has been achieved is one supposed to consider the overtime and idle time issues, and so on.

From the above, it is apparent that the simplex algorithm can be modified to solve goal programming problems: start by finding the optimum solution to the linear program whose objective function is the highest priority (P_1) portion of the original objective function. After this is done, solve another linear program whose objective function is the next highest priority (P_2) portion of the original objective function, and so on.

The interested reader can find a good treatment of the entire topic together with the modified simplex method for solving goal programming models in Lee.[21]

3.7.2.3 Advantages and Disadvantages of Linear Cost Models

The overwhelming advantage of linear cost models is that they generate linear programs that can be solved by readily available and efficient computer codes. Linear programs permit models with a large number of decision variables and constraints to be solved expediently and cheaply. In addition, linear programming lends itself very well to the performance of parametric and sensitivity analyses; this feature can be helpful in making aggregate planning decisions. The shadow price information can be of assistance in identifying opportunities for capacity expansions, marketing penetration strategies, new product introductions, and so on.

As indicated before, the linearity assumptions that are implicit in these models are less restrictive than they appear. First, cost structures might behave linearly within the range of interest of the decision variables under consideration. Second, general convex separable functions can be treated with piecewise linear approximation. Moreover, with some ingenuity certain functions that at first seem to present nonlinear characteristics can be linearized, as indicated in References 11 and 12.

Goal programming models bring in added capability to deal explicitly with multiple, and often conflicting and incommensurable, goals. They also allow one to better handle decision variables whose associated costs might prove difficult, or even impossible, to estimate, such as, for instance, the cost of backorders.

The most serious disadvantage of linear programming models is their failure to deal with demand uncertainties explicitly. However, this can be corrected by introducing safety stocks. Moreover, Dzielinski, Baker, and Manne[10] have reported favorable experiences in using linear programming models under fairly uncertain and dynamic environments.

3.7.3 INTEGRATING THE PRODUCTION PLANNING PROCESS

Production can be defined as the process of converting raw materials into finished products. Effective management of the production process should provide the finished products in appropriate quantities, at the desired times, of the required quality, and at a reasonable cost.

Production management encompasses a large number of decisions that affect several organizational echelons. To understand the role of mathematical models in supporting those decisions and to comprehend where aggregate production planning fits within them, it is useful to classify those decisions according to the taxonomy proposed by Anthony[2] regarding strategic planning, tactical planning, and operations control.

3.7.3.1 Strategic Planning: Facilities Design

Strategic planning is concerned mainly with establishing managerial policies and with developing the necessary resources the enterprise needs to satisfy its external requirements in a manner consistent with its specific goals. Anthony[2] defines *strategic planning* as "the process of deciding on the objectives of the organization, on changes in these objectives, on the resources used to attain these objectives, and on the policies that are to govern the acquisition, use, and disposition of these resources."

In the area of production and inventory management, the most important decisions that can be supported by model-based systems are those concerned with the design of production and distribution facilities. These decisions are on facilities involving major capital investments for the development of new capacity, either through the expansion of existing capacity or the construction or purchase of new facilities and equipment. They include the determination of location and size of new plants and warehouses, the acquisition of new production equipment, the design of working centers within each plant, and the design of transportation facilities, communication equipment, data processing means, and so on.

These decisions are extremely important because, to a great extent, they are responsible for maintaining the competitive capabilities of the firm, determining its rate of growth, and eventually defining its success or failure. An essential characteristic of these strategic decisions is that they have long-lasting effects, thus forcing long planning horizons in their analysis. This in turn requires the consideration of uncertainties and risk attitudes in the decision-making process.

Moreover, investments in new facilities and expansions of existing capacities are resolved at fairly high managerial levels, and are affected by information that is both external and internal to the firm. Thus, any form of rational analysis of these decisions has of necessity a very broad scope, requiring information to be processed in a very aggregate form to allow inclusion of all the dimensions of the problem and to prevent top managers from being distracted by unnecessary operational details.

3.7.3.2 Management Control (Tactical Planning): Aggregate Capacity Planning

Anthony[2] defines *management control* as "the process by which managers assure that resources are obtained and used effectively and efficiently in the accomplishment of the organization's objective." The emphasis of management control is on the resource utilization process.

Once the physical facilities have been decided upon, the basic problem to be resolved is the effective allocation of resources (e.g., production, storage and distribution capacities, work force availabilities, financial, marketing, and managerial resources) to satisfy demand and technological requirements, taking into account the costs and revenues associated with the operation of the resources available to the firm. These decisions are far from simple when we deal with several plants, many distribution centers, many regional and local warehouses, with products requiring complex multistage fabrication and assembly processes, that serve broad market areas affected by strong randomness and seasonalities in their demand patterns. They usually involve the consideration of a medium-range time horizon divided into several periods, and require significant aggregation of the relevant managerial information. Typical decisions to be made within this context are utilization of regular and overtime work force, allocation of aggregate capacity resources to product families, accumulation of seasonal inventories, definition of distribution channels, and selection of transportation and transshipment alternatives. The most common vehicle to communicate the management control decision of the firm is the company's budget.

3.7.3.3 Operational Control: Detailed Production Scheduling

After making an aggregate allocation of the resources of the firm, it is necessary to deal with the day-to-day operational and scheduling decisions. This stage of the decision-making process is called *operational control.* Anthony[2] defines it as "the process of assuring that specific tasks are carried out effectively and efficiently." The operational control decisions require the complete disaggregation of the information generated at higher levels into the details consistent with the managerial procedures followed in daily activities. Typical decisions at this level are:

The assignment of customer orders to individual machines.

The sequencing of these orders in the work shop.

Inventory accounting and inventory control activities.

Dispatching, expediting, and processing of orders.

Vehicular scheduling.

3.7.3.4 The Need for a Hierarchical Decision-Making System

There are significant conclusions that can be drawn from Anthony's classification regarding the nature of a decision support system. First, strategic, tactical, and operational decisions cannot be made in isolation, because they interact strongly with one another. Therefore, an integrated approach is required if one wants to avoid the problems of suboptimization. Second, this approach, although essential, cannot be made without decomposing the elements of the problem in some way, within the context of a hierarchical system that links higher-level decisions with lower-level ones in an effective manner. Decisions that are made at higher levels provide constraints for lower-level decision making: in turn, detailed decisions provide the necessary feedback to evaluate the quality of aggregate decision making.

This hierarchical approach recognizes the distinct characteristics of the type of management participation, the scope of the decision, the level of aggregation of the required information, and the time framework in which the decision is to be made. In our opinion, it would be a serious mistake to attempt to deal with all these decisions simultaneously, via a monolithic system of model. Even if computer and methodological capabilities would permit the solution of a large detailed integrated logistics model, which is clearly not the case today, this approach is inappropriate because it is not a response to the management needs at each level of the organization, and would prevent the interactions between models and managers at each organization echelon.

The basic questions to be resolved when designing a hierarchical system are:

How to partition the decision process into modules or subproblems that properly represent the various levels of decision making in the organizational structure.

How to aggregate and disaggregate the information through the various hierarchical levels.

How to solve each of the subproblems identified by the partitioning procedure.

What linking mechanisms should be used among the subproblems.

How to evaluate the overall performance of the system, particularly with regard to issues of suboptimization introduced by the hierarchical design.

These questions are not easy to answer. Some facts which have to be taken into consideration are:

The organizational structure of the firm that establishes the hierarchical breakdown of responsibilities, identifies the decision makers the system is intended to support, and provides the basis for a preliminary decomposition of the overall decision process.

The nature of the resulting subproblems, which suggest the methodology that might be applicable to solve each of the system modules. Naturally, it is preferable to define subproblems that lend themselves to easy and effective solutions.

The nature of the product structure, which is helpful in identifying ways in which information regarding individual items can be aggregated into families and product types.

The degree of interaction and transfer of information from each of the hierarchical levels of the system. An effective design should facilitate the specification of the constraints that higher-level decisions impose on the lower hierarchical echelons, and the control feedback that is transferred from the lower- to the higher-level decisions. In addition, the feasibility of disaggregation of information should be guaranteed throughout the process, and measures of performance should be available to assess the overall quality of decision making.

For a comprehensive discussion of hierarchical production planning systems, see Reference 14, Ch. 6.

REFERENCES

1. M. Anshen, C. C. Holt, F. Modigliani, F. J. Muth, and H. A. Simon, "Mathematical Models for Production Scheduling," *Harvard Business Review,* March-April 1958, 51–58.
2. R. N. Anthony, *Planning and Control Systems: A Framework for Analysis,* Graduate School of Business Administration, Harvard University, Boston, 1965.
3. H. Antosiewicz and A. J. Hoffman, "A Remark on the Smoothing Problem," *Management Science,* **1**(1), 92–95 (1954).
4. K. R. Baker, "An Experimental Study of the Effectiveness of Rolling Schedules in Production Planning," Graduate School of Business Administration Paper No. 138, Duke University, Durham, NC, October 1975.
5. E. H. Bowman, "Production Scheduling by the Transportation Method of Linear Programming," *Operations Research,* **4**(1), 100–103 (1956).
6. S. P. Bradley, A. C. Hax, and T. L. Magnanti, *Applied Mathematical Programming*. Reading, MA: Addison-Wesley, 1977.
7. L. W. Briskin, "A Method of Unifying Multiple Objective Functions," *Management Science,* **12**(10), B406–B416 (1966).
8. E. S. Buffa, *Operations Management: Problems and Models*. New York: Wiley, 1972.
9. D. Candea, "Issues of Hierarchical Planning in Multistage Production Systems," Technical Report No. 134, Operations Research Center, M.I.T., Cambridge, MA, July 1977.

10. B. P. Dzielinski, C. T. Baker, and A. S. Manne, "Simulation Tests of Lot Size Programming," *Management Science,* **9**(2), 229–258 (1963).
11. C. Haehling von Lanzenauer, "Production and Employment Scheduling in Multi-Stage Production Systems," *Naval Research Logistics Quarterly,* **17**(2), 193–198 (1970).
12. F. Hanssmann and S. W. Hess, "A Linear Programming Approach to Production and Employment Scheduling," *Management Technology,* (1) (1960).
13. A. C. Hax, "Aggregate Production Planning," in J. Moders and S. Elmaghraby Eds., *Handbook of Operations Research.* New York: Van Nostrand Reinhold, 1978.
14. A. C. Hax and D. Candea, *Production and Inventory Management.* Englewood Cliffs, NJ: Prentice-Hall, 1984.
15. A. C. Hax and K. M. Wiig, "The Use of Decision Analysis in Capital Investment Problems," *Sloan Management Review,* **17**(2), 19–48 (1976).
16. A. J. Hoffman and W. Jacobs, "Smooth Patterns of Production," *Management Science,* **1**(1), 86–91 (1954).
17. C. C. Holt, F. Modigliani, J. F. Muth, and H. A. Simon, *Planning Production, Inventories, and Work Force.* Englewood Cliffs, NJ: Prentice-Hall, 1960.
18. L. A. Johnson and D. C. Montgomery, *Operations Research in Production Planning, Scheduling, and Inventory Control.* New York: Wiley, 1974.
19. S. M. Johnson and G. B. Dantzig, "A Production Smoothing Problem," *Proceedings of the Second Symposium in Linear Programming,* Washington, DC, January 1955.
20. R. L. Keeney and H. Raiffa, *Decisions with Multiple Objectives; Preferences and Value Tradeoffs.* New York: Wiley, 1976.
21. S. M. Lee, *Goal Programming for Decision Analysis.* Auerbach, Philadelphia, 1972.
22. S. A. Lippman, A. J. Rolfe, H. M. Wagner, and J. S. C. Yuan, "Optimal Production Scheduling and Employment Smoothing with Deterministic Demands," *Management Science,* **14**(3), 127–158 (1967a).
23. S. A. Lippman, A. J. Rolfe, H. M. Wagner, and J. S. C. Yuan, "Algorithm for Optimal Production Scheduling and Employment Smoothing," *Operations Research,* **15**(6), 1011–1029 (1967b).
24. R. E. McGarrah, *Production and Logistics Management.* New York: Wiley, 1963.
25. R. L. O'Malley, S. E. Elmaghraby, and J. W. Jeske, "An Operational System for Smoothing Batch-Type Production," *Management Science,* **12**(10), B433–B449 (1966).
26. S. C. Yuan, "Algorithms and Multi-Product Model in Production Scheduling and Employment Smoothing," Technical Report No. 22, NSF GS-552, Stanford University, Stanford, CA, August 1967.

CHAPTER 3.8

PRODUCTION FORECASTS

RUSSELL G. HEIKES

Georgia Institute of Technology
Atlanta, Georgia

Predicting future values of a time-oriented variable is essential for efficient operation of a production system. For example, a forecast of demand for a product may be needed to schedule production or adjust inventories of that product. In addition, estimates of quantities such as quality characteristics of a product, costs, cash flow, labor units used, and equipment utilization would be extremely useful. Forecasts of time series such as these are widely utilized as inputs to production planning and scheduling models, inventory control systems, cost control models and other management information systems.

This presentation will focus on forecasting procedures that are based on extrapolation of any pattern that exists in the historical time series of the variable to be forecast. It excludes qualitative techniques where subjective information and judgment relating to an item are used in forecasting. It also excludes techniques where a forecaster seeks to identify a relationship between values of the series being forecast and other factors. These boundaries are selected so that attention is focused on procedures that are applicable when relatively few resources can be expended in forecasting a particular series. This is likely to be the case in a production system with a large number of production items. For high-cost or high-value items for which more accurate prediction is warranted, the forecasting procedures presented here serve as a performance base line for more sophisticated models.

The development of a forecast procedure hinges on the selection of the model used to represent the historical data. A critical element is whether it is assumed that the parameters in the model are subject to changes through time. This element, along with the model form, dictates the parameter estimation procedures required to produce the best possible (if the model assumptions are valid) forecasts. The need for both point and interval forecasts requires that an estimate of forecast error variability be generated.

A large number of forecasting procedures for the class of problems specified above have been proposed in the literature. Some are based on intuitively appealing heuristics, while others are derived from optimization of statistical properties of the forecasts produced. Many of these procedures require relatively simple calculations for producing forecasts. Some examples of the assumed models include:

$$Y_t = \mu_t + \epsilon_t$$

where

$$\mu_t = a \quad \text{or} \quad \mu_t = a + bt \quad \text{or} \quad \mu_t = \alpha\mu_{t-1} \quad \text{or} \quad \mu_t = (a + bt)C_t$$

Some common forecasting procedures will now be reviewed.

3.8.1 FORECASTING PROCEDURES

3.8.1.1 Common Exponential Smoothing Models

Exponential smoothing procedures update estimates of the model parameters recursively each period. Forecasts for future periods are then made using the revised parameter values. The exact form of these updating and forecasting equations depends on the model chosen to represent the series. A heuristic development of these equations for the simpler models is intuitively appealing. Consider the case where the average level of the process is nearly constant over time, or if it is changing it is doing so slowly. Thus the series might be modeled as

$$Y_t = a + \epsilon_t$$

Suppose that at the end of some period T we have available an estimate of the unknown parameter a that was made based on data available at the end of the previous period, $T - 1$. We want to update the estimator, say S_{T-1}, to incorporate the information in the current observation, Y_t. Since S_{T-1} was the forecast value for the current period, it seems reasonable to generate a new estimate by adjusting the old estimate by some fraction of the difference between the forecast and observed values for the current period. Thus the new estimate of a is

$$S_T = S_{T-1} + \alpha(Y_T - S_{T-1})$$

where $0 < \alpha < 1$. Rearranging gives

$$S_T = \alpha Y_T + (1 - \alpha)S_{T-1} \tag{3.8.1}$$

Eq. 3.8.1 defines simple (or first order) exponential smoothing and α is called the *smoothing constant*. Since the model assumes that the mean of the series is constant over time, the forecast for each future period will be the same as the forecast the next period, S_T.

Next consider a model for a process that follows a linear trend; that is

$$Y_T = a + bt + \varepsilon_t$$

If simple exponential smoothing is used to forecast this series it can be shown that the predicted values would lag behind the true values by an amount averaging $(1-\alpha)b/\alpha$. Since this bias depends on the unknown value b, we use the following procedure to estimate b and then adjust the simple exponentially smoothed forecasts accordingly. The procedure consists of the following steps:

1. Apply the exponential smoothing procedure to the smoothed statistics, giving

$$S_T^{[2]} = \alpha S_T + (1-\alpha)S_T^{[2]}$$

This operation is called *double exponential smoothing*.

2. It can be shown that $S_T^{[2]}$ lags behind the true values of the series on average by $2(1-\alpha)b/\alpha$. Thus the difference between the single- and double-smoothed values at time T will estimate the difference between the lags of the two statistics, or

$$S_T - S_T^{[2]} = (1-\alpha)b/\alpha$$

or

$$b = [\alpha/(1-\alpha)]\,(S_T - S_T^{[2]})$$

Consequently the forecast value for the current period is

$$Y_T = S_T + b(1-\alpha)/\alpha$$

$$= 2S_T - S_T^{[2]}$$

To forecast τ periods ahead at time T we use

$$2S_T - S_T^{[2]} + \frac{\tau\alpha}{1-\alpha}(S_T - S_T^{[2]})$$

These two forms of exponential smoothing are widely used in situations where a local (rather than global or constant) trend is expected.

The variance of the forecast errors for single exponential smoothing (assuming the ε's are statistically independent) is

$$\sigma_e^2 = \left(\frac{2}{2-\alpha}\right)\sigma_\varepsilon^2$$

where σ_ε^2 is the variance of the random noise term in the model. Choosing values of α close to zero will minimize σ_e^2, giving the best possible forecasts if the mean of the process remains unchanged. However, if the mean of the process does change, the forecast lags behind that change for a long number of periods when small values of α are used.

It has generally been found that values of α between 0.01 and 0.30 provide a reasonable balance between reduction of σ_e^2 and the ability of the forecasting model to respond to changes in level of the process. The same tradeoff between precision and reponsiveness exists for double exponential smoothing as well; however, the relationship between σ_e^2 and σ_ε^2 is more complex.

3.8.1.2 Regression Models

Model parameters can be estimated in a less heuristic manner via regression modeling techniques. Assume that values of the series $Y_1, Y_2, \ldots, Y_T$ for periods $1, 2, \ldots, T$ have been observed and that the process follows the model

$$Y_t = f(t; b_1, b_2, \ldots, b_k) + \varepsilon_t$$

where the b_i's are parameters to be estimated, and the ε's are independently and identically distributed random variables with mean zero and variance σ_ε^2.

If it is assumed that the model parameters are not changing through time, then ordinary least squares can be used to estimate the parameters. For example, if the model is

$$y_t = a + \varepsilon_t$$

the least squares estimate of a is

$$\bar{Y} = \sum_{t=1}^{T} Y_t / T$$

If the model is

$$Y_t = a + bt + \varepsilon_t$$

then tho least squares estimators are

$$a = \frac{2(2T + 1)}{T(T - 1)} \sum_{t-1}^{T} Y_t - \frac{6}{T(T - 1)} \sum_{t=1}^{T} tY_t$$

$$b = \frac{12}{T(T^2 - 1)} \sum_{t=1}^{T} tY_t - \frac{6}{T(T-1)} \sum_{t=1}^{T} Y_t$$

This approach is generally less acceptable than the exponentially smoothed forecasts, as we are seldom in a position to guarantee that model parameters will not gradually change. In addition, there are slightly greater computational requirements in the least squares regression case, although with current computational facilities this is seldom a critical factor.

To allow the more recent observations to have a greater impact, we can estimate parameters using weighted least squares. That is, we choose parameter estimates to minimize

$$\sum_{t-1}^{T} W_t[(Y_t - f(t, b_1, b_2, \ldots, b_k)]^2$$

where W_t is the weight placed on the tth observation. For example, a moving average forecast of the most recent N periods in a weighted least squares forecast with $W_T = W_{T-1} = \ldots = W_{T=N+1} = 1/N$, and $W_{T-N} = W_{T-N-1} = \ldots = W_1 = 0$. Thus, a moving average forecast for future periods at time T for a process with a constant level is

$$M_T = (Y_T + Y_{T-1} + Y_{T-2} + Y_{T-N+1})/N$$

For processes with a linear trend, the forecast above will lag the process, similar to the lag discussed in the simple exponential smoothing case. Defining a double moving average as

$$M_T^{[2]} = (M_T + M_{T-1} + M_{T-N+1})/N$$

the forecast for period $T + \tau$ at time T is computed from

$$2M_T - M_T^{[2]} + 2\tau(M_T - M_T^{[2]})/(N-1)$$

Alternatively, a frequently used weighting scheme is geometrically decaying weights, i.e.,

$$W_{T-j} = \beta^j, \qquad 0 < \beta < 1$$

By restricting the form of $f(t, b_1, b_2, \ldots, b_k)$ to $\Sigma_{i=1}^{k} b_i z_i(t)$, where the $z_i(t)$ are either polynomial, exponential, or trigonometric functions of t, equations similar to those of the exponential smoothing models in the previous section can be developed. This class of forecast techniques is referred to as *direct smoothing* (also adaptive smoothing or general exponential smoothing). If the $z_i(t)$ are restricted to polynomial terms, we have the case of multiple exponential smoothing.

The difficulty in using direct smoothing is that development of the updating equations requires computation of

$$\lim_{T \to \infty} \sum_{j=0}^{T-1} \beta^j \mathbf{Z}(-j)\mathbf{Z}'(-j)$$

where

$$\mathbf{Z}'(j) = [z_1(j), z_2(j), \ldots, z_k(j)]$$

While this need only be done once (not at each period) and some authors, see References 2 and 11, give the limiting values of numerous terms that are likely to arise, this is frequently a stumbling block in the application of this procedure. Brown[2] gives a good discussion of the computational problems.

The variance of forecast errors is controlled by the selection of β, and there is a tradeoff between precision and responsiveness as was the case in single- and double-exponential smoothing.

3.8.1.3 Seasonal Models

Many of the time series relating to production systems exhibit cyclic or seasonal behavior due to systematic changes in the factors that drive the process to be forecast. Development of forecast procedures for these series depend on the assumed model form for the seasonal as well as the nonseasonal elements.

Several commonly used techniques focus on extracting mean and/or trend components from the observations, and then estimating the seasonal terms from the adjusted observations. These procedures are based on the assumption that if the average of the observations over the number of periods in a season is taken, that average will not be dependent on the seasonal factors. Thus they are based on equal weighting of all observations.

To allow for time-varying model parameters, the direct smoothing approach with trigonometric terms can be used. For example, the model

$$Y_t = b_1 + b_2 t + (b_3 + b_4 t) \sin \frac{2\pi t}{L} + (b_5 + b_6 t) \cos \frac{2\pi t}{L} + \varepsilon_t$$

where L is the number of periods in the season, allows for a trend in the process as well as cyclic terms whose amplitude is growing in time. Note that the cyclic nature is controlled by the frequency of the trigonometric terms. If the seasonality exhibited by the series is not a simple sinusoidal, then terms with higher frequencies will need to be added.

A well-known heuristic due to Winters[18] allows for a completely general seasonal structure as well as a linear trend in the process. The model is

$$Y_t = (a + bt)\, C_t + \varepsilon_t$$

where a is the permanent component, b is the trend component, and C_t is a seasonal factor defined so that over a season of length L we have $\Sigma_{t=1}^{L} C_t = L$. The parameters are updated recursively using

$$a^T = \alpha Y_T/(C_T^{T-L}) + (1 - \alpha)(a^{T-1} + b^{T-1})$$

$$b^T = \beta(b^T - b^{T-1}) + (1 - \beta)\, b_1^{T-1}$$

and

$$C_T^T = \gamma Y_T/a^T + (1 - \gamma)\, C_T^{T-L}$$

where α, β, and γ are smoothing constants between 0 and 1, and the superscripts on a, b, and C_t indicate the time at which the estimate was made. The definition of a has been adjusted in the above equation so that it is the level of the process at time T, rather than at $t = 0$. The updating procedure is intuitively appealing, but is not derived on the basis of optimizing a mathematical or statistical criterion. Several heuristic procedures for generating starting values can be used.

3.8.1.4 Box-Jenkins Models

The procedures discussed so far have ignored the ε_t term in the model. since when it is assumed that the ε's are independent random variables with mean zero, they are of no value in forecasting. However, it has been demonstrated in many cases that the ε's do exhibit some systematic structure. A class of models called the auto-regressive integrated moving average (ARIMA) models, popularized by Box and Jenkins,[4] is useful in forecasting this type of series.

The assumed model for a process with constant mean is

$$Y_t = m + \psi_0 a_t + \psi_1 a_{t-1} + \psi_2 a_{t-2} + \cdots$$
$$= m + \sum_{j=0}^{\infty} \psi_j a_{t-j} \tag{3.8.2}$$

where the a's are independently and identically distributed with mean zero and variance σ_a^2 and the ψ's are model parameters. To overcome the problem of estimating an infinite number of parameters, the model is expressed as

$$Y_t = \mu + \phi_1 Y_{t-1} + \phi_2 Y_{t-2} + \cdots + \phi_p Y_{t-2} - \theta_1 a_{t-1} - \theta_2 a_{t-2} - \cdots - \theta_q a_{t-q} + a_t \tag{3.8.3}$$

The ψ's in the original model form are dependent on the φ's and θ's. It has been found that small values of p and q (usually less than 2) result in Eq. 3.8.3 being an adequate approximation to Eq. 3.8.2 for forecasting purposes in many cases.

Box and Jenkins[4] advocate the following three-step procedure in process modeling: (1) Identification, (2) Estimation, and (3) Diagnostic checking. The correlation of observations k periods apart $k = 1, 2, 3, \ldots$ (called *autocorrelation*) is the focus of the identification procedure. A comparison of the pattern of autocorrelations estimated from process data to the theoretical pattern for various p and q "suggests" an appropriate model (i.e. specific values of p and q). The φ's and θ's are estimated so as to minimize the sum of the squared residual errors, or other selected statistical criterion. This estimation process is usually more complicated than for the previous models discussed, as the model in Eq. 3.8.3 is nonlinear in the parameters to be estimated. Thus, a numerical search procedure is required. Diagnostic checking consists of examining the autocorrelations of the residuals of the fitted model. Lack of correlative structure is indicative that all information has been extracted from the data.

Numerous implementations of digital computer codes to carry out these procedures are available. Enhancements include automatic selection of values for p and q, approximate confidence intervals on estimated parameter values, and automatic generation of point- and interval-forecast values. Even with the availability of these aids, the development and verification of Box-Jenkins models requires considerable skill and time from the forecaster. This may cause it to be less desirable in certain production-oriented applications.

3.8.1.5 Special Techniques

Numerous techniques have been proposed for forecasting series that have special properties such as "lumpiness," boundaries on future values, or relationships to other series being forecast. Wilcox[17] describes a smoothing technique for forecasting demands for inventory control purposes when demands are highly irregular or lumpy. Thomopolous[15] provides a methodology and associated tables when, in addition to having forecast values based on historical data, information or bounds of the demands in future periods are known. An example of this is when forecasting demands and firm orders for some fraction of future periods requirements are in hand at the time the forecast is made.

When additional knowledge of the structure of the time series is available, this information can be incorporated into the forecast procedures. Consider the case where the time series is expected to rise, gradually level off, and then return to zero (for example, demand for spare parts during the useful life of a product). Brown[2] and Thomopolous[15] describe methods for estimating total demand under various assumptions concerning growth decay. Alternatively, consider the case where the series is dropping rapidly and a trend model is being used for forecasting. If it is known that the series cannot have negative values, then some damping of the trend may yield more sensible forecasts.

3.8.2 EXAMPLES

3.8.2.1 Constant Level Processes

The values in Table 3.8.1 illustrate the moving average and simple exponential smoothing forecasts for a process with random deviations about a constant level.

Table 3.8.1. Moving Average and Simple Exponential Smoothing Forecasts

Time Period, t	(1) Observed Series, Y_t	(2) Moving Average Forecast ($N = 4$)	(3) Exponential Smoothing Forecast ($\alpha = 0.1$)
1	50	—	49.78
2	52	—	50.00
3	46	—	49.60
4	51	49.75	49.74
5	54	50.75	50.16
6	49	50.00	50.05
7	47	50.25	49.74
8	51	50.25	49.87
9	48	48.75	49.68
10	53	49.75	50.01

For the moving average forecast, a value of $N = 4$ is arbitrarily chosen. No forecast is computed until the first four values of the time series have been observed. After Y_4 is observed,

$$M_4 = (Y_1 + Y_2 + Y_3 + Y_4)/4 = (50 + 52 + 46 + 51)/4 = 49.75$$

is calculated. This is the forecast for period 5 and later periods at the end of time period 4.

As soon as the observed value for the 5th period is available, the forecasts for future periods will be revised. The new forecast value will be

$$M_5 = (Y_2 + Y_3 + Y_4 + Y_5)/4 = (52 + 46 + 51 + 54)/4 = 50.75$$

The revised forecasts after each period are shown in column (2) of the table.

To generate simple exponential smoothing forecasts, a smoothing constant value of $\alpha = .10$ is chosen, and a starting value $S_0 = 49.75$ (an initial estimate of the level of the process) is selected. After the value of Y_1 is observed, forecasts for future periods are calculated as

$$\begin{aligned} S_1 &= \alpha Y_1 + (1 - \alpha)S_0 \\ &= .1(50) + (1 - .1)49.75 \\ &= 49.78 \end{aligned}$$

This would be revised as each new Y_t is observed, as indicated by the values in column (3) of the table.

3.8.2.2 Trend Processes

The values in column (1) of Table 3.8.2 illustrate a typical trended process. Forecasts are made for period $t + 1$, assuming that data values for periods 1 through t are available using linear regression, double-moving average, and double-exponential smoothing.

It is assumed that the linear regression forecasts are not begun until $t = 5$ so that sufficient data are available to give reasonable stability to the projected trend. The estimated constant and intercept, respectively, are

$$a = \frac{2[(2)(5) + 1]}{5(4)} (27 + 31 + 33 + 40 + 41) - \frac{6}{5(4)} [1(27) + 2(31) + 3(33) + 4(40) + 5(41)]$$

$$= 23.3$$

$$b = \frac{12}{5(5^2 - 1)} [1(27) + 2(31) + 3(33) + 4(40) + 5(41)] - \frac{6}{5(4)} (27 + 31 + 33 + 40 + 41)$$

$$= 3.7$$

Then the forecast for period 6 is

$$\begin{aligned} a + bt &= 23.3 + 3.7(6) \\ &= 45.5 \end{aligned}$$

Table 3.8.2. Forecasts for a Trended Process

Time Period, t	(1) Observed Series, Y_t	(2) Linear Regression Forecast	(3) Double Moving Average Forecasts	(4) Double Exponential Smoothing Forecasts
1	27	—	—	30.70
2	31	—	—	34.46
3	33	—	—	37.87
4	40	—	—	41.99
5	41	45.50	—	45.50
6	45	48.80	—	49.10
7	48	51.90	52.58	52.57
8	52	54.50	53.17	56.14

If we wish to forecast two periods into the future, for period 7, let $t = 7$ and

$$a + bt = 23.3 + 3.7(7)$$

$$= 49.2$$

When the value of Y_6 has been observed, new values of a and b using $Y_1, Y_2, \ldots, Y_6$ would be computed, and the forecasts for periods 7 and later would be revised.

To find the double-moving-average forecasts (we use $N = 4$), first find the single-moving-average forecasts $M_4, M_5, \ldots$. Note that until four single-moving-average values are available, the double-moving-average value cannot be calculated. Thus it is not until Y_7 is available that we have the first double-moving-average forecast. Then

$$M_7^{[2]} = (M_4 + M_5 + M_6 + M_7)/4$$

$$= (32.75 + 36.25 + 39.75 + 43.5)/4$$

$$= 38.05$$

The forecast for period 8 ($\tau = 1$) is then

$$2M_7 - M_7^{[2]} + 2(1)(M_7 - M_7^{[2]})/(4 - 1)$$

$$= 2(43.5) - 38.05 + 2(1)\,(43.5 - 38.05)/3$$

$$= 52.58$$

and for period 9 ($t = 2$) is

$$2(43.5) - 38.05 + 2(2)\,(43.5 - 38.05)/3$$

$$= 56.22$$

When Y_8 is available, M_8 and $M_8^{[2]}$ will be computed and the forecasts for periods 9 and later can be revised.

To begin the double-exponential-smoothing forecasts, values of S_0 and $S_0^{[2]}$ are needed. When estimates of a and b are available (for example, as from the calculations done above) the following equations can be used.

$$S_0 = a - [(1 - \alpha)/\alpha]b$$

$$= 23.3 - [(1 - .1)/.1]3.7 = -10$$

$$S_0^{[2]} = a - [2(1 - \alpha)/\alpha]b$$

$$= 23.3 - [2(1 - .1)/.1]3.7 = 43.3$$

Using $\alpha = .1$, the smoothed values at $t = 1$ are

$$S_1 = (.1)(27) + (1 - .1)(-10)$$

$$= -6.3$$

$$S_1^{[2]} = (.1)(-6.3) + (1 - .1)(-43.3)$$

$$= -39.60$$

The forecast value for period 2 ($\tau = 1$) after Y_1 is observed is

$$2S_1 + S_1^{[2]} + \frac{(1)(.1)}{.9}[S_1 - S_1^{[2]}]$$

$$= 2(-6.3) - (-39.60) + (.11)(-6.3 - (39.60))$$

$$= 30.7$$

The one-period-ahead forecasts for each value of t are shown in column (4) of Table 3.8.2.

3.8.2.3 Winters Seasonal Model

We assume that several years of historical data on a monthly basis are available for a process that follows a cyclic pattern with a periodicity of 12 months. Analysis of the historical data suggests that the seasonality is superimposed on a trend process. The initial estimates of the model parameters are given as

$$\begin{aligned} a^0 &= 154.7 & c_6^0 &= 0.520 \\ b^0 &= .62 & c_7^0 &= 0.755 \\ c_1^0 &= 1.310 & c_8^0 &= 0.770 \\ c_2^0 &= 1.090 & c_9^0 &= 0.895 \\ c_3^0 &= 1.062 & c_{10}^0 &= 1.129 \\ c_4^0 &= 0.896 & c_{11}^0 &= 1.444 \\ c_5^0 &= 0.792 & c_{12}^0 &= 1.355 \end{aligned}$$

To forecast for the first period of the upcoming year we use

$$[a^0 + (b^0)(\tau)]C_1^0 = [154.7 + .62(1)]1.31$$

$$= 203.5$$

Note that the trend was projected for τ periods rather than $t + \tau$ periods, due to redefining a^T as the level of the process at T rather than $t = 0$. To forecast two periods into the future, we have

$$(a^0 + b^0(2)]C_2^0 = [154.7 + (.62)(2)]1.09$$

$$= 170.0$$

After the value of the first period is observed (suppose $Y_1 = 205$) we update the model estimates as follows.

$$a^1 = \alpha(205)/1.31 + (1 - \alpha)(154.7 + .62)$$

$$b^1 = \beta(a^1 - a^0) + (1 - \beta)b^0$$

and

$$C_1^1 = \gamma(205)/a^1 + (1 - \gamma)C_1^0$$

Note that in the general equations previously given for updating the estimates above, the superscript on the C's was $(T - L)$, which has a negative value during the early periods. We use the C_t^0 values for these cases until $(T - L) > 0$. Assuming values of α, β and γ would allow us to continue forecasting and updating as additional Y_t values become available.

Conclusion

A variety of procedures is available for systematic extrapolation of time series. Several of these are sufficiently simple so that even in situations where a large number of series must be forecast, the resources required are minimal. In cases where additional effort may be justified, these basic procedures provide a baseline against which to evaluate alternatives.

REFERENCES

1. D. J. Bamber, "A Versatile Family of Forecasting Systems," *Operational Research Quarterly*, **20**, 111–121 (1969).
2. G. E. P. Box and G. M. Jenkins, *Time Series Analysis, Forecasting, and Control*, 2nd ed., San Francisco: Holden-Day, 1976.
3. R. G. Brown and R. F. Meyer, The Fundamental Theorem of Exponential Smoothing, *Operations Research*, **9**(6), 673–685.
4. R. G. Brown, *Smoothing, Forecasting and Prediction of Discrete Time Series*. Englewood Cliffs, NJ: Prentice-Hall, 1962.
5. D. R. Cox, Prediction by Exponentially Weighted Moving Averages and Related Methods, *Journal of the Royal Statistical Society*, **B23**(2), 414–422 (1961).
6. D. B. Crane and J. R. Crotty, "A Two-Stage Forecasting Model: Exponential Smoothing Multiple Regression," *Management Science*, **13**(2), 501–507 (1967).
7. P. J. Harrison, "Exponential Smoothing and Short-Term Sales Forecasting," *Management Science*, **13**(11), 821–842 (1967).
8. J. O. McClain and L. J. Thomas, "Response-variance Tradeoffs in Adaptive Forecasting," *Operations Research*, **21**(2), 554–568 (1973).
9. E. McKenzie, "An Analysis of General Exponential Smoothing," *Operations Research*, **24**(11), 131–140 (1976).
10. D. C. Montgomery, "Adaptive Control of Exponential Smoothing Parameters by Evolutionary Operation," *AIIE Transactions*, **2**(3), 268–269 (1971).
11. D. C. Montgomery and L. A. Johnson, *Forecasting and Time Series Analysis*. New York: McGraw-Hill, 1976.
12. S. Makridakis, S. C. Wheelwright, and V. E. McGee, *Forecasting: Methods and Applications*, 2nd ed. New York: Wiley, 1983.
13. S. M. Pandit and S. M. Wu, "Exponential Smoothing as a Special Case of a Linear Stochastic System," *Operations Research*, **22**(4), 868–879 (1974).
14. C. C. Pegels, "A Note on Exponential Forecasting," *Management Science*, **15**(5), 311–315 (1969).
15. Nick T. Thomopoulos, *Applied Forecasting Methods*. Englewood Cliffs, NJ: Prentice-Hall, 1980.
16. D. W. Trigg and A. G. Leach, "Exponential Smoothing with an Adaptive Response Rate," *Operational Research Quarterly*, **18**, 53–59 (1967).
17. J. E. Wilcox, "How to Forecast Lumpy Items," *American Production and Inventory Management*, 51–54 (1970).
18. P. R. Winters, "Forecasting Sales by Exponentially Weighted Moving Averages," *Management Science*, **6**(3), 324–342 (1960).

CHAPTER 3.9

PRODUCTION AND PROJECT SCHEDULING

RANDALL P. SADOWSKI
CATHERINE M. HARMONOSKY

Purdue University
West Lafayette, Indiana

3.9.1 DEFINING PRODUCTION SCHEDULING

The ANSI Standard Z94 defines a production schedule as:

> *A plan which authorizes the factory to manufacture a certain quantity of a specified item. Usually initiated by the production planning department.*[1]

Production scheduling actually implies a variety of activities, depending upon a person's manufacturing experience and the particular production environment, leading to a production plan. Successful production scheduling requires careful integration of many factors. Figure 3.9.1 illustrates the breadth of the interaction, ranging from the marketplace to the shop floor.

Basically, scheduling consists of two working levels, exemplified by the department title, "Production Planning and Scheduling," so readily used in industry. One level is an aggregate "planning" of total item quantities for production during given time periods. The second level is the translation of the plan via detailed "scheduling" into specific job sequences suitable for shop-floor implementation. Developing a successful production schedule requires integrating these two levels by actively balancing system requirements vs. capabilities over time.

The aggregate planning stage is often a combination of shoploading or capacity planning and material requirements planning (MRP). Shoploading, capacity planning, and MRP are discussed in depth in other chapters. Unfortunately, this scheduling level does not typically consider system capacities or inherent resource constraints. Consequently, the imposed production load is often inconsistent with true system constraints, inhibiting the possibility of an effective production schedule that will meet customer demands and company objectives.

Regardless of the degree to which the aggregate plan anticipates and incorporates actual shop-floor load conditions, the shop-floor scheduling and control activity must convert the overall plan into a realistic, detailed production schedule. The conversion is most successful when detailed schedules are developed in conjunction with aggregate planning, considering potential bottlenecks and constraints. When aggregate and detailed scheduling are disjoint, as is actual practice in the majority of applications, shop-floor scheduling becomes very difficult and is most often based on past experience with similar situations, rather than consideration of future system conditions.

Ideally, integration of aggregate and detailed scheduling activities allows development of a production plan that is a detailed schedule, effectively utilizing resources at the shop-floor level. This schedule would detail the release time of all jobs to the system and provide, in advance, the sequence of jobs at all machines and workstations within the manufacturing environment. It would allow not only for the effective utilization of manufacturing resources, but also provide advanced notice of the system's ability to meet due dates and effectively control work-in-process. Unfortunately, scheduling theory has not yet evolved to development of detailed schedules for most practical problems.

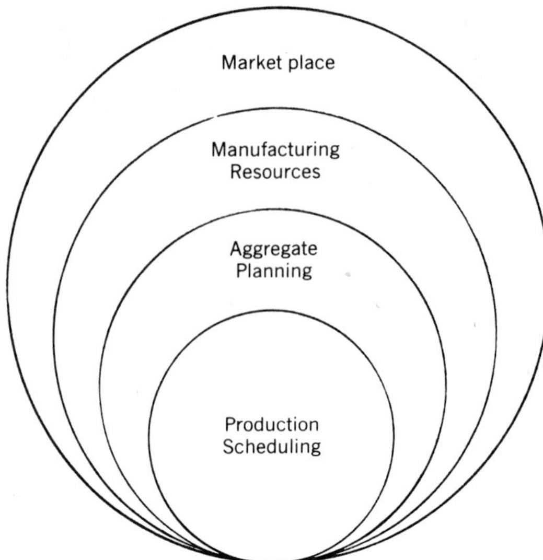

Fig. 3.9.1. The production scheduling environment.

3.9.2 THE SHOP-FLOOR ENVIRONMENT

3.9.2.1 Job Releasing, Dispatching, Expediting

The inability to develop a detailed, realistic schedule often results in a production plan resembling a wish list. The actual implementation of such a plan is often achieved through three different functional activities at the shop-floor level: job releasing, dispatching, and expediting. The first activity involves the actual release of jobs to the system. As these progress to their first or subsequent operations, the dispatching function takes over.

Dispatching involves selecting and sequencing available jobs to be run at an individual workstation and assigning necessary resources to complete those jobs. Dispatching is a localized decision process, frequently based on some type of internal priority associated with each job in the system. This priority can be the desired due date or the perceived criticality of the job. The success of dispatching depends upon the sophistication of support systems providing current information and the ability of the shop-floor personnel to anticipate the impact of their localized decisions on subsequent operations.

Due to localized decision-making techniques, problems often arise, such as jobs within the system exceeding due dates, which were not successfully anticipated. These problems are often resolved by an expediting activity that involves rushing or chasing production jobs through the system in order to meet demands. Ideally, expediting should be limited to instances where a job entered the system with a shorter than normal lead time due to unavoidable circumstances. In practice, the expediting activity is frequently used for jobs that incurred delays at congestion points within the manufacturing environment.

3.9.2.2 Bottleneck Control

Congestion points, or bottlenecks, primarily occur when manufacturing resources required in a given time period are unavailable. Specific resources are quite varied, but can be summarized in one of the following areas: raw materials, machines, workstations, workers, tools, fixtures, jigs, and storage areas. Sometimes it is possible to reallocate excess resources, such as manpower, to allow expediting key jobs through the system. For other resources, such as machines, it may not be possible to obtain the extra resources required to meet demands.

In many environments, it is possible to predict where bottlenecks might occur. Frequently, system control points are established to allow effective allocation and utilization of key resources at these bottleneck points. Such a control procedure is not as efficient as building a total schedule, but it does allow for the intelligent use of available resources at points within the system where problems are most likely to occur.

3.9.2.3 Resource Control

Effective resource control requires understanding both the requirements and availability of key resources. This becomes almost impossible if accurate and timely information regarding these resources is not available. Such information consists of time estimates obtained from standards files, as well as the current status of these resources.

In recent years, many industries have become very sophisticated in their ability to provide key information to the shop floor to support resource control decisions. Numerous software packages and systems have been developed that greatly assist shop-floor control and substantially impact the organization's ability to plan and achieve the production schedule successfully. But implementing a software system usually requires massive investment and extremely long implementation time. Although the introduction of these systems has greatly enhanced the manufacturing environment's capability to adapt to sudden scheduling changes and resource reallocations, they cannot yet successfully plan and implement the production schedule on a continual basis.

3.9.3 SYSTEM EVALUATION

3.9.3.1 Performance Measures

Because production scheduling is an integral part of any manufacturing system, it is desirable to have some means of assessing its effectiveness. However, the success of production scheduling is often difficult to measure. The overall objective of a manufacturing system is profitability, but this objective must be translated into a specific set of performance measures that accurately evaluate the different system functions. The principal performance measures used to evaluate production scheduling are maximization of customer satisfaction and minimization of cost. Customer satisfaction is a measure of how well planned due dates for delivery to an internal or external customer are met. Minimizing cost is an all-encompassing measure implying effective utilization of the required resources and a minimization of work-in-process.

Unfortunately, in practice these two concerns are often contradictory. To achieve 100% customer satisfaction, a production facility could be designed that is grossly overcapacitated and allows high inventory levels providing the greatest amount of flexibilty in dispatching. However, this greatly increases costs. On the other hand, costs can be minimized by exactly capacitating the system and keeping work-in-process low. This often results in the inability to meet customer due dates. In recent years, a new measure of performance has emerged—quality. It is apparent that the manufacturing environment must also produce a quality product that is consistent with customer desires. This has imposed additional constraints and concerns on the manufacturing floor.

3.9.3.2 Surrogate Measures

Due to the difficulty in directly relating profitabilty to shop-floor activity, the principal performance measure is often based on number of items or value completed in a given time period, typically a month. This results in the supervisor vainly trying to produce the critical or expedited products in the first two or three weeks, and then producing only orders with a large number of units, pounds, or high dollar value in the last week to meet the required quota. The resulting system performance provides poor customer satisfaction, high inventories, and frequently results in poor resource utilization.

Since the work-in-process is ordinarily measured only at the end of the time period, the impression may exist that work-in-process is maintained at a reasonable level. Actually, it often becomes obvious that congestion on the manufacturing floor during the actual time period is much higher than is reflected in the ending inventory values. Such a procedure may provide a means to control the throughput of a given department, but it does not globally control the specific parts being produced. Consequently, the high-valued items produced at the end of a month to meet a quota may not be the items required by the next operations. If items are due in an assembly operation, a low-valued item not produced may result in the subsequent assembly operation being delayed and effectively increase the work-in-process of the overall system. When items are produced in advance of their required times, they also add to the overall work-in-process. The net effect, often, is that the true work-in-process for the entire system may actually increase when departments are scheduled individually based on localized evaluation criteria.

3.9.4 PRODUCTION TYPES

The ability to successfully develop and implement any production schedule is dependent on many factors, most importantly, the type of production system. There are numerous ways to classify industrial systems, but the majority of manufacturing operations may be categorized as *product shop* or *process shop*. Both classifications imply certain characteristics regarding the production facility layout and the product type.

A product shop is frequently referred to as a production line or flow shop, where machines or workstations are arranged so that products follow a logical sequence of operations in a straight path from start to finish. Its production is usually limited to a small range of products requiring similar or identical manufacturing operations. A process shop is also called a *job* or *intermittent* shop, and is structured to allow production of a wide variety of different products with different requirements. Typically, similar machines or workstations are grouped together to facilitate this diverse production activity.

The remaining manufacturing operations are generally classified as *model* or *special* shops. These systems usually produce one-of-a-kind, specially designed items that are often quite diverse. This classification also includes special assembly systems that are predominant in the defense industry. Project scheduling may be used in these systems and is described later in this chapter, Section 3.9.11.

Although scheduling any system is difficult, product shop scheduling is better understood, and hence appears somewhat easier to accomplish. Examples of this flow-like production can be found in the petrochemical, paper, and steel industries. The planning process is slightly simplified because all or most items follow an identical path through the system. An assembly line may be considered a specialized product shop. This is scheduled by assembly line balancing, discussed in detail in Chapter 3.11. General product shop systems are scheduled based on releasing jobs at appropriate times and providing inventory buffers at key points to maintain a smooth production flow.

Unfortunately, the majority of manufacturing operations fall into process or job-shop categories, the most difficult to plan and schedule. In reality, most of these systems are not pure job or process shops and tend to fall in a gray area between a product and process shop. Characteristically, they produce a finite number of known, very diverse products, which follow a limited number of flows through the manufacturing system. Although such systems resemble modified flow shops, the criss-crossing of flows and the common use of key resources within the system makes scheduling significantly more difficult than in a pure flow shop.

3.9.5 TRADITIONAL METHODS

3.9.5.1 Manual Methods

Most computerized scheduling methods are based on manual systems that have emerged over the years to control product movement through the system. Manual systems make extensive use of standardized forms to transmit the necessary information for effective scheduling decisions. The scheduling activity is initiated by compiling forecasts for future demand and the current firm orders to create a long-term production plan, usually represented by a master production schedule. Prior to actually creating a detailed production plan, data are gathered to assure that the right raw materials and component parts are available in time for production. This involves obtaining the necessary process and bill of material information and combining the perceived requirements with the inventory records to determine if purchase orders need to be released. These activities are predecessors of material requirements planning and provide a tentative job release schedule.

The output from this phase provides an aggregate production plan, which is converted to a detailed production schedule, often taking the form of a short-term requirements schedule developed at the shop-floor level. Based on the release schedule of the MRP equivalent and the availability of inventory, the shop receives authorization to initiate work. Development of this short-range plan provides production goals for each department for a near-term time frame, typically a week. Depending on the manufacturing environment and sophistication of the manual system, this plan sometimes includes capacity considerations.

Prior to releasing a job to the shop floor, a *shop packet* is developed. Although the exact number and type of forms vary, the packet normally contains routing information of the job, detailing all the operations required and the sequence of these operations, standard operation times, and anticipated setup times. Also included are any special requirements, such as tools, jigs, or fixtures, or information related to inspection or special processes that might be necessary. It also provides information regarding the number required and the due dates dictated by the production plan. A sample job router form is shown in Figure 3.9.2.

Order Number ______________ Date ____________

Page ____ of ____

Part Number ______________ Quantity ____________

Part Description ______________________________

Customer ______________ Ship Date ____________

Oper. Number	Work Center	Description & tooling, etc.	Qty.	Scrap Qty.	Standard Setup Time	Standard Oper. Time	Actual Setup Time	Actual Oper. Time	Due Date	Complete Date

Fig. 3.9.2. Sample job router form.

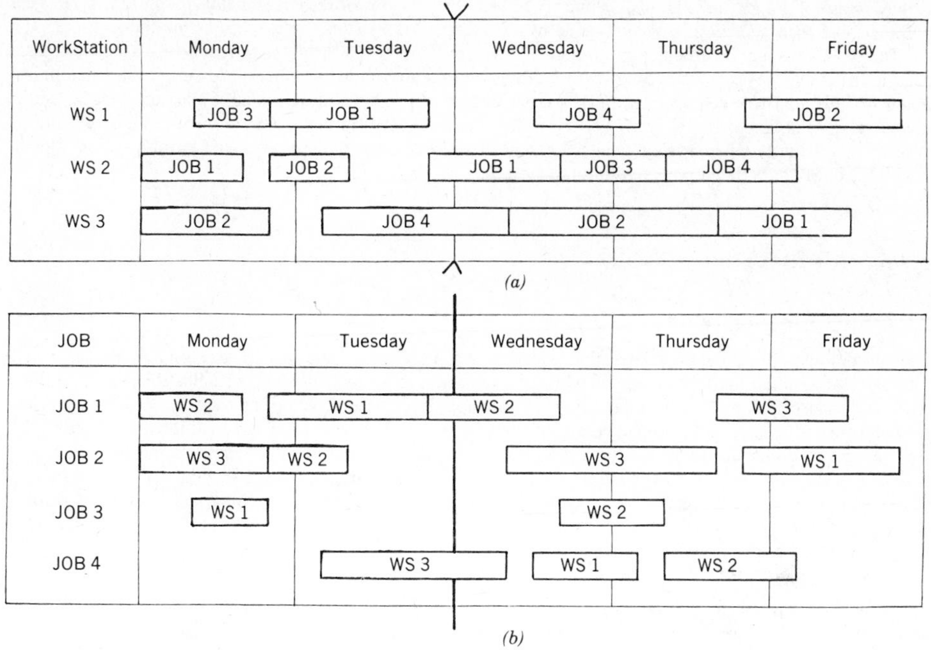

Fig. 3.9.3. (*a*) Resource-based Gantt chart and (*b*) job-based Gantt chart.

A shop packet also contains the necessary forms to initiate work authorizations at individual machine stations, or authorizations required to move the product between stations. Actual processing times are recorded for subsequent comparison to standard times. Additional information, such as scrap and rework, is also noted. Some manual systems include forms allowing centralized dispatching to keep track of the status of each job within the system. Many times, information from a move request or the actual processing times are sent to a central dispatching function at the end of each operation, allowing it to keep track of the location status for jobs currently in the system. This information is the basis for the assignment of scarce resources to the manufacturing system on a day-to-day basis.

3.9.5.2 Mechanical Aids

In a manual system, collection and utilization of information can be a monumental task. Many different techniques have been developed to organize necessary information into a form the decision maker can use to control daily activities. Over the years, many companies have produced mechanical aids to assist the record-handling task. They usually consist of some filing mechanism that allows the user to file jobs and requirements so that key information is easily visualized through simple mechanical methods. Examples are display panels, frequently wall mounted, or filing systems constructed for easy job categorization. The classic approach, used for many years, and still used in small industrial operations, involves the notched card. These cards are produced with prepunched holes on the periphery, which can be notched depending on the category associated with a specific hole. Through the use of a tool similar to an ice pick, the user can easily extract those cards which either do or do not have specific attributes, depending on the design of the sorting system.

Once the information is assembled, the user is faced with the task of actually dispatching or scheduling the activities within the system. Again, many mechanical aids have been developed to facilitate scheduling. Most of these are based on the Gantt chart which was probably the first formalized technique for graphically depicting both anticipated and completed activities on a timeline. (See Fig. 3.9.3.) In its most common form, the horizontal axis of the Gantt chart represents time and the vertical axis represents the resource required to complete the activity, usually machines or workstations. The activities or jobs are scheduled by plotting them as bars on the chart, observing all precedence or routing relationships. Percentage of each activity completed is then indicated, either by shading the appropriate portion of the bar or by placing a caret on the bar. Drawing a vertical line through the current date clearly shows whether any activities are ahead of or behind schedule. A Gantt chart can also be used in reverse fashion, with the vertical axis representing the jobs or tasks to be performed, then plotting the machine or workstation as a bar on the chart.

The variety of mechanisms used in constructing mechanical aids is endless, but their basic function is to display graphically the status of system resources to enable real-time planning or control. Many of them make extensive use of colors to distinguish between various types of jobs or resources and their status; they are often nothing more than a sheet of plywood made by the company for display in a specific environment.

The typical display board contains a series of horizonal slots for inserting precut cards or plastic tabs labeled according to what they represent. Several applications can also be found that employ the use of easily manipulated magnetic boards. These display or control boards are used to develop a visual status of jobs, machine schedules, inventories, or anticipated load vs. capacity within the system. Such techniques are still extensively used in industry, particularly where system congestion tends to occur. They often provide the key information allowing effective control at isolated points within the system.

3.9.6 CURRENT PRACTICE

3.9.6.1 Scheduling Inputs

The scheduling activity requires determining exactly when and where the manufacturing tasks are performed. This activity is initiated at the aggregate level when a master production schedule is established, which considers the overall logistics plan of the manufacturing organization. It is further developed through material-requirements planning, where release and due dates are generated for each job, representing the traditional production plan. But this general plan must still be converted to a detailed schedule, an exact sequence and timetable of all necessary production activities. Practical scheduling techniques at the shop floor facilitate the conversion and can generally be divided into two categories. The first is traditional job dispatching and the second involves development of a finite schedule.

3.9.6.2 Dispatching

The dispatching approach, in contrast to finite scheduling methods that schedule the system prior to job release, delays the detailed timing decisions until resources become available. Attributes of jobs waiting in the queue are examined and a job is selected for processing based on some previously developed priority. Priorities can be based on a job or system attribute or a combination of both. Typical job attributes used to make such decisions are processing times, due dates obtained on the MRP or master production schedule (MPS), or some sort of developed criticality index. Often these indices are based on the relationship of the job to its expected due date, for example, critical ratio scheduling, which develops an aggregate measure indicating the status of the job with respect to anticipated due date. Priorities based on system attributes could incorporate measures of selective machine or worker utilizations as a basis for decision making.

Although a vast amount of literature has been published on job dispatching rules, it is still impossible to identify any single rule that provides the best results for manufacturing job shop operations.[2] The primary advantage of job dispatching rules is their simplicity. The disadvantage is that this form of production scheduling does not provide a global approach to the problem, which may result in optimization of local conditions at the expense of the total system.

3.9.6.3 Finite Scheduling

An optimal finite schedule is the schedule that best meets company goals and is implementable. Although it is not yet possible to develop such optimal finite schedules, heuristic schedules can be generated using a forward or backward scheduling concept. When forward scheduling, the scheduler starts with the known release date of jobs and computes the corresponding completion date of each job based on expected lead times or on a detailed assessment of the available resources. This technique may or may not consider the current status of other jobs that are already in the system. Backward scheduling is essentially the reverse of forward scheduling. Beginning with the due date of each individual job, this method works backward, considering operational times, to determine the start date of jobs. In either case, the expected completion dates for each operation on a job route can be computed.

In most instances, these procedures are used only to develop job priorities, providing estimates of when jobs should be started. However, attempts have been made to utilize forward or backward scheduling to develop detailed finite schedules that present a much more accurate representation of the anticipated shop activity. This is often a very cumbersome procedure and is quite difficult for large systems. But in small systems, it is possible to use a finite scheduling concept, particularly if it is employed with a display or control board, as discussed earlier.

Computerizing forward and backward scheduling has been tried with mixed results. The forward schedule considers the current job status and provides accurate estimates of when job operations can be performed at various workstations within the system. Unfortunately, the resulting completion dates of jobs are often beyond the specified due dates, possibly due to system overload or because the schedule does not effectively utilize the shop resources available. Using backward scheduling provides a good estimate of when the jobs must be started to complete on the due date. But the resulting schedule may conflict with the current shop status, and in some cases, it may indicate that a job should have been started several weeks ago. Naturally, adjustments are necessary to deal with this impossibility.

3.9.6.4 Related Factors

The quality of a finite schedule is obviously very dependent upon the load imposed on the system by the previous aggregate schedules. Consequently, a finite scheduling procedure will not necessarily be able to meet all the due dates imposed by higher level schedules. It can only attempt to utilize the available resources in the best possible way to meet as many due dates as possible. Therefore, many systems still operate on a decision-making process involving dispatching rules utilized on a real-time basis with very little detailed forward scheduling.

System capacity is based on a combination of resources. Capital resources, e.g., equipment, are fixed in the short term. The remaining resources, e.g., manpower and WIP, are more easily adjusted in the short term. The fundamental production scheduling problem is the balancing of these capabilities with requirements. One might assume that as the requirements decrease the scheduling problem becomes easier, since the number of items in the system becomes smaller and the size of the scheduling problem decreases. However, constraints imposed by reduction of controllable resources may make it even more difficult to find a solution that will provide the desired performance. This is particularly true in systems where scheduling decisions are made in real-time based on past experience.

3.9.7 COMPUTERIZED SCHEDULING

3.9.7.1 Production Management Systems

Introduction of the computer into the manufacturing control activity has had a phenomenal industrial impact in recent years. As the sophistication of computers and software increased from early limited use in master scheduling and MRP, computers have become an inherent part of the manufacturing control system. The interaction between the master production schedule, material requirements planning, and resource allocation essentially has replaced the manual production planning technique. If the system is working properly, it provides release and due dates, which are ideally consistent with the aggregate resource availability, for all jobs within the system. This compatibility of requirements to resource availability is the latest extension in computerized scheduling and is currently referred to as manufacturing resource planning, or MRP II.

The production plan resulting from a software system can take several forms, but it is not generally a detailed production schedule that can be directly implemented on the shop floor. Development of the production plan is usually based on fixed lead times, not directly considering potential congestion at the shop floor. A good system does consider the aggregate resource load and indirectly considers congestion.

3.9.7.2 Scheduling with Simulation

One interesting approach that attempts to consider many of the concepts above is the use of forward and backward scheduling involving dispatching concepts. Two computer simulation models of the system are developed. The first models the part flows of the system in reverse. In other words, parts enter their last operation first and in some cases as assembled products. The simulation clock is also run in reverse. Finished parts are then scheduled to arrive to the system at their required due date and proceed backward through the system until they exist as raw materials. If this simulation model is constructed properly, it will operate similarly to an MRP system, which directly considers the capacity of key resources. During the simulation, the time at which each part exits a workstation or exits the system is saved. These times can be thought of as the equivalent latest start times for jobs that have yet to be released to the system. If all of these times are feasible and there are no other jobs in the system, the resulting schedule is implementable. Otherwise, the second simulation model is employed.

The second simulation models part flows in a normal forward fashion. This forward simulation uses the times generated by the backward simulation as estimates of job release dates and operational due dates. Parts and inventory already in the system are also included, and operational due dates are used as job priorities in dispatching decisions at each workstation. As before, the times at which each part enters and exits a workstation are saved and used to develop the schedule. The resulting output from this model provides an implementable schedule, but it does not necessarily guarantee that all due dates are met.

Although this approach has been used rather successfully in some limited environments, it usually requires an extensive simulation model and is prohibitive for manufacturing systems with a large number of jobs that are constantly changing. It is best suited for environments with a limited constant product mix and a finite number of jobs. Ideally, the operation times would be rather large and limited number of alternate routings provided.[3]

3.9.7.3 Commercial Scheduling Systems

Currently there are well over 200 different manufacturing control software packages available.[4] However, only a few claim to develop a detailed implementable production schedule for the shop floor, and many of these have been designed for very specific types of manufacturing environments, thus having limited application.

Systems claiming a detailed scheduling capability are either based on the forward/backward scheduling concept previously discussed, or on the concept of scheduling around bottleneck or control points. Very little information is available to the user referencing the exact procedures followed in these packages, and the limited number of actual industrial applications precludes making generalized statements regarding their capabilities.

Scheduling packages are being marketed with a totally different approach than that used when MRP systems were introduced. Since the MRP concept was easily accessible to anyone interested, the emphasis was on the means of achieving the desired output and relating it to other files and processors a company might require. The available literature on scheduling systems at the shop-floor level does not contain any concepts that can be readily applied to the generalized shop-floor scheduling problem. Consequently, shop-floor scheduling software packages are usually marketed with emphasis on the anticipated benefits with very little, if any, discussion about actual procedures used to attain final schedules. Thus, the intended user can only observe implementations to understand a package's potential value. This marketing approach is very similar to that which has occurred for some time in the area of commercial computerized packages for project scheduling, which are often shrouded in mysterious procedures that appear to give good results for real-world problems.

3.9.8 ALTERNATE SCHEDULING APPROACHES

3.9.8.1 Control Point Scheduling

Scheduling around the bottleneck or control point appears to have merit. Areas in the system that control and constrain the manufacturing flow are isolated, and then production is scheduled based on the capabilities of those bottleneck or control points. In most manufacturing environments there are only a few machines, workstations, or resources constrained to the point of true bottlenecks. If bottlenecks can be predicted for the projected production plan, it seems plausible that realistic schedules could be developed around these points. This practice is quite often observed in manual systems, where control points inhibiting the overall flow of the production process are established, and display or control boards are utilized to schedule those points. Control point scheduling is quite common in the electronics industry, for example, in board build or test departments, where a large number of workstations are fed from a common control point, usually with automatic material handling systems.

More traditional manufacturing firms are also beginning to utilize this concept by establishing centralized storage systems, usually automatic storage and retrieval systems, AS/RS. This provides a central control point at which scheduling decisions are to be made and not left to the discretion of the operator at each workstation. This practice has met with a fair amount of success and at least provides a central point for scheduling. Although simple dispatching concepts or priority rules might still be applied to determine the real-time sequence, improvement is expected by centralization of scheduling decisions at a single point, where common objectives and goals can be pursued.

3.9.8.2 Kanban Systems

The changing business world and the availability of new technology have, and will continue to have, a substantial impact on the manufacturing environment. The development and introduction of Kanban* concepts has created a flurry of activity to adopt just-in-time procedures that minimize work-in-process while simultaneously meeting projected due dates. Kanban is actually an information system with an underlying just-in-time philosophy used to manage the production system.[5] The central idea is the same as that found in traditional MRP systems—produce the right kind of units, at the right time, and in the quantities needed.

In contrast to most control systems, Kanban is a pull system where production is authorized by cards when buffer inventories drop below a certain level. By reducing the number of cards, selective buffer inventories are decreased, and bottlenecks are revealed. These bottlenecks are analyzed in an attempt to increase the production capability until there is no longer a bottleneck. More cards are then removed from the system and the process is repeated with the end objective of reaching zero work-in-process. These improvements are achieved by eliminating waste and reducing setup time to a minimum.

Kanban is applicable only in a limited number of industries where demand is sufficiently constant for production schedules to be frozen for long periods of time, usually a minimum of two weeks. It also requires complete control of all vendor delivery schedules, or buffer inventories, to feed the system. The rigid production schedule requires that a specific number of units be produced each day. This often results in variable length workdays, requiring total worker flexibility and cooperation. Worker cooperation is also integral when a problem arises in the system. The affected worker turns on a red light on a highly visible board, called an Andon board, and the entire production line is shut down until the problem is resolved. During this time, all workers are expected to converge on the problem area to help. The same kind of assistance is expected when changeovers are required.

Fundamentally, the Kanban approach is based on the concept of designing and refining a production system for a very limited number of products so that detailed scheduling is no longer required. However, the success of these systems has generated a vast amount of interest in the concepts of just-in-time production, and numerous attempts are currently being made to modify these concepts for use in the control of more general systems.

---

* *Kanban* is a Japanese word for *card*, and is used in place of *Toyota production system*.

3.9.8.3 Cellular Manufacturing

One approach to overcome the inability to develop implementable finite schedules for entire systems has been to divide the system into smaller elements or departments that are more easily scheduled. Cellular manufacturing, e.g., group technology, functionally creates small flow shops, which process a limited number of similar parts that can be easily scheduled. Ideally each cell starts with a raw material and produces a finished part. Such a structure does not totally eliminate the scheduling problem, but it reduces the complexity, since each cell can be scheduled as if it were a single machine in the overall system.[6]

Cellular systems often work very well upon initial implementation, but experience declining performance over time. One explanation is that these specially designed cells are created to provide production capability for a limited number of common parts. If the product mix changes, the anticipated demand on these cells or centers may change. Thus, certain cells become underutilized while others become overutilized. When this occurs, the manufacturing environment must be restructured to correspond to the new product mix.

3.9.8.4 Automation

Production scheduling is being simplified to some extent by the introduction of more automated equipment. Computer controlled machines, robots, and automated material handling systems do eliminate a large degree of the variability that exists in manual systems, which can render a schedule useless in a short period of time. Theoretically the scheduling problem is simplified, since implementable schedules can be developed based on accurate, predetermined operation times. Unfortunately, automated systems are still subject to scrap, rework, tool breakage, and machine or material handling system downtime. These cause major disruptions and greatly increase the difficulty of system scheduling. This has resulted in many new automated systems being planned, with enormous effort devoted to preventive maintenance and responding to breakdowns within the system.

The introduction of flexible manufacturing systems has drastically altered the philosophy of many manufacturing environments. Flexible manufacturing systems both simplify and complicate production scheduling. Often such systems are designed to produce only a few items, with great flexibility when determining where an operation can be performed within the given system. Other systems are being developed with greater flexibility with respect to the types of products that can be produced, but less flexibility concerning where they can be produced within the system. These automated systems often contain totally computer controlled machining stations and automated material handling systems to deliver work parts to machines. A centralized control facility is usually developed to determine the actual schedule. Interestingly, the increased system flexibility, ordinarily considered a boon to scheduling, has simultaneously complicated the scheduling activities, as it requires the careful monitoring and scheduling of other resources, such as the tooling required to perform a particular operation. Often it is a tradeoff as to which is the most difficult resource to schedule.

A good example is the development of a flexible manufacturing system containing totally flexible machining stations. This allows a large number of parts to be produced at any given station. Although this greatly simplifies the part scheduling activities, it can create complications associated with scheduling tools and fixtures. By taking the same system and restricting the production of certain parts to certain machines, the tool and fixture scheduling problem can be greatly simplified, but only at the expense of the job scheduling problem. Presently, there is no clear-cut obvious direction for flexible systems development.

3.9.8.5 Future Directions

Other new concepts in production scheduling currently being pursued include interactive real-time control and the potential introduction of artificial intelligence and expert systems. It is unclear what impact these concepts will have on production scheduling procedures, but they could provide new avenues for controlling the shop floor. Recently, the concept of computer integrated manufacturing was introduced and should further the capabilities in production scheduling.

3.9.9 ACHIEVING PRODUCTION GOALS

The production scheduling procedure has a phenomenal impact on a firm's ability to meet desired goals and on its profitability. It directly determines and controls resource utilization and the work-in-process at the shop-floor level, elements that are not independent of each other or with other resources in the system. The shop-floor release mechanism directly affects ongoing work-in-process as well as the ability to achieve high resource or machine utilizations.

Alternatively, the firm's objectives constrain scheduling effectiveness, because scheduling is measured by how well subsequent production meets company objectives. The interpretation of objectives during scheduling directly influences schedule development and the ultimate impact the schedule has upon the *actual* system goal. Unfortunately, the true measurement of the system is multiobjective and cannot be accurately translated into scheduling priorities that will achieve all the desired goals. This communication gap can cause serious problems.

In addition, some elements of the manufacturing environment may cause scheduling to have an inadvertently negative impact upon the company goals. For example, control system sophistication greatly increases as more computer systems are introduced. However, many examples exist where sophistication has provided very little benefit due to the manufacturing organization's inability to maintain data integrity. A prime objective of complex control systems is to provide accurate and timely data to support intelligent scheduling decisions. If the data is unavailable or inaccurate, decisions are bound to be inappropriate. All organizational constructs must be considered to facilitate a positive scheduling impact.

It is interesting to note that as early as the 1950s claims were made that various manufacturing companies could easily and successfully generate and implement a production schedule at the shop-floor level that achieved desired goals. In most cases, very little was heard of these systems after their initial introduction. One can only hypothesize that the ongoing results were not good enough to be considered for further discussion or publication. This phenomenon is still occurring today, and it appears that the ability to completely and automatically control and schedule the shop floor is still a vision of the future.

New developments in scheduling are constantly being made. By examining only scheduling articles appearing in technical literature during the 1970s, it is easy to identify more than 500 published articles. Even with this plethora of information available, very little progress has been made in solving practical production scheduling problems within most manufacturing firms. Hopefully, the new approaches, which basically decompose the overall problem to one of developing a schedule for a given production plan, will yield implementable goal achieving results in the foreseeable future.

3.9.10 GENERAL REMARKS ON PRODUCTION SCHEDULING

The topic of production scheduling has recently received renewed attention due to changing business and technical environments. Foreign competition, increased capital cost, new manufacturing technologies, and the need to improve product quality are having a dramatic impact upon the ability of a company to compete. Improved production scheduling techniques, incorporating new manufacturing philosophies, are essential to cope with the changing marketplace.

Unfortunately, the production scheduling problem is combinatorial, making development of optimal solution procedures unlikely. As a result, much emphasis has been placed on developing good heuristic techniques. A great deal of research is currently being conducted in this area and many new concepts are under evaluation in selective manufacturing environments. The next chapter presents basic algorithms and heuristics available for scheduling.

The emergence of high-powered, low-cost computers has provided incentive for developing totally integrated manufacturing systems, further impacting scheduling research. The production scheduling techniques of the future will have to incorporate the global system goals emphasized by computer integrated manufacturing.

Production scheduling cannot be confined to a specific manufacturing area, but encompasses and impacts all levels of the production environment. The emphasis in this chapter has been on the integration aspects of production scheduling at various organizational levels and examining the manufacturing environment's impact on the scheduling task. Many of the newer technologies and concepts touched upon are covered in other chapters. As the manufacturing environment becomes more sophisticated, the art of production scheduling must evolve to a formalized science.

Up to this point, scheduling of the most common manufacturing environments has been discussed. However, there is another environment requiring quite a different scheduling approach, project scheduling. This is most useful for very large scale products, such as ships, and planes, where long manufacturing times are involved. It is also applicable in planning long-term projects at upper organizational levels and can aid integration of those high level plans through to the activities level, resulting in more global system considerations. To complete the discussion of scheduling for all manufacturing environments, project scheduling follows.

3.9.11 PROJECT PLANNING AND CONTROL

The initial concepts of project planning and control emerged during the 1940s in response to the need to better manage and control the complex projects confronted during and after World War II. These concepts were initially applied to defense systems, but have since been applied to numerous nondefense systems in the construction, aerospace and shipbuilding industries, marketing, model shop production, and R & D projects.

Further, many firms are breaking large tasks into smaller projects for convenience. The importance of projects in industry led to development of project management techniques. Successful completion of any project requires careful integration of three key interdependent elements: objectives, resources, and timing. If these key elements are independently determined, project failure is likely. Therefore, the first step in project planning is a clear statement of what objective is to be achieved and the allowable time horizon. Project success is facilitated by having a single main objective. Resources for a project include personnel, funding, data, technology, equipment, and products, some of which may not be readily available or fully developed when needed; therefore, the best project planning is interactive. It starts with a rough plan and proceeds to a final detailed plan, similar to the previously discussed scheduling techniques. Good managers estimate the effects of delays or constraints and modify resources and/or time accordingly.

3.9.11.1 Critical Path Method and PERT

The formalized concepts of project planning and control emerged in the late 1950s with the appearance of CPM (Critical Path Method) and PERT (Program Evaluation and Review Technique). Although the techniques were developed independently, they are both based on the project network concept.

In CPM, the activities or operations necessary to complete the project are shown in a graph called a network, which also shows the order in which activities must be completed. Included with each activity is a single time estimate of how long it will take to complete the activity. Using rather simple computations, the longest route through the network is determined—the critical path. The technique also provides early and late start times for all activities. Knowing the critical path and critical activities can be of great value in project planning and control. If an activity on the critical path is delayed, the entire project will be delayed. A delay in noncritical activities will not necessarily cause the project to be late, unless the delay exceeds available slack. CPM also allows for consideration of time-cost trade-offs.

PERT is very similar to CPM, except that it allows for uncertainty in the time estimates of activities. Each activity is given three time estimates: optimistic, most likely, and pessimistic. These three estimates are used to compute an expected time, which in turn is used to compute the critical path. The initial estimates are used to calculate deviations, which can then be used in the computation of statistical estimates of completion times.

3.9.11.2 Developing the Network

There are several basic steps that must be performed before PERT or CPM can be applied. First, define project objectives and determine the personnel who will be responsible for accomplishing these objectives. Next, define the tasks or activities that must be completed in order to achieve the objectives. In this step, the planner must establish time increments for the tasks or activities, that is, months, days, etc. If the increment is too large, the project may be difficult to control and monitor; if it is too small, accurate time estimates may be difficult. As the planning proceeds, it may be necessary to consolidate several activities into one or conversely, to split one activity into several. Third, develop time estimates for each activity. These may consist of deterministic estimates, as in CPM, or probabilistic estimates, as in PERT.

Finally, the planner must develop the project network by defining all of the interactions or precedences among activities. The precedence relationships must be specified carefully to provide the maximum flexibility in scheduling and should be limited to technological restrictions, at least for the first network. A precedence relationship means that an activity may not begin until all its predecessors are complete. For some activities, this relationship is overly restrictive. In some cases, only a part of the former activity needs to be completed, rather than the entire activity, before the successor may begin. This may be circumvented by breaking the former activity into two or more subactivities and including a precedence restriction between the first of these subactivities and the original successor activity.

Because of the structure of the network, precedence relationships are transitive. Thus, if activity A precedes B, and if B precedes C, this implies that A precedes C, and it is not necessary to include an explicit precedence restriction between A and C. The inclusion of unnecessary precedence relationships will make construction of the network more difficult.

3.9.11.3 Constructing the Network

Once all precedence relationships have been stated, a network consisting of arrows and nodes is constructed. Two different network representations for projects are common: activity-on-arrow and activity-on-node. In the activity-on-arrow format, the activities are represented by arrows and the nodes represent "events," defined as the initiation or completion of one or more activities. Before an activity may be started, all activities that terminate at its start node must be completed. The activity-on-node format represents the activities by means of nodes and the arrows are used to show precedence relationships betwen activities. As in the activity-on-arrow, all predecessors must be complete before an activity may be started. The two different formats are equivalent, and it is not difficult to convert from one to the other. Using the activity-on-node representation makes construction of the network easier, but many people find it more difficult to visualize a project when it is described in this manner.

For an activity-on-arrow network, three rules are generally observed: (1) no two events should have the same number, (2) no two events should be connected by more than one activity, and (3) the network should have one start and one end node. The first and third rules are included to prevent confusion in analysis. The second and third rules appear to limit the types of projects that can be represented. This limitation is removed by the use of dummy activities. Dummy activities have no time delay associated with them; they are used to express precedence relationships only. If the project requires that more than one activity exist between the same two nodes, an additional node, coupled with a dummy activity, can be used to represent the relationship. Dummy activities can also be used to connect all activities with no successors to a single end node and all activities with no predecessors to a start node. The activity-on-node representation is more compact than the activity-on-arrow format; dummy activities are not required. In essence, all the arrows are dummy activities because their only function is to express precedence. Dummy nodes may be required in this representation if it is desired to have one start and one end node for the project.

Examples of the use of dummy nodes and activities for both representations are shown in Figure 3.9.4. Figure 3.9.4*a* shows the use of a dummy activity to prevent activities B and C from having the same start and end node. Figure 3.9.4*b* uses a dummy activity to show a precedence relationship between activities C and E. The activity-on-node format has a dummy node so that the project will have a single end node.

The general concepts and procedures can best be illustrated by considering a simple example of launching a new product. The objective is to be ready for regular production in 22 working weeks. The activity-on-arrow project network and time estimates are shown in Figure 3.9.5. Times for both a CPM and PERT analysis have been included. Figure 3.9.6 contains an activity-on-node representation of the example problem.

The PERT times are given as three estimates:

t_o = optimistic time
t_m = most likely time
t_p = pessimistic time

From these estimates, the expected time t_e can be computed as follows:

$$t_e = (t_o + 4t_m + t_p)/6$$

This provides an estimate of the mean completion time of each activity. For convenience, these times have been made equal to the deterministic times for the CPM analysis. The standard deviation can also be computed for each activity as follows:

$$\sigma = (t_p - t_o)/6$$

These values have also been included in Table 3.9.1.

3.9.11.4 Critical Path Calculations

The initial analysis calls for determining the critical path and is essentially the same for both techniques. This analysis requires a forward pass through the network to establish the earliest possible start and finish times for each activity and a backward pass to establish the lastest possible start and finish times. Figure 3.9.7 gives the computations for both the graphic method, which is suitable for small networks, and the tabular method, normally performed on a computer for large networks.

The computations for the forward pass of the graphic method are given by the circled numbers at each event or node. These circled numbers represent the early finish time for the previous activity. The largest circled values at any node represents the early start time for activities starting at that node. The backward pass values are in the squares. These values represent the latest start time for activities following this event; the smallest value at any node represents the latest ending time for all activities that end at that node. Those values contained in both a circle and a square indicate the critical path times.

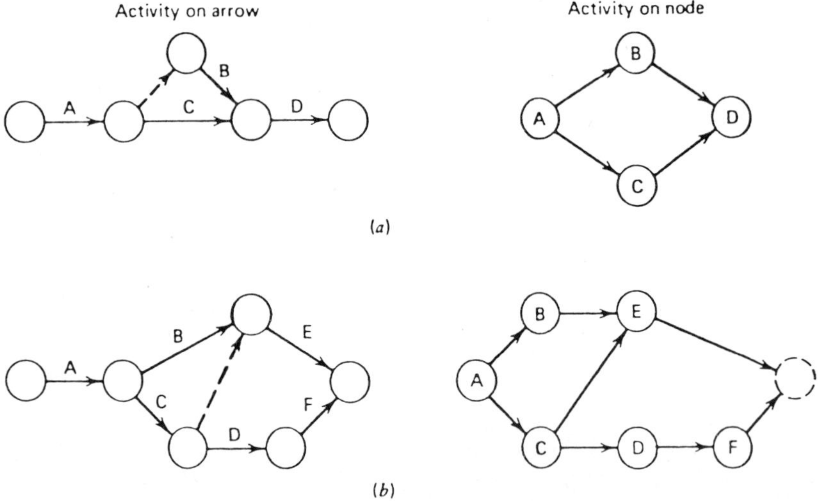

Fig. 3.9.4. Examples of dummy activities.

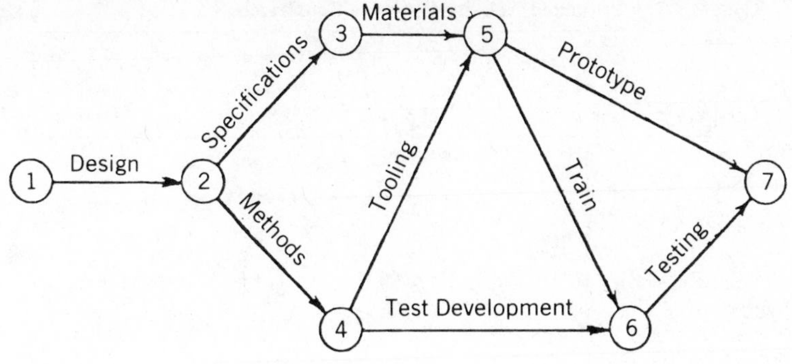

Activity	Arrow	CPM Time	PERT Times				
			t_o	t_m	t_p	t_e	σ
Design	1-2	4	2	4	6	4	0.66
Specifications	2-3	4	3	4	5	4	0.33
Materials	3-5	5	3	5	7	5	0.66
Methods	2-4	3	2	3	4	3	0.33
Tooling	4-5	3	1	2	9	3	1.33
Test Development	4-6	4	3	4	5	4	0.33
Train	5-6	5	3	5	7	5	0.66
Prototype	5-7	8	7	8	9	8	0.33
Testing	6-7	2	1	2	3	2	0.33

Fig. 3.9.5. Example project network.

The equivalent values can be found by the tabular method. The order of the activities is such that all predecessors of an activity are listed before the actual activity is entered. If the network has been constructed so that the start node of each activity has a lower number than its end node, as in the example, then the activities can be entered into the table in increasing order based on the start nodes. Once these values have been entered, all activities that have no predecessors are assigned the same early start time. This value is normally zero, but could be a time representing the actual start of the project.

The difference between the start and finish times, for both early and late times, is the activity duration. In the example, there is a single starting activity, 1–2, which was assigned a start time of zero and a finish time of 4, the activity duration. The early start time of the remaining activities depends on the early finish times of all immediate predecessors, that is, those activities that end at the start node of the activity being examined.

The early start time for an activity is equal to the maximum early finish time of its immediate predecessors. If the activity has a single predecessor, as do activities 2–3 and 2–4, its early start time is equal to the early finish time of the predecessor, activity 1–2, with an early finish of 4. Activities 5–6 and 5–7 each have 2 predecessors (3–5 and 4–5), so their early start time is 13, equal to the maximum of the early finish times,

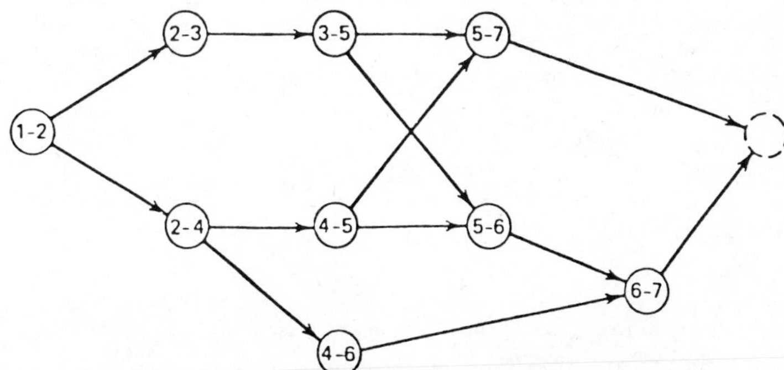

Fig. 3.9.6. Activity-on-node representation of example project network.

Table 3.9.1. Estimated Activity Resource Requirements

Activity	Duration	Resource Requirements		
		Manufacturing Support	Software Support	Consulting Support
1-2	4	2	1	1
2-3	4	1	—	—
3-5	3	2	1	1
2-4	5	2	—	1
4-5	3	3	1	1
4-6	4	2	1	—
5-6	5	2	—	—
5-7	8	2	—	1
6-7	2	1	—	—

13 and 10. The maximum early finish time determines the minimum project duration, or the critical path time. For the example project, a minimum of 21 weeks is required.

The late times are found by a backward pass starting at the bottom of the table, or at the terminating node of the project, node 7. This backward pass is the reverse of the forward pass; the activities are examined in the reverse order, and the late finish time is determined before the late start time.

The first entries are the finish times of all activities that end at the terminating node (5–7 and 6–7). Normally, these times are the same as the critical path time, 21, but a larger value could be used to represent the required completion date of the project. The late start times are then determined by subtracting the respective activity durations from the late finish times. The late finish time for an activity is equal to the minimum late start time of its successors. Activities 5–6 and 4–6 have a single successor (activity 6–7); their late finish times are equal to 19, the late start time of activity 6–7. Activities 4–5 and 3–5 have two successors: activities 5–6 and 5–7, with late start times of 14 and 13, respectively. Thus, the late finish time for activies 4–5 and 3–5 is 13.

The slack or float values are then found by computing the difference between the late and early start times (or the late and early finish times). The critical path is a connected set of activities from the beginning to the end of the network, each of which has the minimum slack value. More than one critical path may exist. If the early finish time of the project is equal to the late finish time, as in the example, the slack for the critical path (or paths) will be equal to zero.

These slack values can be utilized to determine the criticality of paths other than the critical path and are particularly useful for tracking project progress. The critical path for the example is 1-2-3-5-7 with 0 slack. The next most critical path is 5-6-7, with a slack of 1. As the project progresses, the amount of slack may change drastically if any activities are delayed or completed ahead of time. The analysis should be updated with new start and finish times reflecting current status.

3.9.11.5 Statistical Estimates

The times computed in the forward and backward analysis are based on deterministic or expected times, which will seldom occur. A PERT analysis allows the computation of probabilities of completing selected paths on schedule. The initial objective of the example project was to complete the project in 22 working weeks, whereas the critical path time was computed as 21 weeks, or a 50% probability of completing the critical path in 21 weeks. By utilizing the standard deviations calculated in Figure 3.9.5, the probability of completing the critical path in 22 weeks can be computed. Although the individual activity times are assumed to be distributed according to a beta distribution, the path time can be assumed to be normally distributed because of the central limit theorem. The variance of any path can be computed as the sum of the activity variances that make up that path. Thus the standard deviation of the critical path can be computed as follows:

$$\begin{aligned}\sigma_{cp} &= \sqrt{\sigma_{1\text{-}2}^2 + \sigma_{2\text{-}3}^2 + \sigma_{3\text{-}5}^2 + \sigma_{5\text{-}7}^2} \\ &= \sqrt{0.66^2 + 0.33^2 + 0.66^2 + 0.33^2} \\ &= \sqrt{1.108} \\ &= 1.05\end{aligned}$$

Utilizing standard statistical concepts, the probability of completing the critical path in 22 weeks can be computed as

$$Z = (T_c - T_e)/\sigma_{cp}$$

where

T_c = desired completion time (22)
T_e = expected path completion time (21)
σ_{cp} = standard deviation of the critical path (1.05)
Z = number of standard normal deviations of T_c from T_e
Z = (22−21) / 1.05
= 0.952

Using the standard normal tables, the probability of completing the critical path in 22 weeks or less is determined to be approximately 0.83. These results are illustrated in Figure 3.9.8. The probability of completing any other path or partial path in a given time can also be computed using the previous procedure. In addition, it is possible to determine the path duration that would be required for a given probability. For example, if the company wanted to be 95% sure of completing the critical path, they would have to allow 22.7 weeks (Z = 1.65 from the standard normal tables).

One criticism of the PERT statistical analysis is that probability values are calculated for completion of the critical path rather than the project. Unless the critical path is significantly longer than the other paths in the network, it is a poor assumption that the probabilities of completing the critical path and the project are equal. For example, the large variance associated with activity 4–5 could result in a lower probability of completion in 22 weeks for path 1-2-4-5-7 than the critical path. The PERT assumptions result in overly optimistic estimates of the probability of completion of the critical path by a given time. The error in these probability estimates increases as the slack on noncritical paths decreases, as well as when these alternative paths are independent, i.e., they consist of activities not on the current critical path. Consequently, these values should be used with caution.

Several statistical alternatives to surmount these problems have been offered. One is to use a distribution other than beta, for example, gamma, which needs only two time estimates compared to three for beta. Another alternative is to use percentages of the extreme time estimates (t_o and t_p) to increase statistical accuracy.[7]

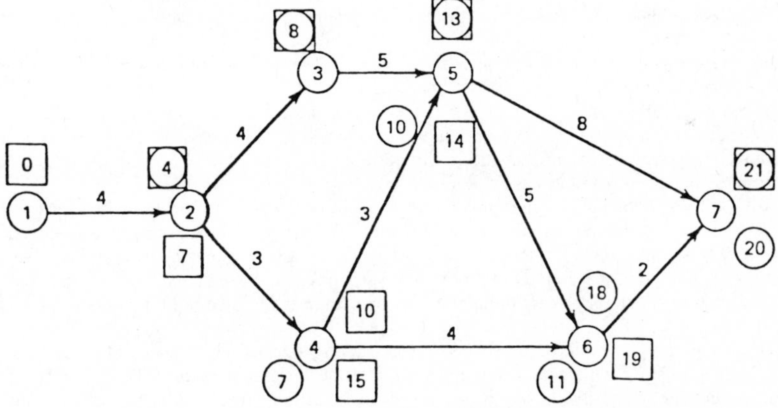

Activity	Time	Early Times Start	Early Times Finish	Late Times Start	Late Times Finish	Slack
1-2	4	0	4	0	4	0
2-3	4	4	8	4	8	0
2-4	3	4	7	7	10	3
3-5	5	8	13	8	13	0
4-5	3	7	10	10	13	3
4-6	4	7	11	15	19	8
5-6	5	13	18	14	19	1
5-7	8	13	21	13	21	0
6-7	2	18	20	19	21	1

Fig. 3.9.7. Forward and backward pass analysis.

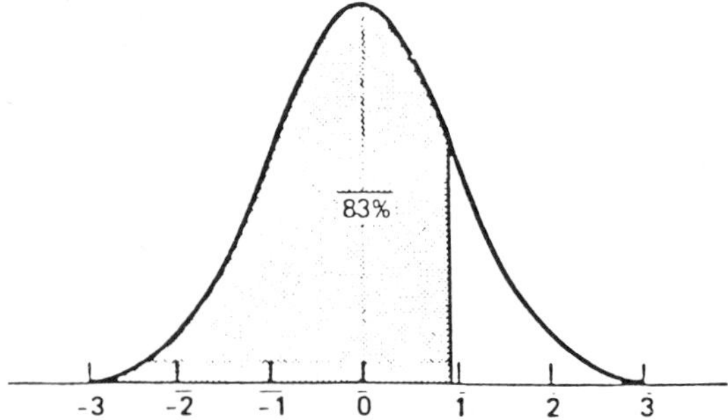

Fig. 3.9.8. Probability of completion.

One other alternative is to employ simulation techniques. With simulation, the user is not restricted to the modified beta distribution for modeling activity durations; any desired distribution type can be employed without increasing the difficulty of analysis. Using time estimates, the project could be simulated several times. The resulting estimates of duration mean and variance will be much more accurate, and the distribution of the mean can be easily obtained. In addition, results apply to the project duration rather than the critical path duration. A criticality index, the probability that the activity is on the critical path, for each activity may also be calculated.

3.9.11.6 Time-Cost Tradeoffs

Using CPM, the user can easily consider time-cost tradeoffs. The objective of this procedure is to reduce project duration by compressing selected activities such that a minimum increase in cost occurs. The user may be able to "crash," or shorten the duration of, selected activities at an increase in cost. To apply this procedure, the user must first identify those activities that can be crashed and the minimum time, or crash time, in which the activities can be completed. At this point the cost associated with these reductions in time must be estimated. Two methods are available—the actual cost for each increment of reduction may be estimated, or the normal and crash-time costs can be estimated, with all intermediate values interpolated by a linear approximation. The two methods are shown in Figure 3.9.9 for activity 5–7. Once the cost estimates for all the activities have been made, the actual project compression can proceed.

If nonlinear cost curves are used, the problem becomes more complex. One method of alleviating the difficulty is to segment each activity into two or more "pseudoactivities," each of which has a linear cost curve. The procedure for time-cost tradeoff that follows is not optimal. If an optimal solution is desired, a network flow algorithm or linear programming may be used.

The basic procedure for linear approximation is to identify that activity on the critical path that can be compressed or crashed at the minimum cost per time unit. This activity is then compressed until the slack on an alternative path becomes zero. In the example, activity 5–7 can be compressed by only one time unit, from 8 to 7, because at that point the time duration of path 5-6-7 has a zero slack, and further reduction of activity 5–7 would not reduce the total project duration. This procedure is repeated until the desired project duration is achieved.

At some point in the procedure, it may become necessary to consider the reduction of several activity times at once in order to achieve a reduction in project duration at the minimum cost. This is the case when the original project has multiple critical paths or when previous crashing creates additional critical paths. At least one activity from each critical path must be crashed for project duration to decrease; some candidate activities may be common to several critical paths.

The cost of crashing is the sum of the costs for each activity. For simple problems with few critical paths, the minimum cost activity or set of activities may be found by inspection. For complex problems, more formalized procedures are required. Once such approach is the use of a linear programming formulation as follows:

$$\min \sum_i \sum_j (a_{ij} - b_{ij}t_{ij})$$

subject to

$$e_i + t_{ij} - e_j \leq 0 \quad \text{for } (i,j) \text{ in network}$$

$$t_{ij} \leq u_{ij} \quad \text{for } (i,j) \text{ in network}$$

$$t_{ij} \geq l_{ij} \quad \text{for } (i,j) \text{ in network}$$

$$e_n \leq T$$

where

a_{ij} = intercept of cost line for activity ij
$-b_{ij}$ = slope of cost line for activity ij
t_{ij} = duration of the activity between nodes i and j
u_{ij} = longest duration for activity ij
l_{ij} = shortest (crash) duration for activity ij
e_i = occurrence time of event i
T = duration of project
n = index for last node in network

The objective function is minimization of total cost. The first constraint prohibits the occurrence of an event until all activities directly preceding it are complete. The second and third constraints restrict activity duration between a lower bound (the crash time) and an upper bound. The upper bound may be the normal activity time or longer, allowing a noncritical activity duration to be extended, at a savings in cost.

If a project is crashed using the basic procedure described earlier, the project cost curve can be easily created by plotting the increasec cost vs. time saved after each reduction. In practice, it is more efficient to use a network flow algorithm to solve the problem and develop the project cost curve.

3.9.11.7 Computerized Network Analysis

The basic critical path scheduling computations for both CPM and PERT may be computerized. This is particularly useful with large networks when updating times to reflect project progress is an essential task, yet tedious and time consuming. Although the computer performs the calculations, the project manager must provide the activity diagram, time and cost information, and decisions, thus maintaining ultimate control. Choosing a program that accurately meets an organization's needs will maximize the real worth of any package. Unfortunately, procedures utilized by these programs are often kept confidential. Thus, packages cannot be judged on methodology, but must be evaluated upon desired features and outputs.

The features of CPM and PERT computer packages will vary, and there are several key points potential users must consider. Capacity, the number of activities the program handles, should be considered based on current and projected needs. Event features include numbering schemes and ability to accommodate multiple start and finish events. Some programs allow network manipulation, such as condensation and revision, while others include cost optimization and control features. Program input (types of data allowed) and output (graphical displays, statistical analysis) must also be considered.[8]

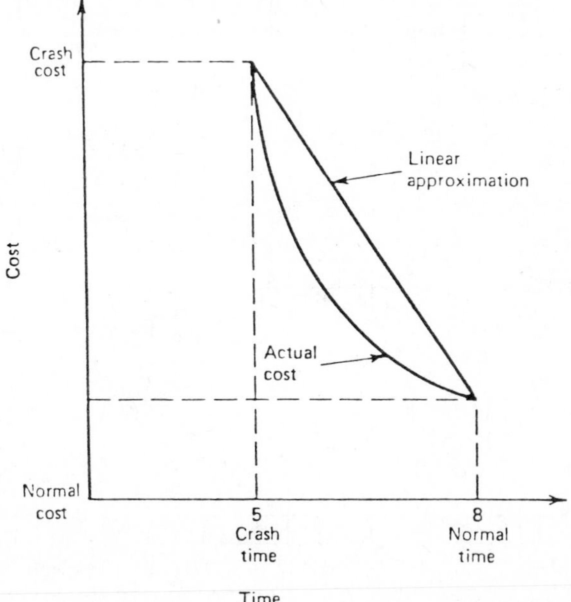

Fig. 3.9.9. Time-cost trade-off curve.

3.9.11.8 Resource Allocation

The CPM and PERT concepts, including time-cost trade-offs and probability estimates, are relatively straightforward and easy to apply. Unfortunately, these techniques do not consider the availability of resources such as manpower and equipment. With the recent advances in technology and subsequent specialization of skills and equipment, resource allocation has become a complex problem. It is not uncommon for a large project to require as many as 40 or 50 different resources.

There are two basic approaches to the resource allocation problem. The first, resource leveling, assumes a project completion date with unlimited available resources and attempts to level the required resources. The second approach, resource constrained, assumes a fixed quantity of available resources and attempts to minimize the project completion date subject to the resource constraints. The real-world problem requires a combination of both approaches, because normally the quantity of resources available is fixed, but it is often possible to access additional key resources for short durations by renting equipment, hiring temporary help, subcontracting, and so on. The total problem often is further complicated by the fact that several projects may be ongoing at the same time.

The initial step of resource allocation is the determination of the types and number of resources required by each activity. These resource requirements depend on the perceived duration of the activity, since the requirements are assumed to be constant for the duration. As with the time-cost trade-off, it is often possible to shorten an activity duration by increasing the resource requirements.

To illustrate the basic concepts, let us use the previous example. Three resource types are required: manufacturing support, software support, and consulting support. The estimated resource requirements for each activity are given in Table 3.9.1. The resource units are given in the number of units per time period. Thus, the first activity requires two units of manufacturing support, one unit of software support, and one unit of consulting support for four weeks.

Prior to proceeding with the development of an actual schedule, it is often useful to determine the relative resource requirements of the project. From the data in Figure 3.9.10, total project requirements are calculated as 73 manufacturing support weeks, 14 software support weeks, and 23 consulting support weeks. If the project is to be completed in 22 weeks, this requires an average of 3.32 manufacturing units, 0.63 software units, and 1.05 consulting units/week. These values can be very misleading, because they do not consider the actual timing of the resource requirements; they do, however, provide a general measure of the project needs.

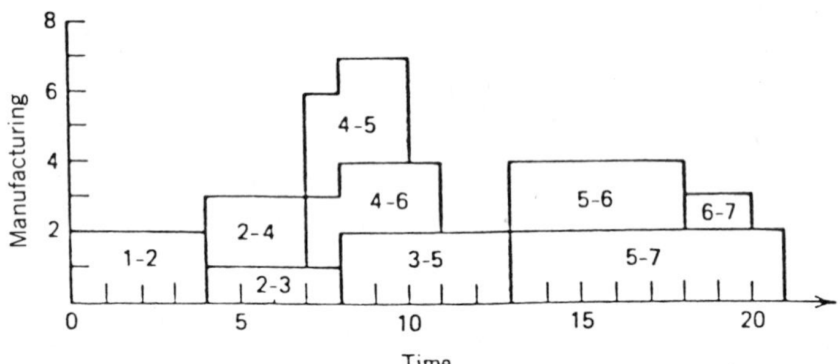

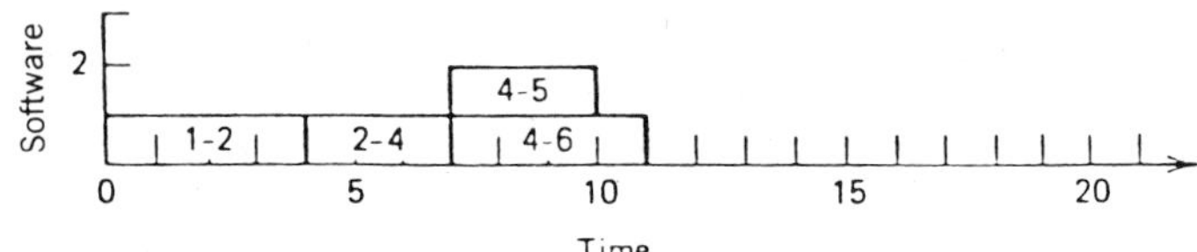

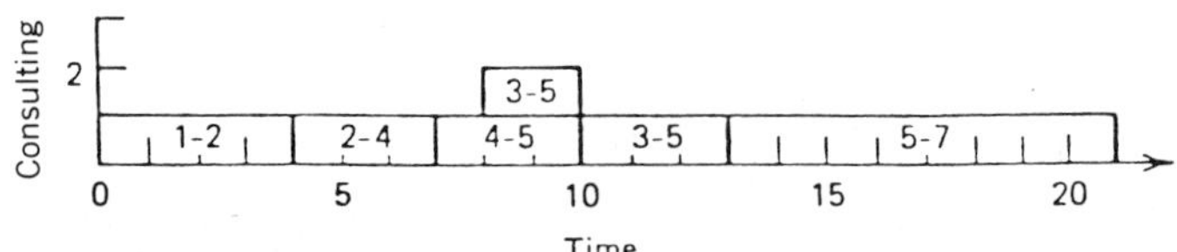

Fig. 3.9.10. Resource requirements for an early start schedule.

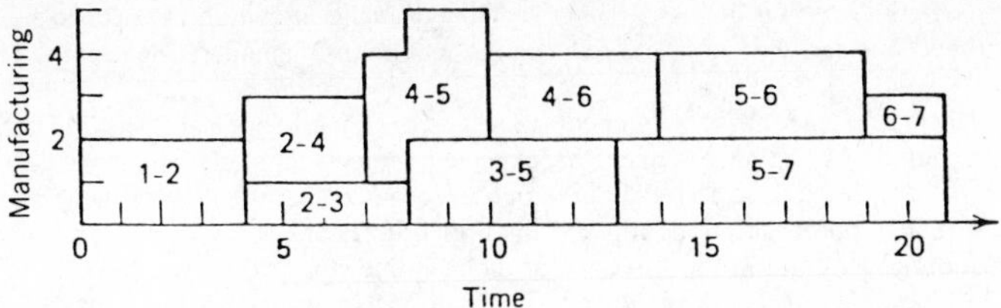

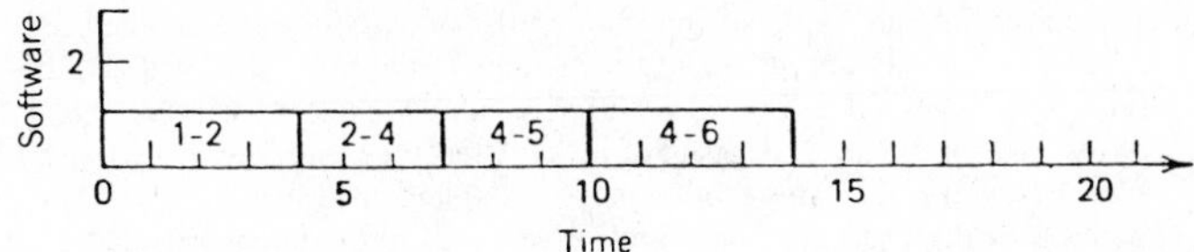

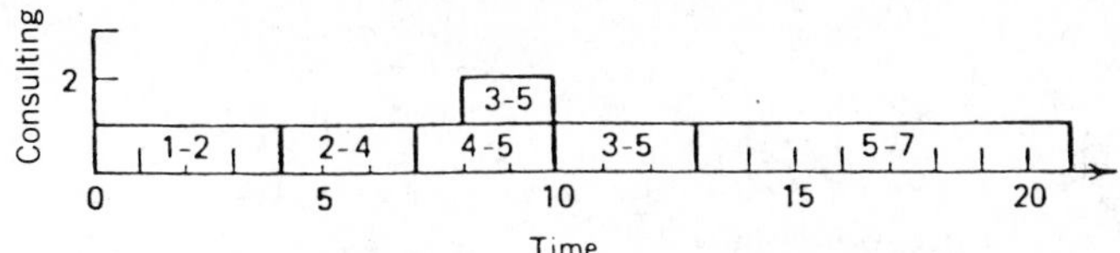

Fig. 3.9.11. Resource leveling schedule.

If the project is small, a Gantt chart of the resource requirements can be constructed to gauge the initial effects of the timing. Scheduling all activities at their earliest start times yields the resource requirements shown in Figure 3.9.10, with the numbers contained in the resource blocks referencing the activity requiring the resource. This schedule requires a maximum of seven manufacturing units, two software units, and two consulting units. The greatest deviation occurs in the manufacturing support schedule, with a range in requirements from two to seven units. Further examination reveals that the simultaneous scheduling of activities 4–5 and 4–6 results in the maximum requirements for both manufacturing and software resources. Resource leveling concepts may be applied to distribute the resource requirements more evenly over time.

3.9.11.9 Resource Leveling

If the project duration is equal to the critical path time, the early and late times determine the general bounds on where each activity can be scheduled. The slack time provides an indication of the maximum amount of movement for each activity. However, caution should be observed when contemplating movements, since the consumption of an activity's slack early in a project will result in successor activities having less or no slack. For example, if the start of activity 2–4 is delayed until time 6, it will not be completed until time 9, and two units of slack are consumed. This reduces the available slack for activity 4–5 from three units to one unit. If the project is to be scheduled for a duration that is longer than the critical path time, the late times and slack must be adjusted. This adjustment is simply a constant, equal to the difference between critical path and the desired duration, added to all the start and finish late times and the slack values.

In applying the concepts of resource leveling to projects with multiple resources, the benefit from leveling one resource may be negated by the effect of rescheduling another resource. If this occurs, it may be necessary to establish priorities for the number of resources or create a weighted performance measure to allow consistent tradeoffs. Naturally, as the number of resources and activities increases, the complexity of the problem grows at a much faster rate.

The application of resource leveling to the example problem for a project duration of 21 weeks yields the results shown in Figure 3.9.11 This rescheduling results in a reduction of the maximum manufacturing support requirements from seven to five and the software support from two to one.

3.9.11.10 Constrained Resources

If the company is limited to only one unit of consulting support, then leveling the resources will not resolve the problem, since two consulting units are still needed for weeks 9 and 10, and the resource constrained approach must be taken. Because 23 consulting weeks are required, the minimum feasible schedule would be 23 weeks. Applying this constraint to the problem and leveling the manufacturing requirements as much as possible yields the schedule shown in Figure 3.9.12. This schedule allows the project to be completed with one consulting unit and one software unit and reduces the maximum manufacturing requirements from five to four.

Even when an acceptable schedule is developed, it must be continually updated. An early activity completion may free critical resources and greatly reduce the expected project completion time; a delay in an activity completion time may have the reverse effect. In either case, resource analysis does provide the project manager with information to make appropriate tradeoffs and intelligent decisions.

Many mathematical programming formulations of the resource constrained and resource leveling problems have been created.[9] They are not useful for most realistic projects because of the combinatorial nature of the problem. However, a great many heuristic, nonoptimizing procedures have been proposed for generating good project schedules.

One class of heuristics ranks the activities and attempts to schedule the highest priority activity first. If precedence or resource constraints are violated, the ranked list of activities is examined until an activity can be scheduled or the list is exhausted. The procedure than steps to the first scheduled activity termination and repeats until all activities are scheduled. Suggested priority rules include least slack, greatest total resource demand, shortest activity duration, greatest demand of a "key" resource, minimum late finish time, and longest chain of following activities. Several of these rules may be implemented in two ways. The priority could be calculated at the beginning of the procedure and never changed, or could be recalculated after each time an activity is scheduled, thereby changing the ordering of the list. Also, these simple heuristics can be combined to produce more complex decision rules.

Numerous computer packages incorporating these or other heuristics are currently being marketed, although little is known about the procedures that they utilize for activity scheduling. Although these packages are generally quite expensive and require large computers, this expense is fully justified in many instances and provides the only reasonable way to manage and control large projects. An alternative is to use several heuristics to generate schedules and choose the best among those schedules. Simulation can be used to evaluate the performance of different heuristics for a particular project when activity times (and possibly some resource requirements and availabilities) are random variables.

3.9.11.11 Other Techniques

Another project scheduling technique originating from military production is line of balance (LOB). The prime concern of this technique is monitoring project progress by measuring actual progress versus schedule dates. There are four basic phases: objective, program, program progress, and comparison of progress to objective.

The objective is stated as number of product units needed/time and is plotted on an objective chart. The program is expressed by a process flow chart (a modified Gantt chart) with time on the horizontal axis and significant steps necessary to meet the objective on the vertical axis. The chart considers lead times and indicates

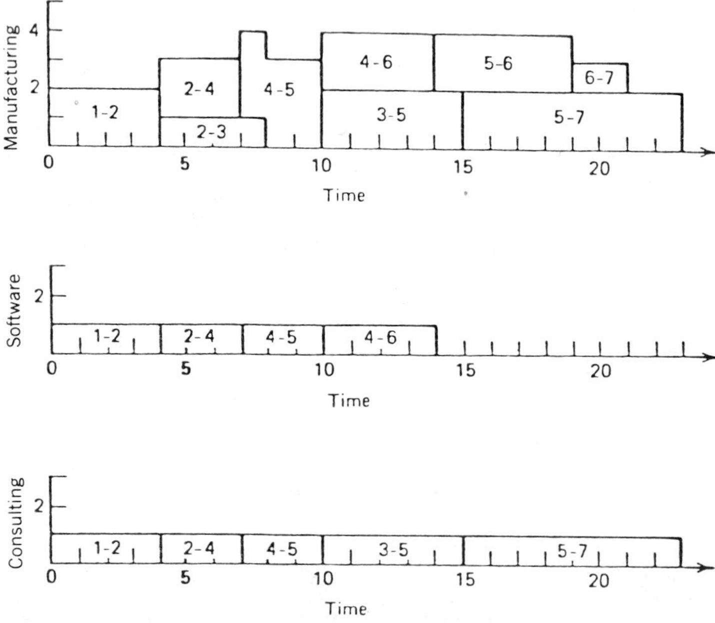

Fig. 3.9.12. Resource constrained schedule.

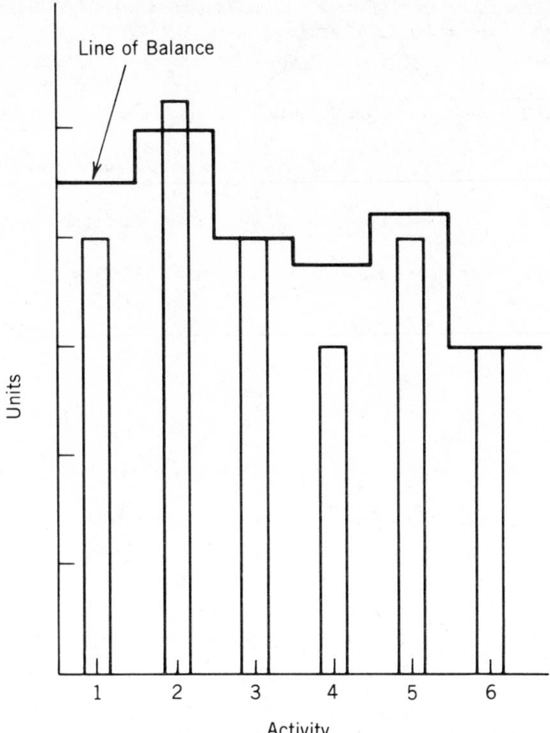

Fig. 3.9.13. Line-of-balance chart.

activity precedence (similar to MRP explosion with due dates). The progress chart is a bar chart with the same units designation as the objective chart. Striking the line of balance involves transferring points for each program step from the objective chart to the progress chart for a given date. As illustrated in Figure 3.9.13, the points graph a step function; a program exactly in phase has a line of balance intersecting each progress chart bar at the top. Bars below the line indicate activities that are behind schedule and bars above the line those that are ahead of schedule. A project manager can quickly note production problems by this highly visible technique.

Another approach, the PERT/cost method, identifies schedule slippages and cost overruns in time for corrective action. From a PERT network and schedule, resource and cost estimates are made per time period. Comparing actual costs to estimates gives project status. Various levels of comparison include overall status for project managers to pinpoint trouble spots, and more specific reports concerning manpower requirements, rate of expenditure, and cost of work. The reports are meant to help managers recognize the effects of applying different levels of resources to different periods of project development.

In addition to single-project management tools, several multiproject scheduling techniques have been developed. One early heuristic method tries to minimize due date slippage using individual project critical path analysis data and is implemented by simulation. Resource Allocation and Multiproject Scheduling, (RAMPS), is a computerized model that attempts to meet due dates under resource constraints. Generally, these techniques and recent research employ simulation and heuristics to develop schedules that attempt to minimize deviation from due dates.

Project scheduling has evolved from its advent in the defense arena to many uses in the private sector. Its main use is on one-of-a-kind projects with long activity durations. With the advances in computing, much larger projects can be monitored and controlled more effectively. Perhaps the greatest contribution of project scheduling is that it requires user forethought when developing an implementable plan leading to successful attainment of project goals.

REFERENCES

1. *Industral Engineering Terminology*, Institute of Industrial Engineers, Atlanta, GA, 1982.
2. J. H. Blackstone, Jr., D. T. Phillips, and G. L. Hogg, "A State-of-the-Art Survey of Dispatching Rules for Manufacturing Job Shop Operations," *International Journal of Production Research*, **20**(1), 27–45 (1982).

3. N. T. Yunk, "Computer Simulation Applied to Shop Floor Scheduling and Control," *1981 Fall Industrial Engineering Conference Proceedings*, Institute of Industrial Engineers, Atlanta, GA, 1981.
4. H. Onari and D. Wechsler, *Survey of Commercially Available Production Management Systems*, R-81-FM-03, CAM-1, 1981.
5. R. W. Hall, *Production Planning and Control in Japan—Driving the Productivity Machine*, APICS Research Report, 1981.
6. T. J. Greene and R. P. Sadowski, "Cellular Manufacturing Control," *Journal of Manufacturing Systems*, **2**(2), 137–145 (1983).
7. G. E. Whitehouse, *Systems Analysis and Design Using Network Techniques*. Englewood Cliffs, NJ: Prentice-Hall, 1973.
8. J. J. Moder and C. R. Phillips, *Project Management with CPM and PERT*. New York: Van Nostrand Reinhold, 1970.
9. E. W. Davis "Project Scheduling Under Resource Constraints—Historical Review and Categorization of Procedures," *AIIE Transactions*, **5**(4), 297–313 (1973).

BIBLIOGRAPHY

Bedworth, D. D., and J. E. Bailey, *Integrated Production Control Systems/Management, Analysis, Design*. New York: Wiley, 1982.

Corke, D. K., *Production Control in Engineering*. London: Camelot Press, 1977.

Graves, S. C., "A Review of Production Scheduling," *Operations Research*, **29**(4), 646–675 (1981).

Greene, J. H., *Production and Inventory Control Handbook*. New York: McGraw-Hill, 1973.

Salvendy, G., *Handbook of Industrial Engineering*. New York: Wiley, 1982.

Spinner, M., *Elements of Project Management: Plan, Schedule, and Control*. Englewood Cliffs, NJ: Prentice-Hall, 1981.

CHAPTER 3.10

SCHEDULING AND SEQUENCING ALGORITHMS

HAMILTON EMMONS

Case Western Reserve University
Cleveland, Ohio

3.10.1 INTRODUCTION

Scheduling may be defined as the allocation of resources over time to perform certain tasks. As a practical matter, such timing and sequencing is a frequent component of managerial decision making. Broad and often conflicting objectives must somehow be translated into specific actions. Frequently, simple heuristic rules are employed without a clear understanding of how they work, how well they work, or whether other policies would work better.

At the same time, a large body of scheduling knowledge exists and continues to grow. Mathematical and computer modeling transform the goals of management into objective functions, and their limitations into constraints. The results of such analysis have not been given wide application heretofore. However, the gap between the simplified idealizations that the researcher constructs and the real world complexity that the manager confronts is narrowing, as increasingly complex models are solved and as our understanding of heuristics develops.

The term "scheduling" can be applied to many different situations. This chapter will focus on *job shop scheduling*. Although the concepts and procedures of job shop scheduling are relevant to hospitals, schools, banks, repair facilities, collection/delivery services, etc., they originated and are used principally in industrial production. In manufacturing, after broader, longer term aggregate planning has formulated master schedules for a period of weeks or months, job shop scheduling makes the detailed, short-term decisions of precisely when and how to do what.

3.10.1.1 Basic Definitions and Notation

To begin with, we state the assumptions and terminology on which this chapter is based. The basic item of work to be performed is called, in various contexts, the *task*, *operation*, or *activity*. A task is characterized by its *processing time* and its resource requirements. A *resource*, or *machine*, is a potential agent for performing work. It may in fact be a machine, or it may be a worker, a truck, or a service counter. There may be several resources, and they may be of various types, where the the type is specified by the kinds of tasks it can process. A task may require one or many resources, of one or several types. Thus, a schedule is a specification of when to do what tasks using which facilities.

Jobs are sets of tasks grouped on the basis of characteristics they share in common. For example, the tasks of a job generally share the same *release date* and *due date*. Of course, the concept of a job arises naturally in practice: it is the basic unit of work from the customer's point of view, as the task is from the manufacturer's point of view.

Usually the main thing tying together the tasks of a job is a *precedence* structure: the requirement that, for technological reasons, certain tasks must be completed before others can be started, thus imposing a "partial order" on the tasks. For example, all possible precedence requirements for a three-task job are illustrated in Figure 3.10.1, where the numbered nodes represent the tasks, and the arrows connect immediate predecessor to immediate successor. In the figure, (a) is the case of *unordered*, *parallel* or *independent* tasks, (b) represents fully *ordered* tasks in *linear* or *chain precedence*, (c) may be described as two *parallel chains* (since a single task is trivially a chain), (d) is the simplest form of an *assembly tree* or *in-tree*, which may be defined as any precedence structure in which each task (except the one "final assembly" task) has one immediate successor, and (e) is similarly a *branching tree* or *out-tree*.

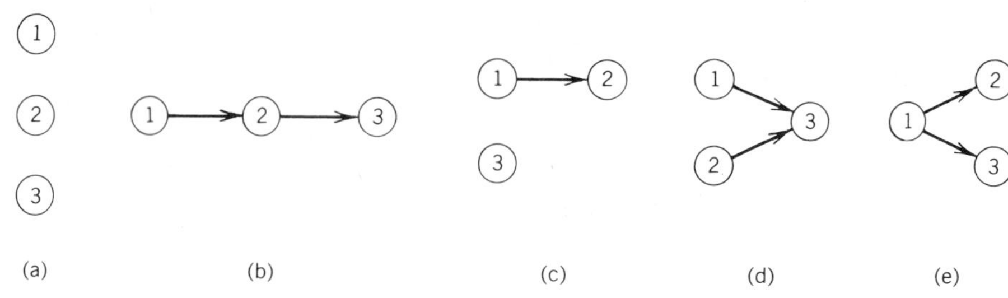

Fig. 3.10.1. All possible orderings of three tasks.

The total work to be performed may be defined in one of two ways. The *static* problem assumes a predetermined, finite number of jobs, each with its own characteristics. The *dynamic* problem models an indefinite stream of jobs arriving for processing. They may be all different and arrive at random times, either singly or in batches. Static problems are most often considered to be fully *deterministic*: all parameters, such as processing times, are assumed to be known constants, either by definition or by accurate estimation. Dynamic problems are naturally *stochastic* or *random*: since jobs are constantly arriving, their characteristics are not known in advance, and we may assume instead a probability distribution for them, based on past experience.

Much of the existing knowledge of scheduling policies is based on the static assumption. While it may seem that the real world is never static, in fact practical operations often function in this mode: arriving orders are collected, and periodically a batch of jobs is scheduled. The alternative is frequent disruptions as new arrivals prompt revisions. We assume static and deterministic settings hereafter when nothing is said to the contrary.

In order to present the ideas and results of good scheduling practice compactly, it is convenient to introduce some terminology and symbols to represent the parameters of jobs and schedules. We have tried to follow common usage wherever it exists. Let:

m = number of machines (unless otherwise stated, one of each type assumed).
n = number of jobs in the static problem.

Job j has the following given descriptors:

r_j = release date, arrival time or ready time, at which job j becomes available for processing.
d_j = due date, by which time job j must be completed, to avoid penalty.
a_j = allowance, or time allowed in shop $= d_j - r_j$.
n_j = number of tasks, for each of which we specify:

- m_{jk} = number of the machines required to perform task k.
- p_{jk} = processing time for task k on machine m_{jk}, including set-up and tear-down times.

p_j = $\Sigma_k p_{jk}$ = total processing time for job j.

The following are schedule-dependent parameters of job j:

W_{jk} = waiting time preceding task k.
W_j = $\Sigma_k W_{jk}$ = total waiting time.
C_j = completion time $= r_j + p_j + W_j$.
F_j = flow time or time in shop $= p_j + W_j = C_j - r_j$.
L_j = lateness $= C_j - d_j = F_j - a_j$. Note that L_j can be positive or negative.
T_j = tardiness = max (O, L_j). We say a job is "tardy" or "late" when it has positive lateness.
I_j = tardiness indicator $= \{1, \text{if } C_j > d_j; 0, \text{if } C_j \leq d_j\}$.

Much of this notation is illustrated in Figure 3.10.2.

We also use the notation Σ_j and $\max_j$ when we intend the summation or maximization to take place over all values of j; usually, from $j = 1$ to $j = n$.

3.10.1.2 Preemption

An important consideration in model definition is the flexibility and speed with which the system can accommodate itself to changes in its environment. For example, when processing of a task ends or a new job becomes available for processing, how much are we bound by earlier decisions and how free are we to adjust to the new situation? One kind of adaptability is *job splitting* or *preemption*: interruption of an in-process task before completion. At one extreme, a decision to start processing a certain job on a certain processor cannot be retracted once made: preemption is prohibited. Think of an oil tanker halfway across the Atlantic. At the other extreme, it may be possible to interrupt the processing of any task freely and with negligible cost. This is often true in small-scale, labor-intensive manufacturing. In between lie the situations where preemption is permitted but at some cost in time and/or money. We shall not discuss further these intermediate cases, but shall speak only of preemption being permitted (without penalty, it is understood) or not.

Clearly, systems with greater flexibility are easier to control: long-term commitments are not necessary. Thus, with preemption allowed, we can make decisions based solely on jobs present and available, without looking ahead; it is not surprising that such *myopic policies* are generally satisfactory, and often optimal. However, nonpreemptive situations are more commonly encountered in practice. Throughout this chapter, preemption is assumed to be prohibited unless otherwise stated.

3.10.1.3 Measures of Performance

Three broad objectives have traditionally been considered for job shop scheduling:

1. Minimize work-in-process inventory.
2. Maximize utilization of resources.
3. Maximize service to customers.

A major reason that scheduling is so difficult in practice is that these goals are usually in conflict: a schedule that performs well by one measure often does badly by another. To investigate the properties of different scheduling rules, and so to gain some insight into how to find a satisfactory compromise of objectives in any given situation, it is necessary to identify specific, quantifiable criteria that reflect the more general goals. Some of the reasonable and commonly used criteria (all of which are to be minimized) are as follows.

Criteria Depending Only on Processing Times

The following objectives, by minimizing delays in various ways, tend to satisfy the objectives of both resource utilization and customer satisfaction. Keep in mind these are quantities to be *minimized*:

1. Greatest completion time of any job, $\max_j C_j$.
2. Greatest flow time.
3. Greatest waiting time.
4. Average completion time of all jobs, $\Sigma_j C_j / n$.
5. Average flow time.
6. Average waiting time.
7. Average weighted completion time, $\Sigma_j w_j C_j / n$.
8. Average weighted flow time.
9. Average weighted waiting time.

In criteria 7, 8, and 9, w_j is a weight indicating the relative importance of job j. Thus, if a completion time of C_j produces a cost of $w_j C_j$, we can interpret w_j as the *linear loss rate*, or *cost per unit time*, of keeping customer j waiting.

Criteria 4, 5, and 6 are special cases of 7, 8, and 9 where all jobs have equal weight. Furthermore, 7, 8, and 9 are equivalent (that is, a schedule that minimizes one minimizes the others) as are 4, 5, and 6, because $F_j = C_j - r_j = p_j + W_j$. The maximal completion time, $\max_j C_j$, is the time to process all jobs from start to

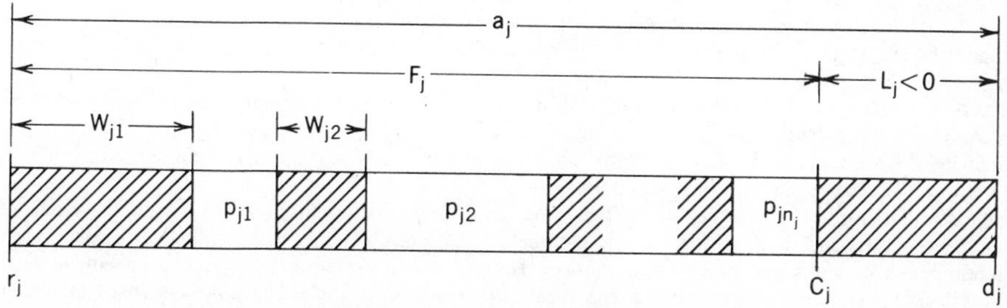

Fig. 3.10.2. Time-related parameters of a job.

finish, and is frequently called the *makespan*. We will denote it M. If all jobs are simultaneously available ($r_j = 0$ for all jobs j), then $\max_j F_j$ and $\max_j C_j = M$ are equivalent. Finally, note that "average" criteria such as $\Sigma_j C_j/n$ are equivalent to "total" criteria such as $\Sigma_j C_j$.

Criteria Involving Due Dates

The following minimization criteria are principally oriented to customer service:

10. Maximal lateness.
11. Maximal tardiness.
12. Average lateness.
13. Average tardiness.
14. Average weighted lateness.
15. Average weighted tardiness.
16. Number of jobs tardy.
17. Weighted number of jobs tardy.

Again, 12, 13, and 16 are special cases of 14, 15, and 17. Also, 12 is equivalent to 4, 5, and 6, as is 14 to 7, 8, and 9 because $L_j = C_j - d_j = F_j - a_j$. We omit the criterion number of jobs late (that is, with positive lateness), since it is equivalent to 16.

Other Criteria

Scheduling objective functions specifically designed to reflect the producer's goals of high utilization of facilities and low work-in-process are not often considered separately. To see why, suppose for a static problem we define machine utilization as {total processing time for all jobs} divided by {total machine time required}. If we assume m machines are tied up until the last task is completed at time M, then utilization is given by $\Sigma_j p_j/mM$. Thus, maximizing utilization is equivalent to minimizing makespan.

Similarly, if we define work-in-process inventory as the average number of jobs on hand during the time necessary to process all jobs, we can show that it is proportional to $\Sigma_j C_j/M$.[1] This means, for a given M, that minimizing the average completion time is an equivalent measure.

Another possibility is to consider multiple objectives. One can minimize a weighted average of several criteria, attempting thereby to balance conflicting goals. Clearly the relative importance given to the chosen objectives is a subjective decision that management must make. Otherwise, one can choose a primary goal that must be satisfied (say, to minimize makespan or to achieve a makespan no greater than some target completion time), and then optimize another criterion (say, minimize average completion time) over the subset of schedules that satisfy the first requirement.

Regular Measures

Finally, it is useful to define a broad class of scheduling criteria, including all those listed above and most other reasonable ones, about which we can sometimes make general statements. *Regular* measures of performance are minimization criteria that depend on the completion times, $C_1, C_2, \ldots, C_n$ of the jobs, and which are nondecreasing functions of every C_j. Equivalently, schedule S is preferred to schedule S' under a regular measure if every job is completed at least as early in S as in S'. Roughly speaking, with a regular measure we are always happy to be able to complete a job earlier, all else being equal.

3.10.1.4 Optimal and Heuristic Algorithms

In this chapter, specific scheduling algorithms will be presented to achieve specified goals under specified circumstances. An *algorithm* is a procedure or sequence of steps. An optimal algorithm will always produce the best schedule; a *heuristic* algorithm usually provides excellent schedules (they may often be optimal, but we cannot be sure), generally with little effort. The catch is that heuristic solutions may sometimes be bad, and again we cannot be sure.

The focus will be on simple rules and policies. In practice such rules are attractive, especially if they appeal to intuition. Of couse, such rules guarantee optimality only under the simplest assumptions. However, their inherent logic often extends to more complex situations and provides a good rule of thumb, even though it may not always produce the very best schedule.

Heuristics are usually satisfactory in practice, since the criterion being minimized is usually a surrogate for one or several broad and ill-defined goals, the parameter values are probably based on imprecise data, and all assumptions are subject to change without notice. Thus, a "good" solution relative to a specific objective function is as likely to serve the purpose as the "best" solution, though it does have the drawback that usually one does not know precisely how "good" it is.

General References

For most of the material to be reported, proofs and full discussions appear in standard texts, three of which[1,2,3] are listed following this chapter. Separate references are provided only for results that are not included in such sources.

3.10.2 ONE-MACHINE SEQUENCING

The question of how best to sequence jobs for one-stage processing by a single facility (whether it be a machine, a worker, a vehicle, or a service counter) is clearly the simplest kind of scheduling problem to start with, and has been extensively studied. Though it may at first appear to be too simple a situation for practical concern, it is frequently encountered. Even in more complex environments, it may be a good approximation, as when a single bottleneck machine dominates a multistage production facility, or when convenient, easily implemented local dispatching rules are considered satisfactory to schedule a large job shop. Furthermore, sequencing policies that are optimal for one machine in isolation often suggest similar approaches that prove useful for multiprocessor scheduling.

In the static and deterministic case, we suppose n jobs are to be scheduled, each job having certain given characteristics, assumed to be known constants, either by definition or by precise estimation. We wish to determine in what order to do the jobs, to achieve some specified objective. We remark that, as long as our measure of performance is regular (recall that this means we are always happier to complete any job earlier) and all jobs are simultaneously available, it is never advantageous to preempt a job or to insert idle time, that is, to hold a machine idle when there is work on hand for it to do. This permits us to restrict our attention to the simplest kind of scheduling policy, the *permutation schedule*, in which it is sufficient to specify the order in which the jobs should be done: start times and completion times then follow automatically.

If all the jobs do not have the same ready time, and still more if the problem involves random elements or a dynamic inflow of jobs, then preemption and delay may be advantageous when permitted.

3.10.2.1 Average Cost Criteria

The following simple scheduling algorith, often called Smith's ratio rule, is recommended when the objective is to give good service to the n customers *on the average*. It is remarkably robust, in that variations and extensions of the idea perform well in a variety of complex situations.

ALGORITHM 3.10.1

Given: n jobs with a known procesing time p_j and a known weight w_j for each job j ($j = 1, \ldots, n$).

Objective: Minimize the average weighted completion time of the n jobs.

Optimal Policy: Sequence the jobs in decreasing order of w_j/p_j.

Besides $\Sigma_j w_j C_j$, as remarked earlier, equivalent objectives that are minimized by this ordering include $\Sigma_j w_j W_j$ and, in case due dates are specified for each job, $\Sigma_j w_j L_j$ (the due dates are irrelevant), but not the more interesting $\Sigma_j w_j T_j$. If we interpret w_j, not as a penalty for delay, but as a *reward* for each unit of waiting time avoided, the rule gives priority to the job for which the reward per unit processing time is greatest.

For example, suppose three jobs (1,2,3) with processing times (4,6,8) and weights (2,1,3), respectively, are to be sequenced on one machine. To minimize the average weighted completion time or equivalent, compute the values of w_j/p_j, getting (½, ⅙, ⅜) so that the optimal sequence is (1,3,2).

If all weights are equal (the unweighted case), the well known *Shortest Processing Time* (SPT) rule results: sequence in increasing order of p_j. Analogously, Smith's ratio rule is frequently called the Shortest Weighted Processing Time (SWPT) rule: sequence in increasing order of p_j/w_j.

If jobs are not all available at the start, but have various values of r_j, preemption becomes beneficial, as when a "heavyweight" job arrives in the middle of processing an unimportant one. Assuming it is possible to preempt without significant loss, the optimal policy in the unweighted case is the logical extension of SPT: at each completion and each arrival time, switch to (or continue with) the waiting job with *Shortest Remaining Processing Time* (SRPT). With weights on the jobs, the corresponding policy is: at each completion and arrival, switch to (or continue with) the job with largest w_j/(remaining p_j). Surprisingly, although a good heuristic, this rule does not always produce the best schedule.

Again, with preemption prohibited, no simple rule can be given to guarantee an optimum. A reasonable heuristic is: At each completion time, choose the waiting job with maximal w_j/p_j to process next.

With precedence constraints, one might hope that Algorithm 3.10.1 could be generalized; for example, by scheduling first the job without predecessors with largest w_j/p_j, removing it, and repeating. This is reasonable but not optimal, as is the similar, but not equivalent policy: schedule last the job without successors having smallest w_j/p_j, remove it, and repeat. Choosing the better of the two schedules so generated is a reasonable heuristic. Other heuristics have been proposed,[4] but optimal algorithms exist only for certain types of precedence constraints: parallel chains,[5] trees,[5] and slightly more general structures.[6]

If processing times cannot be precisely predicted but instead probability distributions are given, the equivalent objective is to minimize the expected value of the total weighted completion times. Again, preemption may be useful (for instance, think of a job that is very likely to be completed quickly, but if not, will take very much longer). However, if preemption is not allowed, the same rule can be applied[7] using the mean values of the processing times: sequence the jobs in decreasing order of $w_j/E(p_j)$. When jobs may be interrupted without cost, one can show under what conditions preemption is never desirable,[8] in which case the same rule obtains; and in any case, show that the following general principle is always applicable: at any time, work on the job that can produce the greatest expected reward rate, for as long as it remains maximal. This policy remains optimal in the dynamic case, with jobs arriving randomly.

If the objective is to minimize the average tardiness or some still more general average delay criterion, then no such simple algorithm exists, even in the unweighted, static, deterministic case. General purpose solution algorithms, relatively complex and time-consuming, have been developed.[9]

3.10.2.2 Maximal Cost Criteria

If we are principally concerned about the worst treated customer rather than the average treatment of all customers, a very general algorithm[10] is available which, while not quite as simple as Smith's ratio rule, is still a straightforward single-pass algorithm. ("Single-pass" means that we successively determine the positions of jobs in a schedule, and once having positioned a job we never reconsider.) Suppose we associate with each customer any nondecreasing penalty function for delay, so that for each job j:

$$f_j(t) = \text{cost of completing job } j \text{ at time } t$$

This could be any of the types of penalities we have considered, such as weighted tardiness:

$$f_j(t) = w_j \max\{0, t - d_j\}$$

or anything else. There could also be different types of penalty functions associated with different jobs.

ALGORITHM 3.10.2

Given: n jobs, with known processing time p_j and known delay cost function $f_j(t)$ for each job j.

Objective: Minimize the largest delay cost suffered by any job.

Optimal Policy: Schedule last a job that is least costly there. That is, choose a job k such that

$$f_k(M) = \min_j[f_j(M)]$$

to be processed last, where $M = \Sigma_j p_j$ is the makespan. Now remove that job from consideration, and repeat the procedure.

For example, suppose three jobs (1,2,3) with processing times (4,6,8), weights (3,2,7), and due dates (8,13,10) are to be sequenced so as to minimize the largest weighted tardiness of any job. The makespan is 18, at which time the jobs would have tardiness (10,5,8) and weighted tardiness (30,10,56). Job 2, being least costly, is scheduled last. Removing it, the makespan for the remaining jobs is 12, at which time the weighted tardiness of jobs 1 and 3 are (12,14), so job 1 is last. The optimal schedule is (3,1,2), giving delay costs of (0,12,10) with maximal cost of 12; no other schedule can do better. Note that it is not necessarily the last job scheduled that has the highest cost.

It is not hard to see that, in the special case where the cost associated with each job is its lateness, or tardiness, a least costly job to put last is one with the latest due date. This leads to the *Earliest Due Date* (EDD) rule: sequence the jobs in increasing order of d_j, which minimizes the maximal lateness, or tardiness.

Another easily derived special case is the objective of minimizing the maximal weighted completion time. The optimal schedule, which surprisingly is independent of processing times, is to sequence in order of decreasing w_j. This has been whimsically titled the VIP schedule.

Algorithm 3.10.2 is valid with or without preemption (since preemption is never desirable anyway). If an arbitrary precedence structure exists among the jobs, the algorithm need only be modified to read: Schedule last a job *without successors* that is least costly there; remove that job and repeat.

A somewhat more complex, but still easily implemented, extension of this procedure is available[11] to accommodate arbitrary job-arrival times, provided preemption is permitted. Another variation[12] deals efficiently with random processing times, when the goal is to minimize the greatest expected delay cost.

3.10.2.3 Number of Tardy Jobs Criterion

If the objective is to complete as many jobs as possible before their due dates, the following algorithm handles the static deterministic case.[13]

ALGORITHM 3.10.3

Given: n jobs with known processing time p_j and known due date d_j for each job j.

Objective: Minimize the number of tardy jobs.

Optimal Policy:

1. Arrange the jobs in EDD order (that is, in increasing order of due dates).
2. Compute completion times in the (remaining) EDD schedule as far as the first late job. If no job is late, go to step 4. Otherwise, suppose the ith job is the first tardy one.
3. Find the longest job among the first i jobs, and remove it from the EDD schedule. Return to step 2.
4. An optimal schedule is given by the remaining EDD schedule followed by all removed jobs (which will be the tardy ones) in any order.

For example, suppose four jobs (1,2,3,4) have processing times (2,5,4,6) and due dates (4,8,10,15). The completion times in the EDD schedule are (2,7,11,17) so the first late job is job 3. Since job 2 is the longest of the first three, we remove it and repeat the procedure with the remaining three jobs. The completion times are now (2,6,12), with all jobs on time. Thus, an optimal schedule is (1,3,4,2), with one job tardy.

Finally, it is easy to see that if all jobs have a common due date, then the SPT schedule minimizes the number tardy.

3.10.2.4 Primary and Secondary Criteria

Often it happens that several different schedules achieve some chosen objective. Since we generally have several goals in mind, it then is desirable to select another objective, and choose the best schedule relative to the secondary measure among those which satisfy the primary one. There are several situations in which such a "crème de la crème" solution can be found efficiently. First, suppose the primary objective is to assure that all jobs in some subset of very important jobs are processed by their due dates. Let us call the select set of jobs E. Of course, it may not be possible to achieve this, in which case the algorithm will be unable to proceed at some point.

ALGORITHM 3.10.4[14]

Given: n jobs, each with known p_j and d_j; and a subset E of jobs.

Objective: Minimize the number of tardy jobs, provided that all jobs in E are not tardy.

Optimal Policy: Use Algorithm 3.10.3, modifying step 3 so that the job chosen for removal is the largest job not in E among the first i jobs.

ALGORITHM 3.10.5[15]

Given: n jobs, each with known p_j and d_j, and a subset E of jobs.

Objective: Minimize the average completion time, provided that all jobs in E are not tardy.

Optimal Policy: Schedule last the longest of the jobs that either are not tardy there or are allowed to be tardy (that is, are not in set E). Remove that job, and repeat.

We may remark that the schedule produced by this algorithm resembles the SPT schedule modified to assure satisfaction of the primary objective.

ALGORITHM 3.10.6

Given: n jobs, each with known p_j and d_j.

Objective: Minimize the average completion time, provided that the number of tardy jobs is minimum.

Heuristic Policy:

1. Use Algorithm 3.10.3 to find a smallest set, E', of jobs to be tardy.
2. Define the set E as all jobs not in E', and use Algorithm 3.10.5.

Though it can be shown by counterexample that this does not guarantee optimality (for example, try the three jobs with processing times (3,1,4) and due dates (3,4,5)), average flow times of only one or two percent over the minimum can be expected.[15]

ALGORITHM 3.10.7[16]

Given: n jobs, each with known p_j and nondecreasing function $f_j(t)$, denoting the cost of completing job j at time t.

Objective: Minimize the average completion time, provided that the largest delay cost of any job is minimum.

Optimal Policy:

1. Use Algorithm 3.10.2 to find the minimal attainable value c of the largest job cost.
2. Schedule last the longest of the jobs that may be last without violating the primary objective; that is, the longest job for which $f_j(M) \leq c$, where $M = \Sigma_j p_j$ is the makespan.
3. Remove that job, and repeat step 2.

The primary objective in such algorithms need not be to minimize some criterion, but only to achieve some satisfactory level. Thus, there may be only one or a few schedules that achieve the smallest possible value, c, of the primary objective, and the best we can do for the secondary criterion may be unsatisfactory. We can relax this constraint to a larger value c', thus giving greater flexibility in step 2 and a better outcome with respect to the secondary objective. Several values of c could be tried until an acceptable compromise between the two criteria is achieved.

For example, if three jobs have processing times (4,6,8), weights (3,2,7), and due dates (8,13,10), we saw that for a primary objective of minimizing the maximal weighted tardiness, Algorithm 3.10.2 gives schedule (3,1,2) with minimax cost of 12. If we insist that no job have higher penalty than this, Algorithm 3.10.7 produces the same schedule, with average completion time of (8+12+18)/3 = 12.7. If we relax the primary criterion and allow the weighted tardiness of a job to go as high as 14, Algorithm 3.10.7 gives schedule (1,3,2) with average completion time of 11.3. No further improvement can be obtained until we allow weighted tardiness of 56 or more, when we obtain the SPT schedule (1,2,3), with average completion time of 10.7.

All these algorithms for a dual criterion have assumed a static and deterministic environment. While few of them remain optimal under more complex assumptions, extensions to allow for weights, release dates, precedence constraints, or stochastic parameters can be constructed using the ideas of this section. While mostly untested, such heuristics should perform well.

3.10.3 PARALLEL PROCESSOR SCHEDULING

Suppose now that each job still consists of one task, but we have $m > 1$ machines, which can all be processing jobs simultaneously, and any machine can process any job. Though the machines are interchangeable, they may not be identical with respect to processing speed. If we define

$$p_{jk} = \text{Processing time of job } j \text{ on machine } k,$$

we may have:

a. *Identical* machines: $p_{jk} = p_j$
b. *Uniform* machines: Each machine k has a speed or efficiency factor, s_k, such that the base processing time p_j of each job assigned to it is scaled up or down proportionately: $p_{jk} = p_j/s_k$.
c. *Unrelated* machines: arbitrary p_{jk}.

If preemption is permitted, it is generally assumed that an interrupted job may be resumed later without loss on the same or a different machine, but that each job may be on only one machine at a time. Another possibility that now arises is that a job may be subdivided and simultaneously be in process on several machines, as when each job actually consists of a large number (a "lot") of small separate items such as washers to be stamped out. For any regular measure, the latter case can be reduced to an equivalent one-machine problem in case it is practical to divide up each job into m parts, so as to occupy all the machines simultaneously for the same length of time. No more will be said about this assumption; we will assume hereafter that a job can be in only one place at a time.

3.10.3.1 Average Cost Criteria

Identical Machines

The only average cost criterion for which a simple optimal algorithm can be given is average completion time (and its equivalent criteria: average waiting time and average lateness). In the basic case (static and deterministic, with jobs ready simultaneously), the SPT rule for one-machine sequencing generalizes easily. The following algorithm, which makes no use of preemption, remains optimal even if preemption is permitted.

ALGORITHM 3.10.8

Given: m identical parallel machines; n jobs, each a single task of length p_j.

Objective: Minimize the average completion time.

Optimal Policy:

1. Schedule the m longest jobs last, one on each processor.
2. Schedule the next m longest jobs next last, one on each processor (the particular assignment of job to machine is irrelevant).
3. Continue scheduling m jobs at a time until the last m or fewer jobs, each of which should be scheduled first on some machine.
4. Process all jobs in the specified order as early as possible.

Since each group of m jobs can be rearranged separately and arbitrarily over the machines, there are many equally good schedules, which provides the opportunity to apply a secondary criterion. Otherwise, it may be convenient to use the following rule, which always produces one of the optimal schedules:

Alternative Policy: Begin by assigning the m shortest jobs to the m machines, and thereafter, whenever a machine completes processing a job, assign to it the shortest remaining job.

This may be called *multimachine SPT* scheduling.

If release dates r_j are given and preemption is permitted, the SPT rule extends to a SRPT rule: Each time a machine becomes idle or a new job arrives, switch to (or continue with) the m ready jobs with shortest remaining processing times.

As usual, the policies that work well in simpler situations suggest rules of thumb for more complex environments. For example, a sensible heuristic to minimize $\Sigma_j w_j C_j$ is to extend the multimachine SPT rule to a *multimachine SWPT* policy: Whenever a machine completes a job, assign to it the remaining job with largest w_j/p_j. No extravagant claims are made for such a heuristic. It is simply a common-sense dispatching rule, which should be satisfactory for most practical purposes.

Uniform Machines

If the machines are uniform and all jobs are simultaneously ready, Algorithm 3.10.8 can be generalized as follows for the nonpreemptive case.

ALGORITHM 3.10.9

Given: m uniform parallel machines, n nonpreemptable jobs.

Objective: Minimize the average completion time.

Optimal Policy: Associate with the jth last position on machine k a cost factor j/s_k. Now schedule the jobs one by one in decreasing order of processing time, building the m sequences backward according to the rule: schedule the next largest job to the unfilled position with next smallest cost factor.

Note how this reduces to Algorithm 3.10.8 if all $s_k = 1$. For example, with two machines having processing time factors $s_1 = 1.75$, $s_2 = 1$ (machine 1 is faster, taking $4/7 \approx 0.57$ times as long for any job), the cost factors for each position on machine 1, starting at the end of the schedule, are (0.57, 1.14, 1.71, 2.28, . . .), and on machine 2, (1,2,3,4, . . .). Hence, if the jobs are numbered in LPT (Longest Processing Time) order: $p_1 \geq p_2 \geq \ldots$, then job 1 goes on machine 1, job 2 on machine 2, jobs 3 and 4 on machine 1, job 5 on machine 2, jobs 6 and 7 on machine 1, etc.

Algorithm 3.10.9 is not optimal when jobs can be split. Instead, the following variation on the SPT theme can be used.

ALGORITHM 3.10.10[17]

Given: m uniform parallel machines, n preemptable jobs.

Objective: Minimize the average completion time.

Optimal Policy: Schedule each job successively, in SPT order, so as to complete it as early as possible.

Thus, if we number the jobs in SPT order ($p_1 \leq p_2 \leq \cdots \leq p_n$) and number the machines in decreasing order of speed ($s_1 \geq s_2 \geq \cdots \geq s_m$), then: (a) Job 1 is scheduled on machine 1 (the smallest job on the fastest machine) without preemption; (b) job 2 is schedule on machine 2 at first, but is transferred to machine 1 as soon as it becomes available; (c) generally, job j starts on the first available machine and switches to faster machines as they become available.

Including release dates is not so easy with uniform machines if jobs are not preemptable, since Algorithm 3.10.9 cannot be converted into a forward-acting SPT-type procedure. However, with preemption allowed, Algorithm 3.10.10 can readily be modified to accommodate new arrivals.

Unrelated Machines

If the machines are unrelated, the problem without release dates can be formulated as an *assignment problem*,[18] a special form of linear program for which very efficient solution methods are known.

3.10.3.2 Maximal Cost Criteria

Identical Machines

One of the most studied problems in job shop scheduling is that of scheduling on parallel machines to minimize the makespan. As usual, this is easier if preemption is allowed.

ALGORITHM 3.10.11

Given: m identical parallel machines; n preemptable jobs.

Objective: Minimize the makespan.

Optimal Policy:

1. Compute the shortest possible makespan:

$$M = \max\{\Sigma_j p_j/m,\ \max_j p_j\} \tag{3.10.1}$$

2. Schedule the jobs so as to complete all work by M, as follows: Fill each machine successively up to time M with jobs chosen in any order. When a job exceeds M, preempt it at M and schedule the remaining part on the next machine.

Without preemption, no simple algorithm can guarantee to minimize the makespan. The following gives satisfactory schedules.

ALGORITHM 3.10.12

Given: m identical parallel machines; n nonpreemptable jobs.

Objective: Minimize the makespan.

Heuristic Policy: Use *multimachine LPT* sequencing: Successively schedule the longest remaining job on the next available machine.

When scheduling on parallel processors with precedence constraints to minimize makespan, various procedures have been proposed for preemptive and nonpreemptive tasks, special precedence structures, etc. A common thread running through these algorithms is the idea of *Critical Path Scheduling* (CPS). To describe it, define the *length* of a chain of tasks as the total processing time of the tasks in the chain; and the *level* of a task in a precedence network as the length of the longest chain starting with that task. CPS is a scheduling procedure that works as follows.

CPS Policy

1. Assign a priority index to each task equal to its level.
2. Successively schedule the task with the highest priority among those whose predecessors are all complete to the next available machine.

This policy appears to be an excellent heuristic to minimize makespan under most conditions, and in some cases is optimum. For example, CPS is an optimal policy[19] if the precedence structure is restricted to trees or (with arbitrary precedence) if there are just two machines, plus in either case preemptable tasks (each of which is divided up into a chain of unit length task before indexing). It is also noteworthy that, without precedence constraints, CPS reduces to LPT.

Uniform Machines

In the deterministic, static all-jobs-available case, two efficient algorithms to minimize makespan are known when preemption is allowed, but both are somewhat complicated. To begin with, the following generalization of Eq. 3.10.1 gives the minimal attainable makespan, which is of interest in itself. Assuming the jobs and machines are numbered such that $p_1 \geqslant p_2 \cdots \geqslant p_n$, $s_1 \geqslant s_2 \geqslant \cdots \geqslant s_m$:

$$M = \max\left\{\frac{p_1}{s_1}, \frac{p_1 + p_2}{s_1 + s_2}, \ldots, \frac{\Sigma_{j=1}^{m-1} p_j}{\Sigma_{j=1}^{m-1} s_j}, \frac{\Sigma_{j=1}^{n} p_j}{\Sigma_{j=1}^{m} s_j}\right\}$$

One algorithm[20] that fits all the jobs into this space is a generalization of LPT scheduling that, roughly speaking, schedules largest jobs on the fastest processors, and whenever two or more jobs reach a point at which they have the same remaining processing time, they share equally the corresponding machines. Detailed implementation is complex, but the general logic may be used heuristically.

Without preemption, the following variation of multimachine LPT generalizes Algorithm 3.10.12.

ALGORITHM 3.10.13

Given: m uniform parallel machines; n nonpreemptable jobs.

Objective: Minimize the makespan.

Heuristic Policy: Successively schedule the longest remaining job to the machine where it will be completed soonest.

Note that we do not assign a job to the first available machine; the choice of machine depends on when it will become available and on its speed. Thus, *inserted idle time*, or *unforced idleness*, will occur in this algorithm. This is the first time an algorithm has employed *delay scheduling*, though it may be noted that even with identical machines, it can be desirable to insert idle time when there are precedence constraints.[21]

3.10.3.3 Concluding Remarks About Parallel Processing

While it is not possible to give simple optimal algorithms for the more complex objectives, the following evident principle is often useful. Once we have decided which jobs to process on which machines, the jobs may be ordered on each machine according to one-machine policies. This two-phase approach lends itself to multicriterion scheduling. If, for example, in the basic nonpreemptive case we wish to minimize the makespan and as a secondary objective minimize the average completion time, we can use *Reverse LPT* scheduling: Use Algorithm 3.10.12 to partition the jobs over the machines, and then reschedule each machine by SPT. If the jobs vary in importance and $\Sigma_j w_j C_j$ is the secondary criterion, then WSPT should be used, and so on.

As a final summation, we remark that with parallel machine scheduling (as was the case with one machine), giving shorter jobs priority over longer ones tends to minimize *average* customer dissatisfaction. On the other hand, to minimize *maximal* delays, thus enhancing resource utilization and keeping the loudest complaints within bounds, the contrary is true: use LPT scheduling (which becomes Critical Path Scheduling with precedence constraints). This is an excellent example of why scheduling can be such a vexing exercise: Two eminently reasonable goals are achieved by precisely opposite policies.

3.10.4 SERIAL PROCESSOR SCHEDULING: THE FLOW SHOP

Now suppose jobs consist of several tasks or operations. In the simplest case, the tasks of each job have chain precedence, and furthermore there is a natural ordering of the machines so that all jobs pass from one machine to another in the same prescribed order. Unless otherwise stated, we assume that there is one machine of each type, and that all tasks are nonpreemptable.

First, note that it is not always best to use a *nondelay* or *permutation* schedule, one that maintains the same job order on all machines. Instead, it may be preferable to allow jobs to pass each other. For example, in a four-machine flow shop suppose we wish to complete two jobs as quickly as possible. If the processing times for jobs 1 and 2 are (1,10,10,1) and (10,1,1,10), respectively, it is easily seen by sketching Gantt charts that either permutation schedule requires a minimal makespan of 32, whereas by scheduling job 1 first on machines 1 and 2, then switching and processing job 2 first on the last two machines, the total time required is only 24.

However, in a flow shop with just two machines, any regular measure can be minimized while maintaining the same job order on both machines. For the makespan objective (the one usually considered), the same is true with three machines, though our example has shown that this is the limit and that in larger flow shops the best permutation schedules can be far from optimal. Despite this, many of the published results consider only permutation schedules and sometimes fail to make clear this limitation. Of course, in some situations job passing may not be permitted by management or may not be possible due to the manufacturing systems design.

3.10.4.1 The Two-Machine Flow Shop

We first present one of the best known results in scheduling theory, known as Johnson's rule, which minimizes makespan in the two-machine flow shop. Recall that p_{jk} denotes the processing time for the kth task (the one that goes on machine k, in a flow shop) of job j. The following is one of several equivalent statements of the rule. It solves both the preemptive and nonpreemptive cases, since preemption is never advantageous.

ALGORITHM 3.10.14

Given: Two serial machines; n jobs with task lengths p_{j1} *and* p_{j2}.

Objective: Minimize the makespan.

Optimal Policy: Schedule first the jobs that are shorter on machine 1 than on machine 2 (that is, jobs j such that $p_{j1} < p_{j2}$) in increasing order of p_{j1}, followed by the remaining jobs in decreasing order of p_{j2}.

Jobs Lags Between Machines

We have assumed above that a job can start on machine 2 as soon as it is finished on machine 1, but no sooner. Now define a *lag*:

y_j = minimal additional time required between ending job j on machine 1 and starting it on machine 2.

Of course, there may be some delay anyway, due to other jobs blocking machine 2, but now we introduce a time that must elapse even if machine 2 is idle.

There are many situations in which lags may occur in practice. The delay may be due to:

1. Transportation between machines.
2. A chemical process (e.g., paint drying) that must run its course.
3. *Nonbottleneck* machines.

Suppose, for example, that between two one-machine stations there are other machines or machine groups through which the jobs must pass, and each intermediate station has capacity to handle many jobs simultaneously. There may be parallel machines, or just one (such as a large oven) that can handle several jobs together. Thus, the first and last stations are bottlenecks, but jobs can move through all others without mutual interference. In such a case, y_j would be the processing time required for job j to traverse the intermediate machines.

Sometimes y_j may be negative, so that a job can start on machine 2 before it is completed on machine 1. Such *lap phasing* or *lap scheduling* can occur when each job consists of producing a lot of small items, or when, for instance, painting on a construction project can begin some time before carpentry ends.

Algorithm 3.10.14 can be easily extended to handle this situation.

ALGORITHM 3.10.15

Given: Two serial machines; n nonpreemptable jobs with task lengths p_{j1} and p_{j2} and a lag y_j (positive or negative) between machines.

Objective: Minimize the makespan.

Optimal permutation policy: Define new processing times $p'_{j1} = p_{j1} + y_j$ and $p'_{j2} = p_{j2} + y_j$, and apply Algorithm 3.10.14.

Note that this is optimal only if we restrict ourselves to permutation schedules: the same job order on both machines. With lags introduced, there could be better schedules involving job passing.

3.10.4.2 The Three-Machine Flow Shop

Even though we can still confine our attention to permutation schedules when seeking to minimize the makespan, the three-machine problem is already too hard for a general optimal algorithm. There are many special cases for which simple procedures have been proposed; almost all are situations in which the middle machine turns out to be a nonbottleneck stage: no job ever has to wait for another at machine 2. However, experiments have shown[22] that only one of these situations is at all likely to arise in practice (unless the shop has a special structure, such as constant processing times in one of the stages). The following condition was found to hold in almost half the randomly generated test problems.

Define a two-machine problem with $p'_{j1} = p_{j1} + p_{j2}$ and $p'_{j2} = p_{j2} + p_{j3}$, and solve it using Algorithm 3.10.14. Suppose this produces a makespan of M, while the same job sequence results in a makespan of M' for the original three-machine problem. If $M' = M + \Sigma_j p_{j2}$, then the sequence is optimal.

Incidentally, the artificial two-machine problem defined above gives reasonable schedules for the three-machine problem even when they are not optimal.

3.10.4.3 The *m*-Machine Flow Shop

The following heuristic has been found[24] to outperform the several others that have been proposed.

ALGORITHM 3.10.16

Given: m serial machines; n nonpreemptable jobs with task lengths p_{jk} $(k = 1, \ldots, m)$.

Objective: Minimize the makespan.

Heuristic Policy: Repeat the following procedure for each value of $i = 1, \ldots, m$: define $p'_{j1} = \Sigma_{k=1}^{i} p_{jk}$ and $p'_{j2} = \Sigma_{k=m-i+1}^{m} p_{jk}$, and apply Algorithm 3.10.14. Then select the best of the $m - 1$ schedules so generated. This solution can be further improved by interchanging adjacent pairs of jobs as often as is beneficial, a tedious but easily computerized process.

3.10.4.4 Concluding Remarks About Serial Processing

Many search-type algorithms using integer programming, dynamic programming, or branch-and-bound have been proposed to solve other variants of the flow shop problem. All require complex and time-consuming computer programs, and are useful only for small problems.

The general insight to be gained from Johnson's rule and its extensions is: schedule the jobs so as to (a) get all the machines to work as quickly as possible, thus start with jobs that pass quickly through the first few machines; (b) keep the machines working as steadily as possible, hence jobs that have increasing times on successive machines should be scheduled early to give the later machines a backlog; (c) let each machine complete its work as nearly synchronously as possible; therefore put at the end those jobs which tend to have decreasing times so that once the early machines are done with them, the remaining work is soon over. With such common sense, sensible decisions can be made in uncertain and fast-changing environments.

3.10.5 THE GENERAL JOB SHOP

In the setting of a general job shop, where different jobs may have to follow different routes through the machines, there is only one situation in which a simple optimal rule can be given. The following result extends Algorithm 3.10.14 (Johnson's rule) to the two-machine job shop in which all jobs have one or two operations.

ALGORITHM 3.10.17

Given: Two machines; n jobs which belong to one of four sets:

J_1 = {jobs requiring machine 1 only};

J_2 = {jobs requiring machine 2 only};

J_{12} = {jobs requiring first machine 1, then machine 2};

J_{21} = {jobs requiring first machine 2, then machine 1}.

Objective: Minimize the makespan.

Optimal Policy: On machine 1, schedule first the jobs in J_{12} in the order dictated by Algorithm 3.10.14. Similarly, on machine 2 schedule the jobs in J_{21} first, using Johnson's rule. Thereafter, any nondelay sequence can be followed on both machines.

A great deal of effort has gone into the general static job shop problem. Highly complex branch-and-bound and mathematical programming algorithms have been developed just to handle the case where there is only one machine of each type, and where each job is *serially* routed (i.e., a *linear* or *chain* precedence structure). Since they convey little insight and lack immediate applicability, they will not be discussed here. We shall instead survey the broad literature on simulation studies of dynamic job shops.

3.10.5.1 Simulation of the Dynamic Job Shop

Because the general job shop scheduling problem is so difficult, research has concentrated on heuristic solution methods. To evaluate and compare different algorithms, it has been necessary to resort to computer simulation.[25]

A number of drawbacks to this approach should be kept in mind. Among the large number of such studies in the literature (and that is the first drawback: the endless variations that can be investigated), the wide variety of assumptions raises doubts about the transferability of results. Fortunately, a recent review[26] of the effects of different arrival rates, shop sizes, and ways of setting due dates concludes that these are not major determinants in scheduling rule performance. The total load carried by the shop (as measured, say, by average machine utilization) is a significant factor. In general we will concentrate on policies that are effective with moderate to heavy loads: lightly loaded systems perform well without careful management.

A further complication in summarizing simulation results arises because each researcher compares a different set of policies, and different studies sometime report conflicting conclusions. Survey articles often report a bewildering profusion of separate findings. We shall instead present the few key policies that have been widely reported to have desirable characteristics.

Having made these concessions to computer simulation, however, there are compensations. It now becomes quite possible to investigate a dynamic job shop with random processing times, and other real-world complexities can also be incorporated. Conclusions drawn with these most general assumptions will of course, carry over to static or deterministic situations.

3.10.5.2 Measures of Performance

As before, the objectives of prime concern will differ somewhat from one shop to another. As a measure of overall throughput directly related to machine utilization and work-in-process inventory, mean flow time is the most popular criterion to be minimized. We will again encounter the trade-off between attaining overall average good performance and taking care of the extreme or worst case. Thus, a rule may have excellent mean-flow-time characteristics but may produce a wide range of flow times. Incidentally, variance of flow times is not an adequate measure of flow time variation because a rule that produces occasional extremely long flow times may still have small variance.

To give attention to the separate needs of individual customers or orders, due dates should be considered. A distinction is sometimes made between externally set due dates, as when producing to customer orders, and internal due dates, when manufacturing to stock. The former tend to be less negotiable and to bear less relationship to the size of the job. Nevertheless, few studies have shown any major effect of this factor on the performance of scheduling rules, so we shall not discuss it further. The particular due-date-related measure generally emphasized is job tardiness.

In sum, the ideal scheduling policy would achieve low values of flow time mean and range, and tardiness mean and range. We may in addition consider shop-related measures such as average machine utilization or mean and maximal queue length, though these are often omitted, presumably because mean flow time is considered an adequate surrogate.

3.10.5.3 Dispatching Rules

For practicality and simplicity, attention has been focused on the simplest type of heuristic algorithm, *dispatching rules*. Dispatching, which is widely and successfully used, means making scheduling decisions one by one as the need arises, based solely on currently available information. Typically a dispatching rule assigns a priority index to each job waiting at a given station or machine, and when a machine next becomes free, the job with the highest priority is selected. When we give examples of priority indexes hereafter, *the smallest index values will correspond to the highest priorities*. Let

$$I_j = \text{priority index for job } j$$

Dispatching rules, of which an enormous number and variety have been proposed, may be classified in various ways. *Local rules* use only information about the jobs currently at the station. Examples are:

SPT (sometimes denoted SI or SIPT in this environment), *Shortest (Imminent) Processing Time*: $I_j = p_j$, where

$$p_j = \text{processing time required by job } j \text{ at this station}$$

LWR, Least Work Remaining: $I_j = P_j$, where

$$P_j = \text{total processing time for all remaining tasks of job } j.$$

On the other hand, *global rules* are based on the entire shop status, and may even involve forecasting. An example is:

WNS, Least Work at Next Station: I_j is the total queueing time job j has in prospect at the next station. Thus, high priority is given to a job that can be passed on to an idle machine; low priority if the next station is already overloaded. This rule can be elaborated to forecast the next-station congestion at the time job j would arrive.

Dispatching rules may also be classified as static or dynamic. With a *dynamic rule*, I_j may change over time. An example:

TSPT, Truncated SPT: $I_j = p_j$, as long as job j has not been waiting at this station longer than some maximal time. If its wait exceeds this limit, it is given top priority.

A job's priority index is fixed throughout the shop with a *static rule*. An example is the EDD rule, with $I_j = d_j$. Finally, a rule may be locally static and globally dynamic, as are SPT, LWR and the simplest of all dispatching rules, FCFS: First Come First Served.

Most of the rules above have been mentioned to give some idea of the abundance and variety of proposed policies, and will not be mentioned again.

3.10.5.4 Recommended Dispatching Rules

From the profusion of reported results, and with special reference to a recent exhaustive study,[27] we now summarize the emergent patterns in the choice of dispatching rules.

Scheduling without Due Dates

If there is one conclusion that all investigators agree on, it is that the *SPT rule is preeminent for minimizing mean flow time*. It has consistently been found superior to all other rules in many studies for this fundamental throughput measure as well as for the corollary measures of facility utilization and work-in-process inventory.

Surprisingly (since it ignores due dates), it also performs well for mean tardiness, especially when the shop is very heavily loaded or when due dates are very tight, in which case it has been reported[2] to outperform good due-date based rules like S/OPN (see *Scheduling against Due Dates*). It has excellent robustness in the face of inaccurate data.

The major shortcoming of the SPT rule is that by constantly hurrying the short tasks ahead, the occasional job with an exceptionally long task will experience intolerable delays. We will now mention one obvious but not highly recommended way to tackle this. Other more successful ways will be mentioned later.

Expediting

It is common practice in many job shops to interrupt the normal schedule occasionally and give some job special priority, in case excessive delay or unexpected demand give it sudden urgency. To maintain the advantages of SPT without the occasional extreme delay, one might set an upper bound on the waiting time at a station, and *expedite* any job that exceeds it. This rule, called *truncated SPT*, does not turn out to be a good strategy: average-time performance deteriorates sharply, even for a very loose bound.

In general, expediting is a dangerous policy, especially if used informally. In times of stress, a few expedited jobs introduce inefficiencies and confusion and stimulate additional emergencies. A snowballing effect can ensue, resulting in a "red tag" on every job in the shop.

Scheduling against Due Dates

While SPT is an adequate rule with respect to mean tardiness, others are significantly better. The following well-known rule is among the best, and with a slight modification, it performs even better. Both produce low average flow times as well.

S/OPN, Slack per Operation or *Slack Time*: $I_j = s_j/n_j$, where

t = present time

s_j = slack for job $j = (d_j - t) - P_j$

n_j = number of remaining tasks for job j, including the present one.

MS/OPN, Modified Slack per Operation: $I_j = s_j p_j / P_j$.

Note that if the present task has a processing time that is about average among the n_j remaining tasks of job j, then $p_j/P_j = 1/n_j$ and the two indices are equal. If p_j is unusually short, then MS/OPN will give it higher priority, and vice versa. Thus the second index has a hint of SPT added to S/OPN, and gives shorter mean flow times without sacrificing the excellent performance of S/OPN against due dates.[27]

Three other dispatching rules are worthy of mention. It will probably never be possible to give a single best policy to follow in all cases; it remains necessary to list several, and trust to the user's judgment and experimentation to choose one that suits his or her needs.

A dispatching rule that has been well accepted in industry but has received little attention in the research community is the following:

CR, Critical Ratio: $I_j = R_j$, where

$$R_j = (d_j - t)/P_j$$

It is instructive to compare this with job slack, $s_j = d_j - t - P_j$, which is the difference rather than the ratio of the same two numbers. Thus, a negative s_j corresponds to an R_j less than 1.

Simulations[29] indicate that the Critical Ratio rule performs slightly better than the Slack Time rule for both mean flow time and mean tardiness criteria, with somewhat smaller variances.

SPTS,[30] *SPT combined with S/OPN*: $I_j = \min\{p_j + X, s_j/n_j\}$.

This is a kind of truncated SPT. The number X is a control parameter, the same for all jobs, that should be set to a value, positive or negative, to regulate the proportion of the time that processing time dominates slack time: about 90–95% is recommended. Effectively, SPTS is an SPT rule with occasional expediting of urgent jobs, but the original study reports that the performance relative to SPT is very comparable, with variances reduced.

SPTR,[31] *SPT combined with CR*: $I_j = \max\{p_j, R_j p_j\}$.

In the most recent comparison of dispatching rules,[25] SPTR was found to perform best of the twenty rules tested against tardiness measures, and among those which had good tardiness characteristics it gave the lowest mean flow times, not much worse than SPT. It should be added that MS/OPN was in a virtual tie with SPTR, and also that while SPT, S/OPN, and MS/OPN were in the study, CR and SPTS were not.

Since the index for SPTR can be rewritten: $I_j = p_j + \max\{0, s_j\, p_j/P_j\}$, we can interpret it as follows. For jobs that are sure to be tardy ($s_j < 0$), use SPT. If a job has some slack, its priority is diminished by adding its modified slack per operaton (see the MS/OPN rule) to its index.

More Complex Policies

Sometimes it is recommended that additional information be added to improve the performance of scheduling procedures. Of course, such refinements come at the cost of extra data collection and processing. For example, if production is for the purpose of replenishing stocked items, then inventory levels and MRP systems could be used to update due dates dynamically. However, simulation results[29] suggest that adjusting due dates does more harm than good. Presumably, altering job priorities as the work progresses through the shop, for reasons unrelated to queue lengths or processing times, leads to confusion and inefficiency.

Another kind of information that could be used is the anticipated queue waiting time at the different machines, which can be used to calculate

$$W_j = \text{estimated future waiting time for all remaining tasks of job } j$$

We can now replace the total remaining processing time of a job, P_j, by $P_j + W_j$ which is a more complete assessment of the remaining time needed to complete the job. Then, slack could be replaced by *modified slack*

$$s_j^* = s_j - W_j$$

and the critical ratio can be similarly refined, giving a *modified critical ratio*

$$R_j^* = \frac{d_j - t}{P_j + W_j}$$

It was found[29] that very little improvement resulted from using this extra data in either the Slack Time or the Critical Ratio rules, partly because average queue length data does not accurately forecast the queues that will actually be encountered in a constantly changing system, and partly because the delay a job will experience at any station depends heavily on its own priority, not merely on the congestion it encounters there.

3.10.5.5 Concluding Remarks About General Shop Scheduling

In the general dynamic job shop, optimality can no longer be hoped for, or indeed readily defined, and we focus on simple practical dispatching heuristics. The SPT rule remains extraordinarily versatile and effective at moving the mass of work along, although at the expense of a few long-delayed orders. Due dates are best satisfied by some form of slack time indexing, and the overall superior dispatching rules are relatively simple combinations of the two concepts.

REFERENCES

1. R. W. Conway, W. L. Maxwell and L. W. Miller, *Theory of Scheduling*. Reading, MA: Addison-Wesley, 1967.
2. K. R. Baker, *Introduction to Sequencing and Scheduling*. New York: Wiley, 1974.
3. S. French, *Sequencing and Scheduling*. New York: Wiley, 1982.
4. H. J. Weiss, "A Greedy Heuristic for Single Machine Sequencing with Precedence Constraints," *Management Science*, **27**, 1209–1216 (1981).
5. J. B. Sidney, "Decomposition Algorithms for Single-Machine Sequencing with Precedence Relations and Deferral Costs," *Operations Research*, **23**, 283–298 (1975).
6. E. L. Lawler, "Sequencing Jobs to Minimize Total Weighted Completion Time Subject to Precedence Constraints," *Ann. Discrete Math.*, **2**, 75–90 (1978).
7. M. H. Rothkopf, "Scheduling Independent Tasks on Parallel Processors," *Management Science*, **12**, 437–447 (1966).
8. K. C. Sevcik, "Scheduling for Minimum Total Loss Using Service Time Distributions," *J. Assoc. Comput. Mach.* **21**, 66–75 (1974).
9. A. H. G. Rinnooy Kan, B. J. Lageweg, J. K. Lenstra, "Minimizing Total Costs in One-Machine Scheduling," *Operations Research* **23**, 908–927 (1975).
10. E. L. Lawler, "Optimal Sequencing of a Single Machine Subject to Precedence Constraints," *Management Science*, **19**, 544–546 (1973).
11. K. R. Baker, E. L. Lawler, J. K. Lenstra, A. H. G. Rinnooy Kan, "Preemptive Scheduling of a Single Machine to Minimize Maximum Cost Subject to Release Dates and Precedence Constraints," *Operations Research*, **31**, 381–386 (1983).

12. T. J. Hodgson, "A Note on Single Machine Sequencing with Random Processing Times," *Management Science*, **23**, 1144–1146 (1977).

13. J. M. Moore, "An *n* job, One Machine Sequencing Algorithm for Minimizing the Number of Late Jobs," *Management Science*, **15**, 102–109 (1968).

14. J. B. Sidney, "An Extension of Moore's Due Date Algorithm," in S. E. Elmaghraby, Ed. *Symposium on the Theory of Scheduling and its Applications*, Lecture Notes in Economics and Mathematical Systems 86, Berlin: Springer, pp. 393–398 (1973).

15. H. Emmons, "One Machine Sequencing to Minimize Mean Flow Time with Minimum Number Tardy," *Naval Research Logistics Quarterly*, **22**, 585–592 (1975).

16. H. Emmons, "A Note on a Scheduling Problem with Dual Criteria," *Naval Research Logistics Quarterly*, **22**, 615–616 (1975).

17. T. Gonzalez, "Optimal Mean Finish Time Preemptive Schedules," *Technical Report 220*, Computer Science Department, Pennsylvania State University, 1977.

18. W. A. Horn, "Minimizing Average Flow Time with Parallel Machines," *Operations Research* **21**, 846–847 (1973).

19. R. R. Muntz, E. G. Coffman, Jr., "Preemptive Scheduling of Real Time Tasks on Multiprocessor Systems," *J. Assoc. Comput. Mach.*, **17**, 324–338 (1970).

20. E. C. Horvath, S. Lam, and R. Sethi, "A Level Algorithm for Preemptive Scheduling," *J. Assoc. Comput. Mach.*, **24**, 32–43 (1977).

21. R. L. Graham, "Combinatorial Scheduling Theory," *Mathematics Today*. New York: Springer-Verlag, 183–212 (1978).

22. A. J. M. Smits and K. R. Baker, "An Experimental Investigation of the Occurrence of Special Cases in the Three-Machine Flowshop Problem," *International Journal of Production Research*, **19**, 737–741 (1981).

23. R. J. Giglio and H. M. Wagner, "Approximate Solutions to the Three-Machine Scheduling Problem," *Operations Research*, **12**, 305–324 (1964).

24. D. G. Dannenbring, "An Evaluation of Flow-Shop Sequencing Heuristics," *Management Science*, **23**, 1174–1182 (1977).

25. J. E. Day and M. P. Hottenstein, "Review of Sequencing Research." *Naval Research Logistics Quarterly*, **17**, 11–39 (1970).

26. J. H. Blackstone, D. T. Phillips and G. L. Hogg, "A State-of-the-Art Survey of Dispatching Rules for Manufacturing Job Shop Operations," *International Journal of Production Research*, **20**, 27–45 (1982).

27. G. R. Bitran, M. Dada and L. O. Sisson, "A Simulation Model for Job Shop Scheduling," Working Paper #1402-83, Sloan School of Management, M.I.T., Cambridge, MA (1983).

28. R. Rochette, and R. P. Sadowski, "A Statistical Comparison of the Performance of Simple Dispatching Rules for a Particular Set of Job Shops," *International Journal of Production Research*, **14**, 63–75 (1976).

29. W. L. Berry and V. Rao, "Critical Ratio Scheduling: An Experimental Analysis," *Management Science*, **22**, 192–201 (1975).

30. M. Oral, and J. L. Malouin, "Evaluation of the Shortest Processing Time Scheduling Rule with Truncated Process," *AIIE Transactions*, **5**, 357–365 (1973).

31. K. R. Baker, and J. W. M. Bertrand, "A Dynamic Priority Rule for Scheduling Against Due Dates." *Journal of Operations Management*, **3**, 37–42 (1982).

CHAPTER 3.11

ASSEMBLY LINE BALANCING

JOHN D. LORENZ
DAVID W. POOCK

GMI Engineering and Management Institute
Flint, Michigan

3.11.1 PRODUCT ASSEMBLY

The manufacture of most products includes not only the fabrication of the individual component parts of the product being produced, but also the fitting together, or assembly, of those component parts into the final product configuration. This assembly activity may be performed in its entirety by an operator at a single assembly station, (one-man build), or it may be performed in a progressive fashion by a number of operators manning a series of assembly stations, (progressive assembly line), where each station performs a portion of the labor required to complete the overall assembly of the product.

The progressive assembly line is based on the principles of interchangeable parts and the division of labor. The ability to fabricate interchangeable parts allows random selection of component parts for assembly, as opposed requiring selective assembly. The division of labor in product assembly is a determining factor in selecting the amount of time that the product spends at each operator's workstation, called the *assembly line cycle time*. In addition, the division of labor ultimately determines the number of workstations to be provided along the progressive assembly line. The nature of the overall assembly task itself imposes an ultimate limit beyond which the division of labor can proceed no further. This ultimate subdivision of assembly work results in the irreducible work element (or *minimum rational work element*), which for practical purposes cannot be further subdivided. The largest such irreducible work element in the overall assembly task effectively dictates the lower limit of feasible cycle times. The upper limit on the cycle time is, of course, the one-man build situation in which a single operator completes the entire assembly task with no division of labor. Modern mass production has found industry typically selecting cycle times relatively close to the lower limit, employing random selection of component parts for assembly as the product moves past fixed workstations at a uniform speed.

Once the extent of the division of labor has been determined, industry is then faced with the problem of allocating the irreducible work elements as equally as possible among the various workstations along the progressive assembly line. This assignment problem, or *assembly line balancing*, is a critical factor in the overall design of the assembly system. The uniform distribution of work elements among the operators along the assembly line not only prevents severe overloading or underloading of operators, but also is a major factor in ensuring high product quality and a smooth flow of finished assemblies.

3.11.2 ASSEMBLY LINE BALANCING

Assembly line balancing is the procedure of assigning work elements to assembly operators so that they are apportioned among the operators as evenly and as compactly as possible without violating any precedence restrictions. The goals of assembly line balancing are:

1. Provide balanced work assignments for all operators such that some operators are not more heavily loaded than others.
2. Provide work assignments that do not fluctuate depending upon the model being assembled in a mixed-model environment.
3. Obtain high labor efficiency by minimizing the total idle time of the operators along the assembly line.

4. Eliminate any severe work overload conditions that may adversely affect product quality and result in subsequent scrap or rework.

The procedure is begun by subdividing the total amount of work required for the overall assembly of the product into its minimum rational work elements, the indivisible elements of work beyond which assembly work cannot rationally be subdivided further. Minimum rational work elements contain the minimum practical number of motions making up a component of work that can be transferred readily from one assembly operator to another. Therefore, a minimum rational work element must be:

1. *Transferable.* It may be performed by only one operator, but it must contain all the motions necessary to be a complete activity so that it may be performed by any assembly line operator.
2. *Minimum sized.* It must be sufficiently small to allow for a good balance.

Assembly line balancing involves the assignment of the minimum rational work elements to the workstations along the progressive assembly line. A workstation, generally manned by a single operator, is an assigned location along the line where a specified amount of work is performed. The product typically moves past the the workstations at a continuous and uniform speed, which is based on the assembly line production requirement. This line speed, in turn, determines the elapsed time between successive units moving along the assembly line. The elapsed time between units dictates the assembly line cycle time, which represents the amount of time that a unit is available for work at each workstation. Salveson's[1] definition of the assembly line balancing problem is:

to minimize the total amount of idle time or equivalently to minimize the number of operators to do a given amount of work for a given assembly line speed.

It is apparent then, that given an assembly line speed and its resultant cycle time, any assembly line balancing procedure should attempt to assign as many minimum rational work elements as possible to each workstation without exceeding the cycle time for any workstation.

Balance delay time is the total amount of idle time of the workstations along the assembly line resulting from the unequal distribution of work content among the workstations. This total idle time resulting from an imbalance in station work content times is expressed as *balance delay*, which is defined as the ratio of the total idle time at all stations along the assembly line to the total time the unit is available for work at all the stations on the line. The following formula is used to calculate the balance delay d (or percent of imbalance) for a given assembly line.

$$d = \frac{\sum_{k=1}^{n} I_k}{nc} \times 100$$

where

k = workstation number, $1 \leqslant k \leqslant n$

I_k = idle time at station k

c = cycle time

n = number of workstations required

The idle time at any workstation along the line, I_k, is the difference between the cycle time (i.e., the amount of time available to perform assembly work at the workstation) and the station work content time, which is the total time required to complete the work elements actually assigned to the station. The idle time at any workstation can be calculated as

$$I_k = c - \sum_{i=1}^{j} t_i$$

where

t_i = duration time for element i

i = work element number for work elements assigned to station k, $1 \leqslant i \leqslant j$

j = number of work elements assigned to station k

Balance delay can then be calculated as

$$d = \frac{\sum_{k=1}^{n} (c - \sum_{i=1}^{j} t_i)}{nc} \times 100$$

$$= \frac{nc - \sum_{k=1}^{n} \sum_{i=1}^{j} t_i}{nc} \times 100$$

But since the total work content time T is the total duration time required to perform the aggregate amount of work required to completely assemble the unit, then

$$T = \sum_{k=1}^{n} \sum_{i=1}^{j} t_i$$

and balance delay can be expressed as

$$d = \frac{nc - T}{nc} \times 100$$

For a given cycle time c and a given total work content time T there exists a minimum number of workstations, $n_{\min}$, which will minimize the value of balance delay for that assembly line. The theoretical minimum number of workstations (or operators) along an assembly line, $n_{\min}$, is the smallest integer value that is greater than or equal to the ratio of cycle time to total work content time. Therefore,

$$\frac{T}{C} \leqslant n_{\min} < \frac{T}{c} + 1$$

$$n_{\min} = \text{an integer}$$

The functional relationship between cycle time and the theoretical minimum balance delay attainable with $n_{\min}$ operators at that cycle time is given by:

$$d_{\min} = \frac{n_{\min} c - T}{n_{\min} c} \times 100$$

This minimum balance delay (in percent) is a function of cycle time, and can be plotted in graphical form as such.

Perfect balance is said to occur when balance delay is zero. Therefore, for perfect balance

$$d_{\min} = \frac{n_{\min} c - T}{n_{\min} c} \times 100 = 0$$

In order to have zero balance delay, there must be no idle time along the assembly line.

$$n_{\min} c - T = 0$$

Therefore, perfect balance (i.e., zero balance delay) can occur only when the total work content time T is evenly divisible by the cycle time c.

$$n_{\min} = \frac{T}{c} = \text{an integer}$$

In actual practice, perfect balance is rarely attained because of two problems encountered in assembly line balancing, as noted by Kilbridge and Wester.[2] The first, the cycle-time problem, arises because of the functional relationship between the cycle time and theoretical balance delay. Perfect balance is attainable only when the total work content time is evenly divisible by the cycle time. Therefore, attainment of perfect balance is limited to a relatively small number of cycle time values: those values for which the ratio of cycle time to total work content time is an integer. Attempting to actually balance the assembly line to the theoretical minimum number of operators introduces the second problem: the line balancing problem. Even though it may be possible theoretically to attain perfect balance at a given cycle time by balancing the line across the theoretical minimum number of operators, in actual practice a greater number of operators may be required due to precedence restrictions and facility restrictions, among others. This inability to balance the assembly line to the theoretical minimum number of operators and the necessity of providing a greater number of operators, with a resultant increase in balance delay, is the line-balancing problem, which will be discussed later.

Consider, for example, a task consisting of 9 minimum rational work elements for the assembly of an automobile backup-light assembly. The work element duration times range from 0.048 to 0.112 min, with a total work content time of 0.802 min being required to complete all 9 work elements. In initiating the task of balancing the assembly line for the production of these backup-light assemblies, it is apparent that the practical maximum cycle time to be considered would be 0.802 min, the one-man build situation in which the complete assembly task would be completed by a single operator. At the other extreme, the practical minimum cycle time to be considered would be 0.112 min, the maximum individual work element duration time. Selecting a cycle time of less than 0.112 min would, of course, result in the inability to complete the longest duration work element within the allotted cycle time.

Plotting the theoretical minimum value of balance delay as a function of cycle time (within the range of practical cycle time values) results in the sawtooth line shown in Figure 3.11.1. The theoretical balance delay value increases as the cycle time increases within the range of cycle times for which a given number of operators is required, up to the cycle time at which one less operator is required. At that point, the balance delay value drops to zero (i.e., perfect balance), and then again increases with increasing cycle times until the cycle time is reached where, once again, the theoretical minimum number of required operators is reduced by one.

Plotting the balance delay function is useful in the subsequent selection of an appropriate cycle time, since the cycle time selected dictates the minimum balance delay value that is theoretically attainable. A cycle time should be selected that, in conjunction with the given number of operators, results in the lowest total amount of idle time, and therefore minimizes balance delay across the entire assembly line. For example, in selecting the cycle time for the automobile backup-light assembly, it is apparent that a cycle time of 0.208 min (balance delay = 3.61%) is preferable to a cycle time of 0.209 min (balance delay = 4.07%) provided that upon the subsequent assignment of work elements, the assembly line can, in fact, be balanced using the theoretical minimum of four operators in each case. Reduction of the cycle time to 0.201 min reduces the theoretical minimum balance delay to 0.25%, because the theoretical minimum number of operators remains at four. On the other hand, reducing the cycle time further, to a value of 0.200 min, results in the requirement of a theoretical minimum number of five operators, resulting in a minimum balance delay of 19.80%. In decreasing from 0.201 min to 0.200 min, the cycle time has been decreased beyond the minimum value for which a theoretical minimum number of four operators is required. Perfect balance is theoretically attainable only at those feasible cycle times for which the ratio of the total work content time T to the cycle time c is integer-valued. In this

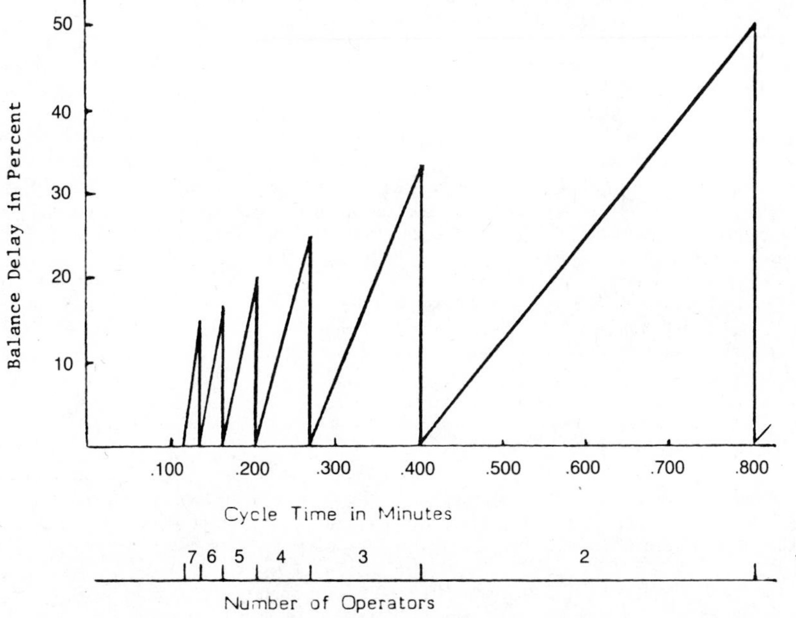

Fig. 3.11.1. Balance delay function for backup-light assembly.

example, because the range of practical cycle times is from 0.112 min (n_{min} = 8 operators) to 0.802 min (n_{min} = 1 operator), the cycle times for which perfect balance may be attained (based on the cycle time problem only) are:

n_{min}	$T/n_{min} = c$
7	.802/7 = .1146 min
6	.802/6 = .1337 min
5	.802/5 = .1604 min
4	.802/4 = .2005 min
3	.802/3 = .2673 min
2	.802/2 = .4010 min
1	.802/1 = .8020 min

In summary, balance delay represents a measure of the total idle time along the assembly line resulting from the unequal division of work among the workstations on the line. Since one of the goals of assembly line balancing is to minimize idle time on the line, line balancing seeks to achieve the lowest possible balance delay. In attempting to balance an assembly line with minimal balance delay, one of three possible situations can arise: (1) given a specified cycle time, minimize the number of workstations required; (2) given a specified number of workstations, minimize the cycle time provided at each workstation (i.e., maximize output); or (3) if neither the cycle time nor the number of workstations is specified, determine the values of the cycle time and the number of workstations for which balance delay will be minimized. Because of the relationship between cycle time and the theoretical minimum number of operators required, and ultimately the relationship between cycle time and balance delay, it is clear that indiscriminant selection of the cycle time can have a severe impact on the total idle time on the assembly line. Once again, however, note that the cycle-time problem addresses only the necessary condition for perfect balance, namely, that balance delay is zero. Selecting a cycle time such that $n_{min} = T/c$ is an integer is necessary but not sufficient for achieving perfect balance. Whether perfect balance is, in fact, attained at that value of cycle time can be determined only upon assignment of the work elements to the operators. If the assembly line can be balanced with n_{min} operators at a cycle time c such that

$$n_{min} = \frac{T}{c} = \text{an integer}$$

then none of the n_{min} operators will have any idle time, and perfect balance results.

One of the major reasons why the line-balancing problem eliminates many of the theoretically feasible perfect balances that result from consideration of the cycle-time problem only is precedence restrictions. Precedence restrictions occur because the final configuration of the product being produced is an assemblage of a number of component parts. Because of the design of the final assembly and its component parts, in most cases certain components must be assembled before other components. Similarly, certain work elements must be performed before other work elements. Hence, the assignment of work elements to operators along the assembly line must be carried out in conjunction with the precedence restrictions of the product being assembled. Progressive assembly of the product by operators along the line must be consistent with the precedence relations inherent in the overall design of the product.

The most common means of presenting the precedence relations associated with the assembly of a product is the *precedence diagram*. Prenting and Battaglia[3] state that "the purpose of a precedence diagram is to convert the actual assembly line situation into a diagrammatic representation that completely describes the work elements for purposes of line balancing." An underlying assumption in the construction of a precedence diagram is that most products can be assembled in a variety of different ways. It is assumed that a number of different sequences of work elements would result in an acceptable assembly. Therefore, the role of the precedence diagram in assembly line balancing is to present all feasible ways of ordering the work elements. The precedence diagram must be constructed such that all conceivable paths through the network (i.e., possible sequences of work elements) will, in fact, result in an acceptable assembly. Hence, all paths through the diagram must be buildable. In addition, the precedence diagram should be constructed so that all conceivable sequences of work elements are presented. In actual practice, however, many theoretically possible sequences of work elements are prohibited because of individual precedence restrictions.

The work elements are diagrammed in such a manner as to reflect all precedence restrictions imposed on their ordering. Each restriction that is imposed will, in general, reduce the number of work element sequences that can be considered, which will, in turn, make line balancing more difficult and result in higher balance delay for the assembly line. Therefore, it is essential that each restriction be questioned to ensure that it is necessary.

The construction of a precedence diagram usually proceeds from left to right, and utilizes the activity-on-the-node network diagramming convention, in which the nodes of the precedence diagram represent the work elements and the arrows connecting the nodes show the precedence relations. Each work element is drawn as far to the left as possible; those nodes shown in the leftmost "column" of the diagram representing those work elements that need not be preceded by any other work element. Those nodes shown in the next column to the right represent those work elements that must be preceded only by those work elements in the leftmost column (already shown on the diagram). Arrows are then drawn from the work elements in the leftmost column to work elements in subsequent columns to the right that they must precede. This procedure is continued until all work elements are shown, with their appropriate precedence relations, on the diagram.

In general, a "vertical" precedence diagram (i.e., a small number of columns with a large number of work elements per column) represents the best situation for line balancing. Because vertical precedence diagrams exhibit fewer precedence restrictions in assigning work elements to operators on the line, a greater number of possible work element sequence assignments are available for consideration. On the other hand, a "horizontal" precedence diagram (i.e., a large number of "columns" with only a small number of work elements per column) represents the worst situation, since the large number of precedence restrictions that resulted in the horizontal diagram restrict the number of possible work element assignments. The extreme example of a horizontal precedence diagram, and one that must be avoided, is the situation that arises when one mistakenly considers a network diagram representation of a single build sequence or an assembly routing to be a precedence diagram. Such a diagram would, of course, have only one work element per column, with each element connected to the next by an arrow, and only one possible work element sequence would be shown. A true precedence diagram, however, shows all possible work element sequences, not merely the sequence presently being followed.

As an example of the construction of a precedence diagram for line balancing, consider again the automobile backup-light assembly discussed earlier. A sketch of the assembly, identifying its component parts, is shown in Figure 3.11.2. As noted earlier, the total work content required to complete the assembly of the backup light includes 9 minimum rational work elements, as summarized in Table 3.11.1. Each work element is identified by an element name: a unique numeric value used merely to provide ease of identification in constructing the precedence diagram and in subsequent line balancing. A brief description of each work element is followed by the duration time (in minutes) for that particular minimum rational work element. The immediate predecessors of each work element are then listed. Immediate predecessors are those work elements that must be completed immediately before the work element in question can be performed. Only the *immediate* predecessors, not all predecessors, are listed. Note, for example, that because element 60 must be preceded by both elements 40 and 50 and because element 40 must be preceded by element 30, element 60 cannot be performed until elements 30, 40, and 50 are completed. Therefore, element 60's predecessors are elements 30, 40, and 50, but element 60's *immediate* predecessors are elements 40 and 50 only.

From this information, the precedence diagram in Fig. 3.11.3 is constructed. The element names are shown at each node of the network diagram and the arrows connecting the nodes reflect the precedence relations among the various work elements. The duration time for each work element is shown immediately below the node for ease in subsequent line balancing.

The precedence diagram graphically shows all possible combinations of work elements, and therefore exhibits all possible assembly build sequences. The construction of the precedence diagram forces one to evaluate objectively any restrictions on the work elements and not to concentrate on a single build sequence based on restrictions that, in many instances, are simply not necessary. In essence, the precedence diagram presents for evaluation all feasible work element sequences, not merely the single assembly build sequence that is most familiar.

The effect that precedence restrictions have on balance delay is observed in the second of the two problems noted earlier: the line balancing problem. If a cycle time is selected such that the total work content time is divided equally by the cycle time, then balance delay can theoretically be eliminated and perfect balance may be attainable. If, however, it is not possible to balance the assembly line to the predicted theoretical minimum number of operators because of precedence restrictions, the distribution of work element duration times, the relationship of the magnitude of the cycle time to the magnitude of each of the work element duration times, etc., then idle time will be present at the various stations along the line, balance delay will not be eliminated, and perfect balance will not be attained. This is the situation referred to as the line-balancing problem. Even

Table 3.11.1.

Element Name	Element Description	Duration Time (min)	Immediate Predecessors
10	Apply glue to housing	0.104	
20	Staple outer gasket to assembly	0.105	10
30	Stake socket to housing	0.102	
40	Assemble bulb to socket	0.100	30
50	Assemble inner gasket to housing	0.053	
60	Position lens to housing	0.048	40, 50
70	Position connector to lead	0.081	30
80	Drive 2 screws to housing	0.112	60
90	Inspect and pack	0.097	20, 70, 80

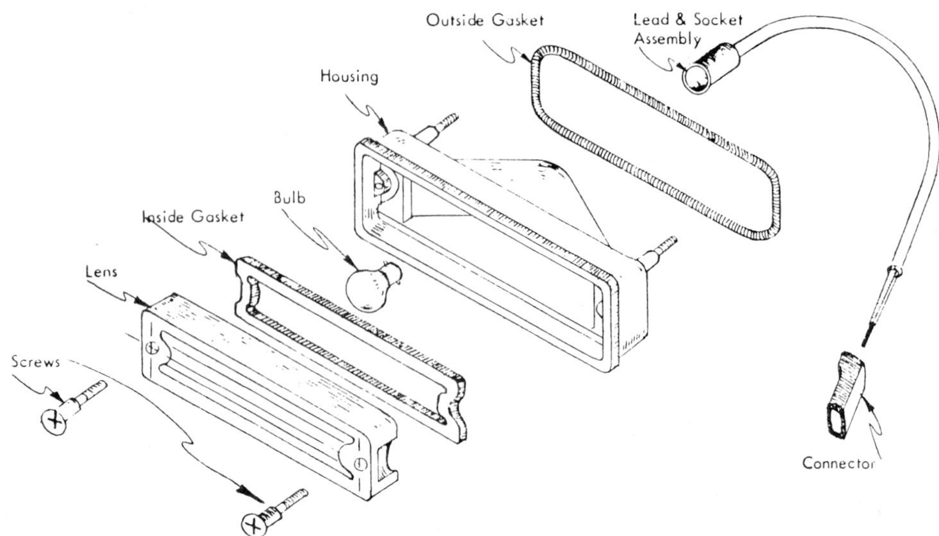

Fig. 3.11.2. Automobile backup-light assembly.

though the cycle-time problem may have been solved, resulting in the selection of a cycle time that theoretically could produce a perfect balance, the line-balancing problem can prevent the actual attainment of that perfect balance.

3.11.3 LINE-BALANCING METHODS

In addressing the line-balancing problem, the packing analogy posed by Kilbridge and Wester[2] is appropriate.

> *The problem of line balancing is to choose that cycle time which will result in the least balance delay and to apportion the assembly work to the stations as compactly as possible. This is to be done without exceeding the cycle time at any station and while heeding given restrictions upon the ordering of work elements. The key relationship in this process is considered to be that between the nature of the work and the length of the cycle time chosen. The nature of the work is defined, for purpose of this analysis, in terms of two characteristics: the size and distribution of indivisible work elements. The balancing problem can be viewed as a form of the "packing problem," in which a given set of equal time intervals is to be filled as fully as possible with work elements having various time lengths and drawn from a known distribution.*
>
> *Considered as belonging to the class of packing problems, the balancing problem is analogous to that of packing a given number of equal-size boxes with blocks of varying sizes. The boxes represent the set*

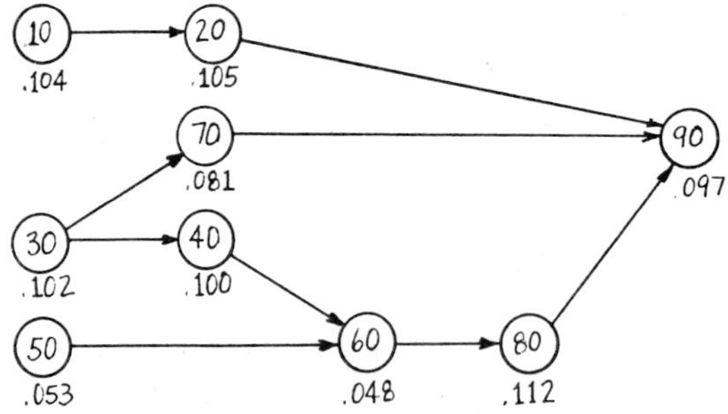

Fig. 3.11.3. Precedence diagram for automobile backup-light assembly.

of equal time intervals (the cycle times of the workstations) and the blocks represent the set of element times (the distribution of work elements). The problem is to pack the boxes with blocks as fully as possible so as to minimize the total of voids left in all boxes.

There are restrictions imposed on certain blocks as to which box they must go into, and where in that box they must fit. Many blocks, however, can be placed where they fit best. The degree of difficulty experienced in optimally packing the boxes is thought to depend on three factors: (1) The sizes, and distribution of sizes, of the blocks; that is, how many large ones there are compared to small ones, and how large the large ones are. In general, other things being equal, the higher the ratio of large to small blocks, the more difficult the packing. (2) The relationship of the size of the boxes to the sizes of the blocks. If the blocks are on the average, very small with respect to the boxes, the packing is relatively easy. Of course, no block may be larger than the boxes. (3) The number of blocks that must go into a specified place in a specified box, and the sizes of these restricted blocks. The more such restricted blocks there are, the more difficult the packing.

Faced with this packing problem, one would normally start with the restricted blocks first, placing them where they must go. This gets them out of the way and gives a clearer view of the remaining choices. Next one would normally take the largest remaining blocks and fit them into the boxes before any larger spaces are wasted by being unnecessarily occupied by smaller blocks. Gradually smaller and smaller blocks would be used until finally the smallest blocks are disposed of by distributing them over and around and between the previously packed blocks. At this point it is convenient to have a large number of very small blocks to occupy gaps between larger blocks and to fill the remaining voids at the tops of the boxes.

In the balancing process described by this analogy, perfection in the distribution of work is seldom achieved. . . .

The line-balancing problem was first recognized in 1954 when Benjamin Bryton,[4] in his master's thesis at Northwestern University, developed what he called a "convergence procedure," which essentially interchanged work elements among stations until the station work content times converged to a common value, thus minimizing idle time for a fixed number of stations. Since Bryton's initial efforts, a number of line-balancing methods have been proposed in the literature, ranging from rigorous mathematical techniques to heuristic procedures.

Salveson[5] suggested that a linear programming model encompassing all possible combinations of work element assignments would be appropriate for addressing the line-balancing problem. Salveson noted, however, that such a matrix would be "enormously large" and for line-balancing problems of practical size, computationally impractical.

An outgrowth of Salveson's work is the application of the branch-and-bound concept to assembly line balancing as discussed by Daellenbach and George.[6] The *branch-and-bound* method is a solution strategy that has emerged over the past decade as one of the major practical tools for the solution of real-life optimization problems. The attractiveness of this method stems from its ability to eliminate implicitly large groups of potential solutions to a problem without explicitly evaluating them. Branch and bound, however, is a strategy, not an algorithm, and the strategy must be merged with the structure of the specific problem at hand, namely balancing an assembly line, to form an implementable solution algorithm.

Application of the branch-and-bound concept involves the construction of a solution tree describing all possible solutions (i.e., all possible work element assignments consistent with precedence restrictions) to the line-balancing problem. A guided search through the solution tree is then conducted to arrive at the best solution. While the branch-and-bound technique does not guarantee an optimal solution, since all possible work element assignments are not evaluated explicitly, the approach does result in good line balance with minimal balance delay in most cases, and with greatly reduced computational time. The computational time is a function of the number of work elements and possible workstations; the greater the number of work elements and possible workstations, the greater the number of theoretically possible work element assignments, which must be either explicitly evaluated or implicitly eliminated. Computational time is also affected by the number of precedence restrictions, since precedence restrictions tend to reduce the number of work element assignments that are theoretically possible. The computational time and effort required to formulate and solve the problem generally limits the application of mathematical technique such as the branch-and-bound approach to simplified or theoretical cases. When applied to practical line-balancing problems of realistic size, such approaches generate an impractically large volume of computation. Therefore, rigorous mathematical techniques are generally of academic interest only, while most real-world line-balancing problems are solved either by using one of the heuristic procedures or simply by trial and error.

The approach to solving the packing problem as suggested by Kilbridge and Wester provides the logic for a relatively simple heuristic approach to line balancing which usually provides a good balance; namely, the application of the largest-candidate rule. Using the largest-candidate rule, work elements are assigned to stations, beginning with the first station on the line, by selecting from those elements that are eligible for assignment in descending order of duration time. For example, consider the problem of balancing the assembly line for producing the automobile backup light at a cycle time of 0.208 min. The largest-candidate rule would be applied by first ranking the minimum rational work elements in order of their duration time. (See Table 3.11.2.)

Table 3.11.2.

Rank	Element Name	Duration Time (min)	Immediate Predecessors
1	80	0.112	60
2	20	0.105	10
3	10	0.104	
4	30	0.102	
5	40	0.100	30
6	90	0.097	20, 70, 80
7	70	0.081	30
8	50	0.053	
9	60	0.048	40, 50

Continuing the application of the largest-candidate rule, consider station 1. Initially, the largest work element eligible for assignment (because of precedence restrictions) is element 10, which has a duration time of 0.104 min, leaving 0.104 (0.208 − 0.104) min of unassigned cycle time. The largest work element eligible for assignment next at station 1 is element 30. (Element 80 is not eligible for assignment because its precedence restriction has not been satisfied. Element 20 is not eligible because its duration time of 0.105 min exceeds the remaining 0.104 min of unassigned cycle time.) The assignment of elements 10 and 30 to workstation 1 result in a cumulative station work content time of 0.206 min, with 0.002 min of unassigned cycle time remaining. Because none of the remaining unassigned work elements have a duration time as small as 0.002 min, assignment of work elements to station 1 is complete, and 0.002 min of idle time remain at that station. The application of the largest-candidate rule continues with consideration of subsequent workstations until all work elements are assigned. Strict adherence to the largest-candidate rule for balancing the backup light assembly line results in assignments as shown in Table 3.11.3.

Application of the largest-candidate rule to the assembly of the backup light at a cycle time of 0.208 min requires five operators, and results in a balance delay of:

$$d = \frac{(5 \times 0.208) - 0.802}{5 \times 0.208} \times 100$$

$$= 22.88\%$$

Note that consideration of only the cycle-time problem would indicate that a theoretical minimum number of four operators is required, resulting in a possible balance delay of only 3.61%. However, in this example, the line-balancing problem has resulted in the requirement of an additional operator and greatly increased idle time (as expressed by the balance delay value) along the assembly line. It should be noted that in this example, using the largest-candidate rule to balance the assembly line at a cycle time of 0.209 min (i.e., a production rate of 287.08 units per hour as opposed to 288.46 units per hour at a cycle time of 0.208 min) results in a balanced assembly line actually requiring only four operators, with a resultant balance delay of 4.07%. If the production requirement is such that the slightly slower line speed (cycle time = 0.209 min vs. 0.208 min) can be tolerated, a sizable reduction in direct labor cost would result. On the other hand, if no line speed reduction beyond the 0.208 min cycle time can be tolerated, consideration of the functional relationship between cycle time and balance delay would argue for selecting an even lower cycle time (i.e., higher line speed) for which no additional operators are required. Inspection of the work element assignments that result from balancing the assembly line at a cycle time of 0.208 min indicates that all stations have at least 0.002 min idle time. Therefore,

Table 3.11.3.

Work Station	Element Name	Immediate Predecessors	Duration Time (min)	Cumulative Station Time (min)	Unassigned Cycle Time (min)
1	10	—	0.104	0.104	0.104
	30	—	0.102	0.206	0.002
2	20	10	0.105	0.105	0.103
	40	30	0.100	0.205	0.003
3	70	30	0.081	0.081	0.127
	50	—	0.053	0.134	0.074
	60	40, 50	0.048	0.182	0.026
4	80	60	0.112	0.112	0.096
5	90	20, 70, 80	0.097	0.097	0.111

Table 3.11.4.

Element Name	Duration Time (min)	Predecessor for: 10	20	30	40	50	60	70	80	90	Positional Weight
10	0.104		i							s	0.306
20	0.105									i	0.202
30	0.102				i		s	i	s	s	0.540
40	0.100						i		s	s	0.357
50	0.053						i		s	s	0.310
60	0.048								i	s	0.257
70	0.081									i	0.178
80	0.112									i	0.209
90	0.097										0.097

the cycle time may be reduced by 0.002 min to 0.206 min (291.26 units per hour) resulting in a reduction in balance delay from 22.88 to 22.14% without rebalancing the line. Further reductions in balance delay by rebalancing the line at still smaller cycle times may be possible.

Whereas the largest-candidate rule assigns to stations those work elements with the largest duration time first without violating precedence relationships or exceeding the station cycle time, Helgeson and Birnie[7] have proposed a heuristic technique that serves as a logical extension of the largest candidate rule. Their technique, the *ranked positional weights* method, considers not only the duration time but also the position in the precedence diagram of each work element as it is assigned to a station along the assembly line. This method assigns work elements on the basis of a positional weight for each element, which consists of the duration time of the work element itself plus the sum of the duration times of all work elements that follow it in the precedence diagram. In essence, the ranked positional weights method attempts to assign first those elements that have a large number of work elements dependent upon their completion. In so doing, it considers each work element's position within the precedence diagram, as well as its duration time, in determining the sequence in which work elements are considered for assignment to the various workstations.

The first step in using the ranked positional weights method is the calculation of the positional weight for each work element. The positional weight matrix shown below lists the element name and duration time for each work element. The work element precedence relationships are shown in the center of the matrix, using a modification of the convention adopted by Wild[8]; namely, immediate successors are indicated with an i, while successors (actually subsequent successors as opposed to immediate successors) are indicated with an s. The positional weight is then calculated by adding the work element's duration time to the sum of the duration times of all succeeding work elements. For example, the positional weight (pw) for work element 10 is:

$$\begin{aligned} \text{pw} &= \text{element 10 } (0.104) \\ &\quad + \text{element 20 } (0.105) \\ &\quad + \text{element 90 } (0.097) \\ \text{pw} &= 0.306 \end{aligned}$$

The positional weight is, therefore, a measure of both the size of the work element and its position in the sequence of work elements (Table 3.11.4). The work elements are then ranked in order of their positional weights (Table 3.11.5).

Table 3.11.5.

Rank	Element Name	Duration Time (min)	Positional Weight	Immediate Predecessors
1	30	0.102	0.540	—
2	40	0.100	0.357	30
3	50	0.053	0.310	—
4	10	0.104	0.306	—
5	60	0.048	0.257	40, 50
6	80	0.112	0.209	60
7	20	0.105	0.202	10
8	70	0.081	0.178	30
9	90	0.097	0.097	20, 70, 80

In using the ranked positional weights methods for balancing the automobile backup light assembly line at a cycle time of 0.208 min, work elements are assigned to workstations in order of decreasing positional weight without violating any precedence constraints and without exceeding the cycle time at any station. With the exception of the use of positional weight as opposed to duration time as the basis for assignment, work elements are assigned in the same manner as they are with the largest-candidate rule. Strict adherence to the ranked positional weights method for balancing the backup light assembly line results in the assignments shown in Table 3.11.6.

Application of the ranked positional weights method to the assembly of the backup light at a cycle time of 0.208 min requires only four operators, and therefore results in a balance delay of 3.61%. Recall that application of the largest-candidate rule to this example at a cycle time of 0.208 min required five operators and resulted in a balance delay of 22.88%. In this example, the ranked positional weights method proved to be a better technique for balancing the assembly line than the largest-candidate rule. Similar results should not be expected in all cases. Even though the logic behind the ranked positional weights method may appear to be more sound than that underlying the largest-candidate rule, some examples of assembly systems obtain better line balancing results from the ranked positional weights method, while other examples obtain better results from the largest-candidate rule. It should also be noted that while both methods should produce good balances, no attempt is made in either technique to arrive at an optimal solution.

While both the largest-candidate rule and the ranked positional weights method may prove computationally practical for manual balancing of assembly lines of realistic size, emphasis in industry is shifting toward the use of computerized line-balancing systems. Computerized line-balancing systems produce solutions quickly, and once the required information is gathered and input for the initial balancing of the line, any subsequent rebalancing that may be necessitated by changes in the product design or in production requirements is performed easily. Computerized line-balancing systems also permit the user to obtain a number of alternative work element assignments at different cycle times, thus permitting the selection of the "best" cycle time, which provides the "best" solution to the line balancing problem. Hence, the most efficient use of assembly operators may be obtained while maintaining the desired production output.

One such computerized line-balancing system is the CALB program. CALB (Computer Aided Line Balancing) is a line-balancing computer program developed by the Advanced Manufacturing Methods group at the Illinois Institute of Technology Research Institute.[9] The CALB program uses essentially the same input data that would be required for manual balancing of the assembly line. The information required by CALB includes:

1. The minimum rational work elements, and their respective duration times, for the assembly of the product.
2. The precedence relationships imposed by the design of the product.
3. Any restrictions imposed upon the assignment of work elements because of plant layout, conveyor fixturing, tool design, labor agreements, and/or management practice.
4. Line speed or production schedule, expressed as a range of permissible cycle times. The upper limit on the cycle time range represents the maximum time available to any workstation to perform work on a unit of output (and therefore dictates the minimum acceptable level of production output), while the lower limit represents the minimum cycle time that management considers an acceptable workload (and therefore dictates the maximum level of production output that can be attained from the line). The same cycle time value may be specified for both the upper and lower limits of the cycle time range. In doing so, a single cycle time value (as opposed to a range of permissible cycle times) will serve as the basis for balancing the assembly line. Specification of a single cycle time deprives the CALB program of the ultimate selection of a cycle time from the range of permissible values.

Table 3.11.6.

Work Station	Element Name	Immediate Predecessors	Duration Time (min)	Cumulative Station Time (min)	Unassigned Cycle Time (min)
1	30	—	0.102	0.102	0.106
	40	30	0.100	0.202	0.006
2	50	—	0.053	0.053	0.155
	10	—	0.104	0.157	0.051
	60	40, 50	0.048	0.205	0.003
3	80	60	0.112	0.112	0.096
	70	30	0.081	0.193	0.015
4	20	10	0.105	0.105	0.103
	90	20, 70, 80	0.097	0.202	0.006

Based on this information, CALB assigns work to a station as dictated by the precedence relationships and as limited by the other restrictions imposed on the balance. CALB uses several simple rules to determine what work elements to assign to a station when it is given a choice in selecting work elements to be assigned. Those rules follow essentially the same type of logic exhibited by the largest-candidate rule and the ranked positional weights method discussed earlier. The heuristic approach that CALB follows considers first the position of the work element in the precedence diagram, and second the duration time of the work element. In assigning work elements to a station, if more than one work element is eligible for assignment (i.e., without violating precedence relationships or exceeding the unassigned cycle time remaining at the station), the work element with the greatest number of immediate successors (based on the precedence relationships) is selected. This tactic tends to open up the precedence diagram for subsequent work element assignment by generating the maximum number of work element candidates for the next selection. If several work elements are eligible for selection and each has the same number of immediate successors, the work element with the largest duration time is chosen for assignment. These rules are followed as long as the accumulated sum of the work element duration times (i.e., station work-content time) at a station does not exceed the assembly line cycle time.

When the work-content time for a given station reaches a value that is greater than or equal to the minimum cycle time but less than the maximum cycle time (as specified by the range of permissible cycle times), CALB creates a new station and begins to assign work to that station by following the rules discussed previously. If the station work-content time is less than the minimum cycle time, CALB continues to try new balance combinations. CALB explores as many feasible solutions as necessary to determine a set of work element assignments that are within the permissible range of cycle times. CALB will, if necessary, exhaust all feasible sequences before accepting an underloaded station. CALB will not, however, allow the sum of the duration times of the work elements assigned to the station to exceed the maximum permissible cycle time. Therefore, while CALB will reluctantly accept an underloaded workstation, it will not permit a station to be overloaded.

CALB's assignment of work elements to a station is not based solely on precedence relationships and limitations imposed on the cycle time. CALB further restricts the assignment of work elements on the basis of a number of types of restrictions that permit the program to simulate actual assembly plant floor conditions. Restrictions that accurately describe the assembly plant environment may be entered into CALB, and then, based on those unique conditions existing in that plant for that particular assembly task, the user is able to determine the manner in which CALB assigns work elements to the stations. These restrictions may be prioritized by the user such that lower priority restrictions may be relaxed if the restrictions appear to be interfering with CALB's ability to balance the assembly line. The restrictions may be used to (1) specify the station(s) to which a work element is or is not to be assigned; (2) attempt the assignment of work elements to stations in which required tooling or operators with specific skills are available; (3) prevent the assignment of an excessive number of fatiguing tasks to an operator; (4) group (or prevent the grouping of) certain work elements together at any station; etc. Other restrictions may be used to attempt to assign work elements to a station in which the operator is positioned properly with respect to the assembly line. If the operator must change position to complete the work elements assigned to his or her station, the program adds to the station work-content time the time required for the operator to change position. Similarly, restrictions may be used to attempt to assign work elements to stations in which the product is properly oriented with respect to the position of the operator. If the product must be repositioned in order for the operator to complete the work elements assigned, the program adds to the station work-content time the time required to reposition the unit. A number of other types of restrictions are provided to control the logic behind the assignment of work elements. It is this provision for a number of different types of restrictions to permit the simulation of actual assembly plant conditions that distinguishes CALB from other line balancing approaches, and thus makes CALB especially well suited to balancing real-world assembly lines.

Using CALB to balance the assembly line for the backup light example discussed earlier would require input data as shown in Table 3.11.7. The work element definition consists of the work element name, description, duration time, immediate predecessors, and the names of any other restrictions that would apply to that work element. Note that the duration time is entered as an integer, with the units being arbitrary but consistent with

Table 3.11.7.

Name	Element Description	Time	Predecessors	Restrictions
10	Apply glue to housing	104		
20	Staple otr gasket-assembly	105	10	
30	Stake socket to housing	102		
40	Assemble bulb to socket	100	30	
50	Assemble inr gasket to housing	53		
60	Position lens to housing	48	40, 50	
70	Position connector to lead	81	30	
80	Drive 2 screws to housing	112	60	
90	Inspect and pack	97	20, 70, 80	

the units for the cycle time. No restrictions other than precedence relationships were specified for the assembly of the automobile backup light.

The remaining input information required by CALB deals with the specification of a cycle time, or range of permissible cycle times, to which the assembly line will be balanced. A single cycle time value, 0.208 min, will be input as 208 consistent with the units specified for the work element duration times, namely thousandths of a minute. The line definition statement required to specify the assembly line cycle time is shown below.

```
/CYCLE/208,208
```

The output from the CALB program consists of two types of reports. The first type merely presents an overall summary of the line-balancing attempt, without going into any detail regarding the work elements assigned to the various stations along the assembly line. This summary report gives the number of stations required, the number of operators required, the total number of work elements required to complete the assembly task, total idle time across all operators on the assembly line (= total available time − total time assigned), total available time (= cycle time × number of operators), total time assigned (= total time for all work elements = total work content time), average time per operator (= total time assigned/number of operators), cycle time limits, cycle time of line (cycle time selected by CALB within cycle time limits), and percent balance delay (= total idle time/total available time).

The summary report produced by the CALB program in balancing the backup-light assembly line is shown below.

```
          BACKUP LIGHT ASSEMBLY
NO. OF STATIONS                5
NO. OF OPERATORS               5
NO. OF ELEMENTS                9
TOTAL TIME ASSIGNED          902
TOTAL IDLE TIME              228
TOTAL AVAILABLE TIME        1030
AVG TIME/OPERATOR           160.40
CYCLE TIME LIMITS            208 to 208
CYCLE TIME OF LINE           206
PCT BALANCE DELAY             22.14
```

The second type of report generated by the CALB program gives the details of the work element assignments to the various stations along the assembly line. Information provided by this report includes the total work time per station (i.e., station work-content time) as well as detailed listings of the work elements assigned to each of the work stations. The work element assignments for the backup light assembly line are shown below.

```
*DESCRIPTION                *PREDECESSORS-RESTRICTIONS*   NAMES*        TIMES*STATION 1
STAKE SOCKET TO HSG                                          30           102
APPLY GLUE TO HSG                                            10           104
                                                          TOTAL TIME      206  STATION 1
*DESCRIPTION                *PREDECESSORS-RESTRICTIONS*   NAMES*        TIMES*STATION 2
STAPLE OTR GASKET-ASSY           10                          20           105
ASM BULB TO SOCKET               30                          40           100
                                                          TOTAL TIME      205  STATION 2
*DESCRIPTION                *PREDECESSORS-RESTRICTIONS*   NAME*         TIMES STATION 3
PSN CONNECTOR TO LEAD            30                          70            81
ASM INR GASKET TO HSG                                        50            53
PSN LENS TO HSG                  40 50                       60            48
                                                          TOTAL TIME      182  STATION 3
*DESCRIPTION                *PREDECESSORS-RESTRICTIONS*   NAME*         TIMES*STATION 4
DRIVE 2 SCREWS TO HSG            60                          80           112
                                                          TOTAL TIME      112  STATION 4
*DESCRIPTION                *PREDECESSORS-RESTRICTIONS*   NAME*         TIMES*STATION 5
INSPECT & PACK                   20 70 80                    90            97
                                                          TOTAL TIME       97  STATION 5
```

The heuristic assignment rules followed by CALB when applied to the automobile backup light line-balancing problem at a cycle time of 0.208 min requires five operators. This compares with five operators being required by the largest-candidate rule, but only four being required when the ranked positional weights method is used to balance the line. Note also that because the largest station work-content time is 0.206 min for station 1, CALB assumes that the line speed will be adjusted so that a cycle time of 0.206 min, and not 0.208 min as specified, will be utilized. Balance delay calculations are based on the provision of a 0.206 min cycle time. Interestingly, using the CALB program to balance the backup light assembly line at a specified cycle time of 0.205 min requires only four operators and results in a balance delay of only 2.20%, which seems contrary to expectations based on the functional relationship between cycle time and balance delay. Reductions in cycle time usually do not result in reductions in the number of operators required. In this particular example, however, the line-balancing problem in conjunction with the assignment rules to which CALB adheres in balancing the line, do in fact result in a reduction in the number of operators from five to four when the specified cycle time is reduced from 0.208 min to 0.205 min.

CALB does not guarantee optimal solutions; however, good balances with low balance delays are usually obtained with minimal computational time. Regardless of whether one of the manual heuristic line-balancing methods or a computerized approach such as CALB is utilized, if that approach is followed inflexibly, substantial balance delay may result. Therefore, Wild[8] suggests that the objective of line balancing should be to:

> *Assign work elements to stations to obtain an output rate in the range X–Y and to minimize the balancing loss and hence the number of stations on the line.*

Therefore, a trial-and-error approach should be adopted in which the line-balancing technique selected is used to obtain line balances for several cycle times within the range of permissible cycle times. The work element assignment that results in the lowest balance delay value would then be selected. Needless to say, the speed with which a computerized line-balancing approach can rebalance the assembly line at different cycle time values argues strongly for its use as opposed to the use of manual methods.

Discussion to this point has centered on the balancing of single-model assembly lines. The balancing of mixed-model lines is relatively straightforward, being a logical extension of the single-model line-balancing problem. In mixed-model assembly lines, the various models of the product being assembled are intermixed as they move down the line. The various models are not batched together, but rather progress through the workstations in some arbitrary order. Therefore, the workstations along a mixed-model assembly line must be prepared to accept virtually any sequence of models. For this reason, mixed-model line balancing does present some unique problems associated with developing work element assignments consistent with a production schedule that calls for different production rates of a number of different models. Therefore, mixed-model line balancing actually requires the solution of two separate but related problems: the line-balancing problem itself, and the problem of determining the best sequence of models to result in the best line balance.

Because of their relatively greater complexity, most mixed-model line-balancing problems are solved by using a computerized line-balancing technique. In using CALB, for example, to balance a mixed-model assembly line, the additional data required (over that required for balancing a single-model line) includes information regarding the number of units of each model to be produced per shift (i.e., the model mix) and an indication of which model(s) require the performance of a given work element. The CALB program then attempts to balance the assembly line, given the specified model mix, on the basis of the work element requirements of an entire production shift as opposed to balancing the line on a cycle time basis as with single-model lines. Precedence relations are adhered to, as are the other types of restrictions provided for by the program. The user may dictate the maximum and minimum station work-content times for each model at each workstation, or may simply specify the maximum and minimum values for the average station work-content time at each station, recognizing that the operator may be slightly overloaded on some models but slightly underloaded on others.

REFERENCES

1. M. E. Salveson, "The Assembly Line Balancing Problem," *Transactions of A.S.M.E.*, **77** (August 1955).
2. M. D. Kilbridge and L. Wester, "The Balance Delay Problem," *Management Science*, **2**(1) (1961).
3. T. O. Prenting and R. M. Battaglia, "The Precedence Diagram: A Tool for Analysis in Assembly Line Balancing," *The Journal of Industrial Engineering*, **15**(4) (1964).
4. B. Bryton, *Balancing of a Continuous Production Line*, Unpublished M.S. Thesis, Northwestern University, 1954.
5. M. E. Salveson, "The Assembly Line Balancing Problem," *The Journal of Industrial Engineering*, **6**(3) (1955).
6. H. G. Daellenbach and J. A. George, *Introduction to Operations Research Techniques*. Boston: Allyn & Bacon, 1978.
7. W. B. Helgeson and D. P. Birnie, "Assembly Line Balancing Using the Ranked Positional Weight Technique," *The Journal of Industrial Engineering*, **12**(6) (1961).
8. R. Wild, *Mass-Production Management*. New York: Wiley, 1972.
9. *Computer Aided Line Balancing (CALB) User's Manual*, Advanced Manufacturing Methods Program of IIT Research Institute, Chicago, IL, 1982.

BIBLIOGRAPHY

Bollenbacher, R. L. "Line Balancing Made Easier," *Industrial Engineering* (November 1971).

Dar-El, E. M., "Mixed Model Assembly Line Sequencing Problems," *Omega*, **6**(4) (1978).

Jackson, J. R., "A Computing Procedure for a Line Balancing Problem," *Management Science*, **2**(3) (1956).

Kilbridge, M. D. and L. Wester, "A Heuristic Method of Assembly Line Balancing," *The Journal of Industrial Engineering*, **12**(4) (1961).

Kovach, J., "CALB—Systemic Assembly Management," *Assembly Engineering* (September 1969).

Mariotti, J. J., "Four Approaches to Manual Assembly Line Balancing," *Industrial Engineering* (June 1970).

Moodie, C. L., "Assembly Line Balancing," *Handbook of Industrial Engineering*, New York: Wiley, 1982.

Moodie, C. L., "Customized Assembly Line Balancing," *Industrial Engineering* (August 1973).

Nawara, G. M., A. K. El-Kharbotly, and A. M. Ali, "An Extended Review on Numerical Methods for Assembly Line Balancing," 2nd International and 5th National Conference on Computers and Industrial Engineering, Orlando, FL, 1983 (unpublished).

Prenting, T. O. and N. T. Thomopoulos, *Humanism and Technology in Assembly Line Systems*, Rochelle Park, NJ: Hayden, 1974.

CHAPTER 3.12

PRODUCTION ACTIVITY CONTROL

ROBERT B. VOLLUM

R. B. Vollum & Associates
Huntingdon Valley, Pennsylvania

The factory is the focal point of asset management. Not only does the responsibility for plant equipment rest there, but the shop floor is also the prime mover/inhibitor of company cash flow. The manner in which a product is moved and controlled determines the level of work-in-process inventory and, therefore, the rate at which money turns over. Cash flow is frequently the dominant factor in a company's success, often surpassing profit or loss in short-term significance. The key to a company's rate of flow is productivity on the shop floor.[1] *Productivity* is generally defined as the measurement of units or value of output vs. a comparable measurement of input. In a more global sense, it can be defined as any action that moves a company closer to its goal of making money.[2]

Of the problems plaguing the shop floor, none is more troubling than inventory shortages. Industry spends hundreds of millions of dollars each year on premium freight and factory expediting to keep production lines moving. That is cost incurred over and above normal freight expense, caused essentially by poor scheduling, poor visibility, and poor control.

When shortages become excessive, companies tend to focus production activity on the problem of meeting shipping bogeys (targets). The resultant month-end push, which frequently accounts for most of the measured output, is concentrated on those orders for which material can be made available, regardless of commitment dates. Material and jobs that are targeted for shipment are "dragged" through the system and resources are diverted to their use. Those orders are typically heavily expedited, with known shortages the subject of daily review by a broad base of management. As a result, the bogey may be met but the mix will differ from the plan. A large percentage of commitments may be missed, while other orders ship early. The aftermath is usually confusion at best, but more likely costly chaos. Work-in-process inventory is glutted with partially completed jobs that cannot move because they contain shortages, and stockroom inventory is inflated with mismatched material that won't complete anything.

Foremen, the backbone of production, are paid to train, supervise, and motivate factory workers to produce quality product at minimal costs, on schedule. In reality, the majority of their time may be consumed chasing parts. That may mean searching through stock shelves or sorting through recent receipts on the dock or expediting jobs through preceding operations. It may also mean expediting purchasing or sitting through shortage meetings or just plain stealing materials from other jobs. Chances are it means all of those things.

3.12.1 CONTRIBUTORS TO PRODUCTIVITY INCREASES

What can executives in manufacturing companies do to improve productivity for their firms and for the nation? Recent studies have concluded that from among applied labor, capital investment, and new technologies, the greatest opportunity (approximately 60%) comes from the application of new technology.

Capital investment in machine centers, robotics, and the like is obviously necessary to keep pace with demands for competitive capacity. But it is the technological advancements of the computer age that presage the new dawn of achievement.

How do we distinguish between capital investment and technology when considering the computer? Very simply. The *money* spent for the acquisition of a computer is capital investment. The *use* of the computer for control of machine centers, for Computer Aided Design/Manufacture (CAD-CAM), for communications, for manufacturing planning and control, for decision support, etc., is applied technology.

The factory floor is an area in many firms that is not taking full advantage of the information processing power of the computer. Ironically, it is the area with the greatest potential for return. Labor, material, and overhead, which make up cost of goods sold, typically represent 50 to 80% of a company's sales dollar. The application of state-of-the-art planning and control technology to the maximization of total factory throughput and to the minimization of those costs, while optimizing customer service, makes good business sense.

Material requirements planning (MRP) changed the face of manufacturing control from a static, backward perspective to one of dynamic projection. It added the elements of timing and relativity to the manufacturing equation. Specific targeted needs became the action triggers, replacing the generalized historical averages that drive "Order Point" systems.

However, material requirements planning does not consider plant capacity in its generation of factory orders. It establishes a single, economical lot size for each part requirement and makes the systemic assumption that sufficient capacity is available for timely satisfaction of the demands. It establishes relative priorities among orders by determining the required completion dates and critical starting dates with no regard to resource contention or order sequencing. In essence, MRP provides broad milestone controls, but no schedule detail. The interpretation of MRP output into factory load by work center by date and the detailed scheduling of each order through its required operational steps require logic normally contained in capacity planning and shop-floor control programs. It is through these routines that the labor, material, and facilities are matched to scheduled orders. Shop-floor control programs typically also contain the logic to dispatch work to machines and to record performance results.

3.12.2 SHOP-FLOOR ENVIRONMENT

There are no easy answers to factory problems. Manufacturing executives and managers spend enormous amounts of time and energy sifting through alternatives to find reasonable solutions from data at hand. There are no "free standing" decisions. Every action has an impact on myriad other factors, which in turn react with still other related conditions or forces as a result of the initial action. It can become an almost unending iteration of change. The overriding question is "What am I going to do?" The machine broke down; the parts have not arrived; a person possessing critical, unique skills is out sick today, etc. What am I going to do?

More so than in many other parts of a manufacturing company, management in the factory needs immediate access to data. It also needs interactive *what if*? capability. The essence of shop operations is change. Management cannot wait for the next ballpark schedule to come out of the weekly MRP run to make decisions. Long-range plans and master schedules have their place, but the foreman deals in the short term. He needs to know what to do at 2:00 pm on Tuesday on vertical boring mill number 1 when the scheduled job gets delayed upstream.

Shop-floor management personnel must have intimate control of everything that happens between the scheduling cracks. Their lives are filled with changes, conflicts, alternatives, and priorities. They continuously manipulate people, material, facilities and jobs. To do this effectively, they must have rapid access to up-to-the-minute information concerning availability, utilization, schedules, delays, etc. In short, they must have control of everything necessary within the four walls of the plant to execute current plans.

3.12.3 RELATIONSHIP TO PLANNING SYSTEM

Manufacturing planning and control systems are generally implemented in a series of steps, with the production activity control portion usually well down the line. This is unfortunate, because beyond the development of certain prerequisite information and programs, execution systems can proceed to implementation simultaneously with the planning systems. The objective should be to place factory management in control of the action at the earliest possible time. Until production activity control is working, management is basically in a reactive position, forced to respond defensively, and often blindly, to the vagaries of change.

A production activity control (shop-floor control) system must receive information from a planning system about orders that have been released for processing. It must also have direct database access to all pertinent information about facilities, people, operations, and material. This can be accomplished within a fully integrated system, or it can be done distributively. It is not essential for the planning and execution systems to reside on the same computer. The key point is that factory people must deal directly with a system that can identify needed action and can process and react to real-time updates. If the planning and execution systems are remote from each other, they must pass data back and forth regularly. In most situations, it is not necessary for both systems to receive simultaneous updates.

3.12.4 MAJOR FUNCTIONAL AREAS

A shop-floor control system must address the following four areas:

Order management and control.

Material movement and control.

Labor activity reporting and control.

Resource management.

The focus is on order management and resource management. Emphasis must also be on the control of material and the proper utilization of labor. Capabilities must exist for order scheduling, order release, input/output control, priority control, order monitoring, material control, time and attendance reporting, work planning and assignments, production reporting, and performance monitoring.

3.12.5 ORDER SCHEDULING

Realistic scheduling is the cornerstone of effective shop-floor control. To be realistic, a schedule must reflect the orderly arrangement of load to fit existing capacity and time constraints. In other words, it must be doable. To maintain realism, factory managers must have the facility to make real-time adjustments and *what if* simulations. They should be able to "see" the projected impact of changes or impromptu demand on resource loads, and to interpret the impact of those adjustments on related schedules. And they should be able to *manage* the schedule, directly, to best advantage. The planning system must produce the schedule targets, but fine-tuning and short-term course changes should be initiated on the factory floor.

Of the scheduling techniques in common use, the MRP-based infinite-load backward scheduling with progressive lead-time offsets from planned completion dates is currently the most widely used in discrete manufacturing. Finite forward scheduling and simulations of machine loading utilizing sophisticated algorithms are newer, more advanced techniques that are rapidly gaining broad acceptance. They focus on the simultaneous control of all factory constraints and on the dynamic calculation of variable process batches and transfer batches to control queues and product flow. Repetitive and process flow operations are generally controlled through cumulative run schedules. However, the Japanese techniques of just-in-time manufacturing are receiving a significant amount of attention throughout industry today, particularly in repetitive manufacturing operations.[3]

Backward scheduling systems (MRP) consider all resources to be of equal schedule importance. No distinction is drawn between critically constrained resources (bottlenecks) and nonconstrained resources. Usually, only one kind of resource (machines, people, tools, fixtures) is scheduled. Backward scheduling calculates the order-start date by subtracting a standard, preassigned lead time from the specified order due date. The start and due dates of each operation on the job are similarly calculated. The detailed operation schedule dates provide the basis for determining the capacity required to manufacture the part.

Capacity planning programs analyze the load placed on each work center by time period throughout the production facilities, as a result of the production orders generated by the planning system. Loads are usually displayed in reports and/or on-line CRT screens as total scheduled standard hours of setup and processing time vs. total standard hours of planned capacity. Load may also be shown as a percentage of demand against available capacity. The orders are calculated to arrive at each work center in accordance with the preset lead times assigned either to the part or to each operational process. The capacity planning process is most often done in a batch computer run, loading all orders at the same time. However, there is justification for the use of on-line updating to test for available capacity as an order is being released. If adequate capacity exists, that capacity can be reserved. If capacity is not available, alternative resources can be checked.

Whenever there is an overload condition, the foreman should have access to each order that is scheduled into the overloaded period. In that way, he can work with the production planners to alleviate the overload. The outputs of capacity planning can be used to plan overtime, allocation of skills, and utilization of work centers.

3.12.6 LEAD TIMES AND QUEUES

Backward scheduling fixed-batch systems such as MRP assume lead times to be standard. Finite forward scheduling systems with dynamically determined variable batches assume lead times to vary as a result of the actual schedule. Manufacturing lead time is the sum of the lead times for every operation required to make the product. Operation lead time consists of queue time before processing, setup time, run time, wait time after processing, and move time (see Fig. 3.12.1). Experience has shown that queue and wait time account for 90% or more of operation lead time in discrete batch manufacturing. Varying the size of the queues varies the level of work-in-process inventory and directly affects the cycle time (manufacturing lead time) of an order. The size of the queue is determined by the total of setup and run hours required by the jobs currently waiting for the resource to become free.

The first step of queue control is to determine the desired optimal queue (Fig. 3.12.2) in each work center. These queues establish the buffers necessary to prevent unscheduled downtime by absorbing the random fluctuation in work arrival. They are most often set empirically, although some computer systems will establish them dynamically.

Each work center should be scheduled so that backlogs are controlled to the optimal level. When work centers queues grow, work-in-process inventory is inflated, lead times increase, and order completion dates may be delayed. Conversely, when queues at critical resources shrink, downtime may occur, adversely affecting productivity. The first objective of a complex shop, therefore, is to control actual queues to an optimal level as a means of simultaneously achieving high productivity, performance to schedule, and minimum inventory investment.[4]

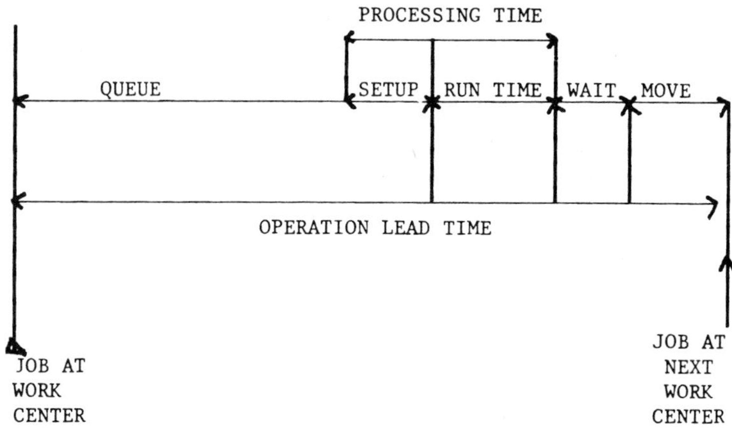

Fig. 3.12.1. Operation lead time.

Work center queues vary in size based on the relationship between output, or work completed, and input, or work arriving, and are usually measured in standard hours. Input is primarily dependent on the rate at which orders are released. Output is dependent on the capacity and utilization of the work center. Queue control is achieved by managing output vs. input. An effective production planning and control system provides actual vs. optimal queue information and output vs. input data to achieve this control.

3.12.7 INPUT/OUTPUT CONTROL

Input/output (I/O) planning and control is a technique used to regulate the rate of job order throughput and shop lead times by managing the value of work-in-process existing in a shop. I/O control can be applied to individual work centers and to the factory as a whole. The focus is on the management of queues; the direct impact is on lead times, inventory turns, and cash flow.

I/O measures the actual flow of work in and out of a work center over a specified period of time. Measurements should be in the same terms as in the capacity plan (standard hours, units, feet, tons, etc.). The essential measurements are:

1. *Planned output (capacity).* The work expected from the work center during the measured period.
2. *Actual output.* A measurement of work completed and moved from the work center.
3. *Planned input (load).* Original measurement includes both existing work and new load projected to arrive during the time period. Subsequent quantities are new load only.

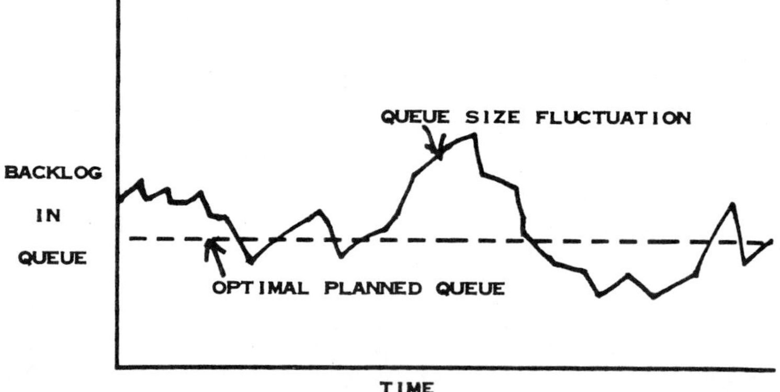

Fig. 3.12.2. Average queue distribution.

4. *Actual input.* The actual quantity of work moved into the work center.
5. *Cumulative deviations (both input and output).* Calculated by subtracting the actual from the planned, plus deviation from the previous periods.
6. *Work-in-process calculation.*

	Week Ending (Day)					
Input	230	237	244	251	258	265
Planned	500	500	500	500	500	500
Actual	490	500	490	515	500	490
Cumulative deviation	−10	−10	−20	−5	−5	−15

	Week Ending (Day)					
Output	230	237	244	251	258	265
Planned	525	550	550	550	550	525
Actual	500	560	540	550	575	490
Cumulative deviation	−25	−15	−25	−25	−0	−35

	Week Ending (Day)						
	223	230	237	244	251	258	265
Work-in-process	750	740	680	630	595	520	520

Fig. 3.12.3. Input/output report. Example of controlled reduction of work-in-process through I/O measurement.

Input/output control is a straightforward, effective means of maintaining a balanced flow through the shop. It applies to all types of production. It can be particularly valuable in job shops that have potential bottleneck operations and/or erratic planned order releases.

Lead times and WIP investment are the primary targets of I/O control. Altering either capacity (output) or load (input) directly impacts the size of the queue. Figure 3.12.3 illustrates a controlled reduction of WIP in queue from 7.5 to 5.2 days against a planned reduction to 5.0 days.

3.12.8 PRODUCTION SPLITTING AND OVERLAPPING

Order lead time can be reduced by splitting one operation and performing it on more than one machine concurrently or by overlapping operations with each other.

If multiple machines and sufficient tools are available, a job can be divided among machines. This procedure is called *splitting* (see Fig. 3.12.4). More than one piece is produced at a time, thus reducing the run time portion of lead time. A common method of determining the economics of a possible split is to calculate the run time/setup time ratio of a job. Depending on how the results compare with preset management values, the load may split among all the possible machines, fewer machines, or not at all. Operation splitting (parallel machining) requires that available resources exist in the work center—multiple-machine capability where machining is the limiting constraint, or sufficient human resource where manpower is the limiting constraint.

Manufacturing orders in process may also be split when it becomes apparent that waiting for the entire quantity to be processed will cause higher-level requirements to be late. When a portion of the order is needed to meet firm higher-level demands and the remainder can be delayed, the required quantity may be split off from the original order to allow completion on the required date. Extra setups are inherent in splitting orders or in splitting among multiple machines.

In overlapped production, the completed portion of a job is sent ahead for processing at a succeeding work center before the entire lot is completed at the current operation (Fig. 3.12.5). Overlapping reduces the lead time of an order by trimming the effect of batch process time and queues between operations. Overlapping requires that the order be closely controlled.

3.12.9 ORDER RELEASING

Timely control of the releasing process is one of the keys to the effective management of work-in-process inventory levels. Management must be able to control the flow and timing of jobs, and it must be able to direct critical materials to priority orders.

Most commonly, the system planning logic will cause specific material reservations to be made after an order has moved into the scheduled release zone. Control will then pass to the production activity control

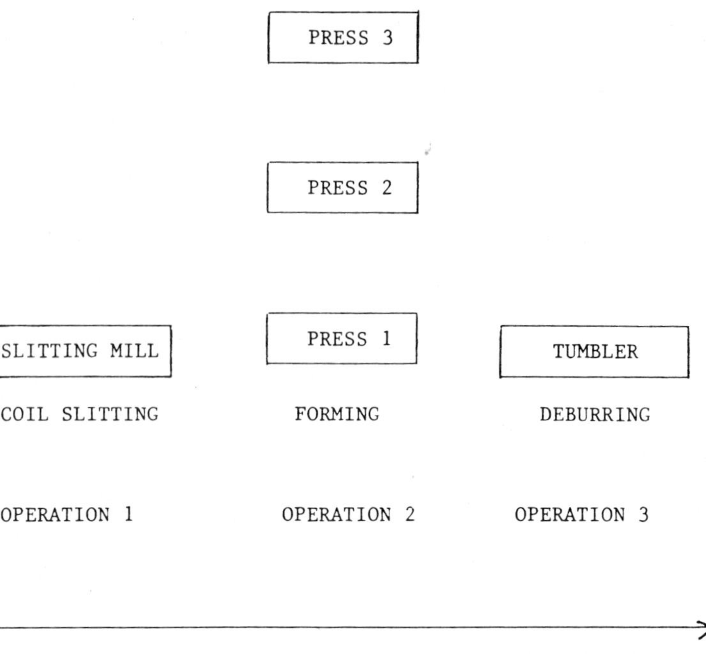

Fig. 3.12.4. Split production.

module. In a totally integrated system, one order release date will be common to both the planning and execution modules.

If production activity control is distributed to a dedicated computer, there may be progressive release dates. The planning system will consider the order to be released as soon as it is consigned to the factory. Shop management, however, may elect to hold the order in queue until it is actually moved to the first operation. In that case, the order status in the production activity control system will remain "unreleased" until the order moves.

The order releasing process should trigger the creation of shop paper. Operation list travelers, material pick lists, drawing lists, and turnaround reporting documents are typical of the type of documents included in a work-order packet. The method of data input employed dictates how the documents are prepared. In a mechanized system, they may be produced for punched card, bar code, magnetic strip, optical character recognition, or CRT keyboard introduction. Most of the older factory data collection systems employed prepunched cards as the input medium. Many current systems are using the other newer techniques, as well as direct machine-to-computer conversations.

3.12.10 PRIORITY CONTROL

Once an order has been released to production, it remains in continuous competition for manufacturing resources until it is finally completed. This is particularly true in discrete job shop operations where orders move progressively from one work center to another. At each stop, an order becomes part of a queue of orders waiting their turn for processing.

Although average order lead time is influenced by work center queue sizes, the completion of any specific job in a work center is a function of two basic factors:

Priority of the job.

Condition of that work center when the job arrives.

A great number of prioritizing rules, rules of thumb, and schemes have been developed over the years to sequence work in the shop. They include efforts to control to due dates, to maximize the number of orders processed, to maximize the value of orders shipped, to treat orders nonpreferentially or randomly, and to establish urgency quotients derived from calculations involving processing times and lead times. Fogarty and Hoffman[5] and Green[2] detail many of the most common rules.

Computer systems generally calculate dynamic priorities for all orders in the shop each time the dispatch list is run, based on the latest update information. Two of the most commonly used priority rules are critical ratio and slack time ratio. With both of these rules, the order with the smallest calculated value receives the highest priority.

$$\text{Critical ratio} = \frac{\text{Time remaining to due date}}{\text{Processing time remaining}}$$

$$\text{Slack time ratio} = \frac{\text{Time remining to due date} - \text{processing time}}{\text{Number of operations remaining}}$$

Another rule that is gaining significant support is the Buffer Time rule. *Buffer time* is defined as the difference between the latest start date and the earliest start date. Remaining order lead time and an externally assigned priority number are also included in this calculation.

Buffer Time External Priority Remaining Order Lead Time

The three elements are written next to each other to build the priority. As with critical ratio and slack time ratio, the order with the lowest number has the highest priority.

3.12.11 ORDER MONITORING

Since the factory is such a dynamically changing environment, the control of change becomes the keystone of productivity. There will always be problems: production mistakes do occur, requiring rework, and there is a recurring need to vary batch sizes, split orders, and overlap operations. Additionally, the exigencies of business will sometimes dictate cancellations, schedule changes, quantity changes, engineering changes, and process

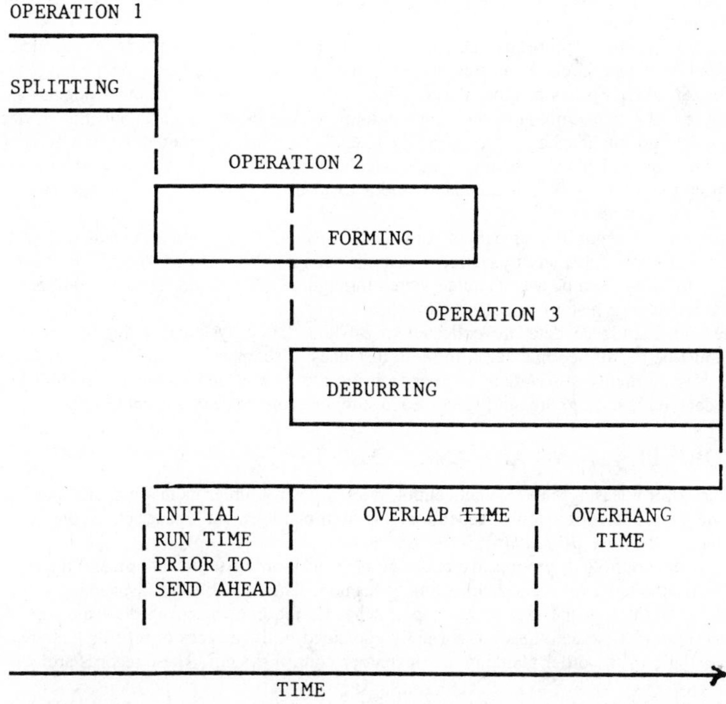

Fig. 3.12.5. Overlapped production.

changes. In other words, the factory floor becomes the final solution for a host of problems, many of which originate elsewhere, and time is often the most pressing enemy.

Problems that occur during production usually demand immediate attention. Frequently there is no time to wait for the next order planning run for a revised plan. In those instances, management must have CRT terminal access to the production activity control system, which must have the capability to respond to direction in real time. The planner or foreman or manager must be able to make things happen, *right now,* interactively, and all the loose ends must be identified for further action or be tied off by the system automatically. For example: if an order must be split with part of it proceeding per schedule while the balance follows an altered routing for rework, the database must be updated; schedules for each resultant new order must be calculated; additional documentation must be created; cost records must be updated; dispatch sequencing must be repositioned. To be most effective, management must have access to up-to-the-minute information about current orders as well as to the history of closed orders.

3.12.12 MATERIAL CONTROL

The two elements of cost that are directly assignable to work orders, and are therefore controllable, are material and labor. Material control has most generally been considered to be an inherent part of the planning system, but the detailed control of in-house inventory is the province of production activity control. While the results of all dock-to-dock inventory transactions must be known to the planning system for inclusion in net requirements calculations, and while material availability must be determined by the planning system before an order is released to production, the minute-to-minute activity control must reside at the operating level.

The production activity control system should direct and monitor the use and movement of inventory "within the four walls." It should have control over the transactions to report issues, receipts, transfers, cycle counts, and adjustments, and it should maintain stock-status information. Since there will always be slips between the plan and reality, the system must also maintain control of shortages and reservations.

3.12.13 TIME AND ATTENDANCE REPORTING

The employee record contains information about an employee that is relative to job performance and provides the basis for labor reporting for both payroll and cost accounting purposes. It identifies the employee by name, identification number, labor grade, pay rate, work skills, shift assignment, department assignment, team assignments, and any other particulars deemed to be pertinent. Performance information about specific job assignments can also be included in this record.

Most mechanized systems capture attendance data through time and attendance terminals located at employee entrances to the plant. Clock in/clock out transactions can be validated directly against the employee record in the database for security purposes or other reasons, if a company so chooses. Another popular approach is to have the employee clock in and out at his/her work station on a multipurpose data collection device. Whatever the approach, time and attendance reporting, as an integral part of the production activity control system, permits direct comparisons between the employee's calculated time at work and the time reported against work assignments throughout the day. When attendance and activity data are integrated, discrepancies can be caught and corrected on the spot.

Time and attendance reporting, as part of a real-time system, permits management immediate access to information about employee availability. Direct questions such as "Who is here today?" by name, and "Who is absent today?" by name, can be asked and answered through a CRT terminal screen. Additional information about individual attendance history can also be obtained.

When time and attendance data are collected in off-line, batch systems or through manual reporting, reconciliation must be performed manually. In these instances, discrepancies will be recognized well after the fact, requiring post-mortem investigation in order to bring the payroll and accounting records into line. The details of these delayed reportings are rarely received in sufficient time to be of any value to factory management.

3.12.14 WORK PLANNING AND ASSIGNMENTS

Order scheduling, order releasing, input/output control, priority control, order monitoring, and time and attendance reporting provide all but one of the tools needed for shop floor management to take active control of the factory. The missing piece is *dynamic dispatching*.

Dispatching is the controlled, progressive releasing of factory orders (work-in-process) through all required operations at each affected work center, in a timely fashion. The objective of dispatching is to provide the optimal flow of work through the factory to support schedule requirements with the least possible cost. The data required to support these activities are normally contained in three types of records that are continuously interfaced to the logic of the order planning and inventory control system. These records are:

1. Manufacturing order records, which contain order due dates, cost data, and order priority data.

2. Routed work-in-process records, which contain released manufacturing order schedule data and are updated by shop floor reporting to maintain job status in relation to progress, quantity, and time.
3. Work center records, which contain capacity, actual load by time period, and performance data (e.g., efficiency, percent of work on schedule, and average input and output).

Current state-of-the-art software usually supports interactive work dispatching on CRT screens. Management can view the up-to-the-minute status and priorities of jobs at, or due into, each work station. Foremen can gauge job arrivals through review of performance data of jobs in upstream work centers.

A well-constructed database with proper communications software gives management the browsing and response capabilities needed to manage the workload effectively. The entire organization can be viewed on-line, either broadly or in specific detail. For example, departmental organization, machine status, employee status, and job skills can be matched readily with critical work running on a temperamental machine; or a quick analysis can be made of the plant-wide status of facilities needed for an upcoming major job.

As computers become more and more integral to the control of shop operations, the tendency toward centralized dispatching is increasing. In either a process or a repetitive, discrete environment, centralized dispatching is usually employed, since the problem is more one of keeping the pipeline continuously flowing than of choosing between viable alternatives. So long as the pipe stays open and the material is available, the work will progress. Discrete batch manufacturing must solve the problem of distinctly different jobs competing for the same shop resources while following divergent paths through a network of dissimilar operations. The problem is further compounded by the variables of time and urgency. Real-time feedback permits up-to-the-minute, centralized status maintenance and activity control with decentralized visibility of the results through CRT terminals. Increasingly, job shops with this automated facility are using the computer to maintain job control data centrally, while permitting the shop departments to dispatch locally, exercising on-site human judgment.

The two controllable commodities within factory operations that determine the rate of flow, and the specific order sequence of flow are (1) operations lead times and (2) order sequencing. Lead times are manageable through the manipulation of lot sizes and queues.

Jobs are sequenced for processing in accordance with their relative urgencies through priority control.

3.12.15 PRODUCTION REPORTING

One of the most significant elements in the move toward effective use of the computer in the shop is the acceptance of automated data collection for production reporting by the rank and file workers. In the early stages of this movement, workers generally felt threatened, with resulting widespread work stoppages and sabotage. Fear of the unknown and suspicion of sinister intent were underlying causes. These fears and suspicions, although not entirely dispelled, have greatly diminished. Today, joining the electronic age has become an exciting prospect for much of the work force. This is particularly true of the younger members who have grown up with an understanding of the value and inevitability of computers.

Data collection is the window into the system, but it is only that. Automating production reporting through electronic devices, such as "seeing eye" bar code wands, provides an interactive link between the worker and the application software of the production activity control system. Since much of the information is precoded in the system, entry is much easier, faster, and less prone to error than in manual systems. Most important, records can be updated directly, in real time.

The production activities that employees must report are those dealing with job performance. Material movement, job/machine setup, part and assembly production, scrap/rework, work interruptions, and nonproduction activities such as union business and idle time are some of the types of reporting required. A single input should update both labor and transaction information, with the system automatically calculating elapsed time. These reports provide the labor input for payroll and, also the job status information essential for order tracking and control. Additionally, management is provided with the up-to-the-minute details required for work center control and quality measurement.

As operations are completed, history information can be developed. Historical data can be organized to analyze the following:

Actual time vs. standard for efficiency calculations.

Labor reporting to identify areas needing additional supervision or training.

Work-in-process actual time for subsequent standard variance analysis.

Actual start and completion compared to standard manufacturing process time to analyze lead times.

Scrap and rework data.

Information to monitor production processes and validate inventory ordering rules.

Work-in-process control should interface continuously with order planning and scheduling and inventory control.

3.12.16 PERFORMANCE MONITORING

For management to control operations properly, it needs factual and timely information. Therefore, the system must not only process data; more important it must organize and present information in a manner suitable for use in decision making. Management runs a factory from conclusions drawn through the applied analysis of efficiency, utilization, cost and activity information. In short, it strives to measure and control productivity.

A recent data processing development that is starting to have a dramatic impact on the manufacturing industry is nonprocedural language, a technique that has been dubbed, "programmerless computing." This technique permits management to organize the data into ad hoc reports through conversation with the computer in simple English language commands. Reports can be organized and reorganized on the spur of the moment from any data resident in the database without data processing intervention.

While it is crucial that information be provided for on-line browsing, it is also important that key data be available in hard copy reports for management perusal. The list of valuable reports is lengthy. Select examples are:

Inventory discrepancy report.
Material reservation report.
Open work order report.
Shortage report.
Scrap report.
Activity status report.
Work center analysis report.
Work center capacity report.
Employee report.
Attendance report.
Dispatch list.
Input/output report.

In summary, a production activity control system provides the facilities needed to effectively manage people, material, work orders, time, and cost in the execution of the plan. The payoffs are clear. Rewards will come in the form of controlled levels of work-in-process, optimal customer service, responsive control of the shop, high productivity and "bottom line" performance.

3.12.17 JUST-IN-TIME MANUFACTURING

Recently, the Japanese system of just-in-time (JIT) manufacturing has received considerable attention throughout the industrial world. Since *just-in-time* is a broad philosophy of management and operation, it will be explored only briefly in this chapter. Interested readers are encouraged to pursue the in-depth treatments provided in the references cited in the bibliography at the end of this chapter.

A technique often used to describe JIT is to focus on the fourteen elements that are typically embodied in the approach (see Table 3.12.1). Many of the elements are cultural in nature and are impractical to duplicate in Western society. Others involve establishing an ongoing program and attitude of improvement in almost every area of the business. The drive is to improve continually until every last element of waste of all kinds is eliminated. Simultaneous attacks are made on product defects, inventory levels, process control, setup times, and cycle times. The approach is basically a systematic, manual, trial-and-error tightening of the controls. As problems surface, they are resolved, thereby creating the potential for further improvement.

Table 3.12.1. Elements of Just-in-Time

Elimination of Waste	Respect for People
Focused factory networks	Lifetime employment
Group technology	Company unions
Quality control	Attitude toward workers
Just-in-time production	Automation/Robotics
Uniform plant loading	Bottom round management
Kanban system	Subcontractor networks
Minimized setup time	Quality circles

The Japanese experience has been very successful where the techniques fit and the efforts have been applied. The Japanese have found that the industries most suited to JIT are those whose production processes are repetitive in nature. It has also been their experience that the road to achieving good just-in-time production is lengthy, often measured in tens of years.

3.12.18 OPTIMIZED PRODUCTION TECHNOLOGY (OPT)

OPT is an emerging technology that is similar to just-in-time in its emphasis on controlling the flow of production through the plant rather than on balancing capacity, and in its focused attention on bottlenecks. However, it differs significantly from JIT in that it is a computerized modeling and simulation technique. OPT is capable of predetermining schedules that will maximize throughput, minimize inventory, and minimize operating expenses for any set of operating condition constraints. It dynamically calculates variable batch sizes, both process and transfer, inventory buffer sizes, and buffer positioning. It finitely schedules all critical resources and infinitely backschedules all other resources to meet the critical needs. OPT simultaneously considers all system-wide constraints to produce globally rather than locally optimized schedules.

OPT, as a philosophy, embodies nine basic rules of manufacturing that differ significantly from traditional rules.

1. Balance flow, not capacity.
2. The level of utilization of a nonbottleneck is not determined by its own potential but by some other constraint in the system.
3. Activation and utilization of a resource are not synonymous.
4. An hour lost at a bottleneck is an hour lost for the total system.
5. An hour saved at a nonbottleneck is a mirage.
6. Bottlenecks govern both throughput and inventories.
7. The transfer batch may not, and often should not, be equal to the process batch.
8. The process batch should be variable, not fixed.
9. Schedules should be established by looking at all of the constraints simultaneously. Lead times are the result of a schedule and cannot be predetermined.

Just as with just-in-time, OPT users have experienced significant results. The bibliography at the end of this chapter contains several references for in-depth material on Optimized Production Technology.

REFERENCES

1. Robert B. Vollum, "In Control with Shop Control," *Production Magazine,* Dec. 1983.
2. Eliyahu M. Goldratt, "Cost Accounting: The Enemy Number One of Productivity," *Proceedings,* American Production and Inventory Control Conference, 433–453 (1983).
3. Robert W. Hall, *Zero Inventories,* Homewood, IL: Dow Jones-Irwin, 1983.
4. William E. Sandman and John P. Hayes, *How to Win Productivity in Manufacturing,* Dresher, PA: Yellow Book of PA., 1980.
5. Donald W. Fogerty and Thomas R. Hoffman, *Production and Inventory Management,* Cincinnati, OH: South West Publishing, 1983.
6. James H. Greene, *Production and Inventory Control Handbook,* New York; McGraw-Hill, 1970.

BIBLIOGRAPHY

APICS, *APICS Shop Floor Control Reprints,* Washington, DC: The American Production and Inventory Control Society, 1973.

Baker, Eugene F., "Flow Management the 'Take Charge' Shop Floor Control System," *APICS 22nd Annual International Conference Proceedings,* 169–174 (1979).

Baker, Kenneth R., *Introduction to Sequencing and Scheduling,* New York: Wiley, 1974.

Bechte, Wolfgang, "Controlling Manufacturing Lead Time and Work-In-Process Inventory by Means of Load-Oriented Order Release," *APICS 25th Annual International Conference Proceedings,* 67–72 (1982).

Berry, William L. and V. Rao, "Critical Ratio Scheduling: An Experimental Analysis," *Management Science,* **22**(2), 192–210 (1975).

Fox, Robert E., "MRP, KANBAN, and OPT—What's Best?," *APICS 25th Annual International Conference Proceedings,* 482–486 (1982).

Fox, Robert E., "OPT vs. MRP—Thoughtware vs. Software," Part I, *Inventories and Production*, **3**(6), November/December 1983.

Fox, Robert E., "OPT vs. MRP—Thoughtware vs. Software," Part II, *Inventories and Production*, **4**(1), January/February 1984.

Fukuda, Ryuji, *Managerial Engineering*, Stamford, CT: Productivity, Inc., 1983.

Garwood, Dave, "Delivery as Promised," *Production and Inventory Management*, **12**(3) (1971).

Goldratt, Eliyahu M., "Optimized Production Timetable: A Revolutionary Program for Industry," *APICS 23rd Annual International Conference Proceedings*, 172–176 (1980).

Goldratt, Eliyahu M., "The Unbalanced Plant," *APICS 24th Annual International Conference Proceedings*, 195–199 (1981).

Goldratt, Eliyahu M., "100% Data Accuracy—Need or Myth?," *APICS 25th Annual International Conference Proceedings*, 64–66 (1982).

Goldratt, Eliyahu M., *The Goal*, Croton-On-Hudson, NY: North River Press, (1985).

Griffin, Kenneth R., "Job Shop Scheduling," *Production and Inventory Management*, **12**(3) (1971).

Jacobs, F. Robert, "OPT Uncovered: Concepts Behind the System," *Industrial Engineering*, 32–41 (Oct. 1984).

Jones, William J., "The Integration of Shop Floor Control into the Materials System," *Proceedings*, American Production and Inventory Control Conference, 133–141 (1978).

Lankford, Raymond L., "Short-Term Planning of Manufacturing Capacity," *APICS 21st Annual International Conference Proceedings*, 37–68 (1978).

Monden, Yasuhiro, *The Toyota Production System: A Practical Approach to Production Management*, Industrial Engineers & Management Press, Institute of Industrial Engineers, Atlanta/Norcross, GA, 1983.

Ohno, Taiichi, *Toyota Production System*, Tokyo: Diamond Company, 1980.

Perreault, Alphedor, "The Bottom Line of Shop Floor Control Begins with a Good Data System," *Proceedings*, American Production and Inventory Control Society (1969).

Putnam, Arnold O., R. Everdell, D. H. Dorman, R. R. Cronan, and L. H. Lindgren, "Updating Critical Ratio and Slack-Time Priority Scheduling Rules," *Production and Inventory Management*, **12**(4) (1971).

Shingo, S., *The Toyota Production System*, Systems Management Association, Osaka, Japan: Shinsei Printing, 1981.

Schonberger, Richard J., "Just-In-Time Production Systems: Replacing Complexity with Simplicity In Manufacturing Management," *Industrial Engineering*, 52–63 (Oct. 1984).

Schonberger, Richard J., *Japanese Manufacturing Techniques: Nine Hidden lessons in Simplicity*, New York: Free Press, 1982.

Wassweiler, William R., "Fundamentals of Shop Floor Control," *APICS 23rd Annual Conference Proceedings*, 352–354 (1980).

Wight, Oliver W., "Input/Output Control: A Real Handle on Lead Time," *Production and Inventory Management*, **11**(3) (1970).

Wautuck, Kenneth A., "The Japanese Approach to Productivity," *APICS 26th Annual International Conference Proceedings* 662–669 (1983).

CHAPTER 3.13

FACTORY FLOOR INFORMATION SYSTEM

GLENN C. DUNLAP

Wandel-Goltermann
Research Triangle Park, North Carolina

DAVID W. JUNG

Arizona State University
Tempe, Arizona

Factory Floor Information System (FFIS) is *the method by which accurate and timely information is communicated to, from, and within the factory floor* in order to optimize the primary goal of manufacturing, which is to create part geometry in required quantities, on schedule, while minimizing cost and maximizing quality.

This chapter will cover the key issues of an FFIS implementation:

Design
Logical model
Control instructions
Data capture
Communications

Before the technical details of FFIS implementation are discussed, a short story will be used to dramatize how a factory supervisor may use the advanced information processing capabilities of the FFIS to solve problems in the factory of the future.

It was late in the afternoon when Pete, the dayshift factory supervisor spotted Joe, the material handler, moving steadily on a course toward one of the many automated machining centers in Cell E of the factory.

With a slight limp, Pete walked slowly toward a position that would intersect the course the material handler was making. As Joe approached, Pete cupped his hands so that he could be heard above the whine, clank, and hiss of the production machines.

"Where are you going with those four-four-ohs?" Pete asked.

"I have instructions to deliver these parts to machine BG09," Joe responded.

"Heck, those instructions must be out-of-date. I just got a call on my radio terminal that machine 9 is down for maintenance, and I can see the flashing light from here," said Pete, his brow beginning to furrow. "Wait here for further instructions."

"Very well, I will wait for further instructions," Joe acknowledged.

Pete returned to the supervisor control room, approached the new advanced FFIS workstation, and with a few keystrokes, quickly accessed the information he needed on the status and history of machine 9 and the parts Joe was carrying. "Just as I thought," he muttered, "it went down ten minutes ago due to some type of power supply failure, and a tool broke."

Further research indicated that the five NGP789440 parts belonging to the material handler were due in assembly that evening. "Let me see if I can piece together this data" thought Pete. "The power supply failed, so the scheduling system automatically moved the three-one-five job to machine 33, which was idle, and assigned a setup specialist to get it going again. Then it assigned the four-four-oh job to machine 9. Maintenance forecasts repair completion in one hour, and quality control has been requested to submit a discrepancy report for the part that was on the machine when it failed. Let's see. I still need some new tooling for this new job. Hey, maybe the material handler has that, too!"

Pete walked back over to where the material handler was waiting. There was tooling on the pallet also. "Why didn't you tell me that you had replacements for the broken tool, too?" he demanded.

"You didn't ask" responded Joe calmly and logically.

"Wise guy," chuckled Pete. "OK, proceed."

"10-4," acknowledged the material handler.

Just at that time, a woman in an orange jumpsuit came up to the control room carrying a black box and some tools.

"Hey Doc, glad to see you," Pete cheerfully exclaimed. "Need to get ol' number nine back in business. The system shows that you'll be done in about an hour. Is that true"?

"Yeah, I think so," answered Doc, the maintenance person, as she pressed a few keys on the FFIS workstation keyboard. The display brought up an isometric view of machine 9, and by touching the screen, she quickly zoomed into the geometric view of the power supply.

"See there," she said, "completely modularized. All I have to do is yank the bad module out and put this in. I'll tell you though, we go through a lot of these things."

Pete said, "As a matter of fact, we're supposed to get wired up pretty soon so that we can tell when these power supplies are about to go. Supposedly, we'll automatically get a call on our radio terminals, and the tools will retract automatically. Would have saved us some money this time, about . . . " he hesitated as he accessed another program on the workstation, ". . . fifteen-hundred bucks, since that bad part can't be reworked. Plus my time and your time! Hopefully, purchasing can do something to solve the problem."

"Speaking of time, I've got to transact in so they'll know where I'm at," she said. "I'll be back for the signoff." Pete watched as she entered the data required on her own radio terminal in just a few seconds. In the distance, the material handler depalletized the load and sped off toward the factory floor warehouse area.

Pete thought, "Well, I'd better get ready to set that job up. The second shift setup specialist will be here soon. Got all the hardware he'll need. Let me check to make sure the part program is set and just review the geometry." With lightning speed, he accessed a view of the four-four-oh and saw that all was well for setting up operation 250. As he looked through the detailed processing instructions, a pleased look came over his grizzled face. "Well, what do you know, the engineers followed our advice about those changes we suggested. Sometimes I think those quality circle meetings actually do some good."

A familiar pleasant voice emanated from the workstation, "Time for your shift passdown report for the second-shift factory supervisor, please." With a sigh of relief, Pete moved closer to the microphone and pressed a function key on the workstation. Then he quickly hot-keyed through graphical views of each of the factory cells, looking for flashing condition lights that would indicate a problem. When he was satisfied that he knew the status of the factory, he spoke.

"Hey, good buddy, everything is under control at the moment. Just need to follow up on the BG09 problem. You'll see what I mean. Have a good shift. Oh, one more thing, watch out for those smart carts . . . the programmers have that new one, Joe, making wisecracks now!"

3.13.1 FACTORY FLOOR INFORMATION SYSTEM (FFIS) DEFINITION

Many respected authorities have made the point that manufacturing is "in the ultimate analysis, a series of data processing operations" and that "all of manufacturing involves creating, storing, transmitting, analyzing, and modifying" information.[1] The short story that opened this chapter was written to dramatize these "data processing operations," particularly those concerning "analysis," in order to show how important it really is to have accurate and timely information when trying to resolve factory problems. Although most factory supervisors would like to have an advanced FFIS workstation, any factory supervisor will tell you that the FFIS described in the story is not available today.

Can such an FFIS be achieved? This chapter will present current methods able to improve the way that information is created, stored, transmitted, analyzed, and modified at the factory floor level, and will reference future trends that may lead to the availability of the FFIS described in the story.

Information is created by a variety of sources in the factory, so it is necessary to understand and emphasize the role that each factory person/component plays in the total scheme of information processing in addition to normal production roles. For example, what is the information processing role of the material handler, whether it be a forklift driver or an automated guided vehicle? In either case, the material handler must be able to provide status and event information concerning the load being carried and its destination.

It is important to give the information creators, such as the maintenance person, quality person, engineer, and toolcrib person, the right tools for information processing. Methods of improving information creation techniques, such as the radio terminal described in the story, will be discussed in sections on control instructions and data capture.

The storage, analysis, and modification of information implies the need for FFIS computers and computer software. While there is no exact computer configuration for every FFIS, powerful intelligent workstations for factory users will be an essential part of the future FFIS network (Fig. 3.13.1). Since automated equipment

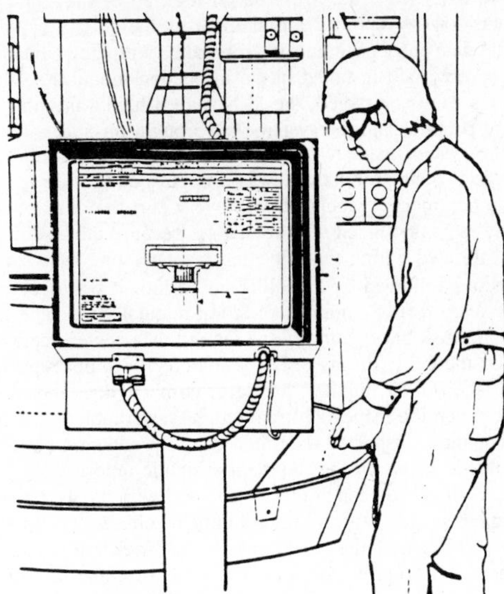

Fig. 3.13.1. The FFIS workstation of the future will provide access to geometric and status information from all computer databases attached to the FFIS network.

will have on-board computers, the tendency will be to distribute computer storage and analysis throughout the factory, instead of taking a centralized approach.

This problem of the distributed nature of FFIS computers and the requirement for the transmittal of information implies the need for a communication network and communications software. The section on communications will provide some alternatives.

Returning to the definition that opened this chapter, communication of information to, from, and within the factory floor is the essence of the FFIS. In addition, the FFIS must be thought of as a value-added service that increases the productivity of the information creators, and increases the aggregate performance of the computer and communications hardware and software that are included in the FFIS network. In addition to user workstations, the FFIS is likely to require a dedicated operations workstation or minicomputer to incorporate these network management and value-added service features.

In the story, Pete, the factory supervisor, was able to resolve a typical problem and avoid overtime through the use of his workstation and the communications network that accessed the information he needed. He was able to use a single workstation for this task, rather than several terminals each attached to a different system. He also avoided making several phone calls or arranging several impromptu meetings in order to obtain the information he needed.

These examples of real productivity gains represent the fundamental purpose and justification for the FFIS of the future. Other examples of value-added functionality will be pointed out in the sections that follow.

3.13.2 FFIS DESIGN ISSUES

FFIS designs are typically deficient because of independent subsystems, inaccessible/incomplete/inconsistent data, and poor peripheral visibility of all of the factory systems that require information.

Analyzing all information sources and roles in the factory makes the architecture of the FFIS more complete. This method should replace the more common approach of identifying the FFIS from a specific factory group's point of view, such as a work-in-process tracking system, a labor data collection system, a machine utilization tracking system, a network of programmable controllers, or a distributed numerical control system, all of which are subsystems of the overall FFIS.

The following example demonstrates how a subsystem designer focused closely on the problem to be solved but did not account for the critical requirements for interfacing to other subsystems. A member of the factory MIS staff at a well-known automotive company received a charter to computerize the quality department's method of recording, retrieving, and reporting factory floor in-process and final inspection results. The forecasted payback of the system was better visibility of repetitive discrepancies and the elimination of paperwork, since all inspection results would be entered through CRT keyboards.

During the course of the design effort, the MIS analyst took all of the steps necessary to implement the system. He planned the formats for the reports with approval by the quality department manager and created the quality subsystem menu and data entry programs in accordance with the manual discrepancy forms currently being used. The CRTs were ordered and installed in the inspection modules scattered throughout the factory area. Finally, when the logistics were completed, the MIS analyst held a briefing for management and a one-hour training session for the inspectors, and the system was brought on-line.

It did not take long for the system inadequacies to surface. The first complaints came from the inspectors—since they were not typists, their productivity was dropping due to the large amount of key entry that was required for recording the comments associated with every part that was scrapped, reworked, or sent to material review. The problem was compounded at shift change, because the host mainframe to which the CRTs were directly networked was saturated during that time and response time was slow.

After a few heated discussions followed by a cooling-off period, it was decided that the inspectors would return to filling out the discrepancy forms' comments section manually, and the forms would be rounded up at the end of each shift for key-to-disk batch entry by the factory data center operator. However, the inspectors would continue to enter data on the CRT for any orders that were 100% discrepant-free. The data key entered by the inspectors included the shop order number, inspector number, date and time of inspection, inspection code, operation, and quantity. When the inspection transaction was complete, the quality database on the host system was updated, and the quantity inspected data field of the production control database was updated so that it could be used for the official count on the shipment schedule report.

Some of the middle managers who had been briefed were upset with the overall design of the quality system. The manufacturing engineering manager saw the opportunity to obtain real-time accurate data to solve the quality problems while they were fresh, rather than waiting for weekly reports. The machine tool planning engineer was upset because there was no provision for entering the machine ID from which the discrepant parts had come. Although the data was filed in the computer instead of a filing cabinet, it was still of no use to him, because he could not cross-reference repetitive discrepancies by machine and machine type.

The inspectors continued to complain because for years they had been required to transact the shop order number, inspector number, inspection type, operation, machine, and quantity at the end of each inspection for time and attendance reporting. Now they were required to do so twice. From their point of view, the transacting stations, which were located within 10 feet of each quality CRT, were easier to use, and automatically recorded the date and time of the transaction.

As the reader may guess, there was no happy or rational end to the story. The decision was made to obtain the time and attendance labor data from the CRT and not require the inspectors to use the transacting stations. The capability to extract information about repetitive quality problems on particular machines or machine types was never realized. There was no direct extraction of in-process inspection results from automated devices.

The problems of overlapping systems, redundant hardware, dissatisfied users, and inadequate provision for peripheral uses of information are found in many environments, but on the factory floor such problems are the rule rather than the exception.

Factory managers must develop a comprehensive plan for the evolution of the overall FFIS, in the same way that corporate managers would develop a plan for the evolution of the CIM system.

First, there must be an FFIS architecture, which takes into consideration both the logical nature of information flow in the organization and the physical nature of data systems and machines.

Second, factory managers must identify factory personnel who will participate in a design task force, rather than simply handing off the design process to the MIS group. During each step of the design process, all of the potential factory-floor users must be given a chance to contribute, so that total information management at the FFIS level is achievable. The task force should have the following responsibilities:

1. Determine the current status of FFIS information usage, including conventional methods of paper forms, meetings, or phone calls. Analyze how the factory workers and components exchange this information.
2. Examine those components that have already been automated to see if information can be "bled off" for use in other systems. The example above showed how one of the basic computerized systems in many shops—the labor transacting system—was capable of providing data in the design of the new quality system.
3. Determine the overall inadequacies of the FFIS and those systems that have been neglected. Redundant systems must be consolidated, and obsolete systems must be scrapped or reworked.
4. Ensure that all advanced manufacturing methods will integrate into the FFIS properly.

Finally, an implementation road map must be developed, in the same way that schedules are developed for implementation of ad hoc islands of automation.

3.13.3 FFIS LOGICAL MODEL

The FFIS architecture requires both a logical and a physical model. The logical model must be developed independently of the physical model to ensure that the limitations of physical systems do not distort or detract from the overall FFIS design objectives.

Figure 3.13.2 presents a logical hierarchical model of the FFIS, and the major applications that it ties together. Site-specific enhancements to this model would include a more detailed diagram of information flow and the database structure. This type of exercise may result in the identification of many specific information-transfer applications for some companies, so it can be seen that completion of the model is not a trivial task.

The logical FFIS model corresponds in many ways to the hierarchical approach being used by many research and manufacturing facilities, including the National Bureau of Standards.[2] The model may include interfaces to nonproduction systems, such as safety and security, also.

As shown by the arrows in the model, the primary purposes of the FFIS are to:

1. Feed forward control instructions to cell, workstation, and equipment level systems. For example, engineering machine control executable programs would be stored in an engineering computer library, but the FFIS would have the responsibility for ensuring transmittal to the machine control.
2. Feedback information to higher-level systems at the corporate, facility, and factory levels. For example, a factory worker event transaction would be transmitted on a batch basis to finance for cost control, and to resource management for realtime dynamic scheduling.

However, the FFIS is not restricted to transmission of information. The intelligent FFIS can integrate the functionality of diverse applications to establish value-added services. For example, if the factory worker event transaction is a "Start Setup," the FFIS can automatically trigger the feed-forward of machine control instructions for the machine indicated in the worker transaction, and automatically update the machine load database.

The following sections will describe the factory systems that the FFIS supports. More detailed information on these systems is also available in other chapters of this handbook. The reader is encouraged to think of ways to improve the logical model and to develop other examples of value-added services.

An overall CIM model would show that the higher-level systems also require internal information management and interface with one another without benefit of the FFIS; however such a model is beyond the scope of this chapter.

3.13.3.1 Resource Management

The primary focus of the FFIS must be to satisfy the requirements of resource management. Resource management relies on an accurate picture of the manpower, machines, and materials (parts and tooling) in the factory.

Resource management consists of both realtime production activity control and longer-term manufacturing resource planning applications. The requirements of the FFIS for handling the information destined for these applications differ. Real-time applications require immediate transmittal of discrete state or event transactions on a one-by-one basis, whereas the daily or weekly requirements of longer-term applications are most likely

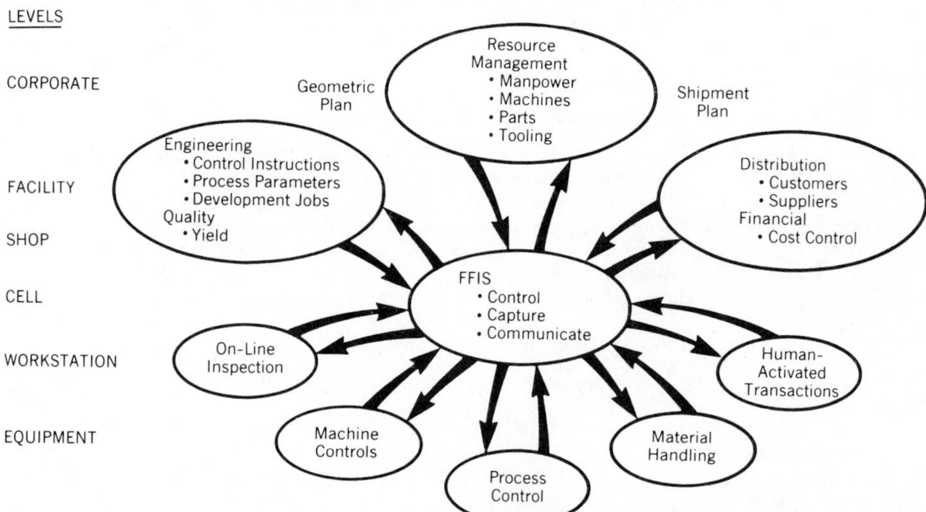

Fig. 3.13.2. Logical model of the FFIS. Instructions are fed forward to factory-level applications, and status information is fed back to higher-level applications. The FFIS performs the communication function, but also adds value to the information by integrating applications.

to be accommodated through batch file transfers. An example of a realtime resource management application is the transmittal of a status message from a failed machine to maintenance. An example of a longer-term application is the weekly summary of the machine's utilization transmitted in graphical form to the facility superintendent or general manager. Both applications will make use of the same original transaction.

The difference between the real-time needs of just-in-time systems and the batch file transfer needs of manufacturing resource planning systems is another example of the need to analyze the timeliness of information transmittal. The FFIS can accommodate both methods.

Understanding the real-time needs of resource management begins with understanding the geometric plan and shipment plan events that require immediate interaction between production support personnel. The story that introduced this chapter showed a typical event that required interaction by the factory supervisor and maintenance person, as well as the material handler. Other production support personnel who require accurate and timely information include manufacturing engineers, machine operators, tooling specialists, toolcrib operators, and inspectors.

Effective control of real-time events almost always requires the interaction of two or more of these production support groups. The ability of production support to solve real-time problems is a function of the communications between groups. In defining the communications problem statement, these factors need to be taken into consideration:

Production support is required as long as production is taking place, which can be up to 24 hours a day, 7 days a week. This requires a continuity of communication across shift boundaries.

Production support needs to be extremely mobile. Although workstations, terminals, and telephones are required at the factory office level to perform the planning function, real-time event communication is usually triggered through the use of radio terminals, which can locate the production support person in the office, on the factory floor, or elsewhere.

Understanding the needs of the batch resource management systems requires an analysis of the reports and database updates required by higher-level systems. Manpower, capacity, and activity planning and measures of performance and productivity require the same event transactions generated for real-time control, but usually in an aggregate or massaged form on a daily or weekly basis.

3.13.3.2 Distribution and Financial Systems

In addition to manufacturing resource planning and production activity control, the factory shipment plan normally includes distribution and financial applications.

The FFIS supports the distribution system by providing the status and location of work orders for customer support personnel. Otherwise, customer support would be required to search for jobs in the factory or make a phone call to a production support person. The FFIS is the source of discrepancy information that can be used by purchasing to determine repetitive problems with suppliers. If just-in-time methods are used to generate material replenishment from suppliers, the FFIS is responsible for determining the most accurate and timely method of transmitting the information to the supplier.

The FFIS has been a traditional source of information for the factory financial systems. In fact, the roots of many FFIS can be traced to the systems established in order to track time and attendance for labor costing. Of course, the opportunity now exists to target costs in much greater detail and track added value of goods through the entire production process.

3.13.3.3 Engineering and Quality Systems

Whereas the above systems are primarily concerned with the information requirements for the shipment plan, engineering and quality systems are primarily concerned with the information requirements and control instruction generation demanded by the geometric plan.

Engineering also closely follows the progress of development jobs through the factory, so that prototype processes can be characterized properly. Usually, engineering will specify parameters that must be charted or logged in order to determine if the geometry is being created as envisioned by the designer.

Quality information requirements include final and in-process inspection results, and traceability of parts.

3.13.3.4 Factory Floor Workstations and Equipment

The logical model also includes the workstations and equipment that would interface to an information system on the factory floor:

Machine controllers

Process control sensing and actuating devices

Material handling equipment

In-process inspection and test equipment

Factory worker activated transacting devices and workstations

These components will be discussed in more detail in the following sections.

3.13.4 FFIS CONTROL

The role of the FFIS in factory-floor control is to ensure that the instructions generated by the higher-level systems are distributed to the factory floor in the most effective manner. This includes instructions from both the shipment and geometric plans, and instructions can be in the form of text, graphic, or machine-control executable programs.

In many of today's job shop factories, each shop order is accompanied by a packet of information containing blueprints, shop-routing signoff sheets, and detailed geometric processing instructions. The volume of paperwork generated in many cases is astounding, and there is always doubt as to the validity of the data contained in the paperwork, since it may have been generated weeks or months prior to the time that the operation actually begins.

In the future, most of this paper-based information will be accessed on a real-time basis through advanced workstations that allow view-only graphic display of geometric information as well as textual processing instructions. These workstations will provide access to most of the information contained in shop order packets, such as dimensional stacking diagrams, setup instructions, tooling specifications, and other technical data.

Control instructions from the shipment plan usually find their way to the factory floor in the form of computer printouts, such as machine loading reports from the scheduling system, and performance reports from the time and attendance system. Typically, computer printouts used in the factory environment are the result of overnight batch runs, so the information conveyed is not timely. Instead of being used purely as information tools, batch reports are often used as worksheets by factory-floor users. In the future, schedule and performance information can be generated at the workstation level for immediate access, either on a CRT display or industrial printer.

3.13.4.1 Distribution of Computer Control Instructions

Just as with any other production tool, factory computer control instructions require generation, storage, and a delivery mechanism. Before discussing the FFIS role as the distribution mechanism, an overview of the various types of computer control instructions used in industry to control process or machining steps will be presented.

> The term *Computer Numerical Control* (CNC) is used in both the metal-working machine tool industry and in the electronics industry to describe a machine that has intelligent controls. The speed and movement of the machine are programmed off-line in high-level languages at a part programming workstation. Afterward, the high-level language instructions are post-processed into machine control instructions for the specific machine the instructions are designed for. Conventionally, the post-processed instructions are punched onto mylar tape or perhaps written to floppy disk. Regardless of the medium, the instructions are treated as any other shop tool, stored in the tool crib until needed, and then kitted and hand-carried to the machine and mounted. Normally, the operator at the machine has the capability of overriding the programmed instructions.
>
> In the semiconductor wafer fab, the term *recipe* is used to describe the high-level programming instruction used to control diffusion furnaces and other equipment. Instead of servo or transducer cutter control, the recipe is used to load/unload wafers from the furnace, and to control the flow and concentration of the diffusion gases. As with the machine tool, the instructions can be programmed off-line or at the machine control itself.
>
> In all types of industries, Programmable Logic Controllers (PLCs) are used to control machine sequences and process logic. The type of higher-level programming is known as *ladder-diagram logic*, which is a result of the PLC's introduction as a replacement for the electrical relay. Because of the PLC's modularity, however, it is now being used for data acquisition, sensing/actuating, and many other functions.

Other special industry machines, such as test equipment for printed circuit boards, have unique high-level languages or perhaps make use of a general-purpose microprocessor and standard programming languages. In addition, process control systems, material handling systems, and inspection systems all have unique control instruction generation requirements.

One may think that industrial robotics is a special case when evaluating computer control instruction requirements, but in fact, stand-alone third generation robots that perform process steps have nearly the same requirements as CNC machines, except that the robot may have more axes.

Although a variety of high-level machine control types exist, they all require a generic solution:

1. The creation of the program.
2. The post-processing and qualification of the program.
3. Storage of the source and post-processed executable program image.
4. Transport of the post-processed image to/from the machine controller.

The FFIS has the responsibility for the control instruction transport, whether the control instructions are on some type of off-line storage medium, or are conveyed to the machine control electronically. From the summary above, it can be seen that the FFIS must be able to accommodate the needs of a variety of controller types including CNC, PLC, robot, and material handling controllers. Consolidating these separate distribution systems is an advantage of the FFIS.

Just as with graphics, it may be necessary for the FFIS to transport machine control instructions back to the programmer when the machine operator overrides the program to correct problems.

In fact, all factory systems require this capability of closed-loop control based on effective transport of control instructions and data capture.

The National Bureau of Standards has implemented such a closed-loop control system based on a hierarchical structure where each controller takes commands from only one higher-level system, but directs several others at a lower level[2]. The NBS approach is to implement controllers as state machines that operate on a predefined cyclical basis. State tables and variables are updated periodically, and are used to generate processing instructions on an as-needed basis. The availability of state data on the machine controller also provides a mechanism for data capture, as will be discussed in Section 3.13.5.

This type of hierarchical structure will become more important as the complexity of manufacturing systems increases, and machine tools and robots become totally integrated with material handling systems in flexible manufacturing cells. In the future, there will still be stand-alone CNCs and process control systems to which control instructions can be transmitted without affecting other automated systems, but more and more engineers and programmers will confront the problem of sending executable programs to CNCs, PLCs, robots, and material handling controllers that are completely integrated and expect information from each other for the performance of a task. The FFIS will play a major role in this complex integration of machine controls, serving as the integrator.

3.13.5 FFIS DATA CAPTURE

Proper management of resources demands that an *accurate* picture of the entire factory be made available to decision makers on a *timely* basis. A decision on how to solve today's problems with yesterday's data or data from the batch reports that were printed overnight is basically a forecasting environment. While this is necessary in some situations, it is certainly less than desirable when today's data are available for today's problems.

This emphasis on accurate and timely data is what leads to the concept of data *capture*, rather than data *collection*. The quality system example demonstrated the difference between capture and collection; the data had lost value to manufacturing when it was rounded up on forms and key-entered in batch mode. In a dynamic manufacturing environment the data capture system must be responsive and versatile. Data on events must be gathered as they happen and distributed to all databases that need it.

3.13.5.1 Overview of Data Capture Methods

In a study conducted for the U.S. Air Force Integrated Computer Aided Manufacturing (ICAM) program,[3] it was shown that from a human factors and accuracy standpoint (Fig. 3.13.3), it is best to capture data automatically, next best by voice, then through human-activated methods.

3.13.5.2 Automated Data Capture Methods

One of the primary methods of capturing data automatically is from existing machine, facility, or process controllers. Controllers (Section 13.4) may or may not be designed with data capture in mind.

For example, the CNC or diffusion furnace programmer does not usually have to worry about the sensing and actuating that the high-level controller requires in order to monitor positions or initiate actions before processing additional instructions. The low-level sensing is carried out by transducers and switches and the actuating by solenoids, motor starters, control valves, and electromechanical or pneumatic servos. Thus, a closed-loop sensor/actuator system is driven by the machine controller transparently to the programmer.

However, sometimes this high-level programmability has its drawbacks. Consider the CNC controller. Despite its sophistication, the CNC controller normally cannot send signals to indicate that it has completed a part! This task can be done in some instances by sensing whether the program end-of-cycle circuit is high or low; as appropriate, a signal can be sent to the master scheduling and manufacturing control databases to update the parts-completed quantity. Another example of a user-installed data capture device is a sensor that indicates that a key component of the machine has broken down.

In many cases, controllers are required solely for the purpose of data acquisition, or sensing and actuating is done independently of a special controller. The output of the sensor is normally a small amount of digital or analog data, and the actuators listed above can control machines, processes, facilities, and perhaps trigger high-level events.

A Process Control system or a Programmable Logic Controller (PLC) are devices usually used when it is necessary to interface with sensing devices and also communicate with higher-level systems. The PLC can be used to maintain state tables at the equipment level, so that critical process parameters can be monitored to fractions of a second, process values can be accumulated, and event status fed back when alarms are required or preset limits exceeded.

A partial list of sensing devices includes:

Flow rate monitors.

Temperature monitors.

Environmental alarms.

Limit switches.

Photoelectric switches.

Inductive proximity switches.

Power transducers.

As previously indicated, machine components, BCD indicators, and operating panels also provide a wealth of usable sensory data, if they can be tapped. The process engineer is normally responsible for determining machine or process sensor/actuator needs, and whether a programmable controller or other intelligent device is needed for sequencing or logic control. In a like manner, the facilities engineer would determine environmental sensor/actuator needs.

Automated Material Handling

The FFIS is responsible for the transport of high-level control instructions and low-level sensing/actuating to/from automated material handling components.

Major material handling components include wired and wireless Automatic Guided Vehicle Systems (AGVS), in-floor towlines, carousels, and overhead conveyor systems, which are usually scheduled and controlled by special-purpose minicomputers. These systems connect raw-material warehouses, work-in-process warehouses, and tool cribs with workstations. AGVS vehicles or "Smart Carts" as they are sometimes called, pick up materials at the warehouse, thread their way through the factory, and stop at designated points to pick up and drop off material. These points are determined by communication with the controller.

More detailed information about material handling equipment may be found in Chapter 4.6.

At the workstation level, transfer equipment is used to unload the guided vehicle or overhead conveyor. In some instances, gravity-fed or servo-driven conveyors are then used to transfer materials between workstations of the FMS cell. These conveyors depend on the ability to detect the presence of material and control its movement through the use of sensor/actuator devices such as photoelectric and limit switches.

Automated part identification methods required by the FFIS include:

Attaching to the part or the container a bar code that can be decoded by an off-surface laser bar-code scanner.

Attaching an RF or microwave generator that can be decoded/encoded by a transceiver.

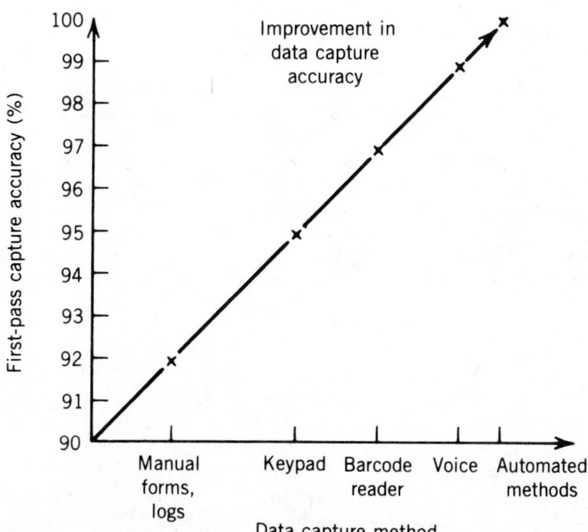

Fig. 3.13.3. First-pass data capture accuracy as a function of possible data capture methods. Human factors data capture accuracy improves quantifiably; actual accuracy percentages depend on site-specific equipment and users. Human factors ease of use also improves as the method becomes more automated.

Laser-scribing the part and detecting the scribe with an image processor.

Recognizing objects via vision systems.

Using scales that weigh parts and use a factor to determine how many are present.

Since the identifying code is often removed during part processing, a means of re-identifying the part needs to be incorporated in the material handling system. This will require the downloading of data to the inscribing device from the FFIS.

In-Process Gauging and Test

Just as with machine control, a variety of equipment is used in different industries for in-process gauging and test. Some of it is microprocessor controlled and some is sensor-actuator oriented. These automated methods are designed to replace time-consuming and error-prone manual methods of iterative off-machine gauging and test.

For example, in metal fabrication industries, Coordinate Measurement Machines (CMM) are high-level program controlled workstations used for high-precision inspection of parts with a number of tolerances to check. Laser micrometers, diode arrays, and tool probes are used right at machine tools for dimensional measurement and mathematical averaging, and are essential when noncontact measurements are required. The digital or analog output from this equipment can be used for machine control.

In the semiconductor wafer fab, nanoline microscope controllers perform a similar type of critical dimension checking and value averaging.

In many industries, image processing is replacing optical comparison (backlighted blueprint measurement) as a means of automated inspection. Special cameras are used to obtain images of a part in very high detail (for example, 1 K × 1 K picture elements or pixels), the image is processed by a dedicated computer compared to files of stored images (perhaps communicated from a CAD system), and data provided to indicate the pass/fail status of the part.

In-process testing techniques commonly used throughout industry are designed to provide statistical time-scaled analysis of in-process parameters to determine whether the product passes or fails, and in some instances (for example, ball bearings or resistors) to assign a product class value based on the quality or properties of the product. Often, these statistical analyses are recorded manually or in an analog manner on strip charts. This makes it difficult to maintain accurate historical records. Therefore, because they are able to store data, dedicated microcomputers are used for many in-process testing functions. The quality engineer is responsible for the evaluation and implementation of these systems.

3.13.5.3 Voice Systems

As Fig. 3.13.3 shows, data capture by voice is the next best method following automated methods. For both control and data capture, voice is the most natural way of transporting information for factory floor personnel. It is difficult to imagine operating a manufacturing plant, especially a large one, without telephones and radio pagers. The FFIS should be implemented in such a way as to reduce the number of needless telephone calls and meetings by providing status information in a more effective manner; however, there are times when a phone call or meeting is absolutely necessary.

The telephone itself is underutilized as a data capture tool. The touch-tone keypad provides a mechanism for entering small amounts of data in conjunction with a voice prompting program. For many applications, the standalone telephone may be useful as a low-cost input device.

The radio pager is another voice-oriented device that is gaining in sophistication. Even with just the basic feature of voice broadcast, the radio pager is an indispensable information tool in the factory. There are advantages to having unidirectional transmission. For example, the maintenance person working on a machine needs to hear a message about the status of a spare part, but doesn't want to have to drop his tools in order to answer the call and then spend five minutes chatting. However, the next decade will be the age of the radio frequency terminal, which can queue messages and provide readouts on demand, in addition to being equipped with keyboards or voice communications with total portability.

As the sophistication of microprocessor-based voice systems grows and the limitations of CPU time and memory to support applications of voice systems become less severe, there will be more integration of voice and data to provide new services. Already, telephones are being integrated with computer workstations to simplify data communications, and radio pagers have some capability for transmission/display of data. In the future, machines will have voice synthesizers as standard equipment, and voice-store-and-forward will be a natural way of appending a message to a graphic display or a shift passdown report.

The limiting factors of this voice development will be technical capability and, more important, the environment. The factory environment has a tremendous amount of background noise that inhibits voice information processing. With current voice recognition systems, the only application that has been truly successful is inspection, because the environment is normally quiet, and inspectors need hands-and-eyes-free data input capability.

3.13.5.4 Human-Activated Data Capture

In many respects, the human being is the most flexible manufacturing component, though not the fastest or most quality reliable. Even in "ghost shift" automated factories, humans are responsible for setting up machines, kitting tooling, a variety of inspections that cannot be automated, and maintaining the machinery. People are often put into service as a backup to broken-down equipment.

Part of the job function of the factory worker is to record event status data. This data entry function has traditionally been required for time and attendance, work-in-process inventory moves, and engineering-specified logs. In many cases, the automation of human transactions has been neglected, and workers enter data manually (forms later rounded up for batch keypunch) or through CRTs. Figure 3.13.3 shows the decrease in accuracy of information when these methods are used.

Often, critical data that workers could enter, such as tool location and maintenance status, has not been tracked. Tools are allowed to wander like gypsies through the factory, causing hectic searches when critical parts must be run. To maintain the quality of the transacted data, and maintain the factory floor user's acceptance, the following criteria should be taken into consideration:

No lines at data input stations.

Short walk to input stations.

Table 3.13.1. Partial List of Event Codes

Group/Code	Description
Setup	
10	Standard setup
17	Nonstandard setup—Material problem
18	Nonstandard setup—Tooling problem
19	Nonstandard setup—Machine tolerance problem
Production	
20	Standard production—Cutting
21	Standard production—Not cutting
22	Standard production—Tool change
25	Nonstandard production—Rework
26	Nonstandard production—Material problem
27	Nonstandard production—Tooling problem
28	Nonstandard production—Machine tolerance problem
Material handling	
30	Start material move
31	Stop material move
32	In transit
33	Job on hold
34	Job quantity audit
35	Tool issue
36	Tool broken or out-of-spec
37	Searching for lost tools
38	Kitting tooling
Inspection	
40	Standard inspection—Inspect 100%
41	Standard inspection—Inspect as specified
42	Standard inspection—Courtesy inspection
43	Re-inspection due to work
44	Re-inspection due to handling damage
45	Nonstandard inspection—Faulty gauge
46	On-line inspection—No discrepancy
47	On-line inspection—Discrepancy detected
Indirect functions	
51	Quality circles
52	Out sick
53	Tardy
54	Leave early
55	On-the-job training
56	Personal time
Dispatches—Warnings	
60	Maintenance required—Electrical
61	Maintenance required—Electronic
62	Maintenance required—Mechanical
63	Mfg. engineer required
64	Tool runner required
65	Process routing circumvention
66	Tool failure
67	Machine failure—Breakdown
Work Stoppage—Idle	
70	Waiting for parts
71	Waiting for maintenance
72	Waiting for manufacturing engineer
73	Waiting for tooling
74	Waiting for inspection
75	Waiting for production control
76	Waiting for material handling
77	Waiting for machine operator
78	Waiting for machine
Maintenance Repair	
80	In-progress
81	Suspended—Need spare parts
82	Suspended—Another skill required
83	Suspended—End-of-shift
84	Suspended—Need more information
85	Complete—Need supervisor signoff
86	Cancelled
Manufacturing Engineering	
90	Helping set up new process
91	Monitoring critical process
92	Testing control program
93	Evaluating producibility problem
94	Performing time study—Standards evaluation

Consolidate data requirements.
Simplify transacting procedures.
Maintain system and device uptime.
Provide alternative input stations.
Short response time.
Users are not typists; do not require extensive keyboard entry.
Provide adequate training.

The type of input device to use is a function of the data being entered for the transaction types, the portability required, the environmental factors, and the frequency of input. The most desirable device to use after automated methods and voice (as shown in Figure 3.13.3), is a hand-held bar code or magnetics reader, which decodes pre-encoded media such as worker badges, work order traveler cards, machine identifiers, and tool IDs. Bar-code and magnetics media can be used to construct "soft" template-like menu sequences and keypads, thus eliminating an expensive piece of hardware. Both bar-code and magnetics have relative applications and tradeoffs. The choice of bar-code depends on the need for off-surface scanning, and the choice of magnetics medium depends on the severity of "environmental conditions," and "density level."[4] In either case, the user should be provided with a pistol-grip device, which is suitable for both left- and right-handed people.

Hand-held readers are more accurate than keyboard input. The following equation shows how the Accuracy (A) of a series of data input operations can be improved by reducing the number of data entry strokes (n) to one by using a hand-held reader.

$$A = (S)^n \qquad \text{where } S = \text{first-pass success rate}$$

Here is an example for a nine-character job ID:

With a keypad, $A = (.95)^9 = 63.0\%$.

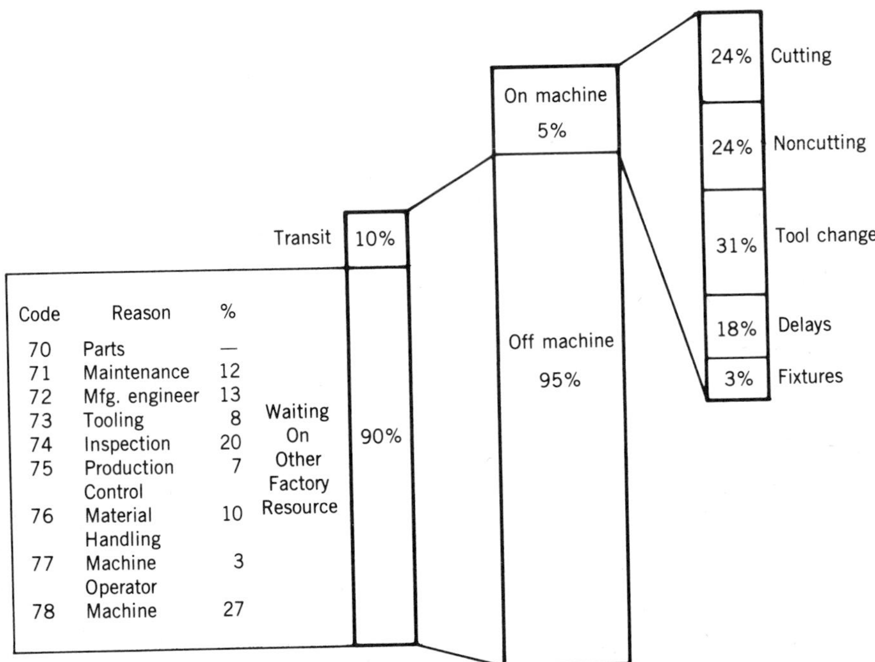

Fig. 3.13.4. Production time breakdown for a typical part in the factory, with an emphasis on the reasons the part waits for processing. As an example, if a part takes 1000 hours to process through the factory, it will be on the machine only 50 hours (5% of 1000) and cutting for only 12 hours (24% of 50) of that time. By comparison, the part will be waiting for processing 855 hours (90% of 1000), and waiting for inspection of material review for 171 hours (20% of 855) of that time. Percentages of waiting time are for demonstration purposes only; actual numbers will depend on site-specific usage.

With a hand-held reader, $A = (.95)^1 = 95\%$.

This example assumes that the user has a first-pass success rate of 95% for both the keypad and hand-held reader. The actual first-pass success will vary from user to user, of course.

Just as with voice processing, environmental problems must be overcome in order to implement bar-code or magnetic systems properly. In metalworking factories, the media must be protected from the hostile environment by lamination. In the semiconductor wafer fab, it is the environment that must be protected by lamination from the "dirty" bar-code or magnetic media.

3.13.5.5 Event Transactions and Codes

Whether event status data is captured automatically or from a human-activated device, a classification of transactions and codes is necessary. The state machines described in the previous section provide a mechanism for generating transactions; however, discrete event transactions are also required. For example, one would not normally track a machine operator's status by physiological monitoring; instead, transactions are made when a significant event occurs.

As shown in the quality system example, the data that are typically captured in an event transaction include the machine/location, employee, shop order, event code, operation, quantity, date, and time.

Some of the other events that require transacting include:

Arrival time (of person or machine) at work site.

Machine setup start.

Machine setup complete.

Production start.

Scheduled work interruption.

Unscheduled work interruption.

Production finish.

Start of material movement.

Completion of material movement.

Note that most events require a *start* transaction, as well as a *stop* transaction. The start transaction is absolutely essential for proper control and accurate depiction of resource utilization.

The key data attribute for the FFIS is the event code. Table 3.13.1 presents a typical list of event codes used in an FFIS.

Event codes are the key to an FFIS because they provide the means to develop algorithms for value-added services. For example, when a code 70 (work stoppage) is detected by the FFIS via a state machine output or discrete event transaction, control actions can be taken immediately at the factory floor level, and higher-level system databases can be updated to reflect the current status of the machine, people, and parts.

A properly designed event-code table facilitates the transition from human-activated data capture to automated data capture, since the control functions that are triggered by the event code are the same. The event-code table also leads to an improved awareness of the detailed reasons for production anomalies. The often-published 5% on-machine–95% off-machine performance statistics shown in Fig. 3.13.4 can actually be extrapolated for any manufacturing facility by designing an FFIS keyed on events.

3.13.6 REALIZATION OF THE FFIS THROUGH COMMUNICATIONS

Up to this point, the logical model, control instruction methods, data capture methods, and event-code table for the FFIS have been identified. It is apparent that there is a variety of physical hardware and software at each site-specific implementation of the FFIS that must be tied together to make the aggregate of all of the distributed intelligent systems more productive. This variety of different vendor systems, each with its own operating systems, database and transaction managers, and communications hardware and software is just as imposing to the FFIS architects as the multitude of applications and databases, each with its own data format, referenced at the beginning of Section 3.13.3.

Once again, one may ask, is the FFIS feasible?

The answer is yes, if current efforts to standardize factory communications systems succeed. A well-designed and open communications network with network management and intelligent information processing application nodes is essential to the FFIS.

At the cell level, hierarchical communication between controllers is appropriate, but as Fig. 3.13.5 shows, an open-system connection is more desirable for the factory-wide FFIS. This statement disputes the pure hierarchical theory proposed by the NBS; however, note once again that there is a differentiation between the logical information processing model, for which a hierarchical approach is appropriate, and the physical model, for which a hierarchical approach has limitations. These include dependence on single-vendor network solutions,

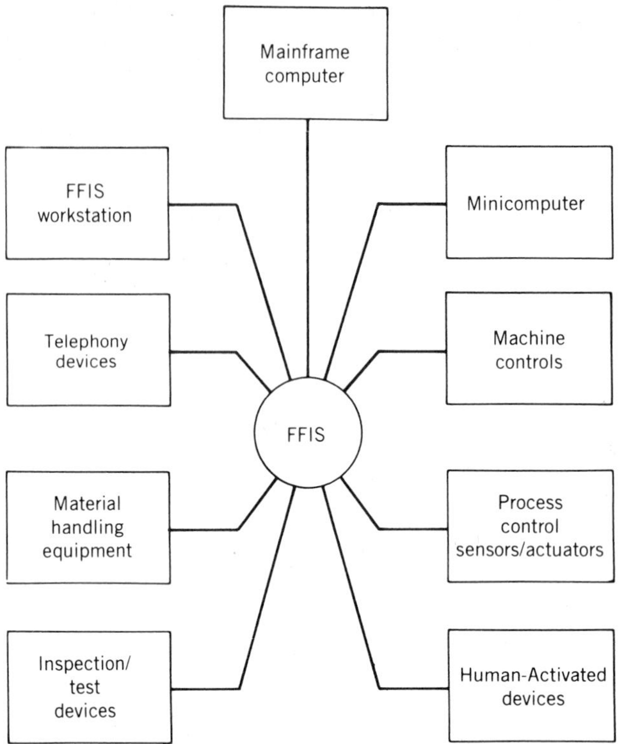

Fig. 3.13.5. The FFIS physical network model. Open compatibility between vendor systems optimizes communications between differing systems. The FFIS plays the role of network manager and protocol conversion.

higher cost of communication hardware and software, and inflexible constraints for integration of future applications.

The FFIS network requires flexibility so that it can handle the needs of all information transport, whether via data, voice, or video. The network must be able to handle reliably the different types of data transport, either the high-volume megabit requirements of graphics and image processing, or the low-volume high-speed requirement of sensing/actuating. Environmental factors need to be considered when choosing factory-hardened communications processors and cabling, just as with data capture devices.

Intelligent processing nodes on the network should be able to transport information to any and all users who need it, instead of to a single host in a hierarchical structure. These intelligent FFIS workstations must be able to access databases from all systems, as described in the story that opened this chapter. Thus, when an unscheduled event occurs, the information about all resources is literally at the fingertips of the production support personnel who need to respond. The FFIS workstation also leads to the possibility of implementing value-added services based on a well-designed state and event code table utilized by all of the distributed systems.

This chapter will not digress into the fundamentals of communications, but a partial list of recommended periodicals is included in the bibliography. FFIS communications is probably the single most important concept for the FFIS designer to understand. It is a dynamic subject and the reader should make an effort to follow the progress of communication standards implementation. In particular, the MAP protocol technology will provide the best avenue for resolving the problems of differing factory hardware and software. In the future, there will also be better standards for operating systems, and neutral data formats will provide a means of communicating between different vendor systems.

The task of providing connectivity and value-added services will not be trivial, of course. The FFIS network may be feasible but not cost-effective for all factories. Each factory will require several man-years of effort to establish the FFIS links, and each application that is added to the network on an ongoing basis may require man-months of effort. Vendors of factory-wide FFIS networks will invest perhaps hundreds of man-years in developing user tools that make the task of implementing the FFIS less expensive and time-consuming, and thus more attractive to their customers.

The FFIS design requires a communication expert to perform the following functions:

Conduct a traffic analysis of all FFIS information transport in order to construct a site-specific table like that shown in Table 3.13.2.

Determine peak-loading periods (Fig. 3.13.6).

Determine the appropriate network requirements and select vendors for communication products, such as protocol converters, Local Area Networks (LANs), LAN gateways, cabling, telephone system interfaces, and data switches.

3.13.7 SUMMARY

Ultimately, the FFIS must be designed with the factory production support personnel in mind. There seems to be polarization of thinking; people believe that there are only two types of information systems, very basic time and attendance devices, or miraculous flexible manufacturing systems. In fact, there is a lot of room for improvement in all factory information systems, and some of the improvements can be made without a major capital investment. For example, forming quality circles to analyze information requirements and upgrade event-code tables only requires a little patience, a lot of persistence, and some good training.

Even in the most highly automated plants, the factory worker plays a major role. For example, in GM's Orion assembly plant, there are over 1100 computer-controlled devices. However, there are also over 3200 hourly workers.[5] Assuming that each worker makes at least 10 event transactions/day, 32 K transactions are made in a single day. It makes sense to try to optimize the methods and results of all that information transport.

By providing the type of FFIS in the scenario that started this chapter, cost-justifiable improvements can be found. According to Dick Green, Editor-in-Chief, *Tooling and Production* magazine (quoting from a subscription newsletter offered by Harbor Research Corp.), "indirect labor, used for moving material and information around our factories, is, in effect, our current data communications network."[6]

If this traditional approach is replaced by the advanced FFIS, much of this indirect labor can be reduced or redirected to more productive pursuits. Accurate and timely information allows the evaluation of many alternatives to production support problems, and leads to fast and correct decisions, as shown in Fig. 3.13.7. The fallout from this improved decision-making process leads to prospective gains in direct labor savings, since there will be less machine operator time charged to waiting, fewer job teardowns and setups because of poor scheduling, and less overtime required to run critical parts. Additional benefits will be found in quality improvements, throughput, inventory reduction, and capital utilization.

Overall, the enterprise effectivity measurement will improve, and production supervisors, like Pete, will continue to have a role in the factory.

Table 3.13.2. Sample of Minimum Data Rates Required for Typical Factory Transmission Types[a]

Transmission Type	Length of Transmission (bits) (L)	Speed Required (seconds) (S)	Minimum Data Rate (bits/s) (MDR)
Sensor scan (32-digit field)	32	1	32
100 Simultaneous sensor scans	3200	0.1	32,000
Time and attendance transaction (64 char record)	512	4	128
Job complete transaction (256 char record)	2048	4	512
50 simultaneous transactions	102,400	4	25,600
Terminal screen		1	19,200
(80 × 30 char)	19,200	8	2400
Executable machine control file download (50 K char file = 410 ft. tape)	400,000	10	40,000
Batch scheduling file download		10	345,600
(432 K char. floppy diskette)	3,456,000	60	57,600
Full color graphics terminal screen (1000 × 1000 pixels)	8,000,000	2	4,000,000

[a] Using the equation MDR = L/S the traffic analyzer can determine what type of communications systems to use for each subsystem of the FFIS, or given the type of communications systems, approximate how long the transmission will take. For example, time and attendance transactions are short in length and a delay in the communications line of 4 seconds is acceptable, compared to the tenth-of-a-second maximum delay acceptable for process control sensor scans. As one progresses through more extensive transmission lengths and encounters simultaneous transmissions from multiple users and demand for faster response times, the minimum required rate or bandwidth increases. The aggregate of all transmission types ultimately determines how much total traffic the FFIS communications system must support.

Note that these simplified examples only represent transmission speeds and raw data lengths. End-to-end delays will increase due to error-detection algorithms, computer processing time, and input-output requirements. Data lengths will increase due to header information and standard packet sizes.

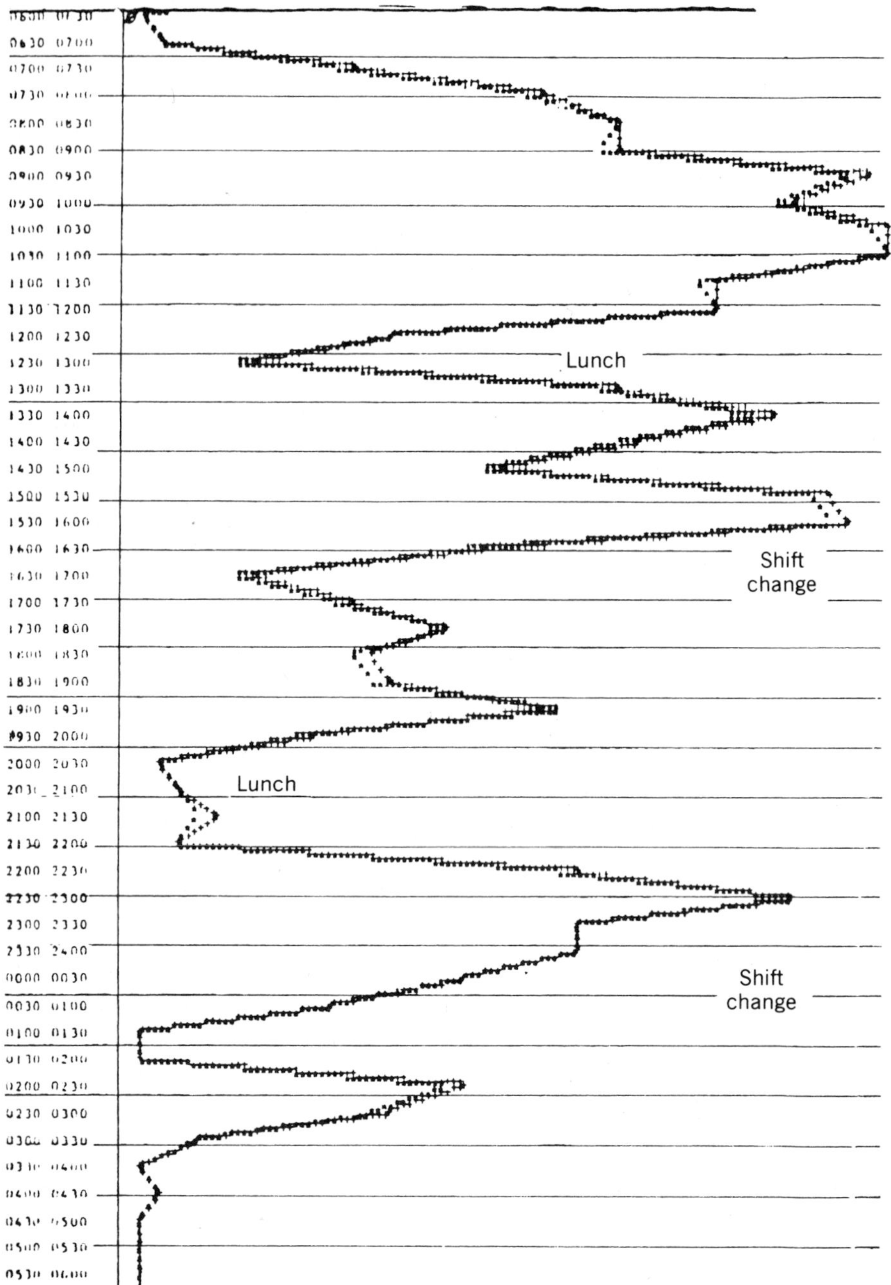

Fig. 3.13.6. Typical factory network load curve for a 24-hour period (courtesy of ITT Standard Electric Lorenz (SEL), Gunzenhausen, West Germany).

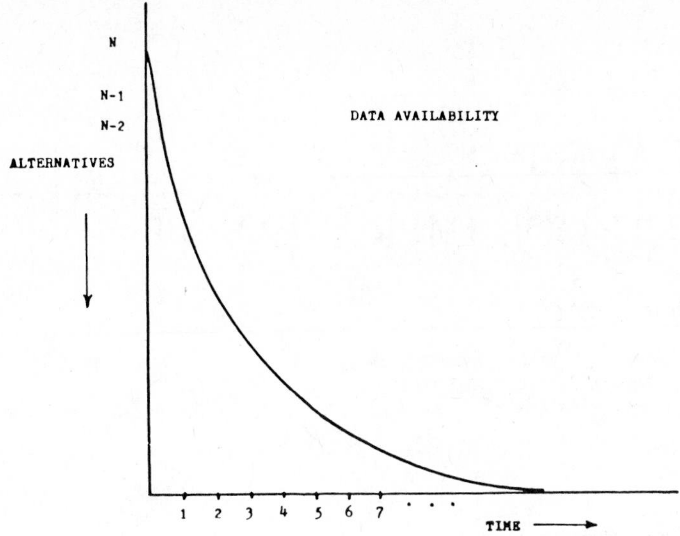

Fig. 3.13.7. Production support solution alternatives as a function of time. In the dynamic environment of the shop floor, resource status changes from minute to minute, and factory supervision is responsible for making the correct decision for allocating resources, perhaps with the help of a computer tool. If all data is immediately available, a number of alternatives (*N*) are possible, but as time goes by, the "window" of opportunity for some alternatives may disappear. For example, when a machine breaks down, there may be five possible alternatives to rescheduling critical parts that are running on the machine—reassigning the operator, etc. As time goes by, machines and operators may be preempted by other scheduled parts, so that by the time the decision is made, there may be only one nonoptimal solution left. This solution may be a teardown of an existing setup, which will result in a resetup and operator overtime at a later date.

REFERENCES

1. Don Hegland, "Integrated Manufacturing and the Factory of the Future," *Production Engineering*, **7** (February 1984).
2. Charles McLean, Mary Mitchell, Edward Barkmeyer, "A Computer Architecture for Small-batch Manufacturing," *IEEE Spectrum*, **59** (May 1983).
3. *Shop Floor Information System*, ICAM Report #AFWAL-TR-82-4147, June, 1982. (Can be obtained by writing to Librarian, AFWAL/MLTC, ICAM CM Library, Wright-Patterson Air Force Base, Ohio, 45433).
4. Seetharama L. Narasimhan and Russell C. Koza, "Automatic Identification Systems Serve as Integrators in the Factory of the Future," *Industrial Engineering*, **58** (February 1985).
5. Al Wrigley, "New GM Car Body Plant Boasts 1,000 Computers, 157 Robots and 22 AGVs," *American Metal Market: Metalworking News* (April 9, 1984).
6. Dick Green, "Indirect Labor—the Bane of Automation," *Tooling and Production*, **5** (June 1984).

BIBLIOGRAPHY

Bedworth, David D., Berry, Gayle L., Jung, David W., and Young, Hewitt H., "Design of a Semiautomated System for Capturing and Processing Shop Floor Information," *Journal of Manufacturing Systems*, **2**(1) 39–52 (1983).

Bowerman, Robert, "Choosing a Local Area Network," *Interface Age*, **56** (July 1983).

Harrington, Joseph Jr., "Managing Manufacturing—Yesterday's Systems, Tools Are Obsolete," *Production*, **84** (August 1982).

Jones, Pam, "Drawing the MAP," *Manufacturing Systems*, **12** (March 1985).

Rauch-Hinden, Wendy, "Special Report. Factory Networking," *Systems & Software*, **28** (December, 1985).

Sadowski, Randall, P., "History of Computer Use in Manufacturing Shows Major Need Now is for Integration," *Industrial Engineering*, **34** (March 1984).

Singer, Larry M., "Designing Factory Data Collection Systems," *Manufacturing Systems*, **31** (February, 1986).

Tanenbaum, Andrew S., *Computer Networks*, Englewood Cliffs, NJ: Prentice Hall, 1981.

"Networking for Control" (10-Part Series—various authors), *Chilton's I&CS: The Industrial and Process Control Magazine*, **58** (2–13), 1985.

CHAPTER 3.14

METHODS IMPROVEMENT

KENNETH KNOTT

The Pennsylvania State University
University Park, Pennsylvania

3.14.1 INTRODUCTION

Methods engineering, one of the cornerstone industrial engineering techniques, is no more than a structured approach to problem solving. An examination of some of the approaches recommended for systems design shows little difference from the concepts used for many years under the name of "Methods Engineering."[1-5] In spite of the minor variations in detailed terminology there seems to be little difference in the overall philosophy of the leading writers on methods improvement. Furthermore, there seems to have been little change in the overall philosophy since the early 1950s. Perhaps this is due to the success of this well-tried approach.

3.14.1.1 The Problem Solving Approach

Two fundamental points relative to the methods engineering approach to problem solving are so important that they must form the basis of any discussion on the subject. What is more, a regular reaffirmation of their importance throughout the discussion is not out of place.

First, it is essential to recognize that the fundamental objective of any methods improvement project is to *install* the improved method. Without the installation, any recommendation is merely academic.

Second, it is necessary to recognize that although the methods engineering approach is presented as a formalized procedure, it must be regarded as being dynamic.

Both points will be raised again in subsequent discussions.

The Methods Engineering Approach

The methods engineering approach, summarized in Fig. 3.14.1, consists of three broad phases:

1. *Select* the appropriate problem.
2. *Understand* the problem.
3. *Innovate* a solution to the problem.

Six factors associated with *install* have an immense influence on the success of any project.

Many writers summarize the procedure in terms of the six steps listed at the top left-hand corner of Fig. 3.14.1. The role of these six steps in the overall procedure is noted in the figure.

3.14.1.2 Install—The Fundamental Problem

The problem solver becomes an integral part of the problem immediately upon arriving in the problem environment and remains so even after a solution is implemented. Five factors that have significant effect upon the success of any change in method have been identified in Fig. 3.14.1.

While there exists a sensitive interaction between these five factors, it is convenient to consider them separately, as below.

The Scale of the Project

Changes in methods of operation always have effects outside the immediate area of application. Two examples will illustrate such effects.

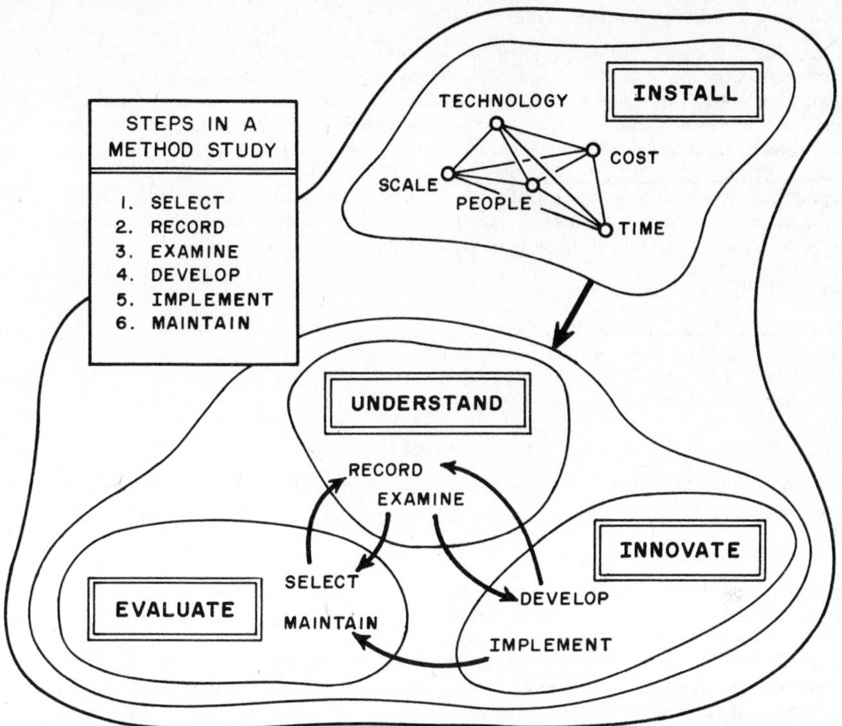

Fig. 3.14.1. A systems map of the methods engineering approach. *Source*: Ref. 6.

Consider the case where the object of a change in method is to speed up the rate of work flow through a workstation. Whether the workstation is in the manufacturing, clerical or service areas, unless the appropriate backup service is provided to assure sufficient input to the work area, potential gains can be significantly reduced or even lost. Furthermore, these effects can reflect on both the previous and subsequent operations. These potential effects must be anticipated, their results estimated, and appropriate accommodations made.

Where the economic considerations of a method change are a prime consideration, there may be some doubt about whether a particular solution should be used. However, the objections are frequently overcome by taking a broader view and identifying points at which similar solutions can be used with advantage.

The Effect of Time on the Success of a Methods Improvement Project

Possibly due to the association that traditionally exists between methods engineering and work measurement, it is common to assume that the times referred to here are the times necessary to perform the tasks being considered for an improved method. While these are important, other aspects of time must also be considered. The total project time, for example, can affect both the economics and acceptability of a project.

Take as an example the case where, for some reason, a change in the method of operating a facility has been delayed. If this delay were to be viewed as unreasonable by those directly affected by the change, the eventual introduction of the new method could meet sufficient resistance to ensure its failure.

Therefore, the influence of all aspects of time and timing must be carefully evaluated at every stage of a methods improvement project.

Costs and Methods Improvement

Cost reduction is probably the most common reason for improving operating methods. However, it should not be considered the only one, since improving the quality of working life has an important role as a stimulus to improvement of methods.

As with the time factor, a broad view of costs is necessary in the methods improvement project. The total economics must be considered, not only the cost per piece. The "one best method" is an elusive concept. Therefore, when the economics of the method are used to evaluate the success of a methods improvement, as soon as the losses resulting from the delay in introducing a new method together with the costs of any further improvements exceed the potential savings, the search for improvement should be abandoned and an installation made.

Methods as a Factor in Methods Improvement

To illustrate the influence of methods on methods improvement, consider the case of a small company that in 1977 investigated the use of a small computer for controlling inventory, work scheduling, work measurement, accounting, and payroll. In spite of the considerable benefits that could have resulted from such a change in operation, the investment was prohibitive so far as the company's limited resources were concerned. By 1981, the developments in computer technology and the associated reduction in the cost of equipment resulted in the original concept now being feasible, with only minor modifications in the methods used.

Rapid changes in technology mean that those involved in methods improvement are faced with technical obsolescence at an ever-increasing rate. Solutions that were apparently impossible a short time previously, become essential almost overnight, because of some technical development.

People—The Unpredictable Factor

The factor "people" has been shown at the center of the installation factors since it is undoubtedly the most sensitive—and the most unpredictable. The methods engineer must be continually alert to the feelings of the people involved in the problem area. Every person in the organization who does not support a new method will, in some way, reduce the potential of the improvement.

3.14.1.3 Evaluation in Methods Improvement Projects

Evaluation is an integral part of any methods improvement project and can be viewed in any one of three different ways, as described below.

Evaluation for Selecting a Project

In any organization there is invariably a greater demand for problem solving resources than availability. Since the productivity of these resources must be optimized, a careful evaluation of the potential benefits of each competing project must be made during selection.

The five installation factors referred to earlier provide a guideline for this type of analysis, since the economic advantages may be nullified by the timing of the project, the scale, the imminence of some technological development, or the attitudes of the people involved. It is not unusual to encounter a case where the choice of one project over another generates strong criticism. Such criticisms often arise because someone feels that his or her "pet" project has been passed over. While it may be possible to show, by using one measure, say the economics of the project, that the wrong choice was made, a broad view of the situation must be taken and recognition made of the fact that business considerations might preclude making public every factor involved.

Evaluating the Limit of Innovation

Methods engineers are notorious for looking for the "best method." Indeed, this well-meant enthusiasm can result in a delay in the introduction of an acceptable method and, consequently, a potential reduction in the effectiveness of the project. There exists a "limit of innovation," where any further improvement has a marginal or even negative value. As the methods engineer proceeds through the project, it is necessary to carry out evaluations to determine if this limit of innovation has been reached.

The evaluation of the limit of innovation should attempt not only to determine when the limit has been reached, but also to identify the reasons.

Evaluation to Maintain the Best Method

With the passage of time, operating methods will inevitably change. This natural "drift" in the methods can have either a good or bad effect on productivity. The operating methods must therefore be subjected to continual scrutiny. *Maintain* is recognized as one of the steps in the methods engineering approach, and many writers indicate that it should be achieved by reviewing the methods at regular intervals. It will be found that the daily production returns can provide an excellent vehicle to carry out those reviews and to trigger any necessary detailed review of the methods. This will, however, almost certainly require a reliable work measurement base.

Change in work methods will result in a change in the work load of an operator. If such a change is not detected or recorded, the seed of an industrial dispute has been sown. A detailed description of the method is an essential part of setting a new method.

3.14.1.4 Understanding the Problem

Understanding a problem for the purpose of improving a method requires a careful study of what is being done. Developing this understanding takes place in two steps, as follows.

Recording the Present Method

First, it is necessary to record the present method being used to perform the task. It is important to note that this is what is actually being done, not what someone thinks is being done or should be done. Record the *method that is actually being used.*

In the process of understanding the present method, those people who are likely to be affected by any change should be involved as early as possible in the improvement process. This ensures that invaluable local knowledge is immediately available to the methods engineer and, further, the operator's essential goodwill is used throughout the project.

It should be noted, however, that there are potential dangers in using this information source. While there are very few people who, if approached in a proper manner, would deliberately give incorrect or misleading information, the information given in good faith is sometimes incorrect or misleading. In view of this, it is necessary for the methods engineer to develop intrapersonal skills that will allow the same question to be asked in several different ways without causing any offense. Unless the answers to the same question can be reconciled, the analyst knows that the information is incorrect.

Whenever possible the methods engineer should personally observe what is actually being performed so that the information obtained from different sources can be verified. This does not in any way imply that where this cannot be done, for example in the development of a new product or facilities, the methods engineering approach is invalid. On the contrary, it is here that the approach should be followed vigorously.

Examining the Present Method

In order to understand the problem fully it is necessary to carry out a critical evaluation of the present method. Several approaches appear in the literature, but it is proposed to consider that which is usually referred to as the *questioning technique*.[8]

This questioning technique centers around the use of the questions what, who, how, when, why, and where, which are asked with respect to:

1. The purpose of the operation.
2. The place where the operation is performed.
3. The sequence in which one operation is done with respect to another.
4. The person who performs the operation.
5. The means used to perform the operation.

The format of this questioning technique is shown in Table 3.14.1. It will be seen that, as part of the questioning technique, the analyst is able to generate ideas for improvement by investigating the possibilities of *eliminating*, *combining, and simplifying* tasks and operations.

Table 3.14.1. A Summary of the Questioning Technique Approach to Methods Improvement

PURPOSE of the activity What is done? Why is it done? What else might be done? What should be done?	ELIMINATE unnecessary parts of the job
PLACE where the activity is carried out Where is it done? Why is it done there? Where else might it be done? Where should it be done?	
SEQUENCE in which activities are undertaken When is it done? Why is it done then? When might it be done? When should it be done?	COMBINE whenever possible or REARRANGE the sequence of operations for more effective results
PERSON by whom the activities are undertaken Who does it? Why does that person do it? Who else might do it? Who should do it?	
MEANS by which the activities are performed. How is it done? Why is it done that way? How else might it be done? How should it be done?	SIMPLIFY the operation

3.14.2 THE TECHNIQUES OF METHODS ENGINEERING

To aid the systematic approach to problem solving dictated by the methods engineering approach it was natural that a set of suitable techniques would develop. Most of these are concerned with methods of recording in-information, either of the present method or the proposed method. Most proved to be useful for both recording and analysis. A brief description of some of the most widely used techniques follows.

3.14.2.1 Process Charts

A *process chart* uses a set of symbols to portray a process diagrammatically, thereby helping a person to visualize the process in examining and improving it.

Process charts may appear in one of several different forms; however, only the two most common are considered here.

Process Chart Symbols

An American Society of Mechanical Engineers (ASME) committee proposed a set of five process-chart symbols. The recommendations of the committee were approved in 1947. A summary of these symbols and their definitions is given in Fig. 3.14.2.

Outline Process Charts

A sample outline process chart is shown in Fig. 3.14.3. In the outline process chart the only symbols used are those for *operation* and *inspection*.

This type of chart can serve two purposes. First, it will provide the analyst with an overall view of the process being studied, enabling the analyst to structure the whole study in a logical and effective manner. In addition to this, the lack of detail provides the analyst with an idea of the maximum improvements that can be obtained from the present method without drastic changes.

The outline process chart shown in Fig. 3.14.3 is for a simple assembly. In constructing such a chart it is most convenient to select the component that forms the "base" onto which all of the other components are assembled. The outline process chart for this component is then constructed on the righthand side of the sheet. The process charts for the remaining components are then constructed, in turn, joining the other components to indicate the order of assembly. The descriptions alongside the symbols are kept brief and to the point, to reduce possible confusion. Where more detail is deemed necessary, it will appear on other documents.

Flow Process Charts

The *flow process chart* uses all five of the process chart symbols and, consequently, provides much greater detail than the outline process chart. A flow process chart may follow the same general form as that shown for the outline process chart; however, the more common form is for each component to be the subject of a separate chart. This allows a far more detailed study to be made.

A sample flow process chart for a single component is shown in Fig. 3.14.4.

○	**OPERATION** Change the shape, form or characteristics of the component or material.
□	**INSPECTION** Check the quality or quantity.
⇨	**TRANSPORT** Move component from one place to another (unless caused at work station during Operation or Inspection).
D	**DELAY** When conditions do not permit next action to be performed.
▽	**STORAGE** Object is protected against unauthorized removal.

Fig. 3.14.2. Process chart symbols.

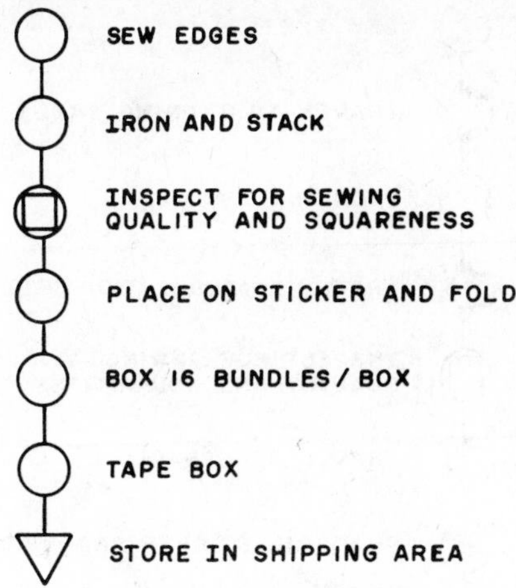

Fig. 3.14.3. Example of an outline process chart.

A Word of Warning

Wherever possible, a flow process chart or an outline process chart, should be constructed based upon direct observation. In addition, it should be noted that this should be direct observation of the man *or* the machine *or* the material. A single chart can relate to only one of these; otherwise, confusion results.

Therefore, three types of flow process charts may be constructed: (1) Flow process chart—man type; (2) Flow process chart—material type; (3) Flow process chart—machine type.

3.14.2.2 The Multiple Activity Chart

A Multiple Activity Chart is a chart on which the activities of more than one subject (worker, machine or item of equipment) are each recorded on a common time scale to show their inter-relationship.[5]

Constructing the Multiple Activity Chart

The activities of each subject are recorded on a separate bar or column against the time scale. An example of such a multiple activity chart is shown in Fig. 3.14.5.

Both the present and proposed methods for the operation are shown in Fig. 3.14.5. Note the way in which any periods of idleness or any inefficiencies in operation appear in the chart.

Multiple activity charts are particularly useful for analyzing and improving operations where there are a number of interrelated activities involving several operators and pieces of equipment. The examples given in the literature seem to emphasize the long-cycle tasks; however, it must not be assumed that they are restricted to this type of operation. In fact, in very short-cycle operations where an operator is interfacing with one or more mechanical devices, the use of a multiple activity chart is almost an essential analysis technique.

The use of multiple activity charts in interfacing automation and robotic devices is obvious.

Measuring the Time Units for Constructing Multiple Activity Charts

The method of determining the times for the various elements to be plotted on a multiple activity chart is governed by the accuracy required. The various techniques available include direct observation by an ordinary wristwatch, a stopwatch, a predetermined motion–time system, photographic, video, or any other means that enables the objectives of the analysis to be achieved. The accuracy of the prediction required and the length of the elements dictate the choice of method of measurement.

3.14.2.3 The String Diagram

Where materials, operators, or equipment move around a work area in a regular or irregular pattern, *a string diagram* is found to provide a simple yet effective means to record and examine the activities.

The string diagram consists of a scale plan or model on which one or more threads are used to trace the path of operators, material, or equipment around a work site. An illustration of a string diagram is given in Fig. 3.14.6.

String diagrams are useful in such situations as the following:

1. When one operator is tending several machines.
2. When several operators are performing a "team" operation on one or several machines.

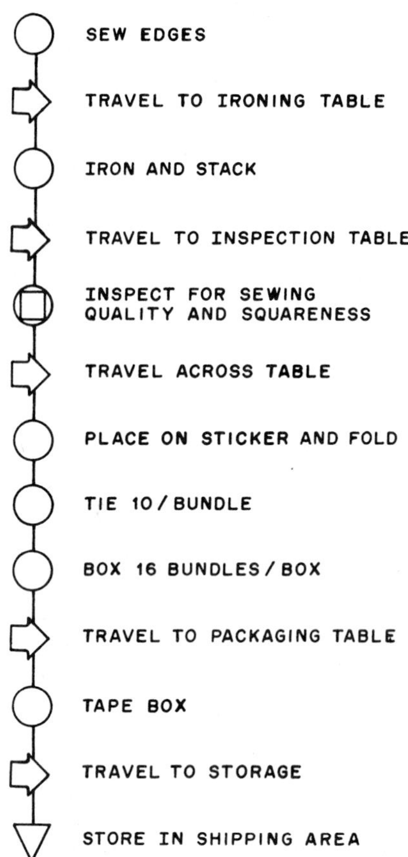

Fig. 3.14.4. Example of a flow process chart.

3. When several alternative layouts are to be compared and evaluated.
4. In restaurants or kitchens.
5. In stores and warehouses.

Constructing the String Diagram

The string diagram, though simple, requires both care and patience to construct. The details may be collected by direct observation or by analyzing records. The actual choice of method depends upon the application. Take for example the case of a manufacturing facility that handles several thousand components and is to be analyzed with the objective of reducing the total material handling in the facility. To use direct observation to collect the data on these components would be impossible. In such a case the data would almost certainly be based upon company records. On the other hand, where a study of the flow of, say, a forklift truck was required, the methods engineer would record the activities of the equipment by direct observation until a representative sample of the work had been obtained.

As components pass through various operations a change in their shape or condition will occur. This may have a significant effect on the unit load that is handled at various stages in a process, and this must be reflected in the string diagram. In a plant manufacturing sheet-metal components, for example, the unit load leaving a shears operation may be a tote pan containing four hundred components. However, to make these four hundred components, five sheets of steel would have to be handled one at a time.

Having collected and, if necessary, classified the data, an accurately drawn scale plan of the work area is prepared. Details of all equipment, doorways, and obstructions are shown. The drawing is then laid out on a board made of some soft material, such as polystyrene, and pins driven firmly into it at points where activities occur or at stopping points. The analyst ties a piece of thread to the pin at the starting point, then, step by step, follows the progress of the subject being studied, stringing the thread from pin to pin.

When the process of "stringing" has been completed it is possible to make an evaluation of the layout based upon the total length of thread used. Similarly, proposed layout changes can be evaluated on the same basis.

The string diagram is particularly useful in explaining the effect of changes in layout without going into technical details.

3.14.2.4 Memomotion Study

The use of time-lapse photography to study operations so that the problem areas could be defined, analyzed, and improved was developed by Marvin E. Mundel.[2] The technique is suitable not only as the basis of methods improvement but has been used successfully for the purpose of setting time standards. The technique has come to be known as *memomotion study*, as observed in Chapter 2.5.

In carrying out a memomotion study, a motion picture is taken of the operation; however, instead of film speed of between 960 and 1440 frames per minute, depending upon the particular application, it may run at between 1 and 60 frames per minute. Some VCRs are now being built with time-lapse capability and will undoubtedly prove to be of great help in improving methods.

Carrying Out a Memomotion Study

A camera with memomotion capacity is set-up so that it cannot be tampered with, and can view all of the area to be studied. After carrying out the time-lapse photography for a period sufficiently long to get a representative study, the results are studied by means of a frame-by-frame analysis.

Sometimes, in analyzing a memomotion study, it is advantageous to run the film also at normal speed, since this will often highlight major bottlenecks or motions of workers coming into or leaving the work area. Operations that appear to be acceptable under normal visual inspection or even using micromotion analysis take on a completely different appearance. In this way, problem areas are detected quickly and effectively.

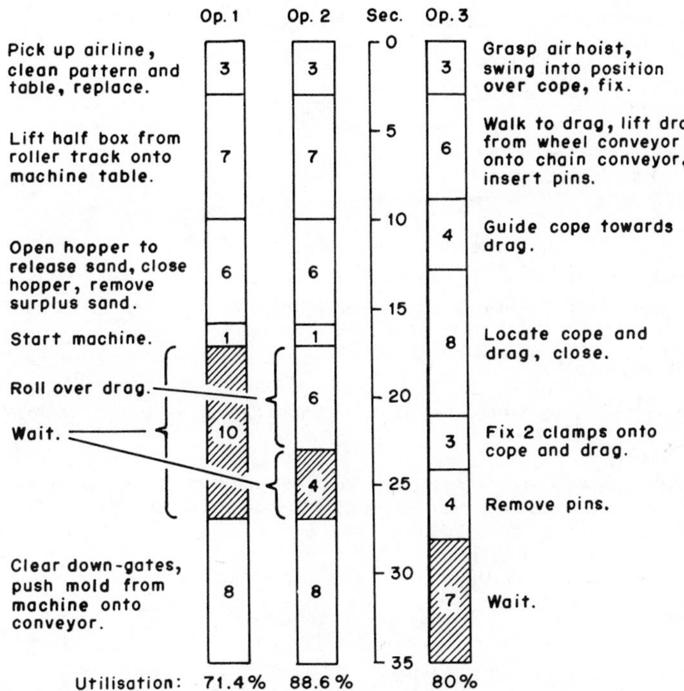

Fig. 3.14.5. Example of a multiple activity chart.

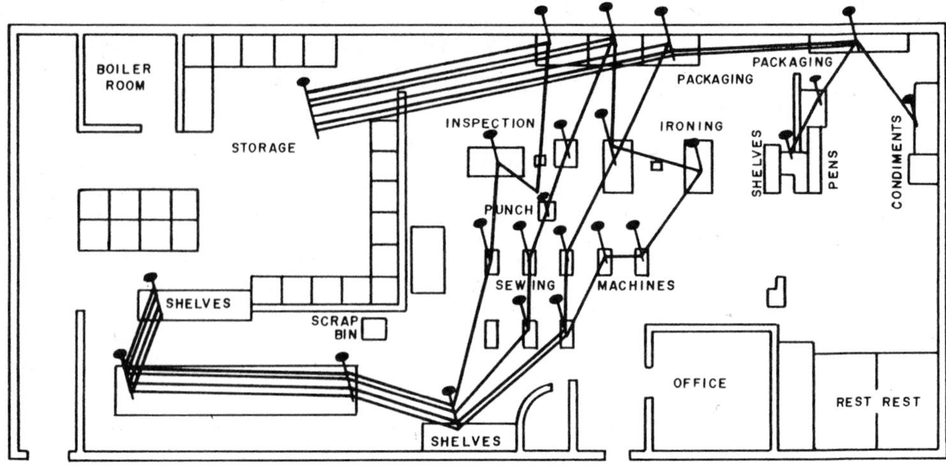

Fig. 3.14.6. Example of a string diagram.

Using the Memomotion Data

The frame-by-frame analysis of the memomotion film or videotape will produce data that can be used as the basis for other recording techniques. String diagrams of the movement of workers around a work area can, for example, be plotted quickly and effectively using the memomotion technique. Where several workers are involved in the operation it is possible to study them either separately or as a whole. The data obtained can be used to construct multiple activity charts and to perform work sampling.

3.14.2.5 Predetermined Motion-Time Systems

A major advance in methods engineering was the development of predetermined motion-time systems. The main use of these systems seems to have been as a means of work measurement; however, their usefulness as a basis for improving methods must not be overlooked.

There are a number of predetermined motion-time systems (PMTS) in common use, all of which seem to have similar characteristics. These characteristics are:

1. Basic elements of motion are recognized and defined.
2. The variables that affect the time to perform these elements of motion have been delineated.
3. The times to perform the various motion elements with different values of these variables have been determined.
4. A shorthand has been developed that allows these motion elements to be described concisely and explicitly.
5. The total time taken to perform a task is assumed to be the sum of the motion elements that describe it.

In several of the predetermined motion-time systems, simplifications of the original systems have been developed. One system, Methods-Time Measurement (MTM), now has a whole "family" of systems, each of which can be combined with the others and has features that make it more suitable for some purposes than others. (For further consideration of MTM, see Chapter 2.5.)

The system, known as MTM-2, is useful in methods engineering and is simple enough so that it can be easily explained.

The MTM-2 System

The MTM-2 system recognizes eleven categories of motion, each of which is very carefully defined. A summary of the way in which these motion categories are recognized is given in Table 3.14.2.

The MTM-2 motion categories use the Time Measurement Unit, which is abbreviated TMU.

1 TMU = 0.00001 hours
= 0.0006 minutes
= 0.0036 seconds

The TMU is expressed at a performance level of 100 on the Westinghouse leveling scale.

Table 3.14.2. A Summary of the MTM-2 Motion Categories

Name of Category	Symbol	Description
Get	GA GB GC	An action of the hand or fingers. The hand reaches to the object, grasps and subsequently releases it. There are three "cases" of Get, A, B, and C.
Put	PA PB PC	An action of the hand or fingers. The hand moves, or moves and positions, the object. There are three "cases" of Put, A, B, and C.
Get weight	GW	The finite time necessary for the muscles of the hand and arm to take up the weight of an object prior to moving it.
Put weight	PW	The weight of an object affects the time necessary to perform a Put. PW is the adjustment for this difference.
Apply pressure	A	Muscular force applied to an object to overcome a resistance, accompanied by little or no movement.
Regrasp	R	The position of an object in the hand is changed, without releasing control
Eye action	E	E may occur in either of two ways. (1) Recognizing a readily distinguishable characteristic of an object. (2) Shifting the eyes from one viewing area to another.
Crank	C	An action of the hand or fingers where an object is rotated more than one half revolution.
Foot motion	F	A motion of the leg, 12 in. or less, or one in which the trunk is not moved intentionally.
Step	S	A motion of the leg, greater than 12 in., or one in which the trunk is moved intentionally.
Bend and arise	B	Lowering of the trunk, and the subsequent arise.

The Distance Variable. Both *Get* and *Put* are affected by the distance of the motion. Since the distance is handled in the same way for each of these motion categories, it is convenient to consider this only once.

The distance for Get and Put is estimated, based on the length of the path generated by the knuckle at the base of the index finger. No attempt is made to estimate the actual distance, but instead to locate this motion within a distance range. These distance ranges and the path they are estimated in are shown in Fig. 3.14.7. No attempt should be made to interpolate within these distance codes.

MTM-2 Category Get. The motion category Get is affected by two variables, the *case* and the *distance*, which are indicated in the Get symbol, an example of which is given in Fig. 3.14.8.

The *case* of the Get may be regarded as the difficulty involved in gaining control of the object. Three cases of Get are recognized in the MTM-2 system. The method used to classify them is illustrated in Fig. 3.14.9.

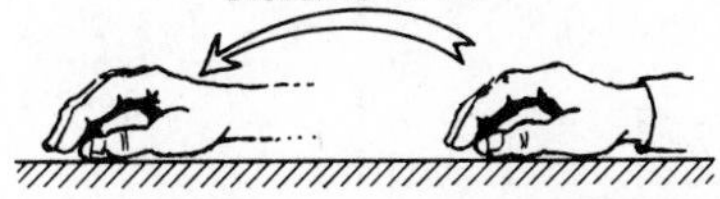

Code	Distance in cm: Over	Up to and including
5	0	5
15	5	15
30	15	30
45	30	45
80	Over 45	

Fig. 3.14.7. Distance codes for Get and Put.

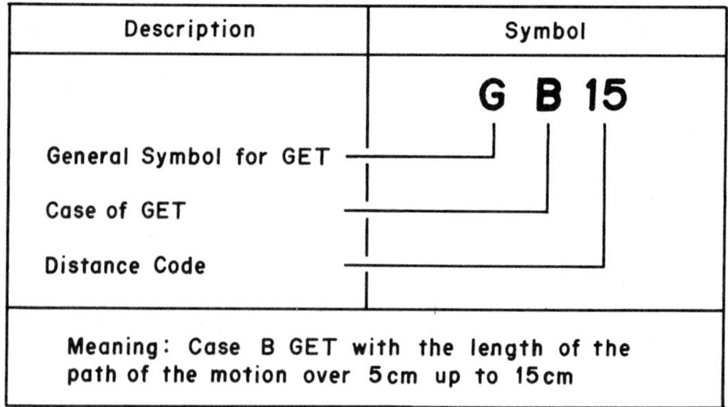

Fig. 3.14.8. The Shorthand used for the category Get.

It is important to note that the decision model embodied in this diagram is the sole guide to the determination of the case of Get. Inexperienced practitioners may try to argue cases on an "intellectual basis." This must be avoided and the decision model used as the sole arbiter.

The TMU values for all of the values of Get are given in Fig. 3.14.10.

The MTM-2 Category *Put*. The form of the symbol used for *Put* is given in Fig. 3.14.11. The variables affecting Put are case and distance.

The *distance* variable is used in the way described earlier. There are three cases of Put, which are classified according to the amount of control required at the termination of the motion. The basis used for this classification is a decision model, which is shown in Fig. 3.14.12. As in the case of Get, there is a temptation on the part

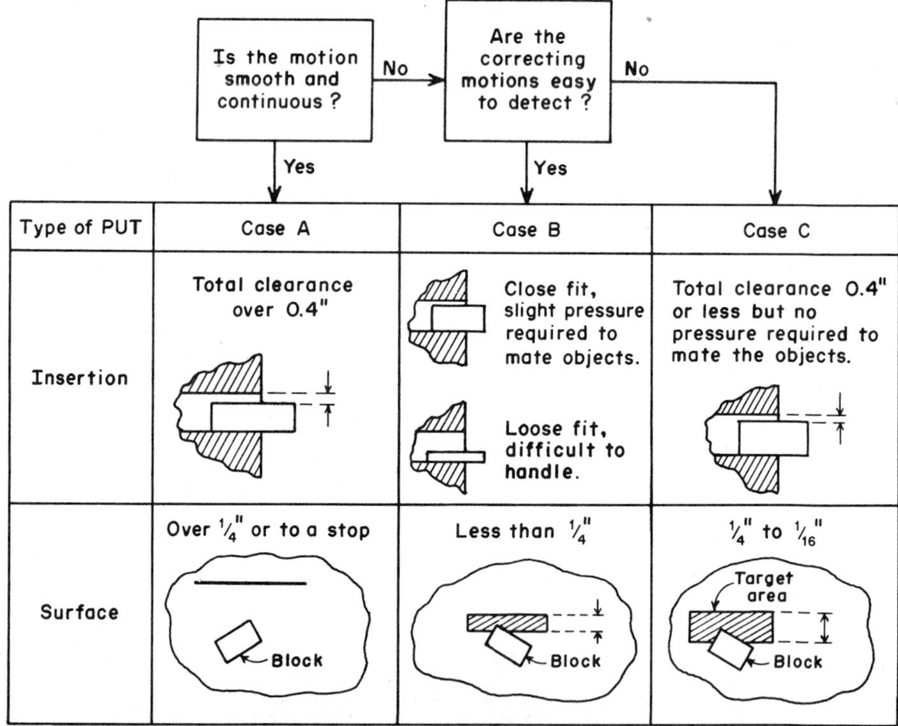

Fig. 3.14.9. Identifying the case of Get.

Category	TMU	Category	TMU	Category	TMU
GA5	3	GB5	7	GC5	14
GA15	6	GB15	10	GC15	19
GA30	9	GB30	14	GC30	23
GA45	13	GB45	18	GC45	27
GA80	17	GB80	23	GC80	32

Fig. 3.14.10. Time values of the MTM-2 motion category Get.

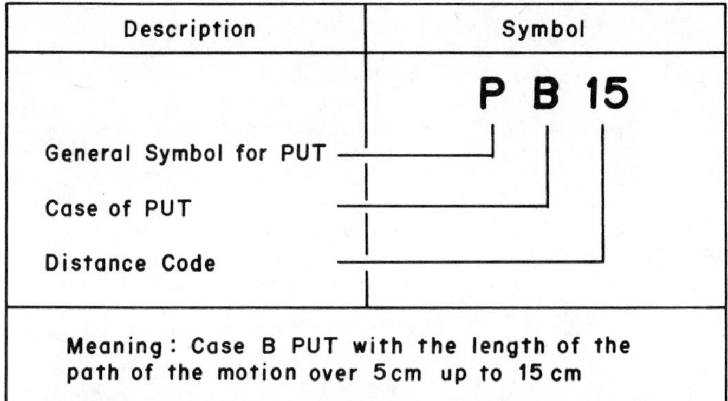

Fig. 3.14.11. The shorthand used for the category Put.

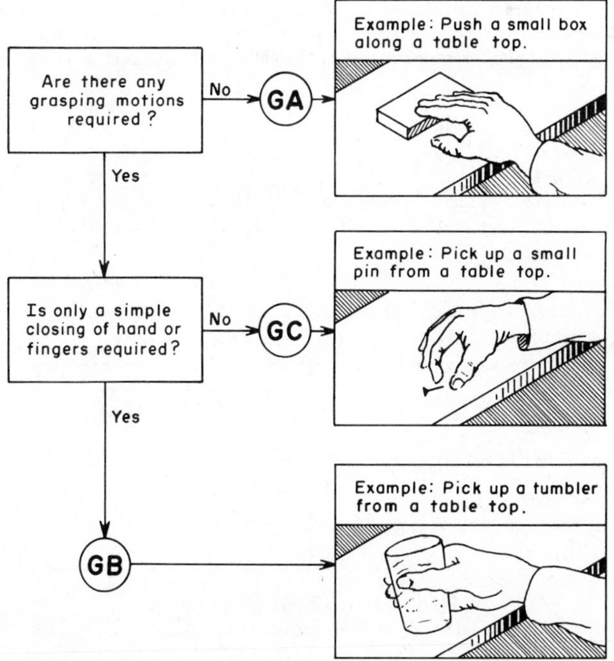

Fig. 3.14.12. Identifying the case of Put.

Category	TMU	Category	TMU	Category	TMU
PA5	3	PB5	10	PC5	21
PA15	6	PB15	15	PC15	26
PA30	11	PB30	19	PC30	30
PA45	15	PB45	24	PC45	36
PA80	20	PB80	30	PC80	41

Fig. 3.14.13. Time values for the MTM-2 motion category Put.

of inexperienced practitioners to rationalize a particular choice. Under no circumstances should this be done, particularly in the case of Put; the decision model must be the basis of choice of the case.

The TMU values for all of the possible cases of Put are given in Fig. 3.14.13.

The Categories *Get Weight* and *Put Weight*. The categories Get and Put, described above, each assume that there is no significant resistance to motion when the motion category is being performed. Clearly this is not always the case and Get Weight and Put Weight are additions made to the total time to account for any extra time required when the weight being handled is significant.

The weight is determined as the Effective Net Weight (ENW), which may be defined as the total resistance to motion per hand. Thus if an operator is to pick up and move an object weighing 12 kg, the ENW would depend upon whether one or two hands were used. If only one hand is used the ENW would be 12 kg; if two hands were used, the ENW would be 6 kg.

The symbol used for the Get Weight category is shown in Fig. 3.14.14 and for the Put Weight on Fig. 3.14.15.

It should be noted that the values of the ENW for the Put Weight category is in steps of 5 kg with only 1 TMU increase for each 5 kg step. Further, where the ENW is determined to be less than 3 kg, the weight is considered too insignificant and the effects of Get Weight and Put Weight are ignored. The value and symbol for Get Weight and Put Weight up to 20 kg ENW are given in Table 3.14.3.

Get Weight can best be described as that finite time required for the hands and fingers to take up any significant weight prior to performing the Put. On the other hand, *Put Weight* is an addition to Put where significant resistance to movement is present. Get Weight may therefore be regarded as the static component of the Put and Put Weight as the dynamic component. It is possible, therefore, for a Put having significant ENW to be performed with a Put Weight but no Get Weight if the component is already under control.

The Category *Apply Pressure*. *Apply Pressure* occurs where a force is applied to overcome a resistance but is accompanied by little or no motion. Apply Pressure has the symbol A and a time value of 14 TMU.

Two restrictions must be noted when the Apply Pressure category is recognized: the maximum distance that may be moved is 0.635 cm and any hesitation must be limited to the operator's minimum reaction time. When the distance restriction is exceeded, the motion must be analyzed as a Put with the appropriate Get Weight and Put Weight. If the minimum reaction time is exceeded, then a process time must be included.

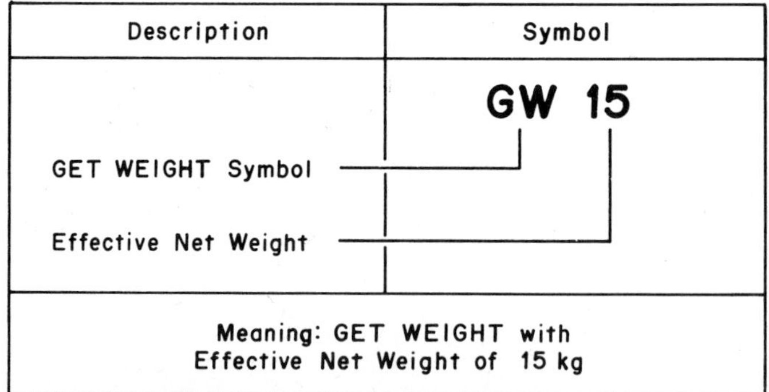

Description
Symbol
GW 15
GET WEIGHT Symbol
Effective Net Weight
Meaning: GET WEIGHT with
Effective Net Weight of 15 kg

Fig. 3.14.14. The shorthand used for the MTM-2 motion category Get Weight.

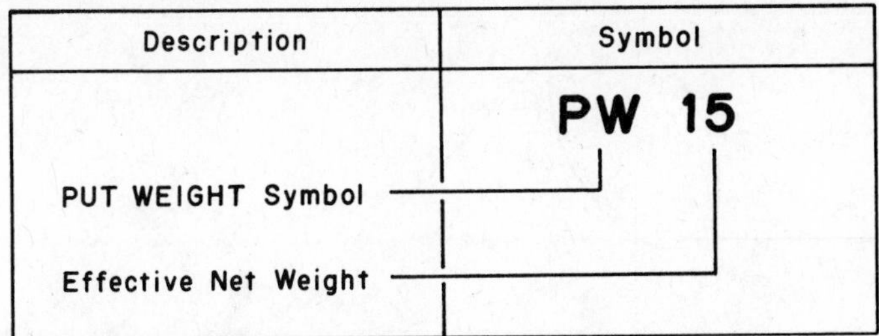

Fig. 3.14.15. The shorthand used for the MTM-2 motion category Put Weight.

The Category *Regrasp*. When an operator holding an object changes its position in the hand, but without relinquishing control, a *Regrasp* occurs. The time allocated to a Regrasp is 7 TMU and its symbol is R.

If an operator is holding an object using two hands and moves, say, the right hand, then the analysis would be a Get rather than a Regrasp, since control was not relinquished. This characteristic of this particular motion category is one which must be fully appreciated by users of MTM-2.

The Category *Eye Action*. The motion category *Eye Action* can occur in either of two ways: (1) A movement of the eyes from one viewing area to another; (2) the focusing of the eyes on a new viewing area. The trainee MTM-2 analyst invariably shows a tendency to allow Eye Actions where they do not occur, so it is necessary to emphasize the rule: Eye Action can occur only where *all other motions cease*.

The observer's Normal Area of Vision (NAV) has an influence upon the analysis of Eye Actions. The basis of the NAV is illustrated in Fig. 3.14.16, but is should be noted that as the distance from the eyes increases, the diameter increases in the same proportion. Thus, if the distance from the eyes increases to 32 in., then the diameter of the circle increases to 8 in.

The analysis of the two types of Eye Action and the influence of the NAV can be illustrated by two simple examples. Assume that two points on an object are to be inspected for some characteristic. If they are located within a 4 in. diameter circle and are 16 in. from the inspector's eyes, then, since the focusing action is associated with the simple binary decision and inspection, the analysis would be a single Eye Action. On the other hand, if the points are more than 4 in. apart, the analysis involves an Eye Action to inspect the first point, followed by a second Eye Action to shift the eyes to the second point and a third Eye Action to inspect the second point.

Table 3.14.3. Symbols and TMU Values for GW and PW

ENW (kg)	Get Weight		Put Weight	
	Symbol	TMU	Symbol	TMU
1				
2				
3	GW3	3		
4	GW4	4	PW5	1
5	GW5	5		
6	GW6	6	PW10	2
7	GW7	7		
8	GW8	8		
9	GW9	9		
10	GW10	10		
11	GW11	11	PW15	3
12	GW12	12		
13	GW13	13		
14	GW14	14		
15	GW15	15		
16	GW16	16	PW20	4
17	GW17	17		
18	GW18	18		
19	GW19	19		
20 max.	GW20	20		

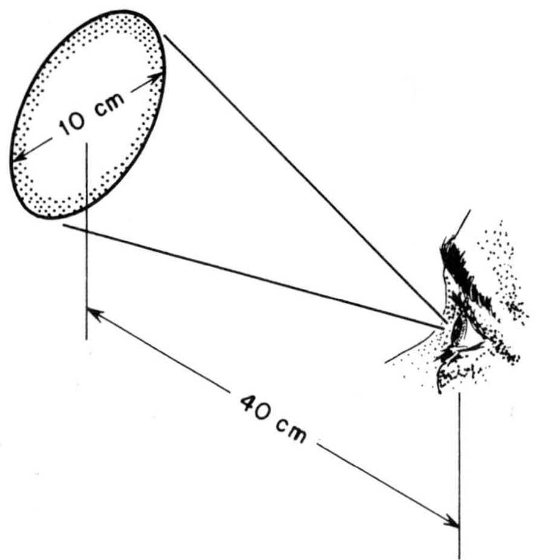

Fig. 3.14.16. The normal area of vision (NAV).

LH	TMU	RH
	30	(PC30 R̸
TOTAL	30	TMU

Fig. 3.14.17. Analysis of Combined Motions.

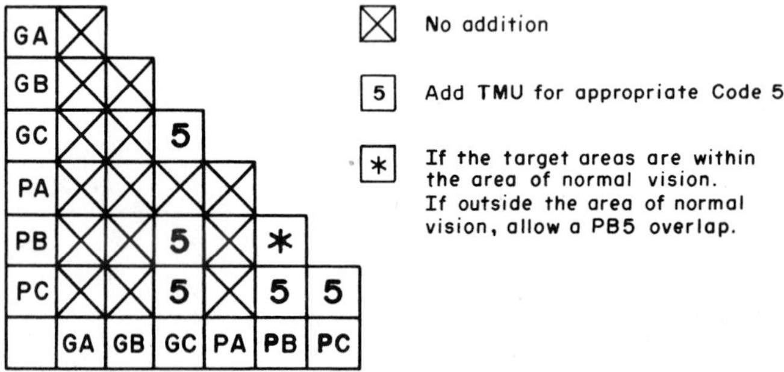

GA	☒					
GB	☒	☒				
GC	☒	☒	5			
PA	☒	☒	☒	☒		
PB	☒	☒	5	☒	*	
PC	☒	☒	5	☒	5	5
	GA	GB	GC	PA	PB	PC

☒ No addition

5 Add TMU for appropriate Code 5

* If the target areas are within the area of normal vision. If outside the area of normal vision, allow a PB5 overlap.

Fig. 3.14.18. Table of Simultaneous Motions.

LH	TMU	RH	Description
(GA15)	23	GC30	Example 1
(G–)	30	GC30	} Example 2
GC5	21		

Fig. 3.14.19. Analysis of Simultaneous Motions.

The Category *Crank*. The MTM-2 category *Crank* has the symbol C and a TMU value of 5. It is the action that generates a cone about the elbow, such at that of an operator turning a handwheel using only one hand. The movement must exceed one half of a revolution, otherwise it should be analyzed as a Put.

The Categories *Step* and *Foot Motion*. *Step* category is listed when there is a lateral motion of the trunk resulting from a movement of the legs. The symbol used for the category Step is S and its time value is 18 TMU. One of the most obvious examples of the category Step is the action of walking, where a Step is allowed each time the foot hits the floor.

A movement of the legs occurring without any lateral movement of the trunk, is given the MTM-2 category *Foot Motion*, whose symbol is F and whose time value is 9 TMU. Foot motion occurs where the operator is operating a foot pedal, the movement occurring as the result of a pivoting action about the ankle or the knee.

The Category *Bend and Arise*. The motion category *Bend and Arise* is given the symbol B and a time value of 61 TMU. It occurs when there is a vertical displacement of the trunk to the extent that an operator can, but not necessarily does, reach below the knees.

Combined and Simultaneous Motions. One of the strengths of using PMTS in methods improvement is the ability to describe and evaluate what happens when an operator is called upon to perform more than one motion with one or more limbs at the same time. MTM-2 classifies this situation as either simultaneous or combined motions.

The combined motion occurs when an operator performs more than one action at the same time with the same limb, for example, pick up a tool and while placing it ready to use, Regrasp it as a matter of convenience. The Regrasp is performed during the Put action, therefore the whole action is said to be a Combined Motion.

In analyzing these combined and simultaneous motions, mention must be made of the Principle of Limiting Motion, which allows only the time for the motion of longest duration. Thus, if a Regrasp is performed as a combined motion with a PC30, only the time for the PC30 is allowed. The analysis would be as shown in Fig. 3.14.17.

Where the operator is required to perform more than one action with more than one limb at the same time, the total action is classified as a simultaneous motion. While simultaneous motions are still subject to the Principle of Limiting Motion, their analysis is somewhat more complex. Depending upon the difficulty of the motions being performed simultaneously, certain additions are made to the total time. A summary of these requirements is given in Fig. 3.14.18.

Two examples of simultaneous motions are given in Fig. 3.14.19. It will be seen that only the time for the limiting motion is given in each case and that a circle is drawn around the motion which is limited out. In the case of the second example, where a GC is performed by each hand, an "overlap" is performed, which is equal to a GC5.

Using MTM-2 for Improving Methods

The whole construction of MTM-2 encourages its use in the improvement of methods. First, the method can be described in terms of the variables that affect the time to perform the task. Second, the limiting motions are very clearly identified. In this way, if it is accepted that any reduction in the time to complete the task represents an improvement in method, the analyst's attention is directed immediately to the areas of potential improvement.

In applying the techniques described above, it is clear that the approach should be disciplined yet inventive. The analyst must not be constrained by what has gone before.

REFERENCES

1. R. M. Currie, *Work Study*, London: Pitman Publishing, 1974.
2. Marvin E. Mundel, *Motion and Time Study Improving Productivity*, Englewood Cliffs, NJ: Prentice-Hall, 1978.
3. Marvin E. Mundel, *Improving Productivity and Effectiveness*, Englewood Cliffs, NJ: Prentice Hall, 1983.
4. B. W. Niebel, *Motion and Time Study*, Richard D. Irwin, Inc., 1982.
5. Wallace J. Richardson, *Cost Improvement, Work Sampling and Short Interval Scheduling*, Reston, VA: Reston Publishing, 1976.
6. Kenneth Knott, *Principles and Practice of MTM-2*, Kenneth Knott Ltd., 1976.
7. E. B. Raybould, and A. L. Minter, *Problem Solving for Management*, Institute of Practitioners in Work Study, Organization and Methods, 1976.
8. *Introduction to Work Study*, Geneva: International Labour Organization, 1976.

CHAPTER 3.15

GROUP TECHNOLOGY

INYONG HAM

The Pennsylvania State University
University Park, Pennsylvania

3.15.1 INTRODUCTION

Group Technology (GT) is generally considered to be a manufacturing philosophy or concept that identifies and exploits the sameness or similarity of parts and operation processes in design and manufacture, and then aims mainly at increasing productivity and reducing production costs. In batch-type manufacturing, each part traditionally has been treated as being unique in design, process planning, production control, tooling, production, etc. However, by grouping similar parts into *part families* based on either their geometrical shapes or operation processes, as exhibited in Fig. 3.15.1, and also, if possible, by forming machine groups, or cells, that process the designated part families, it is possible to reduce costs through more effective design rationalization, manufacturing standardization, higher manufacturing productivity, etc.

The basic concept of group technology has been practiced for many years as part of "good engineering practice" and "scientific management".[1] Applications of group technology concepts are identified by different names and in various forms of engineering and manufacturing functions. Traditionally, group technology practices were limited to productivity improvement in batch-type manufacturing, with different degrees of success, either in the design or manufacturing area. For many years, group technology did not receive formal recognition and has not been rigorously practiced as a systematic scientific technology. In recent years, however, an intensified effort in computer integrated manufacturing (CIM) has stimulated a stronger interest in group technology, since it provides the essential element for the successful development and implementation of CIM through the application of the part-family concept for effective data retrieval and common database systems, for design of manufacturing cells of automated factory systems, etc.

3.15.2 PART FAMILY FORMATION AND MACHINE GROUPING

3.15.2.1 Basic Concept

A part family may be defined as a group of related parts that have some specific sameness and similarities. They may have similar design features, such as geometric shape, and/or they may share similar *processing* requirements. In some cases, however, they may be dissimilar in shape, but related by having all or some common *production* operations. Among many other factors to be considered in grouping part families, the lot size and their frequency of manufacture are important. The greater the similarity of processing requirements and lot frequency, the more effective it is to form the part family. This allows for more practical applications of the group technology concept in forming machine groups, or cells, and in scheduling for optimum sequencing and machine loading.

The grouping of related parts into part families is the key to group technology implementation. The problem that immediately arises is, how are the parts to be grouped efficiently into these families? There are four basic methods used to form part families:

1. Manual visual search.
2. Nomenclatures/functions.
3. Production flow analysis.
4. Classification and coding systems.

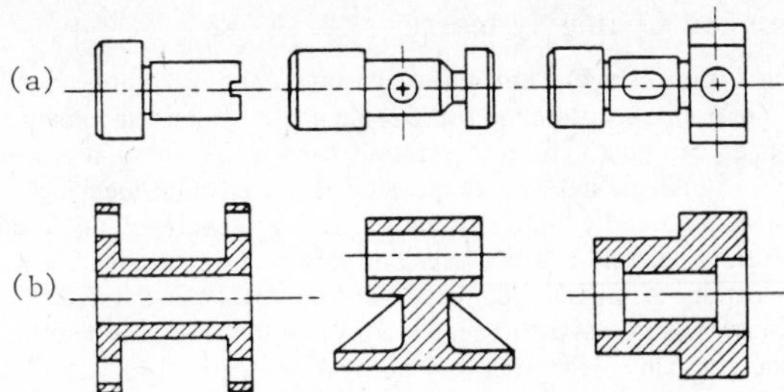

Fig. 3.15.1. Example of part families. (*a*) Similar in shape and geometry, and (*b*) dissimilar in shape but similar in production processes.

The first method is obviously very simple, but limited in its effectiveness when dealing with a large number of parts. The second method is neither scientific nor reliable. In general, the third and fourth methods are more commonly used in forming part families and machine groups for group technology applications. Although production flow analysis is a very practical method, a well-designed classification and coding system based on the specific parameters and code digits of the system is also effective. For better grouping of part families, computer programs using cluster analysis,[2] pattern recognition, etc., have been developed to enhance the conventional methods using production flow analysis and classification and coding systems.

3.15.2.2 Production Flow Analysis

Production flow analysis[3] is a technique for forming part families and/or machine groups/cells by analyzing the operation sequence and the routing of a part through the machines and workstations in the plant. An example is shown in Fig. 3.15.2. To use this method successfully, one should ensure that a company has a reliable data source of routing or operation sheets. Part families can be formed with or without a classification and coding system, since production flow-analysis requires only operation sheets and production data.

(a) Original Data (Before Grouping)

Part # / Machine	1	2	3	4	5	6	7	8	9	10	11	12	13	14	15	16
L	x	x		x		x	x	x	x	x	x			x	x	x
M	x		x	x	x	x	x		x		x	x	x			x
D	x	x	x	x		x	x	x	x	x	x	x	x	x	x	
G	x	x	x		x		x	x	x	x	x		x		x	x

(b) Part Families/Machine Groups (After Grouping)

Part # / Machine	1	7	9	11	4	6	16	2	8	10	15	14	3	13	12	5
L	x	x	x	x	x	x	x									
M	x	x	x	x	x	x	x									
D	x	x	x	x	x	x										
G	x	x	x	x			x									
L								x	x	x	x	x				
D								x	x	x	x	x				
G								x	x	x	x					
M													x	x	x	x
D													x	x	x	
G													x	x		x

Fig. 3.15.2. Example of production flow analysis for forming part families and machine groups (See Fig. 3.15.5.) L = Lathe; M = Milling Machine; D = Drill Press; G = Grinding Machine.

3.15.2.3 Classification and Coding Systems

A classification and coding system is a fundamental of group technology application. "Classification" means the arrangement of items into groups according to some principle or system whereby items are grouped together by their similarities. A "code" can be a system of symbols used in information processing in which numbers or letters or a combination of numbers and/or letters are given a particular meaning.

Many varieties of classification and coding systems[4] are being developed and used around the world. An example of a coded part using a publicly available system[5] is shown in Fig. 3.15.3. Classification and coding for group technology applications is an important and very complex task. Since each company has its own specific needs and conditions, it is necessary to search for or to develop a system that can be adapted to the specific requirements of the company. For group technology applications, a well-designed classification and coding system should be able to group part families as needed, based on specified parameters, and should be capable of effective data retrieval for various functions as required. An example of such part family grouping using a classification and coding system is shown in Fig. 3.15.4.

Many types of general purpose classification and coding systems are available. In addition, many other systems have been developed and used by government agencies and service sectors, for such areas as libraries, museums, office supplies, commodities, insurance, and credit cards. However, these systems are not necessarily designed for group technology applications, though many systems have been used for industrial applications. Bar codes, which are extensively used for commodity classifications, may be used for selective areas of group technology applications, e.g. tooling codes, material codes, etc.

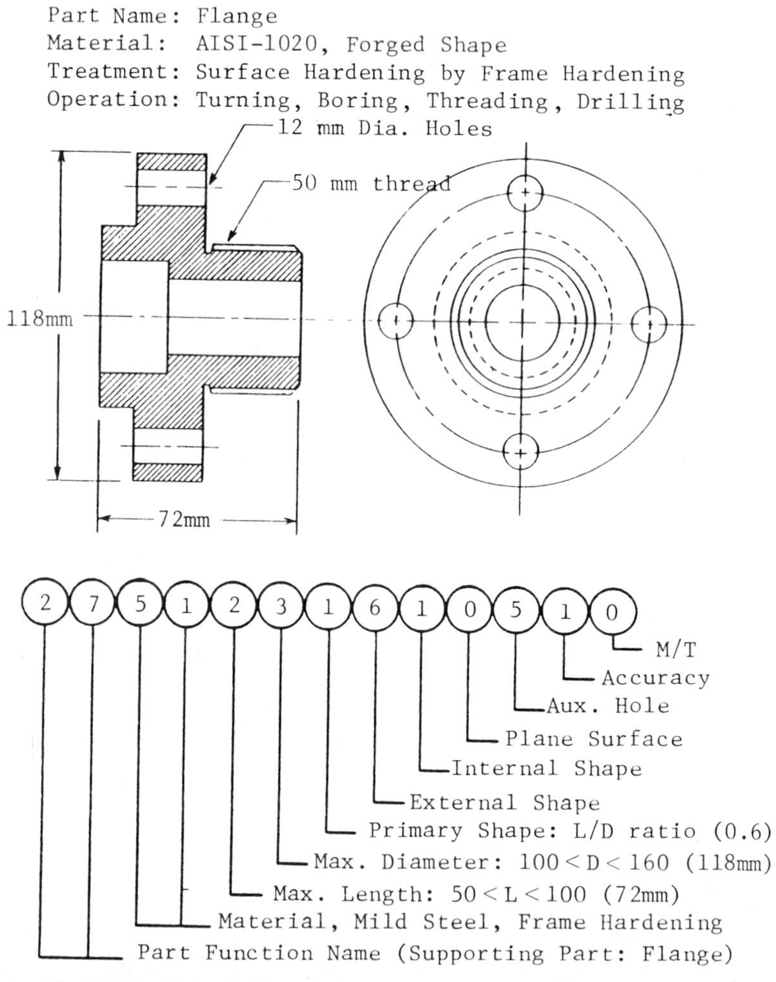

Part Name: Flange
Material: AISI-1020, Forged Shape
Treatment: Surface Hardening by Frame Hardening
Operation: Turning, Boring, Threading, Drilling

12 mm Dia. Holes
50 mm thread
118mm
72mm

2 7 5 1 2 3 1 6 1 0 5 1 0

M/T
Accuracy
Aux. Hole
Plane Surface
Internal Shape
External Shape
Primary Shape: L/D ratio (0.6)
Max. Diameter: $100 < D < 160$ (118mm)
Max. Length: $50 < L < 100$ (72mm)
Material, Mild Steel, Frame Hardening
Part Function Name (Supporting Part: Flange)

Fig. 3.15.3. Example of a coded part using a publicly available coding system.

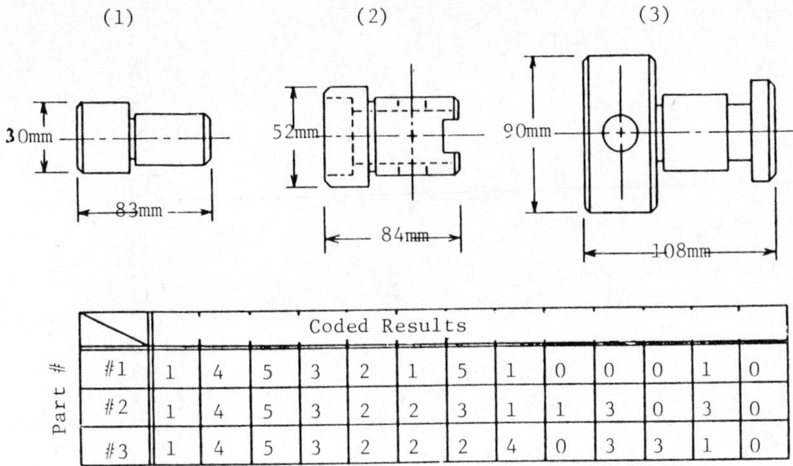

Part #	Coded Results												
#1	1	4	5	3	2	1	5	1	0	0	0	1	0
#2	1	4	5	3	2	2	3	1	1	3	0	3	0
#3	1	4	5	3	2	2	2	4	0	3	3	1	0

Fig. 3.15.4. Examples of coded results of a part family.

A classification and coding system is essential for full exploitation of the benefits of group technology. One should understand, however, that adaptation of a classification and coding system is just the beginning, simply a prerequisite for group technology applications. Installation of a suitable system will provide a powerful and effective data retrieval system, which leads to design rationalization and manufacturing standardization for achieving higher manufacturing productivity.

3.15.2.4 Machine Group/Cell

A group of machines for one or more part families may be arranged to perform all the operations required for the family or families of parts. The machines themselves are arranged in a semiflow line to minimize transportation and waiting. When conventional machine tools are grouped, the result is characteristically very similar to a modern machining center. For automated factory systems, numerical control or computerized numerical control machining centers are grouped, often with robots. These machine groups are sometimes called machine cells, manufacturing cells, manufacturing centers, flexible manufacturing systems, and multistation manufacturing systems.

A conventional functional machine layout and a group layout of machine tools based on the group technology concept are shown in Fig. 3.15.5*a* and *b*, respectively, to illustrate the features of a group/cell layout.

3.15.2.5 Group Tooling

For maximum utilization of tooling, the tooling setup for operations within a part family should be arranged so that all the parts, or as many as possible, can be processed with one group jig fixture and/or one setup. Group fixtures are designed to accept all members of the family, supplemented by adapters that can accommodate some differences between the parts in the family. Instead of designing, fabrication, and using individual fixtures for each part, only one group fixture and the necessary adapters, which are essentially inexpensive, are required. Therefore, it becomes evident how much tooling costs and setup costs can be reduced using group technology.

3.15.3 DESIGN RATIONALIZATION

An important application of the group technology concept using a suitable classification and coding system is in design data retrieval and design rationalization. A classification and coding system provides the following important features for design rationalization:

1. Part family grouping.
2. Retrieval of existing design information for specific needs.
3. Standardization and simplification.
4. Design for economic manufacture.
5. Elimination of duplicate designs.
6. Common database for CAD/CAM (CIM).

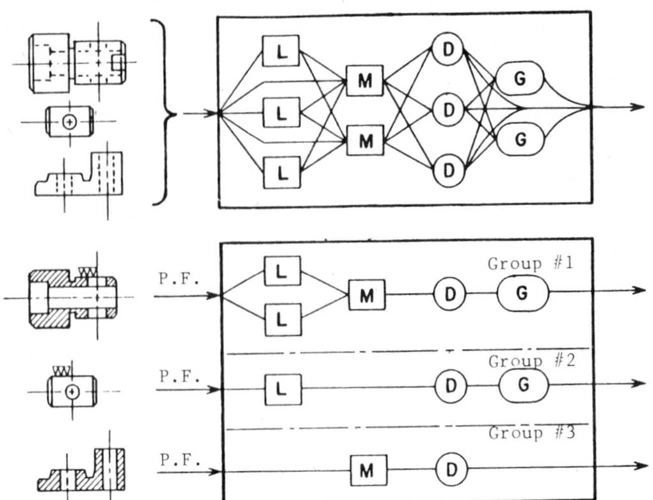

Fig. 3.15.5. Comparison of (*a*) functional layout, and (*b*) group layout. (See Fig. 8.2.2) L = Lathe; M = Milling Machine; D = Drill Press; G = Grinding Machine.

An example of effective design retrieval and design rationalization using the group technology concept[6] is shown in Fig. 3.15.6.

3.15.4 GROUP SCHEDULING

3.15.4.1 Basic Concept

Group technology simplifies general production scheduling, since simple machine group/cell scheduling does not involve an entire shop. Production scheduling associated with the application of group technology is called *group scheduling*.[7] Efficient group scheduling effectively reduces total throughput time and leads to the reduction of work-in-process inventory. Group scheduling has some specific advantages:

1. Optimization for group and job sequence and machine loading.
2. Possibility of flow shop pattern.
3. Reduction of setup times and cost.
4. Economic savings.

When the group/cell layout and part families are properly employed, group scheduling ideally lends itself to optimal sequencing effects. Even though a machine group/cell is not formed, the group scheduling concept can still be applied effectively with the existing layouts. Proper group scheduling is an integral part of group technology and will result in a significant cost reduction.

3.15.4.2 Algorithms for Group Scheduling

Algorithms and mathematical models related to group scheduling have been developed, and corresponding computer programs have also been developed, to schedule jobs of a part family to a corresponding machine group/cell.[7–9] Jobs can be properly sequenced in the family, and the families properly sequenced through the machine groups/cells for the optimum conditions by minimizing total throughput time. A schematic diagram showing the basic models for group scheduling[8] is shown in Fig. 3.15.7.

Optimal Sequencing Analysis

Group scheduling can be analyzed in a multistage manufacturing system. In the case of manufacturing multiple parts (jobs) grouped into several part families, both the optimal group and the optimal group sequences can be determined such that the total flow time (makespan) is minimized. Some of the methods of doing so are the branch-and-bound and heuristic methods.[7,8]

Machine Loading Analysis

The analysis of machine loading for group scheduling is complex, and it is not simple to develop an adequate algorithm for practical applications. However, some mathematical models for machine-loading and product-mix analysis problems for group technology applications are available.[10]

Integrated Applications with MRP

It is important to note that group technology applications directly relate to and influence various planning and inventory activities. The interrelationships between group scheduling and material requirements planning (MRP) should be considered whenever possible. An integrated use of MRP and group technology scheduling provides a viable system for effective production control.[11]

3.15.4.3 Examples of Group Scheduling

Optimum Sequencing for a Single Part Family

To explain the optimum sequencing methods for group scheduling, we will study a simple example using a heuristic algorithm.[8,9] Let us assume that the basic data given in Table 8.2.1 are to be used to find the optimum job sequence processing through four machines for a single part family group that will minimize the total throughput time.

Using the heuristic algorithm it can be easily found that the optimum job sequence is J_2-J_3-J_1. The total throughput times for all possible combinations of this group is as follows:

J_1-J_2-J_3:78 hours
J_1-J_3-J_2:81 hours

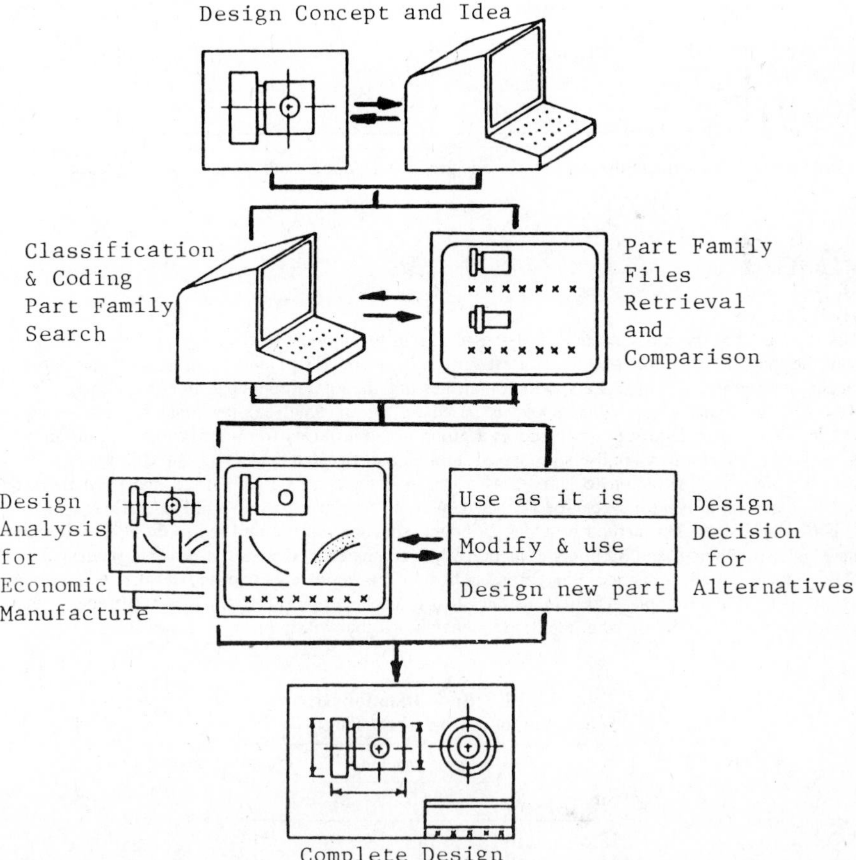

Fig. 3.15.6. Flow diagram for design rationalization using group technology concept.

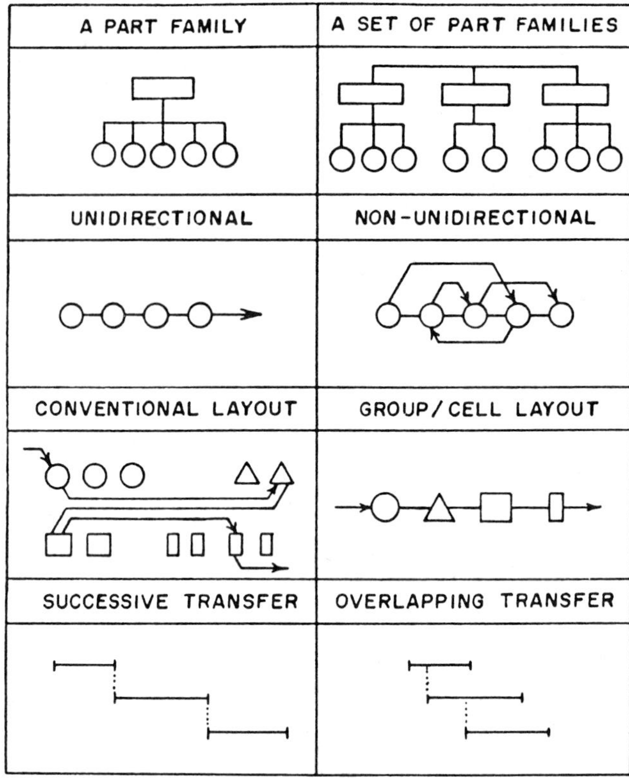

Fig. 3.15.7. Schematic diagram of various group scheduling methods for optimum sequencing.

J_2-J_1-J_3:74 hours
J_2-J_3-J_1:71 hours
J_3-J_1-J_2:72 hours
J_3-J_2-J_1:72 hours

It is obvious that the job sequence (J_2-J_3-J_1) results in the shortest total throughput time. The difference between the longest total time and the shortest time is 10 hours. The Gantt chart showing the optimum job sequencing is exhibited in Fig. 3.15.8, which indicates that the total throughput time is 71 hours.

However, the sequencing method above is based on the so-called unidirectional successive sequencing method in which a part family is processed as a group and transferred from one workstation to another as a batch. However, when a part family is processed through a designated machine group/cell where parts can be transferred from one workstation to another as a part is completed, i.e., unidirectional parallel sequencing method, the total throughput time can be further reduced to 53 hours as shown in Fig. 3.15.9.

This parallel sequencing method provides 29 hours reduction when compared to the optimum sequencing by successive sequencing, and 39 hours reduction when compared to the longest possible sequencing time.

This simple example illustrates that group scheduling is more effective and results in minimum total throughput time. This example also demonstrates that a machine group/cell provides for further reduction in the total throughput time by making parallel sequencing possible with a machine group/cell.

Table 3.15.1. Basic Data for Group Scheduling of a Single Part Family

	Machines			
Job	M_1	M_2	M_3	M_4
J_1	16	14	9	10
J_2	8	12	4	15
J_3	11	15	17	4

Related algorithms of this heuristic method are available for solving the problems for nonuniform, nonunidirectional scheduling where operations or workstations required for the selected part families are neither uniform nor in the same order.[8] In most cases, a real shop presents nonuniform and nonunidirectional problems: the operations required for parts may be in different order, i.e., back and forth.

Optimal Sequencing for a Set of Part Families

Optimization analysis of group scheduling for multiproduct and multistage manufacturing systems with more than one part family[7,8] is demonstrated with the basic data given in Table 3.15.2 and using both the "branch-and-bound" method and the same heuristic method used previously. The solution of this problem by the branch-and-bound method is exhibited in Table 3.15.3. The algorithms for determining optimal group and job sequences to minimize total tardiness are also available. The heuristic method for group scheduling, which is used for optimum sequencing for a single part family, can also be applied to a multiproducts, multistage manufacturing problem. For this example (Table 3.15.2), the solutions for optimal group and job sequences are the same as the solutions by the branch-and-bound method. Although it provides only near optimum solutions, the algorithm for this heuristic method is far simpler and easier to compute, compared to other similar heuristic algorithms and certainly in comparison to the branch and bound method.

Machine Loading Analysis

The analysis of machine loading for group scheduling is a complex problem and it is not simple to develop an adequate algorithm for practical applications. However, some mathematical models for the machine loading and product-mix analysis problems[10] are available.

3.15.5 CAD/CAM APPLICATIONS

3.15.5.1 General Approach

Recently, intensive attention has been given to CAD/CAM systems as computer applications are expanded, and the important role of group technology applications has been recognized. Group technology applications provide a common database for effective integration of CAD and CAM, which leads to successful implementation of CIM. The common database based on the group technology concept plays a critical role in the development and implementation of CAD, effectively creating or modifying engineering product design. In addition, if the database is developed so that it corresponds to groupings of tooling setup, machines, or workstations using group technology concepts, it also provides a basis for computer automated process planning. Computer integrated manufacturing must use group technology applications to be successful.[12]

As CAD/CAM-oriented manufacturing technology develops, more generative and evolutionary systems of group technology should be studied and implemented in all areas of manufacturing, e.g., design, planning, scheduling, tooling, production, testing, assembly, inventory, and control. The ultimate successful implementation

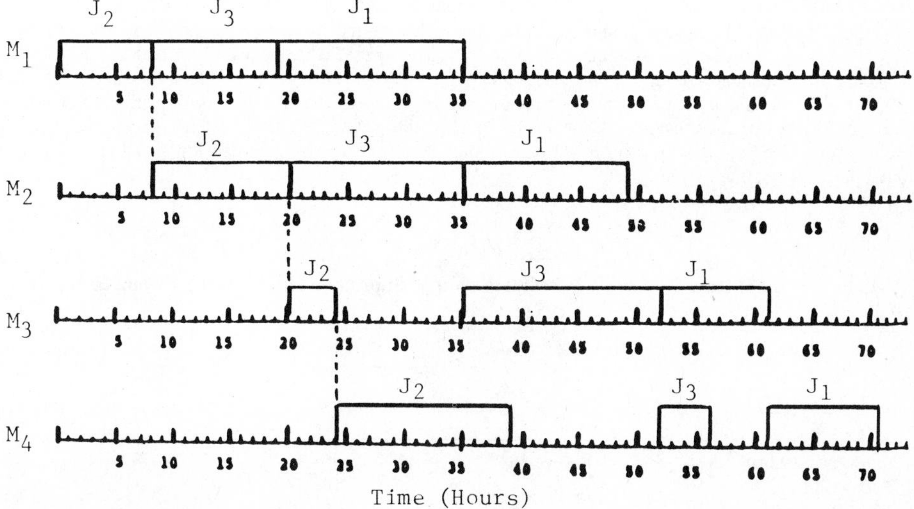

Fig. 3.15.8. Gantt chart for optimum job sequence of J_2-J_3-J_1 with unidirectional, successive sequencing method. (See Table 8.2.1.)

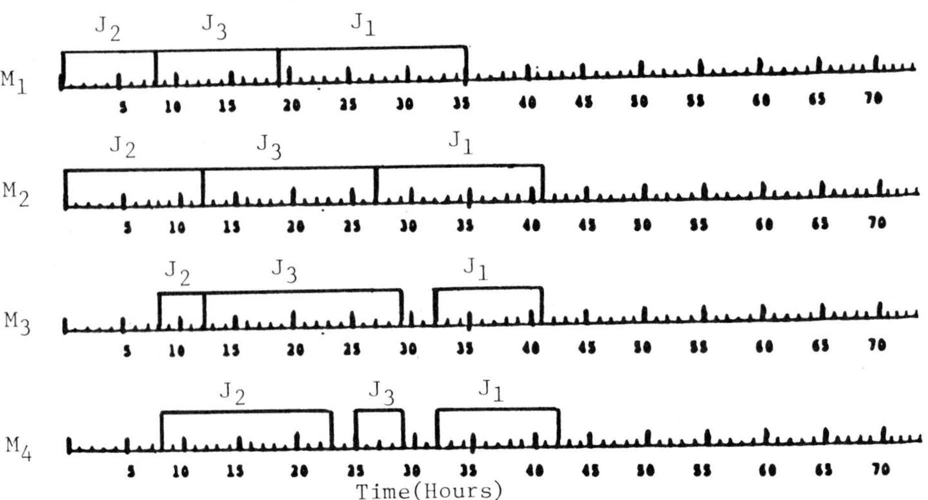

Fig. 3.15.9. Gantt chart for optimum job sequence of J_2-J_3-J_1 with unidirectional, parallel sequencing method. (See Table 8.2.1.)

of CAD/CAM or CIM is certainly based on the balanced development of hardware and software for group technology applications, which provides the essential basis for the further development of CAD/CAM technology.

Group technology is definitely a dynamic and evolutionary technique, whose influence on future manufacturing systems continues to increase. It is clear that the role of group technology is broadened with advancements in theory and application, both for improving productivity in conventional batch-type manufacturing and for proper implementation of the CIM system.

3.15.5.2 Computer Automated Process Planning

For successful implementation of CIM systems, computer automated process planning is essential. Automated process-planning techniques as a basis for a rational and logical approach to the most economical design of components for manufacture are key to achieving optimum manufacturing productivity in CIM systems. The development of a framework of decision-making based on a number of algorithms or logic flow diagrams at each particular decision-making stage is essential, particularly for the retrieval of information on part design specifications, tooling requirements, machining conditions and capabilities, and other pertinent data.

Computer automated process planning can be done either by the *variant* or the *generative* process. Most of the current process-planning systems employ the variant process, which is based on part families and the given database of standard tooling and process routes (e.g., CAM-I CAPP System),[13] as shown in Fig. 3.15.10. The generative technique creates a unique process plan for a particular part. Efforts by various groups have been made to develop generative process planning. Using suitable generative coding systems, geometric modeling, expert systems, etc., along with appropriate optimization logics, it is possible to develop an automatically generative process-planning system.

Table 3.15.2. Basic Data for Group Scheduling on a Multiproduct and Multistage Manufacturing System

	Group											
	G_1				G_2			G_3				
Job		J_{11}	J_{12}	J_{13}		J_{21}	J_{22}		J_{31}	J_{32}	J_{33}	J_{34}
Setup or Processing Time (hours)	S_1	P_{11}	P_{12}	P_{13}	S_2	P_{21}	P_{22}	S_3	P_{31}	P_{32}	P_{33}	P_{34}
M_1	3	3	4	4	7	5	6	4	2	1	9	4
M_2	6	5	6	1	3	5	2	5	2	8	2	3
M_3	4	7	8	7	2	3	5	4	7	4	6	2

Table 3.15.3. Optimal Solution for Minimizing the Total Throughput Time (See Table 3.15.2)

Group Sequence	G_1	G_2	G_3
Job Sequence	J_{13}-J_{11}-J_{12}	J_{22}-J_{21}	J_{31}-J_{33}-J_{32}-J_{34}

3.15.5.3 NC Machines and Machining Centers

Another important application of the group technology concept is software development for numerical control (NC) machining, which is referred to as "Part-Family Programming." Part-family programming is an NC programming method that groups common or similar program elements into a single, master computer program. Therefore, part-family programming increases the productivity of costly NC operations by saving programming time, manpower, and tape proveout time.

NC machining centers are capable of doing the work of several conventional machine tools. They consolidate a number of setups into one, thus having the effect of group layout (cellular or GT layout) of machines. Since NC machining centers are expensive, their use should be optimized through group technology. Where the installation of a machining center is being considered, the maximum benefits will be obtained by considering the place of the whole machining complex that supports and loads NC machine centers in the total factory system. Thus, NC machining centers will generally be operated efficiently under a group technology environment.

3.15.5.4 Automated Factory Systems

The introduction of such new technological innovations as CNC, DNC, machining centers, industrial robots, and microprocessors leads to more integrated applications of group technology for automated factory systems, which are essentially based on hierarchical computer control of automated manufacturing cell systems. Increasing CNC-type minicomputer control of the individual NC machines or machining centers in a cell will in time provide a more economic basis for an overall DNC system of the cells. Eventually it will become profitable to link all the cells or centers in the factory with a large computer, providing an initial basis for overall optimization and automation of the total factory system. The concept for such a system is based on the use of group technology cells, each devoted to the production of a given family of parts or a group of part families.

As the development of CIM leads to more generative design and process planning, classification and coding systems may become an integral part of the total computerized generative system evolving from CIM. The optimum design and effective operation of an automated factory system requires careful analysis of all the requirements that the manufacturing cell systems must satisfy in a group technology environment. Therefore, it is essential to know the part populations and statistics for effective part-family groupings to design optimum manufacturing cell systems as integral parts of the total factory system.

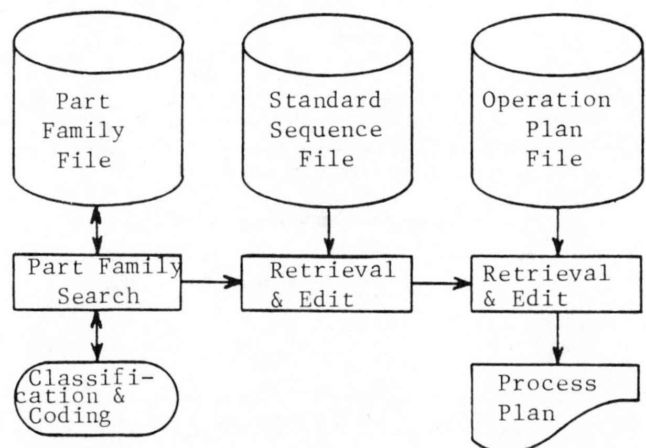

Fig. 3.15.10. Simplified flow diagram of CAM-I CAPP system.

3.15.6 ECONOMIC JUSTIFICATION

Appropriate and successful implementation of group technology will result in many economic benefits through:

1. Improvements in engineering design.
2. Less stock and fewer purchases.
3. Simplified production planning and control.
4. Optimum sequencing and machine loading.
5. Reduction of tooling and setup times and costs.
6. Shorter throughput time.
7. More efficient utilization of production machines.
8. Better quality and high productivity.

When a group technology method is proposed to replace a conventional method, a comparative analysis for economic justification must be made. Many formulae and procedures have been proposed for such economic analyses. Two examples, that is, group tooling costs and group machining costs, are presented.

3.15.6.1 Group Tooling Costs

One of the advantages of group technology applications is the rationalization of tool designs and the reduction of tooling setups, which lead to reduction of tooling costs and production costs as a whole. The cost analysis of group tooling (group jigs and fixtures) in comparison with that of conventional tooling methods becomes essential for the justification of group technology applications in tooling.

(A) CONVENTIONAL TOOLING METHOD

$$C_{tw1} = \sum_{i=1}^{p} C_{w1(i)}$$

(B) GROUP TOOLING METHOD

$$C_{tw2} = \sum_{i=1}^{q} C_{a(i)} + C_{w2}$$

where

C_{tw1} = total tooling costs of conventional methods using p different jigs or fixtures
$C_{w1(i)}$ = cost of jig or fixture i using the conventional tooling method $i = 1, \ldots, p$
C_{tw2} = total costs for group tooling using a group jig or fixture with q different adapters
$C_{a(i)}$ = cost of adapter i, $i = 1, \ldots, q$
C_{w2} = cost of a group jig or fixture
p = number of different jigs or fixtures used (also, possibly, number of different parts to be produced)
q = number of adapters used for the production of a family of parts

(All costs in dollars)

315.6.2 Group Machining Costs

Group machining is one of the most important features of group technology applications. Although group machining is advantageous from various technical points of view, it is still desirable to confirm the advantages of the group machining method over the conventional machining method.

(A) CONVENTIONAL (INDIVIDUAL) MACHINING

$$C_{tm1} = C_o\left[\sum_{i=1}^{n} T_{c1(i)}N_{1(i)} + \sum_{i=1}^{n} T_{s1(i)}\right] + \sum_{i=1}^{n} D_{t1(i)}$$

(B) GROUP MACHINING

$$C_{tm2} = C_o\left[\sum_{i=1}^{n} T_{c2(i)}N_{2(i)} + T_{s2} + \sum_{i=1}^{n-1} T_{sa(i)}\right]$$

$$+ D_{t2} + \sum_{i=1}^{n-1} D_{ta(i)}$$

where

C_{tm1} = total machining cost per lot for conventional machining
C_{tm2} = total machining cost per lot for group machining
C_o = cost per unit time of machining and setup
n = number of parts in the part family
$T_{c1(i)}$ = unit machining time per piece for part i by conventional machining, min/piece
$N_{1(i)}$ = lot size for part i with conventional machining, pieces/lot
$T_{c2(i)}$ = average unit machining time per piece for part i by group machining, min/piece
$N_{2(i)}$ = lot size for part i with group machining
$T_{s1(i)}$ = setup time per lot per part type i for conventional machining, min/lot or part
T_{s2} = setup time per lot (per a part family) for group machining, min/lot or part family
$T_{sa(i)}$ = setup time per adapter type i for group machining, min/adapter
$D_{t1(i)}$ = average depreciation of tooling per lot per part type i for conventional machining, dollars/lot or part
D_{t2} = average depreciation of tooling per lot or part family for group machining, dollars/lot or part
$D_{ta(i)}$ = average depreciation per adapter type i for group machining, dollars/lot or part

3.15.7 MANAGEMENT PROBLEMS

Implementation of a new concept or system calls for a high degree of cooperation between a number of groups or departments in the company. However, in most cases, cooperation is lacking. Close cooperation between design engineering, process planning, production control, inventory control, purchasing, tooling, and the production shops is essential for successful GT implementation. It is common for a great deal of suspicion to follow any change to an existing pattern of life and system. Group technology will require changes not only in working environment but also in thinking.

As the group technology concept has become more accepted, it has also become apparent that benefits obtained through successful GT applications are not determined entirely by its technical characteristics. It adds to job satisfaction by worker participation in decision-making as a group, personalized work relationships, variety in tasks, freedom to determine methods, group production methods, and so on.

REFERENCES

1. H. D. Hathaway, "The Mnemonic Systems of Classification; As Used in the Taylor System of Management," *Industrial Management*, **60**(3) (1920), 173–183.
2. T. Gongaware and I. Ham, "Cluster Analysis Applications for Group Technology Manufacturing Systems," *Proceedings, IX North American Manufacturing Research Conference and 1981 SME Transactions*, May 1981, pp. p. 503–508, Society of Manufacturing Engineers.
3. J. L. Burbidge, *Production Planning*, London: Heinemann, 1971.
4. I. Ham and D. T. Ross, *Integrated Computer-Aided Manufacturing (ICAM) Task-II Final Report*, Vol. 1, Group Technology Classification and Coding, U.S. Air Force Technical Report AFML-TR-77-218, Wright-Patterson Air Force Base, Dayton, Ohio, December, 1977.
5. Japan Society for Promotion of Machine Industry, *Guide Book for Group Technology Implementation* (Japanese). Tokyo, Japan, 1979.
6. A. R. Thompson, "Establishing a Classification & Coding System," Technical Paper No. MS76-276, SME, Dearborn, Michigan, 1976.
7. K. Hitomi and I. Ham, "Group Scheduling Techniques for Multi-production, Multi-stage Manufacturing Systems," *ASME Transactions, Ser. B*, **99**(3), 759–756 (1977).
8. J. Taylor and I. Ham, "The Use of a Micro Computer for Group Scheduling," *Proceedings of the IX North American Manufacturing Research Conference and 1981 SME Transactions*, May, 1981, pp. 483–491, Society of Manufacturing Engineers.
9. V. A. Petrov, *Flowline Group Production Planning*. London: Business Publications, 1968.
10. I. Ham and K. Hitomi, "Machine Loading and Product-mix Analysis for Group Technology," *ASME Transactions*, **100**, 370–374 (August, 1978).
11. N. Sato, J. Ignizio, and I. Ham, "Integrated Applications of Group Scheduling and Materials Requirement Planning (MRP)," *CIRP Annals*, **27**, 471–473 (August, 1978).
12. D. E. Wisnosky, W. A. Harris, and O. L. Shunk, "An Overview of the Air Force Program for Integrated Computer Aided Manufacturing (ICAM)," Technical Paper #MS77-254, SME, Dearborn, Michigan, 1977.

13. CAM-I Training Material for CAPP CAM-I Automated Process Planning, Vol. 1, TM-77-AMP-01, published by Computer-Aided Manufacturing-International, Arlington, TX, 1977.

14. S. O. Mitrafanov, *Scientific Principles of Group Technology* (English translation), J. Grayson, Ed., National Lending Library for Science and Technology, United Kingdom, 1966.

BIBLIOGRAPHY

Arn, E. A., *Group Technology*. New York: Springer-Verlag, 1975.

Burbidge, J. L., Proceedings of International Seminar on Group Technology, International Center for Advanced Technical and Vocational Training, Turin, Italy: 1969.

Burbidge, J. L., "A Study of the Effects of Group Production Methods on the Humanization of Work," final report International Labor Office, Geneva, Switzerland, June, 1975.

Burbidge, J. L., *The Introduction of Group Technology*. New York: Wiley, 1975.

Devries, M. F., S. M. Harvey, and V. A. Tipnis, *Group Technology: An Overview and Bibliography*, Cincinnati: Machinability Data Center (MDC 76-601), 1976.

Edwards, G. A. B., *Readings in Group Technology*, London: Machinery Publishing Company, 1971.

Gallagher, C. C., and W. A. Knight, *Group Technology*, London: Butterworths, 1973.

Ham, I., Group Technology, Ch. 7.8 (pp. 7.8.1–7.8.19) of *Handbook of Industrial Engineering*. New York: Wiley, 1982.

Hyde, W. F., *Improving Productivity by Classification, Coding, and Data Base Standardization; The Key to Maximizing CAD/CAM and Group Technology*. New York: Marcel Dekker, 1981.

Mitrafanov, S. P., *Scientific Principles of Group Technology* (in Russian). Moscow: Mashinostroyenie, 1970.

Mitrafanov, S. P., *Scientific Principles of Machine Building Production* (in Russian). Moscow: Mashinostroyenie, 1976.

Opitz, H., *A Classification System to Describe Workpieces*, Parts 1 and 2. London and New York: Pergamon, 1970.

CHAPTER 3.16

THE TOYOTA PRODUCTION SYSTEM

YASUHIRO MONDEN

University of Tsukuba
Sakura, Japan

The Toyota production system was developed and promoted by Toyota Motor Corporation and is being adopted by many Japanese companies in the aftermath of the 1973 oil shock. Though the main purpose of the system is to reduce costs, the system also helps increase the turnover ratio of capital (i.e., total sales/total assets) and improves the total productivity of a company as a whole.

Even during periods of slow growth, the Toyota production system could make a profit by decreasing costs in a unique manner—that is, by completely eliminating excessive inventory or workforce. It would probably not be overstating our case to say that this is another revolutionary production management system. It follows the Taylor system (scientific management) and the Ford system (mass-assembly line). This chapter examines the basic idea behind this production system, how it makes products, and especially in what areas Japanese innovation can be seen.

Furthermore, the framework of this production system is examined as a unit by connecting its basic ideas or goals with the various tools and methods used for achieving these goals.

3.16.1 BASIC IDEA AND FRAMEWORK

The Toyota production system is a reasonable method of making products, since it completely eliminates unnecessary elements in production for the purpose of cost reduction. The basic idea in such a production system is to produce the kind of units needed, at the time needed, and in the quantities needed. With the realization of this concept, unnecessary intermediate and finished product inventories would be eliminated.

However, although cost-reduction is the system's most important goal, it must achieve three other subgoals in order to achieve its primary objective. They include:

1. Quantity control, which enables the system to adapt to daily and monthly fluctuations in demand in terms of quantities and variety;
2. Quality assurance, which assures that each process will supply only good units to subsequent processes;
3. Respect-for-humanity, which must be cultivated while the system utilizes the human resource to attain its cost objectives.

It should be emphasized here that these three goals cannot exist independently or be achieved independently without influencing each other or the primary goal of cost reduction. It is a special feature of the Toyota production system that the primary goal cannot be achieved without realization of the subgoals and vice versa. All goals are outputs of the same system; with productivity as the ultimate purpose and guiding concept, the Toyota production system strives to realize each of the goals for which it has been designed.

Before discussing the contents of the Toyota production system in detail, an overview of this system is in order (Fig. 3.16.1). The outputs or results side (costs, quality, and humanity) as well as the inputs or constituents side of the Toyota production system are depicted.

A continuous flow of production, or adapting to demand changes in quantities and variety, is created by achieving two key concepts: Just-in-time and Autonomation. These two concepts are the pillars of the Toyota

This chapter is taken with permission of the author and publisher from Monden, Yasuhiro, *Toyota Production System*, Industrial Engineering and Management Press, Norcross, GA: Institute of Industrial Engineers, Inc., 1983.

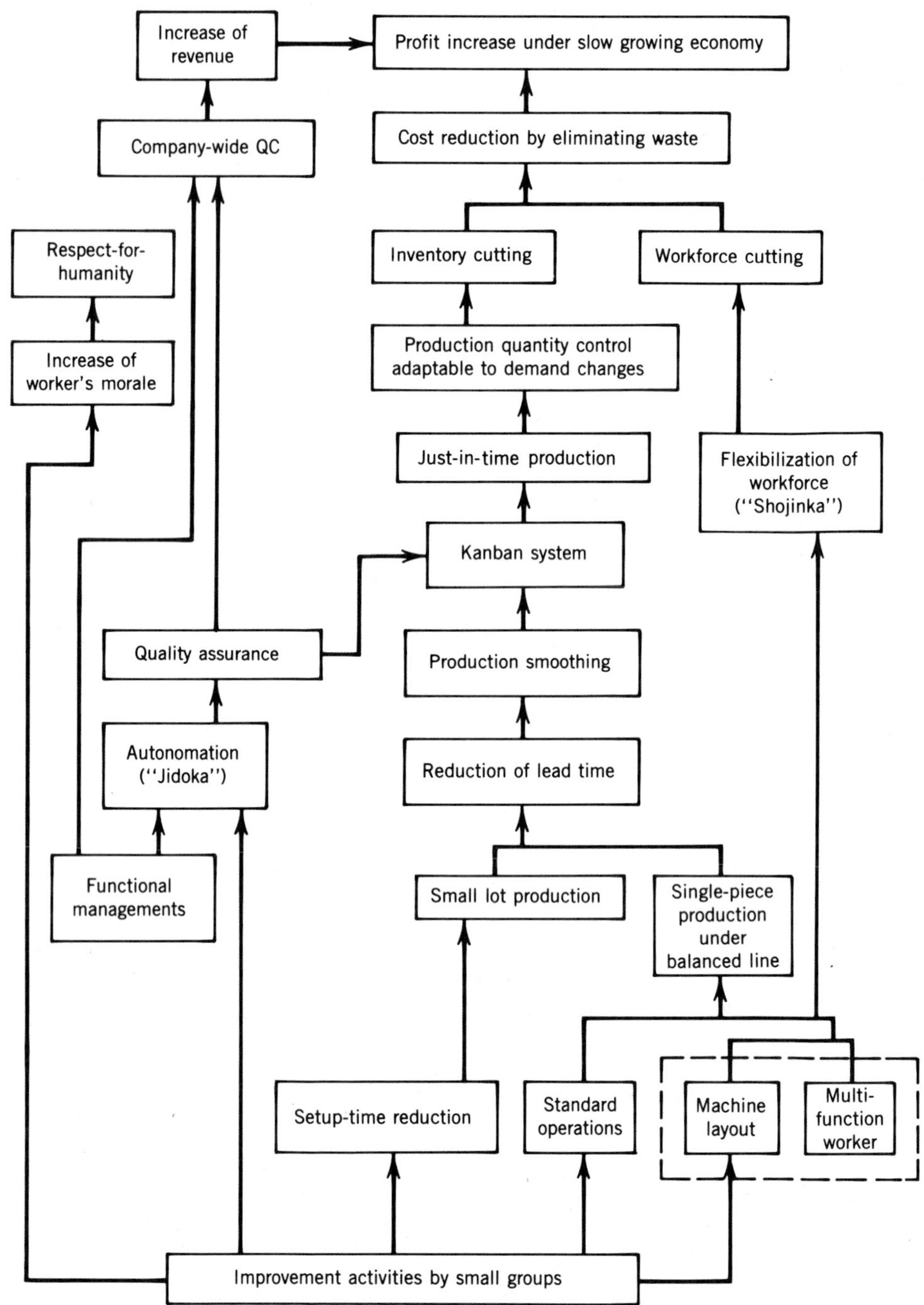

Fig. 3.16.1. How costs, quantity, quality, and humanity are improved by the Toyota production system.

production system. *Just-in-time* basically means to produce the necessary units in the necessary quantities at the necessary time. *Autonomation* ("Jidoka" in Japanese) may be loosely interpreted as autonomous defects control. It supports Just-in-time by never allowing defective units from a preceding process to flow into and disrupt a subsequent process.

Two concepts also key to the Toyota production system include *Flexible Workforce* ("Shojinka" in Japanese) which means varying the number of workers to demand changes, and *Creative thinking or inventive ideas* ("Soikufu"), or capitalizing on worker suggestions.

To realize these four concepts, Toyota has established the following systems and methods:

1. Kanban system to maintain Just-in-time production.
2. Production smoothing method to adapt to demand changes.
3. Shortening of the setup time for reducing the production lead time.
4. Standardization of operations to attain line balancing.
5. Machine layout and the multi-function worker for the flexible workforce concept.
6. Improvement activities by small groups and the suggestion system to reduce the workforce and increase the worker's morale.
7. Visual control system to achieve the Autonomation concept.
8. "Functional Managements" system to promote company-wide quality control, etc.

3.16.1.1 Just-in-Time Production

The idea of producing the necessary units in the necessary quantities at the necessary time is described by the short term Just-in-time. Just-in-time means, for example, that in the process of assembling the parts to build a car, the necessary kinds of subassemblies of the preceding processes should arrive at the product line at the time needed in the necessary quantities. If Just-in-time is realized in the entire firm, then unnecessary inventories in the factory will be completely eliminated, making stores or warehouses unnecessary. The inventory carrying costs will be diminished, and the ratio of capital turnover will be increased.

However, to rely solely on the central planning approach which instructs the production schedules to all processes simultaneously, it is very difficult to realize Just-in-time in all the processes for a product like an automobile, which consists of thousands of parts. Therefore, in the Toyota system, it is necessary to look at the production flow conversely; in other words, the people of a certain process go to the preceding process to withdraw the necessary units in the necessary quantities at the necessary time. Then what the preceding process has to do is produce only enough quantities of units to replace those that have been withdrawn.

In this system what kind of units and how many units needed are written on a taglike card called *Kanban*. The Kanban is sent to the people of a preceding process from the subsequent process. As a result, many processes in a plant are connected with each other. This connecting of processes in a factory allows for better control of necessary quantities for various products.

In the Toyota production system, the Kanban system is supported by the following:

Smoothing of production
Reduction of setup time
Design of machine layout
Standardization of jobs
Improvement activities
Autonomation

3.16.2 KANBAN SYSTEM

Many people call the Toyota production system a Kanban system: this is incorrect. The Toyota production system is the way to make products, whereas the Kanban system is the way to manage the Just-in-time production method. In short, the Kanban system is an information system to harmoniously control the production quantities in every process. Unless the various prerequisites of this system are implemented perfectly (i.e., design of processes, standardization of operations and smoothing of production, etc.), then Just-in-time will be difficult to realize, even though the Kanban system is introduced.

A Kanban is usually a card put in a rectangular vinyl envelope. Two kinds are mainly used: withdrawal Kanban and production-ordering Kanban. A *withdrawal* Kanban details the quantity which the subsequent process should withdraw, while a *production-ordering* Kanban shows the quantity which the preceding process must produce. These cards circulate within Toyota factories, between Toyota and its many cooperative companies, and within the factories of cooperative companies. In this manner, the Kanbans can contribute information on withdrawal and production quantities in order to achieve Just-in-time production.

Suppose we are making products A, B, and C in an assembly line. The parts necessary to produce these products are a and b which are produced by the preceding machining line (Fig. 3.16.2). Parts a and b produced

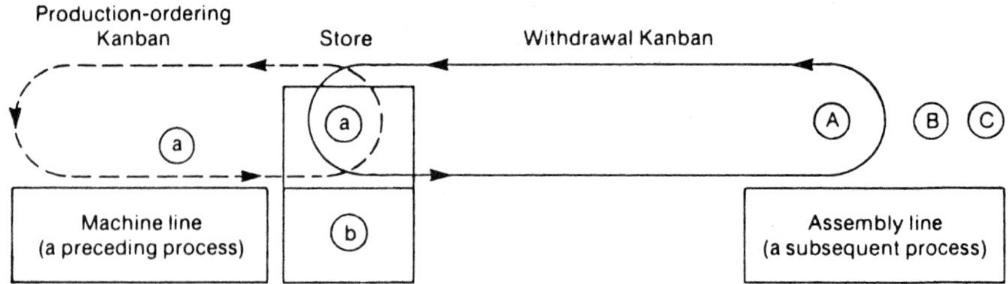

Fig. 3.16.2. The flow of two Kanbans.

by the machining line are stored behind this line, and the production-ordering Kanbans of the line are attached to these parts. The carrier from the assembly line making product A will go to the machining line to withdraw the necessary part a with a withdrawal Kanban. Then, at store a, he picks up as many boxes of this part as his withdrawal Kanbans and he detaches the production-ordering Kanban attached to these boxes. He then brings these boxes back to his assembly line, again with withdrawal Kanbans.

At this time, the production-ordering Kanbans are left at store a of the machining line showing the number of units withdrawn. These Kanbans will be the dispatching information to the machining line. Part a is then produced in the quantity directed by that number of Kanbans. In this machining line, actually, parts a and b are both withdrawn, but these parts are produced according to the detached order of the production-ordering Kanbans.

3.16.2.1 Fine-Tuning Production

Let's consider the fine-tuning of production by using a Kanban. Assume that an engine manufacturing process must produce 100 engines per day. The subsequent process requests five engines per one-time lot by the withdrawal Kanban. These lots are then picked up 20 times per day, which amounts to exactly 100 engines produced daily.

Under such a production plan, if the need occurs to decrease all processes by 10% as a fine-tuning procedure of production planning, the final process in this example has to withdraw engines 18 times per day. Then, since the engine process produces only 90 units in a day, the remaining hours for 10 units of production will be rested by stopping this process. On the other hand, if there is a need to increase production quantities by 10%, the final process must withdraw the engines 22 times per day with the Kanban. Then the preceding process has to produce 110 units, and the additional 10 units would be covered by overtime.

Although the Toyota production system has the production management philosophy that units could be produced without any slack or unnecessary stock by regarding all of the human resources, machines, and materials as perfect, the risk of variations in production needs still exists. This risk is handled by the use of overtime and improvement activities at each process.

3.16.2.2 Smoothing of Production

The smoothing of production is the most important condition for production by Kanban and for minimizing slack time in regards to manpower, equipment, and works in process; it is the cornerstone of the Toyota production system.

As described previously, the subsequent processes go to the preceding processes to withdraw the necessary goods at a necessary time in the necessary quantities. Under such a production rule, if the subsequent process withdraws parts in a fluctuating manner in regards to time or quantity, then the preceding processes should prepare as much inventory, equipment, and manpower as needed to adapt to the peak in the variance of quantities demanded. Also, where there are many sequenced processes, the variance of the quantities withdrawn by each subsequent process may become larger further back to preceding processes. In order to prevent such large variances in all production lines, including the external subcontracted companies, an effort must be made to minimize the fluctuation of production in the final assembly line. Therefore, the assembly line of finished cars, as the final process in the Toyota factory, will convey each type of automobile in its minimum lot size, realizing the ideal of "one-piece" production and conveyance. The line will also receive the necessary parts, in their small lot sizes, from the preceding processes.

In short, production smoothing minimizes the variation in the withdrawn quantity of each part produced at each subassembly, thereby allowing the subassemblies to produce each part at constant speed or at a fixed quantity per hour. Such a smoothing of production can be illustrated by the following example: Suppose there is a production line which is required to produce 10,000 Coronas with 20 eight-hour operating days in a month. The 10,000 Coronas consist of 5,000 sedans, 2,500 hardtops and 2,500 wagons. Dividing these numbers by 20 operating days results in 250 sedans, 125 hardtops and 125 wagons per day; this is the smoothing of production in terms of the average daily number of each kind of car.

During an eight-hour shift or operation (480 min.), all 500 units must be produced. Therefore, the *unit cycle time*, or the average time required to produce one vehicle of any type, is .96 minutes (480/500) or approximately 57.5 seconds.

The proper mix or production sequence can be determined by comparing the actual cycle time to produce one vehicle of any type with the maximum permitted time to produce a specific model of Corona. For example, the maximum time to produce one Corona sedan is determined by dividing the shift time (480 min.) by the number of sedans to be produced in the shift (250); in this case, the maximum time is 1 min., 55 sec. This means that a sedan must and will be generated every 1 min., 55 sec.

Comparing this time interval with the cycle time of 57.5 seconds, it is obvious that another car of any type could be produced between the time one sedan is completed and time when another sedan must be produced. So, the basic sequence is sedan, other, sedan, other, etc.

The maximum time to produce a wagon or a hardtop is 3 min., 50 sec. (480/125). Comparing this figure with the cycle time of 57.5 seconds, it is obvious that three cars of any type can be produced between each wagon or hardtop. If a wagon follows the first sedan in production, then the production sequence would be sedan, wagon, sedan, hardtop, sedan, wagon, sedan, hardtop, etc. This is an example of the smoothing of production in terms of the varieties of products.

Considering the actual manufacturing machines or equipment, a conflict arises between product variety and production smoothing. If a great variety of products are not produced, having specific equipment for mass production will usually be a powerful weapon for cost reduction. In Toyota, however, there are various kinds of cars differentiated in various combinations by types, tires, options, colors, etc. As one example, three or four thousand kinds of Coronas are actually being produced. To promote smoothed production corresponding to such a variety of products, it is necessary to have general-purpose or "flexible" machines. By putting minimum instruments and tools on these machines, Toyota has specified production processes to accommodate the general usefulness of these machines.

An advantage of smoothed production responding to product variety is that the system can adapt smoothly to the variations in customer demand by gradually changing the frequency (times) of lots without altering the lot size in each process, or fine-tuning of production by Kanban. In order to realize the smoothed production, the reduction of production lead time will be necessary to promptly and timely produce various kinds of products. Then the reduction in lead time will require the shortening of setup time for minimizing the lot size.

3.16.2.3 Setup Problems

The most difficult point in promoting smoothed production is the setup problem. In a pressing process, for example, common sense dictates that cost reduction can be obtained through continuously using one type of die, thereby allowing for the biggest lot size and reducing setup costs. However, under the situation where the final process has average its production and reduced the stocks between the punchpress and its subsequent body line, the pressing department as a preceding process must make frequent and speedy setups, which means altering the types of dies for the pressing corresponding to a great variety of products, each frequently withdrawn by the subsequent process.

At Toyota, the setup time of the pressing department had been about 2 or 3 hrs. from 1945 to 1954. It was reduced to a quarter hour in the years 1955–64, and after 1970, it dropped to only 3 min.

In order to shorten the setup time, it is important to neatly prepare in advance the necessary jigs, tools, the next die and materials, and remove the detached die and jigs after the new die is settled and the machine begins to operate. This phase of setup actions is called the *external* setup. Also, the worker should concentrate only on such actions to be taken while the machine is stopping. This phase of setup actions is called the *internal* setup. The most important point is to convert as much as possible of the internal setup to the external setup.

3.16.2.4 Design of Processes

Consider the design or layout of processes in a plant. Previously in this factory, each of five stands of lathes, milling machines, and drilling machines were laid out side by side, and one machine was handled by one worker, e.g., a turner handled only a lathe. According to the Toyota production system, the layout of machines would be rearranged to smooth the production flow. Therefore, each worker would handle three types of machines; for example, a worker might handle a lathe, a milling machine, and a drilling machine at the same time. This system is called *multi-process holding*. In other words, the single-function worker, a concept which previously prevailed in Toyota factories, has become a *multi-function worker*.

In a multi-process holding line, a worker handles several machines of various processes one by one, and work at each process will proceed only when the worker completes his given jobs within a cycle time. As a result, the introduction of each unit to the line is balanced by the completion of another unit of finished product, as ordered by the operations of a cycle time. Such production is called *one-piece* production and conveyance. The rearrangement leads to the following benefits:

Unnecessary inventory between each process can be eliminated.

The multi-process worker concept can decrease the number of workers needed, and thereby increase productivity.

As workers become multi-functional workers, they can participate in the total system of a factory and thereby feel better about their jobs.

By becoming a multi-functional worker, each worker can engage in teamwork, or workers can help each other.

Such a multi-process worker or multi-functional worker concept is a very Japanese-like method. In American and European companies there are many sorts of craft unions in one plant; a turner, for example, handles only a lathe and will not usually work on any other kind of machine, whereas in Japan there exists only an enterprise-union in each company, which makes the mobility of a laborer or the multi-process holding by a laborer very easy. Obviously, this difference may pose one of the major obstacles for American and European companies that might wish to adopt the Toyota production system.

3.16.3 STANDARDIZATION OF JOBS

The standard operation at Toyota is a bit different from the usual operation in that it shows mainly the sequential routine of various operations taken by a worker who handles multiple kinds of machines as a multi-functional worker.

Two kinds of sheets show standard operations: the *standard operations routine sheet*, which looks like a usual man-machine chart, and the *standard operations sheet*, which is tacked up in the factory for all workers to see. The latter sheet specifies the cycle time, standard operations routine, and standard quantity of the work in process.

A cycle time, or tact time, is the standard specified number of minutes and seconds that each line must produce one product or one part. This time is computed by the following two formulas. Initially, the necessary output per month is predetermined from the demand side. Then:

$$\text{Necessary output per day} = \frac{\text{necessary output per month}}{\text{operating days per month}}$$

$$\text{Cycle time} = \frac{\text{operating hours per day}}{\text{necessary output per day}}$$

Each production department will be informed of this necessary quantity per day and the cycle time from the central planning office once in each previous month. In turn, the manager of each process will determine how many workers are necessary for this process to produce one unit of output in a cycle time. The workers of the entire factory then must be repositioned in order that each process will be operated by a minimum number of workers.

The Kanban is not the only information to be given to each process. A Kanban is a type of production dispatching information during the month in question, whereas the daily quantity and cycle time information are given in advance to prepare the master production schedule throughout the factory.

The standard operations routine indicates the sequence of operations that should be taken by a worker in multiple processes of the department. This is the order for a worker to pick up the material, put it on his machine, and detach it after processing by this machine. This order of operations continues for various machines he handles. The line balancing can be achieved among workers in this department since each worker will finish all of his operations within the cycle time.

The standard quantity of work-in-process is the minimum quantity of work-in-process within a production line, which includes the work attached to machines. Without this quantity of work, the predetermined sequence of various kinds of machines in this whole line cannot operate simultaneously. Theoretically, however, if the invisible conveyor belt is realized in this line, there is no need to have any inventory among the successive process.

3.16.4 AUTONOMATION

As noted previously, the two pillars which support the Toyota production system are Just-in-time and Autonomation. In order to realize Just-in-time perfectly, 100% good units must flow to the subsequent process, and this flow must be rhythmic without interruption. Therefore, quality control is so important that it must coexist with the Just-in-time operation throughout the Kanban system. Autonomation means to build in a mechanism a means to prevent mass-production of defective work in machines or product lines. The word Autonomation (in Japanese, "Ninben-no-aru Jidoka," which often abbreviates to "Jidoka") is not automation, but the autonomous check of abnormals in a process.

The autonomous machine is a machine to which an automatic stopping device is attached. In Toyota factories, almost all machines are autonomous, so that mass-production of defects can be prevented and machine breakdowns are automatically checked. The so-called *Foolproof* ("Bakayoke" or "Pokayoke") is one such mechanism to prevent defective work by putting various checking devices on the implements and instruments.

The idea of Autonomation is also expanded to the product lines of manual work. If something abnormal happens in a product line, the worker pushes his stop button, thereby stopping his whole line. The Andon in the Toyota system has an important role in helping this autonomous check, and is a typical example of Toyota's "Visual Control System." For the purpose of detecting troubles in each process, an electric light board, called *Andon*, indicating a line stop, is hung so high in a factory that it can easily be seen by everyone. When some worker calls for help to adjust his delay of a job, he turns on the yellow light on the Andon. If he needs the line stopped to adjust some problem with his machines, he turns on the red light. In summary, Autonomation is a mechanism that autonomously checks something unusual in a process.

3.16.4.1 Improvement Activities

The Toyota production system integrates and attains different goals (i.e., quantity control, quality assurance, and respect-for-humanity) while pursuing its ultimate goal of cost reduction. The process by which all these goals are realized is improvement activities (called "Kaizen" in Japanese), a fundamental element of the Toyota system. This is what makes the Toyota production system really tick. Each worker has the chance to make suggestions and propose improvements via a small group called a *QC* circle. Such a suggestion-making process allows for improvement in quantity control by adapting standard operations routine to changes in cycle time; in quality assurance, by preventing recurrence of defective works and machines; and, lastly, in respect-for-humanity, by allowing each worker to participate in the production process.

3.16.5 SUMMARY

The basic purpose of the Toyota production system is to increase profits by reducing costs—that is, by completely eliminating waste such as excessive stocks or workforce. The concept of *costs* in this context is very broad. It is essentially cash outlay in the past, present, or future deductible from sales revenue to attain a profit. Therefore, *costs* include not only manufacturing costs (reduced by cutting the workforce), but also administrative costs and capital costs (reduced by inventory cutting) and sales costs. To achieve cost reduction, production must promptly and flexibly adapt to changes in market demand without having wasteful slacks. Such an ideal is accomplished by the concept of Just-in-time, producing the necessary times in the necessary quantities at the necessary time. At Toyota, the Kanban system has been developed as a means of dispatching production during a month and managing Just-in-time. In turn, in order to implement the Kanban system, production must be smoothed to level the quantities and variety in the withdrawals of parts by the final assembly line. Such smoothing will require the reduction of the production lead time, since various parts must be produced promptly each day. This can be attained by a small lot production or one piece production and conveyance. The small lot production can be achieved by shortening the setup time, and the one piece production will be realized by the multi-process worker who works in the multi-process holding line. Standard operations routine will assure the completion of all jobs to process one unit of a product in a cycle time. The support of Just-in-time production by 100% "good" products will be assured by Autonomation (autonomous defects-control systems). Finally, improvement activities will contribute to the overall process by modifying standard operations, remedying certain defects, and finally, by increasing the worker's morale.

Where have these basic ideas come from? What need evoked them?

They are believed to have come from the market constraints which characterized the Japanese automobile industry in post-war days: great variety within small quantities of production. Toyota thought consistently from about 1950 that it would be dangerous to blindly imitate the Ford system (one which could minimize the average unit cost by producing in large quantities). American techniques of mass production have been good enough in the age of high-grade growth, which lasted until 1973. In the age of the low-level growth after the oil shock, however, the Toyota production system was given more attention and adopted by many industries in Japan in order to increase profit by decreasing costs or cutting waste.

The Toyota production system is a unique, revolutionary system; however, there is no problem for foreign companies in adopting this system except for the possibility of union problems (i.e., the multi-function worker). Also, the Kanban system is compatible with material requirements planning developed in America.

American and European companies could adopt this system, but might encounter some difficulties if they used it partially. Many Japanese companies are already using it in its imperfect form as well as its perfect form. The Kanban system and the smoothing of production could be particularly important to American and European companies. To implement the Toyota system perfectly, however, top management must proceed through the bargaining process with their union people. Such a process has often been experienced by many Japanese companies, too.

PART 4

MACHINES: PRODUCTION PROCESSES AND EQUIPMENT

CHAPTER 4.1

FOUNDRY PROCESSES

RAYMOND M. NAGAN AND RAJEEV V. NAIK

Arwood, Corporation
Rockleigh, New Jersey

4.1.1 INTRODUCTION

The foundry industry worldwide produces about 90 million tons of castings a year. Cast irons, steels, and aluminum alloys are, by far, the most commonly cast alloys. In the United States the industry consists of over 4,000 foundries, making it, on a value-added basis, the fifth largest. It employs over 400,000 people, but only 3% of the foundries employ over 500 people. The total variety of castings produced reflect a very wide diversity of processes, ranging from sand castings to die and investment castings.

The preference of casting over shaping parts by other manufacturing techniques is attributed to the very compelling advantages of the process. Some of these are listed below:

1. The most intricate configurations can be cast combining heavy and thin sections and minimizing secondary operations.
2. Most of the operations are adaptable to mass production and automation.
3. Size capability ranges from minuscule parts to huge ones.
4. Many alloys are very difficult or impossible to shape in any other way.
5. Complicated configurations can be cast integrally rather than assembled.
6. There is a vast choice of engineering properties available.
7. Properties are predominently isotropic.
8. Design allows the optimum combination of lightness and strength.

The foundry process, evolving as it has for over 6,000 years, has assimilated a vast array of technologies. Particularly in the last 30 years, the casting process has been transformed into a science from its former state as a craft or semiscience. Through all the advancements, the chief virtue of the casting has been its ability to take in molten form the shape of a mold cavity and to solidify into that shape, providing useful engineering properties.

Common to all foundry processes is the manufacture of a pattern from which a mold is made or the direct cutting of a metal or refractory mold or die into which molten metal is poured or injected. Molds produced from patterns can only be used once. Molds or dies that are machined can be reused thousands of times but are limited in the size, complexity of part, and alloys that can be cast in them.

Alternative manufacturing processes to casting are usually considered for simple shapes that can be produced with little need of secondary operations; this is also true when very simple machining operations can be performed on the rough shape (such as a forging) to make the finished part. Traditionally, castings have taken a secondary position with respect to forgings in many fatigue limited and high toughness applications. This has been due primarily to microporosity or inclusions that can detract from the performance reliability of castings. In recent years many processing and engineering improvements have reduced defect incidence, and design consideration has begun to shift to castings. The advent of vacuum casting, castable superalloys, very high-strength castable steels, degassing, argon-oxygen decarburization (AOD), filtration, and solidification control techniques have resulted in an ever-widening array of critical applications. Hot isostatic pressing of castings and the leap in size capability of disposable pattern processes, such as investment casting, have also opened new vistas.

4.1.2 BASIC OPERATIONS OF CASTING PRODUCTION

Although the various processes utilized in the foundry industry today consist of widely varying sequences of operations, it is possible to conceptualize a general flow sheet, as shown in Fig. 4.1.1. While the die or mold preparation is different for each of the foundry processes, most of the other operations are universal. In this review, the basic or universal operations are described first and are then followed by descriptions of specific molding and casting processes.

4.1.2.1 Pattern, Die, and Permanent Mold

A pattern is a master replica of the exterior and some or all of the interior surfaces of the casting; internal features and cavities are usually formed by cores (sand or ceramic shapes) made in separate pattern boxes called core boxes. Reusable patterns are made in a variety of materials, including metal, wood, and epoxy resin. Expendable patterns are made of wax or polystyrene injected into metal dies. In die casting and permanent mold casting, patterns are not used, as the metal is directly injected or poured into metal molds. The patterns and dies have allowances designed into them to compensate for factors such as thermal contraction of the casting through solidification to room temperature, taper for pattern or casting removal, and machining stock. Amortization of pattern and die equipment must be factored into the pricing of castings in order to determine the true purchase price. Obviously, the pattern or tooling costs are more critical for low production volume parts. Numerically controlled machine tools programmed from CAD/CAM systems are increasingly being utilized in producing high precision dies.

4.1.2.2 Mold Making

Various techniques are utilized in producing molds. They often embody unique operations that impart engineering and economic advantages. Since the quality and precision of the mold are critical to the quality and precision of the castings, mold production must be controlled just as rigorously as if the mold were the end-product.

Mechanization and automation of the mold-making process is one of the fastest developing operations of the foundry industry. For example, in sand foundries, sand formulations are automatically blended, mixed, and transferred to automated molding machines, which mold, clamp the top and bottom mold halves (cope and drag) together, and transfer them to conveyor lines, which transport them to metal pouring stations. In die casting and investment casting, robots are used extensively. Such developments are helping to upgrade casting

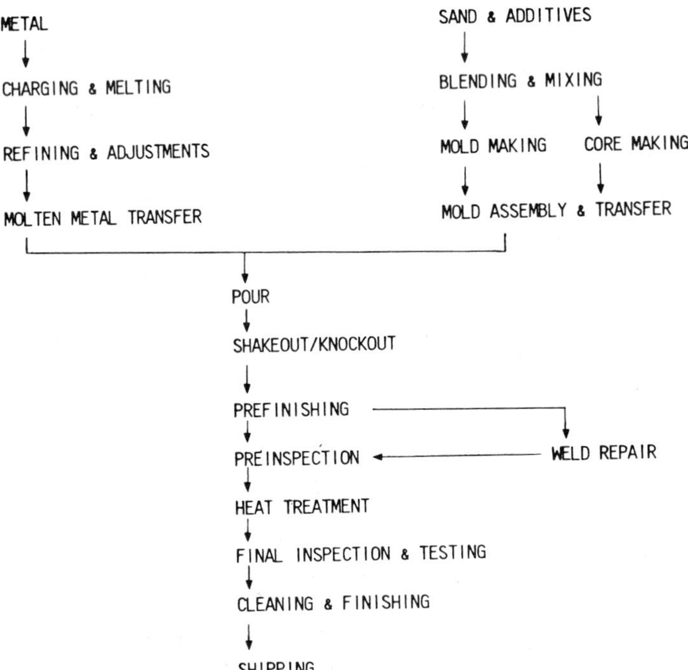

Fig. 4.1.1. General foundry flow chart for sand and ceramic mold casting processes.

Table 4.1.1. Melting Furnaces in Foundry Production

Melting or Refining Unit	Alloys Most Commonly Melted	Treatment in Furnace	Operating Temperature Regime (°F)	Technical and Economical Advantages
Arc	Steels, cast iron	Deoxidation, carburization, desulfurization, dephosphorization	2500–3100	Charge can consist of a wide variety of scrap
Indirect-arc	Stainless steels	Deoxidation	2400–3000	High-production melting
Consumable arc	Titanium alloys	–	3000–3300	Melting reactive metals
Crucible (gas and resistance)	Nonferrous alloys	Degas, grain refinement, modification, deoxidation	500–2500	Low equipment costs
Cupola	Cast iron	Carburization	2500–3000	Low cost, high tonnage, continuous operation
Induction (air)	Steels, cast iron, superalloys copper alloys	Deoxidation, carburization	1800–3200	High rate of melting and versatility
Induction (vacuum)	Superalloys, high alloy steels	–	2500–2900	Casting reactive and oxidizable high-melting-point alloys
Reverberatory	Cast irons and nonferrous alloys	Deoxidation	1300–2800	Large tonnage, intermittent operation

quality while making casting production less labor intensive. The specific mold-making techniques are described in the section on molding and casting processes.

4.1.2.3 Melting, Pouring, and Auxiliary Casting Techniques

The melting and pouring of metal are always rapid sequential operations because of the necessity of maintaining temperature control and the physicochemical condition established in the melting furnace and pouring ladle. Furnace melt charges consist of new metal ingot, scrap, revert scrap, and elemental additions or combinations of these. Control of charge makeup and chemistry is essential to efficiency of melting and conformance to customer specifications. In the melting operation, molten metal treatments such as deoxidation of steel, inoculation of cast iron, and degassing and grain refinement of aluminum alloys are critical to casting quality.

Pouring temperature is a key factor because it affects gas content, chemical composition, and fluidity of the molten metal and the solidified quality (grain size, soundness, and microstructure) of the casting. Controlling mold or die temperature is also important for its effect on filling and rate of solidification. Most pours are performed after the molten metal has been tapped from the furnace into a ladle. A few castings are poured directly from furnace crucibles or tundishes. Each of the types of furnaces listed in Table 4.1.1 embodies technological and economical advantages that favor the use of a particular furnace in melting specific types of stock or alloy families. For example, the electric arc and induction furnaces have become most favored for melting alloys such as low alloy steels and stainless steels above 2500°F; the cupola still remains a heavily used high-production furnace for melting cast irons. Filling the mold cavity with molten metal is usually accomplished by gravity pouring from a bottom-pour, teapot, or open ladle. However, several melt-pour systems or auxiliary systems have been developed that facilitate filling and/or improve the quality of castings (soundness, cleanliness, etc.). Table 4.1.2 lists the techniques commonly utilized in one or more of the various types of foundry operations and the benefits they impart.

4.1.2.4 Gating and Risering Systems

Critical to the quality of all castings are the gating (flow channels) and risering (reservoir) systems, which are designed to fill and feed castings as well as clean molten metal as it flows to the casting cavity of the mold.

Table 4.1.2. Auxiliary Techniques Utilized in Improving Casting Quality and Yield

Technique	Benefit
Vacuum box	Improved filling with lower mold and pour temperatures; cleaner metal
Vacuum assist	Improved filling with lower mold and pour temperatures
Centrifugal	Improved filling and soundness
Vibration	Improved filling and grain refinement
Rollover	Improved filling
Pressure solidification	Improved soundness and metal yield
Upward injection (low pressure)	Improved filling, cleanliness, and soundness
CLA-CLV (proprietary vacuum assist)	Improved filling, cleanliness, metal yield; lower mold and pouring temperatures
Filtering	Improved cleanliness (by retarding filling)

In the economics of casting production, maintaining acceptable quality levels is the first consideration in the design of these systems. Although the ratio of gates and risers weight to casting weight is also an important economical factor, it should not be improved at the expense of maintaining quality levels. Directional solidification, which is the most efficient means for controlling heat transfer to produce sound casting, is promoted by good gating practice: strategic use of risers, chills (heat sinks), exothermic hot topping, and insulation. A sand casting system embodying these principles is shown in Fig. 4.1.2, along with the terminology for the different features of the mold and its gating and risering system.

4.1.2.5 Prefinishing and Finishing

Castings are removed from sand and ceramic mold shells by shake-out, knock-out, and water blast equipment. Permanent mold and die castings are removed by incorporated ejection pins. Next, gates and risers are removed from castings by such devices as hammers, shearing dies, abrasive cutoffs, conventional band saws, and torch cutoffs. Gate stubs and imperfections are removed by chipping hammers, files, grinding wheels, and belts. Finishing is normally completed by sand blast, tumble barrel, grit blast, and vibratory finishing.

4.1.2.6 Heat Treatment

Most alloys require heat treatment in order to homogenize chemistry, relieve stresses, and most importantly to produce the condition and microstructure that will yield the properties sought for in intermediate processing stages (such as straightening); it is also required for application performance. To accomplish these aims, castings can be annealed, solutioned, homogenized, aged, and tempered in a variety of furnaces (e.g., vacuum, hydrogen atmosphere, electric, salt-bath, gas-fired). Depending on the temperature regime of the heat treating cycle and alloy family involved, the protective nature of the furnace atmosphere can be critical to successful production. The cooling stages of these treatments are also critical and can entail slow furnace cooling, air cooling, interrupted quenching, gas, oil, water, and salt quenching, depending on the rate of cooling necessary to achieve proper structure and properties. Also crucial in determining quenching medium and rate is the alloy's susceptibility to quench cracking.

4.1.2.7 Welding

The more challenging and costly a casting is to produce, the more important is the foundry's ability to salvage a defective part. This is usually accomplished by removing defects, usually by grinding, and then welding the resultant excavations. This, of course, makes the weldability of the casting alloy crucial to rejection rate and consequently to the price of the casting. Weldability is also critical because often a casting is a part of a weldment assembly. In such cases, alloys that are very difficult or impossible to weld would not be considered as a casting alloy. However, even weldable alloys may have optimum and marginal compositions within a chemical specification. Weldability is also dependent on the rate of solidification of the casting and the condition in which it is welded.

Fusion welding techniques such as carbon arc, metallic arc, inert gas shielded arc, tungsten, and consumable electrode arc are commonly used in foundries according to the alloys cast.

4.1.2.8 Auxiliary Treatments

Various operations are performed on castings either to restore a surface condition that has deteriorated during solidification (e.g., decarburization restored by recarburization) or to impart improved soundness and properties. These auxiliary treatments include passivation, plating, painting, coating, carburizing, nitriding, chem-milling, shot peening and HIPing (hot isostatic pressing). Most foundries have incorporated one or more of these techniques into their operation.

4.1.2.9 Quality Assurance

Quality assurance embrances four critical functions in the manufacturing process: control of materials, standardization of process controls, inspection of castings, and analysis of metallurgical and statistical results.

Quality control techniques consist of a wide range of nondestructive and destructive tests. The nondestructive tests include visual, dye-penetrant, magnetic particle, radiographic, ultrasonic, eddy current dimensional and surface finish. Mechanical testing procedures conducted on castings, sections cut from castings, or separately cast specimens evaluate hardness, tensile, impact, stress rupture, fatigue, and bending properties. Physical tests include magnetic permeability, coercivity, thermal expansion, and electrical conductivity. Metallographic evaluations are also performed on castings to determine microstructure, inclusion content, soundness, and grain size.

4.1.3 MOLDING AND CASTING PROCESSES

The selection of the right molding process and the method of introduction of metal into the mold cavity are extremely important in producing good castings. The most frequently used casting processes, primarily classified on the basis of these two parameters, include the following:

Sand casting.
Permananet mold casting.
Die casting.
Investment casting.
Plaster mold casting.
Ceramic mold casting.
Centrifugal casting.

4.1.3.1 Sand Casting

Sand casting is the most widely employed casting process and produces by far the largest tonnage of castings. As a production process, it is both very flexible and economical. Most castable alloys can be sand cast, and it is the preferred method of casting unless some aspect of quality or other restrictions prohibit its use. In this process, castings are made by pouring molten metal into sand molds. Molds are typically made by ramming sand mixtures with suitable binders and additive ingredients around a pattern. The pattern is subsequently removed leaving a mold cavity in the sand. Jolting, squeezing, and slinging techniques are primarily used in ramming the sand to produce molds of desired strength. The strength and other related properties of the mold are strongly dependent on the mold constituents and techniques used in making the molds. Sand molding is, therefore, classified mainly on the basis of the binders and molding techniques. The various sand molding processes used in practice are green sand, dry sand, core sand, carbon dioxide, shell, floor and pit, vacuum, and expanded polystyrene molding.

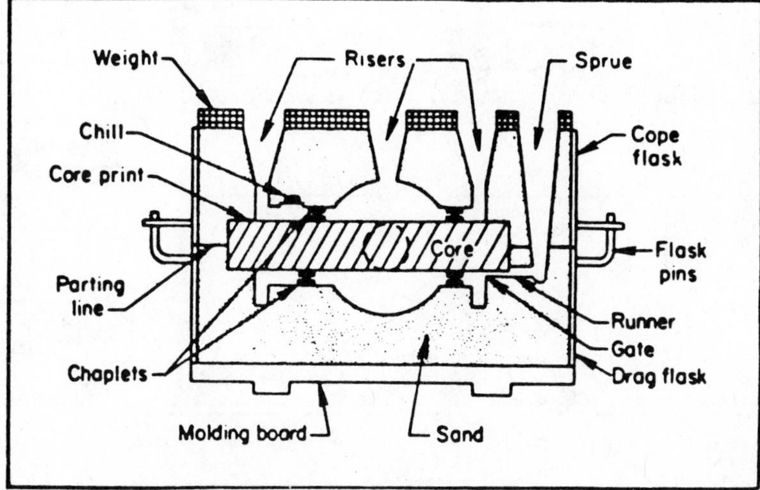

Fig. 4.1.2. Typical sand mold for ferrous and nonferrous castings.[2] Courtesy American Foundrymen's Society.

Green Sand Molding

Green sand molding is one of the oldest and most widely used sand molding processes. The process is highly flexible and economical and is usually the most direct route from pattern to mold cavity ready for pouring. In green sand molding, 4–6% bentonite and 4–8% water are used to produce bonding between the sand particles. The bonding action of clays is due to electrostatic, surface tension, and probably interparticle friction. Usually pitch, wood flour, and other additives are mixed in with the sand to impart special properties to the mold. Green sand can be reused many times by reconditioning it with water, clay, and other ingredients.

There are a few limitations to the use of green sand molding. Castings with thin, long sections cannot be cast in green sand molds because of the poor strength and the chilling action of the mold. Certain metals, such as superalloys and nonferrous alloys, develop defects if poured in green sand molds due to moisture in the mold. The surface finish and dimensional tolerances are relatively poor and may be inadequate for some casting applications. Dimensional variations are strongly dependent on the mold hardness and the metallostatic pressure. Dimensional variations of $\pm 1/64$ in. on small castings and $\pm 3/32$ in. on large castings are typical.

Dry Sand Molding

For dry sand molding, the green sand formulation is modified to impart good strength and related properties in a dry mold. The dry sand molds are made in green condition and subsequently bake-dried. The mold cavities are usually coated with a refractory slurry to provide greater hardness and thermal stability to the mold cavity surface. The molds are dried in air-circulating ovens at 350–600°F or skin dried with torches or radiant heating lamps. Skin dried molds must be poured shortly after drying to avoid moisture penetration into the dried skin. Dry sand molds possess higher strength and most of the advantages of green sand molds. However, dry sand molding is relatively slower and more expensive than green sand molding due to the time-consuming drying operation.

Core Sand Molding

Sometimes large complex casting molds are made entirely out of cores. A number of large and small individual cores are assembled and held securely in position with molding sand and/or suitable clamps. This method is chosen when design does not permit withdrawal of standard patterns from the mold. It is also used in insert molding where metal inserts are held in position in the mold and metal is cast around them. Core boxes are used in place of patterns for making all parts of the mold.

The basic principles of core molding are shown in Fig. 4.1.3. The most commonly used core sand mixtures contain 1–3% core oil or linseed oil as the organic binder. Cereal, bentonite, silica flour, and water are typically added to core sand mixtures to impart special properties to the sand and mold. Core sand molds attain a very high strength when baked between 350–600°F. This is due to hardening of the oil binder. Another variation of core sand molding is the utilization of cement-bonded sands.

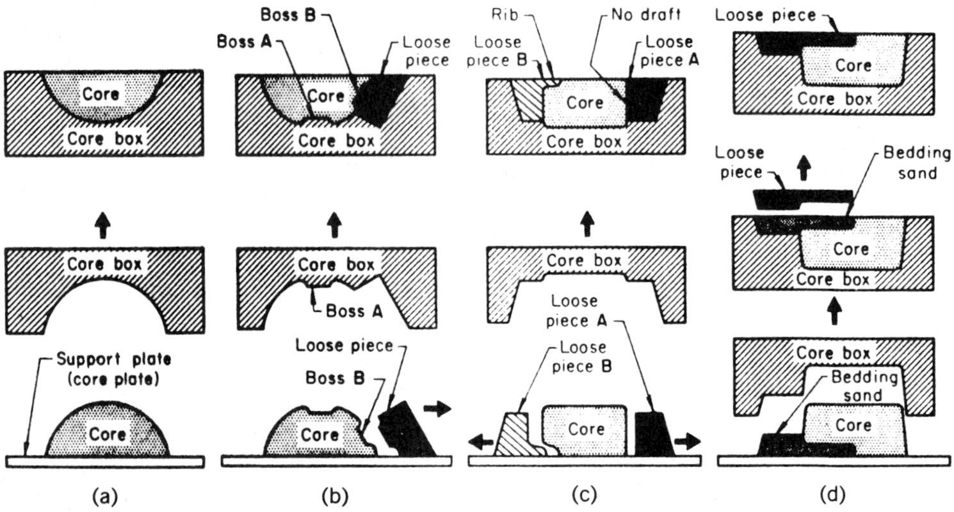

Fig. 4.1.3. Basic principles of core molding.[3] From *Metals Handbook*, Vol. 5, 8th ed., American Society for Metals, 1970.

A process in which the core molds do not require baking is the air-set, or no-bake, process. In this process, a mixture of sand, liquid resins, and catalyst harden with polymerization of the resins. The rate of polymerization is controlled by the amount of catalyst added. The quantity of resin required for good bonding is typically 2–5% by weight of sand and increases with the increase in fineness of the sand. Because of the cost of the core sand binder and difficulties in reusing the sand, core sand molding is costlier.

Carbon Dioxide Molding

The carbon dioxide process is a fast and specialized core sand molding process. The process uses 3–6% sodium silicate (water glass) as a binder. The sand mixture containing sodium silicate possesses good flowability and can be hardened by passing carbon dioxide gas through the mold.

For this process the sand should always be dry (a maximum moisture content of 0.25%) and as clean as possible. The amount of binder required increases with an increase in moisture and a decrease in cleanliness and grain size of the sand. The process requires no baking.

Several additives, such as kaoline, alumina, and invert sugars, are used to improve the properties of sodium silicate-sand mixtures. Carbon dioxide molds easily attain compressive strengths as high as 200 psi when properly gassed. The carbon dioxide process is highly adaptable to mass production and can produce ready-to-use molds and cores in a few minutes. The process is suitable for most casting alloys but is widely used for ferrous and copper base alloys. The carbon dioxide process, however, suffers from mold deterioration with time caused by the absorption of moisture from the atmosphere. The bench life of sodium silicate-bonded sand is much shorter than other core and mold mixtures. Sodium silicate-bonded sands also are difficult to reuse and offer difficulties in shake-out to stronger interparticle bonding.

Shell Molding

In the shell molding process, a mold is formed from a mixture of fine sand and thermosetting resin that is dropped against a heated metal pattern. The heating of the resin causes bonding between sand particles and produces a strong, sturdy shell. After the shell is cured, it is stripped from the pattern. The shell usually constitutes half of a mold and is assembled with its mate and secured together to form the complete mold. Figure 4.1.4 illustrates the shell molding techniques.

Dry sands of 60–140 AFS are generally used in shell molding. As the shells are relatively thin, fine sand does not create permeability problems. The synthetic resin commonly used in shell molding is phenol formaldehyde with about 15% hexamine to produce thermosetting characteristics. The process uses 4–8% thermosetting resin as a binder. Pattern temperatures are maintained between 400–700° F. The shell thickness is usually ¼ in. to ¾ in. and is dependent on pattern temperature and dwell time on the heated patterns. The shell is cured on the heated pattern for 1–3 min. Shell thickness as a function of pattern temperature and duration of contact of resin-sand mixture with the heated pattern is shown in Fig. 4.1.5. After curing, the shell is stripped from the pattern with the aid of ejection pins. Use of silicone parting agents on the pattern considerably improve shell release.

Shell molding is generally used more in making cores than molds. The advantages of shell molding are exceptionally good surface finish and dimensional tolerance and hence decreased finishing and machining costs. The process is production oriented. The limitations of the process are increased cost of patterns, relative inflexibility in gating and risering, and the high cost of binder and the molding operation. The process is also limited to castings of relatively low weight and small size. A tolerance of ± 0.005 in. is possible with metal patterns finished to ± 0.001 in. except across the parting plane.

Floor and Pit Molding

Floor and pit molding are used in the production of large, intricate castings requiring heavy molding work. The castings made by floor and pit molding weigh from a few tons to over 100 tons. Because of the enormity of size and complexity of shape, great engineering effort and control is required in the construction and handling of the mold. Heavy equipment, large amounts of floor space, and many man-hours are generally allocated to these molding operations. Molding is accomplished in large flasks. When the pattern is too large for flasks, the pit molding technique is utilized In pit molding, the pattern is lowered into a large concrete-lined pit and sand is rammed all around it in sequential layers. The pattern usually comprises several pieces, permitting it to be withdrawn from complex features of the mold cavity. The cavity of a pit or floor mold is coated with a refractory mold wash to impart greater dry strength to the mold and better surface finish to the casting. A completed floor or pit mold is dried with dry sand cores in place. Dimensional tolerance of ± ¼ in. for large castings are generally acceptable.

Loam molding is a type of pit and floor molding. Loam is a moist, plastic molding mass containing about 50% sand grains and 50% clay. For this method a substructure to the approximate contour of the casting is made of bricks or wood on which loam is dabbed and worked to proper shape with sweeps. No pattern is required, and shaped steel sweeps or templates are used in generating proper casting contour. The mold is coated with mold washes and dried by forced hot air or torches. The chief advantages of this process are savings in pattern cost and storage. Loam molding, however, is a slow, laborious, and manual process.

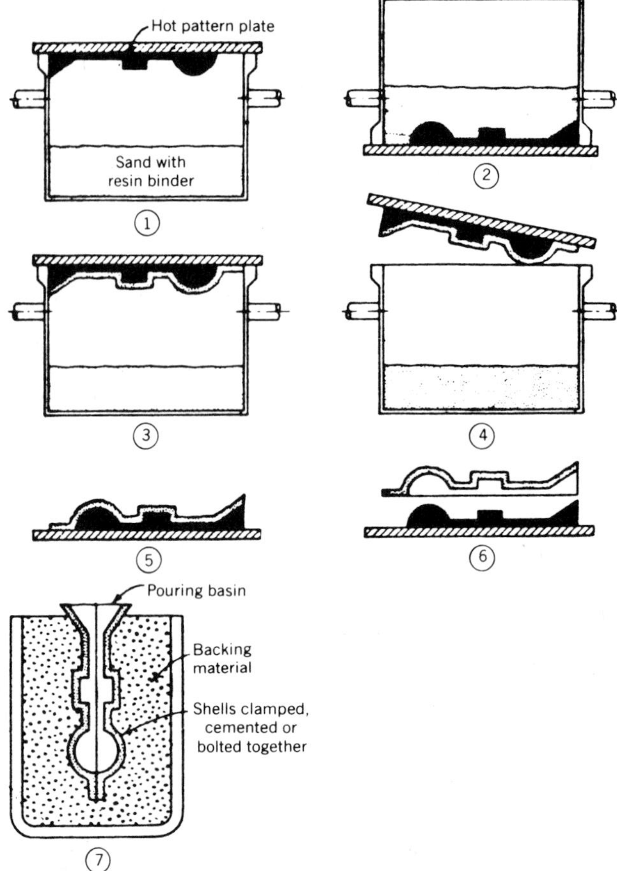

Fig. 4.1.4. Shell molding process.[4]

Vacuum Molding

The vacuum molding process (V-process) employs an innovative molding technique utilizing a thin film of plastic and dry unbonded sand to produce vacuum-sealed molds. The thin film of plastic first is softened under heat, then is drawn around the pattern under vacuum. The flask is then placed over the pattern and clamped in place and filled with unbonded dry sand mixture. The sand mixture consists of coarse (70 mesh) and fine (270 mesh) aggregates to produce good packing. The compaction of the sand is done under vibration. A second sheet of plastic is placed on the top of the flask after excess sand is struck off from the top. The sand bed is evacuated and then the pattern is exposed to atmospheric pressure for easy withdrawal of the pattern. The cope and drag portions of the mold are made in a similar manner and are assembled to form a plastic-lined mold cavity. When the molten metal is poured into the mold, the plastic film melts and diffuses into the sand, leaving a hard glazed surface. After the casting solidifies, the vacuum is released from the flasks to allow the separation of sand from the casting.

V-process molds are more rigid than sodium silicate-bonded molds and produce castings of better surface finish and dimensional accuracy. Because of the plastic film, the pattern life is considerably extended and pattern maintenance costs are lower. The process is suited for castings of varying configurations and weights from a few ounces to 10 tons. The process is suitable for casting ferrous and nonferrous alloys except for superalloys titanium and magnesium alloys. Since there are no binders shake-out equipment is not required. The sand can be recycled and reused many times. Shortcomings of the process are that it is not suitable for high production and castings that require a rapid rate of solidification.

Expanded Polystyrene Molding

In the expanded polystyrene (EPS) casting process (also known as the evaporative casting, cavityless, or lost pattern process), the pattern is evaporated to produce a mold cavity. Low-density patterns are made by expanding

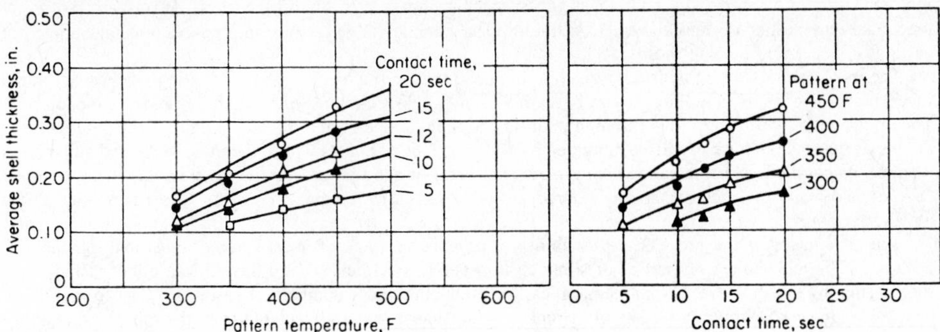

Fig. 4.1.5. Shell thickness as a function of pattern temperature and duration of contact of resin-sand mixture with the heated pattern.[3] From *Metals Handbook*, Vol. 5, 8th ed., American Society for Metals, 1970.

small polystyrene beads in metal molds with the aid of steam, vacuum, or hot air. The gating systems are cut from low-density foam boards and are glued to the pattern. The pattern assembly with the gating system is then coated with a permeable refractory slurry to provide good surface finish and prevent sand collapse during pouring. The coated pattern is encapsulated with coarse sand under vibration in a flask. Coarse, round sand is used to provide maximum permeability for escape of gases during evaporation of the pattern. The evaporative casting process is schematically shown in Fig. 4.1.6.

The EPS casting process produces castings of closer tolerance than those produced by green sand or sodium silicate-bonded molds. The sand requires no binder; hence, no shake-out of the flasks is required. The patterns are extremely light and can be used to produce castings ranging from 20 lb to 20,000 lb. The limitations of the process are high initial pattern die costs and greater care necessary in pattern handling and pouring.

4.1.3.2 Permanent Mold Casting

Permanent mold castings are usually produced in metal molds and sometimes in semipermanent molds made of graphite, silicon carbide, and other refractory materials. The mold cavity and gating system in a permanent

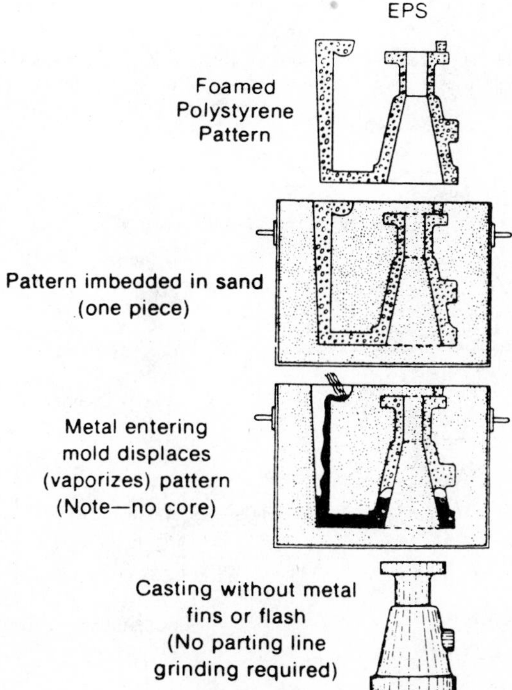

Fig. 4.1.6. Schematic diagram of expanded polystyrene molding.[2] Courtesy American Foundrymen's Society.

mold are usually cast to rough contour and then machine finished to final dimensions. Permanent molds are used in casting most nonferrous alloys and cast irons. The extremely high casting temperatures prohibit commercial casting of steel in permanent molds.

The mold cavity surface is usually given a thin coating of refractory to improve casting ejection and mold life. The thickness of the mold coat can be varied to aid slightly in controlling the mold heat extraction rate. The mold life is significantly lower at high pouring and mold temperatures. Fixed operating temperature ranges are very important and are usually maintained by water cooling channels within the mold walls and automation of mold preparation, pouring, and casting ejection. Metal or sand cores may be used for producing internal cavities. Normally the metal is gravity poured, but in some cases, air pressure of 5–10 psi is applied to the metal after pouring.

The advantages of the process are its ability to produce castings of good surface finish and dimensional accuracy. The mechanical properties of some casting alloys are improved by the chilling effect of the metal mold. The process, however, is not economical for producing small numbers of castings owing to high mold costs. The process is limited to relatively simple casting shapes that are ejectable from the mold. Some alloys cannot be cast into metal molds due to lack of fluidity, high casting temperature, or hot tearing tendency.

4.1.3.3 Die Casting

In die casting, molten metal is injected into a die cavity under high pressure. The pressures used in die casting range from 100–10,000 psi, depending on the type of machines used and metals injected. Two of the principal types of die casting processes in use are shown in Fig. 4.1.7. In the cold chamber process, the metal is ladled into the shot chamber and then injected into the die. Pressures used in the cold chamber process are generally very high. This process is commonly used in casting aluminum, magnesium, copper-base, and other high melting point nonferrous alloys. Ferrous alloys are rarely commercially die cast because of their high pouring temperatures. In the hot chamber, or gooseneck, process, the shot chamber is submerged in the melt. Relatively low pressures of 1,000–2,000 psi are required in the hot chamber operation owing to lower heat losses. Typically, low-melting alloys (e.g., zinc, lead, and tin-base alloys) are cast by the hot chamber process.

The rapid rate of metal injection and solidification make die casting a versatile production process. Production rates as high as 500 shots per hour are possible with modern die casting machines. Because of the high pressures involved, thin sections down to 0.015 in. in small castings can be die cast. The process produces castings with good surface finish and dimensional tolerance (± 0.002 in.). Rapid rate of cooling produces high strength and quality in many die cast alloys. This process is not suitable for short production runs due to heavy initial die costs. It also has limited adaptibility to complex castings of varying thickness, which require effective metal feeding, or casting designs that present difficulties in casting ejection.

4.1.3.4 Investment Casting

Investment casting is one of the oldest processes known to man. It is also known as the lost wax, or precision, casting process. Investment casting uses a specialized mold produced by surrounding an expendable pattern

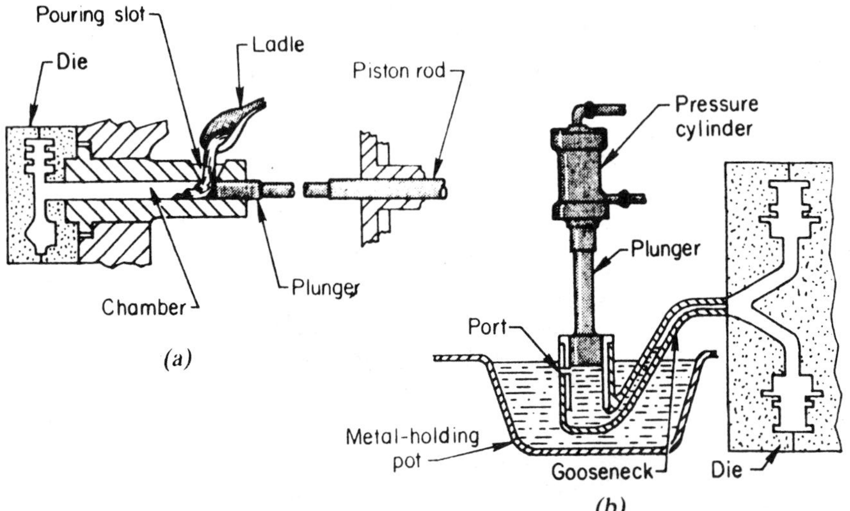

Fig. 4.1.7. Die casting processes: (*a*) cold chamber process and (*b*) gooseneck (hot chamber) process.[2] Courtesy American Foundrymen's Society.

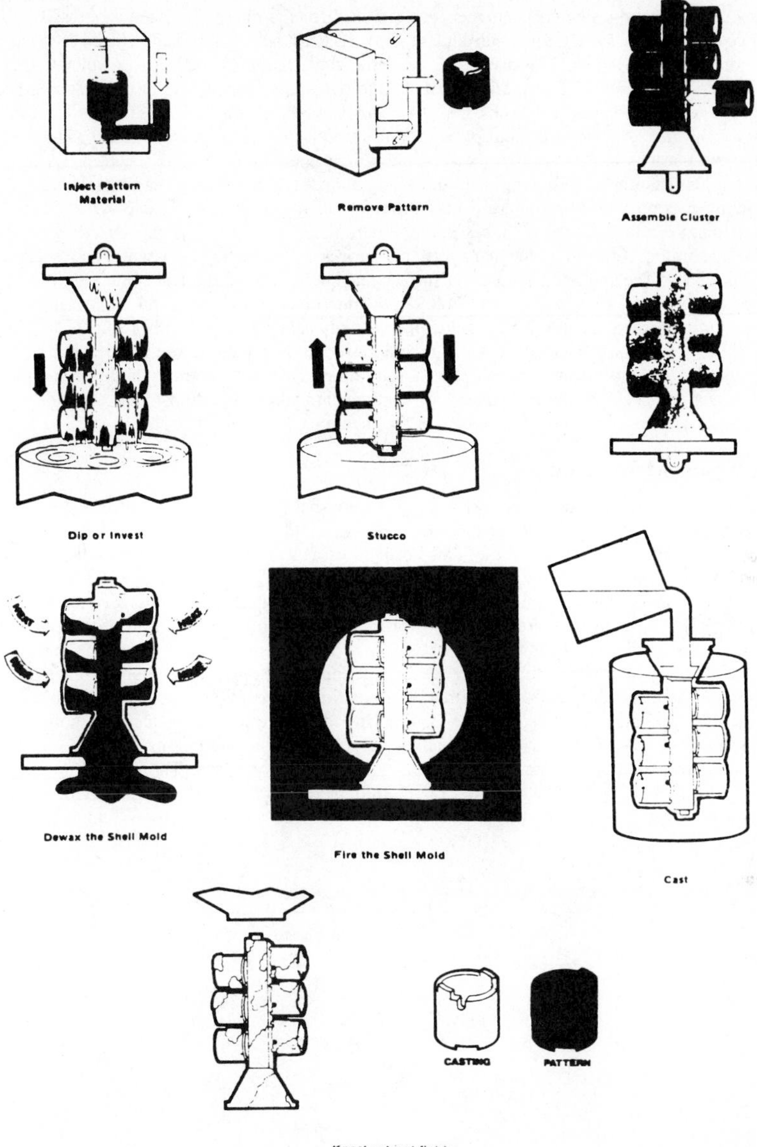

Fig. 4.1.8. Basic production techniques for shell investment casting.[5] Courtesy Investment Casting Institute.

(usually wax or plastic) with a refractory slurry that sets at room temperature. When the pattern is melted or burned out, it leaves behind a mold cavity as precise as the pattern. The molds are fired at elevated temperatures to ensure the burnout of pattern residues, to impart greater strength to the mold, and to extend the fluid life of molten metal filling the mold.

The investment casting process can be subdivided into the shell process and the solid mold process. The two processes differ considerably in precoating of the pattern and the permeability and thermal conductivity of the mold. The pattern in the shell process is always precoated, whereas precoating of the pattern in the solid mold process is generally not required except for casting high melting point alloys. Figure 4.1.8 shows the basic production techniques for shell investment casting. In the shell process, after precoating, the pattern assembly is alternately dipped in a coating slurry and stuccoed with granulated refractory until the shell is built to desired thickness. The composition and viscosity of the coating slurry and the composition and size of the stucco granules vary from the first coat to the last coat to build a composite shell of desired properties. Usually, a more viscous slurry and a finer refractory stucco are used for the initial coat, and progressively less viscous slurry and coarser stucco are used for subsequent coats. Each coat of slurry and stucco is dried prior to the application of subsequent coats. The drying of the initial coats is carried out at a slower rate than later coats and is achieved under controlled temperature and humidity to prevent cracking of the precoat. The total shell building time, varying from three to seven days, is dependent on the number of coats required and the complexity of the part. In contrast, in the solid mold process, a refractory slurry is poured over the pattern contained in a flask. The slurry hardens in air forming a solid mold around the pattern assembly within hours.

The investment casting process is utilized in casting almost any metal that can be melted and poured in air or in a vacuum or controlled atmosphere. Many different casting techniques (Table 4.1.2) have helped make it possible to mass-produce complex shapes that are extremely difficult to produce by other casting processes or even by machining. Owing to the nature of the process, the castings possess good surface finish and dimensional accuracy. The approximate range of surface roughness of an investment casting is 60–125 microinches as compared to 125–250 microinches and 500–1,000 microinches for shell and green sand mold castings, respectively. A dimensional accuracy of ± 0.005 in. is usually achievable in small castings. The process permits close control of metallurgical properties, such as grain size and orientation, and is suitable for castings in the range from ounces to thousands of pounds. The initial tooling costs are generally high for complex and larger castings. The near net shape of the investment cast component makes its end-cost highly competitive with other manufacturing processes because of the reduction in machining and assembly costs.

4.1.3.5 Plaster Mold Casting

Plaster molds are used in producing copper- and aluminum-base alloy castings with good surface finish and dimensional tolerance. The plasters used in making plaster molds consist of plaster of paris ($Ca_2SO_4 \cdot \frac{1}{2}H_2O$) and several ingredients, such as talc, lime, and Portland cement, to control the contraction characteristics of the mold and setting time. Silica or zircon sands comprise up to 50% of the materials for some plaster molds. Water is used to convert the dry mixture into a pourable slurry and for providing water of crystallization. Typically, 140–180 lb of water is required for 100 lb of plaster in the mixture to produce a slurry of good consistency. Optimum mixing of the slurry is necessary to avoid too much air entrapment or premature setting of the plaster. The dry strength of the plaster is strongly dependent on the consistency of the mixture. Plaster is poured onto a metal pattern, with chills already in position and is allowed to set for 20 to 30 min. Setting of plaster involves hydration of gypsum. After setting, the molds are dried to reduce the moisture in the mold. Cores, if used, are secured in the dried mold halves before the halves are clamped together.

Plaster molding usually employs cope and drag or match plate patterns. The Antioch process is a special plaster mold casting technique in which, after the mixture develops an initial set, the patterns are removed and the mold is assembled in green condition. The molds are allowed to stand at room temperature for 6 to 8 hr

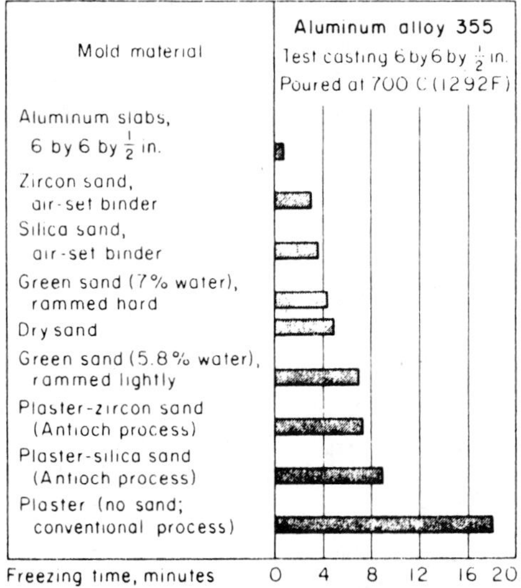

Fig. 4.1.9. Comparison of heat absorption rates for various mold materials, as indicated by freezing time required for identical aluminum alloy test castings.[3] From *Metals Handbook*, Vol. 5, 8th ed., American Society for Metals, 1970.

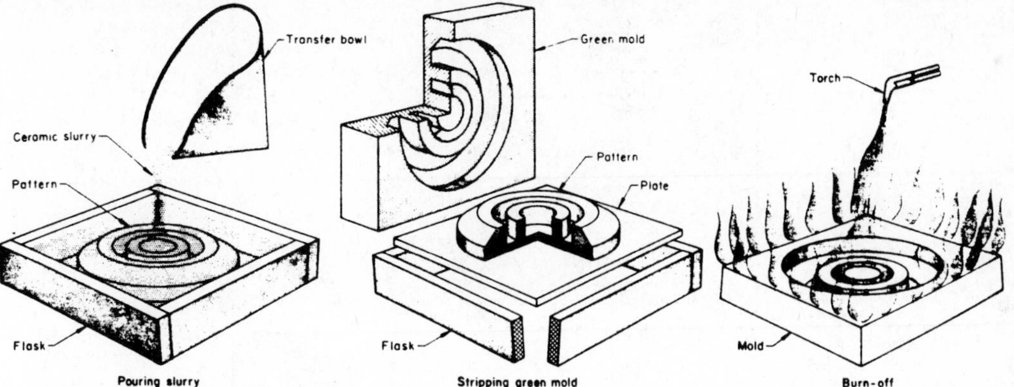

Fig. 4.1.10. Pouring, stripping, and burnoff of a monolithic, all-ceramic mold.[3] From *Metals Handbook*, Vol. 5, 8th ed., American Society for Metals, 1970.

and then autoclaved with a steam pressure of 15 to 20 psi for 6 to 8 hr. After autoclaving, the mold is permitted to remain at room temperature for 12 to 14 hr for rehydration. The rehydrated mold is subjected to a drying cycle at 350–450° F. Drying time depends mainly on the size of the mold and the temperature used. The Antioch process, due to the dehydration-rehydration cycle, produces a very permeable mold of relatively reduced dry strength.

The foamed plaster process offers a means of producing mold permeabilities comparable to those obtained in the Antioch process but without the autoclaving step. In this process, a foaming agent, such as alkyl aryl sulfonate, is added to the plaster mix. A speical method of mixing foams the slurry with many fine air bubbles. The setting and baking procedures for the mold are similar to those in other plaster molding processes. A typical foamed plaster mold contains about 50% porosity and is thermally insulating in nature. Figure 4.1.9 shows a comparison of heat absorption rates for various mold materials. The freezing rate in plaster molds can be considerably improved by embedding chills in the mold.

4.1.3.6 Ceramic Mold Casting

Ceramic molding is most widely used for the production of precision castings considered too large for molding with expendable patterns or production of castings in limited quantity. Ceramic molds are made from fine-grain zircon and calcified high alumina mullite slurries comparable in composition to solid investment slurries. How-

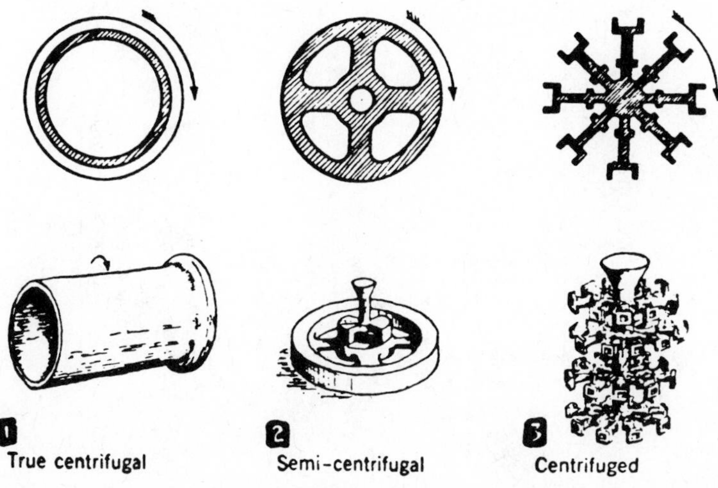

Fig. 4.1.11. Centrifugal casting methods.[6]

Table 4.1.3. A Comparison of Several Foundry Processes with Other Shaping Processes

	Forging	P/M	Weld Assembly	Machined Hog-Out	Sand Casting	Die Casting	Investment Casting
Degree of complexity	Poor	Poor	Good	Fair to good	Very good	Fair to good	Excellent
Relative quality	Excellent	Poor to fair	Excellent	Excellent	Good	Poor to fair	Very good
Mechanical properties	Excellent	Fair	Good to excellent	Excellent	Very good	Fair to good	Very good
Range of alloys utilized	Wide	Narrow	Medium	Medium	Wide	Narrow	Wide
Thinnest section commercially shaped (in.)	0.08	0.03	0.03	0.05	0.10	0.03	0.04
Surface finish and tolerance	Good	Very good	Good	Excellent	Fair to good	Very good	Very good
Size capability	Small to large	Small	Small to very large	Small to very large	Small to very large	Small to medium	Small to medium
Relative cost							
In small quantity	High	Medium	Low	Medium	Low	High	High
In large quantity as finished							
Simple shapes	Low to medium	Low	Medium	Medium to high	Low	Low	Medium
Intricate shapes	High	—	High	High	Low	—	Low

ever, unlike investment molding, permanent patterns and cope and drag molding are utilized in producing the mold.

Ceramic molds can be either all-ceramic (monolithic) molds or composite molds consisting of inexpensive fire clay back-up material with a relatively thin layer of ceramic face coating. Figure 4.1.10 shows the pouring, stripping, and burnoff of an all-ceramic mold. The two most common processes of ceramic molding are the Shaw and the Unicast processes. Both use either all ceramic or composite molds. The processes differ principally in the method of mold stabilization employed and the sequence followed in preparing composite molds. Mold stabilization refers to the treatment given to the mold shortly after the fine ceramic slurry has set but while it is still green. For the Shaw process, mold stabilization is achieved through burnoff of the excess liquid phase. In the Unicast process, the mold is immersed in a liquid hardening bath or a vapor atmosphere to achieve internal stabilization of the mold. Pouring of the fine ceramic slurrry in the Unicast process precedes the preparation of the coarse mold backup, whereas in the Shaw process this sequence is reversed.

Ceramic molds are used in precision casting of both ferrous and nonferrous alloys. The process produces castings of high surface finish and dimensional accuracy. Dimensional tolerances of ± 0.005 in. are easily achievable, rivaling the capability of investment casting. The preference for one process over another is largely dependent on the size of the casting, quantities required, and molding costs involved.

4.1.3.7 Centrifugal Casting

Centrifugal pressures can be applied to advantage in forcing molten metal quickly into molds to prevent premature freezing and to aid slightly in controlling temperature gradients. The castings made by filling the mold under centrifugal forces are referred to as centrifugal castings. Centrifugal casting is subclassified into three categories: (1) true centrifugal casting, (2) semicentrifugal casting, and (3) centrifuging. Figure 4.1.11 shows these three different centrifugal casting methods.

In true centrifugal casting, the casting, such as a cylindrical pipe, is spun about its own axis. No risers and central cores are required, since centrifugal force forms the inner diameter of the casting. In semicentrifugal casting, the casting is spun about its own axis, but risers and cores are needed. Several molds can be stacked along the common axis and cast in one operation. In centrifuging, centrifugal force is used mainly as a mold-filling device.

Centrifugal casting is an intriguing casting process that requires careful balance of metal weight, temperature, and speed of rotation to produce sound castings of high yield. The casting must be ideally suited to the process or it can probably be made to better advantage by other methods. The process is economical for producing tubular objects but generally suffers from relatively high overhead and maintenance costs.

4.1.4 SELECTION OF A SHAPING PROCESS

More than ever before, the design engineer is confronted with an array of ingenious, cost-effective processes for manufacturing parts. Each of the various foundry techniques exists because of the technical and economic advantages it offers in manufacturing a particular shape or family of shapes. These processes also compete with other manufacturing processes, such as forging, powder metallurgy, machining, and weld assembling. Performance, characteristics of manufacturing, and cost of the final component are the decisive factors in the designer's choice of a particular process. In Table 4.1.3 several of the most heavily utilized metal shaping processes are compared in terms of decisive factors. It can readily be seen that given acceptable property levels by all processes, any one factor, such as size capability or final cost of manufacturing, can be the overriding consideration in choosing a shaping process.

REFERENCES

1. "17th Census of World Casting Production [1982]." *Modern Casting*, 22–23 (December 1983).
2. E. L. Kotzin. *Metal Casting and Molding Processes*. Des Plaines, IL.: American Foundrymen's Society, 1981.
3. *Forging and Casting*. Metals Handbook, Vol. 5. Cleveland: American Society for Metals, 1970.
4. M. S. Burton. *Applied Metallurgy for Engineers*. New York: McGraw-Hill, 1956.
5. G. X. Diamond, Ed. *Investment Casting Handbook*. Chicago: Investment Casting Institute, 1968.
6. H. F. Taylor, M. C. Flemings, and J. Wulff. *Foundry Engineering*. New York: Wiley, 1959.

CHAPTER 4.2

MACHINING PROCESSES

J T. BLACK

Auburn University
Auburn, Alabama

4.2.1 INTRODUCTION

The manufacturing process called machining (also metal cutting and chip removal) is actually a large collection of processes designed to remove unwanted material from a workpiece in the form of chips. Machining is primarily used to convert a casting, preform, or block of metal into a desired shape, with size and finish specified to fulfill design requirements. Almost every manufactured product has components that require machining, often to great precision. Therefore, this collection of processes is often the most important of all the basic manufacturing processes (i.e., casting, forming, machining, and joining) because of the value that they add to the final product. In the same light, they are often the most expensive.

The majority of industrial applications of machining are in metals. Although metal cutting has resisted theoretical analysis due to its complexity, considerable progress has been made in the application of these processes in the industrial world.

Machining processes are performed industrially on machine tools. Figure 4.2.1 shows a schematic of four typical machine tools: a milling machine, engine lathe, drill press, and grinder. Workpieces are held in workholding devices, such as fixtures, chucks, jigs, vices, and vacuum chucks. Cutting tools such as horizontal slab mills, turning tools, twist drills, and grinding wheels are used in the machine tools to cut metal. There are four other basic machine tools: shaper/planer, broach, saw, and boring machine. Each of the basic machine tool types have many different versions. Lathes, for example, may be engine lathes, turret lathes, tracer lathes, numerical control (NC) lathes, or automatic screw machines. The basic machine tools are often combined into multiple capability machines known as machining centers. The machining center shown in Fig. 4.2.2 is capable of performing the machining processes normally performed on a milling machine, drilling machine, and a boring mill and is numerically controlled (see section on computer-aided manufacturing) so that the position of the tool with respect to the work is programmed into the machine control unit (MCU). It can move different tools into position, as needed, automatically.

For each of the basic machine tool types, there are many different kinds of workholders, cutting tools, and cutting tool holders, resulting in a rather formidable list of hardware that would fill this handbook (as it has many others). Thus, only the fundamentals of machining are presented here. This chapter will put these processes into perspective, so that the reader understands the problems associated with utilizing these processes in the manufacture of products and integrating them into automated manufacturing systems.

4.2.2 BASIC CHIP FORMATION PROCESSES

The seven basic chip formation processes are listed below with alternative versions in brackets. Each is performed on one or more of the basic machine tools. For example, drilling can be done on drill presses, milling machines, lathes, and some boring machines.

1. Turning (boring, facing, cutoff, taper turning, form cutting, chamfering, recessing, thread cutting)
2. Shaping (planing, vertical shaping)
3. Milling (hobbing, generating, thread milling)
4. Drilling (reaming, tapping, spot facing, counterboring, countersinking)
5. Sawing (filing)

Fig. 4.2.1. Examples of four of the seven basic machine tools used in machining processes.

6. Broaching (internal, external, push, pull, keyway)
7. Abrasive machining (grinding, honing, lapping, superfinishing, polishing)

Turning and shaping are performed with single-point tools; the rest are usually performed with multiple-edged cutting tools. Figs. 4.2.3 and 4.2.4 provide some details of the turning and milling processes.

4.2.3 BASIC FORMULAS FOR MACHINING

For all metal cutting processes, it is necessary to distinguish between speed, feed, and depth of cut. In general, speed (V) is the primary cutting motion that relates the velocity of the cutting tool to the work. It is generally given in units of surface feet per minute (sfpm). Speed (V) is shown in Figs. 4.2.3 and 4.2.4 with heavy, dark arrows. Feed (f) is the amount of material removed per revolution or per pass of the tool over the workpiece, so the units are inches/revolution (cycle) inches/minute, or inches/tooth, depending on the process. Feed is shown by the dashed arrows in the figures. The depth of the cut, given in inches, is indicated by the letter t. Basic equations are needed for the cutting time, or machining time, per piece (CT) in minutes and the metal removal rate (MRR) in cubic inches/minute. The main terms are N for revolutions per minute (RPM), L for

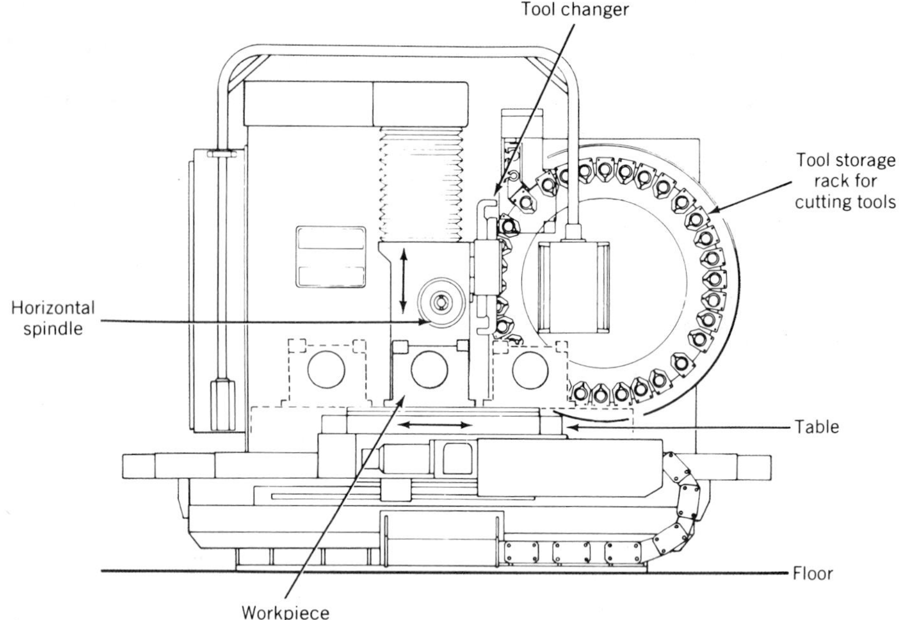

Fig. 4.2.2. Example of a NC machining center that combines the capabilities of a horizontal spindle milling machine, a drill press, and a boring machine.

length of cut, and D for diameter. The basic equations are as fundamental as the processes themselves. In the main, if one keeps track of the units and visualizes the processes, the equations are, for the most part, straightforward.

In the process of turning, for example, the primary cutting motion is rotational, with the tool feeding parallel to the axis of the rotating workpiece. To determine the RPM of the workpiece (an input to the machine tool), the engineer must first specify the recommended cutting speed. Cutting speed is dependent on the material being cut and the cutting tool material. For typical recommendations, refer to the *Machining Data Handbook*. The tables in this book are broken down by process, by material being cut, and by cutting tool material. The same tables also give starting estimates for feeds and upper limits on depths of cut. Having found a recommended speed V, the RPM is determined from

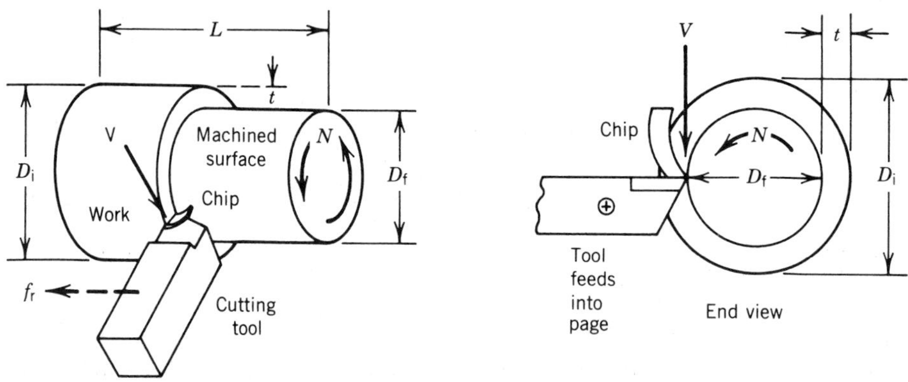

Fig. 4.2.3. Basics of the turning process as usually performed on a lathe.

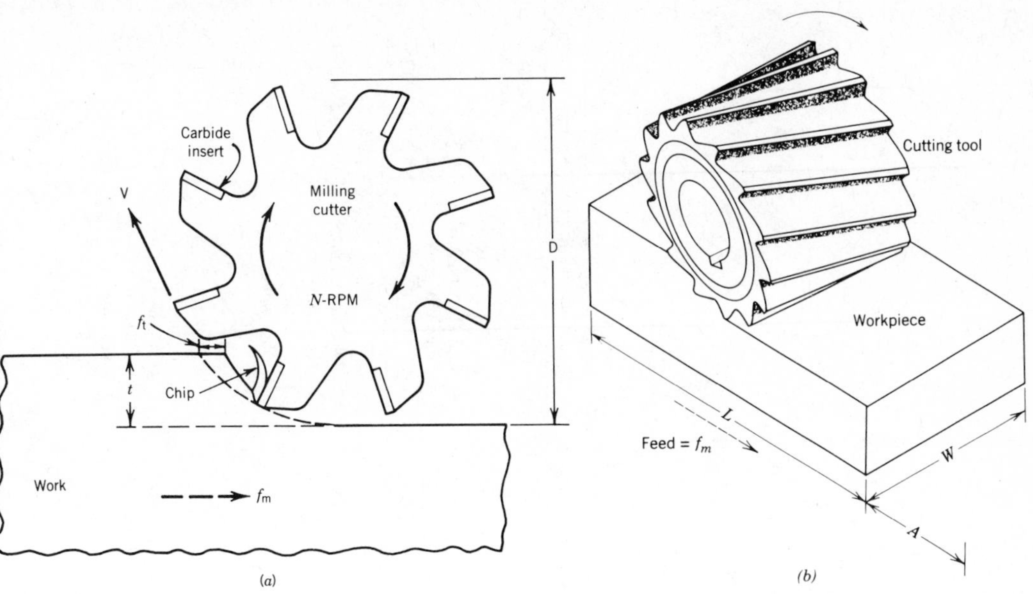

Fig. 4.2.4. (a) Basics of horizontal milling. (b) slab milling pass a workpiece of length *L*.

$$N = 12V/\pi D \qquad (4.2.1)$$

The cutting time is determined from

$$\text{CT} = \frac{L + A}{f_r N}$$

where *A* is an allowance for starting and ending the cut and f_r is the recommended feed in inches/revolution. The rate of material removal, called MRR, is determined from

$$\text{MRR} = \frac{\text{Volume removed}}{\text{Time needed}} = \frac{(\pi D_i^2 - \pi D_f^2)/4}{L/f_r N} \cong 12Vf_r t \qquad (4.2.2)$$

where $t = (D_i - D_f)/2$ (*A* is omitted) and D_i and D_f are initial and final workpiece diameters.

Milling typifies a multiple-tool cutting process in which the tool rotates (mills) at some RPM (*N*) while the work feeds past the tool. RPM is again related to the surface cutting speed for a cutter of diameter *D* according to Eq. (4.2.1). The depth of cut is *t* in inches. The width of cut is the width of the cutter or the work in inches and is usually given the symbol *W*. The length of the cut, *L*, is the length of the work plus some allowance *A* for approach and overtravel (see Fig. 4.2.4*b*). The feed of the table, f_m, in inches/minute is related to the amount of metal each tooth removes during a revolution. This is called the feed per tooth, f_t, as shown in Fig. 4.2.4*a*. Thus

$$f_m = f_t N n \qquad (4.2.4)$$

where *n* is the number of teeth in the cutter (teeth/revolution). For the given work material combination, we again have to select a cutting speed *V* and feed f_t. Note that the cutting velocity is that which occurs at the edge of the mill. The following equations hold:

$$\text{CT} = (L + A)/f_m \text{ min.} \qquad (4.2.5)$$

$$\text{MRR} = LWt/\text{CT} = Wtf_m \text{ in}^3/\text{min., ignoring A} \qquad (4.2.6)$$

Table 4.2.1. The Seven Basic Processes of Machining and Their Major Variations

Process	Raw Material Form	Typical Production Rate	Material Choice	Process Capability	Surface Roughness
1 Shaping (small, flat parts)	Bar, plate, casting	1–4 parts/hr.	Low to medium carbon steels and nonferrous metals best; no hardened parts	±.001–.002 in. for larger parts ±.0001–.0005 in. for small to medium parts	63–250
1.1 Planing (large, flat, surfaces)	Bar, plate, casting	1 part/hr.	Low to medium carbon steels or nonferrous materials best	±.0001 in.– ±.005 in.	63–125
2 Turning on engine lathes (cylindrical parts)	Cylindrical preform, casting, froging	1–10 parts/hr.	All ferrous and nonferrous material considered machinable	+.002 in. on dia. common ±.001 in. obtainable	125–250
2.1 Turning on turret lathe	Bar, rod, tube, preforms	1 part/min.	Any material with good machinability rate	±.003 in. on dia. where needed ±.0100 in. common	125 average
2.2 Turning on automatic screw machine	Bar, rod	10–30 parts/min.	Any material with good machinability rating	.0005 in. and .0010 in. possible .001 in. and .003 in. common	63 average

2.3 Turning on swiss automatic	Rod	12–30 parts/min.	Any good material with good machinability rating	±.0002 in. to common ± .001 in.	63 and better
2.4 Boring (internal turning)	Casting, preform	2–20 hrs./piece	All ferrous and nonferrous metals	±.0005 in.	90–250
3 Milling (Vertical spindle)	Casting, plate preforms	1 part/min. to 1 part/wk.	Any material with good machinability rating	±.0005 in possible ±.001 in. common	63–250
3.1 Milling (Horizontal spindle)	Blank, preforms, rods, plate	1 part/min. to 1 part/wk.	Any material with good machinability rating	±.002 in. or better	63–250
4 Drilling (holes)	Plate, bar, preform	2–20 sec/hole after setup	Any unhardened material; Carbides may penetrate some case-hardened parts	±.002 in. to ±.010 in. common ±.001 in. possible	63–250
5 Sawing	Bar, plate, sheet	3–30 parts/hr.	Any unhardened material	±.015 in. possible	250–1000
6 Broaching	Tube, rod, bar, plate	300–400 parts/min.	Any material with good machinability rating	±.0005 in. to ±.001 in.	32–125
7 Grinding	Plate, rod, wire, bar	1–1,000 pieces/hr.	Nearly all metallic plus many nonmetallic materials	.0001 in. and less	16

The length of approach is

$$A = \sqrt{\frac{D^2}{4} - \left(\frac{D}{2} - t\right)^2} = \sqrt{t\,(D - t)} \text{ in.}$$

It is in this manner that the basic equations for cutting time (also called machining time) and metal removal rate are determined for the various operations. It is worth noting at this point that neither of these basic equations for cutting time included the depth of cut parameter, which must also be selected as an input to the process and does influence the time to machine a workpiece. Depth of cut does enter into the MRR equations.

The machining time usually represents about 20–30% of the cycle time, which is, of course, the inverse of the production rate. The balance of the cycle time is composed of time to load and unload parts; change tools; replace worn or broken tools; adjust or reset the machine tool RPM, feed, or depth of cut setting; and so forth. Table 4.2.1 presents a summary of the basic procedures of machining and some of the major variations.

4.2.4 METAL CUTTING PROCESSES OVERVIEW

In general, the metal cutting processes can be viewed as consisting of independent (input) variables, dependent variables, and independent-dependent interactions or relationships. The engineer or machine tool operator has direct control over the input variables and can specify or select them when setting up the process. The following are descriptions of several input variables in machining.

Material of workpiece. The metallurgy and chemistry of the workpiece can be specified or is known. Quite often, a material is selected for a particular job chiefly because it machines well. Cast iron and aluminum, for example, are said to have good machinability. Other metals machine poorly but are selected to meet other functional design criteria.

Starting geometry (size and shape) of the workpiece. This may be dictated from prior processes (casting, forging, forming, etc.) or may be selected from standard machining stock (e.g., bar stock for screw machines). This variable usually influences directly which machining process(es) will be selected and the depths of cut.

Specific machining processes. The machining processes required to convert the raw material into a finished product must be selected based on the geometry of the part (size, shape, rotational or nonrotational), the required finishes and tolerances, and the quantity of the product to be made.

Tool material. The three most common cutting tool materials in use today for production machining operations are high-speed steel (HSS), carbides, and coated tools. CBN (cubic boron nitride), ceramics, and diamonds are also being employed. Generally speaking, HSS tools are used when large rake angles are needed (see α in Fig. 4.2.5), the tools are complex in design, or cutting speeds will never be very great. The nonferrous tool materials can operate at faster cutting speeds, and they come in a wide variety of grades and geometrics. Selection of the tool material that gives reliable service while fulfilling the

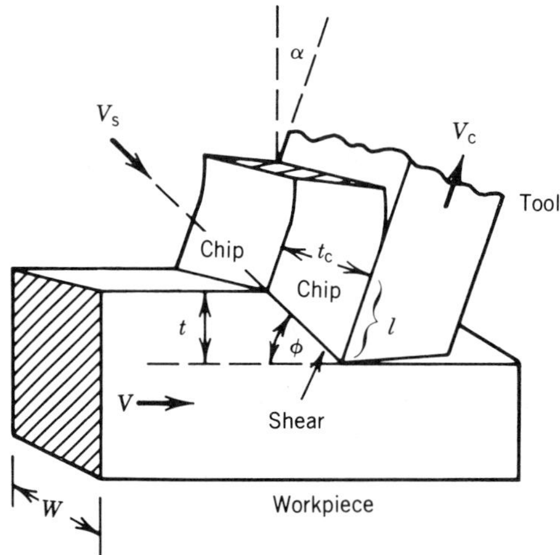

Fig. 4.2.5. Schematic of orthogonal machining. The cutting edge of the tool is perpendicular to the direction of motion (V). The back rake angle is α.

functional requirements has been and still is a bit of an art. The harder the tool material, the better it can resist wear at faster cutting speeds. Retention of hardness at elevated temperatures is a desirable characteristic.

Cutting parameters. For every machining operation, it is necessary to select a cutting speed and a feed. Many factors impinge on the decision maker here because all of the dependent variables are influenced by these two variables and the depth of cut. Proper selection of speed, feed, and depth of cut also depend on what other input variables have been selected; that is, the total amount of material to be removed, the workpiece and tool materials, and the machining process(es) need to be selected before preliminary choices for speed and feed can be made.

Tool geometry. Cutting tools are usually designed to accomplish specific operations and thus the tool geometry (angles) is selected to accomplish specific machining functions. Generally speaking, large rake and clearance angles are preferred but are only possible on HSS tools. Tools made from carbides, ceramics, and other very hard materials must be given small or even negative tool angles, which keep the tool material in compression during machining and thereby avoid tensile failures and brittle fractures.

Workholding devices. The workpieces are located (held in specific position with respect to the tools) and clamped in workholding devices in or on the machine tools. For every machine tool, there are many different kinds of workholding devices, ranging from general purpose vises to specifically designed jigs and fixtures. The workholding devices are the key to precision manufacturing, so the selection (or design and construction) of the correct workholding devices is every bit as important as the selection of the right cutting tool and machine tool.

Cutting fluid. The selection of the right cutting fluid for a particular combination of work material and tool material can mean the difference between success and failure in almost every production machining process. In the main, cutting fluids cool the workpiece, the tool, and the chips; reduce friction (lubricate); carry the chips away from the cutting region; help to improve the surface finish; and provide surface protection to the workpiece.

The dependent variables are determined by the process based on the prior selection of the input or independent variables. Thus, the manufacturing engineer's control over these is indirect at best. The important dependent variables are as follows.

Cutting forces and power. To machine metal at a specified speed, feed, and depth of cut, with a specified lubricant, cutting tool material, and geometry will generate cutting forces and will consume power. A change in any of the variables (except speed) will alter the forces, but the change is indirect in that the engineer does not specify the forces he wants but only the parameters he believes will give him those forces. Forces are important in that they influence the deflections on the tools and workholders, which in turn affects the final part size. Forces also play a role in chatter and vibration phenomena common in machining. Obviously, the manufacturing engineer would like to be able to predict forces (and power) so that he can specify safely the equipment for a manufacturing operation, including the machine tool, cutting tool, and workholding devices.

Size and properties of the finished product. Ultimately, the objective of machining is to obtain a "machines surface" of desired size and tolerance with the desired mechanical properties. Because machining is a plastic deformation process, every machined surface will have some residual plastic deformation left in it. This residual stress will usually be tensile in nature and can interact with surface flaws to produce part failure in fatigue loading situations. In addition, every process has some inherent variability (variation about average size) which changes almost all of the input variables. Thus, the manufacturing engineer must try to select the levels of input variables to produce a product that is within the tolerance specified by the designer and which will have satisfactory surface properties.

Surface finish. The final finish on a machined surface is a function of tool geometry, tool material, workpiece material, machining process, speed, feed, depth of cut, and cutting fluid. Surface finish is also related to the process variability. Rough surfaces will have more variability than smooth surfaces. Often it is necessary to specify multiple cuts—that is, roughing and finish cuts—to achieve the desired surface finish, or it may be necessary to specify multiple processes—that is, follow turning with cylindrical grinding—in order to obtain the desired finish.

Tool wear and tool failure. The plastic deformation and friction inherent in machining generate considerable heat, which lowers the tools' resistance to wear. The problem here is quite subtle but nonetheless significant. As the tool wears, it changes both in geometry and size. The change in geometry can result in increased cutting forces, which will increase deflections in the workpiece and may create a chatter condition. The increased power consumption will in turn increase the heat of the operation and thus accelerate the wear rate. The change in size of the tool means that the size of the workpiece is also changing. Again the engineer has only indirect control over these variables. He can select slower speeds, which will produce less heat and lower wear rates, but the production rates decreases because the MRR is decreased. Alternatively, he can increase the feed or depth of cut to maintain the MRR while he reduces the speed. Increasing either the feed or depth of cut will directly increase the cutting forces; some tool life may be gained, but some precision may be lost due to increased deflection and chatter.

The connections between the input variables and the process behavior is the most important body of knowledge for the manufacturing engineer. Unfortunately, this knowledge is difficult to obtain. Machining is a unique plastic deformation process in that it is unconstrained and operates at very large strains and very high strain rates. The tremendous variety in the input variables results in an almost infinite number of different machining combinations. Basically, there are three ways to deal with such a complex situation.

1. *Experience.* This requires long-time exposure, since knowledge is basically gained by trial and error, with successful combinations transferred to other "similar" situations. This activity goes on in manufacturing every time a new material is introduced into the production facility. It took years for industry to learn how to machine titanium. Unfortunately, the knowledge gained through one process may not transfer well to another even though it appears very similar in its input variables.
2. *Experiments.* Machining experiments are difficult to do, and they are expensive and time consuming. Tool life experiments, for example, are quite commonly done, but, even so, tool life data for most workpiece/tool material combinations are not available. There are just too many combinations. Moreover, the results are not necessarily transferable to the shop floor. Tool life equations are empirically developed from experiments wherein all input variables are kept constant except cutting speed. This experimental arrangement will limit the mode of tool failure to wear. Such results are of little value on the shop floor, where the tools can and do fail from causes other than just wear.
3. *Theories.* There have been many attempts to build mathematical models of the metal cutting process, wherein one can insert the desired values of the independent variables and obtain predicted values of the dependent variables. These models range from crude, first-order approximations to complex, computer-based models using finite element analysis. In the last five years, some modest successes have been reported in the literature wherein accurate predictions of cutting forces and tool wear were accomplished in certain materials. Clearly such efforts are extremely helpful in understanding how the process behaves in general. However, even for these situations, it was necessary to devise two independent experiments, one to characterize the shear strength (τ_s) of the materials in question at large strains and high strain rates and the second to characterize the friction (μ) situation at the interface between the tool and chip. The theory of plastic deformation of metals (dislocation theory) has not yet reached the place where values for τ_s and μ can be predicted from the metallurgy and deformation history of the material.

What can theories for metal cutting tell us about the real world machining? We need to take a brief look at the fundamental nature of these processes in order to respond to this question.

4.2.5 FUNDAMENTAL MECHANISM OF CHIP FORMATION

In this section, the cutting tool geometry is simplified from the three-dimensional (oblique) geometry, which typifies most industrial processes, to a two-dimensional (orthogonal) geometry; the workpiece is a plate, as shown in Fig. 4.2.5. With this 2D model chip formation, the influence of the most critical elements of the tool geometry (rake angle α) and the interactions that occur between the tool and the chip can be more easily examined.

Basically, the chip is formed by a localized shear process that takes place over very narrow regions. Classically called the shear zone or shear plane, this deformation evolves out of a radial compression zone that travels ahead of the shear process as the tool passes over the workpiece. This radial compression zone has, like all plastic deformations, an elastic compression region that converts to a plastic compression region in the workpiece. The plastic compression generates dense dislocation tangles and networks in annealed metals. When this workhardening reaches a saturated condition (fully workhardened), the material has no option but to shear.

The shear process itself is a nonhomogeneous (discontinuous) series of shear fronts (or narrow bands) that produces a lamellar structure in the chips. This is the fundamental structure that occurs on the microscale in all metals when machined and accounts for the unique behavior of the machining process. Shear fronts are very narrow (100–500 A°) compared to the thickness of a lamella (2 to 4 microns) and account for the large strain and high strain rates that typify this process. These fundamental structures are difficult to observe in normal metal cutting but can be readily observed through the use of a scanning electron microscope (SEM) and specially prepared workpieces. Figure 4.2.6 shows SEM micrographs of an orthogonal machining setup wherein the fundamental structure is readily observed. The shear fronts are produced by the movement of many dislocations traveling from the tool tip to the free surface. The lamella represents heavily deformed material that has been segmented by the shear fronts. All metals, when machined, deform by this basic mechanism. The shear fronts relieve the applied stress.

If you have been in a machine shop, you may have picked up a chip with a sawtooth pattern on its top side—the side which did not rub against the tool. This sawtooth pattern was not produced by the shear front lamella structure but by the unloading of the elastic energy stored in the tool and workpiece, which results in chatter and vibration during cutting. The shear front lamella structure can and does exist without any vibration of the tool or work. Therefore, within each sawtooth, there will be many fine shear front lamellas. The geometry of the sawtooth can be changed (even eliminated) by altering the stiffness of the setup or the machine. The

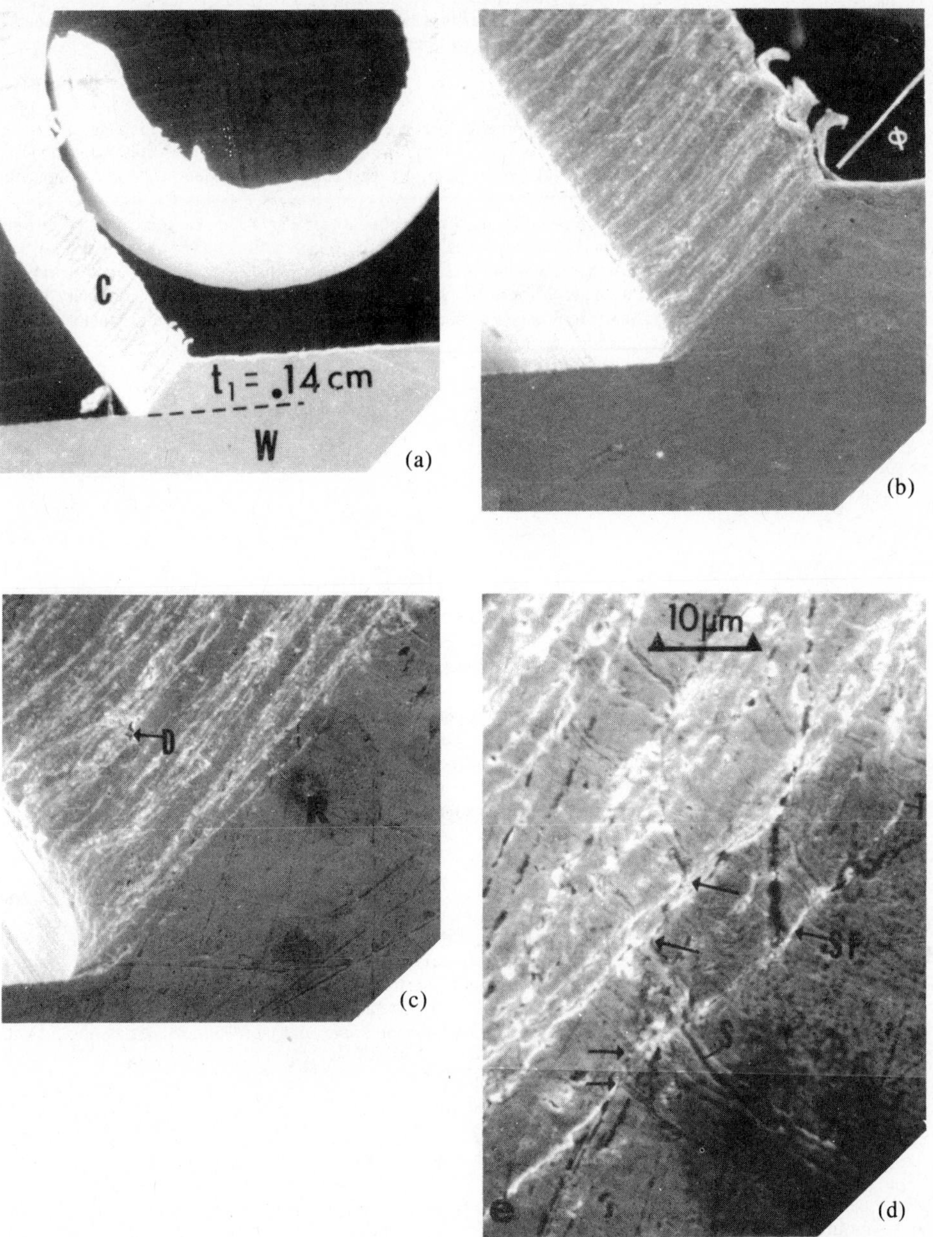

Fig. 4.2.6. Series of SEM micrographs of machined chip (c) that show the shear front lamella structure. (*a*) Polycrystalline gold plate (W) orthogonally machined at a depth of cut of .055 in. (*b*) The gold sheared at an angle of $\phi = 30$ degrees. (*c*) Higher magnification view of the tool tip region, with D being a surface defect opened up by the shear fronts. (*d*) Higher magnification view of region marked R in (*c*), where the shear fronts produced sharp steps (see arrows) in scratches left on the sides of plate when it was polished prior to machining.

shear front lamella structure is fundamental and characteristic to the plastic deformation process itself and is therefore relatively invariant and certainly cannot be eliminated.

4.2.5.1 Orthogonal Machining (Two Forces)

Orthogonal machining setups are used to model oblique (or three-force cutting) processes, which are, of course, what is typically found in processes such as turning, drilling, milling, and shaping. The orthogonal model is an excellent model, however, for gaining an understanding of the behavior of oblique processes without the complications of the third dimension.

Orthogonal machining can be accomplished by machining a plate (see Fig. 4.2.5) or approximated by end-cutting a tube wall in a turning setup. For the purposes of modeling, one assumes the shear process to be a plane, the cutting edge to be perfectly sharp, and no friction contact between the flank of the tool and workpiece surface. The shear process occurs at angle ϕ for a tool with back rake angle α. The chip has velocity V_c and makes contact with the rake face of the tool over length l. Defining the ratio of the uncut chip thickness (t) to the chip thickness (t_c) as r, the chip ratio, we find

$$r = t/t_c = \frac{\sin \phi}{\cos(\phi - \alpha)}$$

or solving this expression for ϕ, we have

$$\tan \phi = \frac{r \cos \alpha}{1 - r \sin \alpha} \tag{4.2.7}$$

There are numerous other ways to measure or compute the shear angle, both during (dynamically) and after (statically) the cutting process has been halted. The shear angle can be measured statically by instantaneously interrupting the cut through the use of "quick stop devices." These devices disengage the cutting tool from the workpiece while cutting is in progress, leaving the chip attached to the workpiece. Optical and scanning electron microscopy is then used to observe the shear angle. High-speed motion pictures have also been used to observe the process at frame rates of as high as 30,000 frames per second. More recently, machining stages have been built that allow the process to be performed inside a scanning electron microscope and recorded on video tapes for high resolution, high magnification examination of the deformation process. The micrographs in Fig. 4.2.6 were made this way. This technique has been used to measure the velocity of the shear fronts (V_s) during cutting, verifying experimentally that the vector sum of V and V_c equals V_s.

For consistency of volume, we observed that

$$\frac{V_c}{V} = \frac{t}{t_c} = r = \frac{\sin \phi}{\cos(\phi - \alpha)} \tag{4.2.8}$$

indicating that the chip ratio (and therefore the shear angle) can be determined dynamically if a reliable means to measure chip velocity can be found. Thus, one could determine ϕ dynamically for a known tool geometry. Therefore, cutting forces can be dynamically predicted, an important consideration in adaptive control machining. Velocities are also important in power calculations, heat determinations and vibration analysis associated with chip formation.

During the cutting, the chip undergoes a shear strain γ of

$$\gamma = \frac{\cos \alpha}{\sin \phi \cos(\phi - \alpha)} \tag{4.2.9}$$

which shows that the shear strain is dependent on the rake angle α and the shear direction ϕ. Generally speaking, metal cutting strains are quite large compared to other plastic information processes, being on the order of 2 to 4 in./in. This large strain occurs, however, over very narrow regions (the shear fronts), which results in extremely high shear strain rates, typically on the order of 10^4 to 10^8 in./in. sec. It is this combination of large strains and high strain rates operating within a process constrained mainly by the workpiece itself and the friction interface (boundary) on the rake face of the tool that results in great difficulties in theoretical analysis of the process.

The properties of the work material (elastic strength, shear strength under compressive loading, strain hardening characteristics, friction behavior, hardness, ductility, etc.) control the chip formation. The cutting parameters (tool materials, tool angles, and edge geometries, which change due to wear, cutting speed, feed, and depth of cut) and the cutting environment (machine tool deflections, cutting fluids, etc.) also influence the chip formation. Further complications arise due to the formation of the built up edge (BUE) on the cutting tool.

Built up edge formation is a dynamic instability produced by the localized high temperature and extreme pressure in the cutting zone. The work material adheres or welds to the cutting edge of the tool similar to a "dead metal zone" in extrusion. Although this material may protect the cutting edge from wear, it radically modifies the geometry of the tool. BUEs are not stable and will slough off periodically, adhering to the chip or passing under the tool and remaining on the machined surface. While BUE formation often can be eliminated or minimized by reducing the depth of the cut, increasing the cutting speed, using positive rake tools, or applying a coolant, it adds greatly to the complexity of the process.

4.2.5.2 Mechanics of Machining

Orthogonal machining has been defined as a two force system, whereas oblique cutting involves a three force situation. Consider Fig. 4.2.7, which shows a free body diagram of a chip that has been separated at the shear plane. The resultant force R acting on the back of the chip is assumed to be equal and opposite to the resultant force R' acting on the shear plane. The resultant R is composed of the friction force F and the normal force N acting on the tool/chip interface contact area (length l times width W). The resultant force R' is composed of a shear force F_s and a normal force F_n acting on the shear plane area A_s. To determine these forces, a third set is needed that can be measured, using a dynamometer, mounted in the workholder (or the tool holder). This set has resultant R'', which is equal in magnitude to all the other resultant forces in the diagram. R'' is composed of a cutting force F_c and a tangential (normal) force F_t. To express the desired forces (F_x, F_n, F, N) in terms of the dynamometer components F_c and F_t and appropriate angles, a circular force diagram is developed in which all six forces are collected in the same force circle. This is shown in Fig. 4.2.8. In this

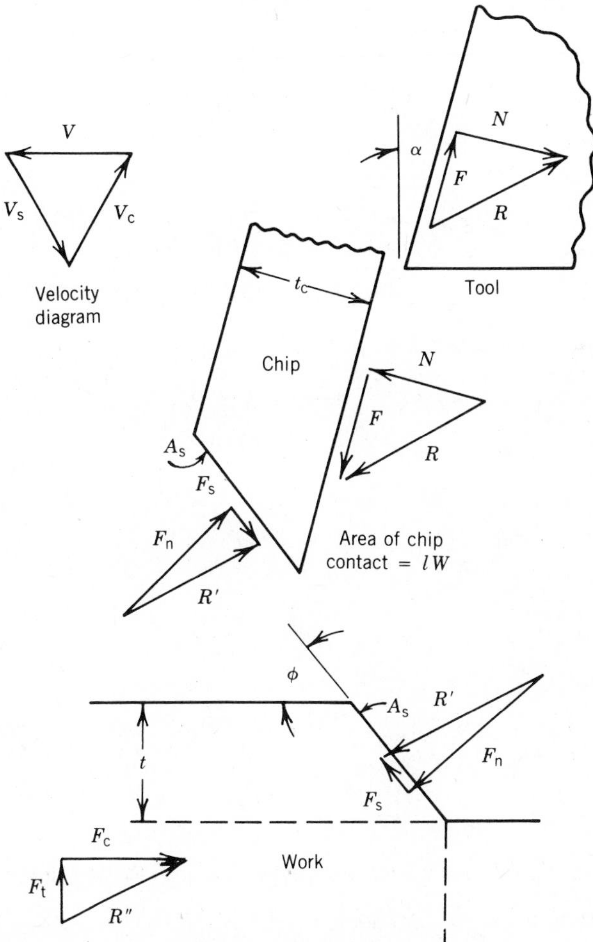

Fig. 4.2.7. Free body mechanics diagram of orthogonal clip formation process showing equilibrium condition in the chip between resultant forces R and R'. A velocity diagram is shown in the upper left.

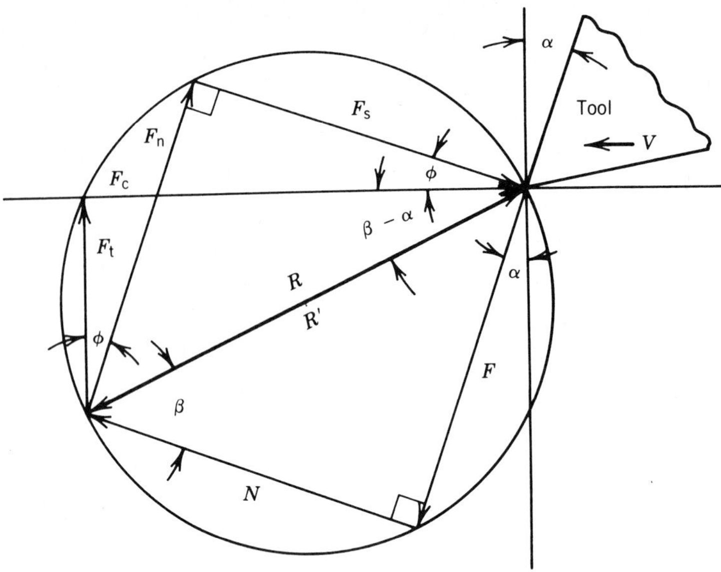

Fig. 4.2.8. Circular force diagram used to derive equations for F_s, F_n, F, and N as functions of F_c, F_t, ϕ, α, and β.

figure, β is the angle between the normal force N and the resultant force R; it is used to describe the friction coefficient, μ, on the tool/chip interface area, which is defined as F/N so that

$$\beta = \tan^{-1} \mu = \tan^{-1} F/N \tag{4.2.10}$$

The friction force F and its normal force N can be shown to be

$$F = F_c \sin \alpha + F_t \cos \alpha \tag{4.2.11}$$

$$N = F_c \cos \alpha - F_t \sin \alpha \tag{4.2.12}$$

where

$$R = \sqrt{F_c^2 + F_t^2} \tag{4.2.13}$$

all in pounds. When the back rake angle α is zero, then $F = F_c$ and $N = F_t$.

The forces parallel and perpendicular to the shear plane can be shown (from the force circle diagram) to be

$$F_s = F_c \cos \phi - F_t \sin \phi \tag{4.2.14}$$

$$F_n = F_c \sin \phi + F_t \cos \phi \tag{4.2.15}$$

both in pounds. F_s is of particular interest, as it is used to compute the shear stress on the shear plane. This shear stress is defined as

$$\tau_s = F_s/A_s \text{ psi} \tag{4.2.16}$$

where $A_s = tW/\sin \phi$.

Where t is the depth of the cut and W is the width of the workpiece, the shear stress is

$$\tau_s = \frac{F_c \sin \phi \cos \phi - F_t \sin^2 \phi}{t\,W} \text{ psi} \tag{4.2.17}$$

For a given polycrystalline metal, this shear stress is a material constant, not sensitive to variations in cutting parameters, tool material, or the cutting environment.

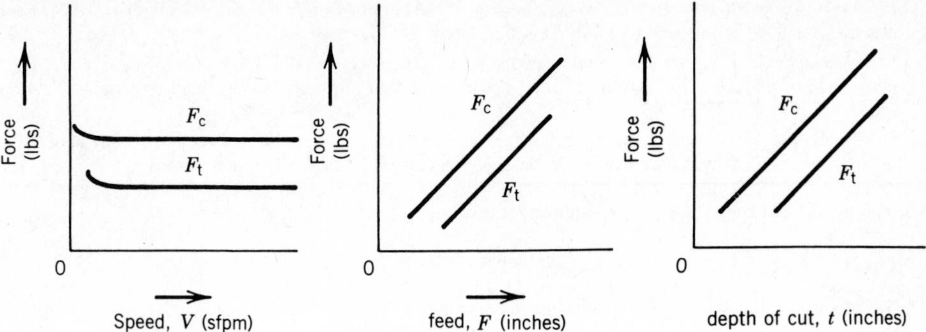

Fig. 4.2.9. General relationship of orthogonal cutting forces to primary cutting parameters speed, feed, and depth of cut.

It is the objective of some metal cutting researchers to be able to derive (predict) the shear stress τ_s and the shear direction ϕ from dislocation theory, but this has not yet been accomplished. Correlations of the shear stress with metallurgical measures such as hardness or dislocation stacking fault energy have been useful in these efforts.

The cutting force F_c is the dominant force in this system, and it is important to understand how it varies with changes in the cutting parameters. As shown in Fig. 4.2.9, the cutting forces typically double when the feed or depth of cut is doubled but remain constant when speed is increased. In addition, the forces will increase (and change direction) when the rake angle is reduced. Now let's take a brief look at the oblique situation with the understanding that what we have learned about the process and F_c from the orthogonal model will hold true for the oblique situation as well.

4.2.5.3 Energy and Power in Machining

In oblique machining, there are three cutting forces, shown schematically in Fig. 4.2.10:

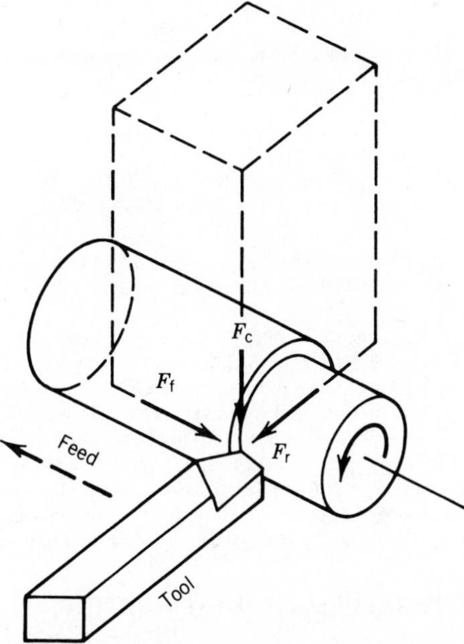

Fig. 4.2.10. Three components of measurable forces acting on a single-point turning tool during oblique machining.

F_c = primary cutting force acting in the direction of the cutting velocity vector. F_c is the largest force and accounts for 99% of the power required by the process.

F_f = feed force acting in the direction of the tool feeds. F_f is usually about 50% of F_c. Only a small percentage of the power is consumed via F_f because the feed rates are usually small compared to cutting speeds.

F_r = radial or thrust force acting perpendicular to the machine surface. This force is typically about 50% of F_f. Power requirements are negligible because velocity in the radial direction is zero.

The energy per unit time P, or power required for cutting, is

$$P = F_c V \quad \text{ft-lb/min } (F_f f_f \text{ and } F_r f_r \text{ negligible}) \tag{4.2.18}$$

The horsepower at the spindle of the machine is, therefore,

$$\text{HP} = F_c V / 33{,}000 \tag{4.2.19}$$

A very useful parameter in metal cutting is called the unit or specific horsepower HP_s, which is defined as

$$\text{HP}_s = \text{HP/MRR (hp/in.}^3\text{/min.)} \tag{4.2.20}$$

In turning, for example, where MRR $\cong 12\, V f_r t$,

$$\text{HP}_s = F_c / 396{,}000\, f_r t \text{ (hp/in.}^3\text{/min.)} \tag{4.2.21}$$

Unit horsepower represents the approximate power needed at the spindle to remove one cubic inch of metal per minute. Specific power factors for some common materials are given in standard machining reference works. Specific horsepower is related to and correlates well with shear stress (τ_s) for a given metal. The major difference is that unit horsepower is sensitive to certain variations in cutting parameters where τ_s is not.

About three-quarters of the energy is consumed in the shear process and one-quarter in tool chip interface friction.

Unit horsepower can be used in a number of ways. First, it can be used to estimate the motor horsepower required to perform a machining operation for a given material. By multiplying HP_s values by the approximate MRR for the process, one obtains the motor horsepower (HP_m). Thus,

$$\text{HP}_m = \frac{\text{HP}_s \times \text{MRR} \times \text{Correction factors}}{E} \tag{4.2.22}$$

where E is a machine efficiency factor that accounts for the power needed to overcome friction and inertia in the machine and drive moving parts. Correction factors are used to account for variations in cutting speed, feed, rake angle, and tool wear.

The primary cutting force F_c can also be roughly estimated according to

$$F_c = \frac{\text{HP}_s \times \text{MRR} \times 33{,}000}{V} \text{ lb} \tag{4.2.23}$$

This type of estimate of the major force F_c is useful in analyzing deflection and vibration problems in machining and in properly designing work holding devices, which must be able to resist movement and deflection of the part during the process.

Increasing the speed, the feed, or the depth of cut will increase the power requirement. Doubling the feed or the depth of cut doubles the cutting force F_c, which in turn doubles the power requirement. Increasing the speed increases HP directly but does not increase the cutting force F_c. This is because the periodicity of spacing of the shear fronts remains constant in a material regardless of the cutting speed.

If speed is doubled, chip length is doubled for the same amount of cutting time. Constant shear front spacing (constant lamella size) means twice as many shear fronts, so energy is doubled. Cutting force F_c, on the other hand, reflects a change in F_s (where $F_s = \tau_s A_s$ and τ_s = shear strength, A_s = shear area). Neither of these two quantities were changed by the change in speed; therefore, F_c remains constant. However, speed does have a strong effect on tool life, because most all the input energy is converted into heat.

4.2.6 HEAT AND TEMPERATURE IN METAL CUTTING

In metal cutting, the power put into the process ($F_c V$) is largely converted to heat, elevating the temperatures of the chip, the workpiece, and the tool. These three elements of the process along with the environment (which includes the cutting fluid) act as the heat sinks. The three main sources of heat, listed in order of their heat generating capacity, are

Shear zone.

Tool/chip interface contact region.

Flank of the tool.

There have been numerous experimental techniques developed to measure cutting temperatures and some excellent theoretical analysis of this "moving" multiple heat source problem. The rate of wear of the tool at the interface can be shown to be directly related to temperature. Because cutting forces are concentrated on small areas near the cutting edge, these forces produce large pressures. The tool material must sustain these properties at elevated temperatures.

The challenge to manufacturers of cutting tools has always been to find materials that satisfy these process needs. Cutting tool materials that soften less at the high temperatures associated with high speeds are said to have "hot hardness," but obtaining this property usually requires a trade-off in toughness, as hardness and toughness are generally opposing properties.

4.2.7 CUTTING TOOL MATERIALS

Obviously, the production rate is a function of the MRR, which depends on V, f, and t. In many machining operations, cutting speed and feed are limited by the capability of the tool material. speeds and feeds must be kept low enough to provide for a minimum acceptable tool life. If not, the time lost changing tools may outweigh the productivity gains due to increased cutting speed (or feed). The materials selected for cutting tools must combine hardness and strength (toughness) with good wear resistance at elevated temperatures. Table 4.2.2 presents a chronological rating of cutting tool materials, clearly showing the typical advances that have occurred in this field over the last eighty years.

The majority of metal cutting done today is performed with cutting tools made from high speed steel, which comes in many grades; carbides made by powder metallurgy techniques; cubic boron nitride (CBN), a man-made material that retains 85–95% of its room temperature hardness at 1800°F; and coated carbides, high speed steels, and ceramics. Coated tools employ a thin, hard, wear-resistant surface layer on tough, shock-resistant carbide, resulting in 50–100% speed increases with the same tool life as uncoated tools. Almost all coated carbide and ceramic tools are designed to be insert type tools, and today coated tools represent about 35% of the carbide tool market.

High-speed steel tools have excellent tensile strengths, while carbides and ceramics tend to be exceptionally strong in compression and far weaker in tension, failing by brittle fracture. This means that carbide tools must be designed and utilized such that they are under compressive loads during cutting. Thus, these tools have small or negative rakes and minimal clearance angles. Some cutting tools, such as twist drills, require large rake angles, as dictated by their basic design. Therefore, these tools are made from high-speed steel (HSS) and must be operated at lower cutting speeds unless coated with titanium carbide or titanium nitride.

4.2.8 TOOL FAILURE

The cutting tools in machining are subjected to high, localized pressure and elevated temperatures as well as mechanical and thermal shocks. They are thus subject to failure, which is the single most critical problem in the application of this process in practice. Tool life (how long the tool lasts) can be classified according to the failure mechanism that causes the tool to die:

Slow death mechanisms. Gradual tool wear on the flank(s) of the tool or wear on the rake face of tool (called crater wear), or both, related to cutting temperature and length of cut

Sudden death mechanisms. Rapid, usually unpredictable, and often catastrophic failures resulting from abrupt, premature death of a tool

Figure 4.2.11 shows a sketch of a "worn" tool, showing crater and flank wear. As the tool wears, its geometry changes. The geometry change alters the cutting forces, the power being consumed, the surface finish, and the dimensional accuracy of the part. The dynamic stability of the process may be altered, as worn tools may chatter in processes usually relatively free of vibration. Wear mechanisms active in this high-temperature environment include abrasive, adhesion, diffusion, or chemical interactions, and any or all of these mechanisms may be operative at a given time in a given process.

Sudden death mechanisms are plastic deformation, brittle fracture, fatigue fracture, or edge chipping. In a particular situation, it is difficult to predict which mechanism will dominate and result in a tool failure. What can be said is that tools, like people, die (or fail) from a great variety of causes under widely varying conditions. Therefore, tool life should be treated as a random variable, probabilistically, and not as a deterministic quantity.

If a machining experiment was repeated 15 times without changing any of the input parameters, and the wear on the flank of the tool was monitored, the outcome might look like Fig. 4.2.12. Tool wear has a random variable nature. Tool failure must therefore be treated as a random variable. In Fig. 4.2.12, the average time to failure is denoted as μ_T and the standard deviation as σ_T, where the tool is declared to be dead when it has

Table 4.2.2. Typical Characteristics and Data for Tool Materials

Tool Materials (year material introduced)	Machining Operation (permissible V for machining steel)	Modes of Tool Wear or Failure	Limitations	Hardness		Typical n Values for $VT^n = C$
				RT	1200°F	
Carbon steels (~1800)	Tapping, drilling, reaming	Buildup, plastic deformation, abrasive wear, microchipping of edges	Low hot hardness, limited hardenability, and limited wear resistance. Low CS, low strength	80 R_A	Mushy	
Low/medium alloy steels	Tapping, drilling, reaming (50 sfpm)	Buildup, plastic deformation, abrasive wear, microchipping of edges	Low hot hardness, limited hardenability and limited wear resistance. Low CS, low strength	80–84 R_A	Mushy	.14–.16
HSS (1907)	Turning, drilling, milling, broaching (100 sfpm)	Flank wear, crater wear	Low hot hardness, limited hardenability, and limited war resistance	84–88 R_A	65–70 R_A	.25
Cemented carbide (1940–1945)	Turning, drilling, milling, broaching (200 sfpm)	Flank wear, crater wear	Cannot use at low speed due to cold welding of chips and microchipping	90–95 R_A	74–84 R_A	.30
Ceramics (1950)	Turning (800 sfpm)	DCL[b] notching, microchipping, gross fracture	Low strength, low thermo/mechanical fatigue strength. Not for interrupted cutting	92–94 R_A 2100 K[c]	88–90 R_A	.40
CBN (1969)	Turning, milling (500 sfpm)	DCL notching, chipping, oxidation, graphitization	Low strength, low chemical stability at higher temperature	~4700 K	Not known	
Diamond-Polycrystal[a] (late 1960s)	Turning (not applicable for milling)	Chipping, oxidation, graphitization	Low strength, low chemical stability at higher temperature	~7000 K	Not known	
Coated carbides (1970) TiC or TiN	Turning (700–900 sfpm)	Flank wear, crater wear	Microchipping at low speed due to cold welding	90–95 R_A	74–84 R_A	.30
Coated HSS TiN	All HSS tools (100–200 sfpm)	Flank wear, crater wear	Thin coating necessary, .0001 in.	80–85 R_A	Not known	.35
Al_2O coated carbide (1980)	Turning (1,000–1,400 sfpm)	Flank wear, crater wear	Not for low-speed work. Not suitable for machining Al or Ti	~93 R_A	~88 R_A	.40

[a] Thin layer (.5–1.5 mm) of fine-grain diamond sinterid to carbide substrate.
[b] DCL = notching of the cutting edge at the depth of cut line.
[c] K = knoop hardness.

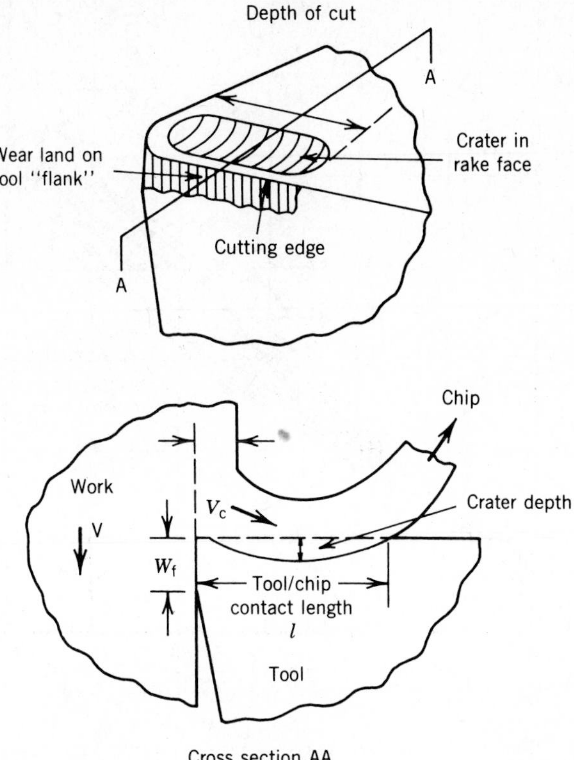

Fig. 4.2.11. Sketch of a worn tool showing various wear elements resulting during oblique cutting.

worn 0.25 in. on the flank. At a given time during the test, say 35 minutes, the flank wear ranged from 0.135 to 0.21 in., with the average being 0.175 in.

Many criteria are used to define tool death. In addition to wear limits, surface finish, failure to conform to size (tolerances), increases in cutting forces and power, or complete failure of the tool can be used. In automated processes, it is very beneficial to be able to monitor or predict the tool wear and failure on-line, so that the tool can be replaced prior to failure wherein defective product may also result. The feed force, for example, has been shown to be a good indirect measure of tool wear.

Once criteria for failure have been established, tool life is that time elapsed between start and finish of the cut, in minutes. Other ways to express tool life, other than time, include

Volume of metal removed between regrinds or replacement of tool
Number of holes drilled with a given tool
Number of pieces machined per tool

4.2.9 TOOL LIFE MODELS

Around the turn of the century, F. W. Taylor and his colleagues developed mathematical models (empirically determined from thousands of cutting experiments) to describe tool life. These were deterministic models and took the form of

$$VT^n = C \tag{4.2.24}$$

where V was cutting speed, T was tool life in minutes to failure, n was a constant primarily dependent on the tool material, and C was an empirical constant dependent on all other input parameters, including feed and work material. Typical values for n are given in Table 4.2.2.

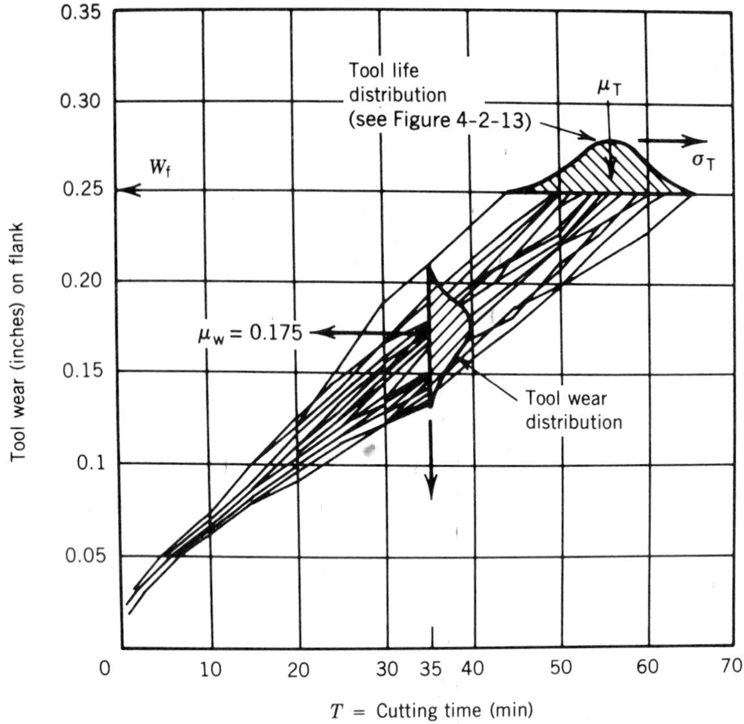

Fig. 4.2.12. Tool wear on the flank displays a random variable nature, as does tool life.

Over the last 80 years, many more sophisticated but still empirical versions of tool life equations have been published. In the main, however, all these equations reflect tool failure by a single death mechanism, usually flank or crater wear. A more realistic approach is to describe tool life probabilistically in the following way. Since T is a function of many variables and is a random variable, one can write

$$T = \frac{1/C^n}{1/V^n} = \frac{K}{V^m}, \qquad m = \frac{1}{n} \tag{4.22.25}$$

where K is now a random variable that represents the effect of all unmeasured factors and V^m is an input variable. The sources of variability include the variation of work material hardness, variations in tool material

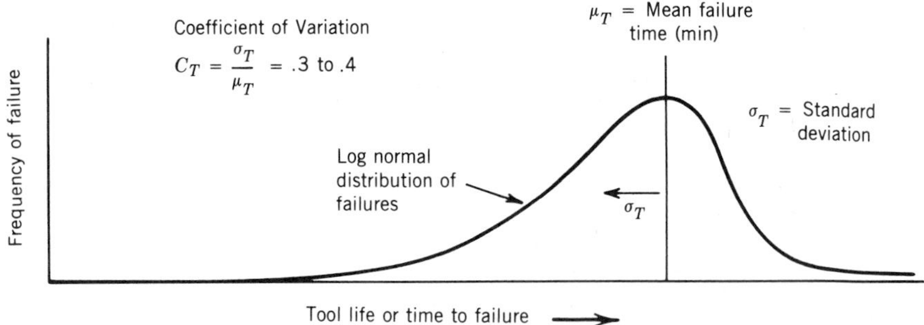

Fig. 4.2.13. Tool life viewed as random variable has a log normal distribution with a large coefficient of variation.

metallurgy and geometry, and changing surface characteristics of workpieces. When all the possible failure modes are considered, one discovers that tool life distributions are usually log normal and have a large coefficient of variation due to the large standard deviation of the life distribution (see Fig. 4.2.13). Now the tool life problem can be viewed from a reliability standpoint, and one can develop viable tool replacement strategies that can be directed toward obtaining production uptimes and superior quality.

It is clear from this research that while many manufacturers of cutting tools have worked toward developing tool materials that have greater tool life (larger μ_T) at high speeds, few have directed their energies toward developing tools that have less variability (smaller σ_T) at all speeds. The reduction in variability would result in smaller coefficients of variation, which now are typically on the order of .3 to .4 for industrial machining operations. Note that this means if you have a tool with a 90-min average life and a 30-min standard deviation, about 16% of the tools will fail in the first 60 min. In automated equipment, early unpredicted failures can be very costly, so the reduction in tool life variability would pay great benefits in quality, productivity, and cost.

4.2.10 PROCESS CAPABILITIES IN MACHINING

The process of machining is utilized to manufacture component parts to match or exceed the specifications called for by the designers of those parts. Designers are responsible for applying the tolerances and specifying the nominal or desired sizes. The manufacturing engineer selects a sequence of processes to manufacture the part. Each of the processes has some inherent capability—that is, some natural level of accuracy (or aim) and precision or variability—that must be known and matched against the functional, specified characteristics of the components being made. In order to do this, a process capability study must be carried out. Using essentially the mathematics of quality control, parts produced by the various processes are measured and analyzed in order to determine the natural or inherent capabilities of the machining process. In machining, this is a rather complex business in that many factors of the process enter into the final result; these include the type, design, age, and control system of the machine tool; the parameters (feed and depth of cut) selected for the machining process; the operator; the workholding devices; the design of cutting tools and toolholders; and the cutting fluids. For example, changing nothing but the work material in a machining operation will alter the process capability, as harder, stronger metals will result in greater cutting forces, more deflections, and greater variability in sizes (dimensions) in the parts. The same sort of thing can happen as tools wear out and become dull, which increases the cutting forces. All in all, process capability is a much neglected and a poorly understood problem that deserves much greater attention than it has received thus far. As processes become more automated, the significance of this activity will become more apparent, and more investigative work will be done on process capabilities. When this happens, we will better understand what are the elements of a process that dominate its process capability. Then we will be able to do a better job matching process capabilities with product design needs, resulting in better quality products and fewer rejected parts. For an expanded discussion of process capability, see Chapter 17, *Materials and Processes in Manufacturing,* 6th ed., and *Quality Planning and Analysis,* 2nd ed.

BIBLIOGRAPHY

DeGarmo, E. P., Black, J T., and Kosher, R. A. *Materials and Processes in Manufacturing,* 6th ed. New York: Macmillan, 1984.

Juran, J. M., and Gryna, F. M., Jr. *Quality Planning and Analysis,* 2nd ed. New York: McGraw-Hill, 1970.

Machining Data Handbook, 2nd ed. Compiled by the technical staff of the Machinability Data Center, Cincinnati, Ohio, 1980.

CHAPTER 4.3

FORMING PROCESSES

RONALD A. KOHSER

University of Missouri
Rolla, Missouri

CHESTER J. VAN TYNE

Lafayette College
Easton, Pennsylvania

The primary goal of any forming process is to transform a mass of selected material into a desired geometric shape with a desired level of precision. This transformation is achieved by exploiting the plastic properties possessed by certain classes of materials. For example, most metals have the ability to flow plastically while in the solid state with no deterioration in engineering properties. In fact, the deformation process can often serve to enhance a metal's properties. The increase in strength and hardness via strain hardening is an example of such an enhancement.

Polymeric materials can also be deformed by processes that are analogous to those used by metal formers. The forming processes applied to polymers, however, are usually performed with a material that is in a semifluid or fluid state. Thus, while the processes themselves are analogous to the solid metal deformation processes, the material behavior is somewhat similar to that of a molten metal. Polymer forming processes, therefore, also bear correlation to the casting processes (e.g., injection molding of plastics can be viewed as analogous to die casting of metals).

The usefulness of a material in our modern civilization is often determined by how easily it can be formed into useful shapes.

4.3.1 CLASSIFICATION OF PROCESSES

There are a variety of ways to classify the deformation processes used for the manufacture of objects. These classification schemes are usually based on a single factor, such as environmental condition, flow behavior during the process, and operational scheme. The classification of metal deformation processes by several methods are presented in Tables 4.3.1–4.3.4. Tables 4.3.1 and 4.3.2 show classification by the two major environmental conditions, namely temperature and stress state. Table 4.3.3 separates processes on the basis of quasi-static or steady state (as in wire drawing or rolling) versus dynamic or constantly changing (as in forging) flow pattern. Table 4.3.4 provides a classification of processes on the basis of tooling. Four of the six processes listed (extrusion, forging, rolling, and wire drawing) are massive deformation processes, wherein there is a considerable increase in the surface-to-volume ratio of the workpiece. Deep drawing and stretch forming, on the other hand, are examples of sheet-forming processes, where the workpiece is deformed into a three-dimensional object without any significant change in thickness. Table 4.3.5 presents typical massive forming processes, and Table 4.3.6 lists a number of sheet-forming operations.

For familiarization, a brief description will be provided for each of the six basic processes of Table 4.3.4.

4.3.1.1 Deep Drawing

Deep drawing, illustrated in Fig. 4.3.1, is a plastic forming process in which a flat sheet of material is formed into a recessed, three-dimensional part with a depth several times the thickness. As the punch descends into the into the die, the material assumes the three-dimensional configuration.

Table 4.3.1. Classification of Deformation Processes Based on Temperature

Process	Temperature
Cold worked	<0.3 absolute melting point of material
Hot worked	>0.6 absolute melting point of material
Warm worked	point of material 0.3–0.6 of absolute melting

Table 4.3.2. Classification of Major Deformation Processes Based on Stress State

Process	Stress State
Indirect compression	Applied tensile stress in one direction; induced compressive stresses by the material and tooling in the other two directions
Direct compression	Applied compressive stress in one direction; induced compressive stresses by the material and tooling in the other two directions
Biaxial tension	Applied tensile stress in two directions; induced compressive stress usually by the material in the third direction.

Table 4.3.3. Classification of Major Deformation Processes Based on Flow Pattern

Process	Flow Pattern
Quasistatic	Single unchanging deformation pattern
Dynamic	Continuously changing pattern during the deformation cycle

Table 4.3.4. Classification of Major Deformation Processes Based on Tooling

Process	Figure	Temperature	Stress State	Flow Pattern
Deep drawing	4.3.1	Hot	Indirect compression	Dynamic
Extrusion	4.3.2	Hot or cold	Direct compression	Quasi-static/Dynamic
Forging	4.3.3	Hot or cold	Direct compression	Dynamic
Rolling	4.3.4	Hot or cold	Direct compression	Quasi-static
Stretch forming	4.3.5	Cold	Biaxial tension	Dynamic
Wire drawing	4.3.6	Cold	Indirect compression	Quasi-static

Table 4.3.5. Typical Massive Forming Processes

Forging	Rolling	Extrusion	Drawing
Closed-die forging with flash	Sheet rolling	Nonlubricated hot extrusion	Drawing
Closed-die forging without flash	Shape rolling	Lubricated direct hot extrusion	Drawing with rolls
Coining	Tube rolling	Hydrostatic extrusion	Ironing
Electro-upsetting	Ring rolling		Tube sinking
Forward extrusion forging	Rotary tube piercing		
Backward extrusion forging	Gear rolling		
Hobbing	Roll forging		
Isothermal forging	Cross rolling		
Nosing	Surface rolling		
Open-die forging	Shear forming (flow turning)		
Orbital forging	Tube reducing		
P/M forging			
Radial forging			
Upsetting			

Source: Ref. 1, Table 2-2.

Table 4.3.6. Typical Sheet Metal Forming Processes

Bending and Straight Flanging	*Deep Recessing and Flanging*
Brake bending	Spinning (and roller flanging)
Roll bending	Deep drawing
	Rubber pad forming
Surface Contouring of Sheet	Marform process
Contour stretch forming (stretch forming)	Rubber diaphragm hydroforming
Androforming	
Age forming	*Shallow Recessing*
Creep forming	Dimpling
Die-quench forming	Drop hammer forming
Bulging	Electromagnetic forming
Vacuum forming	Explosive forming
	Joggling
Linear Contouring	
Linear stretch forming (stretch forming)	
Linear roll forming (roll forming)	

Source: Ref. 1, Table 2-3.

4.3.1.2 Extrusion

In the extrusion process, illustrated in Fig. 4.3.2, the workpiece is compressively forced to flow through a suitably shaped die to form a product with reduced cross section. Basically, the process is analogous to squeezing toothpaste out of a tube. In the most common arrangement (direct extrusion), the workpiece is placed inside a confining chamber. A ram moves forward, first causing the material to conform to the confining chamber. As the ram advances further, the pressure builds until the material flows plastically through the die.

4.3.1.3 Forging

Forging, illustrated in Fig. 4.3.3, is the oldest known metal working process, the modern practice having evolved from the ancient practices of armor makers and blacksmiths. A variety of forging processes have been developed to provide immense flexibility, both in size, shape, complexity, and quantity. The workpiece may be drawn out, increasing its length and decreasing its cross section; upset, increasing the cross section and decreasing the length; or squeezed in closed-impression dies to produce multidirectional flow.

4.3.1.4 Rolling

Figure 4.3.4 illustrates the rolling process, which consists of passing the workpiece between two rolls that rotate in opposite directions, the space between the rolls being somewhat less than the thickness of the entering material. Since the rolls rotate with a surface velocity greater than the speed of the incoming material, friction along the contact interface moves the workpiece forward. The material is squeezed and elongated with a decrease in cross-sectional area. The amount of deformation that can be achieved in a single pass between a pair of rolls depends on the friction conditions along the interface.

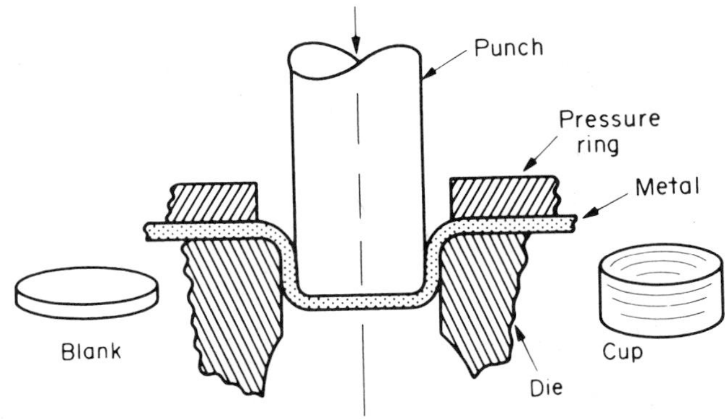

Fig. 4.3.1. Deep drawing. From Ref. 2, Fig. 3.12.

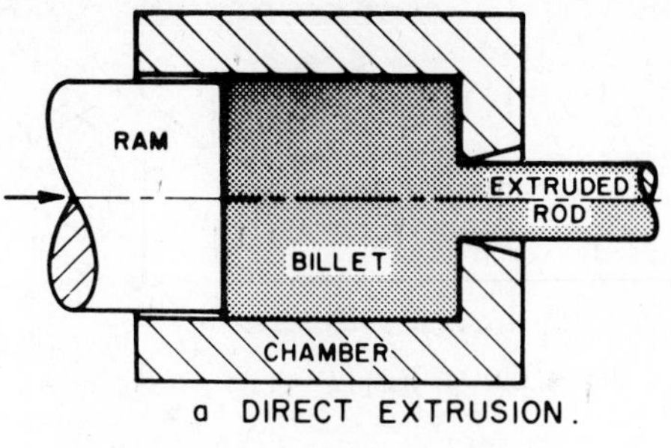

a DIRECT EXTRUSION.

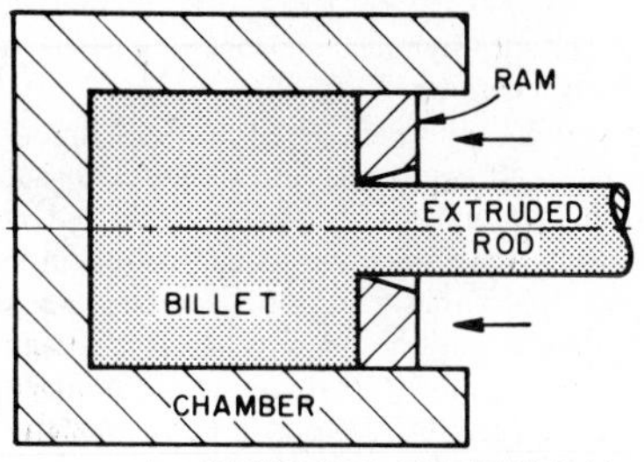

b INDIRECT EXTRUSION.

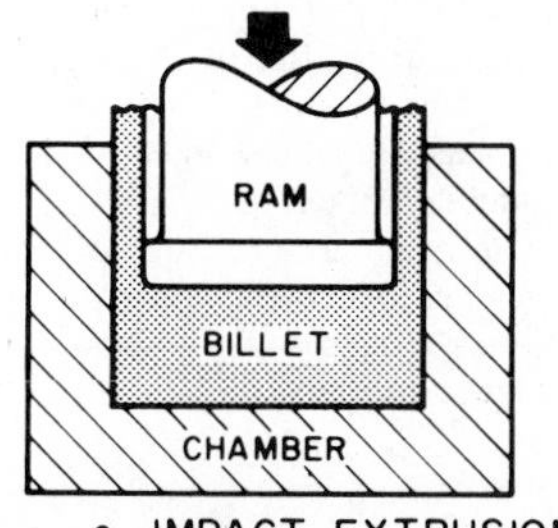

c IMPACT EXTRUSION.

Fig. 4.3.2. Extrusion. From Ref. 3, Fig. 4.3.

4.3.1.5 Stretch Forming

Stretch forming, shown in Fig. 4.3.5, is a process in which the sheet of metal is gripped by two or more sets of jaws that stretch it and wrap it around a formblock. Through proper control of the stretching, most or all of the compressive stresses that accompany other sheet metal processes are eliminated. Consequently, there is very little springback and the workpiece conforms closely to the shape of the formblock.

4.3.1.6 Wire Drawing

Figure 4.3.6 shows the process of wire drawing, in which one end of a wire or rod is reduced or pointed, inserted through a converging die, grasped by grips, and pulled. The workpiece is reduced in cross section and elongated. The reduction in area per pass is small in order to avoid fracture. Therefore, several steps are often required to obtain a desired product.

Returning to the temperature classification of Table 4.3.1, let us now consider deformation processes in each of the three temperature regimes.

4.3.1.7 Cold Working

While many consider forming at room temperature to be cold working, cold working is technically the plastic deformation of materials at a temperature below their recrystallization temperature. In some cases, the working may be done at mildly elevated temperatures to provide increased ductility and reduced strength. From a manufacturing viewpoint, cold working has a number of distinct advantages, and the various cold-working

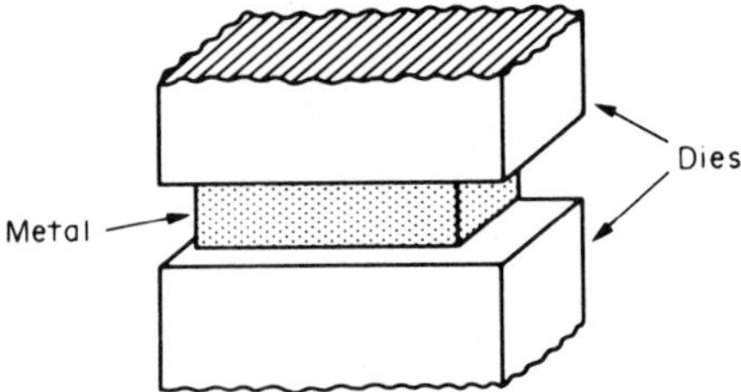

Fig. 4.3.3. Forging. From Ref. 2, Fig. 3.14.

processes have become extremely important. Significant advances in recent years have extended the use of cold working and the trend appears likely to continue.

The cold-working processes are particularly suited for large volume production, when the quantity involved can readily justify the cost of the required equipment and tooling. Considerable effort has been devoted to developing and improving cold-forming equipment. In addition, better and more ductile metals and an improved understanding of basic plastic flow have done much to reduce the difficulties experienced in earlier years. To a very large extent, modern mass production has paralleled, and been made possible by, the development of cold-forming processes. Automated, high-quality production enables the manufacture of low-cost metal products. In addition, most cold-working processes eliminate or minimize the production of waste material and the need for subsequent machining. With increasing efforts in conservation and materials recycling, these benefits become quite significant.

In order to obtain several of the benefits of cold working, the metal must often receive special treatment prior to processing. First, if better surface finish and dimensional accuracy are to be obtained than those produced by hot working, the starting material must be free of existing scale to avoid abrasion and damage to the dies or rolls that are used. Scale is removed by pickling; that is, the metal is dipped in acid and then washed. Second, to assure good dimensional tolerances in cold-worked parts, it is often necessary to start with metal that is uniform in thickness and has a smooth surface.

Another treatment that may be given to metal prior to cold working is annealing. If the cold working is to involve considerable deformation, it is desirable to have as much ductility available as possible. In many cases, annealing is performed after the workpiece has been partially shaped by cold working. Here annealing restores sufficient ductility to permit the final stages of forming to be done without danger of fracture.

4.3.1.8 Hot Working

Hot working is defined as the plastic deformation of metals *above* their recrystallization temperature. Here it is important to note that the recrystallization temperature varies greatly with different materials. Lead and tin are hot worked at room temperature, steels require temperatures near 1100°C (2000°F), and tungsten is still in a cold or warm working state at 1100°C (2000°F). Hot working does not necessarily imply high absolute temperatures.

Plastic deformation above the recrystallization temperature does not produce strain hardening. Therefore, hot working does not cause any increase in yield strength or hardness, or corresponding decrease in ductility. Thus it is possible to alter the shape of metals drastically by hot working without causing them to fracture and without having to use excessively large forces. In addition, the elevated temperatures promote diffusion that can remove chemical inhomogeneities. Pores can be welded shut or reduced in size during deformation, and the metallurgical structure can be altered to improve the final properties.

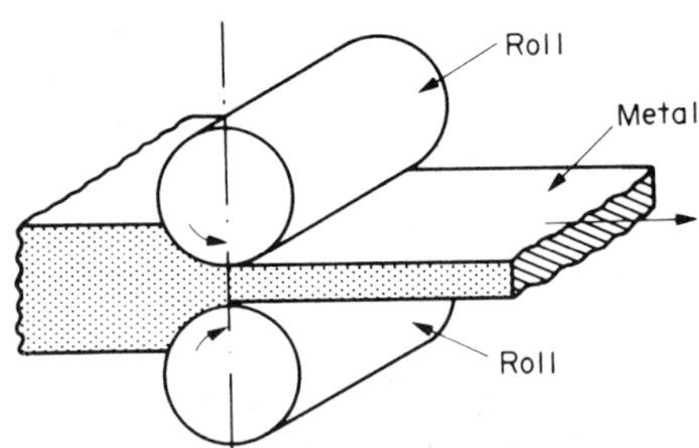

Fig. 4.3.4. Rolling. From Ref. 2, Fig. 3.13.

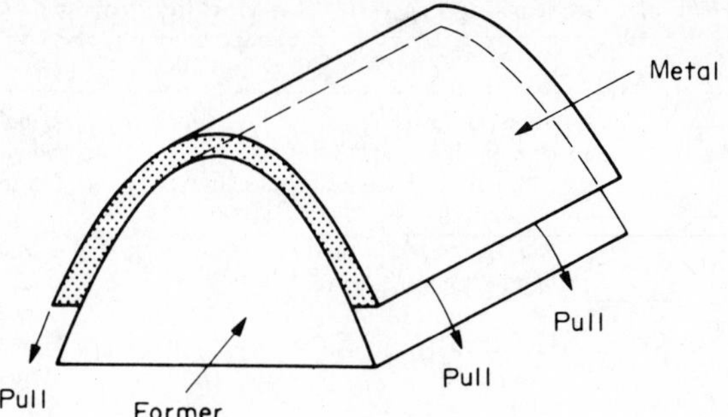

Fig. 4.3.5. Stretch forming. From Ref. 2, Fig. 3.16.

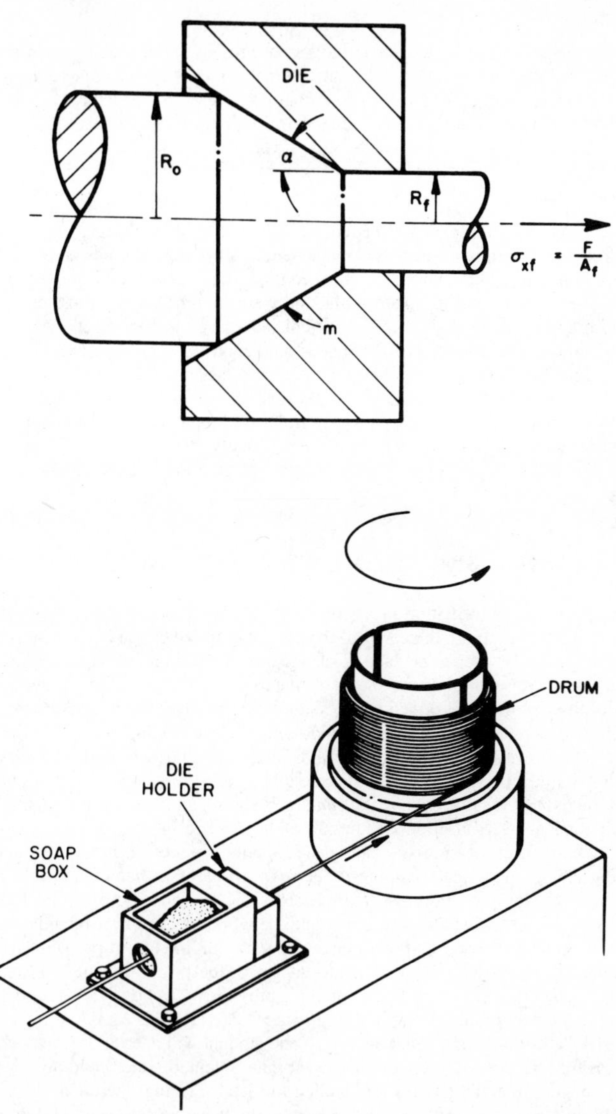

Fig. 4.3.6. Wire drawing. From Ref. 3, Figs. 3.1, 5.2.

Table 4.3.7. Comparison of Cold-Working Process to Hot-Working Process

	Cold Working	Hot Working
Thermal energy required	Small if any	Usually large
Mechanical energy	Usually large	Small
Steps involved	Often several work-anneal cycles required	No intermediate anneals needed
Tolerances in product	Very high	Large variations possible
Surface finish of product	Bright, clean	Dull; scale or contamination possible
Strength of product	Increased	Remains constant
Density of product	Decreased	Remains constant
Residual stresses in product	Often	Seldom
Properties of product	Often directional	Nondirectional

From a negative viewpoint, the high temperatures may promote undesirable reactions between the metal and its surroundings. Tolerances are poorer due to thermal contractions and possible nonuniform cooling. Metallurgical structure may also be nonuniform, the final grain size and structure depending upon reduction, temperature at last deformation, cooling history after deformation, and other factors.

A comparison of hot working with cold working is given in Table 4.3.7.

4.3.1.9 Warm Working

Deformation produced at temperatures intermediate to hot and cold working is known as warm working. Compared to cold working, warm working offers the advantages of reduced loads on the tooling and equipment and increased material ductility. The favorable as-formed properties may well eliminate the need for subsequent heat treatment operations. Compared to hot forming, the warm regime requires less energy consumption (the increased deformation energy is more than offset by reduced energy in heating the workpiece), produces less scaling and decarburization, and provides better dimensional control, less scrap, and longer tool life (while the tools must exert higher forces, there is less thermal shock and thermal fatigue).

Warm working, however, is still a developing field, and there are several barriers to its growth. Material behavior is less well characterized at these previously little-used temperatures. Lubricants have not been fully developed for operation at these temperatures and pressures. Finally, die design technology is not well established for warm working. Nevertheless, the pressures of energy conservation will definitely favor the increased use of warm working.

4.3.2 FRICTION AND LUBRICATION UNDER METALWORKING CONDITIONS

An important consideration in deformation processes is the friction developed between the workpiece and the forming tool or tools. Sometimes more than 50% of the energy supplied by the equipment is spent in overcoming friction. Product quality aspects, such as surface finish and dimensional precision, are directly related to friction. Further, changes in lubrication can alter the mode of material flow during forming and, in so doing, modify the properties of the final product. Production rates, tool design, tool wear, and process optimization often depend on the ability to control friction. Some processes, such as rolling, can only operate with sufficient friction. Friction effects, however, are hard to measure, and, since they depend on such variables as contact area, speed, and temperature, they are difficult to scale down for testing or scale up to production conditions.

It should be recognized that friction during deformation is significantly different from the friction encountered in most mechanical devices. The friction conditions of gears, bearings, journals, and similar components generally involves two surfaces of similar material and strength under elastic loads such that neither body undergoes permanent change in shape. Forming processes, on the other hand, involve a hard, nondeforming tool interacting with a soft workpiece at pressures sufficient to cause plastic flow in the weaker material. Only a single pass is involved as the tool shapes the piece, and the workpiece is often at highly elevated temperature. Figure 4.3.7 shows the change in frictional resistance with variation in contact pressure. At light, elastic loads, friction is proportional to the pressure normal to the interface, the proportionality constant (often denoted as μ) being known as the coefficient of friction. For high pressures, friction becomes independent of contact pressure and is more dependent on the strength of the weaker material. This figure, however, is for unlubricated contact. The presence of a lubricant and variation in type and amount will alter the friction behavior. Nevertheless, it is sufficient to note that friction, lubrication, and wear data obtained under conditions typical of mechanical components will be of questionable value when applied to metal forming operations.

Since the tooling only passes over the workpiece once, any wear experienced by the workpiece is usually not objectionable. In fact, the shiny, fresh surface produced by wear is often desired by customers. Manufacturers

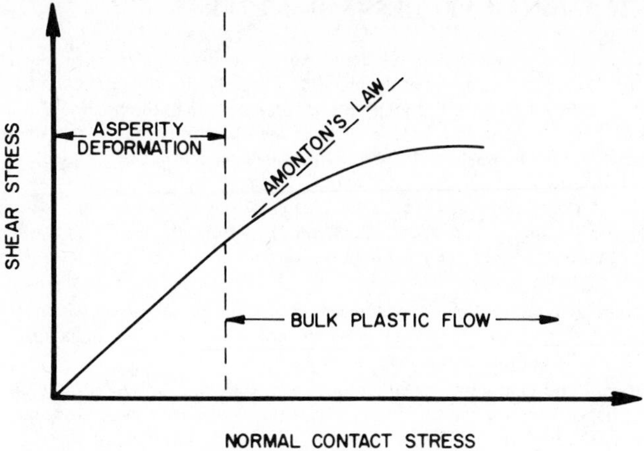

Fig. 4.3.7. Variation in shear stress. From Ref. 3, Fig. 16.8.

failing to produce enough wear, thereby retaining some of the original dull finish, may actually be accused of selling old or substandard products. Wear on the tooling, however, is not desirable. Tooling is expensive, and it is expected to shape many workpieces. Wear here generally means that the dimensions of the workpiece will change. Tolerance control is lost, and at some point the tools will have to be replaced. Other consequences of tool wear include increased frictional resistance (increased required energy and decreased process efficiency); poor surface finish of the product; and loss of production during tool changeover.

Lubrication is of immense importance in forming processes. Lubricants are selected primarily to reduce friction and suppress tool wear. However, there are a number of other aspects that are often considered when selecting a lubricant, and some of these are listed in Table 4.3.8.

In view of the important aspects of lubricants, it is amazing how little science has been applied to their development and selection. Most useful lubricants have been developed on the basis of art and experience. Many alternatives exist, often with proprietary additives or unreported chemistries marketed under a variety of trade names. Selection is often a wasteful, hit-or-miss, proposition. Moreover, when a problem is encountered in a process, lubricant variation is often the easiest and least expensive process variable that can be altered.

Nevertheless, the benefits of friction and lubrication control can be great. For example, if one can achieve full-time fluid separation between a tool and workpiece, the required deformation forces may be reduced 30–40% and tool wear becomes almost nonexistent. Considerable effort, therefore, has been directed to the measurement of friction under general working conditions and for the various forming processes. A science base should be be established that will enable the optimum selection and utilization of lubricants.

Table 4.3.8. Characteristics to Be Considered in Choosing a Lubricant

Reduction of friction
Supression of tool wear
Thermal insulation
Coolant properties
Corrosion resistance
Ease of application
Ease of removal
Toxicity, odor, and flammability
Reactivity with workpiece surface
Thermal stability
Pressure stability
Wetting ability
Cost
Availability

4.3.3 CLASSIFICATION OF PROCESS PARAMETERS

During a deformation process, a complex interrelationship occurs between independent and dependent variables. The independent variables are those aspects of the process that can be directly controlled by the operator. These variables are usually specified during the design aspect and initial setup of the process. Although, theoretically, they all can be varied and are under the direct control of the operator, because of economic considerations, some of these variables are often considered fixed. Typical independent process variables include

Starting material. The operator can specify the chemistry and condition of the material to be deformed, thereby selecting the starting properties and characteristics. While this selection is often based on ease of fabrication, it is often restricted by the need to obtain desired properties in the resultant product and by the availability of raw material.

Starting workpiece geometry. The operator can specify the initial configuration of the material to be deformed. Often this shape is determined by previous processing or by the limitations of raw material and may therefore not be the optimal initial shape for the process.

Tool or die material and geometry. During the design stages of the process, the operator must specify the material and shape of the tooling to be used. Since this is usually a large cost item, the selection must be done with care. The tooling that comes in contact with the deforming material is most critical, since it will have direct control of the deformation and also interact with the deforming material. The surface conditions of the tooling must also receive special consideration. Since success or failure of a process often depends on the tool geometry, mistakes in this realm tend to be very costly. Moreover, attempts to utilize poorly designed tooling through the modification of other process variables generally results in an inefficient forming process.

Lubrication. Friction can account for more than 50% of the energy consumption in a deformation process and occurs primarily at the tool-workpiece interface. Lubricants can help to reduce this loss and provide other useful benefits, such as cooling, thermal insulation, and corrosion resistance.

Starting temperature. The initial temperature of the material can strongly influence its flow behavior during processing and the production rates that can be achieved. If this temperature is not correct, a fractured component, an incorrectly sized component, or a component with incorrect properties may be produced. The lifetime of tooling is also strongly related to the starting temperature of the material.

Speed of the process. The operational speed of most processing equipment can be varied. This speed will influence not only the rate of production but lubricant effectiveness, force and energy requirements, and heating and cooling of the deforming material. Thus, its selection must be done with considerable care.

Amount of deformation. This variable is often specified by the tool geometry selected or the dimensions of the final product. Some processes, though, permit variation in the amount of deformation, and most processes give the operator the choice of single- or multiple-step deformation. Process limitations and the resulting efects on material properties should be considered here.

Table 4.3.9 presents an expanded list of significant process variables.

Once the independent variables are specified, the process then determines the nature and values for a second set of variables, known as the dependent variables. The following are examples of these dependent variables.

Force or energy requirement. To deform a selected material from a given starting shape to a specified final shape with a specified lubricant, tooling geometry, speed, and starting temperature requires a certain amount of energy. This energy requirement can then be directly converted to a power, or force, requirement for the process. A change in any of the independent variables will produce a change in the required energy. However, the effect is indirect. The operator cannot directly specify energy; only the independent variables can be specified. Thus process optimization to reduce energy consumption can only be brought about by judicious selection of the independent variables.

Material properties of the product. While the operator can specify the properties of the starting material, the subsequent deformation, coupled with temperature effects related to the deformation process, will certainly change them. The customer is not interested in the starting properties but the final properties of the product. Thus, while it is often desirable to select starting properties based on compatibility with the process, it is also necessary to know or be able to predict how the process will alter them.

Exit temperature. Deformation generates heat. Hot workpieces cool in cold tooling. Properties depend on both the mechanical and thermal history of the material. Thus it is important to know and be able to control the temperature of the material throughout the process.

Surface finish and precision. Both are product characteristics dependent on the specific details selected in the independent variables.

The nature of the material flow. Generally deformation processes exert external constraints on the material through control and movement of its surfaces. How it flows or deforms internally depends on the specifics of the process and material. Since properties depend on deformation history, control here is vital.

Table 4.3.9. Significant Process Variables

Billet Material

Flow stress as a function of strain, strain rate, temperature and microstructure (constitutive equations)
Workability as a function of strain, strain rate, temperature and microstructure (forming limit curves)
Surface conditions
Thermal/physical properties
Initial conditions (composition, temperature, history/microstructure)
Effects of changes in microstructure and composition on flow stress and workability

Tooling

Geometry of tools
Surface conditions
Material/heat treatment/hardness
Temperature
Stiffness and accuracy

Conditions at Tool/Material Interface

Lubricant type and temperature
Insulation and cooling characteristics of the interface layer
Lubricity and frictional shear stress
Characteristics related to lubricant application and removal

Deformation Zone

Deformation mechanics, model used for analysis
Metal flow, velocities, strain rates, strains (kinematics)
Stresses (variation during deformation)
Temperatures (heat generation and transfer)

Equipment Used

Speed/production rate
Force/energy capabilities
Rigidity and accuracy

Product

Geometry
Dimensional accuracy/tolerances
Surface finish
Microstructure, mechanical and metallurgical properties

Environment

Available manpower
Air, noise and wastewater pollution
Plant and production facilities and control

Source: Ref. 1, Table 2-1.

The customer is only satisfied if the desired geometric shape is produced, with the right set of companion properties, and without surface or internal defects.

4.3.4 INDEPENDENT–DEPENDENT VARIABLE RELATIONSHIPS

In a metal forming process or system, there are independent variables, those aspects of the process over which the operator has direct influence and control, and the dependent variables, those aspects over which the operator desires control but for which his influence is indirect. The dependent variables are determined by the process, based on proper selection of the independent variables. Moreover, if a dependent variable is to be modified, the operator must determine which independent variable (or variables) is to be changed, in what manner, and by how much. Thus, a knowledge of the independent–dependent variable interrelationships is a necessity.

The link between independent and dependent variables is the most important area of knowledge in forming processes. Unfortunately, such links are often difficult to obtain. Deformation processes are complex systems composed of the material being deformed, the tooling performing the deformation, lubrication at surfaces and interfaces, and various other process parameters. The number of different processes and subprocesses is quite large. Various materials often behave differently in the same process. A multitude of different lubricants exist. In fact, some deformation processes may be complex systems of 15 or more independent variables. Moreover, these variables are often interacting. For example, an increase in process speed will tend to alter lubricant effectiveness, reduce time for heat transfer (thereby modifying temperature), and possibly change the apparent strength of the material if it is deformation rate-sensitive.

The ability to predict and control dependent variables comes about in one of three ways.

Experience. This requires a long-time exposure to the process and is limited to the materials, processes, and conditions observed in the past. Extrapolation to new conditions is essentially educated guesswork. In addition, care must be taken to assure transmission of the essential information from individual to individual.

Experimentation. While likely to be the most accurate approach, a direct experiment is both time consuming and costly. For reasons previously discussed, a valid deformation experiment should be conducted under production conditions for size and speed. More often, however, size and speed are reduced when conducting laboratory studies. Heat transfer, lubricant performance, and other significant features tend to be altered. Thus, while the laboratory experiments can provide valuable insight, care should be used when extrapolating their results to different but similar conditions.

Theory. Here one attempts to develop a mathematical model of the process into which numerical values can be inserted for the various independent variables and from which a prediction for the dependent variables can be computed. These techniques are simply applications of the theory of plasticity and vary from crude, first-order approximations to sophisticated, computer-based methods. The solutions may be an equation describing the process and revealing trends or simply a numerical answer based on the specific input values.

Since experience is often limited and may not be properly transmitted and experimentation is generally a rather costly luxury, considerable effort has been directed to the development of mathematical models for predictive use in process design and adaptive use in process control. It is important to note, however, that the accuracy of the models can be no better than that of the input variables. The characteristics of the deforming material must be known for the specific conditions of temperature, strain (amount of prior deformation), and strain rate (speed of deformation) being encountered. Moreover, the same material may behave differently under the same conditions if its microstructure (prior processing history) is different. Microstructure and its effects, however, are difficult to describe in quantitative terms that can be incorporated into a mathematical model. Thus, the characterization of material properties under various conditions is receiving considerable attention at this time, the results being known as constitutive equations.

Friction is another elusive variable, being dependent on pressure, area, surface finish, lubricant, amount of lubricant, speed, and material. Moreover, its value often varies with location within the same forming process and changes with time. Most models, however, incorporate its effects through a single variable of constant magnitude.

While these problems appear to be major barriers to the theoretical approach, it should be noted that the same features will hinder an individual trying to document, characterize, or extrapolate the results of experience or experimentation. In addition, theory can often reveal process aspects that might otherwise go unnoticed, and theory can be quite useful when extending a process into a previously unknown area or designing a new process.

4.3.5 RECENT DEVELOPMENTS IN DEFORMATION PROCESSES

4.3.5.1 Continuous Wiremaking (Continuous Extrusion)

In recent years, there have been a number of innovative developments in the area of wiremaking, generally based on a desire to combine the advantages of wire drawing and extrusion. In a wire drawing operation, the process can be continuous, but, due to the applied tensile stresses, the amount of reduction that can be achieved in a single draw is quite limited. This necessitates the use of a multiple stage operation and multiple tooling in order to achieve a large overall reduction. In contrast, the extrusion process can take very large reductions in a single step, but this is done in a piece-rate, discontinuous fashion. The goal of the continuous wiremaking or continuous extrusion processes is to enable large reductions to be taken continuously. Several of the most recent and innovative processes are listed in Table 4.3.10 and are shown schematically in Fig. 4.3.8.

4.3.5.2 Near Net Shape Forming

Traditionally, manufacturers have relied on several basic techniques, such as casting, forming, and machining, to impart a desired geometrical shape to their products. In recent years, however, stiffer industrial competition, the development of new alloys, shortages of certain materials, and the increase in energy costs have forced the critical analysis and reevaluation of these traditional processing methods. It is becoming very desirable to produce the final product in fewer processing steps and with as little waste as possible. Based on the firm foundation of the traditional processes, several techniques for the manufacture of components to "net shape" or "near net shape" have been and are being developed to meet these challenges.

Table 4.3.10. Qualitative Comparison of Recent Continuous Wiremaking Processes

Process	Figure	Product Length	Reduction Possible	Speed	Power Efficiency	Design Characterisitcs	Dimensional and Surface Quality
Conventional extrusion	4.3.2	Short	Large	Moderate	Moderate	Simple, few moving parts, rugged	Fair
Wire drawing	4.3.6	Unlimited	Small	High	High	Simple, reliable	Good
Extrusion with viscous drag	4.3.8a	Unlimited	Large	Very high[a]	Excessively poor	Complicated, but few moving parts	Good
Continuous-chamber	4.3.8b	Unlimited	Large[b]	Moderate to high	Claimed highest in trade	Many moving parts at high speed, precision, pressure	Good
Conforming (continuous extrusion forming)	4.3.8c	Unlimited	Very large	High	Fair to poor	Simple, few moving parts, rugged[c]	Good
Linex process	4.3.8d	Unlimited	Large	Moderate to high	Fair	(See continuous-chamber extrusion)	Good
Combined hydrostatic extrusion and drawing for fine wire	4.3.8e	Very long (up to 5 $\times 10^5$ ft/load)	Moderate to large	Restricted[d]	Poor but not a factor	Complicated and delicate	Very good
Helical extrusion	4.3.8f	Very long	Sky-high	Moderately high	?	Complicated but rugged	Good
Shaving	4.3.8g	Unlimited	Sky-high	Moderate	Fair to high	Claimed to be very simple[e]	Good
Extrolling	4.3.8h	Unlimited	Moderately high	High	Fair to high	Simple, few moving parts, rugged	Good
Kobe process[f]	4.3.8i	Unlimited	Large	?	Good	Exccessive chamber; gear, seal problems	Good
Melt-extracction process	4.3.8j	Whiskers: limited; wire: unlimited	From melt to wire	High	High	Spooling of product is a problem for whiskers; no problem for wire	Very poor

Source: Ref. 3, Table 5.1.
[a] Limited only by ability to collect product.
[b] Reported 300: 1 in copper at up to 300 kpsi pressure.
[c] Placing die on face of shoe provides rigid die support.
[d] Depends on wire size: for 0.0015-in diameter speeds of 4,000 ft./min. were reported.
[e] Production unit not yet tested.
[f] New process; not yet tested.
[g] Wire is rolled down for better dimensional and surface quality.

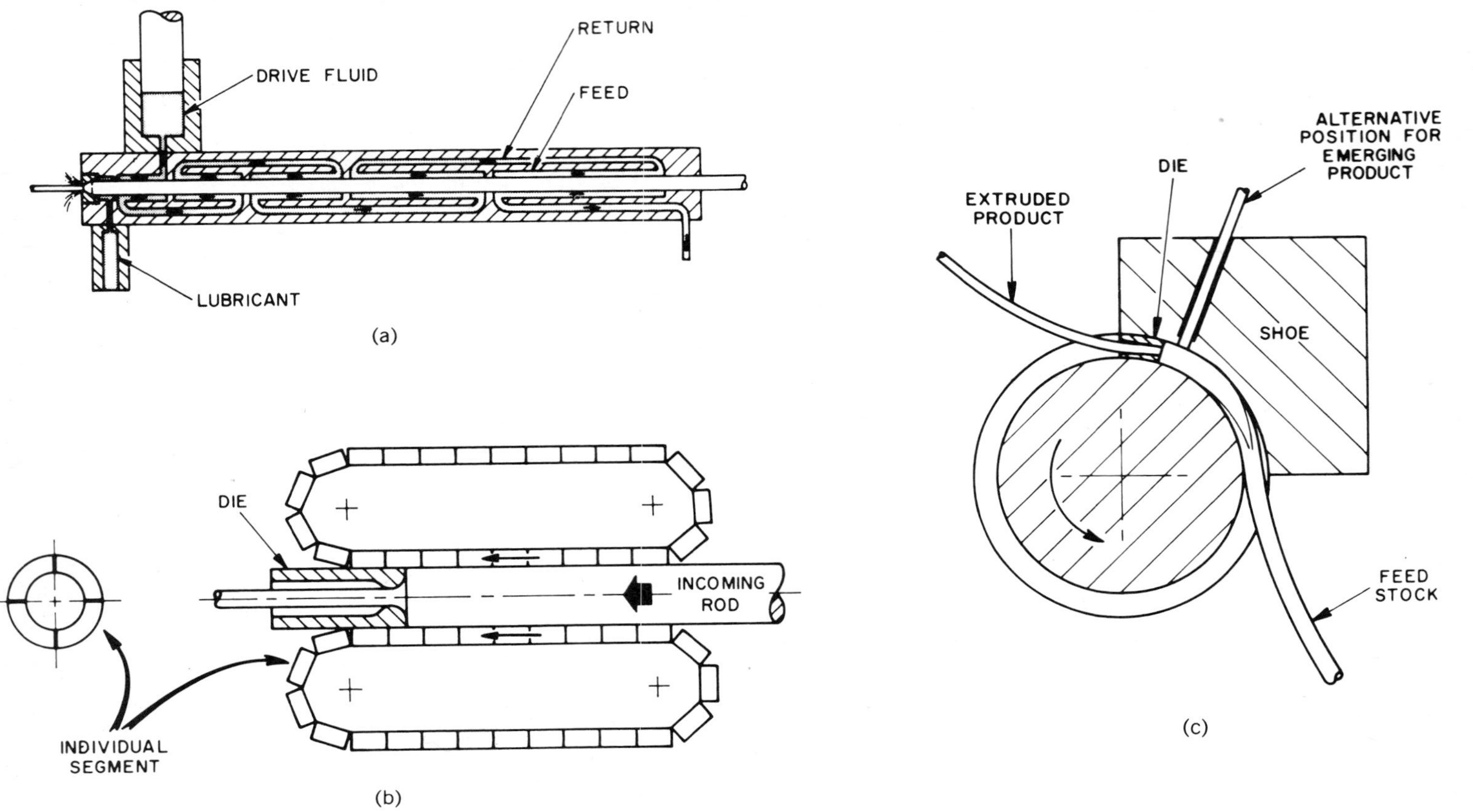

Fig. 4.3.8. (*a*) Extrusion with viscous drag. From Ref. 3, Fig. 5.4. (*b*) Continuous-chamber extrusion. From Ref. 3, Fig. 5.7. (*c*) Conforming. From Ref. 3, Fig. 5.9.

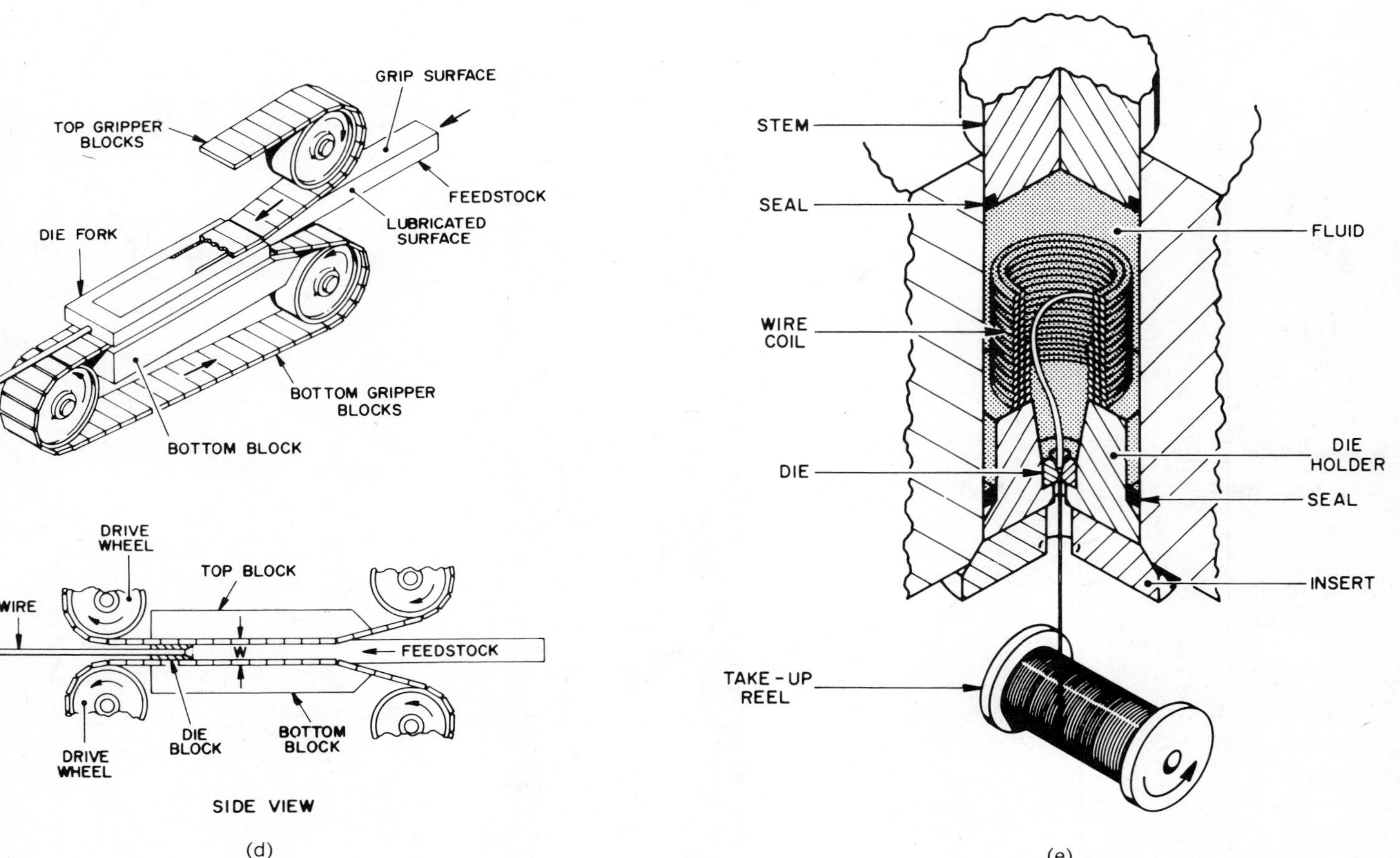

Fig. 4.3.8. (*Continued*) (*d*) Linex process. From Ref. 3, Fig. 5.11. (*e*) Combined hydrostatic extrusion and drawing for fine wire. From Ref. 3, Fig. 5.12. (*Figure continues on p.* **4**.52.)

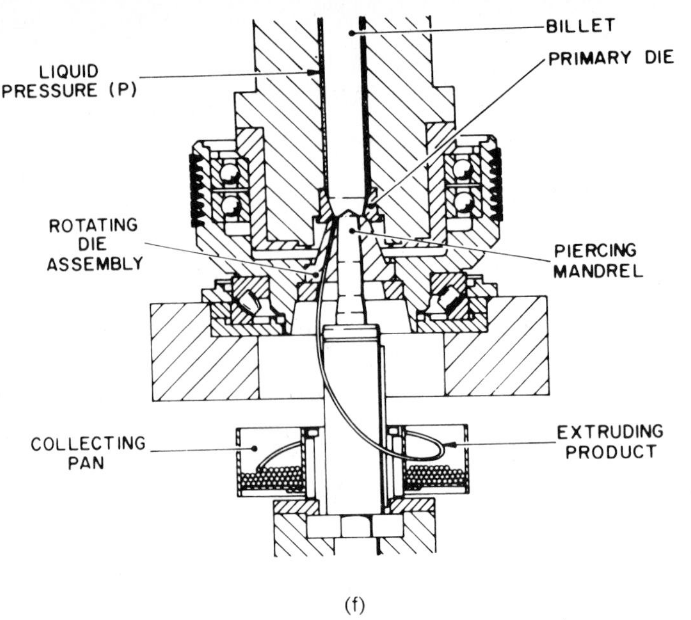

(f)

(g)

(h)

Fig. 4.3.8. (*Continued*) (*f*) Helical extrusion. From Ref. 3, Fig. 5.18. (*g*) Shaving. From Ref. 3, Fig. 5.20. (*h*) Extrolling. From Ref. 3, Fig. 5.21.

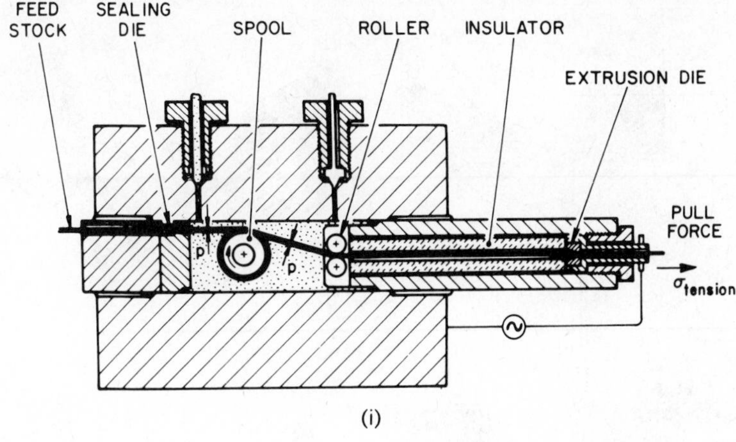

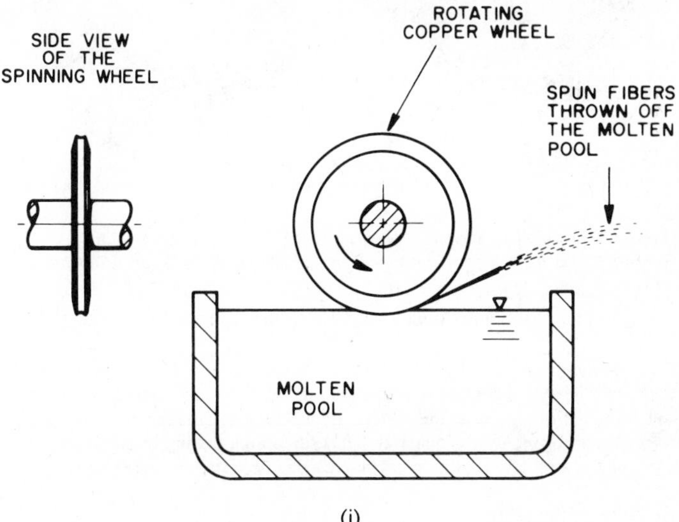

Fig. 4.3.8. (*Continued*) (*i*) Kobe process. From Ref. 3, Fig. 5.17. (*j*) Melt-extraction process. From Ref. 3., Fig. 5.24.

The fundamental incentive for these net shape processes is economic. In spite of the fact that these methods often require extra capital equipment and/or special handling procedures, the overall cost per product can often be substantially decreased by a net shape forming method. These cost reductions, and therefore the justification of the process, can be the result of eliminating or reducing trimming, machining, and final finishing operations; better use of costly or critical materials; a possible substitution of materials; and a possible decrease in total energy consumption.

The process of producing to final net shape is a desirable goal for many metal products. In fact, if one neglects or can eliminate the gates, sprues, and risers, casting can be considered as a net shape process, since useful objects can be produced directly from molten metal. Flashless forging, hot-die forging, isothermal forging, hot isostatic pressing, squeeze forming, and superplastic forming are recent developments that impart a desirable shape in a minimum number of steps. While these processes are developments of different basic techniques, they are all directed to forming a product to net or near net shape.

Flashless Forging

The process of flashless forging has many advantages over the more common forging methods in which a flash is produced (see Fig. 4.3.9). One obvious advantage is that the elimination of the flash reduces the scrap produced. Since no flash is formed, the preform is smaller in size and thus the amount of energy needed to heat the material to the forging temperature is also reduced. A reduction in the number of finishing steps that

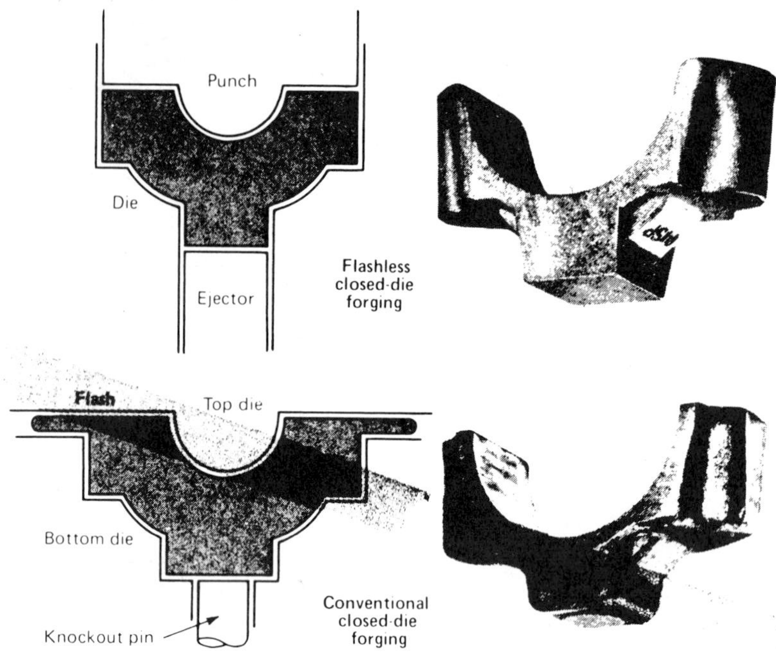

Fig. 4.3.9. Comparison of flashless forging with conventional methods. From Ref. 4, pp. 138–141.

are required in the more traditional processes can also occur. However, in order to use this technique effectively, close and precise control must be exercised over the volume of the blank entering the forging dies. Undersize blanks will prevent complete filling of the die. Oversize blanks may damage the dies or jam the press because there is no escape provided for the excess material. Thus, the initial cutting or cropping of the blanks must be done with precision, and techniques such as fine blanking or fine shearing must be used.

Hot-Die and Isothermal Forging

Heating the dies in a forging press is another technique that can be used to form parts to net or near net shape (see Figs. 4.3.10 and 4.3.11). The high-temperature dies (especially if they are at the same temperature as the

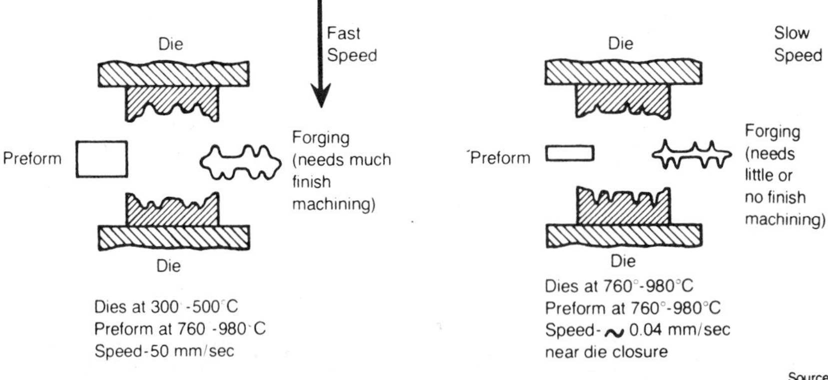

Fig. 4.3.10. Isothermal forging. From Ref. 5, pp. 73–86.

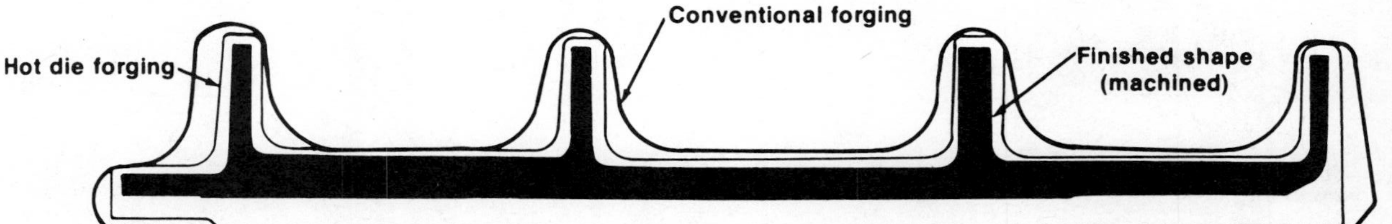

What near net shape is all about. The outermost shape represents conventional closed impression die forging for a 6AL-4V titanium bearing support. The next shape is that produced by hot die forging, followed by the finished machined configuration. Note how close the hot die forging is to final finished dimensions. As cost of material increases, the benefits of near net shape production techniques become dramatic. One could expect even closer to net shape results with isothermal forging.

Fig. 4.3.11. Comparison of hot-die forging with conventional methods. From Ref. 6, pp. 40–43.

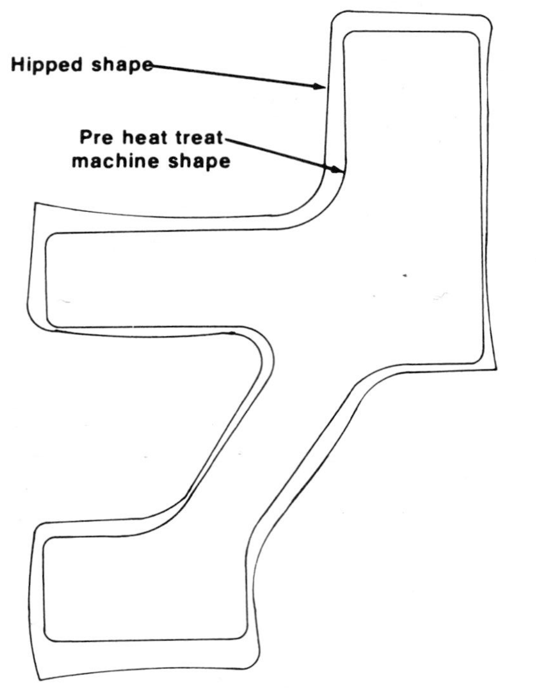

A cross section of a production turbine disc, the HIPed shape is shown by the outside line, with pre-heat treat machined shape inside. Practically no material is removed on at least one surface, and very little is machined away on the rest of the part.

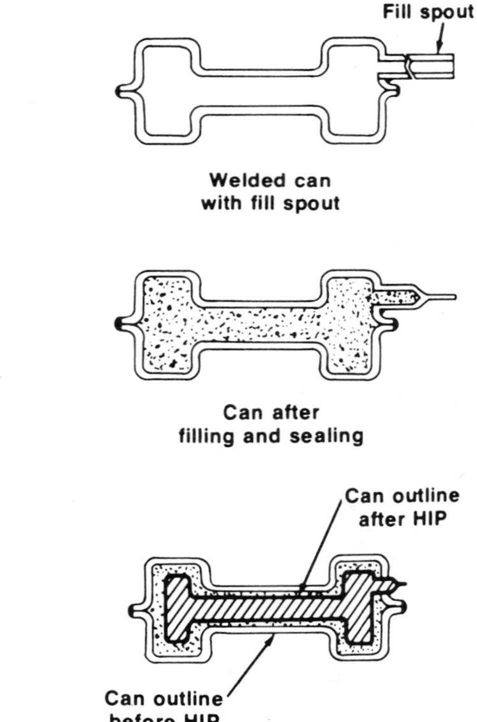

HIPing comprises filling a pre-fabricated metal can with powder in a vacuum, sealing the can and performing the HIP process. Line drawing shows relative container size before and after HIPing.

Fig. 4.3.12. Hot isostatic pressing. From Ref. 6, pp. 40–43.

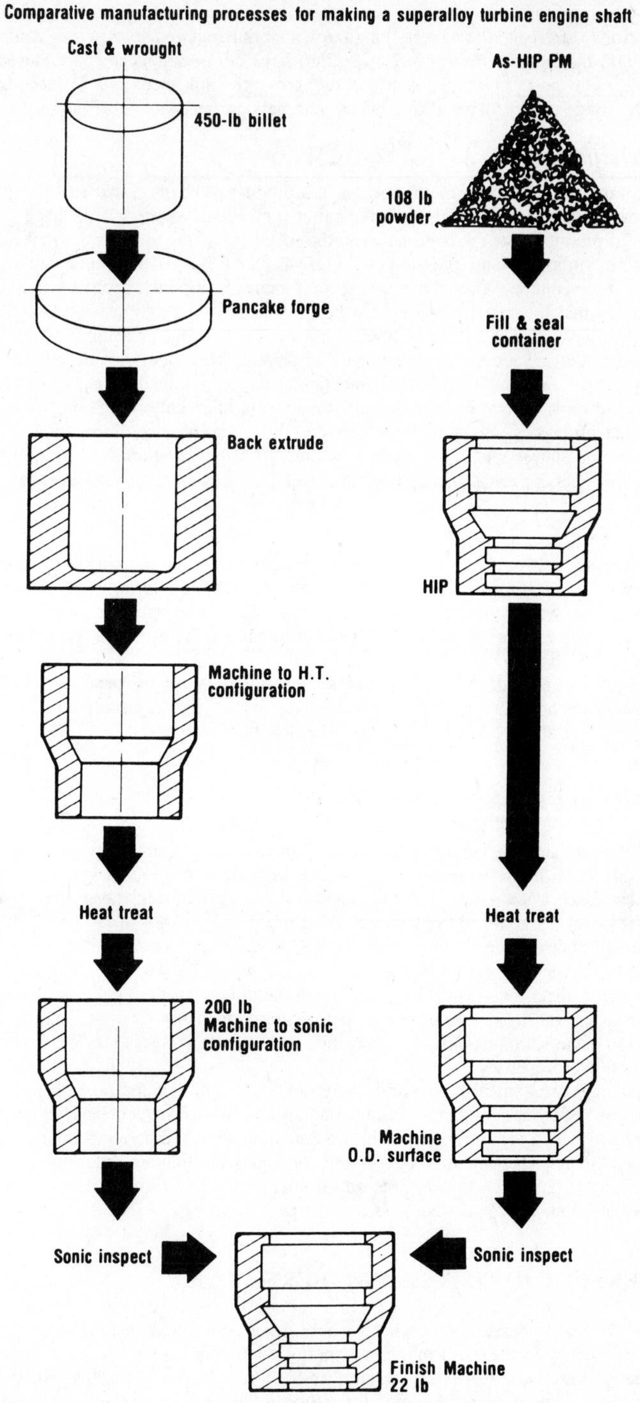

Comparative manufacturing processes for making a superalloy turbine engine shaft

Cast & wrought

450-lb billet

Pancake forge

Back extrude

Machine to H.T. configuration

Heat treat

200 lb
Machine to sonic configuration

Sonic inspect

As-HIP PM

108 lb powder

Fill & seal container

HIP

Heat treat

Machine O.D. surface

Sonic inspect

Finish Machine
22 lb

Fig. 4.3.13. Comparison of hot isostatic pressing with conventional processes. From Ref. 7, pp. 56–61.

metal being formed) allow for easier plastic flow of the material and thus a more complete filling of the dies. This technique can also decrease the excess volume of material required to ensure a complete filling of the critical cavities within the die. By proper preform design, large deformations can be achieved and very precise parts can be made. The cost, however, is high, and presently this technique is reserved for high-priced components usually made of expensive alloys, when materials conservation is of primary importance.

Hot Isostatic Pressing

The forming of complex shapes from metal powder is a production method that has been used for a number of years. With conventional techniques, the powder is first compacted, or pressed, to shape at room temperature and then sintered (to produce a strong bond between the metal particles) by heating to elevated temperature at atmospheric pressure. In hot isostatic pressing (see Figs. 4.3.12 and 4.3.13), the operations are combined by pressing at elevated temperature. While this might seem immensely attractive from a manufacturing viewpoint, one should remember that the metal powder must be protected from reaction with air while it is in a loose condition or low density state at elevated temperature, and the heating, pressing, and cooling operations all must be performed under a vacuum or controlled atmosphere. Nevertheless, the process has a number of attractive features. All of the benefits of powder metallurgy can be retained, along with the added features of 100% full density, isotropic properties, and the elimination of a lubricant during compaction and the resultant improvement in final properties. Many high-temperature alloys can be made to net or near net shape by this technique, with intricate shapes and a minimum of waste. With the development of rapidly solidified powers and other material advances in powder metallurgy, these techniques are likely to take on even greater importance.

Other Processes

Other net shape processes that can be used under certain conditions are hot pressing of scrap material to produce a net shape product; squeeze forming, in which the metal is semimolten during the mechanical forming steps; and superplastic forming, whereby extremely large deformations can be achieved in certain alloy systems with low applied pressure. These techniques are more specialized and can be used only when the metal or alloy has certain unique properties that permit the processing.

In spite of the fact that each of the techniques described above is more expensive or requires more control than the conventional method, the savings in raw material and energy consumption and the elimination or reduction of finishing operations often justify use of the net shape processes.

4.3.6 THERMOMECHANICAL PROCESSING

Under conventional manufacturing conditions, metal forming and heat treatment are generally regarded as being sequential operations. Heat imparted to the metal for hot working is regarded as having served its purpose by providing thermal softening, increased ductility, and recrystallization, and the process is regarded as being complete when the desired shape has been produced. If the properties are to be modified by heat treatment, additional heating and controlled cooling will be required in a separate operation.

With the onset of the "energy crisis" and the rapid and significant increases in fuel cost, attention has been directed to thermomechanical treatments (TMT), wherein metal deformation and heat treatment are performed together. In some cases, the processes of plastic deformation, recovery, recrystallization, precipitation, and phase transformation occur simultaneously and may interact with one another to form new and unique structures with extremely attractive properties.

The object of thermomechanical processing, therefore, is to optimize the properties of the product in the as-rolled or as-hot-worked condition, thereby eliminating the need for subsequent heat treatment. This is achieved by strict composition control of the material in combination with a time-temperature deformation scheme designed to produce the desired structure. Ausforming and control rolling are probably the two most developed of the thermomechanical processes at this time, and it is expected that this family of processes will continue to grow in number and importance as time passes.

4.3.7 HIGH-ENERGY RATE FORMING (HERF)

High-energy rate forming processes, in contrast to bulk deformation and sheet deformation processes, have most of the process energy in the form of kinetic energy within the workpiece prior to the initiation of the deformation. Usually, very high speeds are required in a short period of time, thus HERF processes require techniques that release a large amount of energy for a short duration. The most common sources of energy are chemical, electrical, or magnetic, and the most common processes are explosive forming, electrohydraulic forming, and magnetic-pulse forming.

The advantage of the explosive forming processes is that they can be used on workpieces with a variety of sizes. Usually this method is most suitable for small quantity productions of large parts. Safety is also an important consideration when working with explosives.

Electrohydraulic forming is similar to explosive forming except that the workpiece is smaller and the operation is safer.

Magnetic pulse forming is often used for the final shaping of tubes and embossing sheet metals. It works best with workpieces that have high electrical conductivity.

4.3.8 APPLICATIONS OF CAD/CAM

Computer-aided design and Computer-aided manufacturing (CAD/CAM) are being developed and used for a number of metal forming processes. Here the computational and graphics capabilities of the modern computer are used during several design stages of the forming process. Once a process is adequately modeled, the computer can be used to optimize the process by conducting a series of mathematical (simulation) tests. In addition, the model can also be used for computer control of the process, and, with adaptive control features, can even be used to execute corrective measures to assure production of an acceptable product.

The methodology used in designing and analyzing a forming process generally follows four distinct steps.

First, the constitutive relationships for the material being deformed are obtained. These relationships describe the dynamic flow characteristics of the material and can theoretically be derived from first principles based on a knowledge of the material's composition, structure, and past history. While a large research effort is currently pursuing this approach, most relationships being used today are simply an empirical fit of experimental data generated for the material to be deformed. The dynamic flow model then provides the stress-strain behavior of the material as a function of strain, strain rate, temperature, and other processing variables.

Second, the model for the material behavior is incorporated into a finite element model (FEM) in order to determine the actual flow of the material and the energy required to perform the process being investigated. Specific process variables (such as tool geometry, initial temperature, or initial preform size and shape) can be changed easily in the model and the effect of that change evaluated through determination of a new flow pattern and energy requirement.

Third, the computer model is manipulated so that the proper tool configuration, preform geometry, and temperature are determined that will to produce the final desired shape with the minimum amount of energy required and the minumum amount of processing difficulty.

Since all of the process modeling is done on a computer, the optimum results can then be transferred to a tape for use on a numerically controlled (NC) machine to produce the desired tool geometry. Thus, in the fourth step, forging dies, extrusion dies, wire drawing dies, and so on are machined to the optimum shape determined through the computer modeling effort. This approach eliminates the traditional reliance on trial and error procedures often used during the design stage. While this method is currently being used for the more sophisticated materials and complex geometries, it will eventually be used for almost all materials, processes, and products.

4.3.9 CONCLUSIONS

Through this brief presentation, the reader should sense that metal forming processes are in a state of dynamic change, much of this being in response to the desire to improve productivity, reduce the generation of scrap material, conserve energy, process new materials, and produce new products. New processes are always being developed. Old processes are being improved and modified. Computer design and computer control of processes are now production reality. Thermomechanical processes are being developed. In addition, advances in engineering materials, lubricants, furnaces, materials handling equipment, and other areas all have a direct bearing on metal forming processes. Thus, in the future, there should be a constant evolution of processes and techniques designed to produce new and better products in a more efficient manner.

REFERENCES

1. T. Altan, S. Oh, and H. Gegel. *Metal Forming Fundamentals and Applications*. Metals Park, OH: American Society for Metals, 1983.
2. J. Harris. *Mechanical Working of Metals*. New York: Pergamon Press, 1983.
3. B. Avitzur. *Handbook of Metal-Forming Processes*. New York: Wiley-Interscience, 1983.
4. John T. Winship. "Flashless Forging Is Here." *American Machinist*, 138–141 (June 1981).
5. "How Forging Has Put New Punch into Its Act." *Iron Age*, 73–86 (May 5, 1980).
6. Randolph Gold. "Forging Technologies of the Twenty-First Century." *Precision Metal* (November 1978).
7. John H. Moll. "HIPing the High-Performance Alloys." *Materials Engineering*, 56–61 (November 1981).

BIBLIOGRAPHY

Avitzur B. *Metal Forming: Processes and Analysis*. New York: McGraw-Hill, 1968.

Backofen, W. *Deformation Processing*. Reading, MA: Addison-Wesley, 1972.

DeGarmo, E. P., J T. Black, and R. A. Kohser. *Materials and Processes in Manufacturing*, 6th Edition. New York: Macmillan, 1984. Chapters 13–15.

Eary, D. F., and E. A. Read. *Techniques of Pressworking Sheet Metal*, 2nd Edition. Englewood Cliffs, NJ: Prentice-Hall, 1974.

Hosford, W., and R. Caddell. *Metal Forming—Mechanics and Metallurgy*. Englewood Cliffs, NJ: Prentice-Hall, 1983.

Schey, J. *Tribology in Metalworking—Friction, Lubrication and Wear*. Metals Park, OH: American Society for Metals, 1983.

CHAPTER 4.4

ABRASIVE PROCESSES

S. MALKIN

Technion-Israel Institute of Technology
Haifa, Israel

4.4.1 INTRODUCTION

Abrasive processes include machining and finishing operations that involve the use of abrasives as the cutting medium. The most common abrasive machining process is *grinding*, which is widely used to obtain close tolerances and fine finishes. For grinding, the abrasive grain is bonded into a grinding wheel. Other abrasive processes for producing extremely fine finishes and tolerances are honing, polishing, and lapping. At the other extreme, heavy-duty grinding operations (abrasive machining and snagging) provide rapid material removal with less attention to surface quality and accuracy. Such operations are most widely used for conditioning billets in steel mills. Additional abrasive processes include cut-off, free abrasive machining by particles within a fluid stream, ultrafine slicing and dicing of electronic material, and ultrasonic grinding.

Grinding is commonly used as a final machining operation when the tolerance and finish requirements cannot be held by other prior machining methods (e.g., milling or turning). However, grinding is also applied in some cases as a complete machining process on forgings and castings, thereby saving the time and expense of prior rough machining by other methods. The more widespread use of grinding for complete machining is being accelerated by enhanced precision of forging and casting methods, leading to reduced stock allowance for machining.

In this chapter, the main emphasis is on grinding operations, as these are the most extensively used in production and also the best understood. The same fundamental concepts for grinding are also applicable to other abrasive processes.

4.4.2 ABRASIVE TOOLS

Abrasive tools commonly consist of abrasive grains (grit) held together by a bonding agent in the form of grinding wheels, mounted points, or sticks. *Coated abrasive tools* have a single layer or multilayers of abrasive grains held on a fabric or paper backing by a binder or glue. In some cases, abrasive tools consist only of loose abrasive grains.

Abrasive grains are either naturally occurring or synthetic ceramic materials that are much harder than the materials they cut. Abrasive tools in antiquity were probably mostly sand or sandstone (quartz). Natural abrasives in use today include aluminum oxide (natural corundum and emery), garnet (metallic silicate), and diamond. Modern advances in the abrasives industry can be credited largely to the development of synthetic (man-made) abrasives with controlled properties, shapes, sizes, and chemistry.

The most widely used synthetic abrasive is aluminum oxide (Al_2O_3). Actually, there are numerous types of man-made aluminum oxide-based abrasives, and each has different structural characteristics and chemical compositions that affect the abrasive grain mechanical properties of hardness and toughness. Any increase in grain toughness obtained by altering the structure or chemical composition of the abrasive is usually accompanied by a decrease in hardness. For fine grinding, more friable (less tough) grains are usually preferred to provide sharp cutting edges by brittle fracture. Tougher grains wear less by brittle fracture and are preferred for heavy-duty grinding. Other widely used synthetic abrasives include silicon carbide, diamond, and cubic boron nitride. Again, various types of each grain material are produced with desired properties and shapes tailored to specific applications. Diamond is the hardest known material, and cubic boron nitride is the second hardest. These latter two materials are known as *superabrasives*. Hardnesses of some abrasive materials are listed in Table 4.4.1.

Table 4.4.1. Approximate Indendation Hardnesses of Abrasive Grain Materials

Abrasive	Hardness (GPa)
Quartz	8.1
Garnet	13.3
Aluminum oxide	19.1–21.1
Silicon carbide	24.3
Cubic boron nitride	45–50
Diamond	80–90

Bonded abrasive tools consist of abrasive grains held together by a binding material. Factors that control bonded abrasive tool characteristics include the abrasive grain material, the grain size, the binding material, the relative amounts of abrasive and binder, and the porosity. Figure 4.4.1 shows the standard American system for specifying aluminum oxide and silicon carbon grinding wheels. The symbol A or C indicates aluminum oxide or silicon carbide, respectively, and a manufacturer's prefix is usually included to indicate the particular abrasive type. The abrasive grain *size* is indicated by a mesh number corresponding to the screen sieve (wires per inch) used for sorting the grain: a larger mesh number indicates smaller particles. The wheel *grade* is a relative indication of the binder content; a stronger and harder wheel has more binder (less porosity). The *structure* number indicates the volumetric concentration of abrasive grain material in the wheel; a higher number indicates less abrasive material. The type of binder material specified by a letter is usually followed by a marking that identifies a particular bond material. A typical vitrified grinding might consist of about 50% abrasive grain and 5–10% binder by volume. The remaining volume is porous.

Wheels containing superabrasives (diamond or cubic boron nitride) have a somewhat different system for indicating the abrasive type, grain size (mesh or dimension), bond type, and abrasive concentration. As the abrasive grain is expensive, only a relatively thin layer on the active area of the wheel surface actually consists of bonded abrasive which is attached to a metal or plastic hub. The abrasive grain content is indicated by a concentration number, which is approximately four times the volumetric percentage. Typical concentration numbers range from 50 to 150. Most superabrasive wheels have either a resin or metal binder (matrix), although vitreous bonds are also used to a limited extent. Some metal-bonded superabrasive wheels are manufactured by electroplating the abrasive grain onto a form or hub.

Although abrasive grain materials must be harder than the materials being cut, harder abrasive grains do not necessarily wear less than softer ones. The overriding factor is often the chemical compatibility of the abrasive grain with the workpiece material and environment. Chemical reactions or degradation of the abrasive grain can promote dulling wear or attrition. For example, diamond is not suitable for grinding ferrous materials because it wears excessively by graphitization in the presence of iron at the high temperatures generated at the cutting points. Synthetic cubic boron nitride was developed as an alternative superabrasive for grinding ferrous materials; it wears less and has only about half the hardness of diamond. Attritious wear is usually accompanied by adhesion of the workpiece material to the abrasive, making the wheel appear glazed. Excessive adhesion results in loading of the wheel, whereby the pores between abrasive grains become partially filled up with workpiece material. Glazing and loading cause increased grinding forces and power consumption.

4.4.3 TYPES OF GRINDING OPERATIONS

Most grinding operations can be classified as either *plunge grinding* or *traverse grinding*. Many other grinding operations may not fall into either classification.

Three types of plunge grinding—*straight surface*, *external cylindrical*, and *internal cylindrical*—are illustrated in Fig. 4.4.2, where d_s is the wheel diameter, v_w is the workpiece velocity, a is the wheel depth of cut (downfeed), d_w is the workpiece diameter, and v_f is the radial infeed velocity. Wheel velocities, v_s, are typically about 30 m/sec, although faster wheel velocities even up to 120 m/sec are used in some cases, and somewhat slower velocities may be used for some difficult-to-grind materials. The process is referred to as *high-speed grinding* when wheel velocities of about 60 m/sec or more are used. The maximum allowable wheel velocity may be limited by the bursting strength of the wheel and the capability of the machine. The workpiece velocity v_w is always much slower than the wheel velocity, the ratio v_s/v_w being typically in the range 100–200 for straight surface grinding and about half as much for cylindrical grinding.

For straight surface plunge grinding (Fig. 4.4.2*a*), the depth of cut may be incremented either after each reciprocating stroke of the workpiece motion or every other stroke. In the former case, up-cut grinding strokes with the wheel and workpiece velocities in opposite directions at the wheel-workpiece contact zone are alternated with down-cut strokes with both velocities in the same directions. In the latter case, the grinding will be either up-cut or down-cut, although some continued grinding will occur during the alternate return strokes due to recovered elastic deflection of the grinding system. For cylindrical grinding, the wheel moves into the workpiece

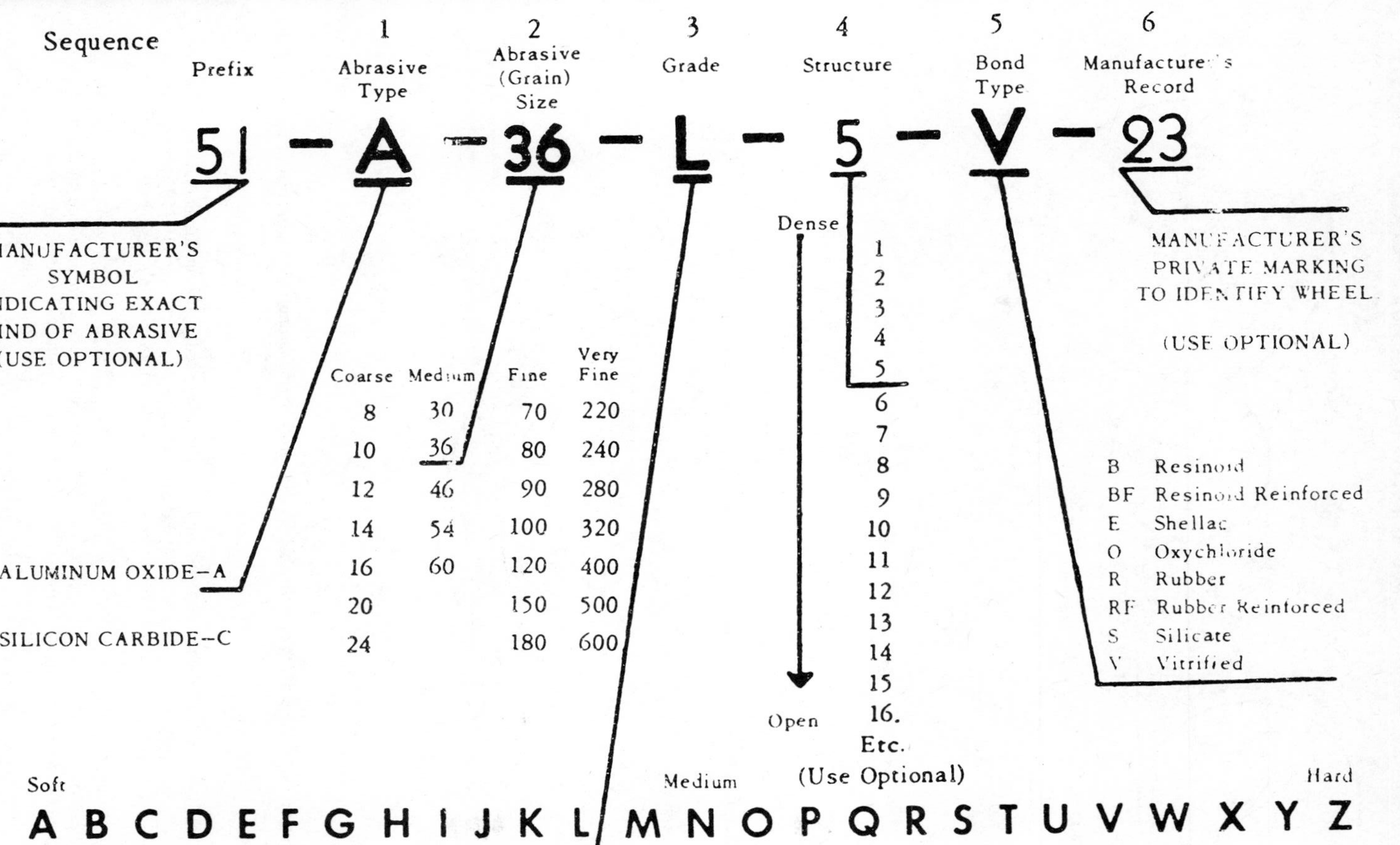

Fig. 4.4.1. Standard American wheel marking system. This material is reproduced with permission from Ameircan National Standard B74.13, 1982, copyright 1982 by the American National Standards Institute. Copies of this standard may be purchased from the American National Standards Institute at 1430 Broadway, New York, NY 10018.

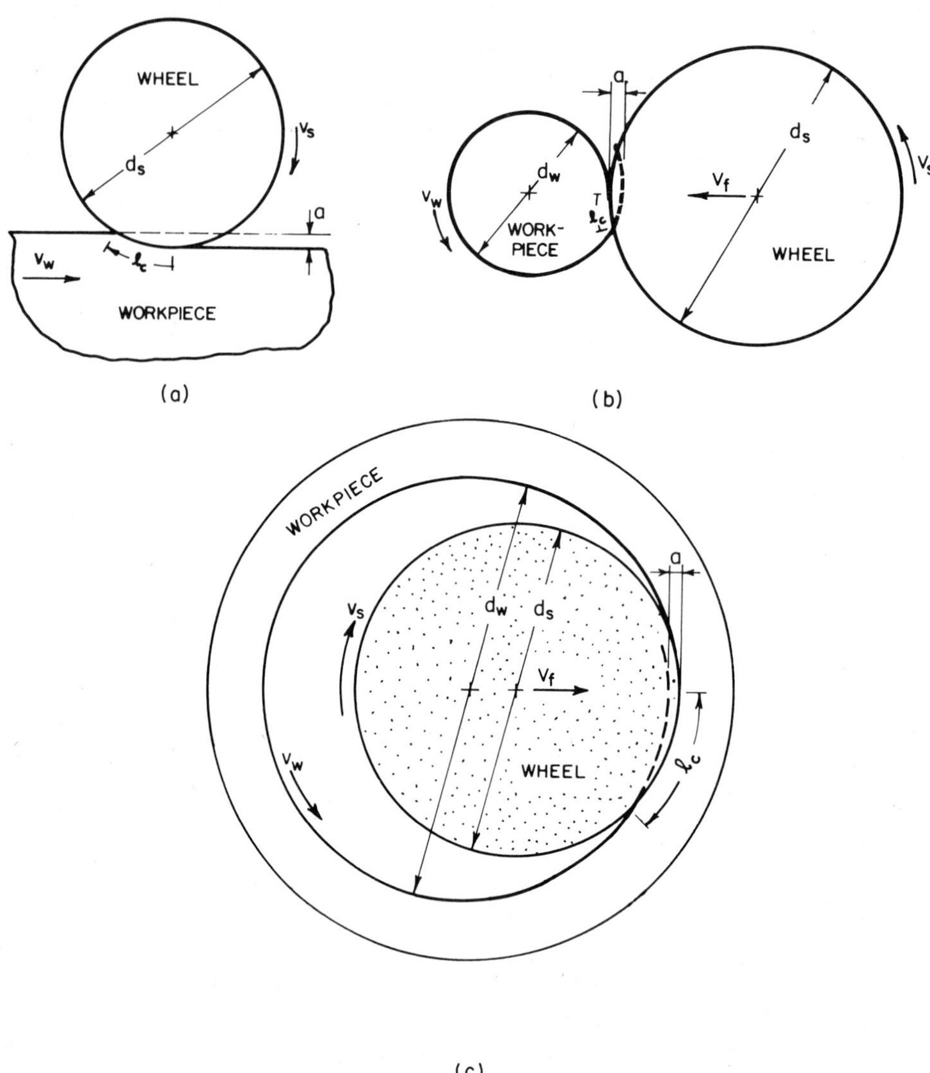

Fig. 4.4.2. Plunge grinding operations. (*a*) Straight surface. (*b*) External cylindrical. (*c*) Internal cylindrical.

at a radial infeed velocity v_f, and the wheel depth of cut is equal to the infeed increment during one workpiece rotation ($a = \pi d_w v_f / v_w$). For straight surface grinding, typical depths of cut are $a \approx 10-50$ μm, and for cylindrical grinding $a \approx 2-20$ μm. For illustration purposes, the relative magnitudes of a are highly exaggerated in Fig. 4.4.2.

A simple cycle for cylindrical plunge grinding is illustrated in Fig. 4.4.3. A *roughing* stage is followed by spark-out with zero infeed and rapid retraction. Other common cycles may have an intermediate finishing stage between roughing and spark-out at a much slower infeed rate than for roughing. For the cycle as shown, the actual infeed lags behind the controlled infeed imparted to the machine, since part of the infeed is taken up by elastic deflection under the contact force developed between the wheel and the workpiece. After an initial *spark-in* transient during roughing, the actual infeed velocity v_f becomes essentially the same as the controlled infeed velocity. Elastic deflection is recovered during spark-out, with grinding continuing at a decreasing rate. Spark-out is also needed to round-up the part from the spiral shape obtained at the end of roughing, and the decreasing removal rate during spark-out is accompanied by an improvement in surface finish. With a stiffer grinding system, there is less elastic deflection, so the spark-in and spark-out transients are more rapid. In general, internal plunge grinders have relatively low stiffness, since the wheel is located at the end of a cantilevered

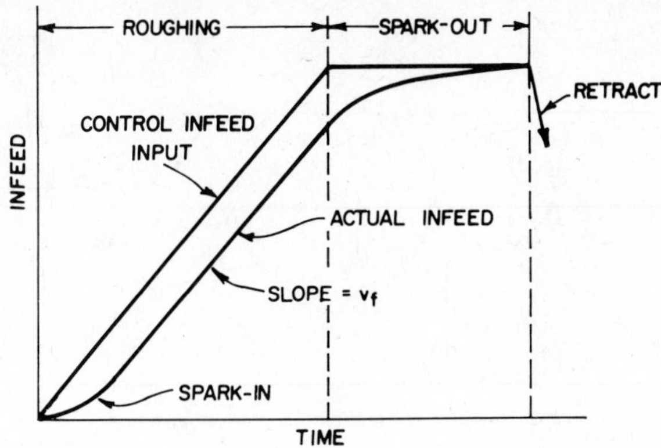

Fig. 4.4.3. Simple cycle for cylindrical plunge grinding.

spindle, and the transients take up a large portion of the grinding cycle. This is less of a problem with *controlled-force internal grinders* for which the normal force rather than the infeed is the controlled (input) parameter.

Traverse grinding is performed by *crossfeeding* (traversing) the workpiece relative to the rotating grinding wheel along the grinding width direction. This type of grinding would apply to straight and cylindrical operations analogous to those for plunge grinding (Fig. 4.4.2). Fig. 4.4.4 illustrates external cylindrical grinding with traverse. In this case, a continuous traverse velocity component v_t of the wheel relative to the workpiece gives a crossfeed width w_t per revolution of the workpiece ($w_t = \pi d_w v_t / v_w$). Essentially the same situation would apply to internal cylindrical grinding. With straight surface grinding, the crossfeed increment w_t is advanced at the end of the stroke while the wheel is out of contact with the workpiece.

In the absence of any wear of the grinding wheel, the total wheel depth of cut a (Fig. 4.4.4) would be removed by the leading edge of the wheel surface over a width w_t, in which case the process would be very similar to plunge grinding a width w_t. Owing to wheel wear, part of the depth a remains behind to be removed by a second wheel area of width w_t adjacent to the first one during the next workpiece rotation. Any wheel wear in this second area leaves behind residual material to be removed by a third width w_t, and so on. Portions

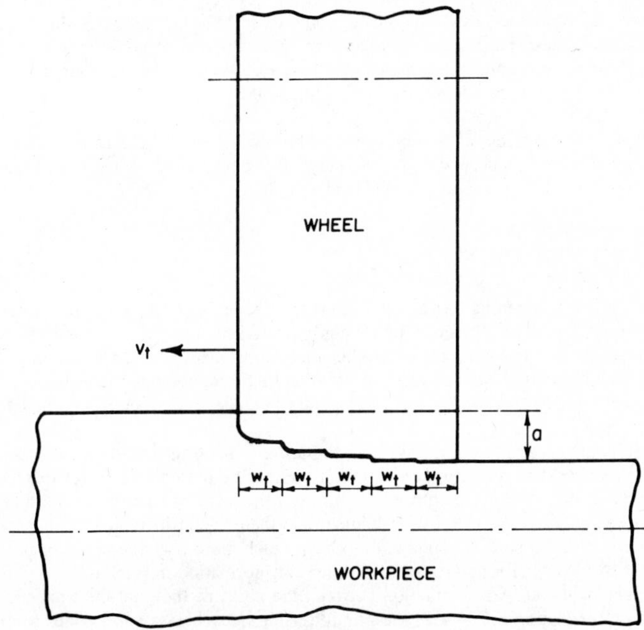

Fig. 4.4.4. External cylindrical grinding with traverse.

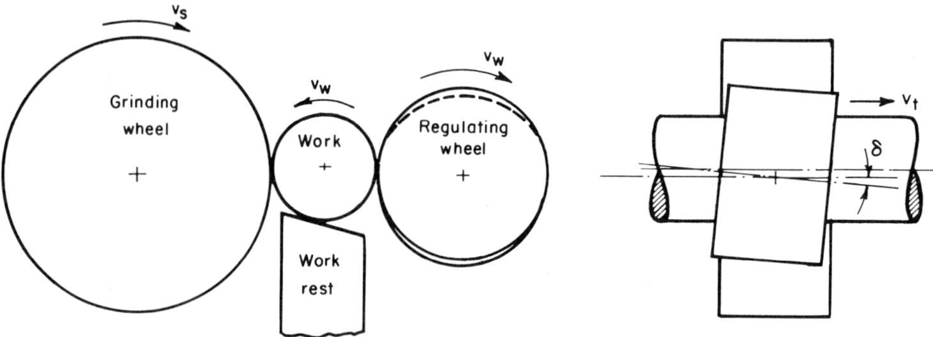

Fig. 4.4.5. Throughfeed centerless grinding.

of the wheel closer to the leading edge cut more material and, therefore, wear more rapidly. If the traverse direction is reversed at the end of the crossfeed path, the same process would occur at the opposite edge of the wheel. The net result would be *crowning* of the wheel, which can lead to a *form error* when traverse grinding to a shoulder.

When feasible, plunge grinding is usually preferred to traverse grinding in production operations, since it provides for simultaneous grinding over a wider area and is easier to control. Traverse grinding, however, is more efficient for *centerless* cylindrical grinding, as illustrated in Fig. 4.4.5. Rather than rotating the workpiece between fixed centers as in regular cylindrical grinding, it is rotated against the grinding wheel by a driven regulating wheel while being supported by a work rest (blade). The regulating wheel is tilted slightly through an angle δ so as to provide a traverse (throughfeed) velocity $v_t = v_w \sin \delta$. In this way, cylindrical parts can be continuously fed through the grinding process.

One widely used type of plunge grinding is *form grinding* for producing parts with cross-sectional profiles. In this case, the wheel is formed with the mirror image of the required shape and the whole profile is ground simultaneously across the grinding width. Form grinding operations with traverse are used for producing helical or threaded components.

There are many other types of grinding operations in addition to plunge grinding and traverse grinding. *Vertical spindle grinding* is especially efficient for producing flat surfaces using the flat face of a grinding wheel rotating in a horizontal plane. The workpiece, or a collection of small workpieces, is mounted on a work table that rotates or translates under the grinding wheel face. *Jig* grinders typically use a small mounted abrasive tool in a vertical spindle to produce accurate holes or contours in a workpiece by rectilinear motion of the work table. Other specialized types of grinding machinery are used for tool grinding and gear grinding. *Abrasive belt grinders* typically have translating continuous (endless) abrasive belts for grinding flat surfaces. With belts containing very fine grit, the process is referred to as polishing.

A relatively new grinding process for high stock removal rates is *creep-feed* grinding. This process is essentially straight plunge grinding with very slow workpiece velocities (creep-feed) and very heavy wheel depths of cut. With this process, a deep slot or profile can often be generated in a single grinding pass. Grinding machines must be specially constructed for creep-feed grinding.

4.4.4 MECHANICS OF GRINDING

The kinematic motions of the grinding tool and the workpiece lead to material removal by interaction of abrasive grains with the workpiece surface. For grinding metals, material removal occurs mainly by a shearing process of *chip formation* similar to what is found with other machining methods such as turning or milling. Similar material removal mechanisms also prevail for fine grinding of brittle nonmetallic materials, including glasses, ceramics, and cemented carbides, although brittle fracture becomes increasingly significant at faster removal rates.

Compared with other machining processes, grinding utilizes much faster cutting velocities, and the cutting points take much shallower depths of cut. Active cutting points are not uniformly distributed over the working surface of the abrasive tool so that the cutting depth is not the same at each point. Furthermore, the cutting tool geometry on abrasive grain tips is not well defined, and there is considerable variation from point to point. On the average, abrasive grain cutting points tend to have much more negative rake angles than other types of cutting tools, so much larger deformations occur during chip formation in grinding.

The *grinding zone* is the area of interaction between the abrasive tool and the workpiece. For the plunge grinding operations illustrated in Fig. 4.4.2, the grinding zone in each case corresponds to the geometrical arc length of contact between the wheel and the workpiece. The contact length l_c can be theoretically expressed by the relationship

$$l_c = (a d_e)^{1/2} \tag{4.4.1}$$

where d_e is the equivalent diameter defined in terms of the wheel and workpiece diameters as

$$d_e = \frac{d_s}{1 \pm d_s/d_w} \tag{4.4.2}$$

The plus sign in the denominator is for external cylindrical grinding (Fig. 4.4.2*b*) and the minus sign for internal grinding (Fig. 4.4.2*c*). For straight surface grinding (Fig. 4.4.2*a*), the equivalent diameter is identical to the wheel diameter d_s. Compared with actual wheel diameters, equivalent diameters are relatively very small in external grinding, thereby leading to small grinding zones, and very large in internal grinding, leading to large grinding zones.

The cutting path of an abrasive grain into the workpiece is illustrated in Fig. 4.4.6 for the case of up-cut straight plunge grinding over a workpiece with width w. As the grain passes through the grinding zone, the depth of cut it takes increases from zero to the maximum value t as it exits the grinding zone. Although the maximum grain depth of cut varies from point to point, the mean value of the maximum grain depth of cut depends on grinding wheel topography and the grinding parameters. In general, the mean value of the maximum grain depth of cut can be written as

$$\bar{t} = K\left[\left(\frac{v_w}{v_s}\right)\left(\frac{a}{d_e}\right)^{1/2}\right]^m \tag{4.4.3}$$

where m is a positive number ($m \leq 1$) that depends on the statistical distribution of cutting point protrusions. This same equation applies to straight surface, external cylindrical, and internal cylindrical grinding, with the equivalent diameter specified accordingly (Eq. 4.4.2). The factor K in Eq. (4.4.3) depends on the cross-sectional shape of the cutting point geometry and on the number (density) of active cutting points per unit area of wheel surface. For a wheel having a larger cutting point density, K is smaller and $\bar{t}$ is also smaller, since there is less material removed per cutting point. Increasing the removal rate by raising v_w or a raises the grain depth of cut,

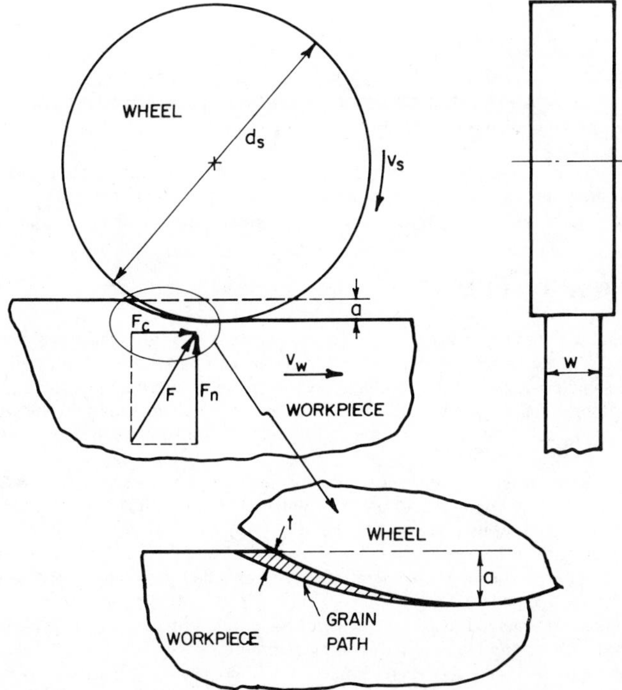

Fig. 4.4.6. Abrasive grain cutting path and forces for up-cut straight surface plunge grinding.

and increasing the wheel velocity v_s or equivalent diameter d_e lowers the grain depth of cut. Typical values of $\bar{t}$ range from about 0.3–4 μm, which is about an order of magnitude less than the corresponding wheel depth of cut (a).

Although material removal occurs mainly by chip formation, the grinding tool-workpiece interactions also involve *plowing* and *sliding*. Chip formation does not commence immediately as the grain enters the grinding zone; the cutting point must first penetrate to a critical depth. Until chip formation commences, the workpiece material is deformed by plowing without any removal. Plowing can also occur with chip formation when part of the material in the grain path is pushed aside into ridges rather than removed as a chip. Grains with very large negative rake angles may only plow the surface. Sliding occurs when flattened areas on the tips of the grains rub against the workpiece surface. Such flattened areas are initially generated by dressing, as described later, and these areas and others can develop and grow by *attritious* grain wear and adhesion of workpiece material. Wheel dulling by the growth of such flat areas is offset to some degree by *self-sharpening* of the wheel due to abrasive grain fracture or whole grain dislodgment from the binder.

The grinding action generates a force F between the wheel and the workpiece (Fig. 4.4.6). This force vector can be divided into a cutting component F_c and a normal component F_n. The cutting component acts tangential to the wheel at the grinding zone and is proportional to the grinding power. The normal force F_n is mainly responsible for the normal elastic deflection, which may be subsequently recovered by spark-out (Fig. 4.4.3). For grinding, the normal force is usually two to four times larger than the tangential force F_c.

The net grinding power is equal to the cutting force F_c times the tangential velocity of the wheel relative to the workpiece. Since the wheel velocity greatly exceeds the workpiece velocity, this grinding power can be expressed as

$$P = F_c v_s$$

A fundamental parameter for grinding, as well as other machining operations, is the *specific energy*. This parameter is defined as the energy consumed per unit volume of material removed. The specific energy is identical to another parameter called the *specific power*, which is defined as the power consumption per unit rate of volumetric removal. The specific energy u (see Fig. 4.4.6) can be written as

$$u = \frac{F_c v_s}{v_w a w} = \frac{P}{Z}$$

where Z is the volumetric removal rate ($Z = v_w a w$). If the specific energy is known, the net grinding power can be obtained by multiplying the specific energy times the removal rate:

$$P = uZ$$

The specific energy for grinding is generally much larger than that for other machining operations. Typical values for production grinding of ferrous metals are $u \approx 20$–80 J/mm^3. Such large energies can be attributed mainly to the sliding and plowing energies in addition to the energy required for chip formation. With finer grinding conditions (smaller $\bar{t}$), the specific energy increases due to increasing magnitudes of the plowing and sliding energies. Furthermore, the portion of specific energy required for chip formation is bigger for grinding than for other machining processes due to the large negative rake angles on abrasive grain cutting points.

4.4.5 DRESSING AND TRUING

Truing is the process of removing material from the cutting face of a grinding wheel so that it runs true. *Dressing* conditions the wheel surface so as to achieve a certain cutting behavior. Both truing and dressing are usually done by the same process, and the combination is commonly referred to as dressing.

Most dressing is done by passing a dressing tool across the face of the rotating grinding wheel. The dressing tool is usually a mounted single-point diamond or a multipoint tool consisting of diamond particles impregnated in a metal binder.

One dressing method used increasingly especially for generating profiles is *rotary diamond dressing*. A rotary diamond dressing tool (roll) consists of an axisymmetrical body with diamond particles impregnated in a metal matrix or electroplated on its surface. The dressing roll must have the same profile as the workpiece; the wheel, then, is dressed with the reverse profile. Since a special diamond roll is required for each profile, its use is justified only for producing many identical parts, or when the required profile cannot be readily obtained by other dressing methods.

Crush dressing is another method to prepare a wheel for profile grinding. The rotating wheel is fed under pressure into cast iron rolls having the required workpiece profile.

The dressing methods described above are applied to grinding wheels containing conventional ceramic abrasives (aluminum oxide and silicon carbide). Multipoint diamond tools are also used for dressing of superabrasive (diamond and cubic boron nitride) wheels, and various other types of rotary diamond and bonded abrasive tools are also used for this purpose. *Brake-controlled dressing* (truing) with vitrified silicon carbide dressing wheels is a specialized treatment applied to superabrasive wheels. In this case, the rotating motion of the dressing wheel being fed into the rotating grinding wheel surface is resisted by a centrifugal brake on the dressing wheel spindle.

Dressing (truing) of diamond or cubic boron nitride wheels is usually followed by *stick dressing*, whereby a bonded tool (stick) containing silicon carbide or aluminum oxide abrasive is fed into the wheel surface. Stick dressing "opens" the wheel surface by removing binder material so the abrasive grains protrude better. Abrasive sticks are also used for periodic cleaning of adhered swarf from the wheel surface.

During dressing, abrasive is removed from the wheel surface by fracture within the grain or by dislodgment of whole or partially fractured grains from the binder. In addition, the abrasive tool can flatten down or dull the abrasive grain tips by a fine wear or polishing action.

The relative severity or agressiveness of the dressing process ranges from *coarse dressing* to *fine dressing*. Coarser dressing indicates faster removal of material from the wheel, and finer dressing indicates a slower rate of removal. With finer dressing, the wheel is duller due to an increased tendency for flattening down the grain tips with less fracture. The fineness of the dressing can be further accentuated by continued dressing action (dressing spark-out or dwell) at the end of the process due to machine deflection recovery without continuing infeed of the dressing tool. Coarser dressing results in a sharper wheel surface. Finer dressing leads to bigger forces due to the increased sliding of the flattened abrasive grain tips against the workpiece. However, finer dressing leads to a smoother wheel surface with more even cutting point protrusion. Because the wheel is smoother, the surface finish transferred by the grinding kinematics from the wheel surface to the workpiece is also smoother. For a given grinding operation, there is an optimum severity of dressing that balances the requirements of good surface finish with the desirability for low grinding forces and power.

During grinding, the wheel wears by attrition, grain fracture, and breaking out of grains at the bond. If attrition prevails, the grinding force progressively increases, so that it becomes necessary to periodically sharpen the wheel by redressing. On the other hand, grain fracture and dislodgment during grinding may lead to deterioration in surface finish, and this will also necessitate redressing. The need for redressing may also be dictated by such factors as loss of form (tolerance) due to wheel wear or machine vibrations (chatter).

4.4.6 GRINDING FLUIDS

Grinding fluid is applied in the vicinity of the grinding zone to cool the workpiece and provide lubrication. Virtually all the energy input to the grinding process is manifested as heat, most of which is conducted to the workpiece. The heat input is concentrated over the restricted area of the grinding zone, leading to high localized temperatures. Cooling by grinding fluids removes heat from the workpiece. Lubrication by grinding fluids reduces energy requirements of the process, thereby reducing the heat input and lowering the temperature.

The two main types of grinding fluids are water-base and straight oils. Water-base fluids consist mostly of soluble oils or emulsions diluted with water; straight oils contain a mixture of various types of oils. Chemical additives are also incorporated into grinding fluids to enhance their lubricity. Various other compounds are added for inhibiting rust, cleaning, and retarding biodegradation.

Water-base fluids are better coolants, and straight oils are better lubricants. In most cases, the cooling action by water-base fluids lowers the bulk temperature of the workpiece but is not sufficient to significantly lower the high localized grinding zone temperature. Straight oils are generally much better lubricants for reducing the grinding energy, thereby providing less heating at the grinding zone. They also tend to be more effective for reducing wheel wear and maintaining finish and form. Although straight oils provide better grinding performance, their use is often restricted due to increased health and fire hazards, which usually necessitate a completely enclosed grinding area and other precautions. Therefore, water-base fluids are much more widely used in production. Synthetic fluids are also being introduced and in some cases are claimed to provide lubrication performance comparable to or even better than that of straight oils.

4.4.7 PRACTICAL LIMITATIONS ON GRINDING PERFORMANCE

Analytical methods have developed for many machining processes in order to determine the optimal operating conditions according to criteria of minimum cost or maximum production rate. Optimal operating conditions are determined as a tradeoff between faster removal rate, which reduces cutting time and its associated cost, and shorter tool life, which results in higher tooling cost. Such analyses are extremely difficult to apply to grinding insofar as the necessary reliable tool life relationships are generally not available. A practical approach to improving the operating efficiency of grinding operations is to progressively increase the removal rate until one or more of the limiting production constraints are reached. In the absence of a production constraint, it is generally possible to grind faster. The most widely encountered constraints for grinding include machine power limitation, machine tool vibrations, thermal damage to the workpiece, excessive wheel wear, poor surface finish, and unacceptable tolerances.

Even when a particular constraint is encountered, it is often possible to alter the grinding conditions so as to introduce a different constraint, thereby making it possible to further raise the grinding rate. However, such action may tend to introduce a different constraint. Therefore, it is especially important to be aware of the production constraints as well as the possibilities for corrective action to relax them. Each of these constraints is briefly described here.

4.4.7.1 Machine Power

The power requirements of the process can be estimated from the magnitude of the specific energy and the removal rate, as described in Section 4.4.4. The net spindle power available on a grinding machine is typically about half its overall rated power, although this will vary from machine to machine.

If the machine is operating at its spindle power limit, any further increase in removal rate requires that the specific grinding energy be lowered. This can be done by using coarser wheel dressing, provided that the surface finish is better than required, since coarser dressing will result in a poorer surface finish as described in Section 4.4.5. Another possibility is to change to a softer wheel, although this can be expected to produce more wheel wear. A third alternative is to use a grinding fluid with better lubricity.

4.4.7.2 Machine Tool Vibrations

Vibrations in grinding operations cause poor surface quality and can limit the production rate. There are generally two types of vibrations: *forced vibrations* and *self-excited vibrations*. Forced vibrations are caused by some external vibration source, such as an unbalanced wheel or other rotating elements, and the chatter frequency coincides with the frequency or some harmonic of the vibration source. Self-excited vibrations occur by a regenerative effect whereby a fundamental instability of the machine is dynamically excited periodically during successive workpiece revolutions. Such *regenerative chatter* builds progressively by the growth of waves (lobes) around the wheel and workpiece peripheries. A stiffer machine is less prone to self-excited vibrations.

Forced vibration can be eliminated or minimized by eliminating or isolating the vibration source. Self-excited vibrations cannot be eliminated in most cases, except by resorting to uneconomically slow grinding rates. However, it may be possible to delay the vibration build-up to an unacceptable level, thereby prolonging the "chatter-free" grinding time between wheel redressings and affording a faster removal rate.

The constraint on the grinding rate imposed by self-excited vibrations (regenerative chatter) may be relaxed by decreasing the cutting stiffness or the wheel wear rate. The cutting stiffness is defined as the ratio of the normal cutting force to the wheel depth of cut, so the cutting stiffness can be made smaller either by reducing the cutting force or by increasing the wheel depth of cut. Force reduction can be achieved as described in Section 4.4.7 for lowering machine power and specific energy. For cylindrical grinding, a larger wheel depth of cut at the same removal rate (same v_f as in Fig. 4.4.2) is obtained by slowing down the workpiece velocity.

With less wheel wear, the regenerative chatter amplitude builds more slowly, since the wheel waviness undulations (lobes) grow more gradually. Wheel wear resistance can be improved by using a harder wheel, but this approach may be detrimental due to bigger grinding forces.

4.4.7.3 Thermal Damage to the Workpiece

Excessive grinding zone temperatures can cause various types of surface damage to the workpiece, such as burning with steels, softening (tempering) of the surface layer with possible rehardening and embrittlement, unfavorable residual tensile stresses and cracks. The possibility for thermal damage can be reduced by lowering the grinding zone temperature.

For plunge grinding operations (Fig. 4.4.2), the peak grinding zone temperature θ can be approximated by the relationship

$$\theta = \frac{1.13\alpha^{1/2} R u v_w^{1/2} a^{3/4}}{k d_e^{1/4}} \tag{4.4.4}$$

where α is the thermal diffusivity, k is the thermal conductivity of the workpiece material, R is the fraction of the total grinding energy generated that is conducted as heat into the workpiece, and u, v_w, a, and d_e are as defined earlier.

The direct way to reduce the temperature is to lower the specific energy u. This can be achieved as described in Section 4.4.7 by using coarser dressing, a softer wheel, or a grinding fluid with better lubricity.

It would appear from Eq. (4.4.4) that another possibility would be to lower the fraction R of heat conducted to the workpiece. For most production grinding operations, about 85% of the grinding energy is conducted as heat to the workpiece ($R \approx 0.85$). This fraction tends to be slightly lower with smaller specific energies, so reducing the specific energy will have the additional beneficial effect of a slightly smaller fraction R. Some of the grinding heat entering the workpiece is subsequently convected out of the workpiece by the grinding fluid, but cooling by the grinding fluid has only a marginal effect on R in most cases. Therefore, cooling by the grinding fluid will have relatively little effect on the peak grinding temperature θ, although it will tend to lower or stabilize the bulk temperature of the workpiece. One notable exception where cooling by the grinding fluid plays an important role is in *creep-feed* grinding (Section 4.4.3), where cooling of the grinding zone is enhanced by supplying copious amounts of fluid under high pressure and by using porous (open) wheels.

According to Eq. (4.4.4), increasing the removal rate with a faster workpiece velocity v_w or larger wheel depth of cut a will raise the temperature. However, because the relative influence of a is greater than that of v_w, the temperature can be reduced while maintaining the same removal rate (same infeed velocity v_f) by increasing the magnitude of v_w, which will give the same proportional decrease in a. Therefore, increasing the workpiece velocity will usually allow for a faster removal rate when thermal damage is the production constraint. This approach may be limited by the greater tendency for self-excited vibrations (regenerative chatter) with faster workpiece velocities.

4.4.7.4 Wheel Wear

Grinding wheel wear can be broadly classified into three categories: attrition or rubbing down of the grain tips, grain fracture, and dislodgment of grain at the binder. All three types of wear occur simultaneously to a greater or lesser degree. Almost all the wear debris can be attributed to grain fracture and dislodgment, but dulling by attrition may be the most significant type of wear insofar as it can cause excessive forces and temperatures (Section 4.4.4). Wear by grain fracture and dislodgment can reduce the flattened dulled area on the wheel (self-sharpening), but this is accompanied by surface finish deterioration and loss of form.

The relative volumetric wheel wear is often expressed by a grindability index called the *G-ratio*, which is defined as the bulk volumetric wheel wear per unit volume of stock removal. For production grinding operations with conventional wheels (aluminum oxide and silicon carbide), typical grinding ratios range from about 10 to 100. Much smaller G-ratios may be obtained for grinding tool steels. With superabrasive wheels (diamond and cubic boron nitride), G-ratios typically range from about 100 to 1,000.

A high G-ratio would seem to be desirable, as it indicates less wheel wear. For many grinding operations with conventional wheels (aluminum oxide and silicon carbide), the wheel cost is relatively insignificant, and more wheel may actually be consumed by dressing than by wear. In this case, the least wearing wheel with the highest G-ratio may not be the best, because it may limit the grinding rate due to production constraint limitations of force or temperature. With superabrasive wheels, the wheel cost is much higher, so the G-ratio becomes a more significant economic factor.

With faster removal rates, the G-ratio decreases, which means that the wheel wear rate increases proportionally faster than the removal rate. This may cause more rapid deterioration in surface finish and form beyond the required limits (see Sections 4.4.7.5 and 4.4.7.6). If larger grinding forces can be tolerated, these production constraints can be relaxed by changing to a slower wearing wheel. This usually means selecting a harder grade conventional wheel or a higher concentration superabrasive wheel. A better lubricant (grinding fluid) can also increase the G-ratio and even lower the grinding forces.

4.4.7.5 Surface Finish

Grinding processes are often selected because of the need to satisfy stringent surface finish and tolerance requirements. For such production grinding operations, typical surface finishes (AA) range from about $R_a \approx 0.15$ μm to $R_a \approx 2$ μm, although the process is also widely applied outside this range. Surface finish requirements may be dictated by the tolerance requirements (see Section 4.4.7.6).

The surface finish generated by grinding is the result of the topography of the active wheel surface interacting with the workpiece under the kinematic motions imposed by the machine. Therefore, the surface finish obtained on the workpiece depends on the roughness of the wheel surface and the operating parameters of the process.

The wheel surface and, consequently, the workpiece surface can be made smoother by using finer dressing conditions. With single-point or multipoint diamond dressing, for example, finer dressing is obtained by decreasing either the radial dressing depth a_d or the crossfeed velocity v_d of the dressing tool across the wheel surface (see Section 4.4.5). The effect of these dressing parameters on surface finish can be approximated by

$$R_a \propto a_d^{1/4} v_d^{1/2} \tag{4.4.5}$$

The crossfeed velocity has a relatively bigger effect than the dressing depth, and this is the dressing parameter that is usually varied on the machine.

A faster stock removal rate generally results in a rougher surface finish. For external and internal cylindrical plunge grinding (Figs. 4.4.2*b* and 4.4.2*c*), the influence of the grinding parameters v_w and a on surface finish can be approximated by

$$R_a \propto (v_w a)^x \tag{4.4.6}$$

Typically $x \approx ½$, although it tends to be smaller for grinding cycles with longer spark-out times.

When the surface finish constraint prevails, it is apparent from the combined effects of Eqs. (4.4.5) and (4.4.6) that the removal rate can be increased while maintaining the same or an even better surface finish by dressing the wheel finer. This approach is feasible only to the extent that larger forces and grinding energies associated with finer dressing can be tolerated.

Smoother surfaces may also be obtained by using grinding wheels with finer grain size. This approach provides only a marginal improvement in surface finish with conventional wheels requiring periodic redressing,

Table 4.4.2. Production Grinding Tolerances (μm)

Cylindrical Grinding	
Diameters	6
Shoulder to shoulder	6
Traverse grind to shoulder	50
Facing runout	6
Corner radii	120
Internal Grinding	
Hole diameter	6
Facing runout	6
Centerless Grinding	
Diameters	3
Parallelism	3
Straight Surface Grinding	
Thickness	8
Flatness	5

and can cause excessive grinding forces due to grain dulling and wheel loading. With superabrasives, the use of fine-grained wheels may be an absolute necessity to obtain the required surface finish. Ultraprecision grinding processes with slow removal rates often depend on the use of extremely fine abrasive grains.

4.4.7.6 Tolerance

Tolerances in grinding can be divided for convenience into *dimensional tolerances* and *form tolerances*. Dimensional tolerances in grinding are mainly concerned with linear lengths or size. Form tolerances in grinding are mainly concerned with cross-sectional shapes.

Linear dimensions of machined components are measured from one finished surface to another or, in the case of diametral dimensions, between two opposite locations on the same cylindrical surface. The roughness of the finished surface can be thought of as a measure of the uncertainty in exactly specifying the location of the surface, so the uncertainty in a linear dimension between two points will depend on the combined surface roughnesses at the measuring points. Therefore, it is generally necessary to have smoother surfaces in order to have less dimensional uncertainty (better tolerance). Accordingly, the surface finish requirement is often the consequence of the dimensional tolerance requirement, and both factors are similarly affected by the grinding conditions. Other factors that adversely affect dimensional tolerances are machine deflection, thermal expansion and distortion of the machine and the workpiece, wheel wear, and vibrations.

For production grinding as well as other machining processes, dimensional tolerances are typically 10 to 50 times as large as the corresponding surface finishes. This ratio will depend on such factors as the machine tool condition, the allowable rejection rate, component interchangeability, and assembly requirements.

The same factors that affect dimensional tolerances also affect form tolerances, since form is also specified in terms of linear as well as angular dimensions. However, the most significant form error is usually caused by grinding wheel wear. This is especially problematic with cross-sectional shapes having sharp radii or deep groves. Better form control requires a slower wearing wheel (larger G-ratio), but this can be expected to cause bigger forces (Section 4.4.7.4).

Table 4.4.2 shows typical tolerances that can be held in production. Larger tolerances will generally allow for faster stock removal rates and more economical production. Much finer tolerances than those in Table 4.4.2 may be achieved by grinding, but this may necessitate special equipment and much higher production costs.

BIBLIOGRAPHY

Bhateja, C. P. "On the Mechanism of the Diamond Dressing of Grinding Wheels." *Proceedings, International Conference on Production Engineering*, Part 1. Tokyo, Japan: JSPE, 1974, pp. 733–739.

Coes, L. *Abrasives*. New York: Springer-Verlag, 1971.

Des Ruisseaux, N. R., and R. D. Zerkle. "Thermal Analysis of the Grinding Process." *Journal of Engineering for Industry, Trans. ASME*, **92**, 428–434 (1970).

Drozda, T. J., and C. Wick, Eds. *Tool and Manufacturing Engineers Handbook*, 4th ed., Vol. 1, Chap. 11, "Machining." Dearborn, MI: SME, 1983.

Farrago, F. T., *Abrasive Methods Engineering*, Vols. 1 and 2. New York: Industrial Press, 1976.

Hahn, R. S., and R. P. Lindsay, "Principles of Grinding." *Machinery* (July–November 1971).

Lewis, K. B., and W. F. Schleicher. *The Grinding Wheel*, 3rd ed. Cleveland, OH: Grinding Wheel Institute, 1976.

Lindsay, R. P. *Precision Grinding Dressing Effects*, SME Paper No. MR 79-321, 1979.

Malkin, S. "Specific Energy and Mechanisms in Abrasive Processes." *Proceedings, Third North American Metalworking Research Conference*. Pittsburgh, PA: Carnegie Press, 1976, pp. 453–465.

Malkin, S. "Burning Limit for Surface and Cylindrical Grinding of Steels." *Annals of the CIRP*, **27**, 233–236 (1978).

Malkin, S. "Grinding of Metals: Theory and Application." *Journal of Applied Metalworking*, **4**, 95–109 (1984).

Malkin, S., and N. H. Cook. "The Wear of Grinding Wheels." *Journal of Engineering for Industry, Trans. ASME*, **93**, 1120–1133 (1971).

Malkin, S., and T. Murray. "Comparison of Single Point and Rotary Dressing of Grinding Wheels." *Proceedings, Fifth North American Metalworking Research Conference*. Dearborn, MI: SME, 1977, pp. 278–283.

Markings for Identifying Grinding Wheels and Other Bonded Abrasives, ANSI Standard B74.13, New York: American National Standards Institute, 1982.

Osman, M., and S. Malkin. "Lubrication by Grinding Fluids at Normal and High Wheel Speeds." *Transactions of the ASLE*, **15**, 261–268 (1972).

Snoeys, R., and D. Brown. "Dominating Parameters in Grinding Wheel and Workpiece Regenerative Chatter." *Proceedings, 10th International MTDR Conference*, September 1969, Manchester, England, pp. 325–348.

Snoeys, R., and M. Maris. "Production Rate Limits in Grinding." *Proceedings, International Conference on Production Engineering*, Part 2, Tokyo, Japan: JSPE, 1974, pp. 70–75.

Snoeys, R., M. Maris, and J. Peters. "Thermally Induced Damage in Grinding." *Annals of the CIRP*, **27**, 571–581 (1978).

Snoeys, R., J. Peters, and A. Decneut. "The Significance of Chip Thickness in Grinding." *Annals of the CIRP*, **23**, 183–186 (1974).

Springborn, R. K., Ed. *Cutting and Grinding Fluids: Selection and Application*. Dearborn, MI: SME (Formerly ASTME), 1967.

Trucks, H. E. *Design for Economical Production*. Dearborn, MI: SME, 1974, Chapter 2.

Zinkak, D., Ed. *Improving Production with Coolants and Lubricants*. Dearborn, MI: SME, 1982.

CHAPTER 4.5

INDUSTRIAL ROBOTS

C. RAY ASFAHL

Fayetteville, Arkansas

The word *robot* epitomizes the public image of industrial automation. This image is only partially correct, since industrial robots are only part of the total automation picture and the public idea of an industrial robot is highly glorified. Despite its shortcomings, however, the industrial robot and the public's fascination with it is propelling industrial automation into factories at a vigorous rate.

Most people think of *Star Wars*'s R2D2* when the word "robot" is mentioned. Such a concept is appropriate because the word "robot" was actually coined on the stage, not in the factory. Robots first appeared in New York on October 9, 1922, in a theatre play entitled *R.U.R*. The creator was Czechoslovakian dramatist Karel Capek, and the word "robot" is a derivative of the Czech word robota, which means "work."[1]

Unlike R2D2 or Capek's robots, most real industrial robots are hardly humanoid in appearance. In fact, a more descriptive term for most industrial robots would be "mechanical arm." Figure 4.5.1 shows a popular model of mechanical arm-type robot, a far cry from R2D2 but much more useful from an industrial standpoint.

The robot in Fig. 4.5.1 meets the Robot Institute of America's definition of *robot*: "A robot is a reprogrammable, multi-functional manipulator designed to move material, parts, tools, or specialized devices through variable programmed motions for the performance of a variety of tasks."[2] Another robot that meets this definition is shown in Fig. 4.5.2. Note that this robot is equipped with an arc welding head instead of grippers to pick up piece parts. A more general definition of robot is provided by Mikell Groover: "An industrial robot is a general-purpose, programmable machine possessing certain anthropomorphic characteristics."[3] Groover's definition does not restrict the concept of robot to manipulator; rather, it leaves open the possibility of other anthropomorphic (humanlike) characteristics, such as judgment, reasoning, and vision.

Not all industrial robots look like mechanical arms. Some robots are enclosed in cubic work envelopes, with the edges of the cubes providing bearing for the robots' movements. This type of robot is sometimes called a gantry-mounted robot.

A key word in the definition of an industrial robot is *programmable*. More than any other, this feature of industrial robots, made practical by the advances of inexpensive microchip circuits in the 1970s, has vaulted industrial robots into the workplace.

Long before the programmable robots came the mechanically fixed manipulators, whose motions are set by mechanical cams installed in the factory. The Japanese robot industry includes the mechanically fixed manipulators within the definition of robot, but this interpretation is debatable. The disparities in robot definitions can distort statistics that compare robot populations in various countries. The mechanical-cam-type manipulators, regardless of whether they are called robots, have an important role in factory automation. These manipulators typify the term *hard automation* as contrasted with *flexible automation* typified by the programmable industrial robot.

4.5.1 ROBOT GEOMETRY

Since robot configurations vary greatly, some classification of robot geometries is in order. The industry has settled on the term *degrees of freedom* to describe the number of ways a robot can move. The form of these movements and the way they are assembled make up the robot *configuration*.

Adapted, with permission of the author and publisher, from C. Ray Asfahl, *Robots and Manufacturing Automation*, Chapter 6, John Wiley and Sons, New York, 1985.

* R2D2 is the name of the personable, fictitious robot that starred in *Star Wars*, a popular movie of the 1970s.

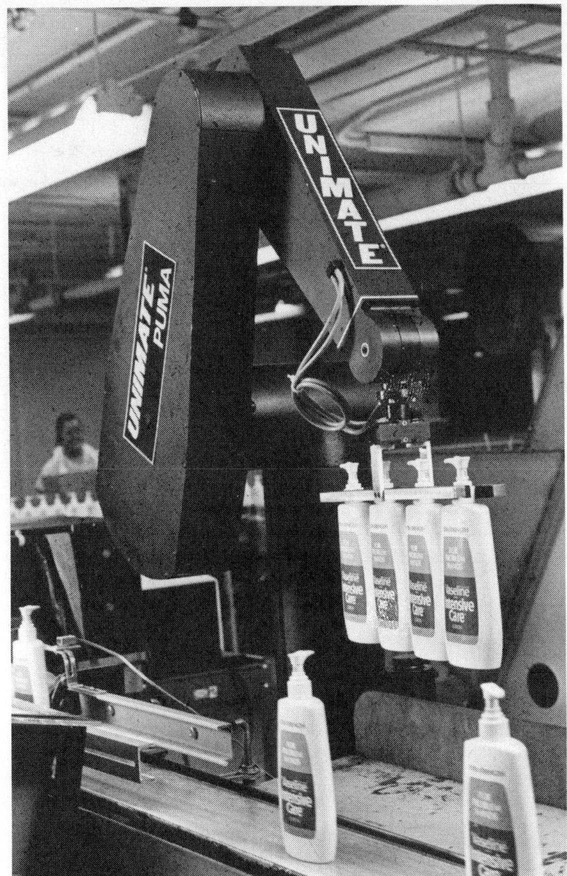

Fig. 4.5.1. Popular model of mechanical arm-type robot. Reprinted by permission of Unimation-Westinghouse, Inc., Danbury, CT.

4.5.1.1 Degrees of Freedom

Every mechanical point on a robot, except in the gripper or tool, at which some form of drive induces motion in a robot part is called a degree of freedom. The motion can be a pivoting or reciprocal motion as is produced by a pneumatic or hydraulic cylinder. Figure 4.5.3 displays a robot with six degrees of freedom:

1. Base rotation.
2. Shoulder flex.
3. Elbow flex.
4. Wrist pitch.
5. Wrist yaw.
6. Wrist roll.

Although there are exceptions, in most robots the degrees of freedom are in series. Thus, the first degree of freedom in the robot of Fig. 4.5.3, base rotation, imparts motion to all of the parts of the robot affected by subsequent degrees of freedom. Conversely, the third degree of freedom (elbow flex), for example, has no effect on the base movement. It follows that the most sophisticated motion in the entire robot is that of the member driven by the highest degree of freedom. Generally speaking, the robot with the most degrees of freedom can produce the most complex movement, but there are other important factors to consider, such as range and quality of motion within a given degree of freedom. This point will become clearer later in this chapter.

Fig. 4.5.2. Industrial robot equipped with an arc welding head. Reprinted by permission of Unimation-Westinghouse, Inc., Danbury, CT.

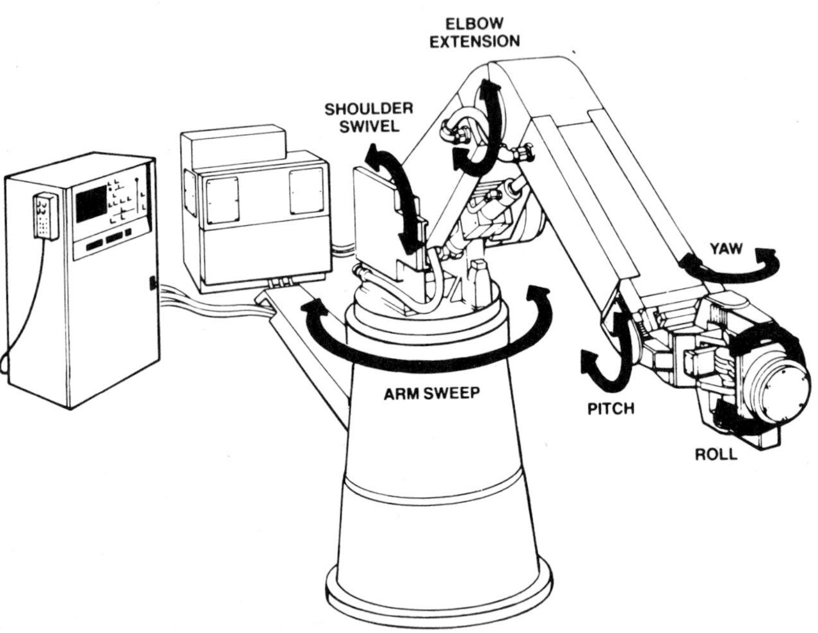

Fig. 4.5.3. Industrial robot with six degrees of freedom. Reprinted by permission of Cincinnati Milacron, Lebanon, OH.

The sequence of the various degrees of freedom and their types of motion determine the physical configuration of the robot. Theoretically, there could be a large number of configurations for a robot with six degrees of freedom. From a practical standpoint, however, almost all robots fall into a few popular configuration categories.

4.5.1.2 Articulating Configurations

Some robots, such as the one in Fig. 4.5.1, actually work like a human arm. Such robots are said to be "articulating." The base rotates in a way similar to a twisting human torso. The shoulder and elbow on most articulating robots pivot on one axis each, perpendicular to the axis of the arm and parallel to the plane on which the base is mounted. The wrist assembly on articulating robots almost always has pitch but may or may not have yaw and roll. The reader may note that the human hand also is very flexible in the pitch axis but has little yaw or roll. The human wrist normally permits yaw of 60° at most. Virtually no roll at all is permitted by the human wrist, but the forearm and even the shoulder can be used together to achieve about 270° roll. Thus, a human can screw in a lightbulb or attach a wingnut, but only in a series of twists and regrasps. Most articulating robots have more roll capability than the human arm.

There is a trick that can produce a yaw-type motion in a five-axis articulating robot that has no yaw. For some applications a 90° roll can be executed and then a pivot on the *pitch axis*. This method is demonstrated in Fig. 4.5.4. Note that the procedure works best for single-point tools, not grippers. If a robot equipped with a gripper is used to attempt a yaw motion by this procedure, the piece part or tool held by the robot may be in the wrong orientation.

A variation of the articulating robot is the horizontal-jointed robot, which simply means that the joint movement in the second, third, and fourth degrees of freedom have their axes arranged vertically so that the joint bends are in a horizontal orientation. A horizontal-jointed robot is shown in Fig. 4.5.5. This type of robot may not resemble a human arm as much as the conventional articulating type but is gaining in popularity due to its usefulness in simple handling tasks.

An even stranger variation of the articulating robot is the polyarticulating trunk robot shown in Fig. 4.5.6. *Trunk* is an apt description of this robot, because it resembles an elephant trunk as it snakes about in a virtually infinite variety of contortions. The advantage of this type of motion is that the robot can extend its trunk into tight workspaces and then orient its tool (usually a paint spray head) into almost any orientation.

4.5.1.3 Polar Configuration

The polar, or "spherical," configuration looks quite different from the articulating configuration but actually is different only in the third axis (third degree of freedom). In place of an elbow, the polar configuration robot has a pneumatic or hydraulic cylinder that provides extension for the arm. A robot in the polar configuration is seen in Fig. 4.5.7. This type of robot is very popular in the automotive industry.

A typical robot motion is to elevate a workpiece along a vertical path while maintaining the orientation of the workpiece. It will be seen later that for the polar configuration robot, this requires simultaneous, coordinated motion in three axes: shoulder, arm extension, and wrist pitch. The articulating robot has a problem of similar complexity, requiring simultaneous, coordinated motion in shoulder, elbow, and wrist pitch.

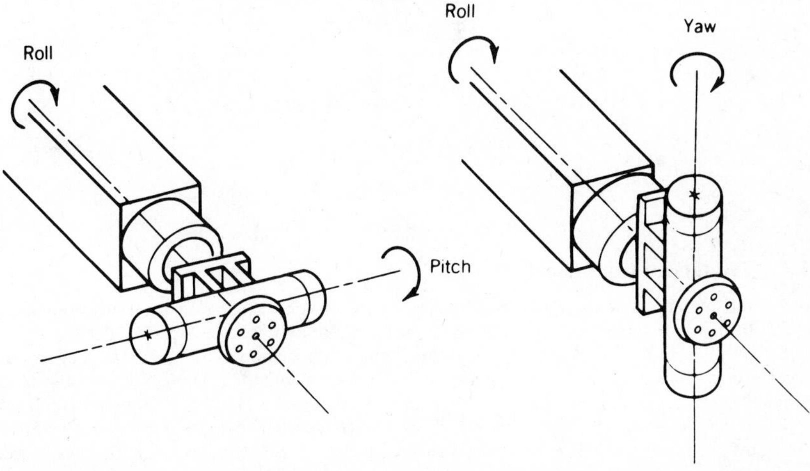

Fig. 4.5.4. To obtain yaw motion, roll 90° and then pitch (feasible for certain tool configurations). (*a*) Roll and pitch axes. (*b*) Roll and yaw axes. Reprinted, with permission, from Asfahl, *Robots and Manufacturing Automation*, p. 142; John Wiley and Sons, Inc.

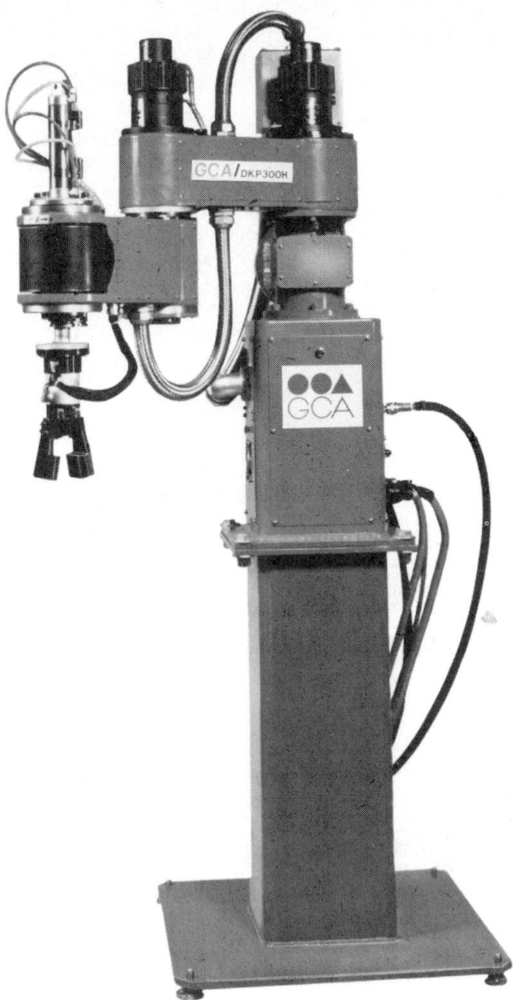

Fig. 4.5.5. Horizontal jointed robot. Courtesy of GCA Corporation, St. Paul, MN.

4.5.1.4 Cylindrical Configuration

A robot of the cylindrical configuration has a vertical reciprocating axis for its second degree of freedom, or base extension. This is usually accomplished by a pneumatic or hydraulic cylinder but may be rack and pinion or chain drive in large robots. A cylindrical configuration robot is shown in Fig. 4.5.8. Axes A, B, C of this robot generate a work envelope in the shape of a cylinder. Cylindrical configuration robots usually have reciprocating motion in their third degree of freedom just as do their cousins of the polar configuration. It is even possible for the cylindrical configuration robots to have *both* base extension and shoulder rotation, giving them an extra degree of freedom between base and wrist. Such robots can have eight degrees of freedom, but this configuration should be considered rare.

The problem of elevating a workpiece or tool in a vertical, straight line is an easy one for the robot of cylindrical configuration. Only one axis of movement is required compared to the three axes of movement dictated by either the polar or articulating configuration. A disadvantage of robots of cylindrical configuration, however, is that they cannot reach around obstacles.

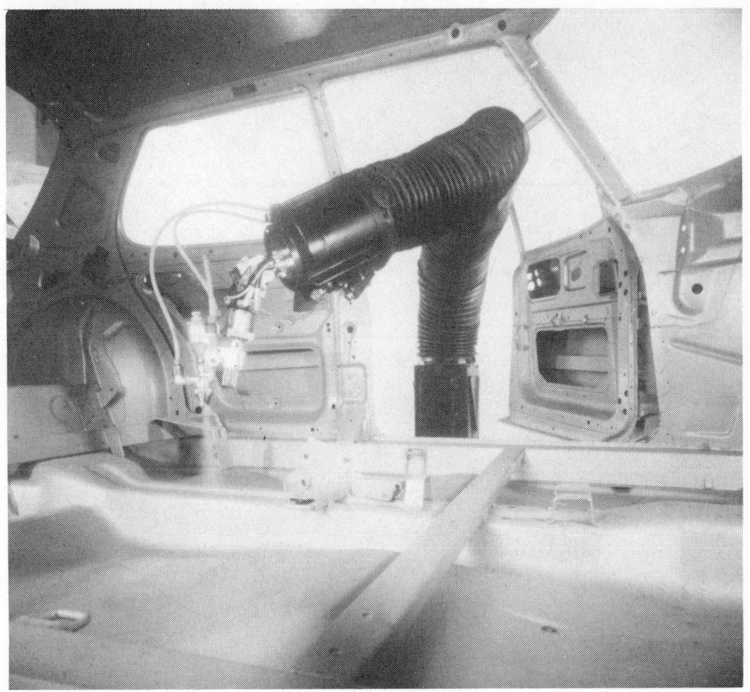

Fig. 4.5.6. Polyarticulating trunkrobot. Reprinted by permission of Spine Robotics, Molndal, Sweden.

Fig. 4.5.7. Industrial robot of polar configuration. Reprinted by permission of Unimation-Westinghouse, Inc., Danbury, CT.

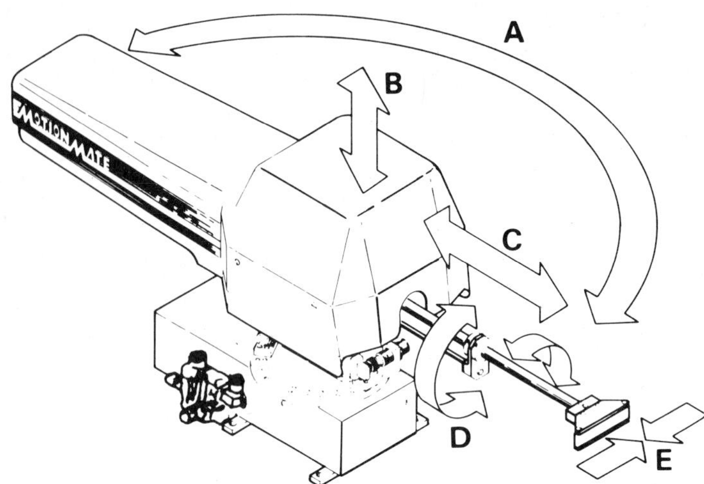

Fig. 4.5.8. Industrial robot of cylindrical configuration. Reprinted by permission of Schrader-Bellows Division of Scovill, Akron, OH.

4.5.1.5 Cartesian Configuration

Machine tool practitioners will feel most at home with robots of the cartesian configuration. This robot is usually gantry-mounted but may be mounted on a track on the floor. The first three axes of a cartesian robot are the familiar x, y, and z axes as in machine tools. The cartesian configuration offers the advantage of rigidity made possible by a boxlike frame to support the robot. Fewer parts on the robot are extended in a cantilevered posture, and even these are only the lighter members close to the tooling. Thus, closer tolerances can be maintained by small cartesian robots, and for very large robots cartesian configuration becomes imperative. Figure 4.5.9 shows a large cartesian robot.

4.5.1.6 Work Envelope

With knowledge of the degrees of freedom of the robot and the physical configuration of the assembly of these degrees of freedom, the user can use geometry and trigonometry to determine the position of the robot *tool center point* (TCP) as a function of the positions of the various robot axes. Figure 4.5.10 shows the computation of the position of the TCP as a function of the axis positions of an articulating robot. Note that each of the angles A, B, and C are measured from a reference parallel to the X axis. Angles A and B are in a counterclockwise direction (positive angles), but angle C is in a clockwise direction and is thus a negative angle. The sine of a negative angle is also negative, so the contribution of the term $(+c \sin C)$ to the formula for Y is indeed appropriate for the wrist angle shown.

The extreme positions of the robot axes describe a boundary for the region in which the robot operates. This boundary encloses what the industry calls the *work envelope*. Figure 4.5.11 displays an example of a work envelope. Work envelope for a robot is an extremely important characteristic that should be considered carefully in every robot acquisition. The size of the work envelope obviously determines the limits of reach and is important from an applications point of view. But also important is the viewpoint of safety. A powerful robot placed in a room that is smaller in some dimension than its work envelope will soon wreck its environment, despite the pride and professional care exercised by its programmers. Even with a hypothetically ideal robot that never malfunctions, there are so many variables to consider in programming, and it is so easy for the user to make an error in some detail, that an accident is inevitable if space for the work envelope is not provided properly.

4.5.1.7 Mobile Robots

The preceding section described robot work envelopes, which are appropriate for most industrial robots, since these robots are mounted either on a fixed base or on a carriage of limited travel on a fixed base. But some robots can walk on legs or move about on wheels. The robot pictured in Fig. 4.5.12 is capable of walking as fast as a human at a brisk pace. It can lift a one-ton payload and can walk with nearly a half-ton payload. In Fig. 4.5.12 the robot not only was capable of climbing into the back of the pickup truck but also of lifting the back end of the pickup and walking a short distance, turning the truck 90° in the process.

Walking robots are not just curiosities or gimmicks for the entertainment world. Many industries have special problems particularly suited for the capabilities of a walking robot. Some examples are underground

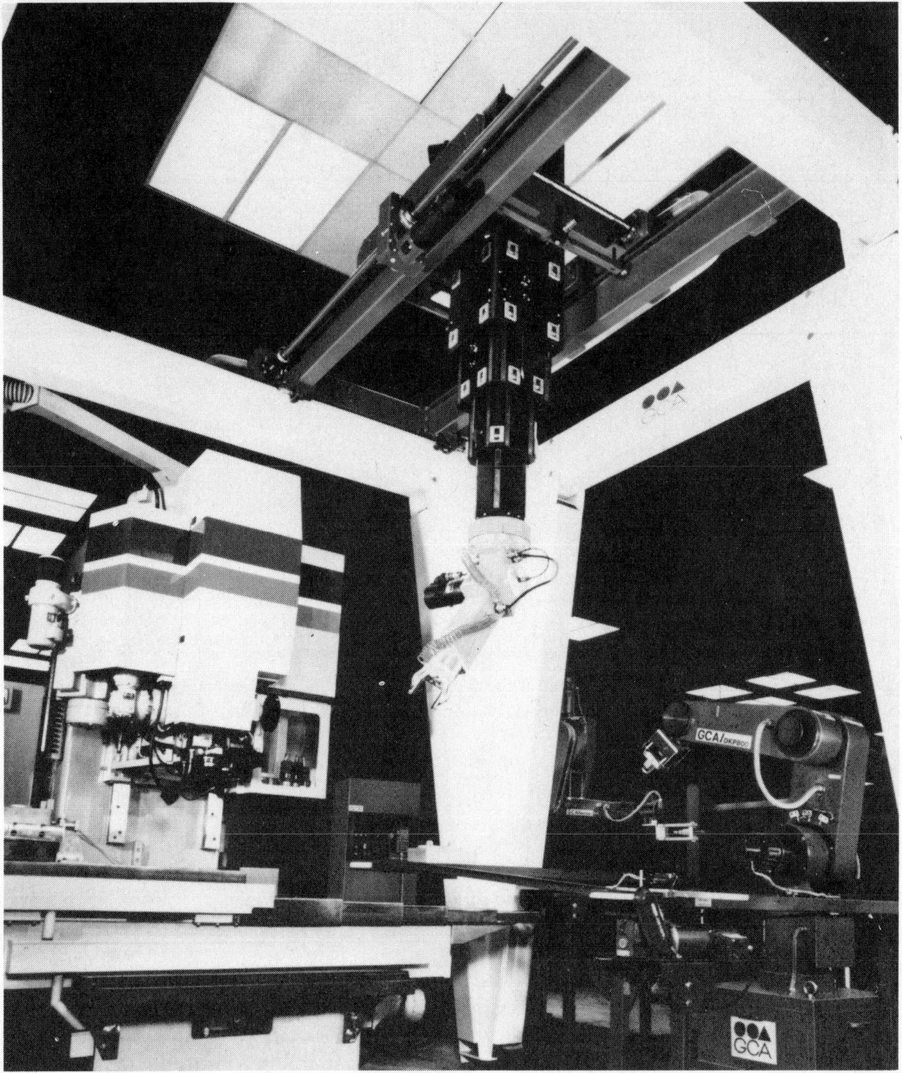

Fig. 4.5.9. Large cartesian gantry-mounted robot. Reprinted by permission of GCA Corporation, St. Paul, MN.

mining, space exploration, nuclear accident cleanup, demolition, fire damage inspection, sea floor exploration, and sentry duty. The walking robots are more complex than wheeled models and can negotiate a more difficult and uneven traveling surface. The wheeled models have their strong points, too, when used as automatic material handling systems and as interoffice mail delivery systems. To emphasize its usefulness, rather than its bizarre appearance, the developers (Odetics, Incorporated) dubbed the machine depicted in Fig. 4.5.12 a "functionoid."[4]

4.5.2 ROBOT DRIVES

The most distinguishing feature used to describe an industrial robot is its power source. The power source usually determines the range of the robot's performance characteristics and in turn the feasibility of various applications, although there is considerable overlap between types. The four principal power sources are hydraulic, pneumatic, electric, and mechanical gear and cam.

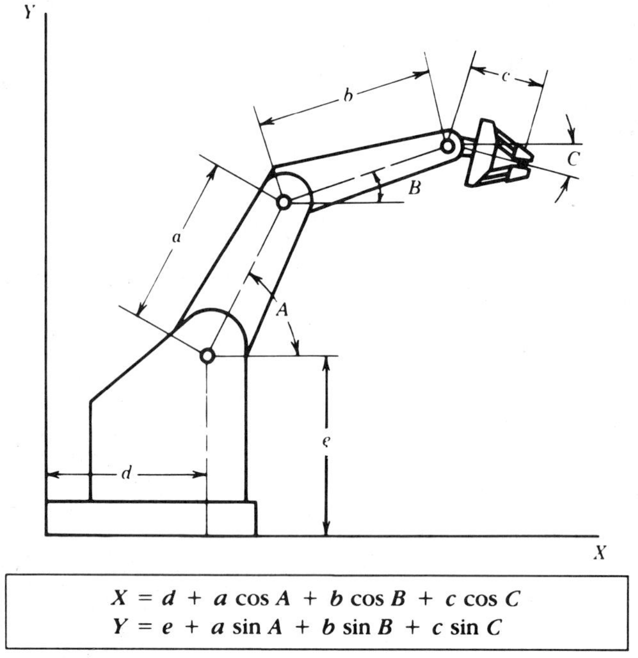

Fig. 4.5.10. Computation of the position of the tool center point (TCP). Reprinted, with permission, from Asfahl, *Robots and Manufacturing Automation*, p. 147; John Wiley & Sons, Inc.

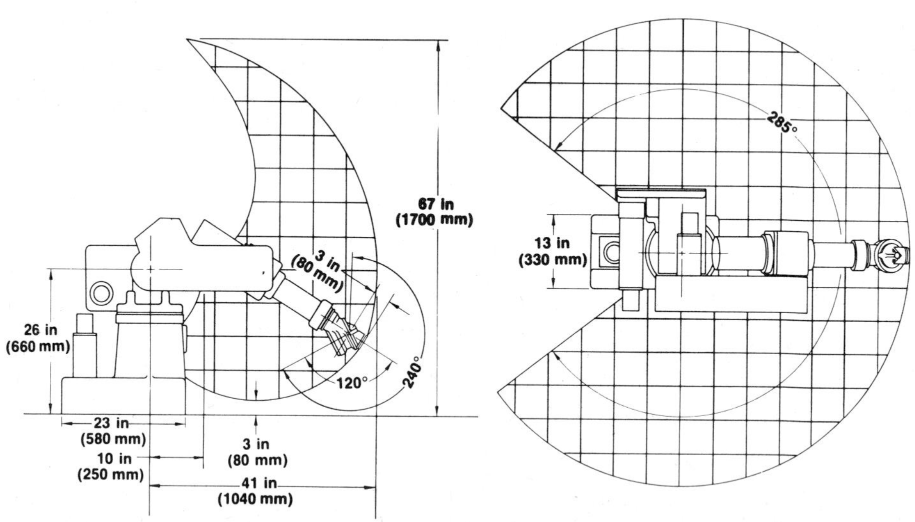

Fig. 4.5.11. Example of a work envelope for an industrial robot. Reprinted by permission of Cincinnati Milacron, Lebanon, OH.

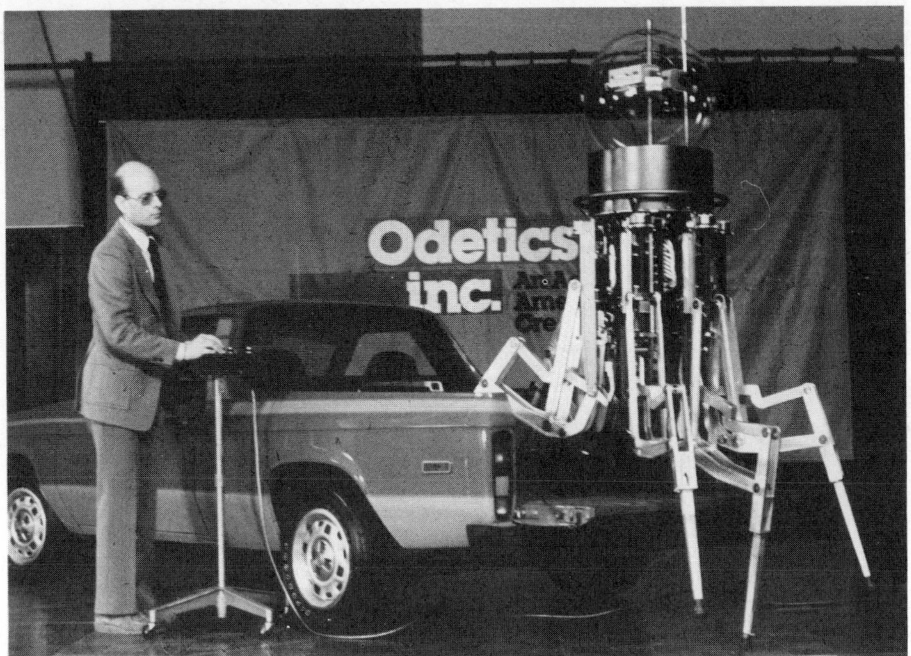

Fig. 4.5.12. A walking robot which can lift a ton and can walk away with a 900-pound load. Photo courtesy of Odetics, Inc., Anaheim, CA.

4.5.2.1 Hydraulic Drives

From a physical standpoint, the most powerful robots are generally the hydraulic models. Hydraulic robots are able to deliver large forces directly to the robot joints and to the gripper or tool center point. Offsetting this advantage is cost, which is usually higher for hydraulic models than for electric or pneumatic models of equivalent capability. Hydraulic models also require a pump and reservoir for the hydraulic fluid in addition to fittings and valves, all designed for high pressure. Figure 4.5.13 illustrates a hydraulic robot that features a payload of 225 lbs.

An important application of hydraulic robots is in spray paint operations. Due to flammability considerations, it may be necessary to employ an explosion-proof robot in paint spray areas, which require equipment that meets National Fire Protection Association (NFPA) standards for Class I, Division I flammable atmospheres.

4.5.2.2 Pneumatic Drives

Some of the least expensive and most practical robots for ordinary pick-and-place operations or for machine loading and unloading are the pneumatic models. The availability of shop air at approximately 90 psi is an obvious advantage. Most factories have compressed air piped throughout their production areas, and this can be conveniently tapped to power a pneumatic robot. The robot in Fig. 4.5.14 is a pneumatic model with a payload of 5 pounds. The configuration for this robot is cylindrical.

Pneumatic robots usually operate at mechanically fixed end points for each axis. Figure 4.5.15 reveals that the mechanical stop is no different from those employed on pneumatic actuating cylinders extensively used in automation long before the advent of robots. The pneumatic robot is really an assembly of several such cylinders, each one representing an axis of motion.

With motion in each axis controlled only at the end points, the reader may be wondering what can be programmable about a pneumatic robot. But remember that timing and sequence are also important, resulting in an infinite variation of possible programmed setups for the pneumatic robot, even without touching a wrench. By further adjustment of the mechanical stops, even more variety can be achieved. Still, a carefully controlled, continuously varying path is impossible to achieve with the typical pneumatic robot.

It should be mentioned here that there is one type of pneumatic robot, certainly atypical in design, that achieves a continuous, controlled motion through the use of *differential dithering*. Differential dithering applies a series of short pulses of compressed air that can act on the robot member in either direction, causing it to follow a continuous path under control. Figure 4.5.16 illustrates a commercially available pneumatic robot that uses the concept of differential dithering to produce a low-cost, continuous, and controlled motion.

Fig. 4.5.13. Hydraulic robot with a payload capacity of over 225 pounds including tooling. Reprinted courtesy of Cincinnati Milacron, Lebanon, OH.

One of the principal advantages of pneumatic robots is their modular construction and their use of standard, commercially available components. This is true of other robots but is especially true of the pneumatic models. This feature presents the possibility of a firm deciding to build its own robots, sometimes at considerable cost savings. Some component suppliers emphasize the "build your own" concept in marketing their products. Any firm that decides to embark on a "build your own" strategy to save hardware costs should also remember to add in the engineering and component procurement costs in addition to the hardware costs.

4.5.2.3 Electric Drives

Electric robots are popular for precision jobs because they can be closely controlled and can be taught to follow complicated paths of motion. One can argue that many hydraulic models can have the same features, but sophisticated motion control is more typical of the all-electric models.

Electric robots can be divided into two groups according to the types of electric motors that drive each of the axes of motion. One type uses stepper motors, which are driven a precise angular displacement for every discrete voltage pulse issued by the control computer. The stepper motor movements can be very precise, provided the torque load does not exceed the motor's design limits. Because of this inherent accuracy, the stepper-motor-type robot is sometimes of the open-loop type; that is, the control computer computes the number of pulses required for the desired movement and dispatches the command to the robot without checking whether the robot actually completes the motion commanded. Unfortuntely, the robot does not always accomplish the commanded motion because it may encounter an obstacle or for some other reason experience slippage in its mechanical linkage from its drive motors to its mechanical members. When this occurs, the open-loop robot unfortunately "loses its way," and its control computer no longer knows the position of the robot's components. When this occurs, the robot may continue into future cycles with a permanent position error that can make its operation completely useless or even destructive. Fortunately, this predicament has remedies, which we shall see later.

The other "species" of electric robot is the DC servo-driven type. These robots invariably incorporate feedback loops from the driven components back to the driver. Thus, the control system continuously monitors the positions of the robot components, compares these positions with the positions desired by the controller, and notes any differences or error conditions. DC current is applied to each motor to correct error conditions until the error goes to zero.

It should be noted that feedback loops can be incorporated into the stepper-motor-type robot also. Optical encoders can be used to monitor the actual angular displacement of the driven component. This information is returned to the control computer, which is programmed to take action to correct any error conditions.

Because it can be used either in an open- or a closed-loop design, one might expect the stepper-motor design to dominate completely its alternative, the DC servo, in the construction of robots. However, the advantage of the DC servo is that it is a continuous device, thereby making possible a smoother and continuously controllable movement.

Of the two basic types of electric robots, the DC servomotor type is the most popular. Figures 4.5.17 and 4.5.18 show examples of the DC servomotor type and stepper-motor type, respectively.

4.5.2.4 Mechanical Gear and Cam Drives

For completeness, we will include the manipulators driven by mechanical gears and cams. These types of manipulators are hard-programmed at the factory and do not meet the "programmable" specification found in most definitions of "industrial robot." For that matter, they are generally electrically driven if traced back to the original power source. But the power is delivered to the components by complicated mechanical linkages. Two principal advantages of the mechanical cam and gear-driven manipulators are low cost and speed.

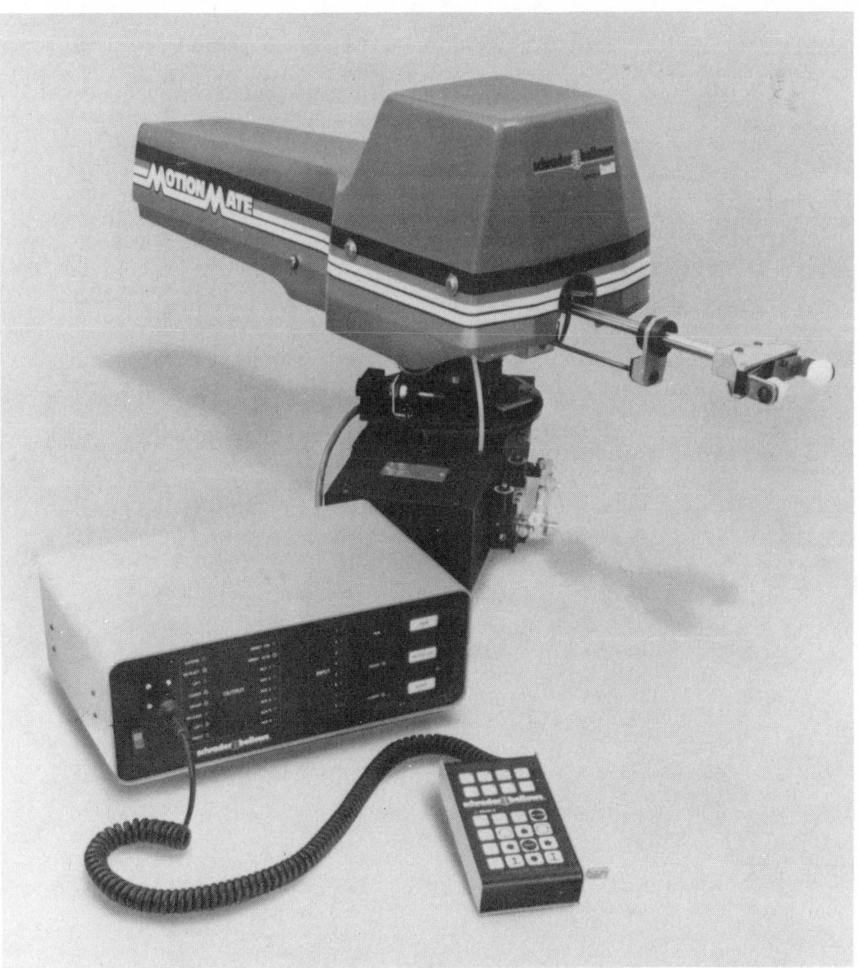

Fig. 4.5.14. Pneumatic-powered robot with a payload capacity of 5 pounds, including gripper tooling. Reprinted by permission of Schrader-Bellows Division of Scovill, Akron, OH.

Fig. 4.5.15. Mechanical stops for axis limits of a pneumatic robot. Reprinted by permission of Schrader-Bellows Division of Scovill, Akron, OH.

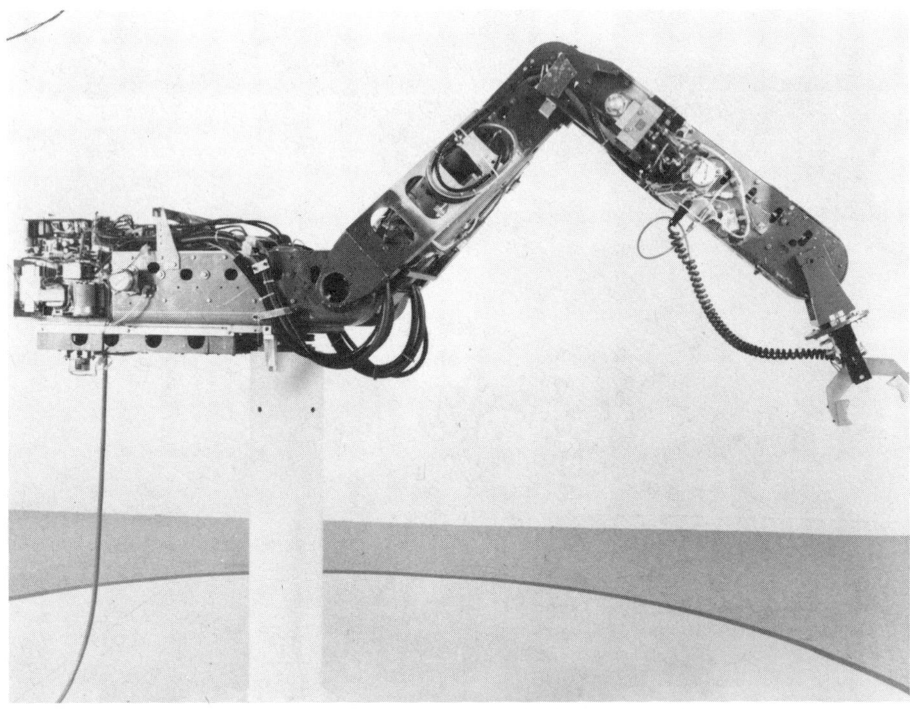

Fig. 4.5.16. Pneumatic robot capable of continuous path control by differential dithering. Reprinted by permission of International Robomation/Intelligence, Carlsbad, CA.

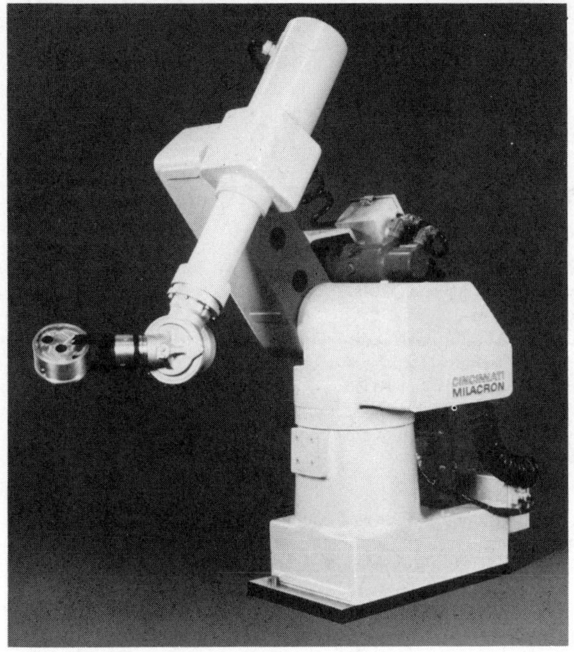

Fig. 4.5.17. Robot driven by electric DC servomotors. Reprinted by permission of Cincinnati Milacron, Lebanon, OH.

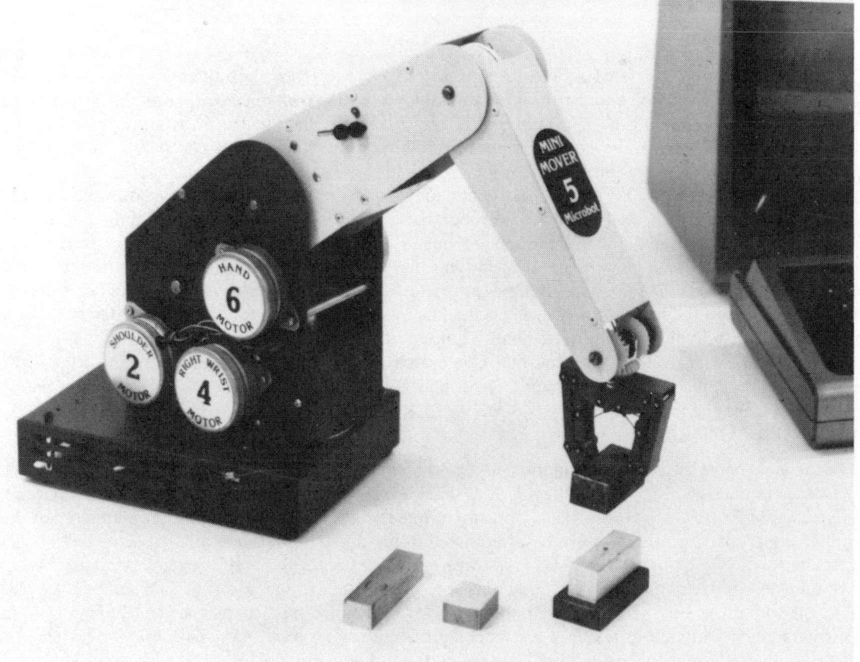

Fig. 4.5.18. Robot driven by electric stepper motors. Reprinted by permission of Microbot, Mountain View, CA.

4.5.3 MOTION CONTROL

The type of drive used may be the most obvious distinguishing feature of the robot, but just as important, although more subtle, is the degree of control possible over the robot motion. This control is affected by the choice of robot drives, as was seen earlier, but it is not completely determined by the drive. Robot users need to know how much motion control they need for their various applications, since the degree of motion control greatly affects the cost of the robot. Three categories of motion control, in order of sophistication, are axis limit, point-to-point, and contouring.

4.5.3.1 Axis Limit

The least sophisticated and therefore the least expensive mode of robot motion control is axis limit, sometimes called "two-position control" because each robot axis typically has two extreme points. In our description of the pneumatic-powered robots, we saw that the extreme points are usually mechanically adjustable stops.

Users of axis-limit robots should have little or no preference for component motion velocities, as these considerations are beyond their control, although a small degree of speed control can be exercised by varying the power source. In addition, pauses can be programmed between robot motions to permit some selectivity in the speed of the overall cycle. The typical application for axis-limit robots is in machine loading and unloading. Axis-limit robots are invariably either pneumatically or hydraulically powered.

4.5.3.2 Point-to-Point

Somewhat better than axis-limit control is point-to-point (PTP) control. In this mode, the user can select any point in space in the robot work envelope, and the robot will move directly to that point. Both the path and speed of movement en route to the destination point are generally uncontrollable. Even if speed is controllable, the robot is point-to-point unless the path en route is also controllable. PTP control is good for component insertion, hole drilling, spot welding, and crude assembly applications. Machine loading and unloading operations to or from a pallet or tray also require PTP motion control.

Point-to-point motion should not be confused with straight-line motion. In general, even a simple straight-line motion between two points cannot be accomplished by a PTP robot. One exception to this generality is a straight-line vertical lift by a robot of cylindrical configuration. But straight-line movements are not an easy task for robots, especially the articulating robots. Simultaneous, controlled movement in more than one axis is *always* required to achieve straight-line movement using a fully articulating robot. To counteract this disadvantage, robot manufacturers have developed computer software routines that handle the mathematical mixing of axis voltages, pulses, or valve openings to achieve straight-line movement upon command.

4.5.3.3 Contouring

As in NC machine tools, the most sophisticated class of robot motion is the full contouring class. Contouring describes motion in which the entire path is continuously or nearly continuously controlled. When the drive is by stepper motor, the control is not quite continuous but still can be classified as an approximation of contouring if there is a feedback loop to the controller and if in addition the controller is capable of varying the *rate* of pulses delivered to the stepper-motor drive.

The difference between PTP motion and continuous path contouring is difficult to distinguish, especially by an observer who did not actually program the robot. By detailed and meticulous programming, the programmer can set up a PTP robot to move in a seemingly continuous contour. Sometimes controller software can provide subroutines to relieve the programmer of specifying the myriad points required to simulate the curvilinear motion with tiny straight-line motions. But programming is not the only problem. The execution time of a PTP robot is dependent on the number of points specified and thus can become too slow to be effective. Continuous contouring motion provides the ability to control not only the *position* of the robot tool but the *velocity* of tool motion in each axis controlled by contouring. Contouring motion control is essential for most spray painting, finishing, gluing, and arc welding operations by robots.

4.5.3.4 Line Tracking

One of the most complex contouring motions is *line tracking*, that is, performing an operation while following alongside a continuously moving conveyor. Line tracking is merely another application of contouring, not a separate class of motion. However, the complexity of line tracking demands intricate programming of the robot controller, especially for robots whose bases are fixed to the floor (true of most robots). Some robots designed specifically for line tracking have a horizontal traverse on a track for the first degree of freedom. The traverse can be adjusted to match the speed of the conveyor, giving such robots a distinct advantage over the fixed-base robots with respect to the task of programming. However, fixed-base robots may be equipped with factory-supplied computer software to which a line-tracking feature can be added conveniently. Therefore, the trade-off between the two types is cost of hardware versus cost of software.

Line tracking has obvious advantages. The product being processed can be transported on a *continuous* conveyor instead of an intermittent one. Continuous conveyors are much simpler mechanically and thus are less expensive and more reliable. With more reliable conveyor operation and quicker repair time, the continuous conveyor keeps idle time at a minimum, maximizing both production machine usage and robot usage.

Particularly suited to line tracking is the robot application to spray painting. Spray painting generally is applied to all sides of a piece part as it is transported by a continuous overhead chain. The robot must be directed at all sides of the part, a feat that can be accomplished most conveniently if the part moves continuously past the work station and if the robot has line-tracking ability. Line tracking is also convenient if multiple operations must be performed by the robot on each part.

4.5.4 ROBOT TOOLING AND APPLICATIONS

A novice in the field can be stunned to discover that he or she has just purchased a robot that can do nothing. Although this chapter frequently refers to robot grippers, most robots do not come equipped with such devices. Remembering that programmability and versatility are hallmarks of the modern industrial robot, the reader will see that the robot manufacturers' strategy is to leave the choice of the end-of-arm tooling to the user. Indeed, many robots are programmed to use a variety of tools or grippers in a single setup, automatically selecting and changing tools according to a prescribed sequence.

4.5.4.1 Grippers

Grippers come in a wide variety of configurations and are often designed by the customer to fit a particular application. Most grippers close upon the part to be picked up, but a large number of them insert their fingers inside the part and then open to grip the part. Many grippers are fashioned to work effectively either way so that the choice is up to the programmer. Figure 4.5.19 illustrates a variety of grippers.

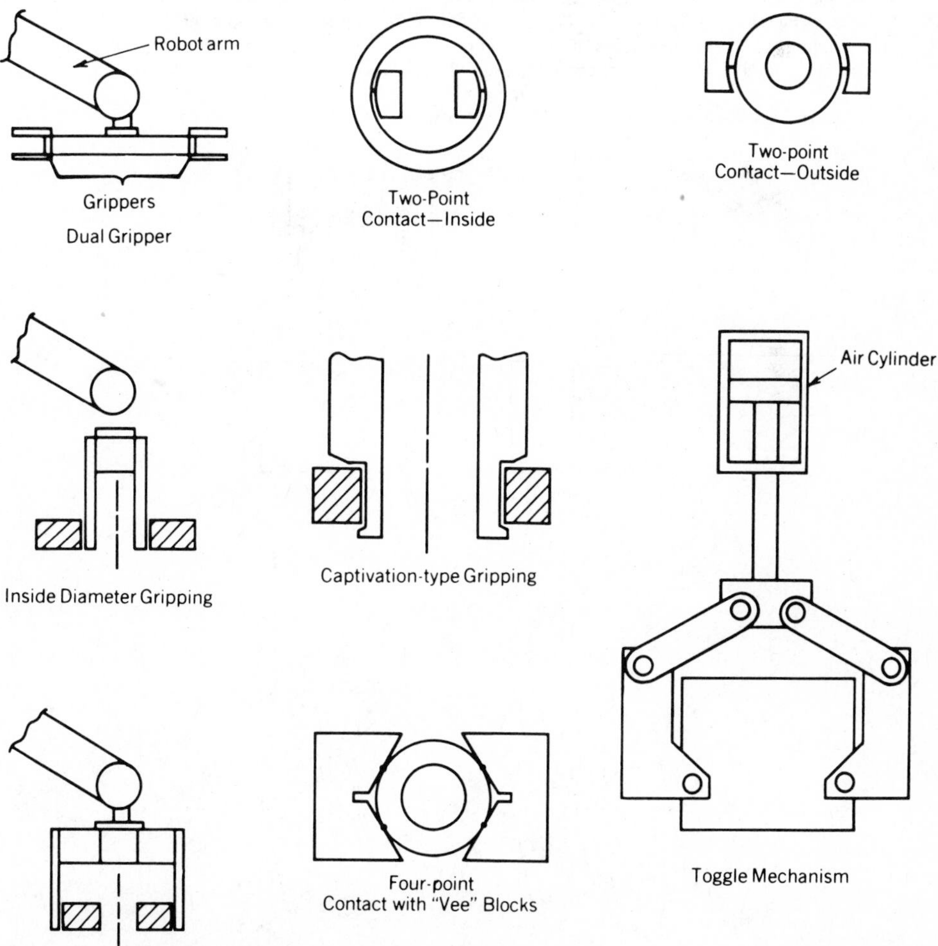

Fig. 4.5.19. End-of-arm tooling for industrial robots. Reprinted by permission of ASEA, Inc., White Plains, NY.

For many applications, a double-handed gripper is more efficient than a single-handed gripper. This is typically true of machine loading and unloading, as the robot is able to both unload and load a given station without moving between stations. This saves much time that would otherwise be wasted in repetitious motion in the principal axis (normally base rotation), and base rotation is typically the slowest axis on the robot.

4.5.4.2 Appliances

Besides the common grippers used for piece parts handling, tool heads of various types can be attached to the end of the robot arm. The wide variety of these end-of-arm tools sets the robot apart from ordinary material-handling devices.

Welding heads are the most common type of robot tool, excluding grippers. Spot welders are the most common, but arc welding robots are growing in importance. Spray painting heads, mentioned earlier in this chapter, are an important type of robot tool. Related to spray painting heads are glue applicators. Both of these tools are useful because of the precision and repeatability of the robot. An unusual tool is a dispenser for electric cable, which is used in the programmed assembly of electrical wire harnesses for aircraft and other large equipment.

4.5.4.3 Part-Compliant Tooling

A tiny misalignment of a piece part or robot tool can result in complete failure of the process and perhaps damage to the product or to the robot hand. A trick to avoid or at least lessen the effect of misalignment problems is to mount the gripper with a flexible connection that allows the robot tool to "give a little" when it encounters the object to be picked up. The industry calls this approach *part-compliant tooling*.

One sophisticated type of part-compliant tooling is called *remote center compliance* (RCC). The concept is illustrated in Fig. 4.5.20. In Fig. 4.5.20*a*, the robot is attempting to insert a pin into a hole, but there is lateral misalignment. The chamfer helps, but with rigid tooling the pin still may not enter the hole. A less rigid tooling still might not deliver the desired result because the lateral component of force at the chamfer will tend

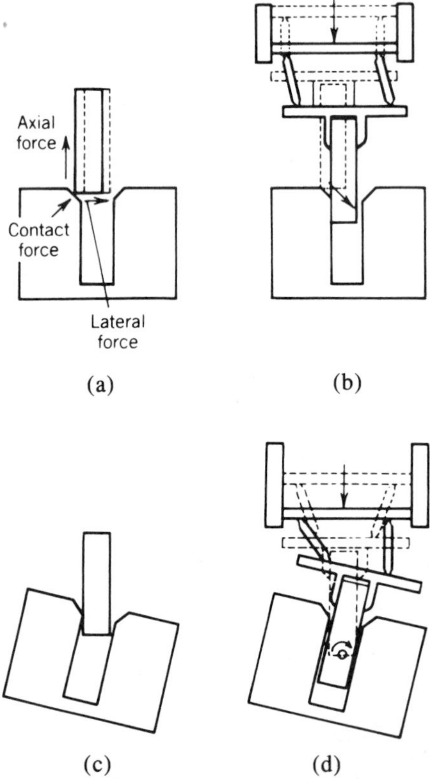

Fig. 4.5.20. Remote center compliance (RCC) tooling to deal with misalignment problems. Reprinted, with permission, from Asfahl, *Robots and Manufacturing Automation*, p. 160; John Wiley and Sons, Inc.

Fig. 4.5.21. Variety of commercially available RCC devices to facilitate close tolerance assembly operations using industrial robots. Reprinted by permission of Lord Corporation, Mechanical Group, Erie, PA.

to rotate the pin about its flexible center of compliance. The assembly in Fig. 4.5.20*b* projects the center of compliance to the leading end of the shaft, which is a remote center of compliance point. The important consequence of this projection is that the pin shifts laterally instead of rotating about its top end.

Figure 4.5.20*c* presents a correct lateral alignment but an error in angular alignment. Note that this time the pin makes contact with the hole at two places. Lateral components of force on the pin are parallel and opposite each other, but these parallel, opposite forces do not act in the same line because the lateral force from the left acts on a point higher on the pin than the lateral force from the right. This causes a moment to act on the pin, and the RCC assembly shown in Fig. 4.5.20*d* permits insertion to occur. The physical appearance of RCC devices is illustrated in Fig. 4.5.21.

An even simpler strategy that works in some cases is to use rubber or nylon in the construction of the fingers themselves so that a soft, compliant touch not unlike that of human fingers is used to pick up the object. Any such flexible pickup method can be considered part-compliant tooling. The principle is so simple and practical that it would seem to be an obvious solution to gripper alignment problems, but a surprisingly large number of robot applications fail because the user or engineer does not think to try this strategy.

4.5.5 PROGRAMMING

It is worth repeating that a key feature of robots is their capability for being reprogrammed for different tasks. This is the feature that was missing from the mechanical manipulators that were seen before the advent of the industrial robot in the 1970s.

Some robots have been programmed by manually inserting pegs in a drum to trip switches according to a desired sequence. Others are programmed using a dialect of a popular general-purpose computer language, such a BASIC. One such dialect is the ARMBASIC robot language, aptly name by Microbot, Inc., its creator and copyright owner.[5] Another robot computer language is VAL, created by Unimation, Inc., for its PUMA series of sophisticated robots.[6]

The vast majority of robots have some form of handheld controller called a *teach pendant* (Fig. 4.5.22). The operator manually moves the robot through the desired motions by pressing buttons on the handheld controller. The pace usually is much slower than the production pace, thereby permitting the operator to program the operation carefully. Later, the program is reset to production speed.

Robot teach-pendant controls are very much like the familiar handheld pendant controls for overhead cranes. But the robot pendant control is much more powerful because it has additional controls for commanding the robot to *memorize* points along the path of motion. Also, robot controls can set timers to synchronize the operation, command the sensing of external inputs, and dispatch output signals to peripheral process equipment.

Moving up to a higher plane of sophistication are the robots that are taught by manually seizing the end of the robot arm and actually pushing it through the series of operations in a dry run of the real production process. The robot can be commanded to remember the path during the teach phase and then commanded to repeat the performance indefinitely. This type of programming is especially good for spray painting and welding applications. Operators skilled in conventional spray painting or welding can teach the robot their skills while simulating an actual manual performance of the job. The robot then merely mimics the actions of its teacher. Of course, the skill acquired by the robot applies strictly to the given application taught, not to spray painting or welding in general.

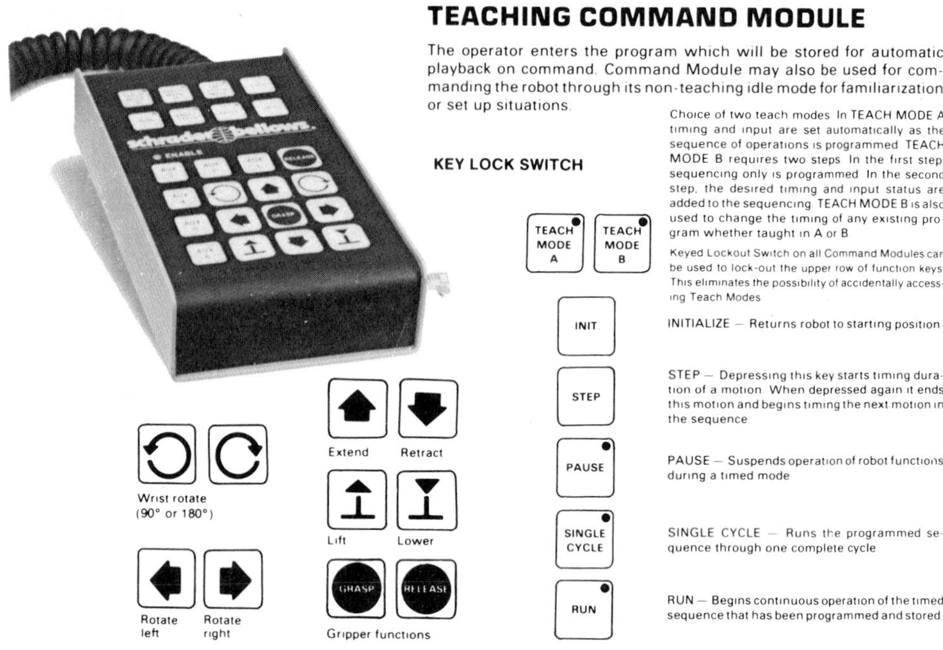

TEACHING COMMAND MODULE
The operator enters the program which will be stored for automatic playback on command. Command Module may also be used for commanding the robot through its non-teaching idle mode for familiarization or set up situations.
KEY LOCK SWITCH
Choice of two teach modes. In TEACH MODE A timing and input are set automatically as the sequence of operations is programmed. TEACH MODE B requires two steps. In the first step, sequencing only is programmed. In the second step, the desired timing and input status are added to the sequencing. TEACH MODE B is also used to change the timing of any existing program whether taught in A or B.
Keyed Lockout Switch on all Command Modules can be used to lock-out the upper row of function keys. This eliminates the possibility of accidentally accessing Teach Modes.
TEACH MODE A
TEACH MODE B
INIT
INITIALIZE — Returns robot to starting position.
STEP
STEP — Depressing this key starts timing duration of a motion. When depressed again it ends this motion and begins timing the next motion in the sequence.
PAUSE
PAUSE — Suspends operation of robot functions during a timed mode.
SINGLE CYCLE
SINGLE CYCLE — Runs the programmed sequence through one complete cycle.
RUN
RUN — Begins continuous operation of the timed sequence that has been programmed and stored.
Wrist rotate (90° or 180°)
Extend
Retract
Lift
Lower
Rotate left
Rotate right
GRASP
RELEASE
Gripper functions

Fig. 4.5.22. Teach pendant control for industrial robot. Reprinted by permission of Schrader-Bellows Division of Scovill, Akron, OH.

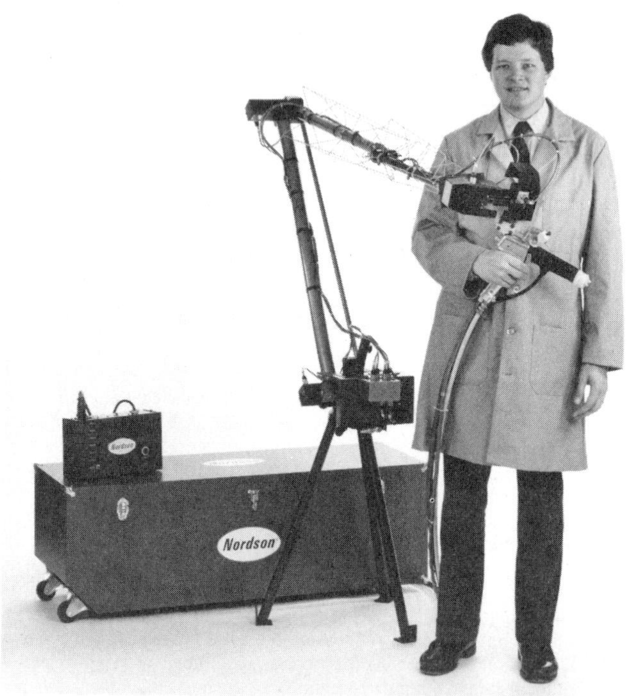

Fig. 4.5.23. Robot training arm for simulating robot motion during the teach mode. Reprinted by permission of Nordson Corporation, Amherst, OH.

A variation is to employ a mechanism that mechanically simulates the robot: a robot training arm, sometimes called a *dummy robot*. Compared with the real robot, the training arm (Fig. 4.5.23) is generally lighter and easier to manipulate by the skilled operator charged with the task of teaching the robot. Spray painting is an ideal application for such mechanisms because the skilled operator must feel as though he or she is actually holding a paint spray gun while teaching the robot. The comparatively light training arm can give that kind of feel to the operator. The training arm transmits its path to the control computer during the teach mode. In turn, the control computer drives the real robot through the same path of motion in the run mode. This chapter has avoided carefully the use of the terms *leadthrough programming* and *walkthrough programming*. Both of these terms are frequently heard among robotics professionals, but their meanings have often been interchanged. Some say that leadthrough refers to teach-pendant control and that walkthrough refers to the manual dry-run mode employed for spray painting operations. Other respected professionals interpret the terms in the opposite manner. The reader is advised to use the terms *leadthrough* and *walkthrough* carefully and to insist on clarification as to whether a teach pendant is used.

The most sophisticated robots are usually programmed using combinations of the teach modes discussed here. Thus, fixed locations can be taught using the teach pendant, and complex paths like arcs and contours can be programmed using computer software such as Unimation's VAL language.

4.5.6 SENSING CAPABILITY

Mechanical manipulation has many applications, but used alone its success is highly dependent on the positioning and orientation of the workpiece. Furthermore, the blind repetition of mechanical sequences can be disastrous if something goes wrong. A robot that can "see" or "feel" its payload and certain aspects of its surroundings has greatly increased utility over its more crude relatives, which operate as deaf, dumb, and blind manipulators.

Sensing capability on a robot can have widely ranging degrees of sophistication in addition to a variety of sensing media. For instance, optical sense capability can vary from a simple photoelectric cell to a complex, three-dimensional vision system. Various sensing categories will now be described, beginning with the simplest and most practical and proceeding to the most advanced systems available: gripper pressure sense, optical presence sense, robot vision, and tactile sensing.

4.5.6.1 Gripper Pressure Sense

The most elementary sense capability on a robot is probably the ability of the gripper to detect grip force between its fingertips. In its simplest form, the grip sensor consists merely of a limit switch that trips when a given preset grip pressure is reached. Such a limit switch is a practical safeguard against overclosure of the gripper in case of either program error or payload dimensional variability. But the advantages go beyond this safeguard: the limit switch can be used to standardize the grip pressure in a gauging operation. Thus, a robot can be used to gauge thickness by simply closing its grippers on an object. This feature of a robot is inexpensive to apply, and accuracy surpassing that of human fingers is easy to achieve.

Sometimes grip pressure and grip closure are in fact the same operation. This can be achieved, for instance, by using "tendon technology," that is, the actuation of the axes by cables leading to motors mounted on the robot base. Figure 4.5.24 shows how one supplier (Microbot) positions the limit switch to sense tension in the cable that controls gripper closure. Figure 4.5.25 charts the relationship between grip opening versus number of stepper pulses on the left and grip force versus number of stepper pulses on the right, illustrating the dual role of the grip closure cable. The thickness gauging capability of this device is rated at a tolerance of ±1/16 inch, but a Microbot Minimover 5 has been observed achieving a much higher resolution than the rating specifies.

The following case study illustrates the calculation of stepper-motor pulses to close upon and grip an object of known dimension with specified force.

Case Study 5.1: Determining Robot Gripper Motor Pulses

A robot gripper is actuated by stepper motor using tendon technology in accordance with Fig. 4.5.25. Suppose the gripper position is open 2½ inches and the command is to close upon an object with a force of 20 ounces. The object is 1⅜ inches wide. How many pulses should be issued to the motor?

Solution

The graph in Figure 4.5.25 intersects the X axis at approximately 895 steps, and full travel from gripper wide open to gripper closed and pinched at a force of three pounds is approximately 976 steps. In order to close upon an object 1⅜ inches wide from a position of open 2½ inches, the motor would need the following number of pulses:

$$\text{Steps to close} = (2^{1/2} - 1^{3/8}\text{in.}) \times \frac{895 \text{ steps}}{3 \text{ in.}} = 336 \text{ steps}$$

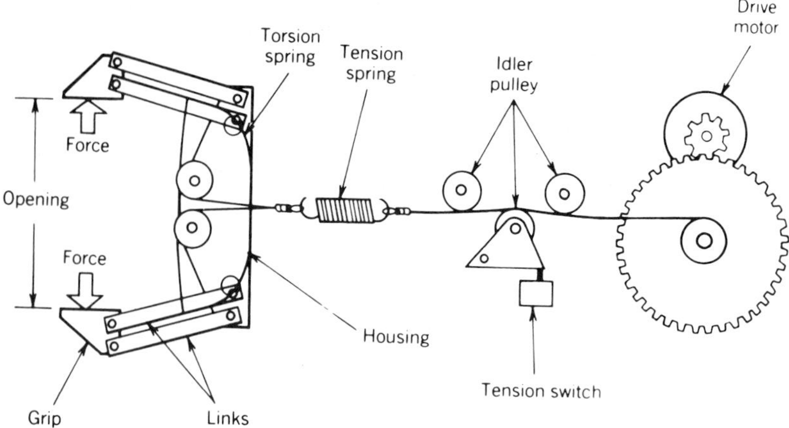

Fig. 4.5.24. "Tendon technology": A cable in the robot arm controls both grip closure and grip pressure. Note the position of the limit switch to detect cable tension. Reprinted by permission of Microbot, Mountain View, CA.

To continue closing to grip with a force of twenty ounces would require

$$\text{Steps to grip} = 20 \text{ oz} \times \frac{1 \text{ lb}}{16 \text{ oz}} \times \frac{(976 - 895) \text{ steps}}{3 \text{ lb}}$$

$$= 34 \text{ steps}$$

The total number of pulses required would be the sum:

$$\text{Total steps} = \text{Steps to close} + \text{steps to grip}$$

$$= 336 \text{ steps} + 34 \text{ steps}$$

$$= 370 \text{ steps}$$

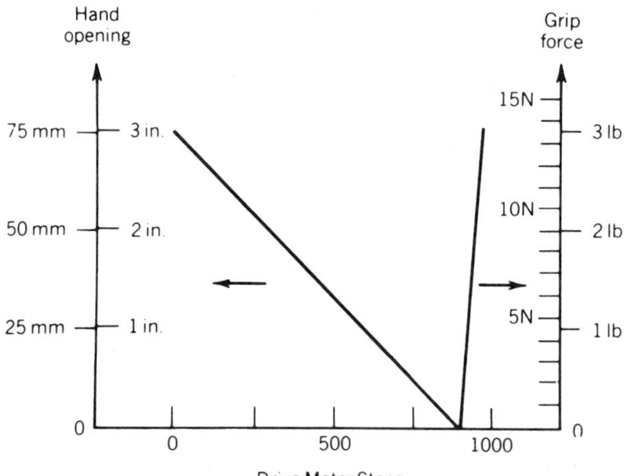

Fig. 4.5.25. Example of the relationship between stepper-motor pulses and grip closure and grip force using tendon technology. Reprinted by permission of Microbot, Mountain View, CA.

4.5.6.2 Optical Presence Sensing

Once a robot has grip pressure sense capability, the next step is to add some sort of presence-sensing mechanism, usually a photoelectric cell. A natural place to mount such a sensor is again on the gripper, to detect that an object is ready to be picked up, but this is by no means the only possible mounting place. In fact, the pick-up point is usually located away from the waiting gripper. In such cases, the only physical connection between the sensor and the robot may be an electrical cable.

Photoelectric sensors also can be placed at various points around the robot work envelope to act as a safety device to stop the robot in the event of an unexpected intrusion into the workspace. In fact, at least one leading manufacturer of robots, Prab, has a policy of including with each new robot a chain-link barrier to enclose the work envelope. The barrier is equipped with an electrically interlocked gate; if the gate is opened, the robot stops.

Photoelectric sensors are the most popular devices for presence sensing, but some others deserve mention. Infrared devices have the advantage of not reading ambient light as a false signal. Radio frequency devices are also a possibility, but these systems must be tuned to trigger at the right time for the right object. Radio frequency devices are affected by the size and conductivity of the object to be sensed, and large variations in objects encountered can cause problems.

4.5.6.3 Robot Vision

Optical presence sensing is a far cry from vision as we humans know it. A great deal of experimentation is being done with pattern recognition, and several robots are capable of recognizing desired objects on a conveyor by their silhouettes. The desired object can be picked up from any orientation and then rotated to a desired orientation before being deposited at the assigned work station.

Robot vision is exciting and shows much promise for the future. But the mainstream of robot applications for at least the remainder of the twentieth century appears to be grounded in more conventional photoelectric presence-sensing devices.

4.5.6.4 Tactile Sensing

The human sense of touch is a marvelous phenomenon, and scientists have a real challenge in developing robot fingers that can actually "feel" the difference between various textures and surface shapes. But William D. Hillis and John Hollerback at MIT have developed a robot fingertip that can "feel" the difference between screws, washers, and cotter pins. Reportedly, Hillis and Hollerback used L'eggs Extra Sheer Pantyhose to provide a matrix of 256 (16 × 16) pressure-sensitive switches to act as data collectors for the computer. The pantyhose material was sandwiched between a piece of conductive silicone rubber and a flexible printed circuit board. The layers were normally separated by the pantyhose, but at pressure points contact was made through the pantyhose mesh, providing a matrix of sensed points. The pattern of the contacts was analyzed by the computer to distinguish objects held by the robot. The pantyhose experiment was a forerunner to modern tactile-sensing systems using electrical contact grids. Helpful for this purpose are piezoelectric materials, that is, materials that emit electric signals when deformed.

4.5.7 PERFORMANCE SPECIFICATIONS

The major features that distinguish highly capable, expensive robots from their cheaper alternatives have already been described. Some general performance specifications that can make a great deal of difference when selecting a robot for a given industrial application are payload, repeatability, and speed.

4.5.7.1 Payload

Payload capacity is an obviously important specification but is not as straightforward to determine as it might seem. Most robots can hold a much heavier weight than they can swing about at maximum speed. Also, the shape of the object held and its surface conditions affect the ability of the robot to handle it efficiently. Payload capacity at arm positions close to the base obviously tend to be higher than capacities at full arm extension.

Some robot manufacturers specify two payload capacities: normal and maximum, static and rates, or static and dynamic. The potential robot user should check carefully to ascertain exactly under what conditions the robot manufacturer is determining the rated payload, especially if only one figure is specified.

4.5.7.2 Repeatability

At this point, the reader should take care to be sure of the difference between "accuracy" and "repeatability." Accuracy is the ability to go to a prescribed point in space defined in terms of *x-y-z* or some other coordinate system. Accuracy is a concern for machine tools. Repeatability is the ability to return to the same spot again and again after that point has already been taught. For industrial robots, repeatability is a more important consideration than accuracy because the robot is usually taught with a teach pendant the first time. The important test is whether the robot can continue to perform the procedure as taught without slipping off target. Engineers are sometimes disappointed to discover that the positioning accuracy of an industrial robot is generally quite inferior to that of numerically controlled machine tools. Robot grippers and tools are generally extended on a much more flimsy framework than is possible for machine tools.

The tightest specification for repeatability is for small pneumatic robots, whose axis positions are stopped mechanically. Some of these robots are rated at ±0.001 in. or tighter. The big standard hydraulic robots, typical in the automotive industry, are rated at ±0.050 in. Spray painting and welding robots may have rated positional tolerances as high as ±0.125 in. or even higher.

There are some tricks that can be used to resolve problems of insufficient repeatability. For instance, a nose cone probe device can be inserted by the robot into a pilot hole just before the operating cycle to assure positioning within a tolerance specified by the customer. The probe insertion is done at the moment when positioning is most critical.

Another strategy, especially useful for open-loop, stepper-motor-controlled robots, is to go to a home position periodically at some point in the cycle to insert a probe and rezero the axis registers, thus reestablishing home position. The user has the option of programming the rezeroing operation to be executed every cycle or every few several cycles, depending on the accuracy desired and the deterioration experienced in each cycle.

In comparing the positioning capability of a robot versus that of the average human, humans can usually position more accurately than most large industrial robots if they try intently enough. But in a repetitive work assignment, a human cannot be relied upon to position intently every time. This explains the superior repeatability of robots over humans. Once programmed, the robot will quickly and consistently go to the same point every cycle. Although humans may be more *accurate*, they are not equipped to operate with as much *repeatability*.

4.5.7.3 Speed

Speed is another characteristic that may disappoint some potential robot users. Pick-and-place cycles used in machine loading and unloading are typically rated at two to three seconds for small pneumatic axis-limit robots. Some of these robots can achieve 1-sec cycles, and cam-operated mechanical manipulators can be even faster. A typical speed for a large, servo-controlled, hydraulic robot is in the range of 50 in./sec.

In many applications, a robot can be slower than its human competitor. Still, use of a robot may be justified on the basis of its higher productivity. This may sound like a contradiction, but it is just another case of the hare and tortoise. The human is the faster "hare," but the robot "tortoise" keeps on working right through breaks, rest period, lunch hours, and even the night.

4.5.8 ROBOT UTILIZATION AND JUSTIFICATION

By now, the reader should be in a position to enumerate several advantages to applying robots in the workplace. Rather than thinking of robots as being superior to human workers, it is useful to consider the human worker superior and deserving of more *meaningful* work than is usually assigned to robots. Robots are effective at boring, repetitive jobs that require little or no intelligence or judgment. Robots are also good for extremely fatiguing, hot jobs or for jobs that must be performed in toxic or otherwise dangerous environments.

Besides the thousands of robot applications in boring or dangerous jobs that are unfit for the superior human, there are other jobs for which the robot has features that are superior to those of humans. Robots have higher repeatability on intricate repetitious tasks. Figure 4.5.26 shows an intricate grinding operation. Each fine edge must be held at a precise angle to the work. It is difficult or impossible for a human to hold the same tool angle while indexing slightly to advance to the next fin, but the precise angles and motions can be programmed easily on the robot, in this case a Unimate PUMA.

Jobs that require handling heavy workpieces may be either humanly impossible or extremely fatiguing and thereby impossible for a human to sustain for a full workshift. For these jobs, robots may be more than desirable; they also may be essential. Many hydraulic models, and some electric ones, have superior lifting strength compared to humans.

Many robot applications present multiple advantages for robots. Die casting is a good example. The work is hot and dangerous, and the workpieces to be handled are often heavy. The same can be said of general foundry work. Welding operations combine the necessity for precision with the problems of exposure to hazards. Robots are excellent candidates for both spot-welding and arc-welding jobs. Like welding, spray painting and other spray finishing operations present hazards but require precision. It may not seem that spray painting is precision work, but left to the discretion of the human worker some units are bound to receive more thorough coverage than others, resulting in variation among the product units. An optimum spray coverage scheme can be exactly duplicated indefinitely by the robot.

4.5.8.1 Labor Resistance

Visions of assembly lines operated entirely by robots conjures up fears in the minds of workers who believe they may be replaced by robots on the job. Such assembly lines are a reality in some rare factories, and it is certainly possible for a robot to replace a given worker on a given job. But to see the robots as a cause of general unemployment is to look at the problem exactly backward. Properly applied, the robot can be used to increase productivity, making the firm more competitive, and thus *preserve* jobs. Applying this rationale to

Fig. 4.5.26. Robot deburrs cooling fins in a heat sink finishing operation, maintaining precise tool angle while indexing slightly for each fin, a difficult job for a human. Reprinted by permission of Unimation-Westinghouse, Inc., Danbury, CT.

the automobile industry, it would be foolhardy for automobile manufacturers in the United States to abandon robots and automation in order to preserve jobs when other automobile-producing countries, notably Japan, would continue producing automobiles as efficiently as possible using robots and automation. Market forces would soon eliminate the less efficient, nonautomated firms. In the words of General Electric's James A. Baker, "Henry Ford didn't invent the automobile; he *automated* its production and in doing so created millions of jobs and the backbone of the American economy."[7]

4.5.8.2 Economic Justification

Robots range widely in cost, from $3,000 to $5,000 on the low end to over $150,000 on the high end, using the early 1980s as a reference point. They are comparable in cost to many machine tools, except for the most expensive NC machine tools, which are more expensive.

Neither the use of machine tools nor robots is difficult to justify when their roles are vital to the feasibility of the process. For instance, some form of robot may be essential when assembling radioactive components in a product.

Economic justification of a robot that is not vital to the process requires more careful consideration of cost factors. If a robot is being considered as a potential replacement for a human operator, the tendency is to capitalize the annual cost of the human operator and compare it with the initial cost of the robot. This is a gross oversimplification of the problem. To begin with, a human operator generally works forty hours per week or less, overtime is costly, and even then the operator has limitations. No worker can work three shifts per day, seven days per week. A robot cannot work continuously either, but a limited amount of maintenance or repair can keep a robot operating three shifts per day, seven days per week. Even with eight hours per week robot downtime, the robot can achieve the equivalent of four human operators if both human and robot operate at the same pace.

The precision of the robot and its ability to repeat a task continually results in cost savings that may be greater than the savings in direct labor costs. The obvious benefit of robot precision is its effect on product quality and the accompanying reductions in scrap and rework costs. Less obvious is the savings in paint, glue, and other materials applied by the robot. The human operator can apply too little paint (a quality problem) or too much paint (both a quality and a material waste problem). The human tendency favors the application of too much paint, the more serious of the two errors.

More subtle economic factors are the support costs for human operators, such as vacation and sick leave, safety equipment, restrooms, cafeterias, parking lots, and even heating and lights. Reportedly, a plant exists in Japan in which essentially all of the operators are robots. The plant operates in darkness, except when a malfunction occurs, in which case a computer controller illuminates only the portion of the plant appropriate for the needs of the maintenance technician. When the problem is eliminated, the computer controller extinguishes the lights one by one as the maintenance worker makes his or her way out of the work area.

Despite their economic advantages, robots also have some economic drawbacks. As with any new equipment purchase, a robot represents a substantial initial cash outlay in return for future annual cost savings. Cash-flow considerations may make a robot investment difficult even if the expected rate of return is high. Obsolescence can also be a factor but not so much now that robots are reprogrammable. Flexible automation equipment including robots are not as prone to obsolescence as special-purpose fixed automation equipment.

Although economics is the common denominator of all robot decisions, some of the decision criteria may be so difficult to quantify that they should be considered separately. Emerson[8] recommends the consideration of three fundamental facets of robot project feasibility: operational, technical, and economic factors. Some managers insist that projects be justified on more than an economic foundation. Even when not required, a sound justification of the operational and technical feasibilities will help to sell the project to management. Managers and decision makers do not want a robot project to go sour by startup problems caused by operational or technical factors such as the following:

Operational Factors

Worker resistance to robots.
Production scheduling infeasibilities.
Interruption of work flow during installation and checkout.
Need for parallel manual production during robot installation and checkout.
Inability to use manual backup to recover from a robot breakdown.
Delivery slippages in robot equipment ordered for installation.
Operator training.
Maintenance scheduling.
Spare parts logistics.
Worker safety.

Technical Factors

Equipment incompatibilities.
Robot reliability inconsistent (either high or low) with reliability of equipment it serves.
Insufficient repeatability of robot motion.
Piece part orientation problems.
Piece part dimensional variations.
Piece parts too fragile.
Problems in interlocking robot with process equipment.
Work envelope conflicts.

Manufacturing automation engineers who can address each of these operational and technical factors when presenting robot projects to management will engender a feeling of confidence that they have done their "homework" and understand the consequences of a decision to purchase and install a robot.

4.5.9 SUMMARY

Real industrial-type robots bear little resemblance to the glamorized, fictional image born of the media, principally the movies. But the real thing has achieved some glamour of its own, mainly due to the technological advances of microchip circuits. The key feature of industrial robots is their programmability or, more precisely, their reprogrammability made possible by their microcomputer controllers.

Robots exercise their degrees of freedom in many different ways, depending on their configuration. The axes of motion may be pivotal, cylindrical, or various combinations of each. The maximum extension of the axes of motion result in the boundaries of the work envelope, an important system criterion. The principal power sources driving the axes of robot motion are hydraulic, pneumatic, and electric. Degree of control over the robot motion is a principal determinant of robot capability and likewise its cost.

Industrial robots usually do not come equipped with grippers or other tooling. The wide variety of tools that can be attached to the end of a robot arm is one of the keys to the robot's great versatility. Another key to this versatility is the ease with which robots are programmed using teach pendants, computer languages, or even mimicry of a skilled human operator.

The usefulness of an industrial robot in the workplace can be enhanced by sensors that provide inputs to the robot's controller. Optical and tactile sensors are the most practical at this time, but voice recognition is a possibility for future industrial-grade robots.

When buying a robot, the engineer needs to consider payload, work envelope, repeatability, accuracy, and speed. Robots can replace human workers on undesirable, fatiguing, or unsafe jobs while at the same time preserving other workers' jobs by making the entire operation more productive. In making economic comparisons, the analyst should consider the robot's ability to work three shifts per day if production volumes warrant, but robot downtime and maintenance cost also must be considered. Economic feasibility is not the sole criterion; technical and operational factors also must be considered.

REFERENCES

1. Baker, James A. "Factory Automation: Industry's Survival Kit." *High Technology* (February 1984).
2. Emerson, Bob. "Robotics: Automation Needs Prompt Different Approaches for Systems Analysis." *Industrial Engineering*, **17**(6), 28 (June 1985).
3. Graves, Gerald R. "Some Considerations for Group Technology Manufacturing in Production Planning." Doctoral dissertation, Oklahoma State University, Stillwater, Oklahoma, 1984.
4. Groover, Mikell P. *Automation, Production Systems, and Computer-Aided Manufacturing*. Englewood Cliffs, NJ: Prentice-Hall, 1980.
5. *Minimover-5 User Reference and Applications Manual*. Mountain View, CA: Microbot, 1980.
6. Russell, Marvin, Jr. "Odex I: The First Functionoid." *Robotics Age*, **5**(5) (September–October 1983).
7. Tanner, William R., "Industrial Robots Today," *Machine and Tool Bluebook* (March 1980).
8. Tanner, William R., ed. *Industrial Robots*, Vols. 1 and 2. Dearborn, MI: Society of Manufacturing Engineers, 1979.
9. *Unimate Puma*™ *Mark II Robot-500 Series*, Vol. II: *User's Guide to VAL*™. Danbury, CT: Unimation-Westinghouse, 1983.

CHAPTER 4.6

MATERIALS HANDLING EQUIPMENT

BRUCE F. BARGER

Executive Editor
Modern Materials Handling
Newton, Massachusetts

4.6.1 INTRODUCTION

Probably no person has done more to demonstrate the importance of materials handling equipment to industry than Henry Ford. When he first used a conveyor to carry automobiles to workstations, he immediately revolutionized mass production.

Yet, despite the obvious impact of materials handling equipment on both methods and productivity, some say that handling does nothing to add to the value of a product, only to its cost. The great majority of managers and engineers alike, however, appreciate the positive value materials handling systems provide in terms of time and space utility. And they are taking advantage of the benefits that a well-engineered operation can provide.

Some people have said that materials handling is "everywhere." And while the immediate reaction is to scoff at such a statement, first consider this widely-accepted definition: "Materials handling involves the movement, positioning, storage, control, and protection of materials throughout a manufacturing operation or warehouse." That certainly doesn't leave out many areas of industry.

4.6.2 PARTS, PACKAGES, AND UNIT LOADS

Think for a moment of what is meant by the "movement and storage of materials." In a factory, whenever a part or product is not being "worked on," it is either being moved to the next production process or it is waiting in storage to be moved to the next process. And in a warehouse, there is literally no operation that does not include materials movement or storage, control and protection.

Consider also the tremendous variety of materials that are processed through a plant or warehouse. Obviously, it is not practical to produce a piece of handling equipment that is specifically designed to handle every part or product in the universe. So families of equipment have been developed for the three groups of materials most common to a warehouse or factory: parts, packages (or containers), and unit loads.

Parts are the smallest materials handling denominator when you have to handle them individually. If necessary, they can be transported one at a time and stored separately. However, there are obvious economies to be gained by putting many parts in a package and handling the packages instead of the parts. One step further, it is more practical to stack several packages or containers on a pallet and to handle palletized loads than it is to handle individual packages.

Thus, with the exception of special item-handling devices and attachments for items of unusual size or shape, most handling equipment is designed to move and store packages and/or unit loads. Of course, because all industries produce many different packages and unit loads, a wide range of handling equipment has "evolved."

The editor expresses his appreciation to the staff of *Modern Materials Handling* for reviewing and proofreading this chapter following the death of Mr. Barger. All illustrations except Figs. 4.6.18 and 4.6.19 are reprinted with permission of Cahners Publishing Company, a Division of Reed Publishing USA.

4.6.3 EQUIPMENT CLASSIFICATIONS

There are several ways of classifying handling equipment. One way is by where the equipment works—on the floor or suspended overhead. Another classification is by whether the equipment is operator-controlled or automated. A third is by the way the equipment travels—over a fixed route or a flexible path.

A fixed-versus-flexible-path classification is particularly useful because it clearly points out the major similarities and differences between equipment types. It also helps a potential user to relate the equipment to the job that is to be accomplished; that is, equipment should be selected to solve a specific problem as well as to ensure damage-free, on-time materials deliveries where they are needed and at minimal cost.

The equipment that is described in this chapter ranges from low to very high levels of mechanization. Equipment for moving materials may include nonpowered floor trucks or powered industrial trucks (both flexible-path) or computer-controlled guided vehicles, conveyors, monorails, or cranes (all fixed path).

Storage equipment includes shelving, racks, and drawer cabinets. Materials are also stored in carousels, unit load and miniload automated storage and retrieval (AS/R) systems, and automatic item and case picking devices.

In all areas of materials handling, there is a strong trend toward greater mechanization, more computer control of equipment, more automation, and more systems integration. However, the purpose of this chapter is not simply to describe the newest technological developments but to highlight the full range of handling and storage equipment from the simplest to the most sophisticated. Descriptions of handling equipment for dry bulk materials and liquid materials have been omitted, but can be found in Ref. 1.

4.6.4 FLEXIBLE-PATH EQUIPMENT

Flexible-path equipment includes operator-controlled, powered, and non-powered floor equipment. *Flexible path* refers to the ability of the equipment to be moved along any route that the operator chooses.

Powered industrial trucks were first developed in the 1920s as the unit load concept was introduced. In the ensuing 50 years, a variety of models have been introduced in the marketplace.

Counterbalanced Fork Trucks

Counterbalanced fork trucks are the most versatile of all fork trucks. They are used both for storage to heights of 20 feet or more and for speedy transportation. Rider models predominate (Fig. 4.6.1), although walkie counterbalanced fork trucks are handy for short-distance hauls and low-level storage (Fig. 4.6.2).

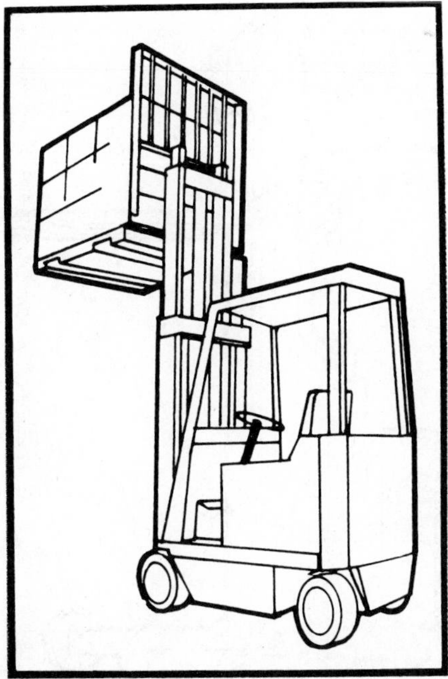

Fig. 4.6.1. Rider counterbalanced fork truck.

Fig. 4.6.2. Walkie counterbalanced fork truck.

Most rider models are four-wheel trucks. Smaller three-wheelers provide greater maneuverability and can operate in narrow storage aisles. They are used more frequently for loading and unloading highway trucks.

Engine-powered or electric trucks are available with cushion (solid) or pneumatic tires and are used both indoors and outdoors. Generally, larger models with load capacities over 10,000 lbs. are diesel or gasoline powered. Lower-capacity trucks are electric (battery powered) or engine powered (diesel, gasoline, or LP-gas).

Narrow-Aisle Trucks

High-lift, narrow-aisle trucks are generally electric. They can maneuver in tight spaces in production areas and are especially useful for storage applications. A narrow-aisle truck has wheeled outriggers that extend to

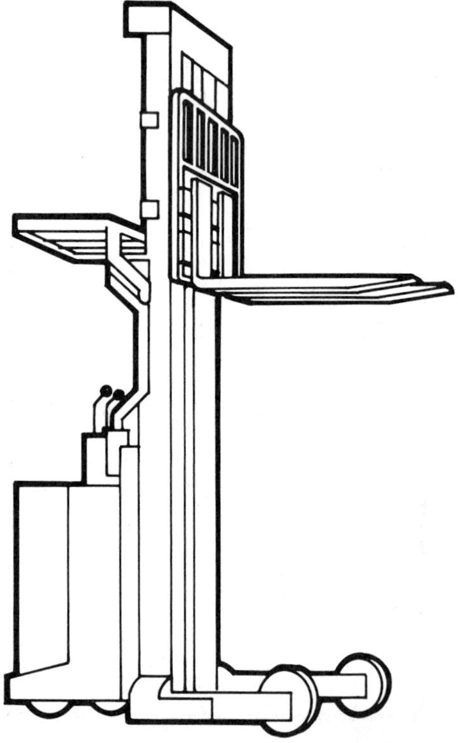

Fig. 4.6.3. Narrow-aisle straddle truck.

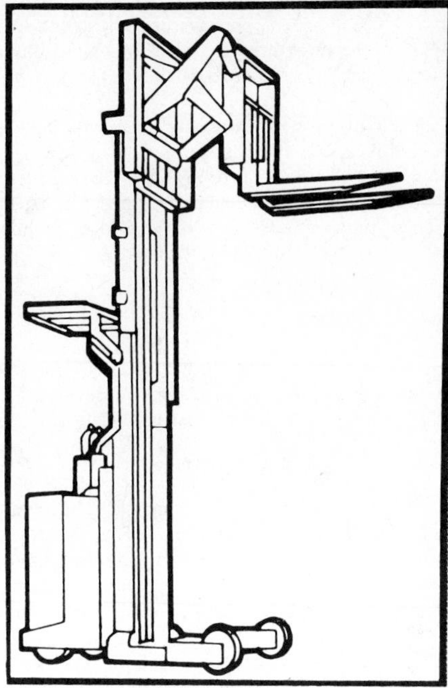

Fig. 4.6.4. Reach truck.

the front at floor level. These outriggers commonly straddle the forks and the loads on the forks. As a result, these trucks have short wheelbases along with ample stability for carrying loads (Fig. 4.6.3).

A short wheelbase truck has an advantage in storage. It can right-angle stack in aisles that are narrower than those required for a counterbalanced truck of the same capacity.

Reach trucks have the special ability to extend forks beyond the outriggers into rack openings (Fig. 4.6.4). They can then retract the loads above the outriggers to carry them. As a result, they can handle a variety of load sizes and also work in narrow storage aisles.

Orderpicking trucks are outrigger trucks with the operator's platform, his controls, and the lifting forks mounted in front of the mast (Fig. 4.6.5). These trucks are designed to work in narrow aisles without turning, thus permitting the operator to pick from shelving, drawers, or racks up to 30 feet high. Usually the forks carry a pallet or a wheeled rack to hold stock.

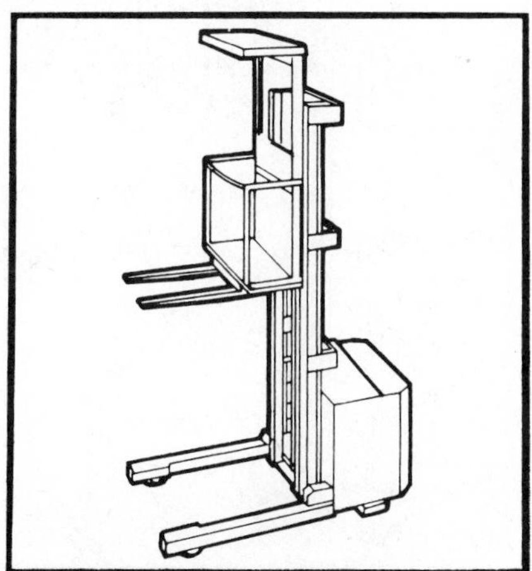

Fig. 4.6.5. Orderpicking truck.

Sideloading trucks do not have to turn in an aisle to stack or retrieve unit loads (Fig. 4.6.6). Some models have a mast that can be rotated 90° to one or both sides for loading/unloading operations. Other sideloading trucks have a fork carriage or "turret" that can be rotated to either side or a shuttle table that moves loads to both sides.

Some sideloading trucks are capable of raising the operator's cab with the load. This is an advantage when stacking or retrieving loads from high levels, as it gives the operator a "bird's eye" view of the load handling operation at all times. It also puts him in position to take inventory or do orderpicking.

Another sideloading truck model is called the sideloader. Sideloader trucks have long bodies and generally extend the forks from the side. These trucks can stack or retrieve loads without turning within an aisle, but only from one side of the aisle. They are used most often for handling long loads, such as lumber, structural shapes, and the like.

Lift Truck Attachments

A wide variety of attachments are available for use with industrial trucks to simplify load picking and positioning. Some of the more common ones include booms, carton clamps, drum handlers, dumpers, fork positioners, push-pull devices to handle slipsheets, rams, roll clamps, scoops, side shifters, and vacuum attachments.

Walkie Pallet Trucks

Walkie pallet trucks are small outrigger-style electric trucks capable of lifting a unit load a few inches off the floor and transporting it from one location to another (Fig. 4.6.7). They are called *walkies* because the operator walks behind the truck and controls its movement and speed through an operator handle.

Many pallet trucks include a small platform on which the operator can stand. These are called *walkie/rider trucks*. Some walkie trucks have load platforms, in place of forks, for moving skids. Walkie high-lift trucks are also available.

Walkies are generally used for short hauls, since their speed is limited to how fast the operator walks. Walkie/rider types may be used for longer hauls, at higher speeds, and for carrying two or more unit loads at once.

Manual Trucks

Nonpowered trucks—manually pushed or pulled—represent a lower level of mechanization and are recommended only for short hauls and auxiliary service.

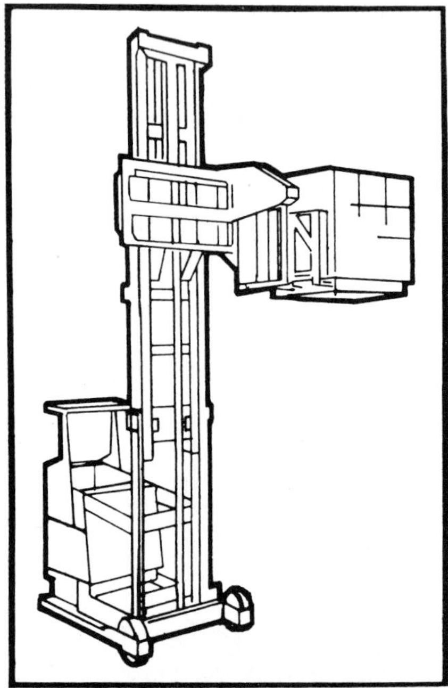

Fig. 4.6.6. Sideloading fork truck.

Fig. 4.6.7. Walkie pallet truck.

Two-wheel hand trucks (Fig. 4.6.8) include designs for specific shapes and types of loads such as bags, cases of beverages, appliances, and drums. They have convenient handles and are equipped to pick up end-supported loads.

Wheeled dollies are also used to support and position drums, appliances, furniture, and the like.

Platform trucks (Fig. 4.6.9) consist of a load-supporting frame or platform with four or more nonpowered wheels or casters. Box sides, shelves, or special superstructures can be added for specific loads.

Industrial trailers and towline carts are customized platform trucks with reinforced platforms and special features. Trailers can be pushed or pulled manually or pulled by a tractor, commonly called a *tug*. Several trailers coupled together form a train. With a tow pin arrangement, you have a basic towline cart. Attach towline carts to a moving in-floor chain and you have a *towline conveyor*.

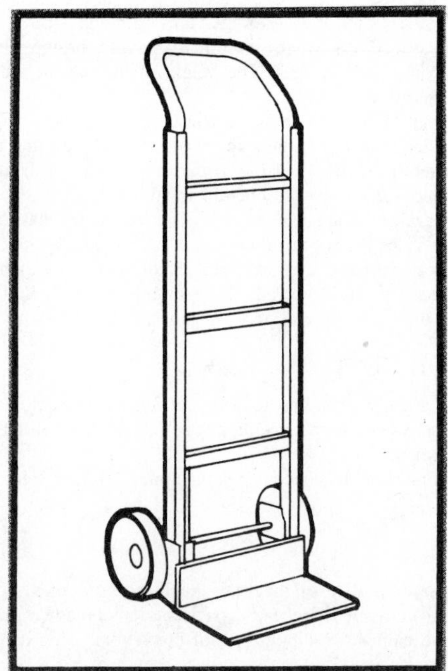

Fig. 4.6.8. Two-wheel hand truck.

Fig. 4.6.9. Platform truck.

Industrial trailers are described by their steering means: caster steer, fifth-wheel steer, and four-wheel steer. Generally, caster-type trailers are recommended for light loads and smooth surfaces. Fifth-wheel steer trucks are suitable for heavier loads and rougher surfaces. Four-wheel steer is best when accurate trailing is essential; however, these trailers are more difficult to maneuver by hand than the others.

Storage Yard Handling

Fork trucks see service in the yard as do other types of flexible-path equipment, including mobile cranes (Fig. 4.6.10), straddle carriers, straddle hoists, and front-end loaders.

Straddle carriers are large, pneumatic-tired vehicles that move over or "straddle" a load. Short retractable arms on the inside of the frame of the carrier extend beneath a load to lift it. The maximum load height is limited by the height of the straddle carrier.

Straddle hoists, sometimes called *straddle cranes*, are designed to lift and move heavy, usually long loads indoors or on unpaved surfaces such as those in a storage yard. Although similar to straddle carriers, the lifting is from the top by means of hoist trolleys, which ride on two parallel bridges. Each bridge is mounted to the top of a pair of tall "legs" with steerable wheels. The height of the legs determines the lifting height. Load capacities of 25 tons or more are not uncommon.

Front-end loaders have a pair of hinged arms which support a bucket (Fig. 4.6.11). The arms extend forward and up to enable a loader to scoop up bulk materials from the ground and load hoppers, conveyors, or highway trucks. Various attachments are available to permit a loader to handle pallet loads or large parts. Pneumatic-tired loaders are faster than those with crawler treads.

Mobile cranes are crawler- or wheel-mounted. They are powered and are classified as flexible-path equipment. A boom does the lifting and provides the reach. Several models of mobile cranes are available, including the industrial truck type, which has a hydraulically powered telescoping boom mounted on a small-wheel chassis. Usually the boom pivots can be raised and lowered. Other mobile cranes include locomotive cranes, crawler-mounted cranes, and cranes mounted on the bed of a highway truck.

4.6.5 FIXED-PATH EQUIPMENT

The one feature that most dramatically separates flexible and fixed-path equipment is the ability of fixed-path equipment to operate automatically without an operator. The path is fixed in the sense that it can be changed only by making physical adjustments.

Four types of fixed-path equipment dominate: conveyors, automatic guided vehicle systems (AGVS), monorails, and cranes.

Conveyors

Conveyors are one of the largest families of materials handling equipment. Usually they are designed by application to handle individual items or loose parts, packages or containers, or unitized loads. As a result, industry has adopted descriptive names for subfamilies of conveyors: parts handling, package handling, and unit load handling conveyors.

Transportation is the basic function of conveyors, but some types can do much more than move products from one point to another. They can accumulate, sort, and provide in-process storage.

Fig. 4.6.10. Mobile cranes. Left, wagon crane; Right, truck crane.

Some conveyors are mounted flush with the plant or warehouse floor to permit trucks and other wheeled equipment to cross over the path. Others are mounted a few feet off the floor for worker convenience. Still others are suspended overhead and, like any type of overhead equipment, leave the factory or warehouse floor unobstructed. Most conveyors are powered, but some are nonpowered and portable.

Conveyors that are mounted above the floor include many types of package and unit load conveyors. These conveyors are usually identified by the load-carrying surface involved: roller, belt, skate wheel, slat, carrier chain. Size and construction determines the load-carrying capacity.

Belt conveyors are typically sized for packages, not unit loads (Fig. 4.6.12). They are also used for sorting in combination with diverters to push or carry packages from the belt to accumulation chutes.

Horizontal and inclined belt conveyors are also offered. With the latter, molded "cleats" keep packages from sliding back. Models with magnetic belts are useful for conveying ferromagnetic parts and scrap. Fixed-position and portable belt conveyors are also available.

Skate wheel conveyors (Fig. 4.6.13) are nonpowered, designed for light loads only, available in fixed-position or portable models, and for horizontal movement only.

Roller conveyors can carry almost any flat-bottom load—case, carton, tote box, or pallet—weighing from less than one pound up to thousands of pounds (Fig. 4.6.14). Transfer sections permit change of direction during conveying. Diverters allow automatic sorting.

Two or more parallel strands of *carrier chain* can move large packages or unit loads. Usually chain conveyors are more cost effective for long hauls than roller conveyors.

Tilting slat and *tilting tray conveyors* provide high-speed sorting. They are especially practical when the items to be sorted come in a wide range of sizes and weights.

Sliding slat conveyors sort by moving product laterally from a delivery lane to several lanes. They can also merge the output of several lanes into one.

Fig. 4.6.11. Front-end loader.

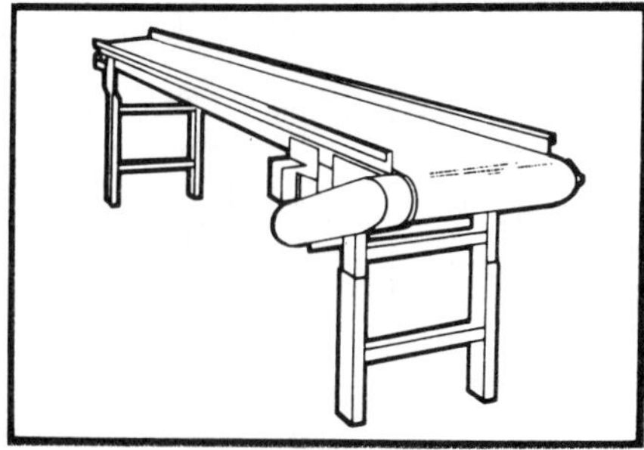

Fig. 4.6.12. Belt conveyor.

Selecting the right type of equipment for sorting requires a careful analysis of different types of conveyors as well as diverting mechanisms and automatic identification equipment.

In the most elementary system, people are employed to identify and push packages off of a roller or belt or skate wheel conveyor into accumulation chutes. A more advanced approach is to replace the people with mechanical diverters to push off the packages. Of course, controls are needed for cycling of the diverters.

As the need for higher sorting speeds increases, the choice often is limited to specialized sorting machines. These may include conveyors with pop-up or *roller diverters* or with "fingers" that sweep the package to an accumulation chute. Tilting tray and slat sorters, of course, are designed specifically for high-speed sorting.

Automatic identification has brought labor savings to sorting operations by eliminating the need for an operator to visually identify each package and identify it for the control unit for the diverters. Scanners or other code readers mounted alongside or above the path of flow can identify machine-readable codes printed on the package or on labels on the package to fully automate the sorting operation (Figs. 4.6.15 and 4.6.16).

Power and free and *trolley conveyors* are two common forms of overhead conveyor. The power and free and trolley conveyors are similar in that both have wheeled trolleys suspended from an overhead rail and they are powered by a moving chain.

All of the trolleys in a trolley conveyor are connected to the chain, so that when it advances, all of the trolleys move forward together (Fig. 4.6.17). Obviously if the conveyor is used for assembly or process operations, all of the operations must be synchronized or special "drop" and accumulation arrangements must be provided.

Fig. 4.6.13. Skate wheel conveyor.

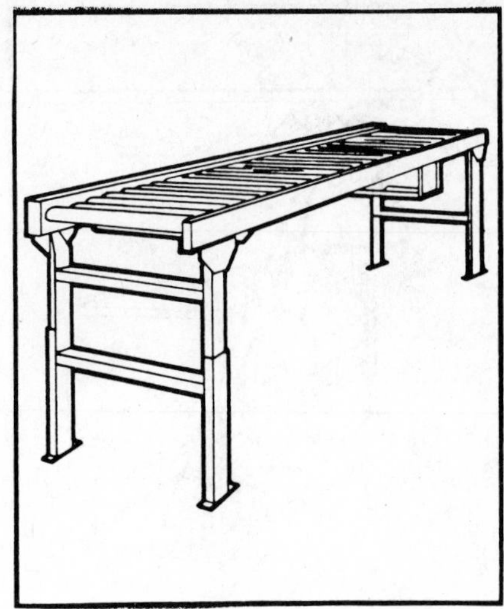

Fig. 4.6.14. Roller conveyor.

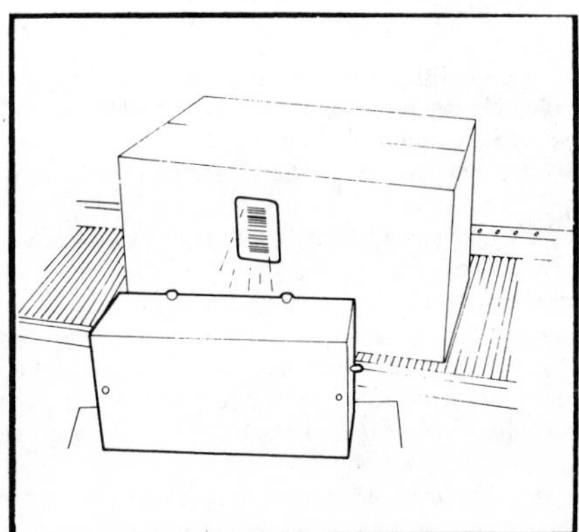

Fig. 4.6.15. Moving-beam scanner.

Fig. 4.6.16. Hand-held scanner.

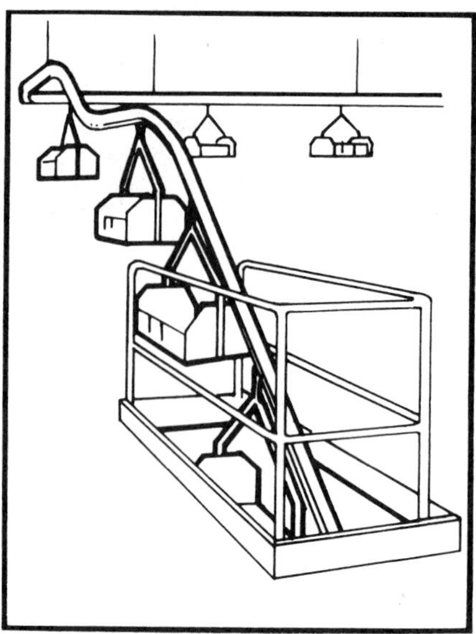

Fig. 4.6.17. Trolley conveyor.

A power and free conveyor (Fig. 4.6.18) is more versatile than a trolley conveyor because the trolleys can be transferred individually from the delivery line to one or more storage lines for accumulation. This capability is made possible by means of a second parallel track from which trolleys are suspended and selectively engaged by pushing elements called "dogs" on the chain-powered track. Trolleys can be switched onto spurs or accumulated as needed.

Carriers that are attached to the trolleys can be fitted with shelves or other superstructures for transporting containers, individual parts, or assemblies.

The conveyor track can be constructed horizontally and with curves to change direction and elevation.

An *inverted power and free conveyor* is a variation of the overhead power and free, but the track is floor-mounted and the trolleys ride on top of the track. Obviously the track provides a barrier that prevents crossover traffic, but the conveyor design permits work to be performed on parts or assemblies carried directly on top of the trolleys. This feature is desirable when it is important that no dirt or grease drip from the chain onto the product being carried.

Car-on-track conveyors (Fig. 4.6.19) resemble inverted power and free conveyors in one respect: individual cars ride on rails and are powered from the bottom. Power is supplied by means of a continuously rotating tube parallel to and located between the rails. By changing the position of a drive wheel on the bottom of each car, relative to the tube, the speed of the car can be increased or decreased or the car can be stopped and positioned very precisely.

A *towline conveyor* (Fig. 4.6.20) is functionally similar to an inverted power and free. However, the track is installed beneath the surface of the floor and does not impede crossover traffic. Loads are carried on wheeled trailers, which are advanced by means of probes that slip into the chain. Switches enable the towcarts to move onto spurs automatically.

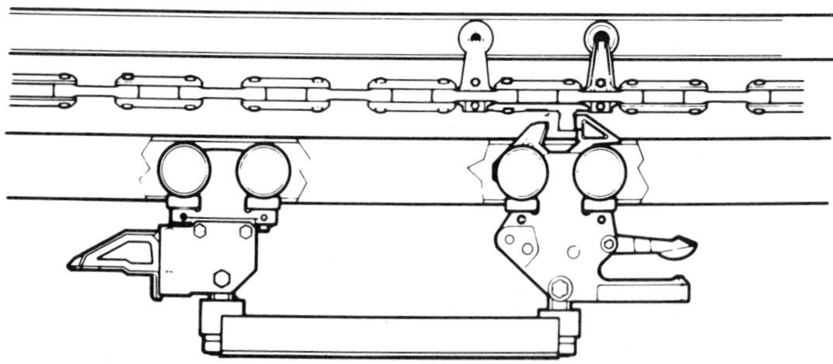

Fig. 4.6.18. Power and free conveyor. Reprinted with permission from Kulwiec, *Material Handling Handbook*.[1]

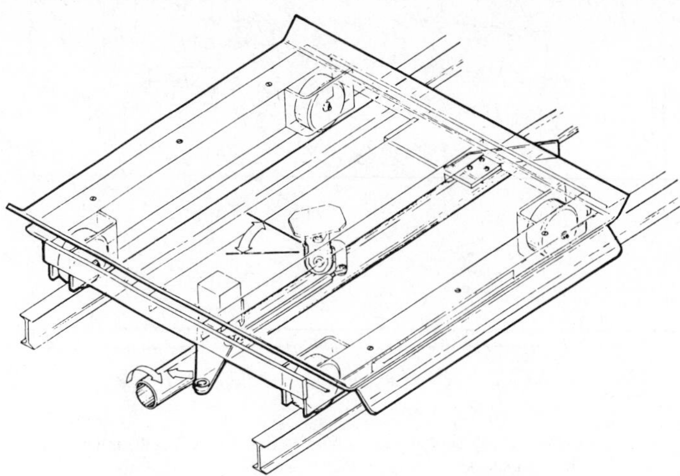

Fig. 4.6.19. Car-on-track conveyor. Reprinted with permission from Kulwiec, *Material Handling Handbook.*[1]

Pneumatic conveyors are practical for transporting documents and small parts through tubes by air pressure. The parts or documents are moved in small cylindrical carriers. This type of conveyor is especially cost effective over long distances. The tubing can be formed to carry the cylinders horizontally, vertically, and around curves.

Automatic Guided Vehicles

Within the past few years, a new category of transportation and storage equipment has been developed: *automatic guided vehicle systems* (AGVS) (Fig. 4.6.21). These are electric vehicles with on-board sensors that enable the vehicle to automatically track along an electrified guide wire located just beneath the surface of the floor in a plant or warehouse. This guidepath can also be painted or taped to the floor.

Sensors on each vehicle detect the field generated by the energized wire or optically detect the path on the floor. Signals are sent to the vehicle's steering mechanism to control its movement. A vehicle can be instructed to start and stop automatically and to follow a designated route that may involve movement across a number of interconnecting loops.

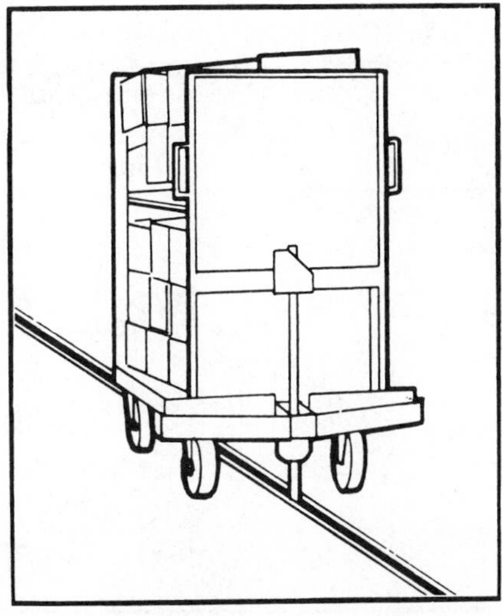

Fig. 4.6.20. Towline conveyor.

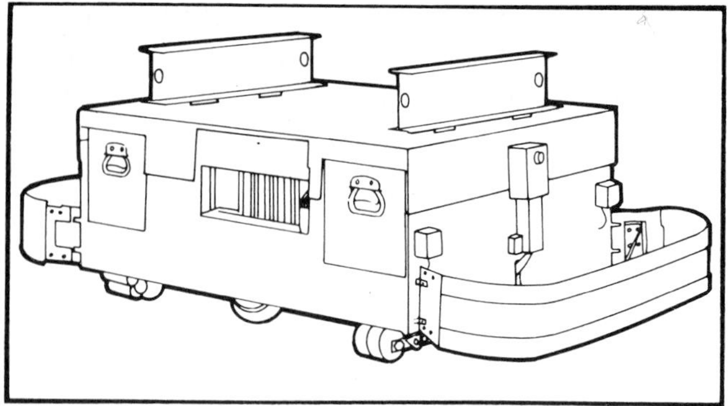

Fig. 4.6.21. Automatic guided vehicle, unit load type.

The concept of steering a vehicle automatically dates back to the 1950s with the introduction of the *driverless tractor train* (Fig. 4.6.22). It followed an in-floor guide wire and towed a number of trailers around a fixed route in a warehouse.

Driverless tractors worked well, but user acceptance came slowly, in part because the technology was quite advanced for the 1950s and in part because it did not eliminate the need for people; fork trucks or other operator-controlled equipment were required to load and unload the trailers. For many prospective users, the "driverless" feature offered little benefit.

Even so, it was obvious that wire guidance could provide safe, automatic steering. It was not long before many orderpicking and other storage trucks were fitted with guidance controls to provide automatic steering through narrow aisles. Today, this remains a recommended application for guidance systems.

In the 1970s, with the development of microprocessors and other programmable controls, the AGV emerged in its current form. Today, there is a family of guided vehicles that includes low-lift pallet trucks, unit load guided vehicles, and high-lift guided vehicles, most if not all with an automatic load/unload capability.

Guided pallet trucks have the ability to pick up a pallet load resting on the floor and move it to another point automatically. They operate with relatively little auxiliary equipment, although the load must be positioned within or next to the vehicle's path.

Unit load guided vehicles have a tilting bed, a roller-conveyor bed, or a hydraulic load platform. They usually interface with a conveyor or a stationary platform that is loaded by a fork truck or other type of equipment.

High-lift guided vehicles (Fig. 4.6.23) look like industrial trucks, but, because of their ability to follow a path automatically, they are truly automated guided vehicles. Side-loading and "turret" models are the most common, some with ability to pick pallet loads from the floor and block stack or store them in racks to heights to 40 feet.

Most applications of AGVs are for load transportation. However, as the concept of flexible assembly becomes popular, automatic guided vehicles will be used more often for assembly operations. There are already installations in which unit load vehicles have been modified to support a specific part of assembly. In Europe, assembly vehicles are being used as the "assembly line" for a number of products, ranging from car engines to entire automobiles.

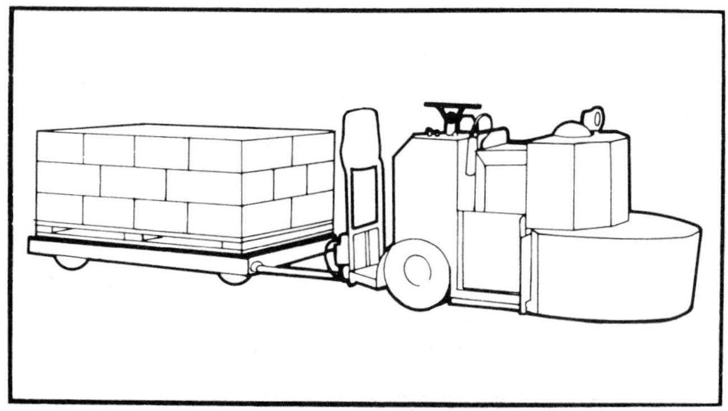

Fig. 4.6.22. Driverless tractor train.

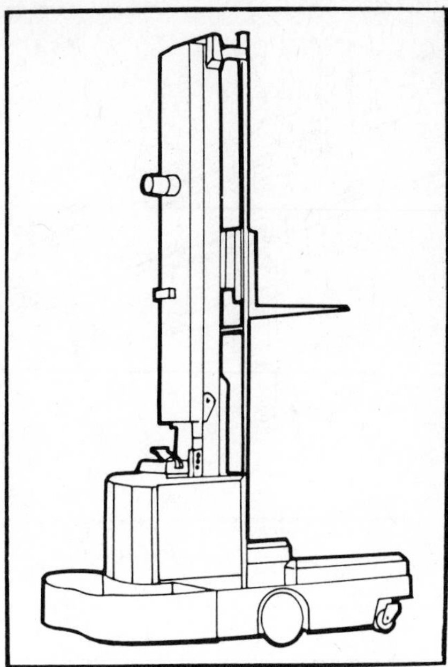

Fig. 4.6.23. High-lift guided vehicle.

Automatic movement of a guided vehicle from the guidepath is a new development now being used on a limited basis only. This technology shows considerable merit. Each vehicle is fitted with a "measuring" wheel or other type of device that calculates the distance the vehicle moves away from the guidepath. Signals are sent to an onboard microprocessor, which keeps track of the distance moved and controls the direction of vehicle's steering wheel. When the vehicle's task is completed, the vehicle returns to the guidepath along the same route.

Hoists, Monorails, and Cranes

Hoists are a basic type of overhead lifting equipment and can be suspended from a rail, track, crane bridge or beam, or wheeled trolley. A hoist consists of a hook, a lifting medium, which is usually a wire rope or chain, and a container for storing the rope or chain. Air motor, electric, and hand-ratched hoists are available in small

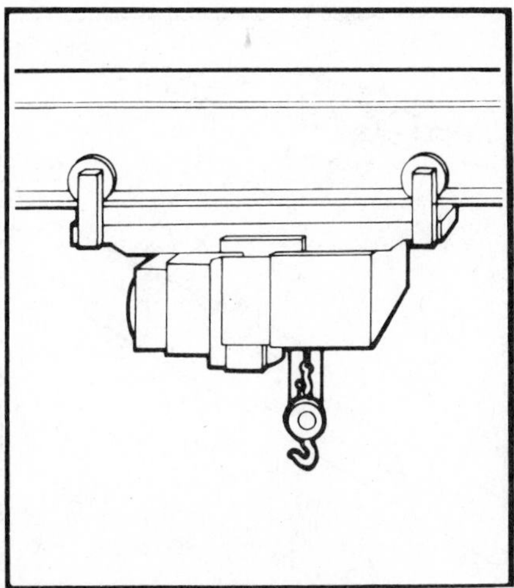

Fig. 4.6.24. Manual monorail.

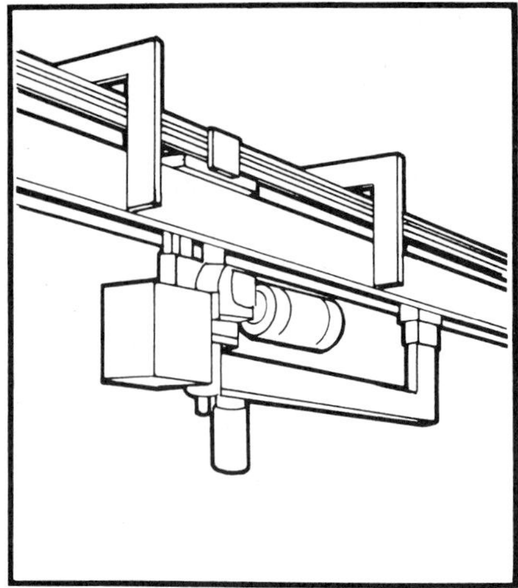

Fig. 4.6.25. Powered monorail.

models. Large-capacity hoists, which can carry 10 tons and more, are electric.

Monorails look something like a conveyor, but each wheeled trolley moves individually along the overhead track instead of being advanced by a chain. Typically, the track is formed with a lower flange, and the wheels of the trolley roll along the top surface of the flange.

Manual monorails (Fig. 4.6.24) are sometimes referred to as rail conveyors when the track is formed of lightweight tubing. The wheeled trolleys ride on the track and are hand-pushed. Powered sections are useful when it is necessary to change direction or automatically switch the trolleys to spurs.

Powered monorail systems (Fig. 4.6.25) include motor-driven trolleys that receive electricity from conductor bars attached to the track. The drive motors are geared to trolley wheels or to a drive wheel of rubber or steel. Drop sections permit a trolley to pick a load from floor-mounted equipment and to transfer a load to another track.

Computer control is the newest development in monorails, providing a means for remotely controlling each trolley.

Cranes are among the oldest types of fixed-path equipment and are the backbone of overhead handling in steel mills, metal fabricating shops, and wherever heavy loads must be moved.

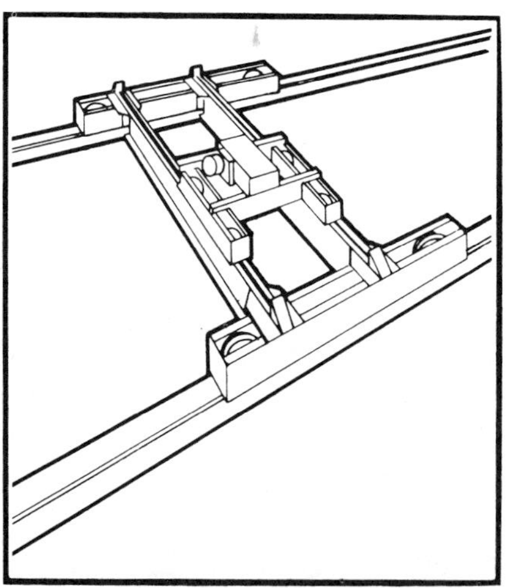

Fig. 4.6.26. Top-running bridge crane.

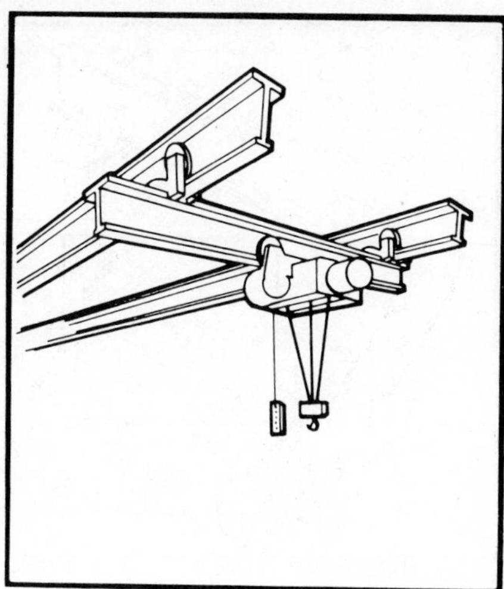

Fig. 4.6.27. Underhung bridge crane.

Not all cranes are suspended overhead, however. Some are mounted on a wall or on a floor pedestal and provide a more limited range of movement; others are mobile or portable.

Overhead cranes consist of a horizontal structural member, or bridge, mounted on wheeled end trucks. The trucks enable the bridge to travel on parallel overhead runway rails. The rails, in turn, are supported by the columns of the building or by uprights.

Loads are lifted and lowered by one or more hoists on the bridge. An operator's cab may be added, or the traveling crane can be operated from the floor through a pendant control or by radio remote control.

A *top-running crane* (Fig. 4.6.26) is one in which the bridge runs on the upper flanges of the runway rails. An *underhung crane* (Fig. 4.6.27) runs on the lower flanges of the runway rails.

Overhead cranes provide one outstanding benefit. They can service the entire area beneath the runway and are not affected by the location of travel aisles or the placement of machines on the floor. Depending on the layout of the plant, adjacent runways can be constructed, each with one or more underhung cranes. This makes it practical to pass loads from one bay of a plant to another.

A *gantry crane* (Fig. 4.6.28) has an overhead bridge supported at both ends by a structure mounted on wheels. The wheels ride on ground-level rails.

Fig. 4.6.28. Gantry crane.

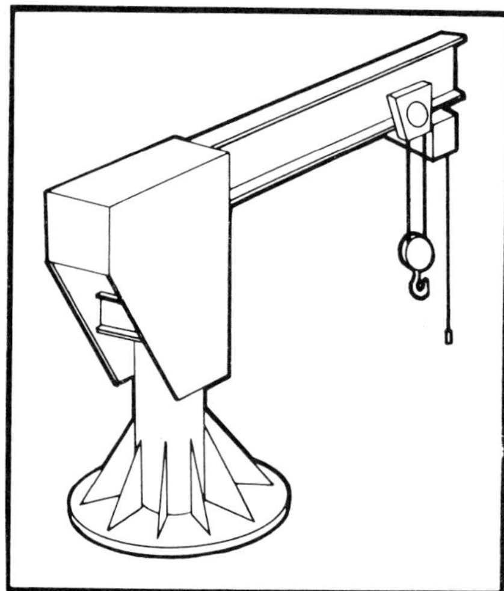

Fig. 4.6.29. Revolving jib crane.

A *semi-gantry crane* has this arrangement at one end of the bridge, with the other end supported on an overhead runway. This is similar to a conventional traveling crane.

Stacker cranes are overhead traveling cranes equipped with a special hoisting mechanism—a telescoping mast that extends toward the floor. The mast can rotate up to 360° and includes forks or other attachments to permit the handling of unit loads and large stock, such as coils of metal and rolls of newsprint.

Large stacker cranes include an operator cab. Small, non-powered models are manually guided and controlled by the operator on the floor. Some have a non-telescoping mast with a lifting carriage for the forks.

Other cranes are mounted permanently to a pedestal or to a rail along one wall. With a pedestal or jib crane (Fig. 4.6.29), the working area is limited to the radius of the pivoting cantilever beam on which a hoist trolley rides. A wall-mounted crane can travel the length of a rail and work as far as the beam and hoist can reach.

Portable cranes (Fig. 4.6.30) are mounted on casters for handling loads in shop areas. Usually they are manually pushed into position.

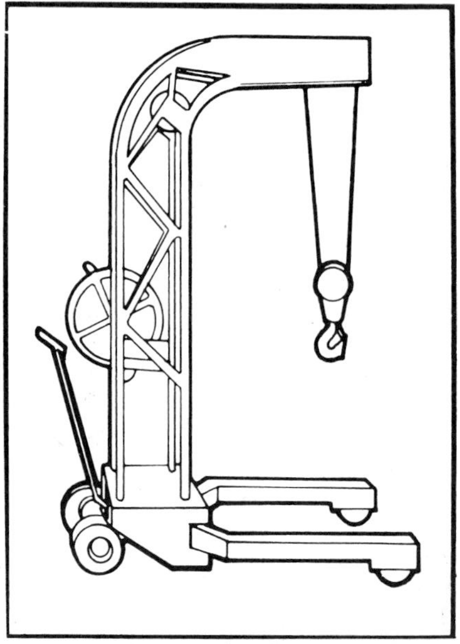

Fig. 4.6.30. Portable cantilever crane with straddle base.

4.6.6 STORAGE EQUIPMENT

The scope of storage "equipment" is as broad as transportation equipment, ranging from portable equipment, such as tote boxes and pallets, to shelving and racks serviced by lift trucks, to huge automated storage systems that tower up to 100 feet or more above the warehouse floor.

Earlier in the chapter we made reference to the economies of handling unit loads. Sometimes that is not practical, and you have to handle parts individually or in containers.

4.6.6.1 Small Containers

Part bins are the basic containers for bulk storage. You can specify metal or paperboard bins in a variety of sizes. Plastic bins are the most popular, though, as they are available in virtually any size, in colors, and formulated from a number of synthetic materials. Generally they are used to store bulk quantities of small parts and hardware.

Tote boxes are one step up in size. Partitions and dividers make it possible to separate parts to prevent damage or to simplify picking.

Many "packages" are corrugated shipping cases or cartons, or overpacks. In manufacturing operations, it is common practice to remove parts from cases at receiving and place them in tote boxes or other shop boxes. Distribution centers, of course, store the cases if they ship in full case quantities. They may also remove the top of the shipping cases and place them in storage when split-case picking is required.

4.6.6.2 Unit Load Containers

Unit loads are designed for mechanized handling, generally by fork trucks or conveyors or AGVS.

The *pallet* (Fig. 4.6.31) is the most common type of unit load "container". Millions of wooden pallets are supplied to industry each year. A far lesser number of metal and plastic pallets are used.

Slipsheets (Fig. 4.6.32) are also used, in some industries, to unitize loads for shipment. These flat, stiff sheets are handled by trucks equipped with special push-pull attachments. Overall, slipsheet usage is by far less common than pallet usage.

When any product or material is put on a pallet, the result is a pallet load. Depending on the type of material being handled, it may be possible to stack pallet loads three or four high without using racks.

Pallet stacking frames (Fig. 4.6.33) are sometimes recommended when the material on the pallet is either "nonstackable" because of its shape or because it would be damaged or crushed by stacking. These are removable metal frames that mechanically clamp to the pallet.

Welded wire containers (Fig. 4.6.34) and *metal shop containers* (Fig. 4.6.35) are other examples of unit loads containers. Welded wire containers may have collapsible sides, which enable the unit to take up less space when not in use. Metal shop containers are often used to hold castings or parts that are hot or oily. They are not collapsible.

4.6.6.3 Racks and Shelving

Tote boxes or cases or unit loads are usually stored in *racks* or *shelving* to get them off the floor and achieve high cube utilization. Good practice calls for storing materials to the ceiling, when practical, to minimize the amount of floor space required for storage.

Shelving

Shelving (Fig. 4.6.36) is inexpensive and comes with adjustable shelves. Walls of shelving 20 feet high or more can be constructed.

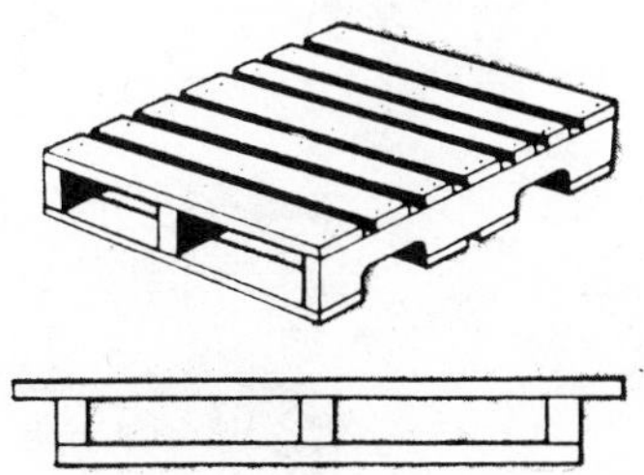

Fig. 4.6.31. Pallets.

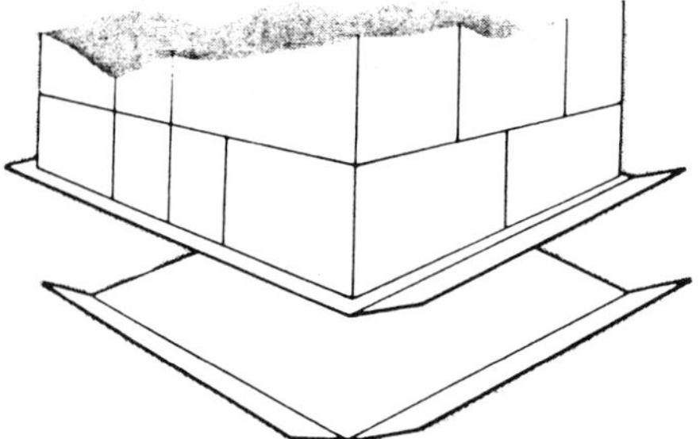

Fig. 4.6.32. Slipsheets.

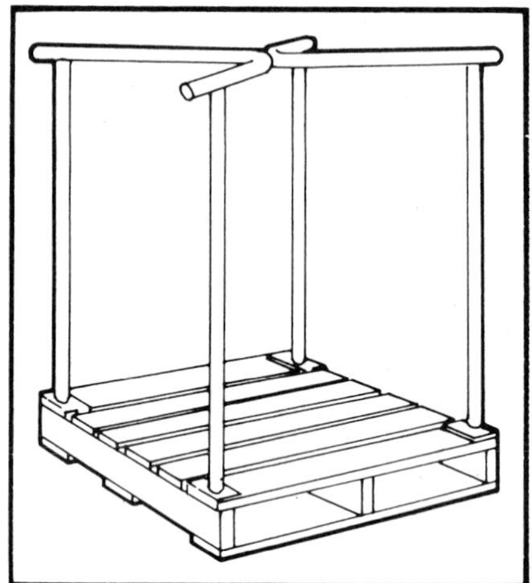

Fig. 4.6.33. Pallet stacking frame

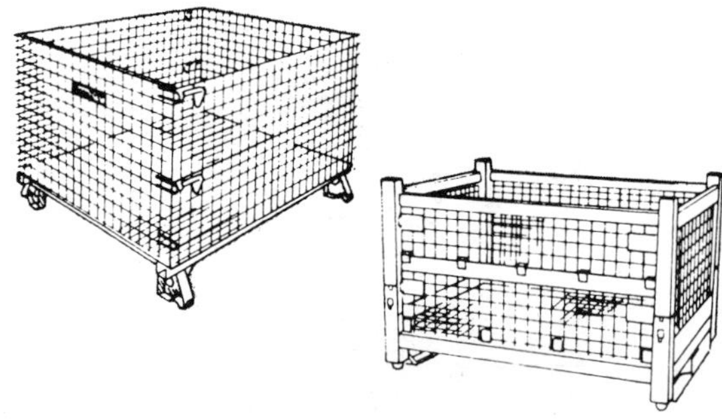

Fig. 4.6.34. Welded wire containers.

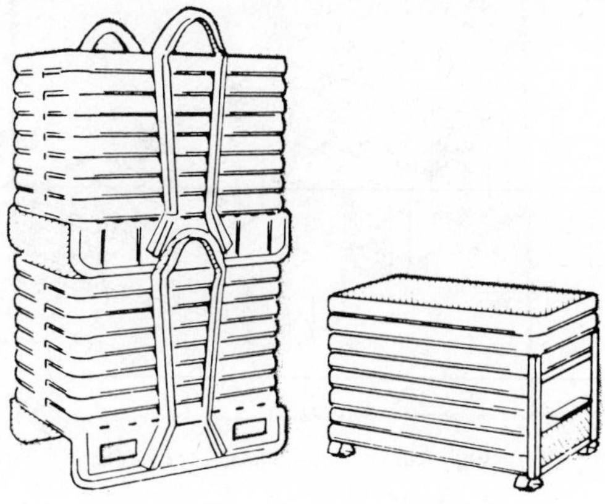

Fig. 4.6.35. Steel containers.

Drawer Storage

Modular drawer cabinets (Fig. 4.6.37) hold individual parts or small boxes and bins or parts. Most drawer cabinets are no higher than 5 feet, so that a stock picker can look into the open drawer to select the right part or tool. Cabinets can be stacked to 20 feet or higher to achieve very high storage density.

Mobile Shelving

Mobile shelving consists of shelving units mounted on wheeled bases. The wheels ride on floor-mounted rails. High-density storage is achieved because the installation does not include a permanent aisle facing each shelving unit. Instead, "aisles" are created as the shelving is moved to the side.

Flow-through Storage

Flow-through or *gravity storage* also requires fewer aisles. This is because each storage opening is a long lane that holds several boxes or cases of the same product.

When the storage structure consists of high walls of gravity lanes, drawer cabinets, or shelving, picking can be performed from orderpicking trucks or man-ride machines that carry the operator through the aisles and lift and lower him to the proper storage opening.

Fig. 4.6.36. Open shelving.

Fig. 4.6.37. Modular drawer cabinets.

Fig. 4.6.38. Computer terminal.

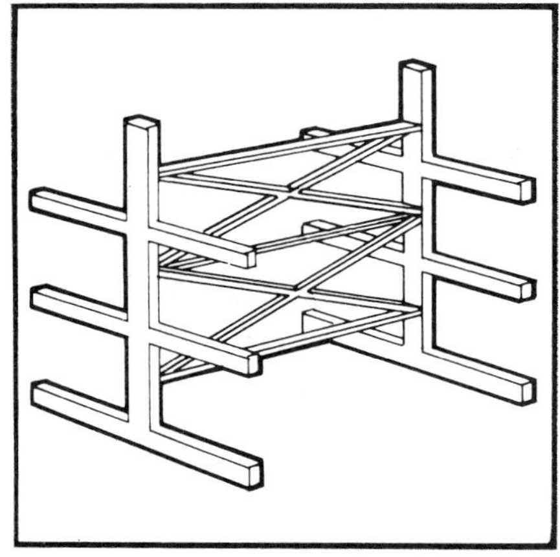

Fig. 4.6.39. Cantilever rack.

Mezzanine Storage

Another way to make good use of overhead space is by using one or more *mezzanines*. Picking becomes a "walk-and-pick" operation.

Orderpicking

Manual orderpicking methods from racks, shelving, drawers, or gravity lane traditionally rely on a printed document to tell the worker what items and what quantities to select. Reading a document involves several different eye movements between the paper and the shelf and product. Since picking errors are likely to occur, ways have been devised to eliminate the need for these documents.

Picking labels, a form of pick list, are affixed directly on the picked case. The operator goes to the indicated storage location and matches a label to each package. Counting is eliminated because there is only one label for each package. Unused labels indicate a product outage.

Displaying picking information on a CRT terminal (Fig. 4.6.38) also eliminates paperwork. These terminals can be mounted on a truck or other picking vehicle or mounted at a fixed pick station. Often the terminals are wired directly to a computer, or, in the case of moving equipment, they are linked by radio frequency signals.

Computer-aided picking systems include a computer-actuated display light mounted at the end of each storage lane or bin. A computer activates the lights to indicate that a pick is required and displays the quantity to select.

Spiral Towers

In some manufacturing operations, *spiral towers* for storage are used to form buffers between production machines. They require little floor space and make good use of the cube. They are also relatively inexpensive, as the stored parts move from the tower by gravity, although a powered lift is needed to elevate the parts to the top of the tower.

When cylindrical parts (parts of rotation) are stored, the load surface in a tower may consist of pairs of rails. For prismatic parts that are fixtured to plastic trays, the load-carrying surface is formed of gravity rollers.

Rack Storage

Racks are designed for unit loads or large parts. For long shapes, *cantilever racks* with horizontal arms extending from columns are popular (Fig. 4.6.39). The arms can be covered with decking to permit the storage of loads of different sizes. It is an approach used by some furniture warehouses, among others.

Pallet racks (Fig. 4.6.40) with post and beam construction represent "conventional" unit load storage. Many pallet racks are one load deep, but if suitable handling equipment is available and the proper product mix occurs, space savings are possible by adopting two-deep storage or by the use of gravity flow lanes.

Drive in and *drive-through racks* are variations of conventional pallet racks. The loads are supported by pairs of narrow "arms" attached to the sides of upright posts. The horizontal distance between arms is sufficiently wide to permit a fork truck to enter a lane to stack and retrieve loads.

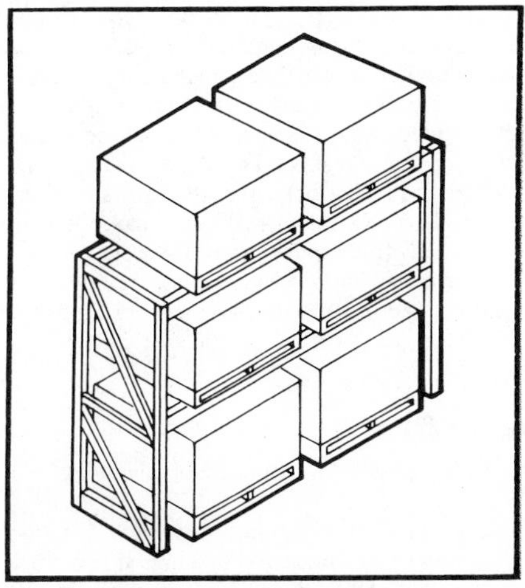

Fig. 4.6.40. Pallet rack.

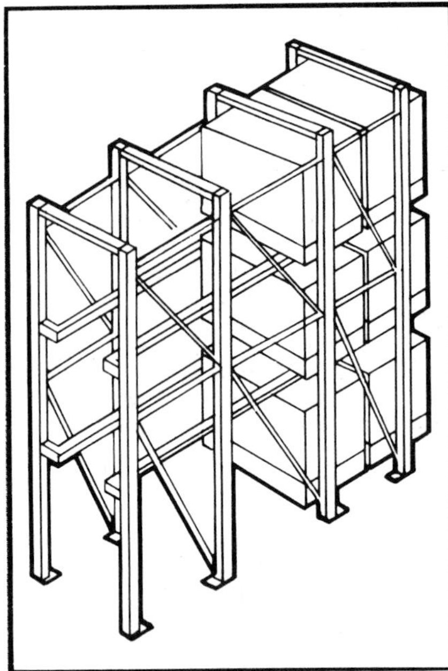

Fig. 4.6.41. Drive-in rack.

Drive-in racks (Fig. 4.6.41) provide lanes into which trucks can be driven to retrieve loads. Because the lanes are closed at the opposite end, truck must back out. With drive-through construction, access can be gained from either end of an open lane.

Gravity racks are recommended if the warehouse has relatively few different products but several loads of each. The storage system consists of long lanes, with input from one end and take-away from the other. Various types of brakes and air-assisted devices are available to control the forward movement of a pallet load through a lane and to prevent the build-up of excessive line pressure that might damage the products being stored.

Mobile racks operate much like mobile shelving. However, because heavier loads are stored on mobile racks, the wheel bases are powered by electric motors. Operation is by push button or by radio frequency signals from a transmitter on a fork truck. Built-in safety devices ensure that a rack will not move inadvertently.

Rack Access

With most storage systems using shelving, drawers, or racks, the put-away function is performed either by an operator walking or working from an orderpicking truck, a man-ride machine or pick car, or by a fork truck or stacker crane. The operator is transported to the stored material to retrieve a load or to pick.

Man-ride machines (Fig. 4.6.42) and pick cars carry an operator into a narrow aisle to allow him to pick or stock materials. Both are rail-mounted and aisle-captive. A man-ride machine has an operator's cab that rides up a rigid mast to permit picking from heights up to 40 feet and sometimes more.

Small models are designed for parts picking only. Picked items are usually accumulated on a multishelf wheeled rack that is carried next to the operator's cab. Larger man-ride machines have a shuttle table that permits storage and retrieval of unit loads from both sides of the aisle.

Pick cars differ from man-ride machines in two ways. First, they have an elevating operator's platform, not a cab that rides up a mast, and thus are restricted to low-level picking. Second, they are equipped with a section of powered conveyor that interfaces with an in-aisle belt conveyor to carry picked goods—usually cases—to an order consolidating area.

4.6.6.4 Automated Storage Systems

Automated storage systems either bring materials to the operator who works at a fixed station or they eliminate the need for an operator.

Unit load AS/R systems are the grandfather of automated storage. They use AS/R machines (Fig. 4.6.43) to store and retrieve unit loads from racks that can be 100 feet high if necessary. Each machine has one or two rigid masts that support a shuttle table. The shuttle moves loads to either side of the narrow aisle through which the machine travels. Control is by computer.

Fig. 4.6.42. Man-ride orderpicking machine.

The first unit load AS/R systems were stand-alone designs. A fork truck brought a load to a pickup/deposit station at the end of the aisle so that it could be transferred to the storage machine. Now, AS/R systems are more likely to be integrated with a conveyor or AGVS or some other transportation equipment. In this way, loads can be stored and moved throughout the plant or warehouse automatically.

These systems are more flexible than they at first might appear. Although the storage machine normally operates in a single aisle, transfer cars are available to move them from one aisle to another. This approach is

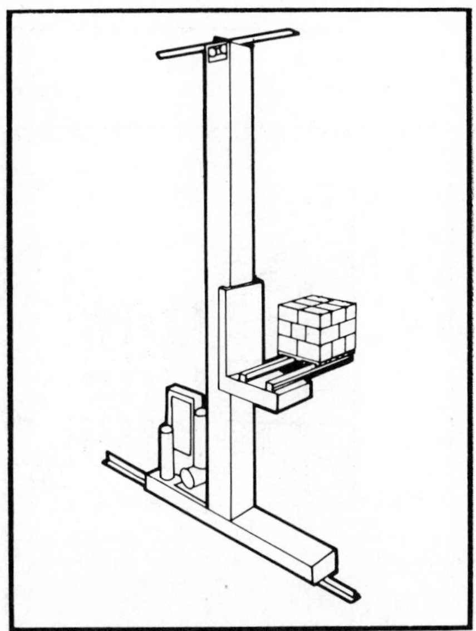

Fig. 4.6.43. Unit load AS/R machine.

Fig. 4.6.44. Mini-load AS/R system.

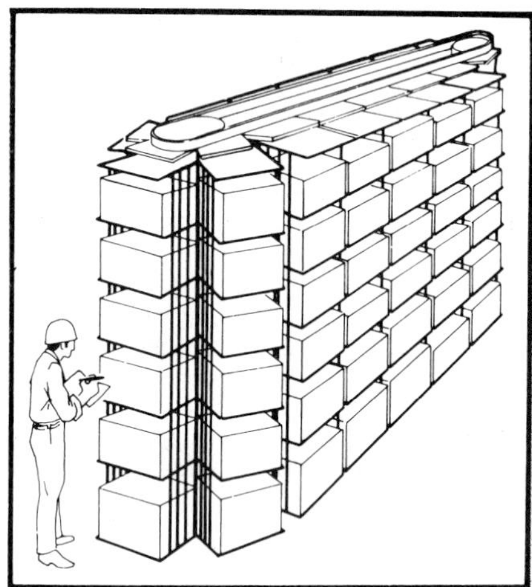

Fig. 4.6.45. Horizontal carousel.

Fig. 4.6.46. Vertical carousel.

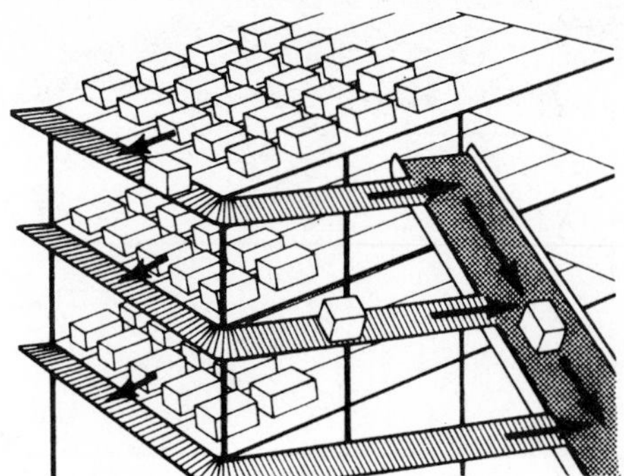

Fig. 4.6.47. Automatic case picking system.

recommended when there is not enough activity in each aisle to justify aisle-captive machines. Most storage machines are designed to handle pallet loads, but special machines have been built to handle coils of metal, engines, car axles, and other automobile components.

Car-in-lane systems provide high-density storage of unit loads that are put in long, horizontal lanes. Immediate access to each stored load is sacrificed in such an arrangement, but this may not be a limitation if there are relatively few different products but at least several loads of each.

The equipment consists of a powered shuttle car that carries individual loads into a lane and retrieves them. Usually the shuttle car is carried on a transfer car or on a man-ride machine to permit it to service a number of storage lanes. Many of these systems are computer-controlled.

Mini-load AS/R systems (Fig. 4.6.44) are scaled-down versions of unit load AS/R systems and are capable of handling bins that can hold up to 500 lbs. of materials. Usually the bins are brought to an end-of-aisle station for picking, then returned. However, some systems are designed so that the end-of-aisle picking station is a conveyor. In this way, bins can be routed automatically to work stations some distance away. Today, most mini-load systems are directed by computer.

Carousels (Fig. 4.6.45) consist of 6- to 10-ft.-tall multishelf bins that are powered by a loop of powered chain either at the top or on the floor. The chain advances the bins to a picking station. Cycling can be controlled by a computer or push button console. Devices that automatically load and unload tote boxes from carousels are available.

When people speak of carousels, they are usually referring to horizontal carousels. But there are also vertical models (Fig. 4.6.46), which require a very small "footprint" on the floor for the amount of materials they can contain. With this equipment, wide trays are supported by two loops of chain that ride around two sprockets, one above the other. By using a push button control, or by means of computer signals, the proper tray can be advanced to an operator pick station.

Automatic item picking equipment operates without human pickers. Individual retail packages, such as bottles of aspirin or tubes of toothpaste, are stored in lanes and released to a take-away belt conveyor. Usually, an entire customer order is filled at one time and carried to packing stations.

Automatic case picking systems (Fig. 4.6.47) operate in a similar manner, but the equipment is much larger. All automatic picking systems are computer-controlled.

4.6.7 AUXILIARY EQUIPMENT

Equipment for the receiving/shipping dock and "plant services" also falls under the materials handling umbrella.

4.6.7.1 Dock Equipment

Dock levelers, dock lifts, ramps, and dock boards provide means by which the plant is connected to a delivery truck.

Dock levelers (Fig. 4.6.48) are mechanized ramps, and, in effect, vary the height of the dock platform so that trucks can be driven from the dock into the bed of a highway delivery truck.

Dock lifts (Fig. 4.6.49) are used when there is no dock per se. They lift a fork truck or other loading equipment to the height of the highway truck bed, then lower the truck to floor or ground level.

Portable dock boards and ramps simply form a bridge between the dock and a railcar or highway truck.

Lights, dock bumpers, dock shelters, and seals (Fig. 4.6.50), and other accessory equipment are also essential to a well-designed dock operation.

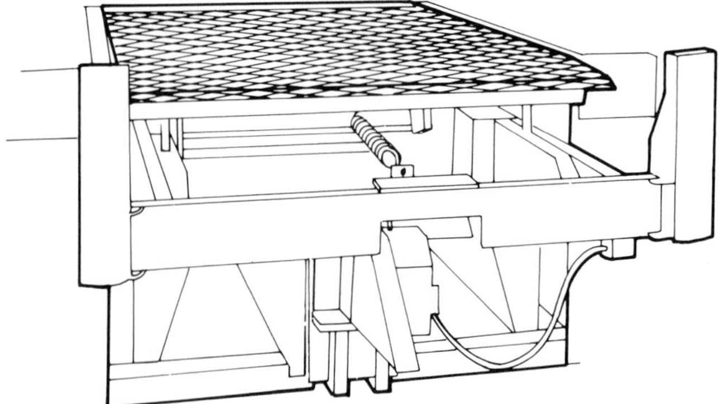

Fig. 4.6.48. Dock leveler.

Fig. 4.6.49. Dock lift.

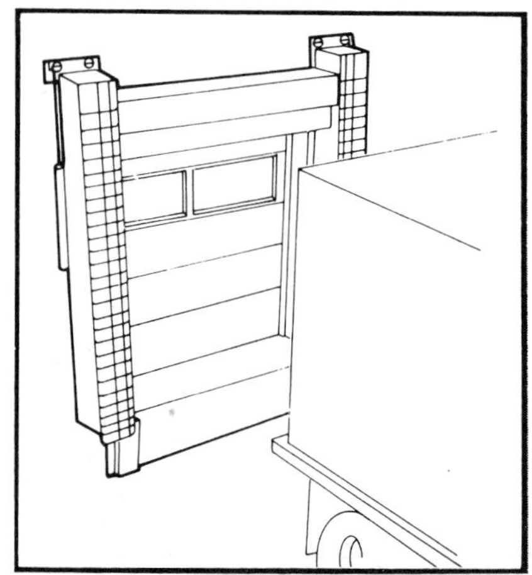

Fig. 4.6.50. Dock seal.

Fig. 4.6.51. Electronic scale.

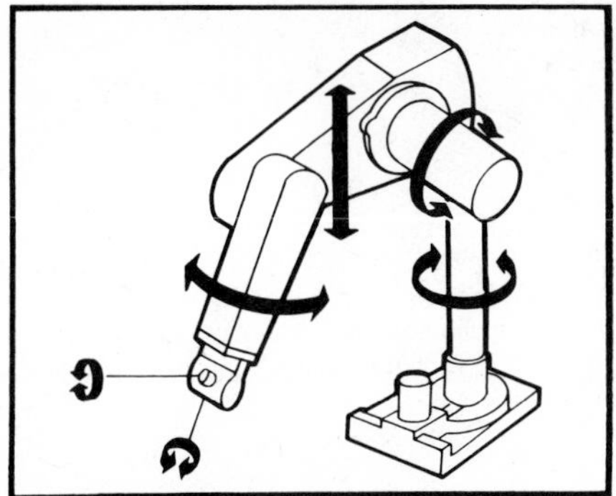

Fig. 4.6.52. Electronic robot.

4.6.7.2 Battery Chargers

Electric trucks have batteries that need periodic recharging. Battery chargers and related equipment, such as battery transfer devices, are necessary whenever you have electric equipment. Other useful truck accessories include battery discharge controls and indicators and hour meters.

4.6.7.3 Weighing Equipment

Weighing equipment has undergone a revolution within the past decade. Mechanical scales are rapidly being replaced by electronic scales (Fig. 4.6.51), which are fast, accurate, and simple to use. Because the equipment is electronic, it can be linked to a computer, printer, or bar code reader to transfer information on a real-time basis. Models include batch, counting, checkweighing, platform, postal computing, monorail, railcar, highway, crane, and lift truck electronic scales. Displays are digital.

4.6.7.4 Industrial Robots

Robots, with their ability to almost duplicate the movements of a human arm, lift loads weighing ounces or hundreds of pounds, and follow computer commands, have properly been called "universal" transfer devices. An electronically programmable robot (Fig. 4.6.52) can pick up a load, move it through a predetermined path in space, and position it precisely to within a few thousandths of an inch on a fixture or work stand, in a machine tool, on a pallet, in a shipping container, or virtually anywhere it is programmed to place the load. Once programmed, it will repeat the same cycle indefinitely or until it receives new instructions.

Manipulator is a word that is used by some people to identify the arm of a robot and by others to describe an operator-controlled positioning device. A manipulator consists of a counterbalanced arm that permits an operator to lift loads weighing hundreds of pounds with little effort. He manually moves the arm to position the part or load in a machine tool or on a pallet or wherever it is needed.

REFERENCE

1. R. A. Kulwiec, *Materials Handling Handbook*, New York: John Wiley and Sons, 1984.

CHAPTER 4.7

AUTOMATIC IDENTIFICATION SYSTEMS

JOHN M. HILL

Logisticon Incorporated
Santa Clara, California

Initial patents covering the use of optical devices for package identification in the material handling environment were issued in Europe in the early 1930s. Now, more than fifty years later, the user community has begun to recognize the potential of identification technology and is moving rapidly to cash in on its application to a host of in-process inventory and flow control problems. This chapter reviews the technical fundamentals, takes a look at current applications, and provides guidelines for assessment of alternative approaches.

4.7 INTRODUCTION

For several years identification systems have been a popular subject for discussion at various professional society and engineering conferences and seminars. One cannot help but suspect that at least a portion of the subject's early popularity had to do with its novelty, its quasi-technical glamour, and its futuristic overtones.

For some this may be true today, although for most the novelty has worn off and continuing interest is based on the demonstrated value of identification systems in a host of manufacturing and warehousing applications where increased productivity is critical.

Most of us would agree that the amount of time a product or component spends in the plant or warehouse directly affects productivity. Further, the degree of manual handling tends to serve as a negative multiplier. There is more than romance, then, to the challenge of moving products faster and with less direct labor dependence. Contemporary material handling systems are seen as primary answers to the productivity question. They tackle time problems and address the key issues of labor shortages and labor-intensive operations. Further, these systems recognize that a simple reduction in time per unit produced may be counterproductive without enhancement of the discipline and control that guards the integrity of receiving, processing, assembly, packaging, storage, and shipping operations. System discipline trades heavily on the timeliness and accuracy of operations feedback. "Morning after" production, shipping, or maintenance reports are historical, not preventive. The best decision making is executed on the basis of events as or before, not after, they occur.

It follows, then, that to achieve greater productivity, material handling systems must provide discipline and control, based not only on plans and performance goals but also on the dynamics of the actual operation. During the past decade, automatic identification systems have emerged as a major source of real-time feedback that is so critical to effective operations management.

What are automatic identification systems? Basically, they are systems that combine machine-readable encoding of goods in process and strategically deployed code readers for purposes of accounting, tracking, or movement control.

For years, the employment of code readers focused on the relatively straightforward control of material flow. During the 1970s, however, and particularly with the advances made in electronics technology, a dramatic redirection of that focus occurred. Automatic identification equipment now plays a primary role in closing the gap between line operations and management with time-critical data on events as they occur. Significantly, and as a by-product of this development, the new systems are playing an increasingly major role in the classical material handling tasks of routing, sorting, consolidation, and so on.

This chapter examines not only how automatic identification adds value to conventional handling system design but also how it can affect selection of the handling technique. Further, suggestions will be made on how innovative integration of an identification system with high-speed material handling can enable users to add capacity and avoid new plant investment. System components, applications, and the growth of usage within the manufacturing environment will be investigated.

To start off with the proper frame of reference, all code reading systems share the following features: a product, part, component, package, pallet, tote box, trolley, barrel, or what have you, whose accurate identification while moving into or through production, warehousing, or distribution will contribute to higher throughput, lower direct labor cost, more efficient handling, increased security, more accurate audit trails, or some combination of these. Then, a code affixed to the product can be automatically read and the product can be identified in terms of what it is, where it came from, where or to whom it is going, or whatever else might be most meaningful to the user. Then, a fixed beam, moving beam, or handheld bar code reader—an OCR, pattern recognition, or vision system—a magnetic stripe reader, or a radio frequency (RF) or surface acoustical wave (SAW) device will read the code within given constraints and, through associated electronics, translate it into meaningful control or information outputs.

The primary variables, of course, are the product, the nature of its identification, and the method of transport through the physical operation. Here, two basic rules apply. First, product movement past an automatic code reader must normally permit line-of-sight reader access to the code pattern within known and controllable parameters. Second, the product must be codable, that is, capable sizewise and surfacewise of being marked, labeled, printed, or tagged with a machine-readable pattern.

4.7.2 CODING

Let's take a close look at coding, which as you will rapidly learn, is both the key to and the Achilles' heel of automatic identification systems. Coding for these systems consists of four major elements:

Code content. The nature and amount of data to be encoded.

Code medium. The material from which the code is fabricated.

Code generation. The method of fabrication.

Code application. How the code is affixed to the product.

In the area of *code content,* a simple distinction may be made between *action codes* and *information codes.* Typically, an action code is directly related to a product's source or destination and, when read, immediately triggers an on-line mechanical response such as conveyor sorting. The information by-product of action codes is normally limited to item counting by source or destination.

Information codes, on the other hand, contain more elaborate data to permit not only control of material flow but also to provide the real-time feedback essential to contemporary production and warehouse management.

The machine-readable format of the code will depend not only on content but also on available coding area, the nature and speed of product flow, and the type of reading equipment selected. We'll examine these considerations in discussing the code readers themselves.

Once the intent and content of the code are established, *code medium* alternatives should be explored. Here, two major questions should be considered:

What are the environmental conditions to which the code will be subjected?

Will the code be affixed to an expendable item, such as a corrugated carton, and read only a few times, or will it be used to permanently identify a reusable product carrier, such as a tote box, pallet, or power and free conveyor trolley?

Most material handling systems are installed in factories, warehouses, or distribution centers where environmental conditions permit the use of optical identification systems. The coding media used in such systems employ either

Alphanumeric characters, bar codes, or unique patterns printed with opaque ink on various backgrounds.

Unique marks fabricated from retroreflective tape or liquid.

Opaque ink codes reflect little or no light as compared to the background surface. This difference in reflection or contrast can be electrically amplified to permit pattern decoding. The best contrast is obtained with good light-absorbing black marks on a white diffuse reflecting background surface, although a variety of opaque inks can be used on lighter, non-white backgrounds. Codes produced with opaque ink are generally less expensive than other alternatives.

Retroreflective media provide a higher level of signal return, typically 200 or more times brighter than most white diffuse reflecting surfaces. They also have the unique property of returning better than 90% of the incident light back to the light source. Retroreflective media are more expensive than opaque ink but have the advantage of being readily distinguishable by the majority of code readers. Further, in tape form, their durability makes them more appropriate for use in the permanent coding of such items as pallets, containers, tote boxes, tow carts, overhead power and free conveyor carriers, or trolleys and storage locations in automated storage and retrieval systems.

For those applications where the coded item is likely to be subjected to extremes of temperature, humidity, or other environmental abuse (e.g., a paint line or a tooling operation where coolants, etc., might contaminate the code), a new family of radio frequency (RF) identification systems is available. These systems employ factory- or field-programmable tags that do not have to be "seen" to be read. Sensed and read with RF signals, RF tags can be packaged in a variety of enclosures specifically designed to meet the environmental demands of a given application. RF systems are also well suited to the permanent coding of the reusable or captive product carriers mentioned earlier.

Similar to RF systems, but still in development, are those based on surface acoustical wave (SAW) technology. Tags consisting of a SAW microchip and small dipole antenna can be read at long range through any nonconductive environmental medium by a low-power hybrid radar and associated reader/decoder.

Code generation and *code application* are so frequently integrated that treatment of them in tandem is appropriate. After environmental considerations, code content and its source play major roles in the selection of media, generation technique, and application method.

An action code, for example, may be no more than a retroreflective mark at a unique vertical position on the side of a carton. Such encoding can be readily accomplished with a manually or automatically applied piece of retroreflective tape. The unique signal return of the medium and its unique location on the carton are all that is necessary to distinguish the carton from others in the mix flowing past the code reader.

On the other hand, a multidigit information code will not be as readily fabricated or applied. Here, the source of code content will play a significant role in selection of the fabrication and application approach. If the data is available ahead of time, there are several alternatives; if not, in other words, if code content is known only hours or seconds before product encoding, the options are reduced.

In the foregoing connection, code generation and application is accomplished either off-line, before product is introduced into the handling or processing system, or on-line, during production or distribution operations. The most common techniques for off-line optical identification systems include

1. *Carton preprinting.* Normally accomplished with opaque ink, this approach incorporates the code pattern with basic packaging graphics during carton fabrication. Code content, typically product identity, is available ahead of time and generation and application are accomplished in one step.
2. *Preprinted labels.* Employing opaque ink, this system offers the greatest flexibility for off-line code generation. Normally produced in roll form, these labels are subsequently applied by hand or with a variety of automatic labelers.
3. *Fixed or adjustable code plates.* Using retroreflective media, these plates can be fabricated for permanent attachment to such product carriers as skids, pallets, or tote boxes, which are used for product transport in captive material handling systems. The fixed plates serve as carrier license plates and permit their tracking by code readers throughout the operation. The adjustable plates are normally used for destination coding (action coding) in automatic routing systems.

On-line optical identification systems include

1. *Bar code printers.* Handset, keyboard, or computer-controlled, these devices can operate on- or off-line. Depending on the model selected, they will create and even apply random, sequential, or batch bar code labels.
2. *Computer line printers.* A variety of impact and non-impact printers are now generating man- and machine-readable orderpicking and product identification labels for code reading systems.
3. *Retroreflective media.* Although retroreflective materials can be converted into a variety of label or permanent code plate formats, their unique signal return permits efficient use in tape or liquid form in many on-line coding situations.

The majority of RF identification systems currently employ factory-programmed tags. Each tag contains a unique multidigit or multicharacter code. For the permanent coding of, say, tote boxes, as long as each code is different, there is no requirement for a given tag to be attached to a specific tote. When the tag must identify a specific item, tag number and item specification are entered into a host computer look-up table at the time the tag is attached to the item.

For on-line requirements, two manufacturers have developed alternative methods for tag-encoding. The first method uses a field programming device. An RF tag is inserted into the device by an operator who subsequently key-enters the desired code. When the entry is validated and displayed, the operator depresses an appropriate key and the code is "burned into" the tag's memory. The second approach uses pulsed microwave signals and complex tag circuitry, which permit tag programming from distances up to three meters.

SAW tags are factory-programmed by lasers that trim an aluminum film pattern of interdigitated transducers (IDTs) deposited on the SAW microchip. The IDTs resemble combs with interlaced teeth and are left in or out by trimming to form binary codes with up to 30 data bits.

Recognizing that the coding system is a critical ingredient of all automatic identification systems, considerable emphasis has been placed on the development of materials and equipment that can accommodate a broad number of specific applications requirements. The other major element is, of course, the code reading equipment.

4.7.3 OPTICAL CODE READERS

There are several types of optical code reading equipment now being used for industrial applications:

Fixed beam readers.
Moving beam readers.
Handheld readers.
Charge couple device vidicon scanners.

4.7.3.1 Fixed Beam Readers

The simplest type of code reader, the *fixed beam reader,* uses one or more individual scanner heads with stationary light sources and light sensors. The term *fixed beam* is used because neither the light source nor associated optics move. The coded product, then, moves into and through the reader's field of view.

There are two versions of the fixed beam reader. The *reflective reader* locates its light source to the side, above, or below the light sensor. The light source illuminates the code surface, while the sensor, through its optics, "looks" at the variations in signal return scattered back from the code area. Reflective readers normally detect opaque ink codes on paper or other diffuse reflecting surfaces.

Reflex or *coaxial readers* illuminate and look at the code area with the same optical system. Here, the light source is focused by a lens onto the coded surface. Varying intensities of light corresponding to the presence, absence, or dimension of the code marks return along the same optical path to the unit's sensor. Reflex (or coaxial) readers are used to detect both opaque ink and retroreflective coding.

When fixed beam readers are used to detect opaque ink codes, the coded surface's presentation to the reader—that is, distance and skew—must be relatively well controlled. If retroreflective coding is used, reading can be accomplished with variations in distance of as much as 30 inches at ranges up to several feet.

Coding for Fixed Beam Readers

As discussed earlier, one of the most straightforward coding techniques is the action code, consisting of a single unique mark at a given height on the side of a carton. A fixed beam reader positioned along a conveyor at the same height can be used to route, sort, or count only those cartons so marked. Naturally, this coding scheme is limited by mark size and carton height.

Increased data content can be handled with a multiple-mark code format. Essentially, such coding is binary in nature; that is, each mark position in the code pattern has a permanently assigned "power of two" weight (i.e., 1, 2, 4, 8, 16, etc.). In one approach, if the mark is present, the assigned position weight is counted; when absent, the weight is not counted. The "value" of the code pattern is determined by adding the assigned weights of all marks present in a given pattern. Determination of mark presence or absence is accomplished in conjunction with the sensing of a parallel clocking track or by timing from detection of the leading edge of the carton.

Another encoding approach accomplishes the same results with wide and narrow marks. Position weight is counted when the mark is wide and not counted when narrow. Code validity is checked by counting the total number of marks in the pattern.

Once the general format of the code is established, the details of code mark size and spacing can be studied. Several factors affect the final selection of dimensions for fixed beam readers, including

The accuracy with which the code can be printed and/or positioned on the product.
The space required to mount the reading heads, side by side or above one another.
Conveyor line speed, vibration, and differences in riding height of products of varying weight.
Light beam size at anticipated maximum reading distance and reader response time.

Typical fixed beam readers can resolve marks as small as .050 in. However, code mark and total pattern dimensions do vary from application to application.

4.7.3.2 Moving Beam Readers

During the 1970s, enormous general interest in bar codes and code reading was stimulated by the grocery industry universal product code (UPC) and its program for automation of the supermarket checkout counter. Most people are now familiar with this 10-digit code printed on virtually all grocery product packaging as well as a host of other items, such as magazines, pharmaceuticals, and cosmetics that are sold through grocery outlets. The code identifies the product and the manufacturer.

When manually passed across the window of a checkout counter code reader, code content is fed to a minicomputer, which transmits the current price back to the cash register terminal. Simultaneously, the computer adjusts inventory balances. Further, most systems tally daily receipts, profile product movement, establish inventory replenishment needs, and generate a variety of other data essential to efficient store operations.

Usage of the UPC is now widespread and growing at a rapid rate. The key to its growth has been, in large measure, a result of the high level of performance of the system scanner, which is a close relative of the industrial moving beam reader. Using, in effect, a flying spot to scan, laser-based moving beam readers will locate, read, and decode miniature patterns with high data content anywhere within a field of view more than 20 inches high by as much as 30 inches deep. Sweeping through this field at a rate of 180 to 600 times a second, the moving beam device takes several looks at each code and compares them before validating and transmitting code content. Additionally, given proper code design, the reader can accommodate wide variations in code alignment and orientation.

Coding for Moving Beam Readers

The moving beam reader is normally employed to read information codes. In this connection, two basic formats are used: the "ladder" code and the "L-shaped" code. Although the former can be read right-side up or inverted, it must pass the reader at or near right angles to the moving beam. The L-shaped code can be read regardless of its orientation.

Moving beam readers also operate through detection of variations in contrast between code marks and background color or carton surface. Using wide and narrow binary coding formats, the devices compare bar thickness within a given pattern to determine pattern data content. For example, a seven-bar, ladderlike code can handle the identification of 128 different items. As with fixed beam readers, each bar in the pattern has a numeric weight (1, 2, 4, 8, 16, 32, 64). When the bar is thick or wide, the weight is counted; when narrow, it is not. If, in the example, the first, third, fourth, fifth, and seventh bars were thick and the balance narrow, the code's value would be 93.

A variety of more elaborate codes are now in use, including those that accommodate the full alpha numeric and/or ASCII (American Standard Code for Information Interchange) character set. Additionally, built-in parity or validity checks are now incorporated in all code formats. Those codes now generally accepted for use in manufacturing, warehousing, and distribution include the following uniform container/transport case symbols:

Interleaved 2 of 5 code. A numeric-only code designated Uniform Symbol Description-1 (USD-1) by the Automatic Identification Manufacturers (AIM) and adopted by the Distribution Symbology Study Group (DSSG) for corrugated shipping container identification.

Code 39. An alpha numeric code designated Uniform Symbol Description-2 (USD-2) by the AIM group and adopted by DSSG for corrugated shipping container identification requiring the full alpha numeric set. Note that Code 39 has also been adopted by the Department of Defense's LOGMARS (Logistics Applications-Automated Marking and Reading Symbols) group for use by defense contractors and DOD activities on all unit packs, outer containers, and selected documentation (Reference Military Standards 1189 and 129H, January 4, 1982, and available through the Naval Publications and Forms Center).

American National Standards Institute (ANSI) bar code. The subcommittee on coding and labeling (MH-10) has developed a standard for bar code for use on shipping cartons, containers, and pallets. It contains all specifications for Interleaved 2 of 5 Code and Code 39.

Other symbols generally used in manufacturing, warehousing, and distribution include the following:

Codabar or USD-4. Permits encoding of the numeric character set (0–9) and six unique control characters. One of the earlier symbols developed, it is still used in a number of retail point-of-sale applications.

Code 128 or USD-6. Permits encoding of the full 128-character ASCII set. It features unique start and stop characters for bidirectional reading, as do the other codes covered earlier. Code 128 also permits variable-length field encoding and provides elaborate character-by-character and full symbol integrity checking.

Code 93 or USD-7. Permits encoding of the full 128-character ASCII set and variable-length fields. The symbol uses two powerful check digits to minimize the possibility of reader error due to printing anomalies.

Code 11 or USD-8. Permits encoding of the numeric set and a unique start and stop character. An extremely compact symbol, Code 11's name is derived from the fact that it is capable of encoding 11 different characters.

Each of the symbols described here is explained in detail in the Automatic Identification Manufacturers' manual. The manual also includes information on the DSSG, DOD, and ANSI standards discussed.

4.7.3.3 Handheld Code Readers

In many applications, an element of vulnerability exists between the mechanized system and shipping or receiving operations when product movement is under direct manual control. In other operations, the flow of product or variations in size and shape may not lend themselves to automatic scanning. The need for accurate product tracking, however, remains critical to cost-effective materials management. To meet this demand, a number of manufacturers have developed handheld code readers (lightpen systems and handheld laser scanners).

The lightpens incorporate both a light source and sensor and are manually drawn across bar codes that may have as many as 10 digits of information per inch. Signals from the pens are decoded by fixed or portable decoding electronics and transmitted directly to on-line computers, or they may be stored on magnetic tape cassettes or solid-state memories for batch transmission to the computer on a routine basis. Fixed-station pen systems are normally deployed at strategic locations central to product movement in nonmechanized operations. Portable units are frequently employed for taking physical inventory, often reading the same bar codes read earlier by automatic scanners as the product moved from manufacturing to warehousing or storage.

The handheld laser scanner's appearance is not unlike that of a contemporary home movie camera. It incorporates a low-power helium-neon laser that is automatically swept from right to left through the scanner window. Activated by a trigger on the device's pistol-like grip, the scanner handles about a 4-inch depth of field roughly 5 inches wide. The scanner offers a number of advantages when either code accessibility or rough or uneven surfaces makes lightpen scanning difficult.

Coding for Handheld Code Readers

Virtually all of the bar code symbols described earlier are suitable for handheld scanners. Generally, bar and space dimensions/thickness can be substantially smaller than those used for moving beam scanners. In all cases, however, final dimensions should be established on the basis of existing standards and supplier assessment of the needs of the actual application.

4.7.3.4 Charge Couple Device/Vidicon Readers

A family of code reading devices that contain no moving parts and are designed to capture, decode, and process either easily printed dot-matrix code patterns or human-readable alphanumeric characters is now emerging as an alternative means of data capture.

Charge Couple Device Scanners

Emerging from a number of federally funded research and development projects in the 1970s, the charge couple device (CCD) belongs to a family of silicon semiconductors capable of image sensing, analog signal processing, and digital or analog memory. They are used for high-speed mail sorting, surveillance, aerial mapping, and, more recently, optical label reading. CCD scanning systems consist of a light source, camera assembly, power supply, and signal decoding and processing electronics. The light source typically consists of a number of high-intensity floodlights focused on the path of labeled item travel. The light is reflected back from the label; it then passes through a high-quality lens and is focused on the CCD chip located within the camera housing. Chip resolution ranges from 128 line image sensors for applications with minimal range or depth of field to 1,024 sensors for greater range and depth of field. The line image sensors scan at a rate of 4,000 times per second, effectively taking several "snapshots" of the encoded item as it passes. Chip output is fed to the system's electronics, which compare snapshots, extract the code, and create output for "action" or "information" purposes.

Top reading scanners are designed to read labels within a 30- by 30-inch field of view at speeds up to 450 feet per minute. Side reading CCD scanners will handle similar speeds but with generally smaller fields of view and restrictions on label orientation. Handheld CCD scanners will read code patterns up to 1¼ inches high by as much as 12 inches long at a range of up to ½ inch.

Coding for CCD Scanners

CCD scanners read what is called a matrix, or presence/absence, code. The code is printed with standard printer slugs such as those used on standard IBM-1403 chain printers. The code uses the presence or absence of a mark in a group of marks to represent a unique number (0–9). Two sets of mark groupings are available: A five-mark pattern in which three of the marks are always present; and, a four-mark group in which two or three of the marks are always present. Each code pattern includes a high-integrity check digit. Further details are available in AIM Publication USD-5. Key features of the presence/absence code include

Ease of printing (in label form or directly on corrugated surfaces), including up to ± 30% latitude on deviations from nominal slug mark dimensions.

Reduced requirements for code contrast (code marks generally need to be only 50% darker than the substrate on which they are printed).

For direct carton printing, red, green, blue, purple, and brown inks as well as black are normally acceptable.

Machine Vision

For years, a number of companies and research organizations have worked on a variety of approaches to automatically reading the human-readable alpha-numeric character set. Until recently, the commercial applications of this technology, known as optical character recognition or OCR, have been confined to document reading and limited use for merchandise tag reading at the retail point of sale. Without tight control of character printing and the reading environment, OCR's performance has not met the criteria established by other automatic identification techniques. A single printing anomaly, such as an ink spot or void, can easily obscure or transpose an OCR character, rendering the label unreadable or likely to be misread. During the past couple of years, however, with the availability of more powerful microprocessors and mounting demand from the robotics industry for "machine vision," a number of suppliers have taken a fresh look at remote character recognition. It appears that they are making substantial progress. The majority of the new systems use an inexpensive vidicon, or television camera, a high-speed microprocessor, and complex firmware that matches images from the camera with memory-resident "pictures" of acceptable characters.

Unlike CCD scanners, machine vision systems do not require high-intensity illumination and can handle characters of varying sizes. Additionally, the systems can distinguish characters printed, etched, or raised on substrates with low contrast. Indeed, machine vision is being evaluated for tire identification where the code is embossed on the side of the tire. Although still in its infancy for classical automatic identification applications, the potential for machine vision in handling systems of the future is undeniable. Indeed, they should be given serious consideration in any assessment of automatic identification alternatives.

4.7.4 RADIO FREQUENCY (RF) SYSTEMS

For years, a variety of major international companies have worked on the development of RF, or microwave, technology for vehicle identification in the transportation market, for railcars, buses, subway cars, and so on. In the early 1980s, new developments in the basic technology shifted the focus to potential applications in industrial process control and material handling where environmental constraints or absence of line-of-sight access to the code preclude the use of optical techniques. Examples include item or carrier identification into or through a paint spraying, baking, or machining operation. In these applications, a machine-readable tag must provide reliable feedback in spite of extremes of temperature or contamination by paint or coolants. RF systems can handle these challenges. They are characterized by three primary components: an identification tag, an antenna, and a reader.

4.7.4.1 RF Identification Tags

The transponder, or tag, consists of a small printed circuit board that contains receiving and transmitting antennae, a small number of discrete components, and an integrated circuit chip for data storage housed in a heavy-duty industrial package. Tag size is determined by the size of the transmitting and receiving antennae, which, in turn, is a function of required range and speed. At limited ranges and speeds, tag size can be reduced to that of a ⅜-inch-thick credit card. On the other hand, tags used to identify buses at ranges of 25 feet or railcars moving at speeds of 100 miles an hour or more are currently housed in a 2½ × 2½ × 7½-inch package. Two basic types of tags are available. A *passive tag* is activated by an external RF signal and responds with a unique 12-character message. This tag is normally preprogrammed at the factory to the user's specification, but it can be field-programmed by the user with a special programming device. Given appropriate enclosure materials and components, these tags can handle temperatures from −40°C to more than 200°C.

A battery-powered tag is available that can receive, store, and transmit up to 8,000 alphanumeric characters of data. In other words, it can actually carry instructions to the station(s) at which the item it identifies is to be processed and can receive and store data along the way, thereby creating an audit trail that can be readily checked at quality control or prior to release for shipment.

4.7.4.2 RF Antennae

The antennae are installed along the line of tagged item movement, transmitting signals to and receiving data from the tags for processing by the system reader. For most industrial applications, antenna construction is quite simple and permits housing in a small, sturdy package. It can be buried beneath inches of nonconductive material (e.g., asphalt, cement, wood, plastic) with no degradation of performance. The antenna emits a conical-shaped RF field. The "capture window" size is determined by the distance from the antenna to the tag path. Low-loss coaxial cables permit antenna placement up to 100 feet from the reader. Depending on throughput rate, up to four antennae can be multiplexed to a single reader.

4.7.4.3 RF Readers

The reader transmits RF signals to and receives data from the tags via the antenna(e). Microprocessor-based electronics decode, store, and compare successive inputs for validation prior to transmitting tag data content to the host computer or local controller.

4.7.4.4 Coding for RF Systems

A variety of high-integrity coding techniques are used by RF system suppliers. One of the more popular codes is analogous to the Interleaved 2 of 5 Code described in USD-1. Another is Manchester II, or Bi-phase L Code, whose parity and error-checking features reduce the probability of undetected error to less than 1 in 10^{24}. RF system suppliers assume full responsibility for all systems components, including factory- or field-programmable tags.

4.7.4.5 RF Systems Tag Costs

In that tag costs are higher than bar code labels costs, RF identification systems are not generally promoted as an alternative to optical systems. They do, however, add a new dimension to automatic identification by offering solutions for applications that cannot be handled optically and for permanent identification of pallets or containers where tag durability can enhance system performance and integrity.

4.7.5 OTHER IDENTIFICATION TECHNOLOGY

4.7.5.1 Surface Acoustical Wave (SAW)

SAW systems employ a reader consisting of a low-power radar transmitter and microcomputer, a bowl-shaped antenna, and the small SAW tags discussed earlier. In operation, the reader transmits radar pulses via the antenna. These pulses are intercepted by the tag's antenna and converted into ultrasonic acoustical waves. The waves pass along the uniquely encoded "teeth" of the tag's IDTs, which modify their amplitude according to tag code content. The waves are then converted back into electromagnetic signals for code transmission to the reader.

The microchips used in SAW tags take up less than .5 square inches of area and may be embedded in a variety of nonconductive substrates. Overall tag dimensions are a function of required range and corresponding tag antenna dimensions, but it is anticipated that generally they will be smaller than the RF tags.

At present, the primary hurdle to commercialization of SAW technology is high volume deposition and encoding of the IDT teeth on the microchip. Given resolution of this challenge, it is anticipated that SAW will become a viable alternative for applications that can use pre-encoded product or carrier ID tags.

4.7.5.2 Magnetic Stripe

Magnetic stripe recording has been used for years in such applications as bank and charge card transaction recording, commuter ticketing, and security access control. To a lesser extent, the technology has been used for shop floor control, failure analysis tracking, work-in-process tracking, and related applications.

Advantages of magnetic stripe recording include high-density data packing (up to 640 characters per card), read/write capability, and readability in spite of card crumpling, wrinkling, or contamination by dirt or oil.

Magnetic stripe cards are read by handheld wands or contact readers; they do require an operator. Card costs are substantially higher than those for bar codes. Accordingly, we do not expect to see more than limited use of the technology in the industrial environment.

4.7.5.3 Voice Recognition

Although not technically a member of the automatic identification product family, voice recognition systems have been promoted and applied in a variety of material handling and process control applications since the early 1970s. It is expected that growth in other applications areas including appliances, toys and games, and security systems will lead to lower unit costs and continued assessment for use in such manufacturing and distribution functions as quality assurance and low-speed product sortation.

There are three varieties of voice recognition systems now available: speaker-dependent, speaker-independent, and continuous speech. Each converts voice input from analog wave forms to digital signals and then matches those signals against various "words" stored in the system's "vocabulary."

Speaker-dependent systems recognize one word (from a 40- to 250-word vocabulary) at a time for a single speaker but must be trained to recognize the speaking style of that speaker. Speaker-independent systems also recognize a word at a time but are not limited to one speaker. Their cost is about four times that of the speaker-dependent system. Continuous-speech systems recognize several words without intervening pauses but must be trained for each individual speaker.

Processing time and accuracy have been the major roadblocks to broader use of voice recognition technology in the real-time plant environment. Because of variations in individual speaker articulation, the systems normally provide feedback to the speaker to verify accuracy of input recognition. Each verification step taken improves accuracy but diminishes system throughput. Accordingly, it is not expected that voice recognition will challenge fully automatic identification techniques but will become an increasingly viable tool for the worker whose hands and eyes are busy performing other tasks.

4.7.6 AUTOMATIC IDENTIFICATION SYSTEM APPLICATIONS

We initiated this discussion with emphasis on the critical contribution that well-deployed automatic identification equipment can make to productivity in material handling applications. Let us take a look at some of the systems now in operation.

Most identification systems applications fall into two categories. The distinction is a function of the primary end-use. As with coding, the first category we define as *action* applications, the second as *information* applications. We shall look at the characteristics of each as well as current examples of their employment.

4.7.6.1 Action Applications

As noted earlier, use for action or control purposes is characterized by coding that bears a direct relationship to a mechanical function within the product handling system, for example, package sorting. In terms of those elements common to code reading systems, the action variety employ the following:

Coding, which typically identifies product by destination within the facility and is generally limited to the execution of a single mechanical task.

Automatic scanners, which are deployed at material handling systems induction or discrete decision points.

Output in the form of signals to on-line controls or data to microcomputers or minicomputers for tracking, sorting, and verifying task completion.

At a major book distribution center, orderpicking is simplified through the use of uniquely positioned strips of retroreflective tape and single-scanner, fixed beam readers. Orders are filled from seven picking areas linked by a common conveyor. Empty cases are delivered along with the packing list to the first zone from which books will be selected for a given order. On completion of his portion of the order, the operator in the first zone consults the packing list for the next designated picking zone. He then affixes a strip of retroreflective tape at a predetermined height on the carton reserved for the designated zone and then places the carton on the conveyor. At each picking zone diverting point, a fixed beam reader is mounted to intercept one of the code mark positions on the cartons. When the reader beam strikes the retroreflective tape mark, the carton will be diverted. At each subsequent zone, the procedure is repeated until the order is completed and routed by express conveyor to shipping. By routing an order only to the zones that hold books for that order, the user has increased his throughput and greatly simplified orderpicking routines.

A large library utilizes a fixed beam code reader in a power roller conveyor system handling tote boxes. The code reader is used to identify and route tote boxes through multifloor buildings. Each tote box carries an adjustable code plate that can be set for 1 of 36 different locations. As requests for literature, books, and so on are received, the appropriate destination is set and the tote box is placed onto the roller conveyor. As the box passes a code reading station, the code pattern is read and, if it is correct for the location, it is diverted for unloading.

Fixed beam code reading is also used in overhead conveyor systems as in the open rail, power and free system handling garments used by a major catalog distribution center. Each of the system's trolleys carries an adjustable code plate that is set for a destination at one of several garment pickup points. Strategically located code readers control trolley movement throughout the picking area until the trolley is full, at which point it is coded for a final accumulation and packaging area.

In another fixed beam reader application, 180 readers monitor 15,000 tote boxes equipped with adjustable code plates to route, sort, and ship 7,000 to 10,000 orders in a replacement parts depot whose inventory fills seventeen acres.

A number of towline conveyor installations now control cart movements with fixed beam readers instead of mechanical and magnetic probe systems common some years ago. An adjustable code unit is mounted on the frame of each tow cart and manually set for a given address prior to cart release. Immediately ahead of each towline spur, a floor-mounted reader looks up and at an angle coinciding with the location of code units on the carts. When the code arrangement on the cart coincides with the location of the scanners in the fixed beam reader, the reader triggers cart diversion.

A number of major dry grocery and frozen food distributors employ identification systems to enhance the efficiency of orderpicking and shipping operations. The code is incorporated in the picking label. It represents one of several shipping docks and can be read right-side up or inverted. Printed on the user's computer line printer as a by-product of administrative order processing, the code is automatically applied to picked cartons by a labeler installed on man-ride picking cars in each aisle of the warehouse.

Picked cartons merge from multiple aisles onto a high-speed conveyor flowing past the scanner on their way to palletizing and shipping. The scanner reads the code through a 6-by-24-inch field of view and provides direct relay output to an analog tracking system that controls sorting. The system has increased sorting accuracy, permitted a 25% increase in throughput, and reduced associated overtime and other operating costs.

In wine distribution, top as opposed to side labeling and a lack of orientation control required the use of an L-shaped omnidirectional code pattern. Preprinted on a portion of the orderpicking label, this pattern represents the shipping dock and can be read regardless of its position on the top of the carton or, for that matter, the carton's orientation on the conveyor. Manually applied, the label is read through an 8-by-24-inch field of view at a line speed of 150 feet per minute. Relay output to a tracking system permits automatic sorting to six shipping docks. The system's accuracy has made a significant contribution to operational security. It has also permitted the increased throughput that leads to lower per unit handling expense.

The world's largest independent hardware distributor doubled the throughput and extended the economic life of its facility through employment of a unique computer-based, action-oriented code reading system and a high-speed sorting line.

Orders picked on nine floors are manually labeled with a computer-generated shipping ticket that incorporates a bar code representing one of 146 different destinations. Once the package is aboard the tilt tray system, the code is read through a field of view 18 inches deep by 24 inches high. Line speed is 315 feet or 70 trays per minute.

The scanner provides code output to a minicomputer, which, on the basis of shaft encoder input on tray travel, tracks the tray to the appropriate tip-up point and dumps the package. The computer records the item count by destination and periodically produces reports for line management.

In addition to the impact of the system on facility capacity, order processing turnaround has been reduced from five days to two. This turnaround improvement has resulted in an increased level of business, which permitted relocation of displaced warehouse personnel to productive assignments in other areas of the operation.

Reviewing the foregoing applications and a variety of others totalling some 30% of the current identification system population, certain advantages and limitations should be highlighted. Advantages of action systems include the following:

Permit the use of a variety of coding alternatives, including the line printers normally used for orderpicking paperwork.

Typically use smaller labels or share space with existing labels and, therefore, have lower associated costs.

May be readily integrated with existing manually controlled routing or sorting systems.

Normally have shorter delivery and installation schedules.

Require lower investment in operation and maintenance training.

Require lower initial investment than information system alternatives.

Action systems have the following disadvantages:

Employ coding generally limited to a single task.

Provide limited information by-product.

Unless computer-based, provide limited expansion or modification capabilities.

4.7.6.2 Information Applications

Information-oriented applications of automatic identification systems are characterized by:

Coding, which provides the opportunity for multiple uses through uniquely identifying the product by type, lot, manufacturing data, build ticket, serial number, purchase order or invoice.

Automatic or handheld scanners, which are deployed to audit and control discrete product movement from receiving to production to storage to shipping.

Output to microcomputers or minicomputers, which match throughput against plan, record, report, and flag exceptions in real time. The computers are also frequently used to control product movement, sorting, storage, consolidation, and shipment through material handling system interfaces.

Where are identification systems now providing this information feedback? Let us start at the receiving dock. A major shoe distributor generates by computer case labels for distribution along with purchase orders to his suppliers. Each label contains a unique eight-digit bar code. All purchase order data, including order number, due date, and case content, such as type of shoe, quantity, and weight, are referenced in computer memory by the unique case code. Suppliers are required to apply these labels to the cases prior to shipment.

On a weekly basis, the business system creates an "open receipts" file that is transmitted to the warehouse minicomputer. The file includes records by serial number of receipts anticipated for the next 60 days.

Upon receipt at the warehouse, cases travel at 175 feet per minute via conveyor past an automatic scanner that reads the code through a 6 × 24-inch field and provides direct output to the minicomputer. The case then moves over a computer-linked, weigh-in-motion scale. The system authenticates the unique case code, verifies case weight, and provides the output for automatic tracking and sorting to storage. Rejects are immediately diverted to an inspection line, and the system produces an exception report identifying the reason for rejection.

Tighter control of incoming inventories has enabled the user to reduce costs associated with short shipments, reject analysis, reprocessing, and associated clerical work. Further, quicker response to demand at the retail level is expected to improve inventory turn. Additional savings have been realized in terms of the higher accuracy and throughput achieved through automatic control of material handling tasks.

In cosmetics distribution, orders are picked and packed in the same area of the facility. Packer instructions are incorporated in a computer-printed document that includes shipping labels for the order. The bar code portion of each label uniquely identifies the consignee. Once sealed and labeled, the cartons flow beneath the scanner at 180 feet per minute. The scanner accommodates a 19-inch variation in carton height.

Code content is fed to a minicomputer that, on the basis of its record of consignee address and carrier availability, tracks and sorts each carton to the appropriate shipping dock. Important by-products of the system are bill of lading preparation and related billing documentation.

A major pet food manufacturer employs a fixed beam reader to accurately monitor finished goods enroute to storage from production by reading product identification codes preprinted on the product cartons. The reader provides direct output to a minicomputer that stores the data for the production of routine reports on throughput by individual product and quantity.

In a soap industry application, moving beam readers read preprinted carton codes for production recording purposes and the automatic control of conveyor line sorting devices. An interesting sidelight here is reader detection of inverted cases. If a case carrying a liquid product is inverted, the system will automatically send it to a manual packing area for inspection prior to shipment.

Little mention has been made of the use of handheld scanners or light pens in the manufacturing or distribution environment. Usage is increasing at the shipping dock, in receiving, at quality control, and in other areas where the nature of the product or product flow does not permit fixed or moving beam reading. For example, in the automotive industry, lightpens are employed in final inspection to verify the proper assembly of engine components and to produce permanent records for EPA compliance.

In energy production, lightpens and a minicomputer keep track of all production moving into and released from vault storage. The pens are also used for periodic inventory auditing to assure the 100% accountability critical to these operations. Inventory taking time has been effectively cut in half.

CCD scanners are used in a variety of retail distribution applications for high-speed package sorting. Since the presence/absence code permits multidigit code printing, many of these systems provide information output that allows the host computer to relieve inventory, maintain or verify item counts by consignees, or other information beyond the simple sorting code.

Vidicon scanners are now being used in a number of pilot applications, including the reading of rubber tire serial numbers, a challenge that has eluded other identification technologies for years. SAW technology is also being evaluated for this purpose.

The primary industrial application of RF identification systems today is in the auto industry, in which high-temperature tags serve as machine-readable license plates for tracking vehicles from the beginning of the assembly line until their release as finished goods. Theoretically, vehicle identification at the beginning of the line coupled with sequential tracking by the facility process control computer would appear to offer the necessary level of control. Practically, however, the undetected removal of a single vehicle from the line—and sequence—could result in robotic installation of four doors on a two-door car or some other equally costly assembly error. Antennae located at each stage of assembly verify vehicle identity prior to initiation of the assembly task.

As noted earlier, magnetic stripe systems have been used for shop floor control and parts tracking on a limited basis. At a major agricultural equipment manufacturing operation, magnetically encoded cards in clear plastic envelopes are attached to parts carriers for work-in-process monitoring. To maximize readability, each card is encoded in multiple tracks. They are read at key transaction points throughout the facility.

The foregoing applications illustrate a number of the areas and disciplines impacted by the use of information-oriented identification systems. Characteristics and additional potential advantages include:

- Code content permitting operations monitoring as well as production, inventory, and shipment control.
- Broader discipline impact and sources of justification.
- Material handling transaction execution verified and recorded not only by event but by discrete item.
- Possibility of direct carton preprinting and permanent carrier labeling as well as RF and other coding approaches.
- Maximum flexibility in terms of accommodating changes to or expansion of code processing or control output requirements, for example, destination reassignment.
- Output readily compatible with business/data processing systems.
- Strong performance in productivity measurement and improvement by manufacturing point, picker, packer, assembly line, and so on.
- Useful in operations demanding a high level of security, such as nuclear fuel production.

To be measured against these advantages are:

- Typically high investment costs.
- Longer delivery and installation schedules.
- Generally but not universally higher coding costs.
- Increased investment in operator and maintenance training.

4.7.7 SUMMARY

Our initial premise contended that automatic identification systems can play a significant role in providing the real-time information and control output so critical to contemporary operations management. This chapter has reviewed the basic elements of these systems and described a number of typical applications.

Few tools surpass the code reader in its ability to draw from the material handling or process environment that data so vital to efficient flow control or production and inventory management. How do you evaluate its suitability for your requirements?

The moving beam reader can locate and read compact, high-density codes almost anywhere on a moving or stationary object. Further, because each pattern is scanned several times, the moving beam device permits the use of less expensive code media.

The fixed beam reader is less flexible in terms of accommodating variations in code placement or orientation and, generally, requires higher quality code media.

The handheld reader has great flexibility given a reasonably motivated operator. Indeed, there should be considerable growth in the use of both wands and handheld laser scanners in the nonmechanized handling environment.

Already in the wings or in limited application industrially are a host of new developments employing CCD, linear arrays, and machine vision as well as RF readers that offer solutions for challenging environments.

One obvious basis for comparing various code readers is price. Moving beam readers normally cost several thousand dollars. Most fixed beam readers are priced below two thousand dollars. Handheld readers including magnetic stripe wands can be purchased for under two thousand dollars but require an operator. CCD, linear array, and machine vision system pricing is well into five figures. Although RF readers and antennae are priced competitively with moving beam devices, RF tags cost substantially more than bar code labels.

Once a decision has been made to proceed with the system design, a specific individual in the organization should be charged with assessing the suitability of automatic identification systems for the particular requirement. Working with other specialists within the organization, the in-house code reading "professional" should develop preliminary data in the following areas:

General application description. Material handling, production, inventory, machine tool, or quality control?

Present approach (if applicable). Product identification technique: clipboard, keyboard, other? Number of personnel and tasks? Type of handling system? Item throughput rate?

System data requirements. In what manner are the products to be identified? Serial or model number? How many digits? A unique code for each product? How many products? If not the foregoing, are products identified by source, destination, or other? Which and how many? Is additional data worth the added cost?

Product coding parameters. Product or package physical features, such as dimensions, minimum and maximum? Other characteristics: irregular shape, background colors, surface composition? Is it reused (e.g., pallet, tote, cart)? Is each package unique to a given product, or is it used for several products? Available coding area on sides? On top? If currently labeling, what does label look like? How is it produced? How is it applied? Any constraints on label size or appearance?

Code reading parameters. Product flow conveyorized? Other? Conveyor width? Speed? Nature of product transport? Desired item throughput? Environment: dusty, damp? Possible contaminants (e.g., paints, coolants)? Temperature?

System Output Requirements. Will the system control product or package flow? If so, what are the diverting or decision points? What are the routine management reporting or other data systems output requirements? As appropriate, what existing data processing capabilities could be employed in the system? Type of computer hardware, size, current peripherals?

Once the foregoing details are assembled and *before* final specifications are established, the in-house automatic identification systems person should consult with equipment suppliers to determine the combination of coding and code reader best suited to the requirements of the system.

Years ago at a material handling seminar, a speaker stressed the importance of totally integrated manufacturing and distribution operations, noting that "the materials handling system of the future must be able to respond to automatically generated material management commands" and that "the response must be fast enough to satisfy the particular timing requirements of the overall operation." Later, at a packaging show, a plant manager warned that success in automating and speeding materials flow has too often been dimmed by the absence of a parallel effort aimed at obtaining real-time information on operation status from the line.

Automatic identification systems offer answers to both challenges by providing the timely, accurate input essential to reliable material handling systems performance and the data collection integrity so important to efficient, total operations management.

The following listing, in chronological order, is by no means exhaustive. It does, however, provide a broad cross section of materials on the various identification systems technologies and a number of their applications.

BIBLIOGRAPHY

Wilcox, R. S. "Diverting by light beams." *Material Handling Engineering* (September 1962).

Wilcox, R. S. "Coding Products with Retro-Reflective Material." *Automation* (February 1965).

"Warehousing Shoes Under Computer Control," *Automation* (July 1965).

Lefer, Henry. "Handling Systems Join as Warehouse Moves In Next Door." *Material Handling Engineering* (February 1967).

"Palletizing Various Chewing Gum Products." *Automation* (May 1967).

"Sorting, Accumulating & Palletizing Cases." *Automation* (June 1968).

Ralston, E. L. "15 Ways to Control Moving Materials." *Control Engineering* (September 1968).

"Selecting Photoelectric Controls: LOW COST CODE READING." *Modern Materials Handling* (November 1970).

"Dispatching Speeds Output: Keeps Pickers Picking." *Modern Materials Handling* (May 1971).

"Automatic Railroad-Car Identification." *Electronics World* (September 1971).

Hill, John M. "How to Read a Label That's Moving 17 mph." *Material Handling Engineering* (Special Issue, Fall 1972).

"Are You Asking Enough of Your Label?" *Material Handling Engineering* (Special Issue, Fall 1972).

Bowlen, R. W. "ABC's of Scanning." *Industrial Engineering* (1973).

"Automatic Identification—How to Speak the Language." *Modern Materials Handling* (March 1974).

"The Emerging Impact of Automatic Identification Systems on Material Handling." Automatic Identification Manufacturers (AIM) Product Section of Materials Handling Institute (MHI), May 1974.

"Printers and Readers Handshake for Effective Bar Code Systems." *Computer Design* (September 1974).

"Because Reliable Sorting is a Must." *Modern Materials Handling* (October 1974).

"All About Data Collection Equipment." *Datapro* (1974, 1975, 1977).

"Automated Warehouse Prints Own Labels for Maximum Flexibility and Minimum Cost." *Rubber Age* (March 1975).

"Bar Codes for Data Entry." *Datamation* (May 1975).

"Scanner Sets-Up Computer for Conveyorized Sorting." *Modern Materials Handling* (November 1975).

Wurz, Albert. "Automatic Identification: A New Code Symbol for Shipping Containers." *Material Handling Engineering* (Special Issue, Fall 1975).

"Automatic Identification Systems for Material Handling & Material Management." AIM Product Section of MHI, March 1976.

"Source Data Automation." *MiniMicro Systems* (May-June 1976).

Overbeke, James E. "Bar Codes on the Assembly Line." *Industry Week* (June 14, 1976).

Benson, James R. "The Intelligent Warehouse." *Datamation* (September 1976).

Hill, John M. "Automatic Identification in Mechanized Systems—The Standardization Challenge." *Material Handling Engineering* (Special Issue, Fall 1976).

"Kresge's Sorting System—120 Cases per Minute." *Modern Materials Handling* (October 1976). (Voice recognition)

Deisenroth, Michael P., and Colin L. Moodie. "An Introduction to Automatic Identification Systems in an Industrial Discrete-Part Manufacturing Environment," School of Industrial Engineering, Purdue University, December 1976.

Mishra, D., and P. Frymire. "Design of Automatic Identification Systems for Manufacturing Control." *Proceedings*, 1977 AIIE Annual Conference.

Percival, Don R. "Photoelectric Sensors for Automated Material Handling." *Proceedings*, Material Handling Systems and Controls Conference, February 1977 (available from MHI).

Bowlen, R. W. "Bar Codes for Industry." *The Platemaker's Bulletin* (April 1977).

Mara, Charles E., and John M. Hill. "Monitoring Assets-In-Process." Presentation to Ford Motor Company's Quinquennial Material Handling Seminar, April 1977.

"A Near Ultimate in Receiving Systems!" *Modern Materials Handling* (September 1977).

Israelsky, E. W., and O. O. Gruenz, Jr. "Human Limitations in Using a Portable Light Pen Device for Manually Scanning Bar Code Patterns." Human Factors Society, October 1977.

Hill, John M. "Warehousing and Order Selection." *Proceedings*, 1977 National Material Handling Forum (available from MHI).

"Avoiding Errors in Shopfloor Data Entry." *Production Engineering* (October 1, 1977).

"The Computer Listens," *New Engineer* (December 1977).

Serchuk, A. "Lasers Open New Packaging Horizons." *Modern Packaging* (January 1978).

Nelson, Ben. "Why Not OCR-A? Some Other Uses." *OCR Today* (February 1978).

"Symbology Study Group Presents First Results of Test Programs." *Paper Board Packaging* (March 1978).

"Photoelectric Sensors: How to Get Top Performance!" *Modern Materials Handling* (March 1978).

"Almost 100% Accuracy in a Case Sorting System." *Modern Materials Handling* (April 1978).

"Scanners and Computers: The Four Jobs They Do Best." *Modern Materials Handling* (July 1978).

Wilderman, Michael B., and Thomas D. Windham. "A Survey of Optical Bar Code Technology Applicable to Air Force Activities." Wright-Patterson AFB, Ohio, School of Systems and Logistics, September 1978.

"Sorting System Provides Accurate Handling for Avon Distribution Center." *Food, Drug & Cosmetic Manufacturing* (January 1979).

"Automatic Identification Equipment" (5 presentations). *Proceedings,* 1979 Automated Material Handling & Storage Systems Conference, March 1979 (available from MHI).

"Identification Systems & The Material Handling Information Blitz." *IMMS Pacesetter,* 3rd Quarter, 1978, and May 1979.

"A Computer Helps Us Label 40,000 Parts a Day!" *Modern Materials Handling* (June 1979).

Struble, Lee, Al Seiferheld, and Bruce Blomgren. "IBM's Showcase for Incoming Materials." *Modern Materials Handling* (June 1979).

"Photoelectric Systems Regulate Conveyor Traffic." *Modern Materials Handling* (Midyear 1979).

Krauss, Dale. "Down-to-Earth Scanning." *Production Engineering* (July 1979).

Andersson, Edmund P. "How Scanners Add Depth and Versatility to Computer-Based Systems." *Material Handling Engineering* (July 1979).

Farrell, Jack W. "Cost/Performance Measurement." *Traffic Management* (September 1979).

Andersson, E. P. "Automatic Identification Enters A New Era." *Proceedings,* 3rd International Conference on Automation in Warehousing, SME, Dearborn, Michigan, November 1979.

"Bar Code Scanner Inputs Data at High Speed." *Design Engineering* (December 1979).

"ITT Courier: The Quest for Perfect Shop Floor Control Moves Ahead." *Material Handling Engineering* (February 1980).

"Photosensors: The 'Eyes' of Material Handling Systems Can Now 'Think' for Themselves." *Material Handling Engineering* (February 1980).

"Bar-Coded Inventory Saves Us $100,000 a Year!" *Modern Materials Handling* (February 1980).

"Proof at Last! We CAN Print a Scannable Symbol on Corrugated Cases." *Modern Materials Handling* (March 1980).

"How Portable Terminals Boost Inventory Accuracy." *Modern Materials Handling* (April 1980).

Bushnell, Rick. "Moving Beam Scanners: Current Applications & Future Solutions" Charles E. Mara. "New Technology in Bar Code Printing Techniques"; Dean C. Percival. "Fixed Beam Code Reading." *Proceedings,* 1980 Automated Material Handling & Storage Systems Conference, April 1980 (available from MHI).

An Introduction to AIM, April 1980 (available from MHI).

"Packaging: The Major Trends." *Modern Materials Handling* (May 1980).

"Image Recognition—New for High Speed Sorting!" *Modern Materials Handling* (May 1980).

Hill, John M. "Computers and AS/RS Revolutionize Warehousing." *Industrial Engineering* (June 1980).

Nelson, Ben. "Coding Can Boost Productivity in Manufacturing & Distribution Process." *Industrial Engineering* (June 1980).

"Don't Give Up on That Multi-floor Warehouse." *Modern Materials Handling* (August 1980).

"In This Warehouse—Control Like Never Before." *Modern Materials Handling* (August 1980).

Hill, John M. "The Growing Significance of Automatic Identification in Package & Unit Load Handling." *Proceedings,* ASME Century II—Emerging Technology Conference, August 1980.

Hill, John M. "Controlling P & I with Contemporary Material Handling Systems." *Industrial Engineering* (September 1980).

Quinlan, Joe. "New Bar Codes for Corrugated." *Material Handling Engineering* (September 1980).

"ITT Courier: Implementing the Master Plan—a Day at a Time." *Material Handling Engineering* (October 1980).

"Fixed Beam Code Readers: Low-Cost Routing and Sorting." *Modern Materials Handling* (October 1980).

"Machines Identify, Measure and Inspect, Yet Fall Short of Vision." *Control Engineering* (November 1980).

"Pack Expo 1980—What the Editors Found." *Modern Materials Handling* (December 1980) (distribution symbol).

"Practical Pattern Recognition Approaches to Industrial Vision Problems." *Assembly Engineering* (December 1980).

"Sortation Enters the Electronic Age." *Material Handling Engineering* (February 1981).

"IBM's Lexington Warehouse—A Marvel of Computer Control." *Modern Materials Handling* (February 1981).

"Double Sorting: The Way to a New Level of Customer Service." *Modern Materials Handling* (February 1981).

Andersson, Edmund P. "Industrial Bar Code Applications Help Control Inventory, Verify Assembly." *Industrial Engineering* (April 1981).

"Pick Car's Travelling Labeler Puts Bar Codes on Cases—at Random, Too." *Packaging Engineering* (May 1981).

Hill, John M. "Sortation/Automatic Identification." *Proceedings,* 1981 National Materials Handling Forum (available from MHI).

"Recommended Practices for Uniform Container Symbol/UCS Transport Case Symbol/TCS." Distribution Symbology Study Group, September 1981 (available from AIM/MHI).

"–the new industrial language." *Material Handling Engineering* (December 1981).

Harmon, Craig K. "Automated Identification Systems." *Proceedings,* 1982 Cedar Rapids Industrial Productivity Conference, June 1982 (Available from the Society of Manufacturing Engineers, Dearborn, Michigan).

"Homing In On The Range—or Factory." *Business Week* (September 20, 1982).

Proceedings, Scan-Tech '82 (November 1982) (available from AIM/MHI).

Automatic Identification Manufacturers Manual, January 1983. Updated periodically. Contains AIM member listing, papers on various code reading systems, uniform symbol descriptions, recommended practices for UCS/TCS, article reprints, glossary, and bibliography (available from AIM/MHI).

"Magnetic Striping: Shop Floor Control/Failure Analysis Tracking." *P & IM Review* (February 1983).

"Bar Code and Voice Recognition Ease Data Entry Problems"; "European Firms Eye U.S. Vision-System Market"; "Data Collection Devices Play Key Role in Automated Factories." *Mini-Micro Systems* (June 1983).

"6 of the Newest Identification Techniques." *Modern Materials Handling* (August 1983).

"Bar Code Guide and Survey Detail What's Available, How It Can Smooth Your Operation." *Industrial Engineering* (September 1983).

Hill, John M. "Radio Frequency (RF) Systems." *P & IM Review* (September 1983).

"Automatic Identification." *Modern Materials Handling* (September 1983).

Proceedings, Scan-Tech '83, (September 1983) (available from AIM/MHI).

"The Latest Trends in Automatic Identification." *Modern Materials Handling* (December 1983).

"How Magical Can a Wand Get?"; "How to Print Bar Codes and Make Sure They're Scannable." *Material Handling Engineering* (December 1983).

"Automatic Data Gathering Anywhere on the Shop Floor"; "Use Advanced Technologies to Automatic Manufacturing." *Modern Materials Handling* (January 1984).

"Machines That Can See: Here Comes a New Generation." *Business Week* (January 9, 1984).

"Source Marking Slashes Errors, Speeds Throughput." *Modern Materials Handling* (January 1984).

"It Takes Vision to Sell Vision." *Electronic Business* (February 1984).

"Basic Elements of Bar Code Scanning and Printing." *1984 Material Handling Engineering Handbook & Directory.*

Periodicals

Bar Code News, bi-monthly publication of North American Technology, Peterborough, New Hampshire.

Scan Newsletter, monthly publication, Great Neck, New York.

Standards

Recommended Practices for Uniform Container Symbol/UCS Transport Case Symbol/TCS, September 1981.

Standard Symbology for Marking Unit Packs, Outer Containers & Selected Documents, MIL-STD-1189, January 4, 1982.

Standard for Shipment and Storage, MIL-STD-129H, January 4, 1982.

Final Recommendations of the Health Industry Bar Code Task Force, September 1983. American Hospital Association, Chicago, Illinois.

American National Standard Specifications for Bar Code Symbols on Transport Packages and Unit Loads, ANSI MH10.8-1984.

CHAPTER 4.8
MAINTENANCE

W. COLEBROOK COOLING

Consultant
South Orange, New Jersey

It is appropriate that a section on maintenance be included in this handbook. Maintenance is a function that has impact on normal customer requirements of quality, delivery, and cost no matter what the product, whether chemical, machined, fabricated, or assemblies.

Maintenance requirements are different in various facilities, but there is a commonality when expressed in broad terms. This chapter emphasizes that two functions, production and maintenance, must work together rather than operate as adversaries. Production as the customer and maintenance as the vendor have the same missions and obligations to the facility: quality of service, timeliness of service, and cost of service. The missions will be enhanced when each function complements the other.

4.8.1 ORGANIZING FOR MAINTENANCE EFFICIENCY

The motto "people working together" illustrates appropriately the necessary relationship between production and maintenance. Just as a working relationship develops between vendor and customer, the maintenance organization should be designed so that production (customer) and maintenance (vendor) develop a continuing working relationship. One fact that creates the need for this relationship, which enhances communications and cooperation, is that production priorities for maintenance work are going to change, sometimes on a daily basis, as a result of operating needs. Production must be disciplined to submit work requests in time to plan and to avoid changing work priorities. Maintenance must also be flexible in order to meet the requirements of production. Both functions must work together to facilitate the missions of the facility.

To ensure a good production-maintenance relationship the ideal organization must organize maintenance by geographic area rather than craft lines. With craft lines, the production supervisor deals with a number of maintenance supervisors. This often leads to questions about who is in charge or who is responsible. When maintenance supervisors are assigned to areas, departments, or some other boundary, production and maintenance supervisors deal with each other on a continuous basis, and the working relationship is established. Daily discussions between the two functions will establish priorities, and more will be accomplished.

The maintenance mechanics should be assigned to geographic areas and report to the maintenance supervisor of the area or production department. The needs with regard to numbers of workers and skills required for each area must be considered in this division of mechanics.

It must be anticipated that maintenance supervisors normally have expertise in specific trades; therefore, each must be available to assist other maintenance supervisors. Also, the division of mechanics must be flexible. As needs arise for additional man-hours or skills in other areas, mechanics will be moved on a temporary basis for their area.

4.8.2 MAINTENANCE BUDGETING

Facilities that place the entire maintenance budget in the maintenance function are in error. The production managers should have the monies for maintenance in their budgets. These managers are responsible for the equipment and certain items in their area of work and thus should have primary interest in maintenance requirements and the cost of maintenance. Maintenance should have monies for buildings, grounds, utilities, maintenance, and so forth but not for items for which production has responsibility. This division should encourage mutual interest in controlling costs and improving maintenance effectiveness as follows:

Nuisance work requests will be reduced.

Production will avoid practices that result in excess repair costs.

Maintenance needs will be anticipated, resulting in time to plan and schedule.

Production and maintenance efforts will be coordinated to reduce delays.

4.8.3 MAINTENANCE FEATHERBEDDING

There is more featherbedding in maintenance operations than any other production function. Restrictive maintenance work rules and condoning poor work habits seriously detract from the efficiency of the facility. Instances of featherbedding increase maintenance costs by:

Adding nonproductive hours to the skilled craftsman's day (travel, job interference, crew balance).

Failing to utilize all the skills and abilities of individual craftsmen.

Causing craftsmen to be lax in their commitment to time and improving their skills.

Discouraging the initiative of individual craftsmen.

Wasting supervisory time.

Complicating and increasing the cost of management controls such as planning and scheduling.

Featherbedding increases production costs by:

Adding to the cost of maintenance.

Increasing equipment downtime.

Failing to use all the abilities of production employees.

Increasing paid hours not worked (waiting for maintenance).

Loss of product during equipment downtime.

In facilities that have union representation, most of these losses occur at the bargaining table. Management must develop a "stiff upper lip" and learn to say no to practices that dull their competitive edge. In the interest of the production mission, production must support maintenance at the bargaining table by promoting unskilled maintenance on equipment, eliminating restrictive craft lines, and ferreting out arbitrary work rules.

4.8.3.1 Unskilled Maintenance on Equipment

Provisions should be made for equipment operators to complete unskilled work on their own equipment. This would reduce job interruption and travel costs for maintenance, aid in planning and scheduling maintenance, relieve skilled craftsmen for skilled work, reduce equipment downtime, and reduce production employee's downtime.

The author has negotiated several labor agreements with international unions that include a provision that "production employees shall complete unskilled maintenance work on their own equipment." This was not the easiest proposal for the parties to reach agreement but worth the effort. Production employees liked the provision and maintenance employees finally accepted the provision.

After reasonable time following agreement, a semiautomated chemical processing operation confirmed that production employees "routinely" completed the following maintenance tasks:

Replacing pumps.

Replacing hoses.

Replacing carriers.

Replacing valves.

Replacing predryer boxes.

Replacing belts.

Replacing carrier arms.

Routine lubrication.

Break and clean plugged lines.

Minor conveyor repairs.

Replacing filters.

PVC pipe repair.

Open and clean Wildeir pumps.

The above was a result of, first, management cooperation at the bargaining table and, second, cooperation between production and maintenance in daily operations.

4.8.3.2 Eliminate Restrictive Craft Lines

Cost increases also occur when maintenance is restricted to craftsmen only working in their trade without flexibility to work in other trades. In an automotive industry agreement, there are fourteen craft trade classifications, with a provision stating "none shall be interchangeable." This creates so much featherbedding that the cost manpower wasted (both production and maintenance) can barely be measured. Planning and scheduling hours are costly, and there is an extreme deemphasis on individual pride and accomplishment.

Management can decrease their losses by at least providing that all maintenance crafts shall complete the unskilled work of other trades and the utility employees shall complete all unskilled maintenance work using hand and power tools. The most management can do is to combine the crafts into general maintenance and assign employees to work within their individual skills and abilities. There may be classifications such as general maintenance, electricians, shop machinists, instrument mechanics, and so forth. It is a fact that with fewer job classifications in maintenance, there is the more flexibility, and the more flexibility, the lower the maintenance costs.

4.8.3.3 Miscellaneous Work Rules

Originally, the bargaining units' interest in seniority provisions were to protect the senior employee in instances of promotions, layoff, and recall. Work rule proposals accepted by management at the bargaining table, however, have in many cases further reduced maintenance effectiveness. For example, management has allowed maintenance employees to choose jobs on locations by seniority on a daily basis and has condoned overtime by senior employees rather than by the maintenance crew that started a particular job. Production must insist that these costly practices be ferreted out by maintenance and then must support maintenance at the bargaining table.

4.8.4 PREVENTIVE MAINTENANCE BY PRODUCTION OPERATORS

Preventive maintenance (PM) as applied to equipment maintenance covers defined, periodic inspections of equipment to uncover parts or conditions that will cause equipment breakdowns, reduce equipment efficiency, or reduce the operating life of equipment. PM also includes the replacement or adjustment of minor components, such as belts and filters.

The four cost factors related to PM in the maintenance function are travel, setting PM aside in favor of repairs, depletion of craft skills, and overhead costs in planning and scheduling.

There are three prime considerations, other than cost, in training equipment operators to complete PM work on their own equipment. Production should:

Protect company assets charged to their area.

Keep the equipment running properly.

Assign more responsibility to production employees, thus fostering pride and the other behavioral favorable to increased production.

A great deal of cooperation between production and maintenance is necessary to accomplish this worthwhile objective. Maintenance should

Establish PM requirements, with tasks in sequence and time schedules for tasks.

Establish sign-off method for production to verify PM.

Assist with operator training and instruction.

Review the PM effectiveness with production through a continuing spot check process.

Preventive maintenance is neglected in most facilities. Many facilities have a well-defined program that fails due to a shortage of man-hours caused by the unexpected nonscheduled maintenance work. Placing this PM responsibility in the hands of production employees eliminates this man-hour shortage in maintenance.

4.8.5 EQUIPMENT LUBRICATION

Production supervisors, whose equipment interest is dependable operation, will insure proper lubrication of equipment charged to them. Proper lubrication, both with timeliness and the proper lubricant, will reduce maintenance charges, equipment downtime, lost production capacity.

Unlike PM, in most instances, specially trained maintenance employees should be responsible for equipment lubrication:

Correct lubricants must be used.

Lubricants must be properly applied in the correct amounts.

It is not economical to store lubricants and lubrication equipment at each piece of equipment.

Certain lubricants may be reclaimed.

Lubricants must be kept free of contamination, whether in the application or reclamation process.

Some lubricants may be flammable and require separate storage area.

Lubricants must be properly identified and stored.

There are definite responsibilities for operating employees with regard to lubricating equipment, and supervisors should train these operators to:

Keep equipment and work areas clean of lubricant and dirt.

Observe and report to their supervisors any signs of excess lubricant and leakage.

Apply oil carefully and refrain from turning down grease cups with force, breaking bearing seals, creating risks with slippery footing, and fire risk.

Cooperate with the lubrication specialist.

The production supervisor who is truly interested in proper functioning of his equipment will insist on proper lubrication specifications for each piece of equipment, regularly scheduled lubrication, lubrication sign-off records at each piece of equipment, and a general inspection at regular intervals with the maintenance representative who is the lubrication specialist. This inspection to review all matters pertaining to lubrication should be both a maintenance and operating responsibility.

4.8.6 CORRECTIVE ACTION

Too many production and maintenance managers adopt the "custodial" approach to maintenance; that is, they wait for costs to occur and then make a repair. Five areas must be addressed before the custodial approach can be eliminated:

1. Asset destruction.
2. Poor craftsmanship.
3. Planning and scheduling.
4. Equipment records.
5. Assigned maintenance.

4.8.6.1 Asset Destruction

There are many managers who condone damage that is the fault of the employee and take the loss rather than face the problem with the careless employee. Examples are damage through careless handling of portable power tools and careless driving of in-plant vehicles that damages doors, columns, and so forth, and the vehicle.

Production and maintenance must assume the responsibility of preventing asset destruction by employees. Normal corrective action procedures are necessary. The seriousness and consequences of careless acts are an important consideration. A facility does not have to expose its assets to rampant destruction and its employees to danger before making a discharge.

4.8.6.2 Poor Craftsmanship

Maintenance is responsible for the work of their employees. Thus, production, which pays the bill, should not pay for "callbacks" and should not condone such inconvenience to operations on a regular basis. A method must be established to identify "callbacks," to charge maintenance rather than production, and to correct the workman's performance.

Poor craftsmanship must be judged on an individual basis, by frequency, by actual severity with regard to both cost and safety, and by the potential severity.

4.8.6.3 Planning and Scheduling

Production must insist that there be an effective work order method to facilitate planning and scheduling. Without it, production pays for:

Ineffective work, unauthorized work, and delays, in short a "firehouse" operation.

Missing definitions of work to be done, thereby hindering maintenance planning.

Incomplete equipment records, which prevent supervisors from reprimanding individuals for asset damage and poor craftsmanship.

Incomplete and inaccurate time charges.

It is surprising how many production managers dislike a controlled maintenance force and a clear-cut work order method, assuming that there is more "flexibility" in the "firehouse" approach. The astute production manager complements maintenance by insisting on a fast response control method designed to meet the facility's needs. This manager will anticipate maintenance needs and submit a work order in time to plan, thereby improving the maintenance service in his areas of responsibility.

It is a fact that most maintenance control systems are "overdesigned" by systems experts who believe maintenance can be controlled by paperwork and the computer. In this case, production must support maintenance.

4.8.6.4 Equipment Records

Equipment records are often poorly maintained or completely overlooked. Production must insist that maintenance adopt a simplified method to create and maintain the following equipment records:

A permanent record of basic equipment data is required for continuing reference.

Space parts requirements and storage locations should be documented on equipment records.

Preventive maintenance, while originally established through the supplier's recommendations, should be reviewed based on operating experience.

Maintenance charges, analyzed by repair items, are necessary to justify monies for replacement, restoration, and additional spare parts.

Repairs must be identified by what, when, why, and who completed the repair.

Federal regulations, such as compliance with OSHA, requires records of equipment maintenance for pinch point equipment, such as power presses, man lifts, cranes, and wire rope.

Equipment records are not a luxury but a necessity, and production must insist and then cooperate with maintenance to install and maintain the collection of equipment data.

4.8.6.5 Assigned Maintenance

It is common practice to assign a maintenance mechanic to an area so that in the event that an adjustment or repair is required on operating equipment, the mechanic is available. This assignment is made to avoid loss of production through equipment downtime and reduce downtime charges of operating employees.

Maintenance time is charged to the department. In the custodial approach to maintenance, there is no record of the adjustments and repairs that are made by the mechanic. Production must realize that if it is necessary to keep a mechanic in the area, there must be a number of production delays. The adjustments or minor repairs must be recorded on a shift basis to analyze the causes of the delays by frequency. It is worth the effort to identify the causes of breakdowns so that an engineering effort can be made to improve equipment performance.

The potential benefit is increased production rather than reduction in maintenance hours. Production may still require the safeguard of an assigned mechanic, but the reduction in equipment downtime through a cause analysis is the payoff.

4.8.7 MAINTENANCE IMPROVEMENT

Maintenance is an area that sometimes lacks dedication to improvement. This is not always due to the managers in maintenance but in many instances to managers of production with their attitude and approach to maintenance.

When production works with maintenance, as suggested throughout this chapter, there will be obvious improvements. Managers cannot be satisfied to live with things the way they are and hope everything will work out. Managers must have a continuing goal of "making everything change for the better."

BIBLIOGRAPHY

Cooling, W. Colebrook. *Simplified Low Cost Maintenance Control*. New York: AMACOM, 1983.

Cooling, W. Colebrook. *Corrective Action*. Barrington, IL: TPC Training Systems, 1983.

PART 5

MATERIAL: ASSET MANAGEMENT

CHAPTER 5.1

MATERIAL REQUIREMENTS PLANNING

WILLIAM L. BERRY
WARREN BOE

The University of Iowa
Iowa City, Iowa

5.1.1 INTRODUCTION

Material Requirements Planning (MRP) has been an evolving concept in the planning and control of operations in manufacturing firms. Its early application, in the 1960s, was a bill-of-material explosion technique for determining the time-phased requirements for individual products and a method for launching manufacturing and purchase orders into actual production. Later, in the 1970s, it was recognized that material requirements planning provides a means of determining production priorities, i.e., determining when open (released) orders need to be rescheduled in order to meet changes in customer requirements for products. The concept of MRP therefore shifted from that of inventory planning to scheduling. Gradually, the MRP concept began to represent much more than simply an inventory planning or a scheduling technique for manufacturing. The notion of a "closed-loop MRP system" evolved, which represented a much broader range of manufacturing planning and control functions, including those of master production scheduling, capacity requirements planning, and shop floor control as well as material requirements planning. A closed loop MRP system provides a means of monitoring and obtaining feedback from manufacturing and purchasing operations in order to insure that the manufacturing plans are, in fact, being implemented.[1,10,16]

The evolution of MRP systems has continued in the 1980s, to encompass a broad range of planning and control functions in manufacturing, beginning with the development of an integrated set of marketing, financial, and manufacturing plans at an executive level, i.e., business planning. Next is the implementation of these plans at a detailed level in manufacturing, i.e., in the plant and at vendors. This view of MRP systems, increasingly being referred to as *Manufacturing Resource Planning* (MRPII),[12] involves the use of a formal system for planning and controlling manufacturing operations. Such systems include a manufacturing database, procedures for updating the database to reflect the actual status of operations, and other computer software to aid in both the development of manufacturing plans and their day-to-day execution. The diagram in Fig. 5.1.1 provides a framework for defining the various functions in Manufacturing Resource Planning and their relationship to one another. One of these functions, labeled material requirements planning, involves the development of detailed material plans using time-phased bill-of-material planning techniques similar to those called material requirements planning (MRP) in the 1960s and 1970s. In this chapter our purpose is to define MRP in both its broadest sense, as manufacturing resource planning (MRPII), and in its more limited sense as a material planning technique (material requirements planning). We begin by describing the various functions in MRPII and then turn to the use of MRP techniques for developing time-phased material plans.

5.1.2 MANUFACTURING RESOURCE PLANNING (MRPII)

The basic elements of a manufacturing resource planning system are shown in Fig. 5.1.1. Three of these elements relate directly to the material planning function (material requirements planning): master production scheduling, capacity requirements planning, and shop-floor control. The arrows indicate the flows of information among these elements.[2,10] In this section each of the elements in MRPII is defined, to provide an overall understanding of the various functions that are performed, as well as the relationship of the material planning function to other planning and execution systems.

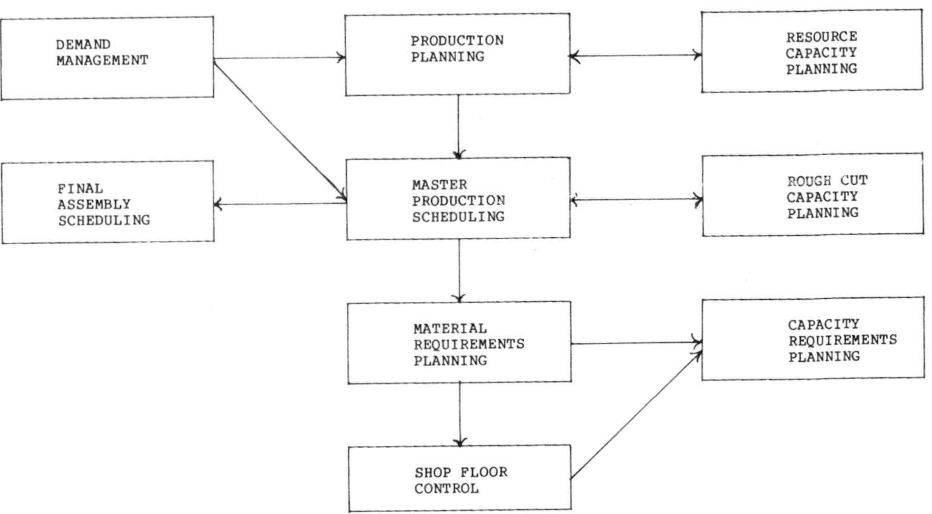

Fig. 5.1.1. MRPII system functions.

5.1.2.1 Demand Management

Demand management represents the forecasting, order entry, order promising, and physical distribution activities in a company. This function includes all the activities that place demands (requirements) for products on manufacturing. These demands may take the form of actual and forecast customer orders, branch warehouse requirements, interplant requirements, international requirements, and service part demand. Clearly, some of these categories of demand will assume more or less importance in particular companies. For instance, the manufacturing output at some firms may be entirely directed at the replenishment of distribution warehouses, while at other firms the production output may be primarily directed toward satisfying individual customer orders. At still other firms, manufacturing may have to satisfy demand from both of these sources.

5.1.2.2 Production Planning

Production planning represents the activities involved in preparing an overall production plan for the business. Such a plan represents a strategic "game plan" for the company, which reflects the desired aggregate output from manufacturing. In some firms the production plan is simply stated in terms of the monthly or quarterly sales-dollar output for the company as a whole, or for individual plants or businesses. In other firms, the production plan is stated in terms of the number of units to be produced monthly in each major product line for the next year. This plan represents top management's "control knob" on the business, and provides the guidelines within which manufacturing is expected to operate. The production plan also represents an agreement among marketing, manufacturing, and finance as to what will be produced and made available for sale to the customers.

5.1.2.3 Resource Planning

Resource planning is the process for determining the long-range capacity needs of the business. This involves translating the long-range sales forecasts and production plans covering the next one to ten years into the required manufacturing facilities. The time horizon considered reflects the lead time required to acquire new facilities. Resource planning often provides the basis for the capital budgeting activities in a company.

5.1.2.4 Final Assembly Scheduling

Final assembly scheduling is the basis for planning and controlling the final assembly and test operations in manufacturing. The preparation and the execution of the final assembly schedule is separate from master scheduling. The master production schedule represents the "anticipated" build schedule for a firm's products, covering the time span between raw-material acquisition and the delivery of the component items to the inventory just before final assembly. The final assembly schedule specifies the "actual" build schedule, beginning at the time when all the components are available for final assembly and ending when the products are shipped to customers. It represents a commitment to the production of specific end-product configurations and serves to accumulate the necessary component parts.

5.1.2.5 Master Production Scheduling

Master production scheduling is the disaggregation of the production plan into a schedule of individual products and the maintenance of this schedule over time. The master production schedule represents a statement of the "anticipated" build schedule for a company's products, and not a forecast of market demand. Typically, this schedule indicates the quantity of specific product configurations to be produced in weekly time periods during the next six months or longer. The product unit selected for master scheduling varies among firms, e.g. end-product items may be used in make-to-stock firms, while a collection of high-level component items from a planning bill of material may be used in make-to-order firms. The master production schedule, in effect, drives the material requirements planning element shown in Fig. 5.1.1, since it supplies the demand information needed to develop the material plan.

5.1.2.6 Rough Cut Capacity Planning

Rough cut capacity planning is the analysis of the master production schedule to determine what critical manufacturing facilities are potential bottlenecks in the flow of production. Such analysis, typically made during the master schedule review or whenever adjustments are made to the master production schedule, focuses on the critical operations in a firm, such as machines that operate on a three-shift basis, critical labor skills, or vendors who supply a key raw material. The rough cut capacity analysis is typically performed less frequently than the weekly shop load analysis performed by the capacity requirements planning function in Fig. 5.1.1.

5.1.2.7 Material Requirements Planning

MRP is the material planning activity for all the components and raw material needed to meet the master production schedule as well as such demands as spare parts and pilot models. Inputs to the material requirements planning function include the master production schedule, bills of material, on-hand inventories, open shop and purchase-order status information, and planning parameters such as lead times, order quantities, safety stock, and shrinkage factors. Using this information, the material requirements planning function explodes the master production schedule to prepare a time-phased plan, which indicates the requirements for subassemblies, components, and raw material items. Open-order and on-hand inventory quantities for such items are "netted out" (deducted) to determine the actual requirements. MRP also determines the due dates for newly released orders as well as any adjustments that are required in open shop and purchase orders.

5.1.2.8 Capacity Requirements Planning

Capacity requirements planning estimates the capacity needs for individual work centers required to meet the material plan produced by the MRP function in Fig. 5.1.1. Inputs to this function include the material plan, the status of open shop orders, and routing and time standards information from manufacturing engineering. Typically, this function provides weekly shop-load information, reported in standard hours, covering a six- to twelve-month time period. Such information is used by manufacturing supervisors in preparing manpower plans and assignments.

5.1.2.9 Shop-Floor Control

Shop-floor control is short-range scheduling and control of plant and vendor operations, involving issuing shop and purchase orders, scheduling, and monitoring the flow of open orders. Shop-floor control reports provide a means of determining which orders to run next at individual machines, assigning operators to machines, and tracking the progress of each manufacturing order through the factory. The planning is at the level of individual machines and operations. Similar functions can be used to schedule vendors and such preproduction activities as engineering and drafting. This is often referred to as an *execution function*; that is, it provides a means of insuring the execution of the material plan as well as a means of obtaining routine feedback concerning any differences between actual operations and the plan.

The diagram shown in Fig. 5.1.1 provides an overall indication of the various functions performed in a manufacturing resource planning (MRPII) system. The concept is to provide a means of translating the company's strategic plans, made at the executive level and involving the marketing, financial, and operating aspects of the business into operational plans that can be executed on the shop floor. The actual implementation of MRPII will differ in individual businesses. Some of these functions may be performed manually, whereas others use computer support. Likewise, there are substantial differences in the time frame covered by each of these functions. Figure 5.1.2 illustrates the approach taken by one firm in implementing MRPII. The diagram indicates the functions included in the MRPII system at a manufacturer of small electrical appliances and the specific linkages between these functions. It also specifies the time horizons involved. The integration between the marketing, manufacturing, and financial planning functions is evident, as is the specific information considered. While the diagram in Fig. 5.1.2 reflects the nature of the small appliance business, the hierarchy of the production planning and control functions presented in Fig. 5.1.1 is readily apparent in Fig. 5.1.2.

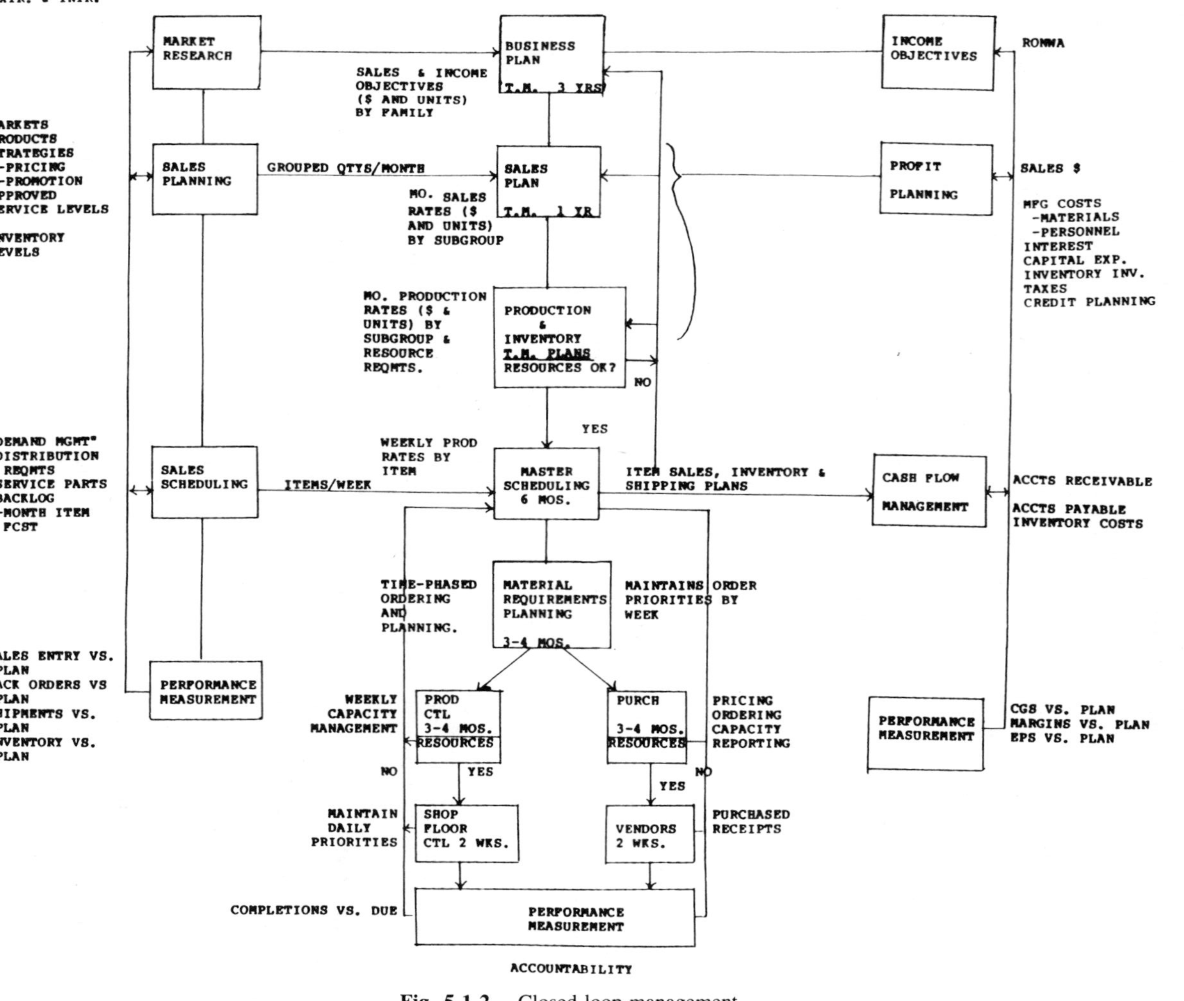

Fig. 5.1.2. Closed loop management.

5.1.3 MATERIAL REQUIREMENTS PLANNING

We now turn to the more restrictive definition of material requirements planning: a method for translating a master production schedule into a set of requirements for individual component items.[8] Three important aspects of this method for developing detailed material plans are discussed in the sections which follow: fundamental MRP concepts, the preparation of time-phased MRP records, and the use of MRP records by an inventory planner for decision-making purposes.

5.1.3.1 MRP Concepts

Three important concepts underly the use of bill-of-material explosion techniques in material requirements planning. These include the principle of dependent demand, the structuring of a manufacturing bill of material, and the use of gross-to-net logic in performing the bill-of-material explosion. Let us illustrate these with an example involving the manufacture of an automobile fan. The bill of material shown in Table 5.1.1 indicates that four component items are required to manufacture this product.

One of the simplest bill-of-material explosion techniques is called "Quickdecking," which is a single-level bill-of-material explosion. For example, suppose that the master production schedule at the automotive fan plant indicates that 500 fans are to be built each week for the next eight weeks. Using the quickdeck bill-of-material explosion approach, 500 blades, armature kits, and motor housings, as well as 1000 hardware kits are required each week. In this case the master schedule quantity each week is multiplied by the usage rate for each item on the bill of material to calculate the weekly requirement for each component. Thus, the demand for these components is derived from the production schedule for an end product, the master production schedule. The demand for manufactured and purchased components is considered to be "dependent demand," since it can be calculated from the end-product production schedule rather than forecast using short-term forecasting methods such as moving averages and exponential smoothing. The concept of dependent demand underlies material requirements planning.

While manual quickdecking has been used successfully for many years by manufacturing firms, it has two limitations. First, this bill-of-material explosion technique does not take into account when the components and raw materials are needed to build the final product. Second, unnecessary inventories of component parts can be accumulated. To illustrate these points suppose that the automotive-fan plant actually builds the product in two assembly stages. In the first stage the armature kit, the motor housing, and one hardware kit are combined to produce a motor, and in the second stage the motor, the blade, and a second hardware kit are combined to produce the completed fan. Each stage requires one week for completion; thus the motor components are needed one week before the fan components. The indented bill of material shown in Table 5.1.2 represents the way the automotive fan is actually built. Here each component is indented below its parent item in such a way that the product structure can be identified.

Now let us further assume that the automotive fan plant currently has the inventory indicated in Table 5.1.3 for each of the fan components. These inventories may build up if they are not considered prior to ordering additional components to meet the master production schedule of 500 fans per week. The use of gross-to-net logic in the current material requirements planning software will prevent the buildup of such unnecessary inventory.

To illustrate the use of gross-to-net logic, we refer to the total requirement for each item above as the "gross" requirement, while the amount to be actually ordered (net of the current inventory) is referred to as the "net" requirement. Note, for example, that only 300 fans need to be built in the first week to meet the master schedule requirement of 500. Likewise, only 150 blades need to be manufactured, since there are already 150 units in stock. Finally, only 150 motor housings need to be manufactured since there are 75 motors and 75 motor housings in stock.

By comparing the net requirement for each component in Table 5.1.3 with the quickdeck requirement of 500 units, it is apparent how quickly inventory could build up if the current stock is not netted out before placing orders for these component items. It is also clear why the indented bill of material, which indicates the timing and manner in which the product is built, is useful in determining the net requirement for the individual component items.

Table 5.1.1. Automobile Fan Bill of Material

Component Item	Usage per Fan
Blade	1
Armature kit	1
Hardware kit	2
Motor housing	1

Table 5.1.2. Indented Bill of Material for the Automotive Fan

Item	Usage Factor
Automotive fan	
Blade	1
Hardware kit	1
Motor	1
Armature kit	1
Hardware kit	1
Motor housing	1

5.1.3.2 Time-Phased MRP Records

The development of an indented bill of material, reflecting the way in which the product is built, and the use of the gross-to-net concept lead naturally to the use of *time-phased records* for individual product components. The time-phased record provides a way of taking into account the length of time required to manufacture a product item in the preparation of the material plan, yielding important reductions in work-in-process inventory and improvements in customer service. In fact, the time-phased record provides a complete summary of the current status of a product item, i.e., the projected requirements, the schedule of open orders, as well as orders planned for release in the future, and the projected inventory level. It is important to understand the information contained in the time-phased record, since it is a basic building block of an MRPII system, and is applied in other MRPII functions such as master production scheduling and distribution planning.

Table 5.1.4 is a time-phased record for one of the components in the automotive fan, i.e., the motor. The gross requirement row represents the total requirements for this item as exploded from the master production schedule for the automotive fan. Since one motor is used on each fan and 500 fans are to be built over the next eight weeks (except for week 1 when only 300 fans are required because of the current finished goods inventory position), the total requirements are as shown in the first line of the table. The second row, scheduled receipts, indicates the present schedule of the open (released) orders for this item that have been placed on either the plant (for an internally manufactured item) or on suppliers. In this case there is one open order for 1000 units due at the start of week 1.

The third row of Table 5.1.4 indicates the current and the projected inventory positions for this item during the next twelve weeks. The figures in this row are calculated by adding any open orders scheduled to be received in a period to the closing inventory from the previous period and subtracting the gross requirement for the period. This process of subtracting the gross requirement each period from the "available" material (scheduled receipts during the period plus beginning inventory) represents the way the "gross-to-net" concept is applied in time-phased records. At some point in moving from the current period (week 1) to future periods in these calculations, the available inventory may not be sufficient to meet future requirements and unsatisfied demand (net requirements) will be determined.* At this point the MRP scheduling logic will begin to create a series of orders planned for future release to the plant (or vendor), taking into consideration the ordering parameters for the item, i.e., the desired lot size, safety stock, and the necessary lead-time offset. These orders are shown in the fourth row of the MRP record—the planned order release row. An example is shown for the motor in Table 5.1.4, where the first planned order release of 1000 units is scheduled to begin production in week 1 in order to satisfy a net requirement of 225 in week 3.

Table 5.1.3. Net Requirements Table[a]

		Requirements	
Item	Current Inventory	Gross	Net
Fan	200	500	300
Blade	150	300	150
Hardware kit	125	300	175
Motor	75	300	225
Armature kit	100	225	125
Hardware kit	—	225	225
Motor housing	75	225	150

[a] The hardware kit appears in two places and its current inventory appears in only one place.

* For example, in Table 5.1.4 the current on-hand inventory of 75 plus the open order for 1000 due at the start of week 1 will be depleted in week 3, when an additional quantity of 225 units is required. A lot size of 1000 is then added to the projected number available at the start of week 3 under the assumption that such an order will be placed at the start of week 1 in order to prevent a stock-out in week 3.

Table 5.1.4. Example Time-Phased MRP Record: Motor[a]

		Week							
Motor		1	2	3	4	5	6	7	8
Gross Requirements		300	500	500	500	500	500	500	500
Scheduled Receipts		1000	0	0	0	0	0	0	0
Projected Available	75	775	275	775	275	775	275	775	275
Planned Order Release		1000	0	1000	0	1000	0	1000	0

[a] Lead time = 2, order quantity = 1000, safety stock = 0.

A careful review of the projected available row in Table 5.1.4 will indicate that the projected on-hand row calculations take into account information from each of the three other rows in determining the anticipated inventory position, i.e., the receipt of both open (previously released) orders and future planned orders, gross requirements, and the present inventory position. Furthermore, if a minimum stock position at the close of each period is desired, i.e., a given safety stock level, replenishment orders will be triggered when the projected stock level drops below the safety stock quantity, as opposed to when the projected inventory level reaches zero.

The planned order release row indicates when each of the future planned orders needs to be released in order to meet the unsatisfied net requirements. The lead-time offset considered in developing the production schedule indicated in this row represents the time required to obtain the item from a vendor or the time required to assemble or fabricate the item if it is obtained from the plant. The lead time to obtain any necessary raw material or components for the item is not considered as a part of this lead time, since it would be considered in the MRP record(s) for those items.

5.1.3.3 Level-by-level MRP Explosions

The time-phased MRP record, a basic building block in an MRP system, contains a summary of the information for an individual product item. In actual practice a company may have as many as 2000 to 100,000 or more items in the product structure to be controlled by the MRP system. We now consider how the information linkages are maintained between the various product items. Two devices are used in linking the MRP records. First, each product item is assigned a level number. Final products or major assemblies that are scheduled and controlled by the master production schedule are assigned the highest level code, level 0, because the bill-of-material explosion process begins with these items. Lower-level components and subassemblies, which are used in the final stages of manufacturing the product, are level numbers starting with 1 and increasingly larger as earlier stages in the manufacturing process are considered. For example, the components used in the final assembly of the automotive fan, e.g., the motor and the blade, are indicated as level 1 items, and the components used to assemble the motor, e.g., the armature kit and the motor housing, are assigned as level 2 items. The level numbers provide a method of determining the exact point in the bill-of-material explosion process (and in the manufacture of the product) at which each item is required.

In performing the bill-of-material explosion, MRP records are first constructed for the level 1 items, using the information in the master production schedule. After all of the level 1 MRP records have been completed, the level 2 MRP records are prepared, and so on, until all the levels in the product structure have been considered. The level numbers are used in order to insure that an MRP record is not prepared for an item until all of the information needed to determine its gross requirements is available.

One problem that is often encountered in performing the bill of material explosion is that some product items are used at several different levels in the product structure. The hardware kit in the automotive fan is an example of such an item. A second bill of material explosion device, "Low Level Coding," is used to insure that the correct linkages are maintained for such items. A low level code is assigned to each item in the product structure. The low level code for an item indicates the lowest level in the product structure at which that item is used. For example, the hardware kit is used at level 1 in assembling the fan, and at level 2 in assembling the motor. In this case, the code assigned to the hardware kit is level 2. The low level code is used to determine when to construct the MRP record for an item in the bill of material. This insures that the gross requirements for an item reflects "all" of the demand for the item from higher level items in the product structure. As an example, the MRP record for the hardware kit shown in Table 5.1.5 has been constructed to illustrate the combining of the demand for this item that is derived from both level 0 (final fan assembly) and level 1 (motor assembly) in determining the total gross requirements. In complex bills of material it is possible for items to be used at several different levels, and it may be necessary to gather requirements from many different items in the explosion process.*

* In addition to the master scheduled demand from higher-level product items, a forecast of spare parts demand for individual items may also be incorporated into the gross requirements row. Often the spare parts forecast is displayed as a separate row in the MRP record, but included in the gross requirements for inventory planning and scheduling purposes.

Table 5.1.5. Low Level Coding for Hardware Kits[a]

Hardware Kits		Week 1	2	3	4	5	6	7	8
Gross Requirements									
from level 0 (MPS)		300	500	500	500	500	500	500	500
from level 1 (motors)		1000	0	1000	0	1000	0	1000	0
Total Gross Requirements		1300	500	1500	500	1500	500	1500	500
Scheduled Receipts		4000	0	0	0	0	0	0	0
Projected Available	125	2825	2325	825	4325	2825	2325	825	4325
Planned Order Release[b]		4000	0	0	0	4000	0	0	0

[a] Order quantity = 4000, lead time = 3, safety stock = 500. [b] Note that when the projected available drops below the safety stock level of 500 in weeks 4 and 8, a planned order is scheduled for release.

5.1.3.4 MRP System Transactions

The MRP records shown in Tables 5.1.4 and 5.1.5 reflect the status of these product items at a given time, i.e., at the start of week 1. If the material plan provided by MRP is to reflect *actual* operating conditions, the MRP database must be updated periodically to reflect the changes in inventory levels, shop orders, and requirements as they occur. This is accomplished by processing "transactions" against the MRP database and "regenerating" the MRP records by performing a bill of material explosion.* Many firms using MRP process transactions against the database daily and regenerate the MRP records weekly as a part of a regular "planning cycle."

As an illustration of transactions processing, suppose that the following transactions occur during week 1 for the hardware kit:

1. 1300 hardware kits are issued to final fan assembly and motor assembly.
2. 800 units on the open order for 4000 hardware kits are scrapped and the balance received into stock.
3. A cycle count of the stockroom indicates an inventory adjustment of -150 hardware kits is required.
4. The planned order for 4000 in week 1 is released to the shop.

After these transactions are processed against the database and the MRP record is updated in the bill-of-material explosion at the start of week 2, the record is as shown in Table 5.1.6. It should be noted that the inventory balance for this item at the start of week 2 is determined by considering the inventory position at the start of week 1 as well as the first three transactions above. Note also that the planned order release in week 1 is now shown as an open order due at the start of week 4.

5.1.3.5 Using the MRP System

Up to this point the discussion has focused on the steps in conducting the bill-of-material explosion and processing the information in the MRP records. It is important to consider the use of this information by the inventory planner in making material planning and scheduling decisions. One important feature of MRP software is the provision for "Exception Messages." These messages direct the inventory planner's attention to those product items where action is required if the plant is to meet the MRP material plan. Such exception notices provide critical information, since the planner may be responsible for controlling 500 to 1500 individual product items. While many types of exception messages can be designed, three are essential: the need to release a planned order, the need to expedite an open order, and the need to reschedule an open order to a later date.

The MRP record for the motor in Table 5.1.4 illustrates conditions under which an exception notice would be issued to the planner to release a planned order. In this case, if the shop is to be given an allowed time of two weeks for assembling the motor, a shop order should be issued for the planned order quantity for 1000 units at the start of week 1 in order to meet the requirement of 500 units in week 3. These 500 motors are needed at the start of week 3, so that 500 automotive fans can be assembled as planned in the master production schedule.

The MRP record for the hardware kit in Table 5.1.6 illustrates a condition where an exception notice would be issued indicating that the open shop order for 4000 units currently due at the start of week 4 must be expedited. The combination of unanticipated scrap on the previous shop order and a negative inventory-cycle count adjustment indicates the need to change the due date on the shop order that has just been released to the shop. Likewise, if the inventory adjustment resulting from the cycle count had been +1300 instead of −150,

* An alternative method for updating the MRP records involves using the "net change" procedure where only those items affected by changes in the database are processed.[8]

the resulting exception notice would have been to reschedule the open shop order due at the start of week 4 to the start of week 5.

The processing of exception notices is an important part of the inventory planner's job. The MRP system enables the planner to focus on those problems that will prevent the plant from meeting the master production schedule and the corresponding commitments to customers. Knowledge of the shop as well as what can and can't be done is important in making scheduling decisions with the MRP system information. Therefore, while MRP system information can aid the planner in making these decisions, his other judgment and coordination skills are important elements in making MRP a valuable tool for planning and controlling manufacturing operations.

Another important aspect of the inventory planner's job is that of monitoring and maintaining the parameter values used in processing the MRP record and building the schedule. Parameters such as lead time offsets, order quantities, safety stock, and shrinkage values need to be evaluated from time to time and adjusted as necessary to reflect changed operating conditions. The setting of such parameters involves analysis and judgment on the part of the inventory planner, and is important in maintaining a valid material plan.

5.1.3.6 Planner Assists

There are several features of MRP software packages that can provide valuable assistance to the inventory planner, one of which is the ability to perform *component availability checks*. For example, prior to releasing the shop order for 1000 motors (Table 5.1.4), the planner can request prerelease information to determine whether all the components required to assemble the order are available in the correct quantities. This feature is useful in insuring that production can, in fact, begin once the shop order is issued to manufacturing.

A second feature is that of *pegging*, that is, the ability to identify the source of the demand shown in the gross requirements row. For example, the total requirement of 1300 units in week 1 of the MRP record for the hardware kit (Table 5.1.5) came from two sources: 300 units from the master production schedule in week 1 and 1000 units from the planned order for the motor in week 1. Such pegging information can be requested by the inventory planner, and can be useful in resolving problems when a gross requirement cannot be met because of scrap, vendor shortages, or for other reasons. This feature enables the planner to determine which higher-level product items will be affected when shortages occur.

Firm planned orders, another useful feature in MRP software packages, enables the planner to fix the quantity and timing of the orders in the planned order row so that the normal MRP bill-of-material explosion process does not alter the timing of these orders. This feature is particularly advantageous in cases where it is not desirable to change the short-term schedule of an item, due to capacity or vendor limitations, or where schedule stability is important.

5.1.3.7 Example MRP Systems

The MRP concepts and techniques for material planning described in this chapter have been illustrated using small examples. We now turn to examples of these techniques from actual practice. Although the example MRP records shown in Figs. 5.1.3 and 5.1.4 contain a substantial amount of information, the concepts used in preparing these records are those covered in this chapter. The records are included to provide examples of the information used routinely by the inventory planners in these firms.

The MRP record shown in Fig. 5.1.3, from the MRP system in the Mocksville, North Carolina plant of Ingersoll-Rand,[3] provides the material plan for item 761 covering a one-year time horizon. Several features of this record should be noted. First, a substantial amount of fixed information about the item is included in the upper portion of the record. Second, the record includes an action-exception message indicating the need to expedite (reschedule in) the open orders for this item. Finally, pegging information is included at the bottom of the record, indicating the source of the requirements placed on this item.

The MRP record shown in Fig. 5.1.4 is from the MRP system at the Viking Pump-Houdaille, Inc. (a subsidiary of Houdaille, Inc.) in Cedar Falls, Iowa.[6] This record provides a material plan for item 2-214-801-100-00 covering the period from 2-21-83 to 10-3-83. Instead of displaying the material plan for this item in a

Table 5.1.6. Transaction Processing Example

		Week						
Hardware Kits		2	3	4	5	6	7	8
Gross Requirements		500	1500	500	1500	500	1500	500
Scheduled Receipts				4000				
Projected Available	1875[a]	1375	−125	3375	1875	1375	3875	3375
Planned Order Release				4000				

[a] The beginning on-hand inventory balance for week 2 is calculated as follows: week 1 on-hand (125) plus the scheduled receipt (3200) less the inventory adjustment (−150) less the week 1 disbursements (−1300) equals 1875.

MATERIAL REQUIREMENTS PLAN

Ingersoll-Rand CURRENT RUN WEEK 07/18/77 ANALYZER N DATE 07/15/77 PAGE 884

PART NUMBER	DESCRIPTION	BALANCE ON HAND	TOTAL SCHED. RECEIPTS	TOTAL REL. ORDERS	TOTAL REL. REQUIREMENTS	S.S.T. CODE	VALUE CLASS	UNIT MEAS.	ANNUAL GROSS REQUIREMENTS	COMMODITY CODE	PRODUCT CODE	ACCT. CODE	BUYER	PRIME VENDOR	ALTERNATE VENDOR	MASTER SCHEDULE	EXPEDITOR	CURRENT ENG. CHNG. NO.
761	GAUGE,DISCH PRESSURE	699	1079	1079	73	B	A	EA	4917	22238	666	132	E	0	0			16279

LEAD TIME PURCH.	LEAD TIME MANUF.	LEAD TIME OVERRIDE	ORDER POLICY	ORDER QTY.-TIME	MINIMUM ORDER QTY.	MAXIMUM ORDER QTY.	MULTIPLE ORDER QTY.	SAFETY STOCK POLICY	SAFETY STOCK QTY.	SCRAP FACTOR	QUEUE/MOVE TIME	RUN TIME	SETUP TIME	STANDARD RUN QTY.	STANDARD COST	LAST ACTUAL COST
20	0	1	D	999999	0	0	0		0	0	T 04WKS	0	0	0	5.350	

DEMAND THIS 4-WEEKS FORECAST	DEMAND THIS 4-WEEKS ACTUAL	SERVICE FORECAST DATE	FORECAST MODEL	FORECAST CODE	MAXIMUM SERVICE ORDER	MANUAL FCST. START DATE	MANUAL SERVICE FORECAST 1ST 13-WKS	2ND 13-WKS	3RD 13-WKS	4TH 13-WKS	5TH 13-WKS	6TH 13-WKS
87	126			AUTO	4	00/00/00	0	0	0	0	0	0

ISSUES	PRODUCTION	SERVICE	INTERCO-COM.	INTERCO-FOR	SCRAP	OTHER	TOTAL
MONTH-TO-DATE ISSUES	96	173	0	0	0	0	269
YEAR-TO-DATE ISSUES	1746	938	0	41	0	366	3262

BIN LOCATION PRIME	BIN LOCATION SECOND	THIRD	FREE ISSUE	LAST COUNT TRANS. DATE
E1E2				06/09/77

ACTION MESSAGE	REFERENCE ENG. CHNG. NO.	OBS. CODE	CURRENT OBS. QTY.	PROJECTED OSMI QTY.	POTENTIAL SURPLUS QTY.	SCRAP LIST CODE	SCRAP LIST QTY.	LAST BOH TRANS. DATE	LAST FILE MAINT. DATE	YTD ADJUSTMENTS NUMBER	TOTAL POS.	TOTAL NEG.
NEED ORDER RESCHEDULE IN	0		0	0	0		0	07/15/77	05/27/77	1	8	42

RUN WEEK	06/27	07/18	07/25	08/01	08/08	08/15	08/22	08/29	09/05	09/12	09/19	09/26	10/03	10/10	10/17	10/24	10/31	11/07
SERVICE		22	22	21	22	21	22	21	22	21	22	21	21	21	22	21	21	21
PRODUCTION	9	93	35	6	13	54	45	18	3	224	49	30	2	180	49	123	89	184
SCHEDULED RECEIPT						500				561								
AVAILABLE	690	575	518	491	456	881	814	775	750	1066	995	944	921	720	649	505	395	190
PLANNED ORDER REL.	133			475				447				483				461		

RUN WEEK	11/14	11/21	11/28	12/05	12/12	12/19	12/26	01/02	01/09	01/16	02/13	03/13	04/10	05/08	06/05	07/03	07/31	10/23
SERVICE	21	20	21	21	21	20	21	20	21	82	81	80	80	79	78	77	227	221
PRODUCTION	1	199	32	1	4	371	6	14	2	365	402	381	307	343	8	8	25	23
SCHEDULED RECEIPT																		18
AVAILABLE	168	51-	104-	126-	151-	542-	569-	603-	626-	1073-	1556-	2017-	2404-	2826-	2912-	2997-	3249-	3475-
PLANNED ORDER REL.		387				422				86	85	84	84	84	83	83	78	

SCHEDULED RECEIPTS

REQUEST DATE	PROMISE DATE	REMAINING QTY.	RECEIVED QTY.	ORDER NO.	D	VENDOR	S	T	R
08/15	09/23	500	0	M52927	1	010400	2	F	1
09/12		561	264	M54169	1	130080	8	P	0
12/31		4	0	MR7065	1	300000	2	P	0
12/31		4	0	MR7102	1	300000	2	P	0
12/31		1	0	MR7006	1	300000	2	P	0
12/31		9	0	MR7124	1	300000	2	P	0

REQUIREMENTS

REQUIRED DATE	REQUIRED QTY.	REFERENCE	ORDER NO.	S	T
02/28	1	R150P1010	36988	1	8
06/14	1	29431612		2	0
06/15	1	65670350		2	0
06/27	1	35132810	86494	1	8
07/06	1	35043926	86514	1	8
07/11	2	35043223	83942	1	8
07/11	52	35002377	83940	1	8
07/15	2	16084812C		1	0
07/15	4	16084818P		1	0
07/18	2	35044296	78995		P
07/18	6	35043447	78992	1	8
07/19	13	35008325	50568		C
08/08	6	35043496	78223		P
09/12	12	35043496	78080		P
10/31	20	35044288	78226		P
10/31	20	35043496	78225		P
11/01	2	2090505 RFL	11111	1	6
01/01	3	35043504	78123		P
01/02	3	35044296	80101		P
01/02	10	35044288	80101		P
03/07	27	35043496	80101		P

REQUIRED DATE	REQUIRED QTY.	REFERENCE	ORDER NO.	S	T
03/07	3	35043447	80101		P
04/02	2	35044296	78122		P

SST Code: Source,Status,Type(stock vs. non-stock, purchase vs. mfg.,ass'y. vs. part,etc.).

Product Code: Indicates primary end item in which this item is used.

Commodity Code: Code to group similar type items.

Master Schedule Code: Indicates whether this item is Master-Scheduled.

Account Code: Financial code for inventory reporting-

Order Policy Code and Qty-Time Code: "D" specifies fixed-order quantity; with 999999 also specified will order a fixed-period quantity.

Lead Time Override: If set on "1" by buyer or analyzer, LT will not be updated by commodity code update and mfg. LT will not be calculated using opn.-by-opn. times.

Eng. Change No.: Number of most recent EC notice currently effective for item.

Ref. Eng. No.: Indicates a forthcoming EC which will affect this item.

Obsolete Code: Indicates if obsolete mat'l was on hand at beginning of calender yr.

PCD-518

Fig. 5.1.3. Sample MRP output. Reprinted with permission of the American Production and Inventory Control Society from *Case Studies in Material Requirements Planning*, p. 99.

```
  EOO     ROP         O/M LTM PTM YTDUSED  LYRUSED     U/C       PASS     FAIL      RFP
  100                   1  35       84        486                  14        9
PLANNING RUN = N * * * * * * * * * * *  ON HAND INV =        17  * * * * * *   PAGE 01
D/Date  S/Date P/F D-TYP    ORDER NO. ITM STS DUEQTY  AV/INV  REMARKS        ORD-AMT
022183          P  EXT      P02508140 001  P       2      15  2-214-801-100-00
022883          P  EXT      P50647000 003  P       1      14  4-2515-2626-502
022883          P  EXT      P50359000 003  P       1      13  4-2520-2626-503
030783          P  EXT      P49543020 004  P       4       9  4-2515-6641-008
030783          p  EXT      P50707000 003  P       2       7  4-2520-2626-008
030783          P  EXT      P49301030 003  P       1       6  4-2520-2646-512
030783          P  EXT      P51178000 003  P       1       5  4-2515-2626-502
031483          P  EXT      P51194020 003  P       2       3
032883          F  EXT      P03800000 003  P       4      1-  4-2515-2626-006
032883  022183  F              231025 000  P     100      99  RELEASE        $5173
041183          F  EXT      P03730000 004  P       1      98  4-2515-2626-017
060683          F  EXT      L0559004R 007  P       1      97
062183          F  EXT      P0250340G 001  P       2      95  2-214-801-100-00
100383          F  EXT      L0559026V 002  P       1      94
```

Fig. 5.1.4. Sample MRP record. Supplied courtesy of Viking Pump Company, Cedar Falls, Iowa.

"horizontal" manner, the plan is displayed vertically. Such a display, referred to as a "bucketless" MRP record, is growing in application—especially in MRP systems that utilize CRT terminals.

It should be noted that the material plan in Fig. 5.1.4 is displayed on a "date" basis instead of using "weekly" time periods as is done in Fig. 5.1.3. Otherwise, the information displayed is similar in both MRP records. The gross requirements, scheduled receipts, and planned order receipts are shown in the "Due Qty" column. The projected available balance is shown in the "AVINV" column, and "Pegging" information is indicated in the "Order No." and "Remarks" columns. Likewise, start dates for scheduled receipts and planned orders are shown in the "S/Date" column. MRP concepts such as "Gross-to-Netting" are incorporated in both Figs. 5.1.3 and 5.1.4.

5.1.4 MRP SYSTEM IMPLEMENTATION

In this chapter we have discussed the technical aspects of material requirements planning systems: the development and maintenance of a material plan that is based on the use of a manufacturing database and computer software. We now turn to the problems of implementing these systems.

Clearly, MRP systems represent a powerful way of improving productivity in manufacturing. Some firms report gains in manufacturing productivity of 15%, reductions in overtime of 50%, reductions in inventory investment of 33%, and reductions in inventory shortages of 80%.[15] However, while the use of information system technology in manufacturing planning and control provides the potential for operating improvements, the implementation failure rate of such systems is considered to be quite high—perhaps on the order of three out of four efforts.[13,14] By failure we mean the nonattainment of the performance objectives established for the system in situations where an investment of $500,000 to $2,000,000 or more is required.

Several reasons for implementation failures in MRP systems have been identified: "technical factors" relating to the implementation procedure, e.g., the accuracy of inventory data; "process factors," relating to the initiation and use of the system, e.g., gaining the active support of top management; and "inner-environmental factors," regarding what may be thought of as the true organizational support for the system, e.g., the organization's willingness to change.[7]

Let us consider some of the technical factors that affect the implementation process. First, procedures need to be installed to insure the accuracy of the manufacturing database in order to provide accurate material plans. This means, for example, the development of accurate bills of material as well as accurate and timely information regarding the status of inventories and open orders. Cycle counting programs, tight control over inventory transactions, and effective procedures for handling engineering change notices are all examples of procedures necessary to insure an accurate database. The objective is to achieve levels of accuracy on the order of 99% for the bills of material and at least 95% for inventory status information.[9]

A second technical factor affecting the implementation process is the selection of the MRP software package itself, including the important step of educating the all users of the system and top management as well, to insure a thorough understanding of how MRPII works *before* a software package is selected. This insures that the needs of the business are considered in the selection of the software package, and that there is a good understanding of how the MRP concepts apply in the company. It is also important that the software package should not be unnecessarily complex and that it contains standard MRP features, e.g., gross-to-netting, and pegging information. These considerations insure the installation of an MRP package that is well understood by the users, has high utility in the particular manufacturing environment, and avoids lengthy delays in obtaining the full benefits of MRP.[4]

Beyond the technical factors in implementing an MRP system, other factors, relating to the way in which the business is managed, can impact the effectiveness of the implementation effort, for example, the management of the implementation process itself. Examples of effective practice in this regard include the development of a formal project implementation plan with defined project milestones and assigned responsibilities. Likewise, the selection of a project manager with substantial manufacturing experience, and a top management oversight committee that conducts regular reviews of project progress against the project plan are critical. Other examples of effective practice include the commitment of top management to provide the necessary resources to staff and support the project, and to do away with the previous planning and control systems once the MRPII system is operating.[9,11]

The managerial processes in a company, beyond those associated with the MRP implementation effort, can also have a major impact on the effectiveness of the MRP system. It is important that top management have a committment to manage operations using formal planning and control techniques.[5] The use of such systems in managing the business often requires significant organizational changes, e.g., new work patterns, new problem solving methods, and often new relationships with other people in the organization. The company must be supportive of these kind of changes. Furthermore, the measures used to evaluate performance in the organization are often not consistent with the use of formal systems for planning and controlling operations. For example, "meeting the plan" on a day-by-day or week-by-week basis and achieving information accuracy represent new goals to be attained in addition to goals such as insuring "efficient" operations. Therefore, substantial organizational change is often required in order to achieve the full benefits of MRP systems.

REFERENCES

1. E. J. Anstead, "How to Define and Plan a System for Integrated Closed Loop Manufacturing Control," *Industrial Engineering*, **15**(9), (September, 1983).
2. W. L. Berry, T. E. Vollmann, and D. C. Whybark, *Master Production Scheduling: Principles and Practices*, American Production and Inventory Control Society, 1979.
3. E. W. Davis, "Material Requirements Planning in a Fabrication and Assembly Environment," Ingersoll-Rand Case Study, Auerbach Publishers, 1982.
4. D. Garwood, "In Search of MRPII Software," *P&IM Review and APICS News* (October, 1983).
5. M. Harwood, "Actualizing the Closed-Loop Concept," *P&IM Review and APICS News* (October, 1983).
6. F. Hansen, "Managing with MRP," The University of Iowa, Presentation, April 20, 1983.
7. J. R. Meredith, "The Implementation of Computer-Based Systems," *Journal of Operations Management*, **2**(1), 11–21 (October, 1981).
8. J. Orlicky, *Material Requirements Planning*. New York: McGraw-Hill, 1975.
9. D. Swann, "MRP: Is It A Myth or Panacea? Key to Answer is Commitment of Management to It," *Industrial Engineering*, **15**(6), (June, 1983).
10. T. E. Vollmann, W. L. Berry, and D. C. Whybark, *Manufacturing Planning and Control Systems*, Dow/Jones/Irwin, 1984.
11. T. F. Wallace, *MRPII: Making It Happen—the Implementer's Guide to Success with Manufacturing Resource Planning*. Brattleboro, VT: Oliver Wight, Inc., 1985.
12. O. Wight, *MRPII: Unlocking America's Productivity Potential*, CBI Publishing, 1981.
13. "MRP Re-Implementation: You Too Can Be Successful," *APICS National Conference Proceedings*, Boston, October 6–9, 1981, pp. 119–120.
14. "The Trick of Material Requirements Planning," *Business Week*, 72D–72J (June 4, 1979).
15. "Productivity Out of MRP—A New Game Plan," *Modern Materials Handling* (January, 1981).
16. Horwitt, E., "MRP Tightens the Front Office/Shop Floor Link," *Business Computer Systems*, 132–145 (October, 1983).

CHAPTER 5.2

MATERIAL HANDLING AND STORAGE

JAMES M. APPLE, JR.
HARVEY V. RICKLES

SysteCon Division, Coopers & Lybrand
Duluth, Georgia

5.2.1 A NEW ROLE FOR MATERIAL HANDLING[5]

Recognition of material handling as an important factor in manufacturing cost is not new. For more than forty years industrial users and material handling equipment suppliers have worked to reduce the cost of the material handling effort. The results were similar to those in the area of manufacturing processes: the efficiency of many individual tasks were optimized.

But in the 1970s, rapid changes in the marketplace began, forcing a more flexible and responsive manufacturing environment. Efficiencies of the individual operations were lost to inefficiencies in the process as a whole. The sum of independently optimized pieces did not add up to total system optimization.

The changes that highlighted system inefficiency include:

1. Higher cost for carrying inventory.
2. Shorter product life in the marketplace.
3. Shorter lead times on product introduction.
4. Rapidly rising costs of labor.
5. Aggressive foreign competition.

For some time, there was, and still is, a focus on the development of *integrated* material handling *systems*. Each handling task was coordinated with those before and after it. Although this approach created smooth-running and efficient handling systems, manufacturing efficiency still did not rise. Manufacturers were still designing inflexible systems that were slow to change, and material handling systems that did not consider manufacturing's overall needs. Finally, at the end of the decade, it became clear that the manufacturing operations and the material handling system must be considered together as an integrated manufacturing system.

Over the last five years the complexity of these totally integrated systems has demonstrated the need for a new approach to planning. Traditionally, the design process has consisted of a series of sequential design decisions, seemingly independently made. As depicted in Fig. 5.2.1, product design decisions are "handed off" to a process designer for decisions concerning the manufacturing processes to be used to produce a product already designed. Typically, little, if any, communication occurs between the product designer and the process designer.

Frequently a designer will create a product or component part with little regard for the production process. The process designer makes independent decisions in selecting the processing technology to be utilized. Often a building is being independently designed to house the manufacturing system; again, with no flow of information regarding facility and equipment requirements. Layout design decisions are then made, following decisions regarding processing and the facility design. Independently, the production planning process occurs and production schedules are established. Finally, consideration is given to the handling, throughput, storage, and control system needed to meet the material flow requirements. Not only does such a process result in frustrations "downstream" from "upstream" decisions that seemingly fail to consider downstream impacts of the decisions, but also there is little chance that the resulting design will meet the firm's overall production goals of optimizing the firm's resources.

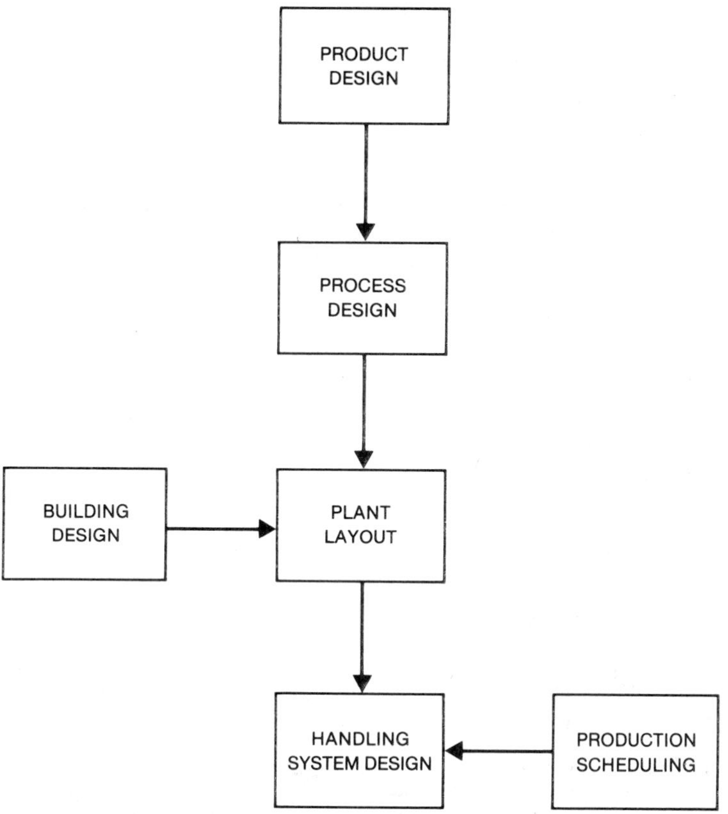

Fig. 5.2.1. The design process.[12]

To achieve the firm's overall goal of production optimization, an *integrated* design of computer integrated manufacturing systems (CIMS) must be accomplished. Ideally, computer integrated manufacturing utilizes a computer network to link together all functions of an enterprise including sales, marketing, accounting, human resources, purchasing, materials management, quality assurance, product design, process design, production design, engineering, maintenance, distribution, etc. Once linked, information can be accessed to fulfill the firm's overall goal.

The best way to coordinate the goals of a firm through CIMS is by using a design team formed with representatives from product design, process design, industrial engineering, facilities design, manufacturing operations, data processing, and marketing. The objective of the team should be to coordinate the efforts of each group; its medium of coordination should be CIMS. Thus, the flow of information required to develop and maintain competitive products can be automated through a CIMS network.

Let us look at an illustration of this network. A customer need is expressed. The perceived need is transformed into conceptual and detail product design by the product design group. The designs are expressed as information. The data generated by product design is entered into the CIMS system, where it is manipulated by the process design group into a process plan. The process plan is communicated in turn to the manufacturing operations group as well as the industrial engineering group to develop the production plan based on information regarding product rates, material requirements, machine loading, demand forecasts, production loading, etc. The information used to develop the production plan would be represented within the CIMS network in the form of historical data, forecasts, drawings, and text. The material handling experts and facilities design group would work with the production plan to design a facility and layouts that would best service production's needs. Through CIMS, the material handling expert would be able to interface with all parties involved to ensure smooth and controlled movement throughout the process. CIMS allows all team members to monitor and advise each group of the impact to their respective areas as changes are made, not after they occur. Thus, CIMS automates the manufacturing process, which may be defined as the transformation of information into physical shapes to satisfy a customer's need.

As material handling system designers interface with other planning functions through CIMS, four major areas of the business have a significant impact on the handling system: (1) marketing, (2) product design, (3) manufacturing and manufacturing control, (4) finance and management. Marketing decisions that affect material handling include those related to unit volume, product mix, packaging, and distribution. Product design factors affecting material handling include changes in materials, component shapes, packaging materials, and product complexity. Manufacturing and manufacturing control strategies that impact material handling include the degree of vertical integration, the type and level of automation, the type and level of control over tooling and work-in-process, lot sizes and production schedules, and the sizes and locations of manufacturing sites. Financial and management considerations include inventory levels, capital availability, labor climate, and productivity goals.

For the material handling plan to be supportive of the overall strategic plan, material handling experts must participate in the development of the overall plan. For example, close coordination is required in developing plans for manufacturing and material handling. The handling/manufacturing interface becomes especially important as the manufacturing plan addresses automatic load/unload of machines, robotics, group technology, transfer lines, flexible manufacturing systems, CNC, DNC, automated guided vehicles, automated storage systems for parts and tooling, and automated scrap removal systems.[3]

The emphasis has shifted from optimizing the efficiency of movement and storage of material to a concentration on the efficiency of the total manufacturing process. Handling systems are consciously designed with excess capacity in order to be responsive to the manufacturing operations. Large handling units are frequently broken down into several smaller loads to provide more scheduling flexibility, convenience at the workplace, and reduced in-process inventory. Differences between the handling requirements for fabrication vs. assembly operations and continuous flow vs. batch processing are accommodated with different kinds of handling and storage systems.

The priorities of the various roles of the handling system have been rearranged. Minimizing movement and storage tasks is secondary to total system optimization. The following list usually guides today's effective system designs.

1. Provide physical discipline of material flow.
2. Provide control and accurate information feedback.
3. Provide responsive material movement.
4. Provide efficient storage.

5.2.2 PRINCIPLES, MEASURES, AND OBJECTIVES OF DESIGN

The environment in which we design material handling systems has changed markedly in the past ten years. Manufacturing process technology has become more sophisticated, automation has become commonplace, and a primary goal in manufacturing system design is the reduction of in-process inventory and total cycle time. At the same time, material handling equipment and approaches to designing material handling systems have changed in order to keep pace. As a background for understanding how and why things are different today, it is helpful to review the objectives and principles of design that have evolved from years of experience by a large number of people.

Twenty years ago the College Industry Committee on Material Handling Education (CICMHE) developed a list of principles to guide the solutions to material handling problems. They were based on the collective experience that had accrued to that point and that matched the current posture toward the planning and managing of manufacturing operations. In 1982, CICMHE adopted a revised set of principles. Comparing these lists gives us real insight into the shift in emphasis from a viewpoint that optimized the material handling functions to one that makes material handling an integral part of an optimized manufacturing or distribution system.

The two lists of principles[13] are followed by observations on the differences. Some are quite obvious, and others more subtle.

Original Twenty Principles of Material Handling.

1. *Planning principle.* Plan all material handling and storage activities to obtain maximum overall operating efficiency.
2. *Systems principle.* Integrate as many handling activities as is practical into a coordinated system of operations, covering vendor, receiving, storage, production, inspection, packaging, warehousing, shipping, transportation, and customer.
3. *Material flow principle.* Provide an operation sequence and equipment layout optimizing material flow.
4. *Simplification principle.* Simplify handling by reducing, eliminating, or combining unnecessary movements and/or equipment.

5. *Gravity principle.* Utilize gravity to move material wherever practical.
6. *Space utilization principle.* Make optimum utilization of building cube.
7. *Unit size principle.* Increase the quantity, size, or weight of unit loads or flow rate.
8. *Mechanization principle.* Mechanize handling operations.
9. *Automation principle.* Provide automation to include production, handling and storage functions.
10. *Equipment selection principle.* In selecting handling equipment, consider all aspects of the material handled—the movement and the method to be used.
11. *Standardization principle.* Standardize handling methods as well as types and sizes of handling equipment.
12. *Adaptability principle.* Use methods and equipment that can best perform a variety of tasks and applications where special-purpose equipment is not justified.
13. *Dead weight principle.* Reduce ratio of dead weight of mobile handling equipment to load carried.
14. *Utilization principle.* Plan for optimum utilization of handling equipment and manpower.
15. *Maintenance principle.* Plan for preventive maintenance and scheduled repairs of all handling equipment.
16. *Obsolescence principle.* Replace obsolete handling methods and equipment when more efficient methods or equipment will improve operations.
17. *Control principle.* Use material handling activities to improve control of production, inventory, and order handling.
18. *Capacity principle.* Use handling equipment to help achieve desired production capacity.
19. *Performance principle.* Determine effectiveness of handling performance in terms of expense per unit handled.
20. *Safety principle.* Provide suitable methods and equipment for safe handling.

Almost 20 years later, a newly named College Industry Council on Material Handling Education (CIC-MHE) revised the listing of 20 principles to read as follows:

1. *Orientation principle.* Study the problem thoroughly prior to preliminary planning in order to identify existing methods and problems, physical and economic constraints, and to establish future requirements and goals.
2. *Planning principle.* Establish a plan to include basic requirements, desirable options, and the consideration of contingencies of all material handling and storage activities.
3. *Systems principle.* Integrate those handling and storage activities which are economically viable into a coordinated operating system including receiving, inspection, storage, production, assembly, packaging, warehousing, shipping, and transportation.
4. *Unit load principle.* Handle product in as large a unit load as practical.
5. *Space utilization principle.* Make effective utilization of all cubic space.
6. *Standardization principle.* Standardize handling methods and equipment wherever possible.
7. *Ergonomic principle.* Recognize human capabilities and limitations by designing material handling equipment and procedures for effective interaction with the people using the system.
8. *Energy principle.* Include energy consumption of the material handling systems and material handling procedures when making comparisons or preparing economic justifications.
9. *Ecology principle.* Use material handling equipment and procedures which minimize adverse effects on the environment.
10. *Mechanization principle.* Mechanize the handling process where feasible to increase efficiency and economy in the handling of materials.
11. *Flexibility principle.* Use methods and equipment which can perform a variety of tasks under a variety of operating conditions.
12. *Simplification principle.* Simplify handling by eliminating, reducing, or combining unnecessary movements and/or equipment.
13. *Gravity principle.* Utilize gravity to move material wherever possible, while respecting limitations concerning safety, product damage and loss.
14. *Safety principle.* Provide safe material handling equipment and methods which follow existing safety codes and regulations in addition to accrued experience.
15. *Computerization principle.* Consider computerization in material handling and storage systems, when circumstances warrant, for improved material and information control.
16. *System flow principle.* Integrate data flow with the physical material flow in handling and storage.
17. *Layout principle.* Prepare an operation sequence and equipment layout for all viable system solutions, then select the alternative system which best integrates efficiency and effectiveness.

18. *Cost principle*. Compare the economic justification of alternate solutions in equipment and methods on the basis of economic effectiveness as measured by expense per unit handled.
19. *Maintenance principle*. Prepare a plan for preventive maintenance and scheduled repairs on all material handling equipment.
20. *Obsolescence principle*. Prepare a long range and economically sound policy for replacement of obsolete equipment and methods with special consideration to after-tax life cycle costs.

The planning principle is now preceded by the orientation principle to recognize finally the need to understand the role of the handling system in the entire manufacturing or distribution process. Appropriately, it is listed first. The planning principle itself has been reworded to indicate that not only should material handling operations be planned, but that planning should look forward to future requirements.

The old material flow principle has given way to a new layout principle, which acknowledges that in some plants a layout must be developed to accommodate a process flow that may not be straight through.

Although the gravity principle still works, cautions for safe and controlled use have been added. The unit load principle has superceded the unit size principle, with emphasis shifted from the movement of as large a load as possible to a load that is sized to match the process and the production schedule.

Automation has yielded to computerization. Today, almost by definition, automation is the computerization of mechanization.

The equipment selection principle, which insists that handling requirements should govern the choice of hardware, has been dropped, perhaps because it seems obvious. It may be to experienced system designers, but evidence in industry clearly points to installations that were based on "gut-feel," or emotion, as they clearly do not match the requirements. The equipment selection principle should have been kept, even if that meant expanding the list.

The now outmoded adaptability principle and the new flexibility principle have similar thrusts, with the latter placing a little more emphasis on the future.

The dead weight principle is dead, as is utilization. Today's systems do seem to have high capacity compared to the loads they carry, and that aspect of the principle is probably outmoded. However, it would be important to recognize the benefits of dual cycle moves, which improve utilization by providing workloads for equipment on its return trip. This concept is normally employed in the design of automated storage and retrieval systems (AS/RS), but is equally useful in standard lift truck storage and for transportation moves.

The old control principle has been reinterpreted in the new system flow principle. In addition to physical control, the usefulness of the handling system for data capture is acknowledged.

The old capacity principle is gone without a replacement. Its important message—to use the handling system to improve the capacity of production equipment—is more important than ever. If it is ignored, the progress of the last five years will be lost.

The new cost principle incorporates the old performance principle and encourages the use of alternative proposed system designs as well as a measure to track day-to-day operating efficiency.

The ergonomic, energy, and ecology principles document the impact of social and economic changes and federal regulations and the need for handling systems to keep pace with progressive attitudes.

In summary, the new principles stress an integrated approach to material handling, emphasizing flexibility, safety, and planning for efficient use of resources.

In addition to the classic principles to guide system design, some quantitative measures have been developed as methods for tracking performance in a number of manufacturing areas. In 1961, James R. Bright of Harvard University authored the *Yale Management Guide to Productivity*, which included seven measures of material handling effectiveness.[6] In 1979, the *Guide* was revised and expanded by John A. White of Georgia Institute of Technology.[11] In the following list from the second edition, where the original measures are repeated, they are denoted by an asterisk (*). Not all measures are applicable to every business. Those that seem particularly appropriate or those that highlight a known trouble area may be selected for independent use.

Resource Utilization

1. *People*.

$$\text{MHL ratio}^* = \frac{\text{Personnel assigned to materials handling duties}}{\text{Total operating work force}}$$

$$\text{DLMH ratio}^* = \frac{\text{Materials handling time spent by direct labor}}{\text{Total direct labor time}}$$

2. *Equipment*.

$$\text{Production equipment}^* \text{ utilization} = \frac{\text{Actual output}}{\text{Theoretical output}}$$

$$\text{Handling equipment utilization} = \frac{\text{Weight moved/hour}}{\text{Theoretical capacity}}$$

3. *Space.*

$$\text{Storage space* utilization} = \frac{\text{Storage cube occupied by material}}{\text{Net usable cube}}$$

$$\text{Aisle space percentage} = \frac{\text{Space occupied by aisles}}{\text{Total space}}$$

4. *Energy.*

$$\text{EUI} = \frac{\text{BTUs consumed/day}}{\text{Cubic space}}$$

Management Control

5. *Materials.*

$$\text{Inventory turnover ratio} = \frac{\text{Annual sales}}{\text{Average annual inventory investment}}$$

$$\text{Inventory fill ratio} = \frac{\text{Line item demands filled/day}}{\text{Line item Demands/day}}$$

6. *Movement.*

$$\text{Movement/operation* ratio} = \frac{\text{Total number of moves}}{\text{Total number of productive operations}}$$

and

$$\text{Average distance/move ratio} = \frac{\text{Total distance traveled/day}}{\text{Total number of moves/day}}$$

7. *Loss.*

$$\text{Damaged loads ratio} = \frac{\text{Number of damaged loads}}{\text{Number of loads}}$$

$$\text{Inventory shrinkage ratio} = \frac{\text{Inventory investment verified}}{\text{Inventory investment expected}}$$

Operating Efficiency

8. *Shipping and Receiving.*

$$\text{RP ratio} = \frac{\text{Pounds received/day}}{\text{Labor hours/day}}$$

$$\text{SP ratio} = \frac{\text{Pounds shipped/day}}{\text{Labor hours/day}}$$

9. *Storage and Retrieval.*

$$\text{OP ratio} = \frac{\text{Equivalent lines or orders picked/day}}{\text{Labor hours required/day}}$$

and

$$\text{Throughput productivity index} = \frac{\text{Throughput achieved/day}}{\text{Throughput capacity/day}}$$

10. *Manufacturing.*

$$\text{Manufacturing cycle efficiency*} = \frac{\text{Total time spent on machines}}{\text{Total time spent in production system}}$$

and

$$\text{JL ratio} = \frac{\text{Number of jobs completed or in process that are late/week}}{\text{Number of jobs completed/week}}$$

Objectives for the design of handling systems that will best serve today's and tomorrow's factories have been outlined as follows:[9]

1. Create an environment that results in the production of high quality products.
2. Provide planned and orderly flows of material, equipment, people, and information.
3. Design systems that can be easily adapted and expanded to match changes in product mix and production volumes.
4. Reduce work-in-process.
5. Provide controlled flow and storage of materials.
6. Integrate processing, assembly, inspection, handling, storage, and control of materials.
7. Eliminate manual material handling at work stations and between work stations.
8. Utilize the capabilities people have above the shoulders, not below,
9. Deliver parts and tooling to workstations and machines in predetermined quantities and physically positioned to allow automatic parts feeding and tool changing at machines.
10. Utilize space most effectively, considering overhead space and impediments to cross traffic.

A number of the objectives listed above are appropriate regardless of the degree of automation employed.

A final emphasis in planning objectives should be placed on controls. As mentioned earlier, one of the primary purposes of the material handling system is to provide physical discipline so that the manufacturing control system can operate effectively. The benefits that we foresee in reducing manufacturing costs will come, not from more efficient physical handling of the material, but rather through the ability to follow the directions of a comprehensive management control system.

To properly design the handling system, requirements must be identified, scheduling intervals and the flexibility of rearranging work within a time frame must be understood, and information requirements must be identified and reported in a timely fashion. The system that will control the material handling operations must be designed to provide an effective interface with existing information control systems. The sequence of performing planning, design, and implementation tasks is important. Consider the following guidelines,[10] which can help to eliminate surprises and redesign that might otherwise result from hasty decisions:

1. Define the scope and objectives for a system before considering alternative solutions.
2. Obtain inputs from operating personnel before finalizing the system design.
3. Generate several alternatives before selecting the preferred system.
4. Collect facts, hard numbers, and/or estimates before economic justification.
5. Develop the materials handling plan before designing the plant layout.
6. Design the plant layout before designing the building.
7. Train personnel before "starting up the system."

At the risk of condensing handling system design to yet another "canned" approach, the following are elements of an integrated handling system:

Container.

Macro handling system between areas.

Micro handling system within work area.

Storage and staging.

Control system.

To produce a complete system, each of these must be designed to be effective in and of itself and yet each must be designed so that the elements work together as a system.

If the manufacturing cycle is long and job priorities may need rearrangement, then order tracking may be required. For some processes, maintaining lot integrity may be a requirement that will change the handling system design.

5.2.3 DESIGN METHODOLOGY

Handling system design, like other system design problems, is aided by some structure to the process. The structure is developed to insure that critical requirements in production are not overlooked, and assist in the development of alternatives from which to choose. Answers to the questions posed in the material handling equation, Fig. 5.2.2, help define the problem and provide data for the selection of the best alternative. Of particular importance is the first question. Why? It serves as a reminder that good process planning may eliminate the need for many material handling moves.

A frequent mistake in designing handling systems is moving forward too quickly. The most successful systems are those that have been thought through carefully with the early involvement of personnel within the organization who can affect the requirements for the system or who will be affected by it. Representatives from the disciplines listed below should be involved during both the planning and the implementation of a new handling system. The material handling or industrial engineer will have full-time responsibility for major systems and will serve as the leader of the planning team. Other members will provide specific inputs related to their area of expertise and will serve as a continuing review panel.

Material Handling System Design Team

Material handling (industrial) engineer

Production engineer.

Production planning, scheduling, material control

Product designers

Facilities

Information

Systems

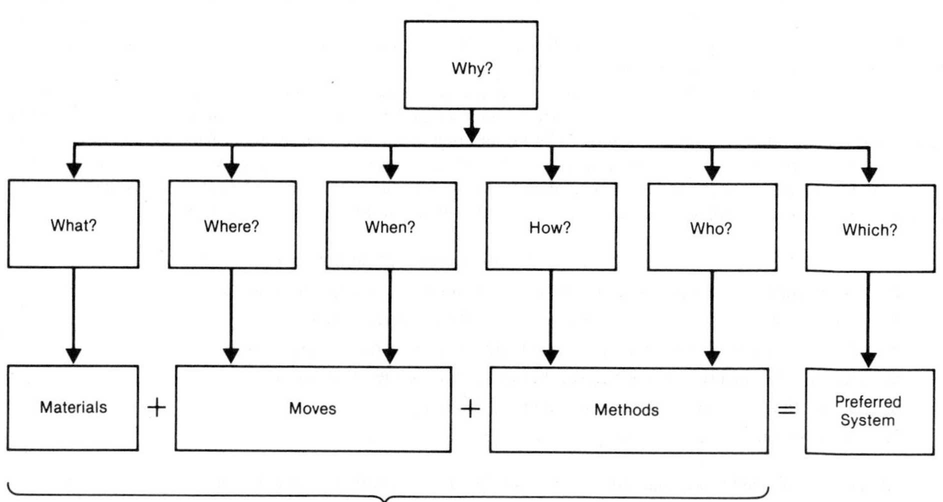

Fig. 5.2.2. Material handling system equation.[7]

Quality and reliability
Operations
  Production
  Support
Personnel, safety, labor relations
Sales, marketing, customer service
Finance

We have found the following series of design steps helpful in trying to coordinate the design of individual system elements:

1. Establish process design.
2. Establish inventory points and levels.
3. Develop block layout.
4. Consider workplace layout.
5. Determine load increment and container load.
6. Evaluate WIP storage methods.
7. Determine horizontal transportation method.
8. Design control system.

No individual step can be done without considering the others. Rather than representing the order of attack, the sequence comes closer to representing the order in which steps must be completed. Frequently, in designing integrated systems, many of the steps are developed simultaneously.

The remaining sections of this chapter will discuss each of these steps.

5.2.4 INTERFACING WITH PROCESS DESIGN

Working in conjunction with manufacturing or process engineers, the manufacturing processes are specified with either a straight-through flow or one consisting of the movement of batches between operations. The straight-through process is one of the key principles of the just-in-time (JIT) philosophy. Success with transfer machining lines and paced assembly lines show clearly that straight-through processes are far more efficient. Too often, if a process is not currently configured in a straight-through fashion, the assumption is made that it cannot be done. However, the productivity gains are too great to let these assumptions prevail.

The following characteristics must exist for straight-through processes:

1. Balance in cycle time between operations.
2. Machines dedicated to process or parts.
3. Reliable process or high machine up-time.
4. Long production runs or high changeover flexibility.

First, look to see if these conditions exist. If not, work with process engineers to see whether it is possible to create these conditions by making minor changes in the manufacturing process. For example, most western tire plants cluster first-stage tire building machines together in one shop, second-stage machines in another shop, etc. Straight-through or JIT principles, require moving these machines into flow lines that orient the machines in process order not process type. *It is important to recognize that there is value in identifying even short segments of the manufacturing process that may be considered as straight-through.* When parts are able to pass from one operation directly to another, handling distances are shortened, the use of containers may be eliminated, and work-in-process inventories are reduced dramatically.

Processes that cannot be arranged for material to flow straight from one to another may be classified as *batch* processes. Their general characteristics are as follows:

1. Machines shared by several parts.
2. Lower volume production.
3. Many different part routings.
4. High set-up time/production time ratio.
5. Storage at next operation.

In general, the manufacturing scene is dominated by batch processes. The propensity to continue in this mode is fostered by a desire to optimize small pieces of the process. Performance measurement systems that emphasize high utilization of direct labor make traditional production supervisors resist changes that would make their operation more dependent upon someone else's. Until performance measurements are expanded to include lower work-in-process inventories and shorter manufacturing times, the value of batch processing will continue to be reinforced. In the meantime, material handling system designers will strive to develop systems that cope effectively with large and complex inventories of work-in-process, double-handling, and long-distance handling of materials.

Work cells resulting from a group technology approach to processing have provided an opportunity to view many traditionally batch flow operations as straight-through flow.

Because the entire manufacturing process may be broken into segments, a typical plant must deal with both batch and continuous flow and also a variety of assembly environments. There are several forms and graphical techniques that display data in a way that helps designers visualize the material flow. Those most commonly used are described and illustrated on the following pages.

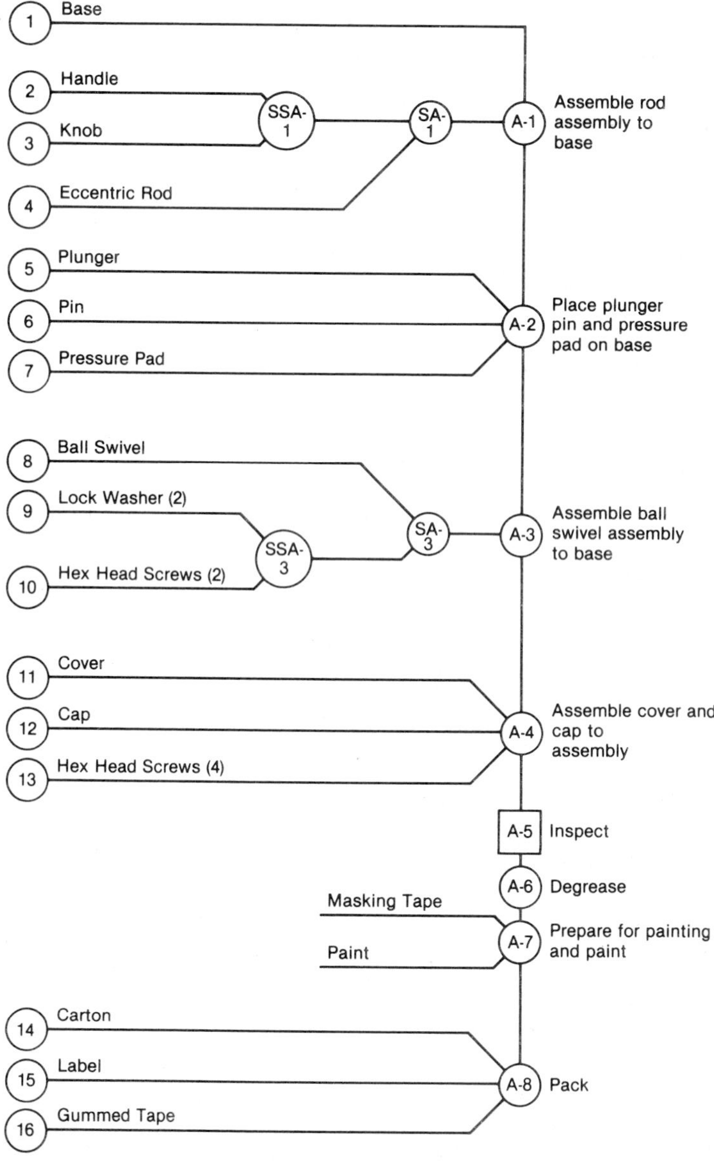

Fig. 5.2.3. Typical assembly chart.[2]

Production routings normally provide the following information:

1. Operation coupling requirements.
2. Production capacities.
3. Operation sequences.
4. Equipment specifications.

In order to use them to track material flow, they must be checked to be sure that they include storage operations, inspection, and off-site processing.

The assembly chart is a graphical representation of the sequence in which parts and subassemblies flow into the assembly of a product. The assembly chart shows:

1. What components make up the product.
2. How the parts go together.
3. What parts make up each subassembly.
4. The flow of parts into assembly.
5. The relationship between parts and subassemblies.
6. An overall picture of the assembly process.
7. The order in which the parts go together.
8. An initial impression of the overall material flow pattern.

A typical assembly chart is shown in Fig. 5.2.3.

Some of the advantages and uses of the *operations process chart* are as follows:

1. Combines production routings and assembly chart for a more complete presentation of information.
2. Shows operations to be performed on each part.
3. Shows sequence of operations on each part.
4. Shows order of part fabrication and assembly.
5. Shows relative complexity of part fabrication.
6. Shows relationship between parts.
7. Indicates relative length of fabrication lines and space required.
8. Shows point at which each part enters the process.
9. Indicates desirability of subassemblies.
10. Distinguishes between purchased and manufactured parts.
11. Aids in planning individual workplaces.
12. Indicates number of employees required.
13. Indicates relative machine, equipment, and personnel concentration.
14. Indicates nature of material flow pattern.
15. Indicates nature of material handling problem.
16. Indicates possible difficulties in production flow.
17. Records manufacturing processes for presentation to others.

An operation process chart is shown in Fig. 5.2.4.

5.2.5 WORK-IN-PROCESS STORAGE AND BUFFER QUEUES

Even with relative success at achieving straight-through operations, there will be points in the process where inventory is needed. Increasing product complexity and production requirements force work-in-process inventories to higher levels. New approaches to production scheduling and shop-floor control are being implemented to move material through the manufacturing system in a more continuous manner. This initiative is helping check uncontrolled inventory growth.

The most important aspect of sizing is to recognize that it is possible to establish an inventory between each pair of operations in the process sequence. The total inventory will be the sum of several unequal increments. In many cases we will establish the inventory between two operations as zero.

The following are places in the process where it makes the most sense to carry inventory:

1. Before universal parts become many specific parts.
2. Prior to equipment shared by several parts.

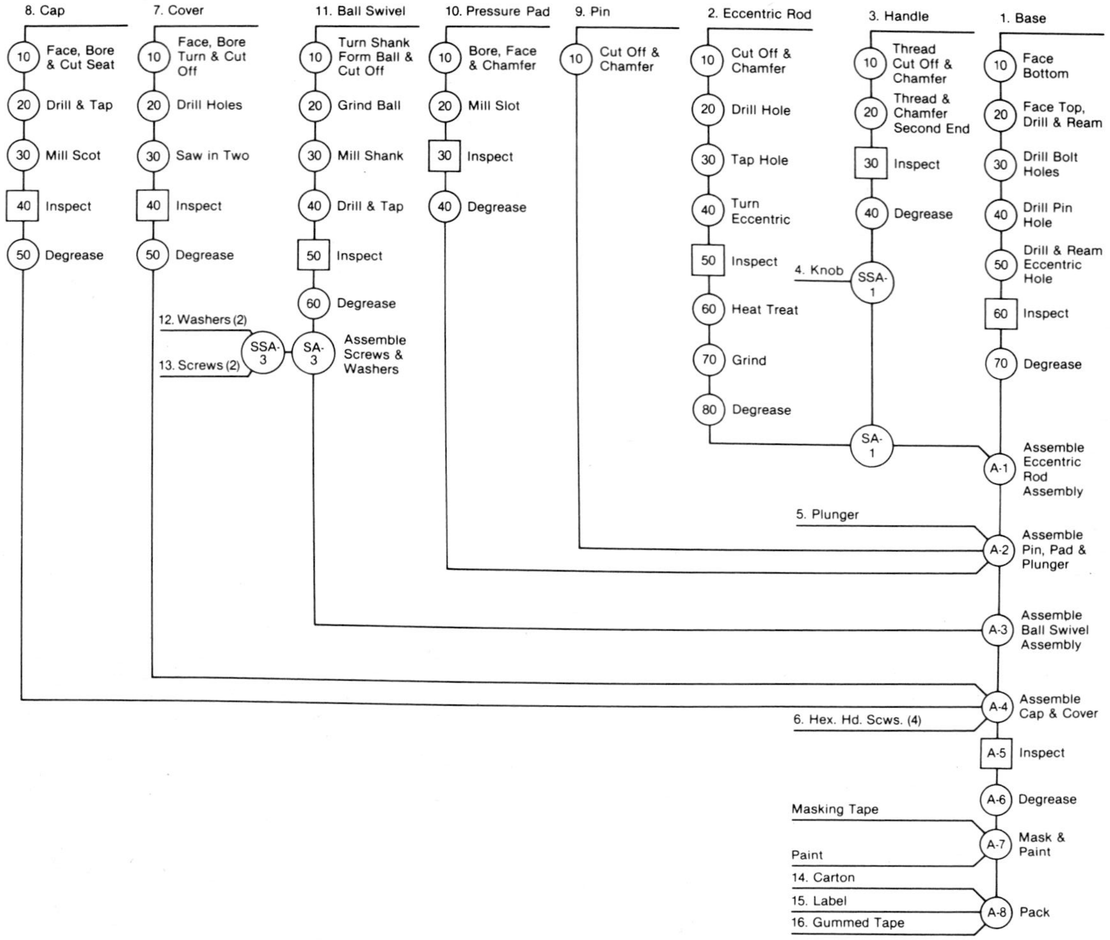

8. Cap
10 Face, Bore & Cut Seat
20 Drill & Tap
30 Mill Scot
40 Inspect
50 Degrease

7. Cover
10 Face, Bore Turn & Cut Off
20 Drill Holes
30 Saw in Two
40 Inspect
50 Degrease

12. Washers (2)
13. Screws (2)
SSA-3

11. Ball Swivel
10 Turn Shank Form Ball & Cut Off
20 Grind Ball
30 Mill Shank
40 Drill & Tap
50 Inspect
60 Degrease
SA-3 Assemble Screws & Washers

10. Pressure Pad
10 Bore, Face & Chamfer
20 Mill Slot
30 Inspect
40 Degrease

9. Pin
10 Cut Off & Chamfer

2. Eccentric Rod
10 Cut Off & Chamfer
20 Drill Hole
30 Tap Hole
40 Turn Eccentric
50 Inspect
60 Heat Treat
70 Grind
80 Degrease

3. Handle
10 Thread Cut Off & Chamfer
20 Thread & Chamfer Second End
30 Inspect
40 Degrease
4. Knob
SSA-1
SA-1

1. Base
10 Face Bottom
20 Face Top, Drill & Ream
30 Drill Bolt Holes
40 Drill Pin Hole
50 Drill & Ream Eccentric Hole
60 Inspect
70 Degrease
A-1 Assemble Eccentric Rod Assembly
5. Plunger
A-2 Assemble Pin, Pad & Plunger
A-3 Assemble Ball Swivel Assembly
A-4 Assemble Cap & Cover
6. Hex. Hd. Scws. (4)
A-5 Inspect
A-6 Degrease
Masking Tape
Paint
A-7 Mask & Paint
14. Carton
15. Label
16. Gummed Tape
A-8 Pack

Fig. 5.2.4. Operation process chart.[2]

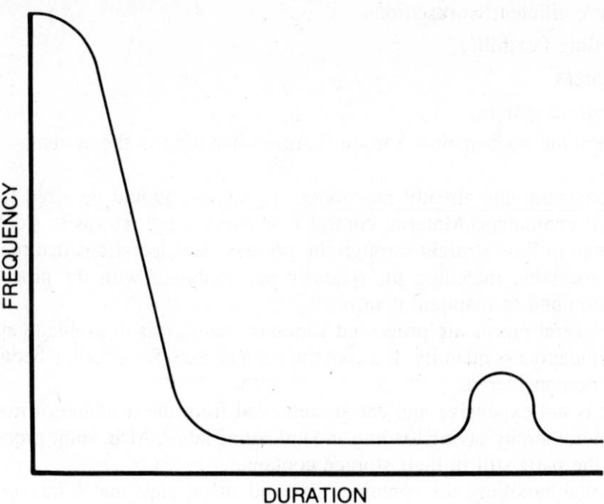

Fig. 5.2.5. Distribution of process downtime.

3. Between operations with widely varying cycle times.
4. Before and after operations with frequent downtime.
5. Prior to operations that increase cube value and handling difficulty and sensitivity.

Downstream scheduling flexibility is improved by storing parts in blank form before specific features are added to make them individual part numbers. Of course, operations that are shared by many parts create inventories related to the scheduling cycle. The size of that inventory is heavily influenced by the concept of economic lot size.

Historically, the primary factor in determining lot size is the set-up time. Although a great deal of effort has been expended to achieve the ultimate "zero set-up time" and a "lot size of one," not enough attention has been paid to the short-term opportunities for reducing current set-up times. Simply measuring and reporting set-up times normally provides incentive to reduce them. Doing basic methods work with the same level of effort that is applied to direct labor operations, usually produces dramatic results. For new machinery it is important to specify changeover time as a part of the performance requirements. The scheduling flexibility gained from reduced set-up times has a significant impact on reducing work-in-process inventories.

Operations with widely varying cycle times normally create inventory requirements. It is particularly noticeable in the context of one- and two-shift operations. However, even when a faster operation has sufficient capacity, we may find that the savings in inventory and handling costs warrant slowing it down to balance with another operation. Because all processes are subject to failure, small inventories must be provided to buffer adjacent operations or suffer the consequences of forced machine downtime.

From a material handling standpoint, it is desirable to completely mechanize the small buffer storage requirements between adjacent operations. The safety stock requirements for complete downtime protection, however, create inventories that are expensive to mechanize. Figure 5.2.5 illustrates the phenomenon that machine downtimes tend to be clustered in either very short or very long durations. The very short downtimes are normally a result of machine jams, broken tools, or adjustment requirements, whereas the long ones represent a more catastrophic occurrence, requiring maintenance intervention. It is the frequently occurring short downtimes that need to be identified and protected with mechanized buffer queues.

If simulation modeling is done to develop inventory requirements in a process line, it is important to use a histogram of real data to characterize downtime. In the case of new machinery, for which there is no history, existing processes that have similar complexity may provide satisfactory data. Sensitivity analysis would certainly be in order.

5.2.6 LOAD SIZE AND CONTAINER DESIGN

Strict adherence to the old unit size principle and a desire to reduce direct material handling costs has left most manufacturing operations with pallet-handling systems between operations. For many plants, pallets or even larger loads *are* still the appropriate load size. But, the current thrust to develop continuous flow processes coupled with a strong drive for automation has led many firms to design handling systems that employ much smaller containers. The benefits of smaller containers include:

1. Compact and more efficient workstations.
2. Improved scheduling flexibility.
3. Smaller staging areas.
4. Lighter duty handling system.
5. The ability to combine both manual and mechanized handling in the system.

In addition to the considerations already mentioned, containers should be sized to permit scheduling of orders in multiples of full containers. Material control is enhanced and returns to stock are minimized when full containers can continue to flow straight through the process. For operations that receive multiple parts per cycle, such as plating processes, matching the quantity per container with the process batch size helps to improve capacity utilization and to maintain control.

In operations where several pieces are processed simultaneously, it is desirable to size the handling unit to match or be a multiple of the process quantity. If a container can be used for handling between several operations, common multiples must be considered.

When a process rack is not expensive and can be removed from the machine, it may be used as a storage and transportation container, thereby saving loading and unloading labor. Also, some processes, such as washing, may be performed with the parts still in their storage container.

In systems using manual handling, the container size and shape must match human capabilities. But even where handling is automated or mechanical assistance is provided, the human factor is still present. Loads in the 50–100-lb range should be avoided in container design. Workers can, and will, at times move the loads manually, resulting in exposure to back injuries.

Many parts require special protection and orientation for handling and storage. We should also recognize the opportunities presented by maintaining part orientation to facilitate automation at the next operation.

Providing surfaces suitable for mechanical handling, stacking, and/or nesting capability and container standardization must all be considered in developing the container system.

The use of standard containers through several or all of the operations in the process eliminates the need for container exchanges. Removing a freshly emptied container from an operation and replacing it with a new empty container of another type doubles the material handling effort serving the operation.

Whenever possible, systems should be developed that "close the loop" in container handling, with equal attention given to planning the return portion of the loop as to the process portion of loop, thus creating self-supporting container systems.

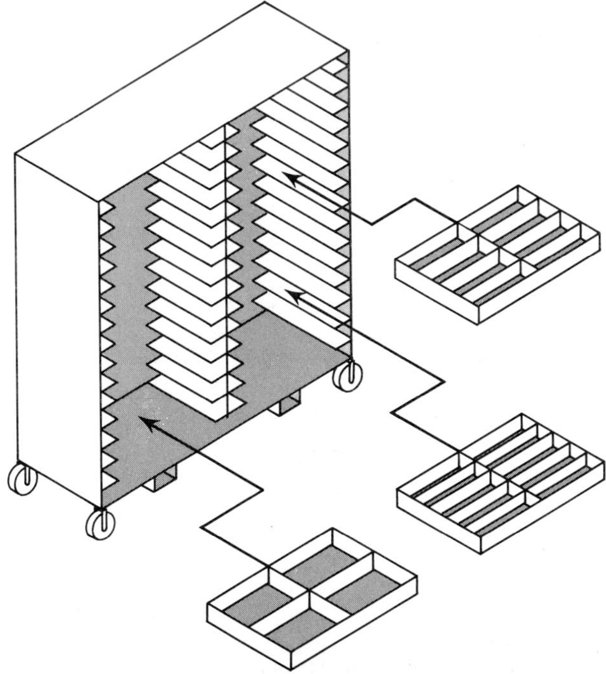

Fig. 5.2.6. Modular containers.

Modular systems may be developed in which smaller containers are circulated within an operation or department, but are combined to create a larger unit load for intraplant transport. Figure 5.2.6 illustrates a family of containers subdivided for part protection and orientation, and a mobile rack that is used to transport them between work centers via an automatic guided vehicle.

5.2.7 LAYOUT AND THE MATERIAL HANDLING SYSTEM

The arrangement of functions and equipment within the plant dictates the scope of the material handling system and in general the size of its workload. Consequently, it is preferable to design the two together. An earlier section discussed the importance of coordinating handling system design with process design. Of course, the flow of the process will have a significant impact on the plant layout, too.

The problems of layout may be divided into three major categories, each having unique requirements.

1. *Plant-wide layout.* This activity involves the arrangement of major functions within the plant. it is most evident in the design of entirely new facilities.
2. *Departmental layout.* A smaller subset of operations related to a specific product or process may be defined as a department, or work center.
3. *Workplace layout.* The workplace is usually the domain of a single worker, although sometimes teams work together in a single workplace to perform a task.

Most frequently the workplace is associated with a single machine or with a station along an assembly line. A growing trend is to combine several machines in a work "cell" so that operations in sequence may be performed without major material handling moves or time delays. Combination of tasks are also used to aid in line balancing.

Another rising trend is the placement of a robot or other fixed automation in a workplace or cell to perform the parts transfer and machine load/unload tasks. Robots at the workplace create a unique set of design requirements for the handling system.

Both plant-wide layouts and departmental layouts are discussed in Ch 7.3 of the *Handbook*, so coverage will be limited to areas directly related to material handling.

5.2.7.1 Plant-Wide Layout

Traditional product-based layouts for high-volume products were such that materials flow in one side of the facility and out the other. Figure 5.2.7 illustrates a typical plant of this type. The flow is unidirectional, the load configuration may change considerably between steps in the process, and thus the material handling systems are dictated by the process needs to and from adjacent functions.

For very-high-volume operations, the handling systems are automated: very efficient, but inflexible. In lower volume or mixed-product environments, the handling systems tend to be manual (often lift trucks) and inefficient.

Figure 5.2.8 illustrates the same flow concept, but arranged in a U so that the flow ends where it began. There are three basic methods to flow material from function to function: straight-through, perimeter, and corridor or spine. Figure 5.2.9 superimposes these flows on the layout.

The handling systems that best serve these flow concepts are quite different. Even in the U-shaped layout, with flow directly from one function to another across a common boundary, the systems are likely to be either dedicated mechanization or manual.

The perimeter flow concept provides an opportunity for unidirectional, high capacity handling. Load sizes will probably be reconfigured as standard containers. Backtracking may be difficult, or interfere with the primary flow. The handling devices might be one of the following:

1. Trolley or power and free conveyor.
2. Overhead roller conveyor with lifts.
3. Towline conveyor and carts.
4. Automatic guided vehicles.
5. Automated powered monorail carriers.

In the facility with the material handling corridor, or spine, load standardization is also most likely. Bidirectional flow permits backtracking, but capacity is limited unless the corridor contains a narrow loop. The loop may take the form of an over-and-under conveyor system. All the other equipment options for the perimeter system apply to the narrow loop. A transfer car or a bidirectional powered monorail carrier could perform the handling tasks, but at reduced throughput levels. They may be appropriate, however, if the loads are especially large and move infrequently.

An interesting variation of the corridor plan might relocate the parts and in-process storage to the sides of the corridor and pass material through the "storage walls" to and from the processes.

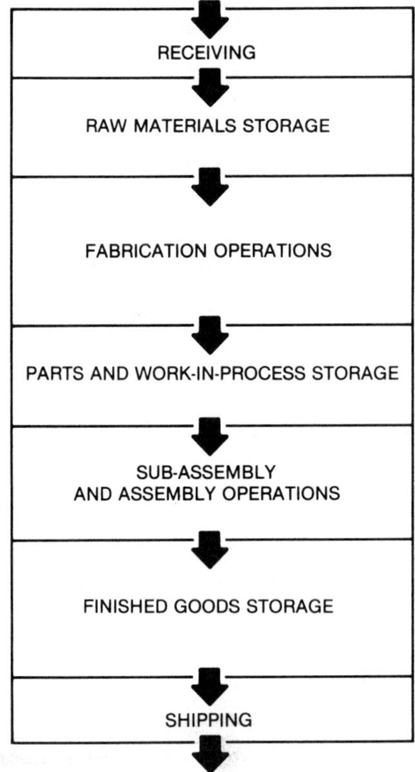

Fig. 5.2.7. Plant layout with straight-through flow.

5.2.7.2 Department Layout

Once an overall plant flow and a block layout have been established, departmental layout should be the next priority in designing a productive facilities layout. A department can be defined as an independent process or a process that is a subset of a larger operation. In many cases a department can be thought of as a microfacility. Thus, many of the same techniques used to analyze plant layouts can be applied at the departmental level too. In this section the discussion will be limited to considerations that should be explored when designing a layout for a department. Traditional analytical techniques are covered in other chapters of this handbook.

Production departments can be classified into three categories,

1. Assembly
2. Process
3. Manufacturing.

In assembly departments, purchased or previously manufactured components are moved through a series of operations to produce either a subassembly or a finished product. Examples of assembly departments range from electronic circuit board assembly, to welding of metal components, to automotive assembly lines. A packaging operation has many of the characteristics of an assembly department.

Process departments produce a new product or material by converting or combining raw materials. The product produced may be used to feed another department or it may be a finished good. Examples of process departments are found in the pharmaceutical, chemical, and food industries. Process departments may also modify a manufactured part, such as plating or heat treating.

Manufacturing departments produce parts. These parts may be a finished product, but more likely they are destined for an assembly department. A manufacturing operation usually alters material physically, by machining or stamping. Most frequently, parts are run in large batches to improve operation productivity. Today, however, many manufacturing operations are being improved to reduce changeover or set-up times, thus making very small batches more economical.

Fortunately, the three classes of departments have many of the same handling systems considerations. The first consideration in determining layout for a department should be what type of process it is. Is it straight-through or batch-oriented? The objective should be to design a department with a straight-through process. A straight-through process will reduce departmental work-in-process and improve flow through the department.

There are several basic flow patterns used within a production department:

1. Straight line.
2. Serpentine or zig-zag.
3. U-shaped.
4. Circular.

Additionally, where a process has been squeezed into an existing facility, or in a job shop where every part does not follow the same flow, a random pattern may exist.

Figure 5.2.10 illustrates the five flows. The straight-line and U-shaped flows are probably the most common in industry today. The flow pattern should be dictated by the process and the best utilization of space that does not hinder the production.

When designing the layout for a department, there are six major material handling issues that should be considered:

1. Interface between department and plant handling systems.
2. Prestage areas for the department.
3. Movement of materials from prestage to work station.
4. Movement of materials at the workstation.
5. Movement of materials after the work has been completed.
6. Outbound staging.

The interface between departmental handling systems and plant-wide handling systems varies greatly. In many cases the plant systems are automated while the departmental handling systems are manual. The layout must be designed to accommodate the needs of both systems at the points of interface between the two. It is

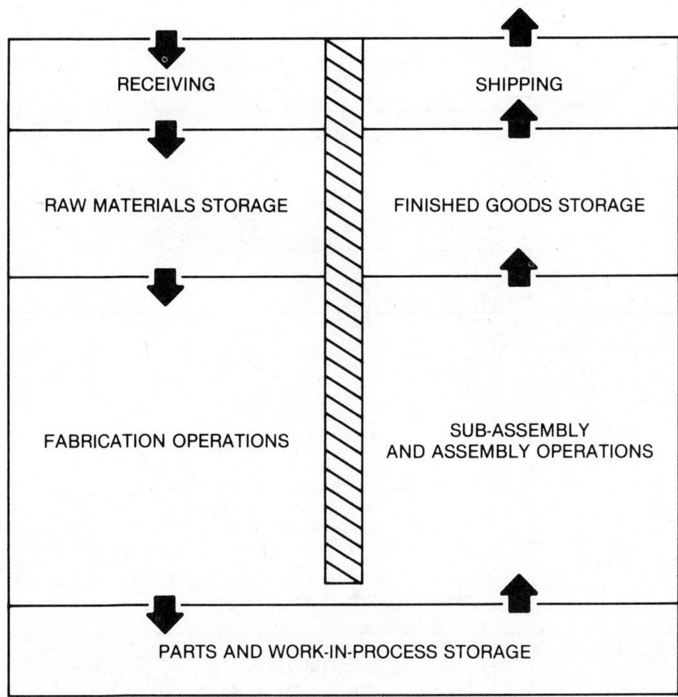

Fig. 5.2.8. Plant layout with U-shaped flow.

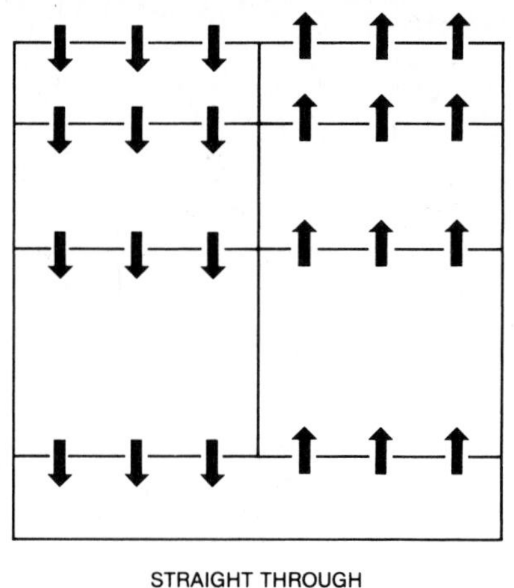

STRAIGHT THROUGH

(a)

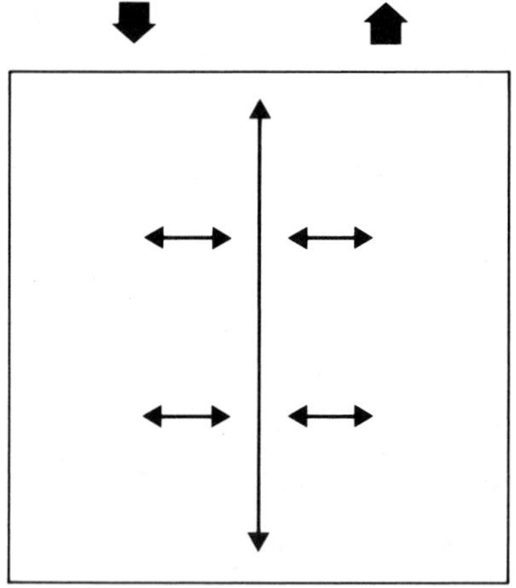

CORRIDOR OR SPINE

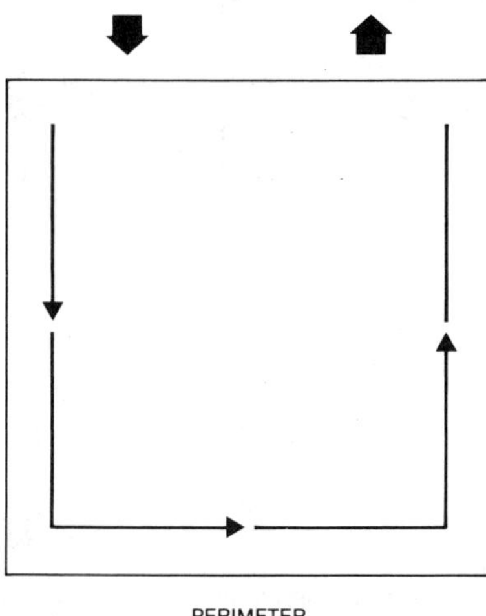

PERIMETER

(b)

Fig. 5.2.9. Alternative material handling system concepts.

also important to recognize that the handling characteristics of raw materials may be much different than those of WIP or the finished goods that come off the other end of the production line.

At the point where the two systems interface, queueing and loading/unloading must be designed carefully. A small queue and automated container or part transfer will help to keep the plant-wide and departmental systems sufficiently independent so that one does not have to wait for the other.

Load elevation, or the level of the container above the floor, becomes critical at the unload point. Within a department, it may be desirable to handle the load at operation working height, but the plant-wide system may perform best with the load close to the floor. A simple lift table as part of the unload station or queueing conveyor can accomplish the change in very little space and at a reasonable cost. It may be either automated or manually operated, as throughput requirements and operator availability dictate.

The prestage area for a department serves several functions, and each will have some bearing on its size and layout. They include:

1. Accumulating the number of parts in a batch-oriented process.
2. Providing a cushion between the department and the warehouse.

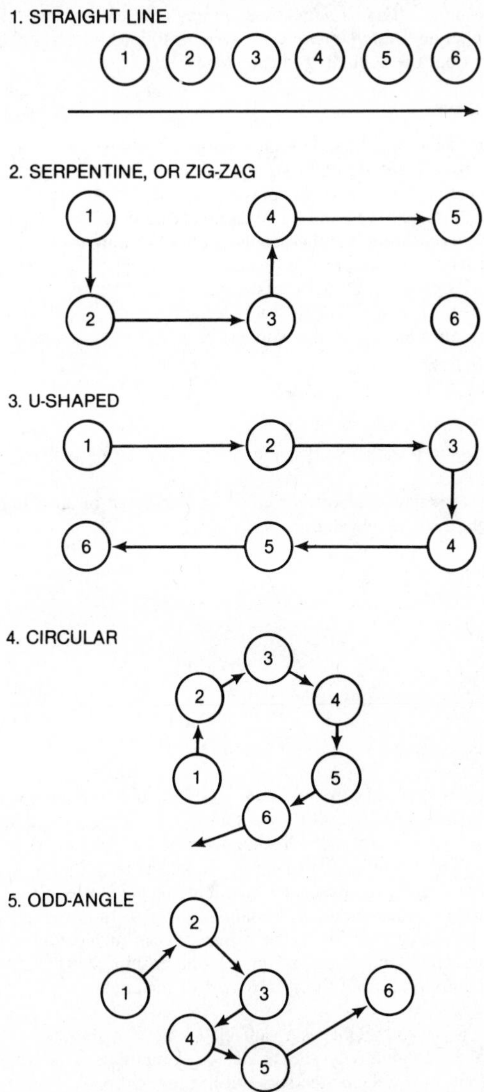

Fig. 5.2.10. General flow patterns.[2]

3. Accumulating a variety of parts arriving for the same assembly job.
4. Providing a backlog of work, readily available to be sure that men and machines will stay busy.
5. Breaking a large load into smaller ones for easier handling within the department.

Regardless of the functions being performed, the objective is to keep this area small and the material under tight control. This might be accomplished with first-in/first-out flow rack or conveyor systems, or it may be done with a random storage system such as a carousel conveyor.

Movement of material from the prestage area to the production line or workstation play an important role in the productivity of a department. The layout should consider the methods of transporting components to each workstation and provide for easy access as well as room for raw materials near the workstation. In some cases it may be possible to prestage or supply components directly to the workstation, thus eliminating one handling move.

Part and container movement within the department may employ any of the flows and methods discussed in the facility layout section; however, careful consideration must be made to see that the handling system does not create barriers to the movement of people.

Once the process has been completed within a department, the finished product of that department must be transported to either a post-staging area (especially if lot integrity must be maintained), a prestaging area for another department, or to a finished goods warehouse. In any case the department layout must be configured in such a way as to allow inter and intra-departmental systems to interface, keeping in mind that the handling systems may have changed from the beginning of the process.

5.2.7.3 Workstation Layout

If bricks are the foundation of a building then the workstation is the foundation of a productive layout. However, it seems that no other area of layout design and material handling is more neglected. Too frequently this area is ignored by the material handling engineers, who may feel that it is out of their domain; similarly, the methods engineers often limit their efforts to improving methods involving only the production task.

Material handling at the workstation can be defined as the handling that must be done after material has been delivered to the workstation and before it is picked up to be removed for the next operation. In general, there are five steps to material handling at the workstation:

1. Preparatory handling of materials at the workstation.
2. Moving materials into the workstation.
3. Manipulating materials within the workstation.
4. Removing materials from the workstation.
5. Transferring materials to the next workstation.

To develop an efficient workstation layout for either an individual or an automated device, the following must be considered for each of the above steps:

1. Manual vs. mechanical handling.
2. Flow
3. Handling distance
4. Floor space
5. Material handling delay
6. Container changes
7. Access requirements

Flow through a workstation should be analyzed and designed in much the same way as it is analyzed and designed for a large facility. First, the overall flow pattern for the workplace should be charted. Many times the pattern is dictated by the production process. To fine-tune the flow, the interrelationships between materials and tasks should be analyzed. The relationships between functions and storage of materials should be ranked in order of importance and quantified by use of the traditional techniques explained in Chapter 7.3.

Handling distances at workstations should be minimized so that time spent transporting materials within the workplace is small. A layout designed with the workstation in mind will be able to minimize distances. Here again some quantitative techniques such as activity relationship charts, which are used in plant layout, can be used to determine the importance of the relationship of materials to machines and personnel within the work area.

Floor space within any production facility is a valuable commodity. However, to use it efficiently does not mean crowding a workstation. Productivity should not be compromised in the name of better floor-space utilization. An efficient layout must balance the space utilization and productive work area.

The majority of material handling delays can be overcome by proper layout design. Disjoining functions through the use of buffers and line balancing will increase throughput and productivity. By providing buffer zones within the layout, delays that occur will not stop production. The purpose of the buffers is to minimize dependence of successive operations, and should be designed for normal downtime, not major malfunctions in the system. A good queue design provides a buffer between an operation and the transportation system to reduce material handling delay, and permits higher utilization of transportation equipment whether it is manual or automated. The key here is that buffers should be as small as reasonably possible without hindering production. In many cases, buffer sizes can be estimated using statistical techniques and data from existing similar production processes.

As raw material moves through a production process, the material handling containers may change in size and unit quantity. The unit load principle of material handling design has long told us that we should move as much material at one time as is feasible. For load transportation purposes, this principle certainly makes sense. However, at the workstation, large containers create long handling distances for operator or robotic movement of parts and require much more floor space. In newer systems, where more frequent delivery may be handled by automated transportation, smaller containers provide many benefits. Positioning within the reach of automated equipment is more easily accomplished, and when manual handling is required it can be done without powered equipment. Container changes are made less disruptive by permitting the operator to manage the input and output queues. Of course, in systems that utilize central storage of work-in-process, the smaller loads will generate a higher frequency of storage and retrieval tasks. Caution should be taken when considering containers for material handling. The container should be designed to protect the product as well as the operator handling it. In particular, operators have a tendency to attempt to lift manually containers in the fifty to one-hundred pound range, thus introducing the worker to potential lifting injuries.

The workplace layout should consider access requirements for the delivery of materials to the workplace, as well as maintenance. By allowing proper access, potential rehandling of materials and machine downtime can be avoided.

A discussion on workstation layout would not be complete without some mention of ergonomics and the principles of motion economy. For a more complete coverage of the subject, see Chap. 2.4.

When designing a workstation, ergonomic data should be used. Also, human factors guidelines should be used in determining human capabilities for lifting and carrying materials. The following eight principles of motion economy apply to the arrangement of the workplace[15]:

1. There should be a definite and fixed place for all tools and materials.
2. Tools, materials, and controls should be located close to the point of use.
3. Gravity-feed bins and containers should be used to deliver material close to the point of use.
4. Drop deliveries should be used whenever safety permits.
5. Materials and tools should be located to permit the best sequence of motions.
6. Provisions should be made for adequate conditions for eyesight. Good illumination is the first requirement for satisfactory visual perception.
7. The height of the workplace and the chair should preferably be arranged so that alternate sitting and standing at work are easily possible.
8. A chair of the type and height to permit good posture should be provided for every worker.

WIP Storage Methods

For those places in the process where inventory cannot be handled in mechanized buffer queues, storage systems must be designed. They may be distributed throughout the plant, or as is becoming increasingly common, they may be centralized to provide an opportunity for high density and higher levels of automation. Centralizing inventories also reduces the total number of load spaces required. Space for surges in inventory requirements may be shared by all the parts and processes. Work-in-process storage systems are most often characterized by handling of full containers, in and out.

In addition to the container design and the number of loads, a major factor in storage system design is the response time requirement. As we try to minimize work-in-process on the manufacturing floor, we develop a need to bring material quickly from the storage system to the operation. The possibility of several demands occurring simultaneously will require a peak throughput capacity considerably above the average demand. Other important considerations in storage systems design include the number of containers in inventory for each part number and the possible need to maintain first-in/first-out inventory control.

Storage Location Assignment

An important factor in the design of storage systems is the method of assigning storage locations to individual parts. There are two basic methods: random, or floating assignment, and dedicated locations. The method selected will have a significant impact on storage cube utilization and on a system's throughput capacity, so it must be chosen carefully.

When an individual part number can be put away in any available storage slot, it may be considered as random location assignment. Storage slots may be classified by size so that there can be a closer match between the cube of the item to be stored and the available locations.

There are two types of randomized storage: (1) true randomized storage, in which all locations have an equally likely chance of having product stored in them and (2) closest-available-slot (CAS). Outgoing items may be picked automatically on a first-in/first-out (FIFO) basis or they may be selected based on the opportunity to complete the pick from a single location. Yet another option is to select based upon an opportunity to empty partially filled locations to make them available for the storage of new items. The CAS type of randomized storage is the most common in today's industrial environment. If the storage levels remain fairly constant and at a high level of utilization, there is little difference between the two types of randomized storage. However, if utilization is low and inventory levels vary, the CAS method of randomized storage will increase throughput in the system.

Dedicated storage based on activity will maximize throughput at the expense of storage space utilization. Conversely, randomized storage will minimize storage space and reduce throughput of the system. Studies have shown that dedicated storage can yield savings in increased throughput of 15 to 50% over pure or true randomized storage. Additionally, studies have shown that dedicated storage can require from 20 to 60% more storage slots than are required for random storage.[13] Thus, the selection of the appropriate storage location assignment method depends upon the importance given to space versus throughput levels.

Some possible automated alternatives for WIP storage are listed below. Complete coverage of equipment alterations may be found in Chapter 4.6.

1. Unit load AS/RS.
2. Miniload and tote handling AS/RS.
3. Carousel conveyors with automated inserter/extractors.
4. Flowthrough lanes with automated sorting and escapement devices.
5. Power and free conveyor systems.

Horizontal Transportation Method

Many factors influencing the horizontal transportation system have already been mentioned. Centralized storage of work-in-process will create more and longer horizontal moves. Smaller containers will further increase the transportation requirements. Queue sizes at the workplace will dictate how quickly the horizontal transportation must move material from one operation to another or from central storage to an operation.

For straight-through flows, it is common to require accumulation of parts or loads within the transportation system. If recirculation of parts is considered, we must also study the impact of the potential waiting time for part arrival and for the availability of empty carriers for part take-away.

Physical characteristics of the handling tasks will affect our selection of viable alternatives. Long distances may justify automated systems. Frequent changes in elevation of floor level may be accommodated with conveyors or monorail systems. Reciprocating lifts are the most likely to be throughput bottlenecks and the most difficult to work around in the event of downtime.

We are normally concerned with physical barriers that the handling system must cross, but it is equally important to recognize that the handling system itself may create a barrier to other traffic. Aisle widths and turning radii are critical factors for towlines and guided vehicle systems.

For any mechanized transportation system, the design challenge is greatest at the load/unload points. The frequency of activity at a particular station and ready availability of an operator help to determine the level of automation. Appropriate queue lengths at the load/unload points will help to achieve higher utilization of the transportation system capacity. A load may be presented at floor level to facilitate inexpensive pallet handling at the station, or it may be presented at a convenient working height to reduce operator effort. Carts have been used as handling containers in both floor running and overhead systems to permit both convenient working height and ease of movement by the operator at the workplace.

5.2.8 SUMMARY

In summary, the design of a material handling system requires careful consideration of many factors. Those that we have found to be essential to successful implementation were described in the chapter. In particular, principles, measures, and objectives of material handling systems design were presented. Additionally, a design methodology was described. The importance of interfacing material handling systems design with process design was also emphasized. Further, due to the increased attention being given to inventory levels in production, we addressed, albeit briefly, the sizing of work-in-process storage and buffer queues. The basic building block of a material handling system, the lot size and container size, was treated. The relationship between the material handling system and the layout of the system was addressed by focusing on the layout of a plant, a department, and a workstation. Finally, issues and alternatives approaches to the physical movement and storage of material were discussed.

REFERENCES

1. James M. Apple, *Material Handling Systems Design*, New York: Ronald Press, 1972.
2. James M. Apple, *Plant Layout and Material Handling*, New York: Wiley, 1977.
3. James M. Apple, Jr., "Strategic Planning for Material Handling Improvements," *Proceedings*, AIIE/MHI Seminar, San Francisco, March 1980.
4. James M. Apple, Jr., and Bruce A. Strahan, "Proper Planning and Control—The Keys to Effective Storage," *Industrial Engineering*, **13**(4), 102–112 (April 1981).
5. James M. Apple, Jr., and John A. White, "Material Handling Requirements Are Altered Dramatically by CIM Information Links," *Industrial Engineering*, **17**(2), 36–41. (February 1985).
6. James R. Bright, *A Management Guide to Productivity*, Yale Materials Handling Division, Yale and Towne, Philadelphia, 1961.
7. Hugh D. Kinney, "Planning an Integrated Maufacturing System," an address to electronics industry executives, Monterrey, CA, May 1982.
8. James A. Tompkins and John A. White, *Facilities Planning*, New York: Wiley, 1984.
9. John A. White, *The Factory of the Future*, Production and Distribution Research Center Report, School of Industrial and Systems Engineering, Georgia Institute of Technology, Atlanta, GA, September 1982.
10. John A. White, "First Things First," *Modern Materials Handling*, **36**(9), 35 (September 21, 1981).
11. John A. White, *Yale Management Guide to Productivity*, 2nd ed., Industrial Truck Division, Eaton Corporation, Philadelphia, 1979.
12. John A. White, "Material Handling Improvements for the Smaller Installation," *Proceedings*, 5th IE Managers Seminar, Atlanta, GA, March 1983.
13. John A. White and James M. Apple Jr., "Long-Range View, Better Systems Integration Needed in Designs for Material Handling," *Industrial Engineering*, **14**(3), 50–58 (March 1982).
14. John A. White and Hugh D. Kinney, "Storage and Warehousing," *Handbook of Industrial Engineering* (G. Salvendy, Ed.), New York: Wiley, 1982.
15. Wesley E. Woodson, *Human Factors Design Handbook*, New York: McGraw-Hill, 1981.

CHAPTER 5.3

MATERIAL FLOW CONTROL

WILLIAM T. SHIRK

AT&T Technologies
Norcross, Georgia

5.3.1 INTRODUCTION

Material Flow Control (MFC) is the systematic regulation[1] of industry's material resources, including raw materials, expense supplies, work-in-process, and finished goods. It controls the dynamic flow of goods through a pipeline starting with procurement and transportation and ending with distribution of finished goods. Its objective is to "*get the right material to the right place in the right quantity and quality at the right time for the right cost*."

MFC is a recent management discipline, which became a top priority in the 1970s because of several trends:

1. Scarcity and higher costs of labor and materials.
2. Foreign competition for markets.
3. Manufacturing specialization, leading to higher inventory values of assemblies and finished goods rather than primary raw materials in the pipeline.

This chapter is introduced with management for effective material flow control, followed by forecasting and inventory control. Material flow strategies are discussed, followed by an overview of the automatic factory.

We will avoid repetition of expert coverage of topics contained elsewhere in this edition of the *Production Handbook*, but will detail new MFC tools that show promise. While higher mathematics are relegated to references, basic models will be illustrated outside the text.

Formal records, tag sets, and accuracy verification, which were standards, are in transition due to new computer power and utility for the dollar. We will emphasize paperless, real-time data systems.

Effective material flow controls are as varied as products or managers. *There is no optimal MFC strategy for all plants*. In the case of Plant C, starting from a patchwork of controls developed to solve specific problems, Manufacturing Resource Planning (MRPII) may be the path to global disciplines and teamwork for most effective MFC strategy. Plant B's high volume, repetitive production [having a material requirements plan (MRP) in place] is best served by stepping up to just-in-time (JIT) strategy. Plant A's strategy would be a distribution resource plan (DRPII). Finally, today's ultimate strategy may be obsolete in the next decade, thus our emphasis on professionalism to keep material managers in the mainstream of this important and rewarding technology.

Most authors[2] define MFC in the context of regulating materials in continuous manufacture, such as metals and chemicals. Tools such as new data systems and just-in-time strategies will help discrete manufacturing emulate continuous manufacturing by pumping and regulating material flows through the inventory pipeline. Some estimates of discrete manufacture place waits and delays at 95% of the cycle. More continuous flows in discrete systems promise large contributions to profitability and customer service.

5.3.1.1 Professionalism

Prospective users may find professional support from many technical societies dedicated to improvement of MFC arts and sciences. We will list five with their major publications.

APICS, The American Production and Inventory Control Society, Falls Church, VA. Publications: *The Production and Inventory Management Review* (monthly), *The APICS Proceedings* (annual), *The APICS Dictionary*.[2]

IIE, The Institute of Industrial Engineers, Norcross, GA. Publications: *Industrial Engineering*, *The IIE Proceedings* (annual).

IMMS, The International Material Management Society, Lansing, MI. Publication: *Modern Material Handling*.

MHRC, The Material Handling Research Center, Georgia Institute of Technology, Atlanta, GA. Membership: Closed to supporting member companies. Publications: Research reports in robotics, factory automation, logistics, warehouse automation issued to member companies. Annual research forum open to public.

MHI, The Material Handling Institute, Charlotte, NC. Publishes: material handling standards, specifications, books, seminars, films, and cassettes.

5.3.1.2 Case Studies

A brief view of the annual results* of three fictitious plants will introduce the reader to basic MFC concepts, jargon, and evaluation techniques (see Table 5.3.1).

		$ × 1,000,000		
Performance Measurements		Plant A	Plant B	Plant C
GCS	Gross cost of sales	$200	$800	$400
I	Value of stock on hand	$ 4	$ 22	$200
Cc[a]	Cost of carrying I	$ 0.8	$ 5.5	$ 60
	% of I = (Cc/I) × 100	20%	25%	30%
	% of GCS = (Cc/GCS) × 100	0.4%	0.7%	15%
TO	Turnovers/yr = GCS/I	50	36	2
W	Weeks stock = 52/TO	1	2	26

[a] Note that Cc can be expressed in dollars or as a ratio.

1. Plant A is a distribution center in a competitive and uncertain market environment. Yet A is a highly profitable venture due to vigorous material controls and warehouse automation programs. These feature:

   AS/RS—Automatic storage and retrieval system

   AGVS—Automatic guided (driverless) vehicles

   AIM—Automatic identification, automatic data entry/scan

   Conveyors/high speed sortation

   Accurate forecasts, schedules

   Powerful dedicated/real time/integrated computers

   A high number of inventory turnovers place A in the enviable position of losing only 0.41% of its gross cost of sales (GCS) to carrying costs.

2. Plant B's automobile assembly operation is producing at near capacity after years of lost sales to foreign competitors. B credits its return to a major share of the market with three goals:

   Maximize quality

   Minimize inventories

   Maximize worker involvement in operating decisions

   An aggressive automation program including smart conveyors/monorails, robots, and an all-out effort to replace lift trucks with automatic guided vehicles should maximize productivity and minimize lead times. Turnovers/year exceeded 36 in 1983, 50 in 1984, and employment of zero inventory strategy in 1986 is expected to produce 100 turnovers/year and make a contribution of more than $3 million per year to plant profitability. The 1985 ratio of carrying cost to gross cost of sales was a slim 0.25%.

3. Plant C dominated its electrical product market and is facing strong competition for the first time in years. C is a technological leader in its field and has a large captive market; however, that market was eroding and sales analysts suggested that a 10% reduction in prices was needed to stem the erosion. The profitability of Plant C was already down to 10% of the gross cost of sales. A 10% decrease in price would have eliminated plant profitability. The owners gave the managers of Plant C two choices:

   Improve profitability from 10% to 30% of the gross cost of sales (adjusted to 1983 level) or

   Consider closing and selling off the operation to competitors.

This chapter will concentrate on the material flow controls needed to achieve the required profitability.

* The fictitious plant performances are selected from the best and worst of several hundred operations viewed through annual reports,[3] interviews, and tours. The study emphasizes the MFC solution. The burden of 15% Cc/GCS is annual 10–12% flows of Plant C's income/profitability in order to maintain excess inventories. In the real world, that might be too large a burden to remain competitive.

Table 5.3.1. Fourth Quarter, 1984, U.S. Industry Average: Average Weeks Stock on Hand [(GCS/*I*) × 52][a]

Fossil fuel	12.5	Chemicals	28.9
Print/publish	12.6	Fabricated metal	29.7
Automotive	16.7	Primary metal	32.3
Wholesales	19.4	Textiles	30.8
Food	21.7	Instruments	35.7
Paper	21.2	Machinery	39.4
Rubber/plastics	24.8	Electrical	40.1
Stone/clay/glass	25.2	Tobacco	46.4
All manufacturing	28.6	Aircraft	86.6

[a] Quarterly Financial Report issued by the Federal Trade Commission.

5.3.2 MANAGEMENT FOR EFFECTIVE MATERIAL FLOW CONTROLS

Townsend[4] lists more than ninety steps to reach corporate objectives. Drucker[5] and Peters and Waterman[6] also demonstrate how corporate cultures must be revised to achieve operating excellence.

5.3.2.1 Goals

Plant C objectives were defined and presented in a format that was understood by the plant population. Standards for this analysis are presented in Chapters 1.6 and 4.1. Figure 5.3.1 displays Plant C cash flows with respect to analysis by ratios and by aggregation of cash flows into four increments consisting of:

(1) Material costs
(2) Direct labor costs
(3) Indirect labor, other costs
(4) Profitability [= GCS − (1 + 2 + 3)]

On the left is a graphic display of current costs cumulated and arranged in descending order, with the activities scaled equally on the horizontal axis. Costs from zero to gross cost of sales are scaled on the vertical axis; 60% of the GCS is materials, 15% is direct labor, 15% is indirect and other costs, and 10% is profitability. This format shows that profitability is being devoured by material costs. Similar analyses can be made for plant operating statistics from production to manpower to inventories to flow charts.[7]

The next chart zooms in on material cost increments and the lowest on carrying costs. The managers now plotted their goals on the left, with shaded areas indicating potential contributions by each increment to profitability. Although materials represent 60% of the gross costs of sales, application of material controls can generate 95% of the increased profitability in this example.

Direct labor contributions were expected through higher sales and improved productivity (higher yields, better flows of materials, automation, computer aided training). Thus, in a "high-tech" shop, trained people could be kept on the job.

5.3.2.2 Materials and Productivity Management (M&PM)

In the past, Plant C concentrated its technical expertise into islands of specialization with small pyramids operating outside of corporate objectives with *umbrellas* of responsibility. The umbrellas not only prevented communications from getting to the pyramids, but also often prevented critical communications from getting out.

Consolidation of the expertise under an M&PM manager has been a key to the success of Plant B. C's managers adapt the same strategy to speed implementation of flow controls to profitability.

5.3.2.3 Personnel

Personnel contributions to this effort are covered in Chapter 1.8, which stated, "The level of the commitment of the workers should match or surpass that of the designers and managers." Additionally, productivity can be improved by:

1. Replacing people with automation.
2. Helping people via training and job aids.
3. Supporting people with human engineered tools and equipment.
4. Selecting and retaining qualified people.
5. Motivating people to work for corporate objectives.
6. Eliminating pointless work and bureaucratic practices.

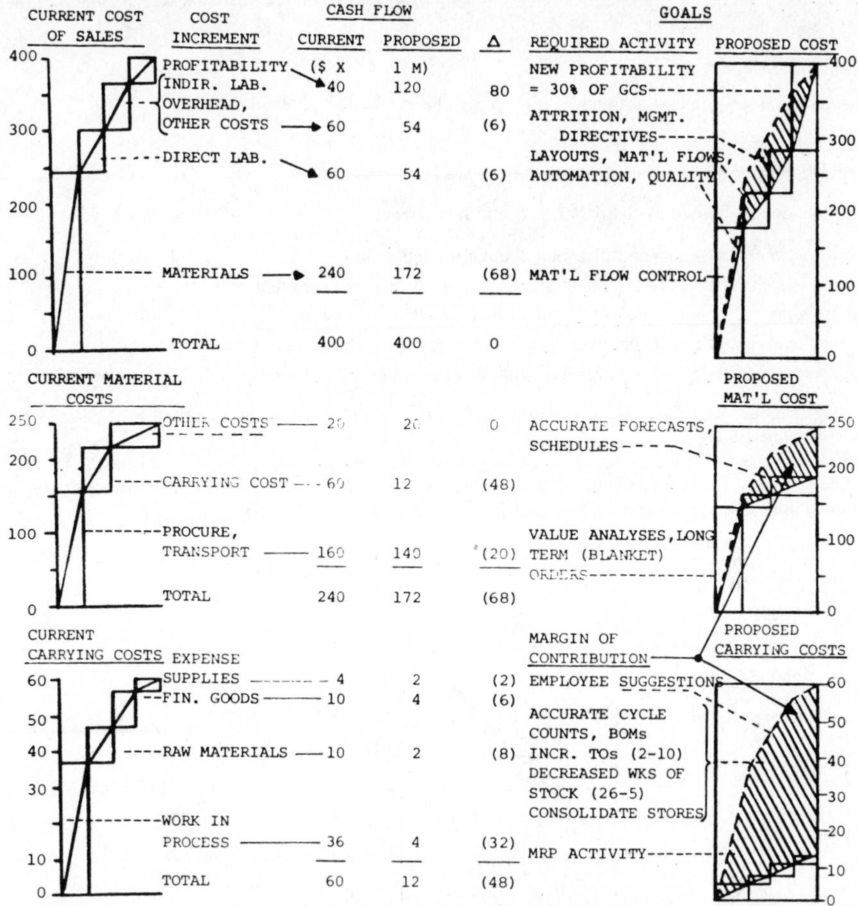

Fig. 5.3.1. Analysis by increment.

5.3.2.4 Implementation

Observation of the proven track record at Plant B, availability of standard software, and availability of consulting expertise to assist in rapid implementation lead to the selection of manufacturing resource planning (MRPII) as Plant C's material flow control strategy.

5.3.3 FORECASTING

5.3.3.1 Objectives

A forecast is Plant C's connection between the external, uncontrollable world and the internal, controllable affairs of production.[8] The functions of forecasting, planning, and scheduling are:

Forecasting is estimating future demand.

Planning suggests a quantity and due date.

Scheduling is the act of creating dates of future events.

The three are often dependent and inseparable, and their applications and feedback become the sources for Plant C's flow controls.

The object of forecasting is to guarantee that demand for final products is satisfied and to maintain a continuous flow of work through the shop.

Top-down forecasting deals with external environmental conditions, such as economic indicators, housing starts, competitor actions, technology, legislation, customer attitudes, and sociology.

Bottom-up forecasts study internal conditions such as management goals, resources, marketability, demand history, lead times, and priorities.

5.3.3.2 Models

Intrinsic/time series forecasting models (Fig. 5.3.2) have two distinct characteristics:

They are based on internal factors.

They are observations related to time increments.

Analysis is done by breaking the data down into five types:

Constant level has a demand distributed about a fixed value.

Trend has a distribution of demand about an upward or downward tendency.

Seasonality presents a repetition of peaks and valleys from year to year.

Cyclic tends also exhibit somewhat regular ups and downs but not necessarily annually.

Random (noise/residual/irregular) distributions have no predictable pattern over a given period of time.

Last period demand forecasts assume that the future will duplicate what happened in the last period. No calculations are needed.

Arithmetic average forecasts assume that the future will look like the *average* for the last time period.

Moving average forecasting is based on the average of a given number of recent observations. The number of observations selected determines how sensitive the forecast is to recent or past data.

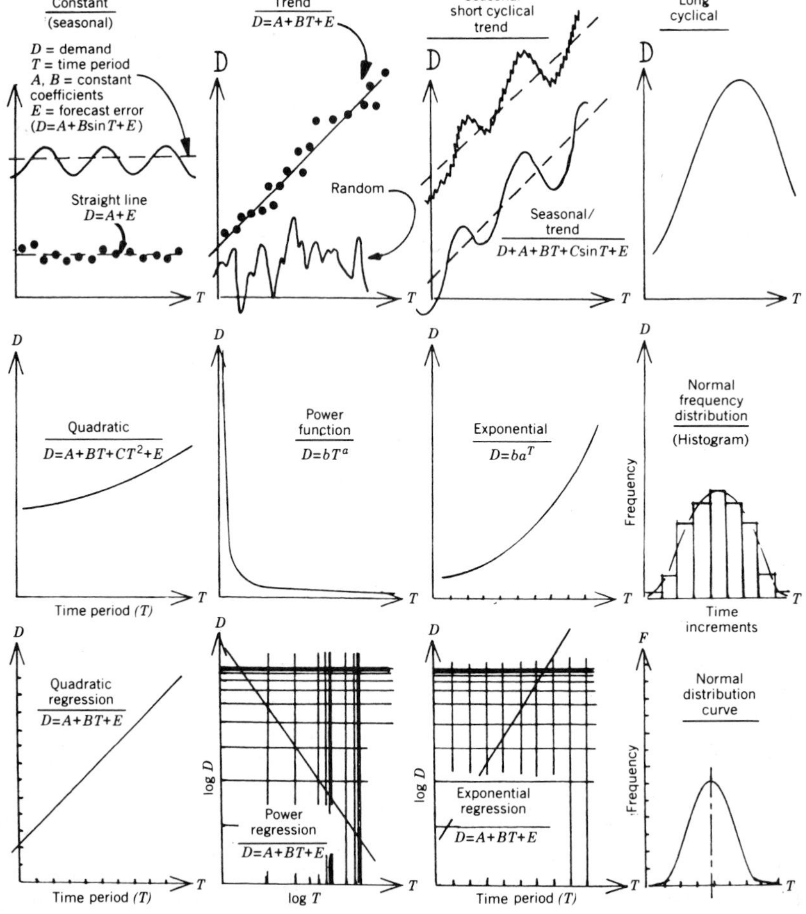

Fig. 5.3.2. Forecasting models.

Exponential smoothing is a forecast in which the latest data are weighted and oldest data are discarded after reaching a certain age. The alpha factor (α) is used to adjust the new and old observations as follows: a high α factor attaches high weight to recent events, while a low α smooths out the recent events in favor of a long view. Assuming that forecasts are more dependent on recent than old events, values for α vary from 0.05 to 0.15.

Seasonal/cyclical models use variations for specific periods.

Regression (rectification) models are a way of simplifying relationships between variables. The objective is to reduce nonlinear to linear relationships.

Polls are the solicitation of opinions about demand per sales, distribution, users, and expert opinions. The Delphi method is an iterative means of maximizing poll accuracy.[10] Questionnaires containing a series of opinions are used to design a second set of questionnaires, and so on, to develop agreement among those polled.

Econometry is the analysis of relationships among variables using simultaneous equations, linear programs, or simulation in order to develop demand forecasts.

Economic indicators are top-down influences on demand, starting with Gross National Product, interest rates, rates of inflation, and other extrinsic (or external) factors such as furniture sales and housing starts, which may have an impact on the overall state of business, rather than individual product forecasts.

Demand is **not** to be confused with sales or production forecasts. Demands are individual product requirements that may or may not become sales. *Dependent demands* are directly related to other demands and are computed from other forecasts.[2]

5.3.3.3 Tracking

Tracking or error feedback, (Fig. 5.3.3) is required to improve forecast accuracy. No forecast is completely accurate, but the more nearly accurate Plant C's forecasts become, the less unsatisfied demand, overtime, overstocks, idle time, and lead time from start of manufacture to finished goods there will be. Tracking also helps select correct forecast models, measures forecast reliability, and can help plan operating requirements.

1. The *tracking signal* to measure bias/skew (deviation from expected data) is:
   Sum of forecast errors divided by the Mean Average Deviation (MAD)
2. The MAD is calculated as follows:
   Mean is the data average.
   Deviation is the absolute (ignores $\pm$) difference of each value from the mean.
   Mad is the sum of the deviations divided by the number of deviations.
3. Demand filter is set to trigger when the demand differs from sales forecasts by a given number of mean average deviations.[2]
4. Least squares correlation measures the deviation of specific data from the line of regression. An absolute value of correlation coefficient of 80–100% is considered a high correlation.[8]
5. Dynamic programming evaluates the available options (models) and selects the best fit by computer.[11]
6. ANOVA (analysis of variance)[12] uses variance, which is the sum of the squares of the deviations from the mean divided by the number of deviations. In other words, the variation is similar to the MAD, except that it averages the squares of the deviations rather than simply the deviations from the mean.

5.3.3.4 Implementation of Forecasting in Plant C

Plant C *establishes goals* of sales, production, and costs of manufacture and *prioritizes* with a PARETO analysis (Fig. 5.3.4). All MFC experts recognize the need to simplify the models that will lead to full-blown control systems, not only to acquire benefits as quickly as possible, but also to test the new ideas on a small population before trying to solve all problems. Needed items not included in the first model will be quickly identified, and the excesses can be jettisoned. The PARETO (or 80/20) view is that a small part of the population accounts for a large part of the activity/cost/sales/labor/volume/,etc. Plant C's engineers find that 5% of the raw material items stored represent 50% of the volume and carrying cost. It is decided that (initially) 10% of the stock items (class A) will be forecasted against Bills of Materials and production plans, while the remainder will be determined historically by exponential smoothing.

Test Pareto model against goals. (See Fig. 5.3.1 and Table 5.3.2.)

Standards for time, unit denomination, and production/sales are now carefully chosen as standards throughout the inventory pipeline.

Final forecasting models are chosen to include all the material forecasts, yields, priorities, other manufacturing steps which could affect shop requirements.

Final forecasting models are also tested and fine-tuned through tracking models.

Data based forecasting begins with applying Delphi-type interactive analysis of all extrinsic sales factors, which can then be used on the intrinsic data. The forecast cannot be any more accurate than the accuracy of

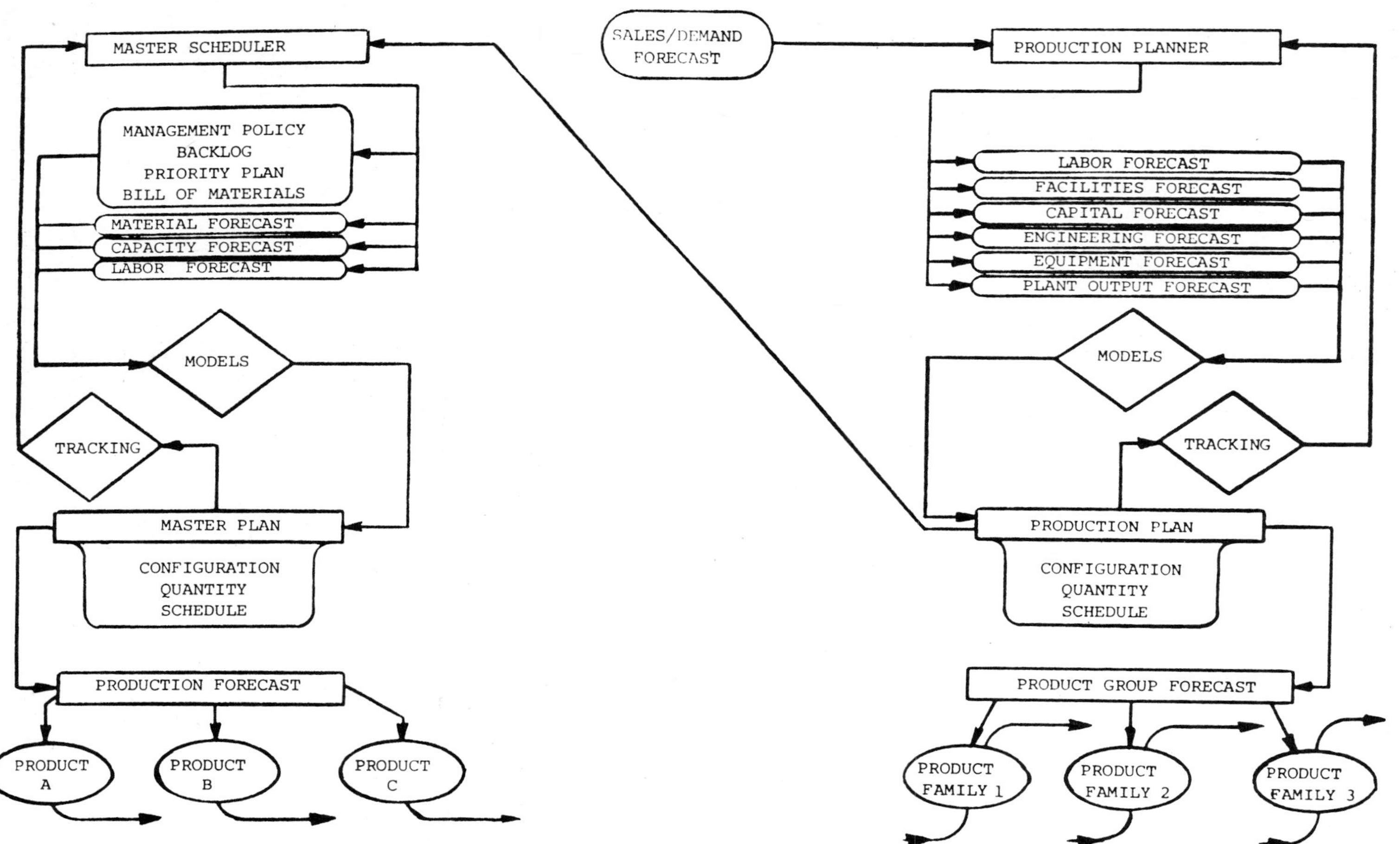

Fig. 5.3.3. Forecasting in plant C's made-to-order MRP system.

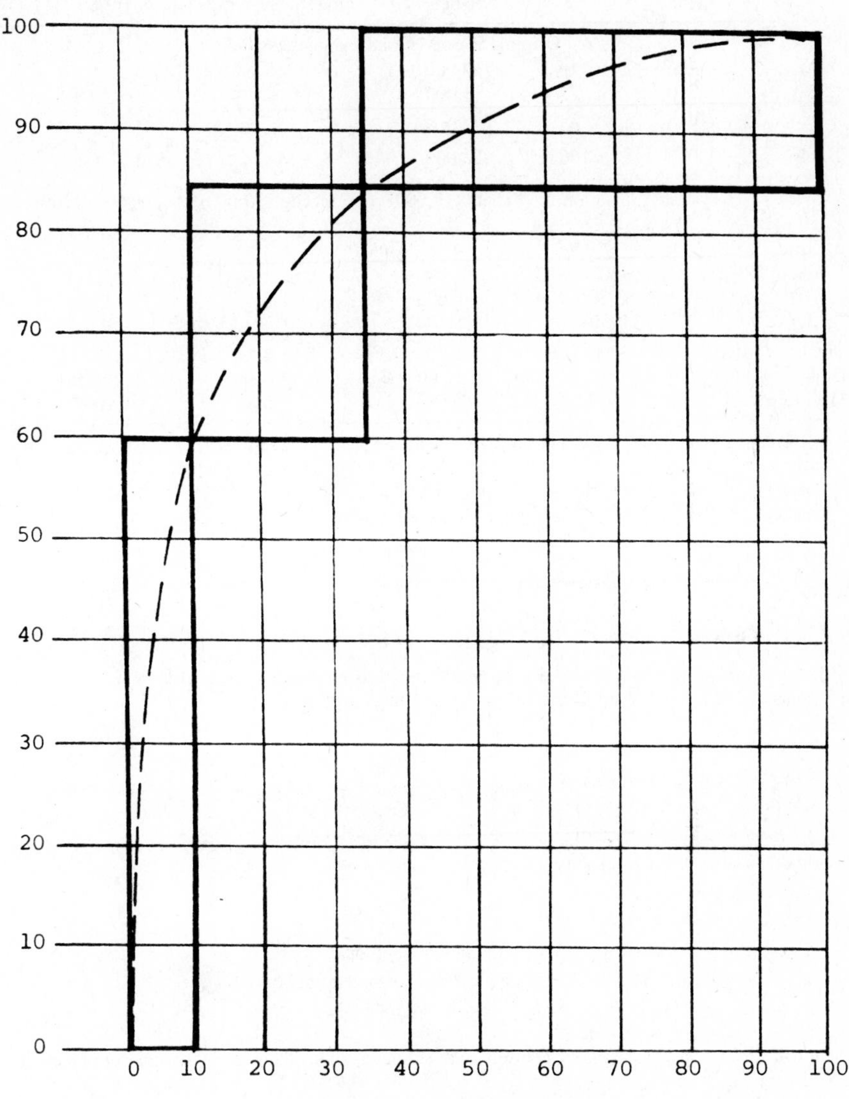

Fig. 5.3.4. Pareto analysis.

historical records. It is unnecessary to apply complex forecasting to any more than 10–20% of the stock. Demand data should be tracked and smoothed.

More intervals (data points) give better results. Histories of at least one–two years are required to identify seasonality.

Most intrinsic data is constant, trend, cyclic/seasonal or a combination of them. Interactive testing of all for the best solution using regression analysis and on-line simulation should be ongoing.

Exponential smoothing can reduce data files of obsolete inventory systems by orders of magnitude. Planning techniques (bills of materials, production plans, etc.) must be included. Lumpy demand patterns should be grouped with other stock to provide aggregate forecasts.

Tracking is the key to accurate forecasts.

Table 5.3.2. Test Pareto Model Against Goals

Pareto[a] Class	Population		Inventory on hand ($ × 1 million)			
	Items	%	Pallets	Investment	Cc @ 30%	WKS Stock[b]
			Original Raw Material Parameters, Plant C			
A	300	10	7,000	$23 M	$ 7 M	35
B	750	25	3,000	7	2	30
C	1950	65	2,000	3	1	13
Totals	3000	100	12,000	$33 M	$10 M	28
	System Utilization = [Pallets (12,000) ÷ Capacity (10,000)] × 100% = 120%					
			Goals[c]			
(A)	(300)		700	$2.3 M	$0.8 M	3.5
(B)[d]	(750)		600	1.4	0.4	6
(C)[d]	(1950)		1,100	1.7	0.8	7
Totals	3000		2,400	$5.4 M	$2 M	5.1
Goals			7,500	$6.6 M	$2 M	
% of Goal			300%	100%	100%	

[a] A,B,C class is determined from standard cost × units on hand for each item.
[b] Uncontrolled production control systems tend to overstock "A" items severely to avoid stockouts.
[c] After implementation, re-evaluation and re-implementations lead to added savings.
[d] B and C items reduced through less rigorous historical analyses.

5.3.4 INVENTORY CONTROL

5.3.4.1 Objectives

Although many profitable businesses now look forward to zero inventories, let us review the inventory policy questions raised daily at Plant C prior to their new material flow controls:

Why are we out of stock?
Why do we have so much stock?
Where shall we put all the stock?
How can we control overflow?
How can we get more capacity?
How can we be sure its there?
How do we know it will get here in time?
How much WIP (work in process) will smooth shop operations?
How much finished goods are required to keep customers' good will?
What will we do with obsolete and stock outside of limits?

Plant C's survival lay in inventory controls with assumed risks. Their inventory control objective can now be stated as:

Have enough stock on hand for maximum customer service and profitability, while minimizing manufacturing and distribution costs.

Without controls, this problem is simply one of optimizing objectives. With controls, the problem becomes one of setting realistic goals and homing in on them with modern MRP or just-in-time data systems.

Whereas the means of *establishing* inventories is forecasting, inventory control uses similar models to *maintain* those levels.

The inventory pipeline for Plant C (Fig. 5.3.5) consists of four general areas of control.

5.3.4.2 Raw Materials

Raw materials begin with vendor and flow into WIP.

In transit raw materials are an investment burden unless materials are on consignment, in which case the suppliers bear all costs until carriers are off-loaded. Plant C finds that, through longer vendor commitments, certain carload deliveries of raw materials can be vendor owned until off-loaded.

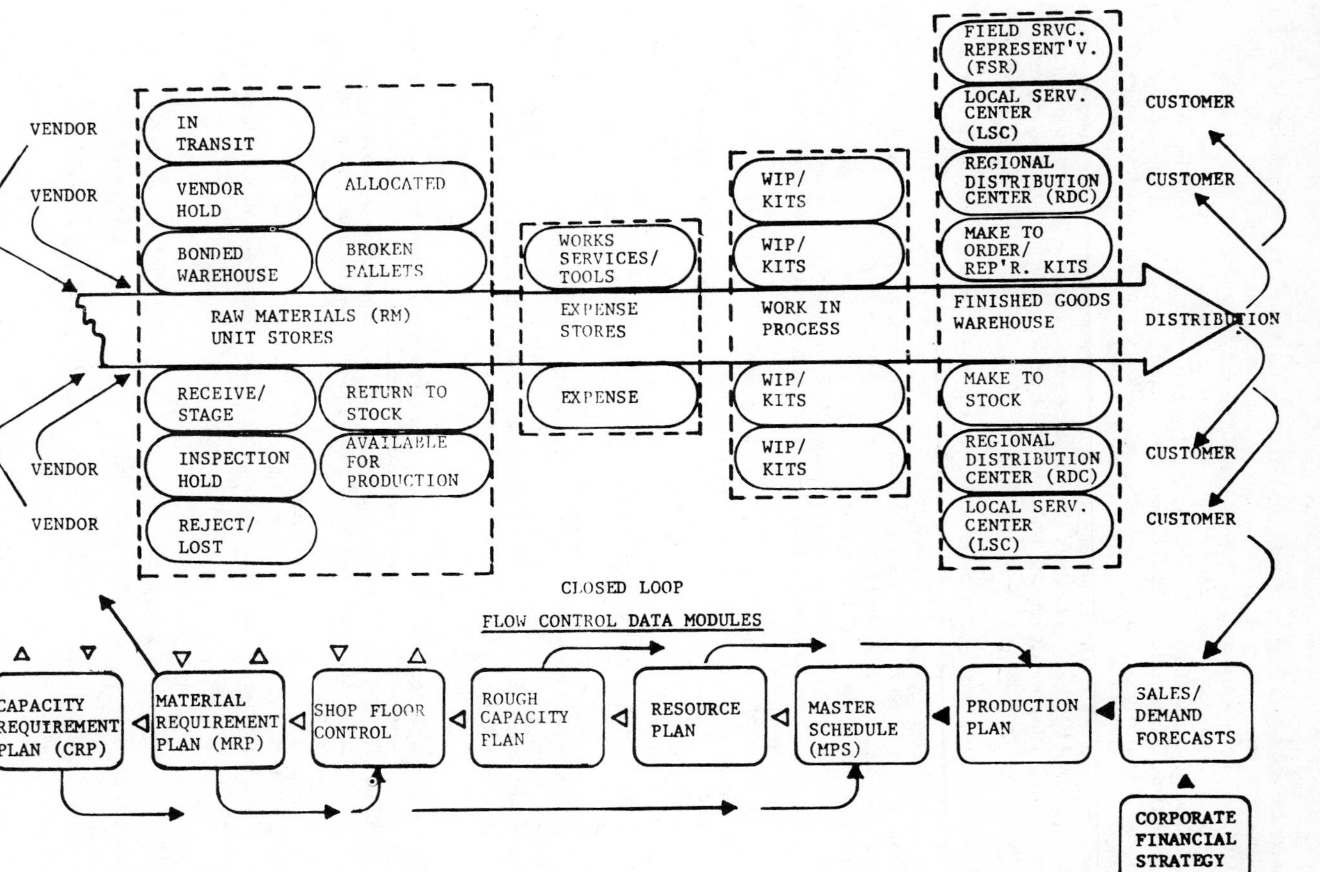

Fig. 5.3.5. Plant C inventory pipeline/flow-control schematic.

Hold for demand is an arrangement (usually on long-term orders) where the supplier maintains goods in his warehouse until release orders are placed, usually by phone or computer. Plant C now employs this for mutual vendor/user benefits, requiring reliable long-term forecasts of one to two years. This arrangement is of interest, especially in the case of long lead times such as those with foreign suppliers.

Hold for bonded warehousing was necessary for speculative inventories, but as a rule, Plant C found them a losing transaction. Warehoused raw materials were often obsolete before required. They were employed to offset the costs of constructing and operating a new warehouse. Plant C now chooses reduced stock levels over those options.

Material in staging was usually not available for production, was not controlled, was subject to damage and was a symptom of a system out of control. Uncontrolled staging often reached levels of 1000 pallets. The secondary staging problem was congestion, which not only degraded productivity, but also crippled dock capacity. New warehouse designs send deliveries directly into staging racks and permit no material on the floor overnight.

Inspection hold/release: Plant C has again reduced inventories by implementing real-time inspection controls for inspection. Materials on hold are now fed directly into the storage system, placed automatically on computer hold, and released by the inspector via the computer.

Rejected materials within the racks are also held by computer until disposition is determined. Rejected materials include goods that exceed shelf life, which Plant C tracks by computer.

Lost materials are the items displayed by Plant C's system as *not available*, and missing from assigned location. In most cases these discrepancies are found during cycle counting or when requested by the system. These items are either incorrectly identified when received, placed in a wrong location, or withdrawn without a requisition through the system.

Inventory hold materials are stock on hand, but allocated or assigned for other purposes and not available for withdrawal. These items may be obsolete and awaiting disposition.

As *in staging* can be a measure of Plant C's material flow efficiency at the dock, *on hold* or *lost* items are measures of storeroom performance. *Units on hold* are itemized on a daily "problem report," which Plant C finds is the only paper document necessary to run warehouses or storerooms. Minimizing the items on the report is the first priority of storeroom managers, inspectors, and operators.

Available for production materials are identified as to quantity, location, and identity, to 99% accuracy. Plant C's manually operated warehouse is equipped with *automatic identification* (AIM) for all stock and stock locations. Plant A found that wanding of bar codes at each transaction and allowing the computer to do record-keeping is the method of data entry that would consistently meet the accuracy levels of automated warehousing. Plant C recognized these advantages and selected AIM with portable wireless terminals for data entry. Each unit load is equipped with a bar code label at receiving, which gives it a unique identity.

*Broken pallets** are usually handled in two ways. One is the *broken pallet area*, which is a second operation operated by equipment designed for stock selection. The second is *broken pallet positions* allocated for this purpose within the unit storage racks. Plant C again borrowed from A in which an automatic storage system simplifies the control system by withdrawing a slave pallet, selecting stock, and automatically returning the pallet to the storage system. Plant C, using a man-up (turret) truck selects partial lots with the pallet remaining in place. The computer entry deducts the withdrawal from the previous pallet balance in a real-time mode.

Returns to stock reduce scrap losses. Originally a formidable control problem, returns were simplified through: (1) using the original ID number; (2) computer matching of quantity required to pallet quantity; (3) re-receiving the material if the original ID is destroyed.

Overflow stock is the material that exceeds planned storage capacity. We previously stated former Plant C raw material levels as:

System capacity:	10,000 Pallets
On hand:	12,000 Pallets
New capacity required:	(12,000/.75) = 16,000 Pallets

Thus, a 60% increase in capacity was required to maintain security and control over assets on hand under current control conditions. Plant C had several options:

Add to existing warehouse @ $500/pallet ($3 M) plus 10% increase in operating costs. (There is no assurance that the new capacity would not be exceeded.)

Procure bonded warehousing for 4500 inactive pallets along with transport services @ $50/pallet/year ($225,000). Stock would be received at Plant C, reloaded on the transport company's carrier, stored off-site, then reloaded on 24 hours notice, redelivered to Plant C, reoffloaded, and then either stored or delivered directly to the shop. Extra handling would add one week to Plant C's inventory level.

Provide unsecured, uncontrolled storage elsewhere (i.e., aisles, in trailers, in cars, and in air bubbles).

* *Broken pallets/units* are unique stock where distribution to the shop is made in less than unit quantities. If part of a unit is withdrawn, the computer adjusts the pallet quantity to a new total, which is the original less the withdrawal. The operator enters old and new quantity as an error check.

Since none of the above options fit the new goals (Fig. 5.3.1), Plant C's management placed inventories under strict control, and using Pareto analysis were able to reduce stock levels to 30% of current rack capacity. Before any new storage capacity is planned, Plant C's managers will ask one question: "Are we doing the best we can with controls?"

5.3.4.3 Expense Supplies

Although, in total, they account for less than 10% of Plant C's carrying costs, expense supplies are 75% of the stock items. They consist of all those stock items that are not part of the finished goods, and were located in more than twenty separately administered storerooms. Improved control and lower costs of operating were gained by consolidation of all storerooms into one raw materials store (unit load, small parts, bulk) and one works services store (bulk, small parts). Elimination of duplicate stock and dual records reduced maintenance carrying costs by 50%. Satellite storage such as shipping containers, lubricants, fuel oils, bulk receivals, and chemicals were assigned to either raw material or work services stores for administration.

The same principles of stock control apply to small parts and unit (pallet) stores. The difference is what constitutes "inactive" stock. Company B, in an effort to reduce works services carrying costs, decided that any stock on hand without activity for more than twelve months would be automatically sold for salvage. In several cases, badly needed repair parts were repurchased from the scrap dealer at premium cost to keep machinery running. Company C established a *plant surplus report*, which would list inactive raw materials and work services stock at each plant. Materials would be placed on that list for three months and be examined by a surplus stock review board, which would determine the best method of disposal and utility to (or burden on) Plant C prior to disposition. In this way, obsolete stock is constantly purged to make space for new commodities.

5.3.4.4 Work in Process

Examination of carrying costs (Fig. 5.3.1) emphasizes that WIP was expected to contribute 40% of the profitability increase for Plant C.

WIP was expected to be cut by a dramatic 89%, from $120M to $13M. The reason for this change is a revised plant-operating philosophy. Previous policy was to send a large part of the production that did not meet finished product requirements to rework storage to be upgraded on a rainy day by operators who would otherwise be idle. The buildup of the rework inventory over several years amounted to $100M. No attempt was made by operators to improve the product, as they were credited with good work as long as the material survived rework specifications. Finally, specifications were being upgraded per customer requirements, so even if the bulk of the rework were completed, it would no longer meet final specifications.

The dilemma was solved when Sales found customers who were willing to buy 80% of the rework ($96M) at about half cost. Carrying costs were reduced by $29M per year, and $10M in out-of-spec materials was scrapped.

Maintaining those levels was done through new emphasis on quality, stabilizing of specifications, and material flow controls. (See Chapter 3.6.)

Other work-in-process consists of raw materials and subassemblies, and are now controlled in Plant C through structured MRP shop-floor scheduling.

The case above is taken from an event that demonstrates the large impact that quality policies can have on material controls.

Kitting procedures exist for the selection of individual parts from a stockroom and sending them to manufacturing as a group or kit. This can be an assembly AS/RS or selection of components by a stock-keeper from designated SKUs. *An SKU* is a stock-keeping unit, or stock/location.[2]

The *kit* can be defined as those components or subassemblies in a container, that (with or without other components or kits) provide the materials for the next manufacturing operation. The kit simplifies the Bill of Materials by loading several components as one unit. Also, several kits may be in the same container, which then becomes a new kit. In the kit assembly area, components and subassemblies are aggregated into kit containers from small-parts storage systems. They can be returned to storage or sent to production.

Plant C finds many advantages to automatic handling and identification of stock in kit form. The complexity of record-keeping is often beyond all but very expert human capability. Second, stock in kit form becomes a high investment if not tightly controlled. Kitting benefits are reduced manufacturing costs (eliminate assembly and waiting time by machinists), reduced handling costs, reduced manufacturing lead times, better parts protection (mechanical/dust/static), adaptability to automation, and simplified process planning.

Kitting has always been a mainstay of Plant C's assembly operations. Kitting is becoming even more important in the manufacturing of electronic components and distribution centers where the kit container can be part of or be the end-product shipping container.

5.3.4.5 Finished Goods

Finished goods can be in two major categories, *service components*, such as spare parts or repair kits, or *finished assemblies* for sale.[13] Whereas stocking of finished goods (such as computers for Plant C, automobiles for Plant B) is done through distributors (Plant A) with their own sales outlets, service components for Plant C are handled (along with direct sales) through Company C's own distribution network. Finished goods made for direct delivery to the customer are made to order.

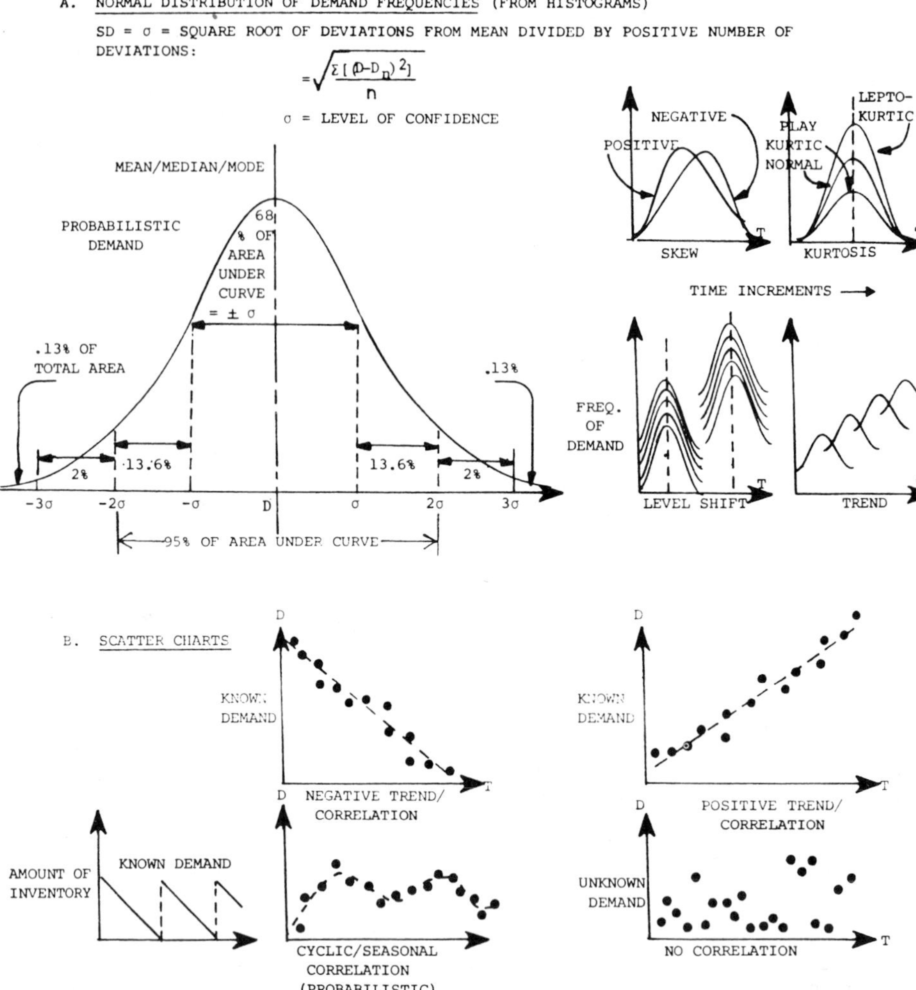

Fig. 5.3.6. Demand models.

As a rule, Company C's plants are becoming more specialized, and finished goods are made to stock. Automated distribution centers are geographically situated to reduce transport costs and lead times. With Plant C as the *fourth echelon*, the Regional Distribution Center becomes the *third echelon* of the dedicated network. The RDC is designed to provide one-day response to local centers for spare parts at a 100% confidence level and complete assemblies at a 95% confidence level (Fig. 5.3.6).[14]

The *second echelon* is the *Local Service Center*, which is stocked to handle all field requests the same day at a 95% confidence level for spare parts, or 65% for finished assemblies. The LSC is a minor repair facility and the RDC handles major repair. The *first echelon* is the *Field Service Representative*, who is Company C's direct contact with the customer. The FSR's van carries enough repair parts to provide a 65% confidence that the customer can make immediate repairs, with same-day support from the LSC for 95% confidence that the spare parts are available.

Company C discovers its products have four distinct life cycle phases:

1. New.
2. Stable demand.
3. Rising sales.
4. Decaying demand.

Replacement parts must be in all echelons prior to distribution of new products. New and decaying demand quantities are determined by market research forecasts of demand, life-cycle tests, and confidence levels at each echelon. Stable demand and rising sales are predicted heuristically, using the forecasting models discussed earlier.

Demand at the first echelon is assumed to be normally distributed, and flows to local and regional centers are determined by economic order quantity forecast.

5.3.4.6 Inventory Models

Inventory models (Fig. 5.3.7) used by Plants A, B, and C for inventory control will be reviewed as material flow control tools:

Static/Certainty

Static. Place one order or manufacture one lot/time period.

Certainty. Buy or manufacture a given lot size (demand is known).

Decision. What is right price?

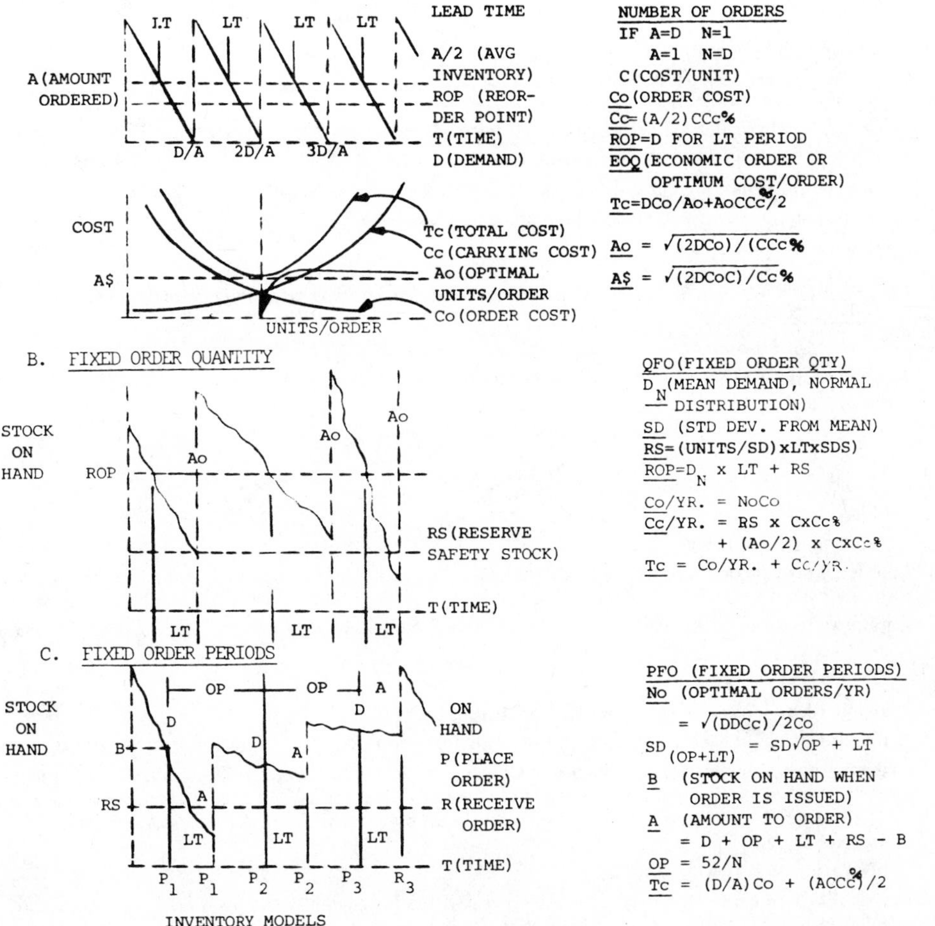

Fig. 5.3.7. Optimization models.

Static/Risk

Risk. There is a probability distribution for the demand.

Probabilities. Determined by market research or forecasts.

Benefit matrix. A chart with demand as columns and options for dealing with demands in rows. Each demand/option has specific paybacks and costs. The *benefit* is the total payback minus the total costs of implementing that option.

Expected values. Expected values of each benefit are determined by multiplying each benefit by its probability of demand to set up the second matrix, the *opportunity matrix*.

Best opportunity. This is selected by adding the expected values in each row and selecting the highest total as the best opportunity or best option.

Static/Uncertainty

Uncertainty. No probabilities of demand are available. Start with benefit matrix.

Decision. Select best option.

Maximin. Select the option with the worst benefit for each option, then select the option with the maximum benefit from the minimum benefits.

Minimax. Applied to costs instead of benefits. Select the option with the minimum cost from the maximum costs (minimax). *Note*: good decision models lead to identical selections of options.

Regret. Employs the benefit matrix to prepare an array of the differences in benefits from the best option, or the *opportunity cost* matrix. This is also known as the *regret* matrix. The maximum regret for each option is selected. The minimum of the maximum regrets are the best options.

Equality. A conversion of uncertainty to risk though assigns equal probabilities to all possible demands.

Dynamic/Certainty (Fig. 5.3.7)

Dynamic. Multiple orders/time period.

D (demand). Known (no over-/understock).

CO (ordering cost). Related to the number of orders placed.

C (cost unit). Can be standard cost or book value.

No (number of orders). Number placed per time period.

LT (lead time). Time between issuing of a procurement order, receipt or delivery.

A (amount). Number of items.

The objective is to find a point between extremes of cost and carrying cost to minimize TC (total cost) or find Ao (optimal or economic order/lot size/quantity, or EOQ).

Carrying costs (Cc). Total annual cost of maintaining stock on hand including:

Capital	Lost opportunity
Taxes	Opportunity
Obsolescence	Damage
Floor space	Equipment
Handling	Labor

Carrying costs can vary from 25 to 35% of the inventory value on hand per year, and has been found in Plant C to vary from 20% for some finished goods to 85% for packaging materials, or 30% overall.

Capital restriction. Inventory investments can be modelled by the use of calculus and simultaneous equations.[15]

Dynamic/Risk—Fixed Order Quantity (QFO) (Fig. 5.3.7)

Demand. Known probability; includes all items with long-term usage, over, under, reserve (safety) stocks.

Order frequency. Determined by fluctuations in demand; based on lead time. When usage reduces stock to ROP (reorder point), order for fixed amount is automatically placed. Two-bin system name indicates two bins required with one bin for safety stock and one for the active stock. Accurate scrubbing for stock records is required.

Order period. Varies with demand.

Decision. Determine Ao (optimal order size) and ROP (reorder point) for a given performance standard (confidence level) in terms of standard deviations.

The requires reserve or safety stock level can be found with the help of basic statistics.[16]

Dynamic/Risk/PFO (fixed period) (Fig. 5.3.7)

Demand. Known probability (normally distributed).

Order frequency. Place a fixed number of orders/year.

Objective. Minimize total costs by determining optimal orders; quantities may vary.

Tests show that PFO requires larger reserve stocks, and as a result is a more expensive model. The following decisions can determine model selection[16,17]:

Select QFO (Two-bin system)	Select PFO (Periodic order system)
1. Annual inventory	Strong cycle counting
2. Fixed order size	Variable order sizes
3. Issue orders any time	Issue orders on schedule
4. Multisource procurements	Single source procurements
5. Random deliveries	Consolidated/scheduled deliveries
6. Low cost/item	High cost item
7. High demand	Low demand

The reader can see conflicting arguments in this idealized chart. Good tracking and flexible data systems can provide a blend of the best of both systems.

Optimization Overview

Although the mathematical logic of optimization is sound, the use of the square root, sum, Lagrangians, simultaneous equations, and statistics places these models beyond the comprehension of many accountants, who are responsible for plant economic policies and routines. Wight[18] makes the claim that few companies can claim large savings through optimization.

> *Getting the right material to the right place at the right time can generate very significant results, even though lot sizes have not been computed scientifically. Getting the right quantity at the wrong time does not accomplish anything.*

On the other hand, Reisman and others[18] remodeled the complete operations of a major manufacturer using these and other sophisticated tools of operations research in 1970. The modern tools of analysis again make O/R attractive to the engineer, if not the accounts manager, who is willing to accept the arithmetic and disciplines of MRP and just-in-time. Plant C still finds EOQ to be a starting point for new manufacture and a valuable tool for evaluating the past.

Modified Box-Jenkins Forecast Model[19]—Mini MRP Program

The purpose of this model was to provide quickly a workable system that utilized Plant C's accurate record system and minicomputers to generate a list of items to be ordered each week based on projected/historic usage, stock on hand, quantities on order, and exponentially smoothed lead time. This system was phased out as MRPII was phased in.

Production schedule. The quantity of each assembly requiring new materials was treated as a final assembly. Assembly production (adjusted for yields) for the next eight weeks was entered each Friday. The next ten months (following the current eight weeks) production was entered on a month-by-month basis.

Projected usage. Every Saturday evening the computer calculated projected usage for each stock item based on (1) assembly shipments and bills of materials for class A items (top 10% of stock on hand by value) or (2) based on historical rates for class B and C items.

Order/reorder points. The computer calculated weekly whether a stock item should be reordered based on whether two weeks plus lead time exceeds weeks of stock on hand and on order.

Overstock. Class A items appeared as overstock when projected usage and historical data indicated an overstock condition.

Understock. When less than one week of stock is available based on projected or historical usage or if stock available plus on order was less than lead time usage.

Lead time calculation. Lead times were recalculated using exponential smoothing each time an item was received.

Reports. Production schedule, order/reorder, overstock/understock, stock status summary.

5.3.4.7 Tool Control[20]

We will now turn our attention to Plant B tool controls, since tools are a large part of their gross cost of sales and setups are a large part of the production cycle.

Typical Plant B machining inputs include raw materials, subassemblies/kits, labor, energy, machine investment, and tooling and fixtures.[21]

1. *Tool/fixture stores* in Plant B fall into three echelons.[22]

   Echelon 1 is the machining cell where securable racks or portable tool magazines are used as storage.

   Echelon 2 is the department dedicated to tool stores. This store has secondary tool recycling and inspection capability, but stocks tools and fixtures unique to the department and serves as a one-day assembly area of common tools, or processes returns to the primary stock.

   Echelon 3 is the main works services storeroom, which houses not only tools but also maintenance and machining supplies. Recycled tools are cleaned, inspected, repackaged, and bar-coded for storage. Worn-out tools are recorded and scrapped to avoid future quality and production delays. Plant B's physical toolroom arrangement is four-foot-high modular drawers, above which are mounted closed shelving or specialized racks three to six feet high (overall 7 to 10 feet). The capacity can be increased with a mezzanine, mobile carriers, or automatic small parts storage. Aisles are 36 in. wide. Operations include one system CRT computer terminal and printer, RF hand-held terminals, bar-code readers, and coded openings and packages. Tool/fixture withdrawals are made by operator security cards (credit-type).

2. *Records* include[23] Tool identification, current, cumulative age, recycle history, reports for damaged tools and operator, vendor information, prepetual inventory, and cycle count routines.

   Straight numeric tools are numbered and the series or stock item number is etched directly on the tool and printed on drawings.

   Classified numeric (Dewey Decimal system) uses six to nine digits for general tool classes.

a. sharp edge	d. impact	g. fire
b. measurement	e. wrenches	h. transport
c. jigs/fixtures	f. holding	i. miscellaneous

   Each class is subdivided by adding as many digits to the key number as necessary.

   Mnemonic uses alpha or literal notion such as:

	M—measuring	
MB—bevels	MG—press gauges	MT—timers
MD—dividers	MM—meters	MY—miscellaneous

   Additional letters focus on specific tools.

3. *Tool marking* is permanently etched plainly and legibly with tool symbols.
4. *Conclusion:* Tooling costs in Plant B represent a large part of the gross cost of sales and require a strong computer integrated tool and flexible manufacturing strategy.

   Just-in-time (JIT) inventory policy embraced by Plant B managers is impacted by tooling. Tool requirements are matched by computer to the Bills of Material and production schedules. The object is to reduce setup times (in many cases) from hours to minutes.

   JIT tooling requires tools arriving in "like new" condition in minimum lead times while maximizing end product quality and profitability.

Other Models

Other models for material flow control employed in Plants A, B, and C are[16]:

1. *Simulation* is the use of artificial data to emulate conditions that may actually occur. The model uses trial and error algorithms to test alternatives to solve complex problems of risk or uncertainty. The development of computer analysis makes this method one of the most useful tools available to the materials and productivity manager. (See Ch. 3.5.) We commend a recent Don Phillips article[24] on simulation in which the utility of the models such as GPSS, SIMSCRIPT, GRASP, SIMULA, DYNAMO, Q-GERTS, SLAM, and GEMS are evaluated in the material flow control environment.
2. *Linear programming* is another optimization tool in situations where constraints are given and functions can be maximized or minimized, such as minimum cost, maximum profit, optimum productivity, and maximum utilization of capacity. Relationships must be linear. The problem is usually stated[25]:

Maximize

$$Z = c_1x_1 + c_2x_2 + \cdots + c_nx_n$$

Subject to

$$a_{11}x_1 + a_{12}x_2 + \cdots + a_{1n}x_n = b_1$$

$$a_{21}x_2 + a_{22}x_2 + \cdots + a_{2n}x_n = b_2$$

$$\vdots$$

a. The objective function (Z) is always a linear sum of terms. Each term is the product of a real number and a decision variable. The decision variable (*x*) is the answer sought as part of the solution. The problem: maximize or minimize Z.
b. Constraints are always linear equalities or inequalities. They are sums of terms that total to be an operating rule or governing process. They represent available resources such as labor, money, materials, and are resource limits.
c. Solutions can be graphic, canonical (a system of simultaneous equations to gain a basic feasible solution), simplex (starts with canonical, and is improved to optimal condition with basic feasible solutions). Other solutions are BIG M, revised and dual simplex and computer solutions (such as MPOS), which easily manages thousands of variables.

3. *Project analysis.* Critical path method (CPM) and program evaluation and review technique (PERT) are forms of network analysis used for deterministic (predictable) problems such as planning of large projects.

PERT assumes activities are stochastic (random variables influenced by time).

5.3.5 APPLIED STRATEGIES

5.3.5.1 MRPII

Material requirements planning (MRP) in Plant C is a system of material flow controls throughout the inventory pipeline (Fig. 5.3.5).

Manufacturing resource planning (MRPII) extends MRP by effectively planning the available resources and simulating production options on the system. Figures 5.3.3 and 5.3.5 show roles of MRPII in closing the information loop from shop activity to master scheduling, in which the corporate and business plans are listed to production, forecasting, material, and capacity plan (see Table 5.3.3).

Implementation of MRP in Plant C requires realistic demands, sound priorities, and accuracies of 95–99% in bills, balances, and routes (Fig. 5.3.8).

MRPII Modules will be briefly defined with emphasis on Material Flow Controls. Fig. 5.3.9 shows MRPII Information Flows.

Table 5.3.3. The MRPII Process[26]

	Closed Loop	Horizon
Plans:	Production resource	Long
	Rough cut capacity (RCCP)	Medium
	Master (CRP)	
	Material (MRP)	
	Bill (BOM)	
	Inventory status	
	Capacity (CRP)	
	Priority	
	Routing	
	Shop-floor control (WIP)	Short

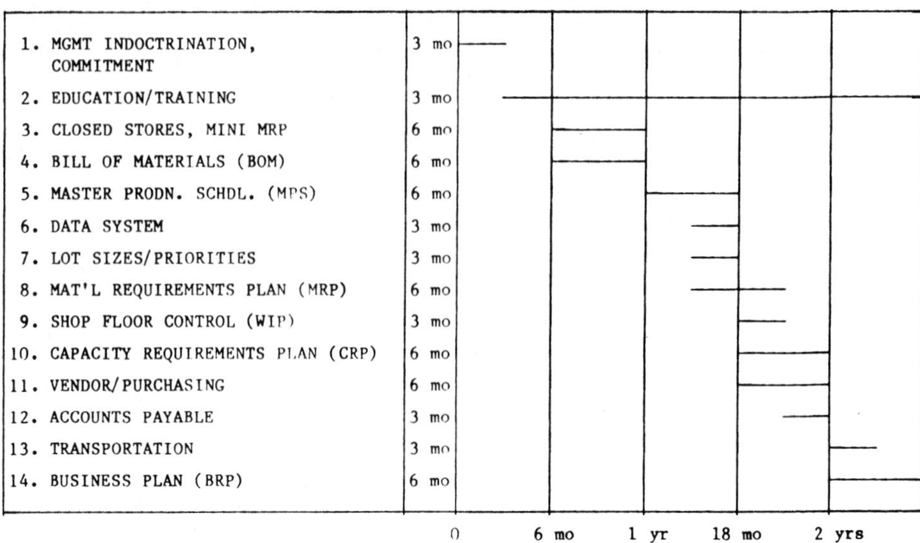

Fig. 5.3.8. Plant C MRPII implementation schedule.

1. *Production planning.* Controlled by the Marketing department, production planning is the main control valve for regulating the inventory pipeline. Fig. 5.3.3 shows the task of planning production rates for production families in terms of standard units or dollars. It has both a long-range (5 year) and a short-range (1 year) *planning horizon.* It is expressed monthly as planned and actual sales, production, and inventory from the following simplified formula:

$$\text{Sales rate} - \text{planned inventory*} = \text{production}$$

   An additional production plan function is to evaluate the impact of sales, capacity, production and inventory changes on the production system.
2. *Rough cut capacity plan.* According to Wight[26] this is a periodic test of key facilities to see if they can meet the production plan prior to preparing the master schedule.
3. *Master production schedule.* Senses the production plan and production performance modules and translates them into timed assembly schedules and resource requirements. The inputs are controlled by the master planner and include monthly demands (hard—3 mo. horizon; soft—12 mo. planning horizon), MRP/CRP feedback, safety stocks, made to stock, release dates according to the following formulas:

$$\text{Periodic forecast (future/actual)} + \text{safety stock} = \text{requirements}$$

$$\text{Periodic requirements} - \text{available (free) stock} = \text{production schedule}$$

   Figure 5.3.3 demonstrates flows of information from resource forecasts through production forecasts with a feedback loop.
4. *Requirements planning* (Fig. 5.3.9) is the synthesis of master schedule, bill of material, and inventory data to schedule and quantify production materials. Independent items can be planned using reorder points, but MRP is required for items or subassemblies when requirements depend on demands for higher (parent) assemblies. Withdrawals of available inventories automatically schedule stock replacements for future demands. It determines material requirements, adds balance on hand (gross requirements) to material on order to determine net requirements or additional material needed and plans order sizes and due dates through lot sizing rules. It subtracts procurement lead time from due dates to provide offset release dates. The bills of material provide for explosion of each final assembly into its individual components of subassemblies and raw materials along with their release dates. Exception messages tell material planners when shop and purchase orders must be released. The orders are given priorities and

* Planned inventory is the sum of the following inventories:

$$\text{Make to stock (allocated)} + \text{make to order backlog} + \text{planned changes } (\pm).$$

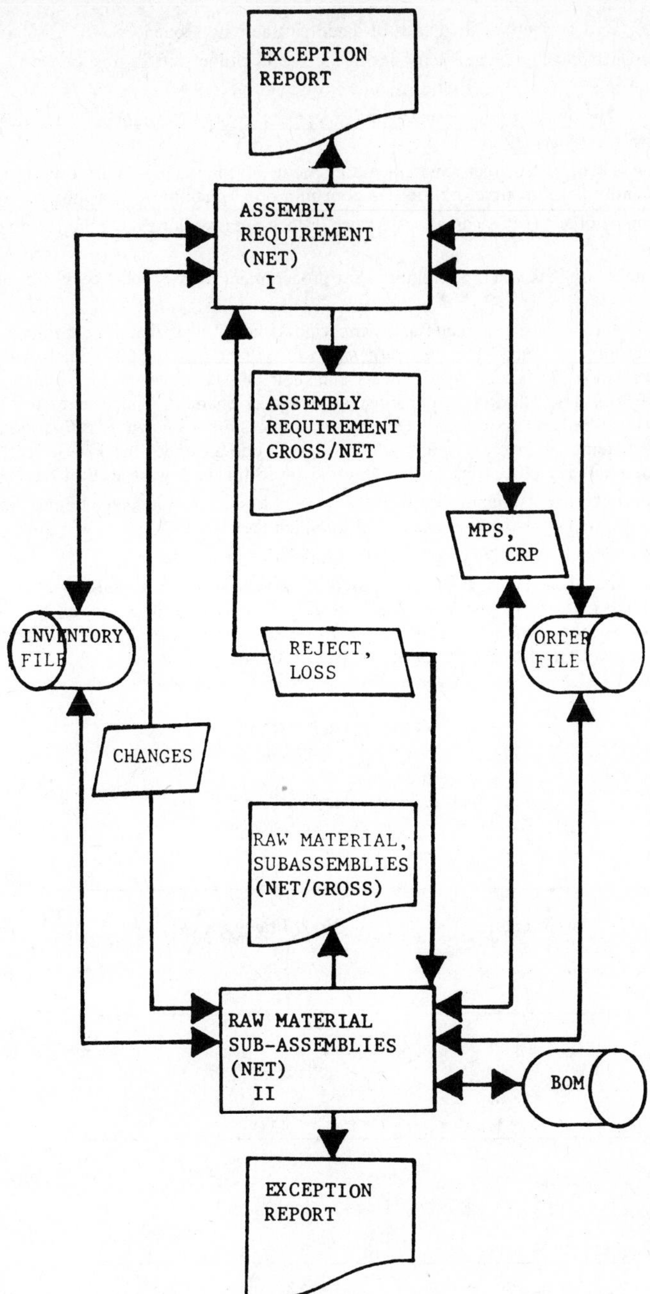

Fig. 5.3.9. Plant C MRP information flow.

approved orders are released as gross requirements. The activities above are known as an *MRP cycle*. At this point, we might compare *optimization* with *MRP flow controls*:

Optimization (reorder point)	MRP
Heuristic/statistics	Plans for future events
Replenishment/usage/lead time	Timed release (OFFSET)
EOQ/safety stock	Dependent/interdependent
Independent BOM	Dependent on BOM

Pegged or activity source lists details of dated part activity sequences.

Bucket reports display pegged activities in the form of time slots or time buckets.

Exception reports display activities required to avoid problems.

Exception sensitivity is programmed into exceptions to prioritize required activities and screen out unnecessary activity.

Lot sizing is the planned order requirement for specific inventory items in the forms of net requirements, order quantity per lead time, multiple max or min order quantities, economic order quantity.

Firm planned orders (FPO) can allocate stock or lock out inventories by computer to guarantee future capacity.

Offsetting determines when net requirement procurement activity must begin for purchase of work orders.

Bills of materials[27] list raw materials, parts, and subassemblies that are consumed in making final assemblies or end items. The *structure tree* for a Plant B automobile seatbelt demonstrates the interdependence of material requirements and their classification into hierarchical levels. Vertical lines are *branches*. The key to planning is linkage or chaining of end items to their components through level-by-level processing. Downward progression is known as *Bill-of-material explosion*. Multilevel items are processed through low level coding, where the lowest level for an item is determined and any processing of the item is delayed till the lowest level of the item is reached.

The *indented* form lists highest levels to the left and lower levels indented to the right of the margin; common use items are listed at each level in which they occur.

Explosion summaries show components and quantities in end items.

Imploded or "where used" lists show assembly levels at which a component is used. Imploded summaries start at the lowest level; *Back flushing* deducts from inventory all the parts (through an explosion) needed to make a specific number of end items; backflushed items are often allocated in a database until they are physically dispatched to the next workcenter.

STRUCTURE TREE

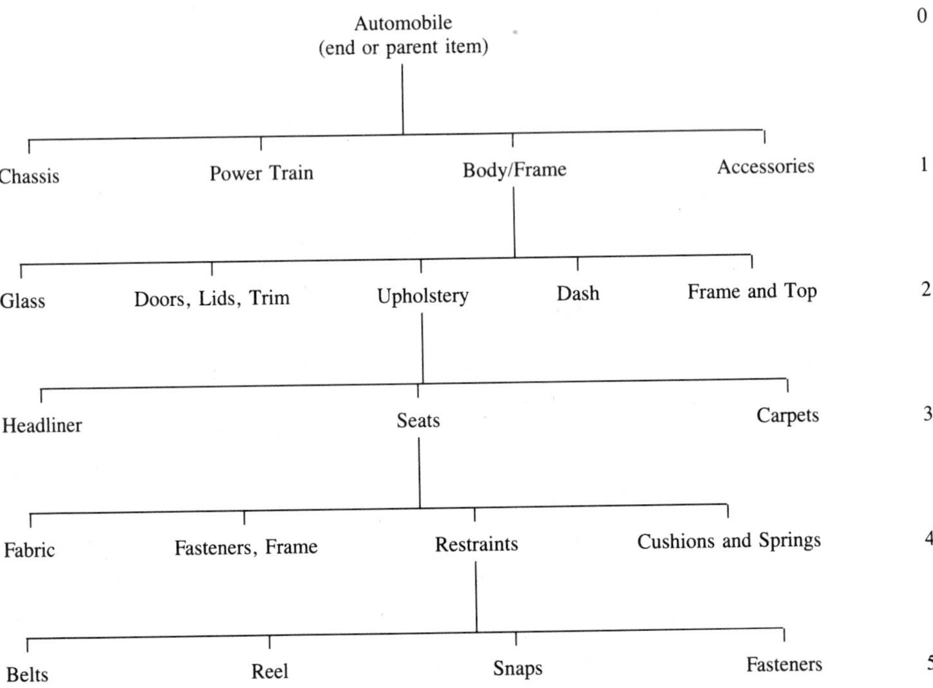

5. *Stores/Tool Control*. Formal data controls have the following advantages over tickets, tag sets, paper reports, and punched cards:

High return on investment.
Accurate records.
Exception reports.
Real time status/values.
Reduced clerical effort.
Less obsolescence.
Allocation/inspection quarantine.
Efficient routing.
Minimized requirements for facilities, space, manpower, stock on hand.
Eliminate paperwork, expediters, annual inventories.

STORES/TOOL CONTROL DATA FLOW

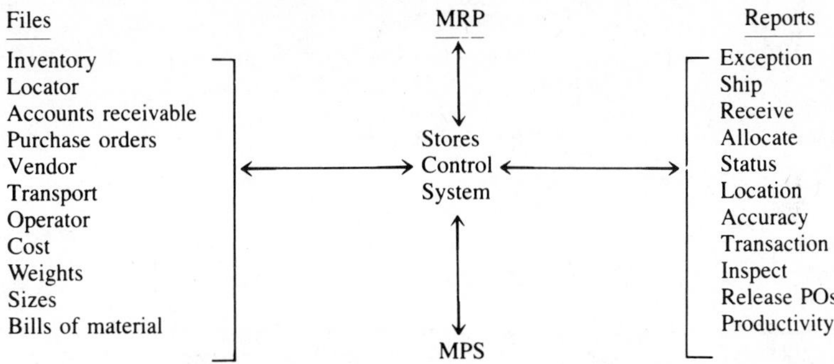

6. *Shop-floor/work in process controls.* Compare planned and actual shop performances and use exponential smoothing to predict realistic cycle times for the next production period. Each production cell makes the following inputs from machine readable travel documents and coded pans, tools, and materials:

Job number	Tool data
Arrival time	Operator
Start time	Routing
Finish time	Machine number
Delivery time	Lot size

Current and expected arrivals are prioritized by computer to improve schedules and similar parts are grouped to reduce setup times.

Quality control employs online tools such as matching to standards, vision systems, automatic test equipment, and computer updates to statistics (SQC) charts.

Priority planning compares planned and need dates, keeps them honest and valid, analyzes backlogs and slack time exceptions and matches planned and current dates.

Validity is tuned to Master Production Schedule.

Integrity is tuned to real-world conditions.

Dependence can be vertical or horizontal.[9]

Real priority relates availability to vertical dependence as follows:

$$\text{Critical ratio} = \text{OD} - \text{CD}/(\text{TR})$$

$$\text{TR} = \text{lead time remaining}$$

$$\text{OD} = \text{order due date}$$

$$\text{CD} = \text{current due date}$$

or

$$\text{Critical ratio} = A/B$$

where

$$A = \text{quantity on hand/order point}$$

$$B = \text{TR/total lead time}$$

Horizontally dependent schedules use fixed order period (PFO) instead of offset order points. If

CR	Job is	Priority
= 1.0	On schedule	Medium
< 1.0	Behind schedule	Higher
> 1.0	Ahead of schedule (has slack time)	Lower

WIP DATA FLOWS

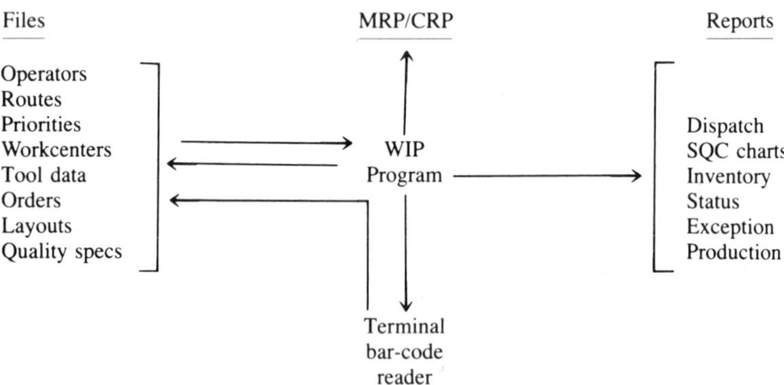

7. *Capacity requirements planning*. Determine shop resources required for short- to medium-range planning horizons by work center and time period.

Decisions may be overtime, rerouting, shift personnel, farm-out, capacity (labor/machines)

Load reports are schedules stated in work load hours by work center.

Load projections or reports include superimposed planned and firm (open) orders in a prioritized and relatively level form.

Infinite forward capacity planning inputs part identifiers, quantities, and totals the output hours beginning with the first job's start date and ending with the due date of the last job.

Finite forward capacity planning inputs work to finite workcenter capacities. Work loads are shifted by computer and smoothed to realistic starting and due dates. Shop orders are released in optimal sequence.

Infinite backward capacity planning starts with the completion date for the last job (due date) as follows:

$$\text{Due date} - (\text{lead time} + \text{setup time} + \text{process time}) = \text{finish time of previous operation}$$

This step is repeated until the start date for the first job is determined.

Cycle (lead) times are taken from the routing for proposed schedules avoid shop floor scheduling conflicts, provide improved resurce utilization, and reduce work queues.

CRP INFORMATION FLOW

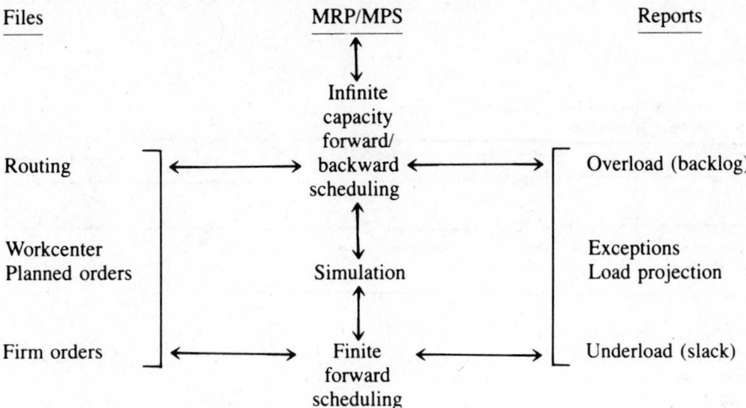

5.3.5.2 MRP Results

MRP results on a broad scale were reviewed through an APICS survey[28] conducted by the University of Minnesota in late 1982. It showed the following MRP user performances:

Class A	= 10%	Class C	= 50%
Class B	= 30%	Class D	= 10%

Average accuracy for BOM, through capacity:

For MRP users = 62%
For non-MRP users = 53%

	PRE-MRP	MRP*
Weeks stock	16	10
Cycle time (wks)	10	6
Meet due dates	60%	90%
Need material	32%	9%
Expediters	10	5

* Expected

CLASSES OF MRP PERFORMANCE[26]

	A	B	C	D
Closed loop priority/CRP plans	uses	inplace	informal	ISD, only
Management usage of MPS	uses	inplace	informal	inaccurate
Game plan	yes	maybe	no plan	ISD
Accuracies (BOM/inventory, routes)	95%	95%	80%	?
Contribution to profitability	10–25%	5–9%	1–4%	?

Benefits to Plant C as a CLASS A MRP user in meeting goals (demonstrated in Fig. 5.3.1) are:

1. Sales increased by 20%.
2. Unit costs to customers reduced by 10%.
3. Profitability 26% of gross cost of sales (up 36%).
4. Inventory turns up from 2 (six months of stock on hand) to 14 (one month of stock on hand).
5. Overtime, lead time (order to due date) down 50%.

MATERIAL FLOW STRUCTURE[9]

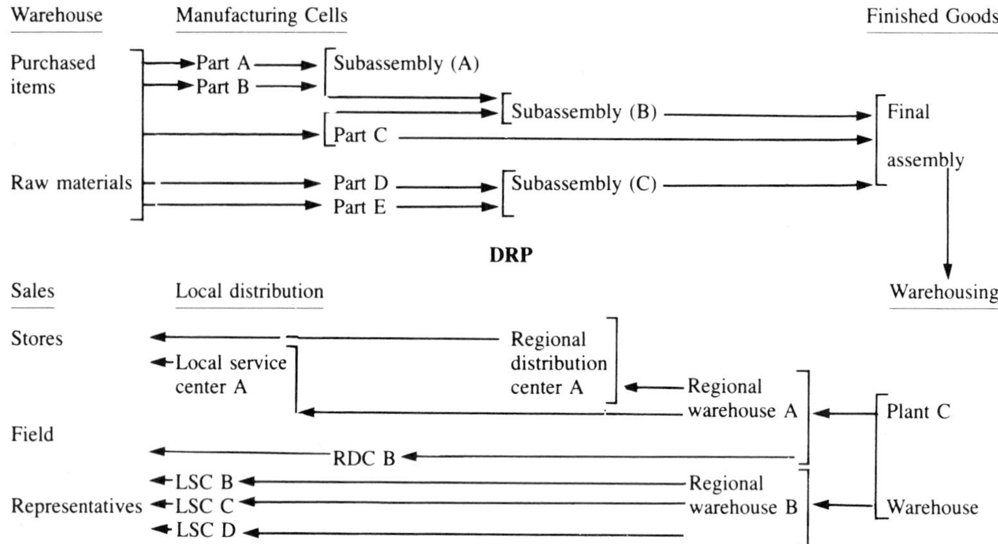

Distribution Resource Planning (DRP)

DRP is MRP applied to finished goods for replenishment and logistics from a master planning schedule (MPS) using time-phased reorders calculated as follows:

RDC forecast per week − [stock on hand + stock in transit − safety stock] = net on-hand balance

Net on-hand balance is expressed as the lead time inventory will last.
Replenishment date is the date when net on-hand balance equals 0.

Net to zero date − lead time for shipment = factory ship date

Resources such as manpower, dollars, space throughout, and transport are developed through MRPII strategies.

Wight[26] claims that distribution's functions are manufacturing dependence, simple forecasts, timing in pace of safety stocks, centralized "*push*" distribution, allocation by computer, and integration with MRPII resource plans.

Optimized Production Technology (OPT) is a product of Creative Output Inc., of Milford, CT.[30] A rough cut capacity plan is used to prepare a detailed process network to identify critical (bottleneck) and noncritical resources. The advantages claimed over MRP are more realistic schedules, fewer shop-floor conflicts, and ease of specific scheduling decisions.

Noncritical resources (serve schedules) start with due dates, assume infinite capacity, and are *backward* scheduled.

Critical resources (split schedules) assume finite capacities and are *forward* scheduled.

OPT/Serve Rules vs MRP[30]

OPT	MRP
1. Balance flows, not capacity.	1. Balance capacity, maintain flows.
2. Bottlenecks is result of other constraints.	2. Manpower potential is self-determined.
3. Utilization/activity of resource are not same.	3. Utilization and activity are same.
4. Time lost at bottleneck is totally lost.	4. Impact is not realized.
5. Time saved at noncritical event is not critical.	5. System is limited by bottlenecks.
6. Bottlenecks decide inventory throughout.	6. Little impact on inventories, temporarily on throughput.

OPT	MRP
7. Transfer not equal to lot size.	7. Lot splitting should be discouraged.
8. One lot (process batch) can be in several stages of manufacture (transfer batches) at the same time. Process batch (lot size) should vary.	8. Lot sizes should be fixed.
9. Schedules should be determined by simultaneous input of all constraints. Lead times cannot be predetermined and are variable.	9. Predetermine lot size; calculate lead time; assign priorities schedule by lead time; adjust capacity by above steps.

OPT STRUCTURE

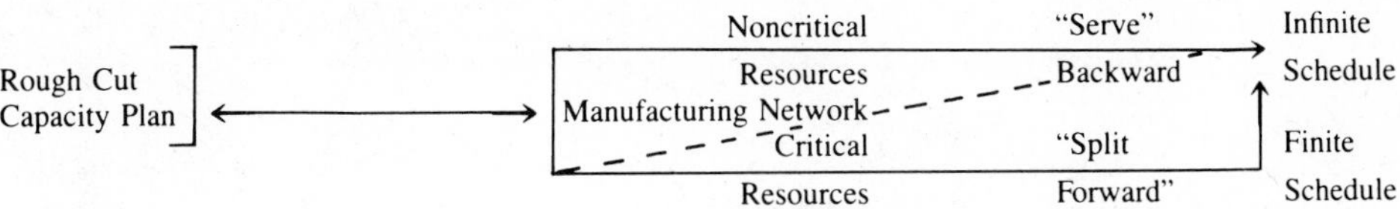

Kanban is a just-in-time production strategy using uniform flow through *Tacto* scheduling, high volume/unit cost through focused factory, employee commitment through quality circles, *Jidoka* (personal accountability for results), reducing setups through group technology, commitment to total preventive maintenance, and synchronized just-in-time production flows (such as in Toyota's Kanban procurement authorizations). The system which has been in development for more than twenty years, has led to breakthroughs in areas of production such as quality and WIP control, and has inspired efforts to develop JIT systems in the United States. Chapter 3.16, "The Toyota Production System," provides in-depth information on TPS. We will attempt to highlight unique TPS features and compare then with MRP.

Tacto time[31] provides for steady flows of work through focused workcenters with monthly schedules, resulting in the same output from day to day and from shift to shift. Uniform economic lot sizes and standard container sizes simplify handling and production leveling. This schedule is passed on to contract vendors one month in advance, to permit their daily deliveries on a JIT basis.

Focused factories are based on the concept that doubling production volume will reduce unit costs by one-third. Similar items are grouped in focused work centers of about 300 people with concentrated manufacture. Nonstandard items are subcontracted to specialists or dependent contractors. Efficiencies are based on lack of fluctuation until next month's schedule change, only (but all of) the materials needed to do the job being on hand, and knowing exactly what to do next. *Maker specialists* provide better quality, lower costs, and higher volumes than their assembly line counterparts.

Kanban cards place demands on preceding operations as the current operation begins. Thus, WIP is pulled through the system.

1. The month's daily schedule is given to final assembly.
2. Final assembly pulls the required parts from the workcenters.
3. Work centers pull parts from their feed centers to manufacture the next day's quota for final assembly, and so on.
4. All workcenters are chained in a pattern that feeds final assembly just in time.
5. Conveyance (Kanban) cards authorize moves.
6. Production (Kanban) cards authorize production.
7. Both cards travel with parts containers and only the information needed to perform the next operation.
8. Users always pick up parts from preceding operations.
9. Pans always contain the same number of parts, and will not be selected without cards in place.
10. Inventory is controlled by the following formula[31]:

$$Y = \frac{D(T_w + T_p)(1 + \alpha)}{a}$$

Y	total cards
D	average production/day

T_w	waiting time/day
T_p	process time/day
a	container capacity $\leq 0.1 \times D$
α	Previous workcenter efficiency (Management Policy)

Planning[32] at Toyota follows the following policies:

	Horizon
Long range:	Three years
Production plan:	One year
Master schedule:	Three months
Production plan cycle:	One month
Long leads/production plan/vendor contracts	Six months
Short-term plan	Three months
Shop schedule	First month
Time buckets	One day
(Includes final assemblies, explosion, daily parts schedules)	
Second month by models/day	

Suppliers or *comakers* on the Kanban system respond to conveyance cards. Parts allocated on consignment with the supplier are considered to be in-house inventory.

Jidoka, or the exercise of the worker's responsibilities, is a major contributor to productivity and quality, since Kanban provides added dignity, quality, and morale. This is the main emphasis driving the Kanban. Workers strive to reduce inventories through minimizing container sizes; operators put in extra individual effort or team up to solve problems that may jeopardize meeting the daily output. When the quota is completed, operators may leave early.

Andon is an electric scoreboard that warns the plant when a workcenter is not producing as planned (quality or productivity). Cycle rates are posted to provide work center incentives.

Exceptions, whether inhouse or vendor, are processed outside the Kanban.

The *benefits of Kanban* are minimum inventories, lead times, cycle times, manpower, rework, setup time, waiting time, and machine down-time. Flexible production schedules and control of abnormality are possible.

Goddard[33] lists the following differences between MRP and Toyota Production System (TPS) strategies:

Objectives/MRPII	Toyota Production Ssystem (TPS)
Inventory/asset	Liability
Lot sizes/EOQ	Immediate needs
Setups/low priority	Extremely rapid
Queues/necessary	Eliminate (correct causes)
Vendors/adversary, multiple multideliver daily	Co-workers, part of team
Quality/tolerate scrap, rework	Zero defects or factory is in trouble
Maintenance/fix when required	Constant/preventative
Lead times/more is better	Short—eliminates expediting
Workers/manage by edict	Manage by term agreement
Outputs/production plan	Leveling
Production/master pr. schedl.	Master production schedule
Matls. reqd./MRP	Kanban cards
Capacity reqd./cap. req. plan (CRP)	Visual
Execute CRP/I/O controls	Visual
Priorities/dispatch reports	Kanban cards
Procurement/purchase orders	Kanban cards, oral orders
Feedback/expected delay report	Andon light
Tools/computer	Cards, lights, visual

Just-in-time production is described by Garwood[34] as a formal strategy for eliminating scrap in manufacture and purchased materials. Vendor lot sizes are reduced by more frequent deliveries as required. Scrap is reduced through quality. Machine down-time is reduced through preventive maintenance and reduction of setup time. WIP lot sizes are minimized and safety stocks and buffers are eliminated. The term *zero inventories* is a step forwrd in MPRII excellence. The trend toward smaller lot sizes (less than 50 units) and wider flexibility leaves conventional discrete manufacturing obsolete. TPS flow controls appear much more adaptable to larger volume, same-kind manufacture. Yet the focused factory has promise even in small lot sizes using the following tools:

Flexible manufacture (FMS), described in Chapters 3.10 and 3.11, is the computer integration of programmable machines serviced by automatic workpart handling systems.[21] Each FMS cell produces a family of workparts, which permits random launching/loading of jobs, reduces manufacturing lead time and WIP, reduces labor and expediting, increases machine utilization, decreases setup time, and provides better management control.

Group technology analyzes similarities in manufacture and part geometry to provide computer classifications into parts families and routing into mchine cells.

Parts classification and coding is the first step in FMS, since it is a part of the initial computer aided design. It provides accurate statistics for machine loading and tool requirements, optimizes setups and increases throughput, optimizes tool and fixture design, and minimizes tool and fixture investments.

Vendor contracts permit more frequent (daily) deliveries of large volume/cost commodities.

Lot sizing minimizes WIP inventories.

Zero defect plan reduces WIP recycled through rework.

Quality circles receive worker input to corporate strategies.

5.3.6 THE AUTOMATED FACTORY

5.3.6.1 Introduction

"The Automated Factory," Part 10 in *Modern Material Handling*,[35] contained a discussion by America's leading material managers and comments by the editor, Miles J. Rowan. We will give a brief overview of those opinions, summarize the new role of formal data based material flow controls, then phase into physical controls and programmable handling.

The *automated system* contains most of the controls described in this volume, integrated into a synchronous, smooth-flowing production system with the following advantages:

1. Improved quality of work life, operator creativity, equipment and manpower utilization, production output, and quality and profitability.
2. Decreased manufacturing cost, shop-floor conflict, inventory cycle time, monotony, and strenuous work.
3. New management shift in commitment from finance to manufacture.

Levels of automation are listed as: conventional, mechanized, automated factory, and automation, all of which have their place, and all of which can make use of advantages of recent technology. The pitfalls are resistance to change and new computer integration into every process, which no longer permits leaving system development to computer professionals. Education and communication are the answer to the first, and a commitment to computer literacy is the answer to the second.

Is automation feasible? Most American workers are cognizant that we are in an international arena. Most understand that unemployment created if industry does not automate is more severe than if it does. In other words, "*Automate or go out of business*." Automation must be carefully and intentionally planned on new production for long-range (5- to 10-year) horizons. According to Kochhar,[9] systems developed without user management and strong user participation have limited success. Blaize Cooke[36] lists the requirements for modern real-time distributed data system success: Thorough documentation of current procedures, logical models of new systems, redundant databases, dedicated computer for each utility function, quick response, backup computer, double (shadow) memories, printers (300 lines/min—shop floor, 1200 lines/min—control), large disks for files. Mainframe computers are not configured for real-time or interrupts, and are not suited for process controls or utility jobs.[37]

5.3.6.2 Computer Integrated Manufacturing System (CIMS)

The computer integrated manufcturing system (CIMS) is conceptualized in Fig. 5.3.10. Most of the data modules have been covered earlier in the MFC section or in other chapters; however, a few points deserve comment.

Architecture is shown in several levels starting with the mainframe, or Corporate Host Computer, which may be located locally or off-site. It contains six months to one year of transaction files (archives) and is accessed by the control system in a batch mode about once per hour.

Control system programs are managed in the local host computer, which performs all tasks in a real-time mode with on-line (real-time) access by *utility computers*. The control computer has a 90-day file capacity and essentially drives the CIMS, thus the Automatic Factory.

Utility processors are dedicated to more specific functions and access each other and the controls through a local area network (LAN). Those processors contain file capacity for up to thirty days of transactions.

Adaptive controls are a new and exciting technology, in which a machine literally "feels" its way through a process by sensing temperatures, pressures, depth of cut, and related conditions. The conditions are fed back into programmable controllers to control machine speeds, feed rates, tool wear, coolant flows, and tool locations. Advantages are precision, quality, tool life, productivity, and machine utilization. They can be combined with vision systems to provide new control flexibility, which may lead to artificial intelligence in Flexible Manufacture.

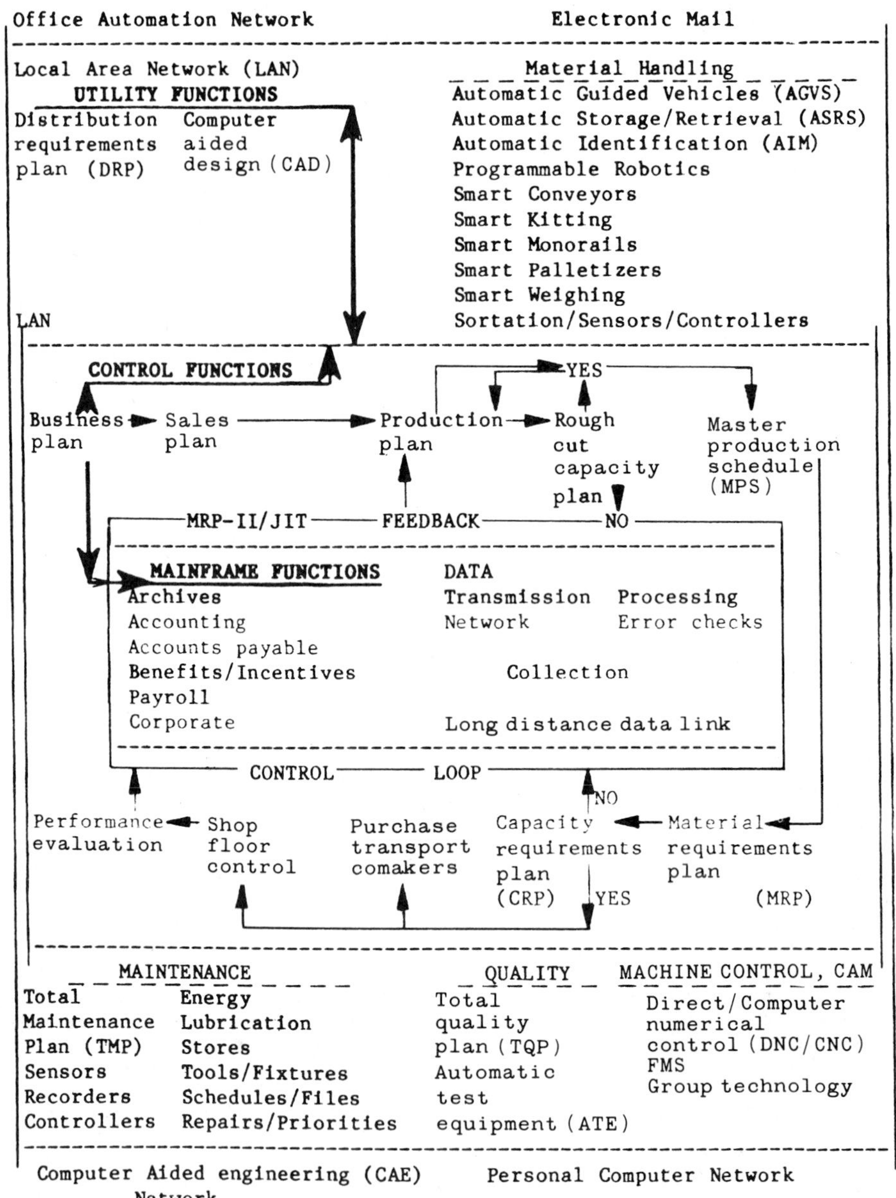

Fig. 5.3.10. Computer integrated manufacturing network.

System protection requires redundancy of computers and files for robust CIMS. The ability of a vendor to emulate local problems off line, to monitor problems on line, and reconstruct data lost in "cockpit" errors or power failures is critical.

Programmable Controllers

Programmable controllers in the Automated Factory appear in many forms. All contain a central processing unit (CPU), I/O system memory, power supply, and software.

The *CPU* contains the operating system, or brain, in the form of chips and printed circuit boards (PCBs). It directs information flows and priorities and performs logical and arithmetic functions. It must be protected from power surges and failure through a power conditioner and battery-powered uninterruptible power supply and grounded for static protection. Dust, smoke, vibration, and heat above 80°F may be detrimental, unless special designs are used.

I/O systems consist of the I/O Bus (a series of printed circuit boards in the processor, which convert signals into operating messages); inputs such as scanners, limit switches and photo cells; and outputs such as motor starters, solenoid valves, lights and microprogrammable controllers (PCs).

Power supply is an uninterruptible source of conditioned 120V power, which is transformed to 12V and 5V dc within the processor.

Software, usually in FORTRAN, Cobol, or C(UNIX) language, tells the foregoing hardware what to do. PCs uses ladder logic similar to relay logic.

Mainframe computers (Fig. 5.3.10) are at the high end of CPU capacities; however, their tasks are not interruptible and must be batched, making them unsuited for real-time controls.

Minicomputers are less expensive, smaller, more accessible, easier to service, and suited to any number of real-time tasks. Tasks can be interrupted, and main memory (on chips) and mass memory (on disks or tapes) can be upgraded by adding boards or disk capability.

Microcomputers, still smaller in size and lower in cost, can be dedicated to complex tasks such as robots or automatic storage systems.

Microprocessors/programmable controllers are dedicated microcomputers with a single chip programmed on a keyboard in Ladder language by electricians and technicians to optimize machine controls with little or no vendor support.

Specifications of controllers should be based on cost, capacity, and task requirements. Memory can be added in inexpensive increments. Maintenance contracts with vendors for dial-in software support and hardware support cost 1–2% of the total system cost each year.

Automatic Identification (AIM)

Automatic identification (AIM) began to show potential as a material flow control with mechanical fingers, magnetic cards, and optical character recognition (OCR). The grocery and distribution trades then adapted the Universal Product Code, a vertical array of strips (bars) that are scanned by a light pen or laser, which converts reflected (blank) segments and absorbed (dark) portions into binary (0/1) signals.

The Department of Defense developed the Logistics Application of Automatic Marking and Reading Symbols (LOGMARS) as MIL STD-129H. The code included 43 typewriter keyboard characters—0–9, a–z, and eight symbols—which are required on all packages sent to the DOD supply system. The code has since been adapted by the automotive and many other industries.

Bar coded data entry can nearly eliminate pricing, tracking, cycle counting, location and kitting errors, while instantly updating data files. Scanning can be through portable (hand-held or truck-mounted) terminals or fixed sensors on conveyors, storage lanes, and machines.

Bar code printing can be done on-site and applied to units as they are launched into the production system, or can be jet-printed while moving. Purchased labels should be checked for readability prior to producing a lot. The Automotive Industry Action Group (AIAG), a nonprofit trade association of major manufacturers, initiated several industry-wide productivity improvement programs including development of standards for bar code symbology specifications and shipping labels.[38] Interface Mechanisms, Inc. has prepared a specification for code 39(tm) alphanumeric bar codes.[39]

CIMS drives the automatic factory. Control systems drive the CIMS. But, to quote Dr. J. C. Majure, "The bar code drives the control system."[19]

Optical character recognition (OCR), though not as accurate as bar coding, can supplement it for human reading. OCR is a required supplement of LOGMARS labels.

Voice recognition is the most user-friendly of the AIM systems, but still has few applications, due to small vocabulary and large computer memory requirements.

Magnetic strips are widely used for work-in-process, due to readability in poor manufacturing environments. Wrinkling, dirt, oil, and other conditions that may destroy bar codes are not harmful.

Transponders are chips that can emit an identifier signal, which can be scanned from several feet away from their source. Applications include identification of rail cars, where signals are not screened by coverings of mud, salt, and ice.

Vision systems are receiving wide attention in traffic control, automatic trailer loading, quality control, and robotics.

Automatic Storage and Retrieval Systems (AS/RS)[36]

AS/RS are flow-control valves in the automatic factory's inventory pipeline despite just-in-time policies that mandate minimal inventories on hand. Under certain conditions, the flow control is needed for the following reasons:

1. Surges in demand or production require controlled reserves.
2. Speculation stocks with long lead times may be allocated for targeted sales that have not materialized.
3. Certain stocks, even in the Toyota Production System, do not fit JIT ordering patterns.
4. Materials on hold for inspection or disposal still need to be under control.

Thus AS/RS will be present, in smaller, but more numerous, loads.

Unit load AS/RSs receive and store a wide variety of palletized merchandise to a high degree of accuracy and control with minimum operator attention. Pallets may be stored one–two deep per lane, with multiple access conveniently located for put-aways or retrievals. Each automatic activity is under computer control through *interfaces*, which may receive from or feed to transporters or directly to workcenters. Interfaces may be transfer (chain or roller) conveyors, robots, or lift/lower automatic guided vehicles. Transfers deal with nonstandardized loads, while robots handle highly repetitive loads of the same shape very efficiently.

Miniload AS/RS is geared to JIT programs by controlling less than unit loads, usually in pans and often in work-in-process lot sizes or kits. Miniloads can automatically deliver kits directly to workstations built into the walls of the systems, and automatically receive processed work into the system as a buffer for the next manufacturing step. In addition to automatic I/O stations, the machinery may include automatic test equipment to monitor quality while storing such parts as printed circuit boards (PCBs).

High-density AS/RS can take several forms where flexibility is not a severe problem. One form is the *car-in-lane*, where a long lane is dedicated to a specific item. The carrier enters the lane to put away from one end. The output end may be serviced by a second (retriever) carrier and the racks may be equipped with conveyors to provide first-in/first-out (FIFO) control. An expansion of the "flow" rack may include many levels and many bays of conveyor-equipped storage on dedicated (slave) pallets. Flow systems are served at one end by an automatic storage machine and at the other by an automatic retrieval machine.

Stacker cranes are designed to span an entire storage area and are computer controlled to a grid on the *x-y* plane. Odd-shaped parts such as airframes and pipe can be stacked within them and can be tracked by means of an onboard terminal in communication with a dedicated computer.

Carousels input and withdraw materials at the same station. The computer rotates each level to present the desired storage container to an automatic I/O loading device. Gutpa[40] has described a smart kitting system using a number of carousels linked by transfer conveyors. Vertical carousels may provide either high-density storage of small parts in a small space or provide access to manufacturing operations on several floors. Rotary racks have shelf levels that rotate independently.

Spiral towers provide work-in-process buffers in a FIFO mode by accepting work into and up a ramp, while discharging to the next work center, at a controlled rate. Small parts can be stored, classified, sorted, oriented, and fed to the next operation through vibrating-bowl spiral feeders.

Computer controls automatically scan identities, route, queue, prioritize, and assign storage locations. Selection and allocation can be done through soft pick lists. Contact can be made with a host computer through on-board terminals or microcomputers.

Smart Kitting

Smart kitting is a process whereby several elements of flow control, including lights, bar codes, weighing, and conveyors combine to perform spectacular throughputs in an automatic mode. In a Plant A system, some 10,000 orders in the form of kits are shipped per day. The kit is the shipping container, and all are the same size, thus, one order may contain more than one kit.

Assembled cartons enter the system at a rate of about thirty per minute. From computer files, a bar-code label is jet-printed and applied to each container. The bar code is matched by computer file with all the information about the order including customer, contents, and total weight and volume for each kit. Sixty percent of the items are manually filled, and 40% are filled from 1024 computer operated magazine dispensers at a rate of up to 2000 items per minute. Dispensers are triggered by bar-code scanners.

Kits are directed to express, manual, or exception lanes. *Manual kitting* occurs when a kit is scanned as it arrives, causing lights to flash on appropriate racks. As a part is selected, the operator presses the light. If another of the same item is needed, the light remains on, until requirements are filled. All kits are check-weighed as they move out of order-filling lanes. Wrong weights divert kits to exception lanes for inspection. Lanes merge for packaging in which contents are heat-sealed into the box for damage protection. Scanners read bar codes, shipping labels are automatically applied, and the kits are lidded and banded for shipment.

Scanners again read bar codes and operate a sorting system that diverts kits into shipping lanes. When a truckload of containers fills a lane, the kits flow into an automatic palletizer, which stacks, stretch wraps the load on a "Slip Sheet" pallet, and automatically discharges the load into a trailer. The kit does not stop moving from the time the container enters the system.

The reader can immediately spot many applications from this example for plants B and C discrete manufacturing systems.

Automatic Guided Vehicle Systems (AGVS)

AGVS are a cost effective and productive means of serving multiple routes and smaller lot sizes for flexible manufacturing/just-in-time systems. They can also interface effectively with automatic storage and retrieval systems, robots, conveyors, machine tools, and elevators.[35] Signals can be transmitted to a central processor from which routing can be monitored and controlled.

Flexible assembly operations may be designed with the work attached to the vehicle or to handle work from station to station, such as automatic machine tools with automatic load/offload stations.

Guidance of AGVS is usually from signals passed through wires buried in the floor (inductive control). *Optical* guidance systems use a painted or taped path, which is scanned by lights (visual or ultraviolet).

Combinations of inductive and optical systems are found in the use of flat conductors taped to the floor with reflective tape. Off-guidepath movement is now possible through wall-mounted sensors, radio controls, measuring wheel, and "dead reckoning." *Dead reckoning* involves *steering the vehicle* to a specific path, then *retracing the path by memory back to the guidepath.*

Lifting and lowering capability increases AGVS utility. Vehicles can be equipped with forks to lift, travel up to 40 feet, and then place a load in storage. Dead reckoning permits the vehicle to leave the guidepath and travel across dock-boards to load and offload over-the-road vehicles, then return to the guidepath.

Scanners can be mounted to read package bar codes and automatically deliver the load to the correct workstation. AGVS is thus becoming a key element in applying material flows in the automatic factory, reliably and flexibly for loads from mail to 25,000 pounds.

Industrial Robots

Robots in the automatic factory either select, place, or process work repetitively and tirelessly, once the initial task pattern is defined.

Robotic systems[35] include dedicated microcomputers, servomechanism arms, and grippers attached to wrists. Arms can be mounted to walls, columns, and floors, and can travel on rails as well as moving carriers (such as AGVs).

Control programs can be CAMS, or programmable teaching pendants. Teach programs record all points, speeds, and motions transmitted to the processor memory and repeat them on demand. Point-to-point control requires entry of a task for each point in space in timed sequence. The stored program routes the robot through the shortest path to each point and directs each activity.

Operating motions are determined by arm mounting and its motion through the work area. *Polar* or *hemispherical* arms are fixed to a pedestal. *Cylindrical pedestals* permit raising, lowering, and rotating the arm on the pedestal. The cylindrical pedestal may be mounted on ways or tracks for cartesian (x-y) travel in addition to up/down/rotation. *Revolute* motion is an emulation of the human arm on a rotating pedestal with a shoulder for raising, an elbow for extension, and a wrist for rotation of a gripper. All the robots can be equipped for retractable arms and gripper rotation in addition to cartesian coordinates (x-y-z); additional degrees of freedom are pitch, roll, and yaw. Up to three arms per machine operating simultaneously are being used for placing, selecting, and processing parts.

Grippers and *end effectors* handle the workpiece using pressure sensors, vacuum, or electromagnets. Most grippers are custom-designed for specific jobs. They are being aided by vision sensors, Lateral Effect Photodiodes (LEPs), Tactile Sensors (arrays of pressure sensors), and other tracking and homing devices.

Power supplies are hydraulic (for less precise and heavier machines), pneumatic, and total electric with lightweight motors at each joint for precision/operations. Combinations are applicable to specific jobs.

Applications include pick and place for loads up to 1000 pounds, painting, welding, palletizing, storage interface, and machine-cell processes. Hazardous radioactive and chemical operations and repetitive or demanding jobs requiring brute force or heavy lifts are potential applications of robotics.

Overhead Fixed Path Transporters (Monorails)

Such transporters use available air space to avoid shop-floor traffic conflicts. The Monorail Manufacturers Association[41] provides descriptive literature and standards.

Smart monorails[36] use on-board microcomputers to communicate with a central computer. Each carrier is independently powered to automatically manage operations such as speed control, interfaces with floor-mounted equipment, lifting or lowering, loading and off-loading.

Track sections that were specially rolled and welded from structural shapes have been redesigned to carry the lighter JIT loads, such as pans, rather than unit loads, such as pallets. They have been equipped with integral controls and safety conductor power supplies.

Controls evolved from magnetic relays to solid-state hard-wired logic to central programmable process to on-board minicomputer to control carrier operations at any control point, commonly called smart monorails.

Advantages are high speeds on straight runs; slow-down for switches, curves, and drop stations; self-powered load/unload/pickup, delivery into flexible manufacturing systems, and redirection while in transit.

Features[35]	Smart Monorails	Chain Conveyors	Power and Free
Overhead path	Yes	Yes	Yes
Travel speeds	High	Low	Low
Variable speeds	Yes	No	No
Programmable carriers	Yes	No	No
Commands in transit	Yes	No	No
Sequencing	Yes	No	Yes
Diversion	Yes	No	Yes
Sortation	Yes	No	Yes
Continuous running	No	Yes	Yes
Buffer/recall	Yes	No	Yes
Individual carrier controls	Yes	No	No
Layout flexibility	Fair	Fair	Poor

Fixed Path Floor Conveyors

There are many forms (roller, chain, slat, live roller, belt, magnetic, fluidized beds, towline, inverted power, and free). We refer the reader to Chapters 3.6, "Material Handling Equipment; and 5.2, "Material Handling," for details.

Controls[35] for floor conveyors are programmable controllers linked to host minicomputers. PCs activate switches, diverters, and drives, act as safety protectors, scan loads, and automatically route work in accordance with automatic identification and keyboard controls. AIM signals are scanned in high-speed package conveyors through sorting devices that divert each package to specific process or transportation lanes at the rate of several hundred per minute.

Inverted power and free conveyors are floor-mounted, upside-down models of overhead transporters. They can accumulate and sort work as programmed, and can be equipped with automatic transfers to feed work into flexible manufacturing workcenters.

Car-on-track systems are becoming the standard for precision car placement. Wheels ride on two tracks, and the cars are driven by a drive wheel making contact with a rotating cylinder. The attitude of the drive wheel on the cylinder determines operating speed, acceleration, and deceleration.

Towlines have been the standard chain-in-floor method for moving heavy loads on fixed paths until automatic guided vehicle systems became prominent transporters. FMS now uses towlines with loops (buffers), interfaces (flight rollerbeds, turntables, or slat conveyors), all of which can be individually programmed to send work to specific jobs. Loops can be programmed to haul waste and balance flows of cars on main track, and monitor car locations by means of digital shaft encoders and proximity switches.

5.3.7 CONCLUSION

Old friends who have served us well in 40 years of material handling such as optimization, Pareto analysis, flowcharting,[7] and Monte Carlo simulation of capacities will still serve to measure the performance of new systems.

Strong managers are needed who can commit to and communicate long-range plans; who are literate in the new productivity and information power and willing to take their reins; who will not relinquish that power to software packages or computer programmers, but will firmly direct programs into user channels. Dr. James Majure proposes that small manufacturers may be able to fit their production to standard software packages, but larger users must tailor software to fit their production.[19]

Older managers directed production with one hand on an expediter, the other on a stack of "hot" lists and their eyes on back schedules. Modern managers have the production plan at their left, the master production schedule at their right, and their eyes on terminals that display responses to rough cut capacity proposals.

Flow control is defined in the APICS dictionary[2] as a system of setting rates, planned introduction of work to meet those rates, and tracking the work to keep it moving—especially in repetitive production. We have shown those principles of repetitive production that can be emulated in discrete production of smaller lot sizes by applying the broad discipline of material flow control, proposed by Dr. Ruddell Reed of Purdue University in the late 1960s. We have reviewed MRP proposed by Wight and JIT under developments by Toyota in the late 1960s. We have demonstrated applications of the bar code, product of a Harvard master's thesis in 1932. These tools have now matured as flow controls and with modern machines and computer power make the Automated Factory a reality.

The automatic factory still has most potential in high-volume, repeatable manufacture.

The benefits of automatic manufacture can be gained in manual operations with automatic identification, and portable terminals, AS/RS, robotics, flexible manufacture, and automatic transport. However, higher priority must be placed on a stronger employee/manager commitment to responsibility and acountability to achieve full integrated-factory potential.

Both manual and automatic systems have a great deal to offer in customer satisfaction and profitability through material flow controls.

Why didn't the Detroit assembly line fulfill flow-control criteria? Lack of flexibility, lack of I/O controls (large just-in-case inventory buffers), and extension of machines to force humans into repetitive machine-like tasks? Large investments required to install, prove-in, and balance assembly lines? The Detroit assembly line *was* the perfect flow-control system as long as every car was black, had two doors, and the employee choices were to work under frustratingly poor conditions or be frustratingly poor.

Entire factory-cities were built around assembly lines for specific car models. Shifts in model designs became plant *and* sociological problems. Enormous investments were passed on to customers who were forced to go to foreign selections. The assembly line became a burden for management, quality, and productivity.

Flexible manufacture, programmable conveyors, robotics, flexible path transport, and quality circles are becoming the new look in Plant B. The material managers now offer the customer thousands of options, provide a cycle time of 36 hours, have less than a week of stock on hand, and can examine paths and simulate automatic guided vehicle traffic on CRTs prior to installing guidepaths.

A reminder in dealing with any material control strategy: *The computer cannot make anything happen, cannot replace good management*, and *no computer is smart enough to out-think people*.

More than $20 billion has been expended by American industry on MRP, with mixed results. Our experience is that computer-*operated* systems, which limit operator/manager control over the job yield poor results. *Integrated* systems, which expand operator/management control, yield spectacular results. Friendly systems must be useful systems for effective material flow control.

REFERENCES

1. R. Reed, *Material Flow Systems*, Schaumburg, IL: IMMS, 1970.
2. T. F. Wallace, *APICS Dictionary 5th ed.*, Falls Church, VA: APICS, 1984.
3. *Corporation Records*, New York: Standard & Poors, 1984.
4. R. Townsend, *Further Up the Organization*, New York: A. A. Knopf, 1984.
5. P. F. Drucker, *Management in Turbulent Times*, New York: Harper & Row, 1984.
6. T. J. Peters and R. H. Waterman, *In Search of Excellence*, New York: Harper & Row/Warner, 1984.
7. R. Muther and K. Haginas, *Systematic-Handling Analysis*, Management and Industrial Research Publications, 1969.
8. R. J. Tersine and J. H. Campbell, *Modern Materials Management*, New York: Elsevier, 1977.
9. A. K. Kochhar, *Development of Computer Based Production Systems*, New York: Halsted/Wiley, 1979.
10. D. S. Ammer, *Materials Management and Purchasing*, Homewood, IL: R. D. Irwin, 1980.
11. B. T. Smith, *Focus Forecasting for Inventory Control*, Boston: CBI, 1978.
12. C. R. Hicks, *Fundamental Concepts in the Design of Experiments*, New York: Holt, Rinehart & Winston, 1973.
13. W. T. Stewart, *Design and Use of Automatic Storage Systems—A Video Tape Indirect Laboratory Experiment in Field Support Service Center*, W. Lafayette, IN: Purdue University, 1982.
14. E. L. Grant and R. S. Leavenworth, *Statistical Quality Control*, New York: McGraw-Hill, 1980.
15. R. Peterson and E. A. Silver, *Decision Systems for Inventory Management and Production Planning*, New York: Wiley, 1979.
16. D. T. Phillips, A. Ravindran and J. J. Solberg, *Operations Research, Principles and Practice*, New York: Wiley, 1976.
17. A. Reisman, *Industrial Inventory Control*, New York: Gordon and Breach, 1972.
18. O. Wight, *Production and Inventory Management in the Computer Age*, Boston: CBI, 1974.
19. J. C. Majure, *Rack Locator System*, Roswell, GA: Majure Data Co., 1984.
20. W. T. Shirk, "Mobile Storage Systems," *Plant Engineering* (May 1983).
21. M. P. Groover, *Automation, Production Systems and CAM*, Englewood Cliffs, N.J.: Prentice-Hall, 1980.
22. R. S. Pressman and J. E. Williams, *N/C and CAM*, New York: Wiley, 1977.
23. R. J. Tersine, *Material Management and Production Systems*, New York: Elsevier, 1976.
24. D. T. Phillips, "Simulation of M/H Systems," *Industrial Engineering* (September 1980).
25. J. E. Biegel, *Production Control, A Quantitative Approach*, Englewood Cliffs, NJ: Prentice-Hall, 1971.
26. O. Wight, *MRP-II, Unlocking America's Productivity Potential*, Boston: CBI, 1983.
27. J. Orlicky, *MRP*, New York: McGraw-Hill, 1975.
28. J. C. Anderson, R. G. Schroeder, S. E. Tupy, and E. M. White, "MRP Systems: State of the Art," *P & IM Review* (4th Quarter, 1982).
29. "How DRP Helps Warehouses Smooth Distribution," *Modern Material Handling* (April 1984).
30. R. F. Jacobs, "Opt Uncovered," *Industrial Engineering* (October 1984).
31. R. W. Hall and J. Nakane, "KANBAN," *Driving the Productivity Machine*, Falls Church, VA: APICS, 1981.
32. D. Nelleman and J. Smith, "Just in Time vs. Just in Case," *P & IM Review* (2nd Quarter, 1982).
33. W. Goddard, "KANBAN VS. MRP," *Modern Material Handling* (November 1982).
34. R. D. Garwood, "Explaining JIT, MRP-II, KANBAN," *P & IM Review* (October 1984).
35. "The Automated Factory," *Modern Material Handling* (February 5, 1984).
36. F. B. Cooke and D. W. Nelson, "Material Control for the Factory of the Future," *APICS Proceedings* (1982).
37. J. A. Martin, *Real Time*, Englewood Cliffs, NJ: Prentice-Hall 1967.
38. *Shipping/Parts Identification Label Standard #B-3-1984*, Southfield, MI: Automotive Industry Action Group, 1984.
39. *Code 39 Alphanumreic Barcode*, Document 601815, Lynnwood, WA: Interface Mechanisms.
40. N. K. Gutpa, "Kitting Matrix," *Industrial Engineering* (January 1982).
41. R. E. Smith, "Robotic Vehicles in the Factory of the Future," *Industrial Engineering* (September 1983).
42. *Specifications for Underhung Cranes and Monorails*, Pittsburgh, PA: Monorail Manufacturers of America.

CHAPTER 5.4

MATERIAL PROTECTION

ALFRED H. MCKINLAY

Consultant
Pattersonville, New York

5.4.1 THE NEED FOR PROTECTION

As products and components move through the production and distribution system, they are subjected to a variety of environmental conditions that may cause damage. Chief among these are shock, vibration, compression, and moisture. To protect against these damaging forces, manufacturers rely on a combination of materials and methods broadly classified as packaging, containerization, or unitization.

5.4.1.1 Environmental Factors Causing Damage

Shock

When an item is handled in the production or distribution system, there is a high probability that it will encounter a number of impact or shock forces. The only question is how much shock it will receive and how often. In their assessment of the common carrier shipping environment, Ostrem and Godshall[1] have captured much of the available data and information relative to measurement of shock input probability. Principal sources of impact are dropping, bumping, rail-car coupling, and poor road conditions such as potholes, bridge transients, and pavement joint misalignment. For smaller items, those which can be carried by one or two persons, by far the greatest potential for damage is an accidental drop or an intentional toss.

Available data on handling show the following general probabilities:

Most manually handled items receive many drops at low heights, while very few receive more than one drop from higher heights.

The heavier an item is, the lower its probable drop height.

The larger an item is, the lower its probable drop height.

Packages are usually dropped on their bases (over 50% of total drops sustained).

Unitized or containerized loads are subjected to fewer drops and from lower heights than single parcels.

A graphical summary of the data described above is shown in Fig. 5.4.1.

In rail-car coupling, packages are subjected to longitudinal shock as well as dynamic compressive force from the lading behind them. Although railroads claim most coupling speeds are below 5 miles per hour (mph), it has been the experience of many shippers that speeds of 6 to 10 mph are frequent, and occasionally as high as 12 to 15 mph. At 4 mph coupling speed, a box car with standard draft gear will transmit a force of 1400 pounds per square foot (psf) against a lading at its end wall. With a 30 in. travel hydraulic cushion draft gear, the same box car can be coupled at over 10 mph before reaching a 1400 psf load.

Transient data from truck shipments over a variety of road conditions have shown a wide dispersion of shock input as a result of potholes, bridges, pavement irregularities, railroad crossings, etc. The worst shock conditions are generated by a lightly loaded trailer. Freight loaded at the rear of the trailer, especially when its wheels are set in a maximum forward position, receive the maximum trailer bed shock input. Precise data are not available, but several studies have recorded numerous shock pulses of 3 to 5 ms (millisecond) durations and peak accelerations of 5 to 15g. Others have noted transient shock pulses which raised freight as much as 6 in. off the trailer floor.

The shock input within the manufacturing plant handling environment is relatively low. Lift trucks, conveyors, and cranes do expose products to varying degrees of shock, but of significantly lower magnitude than those received in over-the-road shipments. One commonly used shock criterion for mechanical handling by cranes or lift trucks suggests that a 9 in. drop on the bottom edges or corners of large packages or unitized loads is the maximum input that one can expect.

Vibration

Vibration is encountered in many places throughout the distribution system and to a lesser degree in the manufacturing process. As described in the Forest Products Laboratory Report,[1] the worst vibration conditions are encountered in highway travel aboard tractor trailers. The input from a trailer bed as it is hauled at 55 mph along a typical interstate highway can easily generate a steady-state environment of $0.5g$ acceleration. From their study, Ostrem and Godshall suggest the vibration envelope curve for highway trailers shown in Fig. 5.4.2. Although it is certainly possible to achieve acceleration levels in trailer vibrations exceeding the $0.5g$ level, normal testing to simulate truck shipping is done at a maximum of $0.5g$.

Rail-car vibration is generally not as severe as for the trailer on a highway, but can produce much higher inputs when extremely poor roadbeds or rolling equipment are encountered. A suggested envelope for vibration input by a railcar is shown in Fig. 5.4.3. The vibrational inputs of aircraft and ocean ships are of significantly less magnitude and generally not considered an important criterion for design. No data have been published on expected vibrational inputs for products moving through the manufacturing process, although it is known that containers of material moving along a roller conveyor system encounter a significant vibrational input.

Compression

Compressive loads can also cause damage to products. The vertical static loads associated with warehousing and storage stacking are the most significant. When containers or packages are stacked one upon the other, the amount of compressive load will vary considerably depending on the height, the stacking strength of the container, and the load-carrying contribution of the contents. With corrugated containers, length of storage time has a significant degrading effect on stacking strength. For horizontal static loads, the major source of damage is the pressure applied by clamp-handling equipment.

In addition to static loads, dynamic compressive loads also result during transportation. These are considerably more difficult to estimate or measure, and little data are available. The longitudinal dynamic loading due to rail-car coupling is the most significant of these compressive forces. Dynamic vertical compressive loads are also a serious problem, especially when heavier products are loaded on light-duty containers in rail and truck shipment. The vibrational input in combination with the heavy load atop a light-duty container can cause significant crushing problems and damage.

Moisture

Moisture, high relative humidity or rain, can have a significant effect on the load-carrying capability of containers and the surface condition of the products within. In the case of corrugated shipping containers, 50% of their normal stacking strength is lost when the relative humidity reaches 90%. As humidity rises even higher, toward 100%, any significant length of time in these very damp conditions can cause almost immediate collapse of corrugated containers.

The effects of high relative humidity and changing temperatures, which produce a condition of condensation within packages, is a major source of much damage, particularly in overseas shipments and unprotected outdoor storage. Moisture trapped within a sealed container when the temperature drops will condense on the container walls and the contents, possibly causing serious corrosion damage. One of the most serious errors is to use a tightly sealed barrier, such as polyethylene film, and then ship a product for a long trip through varying temperature zones. Condensation and then corrosion are almost guaranteed to occur.

5.4.1.2 Determining Product Fragility

To ascertain if one has a product that is susceptible to shock or vibration damage, one must first determine the product fragility relative to those conditions. Test methods have been developed to measure fragility for both shock and vibration.

Shock Fragility

Shock fragility tests are generally run on a shock testing machine. The item to be tested is fastened at the top of a shock table on the machine and the table is then subjected to controlled shock pulses. In a typical test the shock table is raised to a preset drop height and released, and free falls to hit the base of the machine. A shock programmer between the table and the base controls the type of shock felt by the table and the test item mounted on it. The g-level of the shock pulse is controlled simply by adjusting the precharged pressure in the cylinder of the shock testing programmer. A series of drops is then made from varying heights to determine the critical velocity at which damage occurs to the product. Tests are also run in increasing level of acceleration g-level input to determine the maximum or critical acceleration levels. When these two items are known, a damage boundary can be plotted. In cases where the item may be dropped on any of its sides, tests should be performed in each direction of the three axes, and damage boundaries plotted for each, or a total of six damage boundaries. Figure 5.4.4 depicts a damage boundary determined from this method of testing. Shock fragility determination is always made on the product or component itself, not on the packaged product.

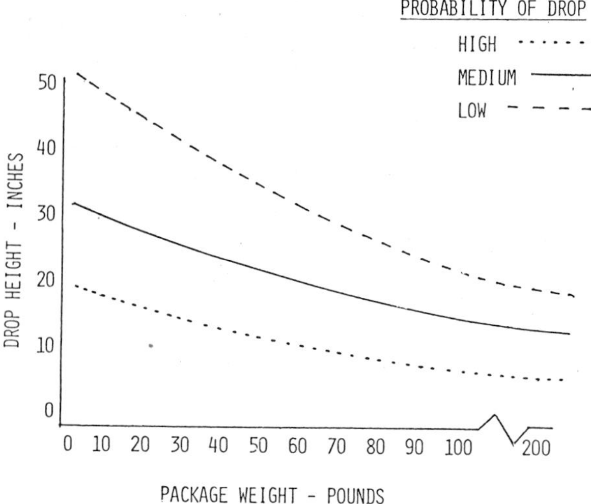

Fig. 5.4.1. Manual handling environment of packages.

Vibration Fragility

Damage from vibration often occurs when some element or component of the product has a natural frequency, which is excited by the environment, such as vibration input through a trailer bed. If this excitation is of sufficient duration, the component accelerations and displacements can be amplified to failure. The response of a component to such vibration is represented in Fig. 5.4.5. At low frequencies, the response of the component of the input acceleration is the same or a one-to-one ratio. Conversely, at high frequencies the response acceleration is much less than the input. At some point in between, however, the response acceleration will be several times the input level. This point, the *natural* or *resonant frequency* point, is the area where the damage is most likely to occur.

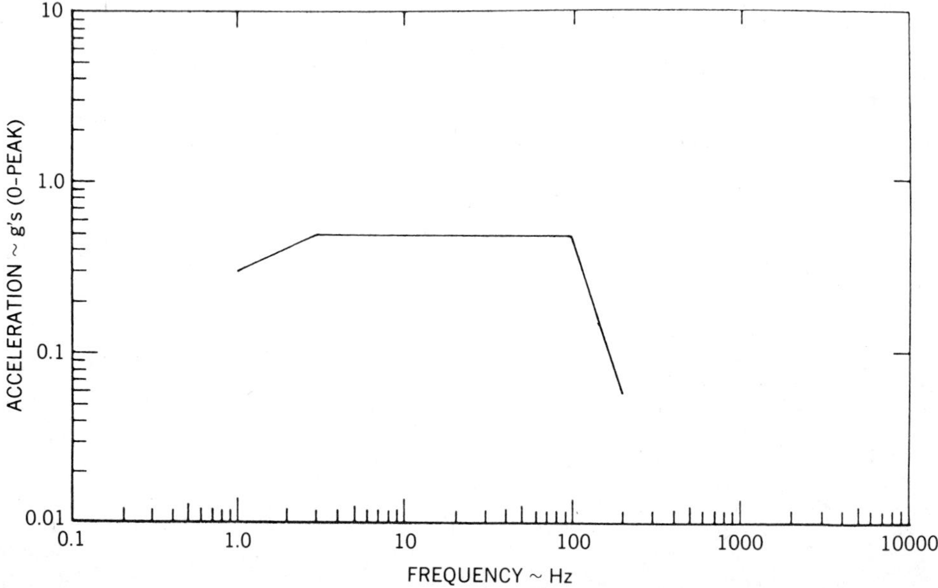

Fig. 5.4.2. Vibration acceleration envelope—trucking environment.

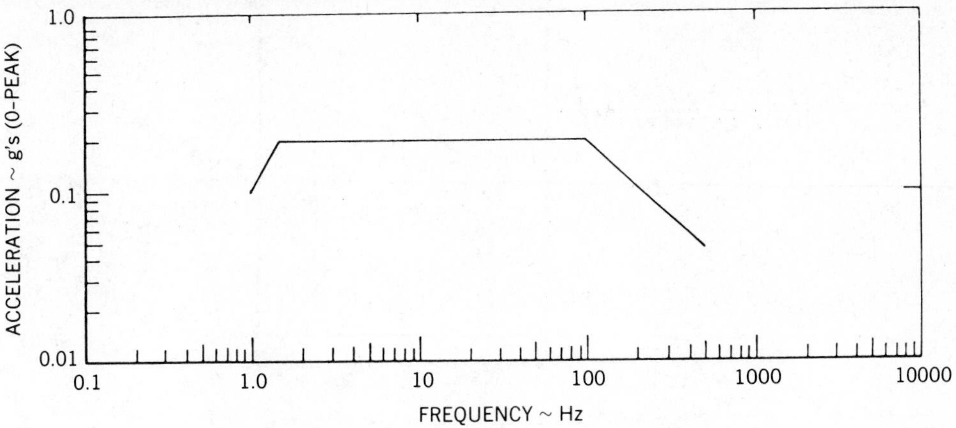

Fig. 5.4.3. Vibration acceleration envelope—rail-car environment.

The resonant frequency point is determined by a test on a vibration machine. The item tested is fastened to the vibration table and then subjected to a vibrational input, typically from 3 to 100 Hz and back down to 3 Hz. As the frequency is slowly being varied between these limits, the item is observed for resonant conditions. As in shock fragility testing, determination for each of the three axes should be done on the product. If in shipping or handling the product is mounted on a definite skid base, only the vertical axis is required for an analysis.

5.4.2 METHODS OF PROTECTING MATERIALS

To protect materials against the environmental conditions that cause damage, three primary methods are employed, depending on the logistical area involved. These include containment of components for intraplant movement, packaging for shipment, and unitizing for finished goods distribution.

5.4.2.1 Containment for Intraplant Movement

As materials move through the manufacturing process, between machines, operation centers, or buildings, they may require protection against damage. Often this protection is provided simply in the form of containment using tote pans. If required, additional protection can be added within the container. Many kinds of containers

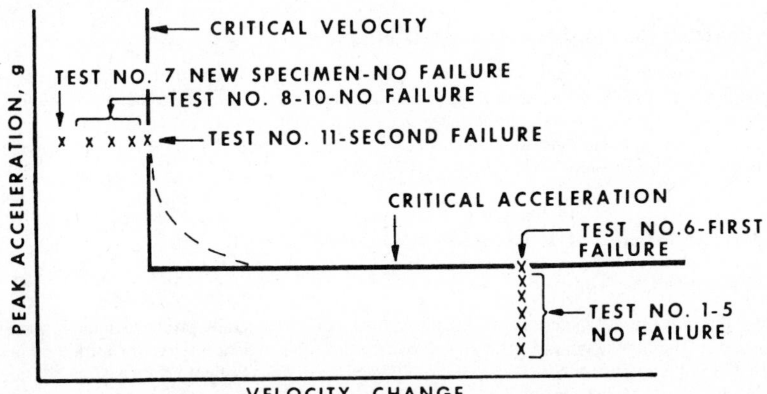

Fig. 5.4.4. Determing shock damage boundary.

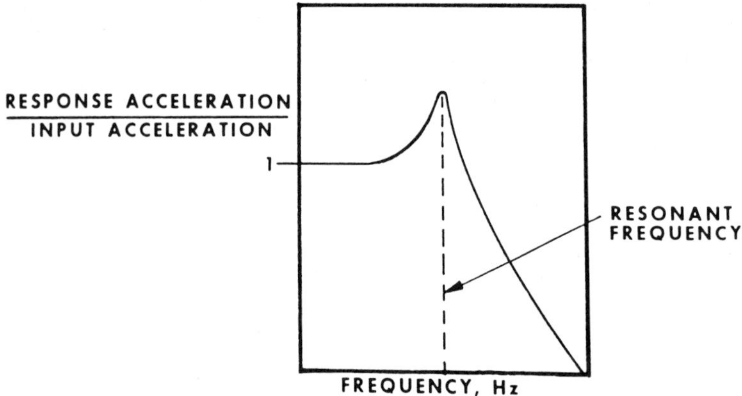

Fig. 5.4.5. Component response to vibration input.

and materials are used, including plastics, metal, wood, fiberboard, and combinations of these materials. In addition to protecting the contents, intraplant containers may provide other benefits, such as:

Reduce handling costs of small components moving between operations.

Become part of the manufacturing process, such as an assembly device.

Provide a device for presentation of parts to automated manufacturing equipment.

5.4.2.2 Packaging for Shipment

Protection of materials in finished goods and distribution is most often provided by packaging. A definition of distribution packaging is "that which provides the basic function of performance in the distribution process." Sometimes it is called "distribution packaging," "protective packaging," or "industrial packaging," all three terms mean the same thing. For this type of packaging, emphasis is placed upon the following major criteria:

1. Protecting the product in distribution.
2. Providing handling capability for the packaged product.
3. Properly utilizing the carrier vehicle and meeting carrier regulations.
4. Providing economical means of packing and sealing.
5. Providing identification of the contents and special instructions for use and handling.
6. Providing customer ease in opening, unpacking, reclosing, reuse, and disposal.

While the materials of containment for intraplant movement are generally reusable, those for packaging in distribution are generally nonreturnable (expendable). Only when arrangements can be made for the customer to easily return packaging materials to the shipper is it possible to use a returnable packaging system. Typical distribution packaging materials include corrugated fiberboard, plastic foams and films, wood, metal, and paper.

5.4.2.3 Unitizing for Finished Goods Distribution

Unitizing is the assembly of a group of items or packages into an appropriate unit for mechanical handling. A unitized load is one in which all of the packages or articles are bound together in one or more units by means of strapping, tying, or gluing. Although the major reasons for unitizing are to reduce material handling costs and to speed the handling operations, the contents of a unit load are generally subjected to less severe handling, therefore increasing the protection of materials.

5.4.3 CONTAINMENT FOR INTRAPLANT MOVEMENT

5.4.3.1 Reusable Containers

The first level of protection for materials moving through the plant is generally provided in the form of containers or *containerizing*. The containers are generally reusable, since the logistic system for intraplant movement is easily controlled and return of containers can be closely monitored. For smaller parts, the containers are generally classified as *tote boxes,* which are available in a wide variety of sizes and materials, such as wood, metal, plastic, fiberglass, or fiberboard. They can be stackable, nestable, and modular in design. (See Fig. 5.4.6.)

When the parts or components are larger, or there are greater volumes to be moved between each operation, skid-based containers that can be handled with lift trucks are used. These containers are generally made of metal, although plastic and wood are also widely used. These larger intraplant containers, as shown in Fig. 5.4.7, are often collapsible to save space for storage and return. Portable stacking racks that fit over wood pallets and metal skids to allow stacking one unit load upon another in the factory while not placing any load upon the components contained therein are also used.

5.4.3.2 Expendable Containers

When it is difficult to control the return of intraplant containers to their original source, expendable containers are often used. The principal material is corrugated fiberboard, sometimes coated with plastic or other materials to provide extra life. The container style is often of the same variety used in packaging, such as half-slotted with cap, telescoping, or double cover. Typically the objective when using an expendable container is to keep its cost as low as possible. Therefore characteristics such as nestability are often sacrificed, since they are expensive.

5.4.3.3 Dunnage

A variety of protective inserts is used within intraplant containers to protect the contents from exterior forces and from bumping against one another within the container. Materials used are metal, plastic, vulcanized fiber, wood, and corrugated fiberboard. The metal inserts are the most expensive but are the only choice when maximum durability is required. Plastic inserts are less expensive and lighter weight, but not quite as durable as metal. Most commonly used are rigid polystyrene and high-density polyethylene. Vulcanized fiber, which has an excellent strength-to-weight ratio, is often used as a divider in containers. Wood inserts can be economical

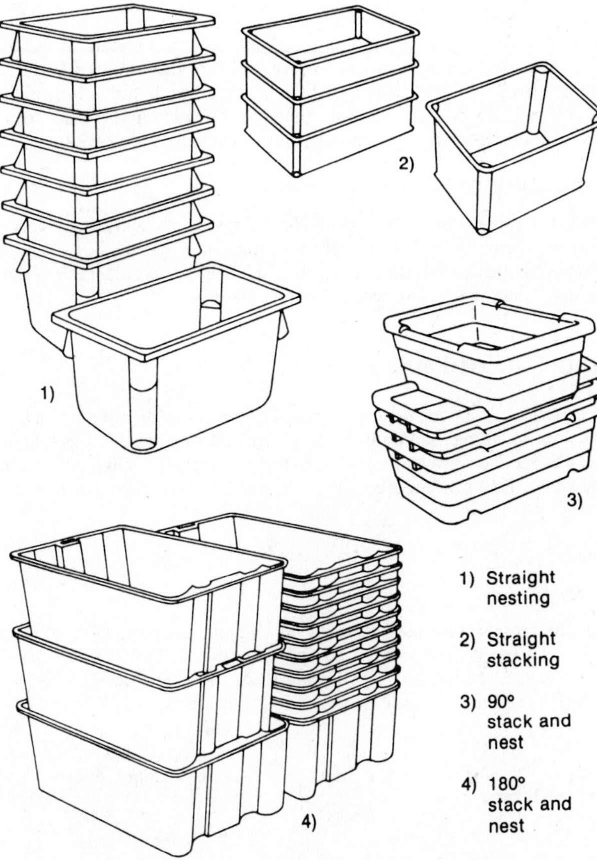

Fig. 5.4.6. Kinds of tote pans.

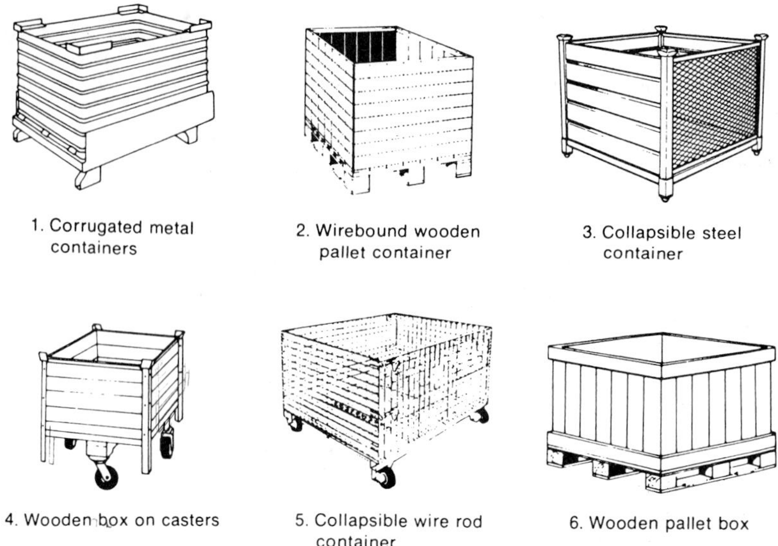

Fig. 5.4.7. Typical large containers for intraplant movement.

for small quantities since no tooling costs are required. Corrugated inserts are the least durable and also the least expensive, used mostly as dividers or as loose sheets placed at the bottom of the containers or between parts. On occasion the corrugated insert is treated to give it grease resistance or longer life. Plastic resilient foam is often used where shock or vibration is a problem in very fragile parts. Polyurethane or polyethylene are the most widely used foams for this purpose. Custom molded trays to fit the exact configuration of contents are often made from vacuum-formed rigid plastics or from expanded polystyrene foam.

5.4.3.4 Vendor Packaging as Intraplant Container

When the volume is sufficient, manufacturers who purchase large components often work with the component suppliers to develop either expendable or reusable shipping containers that can be used as intraplant containers, partway or entirely throughout the manufacturing process. This requires considerable planning and cooperation between supplier and buyer, but can greatly reduce costs for both.

5.4.4 PACKAGING FOR SHIPMENT

Packaging is generally defined as the technique of preparing goods for distribution and can include cleaning, drying, preserving, packing, marking, and unitizing. *Packing* is defined as the selection or construction of the shipping container and assembling of items or packages therein, including any necessary blocking, bracing or cushioning, weatherproofing, exterior reinforcement, and marking the shipping container for identification of contents.

5.4.4.1 Containers

Corrugated Fiberboard

Corrugated fiberboard, the most widely used material for shipping containers today, is generally made from virgin kraft fibers with some percentage of recycled fiber intermixed. Two or more flat sheets of paperboard are glued to one or more center members or corrugating media that have been "fluted" on a corrugator. The combined corrugated board is generally made in one of four flute sizes:

	Number per Linear Foot	Approximate Height (in.) (not including thickness of facings)
A-flute	36 ± 3	3/16
B-flute	50 ± 3	3/32
C-flute	42 ± 3	9/64
E-flute	94 ± 4	3/64

Other combinations of two or three of these corrugating media, generally in B-C, A-C, or A-A-C combinations of flutes, provided a double or triple wall combination for much greater thickness and load-carrying capacity. Rules for both truck and rail shipments require a certain grade of paperboard material to be used for various load-carrying capacities of boxes. The particular carrier rules involved are Rule (Item) 222 for The National Motor Freight Classification and Rule 41 for The Uniform Freight Classification of the railroads. (See Table 5.4.1.)

Many styles of corrugated containers are available. The style chosen depends upon economics, shape of contents, or handling characteristics of the contents. The RSC (regular slotted container) in Fig. 5.4.8 is the most economical and widely used of these varieties. Other common styles are the FOL (full overlap), the 1PF (one-piece folder), the FPF (five-panel folder), and the FTHS (full telescope half slotted). Large two- or three-piece corrugated containers with interlocking caps are often used for bulk materials and for large products such as major appliances.

If one wishes to use a corrugated container for a particular product that is outside of the limits specified in the carrier rules, procedures outlined in the classifications allow one to file for these exceptions.

The various grades of corrugated are generally designated by the *burst strength* specified in the carrier rules. For instance a 200-pound-test box means that the material from which the box is made must have a burst strength of not less than 200 lb/in.^{2} and, according to the rule, can carry no more than 65 lb nor have a maximum united dimensional size greater than 75 in. The grading of board by the burst-strength method permits box manufacturers to develop standard grades. It does leave something to be desired, however, when attempting to select a box that will give the greatest compressive strength for storage.

A better grading of fiberboard strength for users is the *edge crush,* or *short column test,* which can then be used to predict stacking strength in the following formula:

$$C = 5.87P\sqrt{zh}$$

Where

C = the predicted stacking strength in pounds
P = the edge crush value in pounds per inch
z = box perimeter in inches
h = the thickness of the combined board in inches.

Wood

Containers are often made from wood when the product to be shipped is either too heavy or too large for a corrugated fiberboard box, or when production volumes are too low to warrant other types of high-production volume containers. The various types of nailed wood boxes are shown in Fig. 5.4.9. The main difference between them is the construction of the box ends. No carrier rules exist as to the thickness of types of lumber to be used in shipping, although the National Wooden Box and Pallet Association and the Association of American Railroads both publish recommendations on the minimum thickness of lumber to be used for various sizes of boxes and crates vs. the maximum gross weight to be carried. The U.S. Forest Products Laboratory[3] also publishes a *Wood Crate Design Manual* as a guide in construction of wood crates for both domestic and export shipments. Two important things to remember for both wood crates and wood boxes are the three-way corner nailing technique and the use of diagonals for preventing distortion of boxes or crates when they are dropped or subjected to diagonal compressive loads.

Another variety of wooden container used for higher production volumes is the wire-bound box and crate. Using rotary cut lumber of one-quarter-inch thickness or less in addition to wire reinforcements that cover the circumference of the box in several places, these boxes and crates are less costly than a plain wood box or crate, yet have comparable durability and can be stored outdoors.

Cylindrical Shipping Containers

For the shipment of materials such as liquids, semisolids and such solid products as powders and flakes, cylindrical containers are often used. These are sturdy tubes of wood, fiberboard, metal, or plastic, capped at one or both ends, providing an unsupported outer package that can be shipped without further boxing or crating. The fiber drum is the most widely used cylindrical container. As an expendable shipping container, fiber drums are often used for materials that also require a lining in the side wall as a water-vapor barrier or as a coating to prevent contact between the contents and the fiberboard. As with corrugated containers, fiber drums are also strictly regulated by the carrier rules.

Steel cylindrical shipping containers are most widely used for shipping liquid products including hazardous materials. They may be for single-trip shipments of oils, solvents, gasoline, etc., or returnable for the same materials. Steel drums are made in capacities from 15 to 55 gallons in the variety of gauges depending on the weight of the material to be shipped. Lighter gauge pails in capacities of 1–12 gallons are used for shipping smaller quantities of the same materials. When required, linings can be included to protect the contents of the container from contamination or discoloration by the steel. Hazardous materials packaging is regulated by the Department of Transportation Code of Federal Regulations, Title 49.[3]

Table 5.4.1. Truck and Rail Regulations for Corrugated Fiberboard Containers

		Single Wall		Double Wall		Triple Wall	
Gross Weight; Maximum Weight of Box and Contents (pounds)	United Inches; Maximum Inside Dimensions (length, width, and depth added) (inches)	Minimum Combined Weight of Facings (lb/1000 ft^2)	Minimum Bursting Test of Combined Board (lb/in.^{2})	Minimum Combined Weight of Facings Including Center Facing (lb/1000 ft^2)	Minimum Bursting Test of Combined Board (lb/in.^{2})	Minimum Combined Weight of Facings Including Center Facing (lb/1000 ft^2)	Minimum Puncture Test of Combined Board (in.-oz/in. tear)
20	40	52	125				
30	50	66	150				
40	60	75	175				
65	75	84	200	92	200		
90	90	138	275	110	275		
120	100	180	350	126	350		
140	110			222	500		
160	120			270	600		
275	120					264	1100

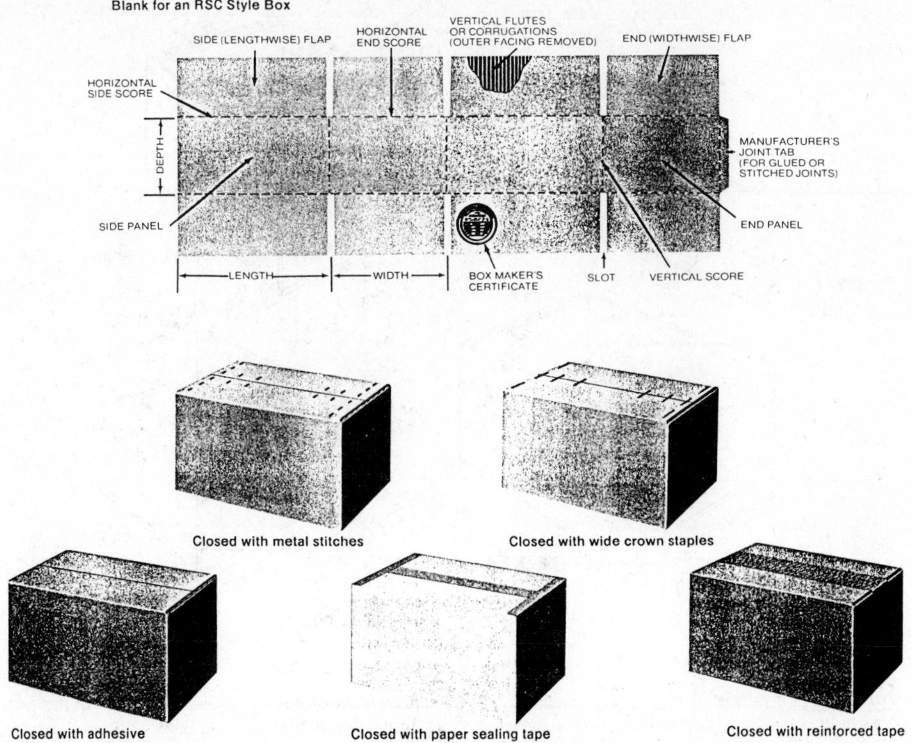

Fig. 5.4.8. Regular slotted corrugated container (RSC).

In recent years plastic cylindrical containers have come into wider use in a variety of sizes from 1 to 55 gallons. Usually made from molded polyethylene, these containers are much more flexible than a rigid steel container. Their chief advantages include flexibility, nontoxicity, light weight, durability, chemical resistance, and reusability.

Bags and Sacks

Another form of shipping container is the bag or sack. The term *sack* generally refers to a heavier duty bag made of flexible materials closed at all sides except one, which forms an opening that may or may not be sealed after filling. The sack may be made of any flexible material, multiple plies of these materials, or a combination of two or more materials such as paper, metal foil, and textiles, any one of which may be coated, laminated, or treated in other ways to provide the properties required for packaging in storage. Sacks are used for bulk shipments of such items as chemicals, fertilizers, cement, grains, feed, etc. The multiwall paper shipping sack, the most widely used bag, is constructed of one or more plies of paper, sometimes in conjunction with one or more plies of plastic. Depending on the method of filling and the product contained, a variety of construction types are available.

5.4.4.2 Inner Packing Materials

Within shipping containers, one may use inner packing materials either to cushion the contents against shock and vibration or to supply blocking and bracing to prevent movement of the contents within the container. This blocking and bracing action is often called *dunnage*.

Cushioning Materials

Many materials are available to cushion products from the shock and vibration environment. The most widely used and least expensive of these is corrugated fiberboard in the form of folded inserts, die-cuts, and pads. Often these corrugated interior packing pieces perform two functions, both cushioning the product and blocking and bracing it within the container. The chief advantage of corrugated is its low cost, while its main disadvantage is the relatively low cushioning performance as compared to other materials that are available.

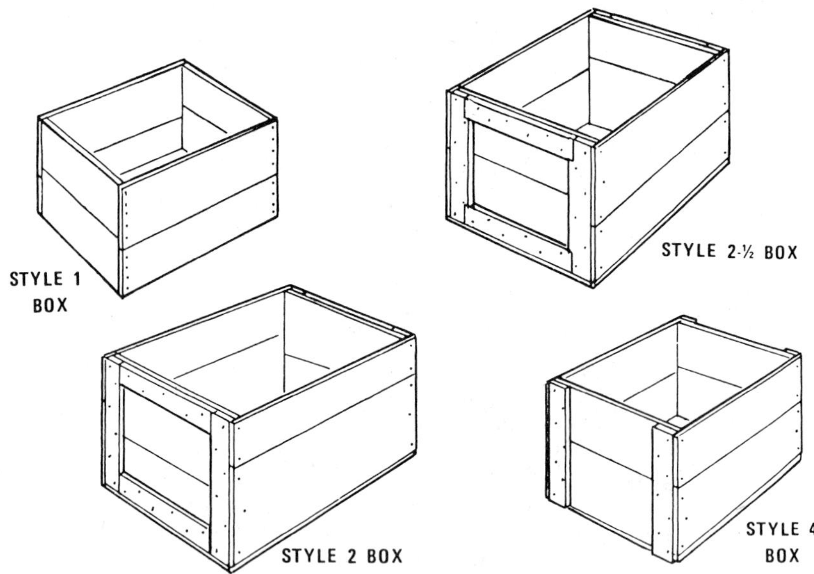

Fig. 5.4.9. Styles of nailed wood boxes.

Expanded polystyrene (EPS) is now widely used as a replacement of corrugated interpacking. In the form of a premolded insert, EPS provides somewhat better shock protection while reducing labor costs in the actual packing operation. EPS is usually used in a density range of 1.2–1.8 pounds per cubic foot (lb/ft^3). In 1984 two new premolded cushioning materials were introduced to the American market. One of them is a polyethylene foam that has better cushioning characteristics than EPS but is considerably more costly, while the other material is a blend of polyethylene and polystyrene offering costs and performance approximately midway between the two materials. Both of these new materials are in limited supply at this time, while EPS is available from a wide number of molders from coast to coast.

When a higher level of cushioning is required, for instance in the range of 15 to 30g, fabricated foams of either polyurethane or polyethylene are often used. The urethanes are best used for lighter products where the static stress on the cushion ranges from 0.03 lb/in.^{2} to 0.2 lb/in.^{2}, while the polyethylene foams are best used in the range of 0.4 to 1.5 lb/in.^{2} Figure 5.4.10 depicts the peak-acceleration–static-stress curves for these materials.

Another variety of foam plastic cushioning is provided by the process known as *foam-in-place*. Two component urethane foam chemicals are mixed in a gun and sprayed into the shipping container to form either a flexible or semirigid cushion around the product. This type of cushioning is best used where there is a wide variety of products entering the packaging or shipping area and the quantity per product is relatively low, precluding the possibility of purchasing premolded forms of cushioning.

Cushioning can also be obtained with special devices such as spring mounts, rubber shear mounts, air-filled "donuts" of rigid plastic, and combinations of wood, foam, and fiberboard materials. The advantage of these specialized cushioning devices is their availability off the shelf, and accompanying well-documented technical information on design and cushioning ability. Their disadvantage is limited source of supply and relatively high cost.

Dunnage—Blocking and Bracing

The most widely used form of dunnage material is loose fill. This may take the form of small particles and pieces of foam plastic, generally EPS or other types of plastic or fiberboard formations. These materials provide some cushioning and do a relatively good job of blocking and bracing low- to medium-weight products within a container. One does have to use care to assure a good distribution of the loose fill material around the product and to be alert to a possibility of settling of the material during long-distance shipments. Other forms of dunnage material used to provide surface protection as well as cushioning and blocking are thin wraps of foam plastic in thicknesses of 1/16–1/4 in. (made from either polyethylene or polypropylene) and polyethylene sheeting with air-encapsulated bubbles. Cellulose wadding is also still used for wrapping and dunnage.

Skin packaging is another form of blocking and bracing used for odd-shaped products, such as household lamps. Here the product is placed on a sheet of corrugated fiberboard and a heated sheet of polyethylene is vacuum formed around the product, holding it down to the corrugated sheet. The skin pack is then placed within a shipping container and held in place with flaps on the corrugated sheet. This type of packing is good for lower volume products where there is a difficult shape to protect or one in which surfaces must be well protected. It is applicable only to small and light-weight products.

Corrosion Protection

To protect ferrous and other materials that may be subject to corrosion, a variety of methods is available. The most widely used method is to coat the corrosion-prone surfaces with a preservative oil. Depending on the application, a number of different preservative oils are available, which may be sprayed, painted, or dip-applied to the surface. The chief drawback of this preservation is the difficulty in removing the oil at the end of the transportation cycle. To reduce the problems of cleanup, materials classified as vapor corrosion inhibitors (VCI) were introduced shortly after World War II. These chemicals give off an invisible vapor which protects any ferrous surface within 18 in. of the emitting chemical. Usually the chemicals are applied to a paper substrate, since this makes it easier to apply in and around the product. It may also be used in the form of small pellets or disks for enclosure within small bags or pouches. The VCI materials have a somewhat shorter life and must be used within enclosed containers. Typically, shelf life of 3–5 years can be expected for a properly sealed container, although water-vapor-sealed liner within the container is not required.

When the highest degree of corrosion protection is required or where preservative oils or VCI cannot be used, the best method of corrosion protection is the use of desiccants and water-vapor-proof sealed barriers. This technique, known as Method II, has been widely used since World War II, principally on electronic components. It is the most difficult to apply, but once properly applied does give the best long-term corrosion protection of any method available.

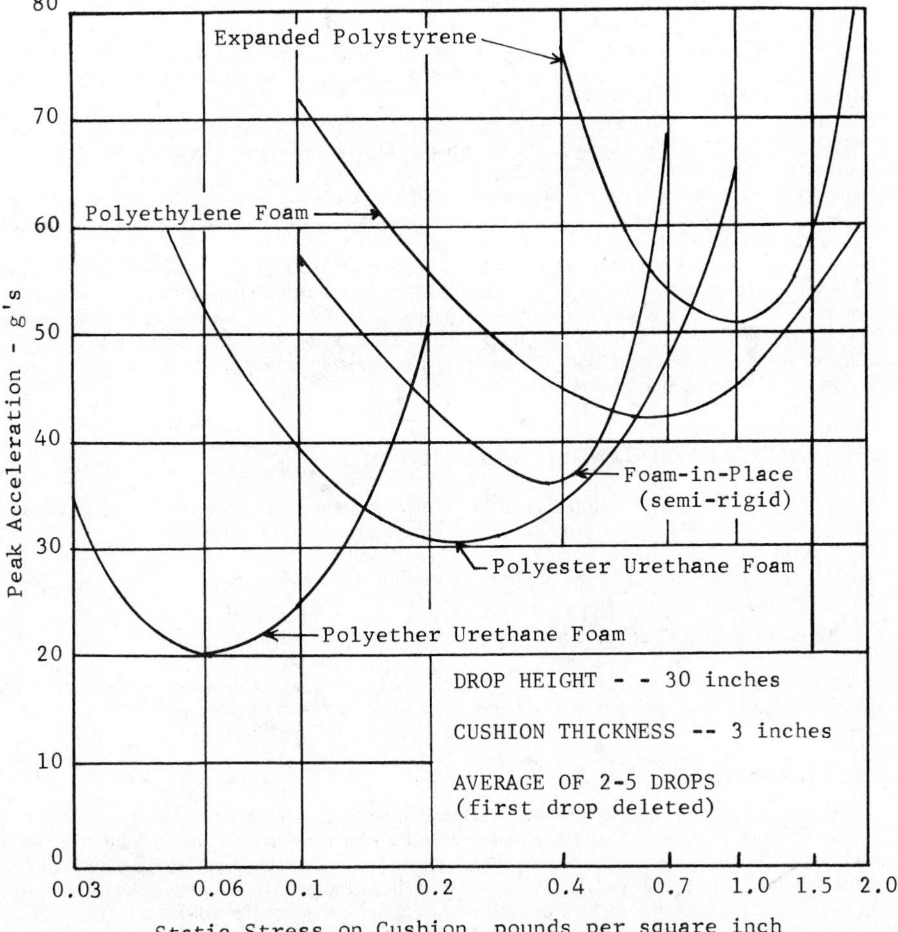

Fig. 5.4.10. Peak acceleration–static stress curves for various cushioning materials.

Electrostatic Discharge Protection

Within the electronics industry, electrostatic discharge (ESD) has become a significant problem, due to the trend toward devices more and more sensitive to lower and lower voltage electrostatic discharge. A number of interior packing materials now available provide ESD protection. These are classified as antistatic, conductive, or shielding, based on their performance in certain specific test procedures. The materials may be treated on the surface or impregnated with conductive substances. Incorporating a foil or metallized layer in a structure by lamination is another means by which conductivity can be achieved.

5.4.4.3 Packaging Design Techniques

The design of packaging should be considered a total systems approach. The following steps are recommended whenever considering any major design or redesign, so that no aspect of the overall problem is overlooked:

1. Define the total distribution system (environment).
2. Know your product.
3. Choose the container and inner packing materials.
4. Design and fabricate prototype packages.
5. Test the prototype packages.

In defining the total system, one must consider the handling within the plant, any warehousing at the plant, the method of shipment, the handling involved at the receiving end, and any storage by the customer or at field warehouses. Each aspect of this total system must have environmental factors assigned to it, such as the degree of shock, vibration, humidity, and temperature that one may reasonably encounter. For the warehousing, one must determine the type and degree of compressive loads that may be applied.

Knowing your product means not only defining the production volumes and dimensions of the product, but also knowing its fragility in terms of shock and vibration, its surface characteristics in terms of its susceptibility to corrosion, electrostatic discharge, etc.

Choosing the proper container and cushioning material will depend on the knowledge of the designer and the range of materials available. The range should be narrowed, once one knows what the environment is (from step 1) and the product characteristics (from step 2).

In the fourth step, one designs and fabricates prototype packages for the various alternatives considered. These prototypes should be as close to production configuration as possible and a sample of at least three of each prototype should be fabricated.

In the last step, test the various prototype packages to determine the degree of protection provided by each one. This testing is described in greater detail in Section 5.4.6. Following all laboratory tests, before integrating into the production cycle, one should always make actual shipping tests via the intended shipping mode to determine final acceptability of any packaging design. Package designers should always keep in mind that packages are only one part of the total distribution system and must fit in with all of the other factors involved such as handling methods, storage equipment, transportation vehicles, and customer facilities.

5.4.5 UNITIZING FOR HANDLING AND STORAGE

When the volume is sufficient, it is often advisable to unitize materials or packages as a means of reducing rough manual handling. Three basic principles for material handling underlie unitization:

1. Handle units that are as large as can be practically lifted and moved.
2. Handle materials as few times as possible.
3. Replace manual labor with mechanical equipment wherever possible.

To design a unitizing system, one must consider material handling and storage equipment to be used, transportation modes, and the packages or products to be shipped. Then a selection from various types of lifting devices and restraining methods is made to develop the most economical unitized load.

5.4.5.1 Lifting Devices

Wood pallets were the first unitizing method developed for the fork truck. Standardization of style and sizes of pallets has been undertaken in the MH-1 Committee of the American National Standards Institute (ANSI).[4] The two most common designs for pallets are two-way entry and four-way entry. *Two-way entry* permits lift truck handling from two opposite ends, while the four-way allows entry on all four sides, either through notches cut in each of the stringers or through a design using blocks as a base for the pallet deck. The six types of pallet generally used in industry are shown in Fig. 5.4.11. For shipping purposes an expendable pallet is usual, unless a general purpose reusable pallet can be guaranteed for return to the shipper. The National Wooden Pallet and Container Association maintains standards for design and construction of the various types of wood pallets.[5]

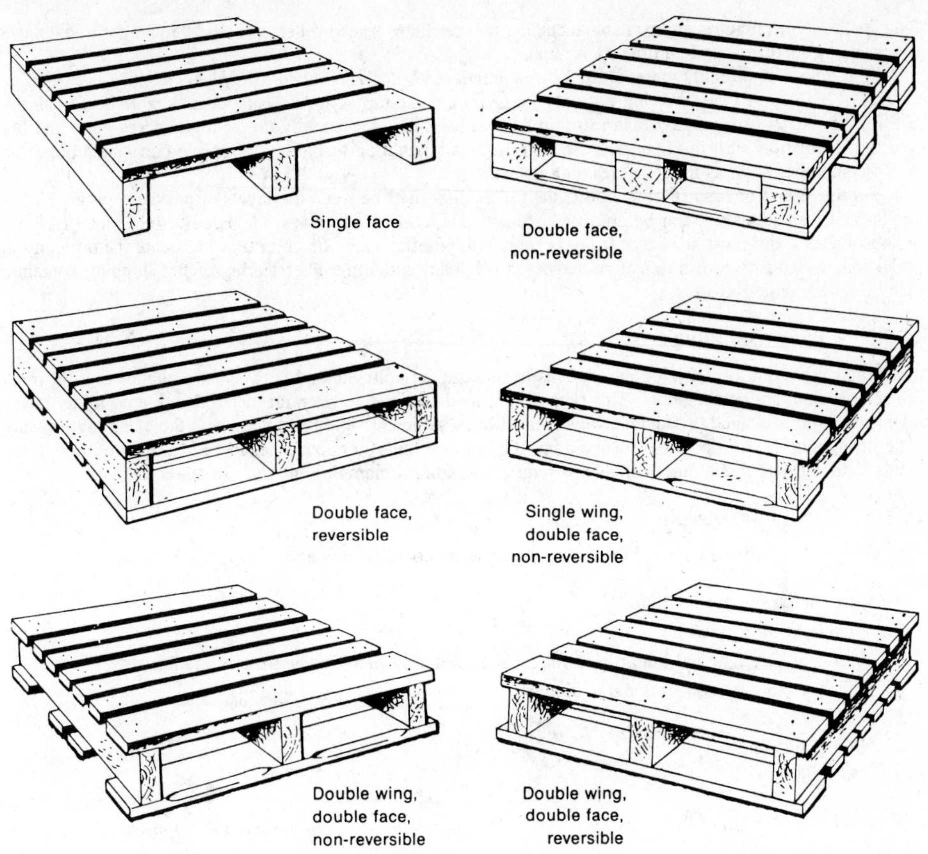

Fig. 5.4.11. Types of wood pallets.

Some pallets are more than a platform; they are part of a complete container. Pallet containers may be used to transport a variety of products, from small parts for assembly to fruits and vegetables from field to cannery. Pallet containers can have wood or fiberboard sides and tops, and they may be either collapsible or noncollapsible. The corrugated "gaylord" box is widely used for bulk storage and handling of raw plastics.

The other lifting device prevalent in industry, especially for unitized handling of food products, is the *slip sheet*. These are handled either with special devices on lift trucks known as "push-pull" attachments or with chisel forks. Slip sheets are usually made of solid fiberboard, corrugated fiberboard, or plastic. Plastic materials are generally used for returnable operations. Slip-sheet materials must have adequate tensile strength, tear resistance, and moisture resistance so that the pull tabs will stay intact during handling. The advantage of a slip sheet for shipping is its light weight and thin profile, which make it possible to load carrier vehicles to full capacity. Their chief disadvantage is the cost of the special handling equipment required. Slip sheets can be used for handling unit loads up to 2000 pounds in weight.

5.4.5.2 Restraining Materials

To stabilize a unit load atop a lifting device such as a pallet or slip sheet, several materials and techniques are available. One of the most widely used is *strapping*, with materials made of either steel or plastic. The plastic strapping materials generally are made of polypropylene, polyester, or nylon. Plastic has the advantage over steel in that it is less expensive, can be heat- or weld-sealed, and will retain a degree of tightness as the package settles. The primary advantage of steel is its high tensile strength and toughness compared to plastic.

Rapidly replacing strapping in recent years, *plastic stretch wrapping* has become a widely used method of restraining unitized loads. For this method, plastic film of polyethylene, polypropylene, or polyvinyl chloride (PVC) is stretched around the sides of the unit load to restrain the packages thereon. The material may be applied in a web of full unit load height or spiral wrapped in a narrower width running up and down the sides of the load. The primary advantage of stretch wrapping over strapping is its ease of application with the proper type of equipment and the more positive restraint of packages from movement during shipment and handling.

Many application techniques are available including prestretching, heated roller, and pass-through type equipment in manual, mechanized, and automated versions.

A third type of restraining material used in some industries is *shrinkable plastic film*. This is usually applied as a bag over the unit load covering not only the sides but the top as well. Using considerable heat to shrink the film, this method of unitized load restraint has not been popular recently due to high energy costs and fire protection regulations. Shrink materials are also used as a bundling technique to replace corrugated boxes on some commodities such as toilet tissues and towels.

A fourth method of load stabilization is the use of *adhesives* between the layers of packages on a pallet or slip sheet. These adhesives can be cold or hot melt. The cost of adhesives compared to other load restraint methods makes it the most attractive from an economic point of view. Its chief drawbacks are the difficulty in application, less positive restraint of packages on the load, and some fiber tearing of the shipping container when boxes are pulled apart.

5.4.5.3 Clamp Handling

An alternative method of unitizing packages is to squeeze the unitized load between two clamps of a lift truck with special clamping attachments. This type of unit load requires no restraint materials nor any lifting device underneath. The packaged product, however, must have sufficient rigidity to withstand the clamping pressure of the lifting device. It is possible for a clamp truck to develop 3000 pounds or more of pressure through the clamps, although regulators are available to reduce and control clamping pressure to lesser amounts.

5.4.5.4 Unit Load Design

When designing a unit load system the following criteria should be a goal:

1. Size and shape are optimum.
2. Materials are low cost.
3. Sufficient strength and integrity to maintain load during handling and transportation.
4. Transportable by conventional equipment.
5. Ease of storage.
6. Minimum weight.
7. Disposability.

With these criteria in mind, the following steps in designing the unit load system are suggested:

1. Determine whether the unit load concept will work. Questions must be answered, such as: is there sufficient volume of product, is there enough movement, can the unit load travel far enough in the distribution process to justify unitization, can the carrier modes handle unit loads, will the receiver accept unit loads, is the product unitizable? Once these questions are answered in the affirmative, then one can move to the second step.
2. Determine the size of the unit load to be used. This will depend on such factors as the types of handling equipment, the type of storage equipment, the size of warehouse storage bays, the size of carrier vehicles, as well as the dimensions of packages to be unitized.
3. Select the type of unit load to be used. Here the choice must be made between a pallet/slip sheet with restraining devices vs. clamp trucking. Selection of the restraining materials and lifting device (pallet vs. slip sheet) must be resolved.
4. Determine the configuration of the unit load and method of building. Here one considers things such as pallet pattern and whether to use column stacking or interlocked layers of boxes. It should be remembered that an interlocked pattern of corrugated shipping containers reduces the overall compression capability of the unitized load as much as 50%. The best way to maximize total compression strength is to be sure that boxes are in straight columns, one corner over the other in each layer. Good stability can be achieved for column stacking by using restraining devices such as stretch wrap or shrink wrap.
5. Test the unit load. The first step is to conduct actual lift truck tests over a defined test course to see if the load can be repeatedly picked up and set down as well as maintaining integrity when turning corners and running over rough terrain. The final test for the unit load should be some actual shipments where loads are sent through the intended distribution system.

5.4.6 TESTING TO VERIFY PROTECTIVE QUALITY

The final step in any design should be a test to determine the adequacy of the design. For material protection, the test procedures have been developed over the years primarily by one group, the ASTM Committee D-10 on Packaging. Since 1914, this committee of volunteer industry experts has developed a great many standard test methods and specifications relative to the measurement of protective quality of packaging and containers.

Although primarily designed for testing packaging for shipment, the same test methods can be used for testing containers for intraplant handling and unit loads for shipping. The packaging test methods are contained in Vol. 15.09 of the *Annual Book of ASTM Standards.*[6]

Until recently the only performance standard for shipping containers was one developed by The National Safe Transit Association and described in their document entitled "Pre-Shipment Test Procedures."[7] For products weighing up to 100 pounds, the test procedure calls for a vibration test followed by a free-fall drop test. For products weighing over 100 pounds, the sequence is a vibration test followed by an inclined impact test. Both procedures include an optional static compression test. The precise method of conducting each test references a particular ASTM test method.

In 1982 ASTM developed a standard practice for *Performance Testing of Shipping Containers and Systems,* their number D4169-82. This practice offers a variety of performance test sequences depending on the particular distribution environment encountered. Each test sequence contains a series of hazard elements in a specified sequence. For instance, for shipment of small packages up to 100 pounds by motor freight or UPS, the typical handling sequence calls for:

1. A drop test to simulate manual handling.
2. A compression test to simulate vehicle stacking.
3. A loose load vibration test to simulate over-the-road travel.
4. A vehicle vibration test to simulate the vehicle input to the package in transportation.
5. Another drop test to simulate manual handling at the receiving end.

Standard D4169-82 contains specific requirements for all the test procedures such as free-fall drop heights, vibration acceleration inputs, and compression design factors.

There is an axiom, "One test is worth 1000 expert opinions." This has been proven many times by experts who have designed "fail safe" packages or shipping container systems and put them through standardized test procedures only to find that they fail. A test should be the last step for any protective system design.

REFERENCES

1. Fred E. Ostrem and W. D. Godshall, "An Assessment of the Common Carrier Shipping Environment," *General Technical Report* FPL22, Madison, WI: U.S. Forest Products Laboratory, 1979.
2. L. O. Anderson and T. B. Heebink, "Wood Crate Design Manual," *Agricultural Handbook #252,* Madison, WI: U.S. Forest Products Laboratory, 1964.
3. "Code of Federal Regulations, Title 49—Parts 100 to 199," Superintendent of Documents, Washington, DC: U.S. Government Printing Office, 1983.
4. "Pallet Sizes," MH-1.2., New York: American National Standards Institute, 1975.
5. "Specifications and Grades for Hardwood Warehouse, Permanent, or Returnable Pallets," Washington, DC: National Wooden Pallet Container Association, 1962.
6. *Annual Book of ASTM Standards, Packaging Test Methods,* Vol. 15.09, Philadelphia: ASTM, 1984.
7. "Pre-Shipment Test Procedures," Chicago, IL: National Safe Transit Association, 1979.

BIBLIOGRAPHY

Fibre Box Handbook, Chicago: Fibre Box Association, 1979.

W. F. Friedman and J. J. Kipnees, *Distribution Packaging,* Huntington, NY: Krieger Publishing, 1977.

"How to Protect Parts in In-Process Containers," *Modern Materials Handling,* December 1980.

Military Packaging Engineering, MIL-HDBK-772, Washington, DC: U.S. Government Printing Office, 1981.

Packaging Cushioning Design, MIL-HDBK-304B, Washington, DC: U.S. Government Printing Office, 1978.

The Packaging Encyclopedia, Cahners, Boston, 1984.

PART 6

MONEY: FINANCIAL MANAGEMENT

CHAPTER 6.1

ECONOMIC ANALYSIS FOR PRODUCTION PLANNING

FERDINAND F. LEIMKUHLER
FREDERICK T. SPARROW

Purdue University
West Lafayette, Indiana

6.1.1 INTRODUCTION

Production economics is the study of how producers combine costly inputs to make profitable outputs. For analytic purposes, production planning can be divided into three stages of decision making as follows:

1. *Short-run input decisions.* How much of the factors of production to use so as to produce a specified output at minimum cost, given a specified plant capacity—the so-called mix problem.
2. *Short-run output decisions.* How much output to make (and input to use) so as to maximize net profit or to achieve some other production objective given a specified plant capacity—the so-called scale problem.
3. *Long-run capacity decisions.* How to schedule long-term investments in production facilities so as to optimize the return on investment—the so-called capacity expansion problem.

These decisions must take into account both the physical capabilities of the production processes used and the market conditions associated with the purchase of inputs, the sale of outputs, and the acquisition of capital funds.

Economic analysis begins with a determination of the minimum-cost way of producing an output in the short run by considering the productivity characteristics of the process and the cost of the inputs. The solution to this mix problem provides a basis for determining how the producer responds to short-run market demand for the output in order to achieve sales objectives, and also how the producer responds to changes in the short-run cost of inputs. Finally, by taking a long-run view of these input-output relationships, a framework is developed for making optimal investment decisions in production facilities. In the first two stages of the analysis, time is treated as an implicit dimension by assuming that input and output activities occur at steady rates and can be adjusted instantaneously. However, when making capacity decisions, the element of optimal timing and construction delays becomes an important explicit aspect of the problem.

The viewpoint taken in this analysis is that of an optimizer, that is, a decision maker who acts so as to achieve a stated objective to the greatest degree possible. Although this goal may not be realized in practice for various reasons, it provides a consistent basis for making comparisons and developing a comprehensive theory of production. This survey of production theory is limited to the more general concepts and models that have been developed. The references listed at the end of the chapter give a thorough treatment of these ideas and include important extensions and applications of the models. The decision rules that emerge from the theory identify the important elements of production decisions, the kinds of data that are needed for decisions, and available methods for processing production information to guide managerial decisions. Simplified numerical examples are used throughout this chapter to demonstrate how analytic methods might be developed and applied in practical situations.

6.1.2 PRODUCTIVITY OF INPUTS

The output rate of a production process depends on the input rates of the factors used and how they combine to produce outputs, expressed mathematically in a production function of the form $q = q(x_1, x_2, \ldots, x_m)$, where q is output rate, and $x_1, x_2, \ldots, x_m$ are input rates for each of the m inputs. The relative efficiency with which a process uses an input is measured by the input's average productivity, that is, output rate divided by input rate, and by its marginal productivity, that is, the rate of change in output rate with a change in input rate.

In the analysis of production systems, the marginal productivity of an input can be represented by the partial derivative of the production function. A production function often used in economic analysis is the Cobb-Douglas function,[3] which has the form

$$q = a(x_1^{b_1})(x_2^{b_2}) \cdots x_i^{b_i} \cdots x_m^{b_m}$$

or

$$\log q = \log a + b_1 \log x_1 + \cdots + b_m \log x_m$$

In this instance, we have

$$q_i' = \frac{\partial q}{\partial x_i} = \frac{b_i q}{x_i} = b_i \bar{q}_i$$

$$q_i' = \text{marginal productivity of input } i \tag{6.1.1}$$

$$\bar{q}_i = \text{average productivity of input } i$$

$$b_i = \text{the productivity coefficient of the } i\text{th input}$$

Note that the logarithmic form of this production function is useful for statistical estimation by means of linear regression methods.

How output changes when all factors are changed proportionally is of special interest. A production process is said to be *homogeneous of degree k* if

$$q(\lambda x_1, \lambda x_2, \ldots, \lambda x_m) = \lambda^k_q(x_1, x_2, \ldots, x_m) \tag{6.1.2}$$

Note that all such functions must pass through the origin. A production process has decreasing, constant, or increasing *returns to scale*, depending on whether k in Eq. 6.1.2 is less than, equal to, or greater than unity, respectively. Production functions with constant returns to scale have marginal productivities of degree zero, which means that input productivities are unchanged for proportionate changes in all inputs. Note further that

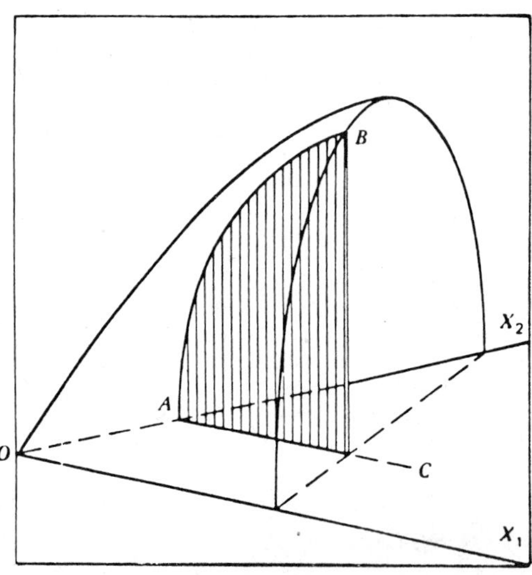

Fig. 6.1.1. Productivity curve $A-B$ when input x_1 changes and input x_2 is constant along expansion path $A-C$.

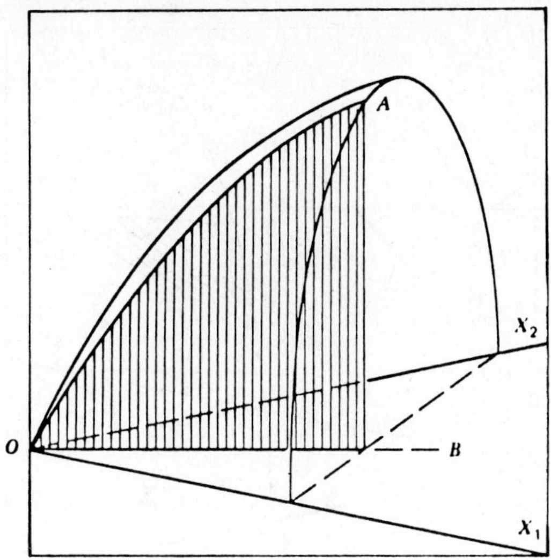

Fig. 6.1.2. Returns to scale along output curve O–A when inputs x_1 and x_2 change proportionally along expansion path O–B.

if one takes the partial derivative with respect to λ of both sides of Eq. 6.1.2, for a constant returns-to-scale function, we obtain,

$$q = q_1'x_1 + q_2'x_2 + \cdots + q_m'x_m \tag{6.1.3}$$

An example of such a process is the following production function

$$\begin{aligned} q &= ax_1^{b_1}x_2^{b_2} = q_1'x_1 + q_2'x_2 \\ &= \left[\frac{b_1q}{x_1}\right]x_1 + \left[\frac{b_2q}{x_2}\right]x_2 \end{aligned} \tag{6.1.4}$$

which implies that

$$\frac{dx_2}{dx_1} = \frac{b_1x_2}{b_2x_1} \tag{6.1.5}$$

The productivity characteristics of this process are plotted for the normal production region in Figs. 6.1.1–6.1.2 for $b_1 + b_2 < 1$. Note that the marginal productivity falls as each input rate is raised individually. These are examples of the economic law of diminishing returns.

6.1.3 SUBSTITUTION OF INPUTS

Those combinations of input rates that yield the same output rate define a constant output curve, or *isoquant*, of a process, as shown in Fig. 6.1.3. The ratio of the marginal productivities of two inputs measures their rate of technical substitution along an isoquant and is equal to the slope of the isoquant times -1. For the previous example, in Eq. 6.1.4, if $b_1 = b_2 = ½$, the marginal productivity ratio q_1'/q_2' equals x_2/x_1, which is the rate of technical substitution of input 2 for input 1. For example, when the process is using 100 units of x_1 and 400 units of x_2, the output level of 400 units of q can be maintained by substituting 4 units of x_2 for 1 unit of x_1. Note that in Fig. 6.1.4, the isoquant is not a straight line—the rate of technical substitution changes as the input ratio changes. When the rate of technical substitution is a positive constant for all input ratios, the inputs are said to be *perfect substitutes*, and when it equals zero, there is no substitution.

An important example of no substitution is in processes where inputs are used in fixed proportions. Such a process is called a linear process with fixed technical coefficients, [1,4] or a Leontief production process, in honor of the Nobel-prize-winning economist who pioneered the use of such processes in describing the input/

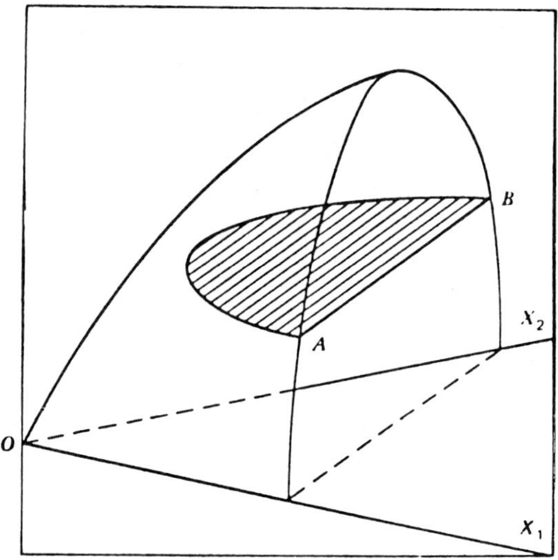

Fig. 6.1.3. Isoquant curve $A-B$ when inputs x_1 and x_2 change so as to keep the output constant.

output structure of economies. The production function of a linear process can be written as follows, where a_i denotes the fixed amount of input i required per unit of output q.

$$q = \text{minimum}\left[\frac{x_1}{a_1}, \frac{x_2}{a_2}, \ldots, \frac{x_m}{a_m}\right] \tag{6.1.6}$$

An example of two linear processes, P_1 and P_2, with only two inputs is shown in Fig. 6.1.5, where the isoquants are L-shaped with corner points on a process vector, that is, the diagonal line from the origin with slope a_2/a_1. Along the process vector the input rate x_i equals a_iq, which is the minimum amount of x_i needed to make q, and the marginal products are zero. If there is an excess of one input, its marginal product is zero, and if there is a shortage of one input, its marginal product is $1/a_i$.

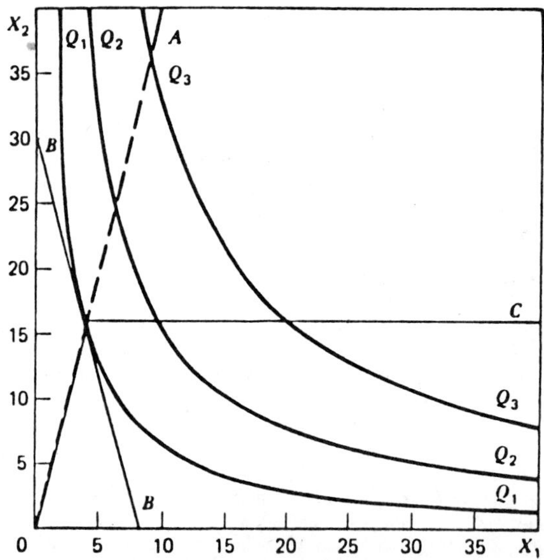

Fig. 6.1.4. Isoquant curves $Q-Q$ for a nonlinear production system with a typical budget line $B-B$, expansion path $O-A$ when inputs are unlimited, and expansion path $O-C$ when input x_2 is limited to 16 units.

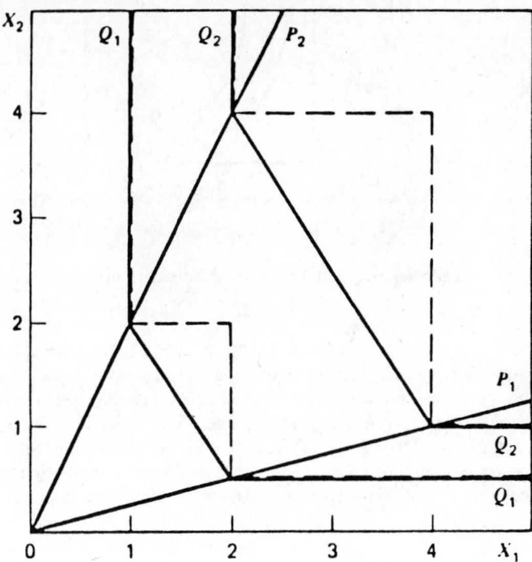

Fig. 6.1.5. Isoquant lines Q–Q for a linear production system with two processes having expansion paths O–P_1 and O–P_2 and used either simultaneously (solid line) or one at a time (dashed line).

When two or more linear processes are used to make the same output, they form a linear production system with coefficients a_{ij} denoting the minimum amount of input i required to make one unit of q by process j. Substitution of processes may or may not be possible. When there is no substitution and processes are used one at a time, the production function has the form

$$q = \text{maximum}_j \ \text{minimum}_i \left[\frac{x_i}{a_{ij}} \right] \tag{6.1.7}$$

That is, the best output obtainable from each process is determined first and then the best of these is used. The isoquant for such a system has a step pattern consisting of segments of the L-shaped isoquants of the processes as shown in Fig. 6.1.5. The case where process substitution can occur by using the processes jointly is discussed in Section 6.1.5.

6.1.4 TECHNICAL CHANGE

Production functions change over time as technical progress increases the productivity of the inputs. Since much of the discussion hinges on the sources of such productivity increases—capital or labor—it will be useful to specialize the production function $q = Q(x_1, \ldots ,x_m)$ to $q = Q(K,L)$ where K = capital and L = labor. Two types of technical progress can be identified: *disembodied* change, where the scientific advance applies equally to all resources in current use, and *embodied* technical progress, where the advance is contained only in new capital equipment. In the first case, the production function can be written as $q = Q(K,L,t)$ where t is the time subscript. Embodied technical progress requires the capital stock to be vintaged by the year of purchase. Thus, the production function must be of the form $q = Q(I_0,I_1, \ldots ,I_t,L)$ where I_t = purchases of capital in period t.

In the case of disembodied technical progress, there remains the issue of its direction—what combination of capital and labor maximizes productivity? Three forms of disembodied progress need to be identified:

Neutral change, where the production function is $q = \alpha(t)Q(K,L)$.
Capital saving, where the function is $q = Q[\alpha(t)K,L]$.
Labor saving, where the function is $q = Q[K,\alpha(t)L]$.

The three can be visualized as alternative ways of shifting the position of the isoquants of a production function, as illustrated in Fig. 6.1.6.

In all three diagrams, shifts in the mixes of capital and labor required to produce a fixed output level are plotted before (t = 0) and after (t = 1), the invention. In the case of neutral change, the amounts of capital

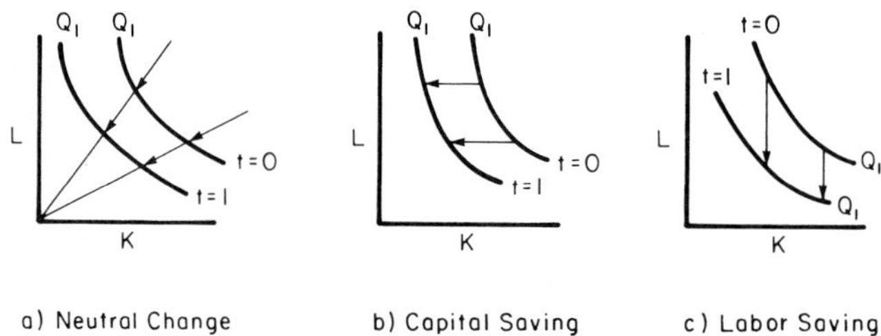

Fig. 6.1.6. Direction of technical change.

and labor are both reduced—the isoquant shifts toward the origin. In the case of capital-saving change, the isoquant shifts to the left, while labor-saving change shifts the isoquant downward. Note that all can be thought of as changing the scales on the axes. The direction of technical change may seem to be an arcane subject until it is realized that it plays a dominant role in labor management collective bargaining. If labor can show that productivity increases were due to the labor force becoming more productive through more education, etc. (see Fig. 6.1.6*c*), then they can argue effectively for wage increases based on increased productivity. Conversely, if management demonstrates that productivity increases are due to the equipment becoming more productive, the case for wage increases is weakened.

6.1.5 LINEAR PRODUCTION SYSTEMS

Where the same output is produced using two or more distinct linear processes, input substitution occurs when different combinations of the processes are used. If q_j denotes the output rate of process j, the sum of the q_j's is the total output (q) of the system, and a_{ij} is the amount of input i per unit of output using process j, then the production function for a linear system is defined by the following linear programming formulation:

$$q = \text{maximum } (q_i + q_2 + \cdots + q_n)$$

where

$$\begin{aligned} q_j &\geqslant 0 \qquad \text{for } j = 1,2, \ldots, n \\ x_1 &\geqslant a_{11}q_1 + a_{12}q_2 + \cdots + a_{1n}q_n \\ &\vdots \\ x_m &\geqslant a_{m1}q_1 + a_{m2}q_2 + \cdots + a_{mn}q_n \end{aligned} \tag{6.1.8}$$

No more than m of the n processes are used at the same time in the linear programming solution.[1,4,5]

An example of a linear production system with one product, two resource inputs, and two linear processes, each of which produce the output, is one with the following production function:

$$\begin{aligned} q &= \text{maximum } (q_1 + q_2)\ ; \quad q_1 \geqslant 0; \quad q_2 \geqslant 0 \\ x_1 &\geqslant 1.00q_1 + 0.52q_2 \\ x_2 &\geqslant 0.25q_1 + 1.00q_2 \end{aligned} \tag{6.1.9}$$

For this system, an isoquant with constant output level q consists of three line segments, as shown in Fig. 6.1.5 and as defined by the following equations:

$$\begin{aligned} x_1 &= a_{12}q = 0.5q, \qquad \text{when } x_2 \geqslant a_{22}q = q \\ x_2 &= a_{21}q = 0.25q, \qquad \text{when } x_1 \geqslant a_{11}q = q \\ x_2 &= \frac{(a_{11}a_{22}q - a_{12}a_{12}q - a_{22}x_1 + a_{21}x_1)}{(a_{11} - a_{12})} \\ x_2 &= 1.75q - 1.5x_1, \qquad \text{when } x_1 \text{ and } x_2 \text{ are otherwise} \end{aligned} \tag{6.1.10}$$

The latter equation defines the isoquants in the triangular region between the two process vectors P_1 and P_2 in Fig. 6.1.5, that is, where the input ratio x_2/x_1 is between $a_{21}/a_{11} = 0.25$ and $a_{22}/a_{12} = 2$ for this example. This is the normal production region of a linear system, since otherwise there are excess amounts of input.

In the normal production region, the marginal products and the technical rate of substitution are as follows:

$$q_1' = \frac{a_{22} - a_{21}}{(a_{11}a_{22} - a_{12}a_{21})} = \frac{6}{7}$$

$$q_2' = \frac{a_{11} - a_{12}}{(a_{11}a_{22} - a_{12}a_{21})} = \frac{4}{7} \qquad (6.1.11)$$

$$\frac{q_1'}{q_2'} = \frac{a_{22} - a_{21}}{a_{11} - a_{12}} = 1.5$$

The constant rate of technical substitution indicates perfect substitution along a linear isoquant. Also, in the normal production region, the output rates are given by the following equations:

$$q = \frac{(a_{22}x_1 - a_{21}x_1 + a_{11}x_2 - a_{12}x_2)}{(a_{11}a_{22} - a_{12}a_{21})}$$

$$= a_1'x_1 + q_2'x_2 = \frac{(6x_1 + 4x_2)}{7}$$

$$q_1 = \frac{(a_{22}x_1 - a_{12}x_2)}{(a_{11}a_{22} - a_{12}a_{21})} = \frac{(8x_2 - 2x_1)}{7} \qquad (6.1.12)$$

$$q_2 = \frac{(a_{11}x_2 - a_{21}x_1)}{(a_{11}a_{22} - a_{12}a_{21})} = \frac{(8x_2 - 2x_1)}{7}$$

These equations are the general solutions to the linear programming problem of Eqs. 6.1.9 in the normal region. Equations 6.1.11 for the marginal products are the optimal values of the corresponding dual variables of the linear programming problem. This analysis can be extended to linear systems with more than two inputs and more than two linear processes.

6.1.6 PRODUCTION COST FOR LINEAR SYSTEMS

The cost of production depends on the technology used and the cost of the inputs. In linear systems, if the cost per unit of each input is fixed, then the cost per unit of output of each process is also fixed under normal conditions. If there is no limitation on the use of inputs and processes, the cheapest process is used, and production cost is equal to the minimum process cost. However, when there are limitations on the amount of input or on the use of a process, increased output requires the use of more expensive processes. The marginal and the average costs of production go up. Under these conditions, a general production cost model is the following linear program, where k_i denotes the cost per unit on input i, c_j denotes the cost per unit of output q_j from process j (exclusive of input costs), and x_i are the available amounts of input:

$$c_q = \text{minimum } (c_1q_1 + c_2q_1 + c_2q_2 + \cdots + c_nq_n)$$

$$c_j = k_1a_{1j} + k_2a_{2j} + \cdots + k_ma_{mj}$$

$$q \leq q_1 + q_2 + \cdots + q_n, \qquad q_i \geq 0$$

$$x_1 \geq a_{11}q_1 + a_{12}q_2 + \cdots + a_{1n}q_n \qquad (6.1.13)$$

$$x_2 \geq a_{21}q_1 + a_{22}q_2 + \cdots + a_{2n}q_n$$

$$x_m \geq a_{m1}q_1 + a_{m2}q_2 + \cdots a_{mn}q_n$$

The cost of producing output q can be found by solving the linear program for all values of q from 0 to the maximum output defined by Eq. 6.1.6. This cost equation will consist of a series of line segments as shown in Fig. 6.1.7.

In the previous example of a linear system with the production function of Eqs 6.1.9, if inputs x_1 and x_2 both cost \$4/unit, then the output from process 1 costs \$5/unit, and the output from process 2 costs \$6/unit.

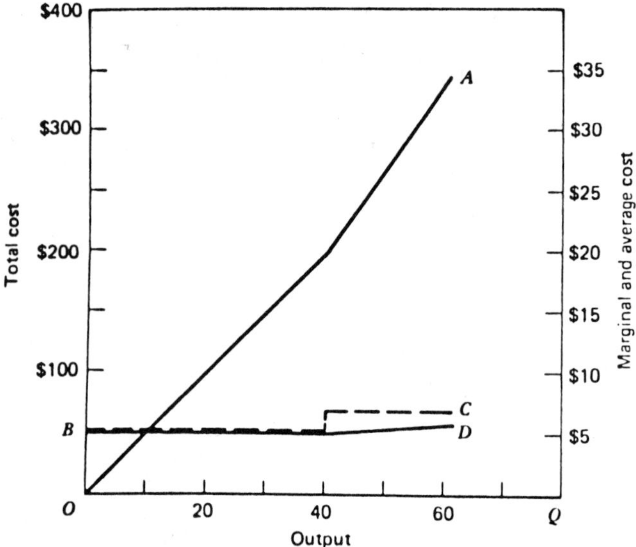

Fig. 6.1.7. Total production cost curve $O-A$, marginal cost $B-C$, and average cost $B-D$ for a linear production system with input limitations.

Furthermore, if there are 40 units of x_1 and 45 units of x_2 available, then the production cost can be found by solving the linear program

$$\begin{aligned} c_q &= \text{minimum } (5q_1 + 6q_2) \\ q &\leqslant q_1 + q_2, \quad q_1 \geqslant 0, \quad q_2 \geqslant 0 \\ 40 &\geqslant 1.00q_1 + 0.50q_2 \\ 45 &\geqslant 25q_1 + 1.00q_2 \end{aligned} \tag{6.1.14}$$

for output rates from 0 to 60. The resulting cost curves are plotted in Fig. 6.1.7, where production cost equals $5q$ for output q between 0 and 40, and equals $7q - 80$ when output is between 40 and 60 units. For output rates up to 40 units, the percentage change in cost is the same as the percentage change in output, but when output is between 40 and 60 units, the percentage change in cost is greater than the percentage change in output.

It is no accident that the cost curve in Fig. 6.1.7 shows diminishing or constant returns to scale; it can be shown that all such curves for linear systems exhibit this property. Indeed, the major drawback of such a formulation is its inability to capture scale economies.

6.1.7 PRODUCTION COST FOR NONLINEAR PRODUCTION SYSTEMS

Given a production function of the form $q = Q(x_1, \ldots, x_m)$ and costs per unit of all inputs, what mix of inputs minimizes the costs of output for any level of output? The problem can be expressed as:

minimize

$$\sum_{i=1}^{m} c_i\,(x_i) = k(x)$$

subject to

$$Q(x_1, \cdots, x_m) = q \tag{6.1.15}$$

where $c_i(x_i)$ = cost of utilizing x_i units of input i.

A general method of finding a minimum production cost is the Lagrangian method used in constrained optimization problems.[1,2] The Lagrangian has the form

$$L(x,v) = k(x) + v[q - Q(x)] \tag{6.1.16}$$

where

x = set of variable input rates, $(x_1, \ldots, x_m)$

v = Lagrangian variable

$k(x)$ = cost as a function of x

$Q(x)$ = output as a function of x (i.e., the production function)

q = some required output level

Those values of the variables that minimize cost usually are found by setting the partial derivatives of the Lagrangian function with respect to the x_i's and v equal to zero and then solving these equations simultaneously.

When the cost function is linear, the *solution* of the Lagrangian function reduces to the following optimality conditions:

$$L(x,v) = k_1x_1 + k_2x_2 + \cdots + k_mx_m + v[q - Q(x)]$$

$$v = \frac{k_1}{q'_1} = \frac{k_2}{q'_2} = \frac{k_m}{q'_m}, \quad \text{when} \quad \frac{\partial L}{\partial x_i} = 0 \tag{6.1.17}$$

$$\frac{k_h}{k_i} = \frac{q'_h}{q'_i} \quad \text{for any pair of inputs}$$

At optimum, the marginal input cost is the same for all inputs and is equal to v, which can be shown to be the minimum marginal cost of production; that is, v equals c'_q when q is being produced at minimum cost. The rate of technical substitution q'_h/q'_i for any pair of inputs along the q-isoquant is equal to the ratio of their unit costs k_h/k_i; that is, the isocost line is just tangent to the isoquant, as shown in Fig. 6.1.4. The locus of the tangency points for all output levels defines the "expansion path" of the system, and defines the optimal mix of inputs to use for all levels of output.

The cost problem can be solved parametrically for q in order to generate a (minimum) cost function, $c(q)$, which gives the costs of production along the expansion path. To illustrate, suppose the production function is Cobb-Douglas of the form $ax_1^{b_1}x_2^{b_2} = q$, and exhibits constant returns to scale—that is, $b_1 + b_2 = 1$—then Eq. (6.1.15) specializes to:

minimize

$$c = c_1x_1 + c_2x_2$$

subject to

$$ax_1^{b_1}x_2^{1-b_1} = q \tag{6.1.18}$$

The Lagrangian conditions become

$$\begin{aligned} q &= ax_1^{b_1}x_2^{1-b_1} \quad \text{(the production function)} \\ 0 &= (1 - b_1)\,c_1x_1 - b_1c_2x_2 \quad \text{(the expansion path)} \end{aligned} \tag{6.1.19}$$

Using Eqs 6.1.18 and 6.1.19, we obtain

$$\hat{x}_1 = \frac{b_1c}{c_1} \qquad \hat{x}_2 = \frac{(1 - b_1)c}{c_2}$$

which, when substituted in Eq. 6.1.19, becomes

$$q = a\left[\frac{b_1c}{c_1}\right]^{b_1}\left[\frac{(1 - b_1)c}{c_2}\right]^{1-b_1}$$

or

$$c(q) = \frac{q}{a}\left[\left(\frac{c_1}{b_1}\right)^{b_1}\left(\frac{c_2}{1 - b_1}\right)^{1-b_1}\right] = \bar{c}q \tag{6.1.20}$$

Note that the constant marginal cost arises from the assumption of constant returns to scale. If $b_1 + b_2 > 1$, then the cost function would have been concave from below; if $b_1 + b_2 < 1$, then it would have been concave from above.

Additional requirements or limitations on the system can be included in the analysis by adding terms to the Lagrangian function. For example, let a production system with a Cobb-Douglas-type production function with $b_1 = 1 - b_1 = ½$ have an upper limit on the input rate x_2 of 16 units, and let there be a fixed cost rate of $48 and a cost of $4/unit of x_1 and $1/unit of x_2. Then, the Lagrangian function and its solution are as follows:

$$L(x,v) = 4x_1 + x_2 + v_1(q - 2x_1^{1/2}) + v_2(16 - x_2)$$

$$q \leqslant 16; \quad x_2 = 4x_1 = q\ ; \quad x_1 = \frac{q}{4}\ ; \quad v_1 = 2\ ; \quad v_2 = 0$$

$$c_q = 48 + 2q\ ; \quad c_q' = \bar{c}_q = 2 + 48/q \tag{6.1.21}$$

$$q \geqslant 16\ ; \quad x_2 = 16\ ; \quad x_1 = q^2/64\ ; \quad v_1 = q/8\ ; \quad v_2 = 1 - q^2/64$$

$$c_q = 64 + q^2/16\ ; \quad c_q' = q/8\ ; \quad \bar{c}_q = 64/1 + q/16$$

The cost functions are plotted in Figs. 6.1.8 and 6.1.9. Note that at optimum, v_1 is equal to the marginal cost of production, and v_2 is equal to the marginal cost of the constraint on x_2; that is, by increasing the x_2 limit, the cost of production would be decreased marginally by an amount equal to v_2. The Lagrangian variables are equivalent to the "shadow prices" of the constraints. In Fig. 6.1.4 the expansion path of the system follows the boundary of the x_2 constraint for outputs of 16 units or more.

In application, engineering production functions seldom contain "economic" variables, commodities that can be bought or sold. Rather, outputs are expressed in terms of engineering variables, which themselves determine the values of the economic variables. To illustrate, consider the problem of constructing a natural gas pipeline of fixed length. If pipeline capacity is expressed in millions of cubic feet per day, then capacity

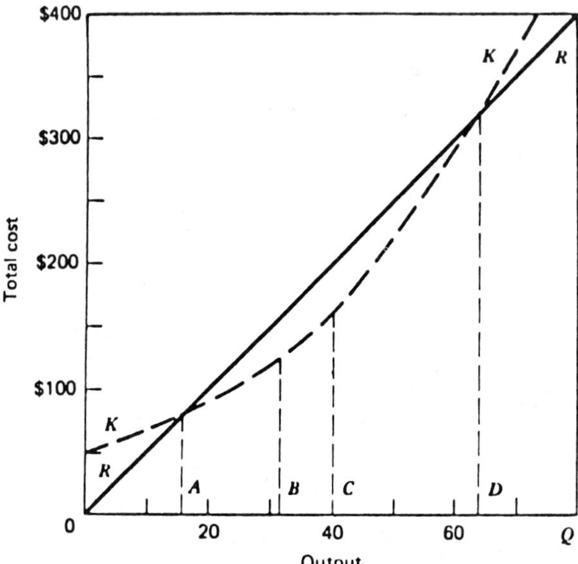

Fig. 6.1.8. Production cost curve $K–K$ for a nonlinear production system, total revenue curve $R–R$ at $5/ unit, and four typical output levels.

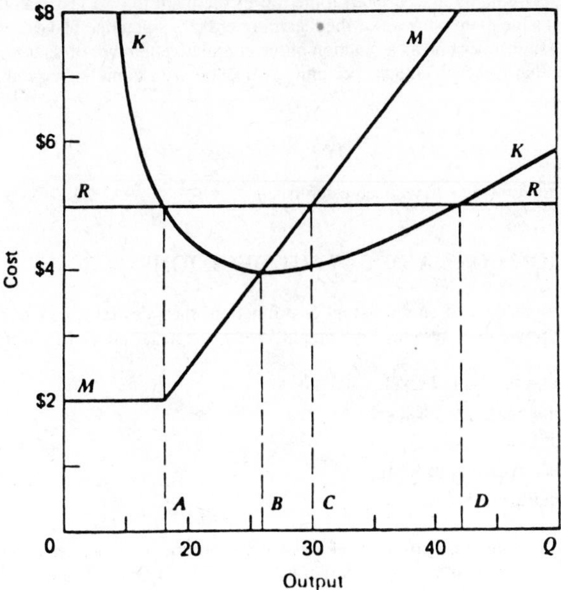

Fig. 6.1.9. Average cost curve K–K and marginal cost curve M–M for a nonlinear production system, with marginal and average revenue curve R–R and four typical output levels.

X is a function of the pipe thickness T, compression ratio R, and inside pipe diameter D, all engineering variables:

$$X = K_1 D^{5/3} T \sqrt{1 - 1/R^2} \tag{6.1.22}$$

The engineering variables, T, R, and D, determine the two economic variables, horsepower and pipe size, according to the following relationships,

$$\begin{aligned} \text{required horsepower} &= \text{H.P.} = (K_2 R - K_3) X \\ \text{required pipe} &= P = K_4 D T \end{aligned} \tag{6.1.23}$$

If c_1 and c_2 are the unit costs of horsepower and pipe, then the cost-minimizing mix of pipe and horsepower is given by the solution to the problem.

$$\min (C_1)(\text{H.P.}) + (C_2)(\text{P}) \tag{6.1.24}$$

$$\begin{aligned} \text{s.t.}\quad \text{H.P.} &= (K_2 R - K_3) X && \text{equations defining economic} \\ P &= K_4 D T && \text{variables in terms of the engineering variables} \\ x &= K_1 D^{5/3} T \sqrt{1 - 1/R^2} && \text{production function expressed in terms of the engineering variables} \end{aligned}$$

Cost functions usually are expressed in terms of total cost per period c, as a function of total output rate per period, q:

$$c = c(q)$$

Of the many other factors that also determine cost, two are so pervasive as to warrant mention: total contemplated output and cumulative output. Usually, costs of production as a function of the rate of production decrease for all rates as total contemplated output and/or cumulative output increase. Thus, we would expect the cost of producing 100 units a day to be reduced if total contemplated output was 100,000 units, rather than 10,000 units, since the scale economies present in constructing a plant that would produce 100,000 units would not be present in a plant designed to produce 10,000 units. Further, given a total contemplated output and a rate per day, we would expect units produced later in the life of the plant to cost less than those produced early in the production history of the plant because of the "learning effect"—practice makes perfect. The literature on progress curves—that is, cost per unit as a function of the cumulative number of units produced—is enormous. Most studies conclude that direct manhours per unit, y, decline with cumulative production, x, according to the relationship

$$y = ax^b > 0 < b < 1 \tag{6.1.25}$$

but more complicated relationships have been established.

6.1.8 PRODUCTION COST AND OUTPUT DECISIONS

Production output decisions depend on the output cost function, the sales revenue function, and the objectives of the managers of the production system. Five output levels of special interest to decision makers are:

A. Minimum break-even output level.
B. Minimum average cost output level.
C. Maximum profit output level.
D. Maximum break-even output level.
E. Maximum sales revenue level.

Level A is important for planning purposes, level B is a usual production objective at the plant level, and level C normally is assumed to be the objective at corporate level. Levels D and E are alternative and appropriate objectives for some production systems. These output decisions are shown graphically in Figs. 6.1.8 and 6.1.9 for the production system in the previous example, with cost curves defined in Eqs. 6.1.21 and with a fixed selling price of $5/unit of output q.

In this example, levels D and E are the same. The conditions for the five output levels can be summarized as follows:

$$\text{A. } \bar{c}_q = \bar{r}_q, \quad \bar{c}'_q < \bar{r}'_q \tag{6.1.26}$$

$$\text{B. } \bar{c}_q = c'_q, \quad c''_q > 0 \tag{6.1.27}$$

$$\text{C. } \bar{c}_q < \bar{r}_q, \quad c'_q = r'_q, \quad c''_q > r''_q \tag{6.1.28}$$

$$\text{D. } \bar{c}_q = \bar{r}_q, \quad \bar{c}'_q > \bar{r}'_q \tag{6.1.29}$$

$$\text{E. } \bar{c}_q \leqslant \bar{r}_q, \quad \bar{r}_q = 0, \quad r''_q < 0 \tag{6.1.30}$$

At break-even level A, average cost $\bar{c}_q$ equals average revenue $\bar{r}_q$ and average cost is changing at a smaller rate, $\bar{c}'_q$, than is average revenue $\bar{r}'_q$. Similar conditions hold at the maximum break-even level D, except the latter condition is reversed. At the minimum profit level C, the marginal cost c'_q and marginal revenue r'_q are equal, and marginal cost is changing at a higher rate c''_q than that of marginal revenue r''_q. Also, to exclude a loss situation, the average cost is not greater than average revenue. In level E, marginal revenue equals zero and is decreasing.

With a fixed selling price, as in the preceding example, the marginal revenue is equal to the selling price, and a profit-maximizing producer supplies that output where selling price equals the marginal cost. For this reason, the marginal cost curve also defines the producer's short-run supply curve, that is, the amount supplied for a fixed price. When the selling price depends on the amount sold, the demand curve, that is, average revenue as a function of output or $\bar{r}_q$, is not a horizontal line. In this case, the profit-maximizing producer still equates marginal cost with marginal revenue, but the supply curve as such is not readily defined. In highly competitive markets, there is a long-run trend for the surviving producers to be producing at the minimum average cost in the industry, with no excess revenue over full cost.

6.1.9 DEMAND AND EXPENDITURE FOR INPUTS

A producer's demand and expenditure for inputs depends on the input prices, the output cost and revenue functions, and the producer's objective, which can be one of the following:

A. Minimize the cost of a given output level.
B. Maximize output with a given cost budget.

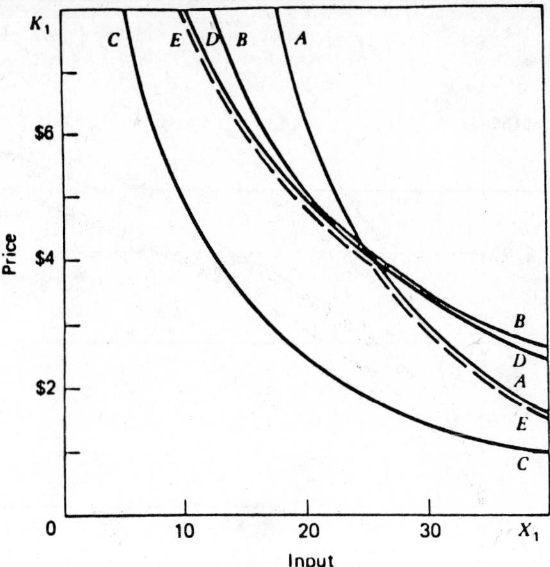

Fig. 6.1.10. Input price-demand curves for a nonlinear production system under five different production objectives.

C. Maximize profit.
D. Maximize the break-even output rate.
E. Maximize sales revenue and avoid a loss.

Objectives, C, D, and E correspond to output levels C, D, and E in the previous section, but not A and B. Input demand curves showing the amount of input needed as a function of its price per unit and the producer's objectives are shown in Fig. 6.1.10. Expenditure curves are shown in Fig. 6.1.11.

The demand curves in Fig. 6.1.10 are derived for a process with the production function of Eq. 6.1.3, a cost of \$1/unit for input x_2, and the following cost and revenue functions:

$$q = 2x_1^{1/2}x_2^{1/2}, \qquad c_x = k_1x_1 + x_2$$
$$c_q = k_1^{1/2}a, \qquad c_q' = \bar{c}_q = k_1^{1/2} \qquad (6.1.31)$$
$$r_q = 4q - 0.02_q^2, \qquad r_q' = 4 - 0.04q, \qquad \bar{r}_q' = 4 - 0.02q$$

For this system, the input curves x_1 (k_1) and the input expenditure curves k_1x_1 for each of the preceding objectives are:

Objective	Demand Curve $x_1(k_1)$	Expenditure k_1x_1
A	$x_1 = q/2k_1^{1/2}$	$k_1x_1 = qk_1^{1/2}/2$
B	$x_1 = c/2k_1$	$k_1x_1 = c/2$
C	$x_1 = 50k_1^{-1/2} - 12.5$	$k_1x_1 = 50k_1^{1/2} - 12.5k_1$
D	$x_1 = 100k_1^{-1/2} - 25$	$k_1x_1 = 100k_1^{1/2} - 25k_1$
E	If $k_1 \leqslant 4$, $x_1 = 50k_1^{-1/2}$	$k_1x_1 = 50k_1^{1/2}$
	If $k_1 \geqslant 4$, same as for objective D	k_1x_1 same as for objective D (6.1.32)

The expenditure curves are plotted in Fig. 6.1.11. In Fig. 6.1.10, demand curve A assumes q is 100, and demand curve B assumes c is \$200. Demand curve A shows only a substitution effect of a price change since the output is constant, but the other curves show both a substitution and an output effect since the change in price also causes a change in output.

When a production process has fixed technical coefficients, that is, when inputs are a fixed proportion of the output, there is no substitution, and price changes can have only an output effect on the input demand.

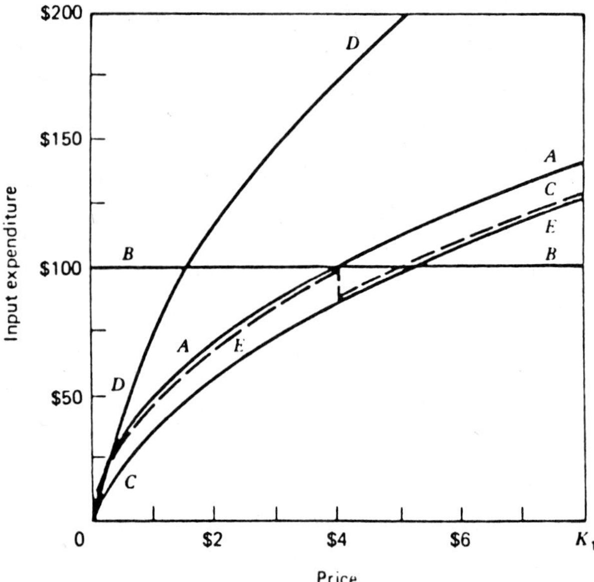

Fig. 6.1.11. Input expenditure curves for input x_1 in a nonlinear production system as a function of the input price k_1 and the production objective.

Input demand curves for such a linear process are shown in Fig. 6.1.12 for each of the preceding objectives. In this example, the revenue function and cost of $1/unit of x_2 are the same as the last example, but the production function and output cost function are as follows:

$$q = \text{minimum}\ (x_1/0.2,\ x_2/1.2)\ ; \qquad c_x = k_1x_1 + x_2$$

$$c_q = (0.2k_1 + 1.2)q\ ; \qquad c_q' = \bar{c}_q = \bar{c}_q = 0.2k_1 + 1.2$$

$$r_q = 4q - 0.02q^2\ ; \qquad r_q' = 4 - 0.04q; \qquad \bar{r}_q = 4 - 0.02q$$

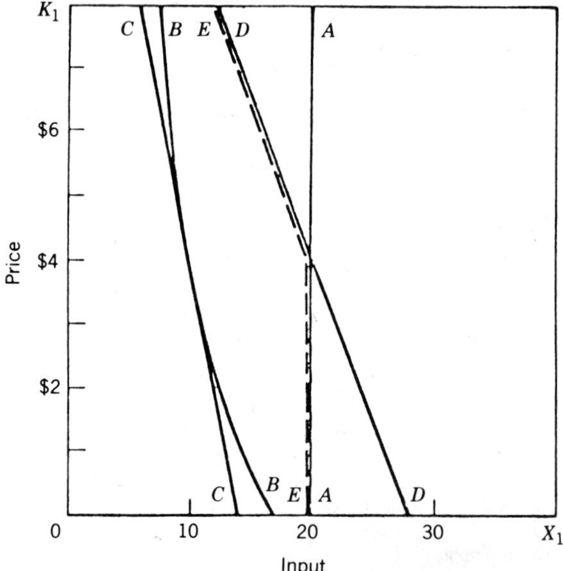

Fig. 6.1.12. Input price-demand curves for a linear production process for different production objectives.

For this system, the input demand curves $x_1(k_1)$ and the input expenditure curves k_1x_1 are as follows:

Objective	Demand Curve $x_1(k_1)$	Expenditure k_1x_1	
A	$x_1 = 0.2q$	$k_1x_1 = 0.2qk_1$	
B	$x_1 = c/k_1 + 6$	$k_1x_1 = 2c/1 + 6k_1$	
C	$x_1 = 14 - k_1$	$k_1x_1 = 14k_1 - k_2^1$	(6.1.34)
D	$x_1 = 28 - 2k_1$	$k_1x_1 = 28k_1 - 2k_1^2$	
E	If $k_1 \leqslant 4$, $x_1 = 20$	$k_1x_1 = 20k_1$	
	If $k_1 \geqslant 4$, same as for objective D	k_1x_1 same as for objective D	

In Fig. 6.1.12, demand curve A assumes q is 100, and curve B assumes c is \$100. The expenditure curves for the input x_1 as a function of price and the different objectives are shown in Fig. 6.1.13.

The price elasticity of demand is the ratio of the percentage decrease in demand to a percentage increase in price; the price elasticity of expenditure is the ratio of the percentage change in expenditure to a percentage change in price. As with other elasticity measures, they are both equal to the ratio of the marginal and average values of the functions. The elasticity for the previous example processes are as follows:

Process and Elasticity	Objective A	Objective B	Objectives C and D	
Nonlinear				
Demand	0.5	1.0	$2/(4 - k_1^{1/2}$	
Expenditure	0.5	0.0	$2/(4 - k_1^{1/2})$	
Linear				(6.1.35)
Demand	0.0	$k_1/6 + k_1$	$k_1/14 - k_1$	
Expenditure	1.0	$6/6 + k_1$	$7 - k_1/7 - 1/2k_1$	

For example, under objective A, minimum cost with fixed output, a 20% rise in the price of x_1 causes a 10% drop in the use of x_1 and a 10% increase in the expenditure on x_1 for the nonlinear system, but in the linear process the use of x_1 is not affected by the price increase, and expenditure on x_1 rises 20%.

Of more general interest is whether the demand is *elastic* or *inelastic*, that is, whether the percentage decrease in use is greater or less than the percentage increase in price. Under objective A, demand is inelastic for the nonlinear case and perfectly inelastic for the linear process. Under objective B, maximum output with fixed budget, demand has unit elasticity for the nonlinear system and is inelastic for the linear process. Under

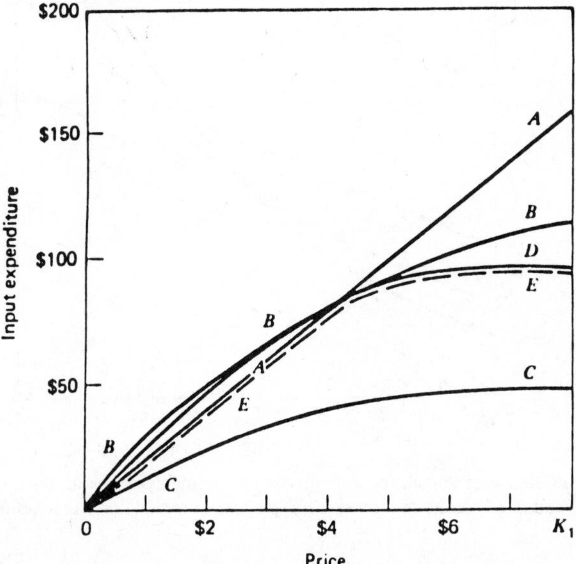

Fig. 6.1.13. Input expenditure curves for input x_1 in a linear production process as a function of the input price k_1 and the production objective.

objective C, maximum profit, and objective D, maximum break-even output, demand is inelastic for low prices of x_1 and elastic for high prices of x_1.

The expenditure on x_1 is inelastic in all cases, except for the nonlinear system under objectives C and D, where expenditure is elastic for high prices of x_1 in both the linear and nonlinear cases. Inelasticity means that the percentage increase in expenditure on x_1 is somewhat less than the percentage increase in the price of x_1, because of adjustments made in other inputs and outputs. Demand and expenditure patterns for linear systems with more than one linear process approach those of the nonlinear system.

6.1.10 SHORT-RUN AND LONG-RUN PRODUCTION COSTS

In the short run, a producer seeks to optimize an existing system that has fixed input limitations and cost characteristics. But in planning for the long run, the producer can specify the system's capacity by considering the *long-run production cost curve*, which defines the minimum cost of output under optimal capacity conditions. The long-run cost curve is the lower boundary, or envelope, of the various possible short-run cost curves.

For example, in Section 6.1.7 and Eqs. 6.1.21, a short-run cost function is derived for a nonlinear production system with a limit of 16 units of input, x_2. Equations 6.1.21 have the general form

$$q \leq s, \quad c(q,s) = k_s s + 2q; \quad c'_q = 2; \quad \bar{c}_q = 2 + k_s s/q$$

$$q \geq s, \quad c(q,s) = (k_s + 1)s + q_2/s; \quad c'_q = 2q/s, \tag{6.1.36}$$

$$\bar{c}_q = \frac{(k_s + 1)s}{(q + q/s)}$$

where s denotes the upper limit for input x_2 and k_2 is the cost per unit time for one unit of "capacity" level s.* For this example, if all parameters are the same as in Eqs. 6.1.21, the short-run total, average, and marginal costs are plotted in Figs. 6.1.14 and 6.1.15 for three different values of s: 16, 25, and 40 units of x_2.

The long-run cost curve can be derived by taking the derivative of the short-run cost function with respect to the capacity variable, setting this equal to zero, and solving for the capacity value that minimizes the short-

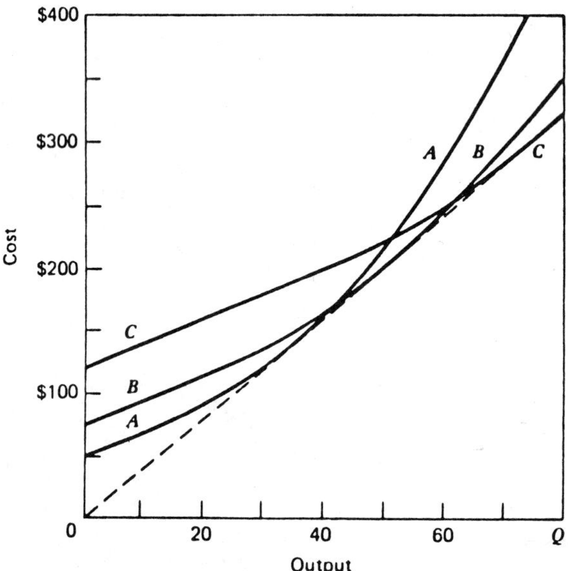

Fig. 6.1.14. Short-run total cost curves for a nonlinear production system with three capacity levels at (*A*) 16 units, (*B*) 25 units, and (*C*) 40 units of input x_2, and the long-run total cost curve (dotted line) for the same system.

* The capacity cost rate is equal to the capital investment multiplied by the continuous capital recovery factor $(\overline{A}/P, i, n) = i(1 - e^{-in})$, where i is the nominal interest rate and n is the life of the investment.

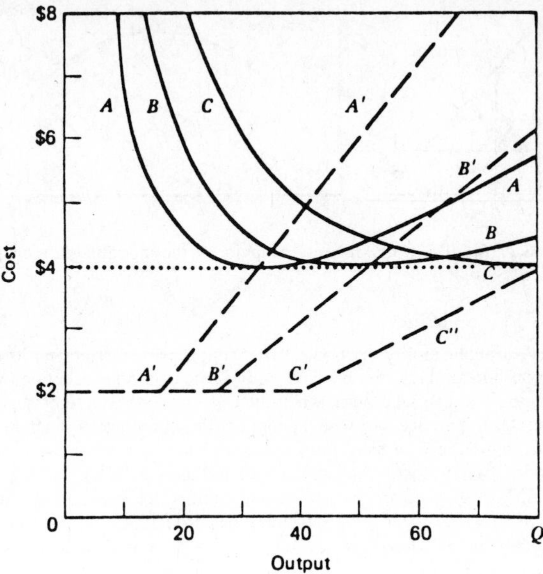

Fig. 6.1.15. Average cost curves (solid lines) and marginal cost curves (dashed lines) for a nonlinear production system with three capacity levels at (*A*) 16 units, (*B*) 25 units, and (*C*) 40 units of input x_2, and the long-run marginal and average cost curve (dotted line) for the same system.

run cost function. Substituting this optimal value of capacity into the short-run cost function will yield the long-run cost function. When this procedure is followed with Eqs. 6.1.36, the optimal capacity level is found to be equal to half of the desired output level. At this value of s, the long-run total, average, and marginal cost functions are as follows:

$$c_q = rq; \qquad c_q' = \bar{c}_q = 4; \qquad s = \frac{q}{2} \tag{6.1.37}$$

These long-run cost functions are plotted in Figs. 6.1.15 and 6.1.16 along with the short-run curves.

The specification of an optimal capacity level depends on the long-run cost function, the revenue function, and the producer's objective. For the preceding example, the optimal output levels and the corresponding optimal capacity levels are derived in order to maximize profit or maximize the break-even output level, when the revenue function is defined as follows:

$$\begin{aligned} r_q &= 6q - 0.02^2_q, \qquad r_q' = 6 - 0.04q \\ r_q' &= c_q', \qquad q = 50, \qquad s_o = 25 \text{ at maximum profit} \\ r_q &= c_q, \qquad q = 100, \qquad s_o = 50 \text{ at break-even} \end{aligned} \tag{6.1.38}$$

Thus, in this example the profit-maximizing producer would prefer a system capacity of 25 units of x_2 and an output of 50 units of q to obtain a net profit rate of \$50. If the producer has to work with a system having a capacity of only 16 units of x_2, the optimal output rate would be about 36 units of q, and the maximum obtainable profit rate would be \$45, or 10% less than optimal. If the capacity were already fixed at 40 units of x_2, the best output rate would be about 67 units of q, with a net profit of only \$40, which is 20% less than is obtainable with an optimal capacity level of 25 units of x_2.

6.1.11 FIRM BEHAVIOR UNDER REGULATORY CONSTRAINT

In the case of industries that exhibit scale economies, breaking up the larger firms in the name of increasing competition and preventing the exercise of monopoly power has an unfortunate side effect: the many small firms cannot produce as efficiently as the large ones. A solution to this dilemma is to maintain the monopoly, but control its prices by allowing it to earn no more than a "fair return" on investment. In return for an exclusive

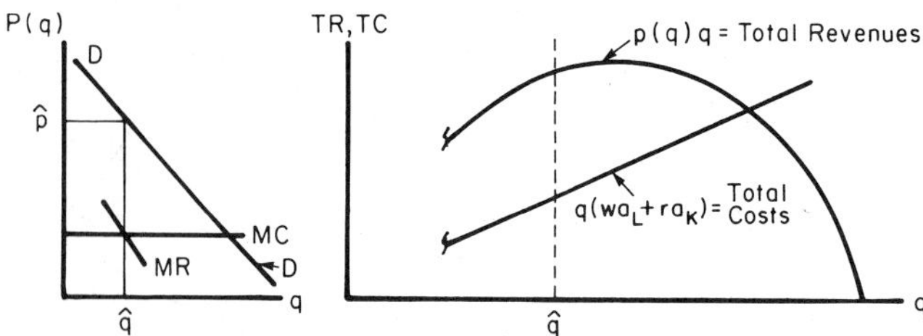

Fig. 6.1.16. Profit maximizing behavior of the unregulated monopolist.

franchise in its territory, a public utility is allowed to charge a price that covers its operating costs *plus* this fair rate. In this way, consumers obtain the benefits of low-cost volume production by the monopolist without the attendant price gouging that can take place when there is only one source of supply.

To see how this operates in practice, suppose the monopolist's production function in its region of operation contained only capital K and labor L in fixed proportions: a_L units of labor per unit of output, and a_K units of capital per unit of output. If labor cost w dollars per unit and the cost of capital was r (an interest rate), then the cost function is simply $c(q) = q(wq_L + ra_k)$. The price at which any output can be sold is a decreasing function of price, $p = p(q)$, which means total revenues are $p(q)q$. Without regulation, what output and price would be chosen by the monopolist? Assuming the monopolist wishes to maximize profit, the problem becomes:

maximize

$$\text{revenues} - \text{expenses} = p(q)q - q(wa_L + ra_K) \tag{6.1.39}$$

The necessary conditions for output maximization are

$$p(q) + qp'(q) = wa_L + ra_K \tag{6.1.40}$$

where $p'(q) = dp/dq$. All this is shown in Fig. 6.1.16.

The problem is that $\hat{q}$ is too low, and $\hat{p}$ is too high. The monopolist profits by restricting his output, allowing him to charge a price $\hat{p}$ well above the unit cost of production, $wa_L + ra_K$. By requiring the monopolist to earn no more than a fair return rate on capital, regulation can increase the optimal output and decrease the price at which it is sold. If s is the limit on the interest return to capital set by the regulatory body, then the imposition of rate of return regulation, as it is known, prohibits those combinations of price and quantity that result in a return greater than s from being chosen by the utility.

The return constraint limits revenues as a function of q to $q(wa_L + sa_K)$, which means those portions of the revenue curve that lie above the line $q(wa_L + sa_K)$ are unobtainable. Figure 6.1.17 shows the impact of rate of return regulation on the behavior of the monopolist. The profit maximizing strategy of the regulated monopolist is now to produce a quantity $\bar{q}$ and charge a price $\bar{p}$. The consumer gets more product at lower cost, yet maintains the advantage of the monopolist's ability to produce in volume.

What if the production function were not of the "*fixed proportions*" type, and capital could be substituted for labor? Certainly, the incentive is there for it to happen, since such a substitution would rotate the limit on revenue constraint counterclockwise, allowing the monopolist to increase profit by charging higher prices and restricting output. This tendency for regulated monopolists to "gold plate"—that is, substitute capital for labor, not to reduce the cost of production, but to increase allowed revenues—has been the subject of much discussion since it was first pointed out.[8] Little, if any, evidence has appeared that indicates that this is a real problem.

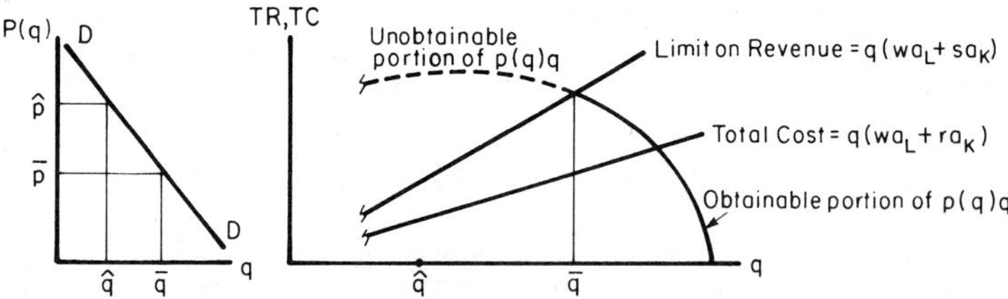

Fig. 6.1.17. Profit maximizing behavior of the regulated monopolist.

6.1.12 EXPANSION OF PRODUCTION CAPACITY

Planning for the expansion of production capacity involves a trade-off between the temporary loss of return on excess production capacity and the economies of scale in building larger production facilities. The initial investment needed for a given level of production capacity frequently can be approximated by a power function[6,7] of the form $k_s q_s^b$, as shown in Fig. 6.1.18. The power b is called the economy of scale factor and is equal to the elasticity of the investment cost function, that is, the ratio of the percentage change in investment cost to a percentage change in capacity q_s. The scale factor 0.6, frequently used in the chemical industry, is called the "six tenths rule." A likely physical basis for the 0.6 rule is the relation between the area (cost) and the volume (capacity) of cylinders (equipment).

When planning the expansion of production capacity in order to satisfy demand, which is expected to grow at a steady annual rate over a fairly long planning horizon, an optimal plan consists of a uniform series of investments at regularly spaced intervals, as shown in Fig. 6.1.19. The amount of capacity added at each interval is equal to the total growth of demand in the interval. When demand is growing at a rate of g units per year, then for each interval of length t years, capacity is increased by gt units at a cost of $k_s(gt)^b$. The equivalent uniform annual cost during the interval t equals $k_s(gt)^b$ multiplied by the capital recovery factor $(\bar{A}/P,i,t) = i/(1 - e^{it})$, and the equivalent present value of capacity cost over an indefinitely long planning horizon is found by dividing the equivalent annual cost by the nominal interest rate, i. Finally, the optimal expansion interval t_o is found by minimizing the present value of the expansion costs, that is,

$$p(t) = \frac{k_s(gt)^b}{1 - e^{-it}} \tag{6.1.41}$$

$$p(t) = 0; \qquad it_0 = b(e^{it_0} - 1) = 0.95 \quad \text{if } b = 0.6$$

When the scale factor b equals 0.5, 0.7, 0.8, and 0.9, the optimal value of it_o decreases with the following approximate respective values: 1.230, 0.676, 0.431, and 0.207. If $b = 1$, $t_o = 0$, indicating that capacity should be increased *continuously* as needed.

Note that the optimal expansion interval depends neither on the cost factor k_s nor on the growth rate g. However, it is inversely proportional to the interest rate; that is, a doubling of the interest rate will cut the optimal interval in half, which implies about 33% reduction in the amount of capital invested in each of the successive capacity expansion projects. The present value of cost is not very sensitive to small changes from the optimal values of the expansion interval. Various applications and extensions of this analysis are discussed in Refs. 6 and 7.

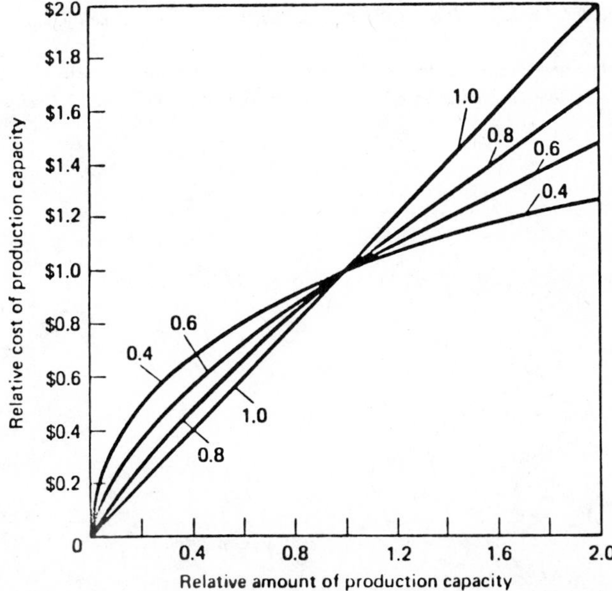

Fig. 6.1.18. Relative cost of expanding production capacity as a function of four different scale factors.

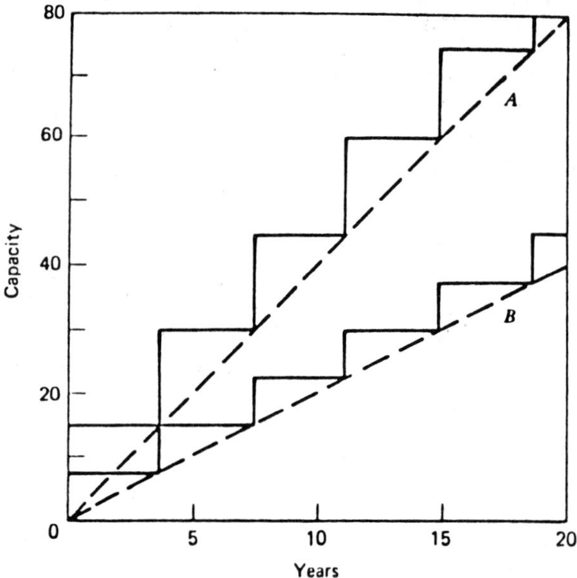

Fig. 6.1.19. Optimal step patterns for expanding production capacity when the scale factor is 0.6, the interest rate is 25%, and the demand for production increases at (*A*) 4 units/year or (*B*) 2 units/year.

REFERENCES

1. W. J. Baumol, *Economic Theory and Operations Analysis*, 4th ed., Englewood Cliffs, NJ: Prentice-Hall, 1977.
2. J. M. Henderson and R. E. Quandt, *Microeconomic Theory*, 3rd ed., New York: McGraw-Hill, 1980.
3. M. Nerlove, *Estimation and Identification of Cobb-Douglas Production Functions*, Chicago: Rand McNally, 1965.
4. R. Dorfman, P. A. Samuelson, and R. Solow, *Linear Programming and Economic Analysis*, New York: McGraw-Hill, 1958.
5. K. E. Boulding and A. W. Spivey, Eds., *Linear Programming and the Theory of the Firm*, New York: Macmillan, 1960.
6. A. S. Manne, Ed., *Investments for Capacity Expansion: Size, Location, and Time-Phasing*, Cambridge, MA: MIT Press, 1967.
7. L. M. Rose, *Engineering Investment Decisions: Planning Under Uncertainty*, Amsterdam: Elsevier, 1976.
8. H. Averch, and L. L. Johnson, "Behavior of the Firm Under Regulatory Constraint," *American Economic Review*, **52**, December 1962.

CHAPTER 6.2

ECONOMIC ANALYSIS

GERALD J. THUESEN

Georgia Institute of Technology
Atlanta, Georgia

6.2.1 INVESTMENT ALTERNATIVES

The process of making investment decisions is based on the fundamental idea that alternative means of achieving stated objectives must be considered. Much experience supports the notion that being able to consider a variety of alternatives leads ultimately to better results than contemplating only one possible course of action. Of course the success of such a philosophy depends on an organization's ability to stimulate the development of good ideas and the accompanying investment alternatives.

It is the purpose of this chapter to consider the methods of evaluating the economic consequences of investment alternatives once they have been identified. To facilitate the description of the economic effects of any investment alternative, a graphical picture referred to as a *cash-flow diagram* will be utilized. This cash-flow diagram will provide a statement of the economic inflows and outflows associated with an investment alternative.

Because it is the *exchange* of cash associated with an alternative that describes its economic characteristics, the cash-flow diagram will reflect both receipts (inflows of cash) and disbursements (outflows of cash). Any receipt received will be represented as an upward arrow (positive cash flow) while any disbursement will be represented as a downward arrow (negative cash flow). These arrows are placed on a time scale so that the timing of the receipts and disbursements is indicated. In addition, the magnitude of the arrows associated with the cash flows will be proportional to the amount of dollars being received or disbursed at each point in time. An example of a cash flow diagram is presented in Fig. 6.2.1. Note that the end-of-period convention is used in preparing cash-flow diagrams, where it is assumed that receipts or disbursements occurring during a period will be concentrated at the end of that period.

The actual cash inflows (receipts) and the actual cash outflows (disbursements) are determined by estimation. This task of estimating the possible future outcomes associated with a proposed investment alternative is the most difficult step of any economic analysis. Unfortunately, there are not standard procedures for assuring accurate estimates of future cash flows. However, a variety of methods exist to assist in the task of cash-flow estimating.[1]

6.2.1.1 Types of Investment Projects

There are many relationships among investment projects that define whether a project is independent or interdependent. It is important to recognize the type of project being analyzed since it will determine how the comparison of the alternatives will be made.

Independent Projects

When the acceptance (or rejection) of a project from a set of projects has no effect on the acceptance (or rejection) of any of the other projects in the set, the project is defined to be *independent*. Although few projects are truly independent, for practical reasons a large group of investment projects will fit this definition. These projects usually assume the characteristics of independence because they are functionally independent; that is, they do not serve the same purpose and there are no obvious dependencies between them. For example, investment proposals to purchase material handling equipment, a security system, and research equipment would be considered independent under most circumstances.

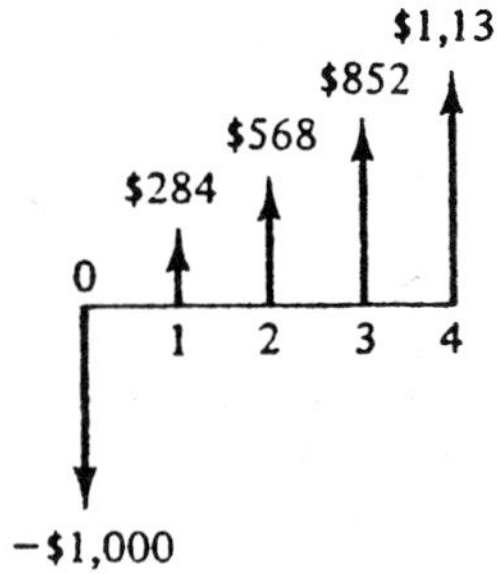

Fig. 6.2.1. Cash flow diagram.

Interdependent Projects

For many projects, the acceptancce of one project will directly or indirectly influence the acceptance of others being considered. This dependency relationship can occur for a variety of reasons.

Projects that are *mutually exclusive* are those projects where the acceptance of any one project precludes the acceptance of any other projects. Projects that are functionally mutually exclusive occur when a particular need must be fulfilled. In this case there are a variety of projects available, each of which will provide the desired function.

Another important dependency relationship that occurs between projects is *contingency*. Thus, if some initial project is undertaken, other auxiliary projects may be possible only if the initial project is completed. These contingent projects are conditional on the implementation of some other project.

In addition to the dependency conditions just discussed, interdependencies may be introduced by having limited funds available for investment. Therefore, the acceptance of some project may utilize such a large amount of the funds available that others must be eliminated from consideration.

6.2.1.2 Expressing Projects as Mutually Exclusive Alternatives

Because of the many possible interdependency relationships among investment projects, it would be difficult to manage the comparison of projects if the comparison methods were based on the nature of interdependency among the projects. Fortunately, interdependent projects can always be rearranged into mutually exclusive alternatives by complete enumeration of all the possible decision options. These options can be viewed as mutually exclusive alternatives, since the idea is to select the single most desired alternative from the set of available alternatives.

Given below is an example of five projects, of which one is independent, three are mutually exclusive, and one is contingent. In addition, there are financial interdependencies among these projects because the funds available for investment are limited. Table 6.2.1 presents the projects and their associated initial investment. Project A is functionally independent of projects B1, B2, B3, and C. Projects B1, B2, and B3 are functionally

Table 6.2.1. Rearrangement of Projects into Mutually Exclusive Alternatives

Project	Initial Investment	Alternative Number	Mutually Exclusive Alternatives					Total Initial Investment
			x_A[a]	x_{B1}	x_{B2}	x_{B3}	x_C	
A	$10,000	1	0	0	0	0	0	$ 0
B1	40,000	2	1	0	0	0	0	10,000
B2	45,000	3	0	1	0	0	0	40,000
B3	50,000	4	0	0	1	0	0	45,000
C	5,000	5	0	0	0	1	0	50,000
		6	0	0	1	0	1	50,000
		7	1	1	0	0	0	50,000
		8	1	0	1	0	0	55,000[b]
		9	1	0	1	0	1	60,000[b]
		10	1	0	0	1	0	60,000[b]

[a] $x_j = \begin{cases} 0 & \text{Project } j \text{ rejected} \\ 1 & \text{Project } j \text{ selected} \end{cases}$

[b] Infeasible alternatives, as the combined investment exceeds $50,000.

Table 6.2.2. Two Alternatives with Different Service Lives

End of Year	Alternative A1	Alternative A2
0	−$140,000	−$200,000
1	−55,000	−10,000
2	−55,000	−10,000
3	−55,000	−10,000
4	−55,000	—
5	−55,000 + 30,000[a]	—

[a]Estimated salvage at end of alternative's service life.

mutually exclusive, while project C is contingent on the implementation of project B2. Additionally, project C is independent of project A, and it is assumed that $50,000 is available for investment.

With this set of circumstances, the possible mutually exclusive combinations are shown in Table 6.2.1. Observe that there are ten mutually exclusive alternatives, three of which are not feasible, as their investment exceeds the funds available. The comparison of alternatives reduces to the comparison of the remaining seven alternatives, each of which must be defined by its estimated cash flows†.

6.2.2 COMPARISON OF ECONOMIC ALTERNATIVES

Since it is always possible to describe investment projects as a set of mutually exclusive alternatives, the question is, "How does one select the best alternative from a set of mutually exclusive alternatives?" Fortunately, it is possible to employ a fundamental decision rule to accomplish this goal.

The fundamental decision rule requires that the differences between the cash flows of two mutually exclusive alternatives be examined to determine the preferred alternative. The proof of this concept is given in Thuesen and Fabrycky.[2]

6.2.2.1 Measures of Investment Worth

To reflect the economic desirability of particular cash flows, a number of measures of investment worth have been utilized. These measures place the cash flows on an equivalent basis, so that the time value of money effects are considered.

Four of the most frequently applied measures of investment worth are presented next. Although these measures have wide acceptance for project evaluation, it is important to be aware of other measures that also can provide useful insights. Discussion of these "other" measures can be found elsewhere.[5]

The four measures defined below require the use of an interest rate that represents the "minimum attractive rate of return," MARR. For a more complete discussion of the application and selection of the MARR, see Thuesen and Fabrycky.[6]

Let F_t be the net cash flow at time t where receipts are (+) and disbursements are (−).

PRESENT WORTH

$$PW(i) = \sum_{t=0}^{n} F_t(1 + i)^{-t}$$

ANNUAL EQUIVALENT

$$AE(i) = \sum_{t=0}^{n} F_t(1 + i)^{-t} \left[\frac{i(1 + i)^n}{(1 + i)^n - 1}\right]$$

FUTURE WORTH

$$FW(i) = \sum_{t=0}^{n} F_t(1 + i)^{n-t}$$

† When large numbers of projects are being considered, these problems can be formulated as linear integer programming problems.[3,4] This formulation allows for the efficient solution of what would require excessive time by complete enumeration.

INTERNAL RATE OF RETURN (i^*)

$$0 = \sum_{t=0}^{n} F_t(1 + i^*)^{-t} \qquad \text{where } -1 < i^* < \infty$$

Incremental Investment Comparisons

The incremental comparison of mutually exclusive alternatives is accomplished by a series of pair-wise comparisons. In order to assure that consistent results occur when applying the four measures of investment worth, it is essential that the mutually exclusive alternatives be ordered in increasing order of the initial year's outlay. In Table 6.2.1 you can see that, based on total initial investment, the mutually exclusive alternatives are already in the proper order. (The alternatives are in ascending order of their initial year outlay.)

The first comparison requires selecting the alternative requiring the smallest initial outlay as the initial "current best." This alternative either must be better than the Do Nothing alternative or Do Nothing must not be feasible.[7] This alternative is then compared to the current "challenging" alternative, which in the first comparison is the one with the next smallest initial outlay. The cash flow representing the difference between the two alternatives is computed, and it is determined if this cash flow is economically desirable. The decision rules to determine the economic desirability of two mutually exclusive alternatives denoted Z1 and Z2 are as follows.

$$PW(i)_{Z2-Z1} > 0, \qquad \text{Select Z2, otherwise select Z1}$$

$$AE(i)_{Z2-Z1} > 0, \qquad \text{Select Z2, otherwise select Z1}$$

$$FW(i)_{Z2-Z1} > 0, \qquad \text{Select Z2, otherwise select Z1}$$

$$i^*_{Z2-Z1} > i, \qquad \text{Select Z2, otherwise select Z1}$$

Once an alternative is selected, it becomes the "current best" and is then compared to the next "challenging" alternative, which is the next one in order of initial investment not considered previously. The pair-wise comparisons continue until all have been compared. Once the last comparison is accomplished, the last "current best" is the economically most desirable alternative in the set of mutually exclusive alternatives.

One note of caution is provided. If the number of sign changes in any of the incremental cash flows being compared is greater than 1, the internal rate of return should not be used. More elaborate tests can be applied to determine if the internal rate of return can be used even if the sign test just mentioned is not satisfied.[8]

Total Investment Comparisons

In many instances, the incremental comparison of mutually exclusive alternatives is cumbersome. Fortunately, three out of the four measures listed, when properly applied to the total cash flow, will yield results consistent with the incremental approach. The different decision rules necessary for the comparisons of mutually exclusive alternatives on a total cash flow basis are as follows.

$$\text{Max or Min}^{\dagger}\ [PW(i)_{Z1}, PW(i)_{Z2}, PW(i)_{Z3}, \cdots,]$$

$$\text{Max or Min}\ [AE(i)_{Z1}, AE(i)_{Z2}, AE(i)_{Z3}, \cdots,]$$

$$\text{Max or Min}\ [FW(i)_{Z1}, FW(i)_{Z2}, FW(i)_{Z3}, \cdots,]$$

It is *not* proper to utilize the internal rate of return when analyzing mutually exclusive alternatives on the basis of their total cash flows. A complete discussion of the inconsistencies involved with the internal rate of return are presented elsewhere.[9]

6.2.2.2 Comparison of Alternatives with Different Life Spans

In most economic comparisons, the lives of the alternatives being considered are not conveniently the same. To assure a fair comparison of the alternatives, a fundamental principle applies: *All alternatives under consideration must be compared over the same time span*. Following this principle is essential so that the economic effect of undertaking one alternative as opposed to another is measured over identical time spans.

Because alternatives must be judged over the same time span, a variety of assumptions are required to place the effects of alternatives within the same study period. Different assumptions are required depending on how the study period selected relates to the actual service life of the alternative.

† If cash flows consist of both receipts and disbursements, the objective is to maximize. If the cash flows represent only costs, the objective is to minimize.

Study Period the Same as Alternative's Life

When this situation arises there is no need to make any adjustment to the cash flow involved. The decision rules discussed would be applied as defined in Section 6.2.2. Thus, the comparison of annual equivalent amounts or present worth amounts will provide the same result.

Study Period Shorter Than Alternative's Life

If an investment alternative has a life longer than the study period, it can be assumed that some value associated with the alternative remains at the end of the study period. For physical assets, this means that as long as the asset has some remaining useful life, the asset has some worth. This worth or *unused* value represents the value of the asset that is yet to be consumed.

Two methods can be utilized to reflect the unused value of an investment alternative. The first is to estimate the actual salvage value that might be received if the asset were sold at the end of the study period. For example, the two alternatives presented in Table 6.2.2 are compared using a three-year study period. It is assumed that the service or revenue to be provided for each alternative is identical for the first three years of service. The negative signs indicate disbursements associated with the initial investment at $t = 0$ and operating costs at the other times. If it is estimated that the actual salvage value for alternative A1 is \$40,000 at the end of three years of service, an annual equivalent comparison for an interest rate of 20% is†

$$\begin{aligned} AE(20)_{A1} &= -\ [(\$140{,}000 - \$40{,}000)\overset{A/P,20,3}{(0.4747)} \\ &\quad + \$40{,}000(0.20)] - \$55{,}000 \\ &= -\ \$110{,}470 \text{ per year} \end{aligned}$$

$$\begin{aligned} AE(20)_{A2} &= -\ \$200{,}000\overset{A/P,20,3}{(0.4747)} - \$10{,}000 \\ &= -\ \$104{,}940 \text{ per year} \end{aligned}$$

By assuming the salvage value at the end of the study period for alternative A1, the alternative now has a life equal to the study period. Using this approach, alternative A1 seems more costly than alternative A2 by \$5,530 per year for the three-year period.

The second approach is to compute annual equivalent amounts based on the service life of each investment. Then the alternatives are compared over the study period using the annual equivalent amounts. For the example in Table 6.2.2, this calculation of annual equivalent amounts at 20% yields

$$\begin{aligned} AE(20)_{A1} &= -\ [(\$140{,}000 - \$30{,}000)\overset{A/P,20,5}{(0.3344)} \\ &\quad + \$30{,}000(0.20)] - \$55{,}000 \\ &= -\ \$97{,}784 \text{ per year} \end{aligned}$$

$$\begin{aligned} AE(20)_{A2} &= -\ \$200{,}000\overset{A/P,20,3}{(0.4747)} - \$10{,}000 \\ &= -\ \$104{,}940 \text{ per year} \end{aligned}$$

Based on this approach, alternative A1 is less costly than alternative A2 by \$7156 per year for three years. We see, therefore, that depending on which approach is utilized, it is possible to have conflicting conclusions. The simplicity of the calculation, and the fact that the salvage value at the end of the study period may not be known, has resulted in widespread use of this second approach by industry.

When an alternative has a service life longer than the study period, an implied salvage value or unused value at the end of the study period can be calculated. This implied salvage value is based on the portion of the annual capital costs of the investment that extend beyond the study period. With P the initial investment and F the salvage value at the end of the investment's life n, the annual capital costs are represented by the capital recovery with return, CR(i).

† For alternative A1, the capital recovery with return formulation is used to calculate the annual equivalents.[5] Also used are the interest factors recommended by the ANSI Z94 Standards on notation.[10] The use of these interest factors and tabulation of the factor values are described in detail in any basic engineering economy text. (See Bibliography.)

$$\text{CR}(i) = (P - F)(A/P,i,n) + Fi$$

If n^* is the length of the study period and n is the service life of the alternative, the implied salvage value at $t = n^*$ is found by solving the expression

$$\text{CR}(i)\overset{P/A,i,n-n^*}{(\qquad)} + F\overset{P/F,i,n-n^*}{(\qquad)}$$

For our example, the implied salvage value resulting from the second approach when $n = 5$, $n^* = 3$ yields

$$\begin{aligned}
&[(\$140{,}000 - \$30{,}000)\overset{A/P,20,5}{(0.3344)} \\
&\quad + \$30{,}000(0.20)]\overset{P/A,20,2}{(1.5278)} \\
&\quad + \$30{,}000\overset{P/F,20,2}{(0.6945)} = \$86{,}200
\end{aligned}$$

If this amount should appear to be unrealistic, then it may be better to attempt the direct assessment of the salvage value at the end of the study period, i.e., use the first approach discussed.

Study Period Longer Than Alternative's Life

When alternatives have lives shorter than the study period, some assumptions must be utilized to describe what occurs from the end of the alternative's life to the end of the study period. It is important to categorize investments in this situation into two distinct types, because different assumptions will be made.

One type of investment is represented by those alternatives that provide identical service each year throughout their corresponding service lives. An example of this case is the comparison of the two alternatives described in Table 6.2.2 for a study period of five years.

Two methods can be utilized to modify the alternatives so they can be compared over equal periods of time. In this example, the time span is five years. The first method requires the explicit estimation of the cash flows that are required for alternative A2 to provide the same service as alternative A1 for years 4 and 5. Suppose that it would cost \$60,000 per year to provide such service for the two years remaining after the original asset had been retired at $t = 3$. The proper analysis gives

$$\begin{aligned}
\text{AE}(20)_{\text{A1}} &= -[(\$140{,}000 - \$30{,}000)\overset{A/P,20,5}{(0.3344)} \\
&\quad + \$30{,}000(0.20)] - \$55{,}000 \\
&= -\$97{,}784 \text{ per year}
\end{aligned}$$

$$\begin{aligned}
\text{AE}(20)_{\text{A2}} &= [-\$200{,}000\overset{A/P,20,3}{(0.4747)} \\
&\quad - \$10{,}000]\overset{P/A,20,3}{(2.1065)}\overset{A/P,20,5}{(0.3344)} \\
&\quad - \$60{,}000\overset{F/A,20,2}{(2.200)}\overset{A/F,20,5}{(0.1344)} \\
&= -\$91{,}661 \text{ per year}
\end{aligned}$$

The conclusion is that alternative A2, with service identical to alternative A1, will save \$6123 per year over the five-year study period.

Another method that can be utilized when identical service is provided by alternatives with different lives, is the *common multiple of lives* approach. Here it is assumed that alternatives having different lives are repeated until a common multiple of lives is reached. The calculations to make the proper comparison are

$$\begin{aligned}
\text{AE}(20)_{\text{A1}} &= -[(\$140{,}000 - \$30{,}000)\overset{A/P,20,5}{(0.3344)} \\
&\quad + \$30{,}000(0.20)] - \$55{,}000 \\
&= -\$97{,}784 \text{ per year}
\end{aligned}$$

$$AE(20)_{A2} = -\$200{,}000\overset{A/P,20,3}{(0.4747)} - \$10{,}000$$

$$= -\$104{,}940 \text{ per year}$$

Here the conclusion favors alternative A1 by $7156 per year for 15 years, the common multiple of the lives. Although the calculation for this approach is simple, it may not be realistic, since the calculations above are based on the (improbable) assumption that the cash-flow patterns for any alternative are being repeated and therefore are identical from cycle to cycle. This approach should be used only when the common multiple of lives is relatively short or it is known that the cash flows will repeat.

Now that methods for analyzing alternatives having identical service have been discussed, the next investments to be considered are those that will neither be repeated nor provide identical service. (A study period longer than the alternative's life is still being assumed.) These *unique* investments have to be described by both their associated receipts and disbursements.

The assumption usually adopted for these *unique* investments is that the cash receipts can be reinvested until the end of the study period at the MARR being used in the calculation. This assumption allows the calculation of the future worth of each alternative at end of the study period. These future worth values can then be compared directly. (This is equivalent to comparing the present worth amounts for each of the alternatives at the MARR.) If the annual equialent methods shown above are applied to unique investments, a different result may occur since the effect of the common multiple of lives assumption is different from the reinvestment assumption.

6.2.2.3 Comparison of Alternatives Considering Inflation

Measuring Inflation Rates

Measuring historical price-level changes for goods or services requires the calculation of a price index. This index represents the ratio of the price of an item at some time to the price at an earlier time. Indices for all types of goods and services are provided by the U.S. Department of Commerce and the Department of Labor.[11–13]

From an index of this type it is possible to compute annual historical inflation rates. Table 6.2.3 presents an example of these rates based on the Consumer Price Index (CPI) for the years 1965–1983.

To find the annual inflation rate from one year to the next, use the expression

$$\text{annual inflation rate for year } (t+1) = \frac{\text{CPI year } (t+1) - \text{CPI year } (t)}{\text{CPI year } (t)}$$

Table 6.2.3. Annual Inflation Rates Based on Consumer Price Index 1965–1985 (base year 1967)

Year	CPI	Annual Inflation Rate (%)
1965	94.5	1.7
1966	97.2	2.9
1967	100.0	2.9
1968	104.2	4.2
1969	109.8	5.4
1970	116.3	5.9
1971	121.3	4.3
1972	125.3	3.3
1973	133.1	6.2
1974	147.7	11.0
1975	161.2	9.1
1976	170.5	5.8
1977	181.5	6.5
1978	195.4	7.7
1979	217.4	11.3
1980	246.8	13.5
1981	272.4	10.4
1982	289.1	6.1
1983	298.4	3.2
1984	311.1	4.3
1985	322.9	3.8

For example, the annual inflation rate for 1982 based on the CPI was

$$\frac{279.3 - 268.8}{268.8} = 0.039 \quad \text{or} \quad 3.9\%$$

Definitions

There are two approaches for the inclusion of inflation effects in investment analyses. One approach is *actual* dollar analysis and the other is *constant* dollar analysis. Because these two approaches will produce the same conclusion, the advantage of one over the other depends on what insight you wish to provide. To understand the connection between the two approaches requires the definition of the following terms. (These definitions are reprinted in part with permission from Thuesen and Fabrycky, *Engineering Economy*, 6th ed., Prentice-Hall, Englewood Cliffs, NJ, 1984, Chap. 5, pp. 113–114.)

Market Interest Rate (i). The market interest rate represents the opportunity to earn as reflected by the actual rates of interest available in finance and business. This rate is a function of the investment activities of investors who are operating within this market. Since astute investors are well aware of the power of money to earn and the detrimental effects of inflation, the interest rates quoted in the marketplace include the effects of both the earning power and the purchasing power of money. When the rate of inflation increases, there is usually a corresponding upward movement in quoted interest rates.

Other names: Combined interest rate, current-dollar interest rate, actual interest rate, minimum attractive rate of return.

Inflation-Free Interest Rate (i'). The inflation-free interest rate represents the earning power of money with the effects of inflation removed. This interest rate is an abstraction. Typically it must be calculated, since it is not generally used in the transactions of the financial marketplace. The inflation-free interest rate is not quoted by bankers, stockbrokers, and other investors and is therefore not generally known to the public. If there is no inflation in an economy, then the inflation-free interest rate and the market interest rate are identical.

Other names: Real interest rate, constant-dollar interest rate.

Inflation Rate (f). The inflation rate is the annual percentage of increase in prices of goods and services.

Other names: Escalation rate, rate of increase in cost of living.

Actual dollars represent the out-of-pocket dollars received or disbursed at any point in time. This amount is measured by totaling the denominations of the currency paid or received.

Other names: Then-current dollars, current dollars, future dollars, escalated dollars, inflated dollars.

Constant dollars represent the hypothetical purchasing power of future receipts and disbursements in terms of the purchasing power of dollars at some base year. This base year can be arbitrarily selected, although in many analyses it is assumed to be time zero, the beginning of the investment.

Other names: Real dollars, deflated dollars, today's dollars, zero-date dollars.

To translate constant dollars (base year $t = 0$) at a time t to actual dollars at the time when the inflation rate per year is f, requires

$$\text{actual dollars} = (1 + f)^t(\text{constant dollars})$$

With this relationship, it is possible to utilize cash flows estimated in either actual or constant dollars and to convert one to the other.

The last important definition is the relationship amont i, i', and f. (Derivation of this relationship is presented in Chapter 5 of Thuesen and Fabrycky, *Engineering Economy*, 6th ed. Prentice-Hall, Englewood Cliffs, NJ, 1984.)

$$(1 + i) = (1 + f)(1 + i')$$

or

$$i' = \frac{(1 + i)}{(1 + f)} - 1$$

This relationship allows for the calculation of any one of these rates if the other two are given.

Actual-Dollar Analysis and Constant-Dollar Analysis

Either actual-dollar or constant-dollar analysis may be utilized to compare investment alternatives. One approach or the other may be considered more appropriate by the nature of the data available or by the type of presentation required. By properly applying the basic principles of these two approaches, the conclusion reached will be the same.

When making calculations that require the translation of dollars at one time to *another* time, the following principles apply.

Actual dollar analysis:

When using actual dollars to compute equivalent values at different points in time use the market rate (i).

Constant dollar analysis:

When using constant dollars to compute equivalent values at different times use the inflation-free rate (i').

To illustrate these principles, suppose our objective is to compare the two alternatives described in Table 6.2.4. The cash flows are given in both actual and constant dollars where i = 20% per year and f = 7% per year. First, the actual dollar analysis gives

$$\text{PW}(20)_{\text{A1}} = -\ \$20{,}000 + \$12{,}000\overset{P/A,20,4}{(2.5887)} = \$11{,}064$$

$$\text{PW}(20)_{\text{A2}} = -\ \$50{,}000 + \$22{,}000\overset{P/A,20,4}{(2.5887)} = \$6{,}951$$

Thus, alternative A1 is more attractive than alternative A2, based on the actual dollar analysis.

Applying a constant-dollar analysis to the constant-dollar cash flows given in Table 6.2.4 requires the use of the inflation-free rate (i').

$$i' = \frac{(1.20)}{(1.07)} - 1 = 0.1215 \quad \text{or} \quad 12.15\%$$

Converting the constant-dollar cash flows to their present worth amounts (also in constant dollars) gives

$$\begin{aligned}\text{PW}(12.15)_{\text{A1}} = &-\ \$20{,}000 + \$11{,}215\overset{P/F,12.15,1}{(0.8917)}\\ &+ \$10{,}481\overset{P/F,12.15,2}{(0.7951)}\\ &+ \$9{,}796\overset{P/F,12.15,3}{(0.7089)}\\ &+ \$9{,}155\overset{P/F,12.15,4}{(0.6321)} = \$11{,}064\end{aligned}$$

$$\begin{aligned}\text{PW}(12.15)_{\text{A2}} = &-\ \$50{,}000 + \$20{,}561\overset{P/F,12.15,1}{(0.8917)}\\ &+ \$19{,}215\overset{P/F,12.15,2}{(0.7951)}\\ &+ \$17{,}959\overset{P/F,12.15,3}{(0.7089)}\\ &+ \$16{,}784\overset{P/F,12.15,4}{(0.6321)} = \$6{,}951\end{aligned}$$

Table 6.2.4. Two Alternatives in Actual and Constant Dollars

	Alternative A1		Alternative A2	
End of Year	Actual Dollars	Constant Dollars[a] (base year, t = 0)	Actual Dollars	Constant Dollars[a] (base year, t = 0)
0	−$20,000	−$20,000	−$50,000	−$50,000
1	12,000	11,215	22000	20,561
2	12,000	10,481	22,000	19,215
3	12,000	9,796	22,000	17,959
4	12,000	9,155	22,000	16,784

[a] Constant dollars calculated using the factor $1/(107)^t$.

Note that these are the same results obtained previously and thus, either analysis leads to the proper conclusion. However, additional insights may result when using a particular approach. For example, it might be that one would desire to know the rate of return earned on alternative A1 when the effects of inflation are considered.

By calculating the rate of return on the actual dollar cash flow you observe the return being earned when inflated dollars are used. For the investment, A1 the rate of return is

$$i^*_{A1} = 47.2\%$$

Calculation of the rate of return on the constant-dollar cash flow gives a more realistic view of the worth of the investment A1. That is, the loss in purchasing power due to inflation is reflected in the constant-dollar cash flow. Based on the constant-dollar cash flow, A1, the rate of return is:

$$i'^*_{A1} = 37.5\%$$

We observe that if actual dollars are used to calculate rates of return, they appear to be greater in magnitude than when using constant dollars. In this instance the use of constant dollars gives a better interpretation of the real return that might be earned. In other cases it may be more useful to utilize actual dollars. Therefore, the analyst should be able to present the results in terms of either actual or constant dollars.

6.2.3 AFTER-TAX ANALYSIS OF INVESTMENT ALTERNATIVES

6.2.3.1 Accelerated Cost Recovery System

Property placed in service during or after 1981 comes under the jurisdiction of the Accelerated Cost Recovery System (ACRS) which defined a different method of recovery for depreciable property than that used prior to 1981. ACRS requires that all eligible depreciable assets be assigned to one of four separate classes of property. These four classes are identified by the recovery period over which the property will be depreciated. A summary of these classes, reprinted with permission from Thuesen and Fabrycky, *Engineering Economy*, 6th ed., Prentice-Hall, Englewood Cliffs, NJ, p. 340, 1984, follows:

Three-year property includes cars and light-duty trucks. Also included is machinery and equipment used in research and experimentation and all other equipment having an ADR life of 4 years or less.

Five-year property includes all personal property not included in any other class. Also included is most production equipment and public utility property with an ADR life between 5 and 18 years.

Ten-year property includes public utility property with an ADR life between 18 and 25 years and depreciable real property such as buildings and structural components with an ADR life less than or equal to 12.5 years.

Fifteen-year property includes depreciable real property with an ADR life greater than 12.5 years and public utility property with an ADR life greater than 25 years.

In these definitions, ADR refers to the Asset Depreciation Range used prior to 1981, where the Internal Revenue Service specified ranges of lives for classes of assets. The ADR life represents the midpoint value of these ranges.[14]

Table 6.2.5. Depreciation Percentage for ACRS Classes

Recovery Year Is	3-Year	5-Year	10-Year	15-Year Public Utility
1	25	15	8	5
2	38	22	14	10
3	37	21	12	9
4		21	10	8
5		21	10	7
6			10	7
7			9	6
8			9	6
9			9	6
10			9	6
11				6
12				6
13				6
14				6
15				6

Table 6.2.6. Straight-Line Recovery Periods for ACRS

Type of Property	Optional Recovery Periods Available
3-year property	3, 5, or 12 years
5-year property	5, 12, or 25 years
10-year property	10, 25, or 35 years
15-year public Utility property	15, 35, or 45 years

When the property is assigned to a class, the depreciation percentage year by year is prescribed. Table 6.2.5 gives these percentages.

The other option available with ACRS is the use of straight-line depreciation for a number of different recovery periods. The taxpayer can select the recovery period based on the type of property. The recovery periods available, as dictated by the property type, are listed in Table 6.2.6. A more detailed description of the federal tax laws are provided by a tax guide.[15]

6.2.3.2 Calculating After-Tax Cash Flows*

The basic philosophy of the comparison of alternatives on an after-tax basis requires that the alternatives be isolated from the other activities within the firm. This approach forces each alternative to be judged on its particular merits, as best as they can be determined. To quantify the after-tax effects of a single alternative it is necessary to convert the before-tax cash flow into an after-tax cash flow. The most common method of accomplishing this conversion is to use the tabular method presented in Table 6.2.7. The example is based on the following information:

First cost	$60,000
Salvage value	None
Annual income	$50,000
Annual operating costs	$24,000
Depreciation method	ACRS (3-year recovery period)
Estimated life	4 years
Effective tax rate	46%

To assure that the tabular calculation procedure is general, positive values are assigned to revenue while costs are assigned negative values. Notice that in Column B the net annual income is found by adding the annual receipts (+) to the annual costs (−), giving net receipts of ($50,000 − $24,000) = $26,000. The initial outlay of the alternative is shown occurring at the present (the end of year 0). If there had been a salvage value at the time of disposal, it would have been included as a cash flow at the end of year 4. This before-tax cash flow represents the net cash flow for all activities except taxes.

Column C presents the depreciation charges as negative amounts. These amounts are costs, but they are *not* cash flows. Their purpose is to allow for the calculation of taxes, which *are* cash flows. Since the recovery

Table 6.2.7. Tabular Calculation of After-Tax Cash Flows

End of Year A	Before-Tax Cash Flow B	Depreciation C	Taxable Income, $B^a + C$ D	Taxes, $-0.46 \times D$ E	After-Tax Cash Flow, B + E F
0	−$60,000			$ 2,400[b]	−$57,600
1	26,000	−$15,000[b]	$11,000	−5,060	20,940
2	26,000	−22,800	3,200	−1,472	24,528
3	26,000	−22,200	3,800	−1,748	24,252
4	26,000	0	26,000	−11,960	14,040

[a] The initial investment does not affect ordinary taxable income.

[b] There is usually a 6% investment credit, since the recovery period is 3 years. If this percentage is used the depreciation base is reduced by one half of the investment credit.[16] However, there is an option that allows the use of the original cost of the asset as the depreciation base if for 3-year-recovery-period assets an investment credit of 4% is applied.[17] To keep calculations simple, the second option has been selected in this example.

* This section (modified) from G. J. Thuesen and W. J. Fabrycky, *Engineering Economy,* 6th ed., © 1984. Reprinted by permission of Prentice-Hall, Inc., Englewood Cliffs, NJ.

period is three years, there are depreciation charges for those three years rather than over four years, the estimated life.

The taxable income for each year in Column D is calculated by adding the amounts in Column B (excluding the first cost) to the amounts in Column C. The investment of funds has no effect on taxable income, since these are funds on which taxes were paid at the time they were earned. Therefore an alternative's first cost is not considered in the calculation of taxable income.

The taxes for each year shown in Column E are found by multiplying the tax rate of −0.46 times the taxable income. Since the taxable income is a positive value in this example, the taxes are negative amounts, indicating that they are disbursements or taxes paid. If the taxes were positive, this would represent a situation where taxes are being saved. That is, less taxes are being paid with the alternative than without it.

The last step is to add the taxes in Column E, which are cash flows, to the before-tax cash flows in Column B. The intermediate step dealing with depreciation is only to account for its tax effects. Since depreciation is not an actual disbursement of cash, it must not be included in the sum of the other cash flows when finding the after-tax cash flow. If the tax cash flows associated with an alternative are known, the use of Columns C and D is not necessary.

Using the after-tax cash flow from Table 6.2.7, the after-tax rate of return can be determined from the following expression:

$$0 = -\$57{,}600 + \$20{,}940(\overset{P/F,i^*,1}{\quad}) + \$24{,}528(\overset{P/F,i^*,2}{\quad})$$

$$+ \$24{,}252(\overset{P/F,i^*,3}{\quad}) + \$14{,}040(\overset{P/F,i^*,4}{\quad})$$

$$i^* = 17.8\% \text{ (after-tax)}$$

This return compares to the before-tax rate of return found from

$$0 = -\$60{,}000 + \$26{,}000(\overset{P/A,i^*,4}{\quad})$$

where

$$i^* = 26.3\% \text{ (before-tax)}$$

To decide whether this alternative meets the profit objectives of the firm, however, the after-tax rate of return must be compared to the after-tax MARR stipulated by the firm.

In many instances in the evaluation of economic alternatives only the costs of the alternatives are considered. Most often this occurs where alternatives that are intended to provide the same service are to be compared. Since the benefits to be derived are equal for each alternative, it is common practice to eliminate the estimation of associated revenues to reduce the evaluation effort.

It might appear that ignoring revenues for alternatives would prevent the comparison of these alternatives on an after-tax basis. Fortunately, this is not the case. It can be seen that the direct application of the procedures presented in Table 6.2.7 will permit the incorporation of tax effects in the comparison cash flows having equal revenue streams. By following the sign convention just used, (+) revenues, (−) costs, the tabular method will produce tax-adjusted cash flows that can be directly compared as long as (1) their revenues are assumed to be equal, and (2) the firm on the whole is realizing a profit. To see how the procedure operates, examine the cost-only cash flow in Table 6.2.8.

For this example, ACRS straight-line depreciation is used. The estimated salvage for the investment is $5000, and the effective tax rate is 42%. Since no annual operating revenues are being considered, the taxable

Table 6.2.8. Tax Effects for a Cost Flow

End of Year A	Before-Tax Cash Flow B	Depreciation Charges[b] C	Taxable Income, B + C D	Taxes (Savings), −0.42 × D E	After-Tax Cash Flow, B + E F
0	−$30,000				−$30,000
1	−15,000	−$10,000	−$25,000	$10,500	−4,500
2	−15,000	−10,000	−25,000	10,500	−4,500
3	−15,000	−10,000	−25,000	10,500	−4,500
3	5,000[a]		5,000	−2,100	2,900

[a] Salvage value.
[b] Depreciation charges ignore salvage value in ACRS.

income appears as $25,000 in costs each year. That is, the effect of this project is to reduce the firm's profits by $25,000 per year (if revenues aren't considered). Assuming that the firm has sufficient earnings from other activities to offset these costs, there will be a "savings" in taxes of $10,500. By following the sign convention utilized in Table 6.2.7, the taxes in Column E are positive, indicating that $10,500 in taxes would be avoided yearly if the costs in Column B and C were incurred.

The after-tax cash flow is found by adding these tax savings to the before-tax cash flow shown in Column B. By comparing this after-tax cash flow with a similarly calculated after-tax cash flow for a competing alternative, the alternative with the minimum equivalent after-tax cost can be identified. Thus, without any change in procedure, an after-tax comparison can be made for projects having the same revenue stream. Since this comparison does not require consideration of the possible revenue effects of the alternatives, much time and effort may be saved in performing the analysis.

REFERENCES

1. P. F. Ostwald, *Cost Estimating,* 2nd ed., Englewoods Cliffs, NJ: Prentice-Hall, 1984.
2. G. J. Thuesen and W. J. Fabrycky, *Engineering Economy,* 6th ed., Englewood Cliffs, NJ: Prentice-Hall, 1984, pp. 180–182.
3. L. E. Bussey, *The Economic Analysis of Industrial Projects,* Englewood Cliffs, NJ: Prentice-Hall, 1978, pp. 276–326.
4. Hillier, F. S. and G. J. Lieberman, *Introduction to Operations Research,* 4th. ed., San Francisco: Holden-Day, 1986.
5. G. J. Thuesen, and W. J. Fabrycky, *Engineering Economy,* 6th ed., Englewood Cliffs, NJ: Prentice-Hall, 1984, Ch. 6.
6. G. J. Thuesen and W. J. Fabrycky, *Engineering Economy,* 6th ed., Englewood Cliffs, NJ: Prentice-Hall, 1984, pp. 185–187.
7. G. J. Thuesen and W. J. Fabrycky, *Engineering Economy,* 6th ed., Englewood Cliffs, NJ: Prentice-Hall, 1984, pp. 187–188.
8. R. H. Bernhard, "Unrecovered Investment, Uniqueness of the Internal Rate, and the Question of Project Acceptability," *Journal of Financial and Quantitative Analysis,* March 1977, 33–38.
9. G. J. Thuesen and W. J. Fabrycky, *Engineering Economy,* 6th ed., Englewood Cliffs, NJ: Prentice-Hall, Inc., 1984, pp. 206–208.
10. ANSI Z94 Standards, *Industrial Engineering Terminology,* Industrial Engineering and Mangement Press, Norcross, GA, 1984, pp. 41–50.
11. *CPI Detailed Report,* U.S. Department of Labor, Bureau of Labor Statistics, published monthly, available from Superintendent of Documents, Washington, DC.
12. *Producer Prices and Price Indexes,* U.S. Department of Labor, Bureau of Labor Statistics, published monthly, available from Superintendent of Documents, Washington, DC.
13. *Statistical Abstracts of the United States,* published annually, U.S. Department of Commerce, Bureau of the Census, available from Superintendent of Documents, Washington, DC.
14. *Federal Tax Handbook, 1980,* Englewood Cliffs, NJ: Prentice-Hall, ¶ 2036.
15. *Federal Tax Handbook,* Englewood Cliffs, NJ: Prentice-Hall, published annually.
16. *Federal Tax Handbook, 1984,* Englewood Cliffs, NJ: Prentice-Hall, Inc., ¶2050(c).
17. *Federal Tax Handbook, 1984,* Englewood Cliffs, NJ: Prentice-Hall, Inc., ¶2050(k).

BIBLIOGRAPHY

Selected Books

American Telephone & Telegraphy Company, *Engineering Economy,* 3rd ed. New York: McGraw-Hill, 1977.

Au, T., and T. P. Au, *Engineering Economics for Capital Investment Analysis,* Boston, MA: Allyn & Bacon, 1983.

Barish, N. N., and Seymour Kaplan, *Economic Analysis for Engineering and Managerial Decision Making,* 2nd ed. New York: McGraw-Hill, 1978.

Bierman, H., and S. Smidt, *The Capital Budgeting Decision,* 5th ed. New York: Macmillan, 1980.

Blanchard, B. S., *Life Cycle Cost.* Portland, OR: M/A Press, 1978.

Blank, L. T., and A. T. Tarquin, *Engineering Economy,* 2nd ed., New York: McGraw-Hill, 1983.

Bussey, L. E., *The Economic Analysis of Industrial Projects,* Englewood Cliffs, NJ: Prentice-Hall, 1978.

Canada, J. R., and J. A. White, *Capital Investment Decision Analysis for Management and Engineering,* Englewood Cliffs, NJ: Prentice-Hall, 1980.

Clifton, D. S., and D. E. Fyffe, *Project Feasibility Analysis,* New York: Wiley, 1977.

Cohn, E., *Public Expenditure Analysis,* Lexington, MA: D.C. Heath, 1972.

Collier, C. A., and W. B. Ledbetter, *Engineering Cost Analysis,* New York: Harper & Row, 1982.

Dasgupta, A. K., and D. W. Pearce, *Cost-Benefit Analysis: Theory and Practice,* New York: Barnes and Noble, 1972.

DeGarmo, E. P., J. R. Canada, and W. G. Sullivan, *Engineering Economy,* 6th ed. New York: Macmillan, 1979.

De La Mare, R. F., *Manufacturing Systems Economics,* East Sussex, England: Holt, Rinehart & Winston, 1982.

Fabrycky, W. J., and G. J. Thuesen, *Economic Decision Analysis,* 2nd ed., Englewood Cliffs, NJ: Prentice-Hall, 1980.

Fabrycky, W. J., P. M. Ghare, and P. E. Torgersen, *Applied Operations Research and Management Science,* Englewood Cliffs, NJ: Prentice-Hall, 1984.

Goicoechea, A., D. R. Hansen, and L. Duckstein, *Multiobjective Decision Analysis with Engineering and Business Applications,* New York: Wiley, 1982.

Grant, E. L., W. G. Ireson, and R. S. Leavenworth, *Principles of Engineering Economy,* 7th ed. New York: The Ronald Press Company, 1982.

Griffiths, R. F., *Dealing with Risk—The Planning Management and Acceptability of Technological Risk,* New York: Wiley, 1981.

Hertz, D. B., and H. Thomas, *Risk Analysis and Its Applications,* New York: Wiley, 1983.

Holloway, C. A., *Decision Making under Uncertainty Models and Choices,* Englewood Cliffs, NJ: Prentice-Hall, 1979.

Humphreys, K. K., and S. Katell, *Basic Cost Engineering,* New York: Marcel Dekker, 1981.

Ignizio, J. P., *Goal Programming and Extensions,* Lexington, MA: D.C. Heath, 1976.

Jelen, F. C., *Cost and Optimization Engineering,* New York: McGraw-Hill, 1983.

Jeynes, P. H., *Profitability and Economic Choice,* Ames, IA: Iowa State University Press, 1968.

Jones, B. W., *Inflation in Engineering Economic Analysis,* New York: Wiley, 1982.

Kaplan, S., *Energy Economics, Quantitative Methods for Energy and Environmental Decisions,* New York: McGraw-Hill, 1983.

Keeney, R. L., and H. Raiffa, *Decisions with Multiple Objectives: Preferences and Value Tradeoffs,* New York: Wiley, 1976.

Mallik, A. K., *Engineering Economy with Computer Applications,* Mahomet, IL: Engineering Technology, Inc., 1979.

Marsh, W. D., *Economics of Electric Utility Power Generation,* New York: Oxford University Press, 1980.

Marston, A., R. Winfrey, and J. C. Hempstead, *Engineering Valuation and Depreciation,* Ames, IA: Iowa State University Press, 1963.

Morris, W. T., *Engineering Economic Analysis.* Reston, VA: Reston Publishing Company, 1976.

Newman, D. G., *Engineering Economic Analysis,* 2nd ed., San Jose, CA: Engineering Press, 1983.

Oakford, R. V., *Capital Budgeting,* New York: Ronald Press, 1970.

Ostwald, P. F., *Cost Estimating,* 2nd ed., Englewood Cliffs, NJ: Prentice-Hall, 1984.

Peters, M. S., and K. D. Timmerhaus, *Plant Design and Economics for Chemical Engineers,* New York: McGraw-Hill, 1980.

Peurifoy, R. L., *Estimating Construction Costs,* New York: McGraw-Hill, 1975.

Raiffa, H., *Decision Analysis: Introductory Lectures on Choice Under Uncertainty.* Reading, MA: Addison-Wesley, 1968.

Riggs, J. L., *Engineering Economics,* 2nd ed., New York: McGraw-Hill, 1982.

Smith, G. W., *Engineering Economy,* 3rd ed. Ames, IA: Iowa State University Press, 1979.

Steiner, H. M., *Public and Private Investments, Socioeconomic Analysis,* New York: Wiley, 1980.

Stermole, F. J., *Economic Evaluation and Investment Decision Methods,* Golden, CO: Investment Evaluations Corp., 1974.

Stewart, R. D., *Cost Estimating,* New York: Wiley, 1982.

Taylor, G. A., *Managerial and Engineering Economy,* 3rd ed. New York: Van Nostrand Reinhold, 1980.

Van Horne, J. C., *Financial Management and Policy,* 6th ed. Englewood Cliffs, NJ: Prentice-Hall, 1983.

White, J. A., M. H. Agee, and K. E. Case, *Principles of Engineering Economic Analysis,* 2nd ed. New York: Wiley, 1984.

Selected Periodicals

Accounting Review
American Economic Review
IIE Transactions
American Journal of Agricultural Economics
Appraisal Journal
California Management Review

Decision Sciences
Economic Journal
Engineering Cost and Production Economics
Engineering News Record
Engineering Economist
Financial Analyst's Journal
Financial Management
Harvard Business Review
Industrial Engineering
Journal of Accountancy
Journal of Business
Journal of Economics and Business
Journal of Finance
Journal of Financial and Quantitative Analysis
Journal of Taxation
Management Science
Public Utilities Fortnightly
Quarterly Review of Economics and Business

CHAPTER 6.3

COST CONTROL SYSTEMS

JOHN R. CANADA

North Carolina State University
Raleigh, North Carolina

6.3.1 DEFINITION AND PHILOSOPHIES

Cost control systems are any procedures used to compare actual costs with planned (anticipated, predetermined, standard, or budgeted) costs. Such systems normally should include determining what variations have occurred, their extent and causes; discovering conditions underlying each cause, and developing or revising policies, plans, methods, and practices for the purpose of eliminating unfavorable conditions.

The essence of any control system is feedback on comparison of actual performance with planned performance, i.e., variance analysis. Variance analysis provides the basis for raising questions, uncovering clues, and directing attention to possible needs for correction/improvement.

Ideally, planned performance is based on flexible budgets or standard costs. Flexible (or variable) budgets/standard costs are those prepared for a range of activity instead of a single level. They provide a dynamic basis for comparison because they are geared to changes in volume.

6.3.2 PRINCIPLES OF COST CONTROL

The following is an outline of cost control principles:*

1. Accounts should be fitted to organization chart so that costs can be segregated by individual responsibilities.
2. Cost accounts by individual responsibilities should be subdivided under uniform classifications to show nature of expenditures.
3. Goals in the form of standards, budgets, and allowances should be set and constantly kept up to date.
4. If cost varies with the rate of activity, variable or flexible budgets and allowances should be developed where justifiable.
5. Standards, budgets, and allowances should be prepared with the cooperation of the person responsible for each cost item and should be agreed to by that person.
6. Variations of actual costs from standard or budget should be segregated and shown in sufficient detail so that responsibility for each variance can be definitely determined.
7. Frequent reports of the costs for which he or she is responsible should be supplied by each person who is responsible for control of any cost element. These reports should emphasize variances of actual costs from standards or budgeted figures.
8. Apportioned or prorated costs over which an executive or subexecutive has *no* control should not be combined in his or her cost reports with the costs over which he or she *does* have control.
9. As an inducement to those responsible for the control of costs, an incentive system of the "savings-sharing" sort should be developed.

A substantial part of this chapter was adapted from L. P. Alford and John R. Bangs, *Production Handbook*, Ronald Press, 1954, and R. Beyer, *Profitability Accounting for Planning and Control*, Ronald Press, 1963. Reprinted with permission, John Wiley & Sons.

* Originally credited to Raymond P. Marple of the National Association of Cost Accountants.

6.3.3 VARIANCES

6.3.3.1 Types of Variance Calculations

Variances from standard cost can be expressed in either absolute or relative numbers. In the first instance, the variance is computed by subtracting actual cost from standard cost. If the actual exceeds the standard cost, the variance is negative, i.e., unfavorable and represents a variance loss; if actual cost is less than standard cost, the variance is positive, i.e., favorable and represents a variance gain. Thus, this method of expressing variances centers the attention of management upon dollar amounts of variation from standard costs.

By the second method, the variance is computed by dividing the standard cost figure into the actual cost figure to obtain the actual cost as a percentage of standard cost. Since standard cost is always the base for comparison, the standard cost is considered 100%. When actual cost has been converted to a percentage of standard, the actual cost percentage can be subtracted from the standard cost percentage (100%). The result, which may be either positive or negative depending upon whether actual cost is greater or less than standard cost, is the *cost variance* expressed as a percentage of standard cost. In contrast with the preceding method, a *relative variation* from standard is thus provided. This method is generally used in connection with basic or measurement standards.

These two methods present complementary aspects of the cost figures; both are required for a complete understanding of the cost variation that has taken place. Variances that are large in terms of dollars are sometimes so small in terms of *percentages* that they pass unnoticed by management if presented in the latter form alone; on the other hand, a large percentage variation may call attention to a substantial deviation from *standard efficiency*, yet the present actual loss in terms of dollars may be small.

6.3.3.2 Analysis of Variances

Since the total difference between the actual and standard cost is composed of variances arising from a variety of causes, it is necessary to resolve this total into its component parts in such a way that the contribution of each causal factor can be isolated. This is accomplished by taking the factors one at a time, while assuming that the other factors are held constant. Thus it is possible to calculate the influence upon cost of each cause of variation.

6.3.3.3 Computing Material and Labor Variances

The method illustrated below represents a mechanical arrangement of the data (Fig. 6.3.1) which makes possible the automatic computation of variances by totals and in detail. Under this method an overall variance is first obtained; this is then broken down in the case of material into usage and price variances.

The basic assumed facts in Fig. 6.3.1 appear in columns 1 to 5 inclusive. Information in columns 1 and 2 is taken from the standard cost card. Column 3 is based on stock ledger cards or other records. Column 4, the standard quantity in the product, is based on production records combined with standard and physical requirements as shown by the standard cost card. Column 5 of course represents actual production figures, taken from the daily or summary production reports. The remaining columns are merely different combinations of the information in the first five.

The same method is used to obtain variances for direct labor (Fig. 6.3.2). Efficiency is represented by a time variance and price by a rate variance. For greater clarity, these are often in turn subdivided. Thus, efficiency might be affected by the presence of learners, and a special allowance must therefore be made so that the foreman is judged only as to conditions under his or her control. Again, the rate variance is affected by the presence of overtime paid for at premium rates. Such overtime should be set forth as a separate variance in order to show the true extent to which basic wage rates have changed.

6.3.3.4 Analyzing Overhead Variances

There are various methods for analyzing overhead variances. Some of these are shown in Figs. 6.3.3 and 6.3.4. The following data are needed to provide the necessary information about variances:

1. Actual overhead is obtained by summing the debits in departmental overhead accounts (departmental expense distribution sheet).
2. Actual number of direct labor hours is compiled from time records.
3. Budgeted overhead is obtained from the flexible budgets by selecting figures corresponding to the actual number of direct labor hours worked.
4. The number of standard direct labor hours in production is found by multiplying the units produced by the standard labor hour content per unit as shown by the standard cost card.
5. The normal overhead rate is obtained from the standard cost card.

(1) Type of Material	(2) Standard Unit Cost	(3) Actual Unit Cost	(4) Standard Quantity (in product)	(5) Actual Quantity	(6) Standard Quantity at Standard Rate	(7) Actual Quantity at Standard Rate	(8) Actual Quantity at Actual Rate	Variances* (9) Overall (6) − (8)	(10) Usage (6) − (7)	(11) Price (7) − (8)
M-1	$1.00	$ .85	500	520	$ 500	$ 520	$ 442	$+58	$−20	$+78
M-2	7.00	7.70	200	205	1400	1435	1579	−179	−35	−144
M-3	2.00	2.20	1,080	1,080	2160	2160	2376	−216	0	−216
Totals					$4060	$4115	$4397	−337	−$55	−$282

* −: Unfavorable variances.
+: Favorable variances.

Fig. 6.3.1. Material cost variances.

Figs. 6.3.3 and 6.3.4 are based on the following assumed data:

	Department A	Department B
1. Standard overhead per hour	$ 1.00	$ 3.00
2. Standard allowed hours for actual production	2,000	800
3. Actual hours run	2,065	790
4. Actual overhead expense	$2,011	$2,870
5. Flexible budget allowances:		
a. For actual hours	$2,032	$2,795
b. For attained capacity:		
Dept. A (2000 h)	$2,000	
Dept. B (800 h)		$2,805

Capacity and Controllable Variances

Capacity and controllable variances represent volume and efficiency variances, respectively. The following facts form the basis for computing capacity and controllable variances:

1. Cleared-in cost.
2. Budget allowance.
3. Actual expenses.

The *cleared-in cost* is the amount charged to work in process, and represents the product of the standard allowed hours for the attained production and the standard hourly rate.

The *budget allowance* is the expense allowed at the attained level of production. It is usually obtained from the flexible budget, by interpolation, if necessary.

Efficiency, Expense, and Utilization Variances. *The expense variance* is the result of spending more or less than the budgeted allowances for indirect materials, indirect labor, etc., at the attained activity level.

The *efficiency variance*, also called *controllable variance*, is the result of using more or less than the standard amount of overhead service. It arises whenever the actual direct labor hours or machine-hours differ from the standard allowed hours.

(1) Operation	(2) Standard Hourly Rate	(3) Actual Hourly Rate	(4) Standard Hours (in product)	(5) Actual Hours	(6) Standard Hours at Standard Rate	(7) Actual Hours at Standard Rate	(8) Actual Hours at Standard Rate	Variances* (9) Overall (6) − (8)	(10) Time (6) − (7)	(11) Rate (7) − (8)
1	$5.00	$5.00	500	525	$ 2,500	$ 2,625	$ 2,625	−$ 125	−$125	$ 0
2	6.00	6.50	1,500	1,540	9,000	9,240	10,010	−1010	−240	−770
3	7.50	8.00	240	230	1,800	1,725	1,840	− 40	+ 75	−115
4	9.50	9.00	560	560	5,320	5,320	5,040	+ 280	0	+280
Totals					$18,620	$18,910	$19,515	−$ 895	−$290	−$605

* −: Unfavorable variances.
+: Favorable variances

Fig. 6.3.2. Direct labor cost variances.

Department	(1) Cleared-in Cost	(2) Allowed Budget	(3) Actual Cost	Variances (4) Overall (1) − (3)	(5) Capacity (1) − (2)	(6) Controllable (2) − (3)
A	$2,000	$2,000	$2,011	−$ 11	$ 0	−$11
B	$2,400	$2,805	$2,870	−470	−405	−65
Totals	$4,400	$4,805	$4,881	−$481	−$405	−$76

Fig. 6.3.3. Overhead variances.

The *utilization variance*, frequently referred to as *capacity* variance, or *volume* variance, is the result of operating more or less than the normal number of hours in any given budget period.

The computation of variances under this method is shown in Fig. 6.3.4. Note that the resulting variances differ from those in Fig. 6.3.3, first, in the kind of information obtained and, second, in the method of computation.

6.3.4 COST REPORTS

6.3.4.1 Management Reports

The well-designed management report should have at least the following characteristics:

1. Amounts should be rounded to the nearest significant digit.
2. The amount of detail shown should be related to the management level at which the report is directed.
3. The results shown should be related to meaningful standards.
4. Where applicable, the impact of actual results upon the profit plan should be reported.
5. Information should be presented in trend form wherever possible.
6. In general, the report should be readable by being concise and avoiding a crowded appearance.

It should be emphasized that reports should direct attention more toward a few exceptions, or variances, from planned performance. Ideally, all managers of a group of people should receive information concerning the performance of each of their subordinates at the lower operating levels, where control over costs is likely to be the primary measure of performance. Both actual results and standards often are more effectively presented in non-dollar units such as people or gallons.

6.3.4.2 Requisites of Cost Reports

The following rules should be observed in the preparation and presentation of reports to attain the characteristics above.

1. *Compactness.* Economizing the managers' time and effort. This means that managers should be presented first with a summary which, by itself, gives them a bird's-eye view of conditions. The *principle of exceptions* may be utilized in the construction of this summary, thus eliminating items that are in line with standards, for these do not require further study or action.
2. *Physical makeup, i.e., the form of the report.* This involves use of descriptive titles, proper dating, the use of clear and concise forms, graphic presentation of results, and good arrangement of data.

Department	(1) Standard Hours at Standard Rate	(2) Actual Hours at Standard Rate	(3) Budget for Actual Hours Worked	(4) Actual Expense	Variances (5) Overall (1) − (4)	(6) Efficiency (1) − (2)	(7) Activity (2) − (3)	(8) Expense (3) − (4)
A	$2,000	$2,065	$2,032	$2,011	−$ 11	−$65	+$ 33	+$21
B	2,400	2,370	2,795	2,870	−470	+30	−425	−75
Totals	$4,400	$4,435	$4,827	$4,881	−$481	−$35	−$392	−$54

Fig. 6.3.4. Overhead variances.

3. *Timeliness.* Managers who have cooperated in setting standards for which they are to be held accountable must have prompt and accurate reports of their actual performance. This might be as often as daily, but normally would be weekly or monthly.
4. *Content.* In general, those items that are controllable by a given manager need to be emphasized in the report. There is no objection to placing in a report information about noncontrollable costs; in fact, to do so may aid managers to acquire a broader understanding of the company's problems. But there must be a strict separation between controllable and noncontrollable items in order that expense control may be positive.

6.3.4.3 Material Cost Reports

It is desirable to separate material cost variances that are the result of purchasing activities from those that are the result of manufacturing activities. The former account largely for price variations, the latter for usage variations. The causes of each of these variations may be subdivided. The causes below are suggestions only.

1. *Price variance sources*:
   a. Changes in market price.
   b. Improper purchasing policies.
      (1) Changes in purchasing policies.
      (2) Wrong quantity.
      (3) Wrong quality or grade.
   c. Errors in recording.
2. *Usage variance sources*:
   a. Changes in design of product, machinery, or tools.
   b. Changes in methods of processing or fabricating.
   c. Spoilage and waste (especially excess spoilage) in production.
   d. Losses in storage of raw materials, finished products, and finished goods through spoilage, theft, waste, etc.
   e. Damage during handling.
   f. Too rigid inspection.
   g. Errors in accounting charges.

Price Variance Reports

Fig. 6.3.5 is a sample report of material price variations. Its purpose is to separate the price-variance factor from the usage factor on both raw and manufactured materials. The usage factor is subjected to later analysis on the basis of controllable efficiencies. This report compares actual against standard in dollars at the time of purchase, including standard and actual unit prices and quantity purchased.

Material Usage Reports

A material usage report would normally express usage standards in normal physical measurement units (as well as, perhaps, in dollars) for each material or product. A daily report might show only those items for which the actual units used exceed standard units allowed. A weekly or monthly report might list complete performance

	Standard Price	Actual Price	Quantity Purchased	Total value at Standard	Total Value at Actual	Price Variations
Raw materials:						
Warp yarn						
10/2 carded	.28	.248	54,136	$15,158	$13,447	$1,711
20/2 carded	.31	.30	10,051	3,116	3,015	101
ETC.						
Spool yarn						
20/2 carded						
30/2 carded						
ETC.						
Bobbin yarn						
80/2 combined						
90/2 combined						

Fig. 6.3.5. Material price variance report.

Scrap Cost by Responsibility—Week Ending April 3, 19 ___

Responsibility Manufacturing Division	Direct Labor	Scrap Cost	Unit Scrap Cost per Direct Labor Dollar	Standard Unit Scrap Cost per Direct Labor Dollar	Variances from Standard Week End 4/3	Variances from Cumulative 8/1–4/3
T. Jones (Foreman)						
Rubber mill—Antenna—1141-X	$	$	$	$	$	$
Rubber mill—Duprene						
Rubber mill—Brake Hose						
Miscellaneous Manufacturing						
Total Manufacturing Division	$ 54,800	$ 3,800	$.069	$.067	$−100	$ −8,700
Material division	$	$	$	$	$	$
Engineering division						
Sales division						
Inspection division						
Plant not determined						
Total Week Ending April 3, 19 ___	$ 54,800	$ 4,500	$.082	$.083	$+55	—
Total Cumulative Car Model 8/1–4/3	$1,500,200	$143,000	$.095	$.085	—	$−17,600

Fig. 6.3.6. Weekly scrap report.

in relation to allowable standards; that is, for each product and material, the report might show the total number produced, total units of material used, standard units, and physical and dollar variances for the period and cumulative to date.

A similar situation exists with respect to scrap and spoilage (Fig. 6.3.6). Note that an allowance for these items is included in the standard. The report, therefore, is intended to focus attention on *excess* scrap and spoilage.

6.3.4.4 Labor Cost Reports

The analysis of direct labor cost variances follows the same general plan as that for direct material cost variances, but the causes of the variances differ. A list of possible labor variance causes follows:

1. *Rate variance sources*:
   a. Wage rate changes.
   b. Change of payment plan, e.g., from piecework to measured day work.
   c. Change in grade of labor used.
   d. Clerical errors.
2. *Sources of time or efficiency variances:*
   a. Selection of workers.
   b. Training of workers
   c. Frequency of change-overs.
   d. Labor turnover.
   e. Incentive wage payment plan.
   f. Working conditions.
   g. Working hours.
   h. Honesty among workers.
   i. Selection of machines and tools.
   j. Changes in design of product.
   k. Changes in machinery, tools, or methods of production.
   l. Adequate accounting or production records.

DIRECT LABOR
WEEKLY GAIN AND LOSS REPORT

FOREMAN: *L.E. Pratt* DEPARTMENT #18 WEEK ENDING: *February 11, 19–*

TYPE OR STYLE	BUDGET			ACTUAL			GAIN PER UNIT	LOSS PER UNIT	REMARKS
	Production	Unit Man Hours	Total Man Hours	Production	Unit Man-Hours	Total Man-Hours			

Fig. 6.3.7. Direct labor, weekly gain and loss report. From Alford and Bangs, *Production Handbook.*

The immediate *control of direct labor cost* is, in most concerns, in the hands of foremen. This requires that they be provided with reports daily or weekly to help them in keeping this element of cost within standard limits. Where the rates paid are determined by a contract with a union or by managers other than foremen, only the usage and selection of the correct grade of labor is changeable to foremen.

Rate Variances

The most obvious cause of rate variances is authorized changes in the wage structure. Such changes should come only as a worker is moved from one job class to another, on the basis of a carefully worked out policy of job classification and salary ranges.

Actually, there are two principal causes of a wage rate variance, aside from authorized wage increases:

1. Employment of high-rate employees on low-rate tasks.
2. Overtime work.

In the former case, an increase in the average hourly rate of a department takes place, although no one has been given a wage increase. In the case of overtime a similar increase takes place, due to the premium wage paid for the excess hours. The presence of both should be checked constantly through reports.

Time Variances

The reasons for excess labor time are more numerous than those for labor rate variances, but they are also harder to discover and control. Consequently, the bulk of labor reports is devoted to the time element.

A comparison between estimated or budgeted unit man-hours and actual unit man-hours, revealing either a gain or loss, is provided for in Fig. 6.3.7. Study of such a report may lead to an investigation of the effectiveness of labor and to a revision of labor policies. It may be found that the right man was not placed in the right job; that a foreman has failed to secure cooperation of workers; that training and instruction of new workers was inefficient and faulty; that wage incentive plans were lacking; or that labor turnover was excessive. Information thus presented and compared has a direct bearing upon individual productivity of workers under the foreman of a department. The report covers a week, but a similar report should be prepared daily. The latter represents standard practice in the case of one automobile manufacturer, where a daily plant report includes a comparison of actual and standard labor costs.

	Standard Cost	Actual Cost	Cost Variations	Cause of Variations			Foremen's Budget	
				Level of Operations	Management Changes	Foremen's Efficiency	Budget Allowance	% Actual of Budget
Productive Cost Centers:								
Yarn Preparation								
Bobbin								
Warp								
Weaving								
Mending								
Bleach								
Dress								
Cutting and Splitting								
Finishing								
Wrap and Label...........								
Stock Room								
Total	$144,864	$138.799	$6.065	$5.150	$1.110*	$2.025	$140.824	98.6

* Denotes red figures.

[Detailed figures purposely omitted.]

Fig. 6.3.8. Analysis of expense variations by cost centers (detailed figures purposely omitted). From Alford and Bangs, *Production Handbook*.

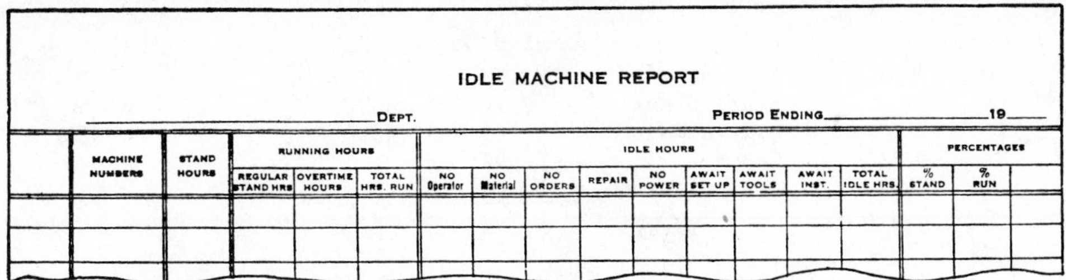

IDLE MACHINE REPORT

_______________ DEPT. PERIOD ENDING_______________19____

	MACHINE NUMBERS	STAND HOURS	RUNNING HOURS			IDLE HOURS									PERCENTAGES		
			REGULAR STAND HRS	OVERTIME HOURS	TOTAL HRS. RUN	NO Operator	NO Material	NO ORDERS	REPAIR	NO POWER	AWAIT SET UP	AWAIT TOOLS	AWAIT INST.	TOTAL IDLE HRS.	% STAND	% RUN	

Fig. 6.3.9. Idle machine time report. From Alford and Bangs, *Production Handbook*.

6.3.4.5 Manufacturing Expenses Reports

Determination of the proper allowance for a given expense is the first step in its control. Once proper allowances are set, control is obtained largely through determining an efficiency variance by comparison of actual expenditures with amounts allowed for a given rate of activity. However, since the expense absorbed differs from that allowed at all but the normal rate of activity, there is also a volume variance.

The expense variances may be classified as follows:

1. *Spending variance* sources:
   a. Using wrong grade of materials.
   b. Using wrong grade of labor.
   c. Failure to get most favorable terms in buying.
   d. Changes in market price.
2. *Efficiency variance* sources:
   a. Waste of materials.
   b. Inefficient labor performance.
   c. Failure to curtail usage of materials and services to correspond with output level.

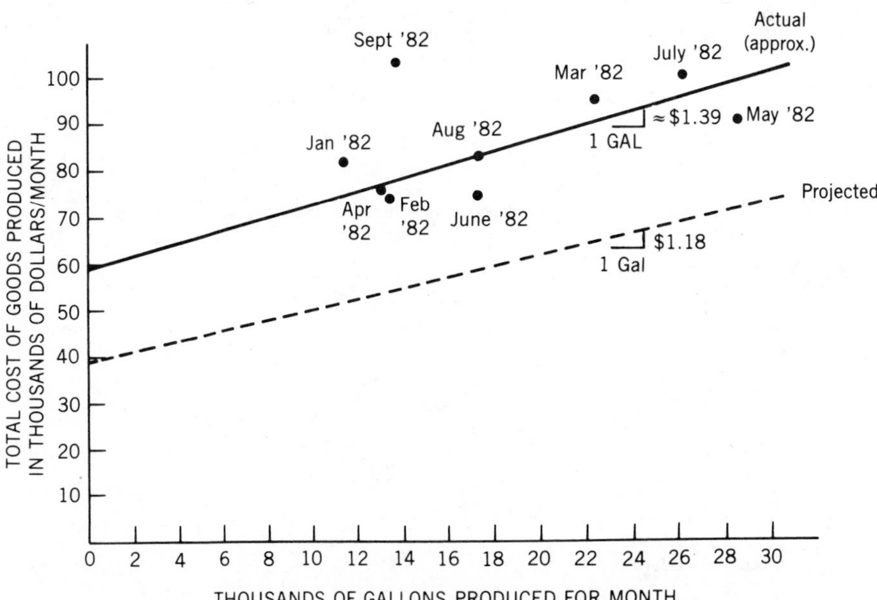

Fig. 6.3.10. Total cost of goods produced vs. gallons/month, January through September 19 __ .

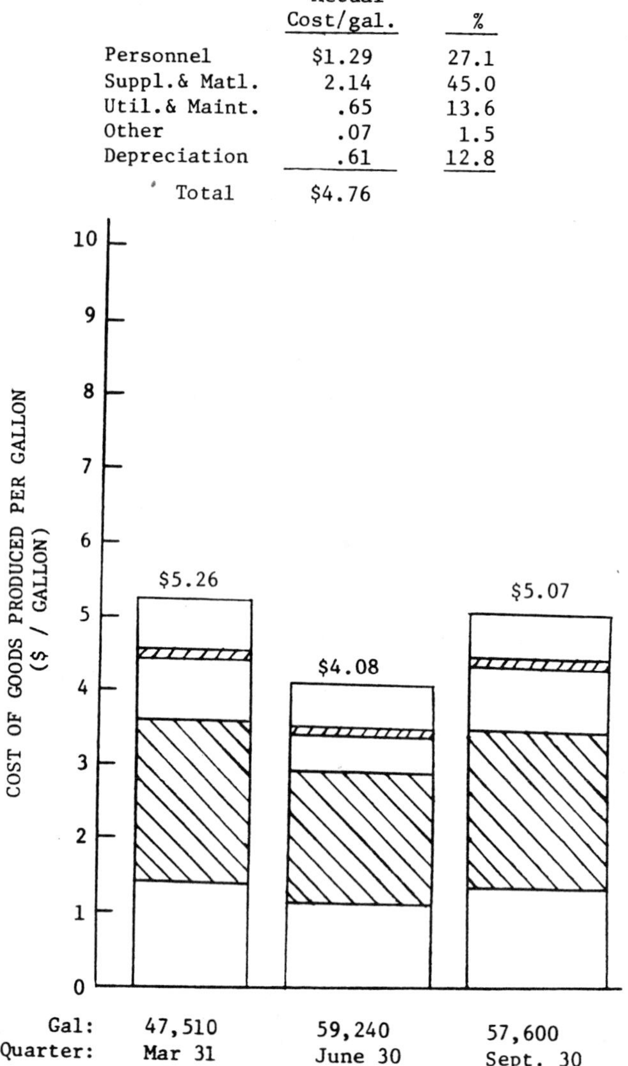

	Actual Cost/gal.	%
Personnel	$1.29	27.1
Suppl.& Matl.	2.14	45.0
Util.& Maint.	.65	13.6
Other	.07	1.5
Depreciation	.61	12.8
Total	$4.76	

Fig. 6.3.11. Cost per gallon produced, categorized for first three quarters of 19 __ .

3. *Utilization variance* sources:
   a. Controllable causes:
      Employees waiting for work.
      Avoidable machine breakdowns.
      Lack of operators.
      Lack of tools.
      Lack of instructions.
   b. Noncontrollable causes:
      Decrease in customer demand.
      Calendar fluctuations.
      Excess plant capacity.

The first group is best studied through budget comparisons, the second and third through departmental cost reports.

Departmental Cost Reports

Figure 6.3.8 illustrates the analysis of expenses by cost centers. Separate reports show the detailed budget for each cost center.

Volume Variance

The difference between budgeted expense and applied expense is a volume variance, which represents idle capacity. The idle time report, Fig. 6.3.9, provides columns for the principal causes of idle time. It gives the number of hours each individual item of equipment was idle; the number of hours it should have run based on standard hours of the shop; the number of overtime hours run; and the percent of total idle time to standard hours and also to total hours actually run. This report is a valuable record for the foreman, as it reflects each week the running condition of the equipment, showing both success and failure of effort to eliminate idle time. As machine rates are figured on normal capacities, there is no idle machine loss in money unless the machine runs less than normal time.

6.3.4.6 Graphical Reporting

Pictorial representations such as graphs provide excellent means of making data more easy to assimilate and highlight. As limited examples, Figs. 6.3.10, 6.3.11, and 6.3.12 show three types of graphical/tabular presentation concerning costs for a small chemical plant.

Figure 6.3.10 shows a plotting of total cost vs. gallons produced by month, together with a straight line to roughly fit the actual data. Also shown is a dashed straight line for the originally projected data. Given the great difference in the two lines, further detail would seem justified.

Figure 6.3.11 shows the cost data on a per-gallon basis divided into five main categories for three-month periods. Note that the width of each histogram bar is in proportion to the gallons provided for each three-month period.

	Projected	Actual	Projected−Actual	Variances of Note % of Projected −100% 0 +100% Un-Favor / Favor
Fixed Cost/Month				
Salaries & fringes				
Administration	$ 9,600/mo	$ 9,700/mo	$−100/mo	
Plant	16,300	13,834	+2,466	
Subtotal	$25,900/mo	$23,534/mo	$+2,366	+9%
Insurance	800	589	+211	
Trucks & cars	2,290	2,623	−353	
Travel subsist	420	1,299	−879	−77%
				−115%
General maintenance & contingency	4,000	7,098	−3,098	−80%
Fixed supplies & materials	5,700	12,235	−6,535	−2%
Subtotal	$13,210	$23,844	−10,634	−21%
Depreciation	$11,000	$11,200		
Total fixed costs/mo	$50,110	$58,578	$−8,468	
Variable Costs/Gallon				
Supplies & materials				−75%
Process A	$0.10/gal	$0.175/gal	$−0.075/gal	−89%
Process B	0.12	0.227	−0.107	−14%
Process C	0.43	0.490	−0.060	−14%
Process D	0.23	0.263	−0.033	
Subtotal	$0.88	$1.155	$−0.275	−24%
Containers	$0.20	$0.13	$+0.07	+35%
Feedstock	$0.10	$0.10	0	
Total variable costs/gal	$1.18	$1.385	$−0.205	−12%

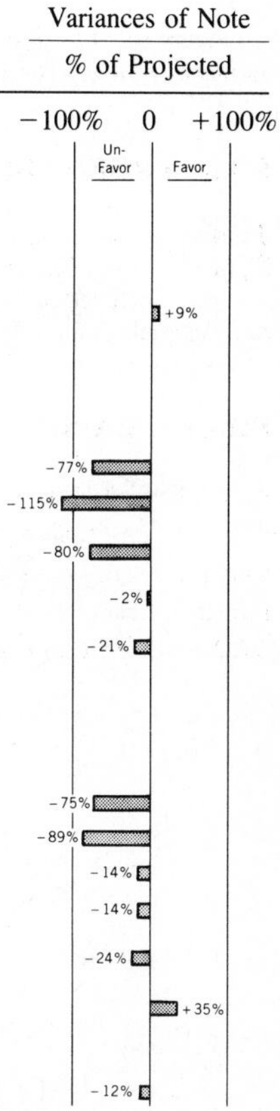

Fig. 6.3.12. Cost variance analysis.

ILLUSTRATING TREND AND EXCEPTION REPORTING

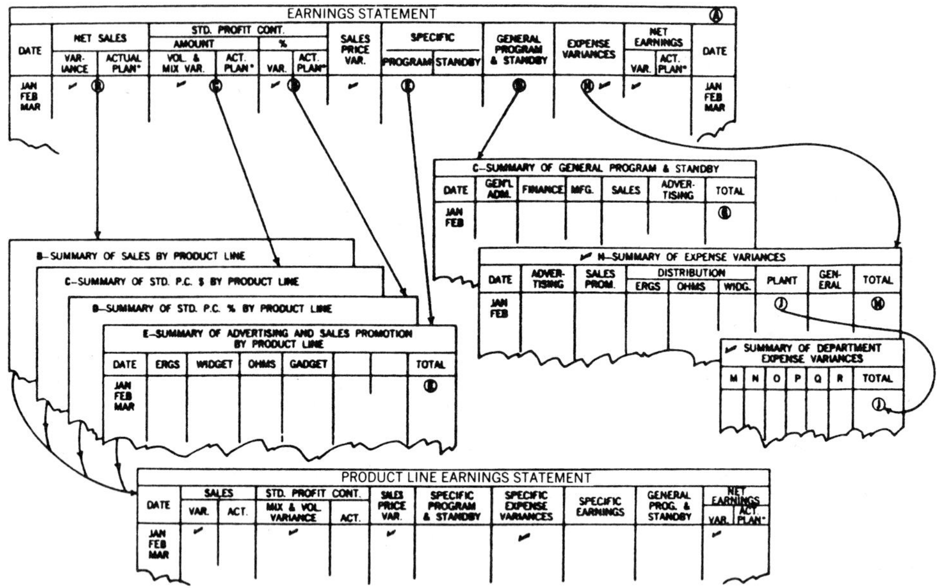

Fig. 6.3.13. Integrated operating statements. From Beyer, *Profitability Accounting for Planning and Control.*

Figure 6.3.12 shows a further breakdown of actual costs and also gives corresponding projected costs and variances of note. The percentage variances are highlighted by horizontal bars in proportion to the amounts.

6.3.5 PROFITABILITY ACCOUNTING AND REPORT INTEGRATION

The logical extension of cost-control systems is to consider also sales and profits by product line over time. Figure 6.3.13 shows how such reports can be broken down and interrelated. Note that the right-hand side generally includes consideration of expense variances, which can be further broken down into overhead, labor, and materials components as described earlier.

BIBLIOGRAPHY

Alford, L. P. and Bangs, John R., *Production Handbook*, New York: Ronald Press/Wiley, 1954.

Beyer, R., *Profitability Accounting for Planning and Control*, New York: Ronald Press/Wiley, 1963.

Gillespie, Cecil, *Cost Accounting and Control*, Englewood Cliffs, NJ: Prentice Hall, 1957.

Horngren, Charles T., *Accounting for Management Control*, 3rd ed., Englewood Cliffs, NJ: Prentice-Hall, 1974.

Shah, Pravin P., *Cost Control and Information Systems*, New York: McGraw-Hill, 1981.

CHAPTER 6.4

UNIT COST ANALYSIS

SAMUEL EILON

Imperial College of Science and Technology
London, England

6.4.1 INTRODUCTION

All managers, and particularly those involved in the production function, are naturally interested in the makeup of the total cost and unit cost (i.e., average cost per unit of output) incurred in the enterprise. One of the preoccupations of these managers is the behavior of the total cost function with respect to output volume, since the size of plant and the level of output may be crucial determinants of the economics of production. Some examples are illustrated graphically in Fig. 6.4.1:

- **(a)** The cost function is linear with a fixed cost element and a constant direct cost per unit, so that the incremental cost per unit declines as volume rises, indicating economies of scale.
- **(b)** The cost function is concave, also leading to economies of scale.
- **(c)** Here the cost function is convex, leading to an increasing unit cost with increased volume, i.e., "diseconomies of scale."
- **(d)** The cost function is concave to start with, exhibiting economies of scale, and then a turning point is reached, beyond which diseconomies of scale occur.

The relationship between total cost and unit cost is given by the simple expression.

$$C = cV \tag{6.4.1}$$

where

$$C = \text{total cost}$$

$$c = \text{unit cost}$$

$$V = \text{output volume}$$

Managers are often interested in analyzing *changes* in total cost and unit cost, so that the causes of change can be identified; alternatively, in planning future activities, it is important to postulate what possible effects changes in the values of certain variables in the system may have on total cost and unit cost, so that appropriate courses of action may be considered. If, in Eq. 6.4.1, the following change takes place from one time period to another,

$$c \rightarrow c + \Delta c$$

(i.e. the unit cost c in the base period becomes $c + \Delta c$ in a subsequent period), and similarly

$$V \rightarrow V + \Delta V$$

leading to

$$C \rightarrow C + \Delta C$$

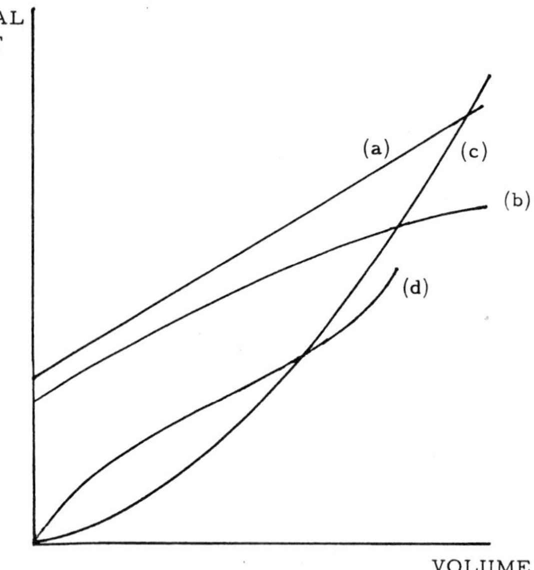

Fig. 6.4.1. Several types of cost models.

where Δc, ΔV, and ΔC are the absolute changes that take place, then it is simple to show with incremental calculus that†

$$C^* = c^* + V^* + c^*V^* \tag{6.4.2}$$

and consequently the relative change in unit cost is

$$c^* = (C^* - V^*)/(1 + V^*) \tag{6.4.3}$$

where

$$c^* = \Delta c/c = \text{the relative change in } c$$

$$C^* = \Delta C/C$$

$$V^* = \Delta V/V$$

By definition, therefore, $V^* > -1$, so that an increase in unit cost is avoided ($c^* \leqslant 0$) if the condition

$$C^* \leqslant V^*$$

is met. Thus, it is possible for the unit cost to decline in spite of an increase in total cost, provided that it is coupled with an even larger proportionate increase in volume.

6.4.2 TYPES OF COST MODELS

Analysis of costs generally needs to be undertaken in some detail, and for that purpose a disaggregation of the cost function becomes necessary. It is often said that a properly managed accounting system should be able to provide the disaggregated information in a form amenable to analysis. However, there are many ways in which cost items can be classified, depending on the conventions associated with any given accounting and managerial control system. Of the many types of cost models that can be constructed, the following four are perhaps most commonly found in practice:

1. The factor inputs model
2. The linear cost model

† The methods of incremental calculus and the major part of this chapter are discussed in greater detail in Ref. 5.

3. The divisional or product cost model
4. The functional cost model

6.4.2.1 The Factor Inputs Cost Model[1]

This model focuses on the major types of inputs to the manufacturing system and is characterized by

$$C = M + W + E \tag{6.4.4}$$

where

M = cost of materials

W = cost of manpower

E = other expenses, including interest charges, depreciation, rent and any other cost not accounted for in M and W

The unit cost then becomes

$$c = c_M + c_W + c_E \tag{6.4.5}$$

where

c_M = unit material cost (i.e., unit cost due to materials) = M/V

c_W = unit labor cost = W/V

c_E = unit cost incurred by other expenses = E/V

It is possible, of course, to disaggregate the cost function further, for example by

$$M = M_1 + M_2 + \cdots$$

where $M_1, M_2, \ldots$ are the costs of raw materials of various types, bought-in components, subassemblies or semifinished products, etc. Similarly, W and E may each be disaggregated into several components if a more detailed analysis is desirable.

6.4.2.2 The Linear Cost Model[3,5]

It is convenient for the purpose of cost analysis to differentiate between *fixed costs* and *variable costs*:

Fixed costs, denoted by F, include all the cost elements that remain unchanged, irrespective of a change in the level of output, V. These costs often cover rent, local taxes (where these are determined by the locality and type of premises used and not by turnover or profit), depreciation of machinery and plant, interest on borrowings to finance fixed assets, salaries and benefits of personnel who must be retained even when output declines, and any other fixed charges on services and facilities. When F is said to be fixed, it is fixed in relation to the output V, that is, when V changes (all other things being equal), the value of F remains unaffected. There may, of course, be other factors that have an effect on F from one time period to another, such as a change in the level of rent, interest rates, fixed manning of machines, office staff, changes in maintenance procedures, or consequences of investment in fixed assets. Any one or a combination of these changes may cause F to change to $F + \Delta F$, the relative increment being $F^* = \Delta F/F$.

Variable costs, denoted by $S(V)$, cover the cost items that are affected by a change in volume, such as direct labor and materials, maintenance of plant (except for maintenance needed even when the plant is idle), sales commissions, depreciation of fixed assets associated with their being in use (i.e., excluding depreciation attributed to the passage of time when the machinery is idle), fuel, and so on.

The distinction between fixed and variable costs is not always easy to make. First, the segregation of cost elements into fixed and variable categories is often dictated by internal accounting conventions, which may have historical roots and may seem arbitrary at times. Second, the fixed cost component tends to change by discrete amounts when the level of activity reaches certain levels, so that in practice F does in fact depend on V and the conventional classification of F as "fixed" is to indicate that it *does not change at the margin* (i.e., when the volume increases or decreases by one unit).

The general fixed and variable cost model then takes the form

$$C(V) = S(V) + F$$

or (6.4.6)

$$C = S + F$$

for short, where S is the variable cost component with the property that $S = 0$ when $V = 0$. The unit cost is

$$c = c_S + c_F \tag{6.4.7}$$

where

$$c_S = \text{unit variable cost} = S/V$$

$$c_F = \text{unit fixed cost} = F/V$$

The linear cost model is a special case of Eqs. 6.4.6 and 6.4.7 when the variable cost component is given by

$$S = sV$$

where the unit variable cost is $c_S = s = \text{const}$, so that

$$C = sV + F \tag{6.4.8}$$

and

$$c = s + F/V \tag{6.4.9}$$

The fixed cost F may be disaggregated into several components, for example interest charges (denoted as J) and advertising costs (denoted as D), which may be sufficiently important to be considered on their own, so that the unit cost function may then be rewritten as

$$c = s + F/V + J/V + D/V \tag{6.4.10}$$

where F here denotes the remaining fixed cost element.

6.4.2.3 Divisional or Product Cost Model

If the enterprise is organized on divisional or product lines, with each division or product constituting a separate cost or profit center, then the cost function is simply the sum of the cost centers:

$$C = C_1 + C_2 + \cdots$$

Such a scheme implies that any central overhead costs (say, at the headquarters office) are allocated to the various business entities and are therefore included in C_1, C_2, etc. Alternatively, the central costs may be unallocated and treated as a separate item (say, F), so that

$$C = F + C_1 + C_2 + \cdots$$

6.4.2.4 Functional Cost Model

Instead of grouping the costs under divisional or product headings, they may be charged to particular functions, such as production, marketing, finance, personnel, and research and development. The model will then take the form of the last equation with C_1, C_2, . . . denoting the costs of the various functions and again with F being the unallocated cost. Thus, there is no fundamental conceptual difference between this and the previous model.

6.4.3 USE OF MULTIMODELS

These models are not, of course, mutually exclusive, and a good accounting system will allow the cost data bank to be organized in such a way that several models may be constructed to serve diverse managerial purposes. For example, models may be constructed in a hierarchical fashion, as indicated in Fig. 6.4.2, where a divisional cost model is first postulated, followed by a breakdown for each division in the form of a cost function of its factor inputs, as in Eq. 6.4.4, or as a linear fixed and variable cost model, as in Eq. 6.4.8. Such a scheme allows a great deal of flexibility, in that it combines a global cost analysis for the enterprise as a whole, incorporating its major constituent parts, with an ability to produce a more detailed picture for whichever part is appropriate.

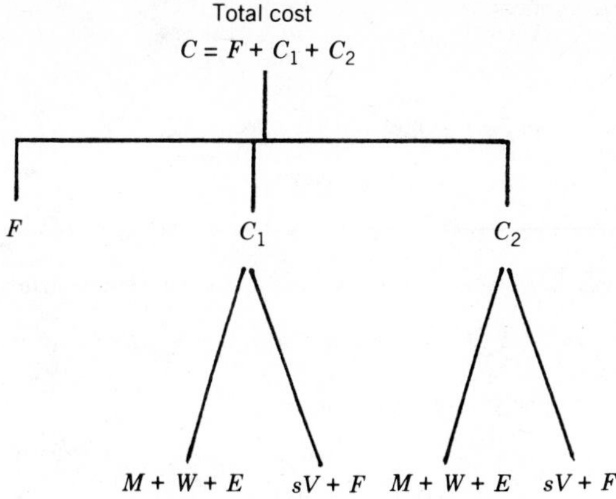

Fig. 6.4.2. A hierarchical cost structure.[5]

The analysis in this chapter is confined to the first two models, which differ fundamentally from each other in their cost structure and purpose. Both models can coexist, as can be seen from the example in Table 6.4.1. The items for the table are listed in three groups, which yield the sums M, W, and E for Eq. 6.4.4, and each item is categorized as fixed or variable (or divided into fixed and variable components), so that the total fixed cost F is determined as the sum of all the entries in its column. Similarly, S is obtained (from which s is ascertained) for Eq. 6.4.8.

6.4.4 ANALYSIS OF THE FACTOR INPUTS COST MODEL

6.4.4.1 The Relative Unit Cost Increment

Applying the addition rule in incremental analysis to the cost function in Eq. 6.4.4 it follows that[1]

$$C^* = f_M M^* + f_W W^* + f_E E^* \tag{6.4.11}$$

where

$$\begin{aligned} f_M &= M/C = \text{materials cost proportion} = c_M/c \\ &= \text{materials unit cost proportion} \\ f_W &= W/C = c_W/c = \text{labor cost (or labor unit cost proportion)} \\ f_E &= E/C = c_E/c = \text{other expenses cost (or unit cost) proportion} \end{aligned} \tag{6.4.12}$$

Table 6.4.1. Data for Two Cost Models[5]

Input	Fixed Cost	Variable Cost	Total
Raw materials		M_1	
Bought components		M_2	
Materials for maintenance	M_3	M_4	M
Fuel and energy	M_5	M_6	
Administration	W_1	W_2	
Production labor	W_3	W_4	W
Other labor costs	W_5	W_6	
Depreciation	E_1		
Interest charges	J_1	J_2	
Rent	E_2		E
Miscellaneous	E_3		
Total	F	S	

and the sum of these cost proportions

$$f_M + f_W + f_E = 1 \tag{6.4.13}$$

Similarly, the effect of changes on the unit cost is derived as[1]

$$\begin{aligned} c^* &= (C^* - V^*)/(1 + V^*) \\ &= (f_M M^* + f_W W^* + f_E E^* - V^*)/(1 + V^*) \end{aligned} \tag{6.4.14}$$

and when the absolute change in volume is small ($V^* << 1$), the approximation

$$c^* \simeq C^* - V^* = f_M M^* + f_W W^* + f_E E^* - V^*$$

may be used. The condition for the unit cost to decline is

$$c^* < 0 \quad \text{if} \quad C^* = f_M M^* + f_W W^* + f_E E^* < V^* \tag{6.4.15}$$

An example for the application of these relationships is given in Fig. 6.4.3, where $f_W = 0.2$ and $f_E = 0.4$ and where the effect of a volume change V^* is shown. The two lines correspond to $E^* = 0$ and $E^* = 5\%$, respectively and the domain to the right of each line signifies a reduction in unit cost ($c^* < 0$) while on each line no change in unit cost takes place ($c^* = 0$).

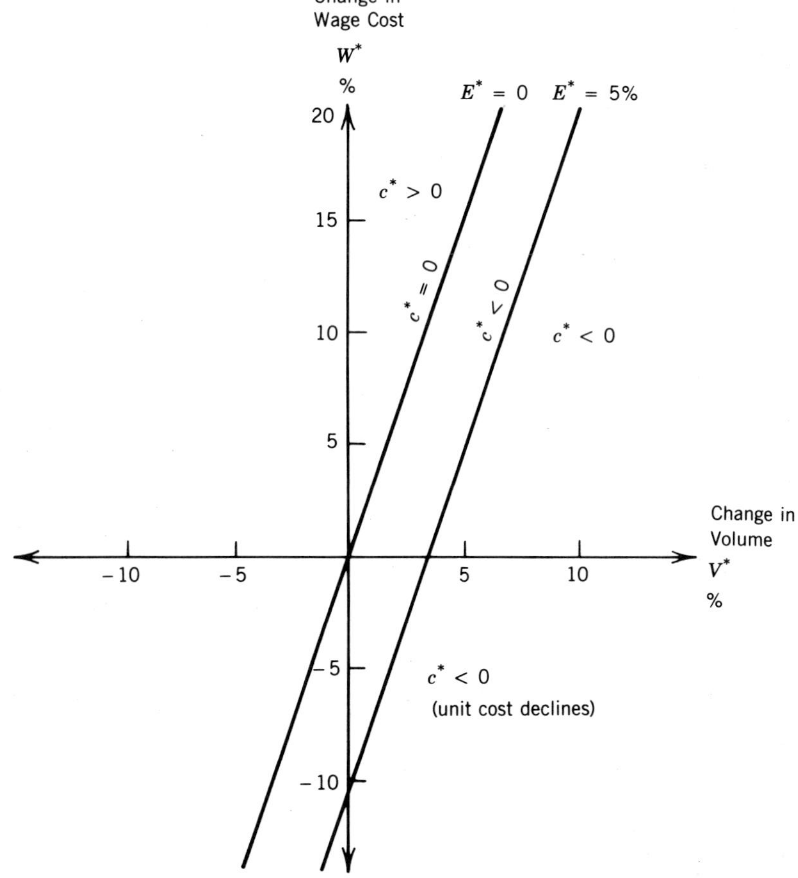

Fig. 6.4.3. The effect of volume change on the unit cost for the factor inputs cost model (for the example $f_W = 0.2, f_E = 0.4$).[5]

Table 6.4.2. An Example

	M	W	N	E	C	V
Period 1	480	640	300	480	1600	1000
Period 2	532	672	306	528	1732	1100
Change (%)	10.8	5.0	2.0	10.0	8.25	10.0

The manpower cost W may be decomposed into several categories of labor, or into

$$W = wN$$

where

$$w = \text{average wage rate}$$

$$N = \text{number of employees}$$

so that W^* in Eqs. 6.4.11 and 6.4.15 may be substituted by

$$W^* = w^* + N^* + w^*N^* \tag{6.4.16}$$

6.4.4.2 Labor Productivity

Productivity is obviously an important factor affecting costs.[2,4,6] There are many ways in which labor productivity may be defined.[5] If the average number of hours per employee remains unchanged, the definition of output per employee may be adopted, namely

$$\pi = V/N$$

from which the relative change in productivity is determined by

$$\pi^* = (V^* - N^*)/(1 + N^*) \tag{6.4.17}$$

so that productivity rises provided

$$\pi^* > 0 \quad \text{if} \quad V^* > N^*$$

and labor productivity falls when the inequalities read the other way. Eqs. 6.4.16 and 6.4.17 may be combined into

$$W^* = N^* + (w^*/\pi^*)(V^* - N^*) \tag{6.4.18}$$

where the relationship between changes in the total employment cost W^*, the wage rate w^*, the volume V^*, and the productivity π^* is given.

Alternatively, it is possible to express the interdependence of all these variables with the change in unit cost by substituting Eq. 6.4.18 into 6.4.14 (for $\pi^* \neq 0$), so that the change in unit cost can be traced to changes in the resource inputs, in the number of employees, in output, and in labor productivity:

$$c^* = [f_M M^* + f_W N^* + f_W (w^*/\pi^*)(V^* - N^*) + f_E E^* - V^*]/(1 + V^*) \tag{6.4.19}$$

6.4.4.3 Example

Consider the example in Table 6.4.2, where data for two periods are given. The total cost changed by 8.2% and the composition of this change is found by Eq. 6.4.11, where the cost components are:

$$f_N = 480/1600 = 0.3$$

similarly

$$f_W = 0.4$$

$$f_E = 0.3$$

so that

$$C^* = 0.3 \times 10.8 + 0.4 \times 5.0 + 0.3 \times 10.0$$

$$= 3.2 + 2.0 + 3.0 = 8.2\%$$

from which it is clear that the contributions of the factor inputs of materials, manpower, and other expenses were 3.2, 2.0, and 3.0%, respectively. The change in unit cost is derived from Eq. 6.4.3:

$$c^* = \frac{C^* - V^*}{1 + V^*} = \frac{0.082 - 0.10}{1.10} = -0.016 \quad \text{or} \quad -1.6\%$$

which shows how the increase in volume compensated for the increase in cost, thereby resulting in a reduction in unit cost.

The change in labor productivity is given by Eq. 6.4.17, yielding $\pi^* = 7.8\%$, so that the change of 5% in the labor cost W^* is attributed to the 2% rise in the number of employees plus 3% due to the second term in Eq. 6.4.18. The breakdown of unit cost is shown in Table 6.4.3. The example shows that the unit material cost increase of 0.8% was offset by the decline of 4.5% in the unit labor cost, while c_E remained unchanged (because in this case $E^* = V^*$), so that the overall cost c declined by 1.6%. The labor productivity increase of 7.8% made a contribution to this decline in unit cost, in spite of a rise in wage rates of 2.9% and an increase in the labor force by 2.0%.

6.4.4.4 Special Cases

1. If $V^* = N^*$, then labor productivity remains unchanged ($\pi^* = 0$), by Eq. 6.4.17.
2. If $w^* = \pi^* \neq 0$, i.e., if wage rates change by the same amount that labor productivity changes, then $W^* = V^*$ by Eq. 6.4.18.
3. If the number of employees remains constant ($N^* = 0$), then $W^* = w^*$ and $\pi^* = V^*$.
4. If the wage rate remains constant ($w^* = 0$), then $W^* = N^*$ and Eq. 6.4.17 becomes

$$\pi^* = (V^* - W^*)/(1 + W^*)$$

5. If material costs and other costs remain unchanged ($M^* = E^* = 0$), then Eq. 6.4.19 is reduced to

$$c^* = [f_W N^* + w^*/\pi^*)(V^* - N^*)f_W - V^*]/(1 + V^*)$$

so that unit cost declines if

$$V^*(1 - f_W w^*/\pi^*) > f_W N^*(1 - w^*/\pi^*)$$

In the event that $w^* = \pi^* \neq 0$, this condition is satisfied when volume increases ($V^* > 0$). And in the case where $w^* < \pi^*$, then a sufficient condition for $c^* < 0$ is given by $V^* > f_W N^*$.

6. If material costs change proportionately to volume, then $M^* = V^*$, in which case

$$c^* < 0 \quad \text{if} \quad W^* - V^* < -(E^* - V^*)f_E/f_W$$

6.4.4.5 Control of c_W

In circumstances where management wishes to ensure that the unit labor cost c_W declines, attention focuses on the following relationship, which is easily derived from the definitions given above:[4]

$$w = \pi c_W$$

or

$$w^* = \pi^* + c_W^*(1 + \pi^*)$$

Table 6.4.3. Breakdown of Unit Cost and Allied Data—An Example

	c_M	c_W	c_E	c	π	w
Period 1	0.480	0.640	0.480	1.600	3.33	2.133
Period 2	0.484	0.611	0.480	1.575	3.59	2.196
Change (%)	0.8	−4.5	0	−1.6	7.8	2.9

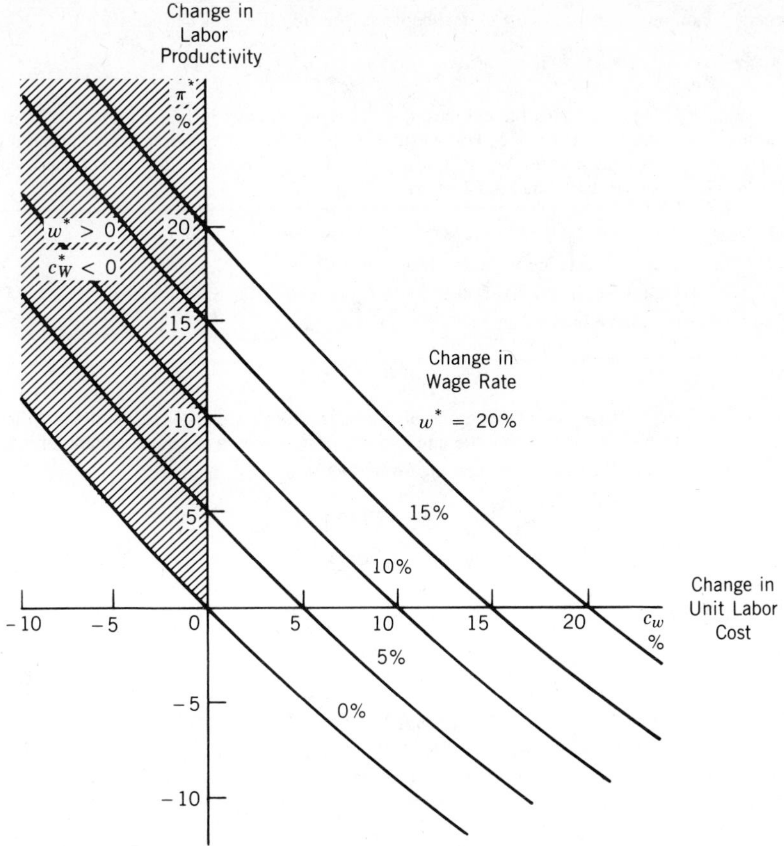

Fig. 6.4.4. Tradeoff between changes in labor productivity (π^*), unit labor cost (c_W^*), and wage rate (w^*).[5]

so that the objective of $c_W^* < 0$ can be achieved if $w^* < \pi^*$. A simple diagram to demonstrate the tradeoff between relative changes in unit labor cost, productivity and the wage rate is given in Fig. 6.4.4, where the shaded area corresponds to the result of a rise in wage rate ($w^* > 0$) coupled with a fall in the labor component of unit cost ($c_W^* < 0$).

6.4.5 ANALYSIS OF THE LINEAR COST MODEL

From the linear cost model in Eq. 6.4.8 and the unit cost in Eq. 6.4.10 it follows that the relative changes in total cost and in unit cost are given by

$$C^* = f_1 (s^* + V^* + s^*V^*) + f_2F^* \tag{6.4.20}$$

and

$$c^* = f_1s^* + f_2(F^* - V^*)/(1 + V^*) \tag{6.4.21}$$

where the cost proportions are defined as

$$f_1 = s/c = S/C$$

$$f_2 = F/C$$

so that

$$f_1 + f_2 = 1$$

and the general condition for the unit cost to decline is derived from Eq. 6.4.21, namely

$$c^* < 0 \quad \text{if} \quad F^* < V^* - (f_1/f_2)s^*(1 + V^*) \tag{6.4.22}$$

The upper bound for F^* to ensure that the unit cost does not rise is shown in Fig. 6.4.5 for the case of $s^* = 10\%$ for various values of f_1 from 0.1 to 0.8. For any given f_1, the value of F^* must lie below the corresponding line to obtain $c^* < 0$. For example, if $s^* = 10\%$ and $f_1 = 0.5$, and if a rise of 20% in volume is envisaged, then the condition for $c^* < 0$ from Eq. 6.4.22 becomes $F^* < 8\%$.

Special Cases

1. If $f_2 = 0$, i.e., all the costs are variable, then $c^* = f_1 s^*$.
2. If $f_1 = 0$, i.e., all the costs are fixed, then $c^* = f_2 (F^* - V^*)/(1 + V^*)$, so that $c^* < 0$ if $F^* < V^*$.
3. If no change in volume is envisaged ($V^* = 0$), then

$$c^* < 0 \quad \text{if } F^* < -(f_1/f_2)s^*$$

This result is useful when an investment in plant and facilities is considered, involving an increase in fixed cost, with the aim of reducing the unit variable cost as shown diagrammatically in Fig. 6.4.6.

4. In the event of both F and s being constant, so that $F^* = s^* = 0$, then

$$C^* = f_1 V^* \tag{6.4.23}$$

and

$$c^* = -f_2 V^*/(1 + V^*) \tag{6.4.24}$$

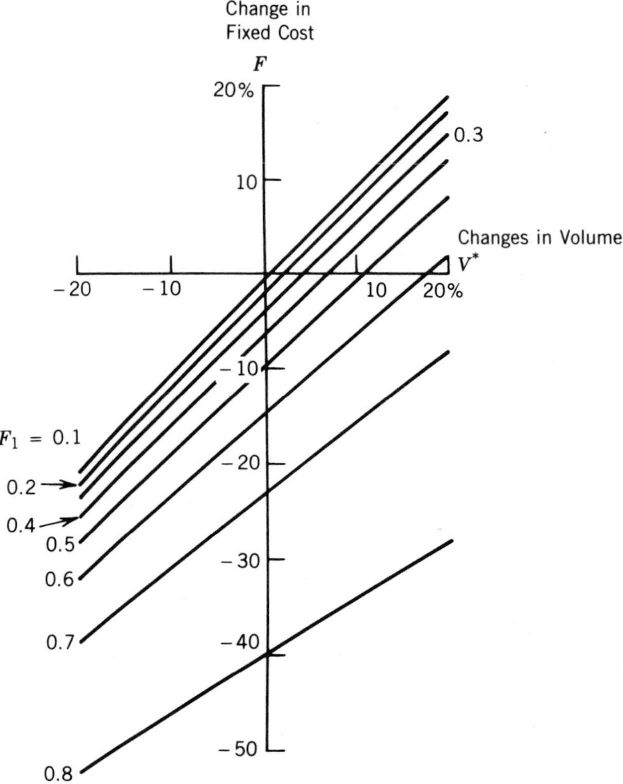

Fig. 6.4.5. Upper bound for the relative increase in fixed cost F^* (when the marginal cost s rises by 10%) to ensure that unit cost does not increase ($c^* \leqslant 0$).

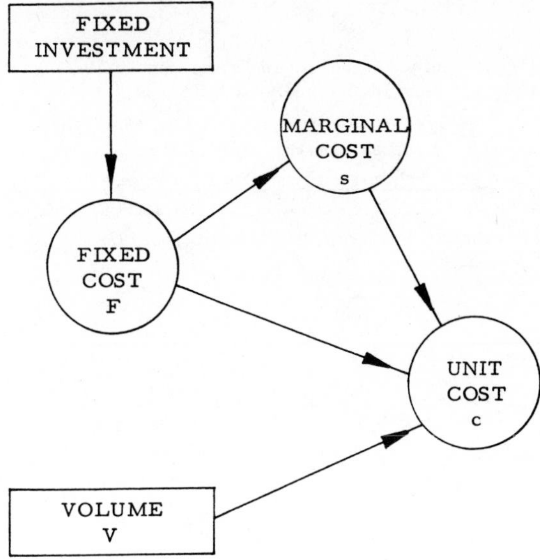

Fig. 6.4.6. Effect of investment and volume on unit cost.[5]

6.4.6 CONCLUSION

The cost models discussed in this chapter provide a practical means of analyzing changes in past performance; they can also be used as a planning tool for computing expected changes in total cost and unit cost for given changes in cost components and/or in volume. The effect of labor productivity and its relationships to unit cost and employee remuneration are highlighted. More elaborate models can be constructed to examine the effect of changes in the number of working hours, the balance between regular hours and overtime, the composition of the labor force, changes in interest rates and the effect of changes in working capital on the cost structure (see[5]).

GLOSSARY

Total cost (C). The sum of all the cost inputs required to produce the output volume, including raw materials, and other purchased goods, manpower (direct or indirect), overheads, and all factory and office expenses, as well as all services.

Unit cost (c). The average total cost per unit of output.

Volume (V). The output of the plant measured in physical terms (number of units, tons, barrels, etc.)

Relative incremental change. The ratio of the absolute increment Δx of a variable to its original value x in the base period. The relative incremental change is denoted as x^*, so that $x^* = \Delta x/x$.

Factor inputs. Resources used as inputs to the factory, such as materials (the cost of which is denoted as M), manpower (the total cost of this factor input being W to cover the workforce of N employees), and other expenses (including all cost items not covered under M and W, as described after Eq. 6.4.11; the other expenses are denoted as E).

Fixed costs (F). The total cost component that is assumed to be unaffected by a change in the level of output V (a more detailed explanation is given in the text).

Variable cost S(V) (or S for short). The direct costs attributed to the output.

Unit variable cost(s). Also called *marginal cost* The direct cost per unit, that is, the increment in total cost when the output volume V changes by one unit. In the linear cost model the marginal cost s is assumed to be constant.

Wage rate (w). The average wage (or remuneration) per employee, i.e., $w = W/N$.

Labor productivity (π). The average physical output per employee, i.e., $\pi = V/N$.

REFERENCES

1. G. P. Cosmetatos and S. Eilon, "Analysis of unit cost, a modelling approach." In *Production Management Systems*," P. Falster and A. Rolstadås, Eds. Amsterdam: North-Holland, 1981, pp. 168–178.
2. S. Eilon, B. Gold, and J. Soesan, *Applied Productivity Analysis for Industry*. Oxford: Pergamon Press, 1976.
3. S. Eilon and G. P. Cosmetatos, "A profitability model for tactical planning," *Omega*, **5**(6), 1977, 673–688.
4. S. Eilon, "Effects of labour productivity and wage rate on costs," *Omega*, **10**(6), 1982, 703–704.
5. S. Eilon, *The Art of Reckoning—Analysis of Performance Criteria*, London: Academic Press, 1984.
6. B. Gold, *Explorations in Managerial Economics*, London: Macmillan, 1971.

PART 7

SPACE: THE PRODUCTION FACILITY

CHAPTER 7.1

FACILITY LOCATION

LEON F. McGINNIS

Georgia Institute of Technology
Atlanta, Georgia

One of the first questions to be resolved when establishing a new production facility is, "Where shall it be located?" The location decision has, potentially, a direct impact on facility acquisition and operating costs, on process configuration, on quality, and on distribution. Land and construction costs vary by geographic region. Variations in energy costs, labor costs, and local industrial base may dictate different process configurations in different candidate locations, e.g., the use of manual assembly rather than sophisticated automation. The quality levels achieved usually depend on the education, training, and experience of the general work force, as well as sociological factors that differ by geographic location. Clearly, distribution costs depend on the availability of transportation and warehousing facilities, as well as the locations of material suppliers and the marketing areas to be served.

Facility location is a major business decision. For this reason, it should be made within the context of a strategic business plan that identifies, over a long-range planning period, what products will be produced, what resources (material, energy, labor, technology, management, and capital) will be required, and locations of the respective product and resource markets. The strategic business plan establishes the requirements for the production-distribution system, and the facility location decisions support the strategic business plan by meeting the requirements in the "best" way.

What is the "best" facility location plan? At least three categories of factors must be considered. The purely economic or cost factors are, of course, important, and include facility costs, production costs, and distribution costs, all of which are affected by the location decision. Although they may be difficult to measure, economic factors are at least quantifiable. Other noneconomic but still objective factors must be considered, e.g., availability of water or skilled labor, and existing governmental regulations. Finally, a host of purely subjective factors exist that, while difficult to define, much less measure, are still important in the facility location decision. Examples of subjective factors include the climate, personal perceptions of the "quality of life" in a location (will the key people be willing to go there?), and the general business climate (will the community be happy to have your new facility?).

Given more than one facility location alternative and more than one decision maker, there will almost surely be disagreement regarding the "best" decision, due to the noneconomic and subjective factors. Thus, ultimately, the facility location decision is another classic example of management decision problems. However, good tools that provide analyses of the economic factors are available to support the facility location decision.

The purpose of this chapter is to review some quantitative modeling and analysis techniques that may be useful in support of the facility location decision. We will discuss "facility location problems" in the very limited context of mathematical models and algorithms for obtaining the corresponding solutions. One should never confuse the solution to one of these mathematical models with the solution to the "real" facility location problem. The obvious question is, "Why bother with these mathematical models if they don't provide a solution to your problem?" The answer is given, along with a discussion of the scope and general form of facility location models, in Section 7.1.1. Subsequent sections deal with particular classes of facility location models. Due to the limitations of space, we will not present the algorithms in great detail, as they may be found in references we will cite. What we will attempt to do is to indicate the magnitude of the data-gathering effort required for each model and the relative difficulty of the solution procedures.

7.1.1 ROLE AND SCOPE OF FACILITY LOCATION MODELS

A mathematical model of a process forces a particular way of thinking about the process, a focus on the constituent parts of the process, and the mechanisms of their interaction. A simple mathematical model, obviously, may not incorporate all parts of the process, may not capture the interaction with full fidelity, and therefore may not fully explain every observation of the process. Even simple models are useful, however, if they provide consistent explanations of some parts of the process.

This is particularly true of facility location models, which provide a framework for collecting and summarizing data describing some aspects of the location problem. They also help to explain the relationship between location and certain cost and performance attributes. Note that facility location models deal only with quantifiable attributes, i.e., attributes that can be measured, recorded, summarized, and predicted with "reasonable" accuracy. Thus, the models provide a concrete measure of the impacts on quantifiable attributes when other factors are used to influence the decision.

In the folklore of management science and operations research, the use of models to structure data gathering and decision making often leads to improved decisions, regardless of the "solution" generated from the models. The implication is that the discipline enforced by the model is as important as the model itself.

Facility location models also can be used to "suboptimize," i.e., to determine the best possible location decision, considering only the quantified attributes incorporated in the model. While such a solution may not be acceptable, it does provide a baseline against which other solutions can be compared, yielding opportunity costs for the complex of unquantified attributes.

The costs incorporated in all the facility location models discussed in this chapter are either fixed or variable. The *fixed costs* are associated with a specified facility or with a specific site, and represent an approximation of the aggregate of the following actual costs:

1. Amortized space and equipment cost.
2. Annual costs for energy, insurance, taxes, etc., not allocated as a variable overhead.
3. Annual costs for labor and supervision not allocated as a variable overhead.

The *variable costs* are associated with a particular production-distribution activity (production at a specific site and delivery at a specific site), and are specified as a cost per unit. The variable costs typically would include:

1. Inbound transportation for material or components.
2. Variable production costs (all costs allocated on a per-unit-produced basis).
3. Variable inventory carrying costs.
4. Outbound transportation costs.

It is very important to note that these are *point estimates* of the costs factors. In reality, significant economies of scale often exist in production facilities, so that the observed "fixed" cost depends on the scale of production.

This is not to say that scale economies are, or should be, ignored in facility location analysis. Experience has shown, however, that directly incorporating scale economies in the cost models usually results in a location model that is too difficult to analyze effectively. Moreover, obtaining a complete description of the costs, including scale economies, is a very demanding task. It is much simpler to obtain one or two point estimates of the "true" cost function, based on current or past production levels. Thus, the use of the simple fixed-plus-variable cost model requires the analyst to make a reasoned judgment regarding the most likely scale of operation of the facility to be located. Implicit in this approach is the need to test the resulting solution, i.e., the projected scale of operation, against the initial assumptions.

All the facility location models reviewed in this chapter take the form of optimization models, i.e.,

minimize

$$f(x)$$

subject to

$$g_k(x) \geqslant 0 \qquad k = 1, \cdots, K$$

$$x \in X$$

where

x is a vector of decision variables

X defines the feasible locations

$f(x)$ is a criterion or objective function

$g_k(x)$ defines a joint constraint on the decision variables

The decision variables are limited to the location of the new facilities (NF), and in some cases, the assignment of requirements, customers, or existing facilities (EF) to the new facilities.

A major division of facility location models hinges on the definition of the set X. If NF locations may be anywhere, then X is the set of all points in the plane, leading to a class of *planar location problems*. If NF locations are limited to points on a network, e.g., a highway network, then X is the set of points at the nodes or on the links of the network, leading to a class of *network location problems*. Finally, NF locations may be limited to a finite set of candidate sites, leading to a class of discrete location problems. The progression, planar–network–discrete, is an important one, since it indicates the direction of increasing model realism, increasing data requirements, and increasing difficulty of solution.

For each of the models presented in the following sections, the discussion will address not only formulation and solution algorithms, but also the number of parameters required, the interpretation of the parameters, and the magnitude of the computational effort required to obtain optimal solutions. Since the models presented will tend to be the simplest in each category, some references to more complex models will be included.

7.1.2 PLANAR LOCATION PROBLEMS

Planar location problems receive their name from the assumption that all facilities may be treated as simple points in the plane, i.e., in Cartesian 2-space. The n existing facilities are represented by the points $P_j = (a_j, b_j)$, $j = 1, \ldots, n$, and the problem is to specify the locations of exactly m new facilities, $X_i = (x_i, y_i)$, $i = 1, \ldots, m$. The facility coordinates may be thought of as longitude and latitude. Several important assumptions are implicit in planar location problem formulations:

1. The geographical region containing the EFs is small enough so that the planar approximation does not distort distances too much.
2. The facilities (EF and NF) may be idealized as points, and any point in the plane is a valid NF location.
3. Variable cost of service can be modeled as a linear function of distance betwen supplier and consumer.
4. The amount of service provided to a specific facility (EF or NF) from a given NF is fixed in advance.
5. All fixed costs are associated with NFs, *not* with locations, and can thus be ignored in the solution procedure.

Assumptions (1) and (2) essentially limit the geographical scope of these models to problems that are neither too large (the region containing the EFs spans a significant fraction of the earth's surface), nor too small (the physical size of the facilities is a significant fraction of the distance between them). Assumption (3) forces a simplified representation of the cost of interaction between facilities, but is probably acceptable in a planning framework. Assumptions (4) and (5) are the most restrictive, especially (4), which can be interpreted as requiring the magnitude of the interactions between facilities to be fixed, independent of their locations. This is probably reasonable if the NFs are each unique, e.g., producing a unique product or service; however, it is not a good assumption if the NFs are essentially identical. Assumption (4) is relaxed in the class of planar location problems referred to as *location-allocation* problems.

Planar distance may be defined using a parametric family of distance measures, or "norms," usually referred to as l_p norms, where

$$d_p(X,P) = [|x - a|^p + |y - b|^p]^{1/p}$$

The l_2 norm measures straight line, or Euclidean distance, which is the minimum possible distance for "real" travel between two points in the plane. Another frequently used distance measure is the l_1 norm, which measures "city block," or *rectilinear* distances, and is the maximum reasonable distance for most "real" travel between two points in the plane. For values of p in the open interval (1,2), the distance measure is less than the rectilinear distance but greater than the Euclidean distance. See Love and Morris[1,2] for discussion of the choice of p.

The choice of distance metric has an impact on the procedure used to solve a planar location problem. In general, there are "good" (i.e., robust, accurate, and efficient) algorithms for planar location problems using the l_1 and l_2 norms, and problems with the l_1 norm are usually the easiest to solve. For $1 < p < 2$, the solution procedures tend to be less efficient, although they remain computationally feasible for most realistic problems.

Another important factor is whether there is only one NF or more than one NF. The multifacility version usually requires a more involved solution procedure, which is more difficult to describe, understand, and code for computer solution. Nevertheless, almost all planar location problems can be solved with a reasonable amount of computational effort.

A third important factor in distinguishing planar location problems is the type of criterion being optimized. We will usually treat the criterion as being stated in terms of cost of interaction. The most common type of criterion is the total cost of all interactions, leading to the *minisum* formulation of the planar location problem. This corresponds to the notion of economic efficiency, or "system optimality." An alternative formulation, however, is stated in terms of the largest cost of interaction, leading to a *minimax* formulation, which corresponds to the notion of economic equity, or "user optimality." While the minimax formulation may have some applications in the location of production facilities, the minisum formulation is more typical, and will be emphasized in this presentation. The minimax formulation is often seen in public-sector problems, such as locating schools, hospitals, or fire stations.

We will begin our presentation of planar location models with the special case of one NF, and use this as a vehicle for introducing the necessary notation and terminology. We will then consider more general m NF problems and the special case of location-allocation problems.

7.1.2.1 One NF

The minisum formulation for one NF is

minimize

$$f(x) = \sum_{j=1}^{n} w_j d(x,P_j)$$

The weights w_j are parameters specified a priori and represent the relative importance or magnitude of the interaction with EF P_j. For example, w_j may have units of [$/mile] and might be defined as

$$w_j = c_j q_j s_j$$

where

$$q_j = \text{average quantity shipped to } P_j \text{ [units]}$$

$$s_j = \text{number of shipments per year [1/yr]}$$

$$c_j = \text{cost per unit-mile when supplying } q_j \text{ units to } P_j \text{ [\$/unit-mile]}$$

In this case, $f(x)$ would be the annual cost of supplying all of the requirements at EFs.

For Euclidean distances, Francis and White[3] describe the HAP algorithm,[4] which is a modification of an algorithm by Weizsfeld.[5] The algorithm requires the following definitions:

$$d(X,P_j) = [(x - a_j)^2 + (y - b_j)^2 + \delta]^{1/2}$$

$$g_j(X) = w_j / d(X,P_j)$$

$$G(X) = \sum_{j=1}^{n} g_j(x)$$

$$r_j(X) = g_j(X)/G(X)$$

$$\nabla_f(X) = \sum_{j=1}^{n} g_j(X)(X - P_j)$$

$$\lambda_f(X) = f(X) - \nabla_f(X)^T X + \min \{\nabla_f(X)^T P_j \mid j = 1, \cdots, n\}$$

The HAP algorithm may be specified as follows:

1. $\Delta \leftarrow \varepsilon$
2. $X^{(1)} \leftarrow \Sigma_{j=1}^{n} [w_j / \Sigma_{i=1}^{n} w_i)] p_j$
3. $t \leftarrow 1$

WHILE $\Delta \geqslant \varepsilon$

4. $X^{(t+1)} \leftarrow \Sigma_{j=1}^{n} r_j(X^{(t)}) P_j$
5. $t \leftarrow t + 1$
6. $\Delta \leftarrow [f(X^{(t)}) - \lambda_f(X^{(t)})] / \lambda_f(X^{(t)})$

ENDWHILE

Note that $X^{(t)}$ and $v_f(X)$ are vectors.

The parameter δ is a "perturbation," a small number, e.g., 10^{-5}, used to prevent any distance from going to zero (see definition of $g_j(X)$). The function $\lambda_f(X)$ always provides a lower bound on the optimal solution value, $f(X^*)$, and ε is an optimization tolerance parameter. When the algorithm terminates, the current solution, $X^{(t)}$ is guaranteed to be within $100\varepsilon\%$ of the optimal solution value.

The HAP algorithm appears to perform well in practice. With a little care, it can be coded in BASIC for use on a microcomputer, and can solve problems with $n = 50$ in a reasonable amount of time (minutes).

For rectilinear distances, the problem is even easier to solve, since it decomposes naturally into two independent problems, one for the x-coordinate and one for the y-coordinate. The basis for the algorithm is the "median condition" discussed by Francis and White.[3] For the x-coordinate problem, the optimal solution is determined as follows:

1. $W \leftarrow \Sigma_{j=1}^{n} w_j$
2. Sort EF into ascending order of a_j. Define $[j]$ to be the index of the jth largest EF coordinate.
3. $x^* = a_{[k]}$ where, $\Sigma_{j=1}^{k} a_{[j]} \leqslant W/2 \leqslant \Sigma_{j=1}^{k+1} a_{[j]}$

Steps 2 and 3 are modified in the obvious way to determine y^*. This algorithm is quite simple, and the only significant computation is due to the sort operations. It is easy to code and capable of solving problems with $n > 100$ even on a microcomputer.

7.1.2.2 *m* NFs

The minisum formulation is:

minimize

$$f(X_1, \cdots, X_m) = f_1(X_1, \cdots, X_m) + f_2(X_1, \cdots, X_m)$$

where

$$f_1(X_1, \cdots, X_m) = \sum_{i=1}^{m} \sum_{j=1}^{n} w_{ij} d(X_i, P_j)$$

$$f_2(X_1, \cdots, X_m) = \sum_{i=1}^{m-1} \sum_{k=i+1}^{m} v_{ik} d(X_i, X_k)$$

In the multifacility problem, not only are there weights w_{ij} associated with distance from NF to EF, but also weights v_{ik} associated with distances between NF. As in the single facility formulation, the weights are defined a priori. Note that if all $v_{ik} = 0$, i.e., there are no interactions between NF, then the problem decomposes into m single facility problems. Also, if the weights are the same for all of the smaller single facility problems, they will all have the same solution.

For Euclidean distances, the HAP procedure generalizes in a straightforward way. Define:

$$g_{ij}^1(t) = w_{ij} / d(X_i^{(t)}, P_j)$$

$$g_{ik}^2(t) = v_{ik} / d(X_i^{(t)}, X_k^{(t)})$$

$$G_i(t) = \sum_{j=1}^{n} g_{ij}^1(t) + \sum_{k \neq i} g_{ik}^2(t)$$

$$r_{ij}^1(t) = g_{ij}^1(t) / G_i(t)$$

$$r_{ik}^2(t) = g_{ik}^2(t) / G_i(t)$$

The multifacility HAP algorithm is:

1. $$\Delta \leftarrow \varepsilon$$
2. $$X_i^{(1)} = \Sigma_{j=1}^{n} [w_{ij} / \Sigma_{j=1}^{n} w_{ij})] P_j$$
3. $$t \leftarrow 1$$

WHILE $\Delta \geqslant \varepsilon$

4. $X_i^{(t+1)} \leftarrow \sum_{j=1}^{n} r_{ij}^1(t)P_j + \sum_{k\neq i} r_{ij}^2(t)X_k^{(t)} \qquad i = 1, \cdots, m$

5. $t \leftarrow t + 1$

6. $\Delta \leftarrow f[X_1^{(t-1)} \cdots, X_m^{(t-1)}] - f[X_1^{(t)}, \cdots, X_m^{(t)})$

END

Step 2 simply initializes the NF locations; other initialization methods may give better computational results. Also, the stopping rule is based on the rate of change in the objective function, so we cannot guarantee the accuracy of the solution value. The multifacility version of HAP also appears to perform well in practice. For large problems, e.g., $m > 10$, $n > 100$, the solution time using a microcomputer implementation may be excessive, although still feasible.

For rectilinear distances, the multifacility problem decomposes naturally into two smaller independent optimization problems, one for the x-coordinate and one for the y-coordinate. An elegant solution procedure for these smaller problems was given by Picard and Ratliff,[7] using a network-based approach. To apply their algorithm for the x-coordinate, suppose the EF are renumbered, if necessary, so that $a_i < a_{i+1}$, $i = 1, \ldots, n - 1$, defining $n - 1$ *intervals* (a_i, a_{i+1}). The algorithm examines each interval and determines which of the NF should lie to the left of the interval, and which to the right (if $a_i = a_{i+1}$, the "interval" is not considered). At the end, every NF will have been assigned uniquely to one of the EF coordinates.

The evaluation for interval k employs a max-flow network. (See, e.g., Bazarra and Jarvis.[8]) The network contains a maximum of $m + 2$ nodes—a source node, a sink node, and a node for each NF, X_i, not yet excluded from this interval (i.e., already shown to have $X_i \leqslant a_k$ or $X_i \geqslant a_{k+1}$). Arc capacities are determined from weights, w_{ij} and v_{ik}. A max-flow, min-cut algorithm applied to this network partitions the NF into two sets, those which will be strictly to the left of a_{k+1}, and those which will lie strictly to the right of a_k. Picard and Ratliff[7] present an example illustrating all the steps in the procedure.

Their algorithm could be implemented in BASIC for use on a microcomputer. The computational effort for a given interval is of the same order of magnitude as solving a PERT/CPM problem of similar size, and many PERT/CPM codes have been written for microcomputers.

7.1.2.3 Location-Allocation

The location-allocation model relaxes the assumption that all the facility interactions are fixed a priori. One mathematical formulation is:

minimize

$$\sum_{i=1}^{m} \sum_{j=1}^{n} w_{ij} Z_{ij} d(X_i, P_j)$$

subject to

$$\sum_{i=1}^{m} Z_{ij} = 1 \qquad j = 1, \cdots, n$$

$$Z_{ij} \geqslant 0 \qquad X_i \text{ unrestricted}$$

where

w_{ij} = cost per mile to service the total requirement at P_j from facility i
Z_{ij} = fraction of total requirement at P_j to be supplied from facility i

Note that the magnitude of the interaction between X_i and P_j is now a decision variable. Also note that there is no limit on the capacity of a NF. We can infer, then, that in an optimal solution, each EF will be assigned to exactly one NF. The optimal solutions to the location-allocation problem are quite different in nature from the optimal solutions to the multifacility problems considered earlier.

If the NF have limited capacity, then the following constraints are added to the formulation:

$$\sum_{j=1}^{n} r_j Z_{ij} \leqslant S_i \qquad i = 1, \cdots, m$$

where

r_j = quantity required at P_j

S_i = capacity of facility i

We still cannot associate site-specific costs with the NF.

While there are some optimization algorithms for location-allocation formulations, they tend to be unreliable in terms of the required computation time. For this reason, heuristic procedures are typically used. These heuristics iterate between the location decisions, the X_is and the interaction decisions, the Z_{ij}s, until no change in either is observed.

The general form of the location-allocation algorithm is:

0. Initialize

$$X_1^{(0)}, \cdots, X_m^{(0)} ; t \leftarrow 0$$

REPEAT

1.

$$c_{ij} = w_{ij} d(X_i^{(t)}, P_j), \text{ all } i \text{ and } j$$

2. Solve allocation subproblem.

minimize

$$\Sigma_i \Sigma_j c_{ij} Z_{ij}$$

s.t.

$$\Sigma_i Z_{ij} = 1 \qquad j = 1, \cdots, n$$

$$Z_{ij} \geqslant 0 \qquad i = 1, \cdots, m; \quad j = 1, \cdots, n$$

3.

$$w'_{ij} = w_{ij} Z_{ij}, \text{ all } i \text{ and } j$$

4. Solve location subproblem.

minimize

$$\Sigma_i \Sigma_i w'_{ij} d(X_i^{(t+1)}, P_j)$$

5.

$$t \leftarrow t + 1$$

UNTIL

$$(X_i^{(t)} = X_i^{(t-1)} \qquad i = 1, \cdots, m)$$

The allocation subproblem can be solved by inspection, i.e., simply assign P_j to the NF with the smallest c_{ij}. Also, the location subproblem decomposes into a set of single-facility location problems, since no EF interacts with more than one NF. This version of the location-allocation problem can be solved using a microcomputer.

If the NF have capacity constraints, the general form of the algorithm is unchanged, but the subproblems are more difficult to solve. The allocation subproblem can be restated as a standard transportation problem, and solved using specialized network algorithms. The location problem no longer decomposes, and must be solved as a multifacility problem. The newer 16-bit microcomputers are certainly capable of solving this version of the location-allocation problem.

7.1.2.4 Summary

Planar location problems are, by and large, easy to solve. There are good, well-documented algorithms, most of which can be implemented on a large microcomputer with satisfactory results for realistic problems. The planar location model, however, requires some stringent assumptions, in particular, the assumption that any location in the plane is feasible.

For a planar location problem having m EF and n NF, the number of parameters to be estimated is $m(n + 2)$. Of these, m represent cost per unit distance for a specific interaction, and $2m$ represents coordinate locations for EFs.

7.1.3 NETWORK LOCATION PROBLEMS

Any transportation or communication system can be modeled as a *network*, i.e., as a graph consisting of vertices and arcs. In a highway network, the vertices correspond to the intersection of two or more road segments; in a communications network, they correspond to switching centers. In a highway network, an arc represents an uninterrupted road segment joining two intersections; in a communications network, it represents an uninterrupted communications link between two switching centers. Our presentation of network location problems will be in the context of a highway network.

In a network location problem, we have a network N defined by a set of vertices V and a set of arcs A. The vertices are numbered $1,2, \ldots ,n$, and there may or may not be an EF at each vertex. Each arc is uniquely defined by its incident vertices and so may be designated by (i,j), where v_i and v_j are the appropriate vertices. Arc (i,j) has a length, l_{ij}, and may or may not have an orientation (i.e., a one-way street has an orientation, a two-way street does not).

The NF locations are again idealized as points and are restricted to the vertices or points on the arcs. Thus, there are an infinite number of feasible NF locations. In this regard, network location problems resemble planar location problems, and site-related costs cannot be modeled. Network location problems differ from planar location problems, however, in several important respects:

1. The distance metric is no longer an issue, since the arc "lengths" are specified.
2. The distance between two points on the network is defined as the *shortest distance* between them.
3. When two or more NF are located, an EF interacts with only the *closest* NF.

A direct result of (1) is that there is no need for concern over the geographic scope of the problem. Because of (2), we will need an optimization procedure, i.e., a shortest path algorithm, in order to determine distances between arbitrary points on the network. Assumption (3) gives the network location problems a flavor of the location-allocation problem.

There is a special class of networks, referred to as *trees*. A tree network contains $m - 1$ arcs and no *cycles*, i.e., there is a unique path between every pair of vertices. The network location models we describe in this section tend to have very elegant and computationally efficient algorithms for the special care of tree networks. However, we are hard-pressed to find many practical examples of production systems on tree networks. Therefore, we cite the outstanding survey papers by Tansel, Francis, and Lowe[11,12] for details on tree network location models and algorithms.

There are three distinct network location formulations. The *p-median problem* is to locate exactly p NF so as to minimize the total weighted distance from EF to the closest NF. This formulation is analogous to the minisum planar location problem in that the total weighted distance is to be minimized, although it differs in that each EF interacts with only one NF. The *p-center problem* is to locate exactly p NF so as to minimize the maximum weighted distance from any EF to its closest NF. With the exception just noted, the p-center problem is analogous to the minimax planar location problem. The *r-cover problem* is to determine the minimum number of NF so that the maximum weighted distance from any EF to its closest NF is no greater than r. A strong and important relationship exists between p-center and r-cover problems.

7.1.3.1 *p*-Median Problem

The p-median network location problem has a valuable property first shown by Hakimi,[13] i.e., there is at least one optimal solution for which every NF is located at a vertex. Two relevant consequences arise from this property. The first is that the problem is trivial if $p \geqslant n$, since, in this case a NF can be located on top of each EF, resulting in a zero solution value. The second consequence is that the problem can now be viewed as a discrete location problem, since there are only a finite number of feasible locations. For this reason, we will defer discussion of the p-median to the section on discrete location models.

7.1.3.2 One Center Problem

A special case of the p-center problem is the 1-center problem, for which the mathematical formulation is

minimize

$$\{\text{maximum } [w_j d(x,v_j)]\} \qquad j = 1, \cdots ,n$$

subject to

$$x \in N$$

where

$$x = \text{location of the NF}$$

$$d(x,v_j) = \text{minimum distance from } x \text{ to } v_j \text{ via the network}$$

The optimal solution to 1-center is (x^*,d^*), i.e., the location of the NF, and the corresponding maximum weighted distance. No NF location can be found with a smaller maximum weighted distance.

The 1-center problem has an important property that makes it relatively easy to solve. For any optimum location x^* there are two vertices, say v_r and v_s, such that x^* lies on a simple path between v_r and v_s, and

$$w_r d(x^*,v_r) = w_s d(x^*,v_s) = d^*$$

This means that the search for an optimum location can be restricted to a finite set of *intersection points*, i.e., points that are on a simple path between two vertices, and have the same weighted distance to each vertex. Christofides[14] describes a procedure for determining all the intersection points. Note that vertices v_r and v_s may generate several intersection points, but only one in any given arc. Thus, an upper bound on the number of intersection points in a network of n vertices and a arcs is given by

$$\frac{a(n!)}{(n - 2)!\ 2\ !}$$

The algorithm for solving 1-center was first given by Hakimi,[15] and simply requires enumerating all the intersection points. With the proper data structures and shortest path algorithms, it should be possible to solve small ($n \leq 100$) 1-center problems on a large 16-bit microcomputer in a reasonable amount of time, say several minutes.

7.1.3.3 *r*-Cover Problems

The mathematical formulation is

minimize

$$\|X\|$$

subject to

$$w_j d(X,v_j) \leq r \qquad j = 1, \cdots ,n$$

$$X \subseteq N$$

where

X denotes the set of NF locations

$\|X\|$ is the number of NF locations

$d(X,v_j)$ is the minimum distance from v_j to some NF

In the solution to r-cover, each vertex v_j can be uniquely assigned to a NF, namely, the one whose weighted distance from v_j is the smallest (break ties arbitrarily). Suppose X_k^* is the location of the kth NF, and denote by S_k the set of EF assigned to the NF at X_k^*.

An important observation is that if we knew S_k, we could determine X_k^* by solving a 1-center problem on the part of the network associated with vertices in S_k. This means that the optimum locations for r-cover can be restricted to vertices and intersection points.

Suppose the feasible locations are denoted by f_j. For $1 \leq j \leq n$, f_j corresponds to v_j, and for $n + 1 \leq j \leq t$, f_j corresponds to an intersection point. The following integer programming formulation can be used to solve r-cover:

minimize

$$\sum_{j=1}^{t} y_j$$

subject to

$$\sum_{j=1}^{t} a_{ij} y_{ij} \geq 1 \qquad i = 1, \cdots , n$$

$$y_j \ \varepsilon \ \{0,1\}$$

where

$$a_{ij} = \begin{cases} 1 & \text{if } w_i d(f_j,v_i) \leq r \\ 0 & \text{otherwise} \end{cases}$$

$$y_j = \begin{cases} 1 & \text{if location } f_j \text{ is selected} \\ 0 & \text{otherwise} \end{cases}$$

This formulation is a special case of a general class of problems called *set covering problems.*[16] It is a difficult problem to solve in general, although there are reasonably good algorithms for the special case corresponding to r-cover. Solution procedures and modified formulations for r-cover are described in Francis and White.[3] Finding optimal solutions to r-cover will probably require a mainframe computer, although good heuristic solutions to small problems might be possible using a mini- or microcomputer.

7.1.3.4 *p*-Center Problems

The mathematical formulation is

minimize

$$\left\{ \begin{array}{l} \text{maximum } [w_j d(X,v_j)] \\ j = 1, \cdots, n \end{array} \right\}$$

subject to

$$\|X\| \leqslant p$$

$$X \subseteqq N$$

Suppose d^* is the optimal solution value for p-center. If the same data are used, and the corresponding r-cover problem is solved for $r = d^*$, then its optimal solution will have p or fewer NF locations. This relationship between p-center and r-cover is the basis of algorithms for solving p-center.

Christofides[14] and Garfinkel, Neebe, and Rao[17] describe algorithms for solving p-center. Both algorithms employ a search for the optimum *value* d^* for p-center. For each trial value, the corresponding r-cover problem is solved. In both algorithms, solving a particular r-cover problem requires solving a set covering problem, so they are computation-intensive algorithms. For the present, at least, solving p-center requires a large, fast, mainframe computer.

7.1.3.5 Summary

It can be argued that network location models provide a more realistic formulation than planar location models, because "true" distances or travel times can be used and NF locations can be limited to "accessible" sites. The price paid for this greater realism is that the algorithms are more complicated and solutions tend to require greater effort, i.e., solution time.

A network location problem with a arcs and n vertices will require $3a + n$ parameters. The network itself can be described by an "arc list," where each arc is specified by its incident vertices and its length (3 parameters). The EF can be described by a "vertex list," where each vertex has a weight w_j. If the weight is zero, there is not an EF at the vertex.

7.1.4 DISCRETE LOCATION PROBLEMS

The single most important characterization of discrete location problems in general is that the feasible facility sites are limited to a finite set of specified locations. The immediate result is that site-specific costs may be introduced into discrete location models, in contrast to the models for planar and network location problems. While this allows the mathematical models to reflect the true costs more accurately, it also increase substantially the magnitude of the parameter estimation task.

A second result is that the cost of assigning a particular EF to a particular NF need not be proportional to the distance between them. Thus, for example, regional differences in transporation costs can be represented easily in the model. Again, while this improves the accuracy of the model, it requires a substantially greater effort to develop the model cost parameters.

In the basic discrete location model, any NF is capable of serving any EF. One interpretation of this is that there is a single "product," or perhaps a product mix, which is consumed by each EF, or customer, and each NF can produce this product. There are multicommodity location models, e.g., by Geoffrion and Graves,[18] but the required solution methodology is not as readily available.

A major distinction among discrete location problems hinges on the capacity of the NF, i.e., whether or not there are any limits on the number of EF that can be served by a given NF. When there are limits, a *capacitated* model is required, and the problem is typically more difficult to solve than it would be if there were no limits. Where there are no limits, the terms *uncapacitated location problem* and *warehouse location problem* are often used.

7.1.4.1 Uncapacitated Problems

The basic formulation is

minimize

$$\sum_{i=1}^{m} f_i y_i + \sum_{i=1}^{m} \sum_{j=1}^{n} c_{ij} x_{ij}$$

subject to

$$\sum_{i=1}^{m} x_{ij} = 1 \qquad j = 1, \cdots, n$$

$$x_{ij} \leq y_i \qquad i = 1, \cdots, m\,;\ j = 1, \cdots, n$$

$$y_i \in \{0,1\}$$

where

f_i = cost for establishing a NF at candidate site i

c_{ij} = cost to serve customer j from a NF at site i

$$y_i = \begin{cases} 1 & \text{if a NF is opened at site } i \\ 0 & \text{otherwise} \end{cases}$$

x_{ij} = fraction of the requirement at EF_j that is served from a NF at site i.

The first thing to note about this formation is that it addresses a *static* version of the problem. That is, the cost to serve a customer may represent the average monthly cost over a planning horizon of, say, two years. If the market is growing significantly and there are different growth rates in different areas, then great care is required in using this model. A facility configuration that is optimal for the market conditions predicted for five years hence may be terrible for the market conditions predicted for next year. Some dynamic versions of the problem have been addressed,[19] but they seem to be substantially harder to solve. Moreover, the parameter estimation task is magnified many times.

Erlenkotter[20] reports what are probably the best computational results for finding optimal solutions to uncapacitated location problems. His algorithm employs a specialized method for solving the linear programming relaxation, within a straightforward branch and bound framework. Using an IBM 360/91, Erlenkotter was able to solve problems having 25 candidate locations, with solution times of less than 0.1 second. Therefore, problems of this size could also be solved using a suitable microcomputer. Erlenkotter's algorithm was implemented in FORTRAN and is in the public domain.

A variation of the uncapacitated location problem, referred to as the *generalized p-median* problem, was studied extensively by Cornujols, Fisher, and Nemhauser.[21] The generalized p-median problem differs from the uncapacitated location problem in the addition of a single constraint:

$$\sum_{i=1}^{m} y_i \leq p$$

The *greedy heuristic* studied by Cornujols, Fisher, and Nemhauser employs a site "profitability" index:

$$\rho_i(\lambda) = f_i + \sum_{j=1}^{n} \min(0, c_{ij} - \lambda_j) \qquad i = 1, \cdots, m$$

In this index, λ_j can be viewed as a reference cost for serving the jth EF. This method of evaluating candidate NF locations can be traced back to Kuehn and Hamburger,[22] and has been used by many others.

The greedy algorithm successively adds NF location until either the maximum number of locations, i.e., p, has been selected, or the cost increases for adding another location. The reference prices λ_j are modified whenever another NF location is added. A statement of the greedy algorithm is given in Table 7.1.1, where I is the set of NF locations selected, $r(k)$ is the index of the kth location selected, and Z is the solution value. This algorithm is quite simple and can readily solve problems with $m = 25$ and $n = 50$ on a microcomputer.

Table 7.1.1. Greedy Algorithm for Generalized p-Median

$$k \leftarrow 1$$

FOR

$$j = 1,2, \cdots ,n$$

$$\lambda_j \leftarrow \max(c_{ij} \mid i = 1, \cdots ,m)$$

END

$$\rho_{r(k)} \leftarrow \min\ (\rho_i(\lambda) \mid i = 1, \cdots ,m)$$

$$I \leftarrow \{r(k)\}$$

WHILE

$$\rho_{r(k)} < 0 \text{ and } \|I\| < p$$

$I \leftarrow I \cup \{r(k)\}$

$k \leftarrow k + 1$

FOR

$j = 1, \ldots ,n$

$\lambda_j^k \leftarrow \min(c_{ij} \mid i \varepsilon I)$

END

$$\rho_{r(k)} \leftarrow \min(\rho_i(\lambda^k) \mid i \notin I)$$

END

$$k \leftarrow \|I\|$$

$$Z \leftarrow \sum_{j=1}^{m} \lambda_j^1 + \Sigma_{i=1}^{k} \rho_{r(k)}$$

For any given solution, represented by the set I, the functions $\rho_i(\lambda)$ can be used to evaluate the impact of dropping a currently selected location, adding a new location, or simultaneously dropping one location and adding a different location. For adding a new location, simply define λ_j to be the current cost to serve EF_j,

$$\lambda_j = \min(c_{ij} | i \varepsilon I)$$

Then if $\rho_i(\lambda) < 0$ for some $i \notin I$, the total cost can be reduced by adding location i. Similarly, for dropping a location $k \ \varepsilon\ I$, define λ_j by

$$\lambda_j = \min(c_{ij} | i \ \varepsilon\ I\text{-}\{k\})$$

Then if $\rho_k(\lambda) > 0$, the total cost can be reduced by dropping NF site k. Finally, for simultaneously dropping NF site s and adding site t, define:

$$\lambda_j^s = \min(c_{ij} | i \ \varepsilon\ I \text{U} \{t\} - \{s\})$$

$$\lambda_j^t = \min(c_{ij} | i \ \varepsilon\ I - \{s\})$$

Then if $\rho_t(\lambda^t) - \rho_s(\lambda^s) < 0$, the total cost can be reduced by dropping site s and adding site t. As with the greedy algorithm, these procedures are easily implemented on a microcomputer.

The uncapacitated version of the discrete location problem can be considered well solved in practice, since there are reasonably reliable algorithms for solving it, at least for $m \leqslant 100$. Of course, some instances of the problem are very difficulty to solve, but these appear to be the exception rather than the rule. Using an uncapacitated model also seems reasonable when the facility capacities are not fixed by technical factors but are to be determined in the decision-making process.

7.1.4.2 Capacitated Problems

When capacity limits at NF locations are specified a priori, and especially if these limits are binding in the optimal solution, the problem is quite difficult to solve. Moreover, the optimal solution is likely to have an EF, or customer, served by two or more NF locations. This type of solution may be undesirable in practice, where single-sourcing is often required. Nevertheless, there is a large body of research on the capacitated discrete location problem.

The formulation for the capacitated problems is:

minimize

$$\sum_{i=1}^{m} f_i y_i + \sum_{i=1}^{m} \sum_{j=1}^{n} c_{ij} x_{ij}$$

subject to

$$\sum_{i=1}^{m} x_{ij} = 1 \qquad j = 1, \cdots, n$$

$$\sum_{j=1}^{n} r_j x_{ij} \leqslant s_i \qquad i = 1, \cdots, m$$

$$x_{ij} \leqslant y_i \qquad i = 1, \cdots, m\ ;\ j = 1, \cdots, n$$

$$y_i \in \{0,1\} \qquad i = 1, \cdots, m$$

where

r_j = requirement at EF_j

s_i = supply capacity at NF location i

and other variables and parameters are as defined earlier. Note that this also is a static model.

The best reported computational results are given by Van Roy,[23] whose solution procedure consists of two distinct phases. The first is a sort of location-allocation heuristic that obtains very good feasible solutions and usually very tight bounds on the optimal solution. The second phase refines the bounds and, if necessary, employs branch and bound to find or verify optimality. The first phase determines locations by *relaxing* the supply capacity constraints, resulting in a modified uncapacitated problem, which is easily solved. The allocations are then determined from the solution of a transportation problem.

Van Roy's algorithm probably should not be considered for implementation on other than mainframe computers. However, the first phase alone could be implemented on a suitable microcomputer for small problems, say $m \leqslant 25$, $n \leqslant 50$. The reported results indicate that the first phase should give very good solutions.

7.1.4.3 Summary

Discrete location models offer the greatest realism in modeling candidate sites and various cost elements. Unfortunately, it may be impossible to find guaranteed optimal solutions for large capacity constrained problems. Uncapacitated problems, however, do seem to be relatively well solved.

A discrete location problem with n EF and m NF will require mn variable cost parameters, m NF fixed cost parameters, n EF requirement parameters, and m NF capacity parameters. Not only are there more parameters to estimate than for continuous or network models, but they may be more difficult to estimate.

7.1.5 SUMMARY

There are three distinct categories of location models—continuous, network, and discrete. This progression of model types corresponds to a progression toward greater realism in the models, greater effort in parameter estimation, and greater effort in solution. For all three categories, there are certain formulations for which "good" algorithms are available, many of which may be expected to perform adequately on microcomputers. At the same time, for each category, there are formulations for which no "good" algorithms are available.

A fundamental requirement for successfully using these quantitative location models is to recognize their limits. They can only tell us from a very limited perspective what would be a good decision. If we are to use them in decision making, we must have some alternative means, such as experience and good judgment, by which to include all the complex factors not represented in the models. Nevertheless, location models provide a powerful tool for describing and analyzing a large class of facility location decisions.

7.1.6 ECONOMICS

The economic aspects of the location decision represent only one of the important factors to be considered. Nevertheless, at least in the early stages of a location study, it is appropriate to focus on the economics. It is also appropriate to approach the development of a cost database via step-wise refinement, i.e., starting with rough cost estimates and refining them as necessary. This approach lends itself well to a corresponding sequential application of location models, with the models being refined in conjunction with the cost database.

Consider, for example, the problem of locating regional distribution centers. In the early phases of the location study, the key issue is the appropriate number of facilities. Thus a location-allocation model using aggregation of customer locations and relatively crude cost approximations can be used. Such a model is easy to solve and easy to parameterize in order to investigate the sensitivity to particular costs.

The result of the location-allocation phase of the study should identify the appropriate range for the number of facilities, e.g., from five to seven. Now it is possible to begin refining cost models for the facilities, based on the projected scale of operation and a reasonable set of potential facility locations, perhaps corresponding to customer locations. At the same time, the transportation cost estimates can be refined, perhaps by geographic region, or by customer, or in detail, i.e., for specific origin-destination pairs.

At this point, the generalized p-median or warehouse location models can be used to obtain solutions based on the refined database. There are several reasons why the decision maker should recognize the need for, and plan for, an iterative process of data and model refinement. It is expensive to develop highly refined data, so the need should be clearly established. As models are refined, it becomes more difficult and costly to perform sensitivity analysis, so it is desirable to focus the detailed model in the "neighborhood" of the optimum solution. Less refined models are useful in identifying this neighborhood.

Step-wise refinement and sensitivity analyses permit the use of location models to quantify the economic aspects of the location decision. There are also formal, though less quantitative approaches to dealing with the noneconomic and subjective aspects. Several authors [23,25–27] provide detailed lists of factors to consider. Various weighting schemes[26] can be used in conjunction with such a list to identify a set of preferred sites. Brown and Gibson[28] describe a detailed, quantitative approach to combining the economic, noneconomic, and subjective factors in a single model. Parks[29,30] also discusses this problem.

7.1.7 CONCLUSION

Facility location is a major business decision, and should be determined within the context of a strategic business plan. The objective in facility location should be to support the strategic business plan in the best possible way. A host of factors, both economic and noneconomic, will be important in determining the best facility location alternative.

This chapter has focused on the basic quantitative models for supporting the facility location decision. Most of the models address only the economic factors, although some include other objective noneconomic factors and even some subjective factors. The primary value of all of these models is that they help to organize and manipulate large quantities of data about the location problem. As with any sophisticated tool, the user is advised to be careful, and not confuse the model solution with the answer to the problem.

REFERENCES

1. R. F. Love and J. G. Morris, "Modelling inter-city road distances by mathematical functions," *Operational Research Quarterly* **23**, 1972, 61–71.
2. R. F. Love and J. G. Morris, "Mathematical models of road travel distances," *Management Science* **25**, 1979, 130–139.
3. R. L. Francis and J. A. White, *Facility Layout and Location: An Analytical Approach*, Englewood Cliffs, NJ: Prentice-Hall, 1974.
4. J. W. Eyster, J. A. White, and W. W. Wierwille, "On solving multifacility location problems using a hyperboloid approximation procedure," *AIIE Transactions* **5**, 1973, 1–6
5. E. Weiszfeld, "Sur le point pour lequel la somme des distances de n points donnés est minimum," *Tohoku Mathematics Journal* **43**, 1936, 355–386.

6. D. W. Hearn and J. Vijay, "Efficient algorithms for the (weighted) minimum circle problem," *Operations Research* **30**, 1982, 777–795.
7. J. C. Picard and H. D. Ratliff, "A cut approach to the rectilinear distance location problems," *Operations Research* **26**, 1978, 422–434.
8. M. S. Bazarra and J. J. Jarvis, *Linear Programming and Network Flows*, New York: Wiley, 1977.
9. D. J. Elzinga, D. W. Hearn, and W. D. Randolph, "Minimax multifacility location with Euclidean distances," *Transportation Science* **10**, 1976, 321–226.
10. P. M. Dearing and R. L. Francis, "A network flow solution to a multifacility minimax location problem involving rectilinear distances," *Transportation Science* **8,** 1974, 126–141.
11. B. C. Tansel, R. L. Francis, and T. J. Lowe, "Location on Networks: A Survey. Part I: The *p*-Center and *p*-Median Problems," *Management Science*, **29**(4), 1983, 482–497.
12. B. C. Tansel, R. L. Francis, and T. S. Lowe, "Location on Networks: A Survey. Part II: Exploiting Tree Network Structure," *Management Science* **29**(4), 1983, 498–511.
13. S. L. Hakimi, "Optimum distribution of switching centers in a communication network and some related graph theoretic problems," *Operations Research* **13**, 1965, 462–475.
14. N. Christofides, *Graph Theory: An Algorithmic Approach*, New York: Academic Press, 1975.
15. S. L. Hakimi, "Optimal locations of switching centers and the absolute centers and medians of a graph," *Operations Research* **12**, 1964, 450–459.
16. R. S. Garfinkel and G. L. Nemhauser, "Optimal Set Coverings: A Survey," in *Perspectives on Optimization: A Collection of Expository Articles*, A. M. Geoffrion, Ed., Reading, MA: Addison Wesley, 1972.
17. R. S. Garfinkel, A. W. Neebe, and M. R. Rao, "The m-Center Problem: Minimax Facility Location," *Management Science* **23**, 1977, 1133.
18. A. M. Geoffrion and G. W. Graves, "Multicommodity distribution design by Benders decomposition," *Management Science* **20**, 1974, 822–844.
19. T. J. Van Roy and D. Erlenkotter, "A dual-based procedure for dynamic facility location," WP-80-31, International Institute for Applied Systems Analysis, A-2361, Laxenburg, Austria, 1980.
20. D. Erlenkotter, "A dual-based procedure for uncapacitated facility location," *Operations Research* **26**, 1978, 992–1009.
21. G. Cornuejols, M. Fisher, and G. Nemhauser, "Location of bank accounts to optimize float: An analytic study of exact and approximate algorithms," *Management Science* **23**, 1977, 789–810.
22. A. A. Kuehn and M. J. Hamburger, "A heuristic program for locating warehouses," *Management Science* **9**, 1963, 643–666.
23. T. J. Van Roy, "A cross decomposition algorithm for capacitated facility location," presented at ORSA/TIMS Meetings, Washington, DC (1980).
24. J. T. Apple, *Plant Layout and Materials Handling*, New York: Ronald Press, 1950.
25. J. M. Moore, *Plant Layout and Design*, New York: Macmillan, 1962.
26. R. Reed, Jr., *Plant Location, Layout, and Maintenance*, Homewood, IL: Richard D. Irwin, 1967.
27. "Site Selection," *Factory*, **125**(5), 1967, 109.
28. P. A. Brown and D. F. Gibson, "A Quantified Model for Facility Site Selection-Application to a Multiplant Location Problem," *AIIE Transactions* **4**(1), 1972, 1–11.
29. G. M. Parks, "Determining the Number, Location and Capacity of Manufacturing Facilities—One Detailed Case Study and Three Quickies," *Proceedings*, Spring Annual Conference, AIIE, 1977.
30. G. M. Parks, "Location: Single and Multiple Facilities," Chap. 10.1 in Gavriel Salvendy, Ed., *Handbook of Industrial Engineering*, New York: Wiley-Interscience, 1982.

CHAPTER 7.2

FACILITIES PLANNING

JAMES A. TOMPKINS

Tompkins Associates, Inc.
Raleigh, North Carolina

7.2.1 INTRODUCTION

A manufacturer in the Midwest made a significant investment in storage equipment for a parts distribution center. The selection decision was based on the need for a "quick fix" to a pressing requirement for increased space utilization. The company soon learned that the "solution" would not provide the required throughput and was not compatible with long-term needs. Unfortunately, similar scenarios are not uncommon in today's manufacturing sector. Correcting this situation requires an active facilities planning effort. This chapter, in addressing the topic of facilities planning, defines:

1. Facilities planning.
2. The objectives of facilities planning.
3. The facilities planning process.
4. Facilities planning strategy.
5. Facilities planning trends.

7.2.2 FACILITIES PLANNING DEFINED

Facilities planning is defined very differently by different people and companies. One recent definition describes facilities planning as an effort to "determine how an activity's tangible fixed assets best support achieving the activity's objectives."[1] In the production context, facilities planning involves the determination of how the manufacturing facility best supports production. These statements are valiant attempts to define a very broad and complex subject. However, they still leave us grasping for a more tangible definition. A look at what facilities planning encompasses provides a clearer picture.

Facilities planning is not synonymous with, yet is comprised of, facilities location and facilities design. Facilities location is the subject of Chap. 7.1 of this volume. *Facilities location* involves determining how the location of an activity supports meeting the activity's objective.[1] The location of the activity with respect to customers and suppliers and its orientation on a plot of land are the two primary considerations in facilities location. The second component of facilities planning is *facilities design*, which is the determination of how the *components* of an activity support achieving the activity's objectives.[1] These components are:

Structural design.
Plant layout (Chap. 7.3).
Material handling systems design (Chap. 5.4).

Figure 7.2.1 serves to illustrate further the hierarchy of facilities planning.

7.2.3 FACILITIES PLANNING OBJECTIVES

The next step in the understanding of facilities planning is developing a "feel" for its underlying objectives. To best grasp these objectives it is again helpful to break facilities planning into its two components, facilities location and facilities design.

In 1930, W. G. Holmes provided a widely accepted statement of the facilities location objective.[2]

To determine the location which in consideration of all factory affecting delivery-to-customers cost of the product(s) to be manufactured, will afford the enterprise the greatest advantage to be obtained by virtue of location.

The facilities design objective is to:[3]

1. Assist the manufacturing process.
2. Maintain flexibility.
3. Utilize space effectively.
4. Minimize equipment investment.
5. Utilize manpower effectively.
6. Enhance employee safety and job satisfaction.

A list of the objectives of facilities planning would look very similar. However, a look at current motivations for facilities planning efforts best illustrates the composite objective of facilities planning.

Facilities planning is a response to several factors that affect the profitability of a firm. For example, between 20 and 50% of the total operating expenses within manufacturing are attributed to materials handling. Furthermore, it is generally agreed that effective facilities planning can reduce these costs by 10 to 30%. Hence, if effective facilities planning were applied, the annual manufacturing productivity in the United States would increase approximately three times more than it has in any year in the last decade.[1] Productivity is obviously a major factor in a firm's profitability. As such, and as the motivation for bringing new equipment and process technology into the firm, productivity is a major issue in facilities planning efforts.

Facilities planning plays a major role in many other arenas that directly or indirectly affect profitability. In 1970, the Occupational Safety and Health Act (OSHA) became law. With it came the responsibility of facilities planning "to assure as far as possible every working man and woman in the nation safe and healthful working conditions and to preserve our human resources."[4] In addition, energy conservation is becoming more and more both a dollar and an environmental issue. In one large office complex the basic facility design has been altered so that many rooms can be heated by the excess energy that is created by the computing equipment. Finally, facilities planning efforts are also motivated by:

Fire protection requirements.

Community rules and regulations governing noise and air pollution and liquid and solid waste disposal.

Security considerations.

A specific statement of the objective of facilities planning is perhaps an impossibility. However, examination of the objectives of its components, facilities location and facilities design, and a review of the considerations motivating continuous facilities planning efforts allows us to tolerate the following overall statement of objective: Facilities planning aims at that result which allows an enterprise to maximize profits for the services rendered.

7.2.4 THE FACILITIES PLANNING PROCESS

If clear, precise, exacting, measurable, and harmonious objectives could be set forth for it, facilities planning could be referred to as a science. However, the objectives of facilities planning are interrelated in a complex fashion, and often conflicting. Therefore, facilities planning remains very much an art. Nevertheless, a systematic procedure can and should be followed in arriving at a facilities plan that indeed allows an enterprise to maximize both profit and return on investment. The systematic procedure to be followed for the planning of a manufacturing facility consists of the following steps:

1. *Update the products to be manufactured.* The products to be produced and the levels of production must be identified.
2. *Specify the manufacturing process required to produce the products.* Primary and support activities to be performed and requirements to be met should be specified in terms of operations, equipment, and personnel. Support activities are those which support the primary activities and allow them to function with minimal interruption and delay. For example, the maintenance function is a support activity for manufacturing.
3. *Determine the interrelationships among departments.* Quantitative relationships among departments, such as volumes of flow, as well as qualitative relationships should be established.
4. *Determine the space requirements for all activities.* Considerations here include all equipment, materials, and personnel requirements.

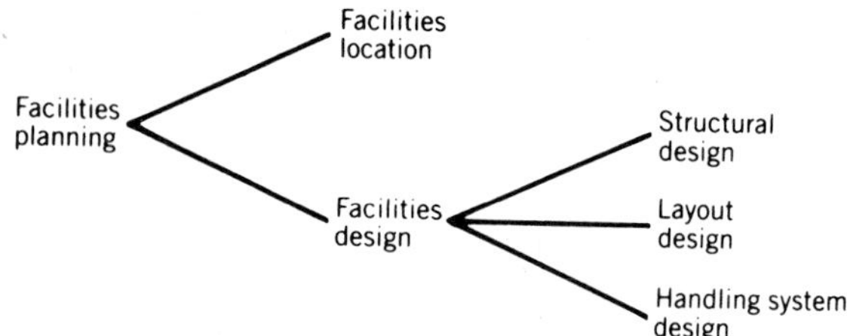

Fig. 7.2.1. Facilities planning hierarchy.[1]

5. *Generate alternative facility plans.* In keeping with the hierarchy of facilities planning, alternative facilities plans should include alternative facility locations, site layouts, and facility designs. Alternative designs should include alternative layout designs, structural designs, and material handling systems designs. Layout alternatives can be generated with traditional techniques or with computerized techniques. *Computer Aided Layout: A User's Guide*,[7] is a comprehensive manual which walks the user through CRAFT, COFAD, PLANET, CORELAP, and ALDEP.

Define the Product(s) to be Manufactured

Specify the manufacturing processes required to produce the product(s)

Determine the interrelationships among departments

Determine the space requirements for all activities

Generate alternative facilities plans

Evaluate the alternative facility plans

Select a facility plan

Implement the facility plan

Maintain and adapt the facility plan

Update the products to be manufactured

PHASE I

PHASE II

PHASE III

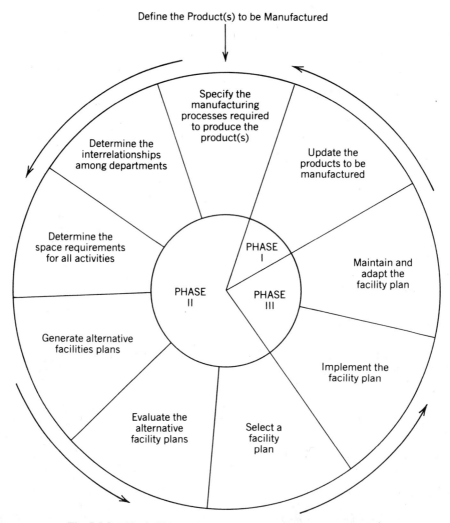

Fig. 7.2.2. The facilities planning process for manufacturing facilities.[1]

6. *Evaluate alternative facility plans.* On the basis of the life-cycle costs, rank the generated plans. For each alternative determine the subjective factors involved and evaluate if and how these factors will affect the facility and its operation.
7. *Select a facility plan.* In this determination, costs are not the only major consideration. The qualitative considerations enumerated in Step 6 should accompany the quantitative considerations in the selection process. The problem is to determine which plan will be the most acceptable in meeting the goals and objectives of the facility.
8. *Implement the facility plan.* Supervising installation of a layout, getting ready to start up, actually starting up, running, and debugging are all part of the implementation phase of facilities planning.
9. *Maintain and adapt the facility plan.* The facility plan must be modified to meet new requirements, capitalize on technological changes in process and equipment, and respond to product design and market changes.

A point that cannot be overemphasized concerning the facilities planning process is that the process must have continuity and it must be continuous, as illustrated in Fig. 7.2.2. It is rare that a facility is not undergoing change, so it is critical that these changes be planned in the context of the overall facility life cycle (Fig. 7.2.2).

7.2.5 FACILITIES PLANNING STRATEGY

Dwight D. Eisenhower is credited with saying "The plan is nothing, but planning is everything." And it is the planning emphasis that distinguishes the activities of the facilities planner from the facilities designer and the facilities locator. The importance of *planning* in facilities planning cannot be stressed enough. The best planners will be the best competitors. Figure 7.2.3 illustrates the cost effectiveness of planning. Successful planning requires effective strategies. A strategy is defined as the art and science of employing the resources of a firm to achieve its objectives.

Two concepts to be considered when developing a planning strategy are contingency planning and strategic master planning. Contingency planning is a *defensive* measure used to guard against a *predictable* future change whose timing is extremely difficult to anticipate. On the other hand, strategic master planning is an *offensive* measure used to capitalize on future *predictable* changes whose timing can be anticipated.[5] Forecasting obviously plays a major role in developing either type of planning strategy. Although not 100% reliable, forecasting provides the best available information concerning the future. In any event, the development of effective strategies for successful planning necessitates an active, as opposed to a reactive, methodology for the facilities planner.

Successful facilities plans and strategies have four common threads. First, successful facility plans consider the major issues which can affect a facilities plan, such as:

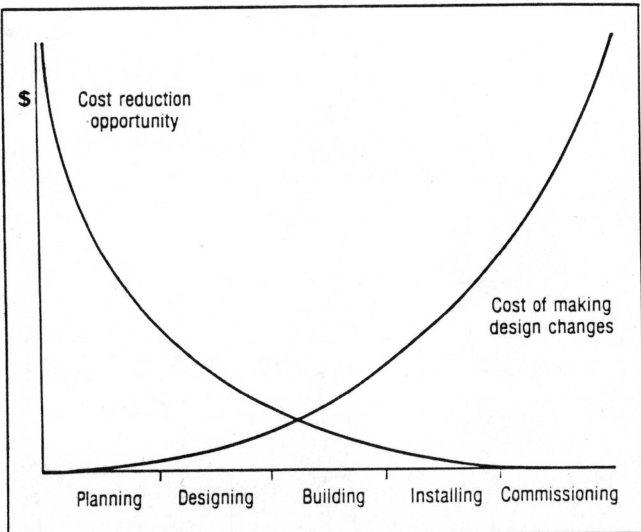

Fig. 7.2.3. Cost reduction opportunities vs. cost of design changes.[1]

23

STRATEGIC FACILITIES PLANNING

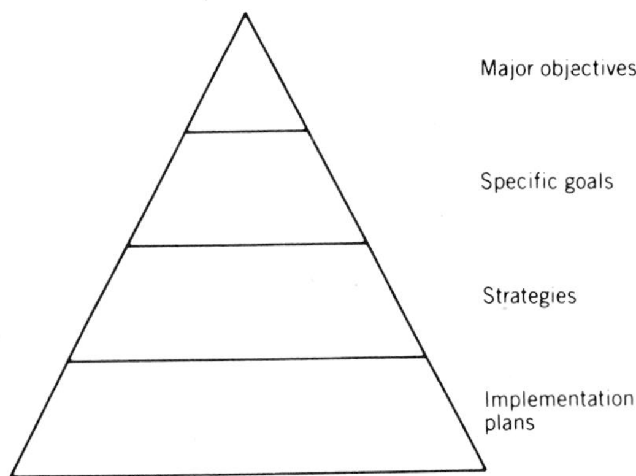

Fig. 7.2.4. Strategies translate objectives into action.[1]

1. Number, locations, and sizes of warehouses and/or distribution centers.
2. Centralized vs. decentralized storage of supplies, raw materials, work-in-process, and finished goods for single and multibuilding sites, as well as single and multisite companies.
3. Acquisition of existing facilities vs. design of model factories and distribution centers of the future.
4. Flexibility required due to market and technological uncertainties.
5. Interface between storage and manufacturing.
6. Level of vertical integration, including subcontract vs. manufacture decisions.
7. Control systems, including material control and equipment control, as well as level of distribution processing.
8. Movement of materials between buildings, between sites, and both inbound and outbound.
9. Changes in customers' and suppliers' technology, as well as the company's own manufacturing technology and material handling, storage, and control technology.

24

DEFINING REQUIREMENTS

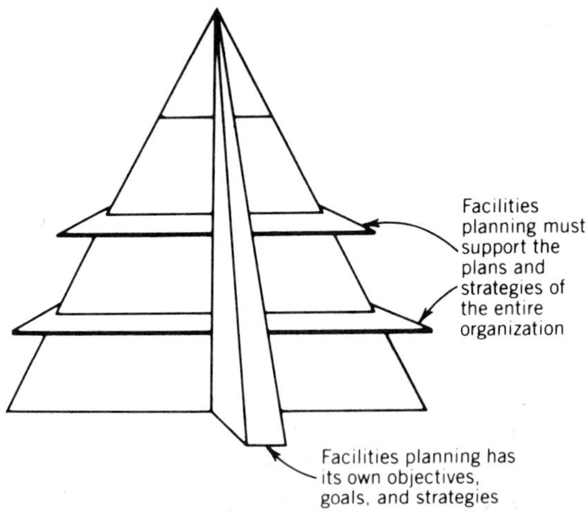

Fig. 7.2.5. Multidimensional impact of material handling strategies.[1]

10. Design-to-cost goals for facilities.
11. Effects of internal functions such as marketing, distribution, and purchasing.

A second commonality of effective facilities plans is that they are action oriented and time phased. Where possible, the plan should set forth very specific actions to be taken to meet a set of manufacturing circumstances. Third, an effective facilities plan should have a defined planning horizon, usually a specific number of years. For example, Texas Instruments has a ten-year strategic plan for each of its plants. Last, an effective facilities plan should consist of a set of formal documents to be taken seriously. Word processors for written material and "friendly" CAD systems for drawings make updating a well-documented facilities plan a reality, and consequently, a responsibility.

Again, the importance of planning cannot be stressed enough. Texas Instruments, a recognized success story in electronics manufacturing, cites its system of strategic planning, OST, as a major reason for its success. The concept of achieving business objectives (O) by developing strategies (S) that translate the business objectives into tactics (T) and the supporting role of facilities planning in those strategies are depicted in Figs. 7.2.4 and 7.2.5 respectively.

John Naisbitt, in *Megatrends*, takes Mr. Eisenhower's comment on planning one step further, "Strategic planning is worthless—unless there is a strategic vision." A vision is a concrete goal that acts to structure every move toward that goal. Naisbitt mentions the example of NASA's strategic vision, to put a man on the moon by the end of the 60s. This vision provided the direction and motivation that established the United States as the world leader in space exploration, a lofty achievement, yet an example of a cloudy strategic vision. In a constantly changing world, strategic planning is not enough. Planning, coexisting with a strategic vision and supported by active strategies, is the challenge that can and must be met by a successful facilities planning effort.

7.2.6 FACILITIES PLANNING TRENDS

As the trend toward automation continues, the facilities planner must plan facilities that respond to this new challenge. It is a common belief that automated facilities are severely limited in their ability to be quickly or economically adapted to any significant changes in facility requirements. However, with a planning approach based on modularity and flexibility, a modern facility can be designed to be more adaptable to future changes than more conventional manufacturing facilities. The key is to plan modularity and flexibility into the facility during the beginning phases of the facilities plan and to continue this emphasis throughout the planning process. Whenever physically and economically possible, modularity and flexibility should be incorporated into the planning of the facility structure, the layout, the production systems, the storage systems, and the material handling systems. The result of such a planning philosophy will be a modern factory, which is both modular and flexible.

Unfortunately, the terms modularity and flexibility are often confused or used interchangeably. Both terms are related to the adaptability of a system; however, they address different aspects of adaptability. *Modularity* is the ability of a system to adapt to a new set of requirements without changing the system's approach. A modular facility can change its size without altering the manner in which it operates. A *flexible* system can adapt to a new set of requirements without being altered at all. *Flexibility* is the ability of a facility to produce different products with no change in the facility, its equipment, or process.

A good analogy to help eliminate the confusion between modularity and flexibility is to think in terms of a photograph album. A *modular* photograph album is an album to which pages could be added or deleted as desired without altering the album covers or means of including pages. A *flexible* photograph album is an album in which a broad range of different-size photographs can be mounted without changing the album in any way. To some extent, the confusion in terminology occurs because systems may be modular but not flexible, flexible but not modular, neither modular, nor flexible, or modular and flexible.

A modular photograph album that is not flexible may be a three-ring binder to which pages could be added, but which will accommodate only one size photograph. A flexible photograph album that is not modular may be a bound album that can hold any size photograph up to a certain maximum. A photograph album that is neither modular nor flexible may be a bound album that will accommodate only one size photograph. Finally, a modular and flexible photograph album may be a three-ring binder to which pages could be added that will accommodate any size photograph up to a certain maximum size.[8]

7.2.6.1 Planning for Modularity

Planning a modular facility implies planning for its expansion. For a facility to be expanded easily and in a manner which will result in the most effective overall facility arrangement, it is absolutely essential that a long-term strategic plan be developed for the site upon which the facility is to be built. Figure 7.2.6 illustates four basic approaches to developing site plans for future expansion. The more traditional approaches to facility expansion are the mirror image, centralized, and decentralized approaches. Typically, facilities that have been expanded via one of these approaches have the inherent problems of poor material flow, major disruption of

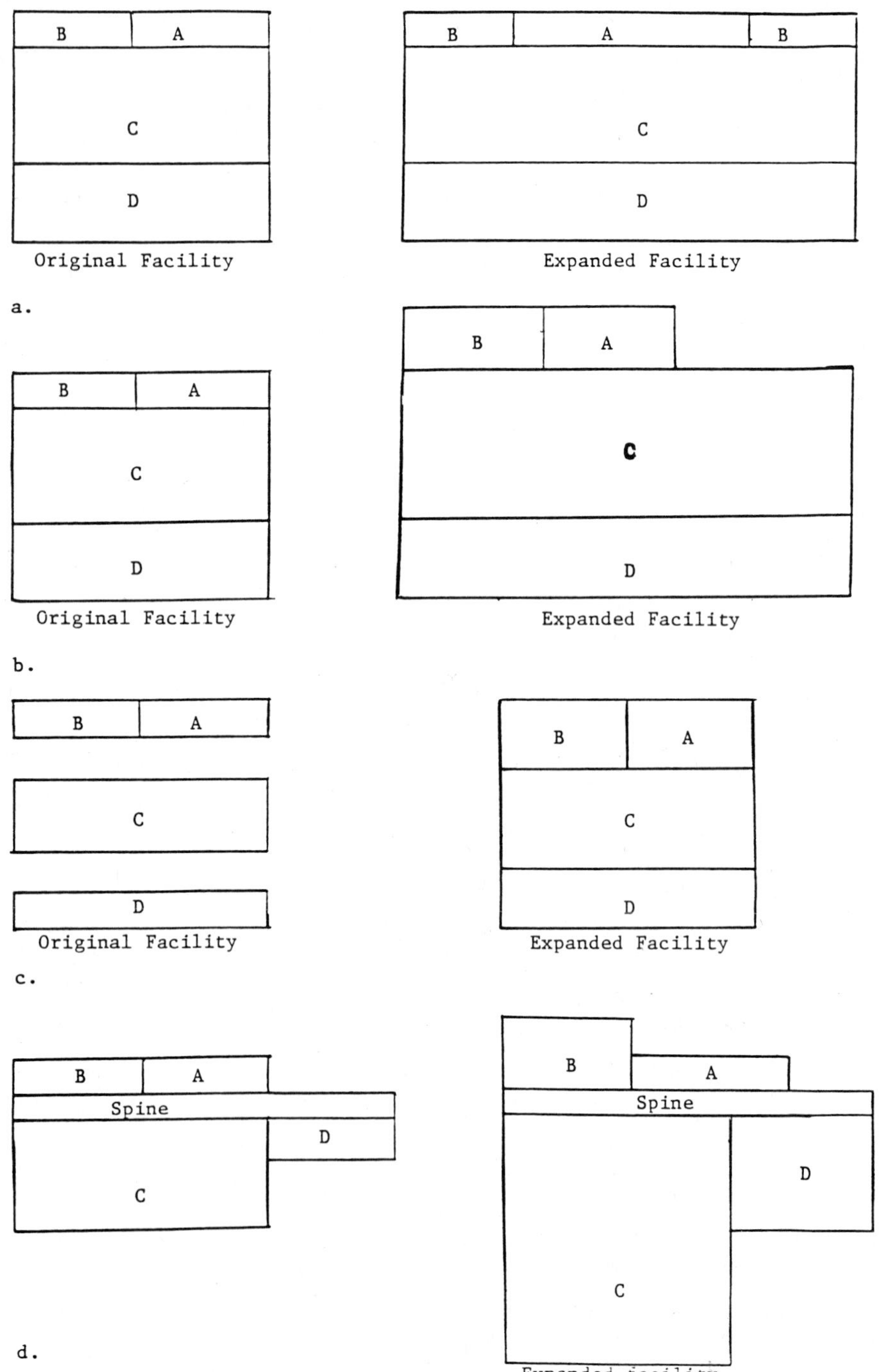

Fig. 7.2.6. Site plans for modular facilities.[1,8] (*a*) Mirror image expansion. (*b*) Centralized expansion. (*c*) Decentralized expansion. (*d*) Spine expansion.

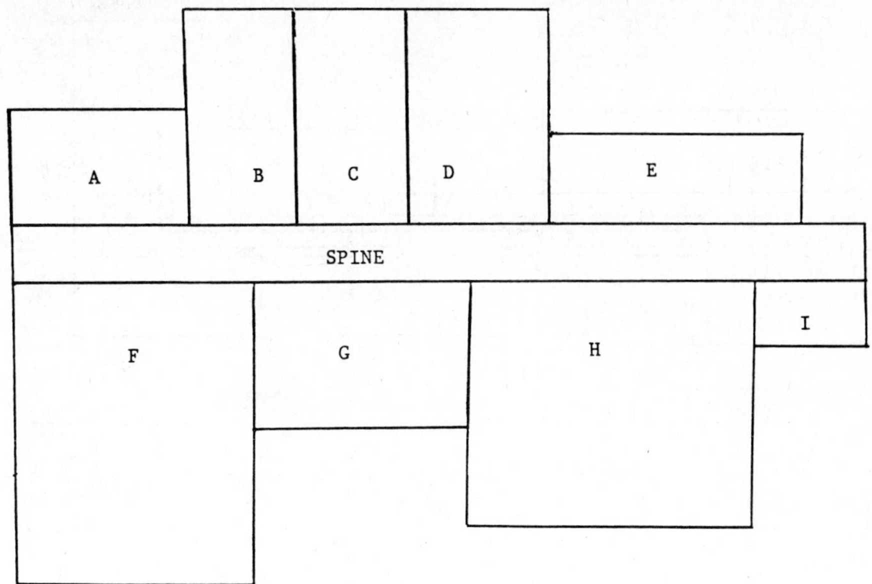

Fig. 7.2.7. Example of a spine facility.[1,9]

all departments during expansion, and uncertainty as to how to expand next. To avoid these problems, the approach to site planning that should be followed is the *spine* concept.

Typically, a spine facility consists of a central spine with the various departments traveling from it. Figure 7.2.7 illustrates a spine facility. All facility services (electricity, compressed air, telephone, etc.) and interdepartmental personnel travel, material handling, and communications are routed through the spine to the various departments. Connecting to the primary systems located within the spine are individual feeder systems that supply facility services, material handling, and communications for the various departments. By locating the primary interdepartment systems in the spine, minimal disruption will occur when the facility is expanded, because no alterations will be required to the primary systems. Only the feeder systems will require expansion. Since it provides the opportunity to expand each department independently without affecting any other department, the spine concept maximizes facility modularity. Figure 7.2.8 illustrates the ease of expansion of a spine facility.

It may also be beneficial to locate material storage in the spine, since this increases the effectiveness of the storage system by avoiding the traditional concept of separate raw materials storerooms, in-process materials storage, and finished goods warehouse. Each of these storage areas performs the same functions: receiving, storing, picking, and shipping material. The only real difference is where the materials are coming from and where they are going to. When all storage is located in the spine, there is no need to make a distinction among raw materials, in-process materials, and finished goods. Consequently, the storage system can be designed based on the physical characteristics of the materials and not their production status. Hence, truly integrated storage systems may be installed.[9]

7.2.6.2 Planning for Flexibility

Establishing a flexible facility implies planning a facility able to produce products in the future with little or no change to its components. Traditionally, facilities have been designed with departments which have been specified either according to the products produced or according to the processes required to produce those products.

Product departments are created by grouping together all processes required to produce a specific product. A classic example of a product department is an automated assembly line. Product departments are usually characterized by high production capacities but low flexibility. As a result, product departments are typically used for high-volume, specialized production.

Process departments are created by grouping similar processes together, e.g., a job shop that has grouped all like machines together. In contrast to product departments, process departments are characterized by low production capacities, but high flexibility. Traditionally, facilities planning involves making a choice between the high production capacity of the product departments and the high flexibility of the process departments. However, the selection of either department arrangement may result in an unacceptable loss of the advantages of the other. The choice between product and process departments does not have to be made. Instead, *group*

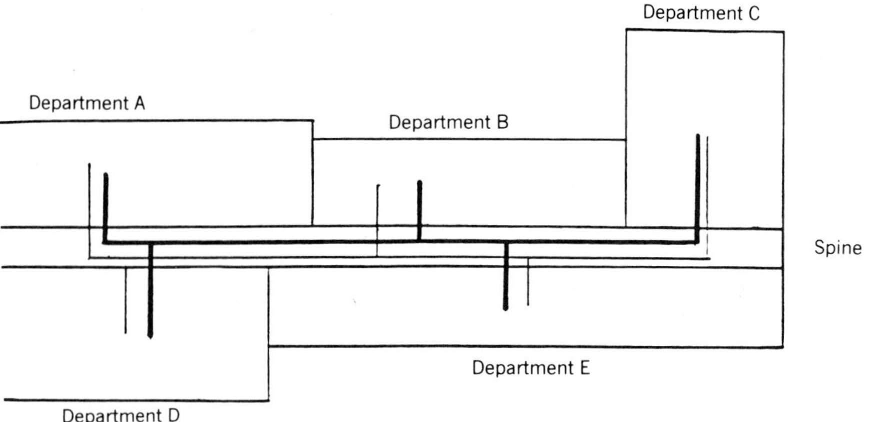

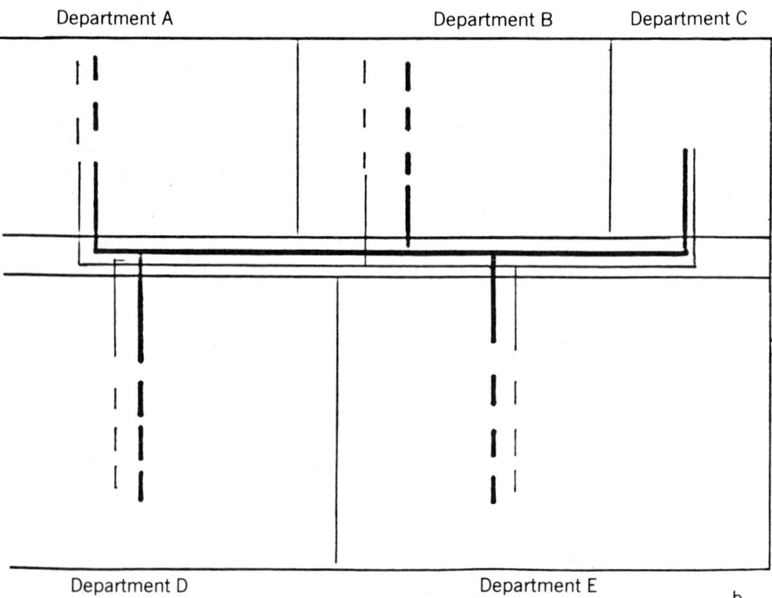

Fig. 7.2.8. Expansion in a spine facility.[9] (*a*) Original facility. (*b*) Expanded facility. Heavy solid line: Material handling and storage system. Light solid line: Facility services and personnel travel patterns. Heavy dashed line: Expanded material handling system. Light dashed line: Expanded facility services and personnel travel patterns.

departments may be created. Group departments are formed by grouping together all processes that are required to produce a specific family of products.

The concept of group departments evolved from the field of group technology. The premise of group technology is that significant efficiencies in manufacturing may be realized by capitalizing on the underlying "sameness" among items. The basis of group technology is the creation of families of similar items by classifying them into categories by shape, material, and/or manufacturing processes. Next, all machines required to produce each family of items are identified and placed into groups. Combining these machines into a group department allows for the efficiencies and high capacities of the product departments, while maintaining the high flexibility of the process departments.[8]

Given the unpredictability of the future, a definite potential for increased efficiencies and flexibility exists via facilities that consist of group departments.

7.2.6.3 Planning for Flexible Manufacturing Systems

Most of the manufacturing in the United States today is *not* high-volume mass production, which is usually associated with automation. Instead it is medium- to low-volume batch production, usually performed in a job shop environment. Thus, it is clear that to improve productivity, the automated factories that are truly needed are those that automate the job shop environment. To develop automated facilities that will be effective for low to medium production levels, it is necessary not only to plan modular and flexible facilities, but also to design flexible manufacturing systems to operate within those facilities. In contrast to conventional fixed automated manufacturing systems, flexible manufacturing systems are designed so that they can be altered quickly and economically to adjust to variations in the production requirements. Specifically, flexibility in manufacturing systems means the ability to produce an assortment of products simultaneously or the ability to quickly accommodate new products or design changes in existing products.

Because it is automated, a flexible manufacturing system has a significant advantage over the semi-automated manufacturing systems presently used for batch production. Flexible manufacturing systems will improve product quality and machine utilization while reducing or eliminating long set-up times, manufacturing cycle times, high inventory levels, and manpower. At the same time, flexible manufacturing systems also offer significant advantages over the conventional automated systems employed for mass production. The biggest advantage is in the total life cycle costs of the manufacturing system. Despite the production efficiency of fixed automated systems, they are limited in application in that they can produce only one type of product at a time. In addition, they require extensive modifications, which can be quite expensive, to accommodate any new product or design changes in existing products. Hence, the justification for fixed automated systems is limited to products that are manufactured in large quantities.

Flexible manufacturing systems, due to their ability to produce multiple products simultaneously, drastically change the economics of automation. *Individual* product volumes are of little consequence as long as the systems' *aggregate* production volumes are high enough to justify the cost of the flexible system. Hence, flexible manufacturing systems can be justified for product volumes previously thought to be too low for automation application.

To date, no universal definition of flexible manufacturing systems has evolved; however, the components of all flexible manufacturing systems are automated production equipment, automated material handling systems for transferring material to and from the production equipment, mechanisms for transferring material between the material handling system and the production equipment, and computer systems for scheduling and equipment control. Automated storage systems also may be included. The goal of a flexible manufacturing system is to establish a totally integrated, computer-controlled, automated production system through the union of computers, material handling systems, and production equipment. This is accomplished by using material handling systems to link independent pieces of production equipment into a manufacturing system and then controlling the system with computers. The goal is to totally control production by having the computer system synchronize material handling and delivery with the production operations.

So far the applications of flexible manufacturing systems have been limited to a few select sections of a plant; however, the flexible manufacturing system concept can easily be expanded to include the total factory. The most logical method to expand the flexible manufacturing system to include the total factory is to develop many small flexible systems for each section or department of the plant and then forming a totally integrated factory-wide system by linking these systems together with facility-wide material handling and communication systems.

One of the most important planning considerations for the development of any flexible manufacturing systems is the selection of its material handling system, for it is the material handling system's capability to provide flexible routing that is the essence of the flexible manufacturing concept. *Flexible routing* is the transportation of different products along various routings to given production equipment and routing a given product over a different path in order to utilize the next free machine. Thus, the flexibility of the automated manufacturing system is directly related to the capabilities of the material handling system.

The material handling system for an automated factory can be divided into two major components, the interdepartmental system and the intradepartmental systems. The interdepartmental system transports material to and from each department or production system, while the intradepartmental system moves materials through the various departments, operations, or production systems. The material handling equipment with potential application for the interdepartmental handling system includes conveyors, self-powered monorail cranes, towlines, monorails, and automated guided vehicle systems. In particular, variable path automated guided vehicles and self-powered monorail cranes can be designed to provide considerably more flexibility than the fixed path conveyors, towlines, or monorails. The vehicles can be easily programmed to follow any of several routings among its travel paths. In addition, the travel paths of these vehicles can be modified easily.

Material handling equipment that merits consideration for the intradepartmental systems because of the flexibility it provides within the departments and production operations, includes conveyors, car-on-track systems, and robots. Of these, the robot is quickly emerging as the most flexible and universal. Most robots can be reassigned to perform any one of a wide variety of tasks automatically by simply changing the robot's software control program and grippers.

The key to control of flexible manufacturing systems is a computer system that incorporates modularity and hierarchical organization. The components of a hierarchical system are a mainframe computer, automated identification equipment, subordinate processing modules, and a cable or fiber optics communications network. The control system divides a manufacturing task into subtasks and assigns them to the subordinate processing modules. Each module controls the material handling and manufacturing operations within its areas. The modules communicate with each other and the mainframe computer via the network communication system. As material is transported through the factory, the material handling system interfaces with the automatic identification equipment, which acts as a "traffic cop" for all material entering and leaving each department. When integrated with the network communication system, these traffic cops provide the information that allows the control system to control totally the flow of materials through the facility. The network communications system acts as a data "highway," and is able to communicate with a wide variety of devices by simply plugging the devices into the highways. Since it can add devices to the network communications system and has hierarchical organization, the control system for a flexible manufacturing system is itself totally modular and flexible.

If the United States is to remain the world leader in productivity, the efficiency of low- to medium-volume production operations must be improved. The automated factory concept of manufacturing promises to provide this improvement. Sound facilities planning is critical to the success of any facility, particularly complex facilities such as automated factories. It is obvious that planning sophisticated automated facilities is very complex and difficult; however, there is a methodical process that provides an organized approach to the planning of all facilities, including those that are automated. To develop a facility that will minimize the total facility life-cycle costs, the facility planner must incorporate modularity and flexibility into the facility design.

7.2.7 SUMMARY

This chapter has described facilities planning in the manufacturing/production context. However, the discussion provided here is applicable in almost any setting. A breakdown of the percent of GNP spent on new facilities (see Table 7.2.1) demonstrates the wide variety of fields open for facilities planning efforts.

The importance and impact of facilities planning motivated the presentation of the material in this chapter. A brief synopsis of that material follows:

1. *Define facilities planning through a composite approach.* Facilities planning is composed of facilities location and facilities design. Facilities design is composed of structural design, plant layout, and material handling systems design.
2. *Establish the objectives of facilities planning.*
   a. Support the organization's mission.
   b. Provide flexibility.
   c. Maximize space utilization.
   d. Promote effective labor utilization.
   e. Minimize capital investment.
   f. Provide for employee safety and job satisfaction.
3. *Describe the facilities planning process.*
   a. Systematic.
   b. Continuous.
   c. Thorough.

Table 7.2.1. Percent of the GNP Typically Expended on New Facilities Since 1955 (by industry grouping)

Industry	%
Manufacturing	3.2
Mining	.2
Railroad	.2
Air & other transportation	.3
Public utilities	1.6
Communication	1.0
Commercial and other	1.5
All Industry	8.0

Source: Ref. 1. Reprinted with permission.

4. *Describe the strategies and importance of the planning process.*
   a. Contingency planning.
   b. Strategic master planning.
   c. Active vs. reactive.

5. *Plan manufacturing facilities to include*:
   a. Flexibility.
   b. Modularity.
   c. Flexible manufacturing systems.

REFERENCES

1. James A. Tompkins and John A. White, *Facilities Planning*, New York: Wiley, 1984.
2. W. G. Holmes, *Plant Location*, New York: McGraw-Hill, 1930.
3. James A. Tompkins, *Facilities Design*, Raleigh, NC: North Carolina State University, 1975.
4. Thomas P. Cullinane and James A. Tompkins, "Facility Layout in the '80's: The Challenging Considerations," *Industrial Engineering*, September 1980, pp. 34–42.
5. James A. Tompkins and Jerry D. Smith, "Optimal Approach to Warehousing Calls for Strategic and Contingency Planning," *Industrial Engineering*, June 1982.
6. John Naisbitt, *Megatrends*, New York: Warner Books, 1982.
7. James A. Tompkins and James M. Moore, *Computer-Aided Layout: A Users Guide*, IIE, Norcross, GA, 1978.
8. James A. Tompkins, "Modularity and Flexibility: Dealing With Future Shock in Facilities Design," *Industrial Engineers*, September 1980, pp. 78–81.
9. James A. Tompkins and John C. Spain, "Utilization of Spine Concept Maximizes Modularity in Facilities Planning," *Industrial Engineering*, March 1983, pp. 34–40.

CHAPTER 7.3

FACILITIES LAYOUT

JOHN J. MARIOTTI
DAVID W. POOCK

GMI Engineering & Management Institute
Flint, Michigan

HARRY K. EDWARDS

University of Michigan
Flint, Michigan

7.3.1 INTRODUCTION

During the facility layout planning process initial decisions that are made by top management focus the efforts of the planners. Most service functions as well as the production departments must have input into the layout process.

For instance, product information may be available in many forms including blueprints, specifications, models, assembly drawings, advance redesign comments (including the expected design life of the models and of each component), exploded views, and physical and chemical properties and prototypes. Production Control might include such items as:

1. Total volumes for the product line and for each model.
2. Schedules for samples, initial production runs, and actual production schedule, including the start-up and acceleration schedules.
3. The expected yearly volumes over the life of the product, and the expected volumes for each part. These are net shipping volumes and do not reflect reject, scrap, and other inefficiencies that occur throughout the procurement, production and distribution cycles.

Financial policies such as return on investment goals, cash flow constraints, and depreciation schedules are necessary. Production policies to be considered include capacity constraints affecting the number and length of shifts, the use of overtime for seasonal variation in customer demand, and the approach to be used for meeting the needs of service volumes for products that will be displaced by new models.

The service policy would probably have to be determined during the layout planning process for the new product line, but management may give some initial direction to reduce the range of alternative solutions. These alternatives might include the following:

1. Purchase service parts from suppliers.
2. Keep the original equipment and use it only for service.
3. Modify the original equipment to produce the new product line as well as service for the original product line.
4. Produce the service volume on general-purpose equipment in a low-volume department.

Typically an industrial or a manufacturing engineering department would coordinate the project and insure that the manufacturing routing is developed. The routing would include estimated costs of labor, material, equipment, tooling, and burden, and would reflect typical efficiency, reject, and equipment utilization rates. Concurrently, suppliers might be asked to submit bids for component parts and assemblies. A comparison would then be made of the cost to make vs. the cost to produce each item.

In creating a routing, each operation description is typically based upon a sketch and a specified workplace design, which includes a description of the duties of the operator, the planned input and output of material, tooling and equipment specifications, and the access space required for service personnel and equipment. The workplace design reflects the overall management philosophy of the enterprise. For example, if all waste is to be eliminated or reduced to a bare minimum as expressed in recent publications describing the Toyota Production System[1,2] then this implies a very detailed specification of each workplace and of clusters of workplaces. The three types of bird-cage layouts shown in Fig. 7.3.1 and the two types of isolated-island layouts in Fig. 7.3.2 are configurations frequently seen in production systems. These are rejected by the Toyota designers for the following reasons:[3]

> *When the entire factory is under this layout, workers are separated from one another and, as such, cannot help each other. It is difficult to attain total balancing of production among the various processes. Unnecessary inventory still occurs among different processes. The mutual relief movement cannot be applied to isolated islands.*
>
> *Since the unnecessary inventory can exist among isolated islands, waiting time of a worker will be absorbed in producing this inventory. Thus, the reallocation of operations among workers to respond to the changes in demand is difficult in this process.*

Once the philosophy of top management has been expressed in terms of policy, the detailed work of the layout planner can begin. One of the very first tasks of the planner is to select an approach to layout. Discussions concerning some of the more prominent approaches to layout, the functions of the layout department, production of the layout, and the generation of alternate layouts (including generation by computer methods) follows.

7.3.2 LAYOUT BY PRODUCT

7.3.2.1 Linear Layout

The linear form of layout (layout by product) is preferred by Toyota over the bird-cage and isolated-island layouts, not only because unnecessary stocking of material between processes can be eliminated, but the double transfer (handling) of material between operations also can be eliminated, since the operator carries or slides the material as he or she walks from operation to operation (Fig. 7.3.3).

In the linear layout, adding or deleting work to or from operator assignments to adjust to changes in demand is more easily done than in other types of basic layout design. In addition, adjacent operators can help each other deal with temporary surges in workload. However, since linear lines are set up independent of other lines, each line typically requires a fractional number of workers, for example, 7.6 workers, and thus 8 workers are assigned. As a result, a series of linear layouts will generally require more workers than other types of layouts.

7.3.2.2 The U-shaped Line

The U-shaped line overcomes some of the limitations of the linear form of layout and is the basic configuration used by Toyota. As shown in Fig. 7.3.4, an operator is assigned work from both sides of the line to allow more flexibility in arranging job assignments to match changes in customer demand. Even with U-shaped lines, the work assignments may lead to a fractional number of operators being assigned to a line. The Toyota response to this problem is to combine several U-shaped lines into one integrated line. In Fig. 7.3.5, six U-shaped layouts are integrated and they include 18, 9, 8, 9, 7 and 10 machines, respectively, for products A through F. In Fig. 7.3.6, the worker assignments are shown for a cycle time of 1 minute (60 units per hour) while in Fig. 7.3.7, the assignment for each worker is shown for a cycle time of 1.2 minutes and the total number of operators has been reduced from 8 to 6. Incidentally, in this example the labor cost was also decreased from .133 hours to .120 hours per unit for a 10% reduction in labor cost. For a comprehensive treatment of the Toyota Production System, see Chap. 3.15.

Advantages of Product Layout

Product layout has several advantages over the alternative layouts:

1. Reduction in work-in-progress, material handling costs, floor space, and production control.
2. Facilitates the use of "product team" style of management.
3. High production rates.
4. Lends itself to the use of specialized handling techniques between operations.

7.3.3 LAYOUT BY PROCESS

The preceding discussion of bird-cage layouts is a fitting introduction to layout by process (also referred to as job shop layout or layout by function). The bird-cage layouts are micro-examples of layouts designed to minimize

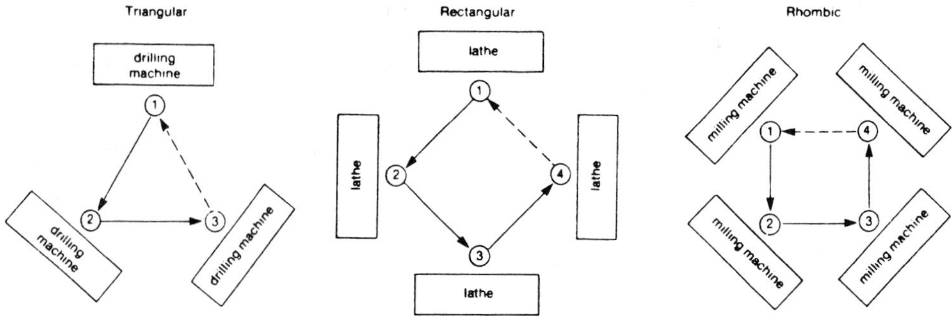

Fig. 7.3.1. Types of bird-cage layouts. *From* Ref. 2.

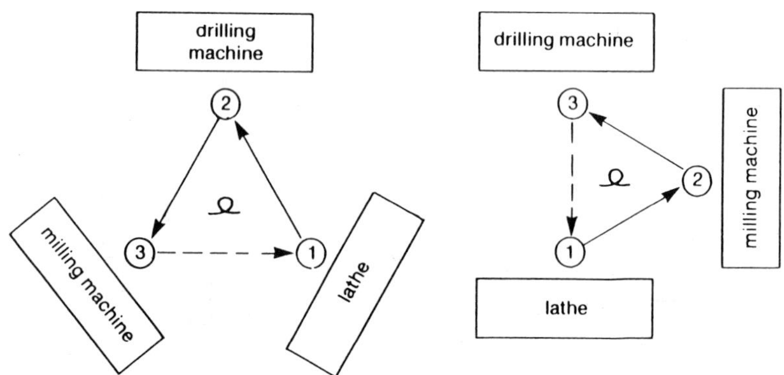

Fig. 7.3.2. Isolated island layouts. *From* Ref. 2.

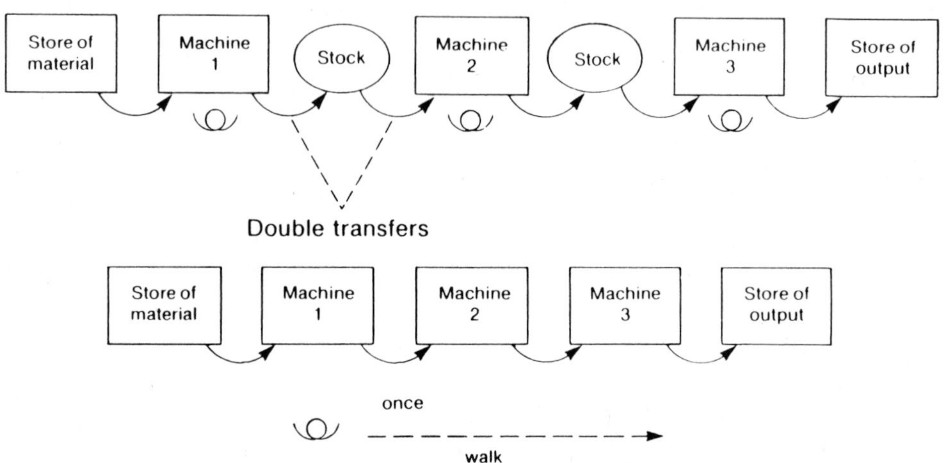

Fig. 7.3.3. Elimination of double transfer. *From* Ref. 2.

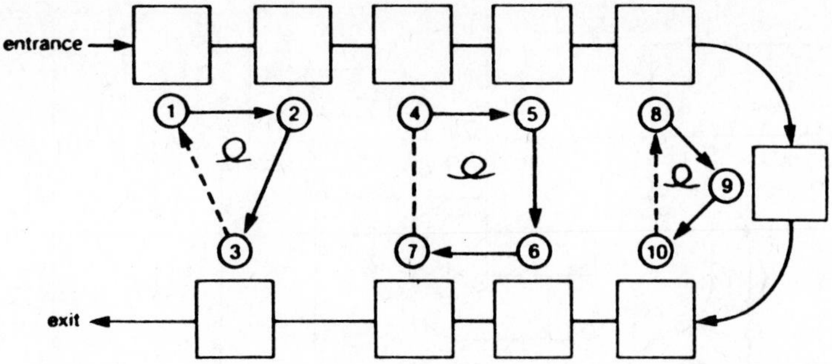

Fig. 7.3.4. Allocation of operations and layout of processes. *From* Ref. 2.

operator cost, and are somewhat unusual. For example, it is more common to arrange lathes in a straight line followed by straight-line clusters of milling and drilling machines, as in the process layout shown in Fig. 7.3.8. In many shops, particularly those with skilled and semiskilled workers, the operators may be restricted not only to operating one *class* of equipment, but may be expected to run only one *piece* of equipment. For example, a lathe operator might not be allowed to run more than one lathe at a time, regardless of the length of the machine cycle and/or the amount of attention required. The layout by process, typically used for batch manufacturing, is found in the textile and shoe industries and the soft trim plants in the automotive industry.

There are many variations in these basic layout systems and it is not unusual to see one or more product type lines set up in a process environment.

Advantages of Process Layout

Process layout also has advantages:

1. Supervisor and operators are very skilled in using the process for a range of products.
2. High utilization of equipment.
3. Few interruptions in schedules because of equipment breakdowns.
4. Lends itself to the use of incentive pay plans.
5. Increased flexibility due to changes in sequence, volume, or manufacturing times.

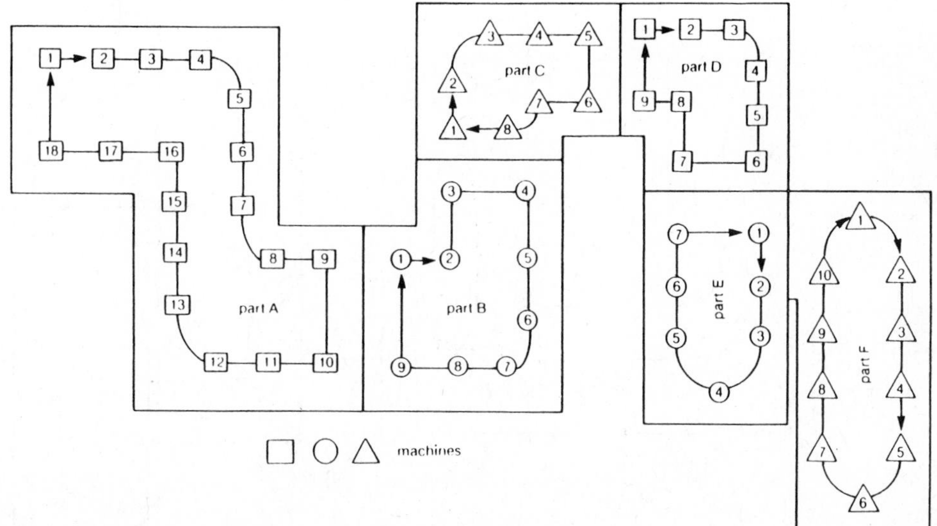

Fig. 7.3.5. Combining six U-shaped lines. *From* Ref. 2.

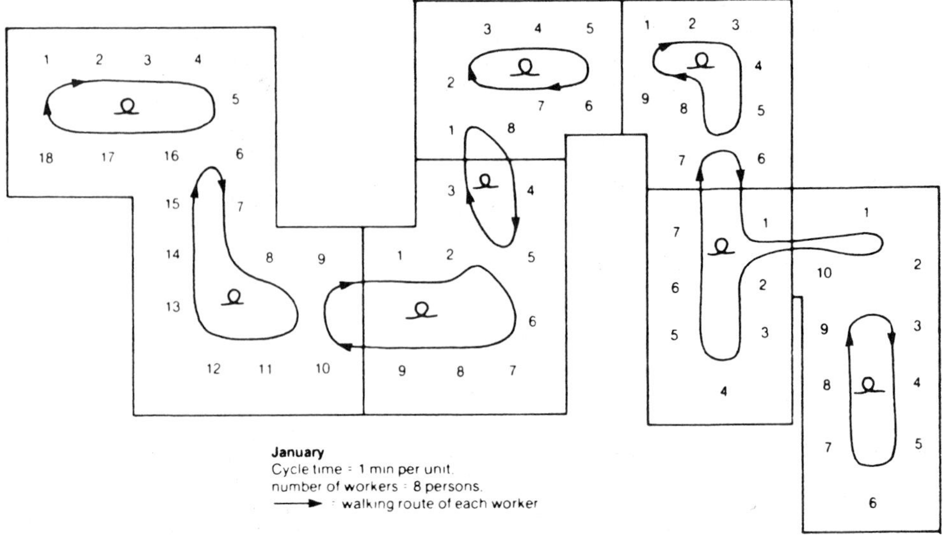

Fig. 7.3.6. Allocation of operations among workers in January. *From* Ref. 2.

7.3.4 LAYOUT BY GROUP TECHNOLOGY

One definition of group technology is "the organization of production in self-contained and self-regulating groups or cells, each of which undertakes the complete manufacture of a family of parts having similar manufacturing characteristics".[4] In actual practice there may be variations in which one or more operations, say painting and plating, will be performed outside the cell. Group technology is a means of bringing the principles and advantages of higher volume production as expressed in product type layouts to small quantity production (Fig. 7.3.8). Groover[5] estimates that manufactured parts in lots of 50 or fewer comprise 25% to 35% of all the manufactured parts produced in the United States and that this may increase to as much as 75% in the years to come.

The benefits of group technology can extend to the entire enterprise, since it usually is accompanied by an analysis and rationalization of part numbers. This may include adopting a classification and coding system that

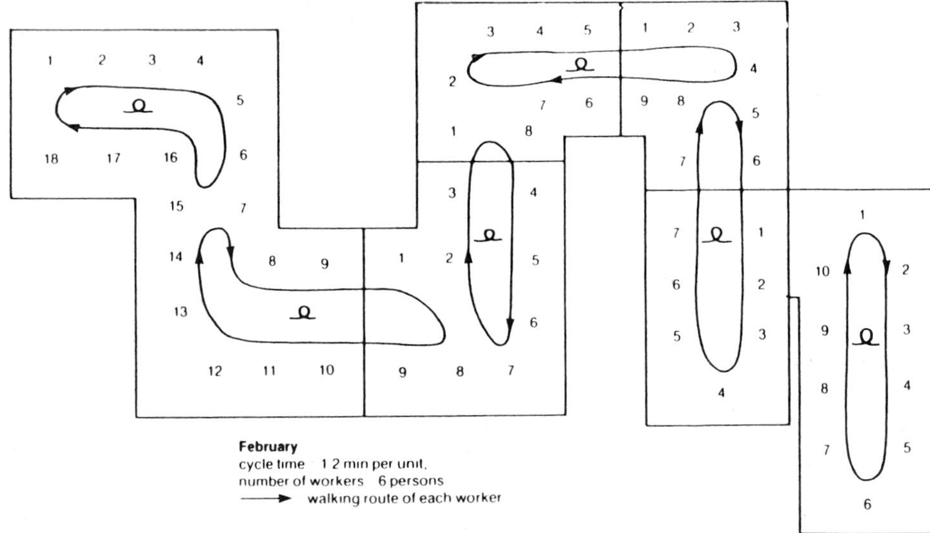

Fig. 7.3.7. Allocation of operations among workers in February. *From* Ref. 2.

PRODUCT LAYOUT

PRODUCT A DEPARTMENT

1 PRESS — 2 MILL — 3 HEAT TREAT — 4 GRIND — 5 PAINT — 6 INSPECT — 7 PACK

PRODUCT B DEPARTMENT

1 CAST — 2 GRIND — 3 MILL — 4 DRILL — 5 BRAZE — 6 PAINT — 7 INSP. — 8 PACK

RAW MATERIALS

FINISHED GOODS STORAGE

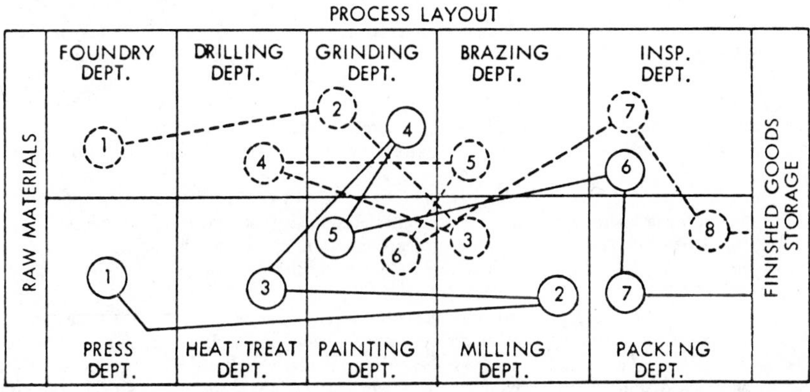

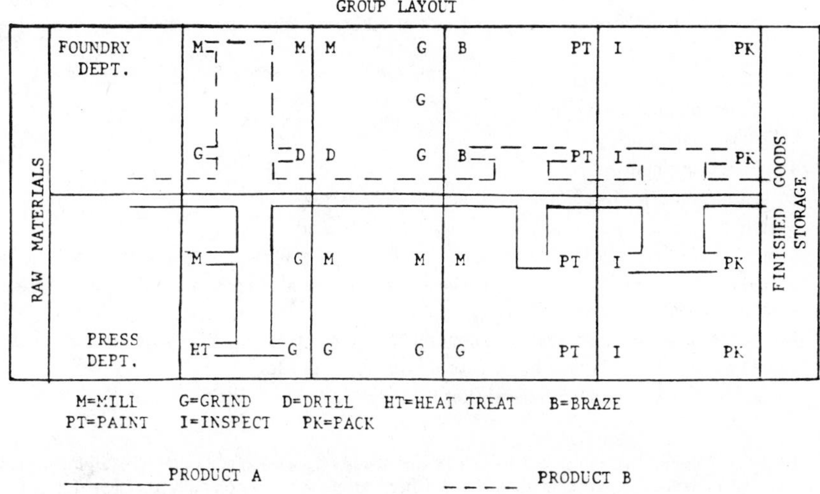

Fig. 7.3.8. Comparison of product, process, and group layout.

can lead to significant improvements in design activities through design data retrieval, rationalization, and design cost reductions.[5–9] For an extended treatment of Group Technology, refer to Chap. 8.2. In recent years, flexible manufacturing systems (FMS) seem to be replacing group technology as the trend-setter in basic layout design for manufacturing. See Chap. 8.10–8.15 for an extended treatment of FMS.

Advantages of Group Technology

Some of the advantages of group technology are listed below.

1. Improved work flow through better control of manufacturing.
2. Improved throughput times.
3. Reduction in work-in-progress stocks.
4. Increased machine utilization, flexibility, and job satisfaction.
5. Reduced absenteeism, paperwork, and set-up costs.
6. Improved environment for solving manufacturing problems.

7.3.5 THE PLANT LAYOUT DEPARTMENT

The plant layout function in large companies is frequently delegated to a specialized group within an industrial or manufacturing engineering department. Over the years, many responsibilities gravitate to the group, sometimes by default.

Since most functions and departments within an enterprise contribute information to the design of a layout, the layout group becomes vulnerable to delays in receiving necessary information and may have to use some loose estimates or not meet project deadlines.

While the major output of a layout group is usually a two-dimensional layout, the group is responsible for other items. The list may include items that in many enterprises are the responsibility of other groups and departments such as production, process, and plant engineering. The responsibilities of the plant layout group may include items from any, or all, of the lists shown in Table 7.3.1.

The lists given in Table 7.3.1 are not exhaustive and it may be obvious that the layout will be affected by most decisions that are made in any enterprise. Few layout decisions can be made without the involvement of a number of other departments. The sheer number of decisions to be made leads to the issuing of procedures to expedite the process. However, the danger of over-reliance on the procedures can sometimes lead to a reluctance to accept change.

The listing shown in Table 7.3.1 gives some credence to the saying, "everything is the business of layout and layout is the business of everyone." Service departments are each unique, but there is much commonality in the decisions and in the policies that come from these departments, which ultimately influence the layout. While most people would initially agree with the statement that employee services should be centralized, exceptions might focus on cafeterias, restrooms, parking lots, time clocks, first aid, employee suggestion office, plant newspaper office, office for counsellors of toxic substance abuse, and the office for employee representatives such as union officials. Some of these are affected by regulations of various governmental agencies as well as by union contracts and historical precedents. In a sense, each of the employee services is unique, but the commonality may become more apparent as a result of asking the following questions.

1. Should the service be offered at all?
2. Should the service be offered to everyone?
   To probationary employees?
   To all regular employees?
   To pensioners? e.g., recreational facilities.
3. Should the service be provided immediately? e.g., first aid.
4. Should it be a centralized or decentralized service? e.g., general stores vs. area cribs.
5. Should the service be dispatched from a central area and/or a decentralized area or should the customer go to the service? e.g., maintenance.
6. Should the service be provided by non-proprietary suppliers? e.g., delivery of products to customers.
7. Should the service equipment be purchased or leased? e.g., fork trucks.
8. Should the service schedule the arrival of the customer? e.g., cafeteria.

Many such questions must be asked to clarify the needs of service facilities for equipment, space, and location. The plant layout department must be sensitive to changes in company policies, for example, the change of scheduling from butt shifts with little or no time interval between day and afternoon shifts to a schedule with a three-hour interval between the two shifts, which greatly decreases the need for parking space.

Table 7.3.1. Responsibilities of the Plant Layout Group

Location and Arrangement of Equipment

Conduct management reviews (2-D and possibly 3-D presentations)
Prepare final 2-D layouts and drawings
Issue work order requests to Plant Engineering
Follow work through Plant Engineering and installation
Coordinate layout projects until they are accepted by the using departments

Layout Standards and Procedures

Kinds, colors, and relative weights of lines
Standards for information to be added, including letters and numerals and where to locate the information on the layout
Grid sheet practices, including content and location of the title block
Specifications of mylar tapes for aisles, rails, and conveyors

Conveyor Systems Design

Determine types of conveyor(s) required
Layout conveyor paths and elevations
Establish load, unload, and work heights, speeds, and capacities
Determine part clearances, guard requirements, and weight to be carried
Design and purchase carriers and related equipment

Office Facilities

Develop office standards
Purchase new furniture
Maintain inventory of used furniture
Order new telephones, and changes to the existing telephone system
Coordinate furniture repair programs
Donate obsolete furniture to organizations
Rearrange people, furniture, and offices
Study standards and make changes to keep old furniture updated to current needs

Workplace Layouts

Prepare workplace layouts for all manufacturing areas including assembly areas
Provide specifications for workplace handling equipment such as tool rails, bridge cranes, work modules.
Provide layouts for general plant equipment.

Employee Facilities

Provide designs for employee parking
Determine restroom requirements
Design cafeteria facilities
Locate time clocks and bulletin boards
Design relief areas, offices, and cribs
Select and purchase office and workplace furnishings
Specify telephone locations

Material Handling Facility Engineering

Determine facility requirements and capacities (storage, shipping, receiving)
Provide battery-charging and truck-repair facilities
Provide engineering assistance for selection and specification of specialized mobile equipment

Planning

Develop advanced plans for rearrangements and addition of facilities for new product programs
Determine future building requirements
Develop plans for future land acquisitions
Obtain cost estimates for new projects and new product studies
Conduct management reviews of facility plans and proposals

Government Agency Interface

Obtain all government permits (zoning, environment, street closings, curb cuts)
Plan and conduct traffic studies
Comply with all safety regulations

(*Table continues on p. 7 • 38*)

Table 7.3.1. Responsibilities of the Plant Layout Group (*Continued*)

Land and Buildings

Layout and general characteristics of most land and leasehold improvements
Layout and general characteristics of most buildings (construction or alteration)

Miscellaneous Layout Items

Integrate minor changes in layouts (requested by Production, Process, Plant or Methods Engineer) with general plans for new layout or rearrangement.
Forward the processing of documents authorizing minor change in layout
Determine location of all signs (exit, danger, aisle, and shelter)
Location of drinking fountains, coat racks, floor matting
Determine the types of intercom and paging systems to be used

Housekeeping Considerations

Location of all trash containers
Determine a painting scheme for aisles and stock barriers

Another example is the change from the use of relief operators on assembly lines to the use of mass relief, in which the line shuts down for a fixed period and may increase the requirement for restroom facilities.

In the past, service areas have been people intensive, but mechanization plus computerization have increased, and should continue to increase, the investment for these services and in many cases reduce the number of people required to supply the service.

7.3.5.1 Office Layout

Creating an office layout is extremely complex, due to the many services required. This is a difficult assignment because it involves groups of people and their conflicting wants and needs as well as the needs of their customers and of their neighboring departments. It usually requires a compromise. Many layout departments have been accustomed to supplying the head of an office group with the grid of the area, templates, and other aids, and requesting that the group supply a layout. The manager usually is advised to involve the group's personnel in the process. It was a small precursor of the emerging practice of involving employees in decision-making processes that affect their lives.

7.3.6 GENERATING ALTERNATIVE LAYOUTS

Engineers analyze both existing and new layouts for improvement possibilities by creating alternative layouts. Since this is still done manually for the most part, several systematic approaches to developing an improved layout are available.[10] Since block (departmental) area diagrams are commonly used, departmental areas must be estimated, initial shapes formulated and relationships among departments recorded.

7.3.6.1 Developing Departmental (Block) Areas

Data Sources and Requirements

The manufacturing routings and production control data are used to organize manufacturing information for use in developing layouts. Using the routings, fabrication charts (Fig. 7.3.9), flow process charts (Fig. 7.3.10), process charts (Fig. 7.3.11), and material handling routing charts (Fig. 7.3.12) are developed. For more information on constructing these charts, see Polk.[11]

As manufacturing routings, and production-control and material-handling data can be retrieved from computers, these charts will be computer-generated or merged with graphics systems to produce block areas, and will be used to develop rough layouts for each department. Many versions might be required before an acceptable layout is designed. To arrive at the block area layouts, space has to be considered for each workstation. The checklist of items for the design of a workstation can be lengthy, and for many work stations it may be unique. The checklist would include such items as raw material, purchased parts, in-process material, rejects, scrap, finished parts, offal and turnings, production equipment, conveyor drops and rises, equipment door swings, cyclic movement of items such as equipment tables, safety items, inspection fixtures, replacement tooling, the operator, maintenance operators, and emergency access space for operators. The list for even a simple operation can be very lengthy.

Estimating Space for Departments

Once a decisions has been made concerning the items needed for workstations, estimating the area required for a department can be done in three ways:

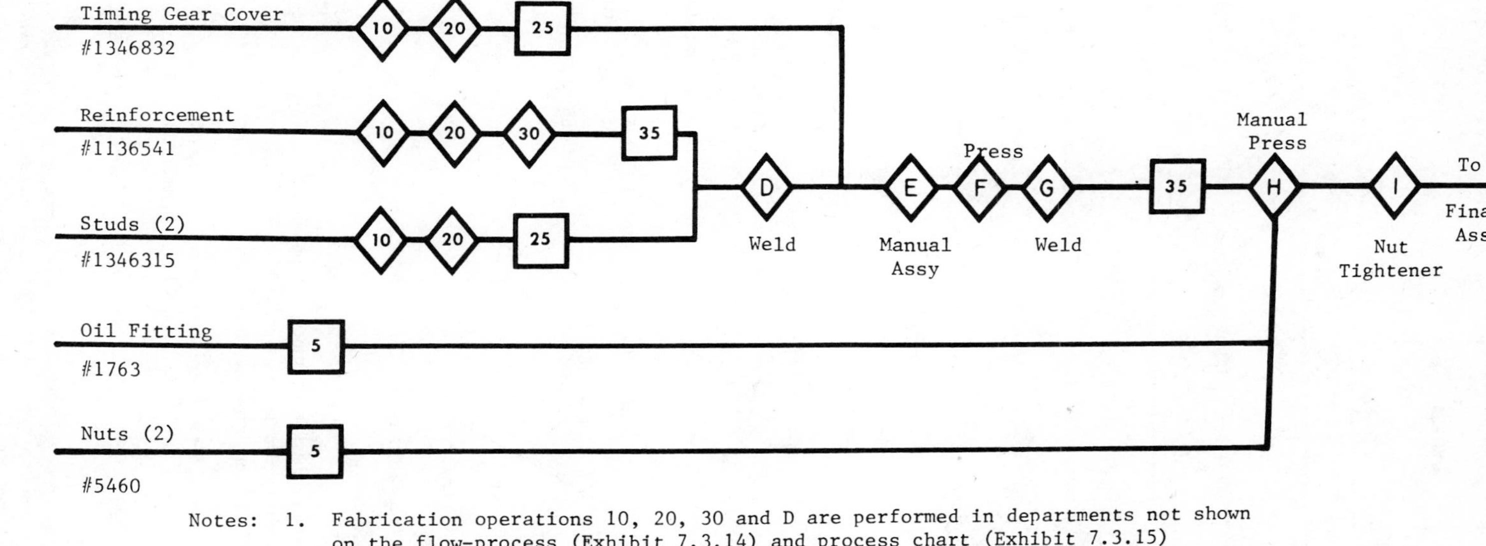

Notes:

1. Fabrication operations 10, 20, 30 and D are performed in departments not shown on the flow-process (Exhibit 7.3.14) and process chart (Exhibit 7.3.15)
2. Fabrication operations E, F, G, H and I are performed in the department in Exhibit 7.3.14 and Exhibit 7.3.15
3. Inspection operations 25 and 35 are after press and assembly operations
4. Inspection operation 5 is inspection of purchased parts
5. Definition of symbols are shown on Exhibit 7.3.15

Fig. 7.3.9. Fabrication chart of timing gear.

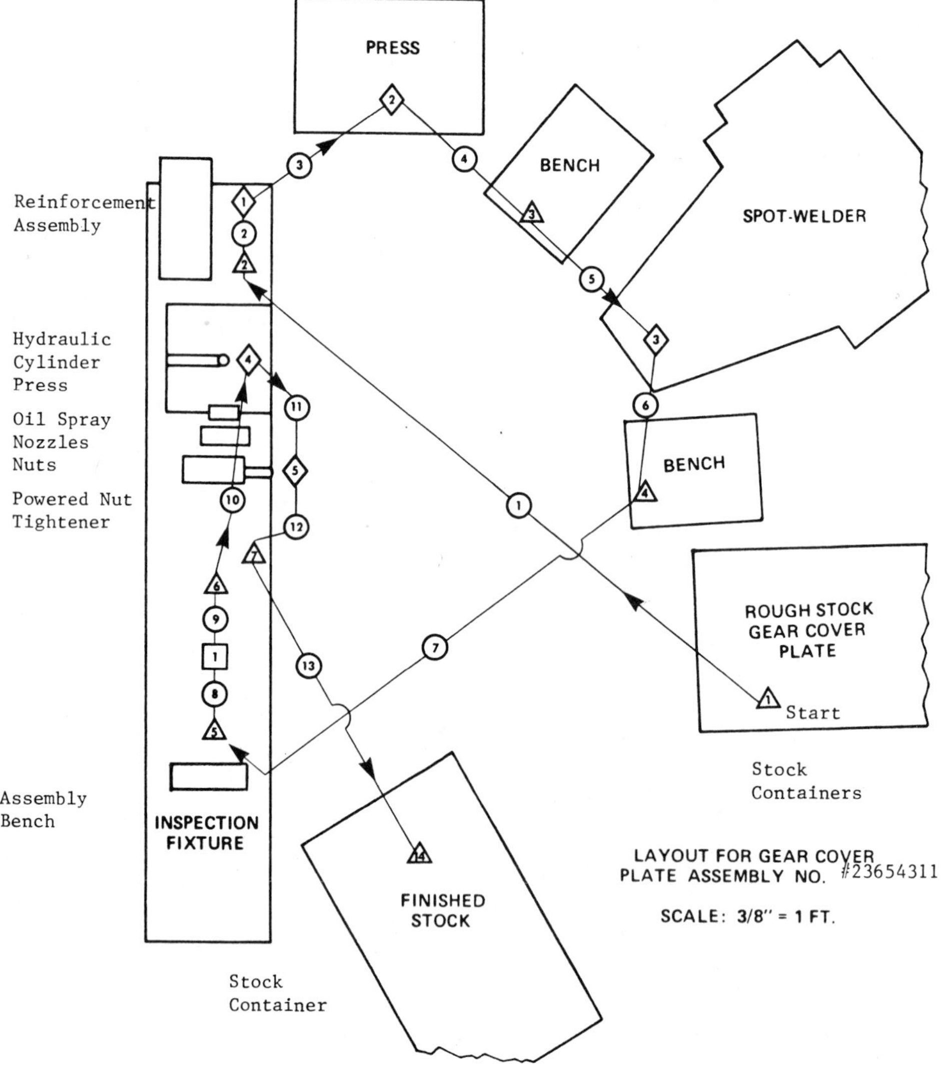

Fig. 7.3.10. Flow-process chart. *Note*: Definitions of symbols are in Fig. 7.3.11.

1. Extrapolate from an existing comparable department by proportionally modifying the length and width.
2. Use templates for each workplace in the department and create a rough layout.
3. Use standard data space values for each workstation.

The first two are commonly used and no further discussion is needed.

The standard space values may be available from proprietary historical records, from vendors or occasionally from a published checklist. Table 7.3.2 is an example of a representative checklist created for a traditional design of large offices and includes some space estimates such as those in the following excerpt from an office layout example:

> *The next phase is to estimate the space required for each work station, group and department until an estimate has been made of the total floor space required. To the normal dimensions of individual work stations must be added an allowance for aisles and access. A rule of thumb is to add 2 ft (0.61 m) to both the length and width dimensions of the work station. The floor space for a workstation requiring 60 ft^2 (5.6 m^2) would be increased to 95 ft^2 (8.83 m^2) by adding the aisles and access allowance.*

The 2 ft added to each dimension provides a medium-density layout. Decreasing the 2 ft (0.61 m) down to a minimum of 1.5 ft (0.46 m) increases the density. Enlarging the allowance over 2 ft (0.61 m)—generally required for adequate separation of work stations in a "pure landscape" layout—decreases the density.

The square footage needed for a single work station is multiplied by the number of work stations to arrive at total square footage required for that type of standard work station. This procedure is repeated for each standard workstation in that particular group or unit. Adding up the total square footage for

		SUMMARY	
Process Description	Timing Gear Cover Plate Assembly	Number of "Fabrications" ◇	5
Object Followed	Gear Cover Plate #23654311	Number of "Moves" ○	13
Name	John Doe Date 8/7	Number of "Stores" △	8
	Present Method ☒ Proposed Method ☐	Number of "Inspects" ☐	1
Remarks:	Normal operation - two operators utilized	TOTAL STEPS	27
		DISTANCE TRAVELLED	57½

Step	Fabrication / Move / Store / Inspect	Description	Distance Moved	How Moved	Parts/Move
1	◇ ○ ▲1 ☐	Rough stock in gondola			
2	◇ ① △ ☐	To Reinf Assem Area	13'	hand	8
3	◇ ○ ▲2 ☐	On table			
4	◇ ② △ ☐	To assem reinf	1½	hand	1
5	◆1 ○ △ ☐	Assem reinf & bolts assys to cover plate			
6	◇ ③ △ ☐	To press for restrike	4½'	hand	1
7	◆2 ○ △ ☐	Restrike assy's			
8	◇ ④ △ ☐	To storage bench	3½'	hand	1
9	◇ ○ ▲3 ☐	On storage bench			
10	◇ ⑤ △ ☐	To spot weld mach.	4'	hand	1
11	◆3 ○ △ ☐	Spot weld reinf & bolt ass'y to cover plate (4 spots)			
12	◇ ⑥ △ ☐	To storage bench	3'	hand	1
13	◇ ○ ▲4 ☐	On storage bench			
14	◇ ⑦ △ ☐	To inspection area	9½'	hand	4/2
15	◇ ○ ▲5 ☐	On inspection bench			
16	◇ ⑧ △ ☐	To inspection operation	2'	hand	1
17	◇ ○ △ ▣1	Gage bolts for width & straighten if necessary			
18	◇ ⑨ △ ☐	To storage area on bench	3'	hand	1
19	◇ ○ ▲6 ☐	On bench			
20	◇ ⑩ △ ☐	To oil supply nozzle area	3'	hand	1
21	◆4 ○ △ ☐	Assem oil spray & crimp to cover - assem nuts to bolts			
22	◇ ⑪ △ ☐	To high cycle wrench	2½'	hand	1
23	◆5 ○ △ ☐	Tighten nuts to bolts			
24	◇ ⑫ △ ☐	To bench for storage	1'	hand	1
25	◇ ○ ▲7 ☐	On bench			
26	◇ ⑬ △ ☐	To finished stock gon.	7'	hand	4/2
27	◇ ○ ▲8 ☐	In gondola			
28	◇ ○ △ ☐				
29	◇ ○ △ ☐				
30	◇ ○ △ ☐				

Fig. 7.3.11. Process chart.

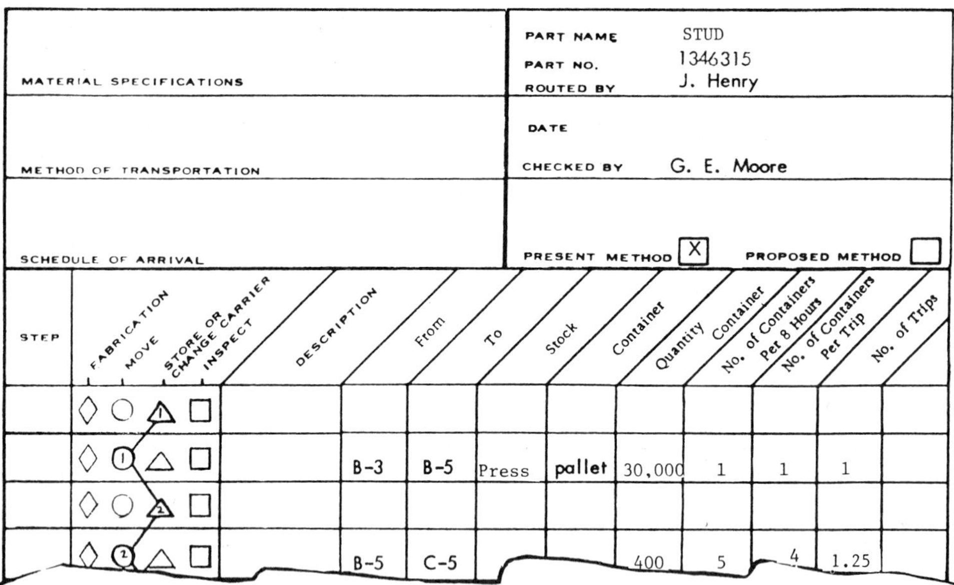

MATERIAL SPECIFICATIONS	PART NAME: STUD PART NO.: 1346315 ROUTED BY: J. Henry
METHOD OF TRANSPORTATION	DATE CHECKED BY: G. E. Moore
SCHEDULE OF ARRIVAL	PRESENT METHOD [X] PROPOSED METHOD [ ]

STEP	FABRICATION / MOVE / STORE OR CHANGE CARRIER / INSPECT	DESCRIPTION	From	To	Stock	Container	Quantity Container	No. of Containers Per 8 Hours	No. of Containers Per Trip	No. of Trips	
	◇ ○ △(1) □										
	◇ ○(1) △ □		B-3	B-5	Press	pallet	30,000	1	1	1	
	◇ ○ △(2) □										
	◇ ○(2) △ □		B-5	C-5			400	5	4	1.25	

Fig. 7.3.12. Material handling routing.

each type of workstation will give the total floor space required for the individual workstations in that group or unit." [12]

The preceding information is used as shown in Fig. 7.3.13. A block layout for a typical office layout is shown in Fig. 7.3.14.

7.3.6.2 Selecting the Layout

After several alternatives have been developed, they are evaluated and ranked. A typical evaluation scheme for an office layout is shown in Table 7.3.3. Additional layouts, in general, should always be generated and evaluated against criteria such as material handling costs or ease of expansion.

7.3.6.3 Manual Development of the Block Area

The building in Fig. 7.3.15 includes six departments and an outside courtyard, which is identified as department 17. This building will be used to illustrate the manual method as well as two computerized methods, CORELAP and CRAFT. The from-to chart (Fig. 7.3.16) includes the quantitative relationships among the departments in trips per hour. In Fig. 7.3.17, the trips below the diagonal, which represent backward flow, have been combined with those above the diagonal, since most computerized techniques do not consider directional flow. This chart could be used as is in the creation of a new layout. However, the trips per hour will be replaced by relationships, since these are required by the CORELAP program. In Fig. 7.3.18 relationship values A, E, I, O, U, X and corresponding values of 6, 5, 4, 3, 2, 1 have been assigned to the trips per hour. The use of letter relationships and similar corresponding values were integrated into a chart by Muther[10] in a historic paper on systematic layout planning (SLP). The assigned weights will not be used in the manual method but will be used in the CORELAP method.

In Fig. 7.3.19 the relationship values are substituted for the trips per hour and the chart is made symmetrical by also placing the values below the diagonal. The values of the rows are added to get the total closeness rating (TCR).

In Fig. 7.3.20 the relationship values are diagrammed, with circles representing the departments. In step 1, only the more important values of 6, 5, and 4 are entered in order to place these departments adjacent to each other. In step 2, the 3s are entered. Now departments 12 and 13, 12 and 16, 11 and 14 are not adjacent. This is scored by taking the product of the relationship value times the nonadjacent distance between the corresponding departments. For example, since the distance between departments 12 and 15 is 2, 1 is deducted from the 2 (an adjustment necessary for a zero distance between adjacent departments) and the result is multiplied by 3. The total score for step 2 is 9.

In step 3, department 12 is relocated, which results in a score of 7. In step 4, department 15 and the 2s are entered. In step 5, a block area diagram with equal areas is created. In Fig. 7.3.21 the exterior courtyard

Table 7.3.2. Representative Checklist for Traditional Design of Large Offices

Desks should face the same general direction.
In open areas desks should be placed in rows of two.
For desks in rows of one, there should be 6 ft (1.83 m) from the front of a desk to the front of the desk behind it.
For desks in rows of two or more, and where ingress and egress is confined to one side, 7 ft (2.13 m) should be allowed from the front of a desk to the front of the desk behind it.
If employees are back to back, allow a minimum of 4 ft (1.22 m) between chairs.
Inside aisles within desk areas should be 3 to 5 ft (0.91 to 1.52 m) wide.
Intermediate aisles should be 4 ft (1.52 m) wide.
Natural lighting should come over the left shoulder or the back of the employee.
From 50 to 75 ft^2 (4.65 to 6.97 m^2) are required for a work space consisting of a desk, shelf space, a chair, and a 2 ft (0.61 m) space allowance on a length and a width.
Desks should not face high-activity aisles and areas.
Desks of employees doing confidential work should not be near entrances.
Desks of employees having much visitor contact should be near entrances, and extra space should be provided.
Desk of the receptionist should be near the visitors' entrance.
Supervisors should be positioned adjacent to the secretaries.
Access to the supervisor's workstation should not be through the work area.
Supervisors in open areas should be separated from their group by 3.3 ft (1 m).
The flow of work should take the shortest distance.
People who have frequent face-to-face conferences should be located near each other.
Employees should be adjacent to those files and references that they use frequently.
Employees should be placed near their supervisors.
Five-drawer file cabinets should be considered in lieu of four-drawer cabinets.
Open shelf filing or lateral file cabinets should be considered in lieu of standard file cabinets.
Four- or five-drawer file cabinets should be considered as a substitute for 2 two-drawer cabinets.
The reception area should create a good impression on visitors, and an allowance of 10 ft^2 (0.93 m^2) should be used per visitor if more than one arrives at a given time.
The layout should have a minimum of offsets and angles.
Large open areas should be used instead of several small areas.
Open areas for more than 50 persons should be subdivided by use of file cabinets, shelving, railings, or low "bank-type" partitions.
Office space should not be used for bulk storage or for storage of inactive files.
Conference space should be provided in rooms rather than in private offices.
Conference and training rooms should be pooled.
The size of a private office will often be determined by existing partitions.
Private offices should have a minimum of 100 ft^2 (9.3 m^2) to a maximum of 300 ft^2 (27.9 m^2).
A 300 ft^2 (27.9 m^2) private office should be used only if the occupant will confer with groups of eight or more people at least once per day.
Related groups and departments should be placed near each other.
Minor activities should be grouped around major ones.
Work should come to the employees.
Water fountains should be in plain view.
Layouts should be arranged to control traffic flow.
Heavy equipment generally should be placed against walls or columns.
Noise-producing workstations should be grouped together.
Access to exits, corridors, stairways, and fire extinguishers should not be obstructed.
All governmental safety codes should be followed.
In planning the office, consider the floor load, columns, window spacing, heating, air conditioning and ventilation ducts, electrical outlets, and lighting and sound.
The scale of the layout should be either ¼ in. = 1 ft (1 cm = 50 cm) or ⅛ in. = 1 ft (1 cm = 100 cm).
Plastic reproducible grid sheets and plastic self-adhesive templates should be considered.

Source: Ref. 18

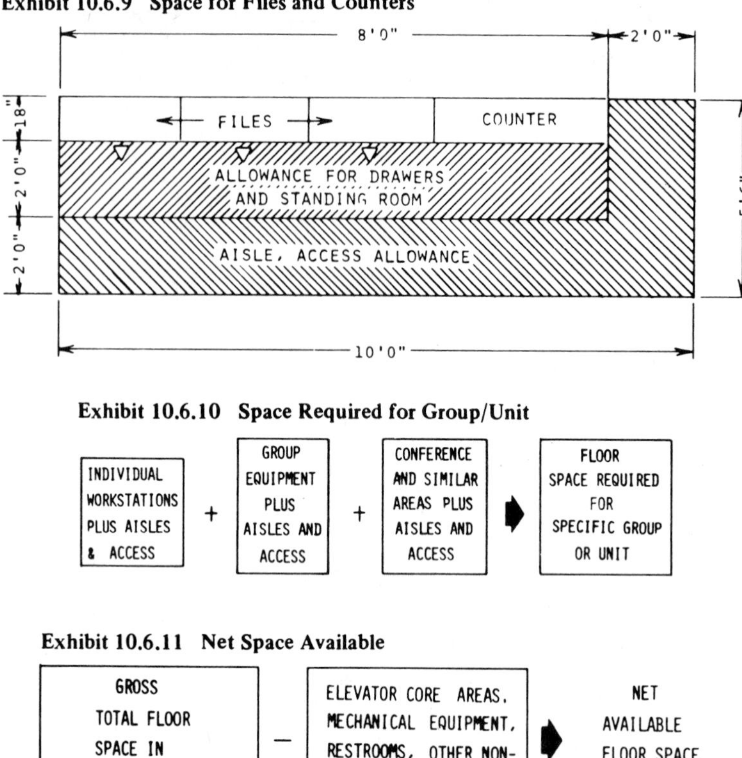

Fig. 7.3.13. Space required for group/unit. Reprinted with permission from Ref. 12.

is added and each department is allocated to the original block area. Department 13 has a very irregular perimeter while each of departments 11, 12, and 14 has a minor irregularity. In step 7 (Fig. 7.3.22), adjustments were made to create straight aisles.

7.3.7 COMPUTER AIDED LAYOUT

The objective of computer aided layout programs is to develop a block area layout of facilities that minimizes the cost of interaction among the blocks (departments). In manufacturing layouts the interaction to be minimized is usually material handling cost, whereas it may be personal contacts in office layouts.

Two major classifications of computer programs for facilities layouts exist, the *construction* type and the *improvement* type. While there are over 22 separate construction programs, the most widely known are CORELAP, ALDEP, and PLANET. There are over ten improvement routines, of which CRAFT and COFAD are most widely known. For an excellent treatment of manual as well as computerized models see Francis and White,[13] Tompkins and Moore,[14] and Tompkins.[15] CORELAP, a construction program, and CRAFT, an improvement program, will be described here in some detail.

7.3.7.1 CORELAP

CORELAP (computerized relationship layout planning), the first computerized construction routine to be developed, requires as input data the relationship chart and the area required by each department. The program selects the most critical department, the one with the highest TCR from the relationship chart, and places it in the center of the layout. Then it selects, in order, those departments that have the highest relationship with the departments already placed.

When two or more departments have the same total score, systematic tie-breaking rules are followed to select the "winner." The departments from Fig. 7.3.23 were added to the layout in the following sequence: 17, 11, 13, 16, 12, 14, 15.

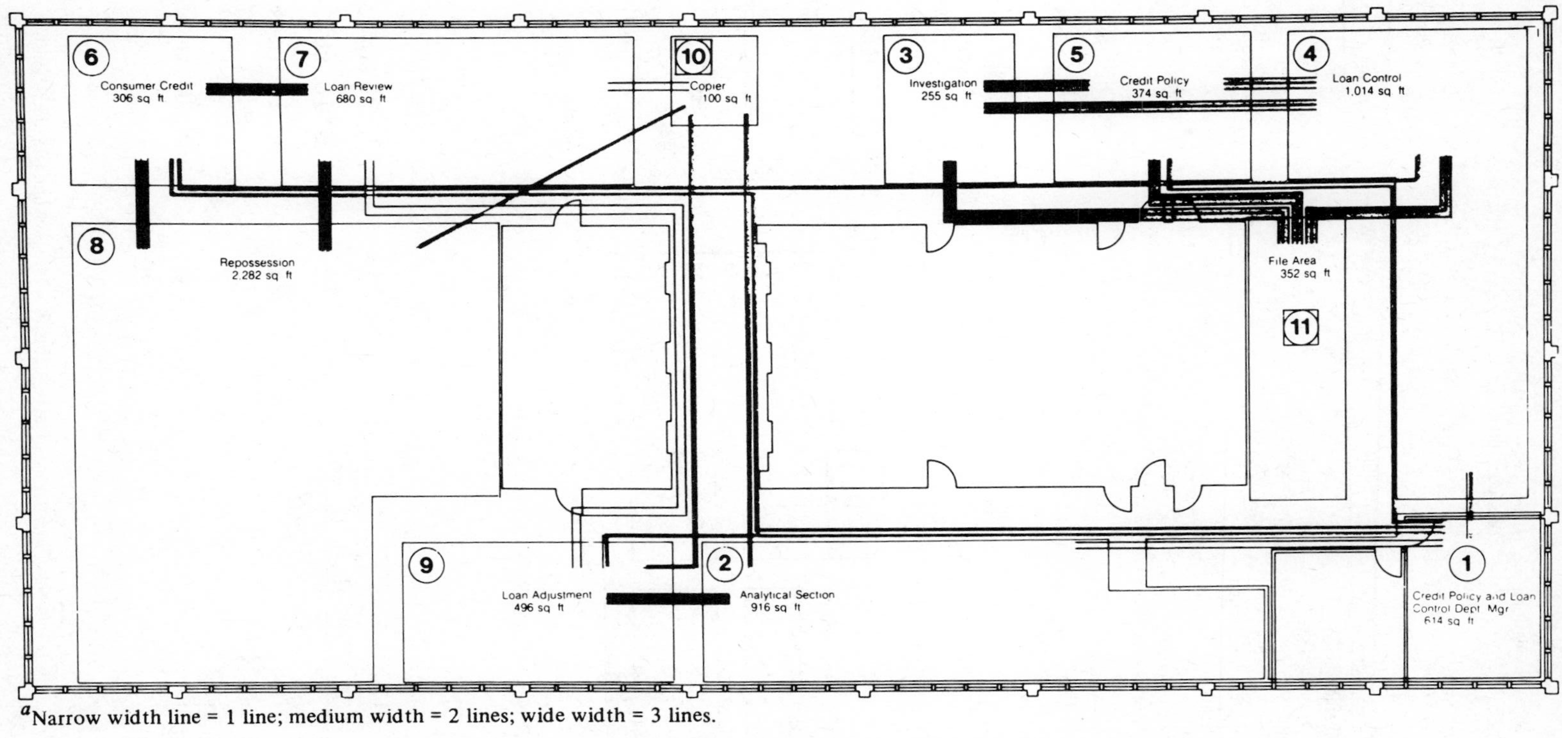

[a]Narrow width line = 1 line; medium width = 2 lines; wide width = 3 lines.

Fig. 7.3.14. Block plan. Reprinted with permission from Ref. 12.

Table 7.3.3. **Evaluating the Block Plan**

Office Layout—Evaluation

Office: Household goods loan   Date: 8-8-80
Project: QWL   Analyst: JJM
Plan: A. Top floor   B. Ground floor

Criteria	Weights	Plant A Rating	Plant A Rating × Weight	Plan B Rating	Plan B Rating × Weight
1. Noise distraction	8	9	72	7	56
2. Visual distraction	6	8	48	7	42
3. Reception	5	8	40	10	50
4. Flow; people, supplies	7	7	49	9	63
5. Convenience of personnel	8	9	72	8	64
6. Flexibility	10	10	100	6	60
7. Aesthetics	7	10	70	6	42
8. Area per employee	3	8	24	7	21
9. Energy costs	4	10	40	5	20
10. Cleaning costs	9	8	72	8	72
11. Maintenance costs	9	8	72	7	63
12. Project cost	10	9	90	6	60
13. Expected productivity	10	10	100	7	70
Totals			849		683

Source: Ref. 18.

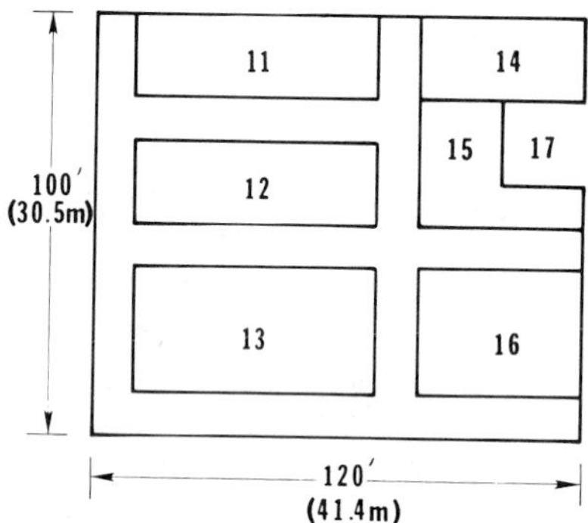

Fig. 7.3.15. Block area layout. Department 17 is an outside courtyard.

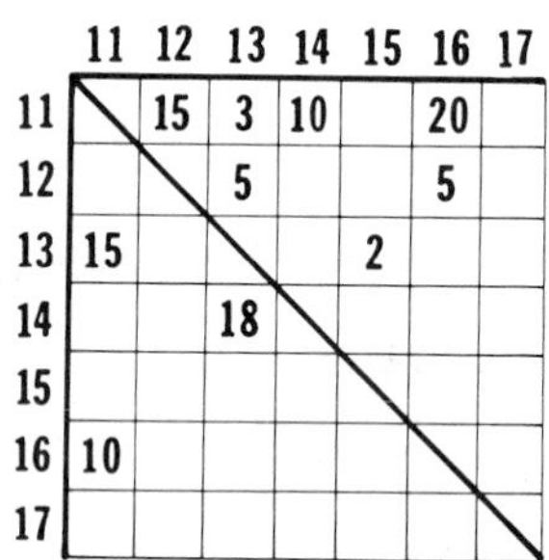

Fig. 7.3.16. From-to chart (trips/hour).

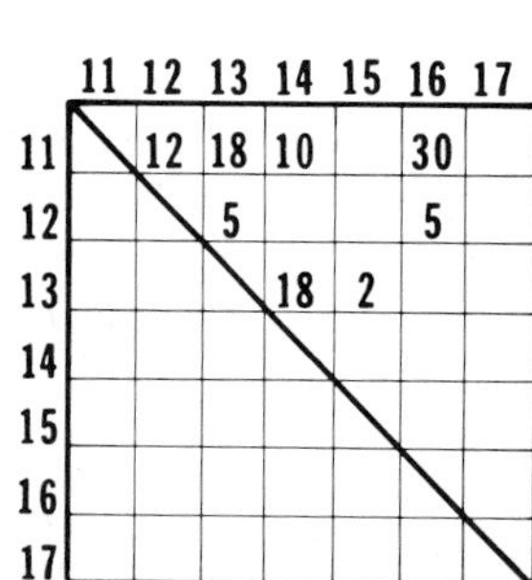

Fig. 7.3.17. All trips combined above the diagonal.

Weights	Relationship Values		Trips/Hr
243	A	6	20 to 30
81	E	5	16 to 20
27	I	4	11 to 15
9	O	3	5 to 10
1	U	2	0 to 5
-729	X	1	
+729		7	For Pre-assigned Departments

Fig. 7.3.18. Assignment of relationship and weights to trips/hour.

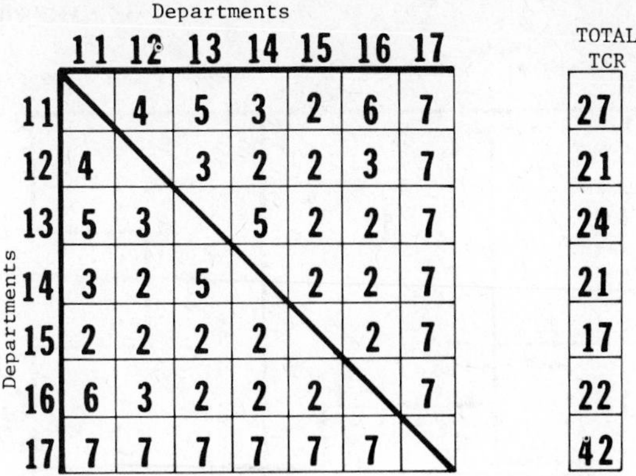

Departments	11	12	13	14	15	16	17	TOTAL TCR
11		4	5	3	2	6	7	27
12	4		3	2	2	3	7	21
13	5	3		5	2	2	7	24
14	3	2	5		2	2	7	21
15	2	2	2	2		2	7	17
16	6	3	2	2	2		7	22
17	7	7	7	7	7	7		42

Fig. 7.3.19. Relationship values between departments.

Step #1 Use 6, 5, 4 Relationship

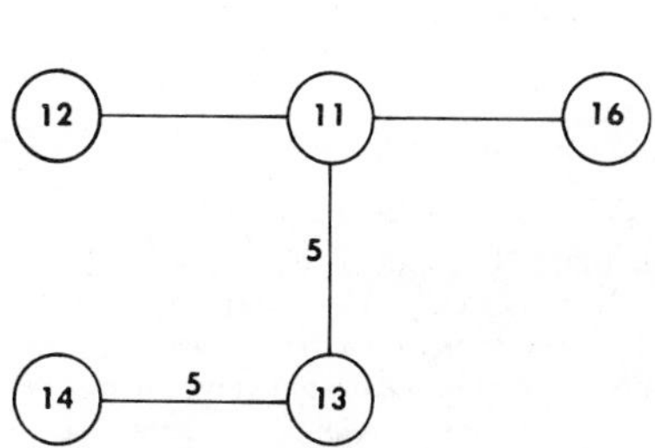

Step #2 Add "3s"

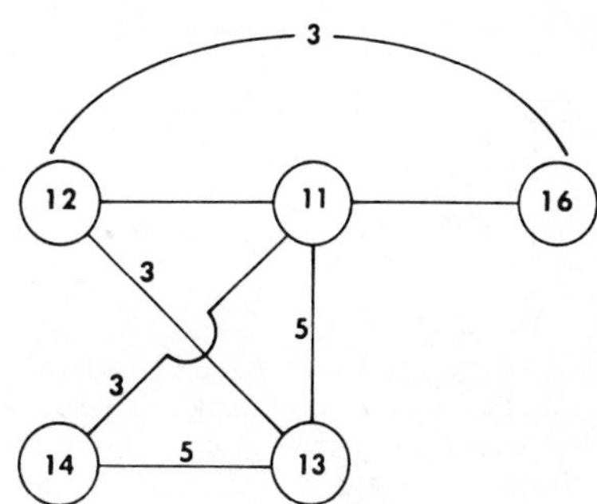

Step #4 Add Dept. 15

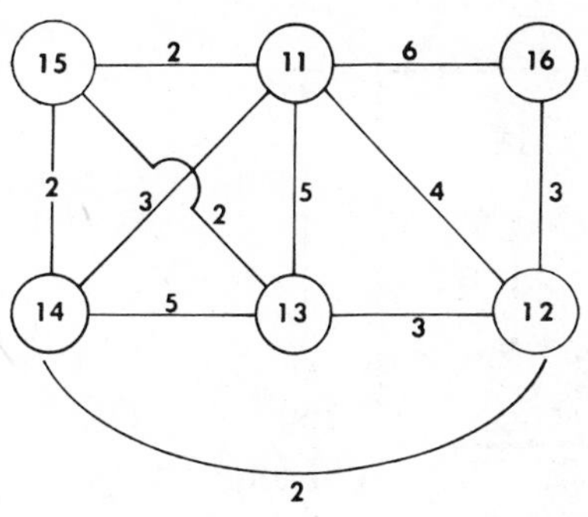

Step #3

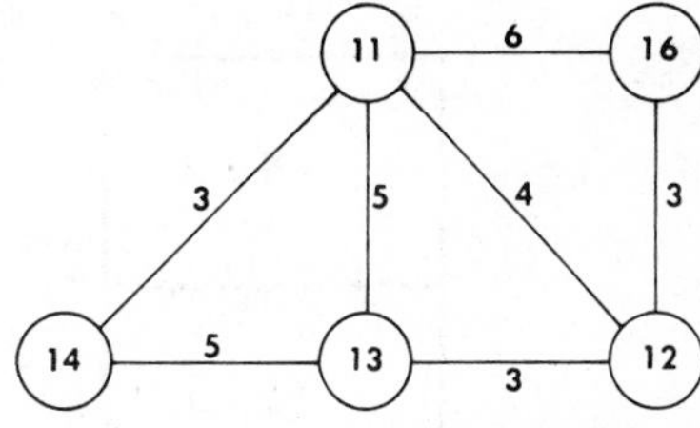

Rearrange to Reduce Number of non adjacent Departments

Step #5 Assume Equal Areas

15	11	16
14	13	12

Fig. 7.3.20. Use of relationship values to create an improved layout.

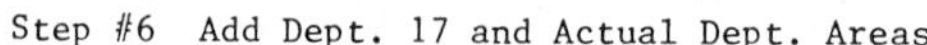

Step #6 Add Dept. 17 and Actual Dept. Areas

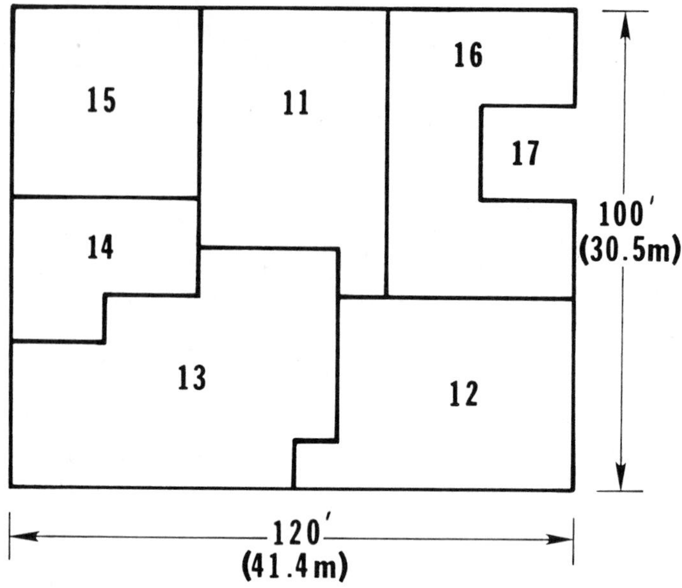

Fig. 7.3.21. Manual layout added department 17 and actual department areas.

This sequence corresponded to the following total closeness relationships (TCR) for these departments: 542, 27, 24, 22, 21, 21, 17. Because department 17 is fixed, the TCR value was calculated by CORELAP as 542. After the sequence is created, the departmental placement weights are used to place the departments as near as possible to departments already located in the layout. The search is made to locate the entering department adjacent to one or more departments with which it has large placement weights. The weights assigned to these departments are listed in the sequence in which the departments were placed: 729, 243, 81, 27, 9, 1, −729.

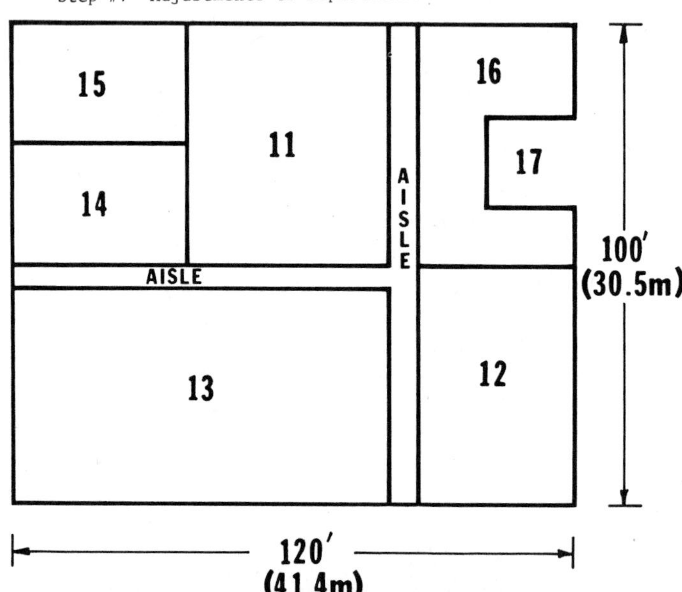

Fig. 7.3.22. Manual layout with aisles added.

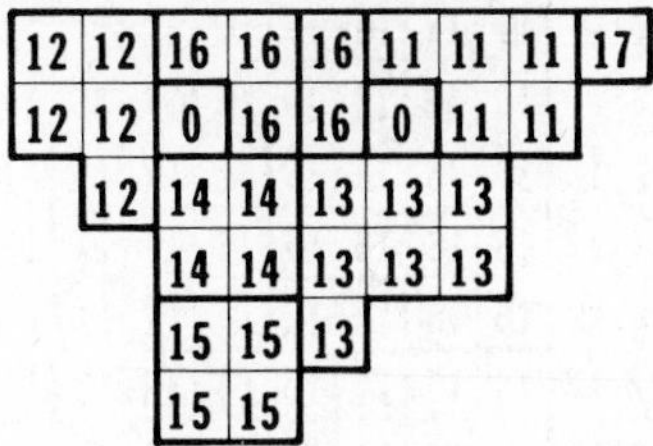

Fig. 7.3.23. CORELAP solution to the sample problem. *Note*: Department 0 represents unallocated area.

The −729 was not used in this example since it applies to an X relationship, which indicates that the departments involved are to be placed at a considerable distance from each other. There are no X relationships in this example.

After the layout is completed, it is scored. The scoring is based on the product of the number of trips between two departments and the number of units of distance between the departments. If two departments touch, the distance is zero and the product is zero. In Fig. 7.3.23 the distance between departments 12 and 16 and 12 and 14 is 0, while the distance between 12 and 15 is 2 and between 11 and 12 is 3. Since the weight between 12 and 15 is 1 and between 11 and 12 is 27, the products of distance times weight is $2 \times 1 = 2$ and $3 \times 27 = 81$, respectively. The total score for the layout is 135.

The user should have some reservations about the results obtained from CORELAP. The perimeter of the resultant layout rarely has a rectangular shape, as is evident in Fig. 7.3.23. A manual reshaping of the layout is shown in Fig. 7.3.24. In addition, not only are the departments greatly reduced in scale, but all departments are not reduced by the same proportion. A manual reshaping expanded the layout from 31 units to the original area of 120 units as shown in Fig. 7.3.25. This example was altered by substituting 1 for each of the placement weights and was analyzed using CORELAP. The resulting layout was identical to the one shown in Fig. 7.3.23. The method of scoring each layout presupposes that a receiving and shipping area for each department exists at the point of contact between two adjacent departments. This, of course, may not reflect the way in which each department is designed.

7.3.7.2 CRAFT

The first computerized improvement facilities design technique was CRAFT (computerized relative allocation of facilities technique), which was presented in 1963 by Armour and Buffa.[16] CRAFT is probably the most widely known, widely used, and written-about computerized facilities design technique in existence. Even though CRAFT seldom yields an optimal solution, in terms of minimizing costs or distances, it does provide fairly good solutions and offers the facilities designer a starting point for improving an initial block area layout.

The input for the CRAFT program includes an analysis control card, the flow matrix data for all the interactions among the departments, the cost of interaction data, and last is the initial rectangular layout which must include all the area within the rectangle.

The Analysis Control Card

In the example problem used for CORELAP and with the manual method, the layout is not perfectly rectangular, a requirement for CRAFT input, and the indentation or courtyard is used as an outside miscellaneous storage area. For CRAFT input the courtyard will be enclosed and identified as department 17. Department 17 will be designated as a "fixed" department in the analysis control card and therefore will remain in place during each iteration of the CRAFT program.

The Unit Area

The first step in preparing a layout for the CRAFT program is to create the unit area. In effect, a grid will be superimposed on the layout. The unit areas must be square and neither the number of rows nor of columns in the grid must exceed 30. Since it is advantageous to use the smallest possible unit area, the largest dimension

12 12 16 16 16 11 11 11

12 12 12 16 16 11 11 11

14 14 14 14 13 13 13 17

15 15 15 15 13 13 13

Fig. 7.3.24. Manual reshaping of CORELAP solution maintaining 31 units of area.

12	12	12	16	16	16	16	16	11	11	11	11
12	12	12	16	16	16	16	16	11	11	11	11
12	12	12	16	16	16	16	11	11	11	17	17
12	12	12	16	16	16	16	11	11	11	17	17
12	12	12	16	16	13	13	11	11	11	11	11
12	12	12	14	14	14	13	13	13	13	11	11
12	12	12	14	14	14	13	13	13	13	13	13
15	15	15	14	14	14	13	13	13	13	13	13
15	15	15	14	13	13	13	13	13	13	13	13
15	15	15	15	15	15	15	15	15	15	13	13

Fig. 7.3.25. Manual reshaping of CORELAP expanded to the original area of 120 units.

of the rectangle should be divided by 30 to create one side of the unit area. This value is then divided into the other side of the rectangle to determine the remaining number of rows or columns. Using the dimensions from Fig. 7.3.18 the calculations would be as follow:

120 ft/30 columns = 4 ft as one side of the unit area

100 ft/4 ft per row = 25 rows

However, for the example in this section only 12 columns will be used in order to produce a reasonable size square-unit area.

120 ft/12 columns = 10 ft

100 ft ÷ 10 ft per row = 10 rows

unit area = 10 × 10 = 100 ft^2

The grid lines are drawn and the department boundaries are modified not only to absorb the aisles but also to insure that all departmental edges coincide with a grid line. Figure 7.3.26 demonstrates this. Each unit area is labeled with its department number.

The from-to chart (flow matrix) shows the number of loads moved per hour among the departments. The CRAFT program will accept a cost matrix, which is then multiplied element by element by the flow matrix. If no costs are available for the cost matrix, a matrix of 1s must be submitted. For both the cost and volume matrixes only 0s may be placed on the diagonal, since CRAFT does not consider intradepartmental moves. The input for CRAFT is entered in the following sequence:

1. Analysis control card
2. Flow matrix

	1	2	3	4	5	6	7	8	9	10	11	2
1	1	1	1	1	1	1	1	4	4	4	4	4
2	1	1	1	1	1	1	1	4	4	4	4	4
3	1	1	1	1	1	1	1	5	5	5	7	7
4	2	2	2	2	2	2	2	5	5	5	7	7
5	2	2	2	2	2	2	2	5	5	5	5	5
6	2	2	2	2	2	2	2	5	5	5	5	5
7	3	3	3	3	3	3	3	6	6	6	6	6
8	3	3	3	3	3	3	3	6	6	6	6	6
9	3	3	3	3	3	3	3	6	6	6	6	6
10	3	3	3	3	3	3	3	6	6	6	6	6

Fig. 7.3.26. The aisles in Fig. 7.3.15 were absorbed by the departments.

3. Cost matrix
4. Layout matrix

THE CRAFT IMPROVEMENT PROCESS

STEP 1. The improvement process beings with the calculation of the centroids of all departmental areas and the rectilinear distances among all departments (centroid to centroid) that have flow relationships greater than zero. This distance matrix is multiplied (by matching elements) by the flow matrix and the products are totaled to give the total move cost (volume-distance product) for the original layout (Fig. 7.3.27).

STEP 2. Those departments that have a move value greater than zero are examined, and if two of them are of equal area, the centroids are exchanged and the revised distance matrix is multiplied by the flow matrix and the cost matrix to get the new total move cost. This is repeated for all pairs of departments that have equal areas, and also for all departments that are adjacent to each other given that at least one of the departments has a move value greater than zero. The process is again repeated for all departments in groups of three, given that two of them have common borders with a third department and given that at least one of the three has a move value greater than zero in the flow matrix.

STEP 3. The interchange that produces the greatest reduction in the volume-distance product is selected, and a new layout is created (Fig. 7.3.28). Step 1 is then repeated and if a reduction in the volume-distance product has actually occurred, a new layout is printed.

STEP 2. The process continues with iterations of steps 2 and 3 until no improvement in the volume-distance product can occur (Figs. 7.3.29 and 7.3.30). The three iterations for the example problem produced a reduction of 38.8% in the total material handling distance traveled.

Comments on CRAFT

The writers have experienced as many as 35 iterations for a layout containing 40 departments and as few as 2 iterations for a layout containing 6 departments. CRAFT has been used for the plot plan of all the property of

LOCATION PATTERN

	1	2	3	4	5	6	7	8	9	10	11	12
1	A	A	A	A	A	A	A	D	D	D	D	D
2	A						A	D	D	D	D	D
3	A	A	A	A	A	A	A	E	E	E	G	G
4	B	B	B	B	B	B	B	E		E	G	G
5	B						B	E		E	E	E
6	B	B	B	B	B	B	B	E	E	E	E	E
7	C	C	C	C	C	C	C	F	F	F	F	F
8	C						C	F				F
9	C						C	F				F
10	C	C	C	C	C	C	C	F	F	F	F	F

TOTAL COST 919.75 TOTAL COST REDUCTION 0.0

MOVEA MOVEB MOVEC ITERATION 0

Fig. 7.3.27. CRAFT computer output, original layout, cost of $919.75. *Note*: Departments 11 to 17 are represented on the computer printout by A to G.

LOCATION PATTERN

	1	2	3	4	5	6	7	8	9	10	11	12
1	A	A	A	A	A	A	A	F	F	F	F	F
2	A						A	F		F	F	F
3	A	A	A	A	A	A	A	F		F	G	G
4	B	B	B	B	B	B	B	F	F	F	G	G
5	B						B	E	F	F	F	F
6	B	B	B	B	B	B	B	E	E	E	E	E
7	C	C	C	C	C	C	C	E				E
8	C						C	E	E	E	E	E
9	C						C	D	D	D	D	D
10	C	C	C	C	C	C	C	D	D	D	D	D

TOTAL COST 693.50 TOTAL COST REDUCTION 226

MOVEA D MOVEB E MOVEC F ITERATION 1

Fig. 7.3.28. CRAFT computer output, iteration 1, new cost $693.50.

LOCATION PATTERN

	1	2	3	4	5	6	7	8	9	10	11	12
1	E	E	E	E	E	E	B	F	F	F	F	F
2	E				E	B	B	F		F	F	F
3	E	E	E	E	E	B	B	F		F	G	G
4	B	B	B	B	B	B	B	F	F	F	G	G
5	B	B	B	B	B	B	B	B	F	F	F	F
6	B	A	A	A	A	A	A	A	A	A	A	A
7	C	C	C	C	C	C	C	A				A
8	C						C	A	A	A	A	A
9	C						C	D	D	D	D	D
10	C	C	C	C	C	C	C	D	D	D	D	D

TOTAL COST 608.68 TOTAL COST REDUCTION 311.0[illegible]

MOVEA E MOVEB B MOVEC A ITERATION 2

Fig. 7.3.29. CRAFT computer output, iteration 2, new cost $608.68.

LOCATION PATTERN

	1	2	3	4	5	6	7	8	9	10	11	12
1	E	E	E	E	E	E	D	F	F	F	F	F
2	E				E	D	D	F		F	F	F
3	E	E	E	E	E	D	D	F		F	G	G
4	C	C	D	D	D	D	D	F	F	F	G	G
5	C	C	C	C	C	C	C	C	F	F	F	F
6	C	A	A	A	A	A	A	A	A	A	A	A
7	C	C	C	C	C	C	C	A				A
8	C		C	C	C	C	C	A	A	A	A	A
9	C	C	C	B	B	B	B	B	B	B	B	B
10	B	B	B	B	B	B	B	B	B	B	B	B

TOTAL COST 562.78 TOTAL COST REDUCTION 356.97

MOVEA D MOVEB C MOVEC B ITERATION 3

Fig. 7.3.30. CRAFT computer output, iteration 3, new cost $562.78, last iteration.

an enterprise as well as for the placement of individual presses in a miscellaneous press department. Tompkins[15] recommends that a number of computer runs should be made for a given problem, using at least three different initial layouts. The layouts usually produce some departments with irregular shapes or infeasible locations.

If one of the overall layouts not only has promise but has departments with irregular shapes or infeasible locations, these should be modified manually and the overall layout be run through the computer again. Additional computer runs can also be made to evaluate the sensitivity of the solutions to changes in the flow matrix.

As with CORELAP, the user should have some reservations about the output from CRAFT. Some of the departments develop very irregular shapes and require much alteration before the block layout is suitable as a basis for a layout that can be implemented. The scoring of the layout presupposes that material flow is between the centroids of departments. This, of course, may not reflect the way in which each department is designed. The restrictions that prevent the input of departments that are not in one cohesive space and that prevent the use of long narrow departments also introduce errors. The improvement process does not insure that departments that have a high amount of material movement between them and that are initially not adjacent to each other, will end up adjacent to each other.

7.3.7.3 COFAD

COFAD, developed by Tompkins,[15] is a modification of CRAFT that allows move costs to reflect equipment data including variable costs, nonvariable costs, and operating characteristics. For each type of handling equipment to be considered, a new from-to matrix is created, since the number of moves required to transport a given amount of material among departments may vary with the equipment. For example, a manually moved platform truck may require many more trips to transport a given amount of material than a fork truck. COFAD assigns the least cost equipment type to each move and then tries to reduce total cost by eliminating a piece of equipment with low utilization. It does so by temporarily reassigning the moves to the next best equipment type and then evaluates the effect on cost. If the cost is reduced, then the reassignment is made permanent.

7.3.7.4 Summary

The use of CORELAP, CRAFT, COFAD, and other similar computer models requires the gathering of data that can be very helpful even if manual methods are to be used. Once the data have been gathered, the transformation of the data for use in the computer models does not require much more effort. The computer models will provide many alternative approaches to the block layout and each will have a score. The scores may be unique to each computer model and thus the results cannot be compared among models without additional computations. In spite of this, the models are worthwhile, since the creativity of the layout analyst should be enhanced by their use.

7.3.8 PRODUCING THE LAYOUT

7.3.8.1 Traditional Methods

The actual production of the layout has traditionally been done manually using either conventional drafting (pen, ink, eraser, T-square, paper, and drawing board) or templates. Generally, the layout is produced in a two-dimensional setting, but occasionally a three-dimensional model is constructed to permit the layout analyst to evaluate overhead clearances and material handling devices that may present problems during the implementation of the layout. Of course, the purpose of producing the layout in such a manner is to bring together all aspects of planning that have been done in order to arrive at such a model of the facility. Material flow, production plans, service and auxiliary activities, individual workplaces, and material handling planning need to be condensed and integrated into single overall plans so that evaluation of alternatives can begin followed by management selection of the layout to be implemented. The actual implementation of the layout is then the final product of the layout analyst's work.[6]

Many layouts are still being produced using the two traditional modes, which are extremely inefficient. The layout model produced by these modes makes it difficult to make adjustments and alterations without using more eraser than pen and ink. Both modes require the layout analysts to do drafting, a poor use of their time.

7.3.8.2 Computer Graphics

Since the middle of the 1970s, the use of computer graphics has had a rapid growth in plant layout departments, since it is more efficient. Some plant layout departments share equipment used by product and tool design departments while others control independent systems.[17–20] Computer graphics systems generally include at least the following:

1. A computer (microcomputer, minicomputer, or a mainframe computer).
2. Computer software.

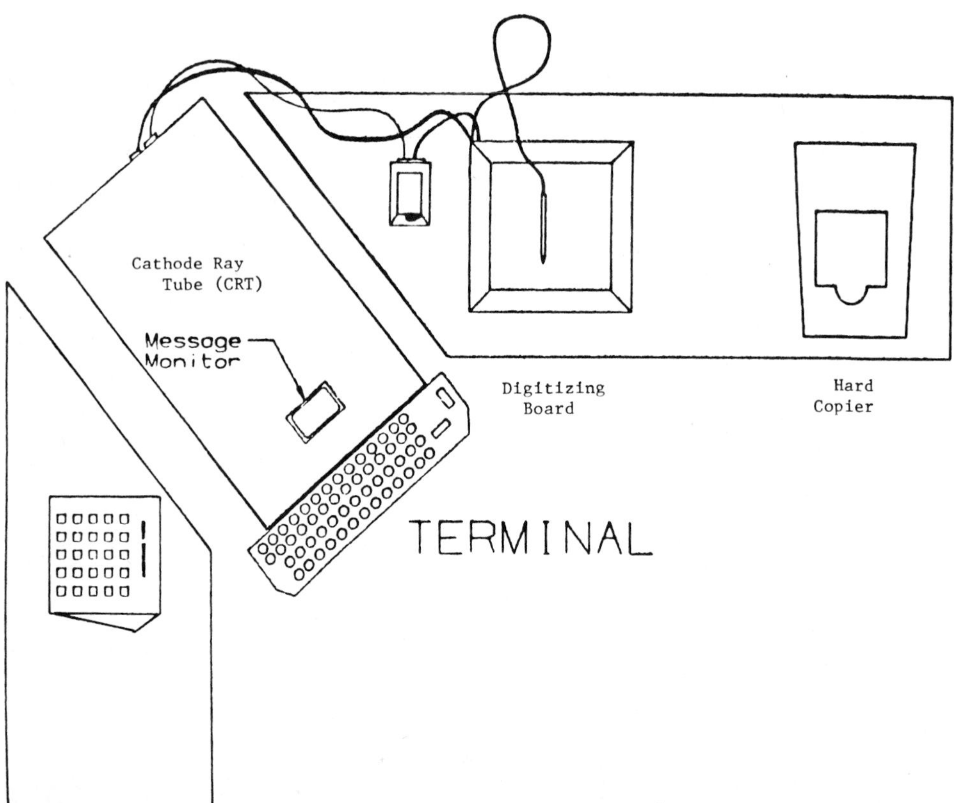

Fig. 7.3.31. Typical graphics workstation.

Table 7.3.4. Assignment of Levels in a Computer Graphics System

1. Foundations, pits
2. Baylines
3. Columns
4. Basic building: walls, restrooms, elevators
5. Aisles
6. Misc. layout: partition, furniture, outlets
7. Machines
8. Monorail conveyors, cranes
9. Material hdlg: gondolas, automated, roller conveyor
10. Platforms, mezzanines
11. Platform equipment and text
12. 6″ and 12″ text
13. Dimensions
14. 18″, 24″, and 5-ft text
15. Area report
16. Title block

3. Computer system support units, e.g., disks and tapes.

4. A visual display unit called a cathode ray tube (CRT).

The computer graphics workstation (Fig. 7.3.31) may also include a digitizing board through which existing layouts are entered, an electronic light pen and tablet, and plotting devices that produce the drawings. The larger, more costly units can support many workstations simultaneously.

The database of the computer graphics systems could include information about many of the items that appear in Table 7.3.1. These can be used to develop various levels of information. Computer graphics systems therefore can have many levels, which in manual layout systems, are called overlays. The user of one system limited to sixteen levels has allocated these as shown in Table 7.3.4. Many systems have 256 levels, and one system has approximately 32,000 levels.[21]

The computer graphics system with its flexibility in forming overlays allows integration of many functions. For example, some have the ability to check for spatial interferences among electrical, hydraulic, handling, and structural systems in a dynamic as well as a static mode. The savings from this can be very large. The zooming, rotating, dimensioning, additional view, and third-dimension creation lead to savings in drafting and implementation time. Designing and analysis time is reduced for analyses that include calculations such as the following:

Perimeter	Weight (given density)
Area	Weight/unit length
Center of mass	Material flow and loads
First moments	Length of a path traveled
Moments of inertia	

These capabilities can be integrated with the part blueprint database to do a preliminary design of material handling containers and packaging, and to refine product design through finite element analyses. By adding yearly volumes from a production control information database, storage and distribution systems can be designed.

Individual workstations for factory and office areas can be created and managed as a database, from which overall plant-layout alternatives can be developed as block area diagrams. Improvement techniques can be used to get better layouts and finally detailed layouts can be developed.

Ergonomic, quality work-methods, and work-measurement analyses can be integrated into a unified job description, simulated, and posted at each workstation for training of supervisors and setup and production operators.

The computer graphics systems can also include standard times to create job assignments for workers who have wide-area responsibilities for functions such as driving tow trucks and fork trucks, housekeeping, preventive maintenance, safety, noise, energy monitoring, and security activities. To create job assignments for wide-area operators which can be readily altered to reflect changes in total volumes, in the popularity of individual models in a product line, and changes in designs, requires a well-planned system. For example, in one existing system, 32 items of information are required to form the data record for calculation of an assignment for a truck driver (Table 7.3.5).

For new buildings, the designs and design changes produced on a computer graphics system significantly improve coordination of subcontractors, which reduces costs and time delays. The architectural prints can then be used as input to the customer's computer and become a prime database for plant, manufacturing, and industrial engineers.[22]

Table 7.3.5. Data Record Estimating Time per Trip for a Fork Truck Record in a Computer Graphics System

1. Boxes per unit load
2. Container type
3. Cycles per shift
4. Cycle type
5. DR's creation date
6. Disposal cost of a unit load
7. Destination department
8. Daily equipment cost
9. Direction of delivery
10. Cycle length
11. Daily labor cost
12. Daily total cost
13. Equipment type
14. Number of unit loads in a stack at source
15. Involvement type
16. Operation ID
17. Part ID
18. Plant ID
19. Number of parts per box
20. Number of unit loads in a stack at destination
21. Routine type
22. Source department
23. Sequence number
24. Number of shifts per day
25. Truck capacity
26. Cycle time
27. Trip type
28. Truck ID
29. Number of unit loads in transport
30. Unit load height
31. Hourly part usage rate
32. Yearly number of working days

The use of computer graphics systems in large companies is more and more being linked to mainframe computers, so database information from many departments will be readily available for the day-to-day design of total systems.

REFERENCES

1. S. Shingo, *Study of Toyota Production System*, Japanese Management Association, Tokyo, 1981.
2. Y. Monden, *Toyota Production System*, Institute of Industrial Engineers, Norcross, GA, 1983.
3. Y. Monden, *Toyota Production System*, Institute of Industrial Engineers, Norcross, GA, 1983, p. 103.
4. B. Wilson, C. Berg, and D. French, *Efficiency of Manufacturing Systems*, New York: Plenum, 1983, p. 151.
5. I. Ham, "Introduction to Group Technology" *Technical Report MMR76-03*, Society of Automotive Engineers, 1967.
6. J. Apple, *Plant Layout and Material Handling*, New York: Ronald Press, 1977.
7. D. Desai, "How One Firm Put A Group Parts Classification System Into Operation," *Industrial Engineering*, November 1981, pp. 78–86.
8. C. Know, "CAD/CAM and Group Technology: The Answer to Systems Interaction?" *Industrial Engineering*, November 1980, pp. 66–72.
9. J. Burbidge, *The Introduction of Group Technology*, New York: Halstead/Wiley, 1975.
10. R. Muther, *Systematic Layout Planning*, Industrial Education Institute, Boston, 1961.
11. E. Polk, *Methods Analysis and Work Measurement*, New York: McGraw-Hill, 1984.
12. G. Salvendy, *Handbook of Industrial Engineering*, New York: Wiley, 1982, Chap. 10.6, p. 13.
13. R. Francis and J. White, *Facility Layout and Location*, Englewood Cliffs, NJ: Prentice-Hall, 1974.

14. J. Tompkins and J. Moore, *Computer Aided Layout: A User's Guide*, Institute of Industrial Engineers, Norcross, GA, 1978.
15. J. Tompkins, "Computer Aided Layout Series," *Modern Materials Handling*, May–November, 1978.
16. G. Armour and E. Buffa, "A Heuristic Algorithm and Simulation Approach to Relative Location of Facilities," *Management Science*, September 1963, pp. 294–309.
17. F. Jacobs, J. Bradford, and L. Ritzman, "Computerized Layout: An Integrated Approach to Specialized Planning and Communications Requirements," *Industrial Engineering*, July 1980, pp. 56–61.
18. J. Zarembski, "Professional Plant Layout System Shows the Way to a Smoother Material Flow," *Industrial Engineering*, April 1981, pp. 122–126.
19. R. Kaplinsky, *Computer Aided Design*, New York: Macmillan, 1982.
20. P. Gheresus and M. Ladha, "Innovation Drawing Applications For Industrial Engineering," *CAD/CAM Technology*, Winter, 1982, pp. 11–14.
21. R. Filley, "CAD For Facilities Planning," *Industrial Engineering*, March 1983, pp. 66–80.
22. H. Edwards, "Computer Graphics: Vital New Facilities Design Tool," *Industrial Engineering*, **12**(9), 1980, 52–54.

CHAPTER 7.4

ENERGY MANAGEMENT

JOHN K. FREUND

Deere and Company
Moline, Illinois

7.4.1 WHY ENERGY MANAGEMENT

Energy management is an important element to controlling production costs and, in fact, can be critical to the ability to perform manufacturing operations at all.

Late in 1973, the five-month oil embargo by OPEC marked the beginning of an upward spiral of energy prices, increased inflation, and shortages of various energy sources. The action by OPEC made manufacturers acutely aware of the nation's vulnerability to geopolitical events and the overall effect of oil prices on the economy. In 1979 similar forces were set in motion when Iran cut back production. In this case a relatively small perturbation, 4% of free-world consumption, resulted in imported oil price increases of 120% in one year.[1]

Large oil price increases have significant economic impacts. For example, a 50% increase in oil prices will add an estimated 3.5% to the level of prices in the United States. Furthermore, an additional 1.5% is added to the underlying inflation rate as higher oil prices move through the economy, causing cost-of-living adjustments in wage contracts, government income redistribution programs, and product price increases.[1] In turn, the resulting loss of disposable income and government policies to control inflation cause increased unemployment and reduced economic growth.

Finally, the delicate nature of the supply-demand balance of oil and natural gas must also be realized. In 1981 the world oil market was changed from a seller's market to a buyer's market by an oil surplus of only two million barrels per day, 1.5% of the world's annual energy consumption and 3.5% of its oil use. Furthermore, the tenuous nature of the world oil supply is accented by a 1981 article in the *Harvard Business Review* by F. A. Lindsay. In this article the opinions of individuals knowledgeable in energy matters, Middle Eastern politics, and national security were quoted. Of those polled, none thought there was less than a 50% chance of a cutoff during the 1980s. Furthermore, many thought the chance that another cutoff could be avoided was less than 10%.[2]

The fragile nature of oil supplies is in turn reflected in natural gas supplies; when oil supplies drop, natural gas is substituted to the maximum extent practicable. In other words, reduced oil supplies can be expected to precipitate gas supply shortfalls and gas supply interruptions, particularly to manufacturers.

In response to the worldwide and domestic energy uncertainties in the decade following 1972, many manufacturers instituted energy management programs. Since the time of "crisis," energy management has evolved to incorporate energy supply, conservation, facility and process planning, governmental affairs, and utility negotiation. Through energy management, the industrial sector is far more fuel efficient. The Department of Energy estimates that industry consumed the same amount of energy in 1981 as it did in 1970, despite the fact that the gross national product (GNP) increased 35%, in real terms. This is in marked contrast to the previous decade, when industrial energy consumption rose 45% while the GNP increased 46%.[3,4]

Industry has made those efficiency improvements for two reasons:

1. Energy management is critically important to mitigating the effects of energy supply interruptions and cost increases induced by national and world situations.
2. Efforts to improve energy efficiency are good investments in their own right. It is common for conservation projects to return any investment costs in one year and payback periods of less than one year are not uncommon. Furthermore, conservation often can be achieved through implementing practices that require no capital investment.

7.4.2 MANAGEMENT IN ENERGY MANAGEMENT

The organization of energy management activities can take many forms and is primarily determined by the existing organizational structure and the energy intensity of the business. The organization can range from centralized, where all energy management activities, such as purchasing, planning, engineering, and policy are concentrated in a single department, to fully decentralized, where energy management activities are performed in separate departments. A survey of 25 major corporations indicated that about half use a centralized organization and about a third are fully decentralized with the others falling somewhere in between.[4]

Regardless of the organizational structure, several elements are keys to a successful energy management program.

Top management support. It is impossible to maintain a successful energy management program without the support of top management. This commitment includes periodic evaluation, review, and direction.

Full-time energy coordinators. An energy coordinator at each manufacturing facility is required. For program effectiveness, energy conservation must be the full-time responsibility of the coordinator. To provide incentive and assure total responsibility, the energy coordinator generates ideas from within the factory, obtains approval, and implements changes.

Established goals. In every endeavor, if success is to be attained, it is essential to establish goals and schedules for their attainment. Either a top-down or bottom-up approach to goal setting can be used effectively. Setting goals has several benefits:

1. Forces all organizational elements to plan ahead.
2. Assigns accountability.
3. Provides a means to measure progress.
4. Increases energy awareness within the corporation.[4]

In addition to setting goals for energy efficiency improvements, recommending energy budgets for buildings and equipment is also appropriate. By establishing "design energy budgets" in terms of annual energy use per unit area per year, architects and engineers are given achievable levels of energy efficiency for the building envelope and equipment. In the cases of environmental energy use, state energy codes may set maximum design energy budgets for new construction. Standards of the American Society of Heating, Refrigeration, and Air Conditioning Engineers (Number 90-75 and 90-80) and the National Conference of States on Building Codes and Standards, Inc. contain feasible and cost-effective energy budget guidelines regarding the annual energy use of buildings and the energy efficiency of lighting and heating, ventilating, and air-conditioning systems.

Complete energy audit. An energy audit, or energy-use analysis, is essential to an effective energy management program. The progression of the energy-use analysis should be from the general to the more detailed. By performing the analysis in increasingly detailed steps, the overall program is initiated immediately. Areas that offer the highest potential will be easily identifiable and efforts can be focused on those high-potential areas. Conversely, starting with a detailed analysis is very time consuming and high-potential opportunities can be masked.

The energy audit starts with the tabulation of the energy use and cost by month for each individual form of energy. Obviously, the more years of data, the greater the degree of accuracy, since unusual weather conditions and varying production schedules can be normalized. Furthermore, with several years of data, it may be possible to make valuable statistical inferences regarding energy use patterns. In particular, if a power plant is part of the facility, the energy going into the property and into the plant can be determined at this point. An analysis of differences in winter and summer energy use also can give valuable insight into the energy used for processes and that used for environmental heating.

The next level of detail in the energy audit is accomplished with a combination of a walk through the facility, and a review of plans and drawings and data on production equipment. The walk-through allows the auditor to get a "feel" of the facility, to witness operational practices, to note anything unusual, to search for obvious conservation opportunities, and actually to take data when conditions allow. Plans and drawings will reveal design details that can be used to estimate the energy used to heat (and/or cool) the building as well as some process energy requirements. Specific data on production equipment is also an excellent source from which to estimate energy requirements of particular processes. Figure 7.4.1 is an example of a summary energy-use analysis of a three million square foot manufacturing facility.

Energy use should be further divided to isolate peripheral energy use from the real process and environmental requirements. *Peripheral energy* can be defined as the energy needed to meet any additional requirements, such as those for occupational safety and health and environmental protection. This approach to the energy-use analysis allows the areas of greatest potential for conservation to be isolated and priorities to be established.

Measurement of results. An accurate, meaningful yardstick must be established to measure results. The measurement tool can take a variety of forms, for example, performance can be measured against a predicted energy use based on variables in production, floor space, and weather, or based on a single variable, such as production or sales. Whatever the technique, the results must be measured and compared to the goals.

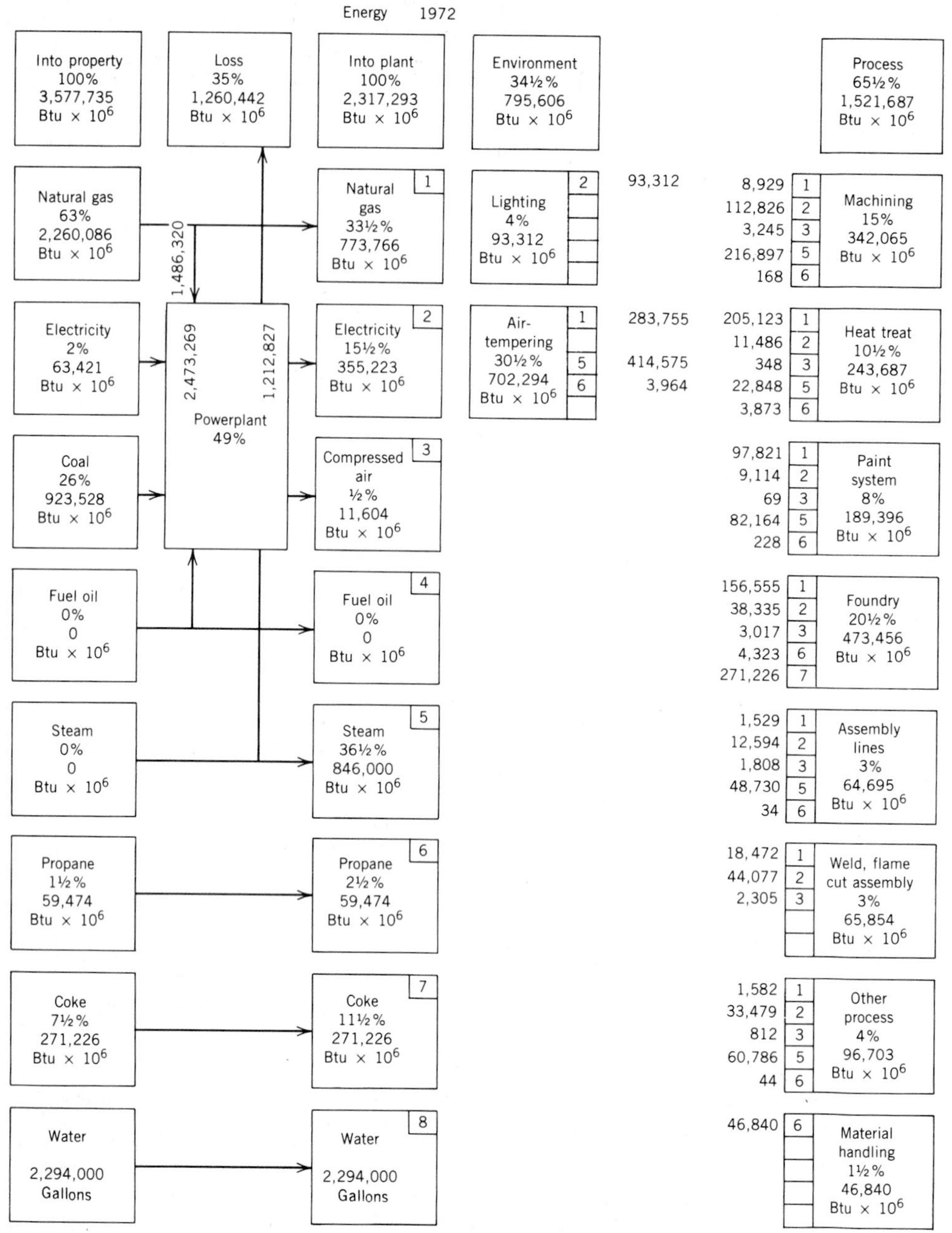

Fig. 7.4.1. Summary of energy audit.

Reporting. Results of the conservation efforts of each facility must be measured and reported periodically to top management. Reports allow programs to be monitored and become tools for recognition, planning and incentive efforts.

The conservation projects that result from adherence to these elements must have objectives beyond conservation. The long-range availability of fuels should also be addressed. Because of the volatile nature of energy supplies, any assessment of long-range fuel availability is likely to be highly subjective. In any case, fuel-use policies coupled with conservation efforts should result in energy mixes at the production facility that have the flexibility of supplying long-range energy requirements with minimum disturbances that might result from changing supply pictures.

7.4.3 ENERGY MANAGEMENT STRATEGIES

The initial emphasis of conservation efforts should generally be on improving administrative procedures and intensifying maintenance operations. Activities in the administrative area could include removing unneeded lighting, turning lights on and off as required, governing equipment operation by functional requirement, not by arbitrary time settings, and calibrating and setting thermostats. Maintenance can encompass a nearly infinite array of activities. Some common maintenance items are detecting and repairing compressed air and steam leaks, and repairing and adjusting burners, particularly in energy-intensive equipment such as boilers and heat treat and forge furnaces.

Once administrative and maintenance aspects of the program have been solidified, additional conservation opportunities must be identified. At this point, alternative engineering paths can be taken. The basic difference can be characterized as a demand or a supply-side approach to conservation. As the terms imply, a demand-side approach would concentrate on reducing energy requirements of the end-use. Conversely, a supply-side approach would emphasize efficiency improvements at the source of energy, such as boilers, and applying heat recovery equipment.

Obviously, the path to follow is highly dependent on the particular circumstances of a production facility. However, it is generally advisable to explore first the demand-side approach to conservation, for the following reasons. First, improvements in supply-side efficiency may be limited if the prescribed maintenance was performed in the first phase of conservation. Second, reduced demand for energy may negate the need for some supply equipment, thus rendering any improvement to supply equipment ineffective. Third, the potential for reductions at the point of use can be great, as well as more cost efficient, than a supply-side approach. Unfortunately, no approach universally produces the best results. However, the importance to a conservation project of addressing energy requirements at the end-use cannot be overemphasized.

7.4.4 CONSERVATION

The previous discussion outlines both the need for organization and the approach of an energy management program. The following section provides a list of conservation activities that should typically be performed. Because, in many cases, a successful conservation program hinges on the inventiveness and imagination of facility engineers, no list of possible projects can ever be considered complete. Furthermore, since conservation opportunities are worthy of treatment in a dedicated handbook, only a selected number of examples will be expanded upon.

Before specific examples are discussed, it may be beneficial to discuss briefly the concept of energy conservation practices and measures and the priority they should have. An energy conservation opportunity that can immediately be achieved through simple changes in operating or maintenance procedures is an energy conservation *practice*. These conservation practices have immediate payback, since no capital investment is required and therefore they should be implemented before capital-intensive solutions are considered. Reduced heating-temperature settings, turning off unneeded lights, and using natural ventilation when outdoor conditions warrant are examples of conservative practices.

Energy conservation *measures*, on the other hand, require capital investment to modify or replace buildings and/or equipment. Consequently, conservation measures have longer payback periods than conservation practices, but typically can yield greater savings in the long run. Examples of conservation measures are purchasing more efficient equipment, adding insulation, and installing higher efficiency lighting.

7.4.4.1 Operations

1. *Reduce operating hours.* Operate heating, ventilating, air conditioning, lighting, water heating, and process equipment by manually or automatically controlling equipment so that it is operated only when needed and that equipment startup periods are governed by process or environmental requirements not by arbitrary time settings.

2. *Reduce ventilation and heated or cooled makeup air requirements.* Building and process ventilation leads to large energy requirements to condition (heat or cool) the air that must come into a facility to replace air that is exhausted. There are numerous ways to reduce the requirements to condition makeup air. The first step is to reduce ventilation to minimum acceptable or required levels. Recommended ventilation guidelines for various types of occupied space are given in assorted government and private publications.[5,6] For that reason and because many ventilation requirements are specified by local building codes and the ventilation associated with manufacturing processes can be mandated by the Occupational Safety and Health Administration (OSHA), specific ventilation rates are not noted here.

   For process exhaust, capturing contaminants at the source ("close capture") is much more energy efficient and environmentally effective than using general area exhaust (such as roof fans). With close-capture techniques, using low-volume exhaust ducted from the source, contaminants are captured before being diluted by and contaminating the surrounding air, thus minimizing the volume of exhaust air required.

   The application of compensating air supplies is another way to reduce the requirement to heat makeup air. The principle of the compensating hood is to use close capture techniques and to introduce unheated makeup air in close proximity to the source of contamination, as shown in Figure 7.4.2. When unheated makeup air is introduced close to the point of contamination, the unheated air mixes

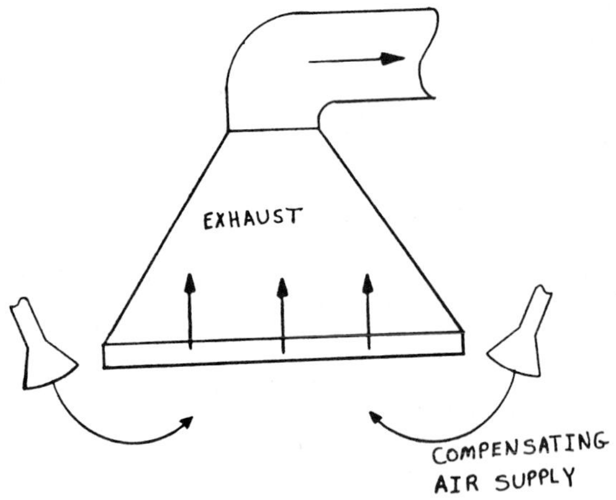

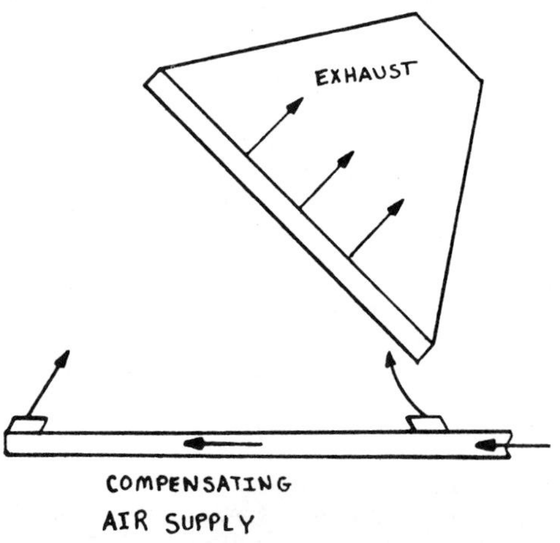

Fig. 7.4.2. Examples of compensating exhaust hood concepts.

with the contaminants and is exhausted; thus a minimum of heated/conditioned air is lost. A secondary benefit of supplying compensating air is that, if properly directed, the makeup air flow directs the contaminants to the point of exhaust, making the system more effective.

Air cleaning with recirculation is also an effective way to reduce exhaust needs. Obviously, recirculation alone will do nothing to improve the in-plant environment. However, recirculating the air through an appropriate cleaning device will improve the in-plant environment and reduce or eliminate the need to heat the recirculated air.

Recirculation has been applied to numerous production environments to reduce energy costs. The key to applying recirculation successfully is capturing the contaminants and using the correct air-cleaning device. Selecting the type of device to remove contaminants from the air can be an involved task. The correct filter to use can be influenced by many factors, including the size of the particulates to be removed.

Dry-fiber air filters are generally effective for relatively large particles (larger than 3μm [microns] in diameter), and some can remove particles with diameters down to 0.5 μm.[7] Absolute filters, typically constructed of a specially treated paper, are very effective for particle sizes of less than 1μm, and can operate under high gas temperatures (0–1000°C). When absolute filters are used, prefiltering the air stream with a more normal filter is advisable to extend the life of the high performance absolute filter media.

Electrostatic air filters are also an effective way to clean the air for recirculation. In an electrostatic precipitator, contaminated air is passed through the corona of high voltage electrodes. As particles traverse the corona, an electrical charge is induced on the particulates. Downstream slightly, the particles subsequently pass through an electrostatic field which separates the charged particles from the air stream and deposits them onto collection plates.

Electrostatic air filters have proven to be very effective for production environments. The uses of electrostatic filters include removing welding smoke, oil mist (such as that resulting from machining operations), dust, and fumes from the air. These types of filters generally perform best for particles having diameters less than 2μm.[7]

3. *Control temperature and humidity.* Heat loss and gain are proportional to the difference in temperature between the outside and inside. Therefore, energy used for heating and cooling can be minimized by controlling space temperatures so that the difference in temperature is as small as possible at all times. Controlling the temperature to different levels during occupied and nonoccupied hours is also important to reduce heating and cooling costs. In the winter, space temperatures should be allowed to drift down at least 5–10°C during unoccupied periods. Obviously, some care must be exercised at the lower temperature setting to protect from general and isolated freeze-up conditions. Conversely, in the summer, if the space is cooled, the temperature should be allowed to drift up. In either case, the controls must be implemented in such a way that when the switch to the unoccupied set-point temperature is made, the heating or cooling systems are not operated to achieve the higher or lower temperature setting.

   Control of humidity levels is also important, since about 5.12×10^6 joules (J) are required to vaporize each kilogram of water (1000 Btus per pound of water). If additional humidity is required, the humidity level should be controlled to a realistic requirement.

4. *Reduce hot water use.* Heating water can be a major energy user in production facilities. About 4.16 MJ are required to heat one cubic meter of water 1°C, (8.3 Btus to heat one gallon of water 1°F), and energy use is directly proportional to the flow rate. Therefore, reducing both the temperature and the flow of hot water will reduce energy costs.

7.4.7.2 Building Shell

1. *Increase energy efficiency of walls, windows, roof, and floor.* Heat loss or gain through the building shell is directly related to the insulating value of the components of the shell, as well as the difference between the indoor and outdoor temperatures. Given that the indoor temperature is controlled in an energy efficient manner, the thermal performance of the building envelope can be improved further by increasing the resistance to heat flow. Typically, the thermal performance of building elements is referred to as the *U value* or the *R value*. The *U value* is a measure of the transmission of heat through a building element, such as the wall or roof in $W/m^2 \cdot K$ or $Btu/(h \cdot ft^2 \cdot °F)$. The R value is the reciprocal of the U value. Lower U values mean better insulating properties.

   The heat lost or gained through walls, roof, and doors can be reduced by adding insulation on the outside or inside. Heat lost or gained through windows can be reduced by installing double or triple glazing or by partially or totally covering the window with insulating panels. Undesired solar heat gained through windows can be controlled by various measures, including shading, reflective films, and landscaping; for example, the use of deciduous trees will reduce summer season solar heat gains while allowing solar heating in the winter. Similarly, unneeded doors can be covered by insulating panels or insulation can be added to the door surfaces or insulating doors installed.

   Heat loss from slab-on-grade floors can be reduced by insulating around the outside perimeter of the floor. The insulating material should be placed vertically around the perimeter to a depth of 0.6 m (2 ft).[8]

2. *Reduce infiltration.* Infiltration is the uncontrolled entry of outside air into a building through such openings as cracks, gaps around windows and doors, etc. Infiltration increases both heating and cooling load of the building and can cause comfort problems.

   Causes of infiltration are varied, and range from wind pressure to unbalanced (exhaust volume greater than supply) ventilation systems. To maintain equilibrium, the air leaving a building through ventilation exhaust or exfiltration that is not replaced by controlled makeup air, will be made up by infiltration.

   To reduce infiltration, exhaust and makeup air supplies should be balanced, preferably by reducing exhaust volumes. To reduce infiltration from other sources, openings in the building shell should be sealed to the maximum extent possible. Caulking cracks around window and door frames and weatherstripping windows and doors are very effective, as is installing vestibules around doors. For planning facilities, orienting the building so that the smallest profile and least number of openings face the prevailing wind is an effective energy planning tool to control infiltration.

7.4.4.3 Distribution Systems

The effective distribution of heated and cooled air, the energy used to distribute air and water, and the energy lost in the distribution process are important energy considerations.

1. *Improve conditioned air distribution.* Distributing heated or cooled air to the occupants of a building is a key to comfort and reduced energy requirements. Therefore, distribution systems must be designed to supply heat to the occupied zones. This can be accomplished by a variety of methods. One way is to install supply air outlets low, i.e., at about the 2–2.8 m (6–8 ft) level, and direct the air downward. In this case (heating), the return air inlets should be mounted near the roof, as shown in Fig. 7.4.3. By using the high return air concept, the stratification (warmest air migrating to roof level) that naturally occurs is used to supply warm air to the heating system.

   Another air distribution concept that takes advantage of the warm air that migrates to the ceiling level supplies air high in the building (at the truss level) at high velocity. To promote circulation of warm air to the occupied zone, return air inlets are mounted near the floor. This air distribution concept is shown in Fig. 7.4.4.

   Supplying a high velocity stream of unheated air at the truss level can also be employed in high-heat-release areas of production facilities. This approach can alleviate unbalanced supply and exhaust systems while reducing heating requirements.

   A third method to use the heat available in stratified air to heat the occupied zone is to simply install ceiling fans, as shown in Fig. 7.4.5. In high-heat-release manufacturing areas, simply moving the warm stratified air down to the occupied zone can significantly reduce heating costs.

   For cooling, the concept of supplying conditioned air in the occupied zone remains valid. However, when in the cooling mode, return air should be picked up low while the warm air at the truss level should be exhausted and madeup with unconditioned air.

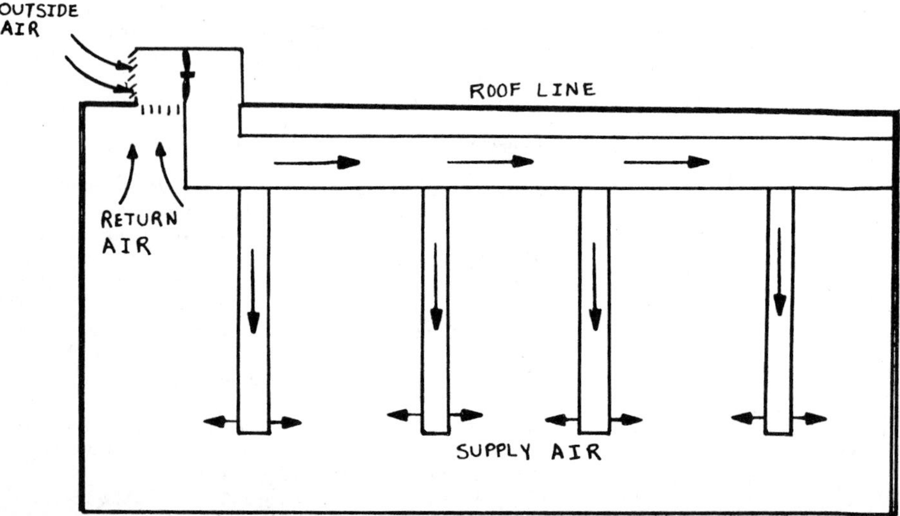

Fig. 7.4.3. Low supply, high return air distribution.

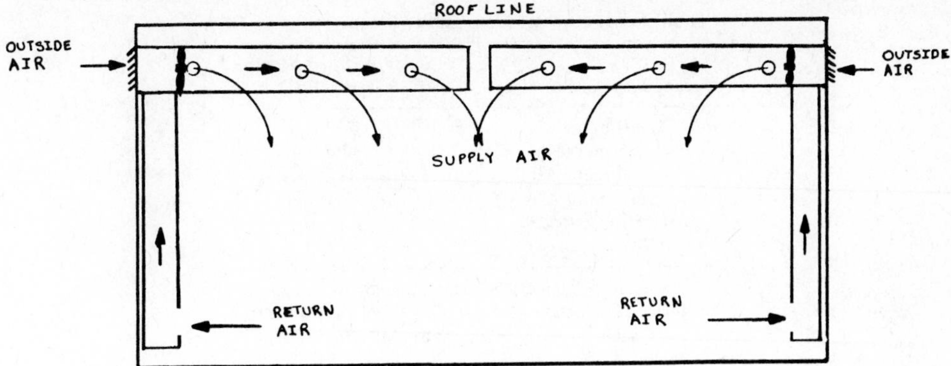

Fig. 7.4.4. High supply, low return air distribution.

2. *Reduce air and water flows.* The horsepower to circulate water and air is typically a hidden but significant cost. Reducing either or both the flow rate and the resistance to flow will yield horsepower savings. Since the power input varies as the cube of the speed, any reduction of air volume, which is proportional to speed, will yield electrical cost savings.

   To reduce fan or pump power requirements, the heating and cooling loads should be reduced first; then the resistance to flow should be reduced and the resulting new flow volume measured; finally, the pump or fan speed to meet the reduced load is lowered.
3. *Insulate ducts and pipes.* To reduce parasitic losses in heated or cooled water and air distribution systems, those distribution systems should be insulated.
4. *Replace steam traps.* Boiler capacity and efficiency are impacted by steam trap performance. When steam traps fail to open and close properly, live steam can escape, thus degrading the performance of the steam supply system. An example of the energy lost through a malfunctioning steam trap at a steam system pressure of 0.86 MPa (125 psi) is shown in Table 7.4.1.

7.4.4.4 Lighting

1. *Reduce lighting levels.* Lighting levels should be reduced to values recommended by various guidelines, such as the Illuminating Engineering Society. Some manufacturers use a guideline of having an illumination of 538 lux (50 footcandles) for office and manufacturing and 215 lux (20 footcandles) for warehousing.

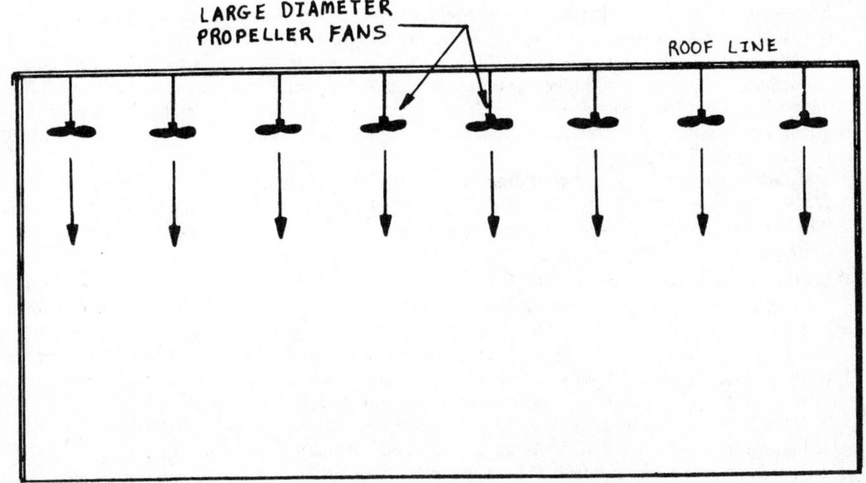

Fig. 7.4.5. Ceiling fans.

Table 7.4.1. Losses from Faulty Steam Traps[a]

Steam Trap (orifice size)	Steam Lost per Month (GJ or MBtu)
12.700 mm (½″)	709
9.525 mm (⅜″)	400
6.350 mm (¼″)	178
3.175 mm (⅛″)	44
1.588 mm (1/16″)	11
0.794 mm (1/32″)	3

[a] For 0.34 MPa (50 psi) steam, the steam loss shown above is reduced by aboud 25%, reduced by 50% for 0.14 MPa (20 psi), and reduced by 75% for 0.034 MPa (5 psi) steam.[8]

Where the work requires additional light, task, not general, lighting is employed. The American National Standards Institute, "Practice for Industrial Lighting," ANSI-A11.1-1973 (June 1973) recommends 323 lux (30 footcandles) for production work areas and 108 lux (10 footcandles) for nonwork areas.

2. *Turn off lights when not needed.*
3. *Improve lighting efficiency.* The efficiency of existing lighting fixtures can be improved by cleaning the fixture and installing higher efficiency lamps. Even greater improvements can be made by replacing low efficiency fixtures with ones having a higher efficiency. The relative efficiency, measured in lumens per watt, of various types of lamps is shown in Table 7.4.2. It must be noted that the efficiencies shown do not include fixture losses.
4. *Utilize natural light.*
5. *Use task lighting.* Rather than providing a high level of general area illumination, reducing general lighting to a safe minimum and providing light at the individual task is an effective way to reduce lighting costs.

7.4.4.5 Electric Power Control

Utility companies generally determine electric costs from a combination of total kilowatt hours used, the short-term (15–30 min) peak demand for electricity, and the power factor of the total facility. Improvements in each of these areas can reduce electric power costs. Previous sections noted numerous approaches to reduce electric power consumption. In this section, some additional items will be discussed.

1. *Reduce consumption.* As noted before, the most effective way to reduce consumption is to shut off equipment when it is not being used. To reduce the consumption of productive equipment, energy efficient motors and transformers should be used. Furthermore, the replacement of oversized motors with motors more closely meeting requirements will increase efficiency.
2. *Reduce peak load.* Electric rates to industrial customers typically take the peak demand for electricity (measured in kilowatts) into account in some fashion. In many cases, a demand charge is applied directly. In other cases, the number of hours of use in a month is used to determine declining block increments in kilowatt hour charges. It is not uncommon for rates to incorporate both schemes. Because the peak demand is measured over a relatively short time span, 15 or 30 min, and the peak typically establishes the demand charge for 12 months, short excursions in demand can contribute markedly to increasing electric costs.

   Therefore, monitoring and controlling electrical demand is important. The demand for electricity can be effectively controlled by *load shedding*. As the name implies, control is accomplished by shutting off large electrical loads when a set peak demand for electricity is approached to prevent the total load from exceeding the peak.
3. *Correct power factor.* Power factor is a measurement of the relative phase relationship of current and voltage in alternating current distribution systems. A power factor of 1.0 indicates that voltage and current are in phase; values close to 1.0 are desirable. Utilities generally penalize companies for power factors less than 0.85. A low power factor is generally caused by a plant operating a large number of partially loaded induction motors. Other inductive devices, such as fluorescent light ballasts and rectifiers used to supply DC power also contribute to a poor power factor.

   A low power factor is undesirable because the losses in electrical distribution and utilization equipment increase while load handling capacity and voltage regulation decrease.

   The most common approach to power factor correction is installing capacitors in the electrical system; using synchronous motors and synchronous condensers is also effective. However, correction by synchronous condensers may not be cost effective. Depending on the electrical distribution system, capacitors can be added to serve individual equipment or a group of equipment. If the capacitors are installed to serve a group, it may be necessary to control the amount of capacitance used as a function of load to prevent a high voltage situation from developing under low load.

4. *Use central control and monitoring.* A central control and monitoring system is a valuable tool to control plant energy use and demand. The ability to monitor and control major energy-using devices such as heating and ventilating units, lights, chillers, air conditioners, and production equipment, provides a tool to more efficiently use plant equipment and personnel.

7.4.4.6 Production Processes

1. *Maintain burners.* For the most part, the aforementioned energy conservation measures apply to increasing the energy efficiency of production processes as well as environmental heating and cooling. The importance of one of those items, burner maintenance and adjustment, to energy efficiency cannot be overemphasized. For energy intensive fuel users, such as heat treat or forge furnaces, repair and adjustment of burners can result in fuel cost savings measured in the hundreds of thousands of dollars annually.[10]
2. *Insulate heated processes.*
3. *Control paint systems.* For manufacturers that paint their product, the energy associated with painting can be one of the largest process energy uses. Unlike other processes in which the majority of energy used goes into adding value to the product, only a small portion of the energy used in painting goes to adding value to the product. A large portion of paint-system energy use, particularly spray painting, goes to satisfying peripheral energy needs—that is, the energy required to negate the impact of process contamination and to meet OSHA, EPA, and other regulations.

An example may best illustrate how the energy requirements of spray-painting operations can be reduced. In this case, an energy audit revealed that a large spray-paint booth was a large energy user and that the majority of the energy used could be categorized as peripheral. Nearly all the energy was associated with the exhaust of 37.75 m^3/s (80,000 cubic feet per minute).

Three reasons existed for exhaust from this operation: (1) an OSHA requirement for an air flow of at least 0.5 m/s (100 feet per minute [cfm]) past the painter, (2) an OSHA requirement that volatile solvent concentrations not exceed 25% of the lower explosive limit (LEL) in the booth and duct work, and (3) a general requirement to contain fumes within the booth.

The OSHA air-flow requirement resulted in the 37.75 m^3/s (80,000 cfm) flow condition, the apparent requirement to protect the painter. However, analysis indicated that 0.94 m^3/s (2,000 cfm) would satisfy both the lower explosive limit and containment needs. Therefore, if the lower air flow could become the actual requirement, the energy needed to heat 36.81 m^3/s (78,000 cfm) could be avoided. For this particular midwest location, an estimated 6.3 TJ (6 GBtu) per year could be saved.

To reduce the exhaust air flow requirements from 37.75 to 0.94 m^3/s, it was obvious that an alternative means of controlling the breathing environment of the painters must be found. In this case, the use of air-supplied hoods was proposed to OSHA as an engineering control. For this example, fresh, conditioned air is supplied directly to the breathing zone of the painters, and the paint system air flow was modified so that 37.75 m^3/s was recirculated in the paint booth and 0.94 m^3/s was exhausted. The operation of this system is depicted in Fig. 7.4.6.

OSHA subsequently approved this control technology and industrial paint systems are using this technology to reduce energy costs. (This control technology is covered under U.S. Patent Number 4,266,504.)

A secondary benefit of reducing paint-system exhaust is that volatile organic solvent emissions can be controlled much more cost effectively. In the example discussed, the 0.94 m^3/s exhaust is ducted to the burners of a nearby prepaint washer, where the solvent in the air stream is incinerated, thus destroying the solvent as well as utilizing the fuel value of the solvent.

Table 7.4.2. Lamp Efficiency Comparison

Lamp	Efficiency (lumens per watt)[9]
Incandescent	15
Incandescent-halogen	30
Fluorescent	80
Low pressure sodium	180
High pressure sodium	120
Mercury	57
Metal halide	80

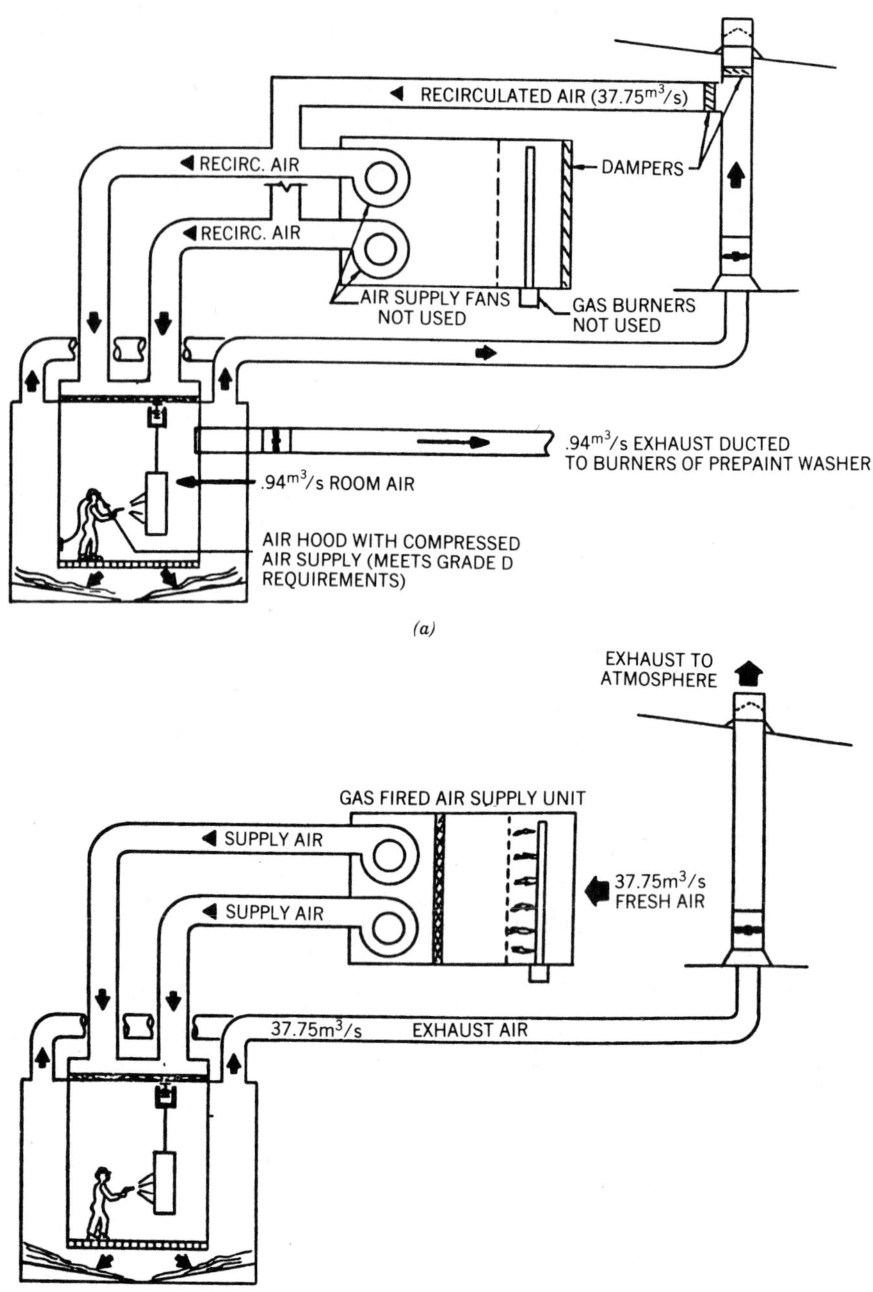

Fig. 7.4.6. Example of (*a*) recirculating and (*b*) nonrecirculating paint booth designs.

4. *Reduce washer temperatures and exhaust.* At water temperatures over 37.8°C (100°F), washers and treatment processes can be the largest paint-system energy user. In one example, a relatively large parts washer with two stages heated to 82.2°C (180°F) used 87% of the total paint-system energy. Reducing the temperature to 65.6°C (150°F) reduced the energy used by 58%, and operating at 48.9 and 32.2°C (120 and 90°F) reduced energy use, relative to operating at 82.2°C, by 79 and 92%, respectively. In other words, increasing the operating temperature from 32.2 to 82.2°C increased the energy requirement thirteenfold.

At these operating temperatures, it was found that the largest heat loss in this particular washer was the latent heat lost in the exhaust of the air-water vapor mixture. At an 82.2°C wash-water temperature, it was estimated that latent heat losses accounted for nearly 92% of total washer energy use. At 32.2°C, latent heat losses are reduced to only 67% of total losses.[11]

Therefore, for energy cost control, it is important that the temperature and exhaust of washers and treatment processes be kept at an absolute minimum.

7.4.5 UNDERSTANDING UTILITY BILLS

An understanding of utility bills is essential to an energy management program. Therefore, elements of electric and natural gas bills are discussed briefly.

7.4.5.1 Electricity

Industrial electric bills typically comprise many cost elements, including customer, demand, and energy charges, power factor penalties, and fuel cost adjustments. Customer charges are periodic charges to cover the cost of billing, meters, meter reading, hookup, and a portion of the distribution equipment. The customer charge is typically a flat monthly free independent of energy use.

The electrical demand is related to the rate at which electricity is used, and is generally expressed in kilowatts (kW) or kilovolt amperes (kVA). Demand charges are designed to recover the cost of installed generation, transmission, and distribution facilities required to meet the greatest demand that the utility must be capable of serving. The demand level used in electricity billing is typically the average demand over a 15 or 30 min period, and the charges are usually in dollars per kW (or kVA). An important element of the demand charge is the application of "ratchet clauses" to the demand level used to calculate the charge. In particular, it is not unusual for utility companies to base the minimum demand charge on the current month's demand or on some percentage (100, 90, or 50%) of the maximum demand of the 11 previous months, whichever is greater. In other words, a demand surge for as little as 15 or 30 min in one month can impact electric bills for an entire year. Therefore, controlling demand is an important aspect of energy management.

Energy charges recover the variable costs of providing electricity, e.g., fuel, purchased power, and operating and maintenance costs of the utility. The energy charge, in cents per kilowatt hour (kWh) can be a flat rate or a block, or stepped rate, in which the charge per kilowatt hour changes in blocks of increased use. In the latter case, the blocks can be explicitly identified as a certain number of kilowatt hours or equated to the number of hours of demand. When the hours of demand method is used, the block level is simply calculated by multiplying the specified number of hours by the demand level.

The energy (or fuel) cost adjustment (ECA) is an additional charge or credit applied to all kilowatt hours used. The ECA reflects changes in the cost of fuels used for generation relative to the costs reflected in the base electric rates, and typically varies month by month.

Industrial electric rates will commonly include a power factor penalty also. Power factor was discussed earlier in this chapter. Typically, penalties are not imposed until the power factor is less than 85%.

7.4.5.2 Natural Gas

Like electric bills, industrial natural gas bills can be comprised of several elements, including customer demand, and energy charges and a cost-of-gas adjustment. Charges for gas are based on the volume used, in 100 cubic feet (ccf) or 1000 cubic feet (mcf) or the energy content of the gas used, in therms (100,000 Btus) or in millions of Btus (occasionally shown as decatherms).

The customer charge is a flat fee independent of energy use. Demand charges in gas sales are usually contingent upon a special gas service contract, in which the customer agrees to purchase a minimum quantity of gas per day on a monthly basis, and is billed for that demand level regardless of use. The utility typically has the right to restrict use to the contract demand level. In addition to the demand charge, contract demand gas rates also incorporate energy charges and purchased gas adjustments. Regardless of the type of gas rate, the energy charge and purchased gas adjustment are simply charges applied to each unit of gas used.

7.4.6 CONCLUSION

Energy management can have an important impact on the profitability of manufacturing companies. That impact will be realized through many channels, including: (1) a direct cost avoidance resulting from increased energy efficiency, (2) higher productivity due to improved working environments, and (3) increased production security through energy security planning. Overall, controlling energy use is a valuable asset to manufacturers now and will continue to be in the future.

Although the preceding discussion outlined key elements in implementing and maintaining an energy management program, additional or more detailed information may be useful to conservation efforts. Energy information is available from many sources, one of which is the utility company, since some utilities have implemented programs to help customers conserve. Another source of information is technical societies. Those societies that have published energy management aids and/or concentrate on energy efficiency issues include the American Society of Heating, Refrigerating, and Air Conditioning Engineers (ASHRAE), the Association of Energy Engineers (AEE), American Institute of Plant Engineers (AIPE), National Electrical Contractors Association, National Electrical Manufacturer's Association, and the Construction Industry Manufacturer's Association. In addition, process-specific professional societies, such as The American Foundrymen's Society, can be excellent sources of information on improving the energy efficiency of manufacturing operations. Last, state and federal agencies can provide information valuable to energy management activities.

REFERENCES

1. Alvin L. Alm, "Energy Supply Interruptions and National Security," *Science*, March 27, 1981, pp. 1379–1385.
2. Franklin A. Lindsay, "Plan for the Next Energy Emergency," *Harvard Business Review*, Sept.–Oct. 1981, pp. 152–168.
3. Energy Information Agency, *Annual Report to Congress 1981*, Department of Energy, *vol. 2: Energy Statistics.*
4. Axel, Helen, *Organizing and Managing for Energy Efficiency*, New York: The Conference Board, 1983.
5. John H. Hagopian and E. Karl Bastress, *Recommended Industrial Ventilation Guidelines*, U.S. Department of Health, Education and Welfare, National Institute for Occupational Safety and Health, Washington, DC, 1976.
6. Committee on Industrial Ventilation, *Industrial Ventilation—A Manual of Recommended Practice, 11th Edition*, American Conference of Governmental Industrial Hygienists, Lansing, MI, 1970.
7. Theodore Baumeister, Eugene A. Avallone, and Theodore Baumeister III, *Marks Standard Handbook for Mechanical Engineers*, 8th ed., New York: McGraw-Hill, 1978.
8. *Architects' and Engineers' Guide to Energy Conservation in Existing Buildings*, DOE/CS-0132, U.S. Department of Energy, Washington, DC, February 1, 1980.
9. Craig B. Smith, *Efficient Electricity Use, A Reference Book on Energy Management for Engineers, Architects, Planners, and Managers*, New York: Pergamon Press, 1976.
10. Mark Pershing, *Paint System Energy Usage*, Deere & Company Study, Dec. 1978.

PART 8

SYSTEM: FACTORY INTEGRATION

CHAPTER 8.1

FACTORIES OF THE FUTURE

JAMES J. SOLBERG
DAVID C. ANDERSON
MOSHE M. BARASH

Purdue University
West Lafayette, Indiana

RICHARD P. PAUL

University of Pennsylvania
Philadelphia, Pennsylvania

8.1.1 INTRODUCTION

Many of the most significant opportunities for improving productivity entail fundamental research that has not even been undertaken yet. . . . The often expressed view that 'the technology is already available—all that remains is to apply it'—is dangerously shortsighted. . . . The science of manufacturing is still in its infancy, when one considers how far we are from reaching the theoretical limits of efficiency.

These excerpts are taken from the report of a study performed by the authors for the National Science Foundation in assessing the manufacturing research needs for future factories. The purpose of the study was to describe the kinds of research required for a highly competitive "factory of the future" to become a reality. This chapter presents the major findings from the study and establishes a foundation for the remaining chapters of the handbook, which treat various aspects of manufacturing systems.

8.1.2 MAJOR THRUSTS DEFINED

A few major themes came to dominate our vision of future factories. Despite the fact that attention was focused on one class of manufacturing (metalworking), these themes were thought to be the central guideposts to progress in all of manufacturing for at least the next decade.

The first major theme is the push for speed. The total manufacturing cycle from design through delivery must be reduced by an order of magnitude. The pressure to eliminate all unnecessary delays is motivated only partly by a desire to provide better service to the customer; *much of the motivation stems from a need for better control and cost reduction*. As we progress in this effort, many problems which are not now regarded as significant (because they are concealed by long lead times) will emerge as serious bottlenecks.

A second major force will be the need for flexibility. The range of capabilities of both hardware and software must be greatly extended in order to accommodate the variations that will become accepted as normal. Equipment must be so versatile that its cost can reasonably be amortized over many different products. *The time and effort involved in set-ups must be reduced to the point where they are not a factor*, so that very small batch sizes can be economic. Fault-tolerance must be built into the system.

The material in this chapter is taken with permission from the *U.S. Army ManTech Journal*, Vol. 10, No. 1, 1985; it is a condensation by the staff of the *U.S. Army ManTech Journal* of a report prepared by the authors and based on a 2-year study they conducted at Purdue University with support of the National Science Foundation.

A third forceful trend will be the transition to more and more machine intelligence. Starting with management of the enormous quantities of data that factories deal with, and gradually moving past data into information, and from there into knowledge, *the entire manufacturing system must escape from its current reliance upon human judgment and interpretation*. It will not be sufficient to capture human expertise in expert systems, because the intuitive methods of people (while often surprisingly effective) are too error prone and arbitrary to provide a basis for good control of tightly coupled, complex systems. Instead, a rational foundation for design and manufacturing decisions must be built from the ground up, based on process models and physical laws. This in turn implies that we have a great deal of research to do just to reach an understanding of basic phenomena. In other words, we will require a science, not just a large collection of experience.

Finally, the integration of technologies is crucial. Advances in narrow, highly specialized fields of research will only be wasted if bottlenecks at the interfaces impede productivity. There is abundant evidence that we face that problem today; that is, factories are unable to gain the full benefit of known technologies because they do not fit together gracefully. Furthermore, the kind of integration that will produce the greatest benefit in the future is not a "paste-together" kind, but a "rebuild-from-common-foundations" kind. The difficulty of getting individuals and institutions to pay serious attention to issues of integration should not be underestimated, *for it is a reversal of the historical trend toward segmentation and specialization*.

8.1.2.1 Obstacles Encountered

Just a few years ago, a major problem was a general lack of awareness that manufacturing industry was facing more than the usual business pressures. The generations of industrial supremacy America had enjoyed was taken so completely for granted that any symptoms of weakness were interpreted as local failures. Now the existence and seriousness of the problem appears to be widely recognized. Dramatic improvements are needed, which suggests revolutionary, expensive, unsettling changes. Some people and some companies may resist these changes, but *the competitive pressures threatening survival are strong enough to overcome much resistance*. Generally speaking, the institutions that can do something about the problem—industrial, academic, and governmental—are ready and willing to act. However, the prospects are still clouded by three major obstacles.

The first major barrier is the sheer difficulty of deciding what to do. Modern manufacturing is technically demanding. Although at one time it called primarily for mechanical skills, today manufacturing requires detailed knowledge of materials, electronics, computers, communications, robotics, and many other advanced technologies. The breadth, diversity, and complexity of modern manufacturing makes it virtually impossible for any one person to comprehend the workings of the entire system. Yet, especially today, the issues are so interrelated that *a systems approach is the only sensible one*. To say that organizing such an effort will be difficult is a gross understatement.

Second, there are not very many people who understand manufacturing broadly enough and deeply enough to make much of a contribution to the effort. This is *the grim heritage of decades of neglect*. Universities have snubbed the area for so long that the talent pool there is perilously small. Even with extraordinary efforts, it will be many years before the necessary higher educational programs can be built.

The third major obstacle to a rapid resolution of our problems in manufacturing is the lack of a scientific foundation for manufacturing knowledge. Aside from a few highly specialized areas, it is generally experience based, often unrecorded, and certainly unorganized.

Despite these difficulties, the research community is responding to the national need. Several innovative programs in multi-disciplinary, joint industry/university research have been formed to deal with these problems directly.

8.1.3 SEEDS OF CHANGE SOWN

In short, America is awakening to the need for a renaissance in manufacturing. After years of complacent self-satisfaction, we have been stirred by foreign competition to recognize our weaknesses. The facts are clear. We no longer hold the commanding lead in technology that once marked our claim as the most advanced industrial nation in the world. More and more, our exports have shifted from manufactured goods to food and raw materials, while our imports have shifted in the opposite direction. In one industry sector after another, market dominance has been ceded to foreign companies.

For some time, observers have seen isolated symptoms, but until recently few have grasped the scope of the problem. Public awareness usually has focused on the difficulties of some one particular industry, such as steel or automobiles. Occasionally it has broadened to wider concerns, such as wage differentials or styles of management or product quality issues. But now, just within the past two or three years, there has emerged a general awareness that something of historic proportions is occurring.

Well before the general public became aware of our worsening competitive position in manufacturing, many leaders in industry perceived systemic weaknesses. In fact, it has been common knowledge in manufacturing circles that our practices have lagged dangerously far behind our potential. Unfortunately, because chronic maladies never attract as much attention as acute crises, little has been done to date to address the larger issues.

Of course, recognizing that a problem exists is a precondition to solving it, so we are quite possibly in a better position now than we were a few years ago.

One important point must be mentioned, however. Although we believe that vast improvements are possible, it is obvious that our standard of living surpasses that of almost any other nation; we also are aware of the role American manufacturing has played in the successful defense of our country throughout its history, and of its role in the national expansion. All of those things strongly suggest that much has been right and continues to be.

We are also aware, however, that the environment in which the manufacturer operates is changing rapidly. And survival requires adaptation—the changes which are under way seem so basic that we must rethink some of our fundamental positions. We cannot continue to operate in our present mode in the climate that we anticipate will exist in just a few years. We do not believe that manufacturing can adapt effectively by making minor modifications to present practice.

So fundamental are the changes that manufacturing will have to face that there must emerge a whole new set of rules for the game—much in the sense that the entire approach to physics had to change as that science moved from the Newtonian era into the present one.

Everyone associated with manufacturing will have to work hard to maintain focus on the real issues. It is altogether too easy to get sidetracked, to give valuable time to strategies which are no more than holding actions and pretenses. But if we can avoid that, the ingenuity of American designers, engineers, and manufacturers will make it possible for American manufacturing to have a strong future.

8.1.3.1 Larger Perspective Required

Despite receiving considerable attention in recent years from a variety of viewpoints, the concept of the future factory remains elusive. Of course, there is intense research activity under way in such areas as Computer Aided Design (CAD), Computer Aided Manufacturing (CAM), robotics, machine vision, and many other technologies employed in production. Modern manufacturing, at least in certain sectors, is far more advanced than the general public realizes. What is missing is the big picture framework that clarifies, in a more than superficial way, how these various technologies can and should fit together.

We believe that manufacturing research stands on the threshold of a very uncertain future. There are literally hundreds of attractive avenues competing for the attention of a small number of capable researchers. Although we certainly never considered it to be our responsibility to dictate policy for the entire research community, we did think it necessary for someone to try to make sense of the whole area and to offer the framework for others to critique. That is what we have done.

For the sake of focus, we limited our attention to the manufacture of products made predominantly of metal, and of medium size (that is, neither so large nor so small that special processing methods would be required). Many of our conclusions would apply just as well to the production of plastic or electronic parts, for example, but we do not address any special issues connected with these. There are, however, important differences between production methods for continuous and discrete products, and we certainly limited our conclusions to the discrete category.

There are several reasons to give primary attention to batch (i.e., intermittent) production, as opposed to high volume mass production. By its very nature, mass production lends itself to very specific, highly specialized hardware and methods of the sort that classical engineering has successfully dealt with in the past, whereas batch production demands something new and different. Furthermore, batch production is more in need of attention, both because it accounts for about four times as much cost as mass production and because the productivity measures are much worse.

For the purpose of prioritizing research needs, it is essential to look past the difficulties faced by industry today. Although many people are justifiably concerned about the ability of American industry to compete in a world economy and are therefore pressing for immediate remedies, we believe that the research community can best serve these interests by understanding the long-term prospects. This belief does not by any means conflict with emphasizing the kind of short-term efforts that could ensure survival; it only declares that we must recognize the difference between temporary, expedient solutions and lasting, fundamental changes.

8.1.3.2 Uncharted Path Ahead

Most attempts to anticipate the future rely heavily upon the detection of current trends. For projections extending only a few years ahead, this is surely the most reliable approach. Long-term projections, however, require a different approach. The systems that emerge over the next twenty years will rely upon *basic advances in technology that no one has even thought of yet*. Under such conditions, it is obviously difficult to predict very many details. Still, it is possible to examine broad historical and economic forces as a clue to the direction of change. Our fundamental premise here is that technological development will, over the long term, respond to needs that are expressed through the marketplace. In other words, we expect that progress will occur most surely in those areas that promise a clear payoff.

The view we have adopted is essentially the same as that which prevails in all Western industrialized nations. Basically, it asserts that any company that can produce a superior product at less cost and in less time will thrive at the expense of other companies. Over the long run, regardless of social policy, labor agreements, or government constraints, the forces of economic competition will prevail. Technologies, companies, or industries that prove to be economically "unfit" relative to the competition will simply not survive. We hold to the notion that the people of the United States, and more generally the population of the world, will be best served by technology that improves competitiveness.

Another clue to long-term progress lies in what economists would call "substitution" effects. When the relative advantages of one technology (or resource, or method, or so forth) shift relative to alternatives, a system will slowly move toward a new point of economic equilibrium, with the advantageous technology displacing the others.

An obviously pertinent example in the present case involves the use of computers in manufacturing. Because of the amazing advances in microelectronics, the relative cost of storing and processing information compared to other cost factors in a system has been greatly reduced. Even if the electronic technology were to freeze at its present level (which, of course, no one expects to happen), it would be many years before the full implications of the lowered cost of computing would be realized. This is only one of many identifiable long term forces working to alter the way manufacturing is done.

8.1.4 THE FUTURE

To speculate briefly on the fundamental question of whether manufacturing can continue to exist in the United States, one chilling scenario for the future is that we will have no factories, but depend upon foreign suppliers for virtually all of our physical goods. Not everyone cares—some people, having no obvious personal stake in manufacturing industry other than as consumers, do not even find such a prospect to be fearsome. Certainly, many of the managers of multinational corporations take the attitude that their plants should be located wherever they can produce most economically. Since *it requires no technological breakthroughs to gain the advantage of lower wage rates*, it can be very tempting to move manufacturing operations offshore. What is now a trickle of losses could become a flood in the near future.

But anyone who looks into easily available statistics on national income will soon realize that, without exaggeration, *manufacturing is the foundation of our economy*. Because so many other of our national priorities depend upon maintaining a healthy economy, the "productivity imperative" is considered to be one of the most critical issues facing our nation.

The factory described, or something like it, will eventually come into being. We hope, but cannot be confident, that it will happen in the United States. The race that is under way among the industrialized nations to achieve the competitive advantages such facilities would give has *no finish line. No one can win.* But any country that falls so far behind that it drops out of the race can fairly be called a loser. What that would mean to life in the United States as we know it is almost too terrible to contemplate.

The significance and difficulty of the competition underscores the importance of having a clear understanding of the goal—something we hope this study will assist in defining.

8.1.4.1 A Future Factory Scenario

In order to present an informal narrative "picture" of a factory of the future, we will track a typical order from first customer contact through delivery. One important point to bear in mind as we go through the steps in some detail is that the entire process will occur much more rapidly than it does today. A single factory order will be fully processed in a matter of hours, as opposed to the weeks that are accepted as normal today.

The range of product are "ordinary" mechanical components and small subassemblies, such as pumps, compressors, motors, gear trains, and the like. We can gain more insight into the methods for complex products by understanding simple ones than by going in the reverse direction.

Customer Interface

We imagine that a customer (who could be either an individual end-user or an organization intending to incorporate this product into some more complex product) has identified a specific need and is seeking a supplier. The steps involved in getting the product to the customer are initiated by a "call." The call may involve vocal person-to-person communication or may not, but will almost always involve computer-to-computer communication. In some cases, the request may be generated by the customer's computer without any human attention. Routine transactions, which occur when the customer knows precisely what he wants and the supplier knows he is able to provide it, would be logged without any human attention on either side. Such simple assignments as finding the cheapest source or quickest delivery are easily handled by fixed rules, and therefore will be handled automatically. Occasional audits by human supervisors will assure that the proper actions are being carried out.

A more interesting scenario occurs when the customer has to "shop" to meet his needs. Perhaps he is unsure whether an advertised product will meet his needs, and must inquire about details. Or perhaps he wants something that is just a little different from the standard product. Maybe he is not quite sure what he wants, and has to learn more about what is available or could be built. In such cases the customer must work through a customer service representative (i.e., a salesperson with sufficient knowledge to deal with the questions). This person may not work for the supplier directly, but rather for a brokerage company specializing in dealing with customer inquiries. A similar function in today's world is provided by travel agents. The reason that this development is likely to occur is that the direct interface with suppliers is going to be sufficiently complicated that occasional users will be at a disadvantage relative to those who do it everyday, which will create an opportunity for intermediaries. Of course, large companies will have their own in-house capability, but the function will be similar.

The first step of the broker's task is to find out what the customer wants. In any product, there are some features that the customer regards as essential, others that may be desirable, and some that are left entirely to the supplier's option. Of course, a different customer or application may necessitate a completely different set of essential features, so no permanent categories can be assumed. The broker must therefore interact with the customer to develop a precise statement of needs.

In most cases, the communication will involve the transmission of pictorial data, not just text.

As the definition of customer requirements is being refined, the broker will be conducting a computer search for suppliers having the probable capability to make the product. For obvious reasons, it is preferable to locate a supplier who has previously made something very similar to what is needed. Similarity in this case refers more to the manufacturing requirements than the visual appearance or function of the product.

Depending upon the nature of the customer's requirements, the need definition stage of the interaction could end in several ways. If the requirements are not particularly demanding, there may be several qualified suppliers available, in which case the broker could proceed to a "request for bid" stage. If the customer becomes aware that his request will be difficult or expensive to meet, he may be willing to modify his requirements.

It is important to realize that suppliers will be much more flexible in meeting the customer requirements than they are today.

Custom Design Work

Suppose that the customer's essential requirements rule out the use of any standard design, even with modifications. In other words, before an appropriate supplier can be identified, a formal design activity must be performed. Again, depending upon the nature of the customer's requirements, several possibilities exist for who will do the design, and how it will occur. One possibility is that the customer will have done his own design work before contacting the broker and will know exactly what he wants. A second possibility is that the product design is so delicate that a specialist must be called in.

Some product families, such as turbochargers or jet engine turbine blades, are so sensitive to subtle details that designing them is a specialized art in itself. Even when designers are supported by a great deal of computer software, relatively few people will possess the expertise to undertake the design of a completely new product of this sort. Consequently, it will make sense for groups of design specialists to assemble in separate companies or divisions of larger companies. There may be only two or three groups in the nation who are capable of doing advanced design work in a certain product category. These groups would serve many production facilities, but may not be closely affiliated with any one in particular.

At the other end of the spectrum, there are some families (e.g., ball bearings) where the design is easy but the manufacturing processes are highly developed.

In the large middle ground between these extremes are the majority of products which involve "common" design principles and manufacturing processes. In general, there are reasons to expect that design and manufacturing will be closely integrated in these cases. The design activity will take place within or in close proximity to the factory where the manufacturing processes occur, and will be heavily influenced by local production capabilities.

Although the actual design of specific products is likely to occur on-site, the development and maintenance of the design tools will be an industry in itself.

When you speak of design today, most people think of engineering drawings. An obvious change is the replacement of the paper drawing by electronic media. However, this is only the most superficial of the changes that will occur.

The shape of a product, including even the most detailed dimensional data (such as tolerances and surface finish), addresses only some of the concerns of the customer. In fact, the customer is usually more interested in the functional performance of the product than what it looks like.

In order to respond promptly to the customer but also to be confident of the answers (for the sake of both the customer and the manufacturer), it will be essential to have extensive computer modeling capabilities on line.

Kinds of computer analysis which are virtually unheard of today will also be performed. Models to aid in the selection of materials, standard tools, and processing methods will assist the designer in making his choices. These models will have access, of course, to the latest information about local plant conditions.

The final result of the design process will be a set of data in a computer database. One expression of this data set could be a realistic picture on a video screen; another could be an engineering drawing. However, there would be no real need to generate either of these.

Until a design is actually implemented as a physical product, it could be considered "disposable." There would be no compelling reason to keep it, because it is rather easily reconstructed from scratch, and conditions at a later time might permit improvements.

The final customer, of course, will see virtually none of this complicated activity, and the designer will be consciously aware of only a small portion of it. The whole process would take place in a few minutes.

Small Factory Orders

A typical factory order will be for only a few units, even if the customer consumes many over the course of a year. There are two strong reasons for this. First, the customer will request only a few at a time in order to keep his own inventory down. Hand in hand with that desire will go a very strong desire for quick response times. So, for example, an order for ten units to be delivered within one day may be followed just a few days later by a similar order for twelve more. Frequent, small orders will be the norm throughout all manufacturing industry.

Second, although it may be possible to accumulate orders from different customers or to fill out production batches with units to stock in anticipation of future demand, there will be little incentive to do so. The production system will be so flexible that setups are no significant factor. In general, the philosophy will be to make only what is immediately needed just before it is to be shipped.

Because the design activity is so tightly integrated to manufacturing, much of the preparatory work which would otherwise delay the processing steps will already have been done. Most of what we now call manufacturing planning or methods engineering will be implicitly included in the design stage. Process planning and part programming will be quite different activities in the future.

Metals to Continue Dominant

Although many new kinds of basic materials—plastics, composities, ceramics, and so forth—are establishing themselves as significant alternatives to metal in certain applications, we do not expect any of them to replace metal as the dominant material in batch manufacturing. In general, metals possess desirable combinations of properties: strength, durability, relatively low cost, and workability. A constantly expanding menu of special alloys offers a rich variety of choices to meet special needs. Sometimes considerations of weight, corrosion resistance, electrical or thermal conductivity, or some other factor will dictate the use of another material, but no other class promises the broad applicability of metals.

Material Handling Significance

A lot of current attention is given to very elaborate systems for workpiece storage and handling. In large measure, the cost and complexity of these systems are merely evidence of failures to control other aspects of the total production system. That is, we try to compensate for ponderous production control and inflated inventories by building bigger and better handling systems.

The average residency time of a workpiece on a machine will be slightly longer than it is today, despite the fact that cutting speeds will be much greater (perhaps ten times as great for routine cuts). The reason for this is that the machines will be so versatile that many consecutive operations will be performed on the same machine. Consequently, the number of trips per workpiece per hour may be somewhat less than it is today.

Incidentally, workpieces are not the only units which circulate in the factory. Tools are just as important. It is not yet known whether they are best served by a separate handling system, or by the same one, but synchronizing the movements of workpieces and tools so that the right combination ends up in the right place at the right time is a challenging research problem with many hidden traps.

Setups on the Fly

Machines will not be shut off to allow time for setups. As a machine changes from one job to the next, all of the necessary tooling will accompany the workpiece into the machine.

Except when absolutely necessary, fixtures will not be specially designed for a single product type. Rather, they will be assembled out of standard elements of various shapes and sizes. The fixture elements and the workpiece will be put together by a robotic assembly machine, working under the guidance of a program that draws data from the part description entered when the product was designed. In other words, it will be automatic.

Processing Changes

Although it is conceivable that some entirely new processing techniques will become important, we think it unlikely. An ideal process would shape a raw blob of metal into any desired form precisely, without waste, without special tools, and with little expenditure of energy. One might think of lasers or some other kind of concentrated energy beam, but no one who is familiar with the engineering aspects of metalworking takes very seriously the notion that the basic processes of today are about to become obsolete. Turning, milling, drilling, grinding, stamping, casting, welding, and the other common methods of cutting, forming and joining will surely constitute the major portion of future work with metal.

The processing portion of the factory by today's standards will seem very small. A dozen or so major pieces of processing equipment, each of which is highly versatile, will be able to perform the duties of many more specialized machines.

You will not see aisles stacked with containers of partially completed workpieces as you do today. In fact, the arrangement of equipment will seem very compact. There may be a few workpieces waiting for an available machine and/or a few machines waiting for work, but the predominant image will be that everything is active.

In many ways, the mechanical part of the factory will seem to be a single machine rather than a collection of machines.

The number of active workpieces at any given time—those which are either in process or partially completed and awaiting subsequent processing—will represent a full order of magnitude reduction from current standards. Similarly, the total start-to-finish span time for any single workpiece will be in the vicinity of twice the total processing time, rather than the 20 to 40 times that is common today.

Human Assembly to Continue

Most products have to be put together at some point. With few exceptions, automated assembly is beyond the capability of currently available mechanical devices, so it is still a labor intensive activity. Indeed, a large percentage of the direct labor content of a typical product is in the assembly operations. Of the common physical operations in a factory, *assembly is undoubtedly the most demanding and will be the last to be automated.*

When you consider the ease with which even a child can put together tinker toys, and the extreme difficulty in getting a robot to perform a similar task, you can begin to appreciate how far we have to go in developing assembly technology.

It will be a serious mistake to even attempt to automate assembly as humans do it. Such a goal is quite possibly unattainable, and even if technically attainable, would almost certainly not be economically justifiable for common factory use. Instead, the effort should first go into altering the product and the environment to make the assembly job easier.

Thus the automation of assembly will be a gradual process, probably taking many decades to reach a point of declining marginal returns. We are speaking here, of course, of general purpose assembly machines, not simply those which are engineered to assemble particular products. Consequently our factory of the future—the jobshop of one or two decades away—is likely to employ a mixture of automated and manual assembly. Easy assembly jobs will be handled by machines working around the clock, while tougher jobs are handled by workers in a conventional work day.

It is worth mentioning in passing that it is not only the final product that requires assembly. Putting together the fixtures and taking them apart, changing the perishable cutters in tools, and certain types of machine repairs all involve assembly operations. Because these are crucial to the operation of the system and because they are susceptible to some degree of standardization, they are likely to be fully automated relatively early.

Quality Assurance Demands

Most American companies today think of quality assurance (or control) in terms of a trade-off between the cost of manufacturing and the cost of warranty service. It is concern for customer satisfaction that forces the issue. In the factory of the future, however, an additional, completely different motive will be in effect. Automated assembly cannot be expected to work without dependably accurate components.

Quality starts with design. Because it will be so important to do everything right the first time, a wide variety of analytical tools will be invoked automatically to prevent the designer from making inadvertent errors, or from specifying tolerances that are either too loose or too tight. Throughout each step of the manufacturing process, inspection will be incorporated either as an integral part of the operation (which is preferable wherever feasible) or immediately before or after. A wide variety of sensing apparatus, much of which does not now exist, will be employed.

A Fault-Tolerant System

We have emphasized repeatedly that the overall production system will be highly integrated, but that does not mean that it will be vulnerable to failures—at least not in the sense that the failure of a single component can bring the whole system to a standstill. Indeed, the versatility of the equipment and the parallelism encouraged by the procedures will provide implicit back-up capability to allow the system to function successfully even when portions are inoperative. A sensible approach, from an economic perspective, is to build a fault-tolerant system.

A System Like a Living Organism

Many people think of advanced automation as precise, regular, and repetitive—as in clockwork. Although this is a seldom attained ideal in mass production, it is not even an appropriate target for batch production. A better image is that of a living organism. The future factory will find it necessary to adapt to a host of short term minor disturbances. The constantly varying product mix, the varying pathways selected, the equipment failures, and many other changing factors all work against any predetermined plan.

There will be an increased pressure on long-term resource management. The factory will be defined by, and its business success limited by, its processing capability, not a product line or any particular market. It will be able to make any kind of product that it has the machinery and tools to make.

It is to be expected that some long term contracts for delivery will be negotiated, when the customer has confidence in both the stability of his needs and the supplier's abilities. Naturally, he will expect a price break. Since many customers will rather hedge than make a long term purchase commitment, a large portion of the business will be of the short-term, low-volume character.

Under these circumstances, a manager will have to keep a close watch on the match between the capabilities demanded by the marketplace and those he is able to offer. This means that the factory must be in a constant state of innovation. There will be many ways to succeed, but standing still will spell certain failure.

Instead of imagining a snapshot image, one should see the factory as a living, evolving, adapting organism.

The Design Function

As crucial as the design phase is in all its aspects, and regardless of the quantity of the product to be produced, the ideal manufacturing system must hold design time to a minimum.

Depending on the complexity of the product, the design phase today can require from months to years. Even minor changes to an existing product can require considerable time. Considering continually shortening product life cycles in our dynamic world marketplace, one sees that shorter design time correlates to greater productivity.

A company's ability to enhance design efficiency has depended, until fairly recently, on the shrewdness and experience of its designers. The design process is a combination of creativity, ingenuity, scientific analysis, and decision-making. One sees that the success of a design depends to varying degrees on the intuitions of the designer. So even the recent attempts to speed some phases of the process through a variety of mechanical means only address part of the problem.

Increasing the efficiency of the design process is a vital ingredient in the overall strategy of a factory of the future. Doing that will require not only mechanical aids, but also a thorough rethinking of what design is and what tools best suit it. In addition, achieving an optimal approach will require solution of some very basic problems in a number of related areas, including such matters as information management, knowledge representation and the fundamentals of part representation.

8.1.4.2 Research Needs

Achieving the design function in the factory of the future will require four main kinds of research. For one, there will need to be major consideration of the database, or *information representation* problem. Along with that, we think there needs to be a major investigation of *what designers do*, how they do it, and how it can be augmented by computers. Much remains to be done in the general area of *graphic and design* capabilities. Finally, *geometric modeling* is really only in its infancy and will require vastly more sophistication and completeness.

In sum, we see as a major research need, both within computer aided design (CAD) itself and between CAD and CAM, a rethinking and restructuring of the database which will render an information system at once more flexible, comprehensive and manageable. This appears to us a fundamental need on which depends progress in many of the other areas.

The use of interactive graphics brings up an interesting point about design. To designers and engineers, the visual sense dominates in many obvious ways. And even as modeling replaces drawing, we are dealing with visual communication. That visual preoccupation, which is a perfectly legitimate one, has taken attention away from several matters of interpretation which are quite important. Shape is only part of the story, after all. Function is an important part of design consideration as well.

Can Creativity Be Computerized?

Since two-thirds of design changes in a given part result from manufacturing problems, *we can no longer afford the cost of an approach which depends so much on trial and error*. The competitive forces at work in the world of metalworking mandate far more efficient communication—far better, far more knowledgeable handling of information.

In any extended discussion of CAD, we must sooner or later confront the notion that design is a creative activity—which implies to many people that it is not amenable to the kind of codification which computers require. *Good designers grasp intuitively matters which are virtually impossible to program*. Even good designers produce bad designs on occasion. To many, that suggests that we cannot automate the design process in a meaningful way. Automation may not be an appropriate goal. Advances in knowledge-based systems and the artificial intelligence field in general have direct applicability and can be applied to this problem. Good design practice is governed by a set of rules, and these computer reasoning systems are driven by rules. They should work together.

More Compact Models

Still in its infancy, geometric modeling seems to clearly offer much promise to designers and engineers. Currently, designers—whatever their mode of operation, automated or not—have to cope with functional matters in very informal ways. In a modeling framework, such information would come readily to mind. And even most current CAD systems offer the designer far more than they offer the engineer. Geometric modeling provides CAD system builders with a robust starting point.

However, even the best current geometric modeling systems suffer from limitations of complexity and size. Virtually all of the systems employ voluminous codes which consume the computer resources. All of the current systems are at best useful in only limited situations. They are unable to produce many shapes, and in most cases fail to provide immediately usable capabilities to today's industry. Clearly, then, a great deal of basic research must address itself to two questions. The first, related to the general database problem, concerns *how to represent the geometric information more compactly*—so it will not consume the entire computer and require inordinately long computation time. The second involves the basic matter of *how to define, mathematically, all of the forms designers need* in the repertoire, and none of the impossible ones.

Governing all of our research in this vital field must be the awareness that the design phase represents a significant part of the expense of getting a part out the door. Everything we can do to expedite that process enhances our competitiveness and increases the profitability of our operation.

8.1.4.3 Robotics in the Future

If we are to design a factory of the future to meet the needs of the batch metalworking industry, we will need some general purpose device which can take the place of operators. An industrial robot of far greater capabilities than any available today will be required. Such automation will only be possible if, at the same time, we develop techniques of design yielding products which are simple to manufacture. If we look at robots available today, we will see that they are unsuitable to meet the needs of automated batch production.

Robots enjoy the dubious distinction of being the most misunderstood elements of the future factory. To many people, robots represent the quintessence of automation—tireless, obedient slaves with all of the advantages and none of the disadvantages of human servants. Some imagine that the future factory will be populated with robot workers performing human functions in human-like ways.

At the other extreme, we have today's industrial robots, which are little more than one-armed mechanical manipulators. They have about the same level of intelligence as a CNC milling machine and somewhat less capability. At least the milling machine has a defined function and performs useful work by itself.

Clearly, the proper image of the industrial robot of the future lies somewhere between the two extremes. Just to clarify what we mean by a robot, it is a computer controlled general purpose "handling" machine that can move end-effectors (tools or grippers) to designated positions within its work region. We regard the important distinguishing feature of a robot to be its multi-functional handling ability.

It should be pointed out that robotics research covers a gamut of topics, some of which are pertinent to factory operations and some of which are not. The field includes, for example, attempts to build walking (legged) robots, which may be important in defense, space exploration, or other "outdoor" applications, but are of little interest in manufacturing. On the other hand, many of the fundamental areas associated with robotics—sensing, actuators, control, computational geometry, computer languages, and artifical intelligence—are the essential building blocks of any form of advanced automation (including machine tools and other devices that are not usually thought of as robots).

We believe that future factories will depend heavily on robots, as we have defined them. However, the industrial robots of today are so limited in ability that considerable development must occur before they will be able to fulfill the role we see for them.

Robot State of the Art

A brief survey illustrates several important aspects of the current state of the robotic art. Of chief significance is the fact that the robots themselves have advanced little since their introduction. They are now faster, more reliable, and computer controlled. But even the very first industrial robots incorporated digital memories and servo loops. There have been few real changes in robot technology. The fact that the repertoire of robot tasks has expanded is due largely to very clever tooling and fixturing.

Almost an obstacle to robot advancement is the ingenuity of the engineers who incorporate robots into their production systems. The robot may be inappropriate for the task at hand, but the engineer finds a way to use it. Not able to modify the robot proper, the engineer creates ingenious end effectors and fixtures. Although the tooling may work, if moves robotics away from flexibility and generality and into the domain of hard automation.

Robots for Batch Production

There are three major problems in attempting to automate batch manufacturing:

1. The short production run, over which the cost of any special purpose automation must be amortized.
2. The set-up time of any form of automation.
3. The space occupied by idle equipment when not in production.

Most automation equipment is expensive to fabricate and takes a relatively long time to design and fabricate. This equipment and engineering cost must be amortized over future production, a period normally measured in years. Once the equipment has been installed, it must be brought into operation, a painstaking, difficult, and time consuming operation.

What is needed to automate batch production is some form of general purpose equipment which can be economically justified by the automation of more than one product. A device is needed in which set-up time is reduced to a minimum, making it economic to use for small batch sizes. A device with these characteristics would obviously resolve our third problem: instead of lying idle and occupying floor space, it would be in use whenever there was a sufficient volume of production in the plant. This form of general purpose automation equipment could be used to manufacture one product now and another next. Its economic justification would be based on total production, not on product-specific production. Such a device would be a true industrial robot.

Clearly, general purpose automation would significantly improve the productivity of the small batch shop. In addition, such an approach to automation would allow the batch shop to incorporate changes in product and radical changes in production. Today's robots, as we have seen, are functionally incapable of performing most of the jobs in batch manufacturing.

Concept Integrated

The logical conclusion of a true robotics approach is a fully automatic batch manufacturing facility in which correctly designed products can be manufactured automatically. The design of products, machines, and robots will be such that their production will be carried out in standard steps for which we can develop and prove correct procedures.

As we pointed out earlier, the economics of batch production require the minimizing of product-specific costs. That is what the robot we propose would accomplish. Accordingly, the costs associated with the robot would amortize over the entire production of the plant, not over the production of a given part.

We are a long way from these goals at the moment. We believe the research we have outlined provides a crucial foundation for the factory of the future. Without it, there can be no true general-purpose equipment. Without that, the automation of batch production is both practically and economically impossible. Our ability to remain economically viable depends in important ways on our ability to improve batch productivity.

8.1.5 MANUFACTURING PROCESS REQUIREMENTS

The manufacturing process in batch work suffers from the general emphasis on mass production that pervades our industry. As a result, *the state of the art is essentially irrelevant to the shape of the future*. Manufacturing in the batch shop, or job shop as we know it, offers little in the way of useful insight. The batch factory of the next century must involve a conceptual discontinuity if a major segment of our industry is to remain competitive or, in some cases, regain a competitive position.

For manufacturing processes to move into the mode which our factory of the future will require, a great deal of development must take place almost simultaneously in at least four general areas:

Automation of manufacturing processes in ways appropriate to batch work.

Process planning.

Production control.

General facility planning.

As a matter of simple fact, our approach to each of these aspects of the manufacturing process will require thoroughgoing rethinking of a number of basic issues.

Because of their fixed-automation background, most people's notions of advanced manufacturing focus on transfer lines and related phenomena.

Basically, a transfer line involves a series of simple steps linked together in a way that makes the result look complicated. In point of fact, the steps are unrelated. There is no real system except in the sense of a concatenation of units—a collection. And the typical transfer line works only because the engineers have planned the work for uniformity.

In mass production, a transfer line can pay off, and when everything is in order the effect is quite impressive. Batch production cannot gain from that approach, however. Typical batch production operations cannot afford the set-up cost. Also, even if a batch manufacturing could establish a flow sequence for one part, others would have different flow patterns. The *variable flow characteristics of batch shops* leads to so much idle time that transfer line automation is completely inappropriate. So job shops still "do it by hand."

8.1.5.1 A Hierarchy of Decisions Advocated

Ultimately, we think the only approach which offers real hope is a generative one. Investigators will have to use models to see how the process ought to take place. And we think that the automatic function should be a high level one. That is, it should specify the part in terms of function, overall strength requirements, and so on, but not in terms of such lower level matters as which machine should make it.

Presently, in very small job shops the shop floor people enjoy a certain latitude in these matters. Whether a support bracket is to be cut, welded, and drilled or stamped and shaped is a low level decision based on such matters as which materials are on hand and which machines are available for use as well as such considerations as strength and function requirements. A part may well be produced one way one month and another way the next. And that is as it should be. That bespeaks a properly flexible system.

For that reason, we advocate a hierarchy of decisions in an automated process planning procedure, because the specific matters of which machine to use have to remain open as long as possible in order that the entire factory is as flexible as possible.

8.1.5.2 Incessant Operation Envisioned

The factory that we envision should be operating almost constantly. When one task ends, the next should begin in minutes or even seconds—instead of the present hours or even days. And, as we have pointed out elsewhere, that operation should involve very high speeds, both with regard to actual machining and assembly and with regard to material handling functions.

In that atmosphere, humans will not be able to keep up with the mechanical aspects of the operation. The entire factory floor may take on much of the aspect of a single machine in present terms—something no one enters during operation.

If that is the case, all the set-up functions which require human intervention will have to take place off-line, away from the machine atmosphere. And if special programming and fixturing requires too much time, parallel teams will have to handle components of the task. Thus, while the machine performs one task, another is made ready and sent in by way of the material handling system. Diminishing that setup effort and time requires that as much as possible be automatic—even the programming.

People will always need to play a role. Economic considerations, however, dictate that the human contribution be relatively high level and minimal. Even current computer operations are often grossly inefficient because humans demand to know the status of things. As the computerized aspects of manufacturing become more and more complex, they near a critical point beyond which the complexity and speed will be so great that people will not be able to make decisions concerning error or process because the machine knowledge, for efficiency's sake, cannot be represented in a form understandable for humans.

8.1.5.3 General-Purpose Material Handling

The automated factory will require special consideration in the area of material handling. Currently, the pallets and fixtures for robot operation are as special purpose as the robots. In the near future, we will need to develop more general purpose packaging so that aspect of the operation will harmonize with the rest.

All of this implies an expanded role for the design phase. In the kind of factory we envision, the designer must specify not only what the part will look like and what it is for, but also how it is to be handled. That is, the design of fixture, container, and pallet become an integral part of the overall design of the part. That is quite a proper outcome, for the question of how to handle the part becomes a major aspect of the process in a genuinely automated system.

This very general discussion of some of the needs of the factory of the future leads to a number of conclusions about needed research. As is the case in robotics, many of these problems require simultaneous work because they involve interdependent aspects of manufacturing. And that underlines the *need for an integrated research approach.*

8.1.6 LOOKING AHEAD

By the end of the century, factories will depend upon a wide range of newly emerging technologies that must somehow fit together into technically feasible and economically justifiable systems. Competitive pressures will force radical changes to improve cost, quality, and flow time. This project was aimed at achieving a comprehensive, coherent view of factories of the future—from an engineering perspective—in order to identify the kinds of research that can best serve future industrial needs. A multidisciplinary team, paying special attention to the interfaces among their areas of expertise, developed a common understanding of a possible future metalworking jobshop and from that extracted several important guidelines for research.

There are opportunities for very substantial improvements in the performance of production systems, provided that certain themes are stressed. *Flexibility, system integration, machine intelligence, and coordination will be critical.* The principal conclusions are that a complete restructuring of the total production process—starting with design—will be necessary and that a firm scientific foundation must be put into place to provide a rational basis for this restructuring.

BIBLIOGRAPHY

This list of references is included as a service to the reader who would like to explore the many perspectives of the future world of manufacturing. This group of books and articles offers a balanced and stimulating entry to the current literature. For an up-to-date glimpse at what is going on in industry, we recommend reviewing the most recent issues of *Automation News, American Metalworking News, American Machinist, Industry Week, and Manufacturing Engineering.*

Abernathy, William J., Kim B. Clark, and Alan M. Kantrow, *Industrial Renaissance*, New York: Basic Books, 1983.

Bloch, E., "Workplace of the Future," *IEEE Transactions on Industry Applications*, **IA-20**, 1984, 8–10.

Bluestone, Barry, and Bennett Harrison, *The Deindustrialization of America*, New York: Basic Books, 1982.

Botkin, J., D. Dimancescu, and R. Strata, *Global Stakes, The Future of High Technology in America*, Cambridge, MA: Ballinger, 1982.

Bylinsky, G., "The Race to the Automatic Factory," *Fortune*, **107**(4), Feb. 21, 1983, 52–64.

Fallows, James, "American Industry: What Ails It, How to Save It," *The Atlantic*, **246**(3), Sept. 1980, 35–50.

Goldhar, J. D., and Jelinek, M., "Plan for Economics of Scope," *Harvard Business Review*, **61**(6), 1983.

Gunn, T. G., "The Mechanization of Design and Manufacturing," *Scientific American*, **247**(3), 1982, 114–131.

Hall, Robert W., *Zero Inventories*, Homewood, IL: Dow Jones-Irwin, 1983.

Hayes, Robert H., and William J., Abernathy, "Managing Our Way to Economic Decline," *Harvard Business Review*, **58**(4), 1980, 67–77.

Hayes, Robert H., and S. Wheelwright, *Restoring Our Competitive Edge: Competing Through Manufacturing*, New York: Wiley, 1984.

Harrington, Joseph, *Understanding the Manufacturing Process*, New York: Dekker, 1984.

Hatvany, J., W. Newman, and M. Sabin, "World Survey of Computer Aided Design," *Computer Aided Design*, **9**(2), 1977, 79–98.

Kanter, Rosabeth M., *The Changemasters: Innovation for Productivity in the American Corporation*, New York: Simon & Schuster, 1983.

Kops, L., ed. *Towards the Factory of the Future*, PED vol. 1, New York: American Society of Mechanical Engineers, 1980.

Melman, Seymour, *Profits Without Production*, New York: Knopf, 1983.

Naisbitt, John, *Megatrends*, New York: Warner Books, 1982.

Noble, David F., *Forces of Production: A Social History of Industrial Automation*, New York: Knopf, 1984.

Office of Technology Assessment, *Computerized Manufacturing Automation: Employment, Education and the Workplace*, Vol. 1., National Technical Information Service, Washington, DC, 1984.

Peters, Thomas J., and Robert H. Waterman, Jr., *In Search of Excellence*, New York: Harper & Row, 1982.

Porter, Michael, *Competitive Strategy*, New York: Free Press, 1980.

Reich, Robert, *The Next American Frontier*, New York: Times Books, 1983.

Reid, K. N., R. Cohen, R. E. Garrett, M. J. Rabins, H. H. Richardson, and W. O. Winer, "Research Needs in Mechanical Systems," *Mechanical Engineering*, **106**(3), 1984, 28–43.

Suh, N. P., "The Future of the Factory," *Robotics and Computer-Integrated Manufacturing*, **1**(1), 1984, 39–49.

Yoshikawa, H., K. Rathmill, J. Hatvany, *Computer Aided Manufacturing: An International Comparison*, Washington D.C.: National Academy Press, National Research Council, 1982.

CHAPTER 8.2

COMPUTER AIDED DESIGN AND COMPUTER AIDED MANUFACTURING

ROBERT E. CROWLEY

Electronic Data Systems
Indianapolis, Indiana

In times past, the engineering and production facilities have been regarded as two individual and disparate groups. This was especially true in most of our larger, more complex organizations. The engineering department was responsible for conceiving and designing products, while the production department was responsible for scheduling and building those products. Sometimes there were interfaces between the two groups, such as during the original conception phase, at drawing release time, and when instigating engineering changes. In general, however, there was little, if any, interdepartmental communication.

8.2.1 TRADITIONAL APPROACH

Traditionally, once the engineering department released a drawing it was more or less forgotten until changes began to flow back from the manufacturing and production departments. When a sufficient number of changes had accumulated, the original drawings were searched out and the engineering changes incorporated. This design-fabricate-alter, then redesign-fabricate-alter, process was iterated almost without end. Carrying this to extremes, the engineering drawings were often used as guidelines, while a large amount of component and assembly design was actually done in the fabrication and assembly areas. Basically, the drawings were changed to fit the system that had been built during the manufacturing process. Hopefully, all these changes would eventually find their way back to the engineering department, though in reality they seldom did.

8.2.2 CONTEMPORARY APPROACH

Today, thanks to computer advances in hardware, software, and communications, we are on the verge of merging these vital engineering and manufacturing functions into an integrated whole. The advances in computing capabilities, coupled with more knowledgeable users, allow us to share information and data beyond the most outrageous dreams of just a few years ago. At present, computer aided design/computer aided manufacturing (CAD/CAM) is a workable, viable concept. Further, it can now be implemented in a well-planned, step-by-step approach, affordable by any who will take the time to learn.

Today's modern computer enables a concept to become a reality even before it hits the shop floor. When a concept is developed in the computer it becomes part of the database. The database, in turn, becomes the information source from which the manufacturing and production departments acquire their data. This same information, although in varying forms, is then carried thoughout the production process. We are no longer concerned with a draftsman's having to interpret the concept from the designer, then the manufacturing and tooling people, in turn, redefining the draftsman's interpretation in order to create what they think the designer wanted in the first place.

8.2.3 THE DATABASE

If we are going to share information, or data, between engineering, manufacturing, and data processing, it starts with the data. But what is a database anyway? *Database* means many things to many people. The data processing department thinks of a database as all employees' social security numbers, along with number of dependents, hourly wage, etc. It could also be the information relating to all our vendors, such as name, address, and amount owed. A database for manufacturing could be all time standards for the milling department, or the information relative to preset tooling, like tool holders and set lengths. In short, a database is a collection of information that is stored in the computer and recalled for processing, optimization, or alteration.

In an engineering department a database is the accumulation of information regarding our product. This may be weight, deformation information, or thermal dissipation. The more information that is stored in our computer, and that we have available for recall, the faster we can change, alter, or improve either existing or future products. The real key, then, to integrating engineering and manufacturing, starts with the evolution of an engineering database. But how does one develop a database for products before the products are built? There are many ways. For one, we can test existing products, either our own or our competitors'. Once we have gathered the information relative to existing product, it is stored in the computer for modeling and simulation in order to improve performance. We can also use our engineering experience to develop customized software, although this is often relatively expensive and time consuming. This information is then used for predicting the product's behavior under varying environmental conditions.

8.2.3.1 Evolution of the Database

Predictive engineering methodologies have evolved during the recent past to the point where design engineers, especially in the mechanical area, can predict, with some degree of accuracy, the manner in which components, subassemblies, assemblies, and even complete systems will react under various external conditions, such as load, force, stress, and heat. Although the knowledge for these predictive techniques has been available for several years, the calculations were so enormous and complex that their solutions were impractical when we were restricted to slide rules and mechanical calculators. Not until the modern computer and its attendant software were developed could we even approach these solutions to extremely complex calculations.

This predictive technique is, in reality, based on a rather simple theory. The design to be analyzed is modeled in the computer and then subjected to load, force, stress, heat, or whatever other environmental conditions may be expected. These conditions are then altered to simulate performance under varying circumstances, predicting how the physical design will behave. One technique for predicting reactions under varying conditions is called the finite element method. Because we can now predict, subject to our given degree of accuracy, the performance of our product in its environment, we should be able to reduce the number of physical prototypes required for the test lab. At a minimum, the first prototype should be much closer to the desired product, which in turn reduces the time frame from conceptual design and drafting to the shop floor.

8.2.3.2 The Finite Element Method

A component or assembly that is to be simulated, or analyzed, is subdivided into a finite number of "elements," hence, the name *finite element method*. These elements can be practically any geometric entity: simple, one-dimensional rods, two-dimensional triangles and polygons, or three-dimensional cubes and polyhedra. Regardless of the shape of the element, however, each contains points, called "nodes," most often equally spaced around its perimeter. Adjacent elements will share common nodes. Each element and its nodes are identified as distinct entities. The creation and identification of each of the elements with its respective nodes is called the finite element mesh, or finite element model.

Using the finite element model as a basis, the engineer can now apply a force at one node and the computer will calculate the distribution of this force throughout the element. Then, because of the shared, or common, nodes between elements, the program determines how this force affects the adjacent elements. These adjacent elements, in turn, affect their adjacent elements and so on throughout the system. In this manner, the engineer can predict how systems, especially complex or oddly shaped ones, will react under varying forces by dividing the whole system into a series of smaller entities. We are, in a sense, doing prototype engineering on the computer model without the need to create a physical model in the laboratory.

8.2.4 THE DATABASE IN DRAFTING

In the late 1960s and early 1970s introduction of turnkey CAD/CAM systems gave drafters a new tool. They were no longer required to draw, evaluate, change, and redraw designs. Instead of erasing and incorporating engineering changes on vellum, they could now recall designs electronically and incorporate these changes more accurately and many times faster. These early systems, however, were primarily drafting systems. Analogous to word processors, the first drawing had to be input from scratch. It was only after the database had been created that real productivity gains were realized. Nevertheless, a revolution was in the making and the drafting department would never be the same.

Today, due to the integration and creation of databases, the drafter is no longer required to input the entire drawing from scratch. With the conceptual engineer's ability to generate the database, the drafter has a starting point. The modern, integrated system allows the drafter to start with the model created by the designer. This model is then dimensioned, toleranced, and notes added, all to company drafting standards. He or she is no longer required to take the output from the designer's computer, which was usually in the form of a huge stack of computer printouts, and regenerate them into input to the drafting system. Elimination of the need to regenerate, or reconstruct, the output of one computer into input to another is where we begin to see truly beneficial productivity increases. This applies not only in the drafting and design areas, but in tooling, manufacturing, production, and QA as well.

8.2.4.1 Graphic Representations

Most of us learned drafting, either in school or on the job, through normal orthographic projections and descriptive geometry techniques. The results are familiar to almost everyone. Visible edges are generally solid heavy lines, while hidden edges are represented as dashed, or broken lines. Section views must be laboriously developed and displayed for clarity, while details and projections are displayed on "Sheet 4 of 6." Although this "wire-frame" method has been in existence for generations, we still create ambiguous designs—designs that look reasonable but are inconsistent and impossible to manufacture. Even attempts at three-dimensional representations cannot always remedy the problem of ambiguity on a drafting board. The paper is two-dimensional, therefore the resulting design must also be displayed in two dimensions.

The early CAD systems, as well as a tremendous number of those sold today, have the potential problem of ambiguity also. Even using three-dimensional databases, as long as the design is displayed with wire-frame techniques, it is possible to create designs that cannot be manufactured. There is an answer, however. Solid modeling, sometimes called volumetric, or three-dimensional modeling, can alleviate, if not eliminate, some of these problems of ambiguity. Though in many cases solid modeling may be difficult to cost justify, both the hardware and software are becoming more price competitive. In a short time solid modeling will probably be within the price range of most companies.

Another advantage of solid modeling is the ability to create photographic-quality representations. Sales brochures and maintenance manuals can be produced before the components have been manufactured. Using the system's color and light source capabilities, lifelike copies are available from the engineering database. We no longer need to rely on artists' renderings or line drawings alone. Further, three-dimensional drawings are now available for shop use. Dissimilar materials, welds, cavities, section cuts, can all be displayed in any view, in three dimensions, and in full color, easing the interpretation requirements for manufacturing personnel. Solid models are also beneficial in eliminating interference and packaging problems. Since two solids cannot occupy the same space, problems are readily identified prior to manufacture and/or assembly.

8.2.5 THE DATABASE IN MANUFACTURING

For many years computer aided manufacturing has meant numerical control. Some of the more sophisticated shops have begun to use the term in reference to shop flow control, and recently it has become associated with the generation of tool paths in a computer aided design system, hence the term CAD/CAM. There is a lot more to computer aided manufacturing than merely shop flow control and numerical control tape generation, however. The integration of engineering and manufacturing technologies can drastically reduce the time a part spends on the shop floor by sharing the database that was created in the engineering department.

Not only is the engineering database utilized in the drafting department, but now the manufacturing, production and quality assurance areas have access to the *exact* data as it was created. No longer do they need to rely on interpretations. Further, when changes or optimizations to the designs are required by these departments, they can now be sent immediately, via data communications, for evaluation by the designer. Changes can be incorporated or rejected before the fact rather than after. Rework, scrap, and overlooked changes on the shop floor potentially can be eliminated. As an aside, "potentially" should be emphasized at this point. Considerable experience will be required before these techniques reach their full potential, as anyone who has ever started up a numerical control operation can attest.

8.2.6 TOOL DESIGN

For an example of the potential an integrated system offers, consider the tool design department. Even with the flexibility in today's turnkey drafting systems, in most instances the tool designers still work from a drawing output by the plotter in the drafting department. Granted, the lines and lettering are clean, neat, and consistent. However, because the turnkey systems are usually limited to the output of engineering drawings, the tool designer still designs the fixture on the drafting board.

Suppose, however, that the tool designer had a terminal connected to the engineering database with which to design the fixture. Suppose, further, that he or she could recall the engineering design electronically on this terminal, and position drill bushings, locators, and clamps without having to physically draw them. Finally, suppose he or she could locate these items in space and build the fixture around the part, freely rotating and moving all components to manufacture the part exactly as it was designed. After all, the same data generated and tested by the design engineer is being used. We are now seeing improvements that will offer truly substantial productivity gains—not only in the engineering and drafting areas, but throughout our factory.

8.2.7 NC-CNC-DNC

It has been written that computers are very fast, highly accurate, but extremely dumb and unimaginative. On the other hand, man is very slow, highly inaccurate, but extremely intelligent and creative. The same analogy can be applied to machine tools. If a machine tool is in good condition it can be fast and accurate but not very creative. Man applies the creativity and the machine tool supplies the speed and accuracy. It is reasonable to expect, therefore, that the further from the actual manufacturing process that we can place the operator, the less chance there is for error in producing a given part.

Numerical control (NC) was a giant step forward in relieving the operator of some of those chances for error. By replacing hand wheels with motors, NC freed the operator of some responsibility in controlling dimensions on the part, as opposed to controlling them via handwheels. The NC machine is considerably more accurate and decisively more consistent than the manually operated machine.

The next giant step forward in the machining process was computer numerical control, or CNC. As a result of the miniaturization of computer hardware, computers were built into the machine controller, resulting in CNC. This enabled the operator to be removed even further from the actual manufacturing process, since there was no need to load and unload tapes between parts or to punch new tapes in order to optimize the program; in addition, the operator was able to communicate with other computers.

Even though, chronologically, direct or distributed numerical control (DNC) was developed prior to computer numerical control, it did not become a truly viable product until after CNC had become a proven technique. The first DNC products tried to bypass the tape reader by transmitting the punched tape image behind the tape reader (BTR) directly from a host computer. Although moderately successful in a few installations, the early DNCs did not have two-way communications and were not readily accepted by industry. It was not until CNCs allowed two-way, computer-to-computer communications that they became a factor on the shop floor.

The maturation of DNC has allowed the operator to become even more removed from the machine tool, further reducing the possibility of human error in the manufacturing process. The more successful DNC systems incorporate a hierarchy of computers. The simplest form would involve a host computer sending machining data to the computers in the machine control unit (MCU), sometimes simply referred to as the "controller." This host computer, depending on the complexity of the part programs, could service several machine tools simultaneously.

In more sophisticated DNC systems, the host computer may pass data to several intermediate, or satellite, computers, which in turn furnish data to the computers in the MCUs. In these hierarchical arrangements, the entire system can expand as new machine tools are brought into service, by the addition of satellites and the expansion of the power of the host, either by increased memory and/or storage or even additional host computer(s). In other words, a DNC system, when properly planned, can grow from the smallest job shop, with one or two CNCs, to the largest manufacturing facility, covering several plants. The expansion capability is practically limitless.

DNC, through evolution from NC and CNC, offers its greatest benefits in removing the operator from the picture almost entirely. The operator becomes a supervisor (similar to the satellite computers) who is no longer concerned with dimensions, feeds, and speeds. The operator is free from turning handwheels and checking dimensions. These error-prone responsibilities are now controlled by the computer, while the operator does what he does best—create.

8.2.8 MACHINABILITY

Another area with considerable room for human error is the selection of optimum feed rates and spindle speeds, which most often are at the discretion of the machinist. Even when the feed rate has been programmed into an NC/CNC system the operator still has control of the manual feed override (MFO) switch and that dial is always easy to find—it is the one surrounded by the silver-colored ring where all the paint has been worn off the control panel!

One of the principal reasons machinability is so difficult for us is the many variables affecting selection of optimum feeds and speeds for any given part. Things like the amount of horsepower available and condition of the specific machine tool, relative to areas like rigidity and worn bearings, and the required surface finish of the part, amount of material to be removed, type of material being machined, tolerance requirements, and cutting tool parameters, all affect feed and speed selection. Further, there are important economic considerations like sacrificing tool life in order to output more parts per hour, vs. fewer parts per hour at a lesser tool cost which must or should be taken into account when selecting feeds and speeds.

Many an argument has developed among operator, set-up man, foreman, and the process planner over what feed rate should be run at a given rpm on a specific part. Prior to the development of computer programs for determining machinability parameters, there were only two methods available: combined experience of the machinist, the set-up man, and others, or preprinted data from handbooks or "slide rules." But these methods are highly susceptible to human error. The handbooks and slide rules, usually written for conservative conditions, meaning suboptimal results, may be good starting points, but still rely heavily on human judgment. While there is nothing wrong with human judgment—it got us where we are today—it does open the door to human error.

Computers, on the other hand, have given us the ability to develop software programs, starting with handbook-type data, that can be altered as required to meet changing variables. For example, the computer can interpolate between depths of cut if the desired depth is not specifically listed in the literature. It then can compensate for insertion of unique conditions applicable to a specific shop or machine tool, thereby customizing the more generalized software. No longer must we wait for years of experience to develop in our employees before they can be expected to produce good parts in minimum time. The computer, with its capacity for memory and recall, allows us to realize faster returns with less experienced personnel.

Further, since we used the computer and the engineering database to determine our cutter path and feed rates, by accumulating this data and adding rapid traverse motions, we know how long the machine should take to produce a part. By inputting these parameters along with cutting tool costs and tool life, we can predict more accurately our production cost per part. Using these data in conjunction with set-up times, batch sizes, raw material costs, overhead, and other pertinent information, we can let the computer determine such things as capacity planning, shop flow, industrial engineering, and parts ordering. Not only are these programs available today, but they are currently being utilized in shops all over the world.

8.2.9 ADAPTIVE CONTROL

Even when using the computer to determine optimum feeds and speeds, there are certain conditions over which we have no control. Hard and soft spots will occur in the piece parts that cannot be predicted. When these are encountered on the machine tool, the feed rate should be slowed down or increased, as conditions warrant, for optimum production. If we are driving a cutting tool up a ramped cut, the maximum amount of material is encountered as the tool enters the workpiece. As the cutter moves up the ramp it encounters less and less material; therefore, the feed and/or speed could be gradually increased as resistance to the cutter decreases.

Adaptive control (AC) is the mechanism by which we vary these cutting conditions. As the name implies, AC is merely a means of adapting, or changing, circumstances to optimize a given situation, and can be applied to many different processes. A thermostat is a good example of AC. As the temperature rises above a set point, either the heat is reduced or the air conditioning started until the temperature drops below another point, after which the heat is increased or the air conditioning turned off. In our perspective, AC monitors one or more cutting conditions, such as heat, force, torque, horsepower, and vibration. As the measured condition(s) exceeds a certain limit the controlled variable(s) is increased or decreased to optimize the cutting process.

One low-cost AC system senses forces at the cutting tool. As the forces on the cutter increase, the feed rate is reduced, and as forces decrease, the feed rate is increased. In spite of its simplicity, this system has documented time savings, when compared to conventional NC and CNC, of greater than 100%, depending on the part, material, and configuration.

In this simple system four sensors are installed in a ring at 90-degree intervals. The ring is installed around the spindle in the machine tool and the sensors are wired to the manual feed override (MFO) switch in the controller. Then, when the machine is in a cutting operation, the sensors detect forces building up at the spindle. These forces can be hard spots in the workpiece, a dull cutter, ramp cuts, or one or more of several other factors. Nevertheless, as the forces increase on the cutter the sensors electronically control the MFO in order to maintain constant force, hence optimum material removal.

Although AC does increase throughput and productivity, one of its biggest advantages is its ability to allow the NC programmer to work to "best" conditions instead of "worst" conditions. For example, if a programmer is facing a casting and the excess stock can vary anywhere from a few thousands of an inch to more than a quarter of an inch, the programmer must assume all castings have the maximum excess. This avoids many broken cutters, but also consumes a lot of wasted time on castings without the maximum excess material. On the other hand, with AC the programmer assumes that all castings have only a few thousands of an inch of excess stock and the AC system allows for machining those with more stock.

8.2.10 PROCESS PLANNING

Computer aided process planning (CAPP) is one of the primary beneficiaries of group technology. Process plans, or routing sheets, have long been developed by manufacturing engineers to direct the flow of a particular part through the production department. Depending on the specific company, these plans may be little more than a "suggestion" as to which machines might be used, or they can be highly elaborate operation sheets, outlining in great detail the methods to be used in the fabrication process. The major goal of the process planner is to develop the optimum flow through the shop and, again, the creation of the plan has evolved into an art form rather than a science.

Today, one of the more common methods for creating the process sheet is to recall a similar part and copy those instructions with a few minor alterations. If a similar part cannot be recalled, it is not unusual to ask one of the more experienced planners for advice. Occasionally it becomes necessary to plan a job from scratch, but that seems to be the exception. In any case, the standardization of routing sheets quite often becomes established by whoever develops the first plan for a totally new part. Other similar parts are then based on this first case. Further, very seldom are existing plans dragged out for optimization; if they produce a good part that is the way of the future for it and similar ones.

The application of computers, however, is gradually creating a scientific method for the evolution of optimal process plans when coupled with group technology and at least portions of the machinability software. With a well-defined GT system, the planner can recall the family of parts to which a part belongs. Then, because all the members of the family have been costed by machinability and industrial engineering software, as previously discussed, we know which routing is the more practical, either from a higher volume, higher cost, or lower volume, lower cost basis. Groover[1] cites an example in which 42 different routings were created for one group of 64 similar parts on 20 different machine tools. Analysis disclosed that all 64 parts could have been built with two different routings, using four machine tools.

Currently, two fundamental techniques are used in CAPP systems. One is a *variant* approach and is a contemporary application of our traditional process planning. In a variant system, group technology and coding and classification techniques are used to categorize parts into families. The families are then grouped by similar manufacturing characteristics so that a "standard" process plan can be developed for each family of parts. Then, when a new part is introduced, the standard plan for that particular family is altered to fit any deviations required to manufacture the new part. The resulting process plan becomes a "variation" of the standard plan, hence the name "variant."

A *generative* approach is closely related to the "expert systems" portion of artificial intelligence. In the generative system, the logic and experience of process planners are captured in the computer and utilized in algorithms, mathematical models, decision trees, etc., to produce optimum plans. The computer actually "generates" the process plan based on part attributes. In such a system a cylindrical part would need to be turned, while a pocket is more apt to be milled, and the generative system creates the process plan accordingly. Needless to say, generative systems are definitely state of the art today and not readily available to smaller shops. Hopefully, however, generative systems are the way of the future due to their abilities to readily take advantage of new manufacturing capabilities as they are implemented in the shop.

8.2.11 FLEXIBLE MANUFACTURING SYSTEMS

So far, we have been discussing those infamous "islands of automation" that are so familiar to all of us. Now we want to start building the bridges to connect those islands. The flexible manufacturing system (FMS) is one of the foremost techniques by which we start pulling all our automation together. FMS utilizes, in one way or another, practically all the previously discussed automation methodologies. The core of FMS is one or more automatic machine tools, either NC, CNC, or DNC, including an automatic tool changer, computer control, and a means of loading and unloading both raw stock and finished parts, generally accomplished through some form of robotics.

By also taking advantage of the other automation techniques, such as machinability and adaptive control, coupled with group technology and process planning, we can start building our unmanned FMS. The CAPP capabilities enable us to determine what machine tools are required to manufacture a family or families of parts. These machines can then be grouped in a "cell" for producing those specific parts. With computer control to monitor the cell, machinability to optimize feeds and speeds, adaptive control to monitor the cutting processes, and robotics to load and unload parts, we have essentially put our operator into the role of supervisor, as opposed to machine attendant.

Once two or more cells are operating, we add means for transferring such components as raw stock, tooling fixtures, tool changers, and finished parts, between these cells by means of computer control, automated storage and retrieval systems (AS/RS), and automatic guided vehicles (AGV). What we are developing is a method for automatically transporting a mixture of parts from the receiving dock to the shipping dock without manual intervention. We have achieved flexibility from our computer system's ability to interchange programs; we have achieved unmanned operations from the automated equipment; and we have generated optimum productivity from our software capabilities in machinability and adaptive control.

Once our manufacturing facility has been fully automated from raw-stock receipt to finished-parts shipment, the next step is the introduction of our design engineering facilities into this information flow. Using the computer's flexibility, the design engineers can load the system with their electronic designs. Further, when the parts lists and bills of material are input to data processing for loading the MRP, purchasing, accounting and scheduling programs, we are quickly approaching the so-called factory of the future or automatic, meaning unmanned or unattended, factory.

8.2.12 LOOKING TO THE FUTURE

One definition of *change* is: "to enter upon a new phase." Perhaps, through engineering and manufacturing integration, we are about to "change" our current methods. Bear in mind, however, that the payback and returns on investment are not as great as we like to see in justifications. Implementation is slow, learning curves are steep. The brutal truth is that we need to accept longer paybacks and smaller ROIs. The good news is that real returns, in both time and profit, are there for those who are willing to make the proper decisions.

REFERENCES

1. Groover, Mikell P., *Automation, Production Systems, and Computer Aided Manufacturing*, Englewood Cliffs, NJ: Prentice-Hall, 1980, pp. 559–560.

CHAPTER 8.3

COMPUTER AIDED PROCESS PLANNING

JOSEPH TULKOFF

Lockheed-Georgia Company
Marietta, Georgia

8.3.1 PROCESS PLANNING

Process planning is that facet of manufacturing engineering concerning the translation of engineering design data into the most efficient method of part manufacture. From process planning flows the plan that translates the engineering concept into a finished part. The Society of Manufacturing Engineers defines process planning as "the systematic determination of the methods by which a product is to be manufactured, economically and competitively." Process plans specify the proper sequence of production operations and the required tools and facilities, using such documents as operations sheets, route cards, and shop orders, all of which transform raw stock into finished parts. As such, process planning requires intimate knowledge of manufacturing processes—machining, forming, painting—plus familiarity with the production capabilities of the specific plant in which a part is to be manufactured.

Process planning, as shown in Fig. 8.3.1, is a key interface with product design and is the lead activity within the manufacturing organization in the general flow of information from product design to the factory floor. Process planners are "knowledge workers," dealing exclusively with information to establish the orderly and efficient "touch" labor functions of tool making, part making, and product assembly.

In the global scheme of manufacturing, it is recognized that any company that manufactures a product composed of discrete parts requires some type of process planning. The *Standard Industrial Classification Manual* (SIC) lists these industries, including, for example, fabricated metal products, machinery, electrical machinery and equipment, transportation equipment, and miscellaneous manufacturing. Embodied within the SIC categories are products such as engines and turbines, farm and garden machinery, construction, mining and material handling, metal working, electrical industrial apparatus, communication equipment, motor vehicles, aircraft and parts, ships and boats, railroad equipment, guided missiles, space vehicles, and many more.

While each of these industries produces different product lines, each product can be decomposed into lower subproduct levels. In the aircraft industry, for instance, a typical lower-level or subproduct breakdown affected by process planning would encompass:

Sheet metal
Nonmetals
Machined parts
Extrusions (nonmachined)
Trim
Tubes
Wiring
Subassembly
Final assembly
Markings and decals
Complete aircraft

The nature of each subproduct design largely dictates the manufacturing process employed. Each subproduct can, in turn, be further decomposed into more specific categories. Inherent in the design of each subproduct

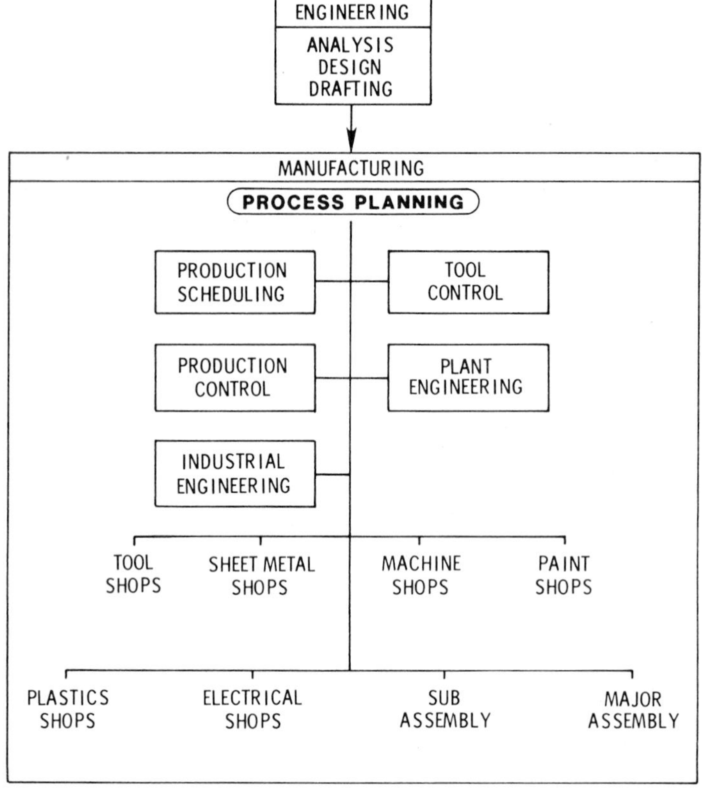

Fig. 8.3.1. Role of process planning within company functions.

are criteria that mandate many of the manufacturing requirements, including facilities and equipment. For example, sheet metal part designs require such equipment as shears, press brakes, hydropresses, routers, and punch presses, while machined parts require machinery such as cut-off saws, tracer mills, vertical mills, lathes, and drill presses. Additionally, many different types of operations can be done on each piece of machinery and in a variety of sequences. Furthermore, each machine usually requires special tooling, uniquely designed for each step in the manufacture of the specific part to be produced.

No matter what the product breakdown, however, the vast spectrum of attendant factory information dictates a step-by-step procedure:

Step 1. Determine systematically all the configuring operations required to produce a part.

Step 2. Identify the ancillary operations required to produce a part.

Step 3. Place all operations in their optimal and logical sequence for the fabrication process.

Figure 8.3.2, an operation sheet, illustrates the completion of these steps. This particular operation sheet is for manufacture of a machined part, bell crank P/N 357544-1. The individual operations with the associated special tooling are exhibited, along with the many machining operations required and the various types of inspection call-outs.

In formulating a process plan, the planner typically must consider the following points when studying the engineering data requirements:

The part number, dash number, and the number of the drawings.

The quantity of parts and the effect on tooling.

The kind, size, shape, and condition of raw material, considering its ability to be machined, formed, shaped, or bonded.

Specific drawing instructions and notes.

Necessary specifications and standards.

```
DPR 77500-10              FABRICATION OPERATION SHEET            08-01-83 09:30

MODEL      PART NUMBER      OPP  M.SER. PLAN  EDIT TYPE MFG TOOLING   DATE  PAGE
C-130        357544-  1                P & S  324             RATE   07-11-83   1

  PART NAME                         GROUP PLANNER NAME/EXT AUTHORITY          STATUS
BELL CRANK                          04.31 CURTIS/DRP 2160  1722-02BA          GENPLN

MAKE FROM NUMBER        CLAD GAUGE/OD WIDTH/WALL LGTH/1ST  LGTH/EAP PTS/BLK INSP
                                .250     1  1/2     2                        QPJM
MATL DESC TYPE MATL  ALLOY CONDITION  SPECIFICATION    MATL SIZE MATERIAL CODE
PLATE     ALUMINUM   7075   T651      QQ-A-250/12        48 X 144  01-0357-020

BOX SIZE: W-    L-     H-       NEST-
REWORK FROM P/N OPP       SUPERSEDED P/N  OPP       SIMO-RELEASE P/N OPP

OP  DEPT CC  LC  TOOL CD TOOL IDENTITY / WORK INSTRUCTIONS    TIME STANDARDS
NO   NO  NO  NO  /MPS NO                                    CODE  SETUP RUN/100

010 1828 35 0622          LAYOUT 1.50  X 2.00               002J    .00     .00

020 1828 35 0380          SAW PER LAYOUT                    308J    .32    2.04

030 1828 35 0627 MPS7253 IDENTIFY M/T A PCT.                223J    .09     .00

040 57                    INSPECT                           REJ      ACPT

                                                         CC TOT     .41    2.04

050 1922 01 0205 JDP-1                  DRILL AND REAM      804J    .41    4.15

                                                         CC TOT     .41    4.15

060 1922 04 0286 MF-1                   PROFILE             289A   1.34    4.28
                          PERIPHERY
                          COMPLETE

                                                         CC TOT    1.34    4.28

070 1922 01 0225          C'SINK 100 PCT X.380 DIA.         804J    .41    2.21
                          NOTE: PART MUST BE C'SINKIN THIS
                          SEQUENCE.

                                                         CC TOT     .41    2.21

080 1922 10 0665 MPS7302 BURR                               290J    .03    2.30

090 1922 10 0696 MPS7257 IDENTIFY TAG AND SEAL              223J    .09     .00

100 57                    INSPECT 100 PERCENT CRITICAL PART REJ      ACPT

                                                         CC TOT     .12    2.30

110 1809 43 0740 MPS1324 A                                  227J    .04     .29

                                                         CC TOT     .04     .29

120 1809 51 0668 MPS7253 IDENTIFY                           223J    .04     .18

130 1809 51 0668 MPS7253 IDENTIFY CAT.                      223J    .00     .09

140 57                    INSPECT HARDNESS                  REJ      ACPT

150 57                    INSPECT 100 PERCENT CRITICAL PART REJ      ACPT

160 57                    INSPECT                           REJ      ACPT

                                                         CC TOT     .04     .27

170                       STOCK
```

Fig. 8.3.2. Operation sheet.

Table 8.3.1. Process Planning Functions

Locating Points for Each Operation
Size and shape of the part
Clamping arrangements
Raw stock or stock in a machined condition
Sequence to machine close tolerances at most opportune time
Close-tolerance holes, machined grooves, recesses, straight machined edges, raw surfaces of castings and forgings
Supports for thin sections
Excess stock for clamping or chucking
Type and Size of Machine for Each Operation
A machine accurate enough to hold tolerances
Methods that will produce desired finishes
A machine that is rigid enough for the job
Number of parts to be produced
Equipment lists for capacities, attachments, and foreign machines
Sequence of Operations
Definition of an operation
Finish specifications, tolerances, locating points, heat treating, roughing operations
Arrangement of all operations in most desirable sequence
Listing enough information to identify each operation
Tooling Necessary for Each Operation
Standard tools (to be used whenever possible)
Special tools required for the project
Types of jigs (drill or weld)
Types of fixtures (mill, lathe, boring, broaching, grinding, saw)
Nonmachining Operations
Assembly
Burring (filing, grinding, belt sanding, tumbling, etc.)
Cleaning (tumbling, degreasing, etching, pickling, sand blasting, etc.)
Heat treating (annealing, normalizing, tempering, and hardening)
Plating (cadmium, zinc, "hard" chromium)
Baking (for plated parts)
Masking (tape, stop-off lacquer, other types)
Surface treatments (anodizing, chromodizing, etc.)
Painting (priming, corrosion resistance)
Stamping (part number and assembly number, rubber-stamp tagging)
Inspection

Heat treatment requirements as they affect machining or forming and sequence.

Finish specifications, such as plating, anodizing, and painting.

Tolerances as they affect equipment, tooling, and sequence of operations.

Machine finishes, since the degree of smoothness sometimes affects processing.

Inspection requirements with consideration given to close tolerance and structural tests.

The part condition required to suit the manufacturing plan to coordinate tooling, hole size, number of holes drilled, holes omitted until assembly.

A mental picture of the completed part, noting the scale, studying all projections, hidden lines, sectional and cutaway views, and checking the next assembly.

Once these areas are taken into consideration, the specific process planning functions can be completed. These functions are shown in Table 8.3.1.

8.3.2 COMPUTER AIDED PROCESS PLANNING

A vast amount of facts and figures evolve in identifying and defining all the activities related to part manufacture. These facts and figures, called data, will become useful information only when they are interpreted, organized, and correlated to show meaningful interrelationships. Since process planning requires consideration of myriad manufacturing aspects, it is one of those activities that can harness the power of the computer to expedite its requirements. In recent years, interest in computer applications has increased as understanding of the computer's capabilities has increased. Process planning procedures that depend exclusively on skilled or trained production specialists are subject to delays, errors, and higher-than-necessary production costs. A key to the development of computer aided planning systems is to organize, rationalize, and standardize the data about parts fabrication, facilities, tooling, and materials into categories and logical relationships. A diagram for describing the functions of a computer aided planning system is shown in Fig. 8.3.3. Since the computer can store vast amounts of factory information and make the many comparisons necessary to achieve an optimum plan, the planner trades the pencil, once used to communicate with the shop floor, for a terminal by which he or she communicates via CRT/computer/printer.

Figure 8.3.4 shows the way in which a computer aided process planning system enhances the flow of information from engineering data to the shop floor. One of the main enhancements is the computer's capability to make the many comparisons necessary to achieve the optimum plan. This particular system, Lockheed's Genplan, is interactive, on-line, and real-time. A blueprint of the part to be process planned is issued by product design. The process planner analyzes the blueprint and assigns a classification code, which is input to the keyboard. The computer creates an approximate process plan. The process planner interacts with the computer to complete a process plan. An interface, computer to computer, allows the newly created process plan to be transferred electronically to the computerized production control functions of scheduling, order writing, and printing of the shop order to the shop floor.

8.3.2.1 Advantages of Computer Aided Process Planning

A Delphi study on manufacturing, conducted by the Society of Manufacturing Engineers (SME) and the University of Michigan, predicts that in the United States, United Kingdom, and Japan, 50% of factories will use computerized process planning by 1990. One reason for this is the growing shortage of experienced process planners. The SME sets the shortage at 50,000 to 100,000 in the United States alone. Competition for these specialists is tremendous as more and more manufacturers emphasize producibility in their plants. Traditionally, it takes five to eight years to develop a machine-part process planner; generally, these people are machinists or mechanics who have risen through the ranks. If the process plan is prepared manually, the process planner is the sole determiner of the plan's effectiveness in achieving the best balance between cost and performance. With the computer, the process planner has the help of the vast amounts of manufacturing knowledge captured in the computer's database. Some of the advanced computerized process planning systems allow novice planners to be nearly as proficient as seasoned planners.

Without computer aided process planning, many opportunities exist to make errors on a single process plan. The computer, however, can have auditing capabilities to check the accuracy of the plan against requirements built into the computer system. For example, in a part using certain types of precipitant hardened steel, an additional 22 steps are required beyond the machining steps. In a fully computerized system, the computer would not release a plan without inclusion of these steps.

The Illinois Institute of Technology Research Institute (IITRI) conducted a survey of both small and large U.S. manufacturers concerning computerized process planning systems for machined parts. Its findings estimated

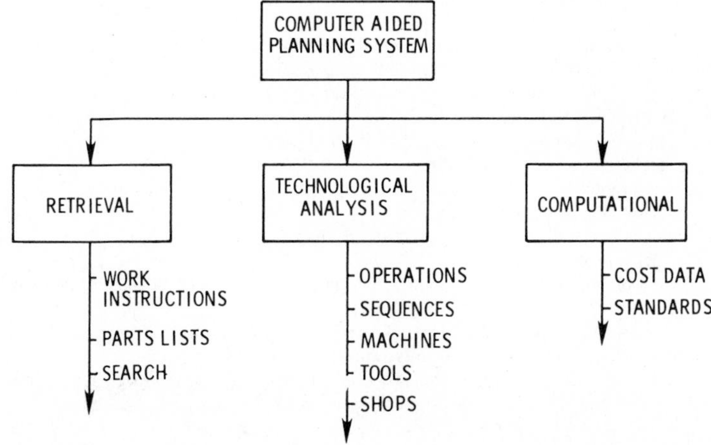

Fig. 8.3.3. Functional diagram of a computer aided planning system.

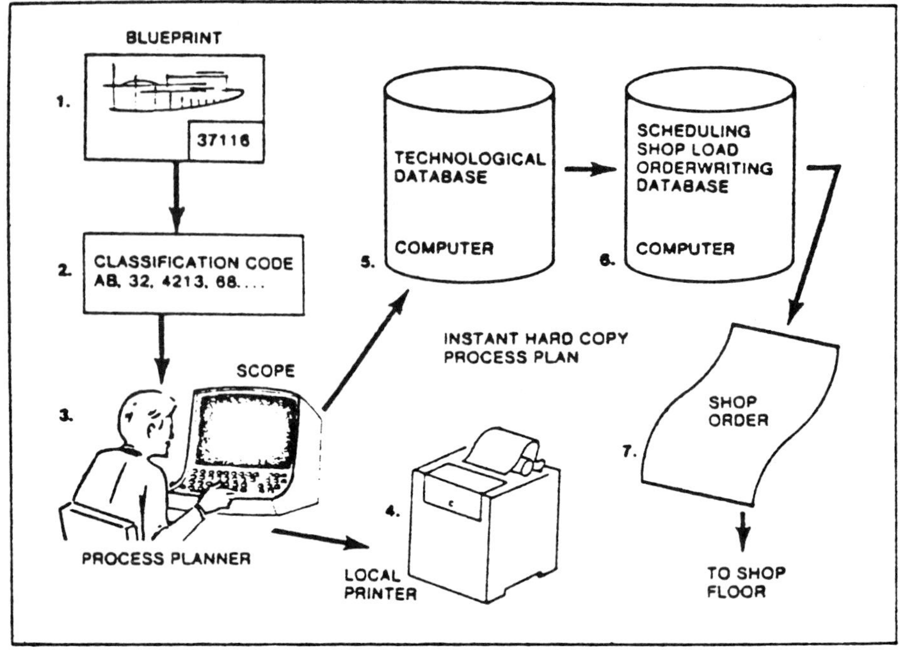

Fig. 8.3.4. Computer-enhanced flow.

labor-hour savings of 58% in process planning, 10% in direct labor, 4% in material, 10% in scrap and rework, 12% in tooling, and 6% in work-in-process resulting from the use of advanced computerized process planning systems for machined parts. The survey also indicated that computer aided process planning benefited many other less quantifiable areas, such as production lead time, machine use, cost-estimating procedures, make-or-buy decisions, production scheduling, capacity planning, plant layout, and quality. In the total cost of a part, process planning accounts for 8%. Computerized process planning, according to the IITRI study, is estimated to reduce this figure by 37.5%.

A brief list of the the advantages of computer-aided process planning is shown in Table 8.3.2.

Table 8.3.2. Advantages of Computer Aided Process Planning

Ability to create or revise process plans rapidly
Reduced novice planner skill and experience level
Immediate access to up-to-date information from a central database
Provision of a basis to introduce group technology into manufacturing
Faster response to engineering changes or changes requested by the shop
Reduced clerical effort to prepare instructions
Use of automatic edits and audits with rules to ensure built-in accuracy
Ability to make mass changes
Ability to print copies remotely
Optimized manufacturing and processing sequences
Selection of the best machine tools to produce high quality parts
Reduction of inaccuracies in manufacturing
Cost savings in process planning, direct shop labor, material, scrap and rework, tooling, and work-in-process inventory
Additional savings in related areas—production lead time, process planning lead time, machine use, product quality, producibility, cost estimating, uniformity, and material handling
Real-time progress and status of process planning work in progress, providing greater management control
Ability to prepare updates and revisions directly on the shop floor through CRTs and printers
Improvement in planning document quality
Standardization of language for work instructions
Technology transfer made possible

8.3.3 NOMENCLATURE RELATING TO COMPUTERIZED PLANNING SYSTEMS

In many computerized planning systems, the planner uses part-classification and coding schemes that enable standard plans to be retrieved. These and other terms need explanation as they apply to most automated planning systems.

8.3.3.1 Classification and Coding Systems

A classification scheme is merely a tool like the computer or the calculator. Its chief use is to group parts into families for definition, retrieval, and analysis. Industry has traditionally used classification and coding systems for many purposes ranging from phone directories to government identification systems, employing color codes, magnetic bar codes, graphic symbols, and the like. In manufacturing, planners classify to create an identification and filing system for the tremendous diversity of substances, objects, concepts, and attributes associated with the manufacturing process. The basic idea of a classification scheme is to allow analysts, in a first step after coding, to get a picture of the total parts spectrum. In the next step, similar parts groups can be put into classes; as a last step, unification can occur. Some computer aided process planning systems use a classification and coding system to retrieve data within the process planning data base.

Types of retrieval code systems include:

Mnemonic. Patterns of words, numbers or letters with a memory device (rhyming pattern, acronym, etc.) for easy recall.

Nomenclature. A naming system.

Numeric. Using numbers.

Alphanumeric. Using a combination of numbers and letters.

One type of coding system, called the Opitz code, is a public-domain code in which the first five digits describe geometric features of a part and the supplementary four digits contain product information such as size, shape of raw material, and tolerances.

The developer of a computerized planning system must either choose one of the available coding schemes like the Opitz code, or develop his or her own method to classify and code his products, and then, by some analytic program, divide those into part family groups.

Whatever scheme the developer chooses, the criteria for any good classification system must include:

Identification and filing capabilities.

Predictive capabilities.

Planning information.

Also, the shorter the code, while still allowing for differentiation in grouping, the more desirable it is for retention, recording, and validation.

8.3.3.2 Part Family

A part family is a collection of related parts that are similar or identical. They may be related by geometric shape and size and require similar manufacturing operations, or they may be dissimilar in shape but related by some common fabricating operations. Parts are said to be *similar* when the type, sequence, and number of operations required to produce the parts are similar. An example of a part family list is shown in Fig. 8.3.5. This part family list is based on the Opitz group technology code and shows parts of similar shape, size, and manufacturing process. The parts shown are from four different aircraft models designed over a 25-year span. The part names vary on the basis of the design function of parts, even though the shapes are similar; therefore, development of part families based on part names is not useful. The grouping of related parts into families is a key to the use of the group technology concept, Section 8.3.3.3.

The methods used to form part families include the following:

Family part matrix. Based on geometric criteria, such as box dimension (length, width, height), slots, holes, flanges, diameters, steps, and radii; and functional criteria, such as tolerances, material, heat treat, and dimensions.

Taxonomy. A classification scheme borrowed from biology, wherein each major division is further divided into a hierarchical arrangement with subprocesses, job classifications, methods, operations, and steps to distinguish and code groups within a subject field.

Production flow analysis. Based on process plan data information using operation sheets as a base of information to identify similarities. (This area is in great need of development in today's manufacturing environment.)

Attribute. An intrinsic quality, character, or characteristic generic to a specific item or group used for identification or grouping.

BUSHINGS AND SPACERS

PART NUMBER	OPITZ CODE NO.		PART NAME
352545-1	10100	0115	BUSHING
352545-2	10100	0115	BUSHING
352545-3	00100	0115	BUSHING
352545-4	10100	0115	BUSHING
352545-5	10100	0115	BUSHING
374582-1	00100	1212	RETAINER
374582-2	00100	1212	RETAINER
375254-1	10100	0115	BEARING
4G13841-103A	10100	0115	BUSHING
1532397-101	10100	1215	SPACER
1532458-101	10100	1115	SLEEVE
JC63-8	00100	1215	SPACER
JF147-6	10100	1115	BUSHING
JL316-1	10100	0112	ROLLER
JL407-1	10100	0115	ROLLER
JL434-30	10100	0115	BUSHING
JL434-33	00100	0215	SPACER
JW90-7	00100	1112	SPACER

Fig. 8.3.5. Part family list.

These types of data retrieval arrays, in combination with the types of retrieval code systems, lead to several types of computerized access methodologies:

Look-up table. Associates process codes with process instruction sets, such as layout, saw, turn, mill, drill, grind, bore, anneal, paint, identify.

Tree structure (process taxonomy). The tree structure or process taxonomy is used to generate process decision tables, and thus plays an important part in the overall process selection method. Plans and subplans are organized in a branching manner such that each component in one subplan contains a reference to a set of more detailed subplans, except when the bottom level is reached.

Decision table. Developed from the process taxonomy. This is done by comparing part characteristics against the process taxonomy in a carefully selected sequence.

One sequence is as follows:

Basic work geometry.

Workpiece size (dimensions, plus volume, perimeter, or cross-sectional area).

Tolerance.

Surface finish.

Material (composition, form, and properties).

Economic lot size.

Production rate.

As each decision table is searched, beginning with the one on basic geometry, certain processes are eliminated, leaving a few. The use of decision table logic permits continued expansion of a system as new processes, equipment, and tooling are developed.

Synthesis algorithm. A part model code that acts against a factory knowledge database to invoke the rules of manufacturing appropriate to the part model.

8.3.3.3 Group Technology

Many computer aided process planning systems are based on group technology principles. Group technology is a very old idea, which takes advantage of similarities in components and processes. In industry, it is used to bring some of the economies of mass production to batch-oriented manufacturing. The concept was first exploited by *S. P. Mitrofanov* who in 1950 began to analyze the equipment and methods necessary to manufacture similar parts. Today, as more and more products come into the marketplace with a decreasing lifetime for each product, a great number of newly designed and fabricated parts and components are being produced; group technology can manage the manufacture of this variety.

One of the basic premises of group technology in factories is that regardless of the number of types of parts that are produced by a company, at some time the different shapes and manufacturing processes of these parts will begin to repeat themselves, although the number of parts continues to increase. The most common approach to the problem of grouping parts with similar shapes is to use a classification and coding system. Since similar parts have the same or a very similar code number, a part family can then be defined as a group of parts with the same characteristics and a common manufacturing method. Figure 8.3.6 shows such a grouping. If a group of parts fits these criteria and is designated as a family, then it is possible to assemble a set of manufacturing instructions—a *standard plan*—that can be used to create a process plan for each member in that family. A standard plan sets up the sequential instructions, including processing requirements, tools, machines, and detailed operating instructions that apply to the part family.

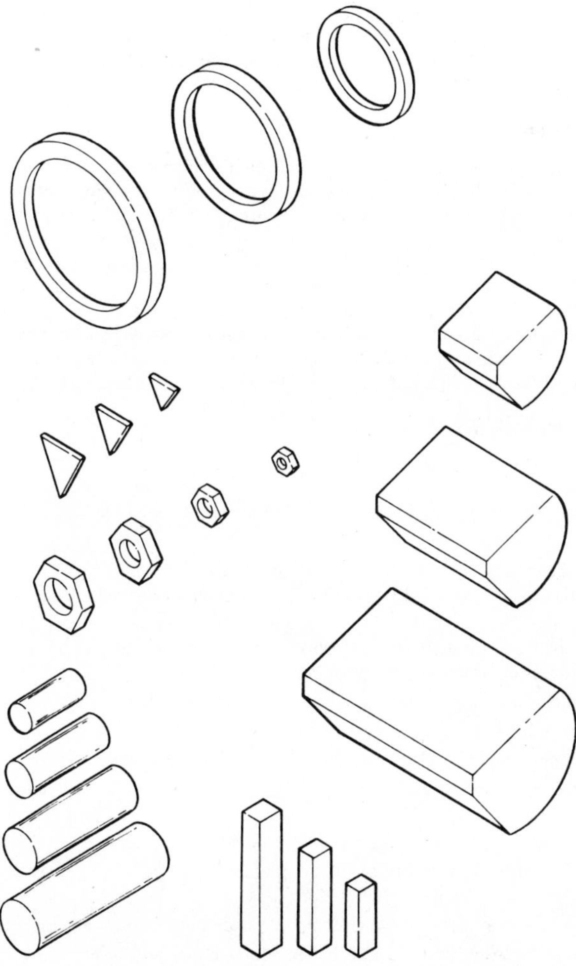

Fig. 8.3.6. Parts family with group technology.

The classification and coding system that allows the computer to group parts into families is the first step required by many systems in moving toward automated process planning.

Two basic group technology coding schemes are in use: hierarchical (monocode) and chain structure (polycode) systems.

Hierarchical systems. These are designed primarily for design and engineering departments. Hierarchical systems group elements in ascending order of specificity according to given principles. Using a numerical code (generally from six to twelve digits), these systems provide the means of coding and classifying drawings for retrieval without unnecessary duplication. In a hierarchical system, the meaning of an individual digit in the part classification number is dependent on the previous digit in the classication number, and does not represent a discrete bit of independent information. The first digit of the number thus defines the type of part, the second clarifies the first digit, the third digit clarifies the second digit, and so forth. Thus, each successive digit identifies increasingly specific subcategories in a family-tree arrangement. In this manner, a large amount of information can be coded into a relatively short number.

Chain structured classification systems. (also called attribute codes). Every digit in a part classification number represents a distinct bit of information, without regard for the previous digit. Every digit in the classification number is a discrete segment of the total part, independent of all others. One digit may be used to define external form, another internal form or shape. By tying the segments together, a complete part can be described, which is internally consistent with the description of all other parts in the system.

8.3.4 TYPES OF COMPUTER AIDED PROCESS PLANNING SYSTEMS

Computer aided process planning systems fall into two categories: variant and generative. *Variant systems*, also referred to as retrieval systems, store standard plans that can be recalled for use of modification. *Generative systems* create new plans on demand, based on the information stored in computer memory.

8.3.4.1 Variant Systems

In some variant systems, standard process plans for each family of parts are developed and filed by a code number, which has been established through group technology principles. Entry of the code number for a proposed new part will cause the system to retrieve the corresponding standard plan. The standard plan can then either be used or modified. Much preliminary work is required to establish the standard plans, categorize the family-of-parts groupings, and establish the classification and coding systems required. Then, the computer, acting more or less as a word processor, helps to assemble and edit an appropriate process for a new part. During the planning stage, various subroutines or canned programs can be called up to help the planner with the many decisions that must be made. All of this takes place in an interactive, conversational mode at a CRT. This approach partially automates the conventional procedure of using an existing plan to produce a new plan. Most computer aided process planning systems are variant.

How to Set Up a Variant System

One step-by-step approach to setting up a variant system is as follows:

1. Analyze a large population of handwritten operation sheets which previously have been used in parts manufacturing.
2. Consider the types of information the worker will need to make the category of parts being analyzed.
3. Evolve modules from a selected population of operation sheets, based on each unique shop as an entity.
4. Examine and analyze operation sheet information from only one shop at a time, rather than the entire fabrication process plan, for the complete part.
5. Establish, standardize, and simplify modules for each process shop.
6. Catalog the modules into a process module code book.
7. Establish a computer database and establish modules in the computer memory.
8. Set up an additional code and listing for single-line operations that do not conform to the configuration of a module.
9. Set up mnemonic codes for identification and retrieval of the modules.
10. Add information about tool orders to the computer data bank to coordinate with the process plan modules through analysis of tool order history files.
11. Add tool order information to the process module code book and to the computer retrieval system.
12. Design a transmittal form or CRT screen input format that will enable the planner to translate the analysis of the engineering drawing from the code book to the transmittal form or CRT input format to the computer.

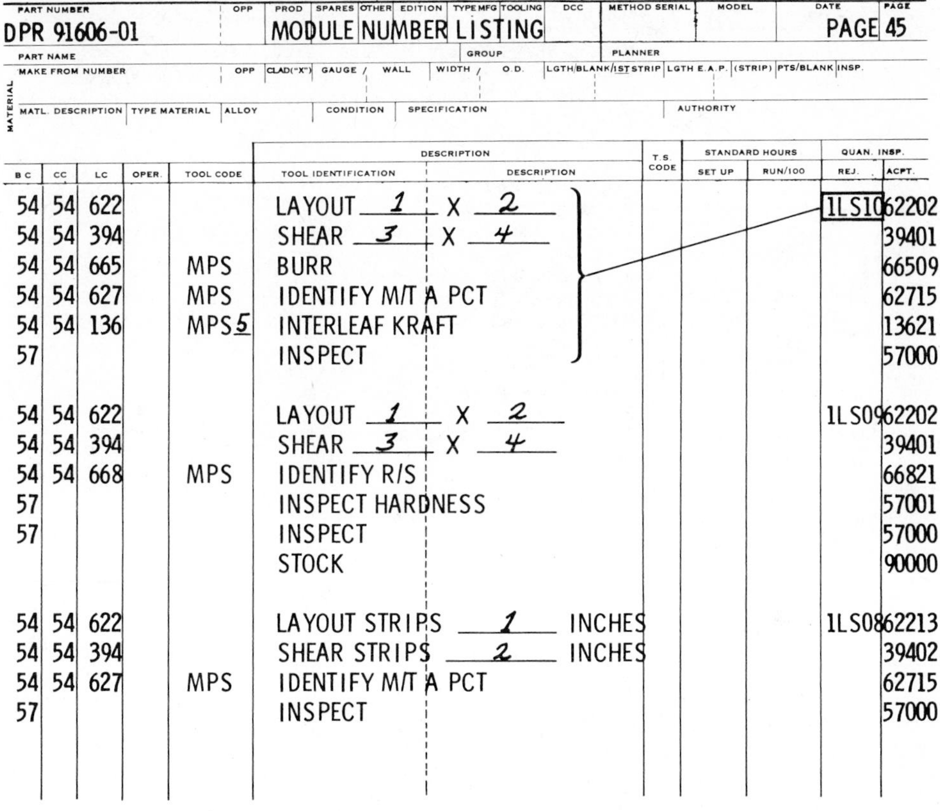

PART NUMBER DPR 91606-01 | OPP | PROD | SPARES | OTHER | EDITION | TYPE MFG | TOOLING | DCC | METHOD SERIAL | MODEL | DATE | PAGE

MODULE NUMBER LISTING — PAGE 45

PART NAME | GROUP | PLANNER

MATERIAL: MAKE FROM NUMBER | OPP | CLAD("X") | GAUGE / | WALL | WIDTH / | O.D. | LGTH/BLANK/1ST STRIP | LGTH E.A.P. (STRIP) | PTS/BLANK | INSP.

MATL. DESCRIPTION | TYPE MATERIAL | ALLOY | CONDITION | SPECIFICATION | AUTHORITY

B C	CC	LC	OPER.	TOOL CODE	DESCRIPTION: TOOL IDENTIFICATION	DESCRIPTION	T.S. CODE	STANDARD HOURS: SET UP	RUN/100	QUAN. INSP.: REJ.	ACPT.
54	54	622			LAYOUT 1 X 2					1LS10	62202
54	54	394			SHEAR 3 X 4						39401
54	54	665		MPS	BURR						66509
54	54	627		MPS	IDENTIFY M/T A PCT						62715
54	54	136		MPS5	INTERLEAF KRAFT						13621
57					INSPECT						57000
54	54	622			LAYOUT 1 X 2					1LS09	62202
54	54	394			SHEAR 3 X 4						39401
54	54	668		MPS	IDENTIFY R/S						66821
57					INSPECT HARDNESS						57001
57					INSPECT						57000
					STOCK						90000
54	54	622			LAYOUT STRIPS 1	INCHES				1LS08	62213
54	54	394			SHEAR STRIPS 2	INCHES					39402
54	54	627		MPS	IDENTIFY M/T A PCT						62715
57					INSPECT						57000

Fig. 8.3.7. Page from process module code book.

13. Program the computer to access these modules in accordance with fabrication logic and produce an operation sheet.
14. Consult with process planners to perform validation exercises to test the system.

Figure 8.3.7 shows a process module for a simple sheet metal part. In addition to the fabrication description steps shown in proper sequence, provision is made for entering part-specific dimensions within selected steps of layout and shear operations. Also shown are the shop and machine identification numbers associated with each process step and the module identification number, 1LS10. The mnemonic code helps the planner identify and memorize information easily. The "LS" stands for Layout and Shear. The "1" indicates that it is a metal preparation operation and the "10" means that this is the tenth module in the LS series. By calling out this code, the planner gets a whole block of associated data. Figure 8.3.8 shows a summary of a fabrication code book's contents and illustrates the structuring of information. Each factory shop is uniquely identified by a cost center number and associated brief descriptor. Keys to module capability for general fabrication processes are indicated by numbers one through eight. Finally, the alpha (two-digit) code identifies specific fabrication processes using the mnemonic code to aid in remembering each process.

Figure 8.3.9 shows the top part of a CRT screen input. On it, the planner fills out the administrative data: the part name, the part number, etc. In the material block, the planner calls out a ten-digit material code. When the material code goes into the computer, it will be translated into a complete raw material description.

Figure 8.3.10 shows the middle section of the CRT screen input format needed to produce fabrication plan details. Illustrated is a string of process module codes depicting the steps for manufacture of a sheet metal part. The codes illustrate the following operations: rout and drill (RD), form brake (FB), degrease/heat treat (DH), and paint booth (PB). Each module code relates to a subset of operations, machines, and tools.

Table 8.3.3 shows additional CRT input format data. Here are indicated many of the relevant factors needed to compute time standards for sheet metal parts. To create the time standards, the computer does many of the computational routines. For the computer time standard on a shearing operation, for example, the planner needs to show such things as thickness, size, and type of material (steel, aluminum, etc.). Different machines have

COST CENTER	FUNCTIONS
15	Mockup-Sheet Metal, Machine Parts, Assembly
33	Sheet Metal-Temporary Tooling, Tooling Not Ready
33	(Saw) Shape, Material preparation-Extru & Bar
41	Shot Peen
42	Finishes and Paint Processing
44	Chemical Milling
45	Penetrant Inspect
48	Plating
54	(Shear) Material preparation - Sheet Metal
55	Rout and Drill - Sheet Metal
57	Brake and Roll Section
58	Drop Hammer and Stretch Form (Sheet Metal), Trim
59	Hot Check and Straightening and Hot Forming Section
60	Punch Press Section
61	Stretch Form (Extrusions) Section
62	Hydro-Form Section
63	Heat Treat and Line-Up
67	Trim and Drill (Extrusions, etc.)
89	Spot Weld Assy. and Processing

KEYS TO MODULE IDENTITY

FIRST CHARACTER (NUMERIC)

1 - Material preparation
2 - Profile and/or hole pattern
3 - Forming
4 - Heat Treat
5 - Finishes and paint
6 - Complete operation sheets
7 - Finish and stock, penetrant Inspect
8 - Sub Assy.

SECOND AND THIRD CHARACTERS (ALPHA) IN MODULES ONLY

BP - Blank and Pierce	PB - Paint Booth
CM - Chem Mill	PD - Profile and Drill
CS - Check and Straighten	PF - Punch - Form
DA - Degrease Age Harden	PI - Penetrant Inspect
DB - Drill - Burr	PL - Shot Peen, Plate
DJ - Degrease - Joggle Hot	PS - Punch - Strippit
DH - Degrease Heat Treat	PU - Punch
DR - Drill - Rout	RB - Rout - Burr
DS - Drill - Saw	RD - Rout - Drill
FB - Form - Brake	RT - Rout
FC - Burr, Finish, Stock	SA - Saw
FD - Form - Drop Hammer	SB - Saw - Burr
FH - Form - Hydro	SD - Saw - Drill
FS - Form - Stretch	SH - Shape
5FS - Finish, Stock	SS - Shape - Saw
7FS - Finish, Stock	ST - Scribe - Trim
FR - Form - Roll	
LSA - Layout - Saw	

Fig. 8.3.8. Fabrication code book contents.

different factors for computing time standards. In this example, all that the planner needs to compute the time standards are the inches required to rout the pin router, the steps involved in hand shearing, the number of holes, the number of inches to be sanded, the box dimension, the quantity of holes, the inches of trim, and the number of strokes. The formulas are stored in the computer to execute the mathematical computations that translate the data into a set of time standards.

Figure 8.3.11 shows a typical process plan produced from the computer system. It calls out administrative information, fabrication instructions, and the setup time and run time standards for each fabrication step, including totals for each shop. Although the computation of time standards can be displayed simply, it can be the most complicated part of the development of computer aided planning systems.

PART NUMBER

PART NAME

PRODUCTION OR SPARES

EDITION NUMBER

DATE

DRAWING CHANGE LETTER

PLANNER NAME

RAW MATERIAL

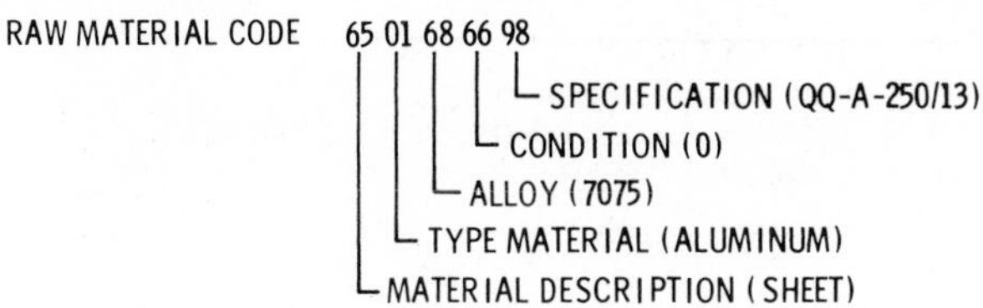

Fig. 8.3.9. Administrative data input.

8.3.4.2 CAM-I CAPP System

The Computer Aided Manufacturing-International (CAM-I) Process Planning Program, a nonprofit research and development corporation dedicated to the advancement of the application of computers to manufacturing problems, developed the CAPP system (Computer Aided Process Planning). More than six versions of this prototype software package designed to provide on-line, interactive creation of process plans based on group technology have been released to date. It is one example of a variant system.

The data structure of CAPP requires six files: part family matrix file, standard sequence file, operation code file, operation plan file, part family setup file, and process plan storage file.

See Fig. 8.3.12, which illustrates the CAPP flow diagram. CAPP operation requires the input of a part classification code into a part family search function, which then interrogates the part family matrix file. If a family match is found, then those data are temporarily stored to allow for the creation of *header data* information.

Header data contain identification information and user-dependent descriptions of the product to be planned. After header data are completed, the system allows retrieval of a standard sequence of user-dependent *operation codes (OPCODES)*. The sequence of OPCODES forms the user-defined standard sequence for the part family. The planner may than edit or modify the data for the particular part being planned.

After the OPCODE sequence has been edited, each operation code may be retrieved from the standard plan data file and edited to be specific for that particular part. The completed information is then stored in the process planning storage file.

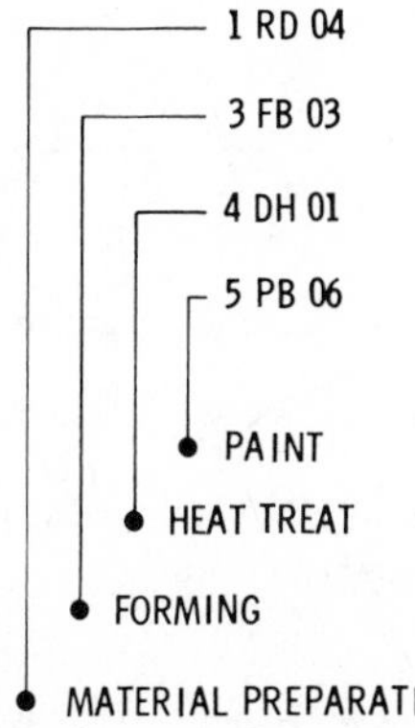

Fig. 8.3.10. Fabrication data input.

Table 8.3.3. Labor Standard Data Input

Cutting (inches)
Mill
Rout
Band saw
Hand shear
File
Shape
Number of Cuts
Table saw
Power shear
Strippet Punch
Number of strokes
Number of punches
Number of adjusts
Power Brake
Number of strokes
Number of bends
Joggles
Number of joggles
Box Dimension
Length, width, height
Holes
Drilled
Laid out

Figure 8.3.13 shows the sequence of events leading to the establishment of the CAPP part family database. As a flow diagram, it illustrates many of the preparatory steps required to install a computer aided process planning system. These steps begin with the selection of a group technology code, familiarization with product parts, and establishment of computer system requirements. The process concludes with standard plans, standard plan retrieval mechanisms, and training for the users of the system.

The CAPP system has the following capabilities:

Part family search.

Part family standardization plans.

Editing.

Instantly accessible storage.

8.3.4.3 Advanced Variant Systems

Advanced variant systems are basically variant but have several quasi-generative features. This type of system works by automatically retrieving existing process plans (based on company experience) that have been stored in the computer. From these, the process planner can make changes, if needed, and then print the final process plan.

The planner can choose one of several approaches to process planning:

1. Creating a plan from scratch using standard process descriptions contained in the system text file. The text can then be assembled and edited for each step of the process.
2. Retrieving an incomplete process plan from the computer and finishing it.
3. Locating an existing process plan by entering the part number, code number, or part family to edit or create a new plan.

The advantage of this kind of system over a basic variant system is that it is possible to start without any of the initial work of classification and coding and without establishing any standard plans. The logic in the software allows the user to back into group technology, establishing codes and plans as the process develops.

```
DPR 77500-10              FABRICATION OPERATION SHEET           08-01-83 09:29

MODEL      PART NUMBER      OPP  M.SER. PLAN  EDIT TYPE MFG TOOLING   DATE  PAGE
C-130       352527-  1                 P & S  316A                  07-11-83    1

  PART NAME                         GROUP PLANNER NAME/EXT AUTHORITY          STATUS
CLIP-DOUBLER INSTL.                 06.21 VAUGHAN           E.J.58776         GENPLN

MAKE FROM NUMBER        CLAD GAUGE/OD WIDTH/WALL LGTH/1ST  LGTH/EAP PTS/BLK INSP
                           X      .051      4          3                        QHPL
MATL DESC TYPE MATL  ALLOY CONDITION  SPECIFICATION     MATL SIZE MATERIAL CODE
SHEET      ALUM      7075   T6        QQ-A-250/13        60 X 180  01-0319-040

BOX SIZE: W-     L-      H-       NEST-
REWORK FROM P/N OPP       SUPERSEDED P/N  OPP       SIMO-RELEASE P/N OPP

OP  DEPT CC  LC  TOOL CD TOOL IDENTITY / WORK INSTRUCTIONS   TIME STANDARDS
NO   NO  NO  NO  /MPS NO                                   CODE  SETUP RUN/100

010 1828 54 0622          LAYOUT STRIPS 3 X 3 3/4  1 PC. +  639J    .24      .00
                          3 1/2 E.A.P.

020 1828 54 0394          SHEAR 3 X 3 3/4  1ST. PC. +     3 639J    .14      .04
                          1/2 E.A.P.

030 1828 54 0627 MPS7253 IDENTIFY M/T A PCT.               223J    .09      .00

040 57                    INSPECT-VERIFY MATERIAL          REJ       ACPT

                                                        CC TOT     .47      .04

050 1807 62 0634 20DM       352527     LR BLANK AND PIERCE  430A    .35      .26

060 1807 62 0500 MPS7302 BURR-BREAK SHARP EDGES            500M    .05     1.40

070 57                    INSPECT                          REJ       ACPT

                                                        CC TOT     .40     1.66

080 1807 57 0428 11A                   PUNCH STRIPPET      336D    .24      .18

090 1807 57 0500 MPS7302 BURR - BREAK SHARP EDGES          502D    .05     2.04

100 1807 57 0603          POLISH BEND RADII                242M    .05      .23

110 1807 57 0425 11A                   FORM--BRAKE         697J    .68      .94
                 MPS7317

120 1807 57 0696 MPS7257 IDENTIFY TAG AND SEAL             223J    .09      .00

130 57                    INSPECT                          REJ       ACPT

                                                        CC TOT    1.11     3.39

140 1809 42 0104 MPS1324 F                                 650J    .06      .67

150 1809 47 0216 MPS1012 1-1                               615J    .07      .54

160 1809 47 0668 MPS7253 IDENTIFY                          223J    .04      .18

170 57                    INSPECT HARDNESS                 REJ       ACPT

180 57                    INSPECT                          REJ       ACPT

                                                        CC TOT     .11      .72

190                       STOCK
```

Fig. 8.3.11. Typical process plan.

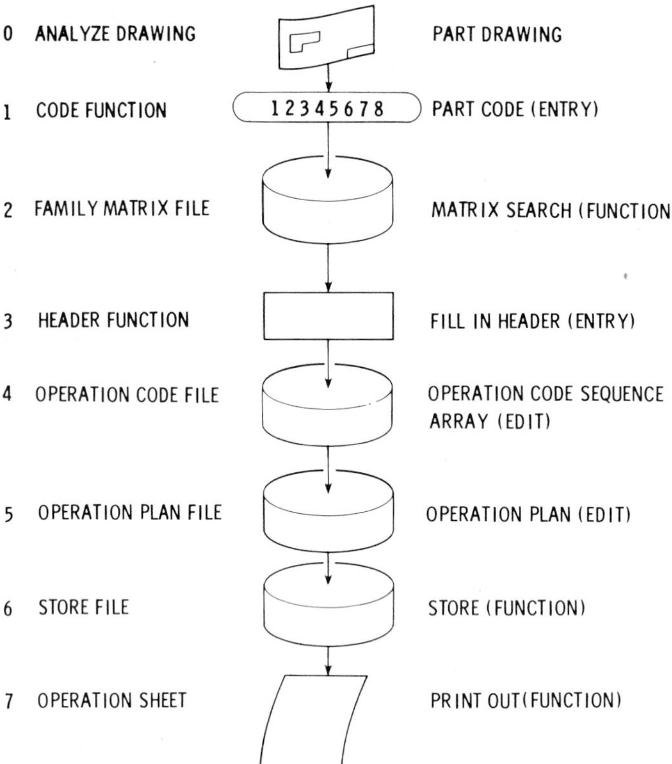

Fig. 8.3.12. CAM-I CAPP flow diagram.

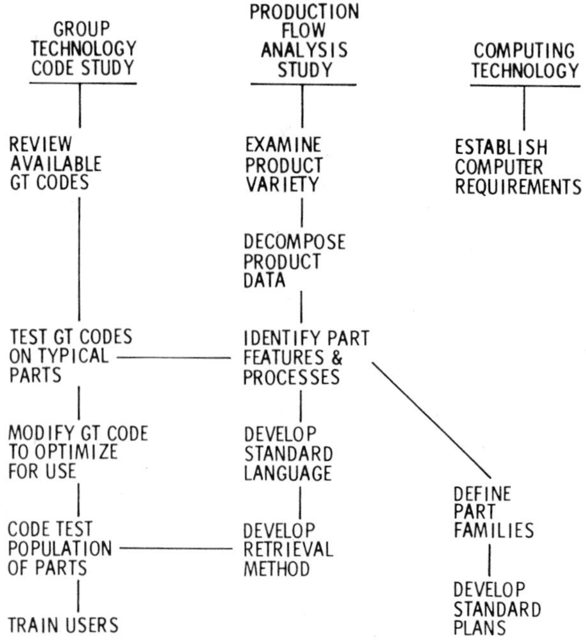

Fig. 8.3.13. Establish CAPP database.

8.3.4.4 Generative Systems

In the generative system, the computer synthesizes a unique part plan using appropriate algorithms that define the various technological decisions that must be made. It is not based on stored standard process plans; it stores equipment capabilities and rules of manufacturing. From these, it synthesizes or generates a specific process plan for a specific part. The synthesis is based on sophisticated decision algorithms, part data, and the manufacturing capabilities available in the shops where the part is to be made. The decision algorithms determine the features of a part and the specific processes required to make that part. The system creates an operation sequence that uses machines and methods available in a particular shop. The planner in such a system performs a monitoring function and only intervenes in some decision conflicts, depending upon part complexity.

In its ultimate realization, such an approach would be universally applicable, allowing any planner to present any part to the system, resulting in the computer's production of a complete plan. Processes would be selected, machines designated, sequences established, and parameters assigned. In actuality, however, decision logic varies from industry to industry and company to company, and theoretical process parameters always need to be adjusted to practical realities. An example from one industry is shown in Fig. 8.3.14, which illustrates the process required of an engineering drawing coded manually by a planner, or automatically via a geometric model and post processor. The code activates the matrix synthesizer software, which includes built-in relationships of manufacturing rules. The matrix synthesizer output acts on a technological information database, thereby providing part-specific and factory-specific information, resulting in a process plan for a specific part.

Most generative process planning systems today are still in the early stages of development and use. Even in these early stages, however, generative process planning systems are capable of part configuration analysis and the instant creation of work instructions for the manufacture of parts. The most sophisticated systems have synthesis capabilities and contain decision logic to capture the manufacturing experience. Generative systems merge the power of the computer with the principles of manufacturing and group technology.

To minimize risks and avoid heavy front-end costs, a generative system must be developed step by step. Using industrial and academic surveys and obtaining substantial user feedback, development can be geared toward a production prototype. This prototype can then be expanded incrementally to address the needs of each of the technological applications with minimum disruption. Such an approach for development provides greater stability and more rapid payback.

The resulting generative system automatically determines the sequence of operations, selects the proper machine tools, and calculates the machining times based on manufacturing logic. Figure 8.3.15 illustrates one

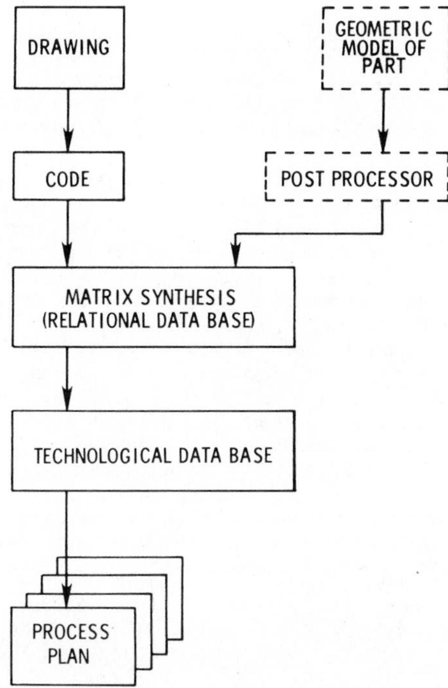

Fig. 8.3.14. Flow diagram (generative system).

Fig. 8.3.15. Planners using Genplan, Lockheed's generative process planning system.

such generative system. With this system, the planner assigns a special part description code. Summarizing the data quickly, the computer evaluates alternatives and makes the best planning decision. When entered by the CRT terminal, the code produces a detailed manufacturing process plan requiring only minor additions. As a result, process plans are consistent not only in methodology but also in sequence, format, and technology.

8.3.5 THE FUTURE OF COMPUTERIZED PROCESS PLANNING

Computerized process planning represents the nucleus of an evolutionary trend toward the automated, integrated factory of the future. In the near future, generative systems will be linked with geometric modeling systems to establish the elusive CAD/CAM link to computer aided design and manufacturing.

Also envisioned are assembly process planning systems with the incorporation of computer aided time standards and computer aided methods to produce assembly tool orders. Process planners will be able not only to create a set of fabrication instructions but also to create assembly job sheets or lists of assembly work instructions given an engineering job assembly drawing and a parts list for a drawing. Using a preplan created by a manufacturing planner as a guide, the planner will be able to analyze engineering drawings, apply group technology codes, and generate assembly work instructions complete with parts lists.

The group technology codes can also describe the assembly conditions and methods needed by manufacturing to meet the engineering and manufacturing requirements. The computer will then analyze these generative codes to synthesize the work instructions via the assembly logic files. Editing can be performed to provide information not included in the generative codes, covering such items as part location number within a product, dimensions, blueprint numbers, tool numbers, production job sheet numbers, item numbers, and the number of mechanical fasteners required. A parts list can be assigned to the appropriate work instructions through a parts list packaging scheme. Final production job sheets will be stored in the computer for time standards to be added, for easy retrieval during revisions, and for release to the production floor.

8.3.6 DEVELOPING TRENDS

8.3.6.1 Expert Systems

Now that state-of-the-art process planning systems are largely or wholly computerized, the future will bring even more computerization. Not only will future systems be able to generate process plans automatically, but they will be able to "think" like an intelligent human being, synthesizing data and making decisions as to the optimum solution to complex problems.

One big bottleneck is the method of accumulating the knowledge required. At present, it is a time-consuming, one-on-one process in which a "knowledge engineer" spends months with an expert to find out how judgments are made. This knowledge is then translated into a computer program, which will be stored electronically in a relational database, a computer archive of relationships between entities. A computer program called the "knowledge-base management system" will update automatically the information in a relational database. A special language will handle interactions between the computer hardware and its stored information. A set of rules will be used to enable the computer to make conclusions when asked a question.

The basis of these decisions will be knowledge engineering and artificial intelligence. Knowledge engineering is the science that captures the data required to make advanced computers and programs operable. Artificial intelligence is the science of making machines accomplish tasks that humans require intelligence to do. Already, Computer-Aided Manufacturing International (CAM-I) has contracted with the United Technologies Research Center (UTRC) and Institut Nationale Polytechnique de Grenoble (INPG) to develop a knowledge-based process planning system. The first stage of the program, completed in June 1984, includes development of an architecture, identification of system functions, and discussion of major design issues. Detailed design of the system, known as XPS-E, began in late 1984.

8.3.6.2 Knowledge Engineering

Knowledge engineering captures the knowledge of shop mechanics, tooling, and engineering—all the knowledge required to make a part—in a computer-sensible way. It is the applied aspect of artificial intelligence that focuses on building expert computer systems, i.e., it is the program or software system to accomplish the task of running advanced computers.

8.3.6.3 Artificial Intelligence

Artificial intelligence requires computers that can think and reason somewhat like human beings and that can understand information conveyed by sight, speech, and motion. The development of reasoning computers that use superspeed symbol manipulation to simulate thought will be based on the brain's inductive thinking powers: the ability to take a series of facts or information and assimilate them into a solution or decision. Already, "knowledge engineers" are building so-called expert systems, which can mimic human expertise in narrowly defined areas. The benefits to be derived from these kinds of computers in the area of process planning are limitless.

8.3.7 CONCLUSION

Because process planning uses the knowledge of many people and processes to make decisions about optimal plans for manufacturing parts, and because it has already benefited from the computer in streamlining process planning, process planning can benefit even more from predicted improvements in computer capabilities, advanced software, and rationalization of factory data. Companies that take advantage of available computer applications, capturing in computerized databases the knowledge required to manufacture parts in their industrial environment, will be in an excellent position to adopt the concepts and applications of expert systems. In so doing, they will be able to stay out front in computerized process planning. Developments in artificial intelligence, knowledge engineering, and expert systems will probably streamline process planning to the point that the computer, when properly activated, will set in motion all the necessary processes required to move automatically from the engineering drawing to the finished part.

BIBLIOGRAPHY

Aerospace Industries Association of America, Inc., *Automated Process Planning Applications*, Project MC 79.7, Washington, DC, Jan. 1979.

Allen, Dell K, "Classification and Coding-Theory and Application." Brigham Young University, Computer-Aided Manufacturing Laboratory, Monograph No. 2, Aug. 1978.

Automatic Software Generation, Artificial Intelligence, and Expert Systems—Special Report, CAM-I Advanced Technical Planning Committee, Aug. 24–26, 1982, Dallas, TX.

CAM-I's Executive Seminar on Coding, Classification, and Group Technology for Automated Planning, Proceedings of a Conference, March 1–2, 1976, San Diego, CA.

CAM-I's International CAM Congress, Proceedings of a Conference, May 14–16, 1974, Hamilton, Ontario, Canada, 1974.

Colding, Bertil, Lester V. Colwell, and Donald N. Smith, *Delphi Forecast of Manufacturing Management*, Society of Manufacturing Engineers and the Univ. of Michigan (Dearborn), 1978.

"Computer Aided Process Planning." IITRI survey, Chicago, IL, 1976.

Dallas, Daniel B., Ed., *Tool and Manufacturing Engineers Handbook*, 3rd ed., Society of Manufacturing Engineers, Dearborn, MI, 1976.

Dunn, Mark S., et al., *Implementation of Computerized Production Process Planning*, Manufacturing Methods and Technology Project 1021. Prepared by United Technologies Research Center for U.S. Army Missile Command System Engineering Directorate, Redstone Arsenal, AL. Report R82-945220-37. Sept. 1982.

Haworth, E. A., "Group Technology—Using the Opitz System," *Production Engineer*, Jan. 1968, pp. 25–35.

Houtzeel, Alexander., "Computer-Assisted Process Planning Minimizes Design and Manufacturing Costs," *Industrial Engineering*, Nov. 1981, pp. 60–64.

Jackson, Richard H., "Automated Process Planning—The Key to Automating Manufacturing Support," SME Technical Paper, 1978.

Mann, Wilbur S., Mark S. Dunn, and Steven J. Pfiederer, *Computerized Production Process Planning*, Prepared by United Technologies Research Center for U.S. Army Missile Research and Development Command, Redstone Arsenal, AL. Report R76-942625-14, Aug. 1977.

Opitz, H. and H. P. Wiendahl, "Group Technology and Manufacturing Systems for Small and Medium Quantity Production," *International Conference on Production Research*, Proceedings of a Conference, Apr. 1970. Birmingham, England.

Peters, J., W. Dumong, and F. Van Dyck, "Group Technology at Work." *American Machinist*, Sept. 16, 1974, pp. 40–43.

"Process Planning," *CAE Annual*, **1**, 1982, 203–207.

Schaffer, George H., "Implementing CIM." *American Machinist*, Special Report 736, Aug. 1981, pp. 151–174.

Standard Industrial Classification Manual, Executive Office of the President, Office of Management and Budget, Washington, D.C., 1972.

Training Manual for CAPP CAM-I Automated Processing Planning, Vol. I of II, TM-77-AMP-01, 1977.

Tulkoff, Joseph, "Process Planning in the Computer Age," *Machine and Tool Blue Book*, Nov. 1981, pp. 74–81.

Tulkoff, Joseph, "Generative Process Planning and Group Technology: Information Control, The Competitive Edge," *Technology Transfer Society Factory of the Future Conference*, Proceedings of a Conference, Feb., 1982, Paris and London.

Tulkoff, Joseph, "Productivity Renaissance through Manufacturing Technology," *Design News*, July 5, 1982, pp. 59–63.

Tulkoff, Joseph, "GT-Based Generative Process Planning," CASA Technical Paper #MS83-915, Publication of the SME, May 1983.

CHAPTER 8.4

COMPUTER CONTROL: SOFTWARE DESIGN

LELAND R. KNEPPELT

Management Science America, Inc.
Winston-Salem, North Carolina

Comments from manufacturing and production personnel indicate that there are crucial problems in the design and operation of computer-based systems. Users do not understand or use much of the output they receive, processing schedules are not met, and data is too old when received. In many cases, a company finds itself inventing problems for the computer to solve based on the existence of the computer rather than the existence of a problem.

In many manufacturing firms, the information systems department is or was embodied within the financial organization. Consequently, the first applications implemented were usually accounting oriented and their orientation was toward automation of clerical tasks. Because of set and often regulated accounting standards, the implementations tended to be straightforward with low risk but low return.

The implementation of systems in support of production and inventory control tends to be of high return, but with an associated high risk. While general techniques such as material requirements planning are evolving into standard practice, they can be used in a variety of ways, depending upon the particular manufacturing environment. Systems supporting the manufacturing environment are more complex and have an orientation toward decision support rather than the automation of clerical functions. Their design and installation requires better communication and an understanding across a broad user base.

The involvement of people in the software design process is critical to a successful installation. The onrushing technology (minis, micros, distributed processing, database, data communications, application software packages) makes many approaches to an effective design possible. However, successful installations have been limited in number due to the lack of recognition of the importance of the human decision-making element in the software design and use.

8.4.1 PROJECT ORGANIZATION

Software in which the ultimate user has taken a direct role in the development has proven historically to be more successful than software that has been designed exclusively by outside sources. The software used in a particular installation may have been purchased from an outside source; however, there is still a design effort for effective utilization within a particular manufacturing environment. A fundamental concept in the organization of a software or systems design effort is that of forming teams within the organization.

The common objective of a software or systems design effort is to make the organization and its people more effective and productive. The organization of teams should allow the manufacturing organization to be conscious of the alternative approaches to the design and use of the software. Some critical questions regarding the project organization include:

Is a key or top management position responsible for the assignment and authorization of the final recommendations?

Does the project organization involve a steering committee and project teams?

What is the composition of the project teams? (Are all impacted user groups involved?)

Is the project planned in clear phases with defined deliverables?

How is the organization informed about the project?

Are special training courses a part of the project?

How will the project be evaluated?

The answer to these questions should provide some prediction on the success of the design and installation effort.

8.4.1.1 Steering Committee

The steering committee provides a format for top management involvement in the design process. Top management needs to communicate to project teams the aims, philosophy, and business goals for the organization. The steering committee is the vehicle for continued participation of top management in defining objectives, providing the means to meet the objectives, and exercising lively and positive control of the project.

The steering committee, which should be presided over by a member of top management, should include senior management whose departments will be involved or impacted in the installation. Their role should be one of providing management insight and operational experience in directing the design effort. In application of new techniques, an outside consultant could be a part of the steering committee.

The general tasks of the steering committee include:

Issuing a directive for the project.

Active participation in all phases of the project.

Assignment of people, providing training and relief from other tasks.

Introduction of project in organization.

Support and authorization of schedules.

Definition of priorities.

Ultimate control on progress, quality, and cost.

Continuation decisions.

Approval and authorization of design.

Review of postimplementation evaluation.

8.4.1.2 Teams on Project

The project team is made up of personnel who constitute the primary users of the resulting design effort. The team members form a working unit to contribute to and to coordinate the accomplishment of the design project. Each member represents the organization in the design process. Team members should be chosen based on a thorough understanding of the organization, its functions, its relationship to other units, its goals and objectives, the informal organization, and any legal or internal policies impacting the organization. Some team members will be full-time on the project, while others need only be available on a part-time basis.

The project team is directly concerned with attaining results from the project. They provide the basic manpower for interviews and investigation of present systems, assigning relative values to requirements and constraints; evaluating alternatives in design and making recommendations, assisting in procedure, forms, method modification and in training programs.

A project team leader or project manager should be assigned. The most capable primary user should be given this position, be relieved from normal duties, and if necessary given suitable training. The project manager should report to the steering committee as well as perform secretarial work for the steering committee. This key individual provides planning, organization, coordination, and assignment of tasks for the project team effort. The project manager should have the skills to encourage input, be a communications link, maintain an awareness of target dates, and generate and sustain active participation and interest. The individual must provide guidance in defining true information and problem-solving requirements. Many users resist changes in current methods.

8.4.1.3 Information Systems Department

A member or members of the project team should represent the information-systems support function. As a part of the project team, the information-systems representative functions as a team member but is also responsible for coordinating the information-systems functions. Participation on the project team should ensure a complete problem definition, with associated information analysis to support the software definition. As actual software development is undertaken, the information-systems representative should coordinate with the team for detailed reviews of the software design as well as supervise coding, testing, debugging, and documentation. The project team should support the systems testing and encourage user activity in the final stages.

8.4.2 SYSTEMS ANALYSIS AND DESIGN

The purpose of the systems analysis and design task is to generate a system specification that will serve as the baseline for further effort. It should clearly provide solutions to all detail problems. The initial step is one of gathering and analyzing data on the existing system and investigating its current use in the environment. The investigation should be carried to a point that will enable the definition of requirements and constraints for the new system. A new system is then defined to meet these requirements and constraints.

8.4.2.1 Problem Definition

Often management initiates a general systems definition of the problem to be solved: "improve profitability." In other cases, the problem may be stated by highlighting a symptom, for example, "I have to know each day where I stand on my production orders." In any case, management, the steering committee, and the project teams must clearly state the proposed objectives of the project. Additional or improved objectives may be defined during the systems analysis and design. However, it is important to document the general objectives and list them in order of priority.

The list of general objectives should then be broken down to a level of detail to allow the measurement of performance. Typical methods of measurement include event logging, attitude survey, rating and weighting, system measurements, and cost/benefit analysis. For each detailed objective, a measurement method, which will be used to predict new system performance as well as cost savings, must be defined.

The main purpose of an explicit statement of objectives is to ensure understanding between management, the steering committee, and the project teams. Once the understanding has been reached, measurement tracking should be put in place to provide a basis for future measurement and comparison. The information from the measurements will assist on critical decisions concerning priorities and project continuation.

8.4.2.2 Study Present Organization

Often, this step in the design process is bypassed, since it is assumed that a detailed study would take too long and that the project team is already knowledgeable on current procedures. Seldom has this proven to be the case in a manufacturing environment. When this analysis has been omitted, the new system presents the wrong type of information, omits required information, allows an existing weakness in the structure of the organization to continue, and provides solutions that do not relate to problem areas.

The intent of this analysis is not to document in detail all existing systems, procedures, operations, and organizations. The analysis should yield sufficient understanding of all relevant present operations to ensure an accurate definition of system requirements. It should serve as a firm basis for the identification of problem areas.

In manufacturing and distribution industries, much of the business is concerned with moving, storing, and controlling material. Consequently, one step in the analysis of the present system is an understanding of the product or material flow. The goal is to become familiar with the production processes in the company's operations as well as to identify specific strengths and weaknesses in the flow of products. Some of the points to cover include: organizational responsibilities of material and production task setting; physical control of inventory points; handling of scrap and rework; and detection of points where trigger actions occur.

A goal in the analysis of product flow is to understand the current information flow, to gain an understanding of documents and other data carriers that currently control the operations. The information flow should be analyzed in relation to the product flow, with particular attention to decision control and feedback points identified in the product flow. Unofficial files, established by personnel for their own use, and verbal information exchanges may indicate deficiencies in the current system.

The study of the information flow is not a trivial effort and it is often difficult to know how to start. One method is to pick the trigger mechanisms to which the organization must respond and follow the information flow associated with the response until all action comes to an end. For example, follow the life of a customer order from receipt in the plant until shipment of product occurs. As the flow is being developed, identify involved departments, obtain copies of all documents and relevant forms, and when possible obtain quantitative data. Be sure to follow not only the mainstream of the information flow but also note where conditional branches occur.

The analysis of the product and information flows, along with the data gathered during their development, provide a basis for evaluating the present system and establishing the requirements and constraints on the new system. When studying the product and information flows, the key areas of interest are their points of contact, since they are normally the key decision-making and control points. Does the information flow provide adequate control, relevant decision support, and accurate feedback?

The analysis steps above should be sufficiently detailed to allow the preparation of detail specifications of the requirements and constraints of the new system. These requirements and constraints should not be based completely on those of the existing system. The aim is to create an entirely separate concept for the new system, not one that is merely a logical extension of the existing system. The basis for the concept of the new system should be developed by research of comparable organizations in the same industry, education on new techniques developed for use in a similar industry, and the potential use of industry standard application software packages.

Organizational, informational, and general requirements and constraints should be defined for the new system. *Organizational* requirements and constraints relate to determining how the existing or future organization will be affected by the new system. They may include recommendations on organization changes. *Informational* requirements and constraints relate to internal or external control information needs, such as legal or corporate policy. *General design* requirements and constraints are those which set limitations or rules for the new system design. They may be expressed as the targeted computer hardware and software configuration or efficiency requirements for the data processing operation.

8.4.2.3 Design New System

The study of the present organization should yield baseline requirements and constraints for the design of the new system. The design of the new system is usually accomplished in two basic steps. First is the definition of "what" the new system will do and provide. Second is the definition of "how" the new system will do it. In actual practice this involves an iteration process between the two steps. The methodology involves building the system top-down by successive refinement. An overall system logical data flow is developed and refined to the detail level. Based on the data flows, the detail of data structures and process logic is defined.

The creation of a logical data flow or information flow is the initial design step for the new system. The basic elements to be shown in the flow are the source or destination of data, the storage of data, processes that transform the flow of data, and arrows which illustrate the flow between the three previous elements. A source or destination of data would be an external entity that either triggers action, such as a customer, or receives system results, such as management. The storage of data might be in any file, accumulation of data might be in a filing cabinet, disk, or microfiche. A process that transforms data is any action point that appends additional data, either through approval or calculations. Finally, connections illustrating the flow between the points are made with arrows and a notation on the reason for flow between points. The objective is to arrive at an overall information flow that can be used to verify the new system from a logical viewpoint as well as be used to begin research on alternative solutions and automation boundaries.

In the information flow, the data storage points have been defined. The next step is to define the data elements present within the data flow. The contents of each data storage point and of the data elements of which they are composed allows the beginning definition of a data dictionary for the new system, which could be developed manually or with one of the many automated data dictionaries. However, at this point the actual data structure within the data stores is not defined.

With each data element in the system defined, the next step is to define what is going on inside each process point. The tools for expressing the process logic include decision trees, pseudocode, structured English, and decision tables. These tools allow definition of the logical decision structure for confirming that the logic expression is correct. For example, in a decision tree the branches of the tree correspond to each of the logical possibilities within a process.

The information flow identified the places where data storage occurred and what data flow resulted between storage points, processes, and sources. The next step is to relate the data flow and the data elements that have been defined in the data dictionary. The goal is to design the logical data stores as simply as possible, given the requirements for immediate access. This is where trade-offs must be made based on the need for on-line access.

8.4.2.4 Data Base Management

The definition of relationship or structure within the data stores should be accomplished with knowledge of the data storage and access technique to be used. Data storage and retrieval could be sequential, random, index sequential, hierarchical, network, or relational. Support for one or more of these techniques is available in many commercial database management software packages. These packages provide physical data processing support for the logical design of the data structure.

Manufacturing systems were the impetus behind the development of database management systems and techniques. The need for a bill-of-material processor was an early requirement, which led to database concepts. Within manufacturing, the need for logical relationships between information on objects which were being controlled has always existed. The most common is the bill of material.

One method of defining required data relationships is to place the groups of data elements within the data stores into one of two categories: (1) data elements which describe or relate to an object, and (2) data elements that describe or relate to the definition of a relationship. For example, customer identification defines an object, the customer. Part identification defines an *object*, the part or material. However, the customer order information describing what has been ordered defines a *relationship*, what part is ordered for what customer. This type of analysis assists in structuring the database.

The use of a commercial database management software package can provide some benefits beyond facilitating physical data storage and retrieval. Most commercial packages also support data security and database recovery, as well as a series of tools for data dictionary, query, screen generation, and report writing.

8.4.2.5 Data Processing Alternatives

Technology is providing many data processing alternatives: minis, micros, and large and medium mainframes. A match must be made between what is currently possible and what is worth doing for the business, as it is being run.

One of the first decisions is the processing mode requirements for a logical activity. Can the processing be done on a batch basis where transactions are captured and the series are processed together? Will on-line support be required? (By on-line, we mean that a direct link is provided for immediate access of information.) Or, does the activity require real-time support? (By real-time, we mean that the system receives the transaction

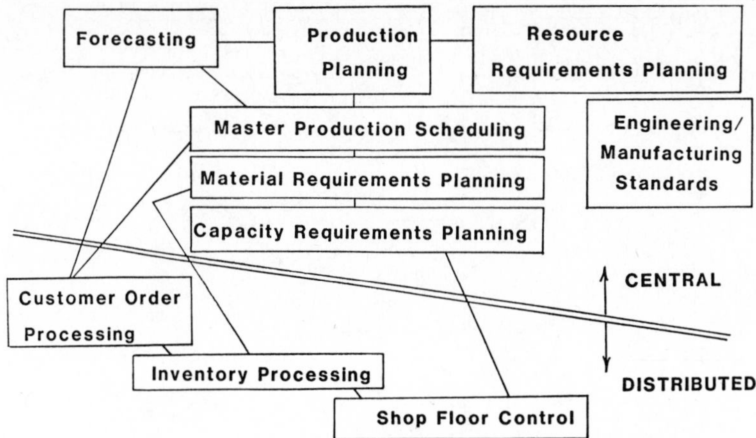

Fig. 8.4.1. A distributed systems approach to manufacturing systems.

data and processes the data fast enough to affect the environment.) An intermediate step between on-line and real-time could be direct update, which receives the transaction data and updates the file but does not affect the overall database.

Another decision relates to computer support. In the past, most computer support was provided through decentralized batch processing. Each function was supported by a stand-alone batch processing capability. While it solved local problems by providing local control, it was expensive due to duplication of equipment, and resulted in slow report consolidation for corporate reporting requirements.

Data processing support then migrated to a centralized approach. This is normally justified on economies of scale (a large computer with cost spread across many users) and on rapid access for corporate reporting. This approach also provides system design standardization and centralized project control. However, it has had only limited success. One reason seems to be the failure of the data processing function to recognize a transfer of some user responsibilities regarding priorities and timeliness. In addition, the centralized approach requires complex mechanisms for bookkeeping, I/O control, and priority within the computer.

With minis and micros, distributed processing has emerged. This approach allocates system functions between multiple computing environments. System functions are placed under local control for determining priorities, timeliness, and data accuracy. A communication link provides rapid access for corporate reporting requirements.

The components of a distributed system are processing sites, communications, database, and system-wide rules. The system-wide rules are based on the approach to distribution of logical processes, which can be based on functional processing requirements or on application needs. For example, a closed-loop manufacturing system may be distributed by accomplishing the planning systems (normally batch) on a mainframe, while the control or feedback applications (real-time) would be done on a mini. Figure 8.4.1 is one example of a functional distribution of applications. In this case, communication of current status from the mini to the mainframe would be done prior to the execution of the planning systems. A continuous direct link between the mini and mainframe is based on the need for immediate access of information in either database.

The distributed processing approach provides some positive features. It makes the user responsible for processing. Consequently, the user has control over scheduling, accuracy, and timeliness. Distributed processing encourages modularity in software design and eases problem identification. It also provides flexibility of technique and location.

Drawbacks of distributed processing include possible duplication of hardware and data and distribution of the support problems. The latter may provide some real management challenges.

8.4.3 MANUFACTURING SOFTWARE REVIEW

A brief review of the major systems and techniques within a manufacturing system will form the basis for describing new software installations.

8.4.3.1 The Planning Systems

Figure 8.4.2 represents a pictorical overview of the planning systems in a closed-loop manufacturing system. Planning starts with the formal definition of the production plan, which is management's primary input to the manufacturing system. Actually, it is a statement of objectives for the plant in terms of output rate and dollars.

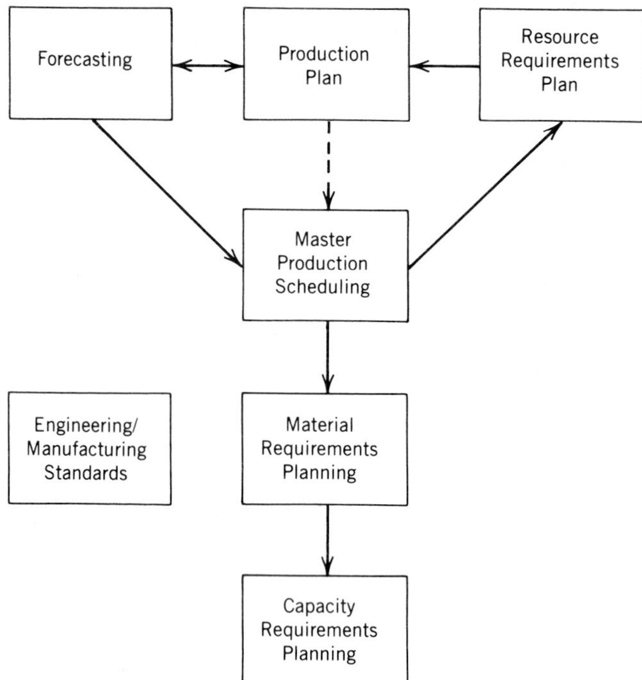

Fig. 8.4.2. Manufacturing planning systems.

The production plan, as well as the master production schedule, are usually based on some form of a forecast. The initial task is to break the production plan down to a master production schedule at the top level of detail planning. Thus, the production plan is a constraint on the master production schedule, since the master production schedule must reflect management's objectives.

The initial check on the master production schedule is to ensure capacity availability at gross level, which is accomplished by using resource requirements planning or some form of rough-cut capacity planning. An overload may require a change in the production plan or a decision to alter the available capacity in long-range planning.

Once the master production schedule has met the planning constraints, it is then used to drive the material requirements planning system. Material requirements planning provides a detailed material plan to support the master production schedule, based on current material availability. It forms the basis for effective priority planning and control during execution of the plan.

Based on the detailed plan from material requirements planning and master production scheduling, capacity requirements planning is done to highlight potential capacity problems at a detailed level. The result may be a change in the overall plan. Capacity requirements planning might also be used as the basis for defining control techniques such as input/output control.

The last system shown in Fig. 8.4.2, Engineering/Manufacturing Standards, is the definition system upon which all planning and control is based. It maintains the material content (bills of material) and capacity usage (standard routing) for every part definition within the plant. It also must reflect future changes in the form of pending engineering changes. This system appears on both Fig. 8.4.2 and Fig. 8.4.3, since both the planning and control systems are based on these essential definitions.

8.4.3.2 The Control Systems

Figure 8.4.3 illustrates the control systems within the closed-loop manufacturing system. The inventory/order processing system provides material availability and timing, which is used in the netting process within the material requirements planning system. Accuracy of inventory balances, order due dates, scrap quantities, and component shortages are essential to an effective material requirements plan. The inventory/order processing system must provide not only for maintenance of this information but also for effective audit capabilities for tracking any discrepancies.

The shop floor control system maintains the detailed tracking of shop orders at the operation-by-operation level. Daily dispatching provides the shop floor personnel with the priorities based upon the detail plan. In addition, the shop floor control information provides current status to the capacity requirements planning system as well as activity tracking to support input/output control.

8.4.3.3 Decision Support Timeframes

The planning and control systems support a wide variety of techniques. It is important to recognize that the overall planning and control process is a continuous cyclic activity. Figure 8.4.4 illustrates the time zones of the levels of material planning and control. In Zone 3, or long-range planning, a "first cut" at developing critical material plans is based on the production plan, using prototype bills of material, "super bills" of family groups. As you move into Zone 2, planning is aimed at breaking down the production plan to the master schedule detail level. This is normally done using a form of super bills of material, since it is still too early to commit to specific configurations of the final product. At some time within Zone 2, the master schedule must be firmed. This point usually is set just beyond the end-to-end lead time for the product or the *planning time fence*. Crossing this fence with a master schedule quantity can trigger material requirements planning and result in procurement of raw material and purchased components.

Once inside the planning time fence, material requirements planning is performed using specific bills of material for the lower levels. Control is accomplished via inventory and order status feedback driven by priorities from the planning systems. Another time fence that comes into play at some time within Zone 1 is the *demand* time fence. This fence is usually set at the point at which the final assembly process begins—the point where the highest stocking level would be pulled. Decisions here must be made on what end item configurations will be assembled either to a customer order or based on a forecast.

On the capacity side of planning and control, Fig. 8.4.5 illustrates the techniques from the perspective of time zones. In the long range, the production plan or master schedule drives resource requirements planning,

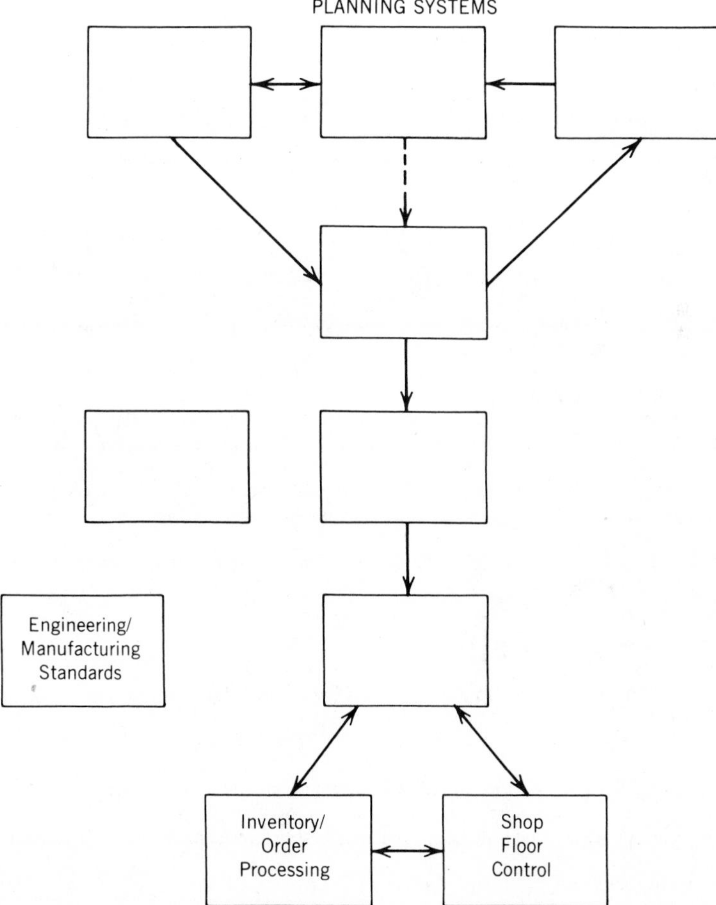

Fig. 8.4.3. Manufacturing control systems.

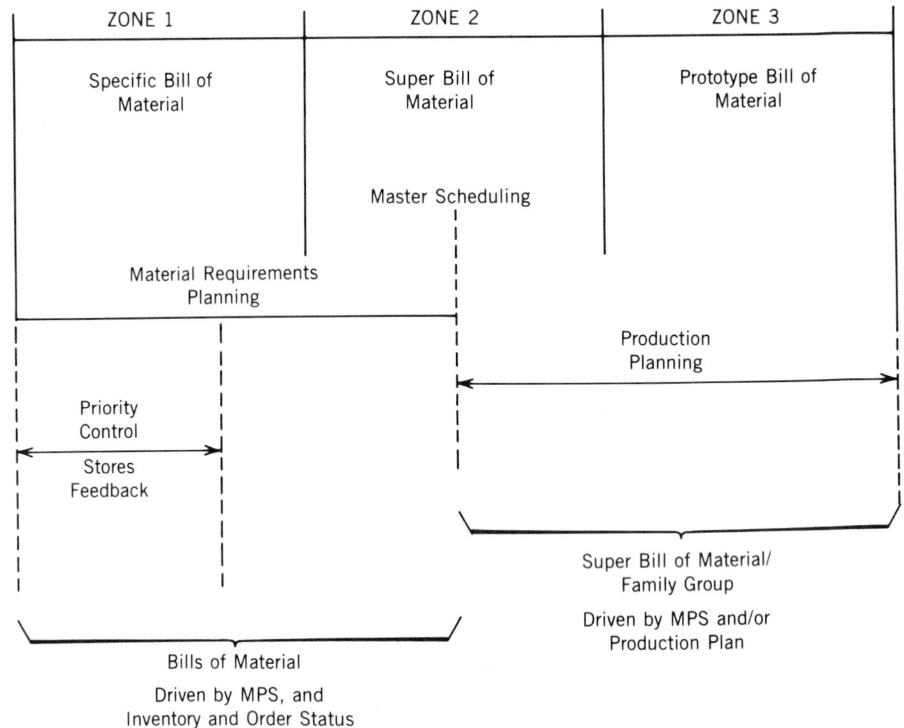

Fig. 8.4.4. Levels of material planning time zones.

using representative routings or bills of labor. The results of this effort should identify potential capacity problems, which may cause a change to the production plan before entry into the material planning systems.

As the time frame moves into Zone 2 and the master schedule is firmed up, capacity requirements planning provides a detailed capacity profile based on the detailed material plan from material requirements planning. Rather than representative routings, capacity requirements uses standard detail routings as well as current work order status from shop floor control.

In the short range, shop floor control provides visibility of capacity problems based on tracking order status at the operation level. This information can be used to generate machine loads as well as input/output tracking.

The review of the closed-loop manufacturing system indicates that the information provided by the overall manufacturing system is aimed at the very essence of manufacturing management at all levels, rational decision making based on the best available data. Success of manufacturing system implementation is usually evaluated using such typical performance measurements as inventory turns and customer service level. Another way of viewing a manufacturing system is that it provides information that conserves other resources through improved decisions.

In brief terms, the fundamentals of manufacturing systems are based on four simple tasks: plan priorities, plan capacity, control priorities, and control capacity.

8.4.4 NEW SOFTWARE INSTALLATION

The majority of functions within a manufacturing system require decision support, rather than automation of clerical functions. This normally means that when the new software is installed it will have an effect on the decision structure of the organization. Some decisions will shift to a higher level in the organization and some will shift to a lower level. The timeframe in which decisions are made may be shortened or lengthened; the level in the organization at which information is accessible and intelligible may change. These are important factors, which must be considered when choosing the approach to use in the installation of new software. Other factors to consider include transaction volume, information restructure requirements, and training.

8.4.4.1 Parallel Processing

Parallel processing in software installation involves executing the transaction activity through both the old and new software. The resultant output from both systems is then compared, to ensure that the new system provides

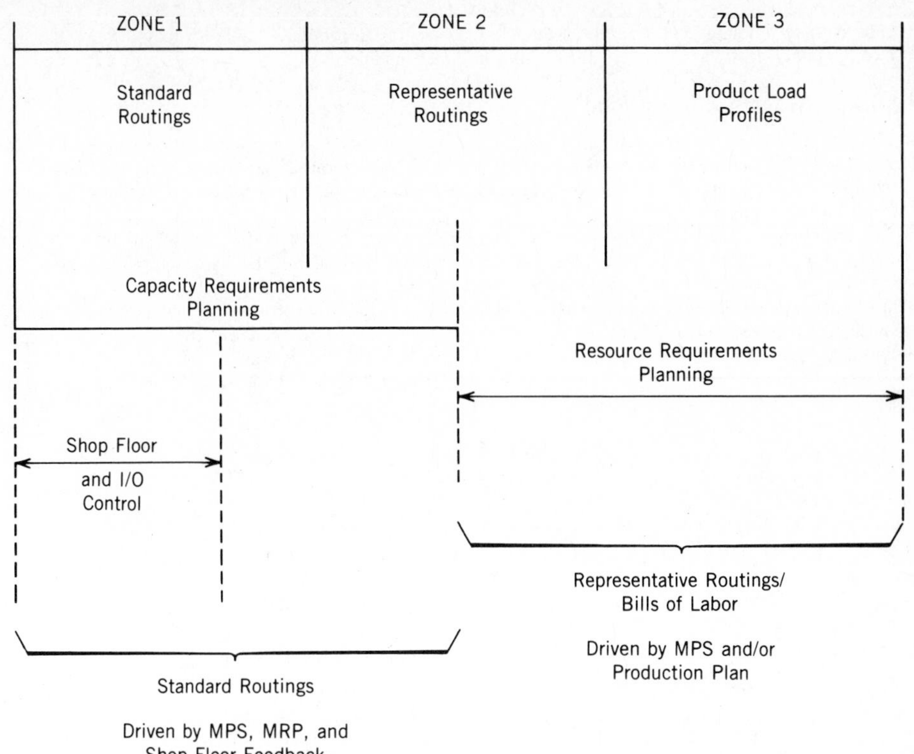

Fig. 8.4.5. Levels of capacity planning time zones.

everything that the old system provides. For accounting applications this approach works well, since much of the output of a system must meet current accounting standards.

Manufacturing systems represent a major opportunity for decision support. The objective is to provide better decision support than was available in the old system. Thus, when we attempt to parallel process the old and new systems, the type of information provided from both systems is often in direct conflict. For example, an order-point-driven inventory system may signal the need to order additional material, whereas a material requirements planning system might project no need for the same material.

Certain functions within a manufacturing system can make use of parallel processing, such as those functions that are more accounting related. An example is the maintenance of on-hand inventory balances in a random location stockroom.

8.4.4.2 Straight Turnover

The *straight turnover* approach to the installation of new software involves turning off the old system and only using the new system. Sometimes called the "cold turkey" approach, this approach prevents the conflicts that arise using parallel processing. However, it requires extensive training for all users of the new system.

At some point, all users must be on the new system. However, a manufacturing system has an impact on most of the organization. In the manufacturing environment, it is difficult to train all users quickly enough to accomplish a straight turnover. It is also difficult to ensure that the system is problem free. When the new software has problems or when information is provided that a user does not understand, the user will revert to what is known, the old system.

8.4.4.3 Phased Turnover

Most manufacturing software is developed as a series of modules, due to the scope of the functional activities. A *phased turnover* approach to software installation involves installing one software module at a time, accomplished in a logical sequence, for example; first the bill-of-material support, then inventory availability support, and finally master production scheduling and material requirements planning. This approach is effective if the overall system utilization has been well conceived.

There are problems with a pure phased turnover. Data established in an initial module might be used to drive logic in a later module. In such cases, there is often limited understanding of the impact. Users often attempt to solve problems with the initial module that need a later module, and also may develop elaborate uses of the initial module, which negates the use of a later module.

8.4.4.4. Conference Room Pilot

The *conference room pilot* approach is more of a quality assurance task than an installation procedure. It is normally used in conjunction with a straight turnover or phased turnover approach to the installation. The conference room pilot approach attempts to capture the benefits of parallel processing—except it recognizes that parallel processing is not realistic.

The initial step is to select a portion of the manufacturing facility or a product line for the pilot operation. The portion selected should have minimum interaction with the remainder of the manufacturing operation. The product line or facility is then converted to the new software (all modules) and all control is accomplished using the new system. The pilot operation will confirm whether the new software is problem free, the training adequate, and will assist in predicting performance under the new system. The term "conference room" implies the physical location of the pilot users, away from their old system and records.

A phased turnover may be required upon completion of the pilot phase due to training and data conversion. Otherwise, a straight turnover may be feasible with timely training.

BIBLIOGRAPHY

Wight, Oliver W., *MRP II: Unlocking America's Productivity Potential*, Williston, VT: Oliver Wight, 1981.

Kneppelt, Leland R., "Implementing Manufacturing Resource Planning/Difficulty of the Task," *Journal of the American Production and Inventory Control Society*, 2nd quarter 1981, pp. 59–77

Hartman, W., H. Matthes and A. Proeme, *Management Information Systems Handbook*, New York: McGraw-Hill, 1968.

DeMarco, Tom. *Structured Analysis and System Specification*, Englewood Cliffs, NJ: Prentice-Hall, 1979.

London, K. R., "Decision Tables," *Auerbach Information Management Series*, 1972

Kahn, B. K. and E. W. Lumsden, "A User-Oriented Framework for Data Dictionary Systems," *Data Base*, **15**(1), 1983, 28–36.

Martin, J., *Computer Data Base Organization*, Englewood Cliffs, NJ: Prentice-Hall, 1975.

Chilson, D. W., and M. E. Kudlac, "Database Design: A Survey of Logical and Physical Design Techniques," *Data Base*, **15**(1), 1983, 11–19.

Martin, J., *Principles of Data Base Management*, Englewood Cliffs, NJ: Prentice-Hall, 1976

Kneppelt, Leland R., "A Simple Distributed Systems Approach to Manufacturing Information Systems," *AFIPS Conference Proceedings*, 1974, pp. 485–490.

Venglarik, James M., "Distributed Processing by Means of Minicomputer Networks," *APICS 22nd Annual Conference Proceedings* (1979), pp. 58–60.

Kneppelt, Leland R., "Real-Time, On-Line, Distributed in the Manufacturing Systems Environment," *APICS 23rd Annual Conference Proceedings* (1980), pp. 18–20.

Wight, Oliver, W., *Production and Inventory Management in the Computer Age*, Boston: Cahners, 1974.

Fogarty, D. W. and T. R. Hoffmann, *Production and Inventory Management*, Cincinnati: South-Western, 1983.

CHAPTER 8.5

COMPUTER CONTROL: HARDWARE DESIGN

JAMES E. HEATON

Oracle, Inc.
Chelsea, Michigan

8.5.1 INTRODUCTION

Computer hardware is a subject that is extremely challenging, especially for a handbook. Because of the dynamic aspects of the subject, it is quite likely that the contents of any chapter treating computer hardware will be obsolete soon after it is written, and more so by the time it is published. Despite the risks of being dated, it is essential that any contemporary treatment of production issues include a coverage of computer hardware; however, the reader is encouraged to challenge the content to ensure its timeliness.

Having made the disclaimer above, it should be noted that in any well-conceived project, a consideration of computer hardware should being with top-down design and application-driven system specifications. Considerable experience suggests that hardware selection must largely be determined by the following.

1. Existing current corporate vendor/brand commitments. These should be honored wherever feasible to reduce spares inventory, operations and maintenance delays, and personnel training costs.
2. Equipment configuration, equipment location and interconnection, and options should be determined by software requirements.
3. Software requirements must be determined by overall system requirements.

This chapter concentrates on these issues and will not address such subjects as project organization and management, application analysis, high-level system design, and software selection and implementation. For consideration of these topics the reader is referred to Chapter 8.4.

This chapter focuses on categories of available hardware, how they can be configured and interconnected, and how best to plan and implement flexible up-datable systems

8.5.2 FUNCTIONAL LEVELS: DEFINED BY INFORMATION FLOW

The traditional approach to computerization envisions a large central staff and equipment complex that attempts to be all things to all users. In reality, this did not work well when originally implemented with the first computer installations of the 1950s and works even less well in the 1980s. Centralized, artificially scarce resources, such as computing power, distort the normal decision-making processes in a line management structure. Extending these distortions to plant floor systems add further confusion and dilution of responsibility to the manufacturing management hierarchy.

If, instead of focusing on computers, one concentrates on operational factors, it is clear that total centralization is only technically essential for airlines reservation systems and other true central database applications. In most other cases the advantages of distributing computing power and databases far exceed the advantages of centralization. This is especially true when the distributed systems match the organizational structure.

In manufacturing, the advantages of decentralization are even more crucial in terms of both employee accountability and motivation. One can identify three general layers of organization and equipment in a modern manufacturing facility, which correspond to three fundamentally different levels of information flow in manufacturing organizations. At one level, groups of people and equipment constitute work cells and/or functional departments. At this level, control equipment interconnection is most economical and therefore most likely to be via brand-proprietary communications interfaces. For example, a department with fifty injection-molding machines, each controlled by a programmable controller, is most likely to have the fifty machines interconnected via a low-level communication network. Likewise a department foreman's industrial-grade personal-computer-based supervisor's workstation is likely to connect to the three or four kinds of equipment in that department via each brand's respective proprietary networks. Thus one personal-computer-like workstation might have one communication network to handle company A's programmable controllers, a second network to handle company B's programmable controllers, and a third network to handle company C's programmable controllers plus other incidental equipment. If and when a machine cluster, a manufacturing cell, or a department is upgraded, the controllers, networks, and workstations are readily replaced within that local area as part of the project.

To interconnect manufacturing cells, machine clusters, and departments, a second level of networking is required. These more general-purpose level two networks are installed as plant information conduits and ideally provide the ability to attach equipment at any randomly chosen location without special procedures or undue complexity. The analogy is that of a power-duct system, into which one can tap for power at any point on the plant floor.

Finally, at the highest level one requires networks to interconnect all major corporate functions, including but not limited to data processing, computer aided design (CAD), computer aided engineering (CAE), computer integrated manufacturing (CIM), materials management, and related functions. This network is the most complex, because it may have to handle simultaneously the transfer of large CAD files and high priority messages about changes in production requirements, customer orders, or other factors.

We will refer to these highest level management functions and networks as Level I, to the intermediate level plant-wide information conduits and management structures as Level II, and to the department or machine-cluster equipment and personnel as Level III. This functional segmentation of operation and computer products is critical to understanding system configuration and equipment selection in industrial automation systems.

The primary responsibility for Level I systems usually rests with the corporate or plant data processing and telecommunications departments. Level II systems are frequently a no man's land sometimes handled by data processing, sometimes by manufacturing. The software selection and management issues discussed in the preceding chapter largely determine the hardware selection and will not be discussed here. The real control requirements start with the Level II networks, Level III networks, and Level III controllers and peripherals.

Before examining specific applications and the associated computer configurations, one must have some insight into the revolutionary changes in available technology and the impact of these changes on proper equipment selection.

8.5.3 HARDWARE/SOFTWARE TRENDS

Taken separately, the recent new technology explosions in hardware, software, and communications seem an orderly development of the immediate past. Taken together, these changes represent a revolution in control, design, implementation, and marketing of industrial (micro) computer based products.

Hardware technology is dominated by three factors. First, the combination of very-large-scale integration (VLSI) technologies and computer aided design will provide increasing use of semiconductor technology as a manufacturing process instead of simply a source of commodity components. This will provide both higher levels of integration and lower power dissipation, unleashing a whole new era of circuit/system designer creativity. These factors represent a qualitative shift in what has become technically feasible over the next decade. In particular, throw-away $400 control computers are now available.

Second, CPUs and memory devices have become true commodities. A few standard CPU designs will dominate "routine" applications. Likewise, standardization of printout and operation of dynamic, static, read-only and electrically erasable, programmable read-only memory (EEPROM) devices has made these into readily available commodities.

Finally, the availability of overwhelming CPU power and memory size has allowed the emergence of more advanced software techniques. C, Pascal, ADA, and possibly Modula-2 represent a new wave of computer programming languages that enhance maintainable design, portability across CPUs, and designed-in real-time capability. ADA and Modula-2, in particular, represent higher level languages that most control engineers will find to be, once they invest the effort to learn them, a major new way of thinking about control.

Portability (the ability to move software from one brand/model of computer to another without rewriting the program) has heretofore been compromised not only by underlying hardware differences, but also by operating system differences. The emergence of UNIX—not so much Bell UNIX, but the conceptual UNIX—provides assurance that a uniform, portable, operating system environment can be implemented across multiple computer brands, models, and architectures of computers.

8.5.4 MAJOR PRODUCT CATEGORIES

In aggregate, these trends offer indefinitely increasing computer power and memory, always implemented with the most cost-effective technology then current, with full protection of vendor and user system and application software investment.

Finally, the Tower of Babel communications chaos is approaching resolution. Ethernet/P802.3, MAP/P802.4, and emerging ISO/OSI upper layer network standardization will allow multiple vendor computers to be networked. This means that several brands of equipment can be effectively integrated into overall control/management systems.

But the realization of the potential offered by these technologies requires products tailored for specific plant floor applications. These products fall into several major categories.

Given the manufacturing company requirements and the available technology, how can these technologies be applied to improve American manufacturing productivity? Basically five complementary classes of products have become available:

Embedded computers
Disguised computers
Command centers
General purpose computers
Industrial local area networks

8.5.4.1 Embedded Microcomputers

Embedded computers are completely hidden from the user. They are programmable for applications (e.g., NC part programming), but not as computers (i.e., they cannot be programmed to perform totally unrelated functions). Thus programmable controllers can be used as robot controllers, but dedicated robot controllers cannot be used as programmable controllers. The "embedded" designation implies a functionally constrained product, programmable in user terms for a limited range of applications. Typical products include:

Computerized numerical control (CNC)
Robot (controllers)
Special equipment (such as gauges, inspection, smart motors, smart instruments, terminals, vision, voice)
Communications Converters (protocol converters, gateways, network interface units)

8.5.4.2 Disguised Microcomputers

Disguised micros are typified by the programmable controller. In fact, an industrial microcomputer is perceived by most buyers as an electronic relay panel, not a computer. This category of disguised computers is particularly important where special application constraints exist:

Where work rules and/or retraining constraints make a visible computer politically undesirable.

Where computer committees (typically data processing oriented and not real-time oriented) review procurements A disguised computer frequently escapes review.

Where a traditional, workable design notation (e.g., relay wiring ladder diagrams, PID loop block diagrams) relates directly to the control problem and general purpose computer languages do not. In this case, which complements the first two, the designer sees a box tailored to his existing conceptual framework. Usually it will be found preferable to a more conventional user-computer interface. As examples, consider:

Visicalc→Accountants
Ladder diagram→Electricians
PID diagrams→Process engineers

In all these cases the simulation of past practice fits the problem better than computer programming languages. In contrast to embedded microcomputers, disguised microcomputers are general purpose in the range of possible control applications. Recent programmable controller product trends have caused a more general recognition that these controllers are really general purpose industrial computers, but many users prefer/need to keep them disguised.

8.5.4.3 Command Centers

The third product category, command centers, consists of vendor proprietary or multibrand networks through which multiple control products are integrated. While possibly programmable in one or more computer languages, the primary programming is interactive specification of the monitoring/reporting/display task to be performed. Typical products are department/process centers including programmable controllers, personal computers, and industrial computers; area centers including large personal and industrial computers; and minicomputers.

8.5.4.4 General Purpose Computers

Fourth, general purpose computers (as seen by the user) will be with us for some time to come, although they will have a declining percentage of the total industrial computer market as they are replaced by the specific application products described above. Used primarily where extensive customization is required, as in large plant-wide systems, these computers plug the gap where preprogrammed solutions do not yet exist. In DNC/FMS systems, general purpose computers are programmed by the vendor/integrator to provide central part program storage and centralized job routing. Future developments will involve dedicated preprogrammed products for most of these functions including super minis, minicomputers, and super microcomputers.

8.5.4.5 Industrial Local Area Networks (ILANs)

The fifth category, industrial local area networks, is the glue that integrates the preceding four product categories. ILANs are unique in two important ways. First, unlike office local area networks, they are designed for the real-time plant floor environment. This means that response time from a message origination at one point to its guaranteed arrival at another point is of greater importance than average network loading. Thus, ILANs frequently have low or very low average message loads in order to assure adequate worst-case real-time response. Other features include ambient electrical noise immunity, guaranteed worst-case delivery, and extensive error checking. In some configurations, physically duplicate networks guarantee delivery even in the presence of cable segment and/or controller failure. Secondly, the diversity of media, controller and command center interfaces, and data rates required on the plant floor assure that there will be no single universal product.

In general, ILANs can be thought of as falling into one of the three management hierarchy levels described above. Level I ILANs are primarily related to higher level management functions and link larger data processing machines and frequently involve interfacility links. Level II ILANs, typically based on IEEE 802.4 broadband cable TV technology, are intended as facility-wide information utilities. These systems can be installed once and expected to meet plant-wide information interconnection requirements for an extended period of time. There is increasing evidence that many companies will provide compatible implementations of the IEEE specification and similar compatible upper layers in the required support software.

Layer III ILANs are intended primarily for extended machine or department level networks. Here the universal ILAN interface cost ($500–$3000) is out of line with the availability of under $500 controllers. Accordingly, many if not most Level III networks will be brand-specific and will accommodate only products of the same brand. Thus a department with three different types of programmable controllers might also have three brand-specific Level III networks hosted from the same department level industrial microcomputer. The industrial microcomputer becomes both the department supervisor's operating console and the link between the various brand-specific Level III networks and the plant-wide standardized Level II network.

8.5.5 LEVEL II/III APPLICATION CONFIGURATIONS

Level III applications involve three different types of control:

- Unit control, for example:
  - Discrete (plastic molding, machine tools, etc.)
  - Batch (dry mixing, batch reactors, etc.)
- Extended control, for example:
  - Paper making machine
  - Conveyors
  - Transfer line
- Facility monitoring/control, for example:
  - Heating, ventilating, air conditioning
  - Energy management
  - Safety/pollution monitoring

In all these applications, dedicated controllers at the lowest functional level are usually best. These are then networked to personal computers (industrial grade where necessary) to provide direct worker/operator and supervisor displays. Generally clusters of like equipment (e.g., a department of identical injection molding machines, a transfer line, and an area-wide conveyor) should use the same brand/model of control equipment and be interconnected via that controller brand network to the personal computer. Where mixed equipment makes up a process (e.g., an FMS cell), mixed brands/models of controllers will be necessary. In this latter case a local, usually panel-mounted, industrial computer (IC) will provide an operator control center as well as supporting multiple different serial protocols to interface with the various local controllers.

In both cases—multiple identical controls on a proprietary network and an IC controller on a heterogeneous equipment set—these department/functional subsystems connect via a plant-wide Level II network to a central data processing computer or to multiple individual department hosts.

To avoid the dilution of line management authority and responsibility inherent in hierarchical central data processing-based systems, the multiple host approach is preferable. In this approach a plant-wide Level II network such as G-NET, IEEE 802.4 broadband, or GM MAPs can be used to allow multiple hosts direct access, on an as-needed basis to all the plant floor ICs and, via the ICs, to individual controllers. Properly configured, the ICs provide the communication interfacing (protocol conversion) necessary to this mixed equipment configuration to function efficiently.

With this approach, numerous additional capabilities can be implemented for relatively low incremental costs. Examples include the following applications.

8.5.5.1 Material Flow Management

Via this system, QQL (quality, quantity, location) of all WIP material can be tracked. Equally important is the fact that process, transit, and idle time can be captured and recorded. Without this data, one frequently does not know where the material flow bottlenecks are.

8.5.5.2 Maintenance Monitoring/Dispatch

Maintenance monitoring uses the network access to all local controllers to identify impending failures as well as to diagnose rapidly actual failures. The logged data is key to identifying maintenance procedure problems, operating problems, and machines in need of overhaul or replacement. The maintenance dispatch function adds the capability to track and prioritize all current assignments for each on-duty maintenance worker. As new alarms arrive, the workers can be reassigned as needed. They usually arrive at the job with the correct tools and parts because of the network-based remote diagnosis.

8.5.5.3 Energy Management and/or HVAC

Dedicated building automation systems have been used for years to control heating, ventilating, and air conditioning equipment for maximum comfort/minimum energy consumption. The industrial systems do these same functions, but also (via the network to the local controllers) minimize process equipment energy consumption. In some systems, other utilities (such as water, sewage, and process chemical flow) are also managed. Some systems produce more than $50,000/month savings.

8.5.5.4 Distributed Numerical Control (DNC)

Instead of independent, free-standing numerically controlled machine tools each equipped with a paper tape reader, newer systems use networked machine tools. Frequently called DNC systems (originally an abbreviation of direct numerical control, more recently of distributed numerical control), these systems use a central computer for part program storage and real-time transmission to various attached NC or CNC controls. The first systems were involved only in part programs and no other functions. Later, more complex, DNC systems use complete computer control of machine tools but no automatic materials handling.

8.5.5.5 Production Monitoring

Production monitoring systems track key operational parameters, via the network, on all production equipment. Typical parameters include total cycles, minimum cycle time, maximum cycle time, mean cycle time, downtime (failure parameters are sent to the maintenance computer), and other parameters designed into each local contoller.

8.5.5.6 Programmable Device Support System (PDSS)

Of all the applications of network interconnected plant floor controllers and workstations, self-maintenance is perhaps the least obvious. Yet as a practical matter, this may be the first, easiest, and most immediately apparent benefit of plant floor ILANs. Imagine a simple system in which a dedicated computer, sometimes referred to as a programmable device-support system, is directly or indirectly networked to all of the plant floor computers, whether they be robots, CNC machines, programmable controllers, microcomputers, or other computer-based plant floor equipment. This support computer stores in its own memory a copy of the correct program for every one of these pieces of equipment and in addition, a copy of the special programs called *diagnostics*, which allow these machines sometimes to identify their own failed components. In addition, this PDSS includes an operator console and the special resources required to create, edit, and test new programs for each of the attached devices. Thus, this becomes a central development, testing, and maintenance point for all the more advanced plant floor equipment.

In the event of a failure, this unit can be used to quickly download, via the network, a replacement copy of the required program when the failed device has been replaced. In the presence of factions not cooperating with inline quality control goals, this system can periodically compare the actual program in each testor with the master copy on its disk. If test parameters have been changed (either through hardware malfunction or local tampering), management can be notified immediately, thus assuring the commitment to proper testing and product quality. From time to time, when equipment functionality must be revised, this tool provides the ability to modify and efficiently test the revised programs with minimum interference with production. Preliminary data from some plants over the last five years suggest that PDSS systems may be able to raise average equipment uptime by 5 to 10%. In some cases, this is equivalent to having installed a facility with 5 or 10% more throughput capacity.

8.5.6 CONCLUSION/RECOMMENDATIONS

Historically, automation projects in the United States have too often been local band-aid attacks on particular problems. Economic and international competitive factors that have become more apparent in recent decades have made it somewhat easier to fund major top-down redesign of complete processes.

But frequently this approach seems too risky. No one really knows how a modern system should be designed. The approach advocated here is:

Top-down design.

Bottom-up implementation.

Layered networks to provide subsystem isolation and interfacing.

These steps allow a learn-by-doing approach to be applied to such projects. In addition, its modular aspect allows future system adjustments to be made without major production disruptions.

CHAPTER 8.6

SIMULATING PRODUCTION SYSTEMS

STEVEN D. DUKET
A. ALAN B. PRITSKER

Pritsker & Associates, Inc.
West Lafayette, Indiana

8.6.1 INTRODUCTION TO SIMULATION MODELING

As the problems associated with production systems continue to grow in size and complexity, the tools of industrial engineering and systems analysis are playing an increasing role in the resolution of these problems. One such tool, computer simulation, is finding wide application for evaluating design options and analyzing control procedures within the production environment. Computer simulation is an analysis tool that can determine the performance of a system given a set of designs or control parameters. With this capability, simulation can be employed at four levels:

1. As an explanatory device to define a system or a problem more clearly.
2. As an analysis vehicle to determine bottlenecks or other critical elements, components, and issues.
3. As a design accessory to evaluate proposed solutions to design or control problems.
4. As a predictor to forecast and aid in planning future developments.

This chapter introduces the concept of computer simulation and the methodology used in performing simulation analysis. Different approaches to simulation modeling are discussed and the major simulation languages available are presented. In addition, the applicability of simulation to production system problems is discussed and examples of actual evaluations are presented.

8.6.1.1 Definition of Simulation

Computer simulation is the process of designing a mathematical/logical model of a production system and experimenting with this model on a computer.[1,2] Simulation encompasses a model-building process as well as the design and implementation of an appropriate experiment involving that model. These experiments permit inferences to be drawn about systems:

1. Without building them if they are only proposed systems.
2. Without disturbing them if they are costly to experiment with.
3. Without danger if the object of an experiment is to determine the limits of performance.

Simulation modeling assumes that we can describe a system in terms acceptable to a computer. In this regard, a key concept is that of "system state description." If a system can be characterized by a set of variables, with each combination of variable values representing a unique state or condition of the system, then manipulation of the variable values simulates movement of the system from state to state. Simulated experimentation involves observing the dynamic behavior of a model over time. Depending on the nature of the model of the system, the observed output processes will be either deterministic or stochastic. Changes in the state of a system can occur continuously over time or at discrete instants in time. Although the procedures for describing the dynamic behavior of discrete- and continuous-change models differ, the basic concept of simulating a system by portraying the changes in the state of the system over time remains the same.

8.6.1.2 Model Building

At the fundamental level, models are simply descriptions of systems. In this abstract sense, our notions about the world are based on models. We all build mental models from which we draw our expectations about our social and physical environment. In science and engineering, these models are formalized with statements in mathematics, logic, and semantics. In order to perform computer experiments with a model, it is generally necessary to have a formal and precise statement of the model. Developing such a model is easier if (1) physical laws that govern the system are known; (2) pictorial or graphic representation of the system can be made; and (3) the variability of system inputs, elements, and outputs is manageable.[3]

The modeling of complex production systems is often more difficult than the modeling of physical systems for the following reasons: (1) few fundamental laws are known; (2) many procedural elements are involved that are difficult to describe and represent; (3) policy inputs that are hard to quantify are required; (4) random components may be significant elements; and (5) human decision making is an integral part of such systems. A simulation approach deals directly with these issues and helps to circumvent many of these difficulties.

Since a model is a description of a system, it is also an abstraction of a system. Model builders must decide on which elements of the system to include. To make such decisions, a purpose for model building should be established. Reference to this purpose should be made when deciding whether an element of a system is significant and hence should be modeled. The success of a modeler depends on how well he or she can define significant elements and the relationship between those elements.

Simulation provides the flexibility to build either aggregate or detailed models and allows differing levels of detail of various system components in the same model. It also supports the concepts of iterative model building by allowing models to be embellished through simple and direct additions.

8.6.1.3 Simulation Modeling Process

As suggested above, the process for the successful development of a simulation model begins with a simple model that is embellished in an evolutionary fashion to meet problem solving requirements. Within this process, the following stages of development can be identified:[2,4]

1. *Problem formulation.* Defining the problem solving objective.
2. *Model building.* Abstracting the system into mathematical/logical relationships in accordance with the problem formulation.
3. *Data acquisition.* Identifying, specifying, and collecting data.
4. *Model translation.* Preparing the model for computer processing.
5. *Verification.* Establishing that the computer program executes as intended.
6. *Validation.* Establishing that a desired accuracy or correspondence exists between the simulation model and the real system.
7. *Strategic and tactical planning.* Establishing the experimental conditions for using the model.
8. *Experimentation.* Executing the simulation model to obtain output values.
9. *Analysis of results.* Analyzing the simulation outputs to draw inferences and make recommendations for problem resolution.
10. *Implementation.* Implementing decisions resulting from the simulation.
11. *Documentation.* The detailed description of the model and its use.

These stages of simulation development are rarely performed in a structured sequence beginning with problem definition and ending with documentation. A simulation project may involve false starts, erroneous assumptions that must later be abandoned, reformulation of the problem objectives, and repeated evaluation and redesign of the model. If properly done, however, this iterative process should result in a simulation model that properly assesses alternatives and enhances the decision-making process.[4]

8.6.2 MODELING APPROACHES

In developing a simulation model, an analyst needs to select a conceptual framework for describing the system to be modeled. The framework or perspective contains a "world view" within which the system functional relationships are perceived and described. This section summarizes the alternative world views for simulation modeling.

8.6.2.1 Discrete Simulation

Discrete simulation occurs when the dependent system variables change discretely at specified points in simulated time, referred to as "event times." The time variable may be either continuous or discrete in such a model, depending on whether the discrete changes in the dependent variable can occur at any time or only at specified times.

As an example of discrete simulation, we will examine the inspection of parts by an inspector at a workstation. Parts come to the inspector, wait for processing if the inspector is busy, are inspected, and then leave the system. Parts arriving when the inspector is busy are queued behind the inspector. To build a discrete simulation of this system, we must define its state and identify the events that can change system status. The system-state

Table 8.6.1. Discrete Simulation of an Inspector

Part Number (1)	Arrival Time (2)	Processing Time (3)	Start Time (4)	Completion Time (5)	Waiting Time (6) = (4) − (2)	Inspector Idle Time (7)
1	3.2	3.8	3.2	7.0	0.0	3.2
2	10.9	3.5	10.9	14.4	0.0	3.9
3	13.2	4.2	14.4	18.6	1.2	
4	14.8	3.1	18.6	21.7	3.8	
5	17.7	2.4	21.7	24.1	4.0	
6	19.8	4.3	24.1	28.4	4.3	
7	21.5	2.7	28.4	31.1	6.9	
8	26.3	2.1	31.1	33.2	4.8	
9	32.1	2.5	33.2	35.7	1.1	
10	36.6	3.4	36.6	40.0	0.0	0.9

variables that must be represented are dependent upon the purpose of the model. If the purpose is to study the effectiveness of the inspector and workstation, then we would need to represent states such as: the inspector's orientation, that is, facing the window or facing the inspection table; the location and state of the inspector's tools; that is, how accurate they are currently; and the inspector's posture, that is, standing or sitting. On the other hand, if the model is intended to study only the efficiency of the inspector in performing tasks, then the representation of the inspector can be aggregated to a single characterization of the time it takes to inspect parts.

For simplicity at this point, we will assume the latter situation. For this case, the state of the system is completely specified by the status of the inspector and by the number of parts waiting for inspection. The state of the system is changed by (1) the arrival of a part and (2) the completion of inspection of a part. To simulate this system we have to generate a stream of part arrivals and their corresponding inspection times—perhaps by sampling from appropriate input probability distributions or reading an input file. Table 8.6.1 summarizes the results of one sample of ten simulated parts.

An event-oriented description of the inspector status and tasks is given in Table 8.6.2. In this table the events are listed in chronological order. A graphic portrayal of the status variables over time is shown in Fig. 8.6.1. The model is discrete because the states of the system change only at certain times. Between these event times, the state of the system does not change. Moreover, when the state of the system does change, it changes

Table 8.6.2. Event Oriented Description of an Operator Simulation

Event Time	Part Number	Event Type	Number in Queue	Number in System	Inspector Status	Inspector Idle Time (min)
0.0		Start	0	0	Idle	
3.2	1	Arrival	0	1	Busy	3.2
7.0	1	Departure	0	0	Idle	
10.9	2	Arrival	0	1	Busy	3.9
13.2	3	Arrival	1	2	Busy	
14.4	2	Departure	0	1	Busy	
14.8	4	Arrival	1	2	Busy	
17.7	5	Arrival	2	3	Busy	
18.6	3	Departure	1	2	Busy	
19.8	6	Arrival	2	3	Busy	
21.5	7	Arrival	3	4	Busy	
21.7	4	Departure	2	3	Busy	
24.1	5	Departure	1	2	Busy	
26.3	8	Arrival	2	3	Busy	
28.4	6	Departure	1	2	Busy	
31.1	7	Departure	0	1	Busy	
32.1	9	Arrival	1	2	Busy	
33.2	8	Departure	0	1	Busy	
35.7	9	Departure	0	0	Idle	
36.6	10	Arrival	0	1	Busy	0.9
40.0	10	Departure	0	0	Idle	

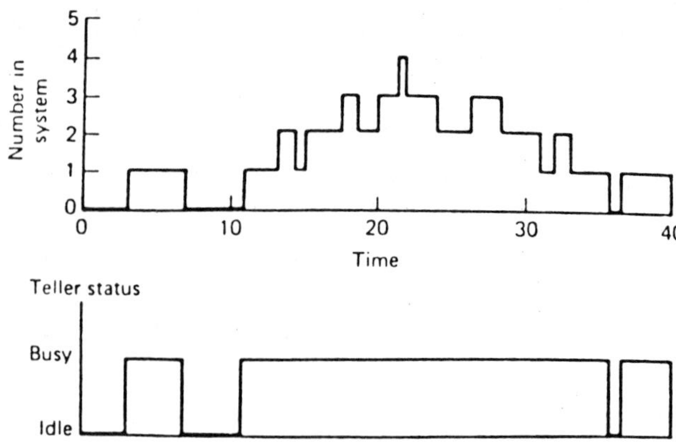

Fig. 8.6.1. Graphic portrayal of inspection process.

in nonzero "jumps." From the data in these figures, performance measures for the system can be calculated. For example, the inspector is idle 20% of the time, the average number of parts waiting and being inspected is 1.45, the average waiting time for parts is 2.6, the maximum wait was 6.9, and the minimum waiting time was 0. This simple example provides a basis for the general discussion of discrete simulation that follows.

The objects within the boundaries of a discrete system, such as people, equipment, and raw materials, are called *entities*. There may be many types of entities, each with its own characteristics or attributes. Although they engage in different types of activities, entities may have a common attribute requiring that they be grouped together. Groupings of entities are called *files*. Inserting an entity into a file implies that it has some relationship to other entities in the file.

The aim of a discrete simulation model is to reproduce the activities that the entities engage in and thereby learn something about the behavior and performance potential of the system. This is done by defining the states of the system and constructing activities that move it from state to state. The state of a system is defined in terms of the values assigned to the attributes of the entities.

In discrete simulation the state of the system (as defined by the modeler) can change only at event times. Since it remains constant between event times, a complete dynamic portrayal of the state of the system can be obtained by advancing simulated time from one event to the next. This timing mechanism is referred to as the *next event approach*, and is used in most discrete simulation languages.

A discrete simulation model can be formulated by (1) defining the changes in state that occur at each event time, (2) describing the activities in which the entities in the system engage, or (3) describing the processes through which the entities in the system flow. The relationship among the concepts of an event, an activity, and a process is depicted in Fig. 8.6.2. An event occurs when decisions to start or end activities are made. A *process* is a time-ordered sequence of events and may encompass several activities. These concepts lead naturally to three alternative world views for discrete simulation modeling. These world views are commonly referred to as *event*, *activity scanning*, and *process* orientations.

In the event-oriented world view, a system is modeled by defining the changes that occur at event times. The task of the modeler is to determine the events that can change the state of the system and then to develop the logic associated with each event type. A simulation of the system is produced by executing the logic associated with each event in a time-ordered sequence.

To create a simulation of an inspector problem using the event orientation, we would maintain a calendar of events and cause them to be executed at the proper points in simulated time. The event calendar would initially contain an event notice corresponding to the first arrival event. As the simulation proceeds, additional arrival events and end-of-service events would be scheduled onto the calendar as prescribed by the logic associated with the events. Each event would be executed in a time-ordered sequence, with simulated time being advanced from one event to the next.

If the modeler employs a general-purpose language such as FORTRAN to code a discrete event model, then a considerable amount of programming effort will be directed at developing the event calendar and a timing mechanism for processing the events in their proper chronological order. Since this function is common to all discrete event models, most simulation languages provide special features for event scheduling as well as other functions that are commonly encountered in discrete event models.[1,5]

Another way to view discrete events is with a process-oriented approach, for example, the process of entities arriving and waiting for processing by a server. The logic associated with such a sequence of conditional activities can be generalized and defined by a single statement, say "queue." A simulation language could then translate the statement into the appropriate sequence of events. A process-oriented language employs such statements to model the flow of entities through a system. These statements define a sequence of events that is automatically executed by the simulation language as the entities move through the process.[6]

The process orientation provides a description of the flow of entities through a process consisting of resources. Its simplicity is derived from the fact that the event logic associated with the statements is contained within the model description.

In the activity scanning orientation, the modeler describes the activities in which the entities in the system engage and prescribes the conditions that cause an activity to start or end. The events that start or end the activity are not scheduled by the modeler, but are initiated from the conditions specified for the activity. As simulated time is advanced, the conditions for either starting or ending an activity are scanned. If the prescribed conditions are satisfied, then the appropriate action for the activity is taken. To ensure that each activity is accounted for, it is necessary to scan the entire set of activities or conditions at each time advance.

For certain types of problems, the activity scanning approach can provide a concise modeling framework. The approach is particularly well suited for situations where an activity duration is indefinite and is determined by the state of the system satisfying a prescribed condition. However, because of the need to scan each activity at each time advance, the approach is relatively inefficient when compared to the discrete event orientation. As a result, the activity scanning orientation has not been widely adopted as a modeling framework.

8.6.2.2 Continuous Simulation

In a continuous simulation model, the state of the system is represented by dependent variables that change continuously over time. To distinguish continuous change variables from discrete change variables, the former are referred to as "state variables." A continuous simulation model is constructed by defining equations for a set of state variables whose dynamic behavior simulates the real system.

Models of continuous systems are frequently written as differential equations, because it is often easier to construct a relationship for the rate of change of the state variable than to devise a relationship for the state variable directly. For example, our modeling effort might produce the following differential equation describing the behavior of the state variable s as a function of time t together with an initial condition at time 0:

$$\begin{aligned} \frac{ds(t)}{dt} &= s(t) + t \\ s(0) &= k \end{aligned} \tag{8.6.1}$$

The simulation analyst's objective is to determine the response of the variable s over a specified time period.

In some cases it is possible to determine an analytical expression for the state variable s, given an equation for ds/dt. However, in many cases of practical importance, an analytical solution for s will not be known. As a result, we must obtain the response s by integrating ds/dt over time using an equation of the following type:

$$s(t_2) = s(t_1) + \int_{t_1}^{t_2} \left(\frac{ds}{dt}\right) dt \tag{8.6.2}$$

How this integration is performed depends upon whether the modeler employs an analog or digital computer.

An analog computer represents the state variables in the model by electrical charges. The dynamic structure of the system is modeled using circuit components such as variable resistors, capacitors, and amplifiers. The

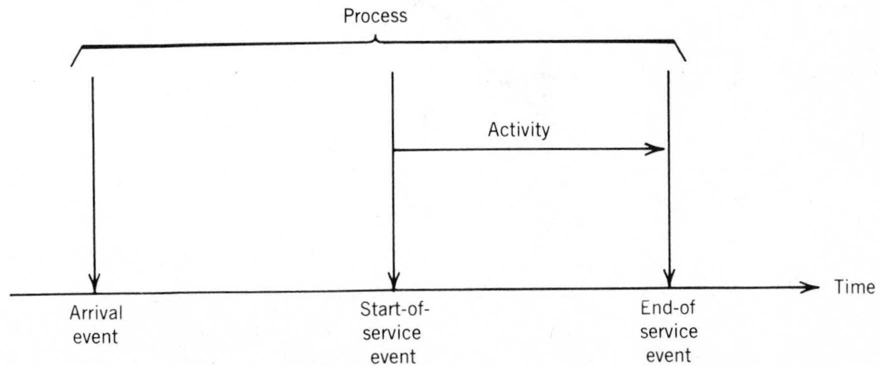

Fig. 8.6.2. Relationship of events, activities, and processes.

principal shortcoming of an analog computer is that the quality of these components limits the accuracy of the results. In addition, the analog computer lacks the logical control functions and data storage capability of the digital computer.

A number of continuous simulation languages have been developed for use on digital computers. A digital computer performs the common mathematical operations, such as addition, multiplication, and logical testing, with great speed and accuracy, and it uses numerical methods to perform the integration operation required in continuous simulation. These methods divide the independent variable (normally time) into small slices referred to as *steps*. The values for the state variables requiring integration are obtained by employing an approximation (normally a Taylor series) to the derivative of the state variable over time. In this situation, there is a trade-off between accuracy of state varible calculations and computer run time. A description of the various numerical integration algorithms can be found in introductory texts on numerical analysis.[7]

Sometimes a continuous system is modeled using difference equations. In these models the time axis is decomposed into time periods of fixed length h. The dynamics of the state variables are described by specifying an equation that calculates the value of the state variable at period $k + 1$ from the value of the state variable at period k. For example, the following difference equation could be employed to describe the dynamics of the state variable s, where r is the rate of change of s.

$$s_{k+1} = s_k + rh \tag{8.6.3}$$

When using difference equations, the essential structure of a continuous simulation model is reflected in the relationship between the rate r projected to period $k + 1$ and the value s of the state variable at period k. Note the similarity between Eqs. 8.6.2 and 8.6.3.

In combined discrete-continuous models, the independent variables may change both discretely and continuously. The world view of a combined model specifies that the system can be described in terms of entities, their associated attributes, and state variables. The behavior of the system model is simulated by computing the values of the state variables at small time steps and the values of attributes of entities at event times.

Two types of events can occur in combined simulations. *Time events* are those commonly thought of in terms of discrete simulation models. In contrast, *state events* are not scheduled, but occur when the system reaches a particular state. For example, as illustrated in Fig. 8.6.3, a state event could be specified to occur whenever state variable X crosses state variable Y in the positive direction. Note that the notion of a state event

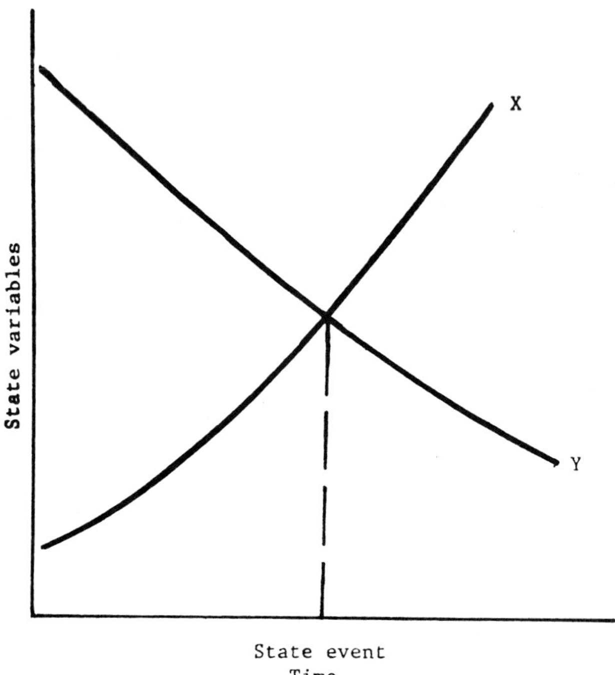

Fig. 8.6.3. Example of state event occurrence.

is similar to that of activity scanning in that the event is not scheduled, but is initiated by the state of the system. The possible occurrence of a state event must be tested for at each time advance in the simulation.

Simulations of production systems frequently, but not always, involve combined discrete-continuous modeling. Systems involving the flow of liquid or the flight of aircraft and systems for which we want to represent the usage of energy are the most common. For such cases, decisions and control inputs (both continuous and discrete controls can be represented) are usually modeled as discrete events and the system responses (levels of materials or positions) are represented as state variables. State events usually trigger control operations. For example, when a crude tank is filled, closing the input valves will be triggered.

The next section discusses simulation languages that have proven effective in modeling production systems.

8.6.3 SIMULATION LANGUAGES

The widespread use of simulation as an analysis tool has led to the development of a number of languages specifically designed for simulation. Shannon[2] has identified the following advantages of using a special-purpose language when performing a simulation study:

1. Reduction of the programming task.
2. Guidance in concept articulation and model formulation.
3. Aid in communication and documentation of the study.
4. Flexibility in embellishment or revision of the model.
5. Provision for the common support functions required in any simulation.

Emshoff and Sisson[8] list the following support functions as requisites for any simulation language:

1. Generation of random variates.
2. Management of the simulated clock.
3. Collection and recording of output data.
4. Summary and statistical analysis of output data.
5. Detection and reporting of error conditions.
6. Generation of standard output reports.

Simulation languages can be categorized as either general or special purpose. General-purpose simulation languages have the capabilities discussed above, but no assumptions are made regarding the characteristics of the systems on which they will be used. Special-purpose simulation languages, on the other hand, have features in addition to those above, which facilitate modeling a particular type of system. As an example, consider a program for solving sets of ordinary differential equations. Such a program would be "general," since it can be used to solve any set of differential equations. On the other hand, a program designed to solve the equations of motion of spring mass systems would have a "specialized" capability because it can address only a restricted class of problems. However, the special-purpose program can address specific analyses relative to spring mass systems and incorporate stylized output reports.

Special-purpose simulation languages have been written to model batch manufacturing operations, safeguard systems, aircraft operations, and other areas. Currently, general-purpose simulation languages are the most applicable to production systems. Numerous discrete, process-oriented, activity scanning, continuous, and combined general-purpose simulation languages exist. Only a few will be discussed here. The General Purpose Simulation System (GPSS)[9] and Q-GERT are process-oriented simulation languages for modeling discrete systems. The principal appeal of GPSS is its nonprogramming orientation. A GPSS model is constructed by combining a set of standard blocks into a block diagram that defines the logical structure of the system. Dynamic entities are represented in GPSS as transactions that move sequentially from block to block as the simulation proceeds. Learning to write a GPSS program consists of learning the functional operation of GPSS blocks and how to combine the blocks logically to represent a system of interest.

The GPSS language* provides almost all the basic simulation functions listed at the beginning of this section. In particular, it has extensive data collection and summary capabilities. Statistics are automatically reported at the end of each run on the utilization of facilities and storages, on queue lengths, and on system sojourn times for transactions. For small, noncomplex problems, GPSS is easy to use. For such problems, the time to build a block design and debug a GPSS model is frequently much less than would be required with a general purpose programming language like FORTRAN. On the other hand, GPSS models usually execute more slowly and hence are more expensive to run. It should be noted that GPSS has a limited capability for generating random variates: apart from the built-in uniform distribution, the user must supply a table for the cumulative distribution function for each random variable to be sampled. However, in the more recently developed version of the

* There are many implementations of GPSS for different computers. GPSS/H is the most advanced and efficient of the GPSS implementations.

GPSS processor, GPSS/H, execution speeds have been substantially improved, and a wider variety of random variate generators has been provided.

Q-GERT[3] is also a process-oriented language. It provides a set of modeling symbols used to construct a network representation of the system to be simulated. Only 10 Q-GERT symbols need be learned by the analyst vs. almost 50 for GPSS. In addition, Q-GERT provides a real-valued clock vs. GPSS's integer-valued clock, and Q-GERT has extensive facilities for random variate generation and statistics collection. One of the most important features of Q-GERT is its convenient interface to user-written FORTRAN insert programs, which the modeler may use to enhance the built-in features of the basic language. Both GPSS and Q-GERT are powerful languages that provide easy, low-cost network model building capability.

The modeling convenience of process-oriented languages, however, has a cost: reduced flexibility and scope of models that can be developed. Process-oriented models are limited to the finite set of modeling constructs that are built into the languages. This limitation is partially offset by the FORTRAN insert capability that Q-GERT and, to a lesser extent, GPSS possess. Discrete event simulation languages, on the other hand, have an essentially unlimited scope, but they provide no built-in process models. SIMSCRIPT II.5 and GASP IV are examples of widely used discrete-event languages. SIMSCRIPT[5] is a complete programming language equivalent in power to FORTRAN or PL/I. SIMSCRIPT is a compiler and has, in addition to conventional language capabilities of computation and input/output, special features for performing discrete event simulations. This means the compiler must be updated for each implementation of SIMSCRIPT. For some computers this has not been done. The principal appeal of SIMSCRIPT as a programming and simulation language is its English-like and free-form syntax. Programs written in SIMSCRIPT are easy to read and tend to be self-documenting.

The discrete simulation modeling framework of SIMSCRIPT is primarily event oriented. The state of the system is defined by entities, their associated attributes, and logical groupings of entities, referred to as *sets*. The dynamic structure of the system is described by defining the changes that occur at event times.

A SIMSCRIPT simulation model consists of a preamble, a main program, and event subprograms. One of the primary functions of the preamble is to define the static structure of the model by prescribing the names of permanent and temporary entities, their associated attributes, and set relationships. For each discrete event, the event name and its corresponding attributes are defined in the preamble. Declarative statements for defining all variable types and arrays are also included. One other function of the preamble is to define variables for which statistics are to be collected. For some models, writing the preamble can be cumbersome.

SIMSCRIPT provides all the standard support facilities previously outlined. Particularly notable are its flexible statement types for creating output reports, although no standard reporting of statistics is provided. Functions are provided for the usual statistical computations and for random variate generation. Since the coding of the event routines is left to the user, the debugging aides in SIMSCRIPT are not extensive; however, the English-like structure of SIMSCRIPT facilitates communication and documentation. Recently, a limited process modeling capability has been appended to SIMSCRIPT.

GASP IV[1] is a discrete-event language that is written in FORTRAN. Event programs and other supporting software are written by the modeler in this widely used and highly portable language. GASP IV provides utilities for all of the normal random variate statistics collection, and the report preparation features of a full-featured simulation language. Models written in GASP IV tend to be more portable than SIMSCRIPT models because of the pervasive presence of FORTRAN compilers, and GASP IV modelers have available to them the vast array of FORTRAN software that is available for graphics, information management, complex numerical procedures, optimization algorithms, etc.

GASP IV was the first widely used language to include combined continuous-discrete features. Using the combined continuous-discrete capabilities of GASP IV, it is possible to simulate the dynamic behavior of an incredibly wide variety of systems. The cost of this flexibility is that the convenience and cost-saving advantages of the process-oriented approach is lost.

SLAM II[4,10] is a simulation language based on Q-GERT and GASP IV that combines process, discrete-event, and continuous capabilities in a single language. Most importantly, it provides well-defined interfaces between orientations. This novel approach permits the modeler to select the most convenient and cost-effective orientation for each subpart of the system. The result is a language with the power and flexibility of GASP IV and SIMSCRIPT and the ease of use of Q-GERT and GPSS.

This important property of SLAM II results because alternative world views can be combined in the same simulation model. Six specific interactions can take place between the network, discrete event, and continuous world views of SLAM II:

1. Entities in the network model can initiate the occurrence of discrete events.
2. Events can alter the flow of entities in the network model.
3. Entities in the network model can cause changes in values of the state variables.
4. State variables reaching prescribed threshold values can initiate entities in the network model.
5. Events can cause changes to the values of state variables.
6. State variables reaching prescribed threshold values can initiate discrete events.

All the simulation languages discussed above can be used for production simulations. However, a simulation language that allows the user a variety of world views, like SLAM II, can be very advantageous for modelers who must address a variety of production systems.

8.6.4 PRODUCTION SYSTEM ANALYSIS USING SIMULATION

A production system is an organized collection of personnel, materials, and machines designed to operate within a predefined workspace. Inputs to the system include raw materials and orders for finished product. Outputs from the system are finished goods. A schematic of a generalized production system is given in Fig. 8.6.4. Management's responsibility for a production system is to apply the proper scheduling procedures and controls so that the system produces the desired goods for the minimal cost.

When performing this task, managers are faced with countless production alternatives ranging from equipment selection to operator scheduling. From a cost perspective, management must monitor or control costs that vary from raw material purchases to investment dollars tied-up in in-process inventory. Because of the complexities of the process and the numerous options available to managers, it is often the case that management can respond to only one objective at one time. Thus, at one time the call might be to minimize order lateness; at another time minimizing inventory is the goal; at still another period, the task may be to maximize throughput. In most cases, management simply does not have the required time or tools to consider multiple objectives simultaneously.

This "local" optimization effect is true not only in the system operations process, but also in the system design process as well. That is, systems are not generally designed with all objectives in mind. Design engineers are requested to develop a system plan to achieve a certain level of throughput. After the design is completed (and found to be too expensive), lobbying over downsized designs by metallurgical, operations, engineering, and finance personnel begins. With this sometimes adversarial environment in the design process plus the overall uncertainty of the finished goods demand in the future, the prospect of optimizing the design relative to overall corporate objectives and subject to constraints on available investment dollars is highly unlikely.

With the increased productivity in the world marketplace, production systems of the future must be more streamlined, more flexible, and more cost effective. Management must take the time to consider more alternatives and to ensure that the design and operating policies implemented are the best possible. Simulation is a decision support tool that allows managers to efficiently and effectively evaluate these alternatives. For this reason, models are becoming mandatory in the production environment.

The power, applicability, and cost effectiveness of simulation methodology is well documented. There exists a wealth of successful applications of simulation modeling to a wide variety of problems for a large number of diverse industries. Selected articles and reports describing some of these successful efforts, including FMS applications, are given in the bibliography at the end of the chapter. See References 11–22.

8.6.5 SIMULATION FOR PRODUCTION SYSTEM DESIGN

Through the use of simulation, system designers can evaluate many different combinations of processing equipment, inventory buffer sizes, material handling devices, inspection procedures, and other equipment/policy concerns. The most effective of these designs can then be "played against" a variety of demand scenarios to test their flexibility. The designers can thus use simulation to account for the complex interactions between components and to develop the best design alternatives. Armed with this information, management can make more informed and thus more cost-effective design decisions.

This type of analysis applies to modification of existing systems as well as to "greenfield" designs. In this case, management normally decides that a system needs expansion to alleviate a bottleneck in some critical

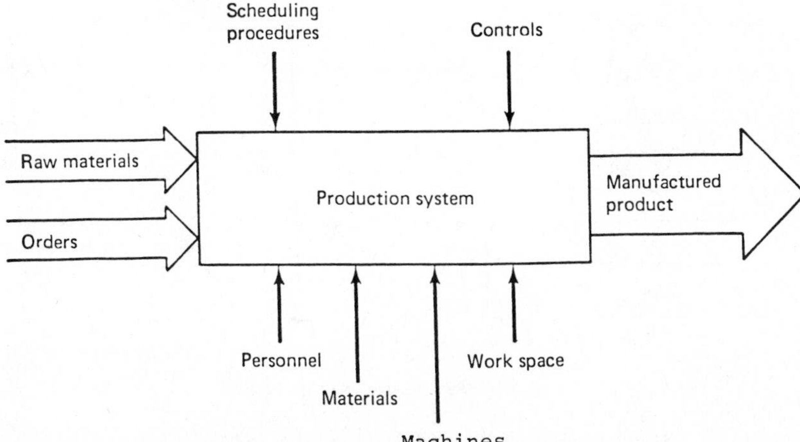

Fig. 8.6.4. Schematic model of a production system.

areas. Here the options are more restricted, because the new processes/procedures must fit into the existing process flow. The new design must be compatible with existing labor, material handling, and safety practices. Because simulation can assess these factors, it assures management that upstream or downstream processes do not prevent the desired benefit from the planned expansion.

The two types of simulation for production system design (new plant and modification to existing facility) are illustrated by the examples below.

8.6.5.1 MLRS System Design[23]

Vought Corporation developed a plan to manufacture the Multiple Launch Rocket System (MLRS) for the U.S. Army. The MLRS vehicle is composed of a tracked vehicle (similar to a personnel carrier) carrying a launch pod container (LPC) holding six rockets. Vought's manufacturing operations would be located in two buildings: the metal parts building and the load, assembly, and pack (LAP) building. A schematic diagram of the proposed facility is given in Fig. 8.6.5. Operations performed in the metal parts building include the fabrication of the necessary detail parts, assembly of the LPC, and production of the motor case. Operations performed in the LAP building include loading the warhead with grenades, joining the warhead and motor case, and loading each LPC with six rockets.

The objectives of the analysis were to evaluate the proposed design by determining the following:

1. The production rate of both the metal parts and LAP facilities as designed.
2. What changes to the metal parts building and LAP facility designs, if any, are necessary to achieve desired production.
3. Storage space requirements for key buffer areas in the manufacturing process.
4. Expected levels of labor and machine utilization.

A model of the proposed system was developed using the SLAM simulation language. This model included a representation of the flow of parts (in batches) through the machines and materials handling equipment within the two buildings. In addition, the model included representation of machine reliability factors, machine repair processes, product quality concerns, inspection processes, and inventory buffering procedures. In both buildings, operation times were calculated by including allowances for set-up, run, PF and D, and overtime. Machine interference and blocking due to in-process storage limitations were also represented in the model.

The model of the LAP building identified a potential bottleneck and was used to verify the redesign of the buffer causing the problem. An illustration of the graphic outputs from the simulation is shown in Fig. 8.6.6 for two operations in series separated by an in-process (buffer) storage area. The models of the two facilities validated production capabilities for a normal rate of production (two 8-hour shifts a day) and surge production rates (two 10-hour and three 8-hour shifts a day).

In April 1980, Vought Corporation was awarded the contract to produce MLRS. The computer modeling analysis was cited by the army as a determinant in awarding the contract to Vought.

8.6.5.2 Billet Preparation Expansion[24]

Based on market projections, the Lackawanna plant of Bethlehem Steel Corporation was anticipating increased demand for its hot rolled bar products. Since the preparation of billets is a critical step in the production of

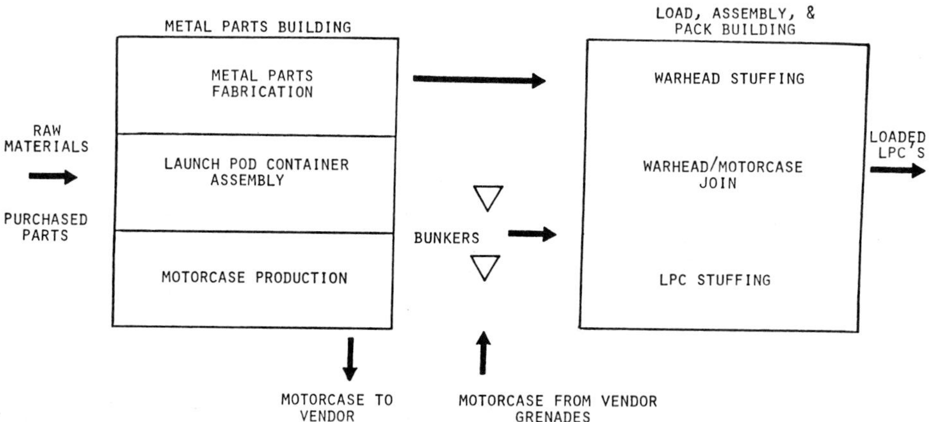

Fig. 8.6.5. Schematic of MLRS production facility.

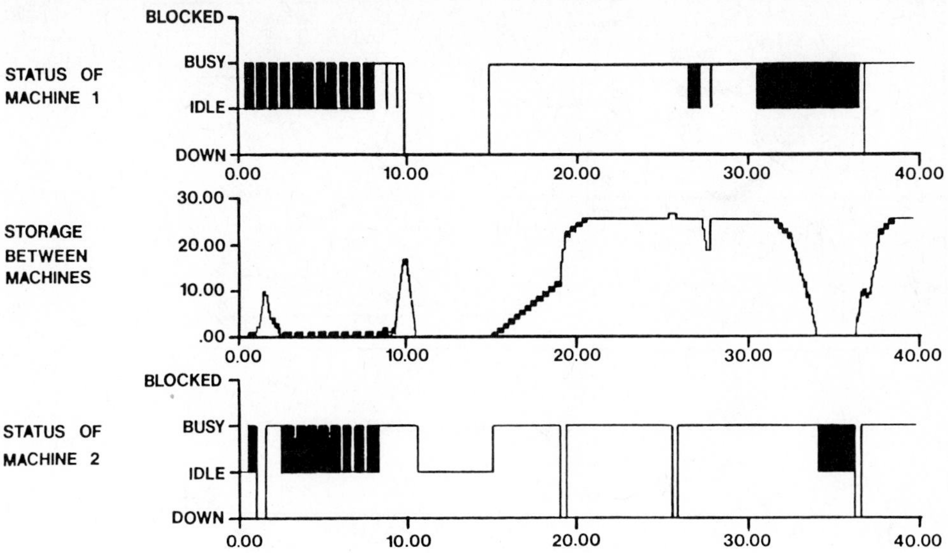

Fig. 8.6.6. Dynamics of machine operations, queues, and buffers for the MLRS production system.

these bar products, Lackawanna management was concerned with the present and future capacity of the billet preparation facilities. Because of the complexity of the physical system, the potential effects of product mix shifts, and the variety of proposed system alterations under consideration, management could not adequately assess the benefits of the designs being proposed. For this reason, a simulation model of the billet preparation facility was constructed. Due to the interdisciplinary nature of the task at hand, Bethlehem management assigned a project team to develop and analyze the model. The team included personnel from plant industrial engineering, corporate industrial engineering, and plant engineering, operations, and metallurgical departments.

A schematic of a major portion of the billet preparation area evaluated in this study is given in Fig. 8.6.7. Not shown here are facilities for gag-straightening, handling of trade billets, and some offloading and storage facilities, all of which were included in the model. Billets are generally moved through the system in "lifts," a set of billets that can be moved at once by an overhead crane. Lifts arrive from the billet mill to one of two hot beds to start the billet preparation process. Each lift must cool before a crane can remove it from a hot bed. After cooling, the lifts are transported to storage racks or specialized cooling/heating furnaces where they cool for 16–30 hours.

After cooling, lifts are moved to temporary storage racks or to one of several required operations. The timing of these movements is largely dependent on the availability of space, crane workloads, and scheduling of facilities. The operations most commonly performed in Building 1 are sonic testing, scarfing, double cutting, visual inspection, marking, and end-cutting (to remove internal defects). To perform these operations, lifts are placed by cranes into a "laydown" area at the north end of the building. As this space is limited, quick turnaround through this area is important. Operations performed in Building 2 are inspection (both visual and ultraviolet light) and grinding. Transfer between the buildings is accomplished via transfer cars or rail.

A model of the system was constructed, including all the processing in the billet preparation area. In addition, the model included machine reliabilities and operating policies. These policies involved a great number of decisions designed to maintain a smooth flow. With this coordination, costly delays to the billet mill (caused by the inability of billet preparation to remove steel from the hot beds) and the bar mill (caused by a lack of billet preparation capacity or late deliveries) can be avoided. Decisions made to provide this control include determining the inventory levels at which billets should be moved between buildings by rail (more costly than using piggyback cars), the number of 8-hour turns per week to schedule for each operation and the sequence of those turns, the number of empty rail cars to request for loading during future turns, alternative routings to be used when desired areas are unavailable, and the assignment of movement tasks to the individual cranes (three cranes in each building). The model also had the flexibility to represent a variety of product mixes (billet sizes and AISI grades) with a wide range of billet qualities.

The model was first configured to represent the existing system. Execution of the model indicated that the desired system capacity could not be achieved because of bottlenecks in the system caused primarily by the grinders and sometimes by the cranes and inspection equipment. Because of these problems, several alternatives for improvement were suggested. The simulation model proved useful in evaluating four equipment configurations in addition to the original system. The first three alternatives involved the installation of an additional grinder in different locations in Buildings 1 and 2. While some additional capacity was realized through each of these

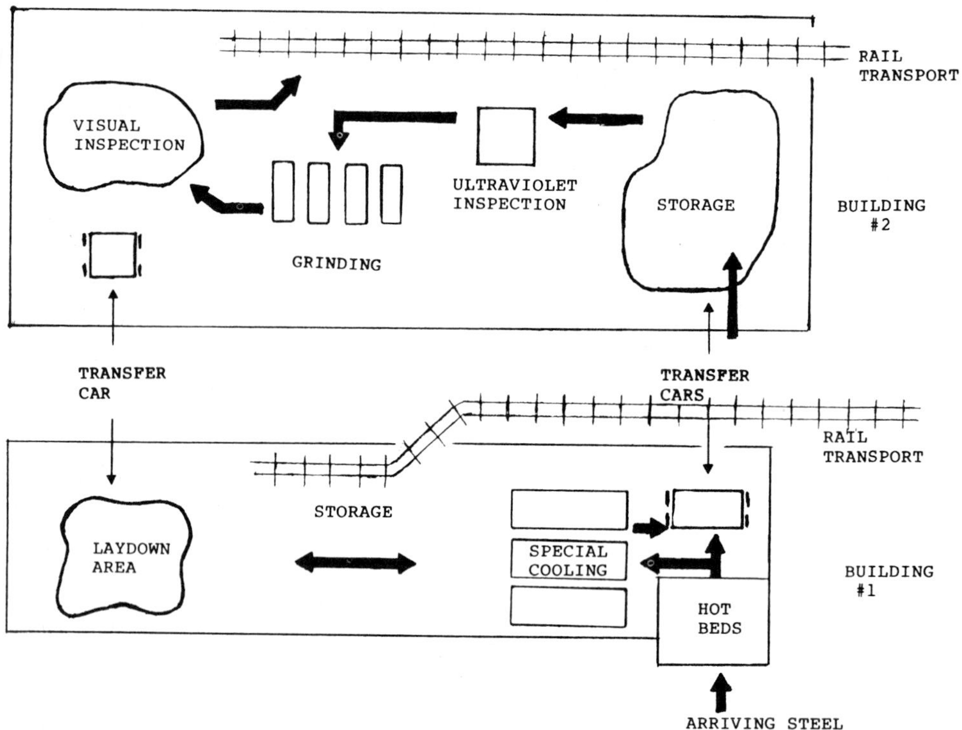

Fig. 8.6.7. Schematic of billet preparation area.

plans, it was not sufficient to meet the demands. The reason these plans failed was crane service. When the grinder was added, the cranes became the bottleneck and only a small increase in production capacity was gained. To alleviate this problem, additional crane capacity was added which, unfortunately, achieved no increase in throughput, but only a major increase in crane interference and congestion (the buildings were not long enough for four cranes).

After these analyses were performed and upon further discussion of the long-term reliability of the existing inspection unit, interest was generated in constructing a model of the system with a new inspection unit in Building 1. This proposed system was radically different from the existing system, as the flow of material was completely reversed, clockwise around the facility rather than counterclockwise. This proposed reverse-flow system included not only an additional inspection unit and grinder in Building 1, but also another crane to serve the new equipment, rearrangement of racks, relocation of transfer cars, and additional material handling equipment to serve the grinders in Building 2. This model confirmed the necessity of an additional crane in Building 1 and demonstrated the feasibility of the entire design. Further, it indicated that fewer transfer cars were actually needed than were initially included in the design, thus avoiding some extra costs.

The problem at hand in this study, as with most system enhancement evaluations, was to design and create a means of analyzing in detail the performance of the system under conditions anticipated in the future. The capacity of this system, and its variants, is limited not simply by individual components, but by the interaction between them. Thus, simple manual methods are not adequate for predicting system performance; simulation methodology is required.

The simulation models developed for this study are highly detailed representations of alternative billet preparation systems. Because the models explicitly account for throughput limitations caused by cranes, train cars, storage capacity, transfer cars, entry and exit tables, cooling time delays, rework, and equipment reliability, the results obtained through them could not be generated manually within any practical amount of time. In this case, then, simulation procedures were very effectively employed to support the system design decision process.

8.6.6 SIMULATION FOR OPERATIONS CONTROL

Simulation can be employed to analyze control strategies for the production environment in the same way that it is used to evaluate alternative designs, that is, a simulation model of the system and the control process is constructed and executed. Different control parameters are supplied to the model for each different execution. The model represents the flow of the system under each particular policy and predicts system performance based on the strategy input. Analyzing a series of simulation runs, the production system manager can determine and implement the most cost-effective strategy.

A schematic of this approach to evaluating scheduling alternatives is given in Fig. 8.6.8. In this figure, a simulation is constructed that will evaluate the performance of the production system. Direct "drivers" to the simulation model include input from a tracking system (to provide the current status of the system to start the simulation model), a set of system performance standards and operating factors (describing how long each particular activity within the system will take for each particular product being processed), the current order database, and a facility schedule (when and which equipment/operators are scheduled at which times). The facility schedule could be manually generated or automatically created by a scheduling algorithm. The simulation model is then executed.

The simulation model produces outputs that describe the ability of the system to meet the schedule in terms of the overall objectives of the organization. These outputs include: utilization of resources, inventory levels, system throughput, and order lateness. An evaluation or review is then performed on these outputs to determine if they are sufficient for the objectives at hand. If not, a new schedule is generated and the procedure is repeated. This process continues until an acceptable schedule is created or until the time for the analysis has ended. At this time, the best schedule generated to date is accepted.

This type of simulation application is effective for the evaluation of equipment schedules, inspection procedures, work order release plans, and other such control options. The example below highlights the use of simulation for control purposes.

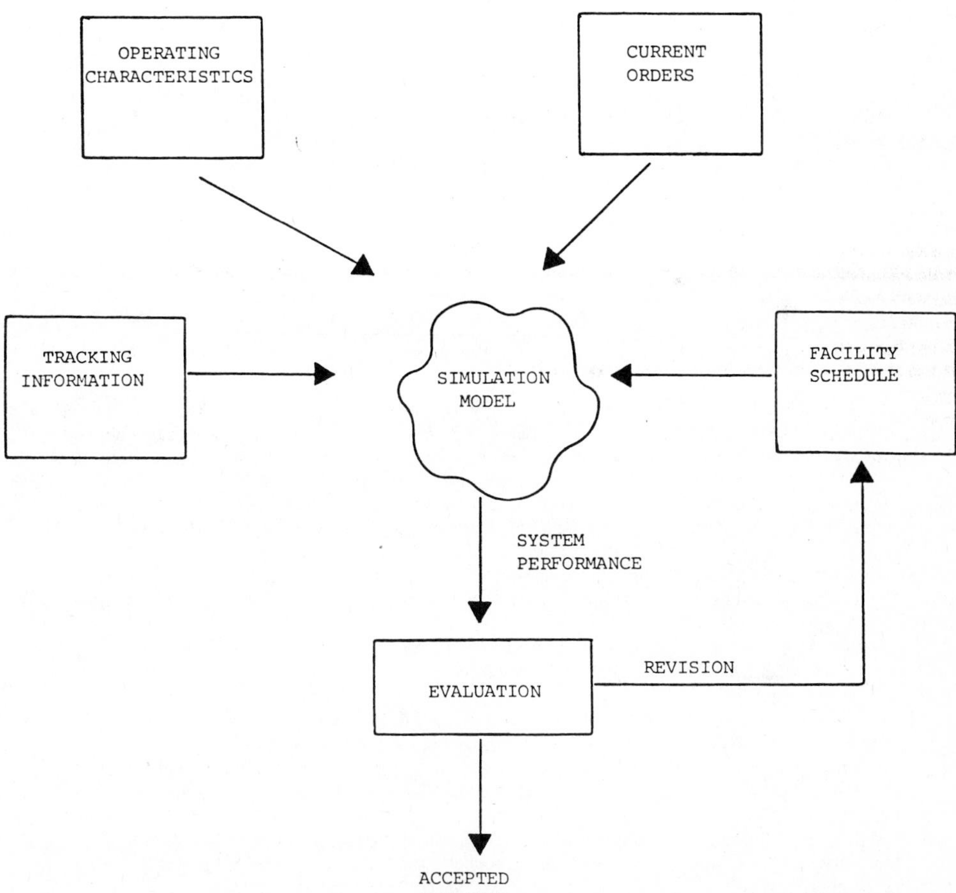

Fig. 8.6.8. The use of simulation for the evaluation of scheduling alternatives.

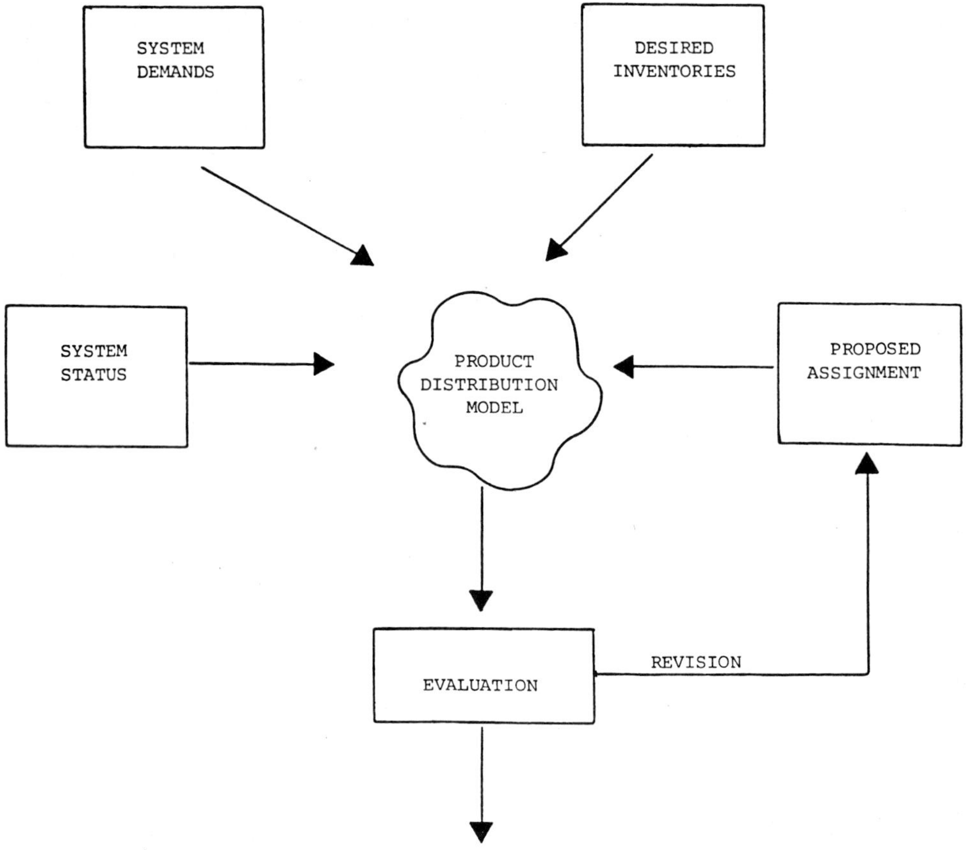

Fig. 8.6.9. Simulation model as an evaluator of a proposed product assignment plan.

8.6.6.1 Distribution Procedure Evaluation[23]

A simulation model of a petroleum product distribution system was developed for a major oil company to generate and evaluate product allocation procedures. The representation of the system by the model included product transportation from refineries to terminals via pipelines, tankers, and barges. Given the physical characteristics of the system and the desired product movements (refinery-to-terminal assignments), the simulation model determines what can be feasibly scheduled, consistent with the following constraints: pipeline flow restrictions including product routing, product cycling, and minimum batch sizes; pipeline shipment time delays; tanker availability; tanker compartment restrictions; barge availability; barge compartment restrictions; refinery production limitations by product group; refinery production limitations by product; refinerey outflow restrictions; blending batch sizes; tanker loading, unloading, and movement delays; and barge loading, unloading, and movement delays.

The simulation model represents the flow of petroleum product through the system and the operating logic required to control this flow. The use of the simulation model as an evaluator of a proposed assignment plan is depicted in Fig. 8.6.9.

Inputs to the model are the initial status of the system, the demands on the system, and the proposed assignment to be tested. The initial status of the system includes the initial levels of inventory of each product at each terminal, tank farm, and refinery; the current status of the pipeline fills; and the current in-transit levels of tankers and barges in this system. The demands on the system include the demands for product at each individual terminal; the desired end-of-period inventory at each terminal, tank farm, and refinery; and the desired end-of-period in-transit inventories for all pipeline, tanker, and barge routes. The proposed assignment is defined by the proposed shipment (volume of each product) from each product source location to each product receiving location.

The model is then exercised to determine whether the proposed assignment can be achieved and how well the system demands are met by this assignment. The outputs from the model include a comparison of the

Table 8.6.3. **Example Outputs from the Assignment Evaluation Simulator**

Pipeline usage report.

Source	Sink	Maximum flow	Actual flow	Desired flow	Difference	Fraction of Time Line: In use	Blocked	Idle
Galveston	JN1	1200.00	987.15	1000.00	-12.85	0.94	0.06	0.00
JN1	Houston	700.00	201.30	200.00	1.30	0.32	0.00	0.68
JN1	JN2	1000.00	785.85	800.00	-14.15	0.94	0.06	0.06
JN2	Buffalo	600.00	97.72	100.00	- 2.28	0.20	0.00	0.80
JN2	JN3	1000.00	688.13	700.00	-11.87	0.94	0.06	0.00
JN3	Dallas	500.00	141.02	150.00	- 8.90	0.36	0.05	0.59
JN3	JN4	900.00	547.11	550.00	- 2.89	0.87	0.10	0.03
JN4	Checotah	300.00	99.72	100.00	- 0.28	0.42	0.07	0.51

Facility inventory report.

Facility	Product	Min inventory Actual	Min inventory Plan	Max inventory Actual	Max inventory Plan	Average inventory	Opening inventory	Closing inventory Actual	Closing inventory Plan	Closing inventory Difference	Product Arrival	Product Outflow	Satisfied demands Actual	Satisfied demands Plan	Satisfied demands Difference
Houston	Regular gasoline	10	3	21	25	15	12	17	16	1	15	0	10	10	0
	No lead gasoline	6	4	9	12	7	7	6	7	-1	4	0	5	5	0
	Heater oil	4	2	12	15	8	4	11	10	1	21	0	14	14	0
	Furnace oil	2	2	5	7	3	5	2	2	0	0	0	3	3	0
	Diesel	3	3	5	10	4	3	4	4	0	6	0	5	6	-1

product flow that was achieved vs. the product flow requested by the proposed assignment; a comparison of satisfied demands at the terminal vs. the actual demand at the terminal; and a comparison of the end-of-period inventories vs. the desired end-of-period inventories. In addition, storage and pipeline utilizations are computed and a list of the actual schedule of batches moved from each source location is presented. Table 8.6.3 provides example outputs from the system.

In this system, there are a large number of individual product types, dozens of terminals, several refineries, and hundreds of different routing possibilities. Given the other restrictions (tanker scheduling, product sequence constraints in pipelines, storage tanker capacities at terminals, etc.), it is impossible within a reasonable time to evaluate manually a proposed allocation plan. For this reason, simulation concepts were employed. By applying simulation, system managers have found that they can accurately predict the flow of product in the system and thus achieve a realistic allocation by using the model. With this procedure, they can now select the best allocation plan from a set of feasible alternatives, a capability that was not available in the past.

8.6.7 SUMMARY

In this chapter, simulation modeling has been defined and its utility with respect to production systems has been discussed. The steps necessary to apply simulation methodology to a production process have also been presented. Modeling techniques based on event-, process-, and activity-oriented approaches to simulation have been discussed along with the major simulation languages that employ each approach. More detailed descriptions of the application of simulation for production system design issues and for operations contol problems have been presented. For each of these application areas, examples of the use of simulation to analyze actual production systems have been examined.

REFERENCES

1. A. A. B. Pritsker, *The GASP IV Simulation Language*, New York: Wiley, 1974.
2. R. E. Shannon, *System Simulation: The Art and Science*, Englewood Cliffs, NJ: Prentice-Hall, 1975.
3. A. A. B. Pritsker, *Modeling and Analysis Using Q-GERT Networks*, New York: Wiley/Halsted, 1979.
4. A. A. B. Pritsker, *Introduction to Simulation and SLAM*, New York: Wiley/Halsted, and West Lafayette, IN: Systems Publishing Corporation, 1986.
5. P. J. Kiviat, R. Villanueva, and H. Markowitz, *The SIMSCRIPT II Programming Language*, Englewood Cliffs, NJ: Prentice-Hall, 1969.
6. W. R. Franta, *The Process View of Simulation*, New York: Elsevier/North-Holland, 1977.
7. B. Carnahan, H. Luther, and J. O. Wilkes, *Applied Numerical Methods*, New York: Wiley, 1969.
8. J. P. Emshoff and R. L. Sisson, *Design and Use of Computer Simulation Models*, London: Macmillan, 1970.
9. T. Scribner, *Simulation Using GPSS*, New York: Wiley, 1974.
10. J. J. O'Reilly, *Technical Reference Manual for SLAM Simulation Program*, West Lafayette, IN: Pritsker & Assoc., 1983.
11. R. W. Swain and J. J. Marsh, III, "A Simulation Analysis of an Automated Hospital Materials Handling System," *AIIE Transactions*, **10**, 1978, 10–18.
12. D. B. Bandy and S. D. Duket, "Q-GERT Model of a Midwest Crude Supply System," presented at ORSA/TIMS National Meeting, October, 1979.
13. M. C. Embury, G. V. Reklaitis, and J. M. Woods, "Simulation of the Operation of a Staged Multi-Process and Limited Interstage Storage Buffers," presented at the 9th International Conference on Systems, Honolulu, 1976.
14. R. J. Miner, D. B. Wortman, and D. Cascio, "Improving the Throughput of a Chemical Plant," *Simulation*, October 1980, pp. 125–132.
15. R. M. Waugh and R. A. Ankener, "Simulation of an Automated Stacker Storage System," *Proceedings, 1977 Winter Simulation Conference*, 1977, pp. 769–776.
16. A. L. Sweet and S. D. Duket, "A Simulation Study of Energy Consumption by Elevators in Tall Buildings," *Computers in Industrial Engineering*, **1**, 1976, pp. 3–11.
17. E. L. Janzen, "Simulation for Materials Handling in a Food Protein Plant," *Proceedings, Winter Simulation Conference*, Miami, December 4–6, 1978.
18. A. Weinberger et al., "The Use of Simulation to Evaluate Capital Investment Alternatives in the Steel Industry: A Case Study," *Proceedings of the 1977 Winter Simulation Conference*, Gaithersburg, MD, December 5–7, 1977.
19. D. J. Medeiros, R. P. Sadowski, D. W. Starks, and B. S. Smith, "A Modular Approach to Simulation of Robotic Systems," *Proceedings, 1980 Winter Simulation Conference*, December 3–5, 1980.
20. D. L. Martin, S. D. Duket, and J. D. Sabuda, "The Future of Simulation in the Mining Industry," presented at ORSA/TIMS Joint National Meeting, April 25–27, 1983, Chicago, IL.

21. V. J. Auterio, "Q-GERT Simulation of Air Terminal Cargo Facilities," *Proceedings, Pittsburgh Modeling Simulation Conference*, **5**, 1974, 1181–1186.

22. David L. Martin and Kenneth J. Musselman, "Simulation in the Life Cycle of Flexible Manufacturing Systems," presented at the ORSA/TIMS Special Interest Conference on FMS, August 16, 1984, Ann Arbor, MI.

23. A. A. B. Pritsker, "Applications of SLAM," *Proceedings, 1981 AIIE Spring Annual Conference*, 1981, pp. 383–390.

24. F. A. DeJohn, C. W. Sanderson, C. T. Lewis, J. R. Gross and S. D. Duket, "The Use of Computer Simulation Programs to Determine Equipment Requirements and Material Flow in the Billet Yard," *Proceedings, 1980 AIIE Spring Annual Conference*, May 11–14, 1980, Atlanta, GA, pp. 402–408.

BIBLIOGRAPHY

Cellier, F. and A. E. Blitz, "GASP V: A Universal Simulation Package," *Proceedings, IFAC Conference*, 1976.

Duket, S. D., and A. A. B. Pritsker, "Simulation and Real Time Factory Control," presented at Society of Manufacturing Engineers Meeting, Southfield, MI, June 20–23, 1983.

Korn, G. A. and J. V. Wait, *Digital Continuous-System Simulation*, Englewood Cliffs, NJ: Prentice-Hall, 1978.

Mitchell, E. L. and J. S. Gauthier, "Advanced Continuous Simulation Language (ACSL)," *Simulation*, **25**, 1976, 72–78.

Pritsker, A. A. B. and C. E. Sigal, *Management Decision Making: A Network Simulation Approach*, Englewood Cliffs, NJ: Prentice-Hall, 1983.

Salvendy, G., Ed., *Handbook of Industrial Engineering*, New York: Wiley, 1982.

Scribner, B., "Operations Research—Out of the Closet and Into the Tool Kit," paper presented at Industrial Engineers Managers Seminar, March 16–18, 1981.

Shannon, R. E. *System Simulation: The Art and Science*, Englewood Cliffs, NJ: Prentice-Hall, 1975.

CHAPTER 8.7

FLEXIBLE MANUFACTURING SYSTEMS

JANE C. AMMONS
LEON F. McGINNIS

Georgia Institute of Technology
Atlanta, Georgia

8.7.1 WHAT IS FMS?

Coincident with developments in computers and automation technology in the past twenty years has been the evolution of a new type of production system. The *flexible manufacturing system*, or FMS, has been developed for the production of related items in small to medium-sized batches. By approaching mass production efficiencies in a more versatile small-batch manufacturing environment, FMS is a highly competitive manufacturing strategy because of its capability to adapt quickly and efficiently to changes in product mix, demand, and design.

8.7.1.1 Definitions

FMS is a manufacturing philosophy based on the concept of effectively controlling material flow through a network of versatile production stations using an efficient and versatile material handling and storage system. Every FMS component contributes to the flexibility of the entire system. Each workstation in the FMS is capable of processing a variety of part types with relatively small changeover times. Linking the workstations, the material handling and storage system can provide timely movement of parts and tools in any desired routing sequence. Typically, there is computer control of part and tool storage and handling, and computer supervision of the workstations. The control system exploits the versatility of the production and material flow functions to give the FMS its flexibility.

Today the M in FMS stands for many types of manufacturing applications, including machining, fabrication, assembly, and even foundry operations. Because the first FMS applications were in the machining environment, many people associate the acronym FMS with flexible *machining* systems. But as demonstrated below, the FMS concept has been developed in many diverse areas since the early days.

Because FMS is a strategic manufacturing philosophy and not simply a technology, the term does not necessarily imply a high level of production and control automation. Typically, the level of automation in an FMS is based on the availability and cost of the automation technology required to meet manufacturing requirements at the time of the system's inception. The rapid development of microprocessor control technology in recent years has led to many totally or partially automated FMS installations. Alternatively, it is projected that many FMS production applications will continue to have manual functions linking "islands of automation" for several years to come.

8.7.1.2 History

The evolution of FMS may be traced to its beginnings in the mid-1960s. One of the earliest systems employing FMS concepts went on-line at Sunstrand Corporation in 1965. This system was composed of eight numerically controlled (NC) machining centers linked by a computer-controlled roller conveyor system.

In 1968 Theodore Williamson conceived and developed the FMS concept as director of R&D at the Molins Machine Tool Company in the United Kingdom. Williamson proposed a system of NC machines to make relatively complex light alloy components for tobacco industry tools. The flexible machining system was named "System 24" because it was intended to operate three shifts a day, with only one of the shifts attended by operators.

The Molins FMS was designed to perform a series of machining operations on a wide variety of parts in virtually any order. After the workpieces were manually loaded on a pallet, they would be moved to machines capable of automatically changing tools and parts as needed. Unfortunately, because the design features of the machines proved too advanced to be reliably realized at that time and because the market for light alloy components did not develop as forecast, Molins discontinued the operation with a $10 million loss.

But the conceptual seed was sown and soon other flexible machining systems followed, in the United States and Japan. Early developments focused on critical hardware, especially in the areas of tooling, fixtures, and material handling equipment. In the early 1970s, the FMS approach was extended in Europe from machining to the development of flexible assembly systems. By 1972 the system scope was expanded with the introduction of microprocessor-based programmable controllers, leading to even more refined control capabilities. Distributed logic came on the scene about 1979.

By 1982 it was estimated that there were 75 flexible manufacturing systems in operation worldwide. Currently there are about 25 multicelled FMSs in operation in the United States, making this country second in installations to Japan, where there are about 50 installations, and ahead of West Germany with 18, France with 8, and the United Kingdom with 5 or 6. Today the international developmental emphasis focuses on both the hardware and control portions as the FMS concept continues to encompass more application areas. This is illustrated by the following examples of FMS installations.

8.7.1.3 Examples of FMS

In this section, several examples of flexible manufacturing systems will be described. These by no means represent the full gamut of FMS applications. Rather, they are intended to indicate the breadth of potential applications. The extensive bibliography at the end of this chapter provides a large number of additional examples.

Traditional Flexible Machining

The FMC Corporation has installed in Aiken, SC, a flexible manufacturing cell for the machining of cast parts for the Bradley Fighting Vehicle, the U.S. Army's high mobility armored troop carrier. Annual production is currently 16 different part types for 672 vehicles, with two of some parts required per vehicle. Additional component types are being considered for future production on this FMS.

The FMC FMS is based on four identical CNC machining centers, each with 90 tool storage capacity and fed by palletized fixtures. Two stations for loading, unloading, cleaning, and orienting parts have a capability for 360° part rotation in the horizontal direction and 90° rotation in the vertical plane. Two automatic work changers (AWCs) provide buffer storage, each with the capacity to store ten pallets as well as provide fixture maintenance and storage. A coordinate measuring machine performs automatic inspections on both completed parts and work in process (WIP). Linking all of these automated systems is a network of three automated guided vehicles.

The system design and its on-line control capabilities give the manufacturing process several types of inherent flexibility. Setup time between different part types has been eliminated, resulting in a very effective response time for changes in part type. Redundancy of the four machining centers' capabilities give the system much greater reliability and many more process flow options. Also, the input/output configuration at each machining center allows for parts to pass each other. This characteristic combined with the temporary buffer space in the AWCs permits much greater routing flexibility. Furthermore, FMC reports that the consistency and quality of production are greatly improved with the FMS.

An Extrusion Trim Cell

The Lockheed, Georgia Company has installed a flexible cell for the machining operations required for sheet metal extrusions of a size that will fit in a 36-inch-long totebox. Typical operations in this cell include drilling and deburring. A wide variety of parts, up to several thousand different part numbers, are processed, and each part type will usually require some special tooling. The operations themselves are typically at manned workstations, although some robotic operations are in the design stage.

The extrusion trim cell employs two microload AS/RS systems, one for tooling and one for WIP. Containers of parts arrive at the cell, are inducted, and have the required tools "married," i.e., placed in the parts container. The parts container is then entered into the WIP AS/RS, where it is automatically routed to the required workstations as they become available, in conjunction with the job priority rating.

This flexible manufacturing system is based in large part on manually operated workstations. The benefits to the system include real-time job tracking, integrated tool control, and, perhaps most important, processing discipline. The jobs are presented to the workstations according to a priority system, and should be processed in that order. Once processed, jobs are automatically returned to the computer-based WIP control system, eliminating the potential for jobs to get "lost" in the shop. Because WIP control is computer based, there are many opportunities for building in intelligence, e.g., using standard times to estimate the completion time for a job, and triggering an audit action for jobs that are significantly later than the estimated completion time.

High-Volume Electronic Assembly

Apple Computers, Inc., has developed a flexible manufacturing system for producing printed circuit boards for the Macintosh personal computer. This is a high volume production environment, since the target is to produce one Macintosh every 22 seconds. The printed circuit board assembly includes both manual and automatic operations. The flexible assembly system is based on two aisles of microload AS/RS. Twenty-two manual assembly stations are around the outside of the two aisles, and eight automated assembly stations are between the aisles. The aisles are 240 feet long and 25 feet high with a 90 foot long mezzanine between them for inspection and corrective action.

As with the Extrusion Trim Cell described earlier, this FMS incorporated a high proportion of manually based workstations. Its primary flexibility is its capability for rapid changes. In particular, the workstations can be reconfigured, removed, or added with relatively little time or expense. Thus, changes in product design, product volume, and perhaps introduction of new but related products can be accommodated readily. Furthermore, the discipline imposed by the system and the real-time tracking of both parts and assemblies is very supportive of the just-in-time philosophy of manufacturing.

Low-Volume Electronic Assembly

Another example of a flexible manufacturing system for printed circuit boards is Westinghouse's Electronic Assembly Plant in College Station, TX. In this FMS, "kits" of parts for a particular printed circuit board are prepared at robotic kitting stations and introduced into a special type of carousel conveyor having independently rotating shelves. The carousel is interfaced to both robotic component insertion and manual stations, and all operations are under complete computer control.

A unique feature of this system is that operations are monitored continuously. The failure of an operation is automatically recorded, and may be used to modify, in real time, the processing at subsequent workstations. In addition, the real-time part monitoring process builds a complete pedigree for each printed circuit board assembled.

This FMS has flexibility of two different types. The first is the flexibility to respond in real time to the result of each assembly operation. The second is the ability to produce a random mix of printed circuit boards in lots of size one. Because of the latter flexibility, the introduction of new products to the FMS is very easy once the kits have been designed and provisioned and the necessary part programs developed for the robotic assembly stations.

Flexible Mechanical Assembly

Starting in 1980, Xerox Diablo in Hayward, CA, phased in a mechanical assembly system for manufacturing letter-quality daisy wheel printers. Currently at the fifth and final stage of development, the actual system output has increased by 500% while floor space has been reduced by 10%. Additionally, overhead cost has been reduced by 50%, manufacturing cost by 40%, and inventory by 30%.

The Diablo printer assembly system has three functional areas in its half million square foot facility. First, the receiving function receives, inspects, and stores raw materials using a bar-coded computer control system. Smaller, high usage items are stored in an AS/RS with a mezzanine for picking and downstairs for put-away, while heavy/bulky/low usage palletized material is stored in high bay narrow aisle pallet racks. When specified thresholds of floor inventory are reached, material is automatically withdrawn for conveyor delivery to manufacturing.

The second functional area is the assembly area, composed of several processes. Printed wiring board assembly has computer controlled WIP carousels linked with automated insertion equipment and assorted test equipment. Electromechanical assembly occurs around three microload AS/RS systems, with kitting stations to feed the progressive assembly lines as well as material interfaces on the cart-on-track-based assembly line.

The assembled and tested printers are automatically transferred to the third functional area for storage, packing, and shipping. Storage occurs in a miniload, with a final quality audit right on the conveyor as the printer moves to the packing area. All three functional areas are completely computer integrated so that overall system performance is enhanced. Flexibility is inherent due to process routing options, generic production equipment, and a great degree of discipline in part and workflow control.

8.7.2 WHY FMS?

Several benefits associated with the FMS concept make it such a competitive manufacturing strategy. Many industry executives feel that because of these advantages, the marketplace is in the midst of a major structural change whose winners will be determined to some degree by decisions reflecting effective FMS design and installation. What are these advantages when compared to conventional manufacturing approaches?

8.7.2.1 "Typical" Justifications

First, one of the most important features of FMS is its quick and effective response to market needs. Because of the inherent flexibility designed into an FMS, meeting demand fluctuations or changes in product mix in a timely fashion is a relatively simple matter of "automatically" readjusting part routings and workstation capabilities as needed. Also, getting related new parts or model changes rushed to the marketplace is facilitated when the FMS is integrated within a CAD/CAM system. Additionally, market competitiveness is enhanced by the shorter lead times required for FMS production. Because the FMS approach minimizes required setup time on machines and part queue times, reductions of product delivery time have been reported up to 75% in some FMSs. This property also makes the FMS concept very applicable to the current just-in-time (JIT) manufacturing trend. All these reasons have convinced many executives that the FMS strategy is required to keep them competitive in their future markets.

Another attraction of the FMS approach is significant reduction in WIP inventory and the corresponding reduction in required plant floor space and carrying costs. The floor space savings can be quite significant, estimated by some at 45–85%. The cost savings associated with reduced inventory are reflected along the entire inventory pipeline, including the supplier. Furthermore, the physical reduction in WIP results in drastically improved flow congestions and the associated enhancement of material control.

A third advantage of the FMS approach is in the area of quality. The automation of work cells associated with FMS yields product of consistent accuracy, and the reduction or elimination of human handling often results in better product. With inspection steps that include real-time feedback, scrap levels are reduced. This translates into raw material savings as well as a reduction in required production repetitions. And for many manufacturers, such as those supplying the defense industry, the ability to consistently guarantee production performance within tolerance is essential to survival. These quality features are important benefits of an FMS.

Another advantage associated with the FMS concept is related to operational efficiencies. Although many systems cannot be justified based on reduced labor costs alone, they can run unattended during breaks and lunch periods as well as significantly longer periods, such as a third shift. It has been reported that some systems save energy costs by running without shop lighting, heating or air conditioning during these times. Additionally, machine utilization can be significantly higher with an FMS.

Finally, an important FMS benefit is a result of the discipline imposed on the manufacturing system. The process of designing and implementing an FMS requires well-thought-through rationalizations for component design and production technique and a corresponding standardization wherever possible. The subsequent product can be produced in more efficient steps, and more effective real-time control can be achieved. Furthermore, this process opens the door for introduction of a spectrum of CAD/CAM benefits.

8.7.2.2 Evaluation Problems

Although the advantages listed above give strong incentive to consider the FMS approach, many of the benefits are intangible or not directly quantifiable using conventional cost accounting procedures. For example, how can the costs associated with making a workstation as versatile as possible be allocated over the projected product lines?

One of the serious questions associated with FMS evaluation is the need for a precise and quantitative measure of system flexibility. Because there are many dimensions to flexibility (see Section 8.7.3.1), the relative flexibility of a particular FMS design is difficult to specify in unconditional and quantifiable terms. The lack of an appropriate flexibility measurement makes it difficult to compare the strategic FMS properties of alternative production system designs.

Additionally, current cost accounting procedures are inadequate to capture the assignment of indirect expenses required to operate and maintain an FMS. For example, with an FMS, labor and energy costs may be reduced while machine utilization is significantly increased when compared to conventional manufacturing systems. But how should these savings be allocated among the varying product mixes and potential future products associated with the FMS? A new approach to indirect expense evaluation is required to properly reflect FMS production costs.

For those advantages for which "hard" numbers can be obtained (in reality these may be "best guess" estimates), a long payback period may be required to justify the significant investment. For example, traditional U.S. systems may be priced upwards of $10 million, unrealistically expensive for smaller firms and difficult for larger ones to justify using conventional return on investment criteria. New criteria and accounting approaches are needed to help justify FMS strategic commitment decisions based on significant intangible benefits in the competitive marketplace.

8.7.3 BASICS OF FMS

From the examples given earlier, it is clear that FMS is a broadly applicable production philosophy, for both fabrication and assembly, and in environments ranging from totally automated through partially automated to completely manual. The key functional elements of an FMS are:

The materials and their processing, including material handling.

Information and communication, including work instructions, performance data, and status information.

Automated planning and control of the movement of materials, tools, and fixtures through the system.

In all three functional areas, the technology is evolving rapidly. What is state of the art today will most likely be obsolete within five years, or at least surpassed by newer, more advanced, technologies.

For this reason, it seems appropriate not to focus on current FMS technology, but rather to emphasize the fundamental concepts embodied in FMS applications. The fundamental concepts that distinguish FMS from both traditional job shop manufacturing and hard automation systems include:

1. Planned but limited flexibility in the processing and material handling equipment.
2. An absolute requirement for maintaining continuous current-status information on both equipment and work in process.
3. A control system capable of responding in real time to contingencies such as tool or machine failure, material, fixture or tool shortage, and changes in quantity or priority requirements.

In addition, most FMS applications have as primary design objectives the reduction of WIP and the associated reduction in response time or manufacturing lead time. These objectives are achieved, however, as a result of flexibility, continuous monitoring, and comprehensive control.

Flexible manufacturing systems vary widely in scope, ranging from a single cell, e.g., a machining center, to an integrated network of cells, spanning operations from fabrication through assembly. In subsequent discussions, an FMS cell will be a collection of workstations or machine tools with an integrated, self-contained material handling and storage system.

The fundamental conceptual elements of an FMS cell are the processing facilities, material storage, material handling, and an interface between the cell and the rest of the manufacturing system. FMS cells are linked with one another, and with conventional processes, by material handling systems that may or may not be under direct computer control.

8.7.3.1 Flexibility

The most flexible machined part fabrication system consists of general-purpose machine tools and skilled machinists. Unfortunately, such systems tend to have poor asset utilization, large amounts of WIP inventories, and long lead times. In an FMS, the goal is to retain enough flexibility to accommodate a range of part types, and at the same time achieve high asset utilization while reducing WIP and manufacturing lead time. Thus flexibility in an FMS is limited by the specific tool and part parameters, which define its capabilities.

Flexibility as an attribute of a manufacturing system is difficult to define concisely. It is much easier to describe the absence of flexibility. For example, lack of flexibility in processing would mean that a machine or workstation could process only a single part type, or perform only a single operation on a variety of part types. The characterization of flexibility will be explored further below.

A central issue in designing an FMS is determining the required or desired degree of flexibility. This design issue is complicated by the lack of good measures of flexibility, and the difficulty of predicting the attributes of parts that may be produced in the future. Obviously, greater flexibility is desirable from the perspective of production management. However, given the technological constraints of flexible automation, greater flexibility also means more complex planning and control problems, and a greater number of "failure modes" for the system.

There are several design and management strategies that preserve the benefits of flexibility without jeopardizing the system through excess complexity. *Group technology* (see Chapter 3.15) techniques can help identify the parts and processes that are candidates for inclusion in the FMS because of similarity of attributes or requirements. *Design standards* (see Chapters 3.1, 3.2, and 3.3) may be difficult to implement, but can significantly reduce the complexity of FMS configuration and control. In addition to design standards, *tooling standards* can reduce the complexity of FMS control and also simplify computer aided process planning (see Chapter 8.3).

Processing/Tooling Flexibility

There are three important characteristics of processing flexibility. First, *operation flexibility* means the ability to perform more than one operation on a given part type. For example, a conventional NC machining center may be able to perform all the machining operations required on, say, a transmission housing casting. Second, *part flexibility* means the ability to perform operations on more than one part type. Again, the NC machining center illustrates this flexibility, since different castings can be machined simply by loading a different part program in the machine controller. Third, and critical for FMS applications, *changeover flexibility* means the ability to change from one part type to another with negligible changeover time. This is also illustrated in the NC machining center, where changeover can be virtually instantaneous, provided that the required tools are available in the tool magazine and that the required part program is available in the machine controller.

Clearly, processing flexibility requires not only flexibility in the workstation or machine tool, but also a companion system for providing the right tools at the right time, and for insuring that the necessary part programs are available. Thus tool control, including not only tool maintenance but also distribution and scheduling, is a critical element in achieving processing flexibility in an FMS. Furthermore, FMS requires rapid changeover between both tools and part programs. For part programs, this usually requires integration between the machining center and a "cell level" control computer, either for part program selection or for downloading the part program.

While machining centers are an obvious illustration of processing flexibility, it is important to remember that processing flexibility can be achieved in assembly operations as well. For example, in the robotic assembly of printed circuit boards, components correspond roughly to tools in fabrication. Rapid component changes can be achieved by using "kits," which present the required components in a condition and orientation to support robotic assembly.

Material Handling Flexibility

For the transport of parts between operations in an FMS, or between FMS cells, three types of flexibility can be identified. (See Chapters 4.6 and 5.2 for details on material handling systems.) *Routing flexibility* means that a particular part can be delivered to any one of a number of alternative stations, depending on both the state of the part and the state of the stations. For example, depending on the result of inspection at a coordinate measuring machine, a part may be delivered to a machining center for the next series of machining operations or to a manual evaluation/rework station. If several identical machining centers are capable of performing the required sequence of operations, the specific destination can be selected to balance their workloads. There are two key features in routing flexibility. One is the simple physical ability to perform a transport operation between a specified origin and a number of potential destinations. The other is the ability to provide the destination decision in real time, rather than as a predetermined part attribute.

An example of a material handling system with very limited route flexibility would be the early generations of the so-called "transporter system," which consisted of a powered conveyor (typically a belt conveyor), with divert arms for removing a container at the appropriate workstation. At a lower level, beneath the transporter conveyor would be a second powered conveyor for returning containers to the dispatch/return station at the end of the system. In this type of system, the route for a given container is from the dispatch/return station to a specified workstation and back. Such systems are used especially in applications where manual assembly of small units is required, e.g., electronics or small power tools.

In a material handling system with total route flexibility, a part can be moved from any workstation to any other workstation. Such route flexibility can be achieved with a very broad range of material handling systems. For example, automated guided vehicles provide route flexibility, as do towline conveyors, smart monorails, car-on-track, and even conveyors. It should be obvious that route flexibility is as much a function of the material handling system's controls as it is the basic material handling hardware.

Path flexibility results from the existence of more than one possible path from a specified origin to a specified destination, combined with controls that permit an intelligent selection from among the alternatives. Path flexibility can be important in system efficiency, since it can be used to avoid congestion and blocking delays. For true flexibility, the path selection must be made in real time as opposed to predetermined on the basis of average congestion.

Configuration flexibility refers to the ease of modifying the transport system, for example, as new centers are added to the FMS, or new parts are added, requiring new station-to-station movements. Note that changing the configuration involves not only the material handling hardware, but the material handling controls as well. Configuration flexibility includes both.

In addition to being characterized by types and degrees of flexibility, material handling systems can be distinguished as either synchronous or nonsynchronous. Examples of synchronous handling systems include continuously moving conveyors and indexing conveyors. Examples of nonsynchronous handling systems include towlines, AGVS, car-on-track, power-and-free, smart monorails, inverted power-and-free, and manually operated carts or vehicles. Generally speaking, nonsynchronous handling systems are better suited to the sophisticated control found in FMS.

8.7.3.2 Tracking and Control

One of the barriers to efficiency in traditional job shops has been the problem of poorly timed resource allocation. Examples include releasing work for which a key machining operation is out of service, for which certain tools or components are not available, or overloading one process, thereby delaying many jobs and incurring excess WIP inventory. This type of resource misallocation occurs in part because the control function does not have complete information on the status of both the facilities and the jobs waiting to be done at the time a control action is required. Only recently has available technology made it feasible to continuously maintain a current status at the individual workstation and job level of detail.

Several key technological elements of continuous monitoring exist in an FMS. At the process level, developments in sensor technology, e.g., vision, tactile, infrared, ultrasound, emf, and fiberoptic, enable the automatic measurement of many part and process attributes that are critical for local control of processes, in addition to being needed for status monitoring. Also, automatic identification via bar codes, RF tags, UV sensitive paint, etc., and suitable readers permit positive tracking of pallets, fixtures, workpieces, and tools at every step of the manufacturing process. (See Chapter 4.7 on automatic identification.)

Simple acquisition of part and process attribute data is not sufficient. To be useful, the data must be processed, communicated, and maintained in a database for use by the FMS control function. Both controller technology and factory networking technology are very active areas of current and near-term development (and are addressed in more detail in Chapters 8.4 and 8.5). Note also that more than just process data is required; the current status of tool availability, pallet and fixture availability, and material availability also must be known. Thus, FMS status monitoring may require integrating data from several somewhat independent and autonomous sources.

It is all too easy to view status monitoring as "just a matter of programming," and therefore not as important as, say, the selection of the process equipment to be incorporated in the FMS. This would be a tragic mistake. One class of failure modes in an FMS is "information failure," i.e., the information on which the control system bases control decisions is in error. This type of failure is often much more difficult to recover from than a simple tool breakage or failure of a pallet to seat properly. Especially in production operations, failure of the status monitoring system can carry much higher costs than failures of specific pieces of hardware.

Currently, and for the foreseeable future, systems analysts and computer programmers are scarce. Thus, not only is the status monitoring system critical to the success of the FMS, but the skills needed for its development and maintenance are in short supply.

8.7.3.3 Comprehensive Control

The current emphasis on computer integration of manufacturing has resulted in a growing appreciation of the importance of the structure, or "architecture," of manufacturing control systems, and in particular, hierarchical control architecture. As an example, the National Bureau of Standards has proposed and implemented a five-level architecture for real-time planning and control in the Automated Manufacturing Research Facility. Table 8.7.1 summarizes the five levels and their functions. The workstation and equipment levels of control deal primarily with the real-time control of processes rather than the management control of the manufacturing system. The next higher level, cell control, is where the selection of control procedures begins to have a significant effect on the performance of the system.

For all practical purposes, the real-time control decisions for an FMS cell are essentially material movement decisions. When a part finishes processing at a workstation, two decisions are required: the disposition of the part, i.e., where to send it, and the activity to which the workstation should be assigned next. When a part leaves a workstation, it may be routed to another workstation or to a storage area. If there are alternative workstations for the next process, then one must be selected. When a workstation completes processing on a part, it becomes available for assignment to another part, if one is waiting, to maintenance, or to idle downtime. Parts that are waiting for the workstation may be in the workstation buffer or in a cell buffer area, and a decision is required regarding which part to *dispatch* to the workstation.

In addition to the routing and dispatching decisions, the cell controller may also have responsibility for the *induction* of parts into the cell, i.e., the selection from among the available candidates for the next part (or part type) to bring into the cell. Viewing the induction, dispatching, and routing decisions as material movement decisions highlights the importance of close coordination between workstation control and control of the material handling system.

The cell-level control decisions should be made in real time, i.e., as parts complete processing, to avoid idle time at the workstations. Also, these decisions require an accurate, current status for workstations and parts in process. Thus, monitoring is another critical cell-level function.

The next higher level, shop control, is concerned with control decisions that can be made on a longer time cycle, or periodically, say, every shift or every day. These decisions address the determination of the set of part types to assign to each cell and the coordination of the flow of parts between cells. The terms *batching* and *balancing* refer to the selection of parts to assign to a cell so that tooling constraints are not violated (the

Table 8.7.1. NBS Hierarchical Control Architecture

Level	Functions
1. Facility	Information management Manufacturing engineering Production management
2. Shop	Task management Resource allocation
3. Cell	Task analysis Batch management Scheduling Dispatching Monitoring
4. Workstation	Setup Equipment tasking Takedown
5. Equipment	Machining Measurement Handling Transport Storage

batch of part types does not require more tooling than can be allocated to the cell) and so that the workstations are reasonably balanced with regard to work load. In making the batching and balancing decisions, shop control should be guided by the master production plan generated at the facility-control level.

An important aspect of the hierarchical control architecture is that each level of the control system provides the framework within which the next lower level of the system must operate. Also, the time span of the decisions tends to decrease at lower levels of the hierarchy, and the scope, either geographically or in terms of the number of operations, tends to narrow.

FMS control problems are difficult, both practically and in theory. At present, there is little in the way of a scientific basis for specifying the control procedures. Thus, however complex and sophisticated they may be, FMS control rules tend to be rather ad hoc. The ad hoc nature of the control rules, together with the unique dynamic behavior in each specific application, makes the comparison of alternative control procedures difficult. Currently, the most reliable approach to selecting between alternative control procedures is making direct comparisons via extensive, detailed simulation analysis. Unfortunately, this can be difficult, time-consuming, and quite expensive, which discourages the search for better control procedures. Although it is not possible to estimate the potential payoff, it should be noted that almost all research on related "pure" scheduling problems seems to indicate that simple control rules do not give the best performance except in very special cases.

8.7.4 GETTING STARTED

The FMS philosophy of production represents a radical departure from conventional methods, and requires the application of sophisticated technologies in an integrated and comprehensive way. Achieving the potential benefits of the FMS approach requires careful preparation, planning, and execution. Jack Bradt, Chairman of the Board of SI Handling, and a leading proponent of FMS and computer integrated manufacturing, recommends the following steps as essential in planning and implementing a corporate FMS program:

1. *Study*. Understand the state of the art in FMS technology and decide if it can be applied successfully in your business.
2. *Commit*. Assign a high level executive to head up the FMS program (several progressive companies have assigned this responsibility at the vice-president level).
3. *Act*. Form an FMS task force with specific charge to:
   a Educate themselves regarding relevant FMS technology.
   b. Educate other key individuals in the corporation.
   c. Develop relationships with external FMS partners who can assist in system planning and development.
   d. Prepare a long-range FMS plan.
   e. Coordinate FMS prototype construction and testing.
   f. Coordinate construction and startup of production FMS.
4. *Set goals*. Based on your business and the state-of-the-art FMS technology, set goals for worker productivity, asset utilization, and product quality; also anticipate impacts on design, engineering, and maintenance functions.
5. *Justify*. Encourage financial justification, but recognize the difficulty of quantifying so-called intangibles, and the importance of "critical mass," i.e., one FMS cell may not be justified by itself, whereas several mutually supportive cells may be justifiable.
6. *Plan*. Develop a strategy and a long-range plan for factory automation, but recognize that implementation will likely be on a piecemeal basis; select a good FMS application to get started.
7. *Prototype*. Developing a prototype based on your people, products, and processes can help to reduce uncertainties and risks, familiarize employees with FMS requirements, identify particular implementation problems, and smooth the transition to production FMS.

These seven steps emphasize the importance of dealing with the FMS philosophy in the context of your own business and corporate environment.

The difficulty of changing production philosophies should not be underestimated. Note that FMS presumes adopting a just-in-time approach to production (see Chapter 3.16). Therefore, full transition from conventional manufacturing to FMS involves the difficulties of implementing JIT, plus the integration of complex FMS technologies. Note also that an FMS "island" may have very little benefit if it is surrounded by a manufacturing system that continues to operate with quality problems, large WIP inventories, and long production lead times.

8.7.4.1 An Action Plan

Once the decision to implement an FMS cell has been made, a specific action plan is needed. The *FMS Handbook* suggests the seven basic steps listed in Table 8.7.2.

In selecting part types for an FMS or FMS cell, bear in mind that some parts may not be appropriate for FMS production. It may be necessary to maintain a small "conventional manufacturing" cell for those parts that do not fit well into an FMS cell. It may also be the case that only a relatively small fraction of all part

Table 8.7.2. Steps in FMS Adoption

1. Select part types and process technologies
2. Design alternative configurations
3. Evaluate alternative configurations
4. Draft request for proposal
5. Evaluate vendor proposals
6. Install and debug FMS
7. Operate FMS

types appear to be candidates for FMS production, resulting in an FMS production system in parallel with the conventional production system.

Redesigning products should be given serious consideration, and a preliminary assessment made of the impact of redesign on the number of parts that are candidates for the FMS. In addition, process selection should consider the potential evolution of the product line, especially if it is a product with a relatively short life cycle. Recognize that by the time the FMS is installed and debugged, many engineering design changes in the product will have been made. Process selection may identify several alternatives (e.g., horizontal vs. vertical machining, or automated vs. robotic insertion) to be considered in more detail in steps 2 and 3.

The FMS *configuration* includes the fixturing, the number and specifications of the various machine tools, the material handling system, and the physical layout. Thus, designing a configuration involves several phases. First, a fixturing concept is needed, based on part geometries and processing requirements, but also considering the material handling system and processing technologies. A rough cut process plan for each part type is used to estimate processing requirements, which together with estimated production requirements, provides the basis for estimating required part cycle times and tool usages. In the second phase, the specific types of process equipment or methods are specified. Based on the previously determined processing requirements, production efficiencies, and other factors, the number of machines or workstations is determined. The final phase involves specifying a material handling system and the physical layout of the FMS. Obviously, a large number of decisions and estimates are involved, and alternative configurations may result from different decisions or estimates.

Performance and cost are two critical measures for evaluating alternative FMS configurations. For most systems, the only reliable method for evaluating performance is a detailed simulation, based on the particular operating rules that will be implemented. (See Chapter 8.6.) The development of detailed simulation models and the data necessary to use them is a time-consuming and often expensive endeavor. Investment analysis would seem to be straightforward, but as pointed out earlier, it is usually very difficult or impossible to assess the secondary impacts of the FMS on overheads, inventories, etc. The evaluation of candidate configurations remains as perhaps the least well understood step in the process of adopting FMS.

In writing a request for proposal (RFP), it is very important to specify performance requirements and the protocol for acceptance testing. There seems to be general agreement that it is best to allow potential vendors to specify the details of the hardware rather than specifying them in the RFP. It may be important, however, to specify the format of the planning and control system, since different approaches to the balancing and batching decisions can lead to quite different system performances.

The procedures used for evaluating vendor proposals will vary from company to company. There are, however, some basic questions to ask in evaluating any proposal. Does the proposed system meet the performance requirements? (How was that established by the vendor?) Are the costs reasonable and complete? (Has the vendor omitted any important cost item?) Does the proposal address any of the important intangibles? Is the project management proposal sound (especially critical if there are subcontractors or multiple vendors)? Does the vendor have enough experience, or is he "going to school" on your project? Is the vendor financially stable? (Will he be around in three to five years to follow up with system modifications?) While it is important to consider the "numbers," i.e., costs, when choosing a vendor, the intangibles are also very important, especially for projects as complex as FMS.

Selection of an FMS vendor is really only the beginning of a successful FMS project. Careful attention should be given to preparing the site, the people, and the production operations that must interface with the FMS. The procedures used to select and train technicians to work in the FMS, along with job specifications, pay scales, and advancement programs, will help to determine the acceptance, and therefore the success of the FMS. Since an FMS will typically include highly sophisticated automated equipment, the requirements for maintenance may be quite different from other factory equipment. Preparation for changes in the maintenance function should be made before the FMS is operational.

8.7.4.2 An Alternative Action Plan

The steps listed in Table 8.7.2 are predicated on having the capability, either in-house or through outside help, to perform a number of detailed complex analyses. Many companies today feel a need to implement at least a limited FMS cell, but do not have the resources (people and time), to do a lengthy preproposal study. In this case, the following alternative to steps 2 and 3 in the table is suggested.

First, develop a detailed database describing the parts to be produced in the FMS, based on the current production method. Included in the database should be the typical lot sizes, the annual requirements, process plans, and standard production hours for the current method. Second, specify how the FMS cell will interface with the rest of the manufacturing system. For example, will parts be required one at a time or in batches? If the rest of the manufacturing system is batch oriented, it may be essential to treat parts in the FMS as batches with hard due dates, even though they may be processed one unit at a time. These are important implications for the FMS planning and control system. Together, they provide an FMS operating scenario for use in system design.

Using the FMS operating scenario, develop a detailed specification for how the FMS should perform, especially vis-à-vis its interface with the rest of the manufacturing system. Include both the operating scenario and the performance specification in the RFP and require that any proposal demonstrate its feasibility with regard to the operating scenario.

This alternative approach to implementing FMS will almost surely limit the benefits that might be obtained, because it does not address product redesign or structural changes in the production system. It is clearly an approach that leads to "islands of automation" and therefore not an ideal approach. Nevertheless, it is an approach that seems to be widely used today as a way to get started in FMS.

8.7.4.3 The Role of Modeling and Analysis

The complexities involved in planning, installing, and operating an FMS can be staggering. Consider that an FMS is often proposed to replace an existing manufacturing process, and may include from a few up to several dozen highly complex, automated, and computer controlled workstations, linked together by an automated, computer controlled material handling and storage system. The FMS may contain hundreds of tools, require constant monitoring of the tools, and a system for automatic tool changing. Anywhere from a dozen to as many as several hundred part types may be produced in the FMS. Perhaps most importantly, the FMS may not produce all the parts required, or may not perform all the required operations.

To describe an FMS completely may require an enormous database containing detailed information on parts, process plans, operation times, workstation capabilities, workstation configurations, material handling system capabilities, etc. All this information is considered, either subjectively or objectively, in making design and control decisions. The role of modeling is:

1. To provide a framework for organizing the large amounts of information describing the problem.
2. To provide standard procedures for manipulating data.
3. To provide guidelines and suggestions for making design and control decisions.

The development of models and methods of analysis for FMS planning and control problems is quite a challenging task, since almost all these problems seem very difficult. Fortunately, in the past few years these problems have received more attention, and some progress has been made in developing useful models.

The remainder of this section will review some of the more important models developed for FMS design and control. The presentation will be organized into three broad categories, corresponding to different levels in the hierarchy of FMS design and control:

1. *Hard system configuration,* i.e., setting the system capabilities; usually the selection of equipment and parts to be processed.
2. *Soft system configuration,* i.e., loading the tools and parts to be processed over some relatively short period, from a day to several weeks.
3. *Real-time control,* i.e., induction of jobs into the system, controlling the flow of jobs within the system, and responding to contingencies such as tool failures, equipment breakdowns, etc.

Rather than a comprehensive review of FMS modeling, this presentation will be an introduction, giving a flavor of the types of models available, the issues important in developing and using models, and the limitations of modeling.

Hard System Configuration

The term *hard system configuration* reflects the setting of limits on the degree of flexibility available in the FMS. The specific equipment, both processing and material handling, is determined by the hard configuration. Hard configuration includes tool holders, fixtures, and pallets, and thus limits the system changes that can be made in the future without major new investments. Note that hard configuration also places limits on the attributes of parts that may be introduced in the future.

There are two distinct classes of models relevant to the hard configuration decision: *normative*, or optimization-based models; and *evaluative* models. Normative models attempt to quantify one or more criteria that are to be optimized, subject to a set of mathematically precise constraints on the decisions. Evaluative models, on the other hand, operate on a set of user-supplied decisions, and attempt to provide an estimate of one or more system performance variables based on an internal representation of the actual system. The utility of both types of models is limited by the accuracy of their mathematical representation of the actual FMS.

An example of a normative model is used by Whitney and Suri[1] in two procedures, called PAMS and PARSE, for analyzing the *part/machine selection* problem. The broad objective in this problem is to select the part types to be produced in the FMS and determine the number and types of machines to be incorporated, so that the total economic benefit from the FMS is maximized. In their model, economic benefit is measured by the annual cost savings for FMS production less the annualized investment cost of the FMS.

The PAMS/PARSE programs are based on a linear programming model of the problem. The decision variables correspond to selections of parts from the list of candidate parts and machines from the list of candidate machine types. Each part type has an annual cost savings for production in the FMS, and there is an annualized investment cost for each machine type. The constraints simply force enough machine time to be available to accommodate all the part types selected for the FMS. An important feature of linear programming models it that the incremental cost impact of a part is independent of the other parts that have been selected.

While this seems straightforward, the model requires quite a lot of detailed data. For example, for each candidate part type, it requires the current variable production costs and also the costs that would be incurred if the part were processed on each of the candidate machine types. These numbers may be difficult to determine.

Evaluative models avoid some of the problems of normative models by allowing the model builder to construct any desired representation of the FMS. Of course, there are tradeoffs involved, since not all representations will be equally easy to work with. Furthermore, evaluative models require the analyst to provide the decisions, so using them to find the best system configuration can be a lengthy process.

Evaluative models generally produce estimates of such performance measures as average machine utilization, part type production rates, average number of parts in queue, etc., based on user-supplied information about the system configuration. For example, the user may have to specify the number of part types selected, the number of workstations selected, the part routing, the distributions of process times at each station, the number of pallets, fixtures, and tools, etc. There are three broad classes of evaluative models, based on the level of detail used to represent the FMS.

Static allocation models are very simple arithmetic models that ignore dynamics, interactions of all sorts, and uncertainties. For example, if an FMS contains a vertical milling machine, then a static allocation model would simply add up all the processing time for vertical milling operations, and divide by the time available per machine to determine the number of machines required. Because static allocation models involve so many simplifications, the performance estimates they provide tend to be overly optimistic. They can, however, be useful in the early development of system concept by rapidly eliminating the clearly infeasible designs.

Queueing network models provide a way to capture some of the time varying performance attributes by considering the dynamics, the interaction of parts through buffers and contention, and the uncertainties in processing times and machine availability. However, queueing network models still involve some simplifications. For example, in a recent paper,[2] Yao describes a simple model that allows limited buffers in front of each workstation, and uses a part routing scheme that is based on the shortest queue, i.e., the next part delivered to a machine is selected on the basis of the shortest machine queue and the urgency rating for all parts that need to go to that machine.

Queueing network models almost always require a number of simplifying assumptions in order to make them analytically and computationally tractable. For example, it is usually assumed that the processing times for each part at each workstation are drawn from an exponential distribution. This means that the variance of the processing times is equal to the mean, and may be a poor assumption for NC equipment, where one would expect the times to be almost constant for a given part type. In Yao's model, e.g., other assumptions include:

There are N pallets, and whenever a part is finished, another part of the same type is immediately loaded.

There are c "carts" or transporters, which are completely independent, i.e., they never interfere with one another.

Whenever a part finishes at a workstation, it is returned instantly to a central material handling (or WIP buffer) area, without the use of one of the c carts.

The time to process a part at the central station and deliver it to its next station is drawn from an exponential distribution.

The service discipline at each workstation is random, i.e., the next job is selected from the queue at random.

Although these assumptions may seem overly restrictive, there is considerable evidence that the queueing network models are fairly robust, in the sense of giving good estimates even when the assumptions are violated. Other queueing network models require similar types of assumptions.

Queueing network models are a very active area of research. Other important references include Suri and Hildebrant[3] on mean value analysis of queues (MVAQ) and Shalev-Oren, et al.[4] on mean value analysis with priority scheduling (PMVA). While queueing network models are relatively easy to work with, require minimal input data, and provide rapid response, they almost always need to be backed up by some more detailed evaluation of system performance, because of the many simplifying assumptions required.

Simulation models allow a very detailed and realistic representation of the system, and can be based on general discrete event simulation languages, e.g., see Schriber[5] on specialized simulation packages. The appeal of simulation modeling is that it allows, at least in theory, an extremely accurate description of the system. On the other hand, such detailed simulation models tend to be expensive to develop from scratch, require large amounts of data, and can be expensive and time-consuming to use. Detailed simulation models remain, however, the best alternative for fine tuning the design of an FMS.

Simulation models can also be developed at a level of detail similar to that for queueing network models, but without the necessity for stringent assumptions about processing times and part handling disciplines. Such simulation models are somewhat easier to build, and can prove useful in determining approximate throughput, sizing and locating buffer storage, and selecting a part release and scheduling procedure. An important element in using this type of simulation model is selecting a format for the results that is appropriate for the design application.

As a final note on evaluative models, it should be pointed out that in most cases the model provides only a "point estimate" of performance, i.e., an estimate that is valid only for the design data provided as input. However, recent work on a technique called *perturbation analysis*[6] provides not only the usual point estimates, but also indicates the effect of marginal changes in performance as the design variables are modified. For example, perturbation analysis applied to a simulation model might give not only the throughput and flowtimes for the given buffer capacities, but also might indicate the magnitude of the change in throughput and flowtime that could be expected if a given buffer size were increased or decreased by one unit. This is a very powerful tool for reducing the amount of time that it takes to find a good design using evaluative models.

Soft System Configuration

It is not unusual to find that an FMS has more part types assigned to it than it can process at one time, because of limits on pallets, WIP, or tool capacity at the machines. When this occurs, it is necessary to determine, over some planning interval, exactly which parts to produce and how to allocate the system resources. The *batching problem* is to select the parts that will be processed over the planning period, and should consider factors such as the requirement for parts (e.g., part runout times), and machine utilization (since low utilization implies low throughput). Once parts have been selected for processing and the required tools are known, there is a *balancing problem*, i.e., assign tools to (configure) the machines and assign part routings to achieve the largest possible machine utilization.

These problems, along with several others, were defined by Stecke,[6] who also gives a model and solution algorithm for the balancing problem. Stecke's balancing model was developed for a machining application, and involved a very elaborate scheme for representing the tool magazine capacity constraint. As a result, the problems were difficult to solve when there were more than a few machines, tools, and part types. An improved solution algorithm, which could accommodate problems with up to 9 machines and 48 operations, was given by Berrada and Stecke[7].

The Stecke model for balancing considers only the workstation processing time imbalance in searching for an optimal solution. That is, it is based on the assumption that other factors, such as part requirements, were considered in selecting the parts to be processed. In contrast, the approach used by Whitney and Gaul[8] explicitly addresses both problems. Their approach uses a probabilistic desirability index and a sequential selection algorithm based on dynamic programming. A potential drawback of their method is that the batching solution may not be feasible. If that happens, they suggest an iterative scheme to refine the batching solution toward feasibility.

An important observation, from the perspective of material handling, is that neither of these approaches considers the number of times that a part is moved. In other words, neither the number of workstations visited nor the number of times that a part revisits a given workstation is addressed. Thus, it is possible that the batching and balancing solutions may induce a large workload for the material handling system. Simulation studies indicate that even small increases in the average material handling system response time can cause significant degradation in the flowtime and throughput performance of an FMS. Unfortunately, the simultaneous consideration of workstation balance and part movement leads to very difficult problems in the general case.

The only work to date which does consider both the balancing and material handling criteria is by Ammons, Lofgren, and McGinnis[9] for a problem in flexible assembly systems. In this problem, there are no technological precedence constraints between operations, so that when a part visits a workstation, it is guaranteed to receive all the operations it needs at that station, and will never revisit it. Their algorithm is based on cluster analysis, modified bin-packing algorithms, and a user selected trade-off between workstation imbalance and total number of transactions by the material handling system. The algorithms were used on a microcomputer to solve problems with 3 workstations, 30 part types, and 90 operations. The solution required only two to three minutes. At this point, the computational limitations are on permissible array size rather than solution time.

It should be noted that the paradigm for batching and balancing implied in these models is that the decisions are made periodically and remain in force until the next decision point. Within this paradigm, the batching and balancing problems are at least approachable as normative optimization models, although they may prove too difficult to optimize in practice. Moreover, there is still a need to *evaluate* the solutions, based on actual real-time control rules, to see if the planned throughput can be achieved. There is, however, another paradigm that has not yet received much research attention. It is conceptually possible to modify the part mix in a dynamic fashion and also to reassign tools dynamically. Both are technically feasible, although the control system implications have not been investigated.

Real-Time Control

In operation, an FMS presents very complex control problems due to the number of active part types, the necessity of coordinating material handling activities with the processing at the workstations, and the possibility of tool failures and equipment breakdowns. For this reason, it would be impractical in most cases to attempt to preschedule all FMS activities, e.g., for a shift, because many changes to the plan would almost certainly be required. Instead, most FMSs incorporate a real-time controller that provides the necessary response to contingencies such as breakdown or variations in completion times of operations.

The essential decisions made by the FMS real-time controller are:

1. When to introduce another part into the system, and what part to introduce.
2. The disposition of parts when they complete processing at a station, i.e., to send them to the next station in their routing or to hold them (if possible) in a central buffer area.
3. When a workstation becomes available, which job to load next on that station.
4. Dispatching and routing of the material handling equipment.

For some systems, some of these decisions may be trivial. For example, if there is only a buffer of one job in front of a workstation, then there is at most one alternative for decision (3).

Although an FMS is conceptually very similar to the traditional job shop in many respects, it is surprising to find that there is very little in the way of models for the FMS real-time control problem from the vast literature on scheduling. This may reflect a predominantly "batch processing" mentality among researchers, which is inappropriate for the real-time control of systems. It may also reflect the fact that traditional job shop scheduling is very concerned with setup times, whereas in an FMS, setup times have been reduced drastically.

Because the influence of setup times has been reduced in an FMS, it is possible to treat certain real-time control problems via optimization models. For example, Kimemia and Gershwin[10] developed a model and procedures for controlling the induction of jobs into an FMS, based on concepts from control theory. The problem they studied had a relatively constant requirement for a fixed set of parts, as might be found in a fabrication shop feeding an assembly line. Later, Akella, Choong and Gershwin[11] enhanced this approach to overcome excessive computational requirements and to improve the solution performance.

A categorization of workflow control problems is given in Han and McGinnis,[12] along with linear programming based control procedures for problems in which the rate of production is to be controlled. These procedures are computationally very simple; in fact, amenable to solution via a microcomputer. This is an important consideration for real-time control of FMSs, since such control is not likely to reside in a main frame "number cruncher."

In some FMS applications, the jobs arrive to the FMS with due dates, rather than having a requirement for the parts to be produced according to a specified consumption rate. Alternatively, the MRP master schedule may specify a required availability date, and the problem is to determine when the job should be released. In either case, there is some existing literature for setting due dates and for scheduling jobs with due dates. In Han and McGinnis[13] the shop operating characteristic curve methodology is used to develop a procedure for setting due dates (or, equivalently, release dates), and a new algorithm is given for the real-time scheduling of jobs with due dates.

The problem of real-time control with due dates is qualitatively more difficult than the problem of controlling the flow of parts to satisfy some target production rate. Thus, at least at this time, it appears that the control algorithms will continue to be dispatch rule oriented rather than optimization oriented.

8.7.4.4 Modeling Summary

There have been significant strides in the development of models for designing and operating flexible manufacturing systems. For design problems, the models tend to be evaluative, based either on some type of queueing formulation or on simulation. For soft configuration problems, a number of useful optimization formulations have been developed, although in some cases the solution procedures are heuristic. Finally, models for the real-time control problem are optimization based, if the problem is to control flow rates, and dispatching rule based if the control problem includes due dates.

The design and control of an FMS presents a set of complex decision problems, often with conflicting criteria and large amounts of detailed information. Models provide a framework for organizing the data, and, properly used, can improve the ultimate system performance.

8.7.5 CONCLUSION

FMS is a highly competitive manufacturing strategy because of its capability to quickly and efficiently adapt to changes in product mix, demand, and design. Inherent FMS flexibility is based on its system structure: versatile production stations with minimal changeover time between part types networked together by a versatile material handling system capable of timely movement of parts and tools in any desired routing sequence. Typically, computer control of part and tool storage and handling exploits the versatility of both the production and the handling systems to give the desired flexibility.

FMS has evolved from a few relatively simple machining applications in the mid-1960s to a wide diversity of applications worldwide today. Example systems range from machining applications and fabrication all the way to assembly processes. Linking these various processes are a diverse set of material handling systems such as towline conveyors, carousels, transporters, microload totestackers, automated guided vehicles, and even manual push carts.

Justifications for FMS may be hard to quantify. Typical justifications may include required flexibility for market competitiveness, reduction of work-in-process inventory, and quality assurance. Additionally, the process of designing and implementing an FMS imposes a discipline on the manufacturing system which provides for further potential benefits.

In this chapter, the key elements of an FMS were described functionally and pertinent characteristics denoted. Flexibilities associated with processing/tooling in the context of operation flexibility, part flexibility, and changeover flexibility were discussed. Material handling flexibilities were presented for routing flexibility, path flexibility, and configuration flexibility. Tracking and control issues were described, and problems associated with comprehensive control were detailed including routing and dispatch decisions, induction, and balancing and batching.

Steps for designing and installing an FMS were outlined and an action plan listed. The role of modeling for the various problems associated with hard configuration, soft configuration, and real-time control was overviewed with examples of specific types of methodologies. The use of these models to enhance overall FMS performance was emphasized.

The ongoing developments in computers and automation technology continue to promote the development of competitive manufacturing strategies. The capabilities and versatilities of the FMS approach make this philosophy especially promising for the pressing demands of the future. The challenge to design, implement, and operate effective and efficient FMS production facilities is the reachable challenge of today.

REFERENCES

1. Whitney, C. K., and R. Suri, "Algorithms for Part and Machine Selection in Flexible Manufacturing Systems," *Annals of Operations Research*, 1985.
2. Yao, D. D., "An FMS Network Model with State-Dependent Routing," *Proceedings of the First ORSA/TIMS Special Interest Conference on FMS*, Ann Arbor, Michigan, 1984
3. Suri, R., and R. R. Hildebrant, "Modelling Flexible Manufacturing Systems Using Mean-Value Analysis," *Journal of Manufacturing Systems*, **3**(1), 1984, 27–38.
4. Shalev-Oren, S., A. Seidman, and P. J. Schweitzer, "Analysis of Flexible Manufacturing Systems with Priority Scheduling: PMVA," *Proceedings of the First ORSA/TIMS Special Interest Conference on FMS*, Ann Arbor, Michigan, 1984.
5. Schriber, T., and F. B. Talbot, "The Use of GPSS/H in Modeling a Typical FMS," *Proceedings of the First ORSA/TIMS Special Interest Conference on FMS*, Ann Arbor, Michigan, 1984.
6. Suri, R., and J. W. Dille, "On-Line Optimization of Flexible Manufacturing Systems Using Mean-Value Analysis," *Proceedings of the First ORSA/TIMS Special Interest Conference on FMS*, Ann Arbor, Michigan, 1984.
7. Berrada, M., and K. E. Stecke, "A Branch and Bound Approach for FMS Machine Loading," *Proceedings of the First ORSA/TIMS Special Interest Conference on FMS*, Ann Arbor, Michigan, 1984.
8. Whitney, C. K., and T. S. Gaul, "Sequential Decision Procedures for Batching and Balancing in FMSs," *Proceedings of the First ORSA/TIMS Special Interest Conference on FMS*, Ann Arbor, Michigan, 1984.
9. Ammons, J. C., C. B. Lofgren, and L. F. McGinnis, "A Large Scale Machine Loading Problem in Flexible Assembly," *Annals of Operations Research* (3), 1985, 319–332.
10. Kimemia, J. G., and S. B. Gershwin, "An Algorithm for the Computer Control of Production in Flexible Manufacturing Systems," *IIE Transactions*, **15**(4), 1983, 353–362.
11. Akella, R., Y. Choong, and S. B. Gershwin, "Performance of Hierarchical Production Scheduling Policy," *Proceedings of the First ORSA/TIMS Special Interest Conference on FMS*, Ann Arbor, Michigan, 1984.
12. Han, M. H., and L. F. McGinnis, "Operating Characteristic Curves for Shop Management and Control," working paper, Material Handling Research Center, Georgia Institute of Technology, 1985.
13. Han, M. H., and L. F. McGinnis, "Workflow Control Models for Computer Controlled Production," working paper, Material Handling Research Center, Georgia Institute of Technology, 1985.

BIBLIOGRAPHY*

"An Aerospace CIM System," *Manufacturing Engineering* (**91**)4, 1983, 67–68.

"An Award-Winning CIM System," *Manufacturing Engineering* (**90**)2, 1983, 49–50.

* In addition to the sources listed as references.

Andel, Tom, "Aerospace Manufacturer Unveils Flexible Manufacturing Components," *Material Handling Engineering* **(92)**5, 1984, 48–54.

Asfahl, C. Ray, *Robots and Manufacturing Automation*, New York: Wiley, 1985.

"Automating Applicance Manufacturing," *Manufacturing Engineering* **(91)**2, 1983, 61.

Behne, Thomas A., "Opel Engine Line Goes Flexible," *Automotive Industries* **(164)**8, 1984, 23–24.

Bergstrom, Robin P., "Automated Pump Assembly Pays Off," *Manufacturing Engineering* **(92)**6, 1984, 44–47.

Bergstrom, Robin P., "FMS: The Drive Toward Cells," *Manufacturing Engineering* **(95)**2, 1985, 34–38.

Bryce, A. L. Graham, and P. A. Roberts, "Flexible Machining Systems in the U.S.A.," *Proceedings of the 1st International Conference on Flexible Manufacturing Systems*, New York: IFS Publications and North Holland, 1982.

Burgam, Patrick M. "Flexible Fabrication Moves in at Hughes Aircraft," *Manufacturing Engineering* **(91)**3, 1983, 56–57.

"Control Key for Harris FMS," *Manufacturing Engineering* **(91)**3, 1983, 60.

"Conveyor Systems for Flexible Assembly Operations," *Modern Materials Handling* **(40)**8, 1985, 67–70.

Cooper, R., and R. Jaikumar, "Management Control of the Flexible Machining System," *Proceedings of the First ORSA/TIMS Special Interest Conference on FMS*, Ann Arbor, Michigan, 1984.

Curtin, Frank T. "Automating Existing Facilities: GE Modernizes Dishwasher, Transportation Equipment Plants," *Industrial Engineering* **(15)**9, 1983, 32–38.

Dallas, Daniel B., "The Robot Enters the System," *Manufacturing Engineering* **(82)**2, 1979, 70–73.

Drozda, Thomas J., "Flexible Assembly System Features Automatic Setup," *Manufacturing Engineering* **(93)**6, 1984, 75–76.

Dupont-Gatelmand, Catherine, "A Survey of Flexible Manufacturing Systems," *Journal of Manufacturing Systems*, **1**(1), 1983, 1–15.

Eversheim, W., and P. Herrmann, "Recent Trends in Flexible Automated Manufacturing," *Journal of Manufacturing Engineering*, **(1)**2, 1982, 139–147.

"Facility Run By 4 Workers," *CIM Technology* **(3)**4, 1984, 15.

"Flexible Machining for High-Volume Boring," *Manufacturing Engineering* **(91)**3, 1983, 59.

"Flexible Machining System to Make Army Vehicle Parts," *Tooling and Production* **(50)**4, 1984, 80–81.

Flexible Manufacturing Systems Handbook, Park Ridge, NJ: Noyes Publications, 1984.

"FMS at GE," *Manufacturing Engineering* **(91)**3, 1983, 66–67.

Hartley, John, *FMS at Work*, New York: IFS Publications and North Holland, 1984.

Hegland, Donald E., "Flexible Manufacturing—A Strategy for Winners," *Production Engineering* **(29)**9, 1982, 41–46.

Holland, James R., Ed., *Flexible Manufacturing Systems*, Dearborn, MI: Society of Manufacturing Engineers, 1984.

Klahorst, H. Thomas, "Flexible Manufacturing Systems: Combining Elements to Lower Costs, Add Flexibility," *Industrial Engineering* **(13)**11, 1981, 112–117.

Klahorst, H. Thomas, "How to Plan Your FMS," *Manufacturing Engineering* **(91)**3, 1983, 52–54.

Knight, J. A. G., "The Latest Developments of FMS in Japan," *Proceedings of the 1st International Conference on Flexible Manufacturing Systems*, New York: IFS Publications and North Holland, 1982.

Knill, Bernie, "Vought Aero Products Reveals New Lessons in Flexible Manufacturing," *Material Handling Engineering* **(40)**1, 1985, 78–85.

Krauskopf, Bruce, "Flexible Manufacturing of Machine Tools," *Manufacturing Engineering* **(93)**2, 1984, 141–142.

Lofgren, Gunnar, "Automatic Guided Vehicles Perform As Production Line Systems," *Industrial Engineering* **(13)**11, 1981, 36–38.

"Mazak's FMS: Making Machine Tool Manufacturing Look Easy," *Manufacturing Engineering* **(91)**3, 1983, 63.

Millar, William G., "Flexible Manufacturing Systems," *Data Processing* **(26)**9, 1984, 30–32.

Mortimer, John, (Editor) *The FMS Report*, New York: IFS Publications, 1984.

Ohmi, T., Y. Ito, and Y. Yoshida, "Flexible Manufacturing Systems in Japan," *Proceedings of the 1st International Conference on Flexible Manufacturing Systems*, New York: IFS Publications and North Holland, 1982.

"Our Flexible Systems Help Us Make a Better Product," *Modern Materials Handling* **(39)**10, 1984, 46–49.

"Planning Dismantles the Barriers: Networks of Flexible Flow," *Production Engineering* **(31)**3, 1984, 44–48.

Ranky, Paul G., *The Design and Operation of FMS*, New York: IFS Publications and North Holland, 1983.

"Robotized Servo Manufacture and Assembly," *Manufacturing Engineering* **(90)**6, 1983, 53–55.

Rohan, Thomas M., "Tooling Up for Flexibility," *Industry Week* **(214)**5, 1982, 34–38.

Rolston, L. J., "Modeling Flexible Manufacturing Systems with MAP/1," *Proceedings of the First ORSA/TIMS Special Interest Conference on FMS*, Ann Arbor, Michigan, 1984.

Rooks, B. W., Ed., *Proceedings of the Second Annual Conference on Automated Manufacturing*, New York: IFS Publications and North Holland, 1983.

Schwind, Gene, "GM's Orion Assembly Plant: Another Step Up the Automation Ladder," *Material Handling Engineering* (**39**)5, 1984, 36–41.

Spur, G., and K. Mertins, "Flexible Manufacturing Systems in Germany, Conditions and Development Trends," *Proceedings of the 1st International Conference on Flexible Manufacturing Systems*, New York: IFS Publications and North Holland, 1982.

Stauffer, Robert N., "Flexible Manufacturing Systems: Bendix Builds a Big One," *Manufacturing Engineering* (**87**)2, 1981, 92–93.

Stauffer, Robert N., "General Electric's CIM System Automates Entire Business Cycle," *CIM Technology* (**3**)4, 1984, 20–21.

Stecke, K. E., "Formulation and Solution of Nonlinear Integer Production Planning Problems for Flexible Manufacturing Systems," *Management Science*, (**29**)3, 1983, 273–288.

"The Coming of the Automatic Factory," *Manufacturing Engineering* (**84**)3, 1980, 69–75.

"The Remaking of a Chrysler Plant," *Fortune* (**108**)10, 1983, 178–185.

"We Achieved Flexible Flow to 250 Assembly Stations," *Modern Materials Handling* (**39**)14, 1984, 48–51.

CHAPTER 8.8

ADVANCED FLEXIBLE MANUFACTURING SYSTEMS

SAMUEL B. KORIN
JOHN C. LUBER

IBM Corporation
Thornwood, New York

8.8.1 INTRODUCTION

8.8.1.1 Definition

The concept of the flexible manufacturing system (FMS) was treated in Chapter 8.7. In summary, its purpose is to provide the ability to manufacture a defined, broad spectrum of product types or component part numbers through programmable tooling, routings, and material handling systems; in such a system, the transition from one product type or part number to another on the manufacturing floor is done automatically, under computer control, generally without human intervention. Conceptually analogous to the achievement of "economy of scale" in mass-production manufacturing, business performance in the FMS is attained through what can be described as "economy of scope."

In this chapter, since the topic of the FMS has already been discussed, consideration of the Advanced Flexible Manufacturing Systems (AFMS) will be treated in the following context.

Advanced means continuing and extending the manufacturing system integration process

1. to effectively accommodate a part-number set sufficiently large that the capability for product variation is not constrained, and
2. to incorporate additional adaptive control features as they become technically and economically feasible.

Examples are distributed microprocessor intelligence applied to conditional routings; self-diagnostic, self-correcting maintenance; and automatic verification of conformance to specification, including automatic calibration and compensation for mechanical wear.

Flexible means

1. expanding the capability of the conventional FMS (which generally fabricates or processes families of parts or assemblies very similar in material content and geometric configuration), and
2. accommodating a broader variety of materials and geometries through the application of greater control intelligence, such as provided by advanced robotics and materials handling systems.

Product flexibility is the prime requirement around which the AFMS will be designed.

To provide the capability to meet this essential AFMS characteristic, a set of manufacturing systems elements must be planned, implemented, and integrated in such a way as to allow the model/part number requirements to be virtually "transparent" to the system and to provide for flexibility in the utilization of materials and processes.

In general, the attributes of transparency and flexibility are made possible through integrated computer control and integrated materials handling, which are prerequisites for an integrated manufacturing system.

8.8.1.2 Scope

No one integrated system configuration can be specified for the design of an AFMS. Nor, indeed, does one set of product parameters exist that would characterize a manufacturing system as an AFMS. The objectives of a particular business, the needs of its marketplace, and the nature of the product itself will determine the degree of flexibility required. But in specifically planning an AFMS, a series of uniquely relevant statements can be made with respect to developing the functional objectives for such a system. They can be addressed in these three categories:

1. Functional objectives: product.
2. Functional objectives: manufacturing system.
3. Functional objectives: the physical facility

In this chapter, as we identify a set of product design and manufacturing system features as objectives for planning and operating an AFMS, we will reference those sections of this volume that discuss those features in more detail.

8.8.2 FUNCTIONAL OBJECTIVES: PRODUCT

8.8.2.1 Projection of Product Parameters

The first consideration must be given to the *functional objectives* for the *product* to be manufactured, as it must be for the design of any manufacturing system. In the context of the AFMS, a prime activity must be the estimation of range(s) for these particular *product parameters*:

1. *Physical size.* Dimensions and weight, so that the functional requirements for the equipment and the "flexible" materials handling systems can be stated.
2. *Volume of part numbers in the system.* Required for the information system.
3. *Types of direct materials.*
4. *Types of indirect materials* (chemicals, solvents, etc.).

Both the parameters themselves (e.g., materials) as well as the ranges need to be projected to the degree possible, considering the extent to which supporting data is available. This must be done in order to establish a reference base for the system design. This will mean making decisions on what the system will *not* handle as well as what it will. It may also be possible to build flexibility into the system for contingent product design considerations. For example, a small change in the functional requirements for a unit of production equipment may permit it to handle a broader spectrum of materials, some of which have potential for being incorporated in a future product design.

8.8.2.2 Design for Automation

The product should be designed for automation, both in the fabrication of component parts and in assembly. This should be the case even if some operations do not support automation, either because of economic considerations, or by their very nature (where the adaptability of the human operator is an asset). Designing for automation will allow for the later introduction of automated operations, if these become feasible in time. More importantly, it will tend to introduce design discipline and practices which will enhance the production of parts in the nonautomated mode. See Chapters 3.1–3.3.

8.8.2.3 Automated Design System

AFMS parts manufacture begins with the use of an automated design system utilizing computer graphics as the basic design tool. Design changes can then be translated *directly* into the data necessary to characterize the manufacturing operation through the automatic conversion or postprocessing of a digital description of a part into a digitized description of the process.

This link between computer aided design (CAD) and computer aided manufacturing (CAM) must be an intrinsic feature of the AFMS; it is essential for the effectiveness of the AFMS that "the slash be removed" from CAD/CAM, that the two subsystems function as a single, integrated system. See Chapter 8.2.

A further example of such a functional relationship is the establishing of the *product routing* from the data in the graphic design system, and the graphic projection of it at the factory operations. See Chapter 8.3.

8.8.2.4 Product Mix Flexibility

While the AFMS is designed to maximize the potential for handling product variation, the operational demands on the system can be decreased through the judicious selection and application of component parts and subassemblies. Product design objectives should be set to maximize the degree of commonality of parts across product types and to emphasize the use of "standard" commercially available parts wherever possible.

8.8.2.5 Product Volume Flexibility

The statement of product functional objectives also requires the determination of the degree of flexibility for *volume variation* to be handled by the system. In a true AFMS, the effect of economy of scope can be realized by looking at the total volume of all the jobs, even as the lot size approaches one. However, in planning the system, the total work content will have to be evaluated for the assumed range of product mix.

The central focus for attaining volume flexibility will be the design of the manufacturing system itself, which will be considered in the following section.

8.8.3 MANUFACTURING SYSTEM FUNCTIONAL OBJECTIVES

8.8.3.1 Process Flow

The first step in defining the functional objectives for the manufacturing system is to state an objective for the *macroflow* of product through the system. This will begin the process of dealing with the need to provide for volume flexibility, as stated in the previous section.

Several approaches can be taken to address this need. If the nature of the manufacturing operation is such that the system architect has opted to arrange the equipment in a *process-centered* format (i.e., groupings of like equipment/tooling), then contingency space should be planned within each process center to permit the addition of tools for an increase in volume in excess of what might be managed through overtime operation.

If management believes that the process flow will be relatively stable through the expected life of the system, and therefore a *serial-flow* format can be utilized to optimize production efficiency, then the manufacturing system is best designed in a series of *modular lines*. These can be incrementally operated to match the demand at any point in time. While the series of modular production units can be an effective configuration, depending on the nature of the particular product or process, in general, for an FMS, the process-centered approach will prove most feasible.

Within the spectrum ranging from purely process-centered to pure serial-flow, there is a combination of the two, driven largely by utilization of the concept of *group technology*. The manufacturing system operation can be optimized by processing similar parts in the same set of tooling. This opportunity depends on the nature of the product and its component parts. In this instance, the AFMS configuration will be a combination of "process centers," each composed of the equipment and tooling set required for the like parts, arranged internally in a form of serial flow that enhances the movement of parts through that center. See Chapter 5.2.

8.8.3.2 Programmable Processing

The feature of programmable processing in the AFMS permits the manufacturing engineer to specify, in the extreme case, unique manufacturing operations for each part and assembly flowing through the system. The constraints on this have already been given consideration in terms of the range of size, shape, and materials that the AFMS will accommodate. Within those limits, one should be able to route a part to any tool/process center, call for the (automatic) selection of the tool/process required by the part, and insert into the data system the dimensional data/process setpoints under which the operation will be executed.

This implies the capability for *automatic tool selection and changing* for the "mechanical" aspects of an operation, and automatic control of indirect materials for the "chemical" aspects of an operation.

The practice of programmable processing will also call for the comprehensive *identification of process variables* before the product is introduced into the production system. In the AFMS environment, where feasible, a manufacturing pilot line can be a valuable tool to stress the process in order to surface process variability against process specs. Conventionally, product pilot lines have been used to stress the product in order to surface product variability against product specs.

When the nature of the operation precludes a pilot line capability, preproduction information should be obtained by processing the parts in the AFMS itself. If new equipment is involved, the acceptance testing period for it can be used for this purpose. Whatever medium is used, the aim is to surface process variables before-the-fact and assure the ability in the AFMS to *identify*, *measure*, *analyze*, and *control* them.

8.8.3.3 Process Control

The AFMS requires a process control system that will react to the variability in specified dimensions and setpoints. Inherent in such a system is the need for *adaptive control*, in which adjustments are automatically made at the operation to compensate for tool wear, etc. Complementary to such real-time control is a *programmable verification system*, which monitors the actual performance of the operation against the variable input parameters. This is done through on-line, automatic measurement of product dimensions and process variables, under computer control. If automatic adaptive control is not in effect in the system, appropriate feedback of data and trends for decision making is provided.

Associated with automatic measurement is the use of *automatic visual inspection systems* in which the conventional human inspection role is performed instead through computer vision, based on the principles of image analysis. The need for automatic inspection systems is driven by the ever- increasing demand of technology for the detection of smaller and smaller defects and, in the AFMS, the need to readily accommodate variability in inspection criteria by part number.

The functional requirement for adaptive control and programmable verification is consistent with the quality goal of "zero-defects," which translates into the assurance at each operation that it was carried out in conformance with the product/process specs.

8.8.3.4 Programmable Assembly

Different model types within a product line are generally produced through the assembly of different components from similar families of parts. Such assembly in the AFMS can be achieved by routing the product to the appropriate *assembly robot*. The programmable nature of the robot permits the selection of the correct parts for the option at its workstation. To provide the most advanced capability and flexibility, the AFMS will intrinsically incorporate applications of *artificial intelligence* for the recognition and orientation of parts in the assembly process. See Chapters 3.2 and 4.5.

8.8.3.5 Information System

The fact that the AFMS must be virtually transparent to the variety of part numbers accommodated by the system will determine the specification of the database capability for the logistics system. *Distributed data processing* will supply the information needed by the finite set of part numbers being processed on the floor at any point in time.

Within the processing domain, the AFMS will provide, along with the part number and routing data for the logistics system, the technical data required for each processing operation, including:

1. The control data to define tool operation, to establish setpoints for process operations, and to select unique component parts for assembly operations.
2. The parametric data to verify that the operation did in fact transform or assemble the part in conformance with its spec.

The technical data specifying how to make the product should reside in the same system as the logistical data that is used to move the part(s) between operations. The data will be transmitted to and from the processing operation via the distributed information system hierarchy. This will permit the direct link of design data, after the necessary translation, into the dimensional/process data required at the operation.

8.8.3.6 Turnaround Time

A functional objective should be stated for the turnaround time performance of the system, i.e., the elapsed time between the release of materials to the floor and the putting of finished goods to stock. The turnaround time is a key manufacturing performance parameter, affecting customer satisfaction, inventory levels, parts obsolescence, parts losses, etc. The objective can usually be expressed as a multiplicative factor of the total "raw process time."

Definitions:

cycle-time = elapsed time between the actual start of material/parts into an individual operation and its completion

raw process time = summation of cycle times for all operations.

A turnaround time equivalent to two to five times the raw process time can be considered a meaningful objective, depending on the business. A *simulation model* can be an excellent tool for understanding and evaluating this objective.

8.8.4 FUNCTIONAL OBJECTIVES: THE PHYSICAL FACILITY

8.8.4.1 Materials Handling

The AFMS will incorporate a materials handling system responsive to a wide range of product/component physical sizes. A means of achieving this is the use of *standard pallets/carriers* for assemblies or large parts, and *standard containers* for component parts. In an AFMS, a prime function of the materials handling system may have to be the use, as well, of *variable fixtures on standard carriers*. These will provide the interface between the array of different parts/assemblies and the defined set of equipment that must process all of them; the fixtures/holders adapt to the parts, the carriers to the processing equipment. An added function of the logistics system, then, is the (programmable) control of the parts handling media. The nature of the products/parts will determine the particular materials handling technology employed (conveyors, automated guided vehicles, etc.). See Chapters 4.6 and 5.2.

8.8.4.2 Work-in-Process Storage

A manufacturing system design consideration essential to the AFMS is the use of *buffer storage (programmable)* integral with the process flow, i.e., resident on the processing floor, and under the control of the floor control subsystem. The multiplicity of parts routings accessing multifunction equipment, inherent in an AFMS-type operation, does require the ability to store work-in-process to a highly variable but controlled degree. The design parameters for such buffers are probably best established, again, through the use of a *simulation model*, which exercises the variability of loading assumed for the flexible manufacturing system. Such buffers, the physical design of which will also be a function of the nature of the products/parts, must be planned integrally with the overall materials handling system. See Chapter 5.2.

8.8.4.3 Physical Product Tracking

A further functional requirement of the materials handling system is that each transfer (move) be considered as equivalent to an operation, in the routing, under programmable control. This provides the basis for the control of the work-in-process and the overall turnaround time, both of which must be directly managed in the AFMS, i.e., specifically planned and positively tracked. Integral, then, in the AFMS is the *automatic tracking* of the physical products/parts via electronic identification and reading, both at each operation as well as between operations. See Chapter 4.7.

8.8.4.4 Facilities Subsystem

The necessity for providing flexibility in the range of materials that can be processed by the AFMS also impacts the design of the facilities subsystem. Comparable to the manner in which the process equipment requires adaptability for potential variability in direct material, the facilities subsystem must permit variability in indirect material and consumables (primarily). Its chief attributes must be flexibility, modularity, and integration into the total manufacturing system. For example, fluid piping from and to centralized plant facilities must be accessible and replaceable if the product/material/process variability call for such change. Contingency capacity for services (power, air, and drainage) must be considered when decisions are made about the functional objectives for the product and the manufacturing system.

8.8.5 THE AFMS AS A "FACTORY OF THE FUTURE"

8.8.5.1 Planning

In addition to the generally understood purpose of a flexible manufacturing system, that is, that it must accommodate a broad spectrum of product types for the marketplace, the "factory of the future" should also have the flexibility for transition to the manufacture of new product designs, with new materials, and to new standards. The need for this type of adaptability is vividly exemplified by the effect of technology changes on the manufacture of such products as watches, computers, airframes, and typewriters. Each of these industries saw a significant shift from mechanical or electromechanical fabrication to electronic and "process-sensitive" manufacturing. While industry in general has not always been able to anticipate the full extent of such shifts, it is necessary to make the effort to look ahead in order to protect as much as possible against premature obsolescence.

The AFMS is a *capital intensive enterprise*; the degree to which the product planners and manufacturing systems architects can envision the potential impacts on the product line and the manufacturing system will significantly influence the success of the business. In this activity, they will be called upon to project not only the changes in product design, but, even more fundamentally, changes in types of materials that will have to be handled, transformed, and assembled. Looking at the other side of the coin, a well-planned and executed AFMS will minimize the exposure to capital obsolescence by providing the greatest opportunity for future product and process flexibility.

8.8.5.2 Application

Possibly the most critical aspect of the AFMS application is the *identification and specification of the limits* (product type, part numbers, etc.), within which the system is expected to operate at the start of the planning process and, with appropriate modification, throughout it.

Given this understanding, all or any of the concepts outlined in this chapter are translatable, using state-of-the-art processing, material handling, and control technologies, into a functional AFMS, as detailed throughout this book. As always, this must be put into the context of the tradeoffs relating to other aspects of the business decision-making process, which may modify the extent to which they are implemented.

The degree to which one would depart from an "idealized" or even technically feasible structure for any given manufacturing system will depend upon business need, the nature of the product family, and economics. Lesser degrees of intrinsic system control will, of course, narrow the window of product/part diversification that the system can handle. The system planning process, through analysis and/or simulation, can be used to identify the impact of such constraints.

8.8.6 SUMMARY

The advanced flexible manufacturing system is intended to provide virtual transparency to product/component part-number requirements in order to meet a business need reflecting market demand. An AFMS can be designed using state-of-the-art manufacturing technologies. Specific technologies that make this possible have been reviewed; a checklist of these functional capabilities for an AFMS is given below.

Design for automation.
Automated design system.
Computer-driven planning and routing.
Group-technology-based process flow.
Direct link, design data to process data.

Programmable processing.
Automatic tool selection and changing.
Adaptive control.
Programmable verification system.
Automatic visual inspection.
Programmable assembly/advanced robotics.
Modular workstations and facilities.

Adaptable materials handling carriers/systems.
Programmable storage systems.
Automatic (physical) part identification and tracking.

Distributed data processing.
Operational simulation.

Appropriate references have been made throughout this chapter to more detailed considerations of these topics.

CHAPTER 8.9

CONTROLLING THE FLEXIBLE FACTORY

JOHN W. MUELLER

Litton Automated Systems
Hebron, Kentucky

8.9.1 INTRODUCTION

The control system for the flexible factory must have the following attributes:

Functional segregation
Modularity
Flexibility
Expandability

Functional segregation simply means that primary factory functions should have their own control systems. Financial systems should be physically separate from material planning systems, machinery control systems should be physically separate from financial systems, etc. *Modularity* means that the control system should be constructed so as to allow new control devices to be added without requiring a re-creation of the system architecture. *Flexibility* means that the system must be designed to allow changes to be made to subsystems already in place without jeopardizing the operation of the business. *Expandability* means that the fundamental control architecture can allow for growth. Growth in this sense means that computers already in place can be subdivided into two or more computers, and that additional control devices can be accommodated.

These functions must be provided for within and at all levels of the control system. The flexible factory control system is neither a patch to nor or a modification of an existing control system, it is a control system based on a philosophy that once established must be adhered to any time a new piece of equipment is purchased or a new requirement is placed on the business.

8.9.2 LIMITATIONS OF EARLY SYSTEMS

Historically, control systems have been designed to meet the needs of a particular equipment purchase. Often additional functions are provided in these equipment controls that logically belong elsewhere. The result of such purchases is a hodgepodge control system that cannot be modified or expanded because it was never designed to be modified or expanded. Worse yet, once in place, a high level of operational dependence is placed on these systems, making it virtually impossible to reconfigure them to be compatible with a proper control philosophy. An example of this type of purchase is an AS/RS control system that incorportes AS/RS control, a partial order entry system, or a purchase order verification system, all in the same computer (Fig. 8.9.1). Order entry is a function that is dependent upon the AS/RS to obtain current stock levels and to cause orders to be filled. Order entry must also address production planning, allocation, and many other business related issues. The order entry function, because it is associated with planning, is changeable. The fundamental control of an AS/RS has a low probability of change and constant operational dependence. It does not make sense to combine these two functions in the same computer. A similar case can be made for the purchase order verification system.

Separation of these two control functions into two computers increases computer hardware costs because two computers are now required, and software costs are also increased because a communications link between the two computers must be established. Communications links require greater discipline and effort on the part of the programmers because they are not as convenient as having everything in one machine.

8.9.2.2 Separating Administrative and Machine Control

A control system design for a limited factory management function that is convenient to programmers will probably not provide the long-range flexibility needed. However, the two-computer approach provides the advantages of a one-time development cost for AS/RS administrative software that can be reapplicable. This idea is of greater value to the user than to the vendor, since the administrative computer defines a technique for doing business. If business techniques are similar from operation to operation within the organization, the user should be able to reuse the software at different sites. The vendor who must deal with a variety of users will probably not be successful in moving the package from customer to customer without change. However, he may benefit by establishing a module that needs only minor change for some reapplications.

This technique of separating administrative control from machine control can be plagued with the same problems as the single computer system if the designers are not careful in assigning functions to the system.

8.9.3 TRANSPORTATION SYSTEM CONTROL

Mechanized transportation systems are the arteries of the flexible factory and undergo constant change. Their control systems must also allow for change.

8.9.3.1 System Developments

Not too many years ago, mechanized conveyor systems were controlled by electromechanical devices, which consisted principally of relays and timers in many variations. These were binary devices, which displayed an on/off condition. Data handling was very cumbersome, difficult, and expensive because of this limitation. Such devices were best suited for the control of machinery, since most machine control is event oriented (load on transfer, transfer raised, load ejected, etc.). Symbols for producing control schematics were developed and became a language for engineers to use in expressing control logic functions to maintenance people.

The advent of solid state electronics brought with it a new set of symbols and a new language and new techniques that to this day change constantly. The lower cost, space advantage, higher reliability, and flexibility of solid state electronics could not be ignored by the machine control industry. Opportunities existed, if a method could be developed to place this technology in the hands of individuals familiar with relay logic. Thus the programmable controller was conceived. It is a solid state microcomputer based device that is programmed using relay logic. The solid state devices in the programmable controller are capable of data handling, and most programmable controllers sold today permit a limited amount of data handling.

In earlier systems two types of individuals programmed transportation systems, controls engineers and programmers. The controls engineers were responsible for the machinery controls, which are usually accomplished in a programmable controller. The programmers were responsible for maintaining the linkages between the material and the data in a computer system. This was the case because of the limited data handling capability of the programmable controller and also because the data were of interest to the programmers in controlling an AS/RS or other highly mechanized systems served by the transportation system. The programmable controller would wait for instructions from the computer before it made a major load-routing decision, thus helping to insure the integrity of the material/data linkages. The computer system also provided a friendly operator interface for recovering from error conditions. This technique was effective but it was also costly, since the programmers would duplicate much of what the controls engineer did. In addition, both parties had to devise techniques to identify the party responsible for design errors. Changes to the transportation system required changes to both the computer and the programmable controller.

The data handling and operator interface capabilities of programmable controllers have improved over the years to the point that most transportation system data tracking can reside in the programmable controller. Contemporary tracking systems are table driven. The computer includes a simple dispatch table containing the data to be associated with the load. To minimize the amount of data handled, it provides the programmable controller with a token or pseudonumber and a destination to link to the location of the load on the transportation system. When a load arrives at its destination, the programmable controller sends the computer a completion message on an interrupt basis containing the token number. Many times changes to the transportation system can be accomplished by changing only the programmable controller.

Microprocessors are also used in machine control where programmable controllers cannot perform the control function needed. The microprocessor allows the controls engineer complete freedom in the design of the machine control system and controller packaging, i.e., he or she can accommodate any type of device that must be electrically interfaced to the microprocessor on space constrained machinery. The engineer is not limited to using the interface devices and packaging provided by the programmable controller manufacturer. The microprocessor allows custom display and keyboard devices as well as special packaging.

The machine control microprocessor has the disadvantage of high initial development cost and high maintenance costs, since most programming is done in assembler language. Assembler language programming is more

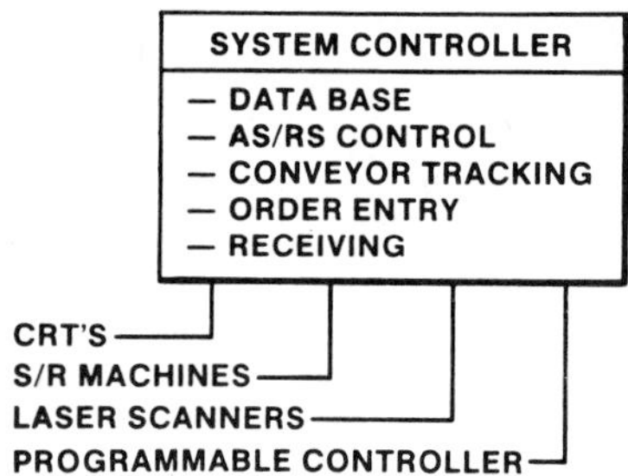

Fig. 8.9.1. Single computer configuration.

tedious, more error prone, and more difficult to debug than other languages. The program must also be constructed in a modular manner to simplify future modification and the training of new programmers.

8.9.3.2 Material Tracking

The mechanized transportation systems of the flexible factory require data to be tracked with the material that they move. They are generally point-to-point dispatch systems with path optimization and rerouting capabilities. There are two techniques of maintaining the relationship between the material and the data:

Keep the data in the memory of the controller and monitor the movement of material along the system.

Apply an optically or magnetically readable tag to the material and allow the data to move with the material.

The first technique requires that all information be memory resident in the controller. The controller monitors proximity sensors located along the network and adjusts pointers to constantly maintain a link between the location of the material and the data. This logic is usually very complex, because the designer must allow for the unexpected, i.e., a person or object or reflection blocking or actuating a photoelectric device. These events, if not properly considered, can destroy the integrity of the tracking system. The usual method of dealing with an unexpected event is to stop the system and require a partial audit to be taken. The controller must employ battery backup or some other technique to prevent the loss of data on a power failure. This control technique almost always requires a visual inspection of the transportation system prior to restart/recovery. The complexity of the control design for these types of systems centers around error recovery.

The second technique, tag applied to load, has the advantage of allowing changes to be made to the transportation system with minimal impact on the control system. Scanning devices located at decision points along the transportation system read the tags and in conjunction with their controller, decide what action to take. Usually it is physically impossible to place all the information concerning the carrier on the tag. Typically the tag contains a unique number and the controller contains a table of the information associated with the carrier, keyed by tag number. The table also contains the destination and the date and time when the carrier

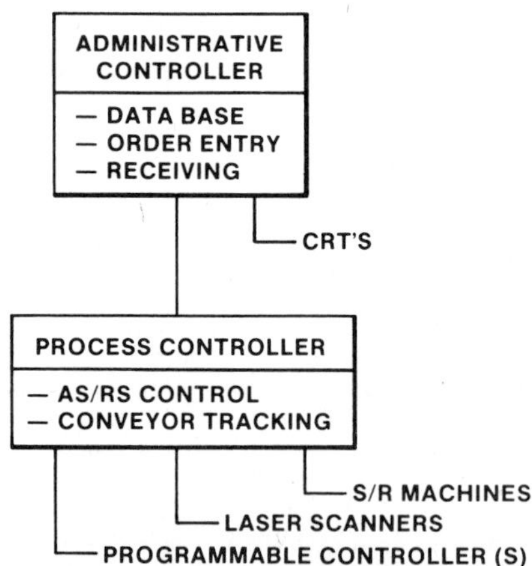

Fig. 8.9.2. Multiple computer configuration.

was introduced to the system. Decision making is simple and fast, requiring only a table look-up. Error recovery is simplified in that complex linkages between the material location and the information do not have to be maintained and resolved. This technique has the disadvantage of requiring expensive scanning devices and identification tags.

The tracking technique eliminates the need for label printers, applicators, and special carriers. However, it requires more programming effort and a longer and more complex fault recovery technique than the tag reading method.

Many systems being implemented today make use of a combination of both techniques. Tracking is generally used in concentrated conveyor intensive areas with many decision points and where tracking errors would not be catastrophic. An example is manned workstations where tracking errors could be visually confirmed. Scanners are placed judiciously to confirm the integrity of the tracking where tracking errors would be catastrophic. An example is the entry point to an AS/RS aisle input dispatch conveyor, where tracking errors could destroy the integrity of the entire inventory system.

Technology is also being developed to permit the data contained on a tag to be changed, i.e., the tag can be both read and written to as it advances along the transportation system. Currrently, these devices are expensive, but as with any solid state device, time and demand will bring the prices down. Such devices may have a large impact on the design of control systems, since it is conceivable that all required information can be carried on the load itself. Likewise, AGVS control system design and data tracking could be simplified by mounting the reader on the AGV and writing the next destination into the tag. The AGV could then route the load using its own local intelligence.

8.9.4 FACTORY CONTROL OVERVIEW

The automatic factory control system is no different than the control system for an efficient manual factory. In fact, a good test of the adequacy of the control system design is its ability to be installed in a manual factory which can then progress toward automatic operations without having to make major control design changes.

The automatic factory must be provided with the same top level administrative and planning functions as the manual factory. The major difference between the two control systems is the timing of decisions and reactions. Manual factories react at human speeds—minutes and seconds—while automatic factories react at machine speeds—seconds and milliseconds. Hence, one would expect to see more control devices at the lower level of the control hierarchy because there is more machinery at that level and an architecture that permits rapid message exchange in the flexible factory control system design.

The factory control system begins with a discipline for organizing the various operations within the factory and ends with a method of gathering and distributing information or integrating those operations. The flexible factory is organized into cost centers or workcenters as most factory operations are managed and measured today. Workcenters have the following common characteristics:

Material enters the workcenter.
Labor enters the workcenter.
An operation is performed.
Material exits the workcenter.

More formally, a workcenter is a group of similar or related activities or a function generally identified as a department, which is also a logical group of activities from a cost collection point of view. Typical workcenters found in manufacturing operations are:

Receiving dock operations.
Receiving inspection.
Receiving stores.
Machining centers.
Minor subassemblies 1,2,3, ...
Major subassemblies 1,2,3, ...
Final assembly.
Test.
Repair.
Button up.
Packaging and accessorizing.
Finished goods storage.
Distribution.
Shipping.

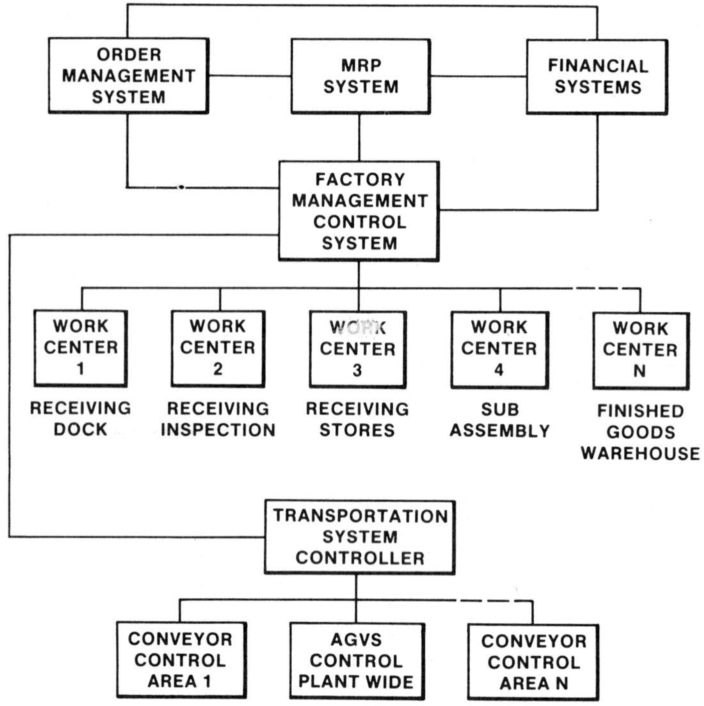

Fig. 8.9.3. Factory control system logical diagram.

The factory control system consists of three major logical elements: work center control systems, a factory management control system, and a transportation control system (Fig. 8.9.3). The factory management control system (FMCS) integrates the work center control systems and provides real-time shop-floor control. FMCS is not an MRP or other planning system, it is an execution system with the responsibility of directing and coordinating the operations at the workcenters.

The underlying design philosophy of the factory control system should be to help plant management personnel to do their jobs better, not to replace them. It should minimize the need for paperwork and paper handling, eliminate the need for expediting, and provide a reasonable level of optimization. Perfect optimization is not possible, because any activity that relies on machinery and people will always be imperfect. There will be material defects, machine breakdowns, emergencies, absences, and learning curves to contend with. This philosophy does not mean that the workcenter supervisor is free from constraints but rather that he or she should be given some latitude to deviate from a theoretically optimal schedule. However, the effects of any deviation must be determined by the control system and, if significant, the approval of plant management must be sought.

FMCS provides the following principal functions:

Workcenter loading.
Job release.
Material dispatch.
Labor statistics.
Factory inventory.
Job tracking.
Interactive shop-floor management.

8.9.5 LOGICAL CONTROL SYSTEMS CONCEPTS

The factory control system has been described as containing "logical" elements. The word *logical* means that the elements are described as physically different systems, though they may be one integrated system. Advances in computer networking technology are driving control systems toward a modular distributed design. The traditional large single computer system that has the capability to perform many tasks simultaneously is being complemented with the concept of multiple smaller computers each running a single or a limited number of tasks.

An emerging form of factory control utilizes networking. Networking technology is much more than creating a high-speed information bus to allow computers to share information with each other. It is the creation of a "logical" single computer that extends itself throughout the factory. The network control software allows for a distributed database, i.e., system files do not have to reside on a single device or within a single computer, they can reside on any device anywhere in the system. The applications program in a computer attached to the network needs only to call for the file by its logical name and logical device number as it would in a large single computer system. The network software will obtain the information. Much of the custom programming effort and all of the protocol required in earlier distributed systems to effect communications between computers is provided in the standardized networking software and solid state devices obtained from the network vendor. Hence, the implementation of distributed systems today from a programmer's perspective is much easier than in the past.

File servers also represent a viable control concept. It is often desirable to centralize many of the major files required to operate a factory into one or more devices for security and backup purposes. These devices are called *file servers*, since they are slaves that make no decisions other than security related decisions. These devices can become large if the database is large, and their disk drives must operate quickly, since they must serve multiple users on the network. They may also contain standardized database software to simplify the development of application programs in the satellite computers. If database software is employed, the server must also contain a fast processor and make use of other techniques such as cache memory to speed file searching. A new generation of file-server computer systems is being created, which are optimized architecturally for providing rapid access to data. However, note that in distributed network applications where access to a central database is required by a number of modules/processors, the file server may become the bottleneck and the network will be I/O bound. If this is the case, the network is no better than a single computer.

Another advanced concept is that of a *floating host computer*. A fully distributed computer system may not contain a single computer that can be identified as the factory manager. Decision making is performed throughout the entire system with the individual decisions producing a resultant "best" decision for the factory as a whole. The concept is similar to the processes used in human group decision making, and borders on the field of artificial intelligence, which is receiving increased attention.

The distributed design has several advantages over the single large machine concept:

Flexibility
Modularity
Lower cost
Lower risk

It is more flexible, because functions can be added by merely connecting new computers to the network. Modularity is obtained by limiting the functions performed by each computer. Lower cost is obtained from the use of microprocessors, standardized networks, and inexpensive peripherals. Lower risk is inherent in the design, because the control system was initially designed to be expanded. A change or addition will have a minimum impact to the existing design, i.e., modules can be added or changes can be made with minimal disruption to current factory operations.

As long as a business continues to grow, a single large computer will never be large enough. As long as large-computer technology provides ever larger computers, this may not be a problem. However, the reality is that the next generation of larger computers may not be available when they are needed, or may not be compatible with the previous generation. The large-computer user will inevitably be faced with the task of dividing the functions performed by the large computer into two or more computers. Applications must be designed to allow for this eventual separation.

8.9.6 WORK CENTER CONTROL

The implementation of a factory control system begins with a philosophy for managing workcenters. The initial computer system design should assume that all workcenters are operated manually and that they will be upgraded to automatic control in the future. The underlying thought is to develop a set of transactions that can initially be performed by a person at a CRT terminal (Fig. 8.9.4) and later emulated by automatic equipment through the same device interface (Fig. 8.9.5). Detailed inventories within the workcenter should be kept on micro- or minicomputers assigned to the workcenters. The computers can transmit summary or detailed reports to FMCS through the workcenter device interface. The use of intelligent block transmitting terminals in place of high-overhead interactive character transmitting terminals will allow an easier upgrade path to automation. The initial implementation of a local area network throughout the factory, employing personal computers as intelligent CRTs in the manually controlled factory, will facilitate the migration to full factory automation.

The workcenter control system is an execution and data collection system. It makes decisions local to the workcenters only and is a slave to FMCS. The functions that this control system provides are:

Fig. 8.9.4. Simple workcenter control system.

Delegation of workcenter activities.

Inventory by location within the workcenter.

Workcell (robotics) control.

Intraworkcenter transporation system control.

Workcenter labor statistics.

FMCS should provide the workcenter control system with a prioritized list of activities that must be accomplished. FMCS should *not* provide rigidly optimized directives that must be accomplished in an exact sequence. It should allow the workcenter supervisor some latitude in choosing tasks to match the capabilities of the human and mechanical resources currently available in the workcenter. Many times people and a specific machine in a group of seemingly identical machines must be matched to jobs. The FMCS directive should be simple, for example, a listing of the current stock levels and the needed stock levels of certain parts in adjacent operations. These simple directives will allow the workcenter supervisor to manage his area in harmony with the rest of the factory.

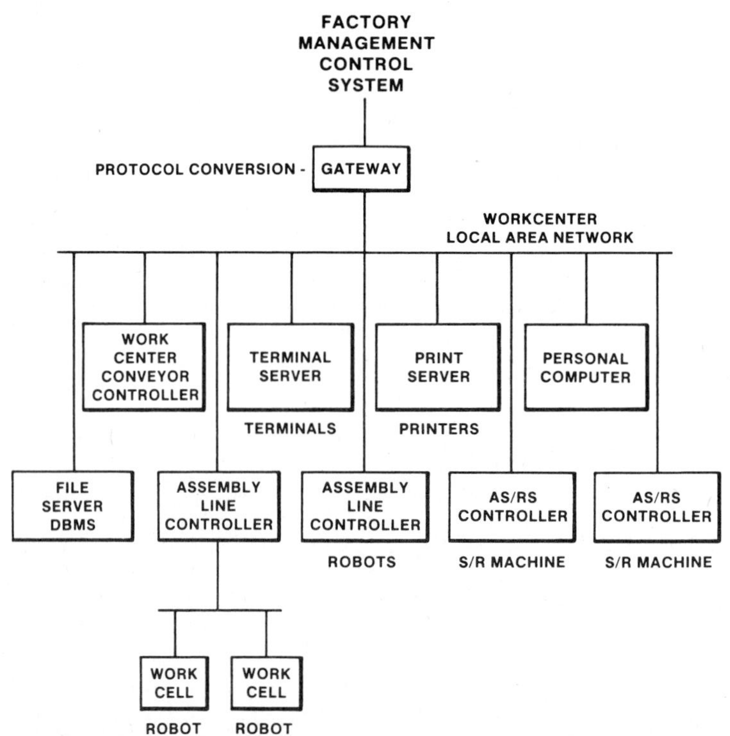

Fig. 8.9.5. Complex workcenter control system.

A detailed inventory by location should also be kept in the workcenter controller, since it is of concern only to the workcenter. FMCS should keep summary reports of workcenter inventories. Information should be kept where it is needed and generated when it is needed. It does not make sense to burden a network with single transactions as the transactions are created, when FMCS will do nothing more than store them until they are needed. It makes more sense to consolidate them at the workcenter and to transmit them in a batch to FMCS when they are needed.

Robotics control should be accomplished in the workcenter. Downloading of programs to robots and material delivery and pickup demands should all be managed by workcenter controls.

All transportation systems within the workcenter should be managed by control systems local to the workcenter. Assembly lines and material delivery systems between a workcenter AS/RS and the robotic work cells fall into this category.

The gathering of labor statistics is the responsibility of a workcenter, since the labor is expended in the workcenter.

Figure 8.9.5 describes a complex workcenter control system for an automated assembly line integrated by its own local area network. This subsystem is interfaced to FMCS in the same manner that the terminal in Fig. 8.9.4 is interfaced to FMCS. The complex system employs a device called a *gateway* to perform the protocol conversion necessary to move information between dissimilar vendor networks. Gateway functions also can be provided by a central workcenter computer. Figure 8.9.5 also shows devices called *terminal servers* and *printer servers*. These devices are usually simple microprocessors that perform a queueing and block transmission function to minimize transactions on the network.

Personal computers should also be accommodated in the network to allow supervisory personnel to obtain and analyze information. These computers may also contain decision-making logic to direct and control the activities of the workcenter. Their uncomplicated operating systems and programming languages make them a logical choice as flexible workcenter operator interface devices. The personal computers are programmed to allow the workcenter supervisor to modify or create a prioritization file that is kept in the file server for access by all other workcenter device controllers. The file server centralizes and consolidates workcenter data files, setup files, production requirements files, and prioritization files. It also receives and provides information to FMCS.

Assembly line controllers coordinate and monitor the activities on the assembly line. They also can download programs to the robots, since most robot controls have a limited program storage capability. The robotic work cell controllers provide direct robot control and also can be interconnected with their own local area network as shown in Fig. 8.9.4. AS/RS and conveyor systems may be employed to deliver parts to and from the robots automatically. These devices are provided with their own individual controllers. The AS/RS controller monitors sensors in the work cells and on the conveyor system and dispatches the S/R machine to provide replenishment material or to pick up completed work from the work cells.

The implementation of the hierarchical distributed control system shown in Fig. 8.9.4 can provide tremendous flexibility by providing a physical design that allows expansion by simply adding more controller modules to the network. However, this design requires careful thought and planning to maximize its inherent flexibility. The functions to be provided by each element and the desired flexibility must be well defined in advance. File expansion must be anticipated and provided for, as well as modifications to the AS/RS racking system and workstation setups. The software design for each controller should be table-, parameter-, or switch-driven, so that change can be accomplished quickly by operating personnel. Detailed files that will always be required to operate a device should be located in the controller for that device. If information contained in the controller is occasionally required by other controllers, a provision should be made to transmit the information to the requesting devices. A simple rule of thumb is to allow the network to execute the same commands and to receive the same information as that provided to an operator at the device controller console terminal. Obviously, commands that require human presence for safety purposes should be excluded.

8.9.7 TRANSPORTATION SYSTEM CONTROL

The transportation system for the flexible factory is controlled as a separate entity. Like a workcenter, it is an execution-only, or slave, system. Its principal responsibility is to move material from workcenter to workcenter in the shortest amount of time. It must maintain an inventory of the material in-transit to facilitate knowledge of the plant-wide inventory. It is *mechanically* integrated with manufacturing operations more than it is *informationally* integrated. The design of the control system must be flexible and modular, employing simple control elements such as programmable controllers.

It should also make use of networking to allow for change or expansion. The need for flexibility is more important in this control system than in any other, since transportation systems undergo constant change and expansion. Architecturally, this control system must be independent of the intraworkcenter transportation control system previously described, to allow changes or modifications to be made at the workcenter level without jeopardizing the entire factory operation. The transportation control system should make use of bar-code readers and look-up tables to eliminate the need for complex interactions with the workcenter control systems. The mechanical design should incorporate closed loops to allow recirculation of material to facilitate error recovery and to accommodate easily the future addition of mechanized workcenters.

8.9.8 INTEGRATION

Now that the three major logical elements (workcenter control system, factory management control system, and transportation control system) of the flexible factory control system have been described, description of the integration of these systems into a single factory control system remains.

Factory operations require three other major control modules, shown in Fig. 8.9.3: financial, order management, and material and capacity requirements planning (MRP). The FMCS described is dependent upon these other systems for direction. It assumes that MRP knows how to do its job and will get it done correctly in the end. MRP is a planning system that considers current resources and future demands. It tells the organization what to order and when to order it, it considers current resources and schedules them to produce a result. Because of the amount of computation required, it is generally run periodically on a batch basis (biweekly or once a month).

FMCS receives the planned directives from the MRP system and issues them to the workcenters. In its most complex form of controlling a job shop, FMCS is analogous to a dynamic PERT scheduling system. It knows the critical paths and issues directives to the workcenters to cause schedules to be met, it copes with the reality of defective parts, late deliveries, and personnel absences as they occur, and attempts to compensate for them. It allocates material to similar workcenters and provides decision-making flexibility to the workcenters by allowing supervisors or workcenter computer sytems to maximize the deployment of their resources on a short-term basis. In its simplest form, controlling a factory producing only one product, it merely allocates material, collects statistics, and maintains a factory-wide inventory.

The demand management system receives and anticipates customer orders. If the factory is run on a pull basis (no finished-goods warehouse), it will allocate the output of the factory, i.e., it will create a mix of customer orders that matches the capabilities of the factory. If the factory is run on a push basis, it will create an inventory that matches marketing projections.

The financial system is the business management system and is concerned with receivables, payables, payroll, product cost acounting, etc.

The information passed between FMCS and the workcenter controllers consists of the following principal elements:

Completions within a workcenter (from workcenter to FMCS).

Material disbursement instructions (from FMCS to workcenters).

Receipt of material at a workcenter (from workcenters to FMCS).

Material consumed within a workcenter (from workcenter to FMCS).

Scrap generated within a workcenter (from workcenter to FMCS).

Labor expended in a workcenter (from workcenter to FMCS).

Emergency demands (from FMCS to workcenter).

Build orders (from FMCS to workcenter).

Inventory (from workcenter to FMCS).

The information passed between FMCS and the transportation system controller is as follows:

Workcenter destination of material to be moved and relevant statistical information (part number, quantity, barcode, date code, lot no., etc.).

Arrivals

8.9.8.1 FMCS Attributes

This section is a logical progression of the attributes of FMCS as the complexity of the factory increases. The examples begin with a highly automated assembly factory producing only one product, on one assembly line, with no variations, in a pull environment (no warehouse). The factory is literally one machine surrounded by walls. The examples end with a very complex batch production facility producing a multitude of products:

EXAMPLE A: ONE PRODUCT, ONE ASSEMBLY MODULE

FMCS attributes:

Maintain a plant-wide summary inventory by workcenter.

Provide a prioritized receipt processing list to receiving inspection and stores.

Log material into and out of workcenter inventories.

Log finished product and scrap produced by the workcenters.

Provide a prioritized build list to the subassembly areas.

Log summary test statistics.

Log repair statistics.

Provide a prioritized list of shipping requirements to packing and shipping.

EXAMPLE B: ONE PRODUCT, ONE ASSEMBLY MODULE WITH VARIATIONS CAUSED BY ACCESSORIES

Additional FMCS attributes:

Provide batching information to accessorizing to minimize setup changes and to meet customer order requirements by shipping unit capacity and schedule.

EXAMPLE C: ONE PRODUCT, SEVERAL ASSEMBLY MODULES WITH VARIATIONS

Additional FMCS attributes:

Distribute and balance material and subassemblies among modules.

EXAMPLE D: MULTIPLE PRODUCTS, ONE PRODUCT PER ASSEMBLY MODULE WITH VARIATIONS (LARGE-SCALE BATCH PRODUCTION)

Additional FMCS attributes:

Obtain time schedule of planned receipts from MRP.

Obtain time schedule of shipping commitments from the demand management system.

Perform dynamic scheduling (analogous to real time PERT charting) at the factory level for assembly line run length planning purposes.

Predetermine WIP and inventory buildup and compare it to available space for batch produced items.

Assess financial tradeoffs, cost of WIP vs. cost of setup.

EXAMPLE E: FULL BATCH PRODUCTION, MANY PRODUCTS, MANY VARIATIONS

Additional FMCS attributes:

Distribute computational burden by providing capability to perform dynamic scheduling within workcenter control systems.

Capability for workcenter to predetermine WIP and inventory build up.

Capability for workcenter to determine costs of various alternates.

Capability for FMCS to determine the best plan for the factory as a whole in a distributed decision-making environment.

The examples above contain the most obvious attributes and are not intended to be an exhaustive listing of all of the FMCS attributes required in each of these situations. However, they should serve as a starting point for the reader.

8.9.9 SUMMARY

In summary, the flexible factory control system is not a mystical commodity. It is a control system that finds its origins in serving an efficient manual factory. The technology required to construct it exists today and will only improve tomorrow. In the past we have started at the bottom by automating workcenters or portions of workcenters and have found these control systems to be difficult to change or to integrate with later systems. We have learned that we must begin at the top with a philosophy and a technique that allows for change and migration to a broader scope of control. The ultimate flexible factory control system must include a network of standardized, predefined data receptacles that allow the accommodation of additions and upgrades by merely changing the device that is plugged into the receptacle.

CHAPTER 8.10

IMPLEMENTING THE FLEXIBLE FACTORY

RICHARD L. ENGWALL

Westinghouse Electric Corporation
Columbia, Maryland

8.10.1 INTRODUCTION AND BACKGROUND

The success of implementing the flexible factory is predicated on doing an effective job of organizing for production to optimally utilize the five Ms—Men, Methods, Machines, Material, Money—and by properly designing the production facility for factory integration. In this chapter, we will first look at why Japanese factories are successful, the nine hidden lessons in simplicity that have been successful in implementing factories within Japan, and how to manage quality so that it becomes not only a source of profit but also assures successful implementation of the flexible factory.

In his article "Why Japanese Factories Work," Robert H. Hayes states that

> *The Japanese have achieved their current level of manufacturing excellence mostly by doing simple things; but doing them very well and slowly improving them all the time.*[1]

In particular, we must "get back to basics" if we are to implement new technology successfully. We need to plan, analyze alternatives, evaluate and select the best alternative(s); train, implement, provide and solicit feedback, and repeat the same cycle again and again, until it is no longer cost effective to obtain further optimization.

> *Japanese managers have never stopped emphasizing basics. To them, every stage of the manufacturing process—from product design to distribution—is equally important.*[1]

So it is with implementing the flexible factory. Each step, in sequence, must be successfully implemented before we proceed with the next step. The Japanese work constantly to improve the product design, manufacturing process, manufacturing support systems, and worker skills through cooperation at all levels. Their ultimate goal is perfect products and error-free operations.

The Japanese consider inventory the "root of all evil." If you minimize inventory you need fewer inventory managers, a less complex inventory control system, and also fewer expeditors, for there is less to expedite. With minimal inventory, when something goes wrong the entire system stops, and the whole organization immediately becomes aware of the problem and has to move expeditiously to resolve it. If one has buffer inventories, these potential problems can stay hidden and may never get corrected.

Furthermore from a quality standpoint, the Japanese treat a "defect as a treasure." Quality, to the Japanese, means error-free operation.

> *Any defect in any part of the manufacturing operation, therefore, becomes a quality problem in management's view—another 'grain of rice' to be pursued and eliminated.*[1]

In Richard J. Schoenberger's recent book, *Japanese Manufacturing Techniques*,[2] there are nine hidden lessons in simplicity that are pertinent to successful implementation. They are:

Lesson 1. Management technology is a highly transportable commodity. The Japanese have had little trouble learning our techniques, and we will have little trouble learning theirs.

Lesson 2. Just-in-time production exposes problems otherwise hidden by excess inventories and staff. It is imperative to continually improve the just-in-time/total-quality-control system.

Lesson 3. Quality begins with production, and requires a company-wide "habit of improvement." Production, not quality control, must have primary responsibility for quality; and everybody, including top management, must participate in project-by-project quality improvement.

Lesson 4. Culture is no obstacle; techniques can change behavior. The Kawasaki, Nebraska, experience is that Western managers and workers begin to behave more like their Japanese counterparts as just-in-time techniques are adopted.

Lesson 5. Simplify, and goods will flow like water. Emphasize simpler plant configurations—break down the barriers between shops.

Lesson 6. Flexibility opens doors. Western workers are over-specialized; Japanese-style labor flexibility is the key to effective resource management.

Lesson 7. Travel light and make numerous trips—like the water beetle. Have your suppliers deliver every day, or more often.

Lesson 8. Seek more self-improvement, fewer programs, less specialist intervention. Industry does not need a lot of improvement programs coordinated or run by specialists; production managers and workers can do it themselves.

Lesson 9. Simplicity is the natural state. Industry is ready to change its ways, and now we know what to do: simplify and reduce, simplify and integrate, simplify and expect results.

Quality improvement has to be one of the key strategies in implementing flexible automation in the "Factory of the Future." By putting an emphasis on quality when designing the product while concurrently designing a manufacturing process that produces quality, and then emphasizing quality with suppliers and employees, one can simultaneously improve quality, improve productivity, and reduce costs. By literally "doing it right the first time," one has the potential for substantial productivity improvement, as well as for improving customer satisfaction. We need to follow the Japanese in implementing their total quality control concept which emphasizes:

1. A goal of continual quality improvement, project after project (rejection of the Western notion of an "acceptable quality level").
2. Worker (not Quality Control department) responsibility.
3. Quality control of every process, not reliance upon inspection of lots for only selected processes (defect prevention, not random detection).
4. Measures of quality that are visible, visual, simple, and understandable, even to the casual observer.
5. Automatic quality measurement devices (self-developed).[2]

The detailed approach toward achieving these quality improvement goals in the design, manufacturing and field use of a company's products should be singular: "Do it right the first time." Identify, analyze, and prevent reoccurrence of deficiencies that add to the costs, impact the schedules, and detract from the reliability of products, be they in the design, the hardware, processes, procedures, management, or any other aspect of the overall cycle.

In summary, two key company-wide objectives should drive the overall general system design and implementation of the "Factory of the Future"

1. *Quality improvement.* This translates to "doing it right the first time" from product design through shipment and subsequent service to the customer.
2. *Just-in-time production.* This translates to reducing and controlling work-in-process inventories from receiving through manufacturing and distributing the end product to the customer.

These two objectives create a manufacturing need for real-time control and feedback of data and material through every step of the manufacturing process.

8.10.2 FACTORY MISSION AND GOALS

The mission of the flexible factory should be to manufacture the end product at the lowest possible total cost, while providing on-time delivery of quality products to one's customers. This must be accomplished cost effectively while maintaining a desirable quality of work life for all employees by providing a participative, challenging, safe, orderly, and open working environment.

The operating philosophy should be to accomplish this mission through a balance of state-of-the-art technology and innovative human resource approaches that will strive to satisfy the technical, individual, and social needs of each employee.

It is the goal of management to enthusiastically facilitate a working environment to establish pride in company membership, avenues for individual and group recognition for performance achievement, adaptability to change to meet ever-changing technology needs, and challenging work content. The job should be reasonably demanding in terms of use of each employee's creative intellect, and should allow the individual some latitude for decision making. The job should satisfy the individual's need to relate what he/she does and how he/she contributes to the mission and objectives of the organization, and to feel that the job leads to some desirable future, although not necessarily promotion. The job should meet the individual's needs to learn and go on learning. An abundance of information demonstrates that organizational effectiveness is superior and motivation is high when these principles are applied. And, most of all, implementing the flexible factory is successful.

8.10.3 TOP-DOWN STRATEGIC PLANNING AND BOTTOM-UP TACTICAL IMPLEMENTATION

In order to implement the flexible factory successfully over time, one must to do a total, top-down strategic plan for what the factory will "look like" over the next ten years. It is essential to "wire" the factory for all the interfaces anticipated and to implement generic, modular technology, systems, and human resource innovations. Assuming that our top-down strategic planning (1) focuses on organizing for production as covered in Section 1 of this book; (2) incorporates the proper methods for designing, improving, and controlling production systems as defined in Section 3; (3) uses the current state-of-the-art machines for implementing current strategy as defined in Section 4; (4) uses our material/asset management resources as identified in Section 5; and (5) properly designs the space for the production facility, as identified in Section 7, we will have designed a factory integration system that can be implemented with a bottom-up tactical implementation.

Top-down strategic planning provides the continuity and long-term focus needed to prepare for the evolutionary process of systems implementation with changing technology and requirements.

8.10.4 GENERAL SYSTEMS DESIGN

The General Systems Design approach (see Fig. 8.10.1) for developing flexible factory automation for the factory of the future incorporates integrated computers, material control/distribution, and process automation. Because of the complexity and length of time needed to evolve such an automated factory, one must plan for the systematic introduction of proven submodules that can ultimately be integrated into an automated real-time system. The key system elements must be designed explicitly to include interfaces for future integration and be generic and modular in concept and design.

8.10.4.1 Integrated Computer Systems

The basic computer system architecture and database design must accommodate evolution into the projected CAD/CAM system of the future. In fact, the initial integration of quality, test, process control, business, and material handling computer subsystems must be in place before a total CAD/CAM System can be implemented. An integrated production planning and control computer network that provides real-time scheduling and tracking of work from initial receipt through shipment of the final product is needed to manage the factory. Automated material handling systems that integrate storage, retrieval, dispatch, and distribution of all work in process must act in concert with the immediate production priorities. At the time of factory startup, new process mechanization should include the state-of-the-art equipment for the processes required to manufacture the needed items. Moreover, these key system elements must be designed explicitly to include interfaces for the future integration of the advanced robotics that will ultimately be used in the factory of the future.

More specifically, the integrated computer systems can be broken down into three logical operating system levels: planning, execution, and direct control. Planning systems are used to prepare for the manufacturing activity and to monitor and analyze results of the manufacturing processes. These systems should include: manufacturing engineering, quality, and finance, industrial engineering, and plant business systems. Execution systems are tactical in nature and support the factory floor with the real-time information required for short-term decision making. These systems should include: manufacturing control, test control, material handling and control, and work management. Direct control systems are normally real time in nature, and control physical processes embedded in the manufacturing processes. These systems should include: automated manufacturing equipment, automated test equipment, and material handling equipment.

8.10.4.2 Material Handling/Distribution Systems

The next key element required to implement flexible factory automation is *material control distribution systems*. For a truly flexible manufacturing system (FMS), centralized work-in-process (WIP) automated storage and retrieval system (AS/RS) is advantageous. Material should be stored automatically in the WIP AS/RS and dispatched to and from work centers under real-time computer control. Transporters, conveyors, or trucks will deliver the highest priority job to the appropriate work center queue, where the operator/robot will remove the

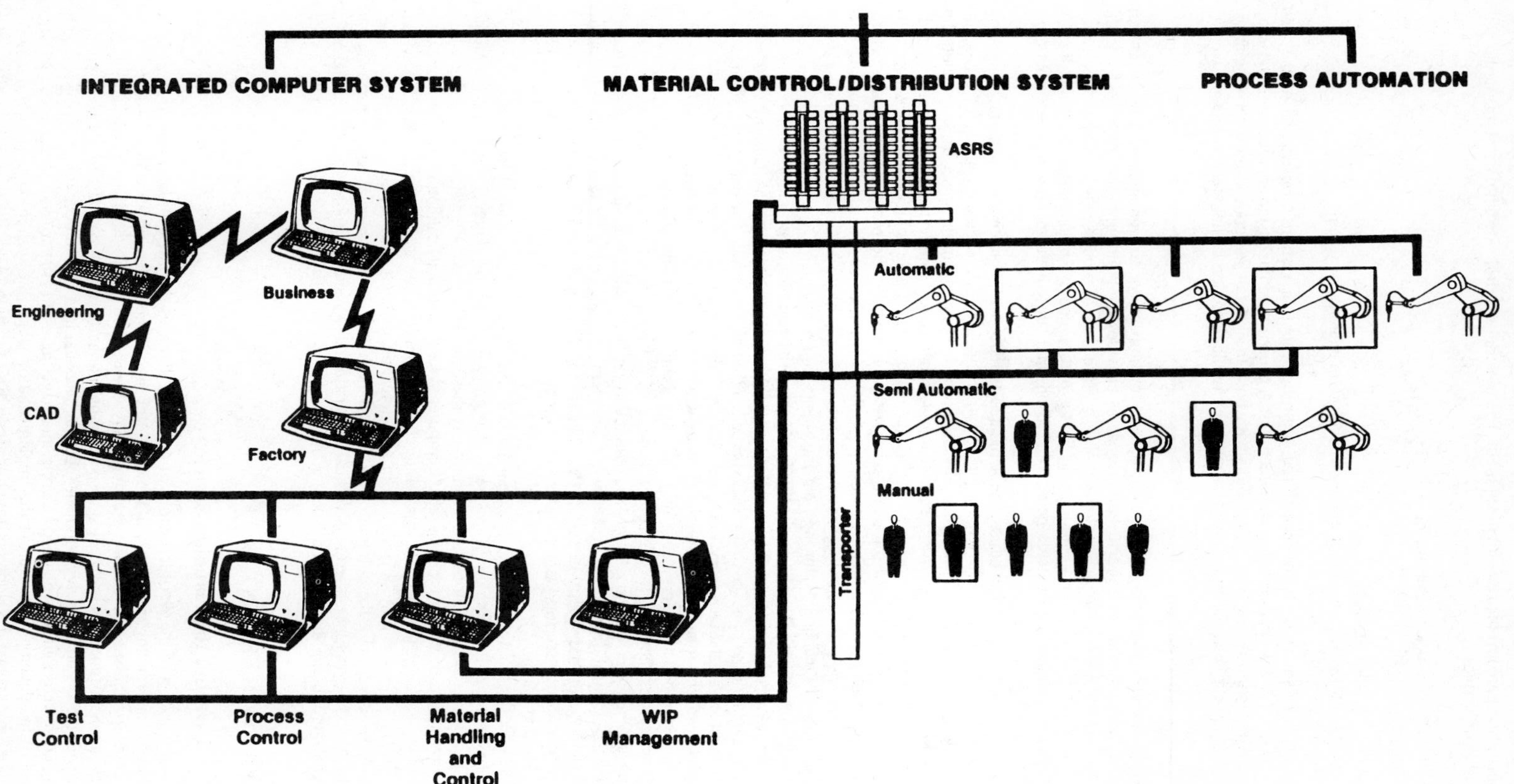

Fig. 8.10.1. General systems design approach.

prior job in queue and perform the operation required. The material control/distribution system should achieve the following key objectives:

Provide dock-to-stock/stock-to-ship material tracking.

Ensure product integrity.

Maintain product identification.

Provide a totally automated, integrated material handling/management system.

Considerable synergism exists between material control/distribution systems, irrespective of the facility, manufacturing process, and/or product being manufactured. A generic modular design approach must be followed to assure maximum flexibility and total integration with both the overall integrated computer system(s) and process automation objectives, and to ease the training and implementation of new technology, as well.

8.10.4.3 Process Automation

The third key element for developing flexible factory automation is *process automation*. The integration of computer systems and material control/distribution systems provides the building blocks required to gradually automate the process for the ultimate technology evolution from manual processes to semiautomatic (operator assist), to automatic (no human interface required). This evolution enables planning for future automation of the robotic loading and unloading of items being worked on, and subsequent incorporation of robotic count, verify, and pick-and-kit functions.

The flexible manufacturing system can be implemented initially with conventional manufacturing process equipment. Distribution of material to the work group (group technology cell) should be optimized for expansion to future workstation requirements. These workcells should be designed from the onset from a sociotechnical standpoint, that is, organized around the human being. Ultimately, as simple tasks can be automated and replaced by robots within the same workcell, the human being will be able achieve job enrichment through vertical integration and by working on the next higher assembly task for the product being manufactured in his/her work group. One should evaluate the design of the workcell to try to achieve an optimum balance between human interactions and technology advances. The workcell should be designed to accommodate both people and future robotics technology through enhancements, not redesign.

At the initial stages of introduction of the FMS, the material handling distribution can be to workcells. However, the long-range objective should be to provide totally mechanized material handling distribution to each individual workstation, as opposed to just the workcell, as initially required. The material handling system's design should anticipate all the necessary key gates, and include a modular/expandable database design with sufficient computer capacity to accommodate the ultimate workstation control. Workstation control is necessary to utilize the planned robotic systems being required for robotic process automation, so one can implement stand-alone subsystems as they are developed and debugged, thus reaping their respective benefits earlier. In addition, the development of a sophisticated FMS is required to verify the subsystems prior to systems integration. The planned material control distribution system is the key element in achieving these objectives. In summary, the ultimate material control/distribution system should be capable of making the entire manufacturing area into a single flexible manufacturing system under total computer/material handling control, irrespective of the degree of process automation.

To implement the flexible factory, it is essential that we first implement the state-of-the-art integrated computer systems and material handling/control systems that are wired "for the future," while incorporating the latest, most cost-effective process automation. Automating these processes takes significantly more development time than the current state of the art for integrated computer systems and material control/distribution systems. Substantial benefits can be gained from incorporating current systems while waiting for process automation to be developed. In fact, this is highly desirable from a sociotechnical standpoint. When a person performs the process while we are developing the automation to replace the human on that menial task, the person can be trained to move up to a next higher assembly (vertical integration) task requiring more skill and intellect. Thus, the person is constantly being driven toward new goals at a pace commensurate with his/her ability to learn.

8.10.5 USER INVOLVEMENT

8.10.5.1 The User and the Planning System

Successful implementation of the flexible factory of the future requires that the user be involved from the time of the initial design concept. Figure 8.10.2, illustrating user involvement, describes the degree of involvement of the system development function and the user functions. Please note that in the planning cycle, the user and the systems development functions have equal responsibility and authority for defining the system. During the development stage, the systems and development function takes the lead role in developing the system to meet the requirements that were mutually planned. During this stage the user is never totally left out of the process, but kept abreast of the development. During the implementation stage, the system ownership of development transfers from the systems and development function to the user function. And, likewise, during the operation

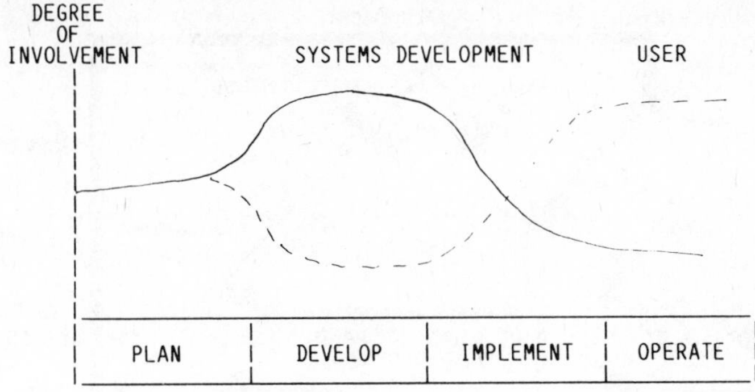

Fig. 8.10.2. User involvement.

cycle, the systems and development function never drops out completely. The user needs to feel that the systems and development function is responsible for assuring reliable and maintainable functionality of the system.

8.10.5.2 Project Development Cycle

Good project methodology tools are required to assure that the requirements, specification, design, generation, and implementation of systems design is under tightly structured design and change control rules. One of the most difficult tasks in implementing systems is the match of system specifications with user requirements. It is essential to remember that system specifications are based on user requirements. Frequent user reviews of the various stages of the project development cycle are required to assure that a system is being designed to meet user needs. It is recommended that user signoff, as well as the signoff by key functional departments involved in operating and maintaining the system, such as the management information system, maintenance department, and tool room, take place at every stage of the project development cycle.

8.10.5.3 Operating Scenario

The systems development function must review the operating scenario prior to doing systems design. In order to develop a system that meets their requirements, the activity being performed by the user and the expected results must be reviewed. Therefore, one needs to look at the total system from a global point of view, which encompasses the hardware requirements for the computer, communications network, applications hardware (including equipment and tools), and peripheral equipment. Similarly, one must consider all aspects of software, including the operating system and communication software, as well as the applications software. One needs to review also the workcells required to perform the activity to be performed, e.g., flexible manufacturing system (machining) vs. conventional machining, or manual assembly vs. automatic assembly. Similar workcells have different requirements for automated material distribution and control and computer-aided test and inspection cells. One needs to look at the data required, the timeliness and control of that data, as well as the configuration management of that data for each type of required workcell. This is in addition to the tools, equipment, and material supplied as input, as well as the material output requirements. The support systems for the flexible factory operating scenarios are based on a product, process, organization, and management philosophy that is different for each function. Such things as quality reporting, financial reporting, production planning, and inventory control reporting, management visibility, receiving and receiving inspection, kitting, as well as tool development, control and distribution are different for different functions being performed. If we are to implement those systems successfully, it is essential to develop the systems for the operating scenarios through a structured methodology that meets the user requirements.

8.10.5.4 Systems Test

Systems test is the single most important function in successful implementation of the flexible factory. The time and effort spent upfront in debugging the system and assuring functionality, reliability, maintainability, and usability pays off in the end. A systems control and quality assurance function should be established to review and audit systems for integrity of systems design, appropriate intrasystems communications, and assurance of development to authorized design. Utilizing the systems development methodology (SDM) defined in Section 8.10.3, all systems projects should be reviewed at the following points of audit: project initiation/requirements, completion of functional design, development of test plans, execution of test plans, and operational turnover. At project initiation, the systems control and quality assurance function should assist in setting project standards per SDM, assist the project team in selection of the user community, and aid in acquiring signoff. The latter includes validating systems bounds, scope, and points of interface, and reviewing and signing off on requirements or project objectives documentation.

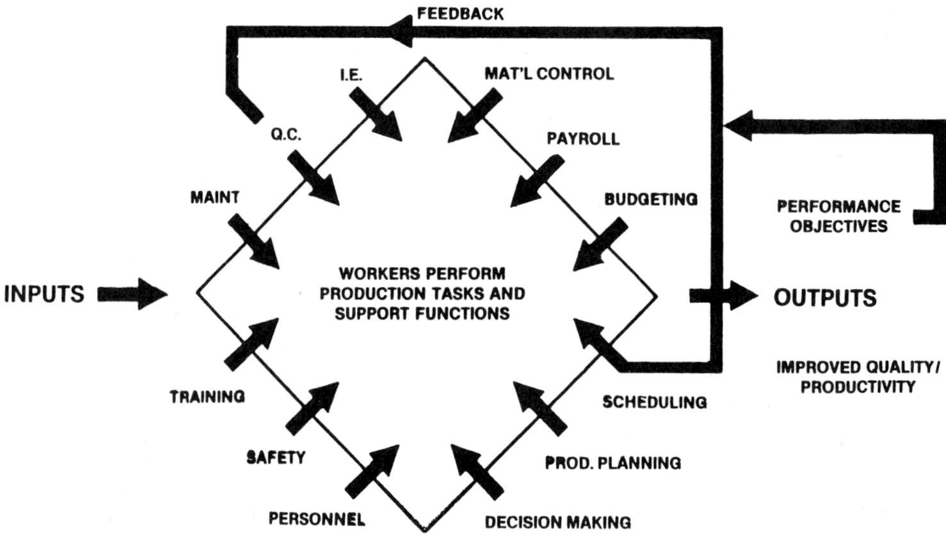

Fig. 8.10.3. Participative design will result in a psychologically enriched workforce.

At the functional design stage, systems test should assist in resolving design issues and defining project interfaces, should review logical database design where appropriate, and review and signoff the system document.

At the point of establishing test plans, this function should review the test plan against the design document for test completeness, define the proper testing environment, the appropriate database, and the technical support needed to perform the test(s). In addition, the proper level of user participation must be secured, and review and signoff of the test plan accomplished.

At the point of operational turnover, the systems test function should review the user's implementation plan/strategy for validity, assist in developing acceptance criteria, defining the proper production environment, securing the appropriate database and technical support for production, and securing user system signoff.

It is essential for successful systems implementation to assure that the systems are adequately tested, implemented, and debugged, so that the user's first interaction with the system is positive. Nothing is more frustrating for the user than to be taken off the production line to learn this new system and to find some bug. The credibility of the system developers comes into question. Too often, companies are interested in getting the system into production before it is really ready.

8.10.6 PARTICIPATIVE WORK DESIGN

Participative work design is a new structured technique that evolves a psychologically enriched workforce by evaluating worker inputs, outputs and feedback loops. Figure 8.10.3 shows this process schematically. Participative work design relates workers' needs to company needs—and improves the quality of work life and productivity.

The same types of input criteria used for a process design analysis are evaluated from a worker's point of view. The inputs are those tools, information, and material that we provide the workers to perform production tasks and support functions that result in improved quality and productivity outputs. When work design is evaluated from a total input/output point of view, performance objectives can provide feedback to provide a closed-loop system. Establishing participative work teams that are integrated with the manufacturing processes enables problem solving and decision making to take place at the worker/team level. Defect cause-and-effect analysis and problem solving will reinforce "the manufacturing operator responsible for quality" concept. At the same time, one can expect a more satisfied work force, which will take pride in its own personal and team efforts.

Lyman D. Ketchum, a consultant, states[3] that a sociotechnical approach to participative work design offers the best way to achieve and sustain success. Participation is crucial to success. First, involve all levels in analyzing what things are like today and how they got that way. Then encourage and permit broad participation in defining a desired future state that is the kind of workplace they will want. Next, encourage broad participation in working out the changes necessary to move the organization toward the desired state.

Broadly stated, the sociotechnical approach pioneered by Eric Trist, Fred Emery, and colleagues in the Tavistock Institute and now being introduced in the factory by consultants like Lyman D. Ketchum, answers six questions:

1. What does the technical system need from people to run well?
2. What does a social system (people) require from a technical system (work) in order to meet peoples' needs (healthy quality of work life)?
3. How well are people meeting the need of a technical system?
4. How well does the technical system meet the needs of the people?
5. What changes in these two systems (social and technical) would improve conditions in both?
6. What must be done to bring about these changes?[3]

Before the employees' commitment to the development of a less selfish, or cooperative, approach to work can be achieved, there must be understanding. Understanding comes from the open expression of skepticism through a process of debate and of analysis. There must be an incentive to dig into old assumptions and ask hard questions, as well as a sufficient level of trust so that individuals will speak candidly to peers, subordinates and superiors. Above all, cooperation comes before any of these other things.

The user must be involved in creating the technology of the flexible factory of the future. People, as well as technology, need to be coordinated. We must create an environment that encourages the employee to replace manual work with more cost-effective automation, while obtaining new work to sustain the workforce while productivity is being increased. This involves further development of peoples' skills and creation of broader job descriptions, new structures and incentives, and a new philosophy of management. A social and technical "marriage" is needed to introduce flexible factory automation in the factory of the future.

People are not programmable; they have needs that require constant fulfillment. Satisfying these needs is important. Some of the workers' principal needs are (1) decision making—some is better than none, more is usually better than less; (2) reasonable use of one's intellect—not too little, not too much, what is optimum varies as one develops; (3) the opportunity to learn and continue learning; and (4) being and staying informed.

8.10.7 TRAINING AND IMPLEMENTATION

Training and developing peoples' abilities properly is essential to successful implementation of technological change. This area is usually overlooked or misunderstood. (The higher the technology, the more complicated the training program.) A detailed, step-by-step user's manual, lesson plans, and hands-on training program that is in the user's language and at the appropriate intellectual level is needed for successful training. In addition, the trainer should be the trainee's supervisor, not someone who comes and goes from any support function. This helps to establish an optimal supervisor-worker relationship.

Before people are trained on the application, they should be trained on how to use the tool. This is too often overlooked. Similarly, the objectives and goals of the job function must be established before orienting the worker to his/her area of responsibility. Then we need to teach the specific application. Moreover, when training relates to and involves the worker (trainee) and supervisor (trainer), successful implementation is more readily achieved. This requires an organized, structured approach to training.

8.10.7.1 Training Plan

More specifically, the social-technical process (see Fig. 8.10.4) describes how to introduce new technology into the workplace. The process begins with the user manual as the prime input to lesson-plan development. It is preferable that the people designing the system do not write the user manual. Instead, use technical writers to translate the Technical Development Specification into a user manual, written at the reading level of the people who will use the system, and broken down into modules, each one covering a specific task to be performed. The user manual then becomes the basis for developing highly structured lesson plans and teacher aids.

The lesson plan should be prepared by people with an education background. One good source is interns who are M.A. or Ph.D. students in education from local colleges and universities. Like the user manual, the lesson plans should be divided into task modules. They should be proficiency based; that is, the trainees learn to perform the task during the course of the lesson, and their proficiency in accomplishing the task is checked by performance. The lesson plans provide the instructors with talking points for introductory and acting narratives, instructions on how and when to use visual aids, and step-by-step procedures for taking the trainees through the tasks contained in each module.

The lesson plans relieve the trainers of the burden of preparing lessons, and ensure that all trainees learn the same material. A detailed, highly structured curriculum is necessary, since most first-line supervisors are not professional trainers. The trainers usually should be trained by their supervisors. This helps pass on "systems ownership" to the user, and further assures continuity of future training on the job. The use of supervisors as part-time trainers is both a necessity and a logical extension of existing organizational relationships.

The supervisors need to be taught how to train their subordinates and/or peers. Many universities, companies, and/or consultants have excellent "Train-the-Trainers" courses. The lesson plans (described above) are used to train the master trainers, as well as the trainers. A systems approach should be used to train the supervisors/trainers. In order to be used effectively as master trainers, the engineers and technical persons who develop

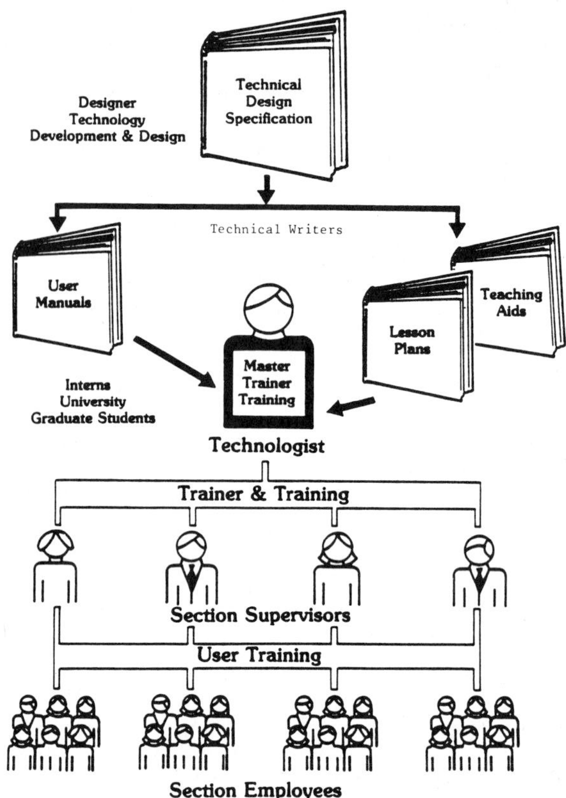

Fig. 8.10.4. Sociotechnical process; introducing a new technology into the workplace.

the system also need to be trained on how to train. The supervisory training program should consist of (1) demonstrations of the system; (2) lectures (based on the lesson plans) and hands-on training to teach proficiency over the entire system; and (3) practice sessions during which the supervisors train each other.

Using the lesson plans developed by the interns, the system developers teach the first-line supervisors how to perform all the tasks involved in operating the entire system. This is in contrast to worker training, which covers only those tasks necessary to perform particular jobs. Using the system developers as master trainers serves a multiple purpose: (1) it provides a system developer an opportunity to review the lesson plan for technical correctness; (2) it provides experience in explaining highly technical subjects to nonexperts in layman's terms; and (3) it provides an opportunity for direct interchange between the developers and users of the system.

As the user manual is being developed, a system developer training facilitator should be assigned to establish communications with the users. Similarly, a user training manager should be assigned where economically feasible. Initial meetings are needed to identify the specific people to be trained, to designate user master trainers, and to identify any prerequisite skills that must be covered in training. The system developer facilitator then develops a training plan for a particular system including (1) scheduling the needed train-the-trainers classes; (2) scheduling all training classes; and (3) negotiating agreement between the developer and user.

As a result of early involvement in the training process, employees find a sense of "ownership" that will motivate them to assure successful implementation. One also finds that changes are made to the initial system being developed to better accommodate the worker. In a recent article, Kelvin Cross of Wang Laboratories stated, "All too often a new automated system is implemented with little or no regard for the kind of job being created. Yet with a little consideration of human factors and motivation, the effectiveness of both the automated process and the workplace can be maximized."[4] In an article, Alice M. Greene states that, "Conversion to the factory of the future is a slow arduous process measured against world time, but not against an individual's clock."[5] Thomas L. Gunn, Manager, Computer Integrated Manufacturing Group, at Arthur D. Little, Inc. predicts that "in ten years nearly all manufacturing employees—from shop floor to top management—will be spending roughly one day in five in seminars and training programs to keep pace with new technology."[6] Training is truly one of the keys to successful systems implementation.

Here again, the more generic, modular, and decoupled the systems design is, the easier and more effective the training and learning task will be. Implementing new technology over time should be in a generic, modular, and decoupled way that enables people to continue to learn and upgrade their skills, as well as to achieve job enrichment and job satisfaction. As a result of this evolutionary implementation of the flexible factory of the future, people will better accept robotic automation, seeing it not as job replacement but as a tool (friend) to help take over a boring task and enable the individual to learn more complex tasks.

8.10.7.2 Plan to Develop Implementation Plan

As indicated earlier in this chapter, planning is integral to successful implementation. Basically, one needs to have a systematic approach to developing an implementation plan. Figure 8.10.5 identifies the work to do (action to take), responsibility, and project schedule dates needed to identify all of the activities and their appropriate interfaces at the system, task, and organizational levels. Then, one must prepare a *gross implementation plan* and, with the users, determine the appropriate level of effort involved in implementing the plan. Similarly, one needs to evaluate the impact of global concerns, common concerns, and system specific concerns involved in implementing the design. *Global concerns* are concerns or issues that affect all systems. *Common concerns* are concerns or issues that impact one or more systems, but are unique to each system. *Systems specific concerns* or issues are those that affect only a particular system.

Both the user and systems development management must identify people availability for systems implementation, and subsequently schedule people accordingly to perform the activities defined above. This *net implementation plan* then needs to be communicated to all parties concerned and monitored and maintained accordingly.

8.10.7.3 Implementation Plan

An implementation plan defines, monitors, and tracks the availability of equipment, software/systems, data (information), material, and people. Of particular importance are those tasks associated with systems testing, training, and preparation for cutover to production. Unfortunately, too many companies think the job is nearly complete when the software system modules have been coded and/or the mechanical equipment modules have been constructed, tested, and debugged. That actually is far from reality. Each system module needs to be tested with the the other modules with which it interfaces, whether software or hardware. The more integrated and complex the systems design, the more complex the testing, debugging, and retesting.

At Westinghouse Defense and Electronics Center in Baltimore, MD, as many as 27 tasks are tracked on each system module(s), as shown in Figs. 8.10.6–8.10.8. The following key definitions apply to the critical tasks for which specific responsibility is assigned and schedule performance tracked:

System requirements agreement. Defines all requirements that a system must meet/perform. Provides a mutual understanding of system agreements between the developer and the user.

System test plan. A development function document that defines all tests required to fully exercise a particular system.

System test. Test to assure accurate performance of all functions that a system provides. The system test plan is used to direct system test.

System integration test. After a system has been fully tested, this test assures accurate data/database interface between a particular system and all systems with which it interfaces.

User test plan. A user developed test plan that exercises the system from a user's point of view and is a condition of systems acceptance.

User acceptance criteria. A user-generated list for each system of milestones that must be completed before the system is considered acceptable; for example, user manual, training plans, and user test.

Training plan. Defines user plans to train personnel for a particular system. This includes training materials, schedules, equipment, etc.

Lesson plans. A development function document necessary to properly train the user for a particular system.

User manual. Defines how to use the system from a user's point of view. This would include such things as screen formats, how to respond to error messages, and report interpretations.

User familiarization. Involves a walkthrough of the system's operations with developer, development functional staff, and functional user.

Development staff system test. Execution of steps identified before successful operation of this step would lead to preacceptance signoff of system being developed.

Functional user test. Same as above, performed or reviewed by functional user.

Operations manual. Defines how the system operates, when the system operates, JCL, backup procedures, reload procedures, data archival, disaster actions, manual procedures in case of shutdown, etc.

Preacceptance signoff. Agreement by the functional user that the system performs to all requirements and is ready to go into production.

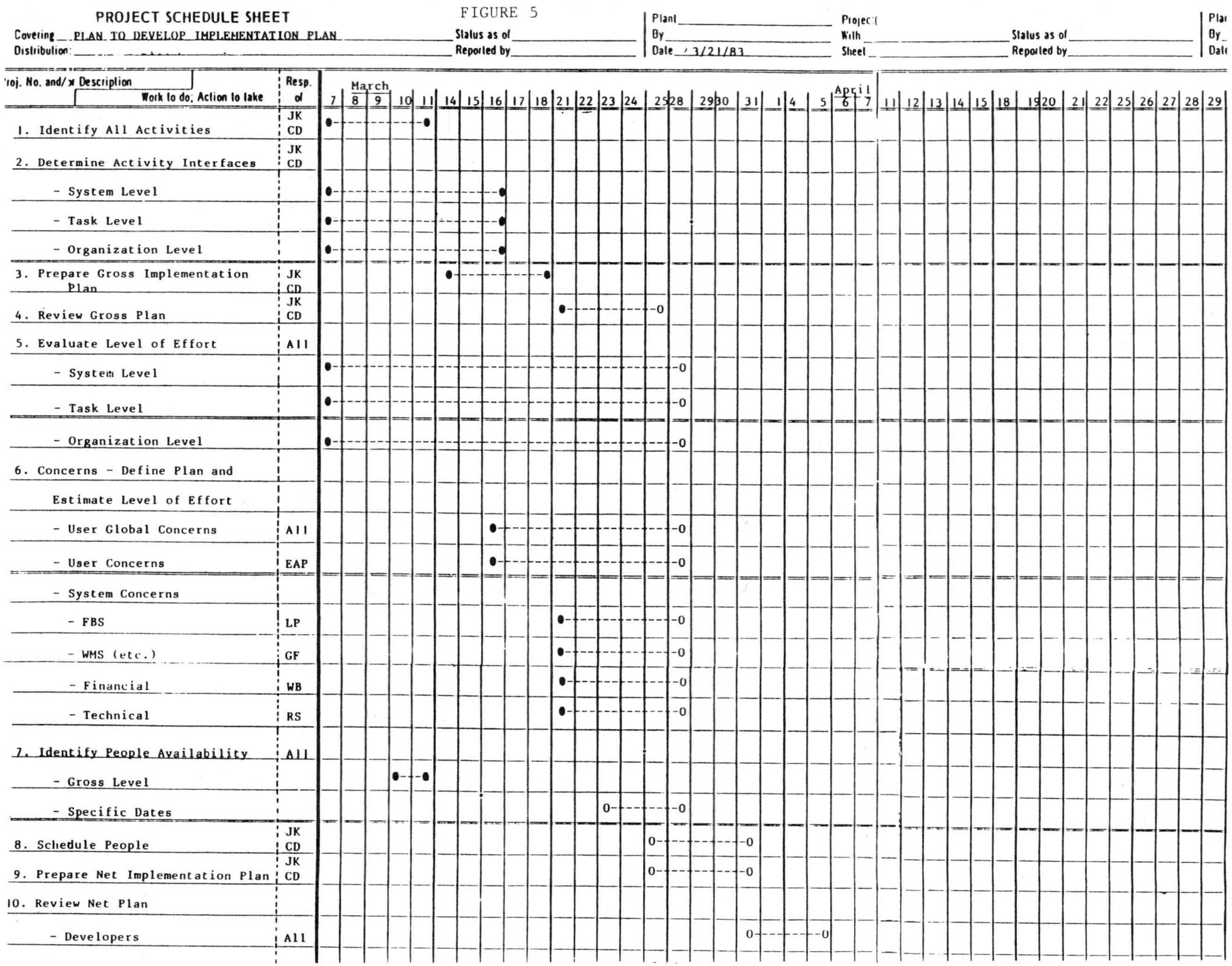

FIGURE 5

PROJECT SCHEDULE SHEET

Covering PLAN TO DEVELOP IMPLEMENTATION PLAN — Status as of ______ Reported by ______ — Plant ______ By ______ Date 3/21/83 — Project ______ With ______ Sheet ______ — Status as of ______ Reported by ______ — Plan ______ By ______ Date ______

Distribution: ______

Proj. No. and/or Description — Work to do; Action to take	Resp. of		March																					April																
		7	8	9	10	11	14	15	16	17	18	21	22	23	24	25	28	29	30	31	1	4	5	6	7	11	12	13	14	15	18	19	20	21	22	25	26	27	28	29
1. Identify All Activities	JK CD	●				●																																		
2. Determine Activity Interfaces	JK CD																																							
- System Level		●							●																															
- Task Level		●							●																															
- Organization Level		●							●																															
3. Prepare Gross Implementation Plan	JK CD						●				●																													
4. Review Gross Plan	JK CD											●				0																								
5. Evaluate Level of Effort	All																																							
- System Level		●															0																							
- Task Level		●															0																							
- Organization Level		●															0																							
6. Concerns - Define Plan and Estimate Level of Effort																																								
- User Global Concerns	All								●								0																							
- User Concerns	EAP								●								0																							
- System Concerns																																								
- FBS	LP											●					0																							
- WMS (etc.)	GF											●					0																							
- Financial	WB											●					0																							
- Technical	RS											●					0																							
7. Identify People Availability	All																																							
- Gross Level					●	●																																		
- Specific Dates														0			0																							
8. Schedule People	JK CD															0				0																				
9. Prepare Net Implementation Plan	JK CD															0				0																				
10. Review Net Plan																																								
- Developers	All																			0			0																	

Fig. 8.10.5. Plan to develop implementation plan.

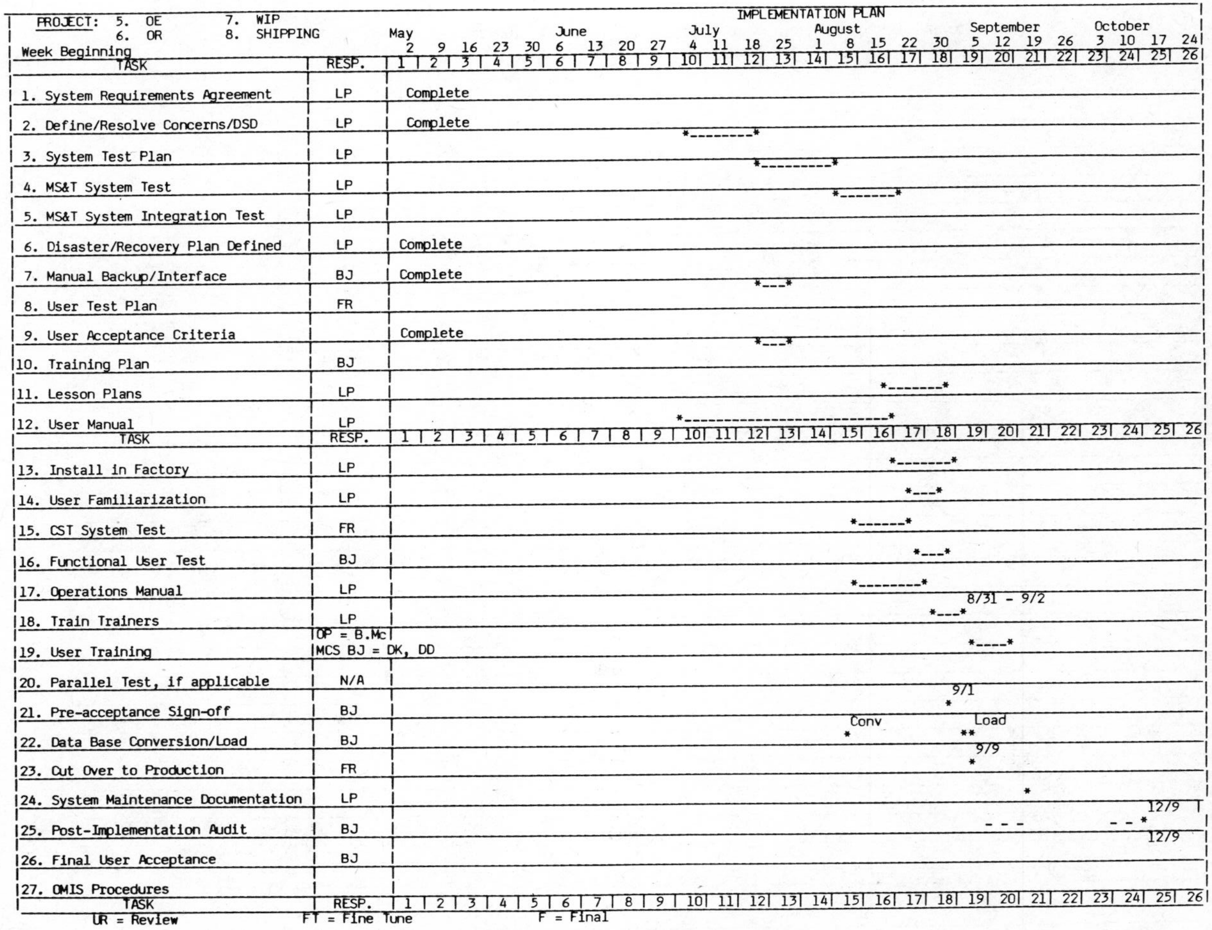

PROJECT: 5. OE 6. OR 7. WIP 8. SHIPPING

IMPLEMENTATION PLAN

Week Beginning: May 2, 9, 16, 23, 30; June 6, 13, 20, 27; July 4, 11, 18, 25; August 1, 8, 15, 22, 30; September 5, 12, 19, 26; October 3, 10, 17, 24 (weeks 1–26)

TASK	RESP.	Schedule (weeks 1–26)
1. System Requirements Agreement	LP	Complete
2. Define/Resolve Concerns/DSD	LP	Complete
3. System Test Plan	LP	*--------*
4. MS&T System Test	LP	*---------*
5. MS&T System Integration Test	LP	*-------*
6. Disaster/Recovery Plan Defined	LP	Complete
7. Manual Backup/Interface	BJ	Complete
8. User Test Plan	FR	*---*
9. User Acceptance Criteria		Complete
10. Training Plan	BJ	*---*
11. Lesson Plans	LP	*-------*
12. User Manual	LP	*------------------------*
13. Install in Factory	LP	*-------*
14. User Familiarization	LP	*---*
15. CST System Test	FR	*------*
16. Functional User Test	BJ	*---*
17. Operations Manual	LP	*-------*
18. Train Trainers	LP	*---* 8/31 - 9/2
19. User Training	OP = B.Mc MCS BJ = DK, DD	*----*
20. Parallel Test, if applicable	N/A	
21. Pre-acceptance Sign-off	BJ	9/1 *
22. Data Base Conversion/Load	BJ	Conv * Load **
23. Cut Over to Production	FR	9/9 *
24. System Maintenance Documentation	LP	*
25. Post-Implementation Audit	BJ	- - - - - * 12/9
26. Final User Acceptance	BJ	12/9
27. OMIS Procedures		

UR = Review FT = Fine Tune F = Final

Fig. 8.10.6. Implementation plan—system modules.

FIGURE 7

INTEGRATED SCHEDULE FOR FACTORY SYSTEMS — IMPLEMENTATION PLAN

TASK	RESP.	PARTICIPANTS	Schedule (weeks 1–26)
1. Define/Resolve Concerns			*---------------*
2. System Test Plan			5-10,17/UR *--*0------0
3. MS&T System Test			5-10 17 *---0-*----0
4. MS&T Integration Test			5,6,7,8 17 *---------0--*----0
5. Fine Tune/Retest			17 5-8 *--------*0-----0
6. Error Recovery Procedures			5-10,17 *------*
7. Manual Backup Plan			5-10,17 *------*
8. CST Test Plan			5-10.17-Integrated *-------*
9. User Acceptance Criteria			5-10,17 *------*
10. Integrated Training Plan			5-10,17 *-----*
11. Training Lesson Plans			5-8/ 9,10,17 *------0-----0---*
12. User Manual Draft			5-8 9,10,17 *-0---------------*-------0
12a. User Review			5-8 9,10,17 *------*0-------0
12b. Rework to Final			5-8 9,10,17 *-------*0------0
13. Install in Factory			9-10 5-8 17 *---* *---* *---*
14. User Familiarization			9-10 5-8 17 *---* *---* *---*
15. CST/User Test			9-10 5-8 17 *---* *-----------0-------0--*
16. Operations Manual			9-10 5-8 17 *---*0------0*--*
17. Train Trainers			9-10 5-8 17 *---* *---* *---*
18. User Training			5-10, 17 *--- --- --- --------------*
19. Parallel Test			17 *-------*
20. Pre-Acceptance Sign-Off			9-10 5-8, 17 *-* *-*
21. Database Load/Conversion			5-10, 17 *-----*
22. Cut Over to Production			#
23. System Maintenance Document			9-10 5-8 17 *----*0-----------0*------*
24. Post Implementation Audit			----+ ----+ ---+
25. Final User Acceptance			*
26. OMIS Procedures			

KEY:
5 - Order Entry
6 - Order Release
7 - WMS - WIP
8 - Shipping
9 - SFDC
10 - SFDC-QC
11 - WMS/Financial Systems

Fig. 8.10.7. Implementation plan—integrated schedule for factory systems.

SUMMARY OF DATES THAT SYSTEMS ENTER PRODUCTION IMPLEMENTATION PLAN																										
SYSTEMS	1	2	3	4	5	6	7	8	9	10	11	12	13	14	15	16	17	18	19	20	21	22	23	24	25	26
1. INVENTORY I - FINE TUNED			*																							
2. MCS-FA			*																							
3. WCR								*																		
4. BOM - FINE TUNED				*																						
5-8. OE/OR/WMS/SHIPPING												*														
9. SFDC												*														
10-14. QC/ECT/MISR/NRS[2]										*																
13-14. NRS STAND-ALONE				*																						
15-16. PAYROLL/PERSONNEL		*																								
17. WMS/FINANCIAL SYSTEMS												*														
18. EDS			INSTALLED-BEING ENHANCED																							
19. TEST ENGR.			INSTALLED-BEING ENHANCED																							
20. MFG CONTROL			INSTALLED-BEING ENHANCED																							
21. CV			INSTALLED-BEING ENHANCED																							
22. MNTCE & CALIB./ASSET CONTROL					*																					
23. PART SUBSTITUTION		*																								
TASK	1	2	3	4	5	6	7	8	9	10	11	12	13	14	15	16	17	18	19	20	21	22	23	24	25	26

Fig. 8.10.8. Implementation plan—summary of dates that systems enter production.

Database conversion/load. Load of active work/status/material into a production database, prior to cutover to production (e.g., work-in-process inventory).

Cutover to production. The activity of going "live" with a system.

Systems maintenance documentation. Commonly called a program maintenance manual, it defines all software in a system by name and function as well as the operating environment. It is used to provide a maintenance programmer with any information necessary to understand the program and describes required maintenance procedures.

Postimplementation audit. The followup review of implemented systems to assure adequate, accurate performance. Normally should occur at 30, 60, and 90 days. Functional user is responsible for defining audit procedures and for performing audit.

Final user acceptance. After postimplementation audit, formal user signoff of system.

An effective systems implementation plan must not only include a detailed plan as defined above, but also requires extensive formal and informal communication. An independent function should be established to develop a systems implementation plan with the key system-development and user personnel. This function then needs to be the focal point of communication during the transition from development to production. The formal plan should include not only the milestone plans (Figs. 8.10.6–8.10.8), but also a published, detailed listing of all personnel associated with the development of the plan or involved with the automated systems development or users. This listing should include telephone numbers, both at business and at home (in case of emergency), and also mail stops.

Further, a chart indicating the responsible parties involved for each system module to be installed should be published. To assist in establishing a common understanding among the various levels of education, experience, and background involved in implementing systems in the factory, a dictionary of terms should be published.

Frequent (normally weekly) meetings should be scheduled to review the current status of systems implementation, including systems functionality, scheduling, resolution of issues and/or concerns for all systems, equipment, material, and people needs. Alice M. Greene pointed out how the Japanese excel in introducing new technology in the workplace: "They offer job security, extensively educate and train the workforce and explain the need for new technology to *everyone*, not just middle management.[5]

The following systems documentation items are also critical to successful implementation: (1) user manual; (2) training material (lesson plans, training aids); (3) systems maintenance manual; and (4) operations manual.

All too often, training is started and the system is rushed into production before all documentation is available. It is difficult to retain what is initially learned while being taught a complex new system without being able to refer back to systems documentation. Systems documentation is another key element that must be in place before successful systems implementation can begin. Likewise, the more generic, modular, and decoupled the systems design is, the better will be the change for successful implementation to begin. It is easier and more effective to learn one module at a time well, than to attempt multiple modules simultaneously. In a recent article, J. Randolph New and Daniel D. Singer of Loyola College state: "As a result of early involvement in a change process, employees may find a sense of 'Ownership' of the change that motivates them to carry it out."[6] Understanding the sources of resistance to change is essential to successfully implementing new manufacturing systems and technology.

As the systems development function works with the user(s) to implement the flexible factory, an open, objective assessment of user concerns must be conducted. It is preferable to establish independent systems implementation, systems control, and quality assurance function to monitor and control successful systems implementation between the systems development and user functions. Concerns whether global, common, and/or systems specific concerns, as defined earlier, should be identified, and responsibility for resolving them should be assigned. Responses should not only be communicated, but also evaluated and assessed from an independent, "consultant" standpoint for assuring resolution of the issues or concerns from both a systems development and users' standpoint. Too often, when systems development thinks they have resolved an issue or concern, it is not perceived in the same way by the user, because of a language/communication barrier between systems development and the user. The independent systems implementation, systems control, and quality assurance function serves as the interpreter and moderator between both functions to assure successful systems implementation.

Formal procedures for tracking issues and concerns must be established, monitored, and published. Reports should identify issues and/or concerns, responsibility for resolution (both systems development, user and/or other function), the response itself, and a schedule for resolution, both required and completed. As issues and/or concerns are resolved, they should be archived and only open items should be printed out in subsequent communication. The actual reporting of key issues and concerns to all parties involved is a critical element to successful systems implementation.

In conclusion, implementation of the factory of the future is evolutionary. The following factors are critical in achieving successful implementation:

1. Establishing a proper mission and goals.
2. Taking the time to perform top-down strategic planning before focusing on bottom-up tactical implementation.

3. Proceeding with evolutionary general systems design by implementing integrated computer systems and material control/distribution systems before introducing process automation.
4. Involving the users not only in the planning and project development stages, but also in the systems test stage.
5. Foremost, getting involved with the real-world operation, both present and future, not just creating an "ivory tower" paper design.
6. Designing the system from a participative work designer sociotechnical aspect.
7. Establishing an independent function from a development and user standpoint for systems control, quality assurance and implementation planning and control. Developing and maintaining a detailed implementation plan as well as comprehensive training plan.

REFERENCES

1. Robert H. Hayes, "Why Japanese Factories Work," *Harvard Business Review*, July–August 1981, pp. 59–62.
2. Richard J, Schoenberger, *Japanese Manufacturing Techniques*, New York: The Free Press, 1982, p. 67.
3. Lyman D. Ketchum, "How To Start and Sustain a Work Redesign Program," *National Productivity Review*, Winter 1981–82, p. 82.
4. F. Kelvin Cross, "IE's Must Address Automation's Effects on Worker Automation and Capability," *Industrial Engineering*, August 1983, p. 75.
5. Alice M. Greene, "Factory of the Future: The Love/Hate Response to Automation," *Iron Age*, February 25, 1983, p. 47.
6. J. Randolph New and Daniel B. Singer, "Understanding Why People Reject New Ideas Helps IE's Convert Resistance to Acceptance," *Industrial Engineering*, May 1983, p. 53.

BIBLIOGRAPHY*

Cover Story, "A Work Revolution in US Industry," *Business Week*, May 16, 1983, p. 100.

Crosby, Phillip B., *Quality is Free*, New York: McGraw-Hill, 1979.

Groover, Mikell P., *Automation, Production Systems and Computer-Aided Manufacturing*, Englewood Cliffs, NJ: Prentice-Hall, 1980.

Wight, Oliver W., *MRPII: Unlocking America's Productivity Potential*, Willison, VT: Oliver Wight, 1981.

* In addition to the sources listed as references.

CHAPTER 8.11

MANAGING THE FLEXIBLE FACTORY

JOHN A. MADDOX

Consultant
Arlington, Texas

8.11.1 INTRODUCTION

When my previous company first made the decision to invest in a flexible manufacturing system back in 1979, and gave me the responsibility to design and create it, I spent many hours contemplating how its management would differ, if at all, from a conventional operation. Four years and several million dollars later, I am still contemplating the same question.

Perhaps the easiest approach to this subject is to understand exactly what flexible manufacturing really means from an operational management point of view.

Let me start by briefly describing the systems we eventually installed and some of the design objectives and criteria we hoped to accomplish.

8.11.2 SYSTEMS INSTALLED

We installed two FMS systems, both in the same facility and under one management. One was for the production of prismatic parts, the other for parts of rotation, or round parts.

In creating these two systems, our approach was a little different from that of most others, inasmuch as we chose to do our own design based on commercially available hardware. This resulted in systems that contained a variety of machine tool types and makes, different electronic controls, and an overall integrated software operating system designed specifically to our requirements.

Both systems, which form the total operation, were tied into and controlled by a central computer. From this host machine all communications to and from the systems were exercised through a network of microprocessors and secondary computers to achieve a totally computer-integrated manufacturing system (CIMS). This included a full DNC system, although we chose a different approach than is usually associated with this acronym. We chose "distributed" numerical control, rather than direct. We downloaded complete and total machine programs, on demand, all at once, rather than block by block. These programs then resided in the memory of the machine controls, all of which were CNC, and had the effect of reducing communications traffic.

The reason for choosing this method, which proved to be correct, was to facilitate the use of a minicomputer for the host, rather than some large mainframe CPU. This approach proved to be simpler, quicker, and more effective than other DNC systems we observed, and considerably less expensive both to buy and operate.

The two systems contained a total of eighteen different machines and their CNC controls that were directly tied to the host computer, plus an additional four pieces of equipment that were manually controlled.

8.11.3 DESIGN CRITERIA AND OBJECTIVES

We established a number of design criteria and objectives that, I am sure, influenced the operation management task. Briefly, these various criteria and objectives were:

To reduce product costs through improved methods and technology and reduced labor content.

To improve product quality through reduced material handling (damage control) and the use of statistical control methods.

To increase product throughput and thereby minimize inventories.

To utilize available and proven hardware technology of current designs, but to avoid custom or specialized design whenever practical.

The end results of these objectives have been an operation that, by comparison to traditional manufacturing, reduced the total number of people by about 65% and lowered costs on an average of 30%.

In some individual cases, costs were lowered by 76% and the labor content reduced by over 80%, but overall the net savings were about one third below previous costs. Additionally, we were credited with being one of the highest quality producers in the company.

One other key criterion worth mentioning here is that we structured our operation around the concepts of a cost center. It was designed to produce a family of parts common to several of our own operating product divisions, all of which are autonomous profit centers. We, alternatively, transferred all production at cost and measured our performance against the savings created for the end user division in the form of a discount below their published costs. This had the effect of baring the soul, because there were no hidden or obscure budgets in which to conceal cost overruns or gains.

8.11.4 PHILOSOPHIES AND CONCEPTS

Managing any operation will always be a byproduct of the philosophies and concepts you believe in, so this is probably a good time to identify those that we established.

8.11.4.1 Level of Automation

As this chapter is concerned only with flexible manufacturing, which implies a high degree of automation, let me start with that. I have never subscribed to the theory of total automation in a manufacturing production environment. To me the most practical and effective approach is a careful and controlled blend of automation and manual (operator controlled) functions.

Automation, per se, is very good at the repetitious and higher volume production, and at maintaining some predetermined level of consistency. Assuming that it is correctly maintained, managed, and programmed, automation will continuously generate product at a constant level of cost and quality. However, the greater the variety of tasks or products demanded from automation, the less effective it will become. This proved to be our first real philosophical lesson.

8.11.4.2 Level of Flexibility

The term *flexible*, in the context of manufacturing, is really an enigma. In terms of variety of either functions or products, the levels of flexibility are limited and costly. The ability to produce a high variety of product will still require similar levels of tooling and fixtures, and each changeover will usually require some form of setup. Admittedly, by such procedures as code and classification and family of parts planning you can improve the organization and scheduling of changeover and variety, but it will still impose downtime. True, some of the machining centers now available are more flexible in the number of different operations and functions they can perform per setup, so you can, in effect, reduce the total number of machines required. However, increased downtime is incurred due to changeover thereby reducing the operating capacity. There is, therefore, a flexibility trade-off.

Our approach was to define our part families carefully by similarity of operation, rather than by shape or size. An example of this is keyways.

Just about all of the 700 different round parts our system was designed to produce (there were originally over 1200 but were reduced through coding and standardization) required keyways. Usually the requirements were for multiple keyways, two or three, and those were generally all different sizes within any given part.

These 700 different parts varied in size and shape from diameters of 1½ to 16 in, lengths from 15 to 120 in., and weights from 25 to 2000 lb. Additionally, the shorter parts tended to be heavier, with the majority having a length-to-diameter ratio of 15:1, and 25:1 not uncommon. They were all precision-ground parts typically requiring tolerance control for both sizes and concentricity of 0.001 in. or better, and surface finish requirements in the range of 5–63 μin.

Since production startup, as luck would have it, our part mix changed considerably from the original plan. As a result of this change we found that those parts originally planned for ran very well through the system, utilizing all aspects of the automation, material handling, and robotics. However, a number of additional parts that were not originally planned for, but now produced, gave us considerable problems.

An interesting phenomenon evolved. Although the layout and configuration for those newer parts was not ideal, causing additional material handling time, we were still able to achieve similar savings. The reason? Our ability through the software operating systems and communications network to adjust schedules and mix changes quickly to maintain both productivity and utilization of the cutting tools.

8.11.4.3 Design Approach

At the onset we recognized that reductions in costs or increases in productivity would not result from faster metal cutting. Figure 8.11.1, which has been widely published and verified, illustrates a different problem.

The net result of this data is to demonstrate that the actual amount of time a part actually spends being productively worked on to produce its final form, is really not much more than 2% of its total life cycle in manufacturing.

If you accept this premise, as we did, then you conclude that any reduction in the 95% portion will translate into a far greater and more immediate savings than in the 2% portion. Put another way, a 30% cost reduction in the 2% portion will yield less than, say, a 10% reduction in the 95% portion. Additionally, any cost reductions to the actual metal cutting and forming operations will be considerably more difficult and costly to achieve.

With this in mind, we designed a system that would nullify the effects of the 95%, by making them transparent, or nonpertinent, to the production cycle. We did this through a number of different approaches and disciplines.

First we designed an operating system that managed capacity planning, scheduling, and dispatch, and the DNC download functions. Additionally it monitored and tracked all production released to the system and provided constant status reporting. Two fundamental criteria were used to structure that system protocol. These criteria were:

1. All production decisions and responses to the system for product already released will be predicated exclusively by demand from the system to maximize the uptime of the machine tools. All material handling movements will take a secondary priority to these demands.
2. All scheduling and dispatching of released product will be prioritized based on setup and changeover. Prior to releasing an order into a specific workcell, the queue of waiting work will be matched against the existing setup of the workcell for compatibility. In the event no exact match exists in the queue, the next order causing the least amount of changeover will be selected.

These two criteria imposed their own disciplines upon our management of the system, namely that we load the system to capacity over a given period, say one week, but then let the operating system create the precise priority in which those order requirements were met.

This had the effect of doing away with expediting, except in breakdown or emergency situations. In does, however, take a high degree of trust and courage to rely on the system and resist the temptation to override it, although, of course, we retained that capability. It is difficult to accept initially, but long term creates the most effective utilization.

8.11.4.4 Preventive Maintenance (PM)

We all believe in PM, and do a lot of talking and explaining its necessities, but then continually ignore it in the name of shipments. You must accept, along with mother and the flag, that PM is necessary, vital, and cannot be tampered with. We actually overcame the problem by building the PM schedules for each machine right into the operating system software. When PM for a particular piece of equipment became necessary, the system simply stopped sending it work until we indicated that the PM had been performed.

At times this is frustrating, and occasionally you will receive calls from irate customers, but in our first year of full production the uptime of the system was slightly over 94% of the productive time said to be available.

8.11.5 THE FLEXIBILITY TRADE-OFF

As stated earlier, flexibility creates a trade-off. It also creates disciplines and procedures that must be followed. Stay within the parameters under which the system operates best and it will perform well; stray outside those parameters and the result will be mass confusion and utter chaos. You have a choice.

In terms of flexibility, you also have a choice. If you want the flexibility of producing a high variety of different part numbers on the absolute minimum number of machine tools, then you can have it if you give up some utilization, productivity, and throughput. But if you want to maximize your utilization, productivity, throughput, and scheduling flexibility, then you must reduce the workload variety.

When we talk about flexible machining or manufacturing systems (FMS) we must recognize the subtle differences between the two terms. Flexibility can be achieved, but in some specific form. It cannot be all things to all people, therefore you need to decide at the onset which direction best suits your requirements. Once chosen, flexible systems actually restrict the amount of flexibility you have in comparison to conventional production shops.

We also found that the ratio of direct to indirect labor changes, a phenomenon that some accounting and financial people find hard to accept initially. Because of the levels of automation and computer control now available, systems require higher levels of support people. You must be capable of quick recovery in the event

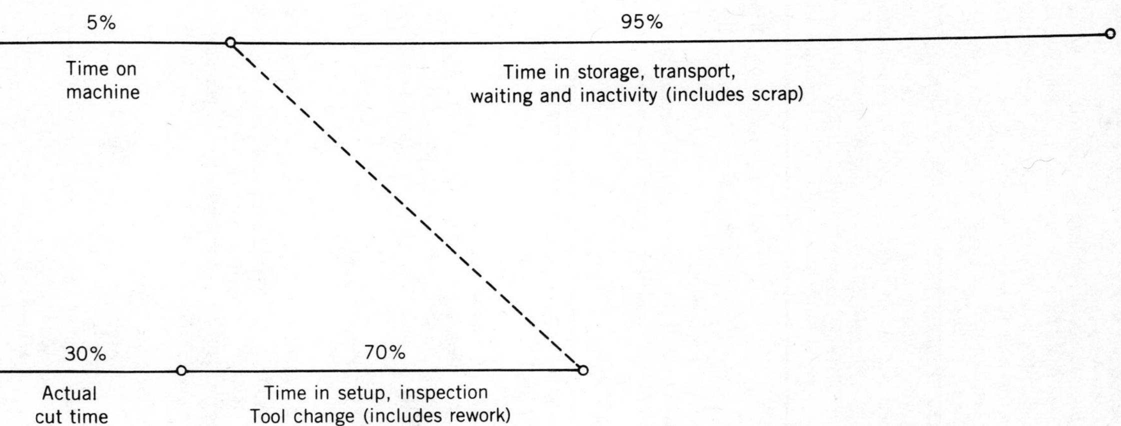

Fig. 8.11.1. Average life of production part (typical batch production shop).

of a breakdown or failure; therefore you need technicians who can jump in and operate *all* the equipment, randomly.

Obviously a very competent maintenance department is also required. We chose to provide all our own maintenance expertise from the traditional "fix it" through the diagnosing and repair of all electronic circuit boards. Given that some machining cells which comprise, say, two large 4-axis lathes, a robot, and material delivery and transport system, can cost over $1 million, you really cannot afford to wait two or three days for the supplier to deliver a $25 P/C board or valve for you. You must have the spare parts in inventory and the required skills to repair it immediately.

Without this capability, you can lose not just valuable product and production time, but recovering from such a loss in terms of getting the whole system back on line can take several days. Again you have a choice. Either accept the risks involved and plan for such losses, or hire and train a highly competent maintenance crew and carry a substantial inventory of spare parts and diagnostic equipment.

No one will like this type of inventory, and you will have difficulty convincing them of its necessity, but operating without it is worse. I compare it to insurance; you pay the premiums in the hopes you will never have to use it, but if you do, you are really glad you have it.

8.11.6 DESIGN OF THE SYSTEM

Because of the approach we followed, which was based on the increase of throughput, we spent considerable time and effort on the problems of scheduling and order release. To fully understand why, let me explain briefly the design of the system.

The system was designed to provide sufficient queueing on-line for the effective balancing and utilization of the machine tools, but not so much that product would sit in work-in-progress (WIP) for long periods. The concept is simple; design a system that provides minimal on-line storage so that order entry rate is governed by shipment rate, with no place for product to get lost.

The total material handling system, depicted in Figure 8.11.2 was an automated, car-on-track palletized railway under computer control. Once all the cars and pallets were full, you could not input more work until you had offloaded shipments. The net effect was to hold down the size of WIP inventories, while increasing the inventory turns threefold over conventional shop operations. I project that ultimately the inventory will turn about 10–12 times annually, presently it is between 5 and 6 times. Conventional operations are running between 2 and 2½ times.

All these advantages are obviously desirable, but as can be expected, they also bring with them some disadvantages. With such highly automated systems and their imposed disciplines and procedures, things can happen quickly.

You must insure that what you release to the system is what you require from the system. If you are wrong, chances are you will be wrong very efficiently. To correct a discovered mistake halfway through the system requires offloading, setting aside, rescheduling and re-releasing the correct orders, and then reentering the incorrect orders at some later date. We found that, once released, we could process significant number of parts in a very short time. What used to take 3–4 shifts of production time in our conventional operations can now be accomplished in 1–1½ shifts.

If you remember the chart recalling our efforts to reduce the effects of the 95%, you will realize how this could happen.

One more example of the inflexibility of flexible systems.

There were, of course, many obvious things we did in preparing for the operational management of the systems. Advance communications to those involved; training programs; careful planning, etc. However, if I think back to the things we did that I now believe had the most significant and lasting impact, I would have to include the following:

Code and classification of all parts.
Family of parts grouping and definition.
Process and tooling standardization.
Preventive maintenance planning.
Interactive scheduling based on load and capacity.
Extensive training.
Open, two-way communications at all levels.
Partnership approach to suppliers and vendors.
Singular understanding of the ultimate objectives.
Commitment to simplicity and excellence.

The last two may seem obvious, but all too often the most well intended and well conceived projects fail as a direct result of their lack.

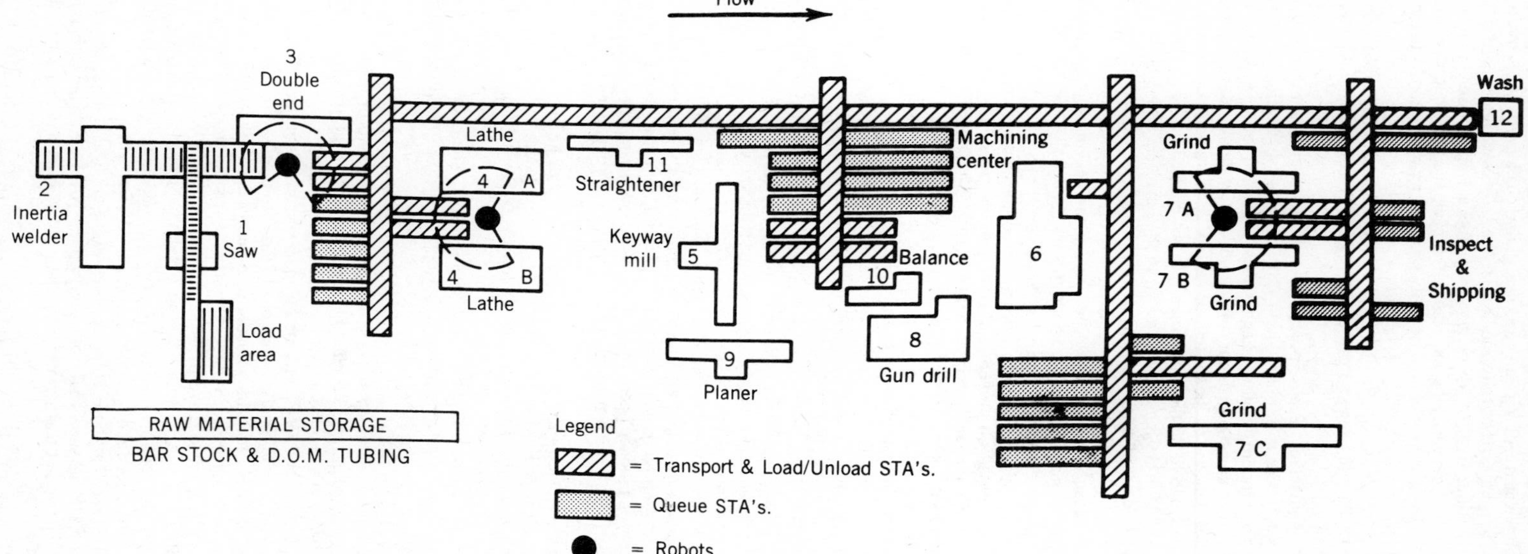

Fig. 8.11.2. Layout of cylindrical-part FMS (not to scale; proportionally correct).

As we have all learned to our chagrin at one time or another, it is the easiest thing in the world to create something that is overly complicated. A far greater and demanding challenge is to keep it simple. Additionally, it has been my experience that unless carefully monitored, the most well intentioned people will eventually drift off on some tangent and lose sight of the original objectives. I am not sure which of these two points precipitates the other, and in fact that may be the classic "chicken and egg" question. We made a concerted effort to overcome these pitfalls, and in so doing probably added at least six months to the design stage of the program, but in retrospect it was a worthwhile investment.

Throughout every conceivable stage of the total project we continually challenged each and every decision and course of action against our fundamental objectives. Did they solve anything? Did they in fact mean anything? Were they relevant?

We consciously reminded ourselves of those objectives and, as a result, spent virtually no time working on things of little value. In fact, I was surprised at how little of what we developed actually ended up being superfluous. Occasionally some modification or changes were required, but no redesigns or major problems.

We particularly extended this philosophy to our software development. We were not concerned, as software people so often are, with the elegance of the system or design, or its ranking in world creativity circles, but rather whether it was good enough to do the job. As soon as we reached that stage, no matter how crude, we stopped working on that module or design and moved on to the next one. Our philosophy was that we could always go back at some later date and augment what we had done, but by then our efforts would be based on some additional knowledge or experience.

We chose this approach because we really had no way, at the beginning, to predict how such a system might react. From an operational management viewpoint it is imperative that you are able to predict with some measure of certainty how the system(s) will behave in a variety of situations. Without this one ability, you will never be completely in control and therefore will have accomplished nothing.

Finally, a personal opinion based on proven, and often painful experience. *There is no such thing as an effective or practical turnkey manufacturing system*. Many people, both users and suppliers alike, believe you can order a system and await its delivery. To some degree this may be true, but remember the word "flexible." You have to assume that changes will be required during the life of the system. Both products and markets will evolve and change. So too will varieties, mix, and volumes.

If you do not have qualified and knowledgeable people on board to react to these changes as they occur, you run the serious risk of making critical mistakes that will haunt your operations, and your reputation, for a long time.

The operations people at all levels should be involved as early as possible in the project's life. *You* will eventually own the system, not the supplier who will quickly move on to other projects, leaving you the problems and shortcomings. This is not to imply any inadequacy on the suppliers' part; they are generally very professional, but systems are very personal. No one understands your problems better than you do yourself. It is very difficult to communicate knowledge and experience built up over decades. Your own people should play a very active role in the design stages to support and maintain the system when installed.

Use the suppliers' expertise in the products they build, but do not allow your problems to be redefined to suit your suppliers' solution or product lines. If necessary, use other suppliers' equipment in specific areas. The most effective manner in which you can do that is to be your own program manager, not always convenient, but ultimately most effective.

Use consultants for design reviews and in specific areas; they tend to be unbiased, and given the overall cost of an FMS, their fees are small. Always remember it is your system, which must solve your problems. You are not building a showpiece for the machine tool industry. What you want is advanced technology, not gimmickery.

This approach provides the greatest assurance that the majority of the knowledge gained in system design will reside with your own people, not the suppliers. That is my principal objection to turnkey projects—they rob you of knowledge.

People management is also a critical area of concern and forethought. In a traditional batch production environment, every machine and operator represents an autonomous island of production whose coordination and scheduling is handled by someone else. In a flexible, or cell environment, the need for coordination between stations or cells must be built right into the operating control procedures.

Communications of schedules and tasks must be far more organized and distributed than usual.

For a system to run smoothly, you require a formal and structured set of operating policies and procedures. Once operational, you will not have the same amount of time in which to react to situations that you may be used to. The operations people must clearly understand their roles and responsibilities when intervention is required. Similarly, they must learn to leave the system alone once it is correctly setup and producing. The age-old habit of continually trying to adjust for every little "blip" observed or perceived must be discontinued. Let the machines and their CNC controls do their job. You will undoubtedly pay a lot of money for them, so let them earn their keep. It requires faith and confidence, but it does work.

At this time I would suggest that you become familiar with statistical control methods at all levels within your operations group, not just at the supervision or quality control level. Learn the various methods of statistical measurement and control charts, and teach your people how to interpret and understand their meanings.

Above all, you must resist the temptation to keep "playing with the train set," and *let it run*. Only with such operating knowledge can you begin to establish the database necessary to make your system predictable, and therefore controllable.

My earlier comments about not seeking perfection also apply to the operational management of your system. The time for comtemplation and deliberation is during the design stage; once installed you need to generate revenue to achieve an acceptable payback. Continually trying to improve your system is both costly and time consuming. Let it alone and save your improvements and observations for the next system you install, because if you do not, the economic returns will become so distorted there may not be a second chance.

Left to its own devices, your system will gradually establish its own operating parameters. You will quickly learn both its upper and lower limits of operating capability, and can then adjust accordingly. The trick is to design a system that basically matches your quality requirements from the onset. Generally all that will be required is a little fine tuning after installation: speed and feed rates, cutting nomenclature, carbide grades or coating, stock removal rates and allowances, etc. These methods are traditional and well understood; it is not necessary to constantly vary them!

A more realistic danger is actually to miss your original quality requirements by not fully understanding them up front. During the design phase of your system, once you have selected a family of parts to be produced on it, a great deal of coordination and cooperation is required between manufacturing and engineering, both in terms of fully understanding the requirements and also what can be done to improve those requirements. Engineering has a significant impact on both the quality and costs of product, but often lacks the experience or knowledge to recognize these areas. The adversary relationship that generally exists between engineering and manufacturing does nothing to relieve this problem, and certainly confrontation is not an answer.

Education on both sides is required for each to fully understand and appreciate the other's problems and constraints, and only then will you be able to achieve the full exploitation of the system. Finding out too late that the system you installed is not capable of achieving the results you require is deadly.

Let your system run in full automation, establish its own natural control parameters, fine tune them to match your individual requirements, and then turn it loose on production with the absolute minimum of supervision. Through the application of good statistical control procedures and methods you will consistently produce product to a predictable level of both quality and cost, and finally return the control of quality and productivity to the hands of management. The less of this you do, the more control management will have to relinquish.

If you do your homework well, manage the program effectively, say your prayers and eat all your spinach, you will end up with a good system that will serve you well. Now all you have to do is have faith, confidence, and trust.

INDEX

Abrasive processes, 4 • 61
  finishing, 4 • 61
  grinding, 4 • 61, 4 • 66
  honing, 4 • 61
  lapping, 4 • 61
  machining, 4 • 61
  polishing, 4 • 61
Accelerated cost recovery system (ACRS):
  ADR life, 6 • 32
  recovery period, 6 • 32
Acceptable quality level (AQL), 3 • 111
Acceptance process level (APL), 3 • 108
Acceptance sampling methods:
  attributes, 3 • 111
  bulk, 3 • 115
  chain, 3 • 115
  continuous, 3 • 115
  skip-lot, 3 • 115
  variables, 3 • 111
Actuator, 3 • 69
Adaptation to uncertainty, 2 • 14
Adhesives, 5 • 86
Advertising and promotion, 3 • 101
Aftertax analysis, 6 • 32
Aiding (*see also* Training)
  adaptive, 2 • 62
  expert systems, 2 • 62
Aggregate, 3 • 116
Aggregate capacity planning, 3 • 125
Aggregate production planning (*see also* Aggregate planning)
  costs relevant to, 3 • 117
  fixed cost models, 3 • 117
  linear cost models, 3 • 117, 3 • 118
  nonlinear cost models, 3 • 117
  operations control, 3 • 125
  quadratic cost models, 3 • 117
  strategic planning, 3 • 125
  tactical planning, 3 • 125
  taxonomy, 3 • 125
  the role of models in, 3 • 117
  when to use, 3 • 116
Algorithm(s):
  decision support system, 1 • 69
  HAP, 7 • 6
  minimum annual cost, 1 • 31
  multifacility HAP, 7 • 7
  optimal and heuristic, 3 • 162
  scheduling and sequencing, 3 • 159
  synthesis, 8 • 28
Allowances, standard time:
  fatigue, 2 • 110
  personal time, 2 • 110
  unavoidable delays, 2 • 110
Alternative world views, 8 • 64
Aluminum oxide, 4 • 61
American (U.S.) firms, 1 • 12
Andon (visual control system), 3 • 255
Angle:
  back rake, 4 • 30
  negative rake, 4 • 68
  rake, 4 • 26
  shear, 4 • 28
ANSI/ASQC Standard A3, 3 • 96
ANSI/ASQC Z1.4, 3 • 113
ANSI/ASQC Z1.9, 3 • 114
ANSI/ASQC Z1.15, 3 • 98
Anthropometric:
  dynamic dimensions, 2 • 74
  static dimensions, 2 • 74
Anthropometrics, 2 • 54
Arcs, 7 • 9
Argon-oxygen decarburization (AOD), 4 • 3
ARMBASIC, 4 • 91
Articulating, 4 • 76
Artificial intelligence, 3 • 80, 3 • 92, 8 • 39
Assembly:
  automatic, 3 • 21
  automated, 3 • 21, 3 • 39
  chart, 5 • 25
  flexible, 3 • 36, 8 • 75
  line, 1 • 6, 2 • 10
  line balancing (*see* Assembly line balancing)
  line cycle time, 3 • 176

Assembly (*continued*)
line layout, 1 • 6
product, 3 • 176
robotic, 3 • 37
rotary machines, 3 • 31
selective, 3 • 36
Assembly line balancing:
branch and bound, 3 • 183
computer aided line balancing (CALB), 3 • 186
defined, 3 • 177
goals, 3 • 176
heuristic procedures, 3 • 183
largest-candidate rule, 3 • 184
methods, 3 • 182
procedure, 3 • 177
ranked positional weight, 3 • 185
Assessment:
subjective, 2 • 63
workload, 2 • 63
Assignable causes, 3 • 103
Attributes:
of flexible factory control systems, 8 • 96
Attritious:
grain wear, 4 • 68
wear, 4 • 62
Ausforming, 4 • 55
Autocorrelation, 3 • 132
Automated:
Advanced Office Controls (AUTO-AOC), 2 • 98, 2 • 117
assembly, 3 • 21
assembly machines, 3 • 39
data capture methods, 3 • 210
guided vehicles (AGV), 8 • 20
factory, 1 • 3, 1 • 8, 5 • 65
material handling, 2 • 93, 3 • 211
storage and retrieval systems (AS/RS), 8 • 108
storage systems, 2 • 93
Automated factory, 1 • 3, 1 • 8, 5 • 65
Automated storage systems:
automatic case picking, 4 • 125
automatic item picking, 4 • 125
car-in-lane systems, 4 • 125
carousels, 4 • 125
miniload AS/R systems, 4 • 125
unit load AS/R systems, 4 • 122
Automatic assembly, 3 • 21
auxiliary orienter, 3 • 27
carousel, 3 • 34
centerboard feeders, 3 • 29
centrifugal feeders, 3 • 29
elevator feeders, 3 • 30
flexible, 3 • 36
general system, 3 • 21
half-centerboard feeders, 3 • 30
in-line free transfer, 3 • 34
in-line indexing, 3 • 34
machines, 3 • 31
magazines, 3 • 22, 3 • 24
nonvibratory feeders, 3 • 28
part orientation, 3 • 26
parts feeders, 3 • 21
pick and place, 3 • 31
robots, 3 • 36
rotary and in-line indexers, 3 • 24
rotary assembly machines, 3 • 31
selective, 3 • 36
semiflexible bowl feeders, 3 • 27
transfer devices, 3 • 23, 3 • 30
vibratory bowl feeders, 3 • 25, 3 • 28
Automatic control, 3 • 68
Automatic guided vehicle(s):
driverless tractors, 4 • 111
guided pallet trucks, 4 • 112
high-lift guided vehicles, 4 • 112
systems, 5 • 68
unit load vehicles, 4 • 112
Automatic identification, 4 • 128, 5 • 67, 7 • 28
applications, 4 • 135
codes, 4 • 129
coding, 4 • 129
fixed beam readers, 4 • 129
handheld bar code reader, 4 • 129
magnetic stripe reader, 4 • 129
moving beam reader, 4 • 129
OCR (optical character recognition), 4 • 129
pattern recognition, 4 • 129
radio frequency (RF), 4 • 129
surface acoustical wave (SAW), 4 • 129
systems, 4 • 128
technology, 4 • 128
vision, 4 • 129
Automatic storage and retrieval systems, 5 • 67
Automation:
continents of, 1 • 8
fixed, 3 • 21
flexible, 3 • 21, 4 • 74
hard, 3 • 21, 4 • 74
hemispheres of, 1 • 8
islands of, 1 • 8
level of, 8 • 120, 8 • 123
planets of, 1 • 8
process, 8 • 110
universe of, 1 • 8
Autonomation (Jidoka), 3 • 254
Auxiliary material handling equipment:
industrial robots, 4 • 127
battery chargers, 4 • 127
dock, 4 • 125
weighing, 4 • 127
Auxiliary orienter, 3 • 27
Average, 3 • 102
Average outgoing quality (AOQ):
with replacement, 3 • 112
without replacement, 3 • 113
Average outgoing quality limit (AOQL), 3 • 112

Average total inspection (ATI), 3 • 112
Average sample number (ASN), 3 • 113
Axis limit motion control, 4 • 88

Babbage, Charles, 2 • 11
Bakayoke (foolproof), 3 • 254
Balance delay time, 3 • 177
Balancing the line, 2 • 12
Batch processes, 5 • 23
Bill of material (BOM), 1 • 6, 1 • 46, 1 • 63
Billet preparation, 8 • 66
Biomechanics, 2 • 74
Bird-cage layouts, 7 • 31
Block areas, 7 • 38
  manual development, 7 • 42
Block diagrams, 3 • 73
Bonded abrasive tools, 4 • 62
Bottleneck control, 3 • 138
Bottom line, 2 • 138
Bottom-up implementation, 8 • 56
Brake-controlled dressing, 4 • 69
Bowl feeders:
  semiflexible, 3 • 27
  vibratory, 3 • 25
Budgets, 1 • 6
Build curves, 1 • 7
Building shell, 7 • 63
Built up edge (BUE), 4 • 28
Business:
  plan, 1 • 46
  requirements, 1 • 4
  requirements capacity, 1 • 6

CAD (*see* Computer aided design)
CAD/CAM, 4 • 4, 4 • 59, 8 • 15, 8 • 108 (*see also* Computer aided design and Computer aided manufacturing)
Calendaring, 3 • 16
Calibration/mastering, 1 • 5
CAM (*see* Computer aided manufacturing)
Capacity:
  deluxe, 1 • 29
  deterministic demand, 1 • 24
  electric power, 1 • 24
  expansion, 1 • 23, 6 • 21
  heavy process industries, 1 • 24
  loading, 1 • 64
  key issues, 1 • 24
  planning, 1 • 23, 1 • 64, 3 • 192
  public services, 1 • 24
  rearrangements of, 1 • 29
  shortages, 1 • 26
  standard, 1 • 29
  technology, 1 • 26
  telecommunications networks, 1 • 24
  utilization, 1 • 26
  water resources, 1 • 24
Capek, Karel, 4 • 74
Capital, 1 • 6
CAPP (*see* Computer aided process planning)
Carousel, 3 • 34
Cash-flow diagram, 6 • 23
Castable:
  steels, 4 • 3
  superalloys, 4 • 3
Casting:
  advantages, 4 • 3
  auxiliary techniques, 4 • 5
  basic operations, 4 • 4
  chills, 4 • 6
  cores, 4 • 4
  design parameters, 3 • 8
  die, 4 • 4
  directional solidification, 4 • 6
  exothermic hot topping, 4 • 6
  finishing, 4 • 6
  gating, 4 • 5
  heat sinks, 4 • 6
  heat treatment, 4 • 6
  hot isostatic pressing of, 4 • 3
  insulation, 4 • 6
  mold making, 4 • 4
  molding, 4 • 5
  pattern, 4 • 4
  pouring, 4 • 5
  prefinishing, 4 • 6
  process, 3 • 16
  processes (*see* Casting processes)
  quality, 4 • 6
  quality assurance, 4 • 7
  risering, 4 • 5
  risers, 4 • 6
  vacuum, 4 • 3
  welding, 4 • 6
  yield, 4 • 6
Casting processes, 4 • 7, 4 • 18
  centrifugal, 4 • 7
  ceramic, 4 • 7
  die, 4 • 7
  investment, 4 • 7
  permanent mold (*see* Permanent mold casting)
  plaster mold (*see* Plaster mold casting)
  sand (*see* Sand casting)
Cellular manufacturing, 3 • 145
Centerboard feeders, 3 • 29
Centrifugal feeders, 3 • 29
Change, changing:
  rationale for, 1 • 68
  world of work, 1 • 67
Charts:
  assembly, 5 • 25
  fabrication, 7 • 38
  flow process, 7 • 38, 3 • 224

Charts (*continued*)
from-to, 7 • 42
material handling routing, 7 • 38
multiple activity, 3 • 225
operations process, 5 • 25
outline process, 3 • 224
process, 7 • 38, 3 • 224
Checklist, 7 • 40
Chemical distribution, 1 • 76
Chip:
formation, 4 • 26, 4 • 68
ratio, 4 • 28
Chucks, 4 • 18
Circadian rhythm, 2 • 85
Circular force diagram, 4 • 30
Clamp handling, 5 • 86
Classification and coding systems, 3 • 237
basic idea of, 8 • 27
criteria, 8 • 27
Clean room, 1 • 76
Closed-loop, 4 • 85
CNC (computer numerical control), 8 • 18
Coarse dressing, 4 • 69
Coatings for vibratory bowl feeders, 3 • 28
Code
Opitz, 8 • 27
Coding and classification, 8 • 20
Coefficient of friction, 4 • 44
Cold heading, 3 • 10
Common multiple of lives, 6 • 28
Common Staffing System, 1 • 55
Compensation plans:
group bonus, 2 • 16
group piecework, 2 • 16
pay by knowledge, 2 • 16
pay by skill, 2 • 16
Scanlon plan, 2 • 16
Competitive:
advantage, 1 • 17
benchmarking, 1 • 39
benchmarking, defined, 1 • 39
challenge, 1 • 39
plan, 1 • 34
planning, 1 • 33
planning concept, 1 • 34
reactions, 1 • 38
weapon, 1 • 11, 1 • 21
Competitive benchmarking:
analysis, 1 • 40
critical success factors, 1 • 41
defined, 1 • 39
implementation, 1 • 41
key questions, 1 • 44
maturity, 1 • 41
organizational outputs, 1 • 43
planning, 1 • 40
process, 1 • 40, 1 • 43
startup, 1 • 41
Complementary computer products:
command centers, 8 • 53
embedded computers, 8 • 53
disguised computers, 8 • 53
general purpose computers, 8 • 53
industrial local area networks, 8 • 53
Component availability checks, 5 • 11
Component design, 1 • 80
Computer:
aided design (CAD), 8 • 10, 8 • 15, 8 • 52
aided engineering (CAE), 8 • 52
aided layout (*see* Computer aided layout)
aided manufacturing (CAM), 8 • 10, 8 • 15, 8 • 52
control, 1 • 7
hierarchy, 8 • 18
host, 8 • 18
integrated manufacturing system (CIMS), 1 • 69, 5 • 16, 5 • 65
numerical control (CNC), 3 • 209, 8 • 18
satellite, 8 • 18
Computer aided:
design (CAD), 1 • 5, 1 • 7, 1 • 76, 2 • 93, 3 • 191, 4 • 59, 7 • 23
line balancing (CALB), 3 • 186
manufacturing (CAM), 1 • 7, 2 • 93, 3 • 191, 4 • 59
manufacturing planning, 3 • 17
process planning (CAPP), 1 • 7
Computer aided layout:
ALDEP, 7 • 20, 7 • 44
COFAD, 7 • 20, 7 • 44, 7 • 53
construction programs, 7 • 44
CORELAP, 7 • 20, 7 • 44
CRAFT, 7 • 20, 7 • 44, 7 • 49
improvement programs, 7 • 44
PLANET, 7 • 20, 7 • 44
Computer aided process planning (CAPP), 8 • 19, 8 • 21
advanced variant system, 8 • 34
artificial intelligence, 8 • 38
CAM-I CAPP system, 8 • 33
establishment of, 8 • 30
expert systems, 8 • 38
future of, 8 • 38
generative, 8 • 30, 8 • 37
Genplan, 8 • 25
information flow, 8 • 25
knowledge engineering, 8 • 38
savings, 8 • 26
types, 8 • 30
variant, 8 • 30
Computer control, 8 • 41
hardware design, 8 • 51
software design, 8 • 41
Computerized:
facilities, 1 • 71
network analysis, 3 • 153

numerical control (CNC), 1 • 68
standard data, 2 • 97
work measurement, 2 • 97, 2 • 113
Confidence-interval, 1 • 80
Configuration:
cartesian, 4 • 78
cylindrical, 4 • 78
polar, 4 • 77
spherical, 4 • 77
Constituitive:
equations, 4 • 48
relationships, 4 • 59
Constraints, 3 • 116
Consumer's risk, 3 • 103
Container:
design, 5 • 27
sizing, 5 • 27
Containerizing, 5 • 76
Containers:
bags, 5 • 81
corrugated fiberboard, 5 • 78
cylindrical, 5 • 79
sacks, 5 • 81
wood, 5 • 79
Contingency planning, 7 • 21
Continuous extrusion, 4 • 48
Continuous wiremaking, 4 • 48
Control:
ability, 3 • 69
activity, 3 • 69
adaptive (AC), 3 • 77, 8 • 19
analog, 3 • 71
attributes, 3 • 69
automatic, 3 • 68
bottleneck, 3 • 138
change, 3 • 100
closed-loop, 3 • 210
computer, 3 • 71, 8 • 41
computer numerical (CNC), 8 • 18
conveyor, 8 • 96
cost systems, 6 • 38
direct digital, 3 • 74
direct numerical (DNC), 8 • 18
display design, 3 • 88
distributed, 3 • 75
distributed numerical (DNC), 8 • 18
dynamic optimizing, 3 • 78
dynamic response, 3 • 69
expert systems, 3 • 80
factory, 8 • 98 (*see also* Factory control)
factory-floor, 3 • 209
factory floor information systems, 3 • 209
feedback, 3 • 69
functions, 3 • 68
fundamentals, 3 • 69
hierarchy, 3 • 74
input/output, 3 • 194
learning, 3 • 80
logical, 8 • 100
loop, 3 • 70
machine, 8 • 97
machine intelligence, 3 • 80
management, 3 • 125
manual, 3 • 76
material, 3 • 100, 3 • 198
measurement, 3 • 100
models, 3 • 68, 3 • 73
motion, 4 • 88
motivation, 3 • 68
multivariable noninteracting, 3 • 80
numerical (NC), 8 • 18
of labor cost, 6 • 56
on-off, 3 • 69
open-loop, 3 • 74
operating characteristics, 3 • 69
operational, 3 • 125
optimal, 3 • 77
optimizing, 3 • 74
priority, 3 • 196
purpose, 3 • 68
queue, 3 • 193
real time, 8 • 86
resource, 3 • 138
rolling, 4 • 55
shop-floor, 3 • 192
strategies, 8 • 68
supervisory, 3 • 74
system, 3 • 68, 8 • 46
system attributes, 8 • 96
systems design, 3 • 68
theory, 3 • 73
transfer function, 3 • 69
transportation system, 8 • 97, 8 • 103
work center, 8 • 101
Control charts, 3 • 103
acceptance, 3 • 108
c, number of nonconformities, 3 • 108
cusum, 3 • 108
individual, 3 • 108
median, 3 • 107
moving average, 3 • 108
moving range, 3 • 108
np, 3 • 109
p, fraction defective, 3 • 108
range, 3 • 103
sample average, 3 • 103
sample standard deviation, 3 • 107
stabilized p, 3 • 109
standard deviation, 3 • 107
u, 3 • 109
Control limits, 3 • 103
Control models:
cognitive, 3 • 76
company systems, 3 • 83
feedback, 3 • 73
feedforward, 3 • 73

Control models (*continued*)
in man-machine systems, 3 • 88
in production, 3 • 81
information systems, 3 • 90
manual control, 3 • 76
of manufacturing systems, 3 • 76
open-loop, 3 • 73
overall plant, 3 • 83
preview, 3 • 76
production, 3 • 84
robot, 3 • 82
Controlled-force internal grinders, 4 • 65
Controller, 8 • 18
Controls, man-machine:
characteristics, 2 • 77
dedicated, 2 • 56
factors, 2 • 76, 2 • 77
general-purpose input devices, 2 • 56
layout principles, 2 • 77, 2 • 78
principles, 2 • 76
systems, 2 • 56
tools, 2 • 56
Conveyors:
belt, 4 • 107
carrier chain, 4 • 107
cart-on-track conveyors, 4 • 110
diverting mechanisms, 4 • 108
inverted power and free, 4 • 110
pneumatic, 4 • 111
power and free, 4 • 108
roller, 4 • 107
skate wheel, 4 • 107
sliding slat, 4 • 107
tilting slat, 4 • 107
tilting tray, 4 • 107
towline, 4 • 110
trolley, 4 • 108
Coolants, 4 • 69
Corrective action, 3 • 101
Corrosion protection, 5 • 83
Corrugated:
containers, 5 • 79
fiberboard, 5 • 78
Cost(s):
actual, 6 • 39
average per unit of output, 6 • 49
control systems, 6 • 38
fixed, 6 • 51
hierarchical structure, 6 • 53
joint set-ups, 1 • 29
labor, 1 • 6
linear model (*see* Models)
material, 1 • 6
matrix, 7 • 50
of quality, 1 • 78
overhead, 1 • 6
reduction, 1 • 6
relative increment, 6 • 53
standard, 6 • 39
total, 6 • 49
transportation, 1 • 31
unit, 6 • 49, 6 • 53
unit labor, 6 • 56
variable, 6 • 51
variance, 6 • 39
Cost control, 6 • 38
principles of, 6 • 38
variances, 6 • 38
Cost function:
fixed charge, 1 • 25
power, 1 • 25
Costs of aggregate production planning:
basic production costs, 3 • 117
inventory related costs, 3 • 117
related to production rate changes, 3 • 117
CPM (critical path method), 3 • 147
Crash time, 3 • 152
Creep-feed grinding, 4 • 66
Criteria:
for clustering tasks and aligning boundaries, 2 • 28
secondary, 3 • 165
Critical path, 3 • 150
Critical path method (CPM), 3 • 147
Critical ratio, 3 • 197
Critical sequence, 1 • 5
Crossfeeding, 4 • 65
Cross-training, 2 • 14
Crowning, 4 • 66
Crush dressing, 4 • 68
Cubic boron nitride, 4 • 61
Cushioning, 5 • 81
Cutting:
fluid, 4 • 25
forces, 4 • 25
tool materials, 4 • 33
Cutting parameters, 4 • 25
Cutting time, 4 • 21
Cycle time, 3 • 254
cognitive system, 2 • 71
perceptual system, 2 • 70

Damage:
compression, 5 • 72
moisture, 5 • 72
shock, 5 • 72
vibration, 5 • 72
Data:
attributes, 3 • 103
base management, 8 • 44
processing alternative, 8 • 44
variables, 3 • 103, 3 • 113
wear, 4 • 44
Data retrieval arrays:
decision table, 8 • 28
look-up table, 8 • 28

process taxonomy, 8 • 28
synthesis algorithm, 8 • 28
tree structure, 8 • 28
Data retrieval codes:
alphanumeric, 8 • 27
mnemonic, 8 • 27
nomenclature, 8 • 27
Opitz, 8 • 27
Database, 8 • 15 (*see also* Data base)
drafting, 8 • 16
engineering, 8 • 16
for products, 8 • 16
for manufacturing, 8 • 16
DC servo motor, 4 • 84
Decision making, 2 • 58
Decision support:
algorithm, 1 • 69
system, 2 • 93, 3 • 125
timeframe, 8 • 47
Decisions:
social, 2 • 10
social system, 2 • 9
technological, 2 • 9
Deep drawing, 4 • 38
Deformation, 4 • 46
Degrees of freedom, 4 • 74
Degassing, 4 • 3
Dehumanizing work, 2 • 12
Deming, W. Edwards, 3 • 98
Deming's 14-point program, 3 • 99
Department:
group, 7 • 26
plant layout, 7 • 36
process, 7 • 25
product, 7 • 25
Departmentalization, 1 • 66
Depreciation, 6 • 33
Design:
analysis, 8 • 65
assurance and change control, 3 • 100
change control, 3 • 100
container, 5 • 27
for automation, 8 • 91
for manufacture, 1 • 78, 3 • 3
packaging, 5 • 84
parameters for casting, 3 • 8
parameters for forging, 3 • 9
parameters for basic manufacturing processes, 3 • 11
parameters for secondary forming and sizing, 3 • 13
product, 3 • 4
requalification, 3 • 100
software, 8 • 41
to assemble, 3 • 21
unit load, 5 • 86
validation, 3 • 100
Design-to-cost:
philosophy, 3 • 48
process, 3 • 40
small lot manufacturing, 3 • 46
Designing:
a training course, 2 • 129
for shaping plastic components, 3 • 12
Diagram:
cash-flow, 6 • 23
Diamond, 4 • 61
Die material, 4 • 46
Direct numerical control (DNC), 8 • 18
Discount:
cost, 1 • 24
rate, 1 • 24
Disembodied change:
capital saving, 6 • 7
labor saving, 6 • 7
neutral saving, 6 • 7
Dispatching, 3 • 138, 3 • 198
Dispatching rules, 3 • 172
Display design:
control models, 3 • 88
Displays, attributes:
display format, 2 • 56
display technology, 2 • 55
display type, 2 • 55
physical characteristics, 2 • 55
Displays, types:
analog, 2 • 56
auditory, 2 • 55, 2 • 56, 2 • 68
digital, 2 • 56, 2 • 76
dynamic, 2 • 55
electronics, 2 • 76
pattern-oriented, 2 • 55
static, 2 • 55
tactile, 2 • 55, 2 • 56, 2 • 68, 2 • 76
variable-oriented, 2 • 55
visual, 2 • 55, 2 • 56, 2 • 68, 2 • 76
visual display terminals, 2 • 68, 2 • 86
Distances:
Euclidean, 7 • 6
rectilinear, 7 • 6
Distributed numerical control (DNC), 8 • 18, 8 • 55
Distributed systems approach to manufacturing systems, 8 • 45
Distribution system(s), 7 • 64, 8 • 70, 8 • 108
Division of labor, 3 • 176
DNC (direct or distributed numerical control), 8 • 18
Dock equipment:
dock boards, 4 • 125
dock bumpers, 4 • 125
dock levelers, 4 • 125
dock lifts, 4 • 125
dock seals, 4 • 125
dock shelters, 4 • 125
ramps, 4 • 125
Drawings, engineering, 3 • 3
Dressing, 4 • 68

Drives:
electric, 4 • 84
hydraulic, 4 • 83
mechanical gear and cam, 4 • 85
pneumatic, 4 • 83
robot, 4 • 81
Dummy robot, 4 • 93
Dunnage, 5 • 77, 5 • 81
Dynamic:
analysis, 8 • 57
dispatching, 3 • 198
programming, 1 • 27, 1 • 29
Dynamometer, 4 • 29

Economic analysis, 6 • 3
Economic comparisons, 6 • 25
alternatives with different life spans, 6 • 26
incremental investment comparisons, 6 • 26
total investment comparisons, 6 • 26
Economic justification of robots, 4 • 96
Economic lot size, 5 • 27
Economics:
of production, 6 • 49
Economies of scale (*see* Scale)
Economy-of-scale, 1 • 24, 1 • 33
Efficiency, 3 • 50
Electric power control, 7 • 66
Electroforming, 3 • 10
Electrohydraulic forming, 4 • 58
Electronics industry, 1 • 73
Electrostatic discharge (SD), 5 • 84
Elevator feeders, 3 • 30
Empirical methods, 2 • 62
Employee:
selection, training, and motivation, 3 • 102
Energy management, 1 • 6, 7 • 58
and/or HVAC, 8 • 55
conservation, 7 • 61
distribution systems, 7 • 64
electric power control, 7 • 66
energy audit, 7 • 59
lighting, 7 • 65
management in, 7 • 59
need for, 7 • 58
operations, 7 • 61
production processes, 7 • 67
strategies, 7 • 61
Energy requirement, 4 • 46
Engineering:
anthropometry, 2 • 73
design, 1 • 62
drawings, 3 • 3
industrial, 2 • 5
knowledge, 8 • 39
manufacturing, 1 • 63
methods, 3 • 220
predictive, 8 • 16
psychology, 2 • 53
Engineers:
industrial, 2 • 97
Environment:
manufacturing, 1 • 72
elements of a manufacturing, 1 • 76
physical workplace, 1 • 72
shipping, 5 • 72
Environmental considerations:
noise, 2 • 82
vibration, 2 • 82
Equipment:
automated storage, 4 • 122
auxiliary material handling, 4 • 125
classifications, 4 • 101
fixed-path material handling, 4 • 106
flexible-path material handling, 4 • 106
material handling, 4 • 100
storage, 4 • 117
Ergonomics:
relation to human factors, 2 • 68, 2 • 94
relation to man-machine systems, 2 • 52
5Ms, 2Ss, 1 • 4, 1 • 5
Error cause removal (ECR) program, 2 • 86
Estimates, 2 • 99
Estimating space, 7 • 38
Ethernet/P802.3, 8 • 53
Euclidean distances, 7 • 6
Event transactions and codes, 3 • 215
Expandability, 8 • 96
Expansion of production capacity, 6 • 21
Expediting, 3 • 138, 3 • 172
Expense, 1 • 6
Expert systems, 2 • 62, 2 • 93, 3 • 80
Explosive forming, 4 • 58
Extrusion, 3 • 10, 3 • 16

Facility, facilities:
advanced FMS, 8 • 94
components, 7 • 18
design, 7 • 18
in manufacturing, 1 • 58
layout, 7 • 30
location, 7 • 3, 7 • 18
planning, 7 • 18 (*see also* Facilities planning)
planning process, 7 • 19
single facility problems, 1 • 26
multilocation problems, 1 • 26
multitype problems, 1 • 26
multifacility problems, 1 • 30
two-facility problems, 1 • 29
5Ms, 2Ss, 1 • 5
Facilities planning, 7 • 18
defined, 7 • 18
flexible manufacturing systems, 7 • 27
group department, 7 • 26
objectives, 7 • 18

planning process, 7 • 19
procedure, 7 • 19
process department, 7 • 25
product department, 7 • 25
strategy, 7 • 21
trends, 7 • 23
Facility layout:
industrial safety considerations, 2 • 37
Facility location, 7 • 3
discrete location, 7 • 12
fixed costs, 7 • 4
location-allocation, 7 • 8
mathematical models, 7 • 3
minimax formulation, 7 • 5
minisum formulation, 7 • 5
network location, 7 • 9
one center, 7 • 10
p-center, 7 • 10
planar location, 7 • 5
p-median, 7 • 10
r-cover, 7 • 10
variable costs, 7 • 4
Factory:
automated, 1 • 3, 1 • 8, 5 • 65
flexible, 1 • 3
layout, 1 • 6
Factory control:
overview, 8 • 98
Factory floor information system (FFIS):
architecture, 3 • 206
control, 3 • 209
data capture, 3 • 210
defined, 3 • 203, 3 • 204
design issues, 3 • 205
distribution and financial systems, 3 • 208
engineering and quality systems, 3 • 208
factory floor workstations and equipment, 3 • 208
logical model, 3 • 206
physical model, 3 • 206
resource management, 3 • 207
Factory of the future:
custom design work, 8 • 7
design function, 8 • 10
fault tolerant, 8 • 9
human assembly, 8 • 9
living organism, 8 • 9
material handling, 8 • 8
materials, 8 • 8
processing changes, 8 • 8
quality assurance, 8 • 9
scenario, 8 • 6
setups, 8 • 8
small factory order, 8 • 8
Factories of the future, 8 • 3
major thrusts defined, 8 • 3
research needs, 8 • 10
Family of parts, 8 • 20
Feasibility:
operational, 4 • 98
robot, 4 • 98
technical, 4 • 98
Federal tax laws, 6 • 33
Feed rates, 8 • 18
Feedback:
user/consumer, 3 • 101
control, 3 • 69
Feeders:
centerboard, 3 • 29
centrifugal, 3 • 29
elevator, 3 • 30
half-centerboards, 3 • 30
nonvibratory, 3 • 28
parts, 3 • 21
semiflexible bowl, 3 • 27
vibratory, 3 • 25
Field performance, 3 • 101
Files:
item master, 1 • 63
product structure, 1 • 63
production router, 1 • 63
safety, 2 • 36, 2 • 37
work center, 1 • 63
Filtration, 4 • 3
Finite:
dressing, 4 • 69
element mesh, 8 • 16
element method, 8 • 16
element model (FEM), 4 • 59
horizon, 1 • 27
set of capacity expansion projects, 1 • 28
Firm:
behavior of the, 6 • 19
planned orders, 5 • 11
Fitt's law, 2 • 71, 2 • 90
Fixed-path floor conveyors, 5 • 70
Fixed-path material handling equipment:
automatic guided vehicles, 4 • 111
conveyors, 4 • 106
hoists, monorails, and cranes, 4 • 113
Fixtures, 4 • 18
Flashless forging, 4 • 53
Flexibility, 2 • 15, 7 • 23, 8 • 3, 8 • 13, 8 • 78, 8 • 96
level of, 8 • 123
material handling, 8 • 79
processing/tooling, 8 • 78
product mix, 8 • 91
product volume, 8 • 92
Flexible:
assembly machines, 3 • 39
factories, 1 • 3
manufacturing system models, 3 • 64
robots, 1 • 5
Flexible factory:
concepts, 8 • 123
controlling, 8 • 96

Flexible factory (*continued*)
design criteria, 8 • 122
implementing, 8 • 106
managing, 8 • 122
mission, 8 • 107
objectives, 8 • 122
philosophies, 8 • 123
Flexible manufacturing systems, 7 • 27, 8 • 20, 8 • 108 (*see also* Flexible manufacturing systems, advanced)
action plan, 8 • 81
adoption steps, 8 • 82
advanced, 8 • 90
basics of, 8 • 77
comprehensive control, 8 • 80
concepts, 8 • 78
configuration, 8 • 83, 8 • 85
definition, 8 • 74, 8 • 90
examples, 8 • 75
getting started, 8 • 81
history, 8 • 74
justification, 8 • 76
modeling and analysis, 8 • 83
real time control, 8 • 86
simulation modeling, 8 • 65
tracking and control, 8 • 79
Flexible manufacturing systems, advanced:
definition, 8 • 90
facilities, 8 • 94
factory of the future, 8 • 94
functional objectives for the manufacturing system, 8 • 92
functional objectives for the physical process, 8 • 93
functional objectives for the product, 8 • 91
Flexible-path material handling equipment:
counterbalanced fork trucks, 4 • 101
lift truck attachments, 4 • 104
manual trucks, 4 • 104
narrow aisle trucks, 4 • 102
storage yard handling, 4 • 106
walkie pallet trucks, 4 • 102
Floating host computer, 8 • 101
Flow:
material, 4 • 46
patterns, 4 • 97
FMS (*see* Flexible manufacturing systems)
Ford, Henry, 2 • 12, 4 • 97
Forecast(s):
production, 3 • 128
sales, 1 • 61
Forecasting, 1 • 61, 3 • 128
objectives, 5 • 41
models, 5 • 42
tracking, 5 • 43
implementation, 5 • 43
Forging, 4 • 3
Foundry processes, 4 • 3
die, 4 • 3
investment, 4 • 3
sand, 4 • 3
Forming processes, 4 • 38
annealing, 4 • 42
classification of, 4 • 38
cold working, 4 • 41
deep drawing, 4 • 38
dynamic, 4 • 38
electrohydraulic, 4 • 58
explosive, 4 • 58
extrusion, 4 • 40
forging, 4 • 40
hot working, 4 • 42
in presses, 3 • 10
magnetic-pulse, 4 • 58
massive, 4 • 38
press, 3 • 10
quasi-static, 4 • 38
recent developments, 4 • 48
rolling, 4 • 40
sheet-forming, 4 • 38
six basic, 4 • 39
stress state, 4 • 38
stretch-forming, 4 • 41
temperature, 4 • 38
theory, 4 • 48
tooling, 4 • 38
warm working, 4 • 44
wire drawing, 4 • 41
Fraction nonconforming, 3 • 102
Friction, 4 • 44
Functional:
analysis, 1 • 70
segregation, 8 • 96
Functionoid, 4 • 81
Future development of control systems, 3 • 91

Gains:
derivative, 3 • 71
integral, 3 • 71
proportional, 3 • 71
Gantry-mounted robot, 4 • 80
Gate, 4 • 95
Gauging, 1 • 5
Geometric modeling, 8 • 10
Gilbreth, Frank B., 2 • 97
Gilbreth, Lillian M., 2 • 97
Glazing, 4 • 62
Goal programming, 3 • 118, 3 • 124
Goals:
price/volume profit and cost, 3 • 40
Goals, operators, methods, and selection rules, 2 • 90
Goodness of fit test, 3 • 110
GPSS, 3 • 49
Graph, 7 • 9
Graphic representation, 8 • 17

Graphical reporting, 6 • 47
Greedy heuristic, 7 • 13
Grinding:
  fluids, 4 • 69
  wheels, 4 • 18
Grinding mechanics:
  chip formation, 4 • 66
  grinding zone, 4 • 66
Grinding operations, 4 • 62
  centerless cylindrical, 4 • 66
  creep-feed, 4 • 66
  external cylindrical, 4 • 62
  form, 4 • 66
  high-speed, 4 • 62
  internal cylindrical, 4 • 62
  plunge, 4 • 62
  straight surface, 4 • 62
  traverse, 4 • 62
  vertical spindle, 4 • 66
Group scheduling:
  algorithms, 3 • 240
  basic concept, 3 • 240
  examples, 3 • 241
Group technology (GT), 1 • 7, 3 • 236, 8 • 19
  and CAD/CAM, 3 • 243
  and CAPP, 3 • 244
  and CIM, 3 • 243
  attribute codes, 8 • 30
  automated factory system, 3 • 245
  chain structure, 8 • 30
  coding schemes, 8 • 30
  hierarchical, 8 • 30
  in functional design, 3 • 17
  manufacturing plans, 8 • 29
  Mitrofanov, S. P., 8 • 29
  monocode, 8 • 30
  polycode, 8 • 30
  premises, 8 • 29
  standard plans, 8 • 29
Group technology economics:
  group machining costs, 3 • 246
  group tooling costs, 3 • 246
Group tooling, 3 • 239

Half-centerboards, 3 • 30
Handling, storage, and shipping, 3 • 101
HAP algorithm, 7 • 6
Hard automation, 1 • 5
Hardware design, 8 • 51
Hardware/software trends, 8 • 52
Herzberg, Frederick, 2 • 13
Heuristic:
  greedy, 7 • 13
Hick's law, 2 • 72
Hierarchical:
  decision making, 3 • 125
  production planning system, 3 • 126
Hierarchy:
  of work units, 1 • 49
High-energy rate forming (HERF) processes, 4 • 58
Histogram, 3 • 110
Historical data, 2 • 99
Hoists, monorails, and cranes:
  gantry, 4 • 116
  hoists, 4 • 113
  manual monorails, 4 • 114
  overhead crane, 4 • 115
  portable crane, 4 • 116
  powered monorail systems, 4 • 114
  semigantry crane, 4 • 116
  stacker crane, 4 • 116
  top running crane, 4 • 115
  underhung crane, 4 • 115
Holistic, 1 • 72
Horizon:
  finite, 1 • 27, 3 • 120
  infinite, 1 • 26
  planning, 1 • 26, 3 • 118
  rolling, 3 • 120
  time, 3 • 119
Horizontal slab mills, 4 • 18
Hot-die forging, 4 • 54
Hot isostatic pressing, 4 • 55
Human-activated data capture, 3 • 212
Human-computer interaction, 2 • 89
Human error, 2 • 87
Human factors, 5 • 35
  relation to ergonomics, 2 • 68
  relation to man-machine systems, 2 • 52
Human/machine interaction, 1 • 69
Human outputs:
  cortical responses, 2 • 54
  electromyographic signals, 2 • 54
  motion, 2 • 54
  speech, 2 • 54
Human performance factors:
  ability, 2 • 55
  acceleration, 2 • 55
  aptitude, 2 • 55
  attitude, 2 • 55
  illumination, 2 • 55
  motion, 2 • 55
  motivation, 2 • 55
  noise, 2 • 55
  organization, 2 • 55
  safety, 2 • 55
  style, 2 • 55
  temperature, 2 • 55
  vibration, 2 • 55
  weightlessness, 2 • 55

Identification, 3 • 101
Illuminance:
  defined, 2 • 79

Illuminance (*continued*)
recommended ranges, 2 • 79
Illumination:
industrial safety considerations, 2 • 27
Implementation, 8 • 113
bottom-up, 8 • 108
Implied salvage value, 6 • 28
Improvement activities, 3 • 255
Industrial engineering, 2 • 97
Industrial Revolution, 2 • 125
Industrial robots, 2 • 91, 4 • 89, 5 • 69
Inflation:
actual-dollar analysis, 6 • 30
actual dollars, 6 • 30
annual rate, 6 • 29
constant-dollar analysis, 6 • 30
constant dollars, 6 • 30
Consumer Price Index, 6 • 29
measuring, 6 • 29
Information processing, human:
automatic control, 2 • 57
cognitive, 2 • 69
decision making, 2 • 58, 2 • 69
elements, 2 • 57
human-computer interaction, 2 • 58
keyboard design, 2 • 87, 2 • 88
manual control, 2 • 57
model, 2 • 90
motor, 2 • 69
perceptual system, 2 • 68
problem solving, 2 • 58
rate, 2 • 87
speech recognition, 2 • 87
supervisory control, 2 • 57
Information systems:
control models, 3 • 90
Information theory, 2 • 72
In-line:
free-transfer, 3 • 34
indexing, 3 • 34
In-process gauging and test, 3 • 212
Inputs:
demand and expenditure for, 6 • 14
productivity of, 6 • 3
substitution of, 6 • 5
Inspection, 3 • 99, 3 • 101
Installation and use, 3 • 101
Integrated:
computer systems, 8 • 108
design, 5 • 16
manufacturing planning, 1 • 65
manufacturing systems, 1 • 66, 5 • 15
systems, 1 • 59
Integration, 8 • 4, 8 • 13, 8 • 104
Intelligence, 8 • 4, 8 • 13
Intelligent manufacturing systems, 2 • 93
Interlock, 4 • 95
Internal coordination, 2 • 14
Intersection points, 7 • 11
Inventory:
carrying cost, 3 • 124
levels, 3 • 116
management, 1 • 62
order techniques, 1 • 62
physical audits, 1 • 62
Inventory control:
expense supplies, 5 • 49
finished goods, 5 • 49
models, 5 • 51
objectives, 5 • 46
order point-order quantity, 3 • 116
raw materials, 5 • 46
tool control, 5 • 53
work in process, 5 • 49
Investment problems, 1 • 25
Investment project types:
independent projects, 6 • 23
interdependent projects, 6 • 24
mutually exclusive alternatives, 6 • 24
Islands of automation, 8 • 20
ISO/OSI, 8 • 53
Isolated island layouts, 7 • 31
Isostatic, 4 • 3
Isothermal forging, 4 • 54

Japanese manufacturing:
counterparts, 1 • 39
factories, 8 • 106
managers, 8 • 106
simplicity, 8 • 106
success, 1 • 3
5Ms and 2Ss, 1 • 3
Jidoka (autonomation), 3 • 251, 3 • 254
Jigs, 4 • 18
Job(s):
classification, 1 • 4
definition of, 2 • 4
design, 2 • 3
deterministic specification, 2 • 12
enlargement, 2 • 13, 2 • 82
enrichment, 2 • 13, 2 • 82
linking, 2 • 11
multiple dimensions of, 2 • 3
myopia, 2 • 12
releasing, 3 • 138
satisfaction, 2 • 7
shop (*see* Job shop)
Job analysis:
approaches, 2 • 80, 2 • 81
cross-boundary interaction analysis, 2 • 25
for training, 2 • 129
loci of organizational stability and instability, 2 • 27
mobility analysis, 2 • 25

responsibility analysis, 2 • 25
task ratings, 2 • 25
transformation flow chart, 2 • 16
unit operations, 2 • 17
variance analysis, 2 • 18
Job design:
decisions for self-maintaining work teams, 2 • 15
definition of, 2 • 4, 2 • 82
impact of uncertainty on, 2 • 4
processes and strategies, 2 • 28, 2 • 29
role of consultants, 2 • 30
systematic factors in, 2 • 6
types of, 2 • 11
man-machine systems, 2 • 61
human factors considerations, 2 • 80
Job shop, 3 • 139, 3 • 199, 3 • 209
scheduling, 3 • 159, 3 • 171
simulation, 3 • 171
Johnson's rule, 3 • 170
Just-in-time (JIT), 1 • 61, 3 • 193, 3 • 200, 3 • 251, 5 • 23

Kanban, 3 • 144, 5 • 63
production-ordering, 3 • 251
withdrawal, 3 • 251
Kelvin, Lord, 1 • 45
Kitting, 5 • 68
Knowledge, 2 • 126
engineering, 8 • 39
workers, 1 • 60

Labor productivity, 6 • 55
Labor resistance to robots, 4 • 96
Ladder diagram logic, 3 • 209
Lagrangian function, 6 • 12
Lamellar structure, 4 • 26
Lathes:
automatic screw machine, 4 • 18
engine, 4 • 18
numerical control (NC), 4 • 18
tracer, 4 • 18
turret, 4 • 18
Layered networks, 8 • 56
Layout(s):
alternative, 7 • 38
bird-cage, 7 • 31
by group technology, 7 • 34
by process, 7 • 31
by product, 7 • 31
departmental, 5 • 29
isolated island, 7 • 31
linear, 7 • 31
material handling issues, 5 • 31
office, 7 • 38
open-type, 1 • 74
organizational, 1 • 70
plant, 5 • 29
plant-wide, 5 • 29
producing, 7 • 54
U-shaped line, 7 • 31
workplace, 5 • 29
Lead time, 3 • 193
Learning:
curve, 2 • 110, 3 • 43
principles of, 2 • 127
Lighting, 7 • 65
Limit switch, 4 • 93
Limiting quality (LQ), 3 • 111
Line:
balance, 1 • 4
balancing, 1 • 68
of balance (LOB), 3 • 156
Linear production systems, 6 • 8
Linear programming, 3 • 118, 3 • 122
Linear systems:
production cost for, 6 • 9
Load size, 5 • 27
Local area networks (LANs), 3 • 217
Location-allocation problems, 7 • 8
Lot acceptance, 3 • 102
Lubricants, 4 • 45
Lubrication, 4 • 44, 4 • 46
Luminance, 2 • 79

Machinability, 8 • 18
Machine, 1 • 3
control unit (MCU), 4 • 18, 8 • 18
group/cell, 3 • 238
interference models, 3 • 51
Machine tools:
boring machine, 4 • 18
broacher, 4 • 18
drill press, 4 • 18
lathe (*see also* Lathes), 4 • 18
grinder, 4 • 18
milling machine, 4 • 18
planer, 4 • 18
saw, 4 • 18
shaper, 4 • 18
Machining:
basic formulas for, 4 • 19
center, 4 • 18
flexible, 8 • 75
mechanics of, 4 • 29
orthogonal, 4 • 28
process capabilities in, 4 • 37
Magazines, 3 • 22, 3 • 24
Magnetic-pulse forming, 4 • 58
Magnetic stripe, 4 • 135
Maintenance:
budgeting, 4 • 143

Maintenance (*continued*)
corrective action, 4 • 146
efficiency, 4 • 143
featherbedding, 4 • 144
monitoring/dispatch, 8 • 55
planning and scheduling, 4 • 146
preventive, 4 • 145, 8 • 124
records, 4 • 147
Makespan, 3 • 162
Malicious obedience, 2 • 12
Man-machine systems (also referred to as human-machine systems, man-computer systems, and man-machine/computer systems)
authority/responsibility, 2 • 53
components, 2 • 68
control models, 3 • 88
design issues, 2 • 61
design methods, 2 • 62
design process, 2 • 59
endogenous factors, 2 • 55
environmental compatibility, 2 • 52
exogenous factors, 2 • 55
information processing, 2 • 53, 2 • 56, 2 • 68, 2 • 87, 2 • 90
input/ouput devices, 2 • 87
input/output requirements, 2 • 53, 2 • 54
levels of issues, 2 • 53
measurement techniques, 2 • 63
overall issues, 2 • 52
overview, 2 • 53
relation to ergonomics, 2 • 55
relation to human factors, 2 • 55
simulation, 2 • 62
system design, 2 • 59
systems perspective, 2 • 52
visual characteristics, 2 • 68
Management:
control, 3 • 125, 5 • 20
inventory, 1 • 62
of design for manufacture, 1 • 78
of product cost, 1 • 80
Manpower, 1 • 3, 1 • 5
Manpower planning, 1 • 7
Manual feed override (MFO), 8 • 18
Manual trucks:
industrial trailers, 4 • 105
platform trucks, 4 • 105
towline carts, 4 • 105
two-wheel hand trucks, 4 • 105
wheeled dollies, 4 • 105
Manufacturing:
automation protocol (*see* MAP)
closed-loop, 1 • 59
control, 1 • 64
control system, 1 • 58, 8 • 47
engineer, 4 • 26
enterprise, 1 • 4, 1 • 10
environment, 1 • 72
Japanese, 1 • 3
management, 1 • 58
planning and reporting system, 1 • 65, 8 • 46
software, 8 • 45
strategy, 1 • 11
5Ms, 2Ss, 1 • 3
Manufacturing facility:
new concepts, 1 • 75
Manufacturing process:
four main classes, 3 • 20
liquid material basic processes, 3 • 6
plastic material basic processes, 3 • 7
selection, 3 • 6
solid state basic processes, 3 • 7
Manufacturing resource planning (MRPII), 5 • 3
capacity requirements planning, 5 • 5
demand management, 5 • 4
final assembly scheduling, 5 • 4
master production planning, 5 • 5
material requirements planning, 5 • 5
production planning, 5 • 4
resource planning, 5 • 4
rough cut capacity planning, 5 • 5
shop-floor control, 5 • 5
Manufacturing strategy:
competitive advantage, 1 • 17
corporate concept, 1 • 16
evaluation criteria, 1 • 16
importance, 1 • 11
Manpower, 1 • 5
MAP/P802.4, 8 • 53
Market elasticity, 3 • 42
Market information, 3 • 101
Master standard data, 2 • 104, 2 • 108
Material, 1 • 3, 1 • 5
control, 1 • 6, 3 • 100
flow, 4 • 46 (*see also* Material flow)
flow management, 8 • 55
handling (*see* Material handling)
procurement, 1 • 7
properties, 4 • 46
requirements planning (MRP), 1 • 6, 1 • 62, 1 • 63, 3 • 192
selection, 3 • 6
Material flow
industrial safety considerations, 2 • 37
Material flow control, 5 • 38
distribution resource planning, 5 • 62
goals, 5 • 40
kanban, 5 • 63
materials and productivity management, 5 • 40
optimized production technology, 5 • 62
personnel, 5 • 40
forecasting, 5 • 41
strategies, 5 • 55
Material handling, 1 • 6, 7 • 27, 8 • 13, 8 • 108
advanced FMS, 8 • 94
design of, 5 • 22

equipment, 4 • 100
flexibility, 8 • 78
industrial safety considerations, 2 • 37
in-plant transportation, 5 • 36
integrated with furniture, 1 • 74
interface with process design, 5 • 23
issues related to layout, 5 • 31
manual limits, 2 • 74
objectives of design, 5 • 21
principles of, 5 • 17–5 • 19
priorities, 5 • 17
role of, 5 • 15
systems design, 7 • 18
Material:
protection, 5 • 72
tracking, 8 • 98
Material requirements planning (MRP), 5 • 3, 8 • 20
closed-loop system, 5 • 3
concepts, 5 • 7
dependent demand, 5 • 7
example systems, 5 • 11
exception messages, 5 • 10
gross-to-net, 5 • 7
low-level coding, 5 • 9
quickdecking, 5 • 7
system implementation, 5 • 13
system transactions, 5 • 10
time-phased records, 5 • 8
level-by-level explosions, 5 • 9
Material storage:
dedicated, 5 • 35
industrial safety considerations, 2 • 37
location assignment, 5 • 35
randomized, 5 • 36
work-in-process methods, 5 • 35
Materials:
cutting tool, 4 • 33
piezoelectric, 4 • 95
Materials flow, 5 • 29 (*see also* Material flow)
Materials handling equipment, 4 • 100
classifications, 4 • 101
Mathematical techniques, 2 • 107
Maximum principle, 3 • 76
Maynard, Harold B., 2 • 97
Maynard Operation Sequence Technique (MOST), 2 • 97, 2 • 104, 2 • 116
Mean, 3 • 102
Measure(s):
of performance, 1 • 45
of productivity, 1 • 45
participative development, 1 • 53
hierarchy, 1 • 45
why?, 1 • 46
of organizational structure, 1 • 55
Measures of investment worth:
annual equivalent, 6 • 25
future worth, 6 • 25
internal rate of return, 6 • 26
present worth, 6 • 25
Measuring:
functional activities, 1 • 55
support activities, 1 • 50
Measurement:
activity, 2 • 63
control, 3 • 100
physiological, 2 • 63
work (*see* Work measurement)
objectives, 1 • 45, 1 • 46
of work, 1 • 45
needs at different levels, 1 • 47
Firm-Level Productivity Measurement System, 1 • 48
Nominal Group Technique, 1 • 53
Overhead Value Analysis, 1 • 55
Common Staffing System, 1 • 55
Median, 3 • 102
Memomotion study, 3 • 227
Memory:
long-term, 2 • 70
working, 2 • 70
Method, 1 • 3
Methods engineering approach, 3 • 220
Methods improvement, 3 • 220
Methods Time Measurement (MTM), 2 • 97, 2 • 104, 3 • 228
Microcomputers:
used for productivity measurement, 1 • 56
Micro-Matic Methods and Measurement (4-M) Data System, 2 • 97, 2 • 113
Micromotion, 2 • 97
MIL-STD-105D, 3 • 113
Minimal critical specifications, 2 • 14
Minimum annual cost algorithm, 1 • 31
Minimum attractive rate of return (MARR), 6 • 25
Minimum rational work element, 3 • 176
Model development:
basic approach, 3 • 50
steps, 3 • 50, 3 • 51
Modeling:
methodologies, 3 • 73
Models (*see also* Control models):
analytic, 3 • 50
approximate, 3 • 50
assembly line with rework, 3 • 65
assembly line with scrapping, 3 • 65
assembly process, 3 • 65
assumptions, 3 • 51
Box-Jenkins, 3 • 132
building, 8 • 57
capacitated, 7 • 12
closed queues, 3 • 62, 3 • 64
continuous location, 7 • 15
control, 3 • 68, 3 • 73
discrete location, 7 • 15
divisional or product cost, 6 • 51
exact, 3 • 50

Models (*continued*)
  exponential smoothing, 3 • 128
  facility location, 7 • 5
  factor inputs, 6 • 50
  finite element (FEM), 4 • 59, 8 • 16
  fixed-work-force, 3 • 118
  flexible manufacturing system, 3 • 64
  flow charts, 1 • 70
  forecasting, 5 • 42
  functional cost, 6 • 51
  goal programming, 3 • 122
  heuristic capacity expansion methods, 1 • 26
  inventory control, 5 • 51
  linear cost, 3 • 124, 6 • 50, 6 • 51, 6 • 57
  location, 7 • 15
  machine breakdown, 3 • 51
  machine interference, 3 • 51
  mathematical/logical, 8 • 57
  minisum formulation, 7 • 5
  minimax formulation, 7 • 5
  multimachine job shops, 3 • 64
  multistage flow line, 3 • 63
  network location, 7 • 15
  normal standby, 3 • 55
  parallel and standby systems, 3 • 54
  parallel machines, 3 • 54
  process charts, 1 • 70
  priority standby, 3 • 55
  production control, 3 • 84
  production system, 3 • 49
  quality, 3 • 65
  regression, 3 • 130
  seasonal, 3 • 131
  sequencing, 1 • 29
  sensitivity analysis, 7 • 16
  simulation, 3 • 50, 3 • 143
  splitting, 3 • 55
  standby machines, 3 • 55
  state-dependent arrival and service rates, 3 • 61
  step-wise refinement, 7 • 16
  task time variability, 3 • 60
  test and rework loops, 3 • 65
  tool life, 4 • 35
  transfer line, 3 • 55
  two machine job shops, 3 • 64
  two-dimensional layouts, 1 • 70
  two-stage flow line, 3 • 63
  types, 3 • 50
  types of cost, 6 • 51
  unlimited in-process inventories, 3 • 60
  variable-work force, 3 • 120
  Winter's seasonal forecasting, 3 • 135
Modified control limits, 3 • 108
Modular:
  assembly machines, 3 • 39
  facility, 1 • 74
  furniture, 1 • 74
Modularity, 7 • 23, 8 • 96
  spine concept, 7 • 24
Molding:
  blow, 3 • 16
  cold, 3 • 16
  compression, 3 • 12
  injection, 3 • 16
  transfer, 3 • 12
Money, 1 • 3
Monorails, 5 • 69
Motion:
  control, 4 • 88 (*see also* Motion control for robots)
  economy, 2 • 97
Motion control for robots:
  axis limit, 4 • 88
  contouring, 4 • 88
  line-tracking, 4 • 88
  point-to-point, 4 • 88
Motion-Time Analysis (MTA), 2 • 97
Motivation, 2 • 126
Moving average, 3 • 102
Moving range, 3 • 102
MRP (*see* Material reguirements planning)
MTM-2, 3 • 228
Multidiscipline groups, 1 • 69
Multimodels, 6 • 52
Multiple launch rocket system (MLRS), 8 • 66
Multi-process holding, 3 • 253
Multi-function worker, 3 • 253

Narrow-aisle trucks:
  high-lift, narrow-aisle, 4 • 102
  orderpicking, 4 • 103
  reach, 4 • 103
  sideloading, 4 • 104
Narrow limit gauging (NLG), 3 • 103
National Bureau of Standards, 3 • 210
  hierarchical control architecture, 8 • 80
National Institute of Occupational Safety and Health (NIOSH), 2 • 74
Needs:
  individual, 2 • 7
  organizational, 2 • 7
  production and technological, 2 • 7
Negative rake angle, 4 • 68
Near net shape forming, 4 • 48
Net present value (NPV), 1 • 6
Network:
  activity-on-node, 3 • 147, 3 • 180
  closed queueing, 3 • 64
  communications, 7 • 28
  constructing, 3 • 147
  flow, 1 • 27, 1 • 30
  open queueing, 3 • 64
  project, 3 • 147
Network location problems, 7 • 9
  p-center problem, 7 • 10

p-median problem, 7 • 10
r-cover problem, 7 • 10
NC (numerical control), 8 • 18
Nine hidden lessons, 8 • 106
Nodes, 8 • 16
Nominal:
design, 1 • 80
group technique, 1 • 53, 2 • 30
part characteristics, 1 • 80
Nonconformances, 3 • 102
Nonconforming product, 3 • 102
Nonconformities, 3 • 102
Nonlinear production systems, 6 • 10
Nonlinear systems:
production cost for, 6 • 10
Nonrandom patterns, 3 • 103
Normal probability paper, 3 • 110
Numerical control (NC), 8 • 18
Numerically controlled (NC) machine, 4 • 59

Occupational:
Safety and Health Act (OSHA), 2 • 35
Safety and Health Administration (OSHA), 2 • 43
specialty, 1 • 67
strain, 2 • 85
stress, 2 • 83
One-man build, 3 • 176
One-piece production and conveyance, 3 • 253
Open-loop, 4 • 84
Operating charcteristic (OC) curve, 3 • 111
Type A, 3 • 111
Type B, 3 • 111
Operations:
control, 8 • 68
process chart, 5 • 25
research, 1 • 23
sheet, 3 • 254
Operating efficiency, 5 • 20
Opitz code, 8 • 27
OPT (*see* Optimized production technology)
Optical code readers:
charge couple device (CCD), 4 • 133
fixed beam readers, 4 • 131
handheld code readers, 4 • 132
machine vision, 4 • 133
moving beam readers, 4 • 131
Optical encoder, 4 • 85
Optimization techniques, 2 • 62
Optimized production technology (OPT), 5 • 62
bottleneck, 3 • 201
rules, 3 • 201
Order, 1 • 7
monitoring, 3 • 197
scheduling, 3 • 193
Order point systems, 3 • 192
Order quantities, 1 • 62
Order releasing, 3 • 195
Organization:
dimensions of, 2 • 3, 2 • 7
matrix structure, 2 • 16
Organizational learning, 2 • 14
Out of control (OOC), 3 • 103
Overhead fixed path transporters, 5 • 69
Overhead value analysis, 1 • 55

Pace rating, 2 • 97
Packages, 4 • 100
Packaging, 1 • 5, 1 • 6, 5 • 76, 5 • 78
design techniques, 5 • 84
Packing, 5 • 78
Pallets, 5 • 84
Parallel and standby systems:
parallel machines, 3 • 54
standby machines, 3 • 55
normal standby, 3 • 55
priority standby, 3 • 55
splitting, 3 • 55
Pareto, 3 • 45
Part family, 3 • 236
attribute, 8 • 27
definition of, 8 • 27
family part matrix, 8 • 27
production flow analysis, 8 • 27
similarity, 8 • 27
taxonomy, 8 • 27
Part orientation, 3 • 26
Participative work design, 8 • 112
Parts, 4 • 100
Parts feeders, 3 • 21
Parts inspection, 1 • 5
Pattern recognition, 3 • 106
Payback, 1 • 6
Payload, 4 • 95
Peer counseling, 2 • 14
Pegging, 5 • 11
Perfect balance, 3 • 178
Perfect substitutes, 6 • 5
Performance, 1 • 4, 1 • 7
and productivity, 1 • 47
measurement, 1 • 45, 1 • 65
measures, 1 • 45
organizational, 1 • 46
specification, 1 • 79, 4 • 95
system, 3 • 50
Permanent mold casting, 4 • 7, 4 • 11
die casting, 4 • 12
investment casting, 4 • 12
PERT (program evaluation and review technique), 3 • 147
PERT/CPM problem, 7 • 8
Philosophy:
business, 1 • 11

Philosophy (*continued*)
company, 1 • 11
defined, 1 • 12
design-to-cost, 3 • 48
Physical constraints, 1 • 6
Pick and place:
transfer devices, 3 • 33
Piece-rate incentive compensation, 1 • 45
Piezoelectric, 4 • 95
Plan:
implementation, 8 • 115
marketing, 1 • 60
production, 1 • 46
sales, 1 • 60
training, 8 • 113
Planning:
capacity, 1 • 23, 1 • 64
competitive, 1 • 33
computer aided process (CAPP), 8 • 33
demand, 1 • 61
horizon, 1 • 26
production, 1 • 61
manufacturing, 1 • 33
materials requirements (MRP), 5 • 5
strategic, 1 • 33, 1 • 59
Planning function, 1 • 6
Planning horizon, 3 • 116
Plant layout, 5 • 29, 7 • 18
department, 7 • 36
Plaster mold casting, 4 • 7, 4 • 14
centrifugal casting, 4 • 17
ceramic mold casting, 4 • 15
Plastic deformation, 4 • 26
Polyarticulating, 4 • 76
Pontryagin, 3 • 77
Position analysis questionnaire, 2 • 80
Power law of practice, 2 • 72
Precedence:
chain, 3 • 159
linear, 3 • 159
tree, 3 • 159
Precedence diagram, 3 • 180
Predetermined motion time study, 3 • 228
Predetermined motion time systems, 2 • 97, 2 • 103
Predictive engineering, 8 • 16
Preemption, 3 • 160
Press forming, 3 • 10
Principle:
maximum, 3 • 76
Principles:
learning, 2 • 127
Probability of acceptance, 3 • 103
Problem solving, 2 • 58
Problem solving approach, 3 • 220
Problems:
capacitated, 7 • 15
discrete location, 7 • 12
generalized p-median, 7 • 13
location-allocation, 7 • 8
network location, 7 • 5, 7 • 9
one center, 7 • 10
PERT/CPM, 7 • 8
p-center, 7 • 10, 7 • 12
planar location, 7 • 5
p-median, 7 • 10
r-cover, 7 • 10, 7 • 11
set covering, 7 • 12
uncapacitated location, 7 • 12, 7 • 13
warehouse location, 7 • 12
Process:
automation, 8 • 110
capability analysis, 3 • 103, 3 • 109
control, 3 • 103, 8 • 92
design, 5 • 23
flow, 8 • 92
layout, 7 • 31
monitoring, 3 • 74
parameters, 4 • 46
planning, 8 • 19
planning and control, 3 • 100
Process plans:
definition of, 8 • 21
factory information, 8 • 22
formulation of, 8 • 22
in aircraft industry, 8 • 21
operation sheet, 8 • 22
Processes:
abrasive machining, 4 • 19, 4 • 61
batch, 5 • 23
broaching, 4 • 18
casting (*see* Casting and Casting processes)
chip formation, 4 • 18
constant level, 3 • 133
deformation, 4 • 38
drilling, 4 • 18
forging, 4 • 3
forming, 4 • 18, 4 • 38 (*see also* Forming processes)
foundry, 4 • 3
joining, 4 • 18
machining, 4 • 18
metal cutting, 4 • 24
milling, 4 • 18
molding, 4 • 7
sawing, 4 • 18
seven basic, 4 • 22
shaping, 4 • 3, 4 • 17
specific machining, 4 • 24
straight-through, 5 • 23
thermomechanical, 4 • 55
trend, 3 • 133
turning, 4 • 18
Processes and strategies:
for designing new jobs, 2 • 28
for redesigning existing jobs, 2 • 29
Producer's risk, 3 • 103

Producing layouts:
computer graphics, 7 • 54
data record, 7 • 55
overlays, 7 • 55
traditional methods, 7 • 54
Product:
cost, 1 • 78
design process, 3 • 4
-development cycle, 1 • 79
layout, 7 • 31 (*see also* Layout(s))
mix, 1 • 33
Product fragility:
shock, 5 • 73
vibration, 5 • 73
Production:
activity control, 3 • 191
capacity expansion, 6 • 21
economics, 6 • 3, 6 • 49
fine-tuning, 3 • 252
flow analysis, 3 • 237
forecasts, 3 • 128
-inventory balance equation, 3 • 121
islands, 2 • 15
levels, 3 • 116
linear systems, 6 • 8
models, 3 • 49
monitoring, 8 • 55
nonlinear systems, 6 • 10
organization, 1 • 66
output, 6 • 14
overlapping, 3 • 195
plan, 1 • 46
planning, 3 • 116
planning process, 3 • 124
rate, 3 • 50
reporting, 3 • 199
routings, 5 • 25
schedule, 1 • 46
scheduling (*see* Production scheduling)
smoothing, 3 • 252
splitting, 3 • 195
supervisors, 2 • 137
system analysis, 8 • 65
system control models, 3 • 68
system design, 8 • 65
system models, 3 • 49
workers, 2 • 136
Production cost(s), 6 • 14
linear systems, 6 • 9
long-run, 6 • 18
nonlinear systems, 6 • 10
short-run, 6 • 18
Production planning, 6 • 3
economic analysis for, 6 • 3
Production process:
homogeneous of degree k, 6 • 4
Production scheduling, 3 • 125, 3 • 137
backward scheduling, 3 • 142
control point, 3 • 144
finite scheduling, 3 • 142
flexible manufacturing system, 3 • 145
forward scheduling, 3 • 142
Gantt chart, 3 • 141
kanban systems, 3 • 145
manual methods, 3 • 140
simulation methods, 3 • 143
Productivity, 3 • 191, 7 • 19, 7 • 28
American Productivity Center (APC), 1 • 48
and performance, 1 • 47
competitive benchmarking premise, 1 • 39
Firm-Level Productivity Measurement System, 1 • 48
impact of industrial safety on, 2 • 36
impact of manpower and equipment on, 1 • 58
improvement, 3 • 98
measures, 1 • 45, 5 • 19
of inputs, 6 • 3
of support activities, 1 • 50
related to profitability, 1 • 48
Profitability, 1 • 48
Program Evaluation and Review Technique (PERT), 3 • 147
Programmable:
controllers, 5 • 66
device support system (PDSS), 8 • 55
logic controllers, 3 • 209
Programming robots:
leadthrough, 4 • 93
walkthrough, 4 • 93
Progressive assembly line, 3 • 176
Project:
development cycle, 8 • 111
planning, 3 • 146
scheduling, 3 • 137
Project network, 3 • 147
Project teams, 1 • 69
Prototype, 1 • 79
Psychology:
engineering, 2 • 53
experimental, 2 • 53
Psychomotor skill, 2 • 71

QC circles, 2 • 86
Quality, 1 • 5
assurance, 3 • 96, 4 • 7
-at-source, 1 • 59
characteristics, 3 • 96
circles, 2 • 86
control, 2 • 86, 3 • 96
cost of, 1 • 78
destructive tests, 4 • 7
human error, 2 • 86
models, 3 • 65
nondestructive tests, 4 • 7

Quality (*continued*)
of design, 3 • 96
of goods, 1 • 78
of production, 3 • 96
of residual effects, 3 • 96
of sales, use, and service, 3 • 96
-of-working-life, 2 • 7, 2 • 30
parameters, 3 • 96
systems, 1 • 7
vendor, 1 • 80
verifying, 5 • 86
Quality costs:
appraisal, 3 • 98
categories, 3 • 98
external failure, 3 • 98
information, 3 • 101
internal failure, 3 • 98
manual, 3 • 100
objectives, 3 • 100
planning, 3 • 100
policies, 3 • 100
prevention, 3 • 98
program, 3 • 100
reports, 3 • 101
system, 3 • 98
Questioning technique, 3 • 223
combining, 3 • 223
eliminating, 3 • 223
simplifying, 3 • 223
Queue control, 3 • 193
Queueing network analysis, 8 • 84
Quick, Joseph H., 2 • 97

Racks and shelving:
adjustable shelving, 4 • 117
flow-through storage, 4 • 119
mobile shelving, 4 • 119
modular drawer cabinets, 4 • 119
Radial compression zone, 4 • 26
Radio frequency (RF):
RF antennae, 4 • 134
RF identification tags, 4 • 134
RF readers, 4 • 134
Range, 3 • 102
Rate:
inflation, 6 • 30
inflation-free interest, 6 • 30
interest, 6 • 30
market interest, 6 • 30
of material removal (MRR), 4 • 21
Real time control, 8 • 86
Recall of product, 3 • 102
Receiving inspection, 3 • 100
Rectilinear distances, 7 • 6
Recurrence file, 3 • 102
Redesign principles, 2 • 30
Regenerative chatter, 4 • 70
Regulations:
general health and safety, 2 • 47
Regulator, 3 • 71
Regulatory constraint, 6 • 19
Rejectable process level (RPL), 3 • 108
Relationship values, 7 • 42
Reliability:
human performance, 2 • 86
Reorder point:
fixed, 1 • 62
variable, 1 • 62
Reproducibility, 1 • 80
Requirements:
injury/illness recordkeeping and reporting, 2 • 43
product, 3 • 41
Reports:
cost, 6 • 41
departmental cost, 6 • 47
labor cost, 6 • 43
management, 6 • 41
manufacturing expenses, 6 • 45
material cost, 6 • 42
material usage, 6 • 42
price variance, 6 • 42
Resource:
allocation, 3 • 154
constrained, 3 • 156
leveling, 3 • 155
utilization, 5 • 19
Resources:
natural grouping of, 1 • 67
Return on investment (ROI), 1 • 6
Returns to scale, 6 • 4
Robot:
applications, 4 • 89
assembly, 3 • 36
classification (Japanese), 2 • 91
control, 3 • 82
definition, Electric Machinery Law (Japan), 2 • 91
definition, Robotics Institute of America (RIA), 2 • 91
economic justification, 4 • 97
gripper, 4 • 93
human issues, 2 • 93
justification, 4 • 96
labor resistance, 4 • 96
payload, 4 • 95
performance specifications, 4 • 95
programming, 4 • 91
project feasibility, 4 • 98
repeatability, 4 • 95
sensing, 4 • 93
social impacts, 2 • 93
speed, 4 • 96
technical feasibility, 4 • 98
tooling, 4 • 89
training arm, 4 • 89

utilization, 4 • 96
vision, 4 • 93, 4 • 95
Robota, 4 • 74
Robotic:
assembly, 3 • 37
feeding, 3 • 37
orienting, 3 • 37
transferring, 3 • 37
Robots, industrial, 5 • 69
articulating, 4 • 76
configuration, 4 • 74 (*see also* Configuration)
gantry-mounted, 4 • 74
manipulator, 4 • 127
mobile, 4 • 80
programmable, 4 • 127
walking, 4 • 70
Roll forming, 3 • 10
Rotary and in-line indexers, 3 • 24
Rotary assembly machines, 3 • 31
Roughing, 4 • 64
Routine sheet, 3 • 254

Safety, industrial, 1 • 4–1 • 6
accident reporting, 2 • 33
biochemical hazards, 2 • 55
bulletins, 2 • 41
campaign, 2 • 36
checklist of unsafe conditions, 2 • 37
classification of hazards, 2 • 38
elements of accident prevention, 2 • 36
eliminating unsafe conditions, 2 • 36
enforcement of rules, 2 • 36
files, 2 • 36
functions, 2 • 34
housekeeping, 2 • 36, 2 • 46–2 • 49
human performance, 2 • 55
management principles, 2 • 35
Occupational Safety and Health Administration (OSHA), 2 • 43
office, 2 • 33
personal protective equipment, 2 • 36
policy, 2 • 33, 2 • 36
potential unsafe conditions, 2 • 38
problem sources to be inspected, 2 • 39
procedures, 2 • 40–2 • 46
radiological hazards, 2 • 55
regular duties, 2 • 36
responsibilities, 2 • 33
robots, 4 • 95
supervisory functions, 2 • 34
training, 2 • 46
Safety stock, 3 • 119
Safety stocks, 3 • 121
SAINT, 2 • 62
Sales and service, 3 • 101
Sample, 3 • 102
Sampling plans:
double, 3 • 111
multiple, 3 • 111
sequential, 3 • 111
single, 3 • 111
Sand casting, 4 • 7
carbon dioxide molding, 4 • 9
core sand molding, 4 • 8
dry sand molding, 4 • 8
expanded polystyrene molding, 4 • 10
floor molding, 4 • 9
green sand molding, 4 • 8
pit molding, 4 • 9
shell molding, 4 • 9
vacuum molding, 4 • 10
Scale:
economies of, 6 • 49
diseconomies of, 6 • 49
returns to, 6 • 4
Scanning electron microscope, 4 • 26
Scheduling, 8 • 69
algorithms, 3 • 159
critical path, 3 • 168
earliest due date (EDD) rule, 3 • 164
flow shop, 3 • 169
job shop, 3 • 159, 3 • 171
master, 1 • 61
multimachine SPT rule, 3 • 167
one-machine, 3 • 163
order, 3 • 193
parallel processor, 3 • 166
performance measures, 3 • 161
serial processor, 3 • 169
shortest processing time (SPT), 3 • 163
terminology for, 3 • 160
Schwab, John L., 2 • 97
Scientific management, 1 • 45, 2 • 12, 2 • 82, 2 • 97, 2 • 125
Segur, A. B., 2 • 97
Self-:
correcting, 3 • 71
sharpening, 4 • 68
Sensing:
hearing, 2 • 54
kinesthetic, 2 • 54
systems, 2 • 54
taste, 2 • 54
touch, 2 • 54
vision, 2 • 54
Sensors:
optical presence, 4 • 95
pressure, 4 • 93
tactile, 4 • 95
Sequencing, 1 • 29
algorithms, 3 • 159
one-machine, 3 • 163
Sequencing, 1 • 29
Servomechanism, 3 • 69

Set point, 3 • 71
Setup time reduction:
external setup, 3 • 253
internal setup, 3 • 253
setup time, 3 • 253
Shadow price, 3 • 124
Shaping, 4 • 3
Shaping plastic components, 3 • 12
Shear:
angle, 4 • 28
plane or zone, 4 • 26
Shelf life, 1 • 6
Shipping environment, 5 • 72
Shop:
job, 3 • 139
model, 3 • 139
process, 3 • 139
product, 3 • 139
special, 3 • 139
Shop-floor control:
functional areas, 3 • 192
Shop floor environment, 3 • 192
Shortest path, 1 • 28
Shortest processing time (SPT), 3 • 163
Shrinkable plastic film, 5 • 86
Signal flow graphs, 3 • 73
Signals:
analog, 3 • 71
digital, 3 • 71
Silicon carbide, 4 • 61
Simplicity, 8 • 106
SIMSCRIPT, 3 • 49
Simulation, 1 • 7, 1 • 70, 3 • 171
activity scanning, 8 • 61
computer, 8 • 57
continuous, 8 • 61
definition of, 8 • 57
discrete, 8 • 58
event, 8 • 60
GASP IV, 8 • 67
GPSS, 8 • 63
Human Operator Simulator, 2 • 62
job shop, 3 • 171
languages, 2 • 62, 3 • 49, 8 • 63
man-machine systems, 2 • 62
modeling, 8 • 57
modeling process, 8 • 57
Monte Carlo, 2 • 62
process, 8 • 60
Q-GERT, 8 • 63
SAINT, 2 • 62
SIMSCRIPT, 8 • 64
SLAM, 2 • 62, 8 • 66
SLAM II, 8 • 64
Skill, 2 • 126
Slack time ratio, 3 • 197
SLAM, 2 • 62, 8 • 66
Sliding, 4 • 68
Slip sheets, 5 • 85
Small containers:
part bins, 4 • 117
tote boxes, 4 • 117
Smith, Adam, 2 • 11
Sociotechnical:
analysis approaches, 2 • 30
system, 2 • 3, 2 • 14
Software design, 8 • 41
Software installation:
conference room pilot, 8 • 50
parallel processing, 8 • 48
phased turnover, 8 • 49
straight turnover, 8 • 49
Solid modeling, 8 • 17
three-dimensional, 8 • 17
volumetric, 8 • 17
Solidification control, 4 • 3
Sound pressure level, 2 • 82
Space, 1 • 3
utilization, 1 • 6
Spark-in transient, 4 • 64
Specific:
energy, 4 • 68
power, 4 • 68
Specification limit, 3 • 103
Specifications for robot performance, 4 • 95
Speeds and feeds, 1 • 5
Spindle speeds, 8 • 18
Spinning, 3 • 10
Squeeze forming, 4 • 53
Stable mixture, 3 • 106
Standard deviation, 3 • 102
Standard operation, 3 • 254
Standard operation routine, 3 • 254
Standard time allowances, 2 • 110
Standards, 1 • 4
Standardization:
of processes, 2 • 12
of products, 2 • 12
of services, 2 • 12
Star Wars, 4 • 74
State of statistical control (SOSC), 3 • 103
Station:
layout, 1 • 5, 1 • 6
location/usage, 1 • 6
replenishment, 1 • 6
Statistical:
capability, 1 • 80
tolerance analysis, 1 • 80
Statistical process quality control, 3 • 103
Statistical quality control, 3 • 102
Steady state optimization, 3 • 77
Stegemerten, Gustave J., 2 • 97
Stepper motor, 4 • 84
Stockouts, 3 • 121
Storage, 1 • 6
automated, 2 • 93

Storage equipment:
automated storage systems, 4 • 122
racks and shelving, 4 • 117
small containers, 4 • 117
unit load containers, 4 • 117
Storage system, 7 • 25
Storage yard handling:
front-end loaders, 4 • 106
mobile cranes, 4 • 106
straddle carriers, 4 • 106
straddle cranes, 4 • 106
straddle hoists, 4 • 106
Stopwatch time study, 2 • 99
Straight-through processes, 5 • 23
Straight-line depreciation, 6 • 33
Strain hardening, 4 • 38
Strapping, 5 • 85
Strategy, strategies:
business, 1 • 11
characteristics, 1 • 13
competitive, 1 • 11
differentiation/value added, 1 • 34
financial, 1 • 11
for job design, 2 • 28, 2 • 29
formulation, 1 • 35
functional, 1 • 14
levels, 1 • 13
low-cost producer, 1 • 34
manufacturing, 1 • 11, 1 • 14, 1 • 60
manufacturing, decision categories, 1 • 15
market niche, 1 • 34
marketing, 1 • 11
product volume/mix, 1 • 34
Strategic:
audit, 1 • 33
business unit (SBU), 1 • 13, 1 • 14
capabilities, 1 • 15
master planning, 7 • 21
planning, 1 • 11, 8 • 108
planning process, 1 • 33
planning unit (SPU), 1 • 14
Stretch wrapping, 5 • 85
Structure:
manufacturing, 1 • 14
Stress:
occupational, 2 • 83
in shiftwork, 2 • 85
String diagram, 3 • 225
Study period, 6 • 27
Subgroup, 3 • 102
Subgroup variation:
between, 3 • 103
within, 3 • 103
Suboptimization, 3 • 126
Substitution of inputs, 6 • 3
Superabrasives, 4 • 61
Superplastic forming, 4 • 38
Supervisory training, 2 • 135
Supplier process capability, 1 • 7
Supplier quality:
control or surveillance, 3 • 100
information, 3 • 100
management, 3 • 100
Surface acoustical wave (SAW), 4 • 135
Surface finish, 4 • 25, 4 • 46, 4 • 71
Synergistic, 1 • 73
System, 1 • 3
distribution, 8 • 70
performance, 3 • 50
social, 2 • 4
social support, 2 • 4
state, 8 • 57
technical, 2 • 4
Systematic layout planning, 7 • 42
Systems:
analysis and design, 8 • 42
automated storage, 2 • 93
integrated, 1 • 59
integration, 1 • 8

Tactile sensing, 4 • 95
Task analysis (*see* Job analysis)
Tasks:
man-machine, 2 • 59
sources of, 2 • 7
Taylor, Frederick W., 1 • 45, 1 • 68, 2 • 11, 2 • 12, 2 • 97
Taylor series, 8 • 62
Taylorism, 1 • 45
Teach pendant, 4 • 91
Technical progress:
disembodied change, 6 • 7
embodied change, 6 • 7
Technique for human error rate prediction (THERP), 2 • 86
Technological:
assessment, 2 • 20
determinism, 2 • 10
Technology dimensions, 2 • 23
Temperature, 4 • 46
Test equipment, 1 • 5
Testing packaging, 5 • 87
Therbligs, 2 • 97
Thermal damage, 4 • 70
Thermoforming, 3 • 16
Thermomechanical treatments, 4 • 55
Time:
and attendance reporting, 3 • 198
horizon, 3 • 116
lags, 3 • 69
measurement unit (TMU), 3 • 228
series, 3 • 128
standard, 2 • 97
standards, 1 • 5
study, 1 • 45, 2 • 98

Tolerances, 1 • 80
  dimensional, 4 • 72
  form, 4 • 72
Tool:
  control, 5 • 53
  design, 8 • 17
  failure, 4 • 25, 4 • 33
  geometry, 4 • 25, 4 • 26
  life, 4 • 33
  life models, 4 • 35
  management, 1 • 5
  material, 4 • 24
  wear, 4 • 25
Tooling, 1 • 5
  part-compliant, 4 • 90
  misalignment, 4 • 90
  remote center compliance, 4 • 90
  robot, 4 • 89
Tools:
  abrasive, 4 • 61
  bonded abrasive, 4 • 62
  controlled abrasive, 4 • 61
Top-down design, 8 • 56
Total facility, 1 • 75
Tote boxes, 5 • 77
Tower of Babel, 8 • 53
Toyota production system (TPS), 7 • 31
  autonomation (Jidoka), 3 • 251
  background, 3 • 249
  cost concept, 3 • 255
  creative thinking or inventive ideas (Soikufu), 3 • 251
  flexible workforce (Shojinka), 3 • 251
  framework, 3 • 249
  Just-In-Time production, 3 • 251
  kanban, 3 • 251
Tradeoffs:
  function versus cost, 3 • 42
  time-cost, 3 • 152
Training, 1 • 4, 8 • 113
  bottom line impact, 2 • 138
  concepts, 2 • 126
  course, 2 • 129
  embedded, 2 • 62
  employee, 2 • 136
  implementation, 2 • 138
  industrial employee, 1 • 125
  supervisory, 2 • 135
Transducer, 3 • 71
Transfer devices, 3 • 23
Transfer line models:
  finite capacity banks, 3 • 56
  infinite capacity banks, 3 • 56
  multiple stage extensions, 3 • 60
  no inventory banks, 3 • 55
  optimal line division, 3 • 60
  three-stage line, 3 • 59
  two-stage line, 3 • 56
Transportation problem, 1 • 31
Trees, 7 • 10
Truing, 4 • 68
Turnaround time, 8 • 92
Turning tools, 4 • 18
Twist drills, 4 • 18
Type 1 error, producer's risk, 3 • 103
Type 2 error, consumer's risk, 3 • 103

Uncertainty:
  human, 2 • 4
  technological, 2 • 4
Understanding utility bills, 7 • 69
  electricity, 7 • 69
  natural gas, 7 • 69
Unit load containers:
  metal shop, 4 • 117
  pallet stacking frames, 4 • 117
  pallets, 4 • 117
  slipsheets, 4 • 117
  welded wire, 4 • 117
Unit load design, 5 • 86
Unit loads, 4 • 100
Unit sales volume, 1 • 79
Unitizing, 5 • 76, 5 • 84
UniVation System, 2 • 97, 2 • 117
Unused value, 6 • 27
U.S. Air Force Integrated Computer Aided Manufacturing (ICAM) program, 3 • 210
User contact, 3 • 101
Utility theory, 3 • 122

Vacuum chucks, 4 • 18
Vacuum waste collection, 1 • 76
VAL, 4 • 91
Value:
  salvage, 6 • 28
  unused, 6 • 27
Variables acceptance sampling, 3 • 113
Variance analysis, 2 • 30
Variances:
  capacity, 6 • 40
  controllable, 6 • 40
  efficiency, 6 • 40, 6 • 45
  expense, 6 • 40
  labor, 6 • 39
  material, 6 • 39
  overhead, 6 • 39
  price, 6 • 42
  rate, 6 • 44
  spending, 6 • 45
  time, 6 • 44
  utilization, 6 • 41, 6 • 46
  volume, 6 • 47
Variant approach, 8 • 20

Variation, 1 • 80
Vendor packaging, 5 • 78
Ventilation:
  industrial safety considerations, 2 • 37
Vertices, 7 • 9
Vibrations:
  forced, 4 • 70
  self-excited, 4 • 70
  sequential to the hands, 2 • 83
  whole body, 2 • 83
Vices, 4 • 18
Visual:
  acuity, 2 • 69
  control system (Andon), 3 • 255
  displays (*see* Displays)
  display terminals, 2 • 68, 2 • 86
  sensitivity, 2 • 69
Vision:
  robot, 4 • 95
Voice recognition, 4 • 135
Voice systems, 3 • 212
Volvo Kalmar plant, 2 • 15

Warehouse, 7 • 25
Waste management, 1 • 76
Wheel:
  grade, 4 • 62
  wear, 4 • 71
WOCOM, 2 • 97, 2 • 115
Wood:
  boxes, 5 • 79
  crates, 5 • 79
  pallets, 5 • 84
  wire-bound box, 5 • 79
Work:
  changing world of, 1 • 67
  envelope, 4 • 78, 4 • 80, 4 • 95
  subdivision of, 1 • 68
Work-factor, 2 • 97, 2 • 103
Work-force levels, 3 • 116
Work-in-process (WIP), 8 • 108
Workpiece, 4 • 70
Work measurement, 1 • 45
  defined, 2 • 97
  result, time standard, 2 • 97
  computerized, 2 • 113
  considerations, 2 • 108
  techniques, 2 • 98
Work sampling, 2 • 104
Workstations:
  industrial safety considerations, 2 • 37
Work teams:
  advantages and disadvantages, 2 • 16
  consensus decision making, 2 • 16
  self-maintaining work units, 2 • 14
Work unit(s):
  defined, 1 • 49
  hierarchy of, 1 • 49
  structure, 1 • 51
Worker motivation, 2 • 15
Working to rule, 2 • 12
Workload:
  mental, 2 • 57
  physical, 2 • 57
Workpiece, 4 • 7, 4 • 24
Workplace:
  design, 2 • 79
  environment, 1 • 72
Workspace layout, 2 • 61
Workstation design, 2 • 61, 2 • 87
Worst case analysis, 1 • 78
Worth:
  annual, 6 • 25
  future, 6 • 25
  present, 6 • 25

Yerkes-Dodson law, 2 • 85

Zero defects, 3 • 37